Quantitative Geology and Geostatistics

Volume 19

More information about this series at http://www.springer.com/series/6466

J. Jaime Gómez-Hernández • Javier Rodrigo-Ilarri
María Elena Rodrigo-Clavero • Eduardo Cassiraga
José Antonio Vargas-Guzmán
Editors

Geostatistics Valencia 2016

Springer

Editors
J. Jaime Gómez-Hernández
Institute for Water and Environmental Engineering
Universitat Politècnica de València
Valencia, Spain

Javier Rodrigo-Ilarri
Institute for Water and Environmental Engineering
Universitat Politècnica de València
Valencia, Spain

María Elena Rodrigo-Clavero
Institute for Water and Environmental Engineering
Universitat Politècnica de València
Valencia, Spain

Eduardo Cassiraga
Institute for Water and Environmental Engineering
Universitat Politècnica de València
Valencia, Spain

José Antonio Vargas-Guzmán
Saudi Aramco
Dhahran, Saudi Arabia

ISSN 0924-1973 ISSN 2215-1834 (electronic)
Quantitative Geology and Geostatistics
ISBN 978-3-319-46818-1 ISBN 978-3-319-46819-8 (eBook)
DOI 10.1007/978-3-319-46819-8

Library of Congress Control Number: 2016962455

Printed on acid-free paper

This Springer imprint is published by Springer Nature
The registered company is Springer International Publishing AG
The registered company address is: Gewerbestrasse 11, 6330 Cham, Switzerland

Preface

The 10th International Geostatistical Congress took place in Valencia on September 2016 following the traditional quadrennial gathering of theoreticians, engineers, and practitioners on geostatistics. From September 5 until September 9, more than 200 experts discussed the latest developments on the many fields in which geostatistics have found application. This book contains 66 technical papers presented at the conference by those participants who chose to submit their paper for the book of proceedings. All of them have been reviewed by at least two experts in their field. The proceedings are divided in seven parts: Theory, Mining, Petroleum, Hydro(geo)logy, Environmental, Big Data, and Health, containing theoretical and applied papers on the subjects of matter. There is also a paper that has been singled out containing a semblance of Professor Danie Krige, the person who gives name to the technique on which geostatistics is built upon, who passed away on March 2013, and on whose memory a special session took place on Wednesday, September 7, 2016.

Acknowledgments

The 10th International Geostatistical Congress would have not been possible without the financial support of a number of institutions and companies. The Organizing Committee is very thankful to the Universitat Politècnica de València; the Polytechnic City of Innovation; the Valencian Regional Ministry of Education, Research, Culture and Sport (ref. AORG/2016/050); the International Association of Mathematical Geology; and Geovariances. Iberia and Geoconnexion are also thanked for their support.

Last, but certainly not least, special thanks are due to the Saudi Arabian Oil Company (SAUDI ARAMCO) for undertaking the exclusive sponsorhip of this Geostatistics Valencia 2016 proceedings book by Springer.

Organizing Committee

grupo de HIDROGEOLOGIA

J. Jaime Gómez-Hernández (Chairman)
Javier Rodrigo-Ilarri
Eduardo Cassiraga
Teng Xu
Julio C. Gutiérrez-Esparza
M. Elena Rodrigo-Clavero
Institute for Water and Environmental Engineering
Universitat Politècnica de València

Scientific Committee

Rachid Ababou, Institut de Mecanique des Fluides de Toulouse, France
Petter Abrahamsen, Norwegian Computing Center, Norway
Denis Allard, BioSP, INRA, France
José Almeida, FCT – Universidade Nova de Lisboa, Portugal
Andras Bardossy, University of Stuttgart, Germany
Joerg Benndorf, Delft University of Technology, Netherlands
Jeff Boisvert, University of Alberta, Canada
Alexandre Boucher, Advanced Resources and Risk Technology, LLC, USA
Bruce Buxton, Battelle Memorial Institute, USA
Jef Caers, Stanford University, USA
José E Capilla, Universitat Politècnica de València, Spain
Eduardo Cassiraga, Universitat Politècnica de València, Spain
Snehamoy Chatterjee, Michigan Technological University, USA
João Felipe Costa, Federal University of Rio Grande do Sul, Brazil
Colin Daly, Schlumberger, UK
Ernesto Della Rossa, Eni, Italy
Clayton Deutsch, University of Alberta, Canada
Jared Deutsch, University of Alberta, Canada
Roussos Dimitrakopoulos, McGill University, Canada
Christina Dohm, Anglo American, South Africa
Peter Dowd, University of Adelaide, Australia
Olivier Dubrule, Imperial College/Total, UK
Xavier Emery, University of Chile, Chile
Jianlin Fu, Universitat Politècnica de València, Valencia, Spain
Henrique Garcia-Pereira, Instituto Superior Técnico, Portugal
J. Jaime Gómez-Hernández, Universitat Politècnica de València, Spain
Pierre Goovaerts, BioMedware Inc., USA
Dario Grana, University of Wyoming, USA
Carolina Guardiola Albert, Instituto Geológico y Minero de España, Spain

Contents

Part III Mining Engineering

Part I
In Honor of Professor Danie Krige

Professor Danie Krige's First Memorial Lecture: The Basic Tenets of Evaluating the Mineral Resource Assets of Mining Companies, as Observed in Professor Danie Krige's Pioneering Work Over Half a Century

W. Assibey-Bonsu

Abstract This paper provides a write-up of the first Professor Danie Krige memorial lecture in 2014, which was organised by the University of the Witwatersrand in collaboration with the Southern African Institute of Mining and Metallurgy (SAIMM) and the Geostatistical Association of Southern Africa, where his wife Mrs Ansie Krige, the SAIMM and Professor R.C.A. Minnitt also spoke. The memorial lecture was presented by his previous PhD graduate student, Dr Winfred Assibey-Bonsu.

During that inaugural memorial lecture, the SAIMM highlighted three activities that the institute would hold going forward, so as to remember this great South African mining pioneer:

- The publication of a Danie Krige Commemorative Volume of the SAIMM Journal
- An annual Danie Krige Memorial Lecture to be facilitated by the School of Mining Engineering of the University of the Witwatersrand
- The annual award of a Danie Krige medal

What follows is both a tribute to his work and a testimony to the great man's deep personal integrity, belief in family, humility and faith in Christ, all of which led him to become not only a giant in the South African mining industry but indeed worldwide.

Figure 1 shows Professor Danie Gerhardus Krige.

The paper was first published by the Southern African Institute of Mining and Metallurgy during the Danie Krige Geostatistical Conference, Johannesburg, 2015.

W. Assibey-Bonsu (✉)
Gold Fields Limited, Perth Office, PO Box 628, West Perth 6872, Australia
e-mail: Winfred.AssibeyBonsu@goldfields.com.au

J.J. Gómez-Hernández et al. (eds.), *Geostatistics Valencia 2016*, Quantitative Geology and Geostatistics 19, DOI 10.1007/978-3-319-46819-8_1

Fig. 1 Professor Danie Gerhardus Krige: 26 August 1919–2 March 2013

1 Introduction

It has been said that 'we make a living by what we receive, but we make a life by what we give'. Professor Krige epitomised this in both thought and deed, where he showed that true success in life does not revolve around material accomplishments accrued as an individual but is defined by that which one does and leaves for others.

It has been the author's privilege to have had an association with Professor Krige for over 20 years, both initially as a student during a doctorate thesis at the University of the Witwatersrand and later with him as mentor, counsellor and 'father figure' for the period that followed.

This paper will cover the two aspects that defined Professor Krige, firstly with respect to his personal life and career, including the achievements of both, whilst the second part will briefly touch on his immense contribution to industry and the world for over half a century, through his pioneering work in ore evaluation, economics and of course geostatistics. Indeed, his passing was recorded in Wikipedia under notable persons, a distinction he shared with renowned persons such as Margaret Thatcher.

2 The Great Man: Professor Danie Krige

2.1 Family and Faith

This memorial lecture would be incomplete without firstly throwing light on some of the things Professor Krige held very dear in his life, taken from his interview in 2012 with Professor R.C.A. Minnitt of the University of the Witwatersrand.

Fig. 2 A 1930 photograph of Professor Krige aged 11 (*front middle*) with family

Professor Krige was born in Bothaville in the Free State and was the youngest of nine children born to a pastor.

Figure 2 shows a 1930 photograph of Professor Krige aged 11 with family.

Professor Krige was a devout Christian, who always emphasised that what made a difference in his life was his belief in Jesus Christ. He also acknowledged that he had been the recipient of gifts of grace from the Creator – 'grace given to him' – drawing attention to six specific areas, in which he could identify the grace of the Almighty at work in his life and career:

The first gift of grace:

It was a tribute to his parents for the practical application of a godly lifestyle, the establishment of a firm foundation and a life philosophy that was modelled by them in every area of life. An example being that even with the limited resources at their disposal, they ensured that seven of the nine siblings received a tertiary education.

The second gift of grace:

The second of the gifts of grace that he acknowledged was the support he had received from his two spouses. He was happily married for 45 years to his first wife (until her death) and thereafter for 20 years to Ansie.

The third gift of grace:

The third gift of grace was the way in which his career developed and the various changes in direction that it took, as his research unfolded.

The fourth gift of grace:

The fourth gift of grace was that when he returned to work at Anglovaal, they began to apply his advanced methods of evaluation on their mines.

The fifth gift of grace:

The fifth gift of grace was that on retirement from Anglovaal at the age of 60, he received the unexpected opportunity of taking up the Chair of Professor of Mineral Economics at the University of the Witwatersrand, which he occupied for the next 10 years. This enabled him to teach and undertake extensive consulting work for mining companies both locally and internationally and was, in his opinion, a great blessing.

The sixth gift of grace:

The final gift of grace was that after leaving the University of the Witwatersrand, he was still able to undertake extensive national and international consulting work, which kept him occupied and young for the following 20 years.

He also acknowledged with deep gratitude that whilst the opportunities presented themselves to him, it was his responsibility to make good use of them and that without these gifts of grace, his life's work would not have been possible.

The photos that follow bear testimony to his strong belief in family values, those same ones he was blessed with as a young boy.

Figure 3 shows Professor Krige and family.

Figure 4 shows Professor Krige with great-grandchildren.

Figure 5 shows Professor Krige celebrating his 90th birthday with Ansie.

2.2 *Career, Achievements and Awards*

Professor Krige matriculated from Monument High School at the age of 15 and, in 1938 at the age of 19, graduated as a Mining Engineer from the University of the Witwatersrand. It was clear early on that he was destined for great achievements.

The two photos in Fig. 6 show the difference between the robe of a university graduate and typical clothes of an underground miner. It provides a perfect illustration of Professor Krige's values regarding theoretical developments aimed at solving practical problems.

Figure 6 shows Professor Krige as a university graduate and also learning the trade, 1939.

2.2.1 Career

Professor Krige worked with Anglo Transvaal on a number of gold mines in the Witwatersrand until 1943 and thereafter joined the Government Mining

Fig. 3 Professor Krige and family

Fig. 4 Professor Krige with great-grandchildren

Fig. 5 Professor Krige celebrating his 90th birthday with wife Ansie

Fig. 6 Shows Professor Krige as a university graduate and also learning the trade

Engineering Department, where he worked for a further 8 years. He spent time studying data and developing mathematical models. He returned to industry as Group Financial Engineer of the Anglovaal Group until 1981, when he 'retired'. He then spent another 10 years (of his 'retirement') as Professor of Mineral Economics at the University of the Witwatersrand.

Professor Krige's seminal papers, published in the Journal of Chemical, Metallurgical and Mining Society of South Africa, led to additional fundamental research in France on 'regionalised variables' by Professor George Matheron and his team. Professor Matheron named the new method of linear estimation of the regionalised variables using a spatial model, 'kriging', in recognition of Professor Krige's distinguished pioneering work.

His 1951 paper, based on his MSc (Eng.) thesis at the University of the Witwatersrand, expounded his pioneering work in geostatistics in more detail. His research and paper covered and assisted with the statistical explanation of conditional biases in block evaluation. It stimulated the use of regression corrections for routine ore reserve evaluations by several mines, and the technique was essentially the first elementary basis of what is now known as kriging. The paper introduced, inter alia, the basic geostatistical concepts of support, spatial structure, selective mining units and grade-tonnage curves. The concept of recoverable resources/reserves in current use is based on what is known as 'Krige's relationship'.

Kriging is currently applied worldwide in the fields of exploration, ore evaluation, environment, petroleum, agriculture, fisheries and other disciplines. His outstanding influence on the worldwide mining industry is visible every day, as shown by the decision-making processes followed by international mining companies.

Over the course of his career, he published some 96 technical papers including the Geostatistics Monograph, the first in the series of the SAIMM. A complete record of all his publications is available in digital format from the SAIMM.

2.2.2 Dedicated Service

As a professional engineer, Professor Krige served for many years on the mining committee of the Engineering Council of South Africa and on the Council of the SAIMM; he was a co-founder of the International Association of Mathematical Geology, Geostatistical Association of Southern Africa, Geostatistical Association of Australia and the Statistical Association of South Africa.

He also served as a director of several companies, on the subcommittee of the South African Prime Minister's Economic Advisory Council during 1967/1968, as well as on various committees of the South African Chamber of Mines. He was a member of the SAMREC Working Committee for The South African Code for Reporting of Exploration Assets, Mineral Resources and Mineral Reserves (SAMREC Code) as first published in 2000.

Amongst all of this, he still managed to find time to (i) design the state aid formula, which assisted a large number of gold mines to survive the period of low gold prices, (ii) establish the original South African uranium contracts and, (iii) in 1955 and in Afrikaans, published probably one of the first papers on risk analysis for new mining investment. He also gave major inputs in the fields of financial analysis and taxation.

Professor Krige was especially committed to the Application of Computers and Operations Research in the Mineral Industry (APCOM). He was South Africa's representative on the International APCOM Council from its inception, served as the Chairman of the International APCOM Council and was the first member, outside of the United States, to be elected to this position. He initiated and was directly involved with all arrangements for the symposia held in South Africa in 1972, 1987 and 2003 and is believed to have attended all APCOM symposia until he was almost 90 years old. In 2003, 2 weeks after a major operation, he managed to convince his medical doctors to allow him to attend the 2003 APCOM in Cape Town, South Africa, where he was a keynote speaker and also presented two other papers.

During his time as a Professor of Mineral Economics at the University of the Witwatersrand, he was responsible for postgraduate courses in geostatistics and mineral economics and supervised many masters and doctoral theses. Both whilst there and afterwards, he presented courses in geostatistics and/or lectured at local South African and also international universities in Australia, Germany, Taiwan, Chile, Russia and China, to name but a few. He also still found the time to provide valuable consultancy work both locally and internationally and also participated in and contributed to many international congresses all over the world.

2.2.3 Achievements and Awards

Over his life time, he was the recipient of numerous awards locally and internationally, too many to mention all. His academic achievements and awards included:

- DSc (Eng.) 1963, University of the Witwatersrand
- DIng (HC) 1981, Honorary Degree, University of Pretoria
- Honorary Doctorate Degree from Moscow State Mining University
- Honorary Doctorate Degree from University of South Africa (UNISA)
- Order of Meritorious Service Class 1, Gold, by South African State President
- The highest award of the SAIMM, the Brigadier Stokes Award, 1984
- Many other merit awards from SAIMM including two gold medals in 1966 and 1980 and two silver medals in 1979 and 1993
- International Association of Mathematical geology – William Krumbein medal, 1984
- One of the highest awards from the American Society of Mining Engineers – the Daniel Jackling Award
- Several awards from APCOM International Council, including the Distinguished Achievement Award, 1989

Fig. 7 Professor Krige receiving the US NAE Award

Fig. 8 Professor Krige receiving the Order of the Baobab in silver

- Elected as Foreign Associate of US National Academy of Engineers (NAE) 2010, the first South African to ever receive this award for his distinguished contributions to Engineering
- Order of the Baobab in silver – awarded by President Jacob Zuma

Figure 7 shows him during the US NAE Award session (Professor Krige fifth from left in front row).

Figure 8 shows him being awarded the Order of the Baobab in silver by the South African State President Jacob Zuma.

3 Professor Danie Krige's Work on Essential Tenets in Evaluating the Mineral Resource Assets of Mining Companies

Although impossible to provide a thorough list in this paper, the author will try to detail at least some of the many principles that Professor Krige brought forth over half a century.

3.1 *Historical Background and Motivation: The Capital Intensiveness of Mining*

The mining industry requires very capital-intensive investments. Figure 9 provides some examples in this regard.

The left graph in Fig. 9 shows that in 2007 Rio Tinto's acquisition of a Canadian aluminium company (Alcan) costs 38.1 billion US dollars. It further shows that the estimated cost for Billiton's Olympic Dam project in Australia was estimated at 27 billion US dollars.

The right graph in Fig. 9 illustrates that in 2007 Gold Fields Limited acquired the South Deep Gold Mine in South Africa at a cost of 2.5 billion US dollars (the equivalent of 22.2 billion rand at the respective 2007 exchange rate). It also shows that in 2011 Newmont's acquisition of Fronteer Gold Inc. costs 2.3 billion US dollars and that Barrick's ongoing development of the Pascua-Lama gold mine in South America was estimated at 8.5 billion US dollars.

Mineral resources and mineral reserves are the fundamental assets of mining companies, and capital-intensive investments are made with respect to these. The

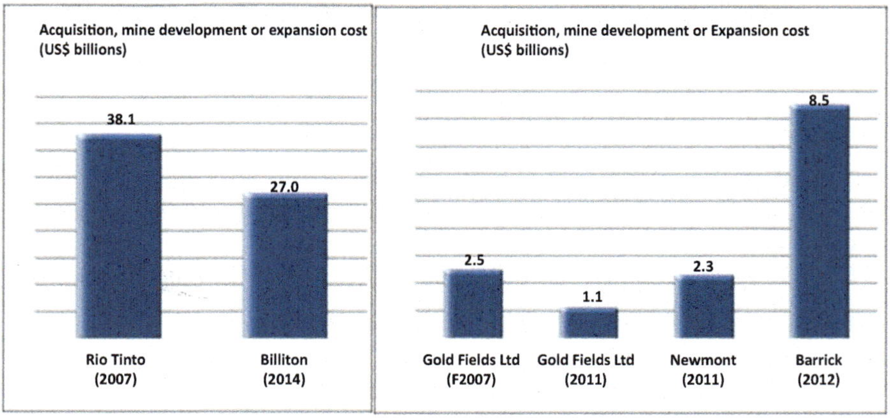

Fig. 9 Shows capital-intensive investments required in the mining industry

strategic objective is to explore, acquire, develop and ultimately mine them, but one critical risk exists in the uncertainty of the estimation of resources/reserves. If, after intensive capital investments, it is subsequently found that the expected mineral resources and mineral reserves were inefficiently estimated or valued, billions of dollars may be lost. Professor Krige's pioneering and research work provides technical solutions to mitigate these technical and financial risks when evaluating these fundamental assets.

3.2 Essential Tenets in Evaluating Mineral Resource Assets of Mining Companies Based on Over Half a Century of Professor Krige's Pioneering Work

3.2.1 Data Integrity

Professor Krige emphasised the critical importance of data integrity as the lifeblood of mineral resource and reserve evaluation. This includes data validation and authorisation, use of standards and blanks with approved laboratories and also database safety and security, which are all critical requirements of the Sarbanes-Oxley Act of 2002 (SOX) that is necessary for compliance with the New York Stock Exchange regulations.

3.2.2 Geology Models

Professor Krige highlighted geology as the foundation of mineral resource and reserve modelling. He emphasised that different ore bodies behave differently and that the main geological characteristics, including lithological and structural features, mode of origin and formation as well as controls of mineralisation, are critical inputs in ore body modelling.

He further highlighted the dangerous practice of subdividing ore bodies, not on geological grounds but directly on grade only, as this can lead to serious biases, particularly where data in one or more subdivisions are insufficient to allow proper geostatistical analysis.

3.2.3 Geostatistics Technology: Technique Selection and Optimal Application

In the field of mineral resource and reserve evaluation, geology and geostatistics are two inseparable sides of the same coin. As highlighted above, on the one side, geology concentrates on the physical features of the ore body, such as structures, source, deposition and type of mineralisation. Geostatistics is the other side of the coin and provides mathematical, statistical and geostatistical models for the analytical sampling data available, in order to introduce efficient evaluation techniques for resource and reserves estimates and to attach confidence limits to these estimates.

Uncertainty is fundamental to all branches of science and to human life itself. It is the reason for the introduction of mathematical and statistical techniques in geology and to the birth of geostatistics over half a century ago.

3.2.4 Frequency Distribution

The initial efforts in applying classical statistical procedures to ore evaluation in South Africa date back to 1919 (Watermeyer) and 1929 (Truscott). It was only in the late 1940s and early 1950s that Sichel (1947, 1952) introduced the lognormal model for gold values and using this model developed the 't' estimator. Departures from the usual lognormal model were largely overcome with the introduction in 1960 (Krige 1960) of the three-parameter lognormal, which requires an additive constant before taking logarithms. However, there were still cases which could not be covered by the three-parameter lognormal, and Sichel et al. (1992) introduced the more flexible compound lognormal distribution, originally developed by him for diamond distributions.

3.3 Spatial Concepts and the Birth of Geostatistics and Kriging

Geostatistics as such did not really originate until the basic concept of ore grades as a spatial variable, with a spatial structure, was introduced in 1951/1952 by Professor Krige.

This arose firstly in his endeavour to explain the experience seen on the South African gold mines for many decades, where ore reserve block estimates consistently showed significant undervaluation in the lower-grade categories and overvaluation for estimates in the higher-grade categories, during subsequent mining, i.e. what is now known as conditional biases, and illustrated below in the form of a simple diagram (Fig. 10). His pioneering work provided the geostatistical explanation of conditional biases, as unavoidable errors resulting from the use of limited data on the periphery of blocks, which were used in evaluating ore reserve blocks. He proposed and implemented corrective measures to eliminate these significant conditional biases. The regression corrections were applied routinely to block estimates on several mines in the early 1950s and represented the actual birth of kriging. The regressed estimate was, in effect, a weighted average of the peripheral estimate and the global mean of the mine section; it was the first application of kriging. It could be called 'simple elementary kriging', being based on the spatial correlation between the peripheral values and the actual grades of the

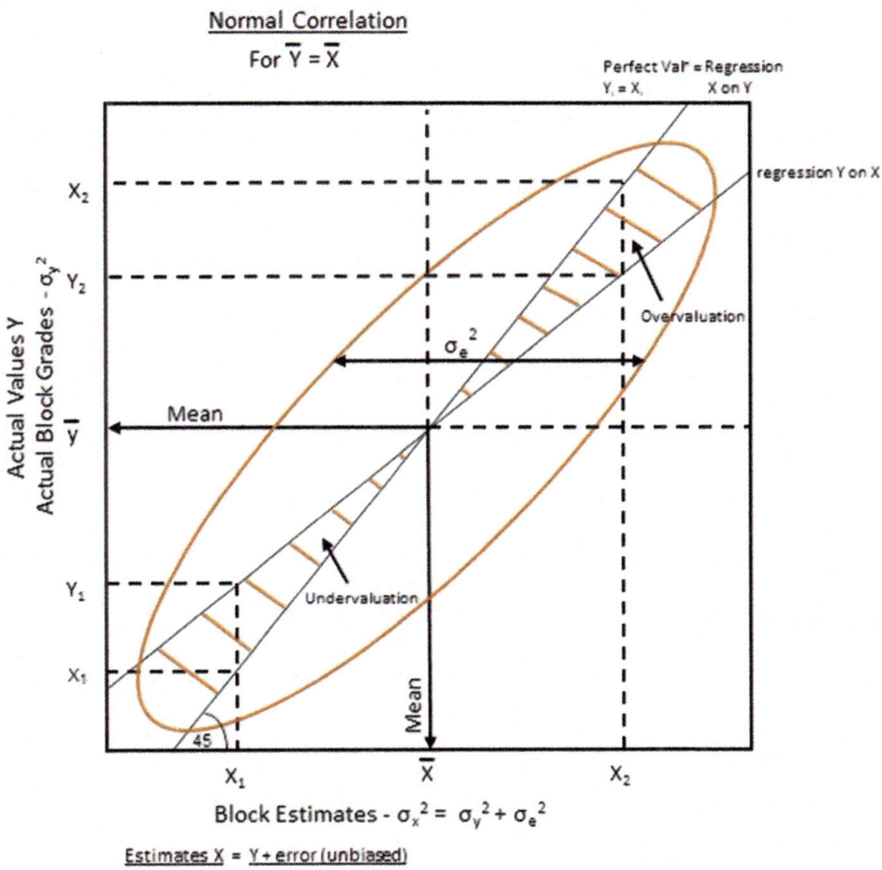

Fig. 10 Conditional biases

ore inside the blocks and giving proper weight to the data outside the block periphery via the mean. In this way, the spatial concept and kriging were introduced. The concept of 'support' is very basic to geostatistics and was first covered by Ross (1950) and further developed by Krige (1951), including Krige's variance size of area relationship.

Figure 10 illustrates conditional biases.

3.4 Spatial Structure and Variograms

Professor Krige's pioneering work in the early 1950s aroused interest worldwide, particularly in France where, under Professor Allais, Professor Krige's papers were republished in French (Krige 1955). One of Professor Allais students, later to become world renowned as Professor Matheron, started the development of the theory of regionalised variables. Matheron also then proposed the use of the variogram to define the spatial structure. This model is an extension and refinement of the concept covered by De Wijs (1951/1953). Professor Krige's regressed estimates were then still called 'weighted moving averages' until Matheron's insistence in the middle 1960s on the term 'kriging' in recognition of Professor Krige's pioneering work.

During 1963 to 1966 (Krige and Ueckermann 1963, 1966), the spatial patterns were defined in far more detail. These studies covered the spatial correlations between individual 'point' sample values, as well as those between regularised data blocks. The corresponding correlograms or covariograms were used on a simple kriging basis for block evaluations. Kriging on a routine basis for ore reserve evaluation was, therefore, already in use on some Anglovaal gold mines more than 50 years ago.

3.5 Conditional Unbiasedness

It is instructive to observe that on the South African gold mines, the improvement in the standard of block evaluations due to the elimination of conditional biases accounts for some 70 % of the total level of improvement achievable today, using the most sophisticated geostatistical techniques. It is for this reason that Professor Krige placed so much emphasis on the 'proper' implementation of the methods to mitigate conditional biases. Thus, it is critical that the elimination of conditional biases is not only the major contributor to the reduction of uncertainty in assessing the mineral resources of mining companies but also an integral and fundamental part of any kriging and mineral resource and reserve assessment process.

3.6 Conditional Biases

The elimination of these biases is basic to ore evaluation and all geostatistical procedures. David, in his 1977 popular geostatistics book ('Geostatistical Ore Reserve Estimation'), emphasised that the elimination of these biases is basic to ore evaluation and all geostatistical procedures. As per David, conditional unbiasedness is 'the key point of Krige's 1951 paper, one of the key points of his 1976 paper but even then, still appeared as a revelation to many people'.

3.6.1 What Contributes to Conditional Biases

Any increase in knowledge and available data relevant to any uncertainty being studied will reduce the level of uncertainty, provided the knowledge is applied properly. Knowledge will never be perfect and data never complete, and therefore uncertainty will never be entirely eliminated. However, any procedure or technique, which does not use all relevant data in order to provide the 'best' perspective on the remaining uncertainty, must not be accepted. Unfortunately, Professor Krige reported that in his worldwide experience, he encountered many cases where practitioners had erred in respect of this fundamental concept. In too many cases, mineral resources and mineral reserves were estimated on limited data, and relevant data was ignored. The use of insufficient data can still be a problem today. In 1950 only the peripheral data for each block was used, whilst now with the use of geostatistics, the data search routine is still often inadequate, even with the complete database available to the computer. This is compounded with often no advanced analysis to determine the minimum search routine required to eliminate the biases and no follow-up studies to record the presence of these biases and the need to eliminate them.

3.6.2 Practical Examples of Outcomes of Conditional Biases

The graphs and tables that follow, some also taken from Professor Krige's historical and practical work, illustrate the effect and outcomes of conditional biases.

Figure 11 illustrates feasibility block estimates versus final production blast-hole averages of an aluminium deposit, showing no correlation between the feasibility block estimates and that observed during production, as demonstrated by the regression trend, which could lead to significant risk in invested capital. Figure 12 illustrates similar conditional biases problems, demonstrates why they are important and shows how they result in misclassification of ore blocks, which lead to levels of profit well below what can be achieved. Figure 13 demonstrates the improved estimates of Fig. 12 analysis that can be achieved through using 'proper' kriging with an adequate search.

More recent practical examples of conditional biases are included in the tables and graphs that follow (Tables 1 and 2 and Fig. 14). Table 1, an open-pit historical mined-out case study, demonstrates that even the latest sophisticated geostatistics method used to estimate recoverable resources can suffer from inherent conditional biases. Table 2 shows the effect of conditional biases over time, from a historically mined-out case study, with consistently large negative percentage errors for tonnes and positive ones for grade, over various time periods and cut-offs. Figure 14 illustrates the financial impact the errors would have, over the respective cut-offs and time periods.

3.6.3 Conditional Biases: Testing Tools

Mineral resource estimation for a new or an existing mine covers two major stages:

- At the initial or first stage, the data is limited and is obtained either from a broad drill hole grid or from an initial main development grid.

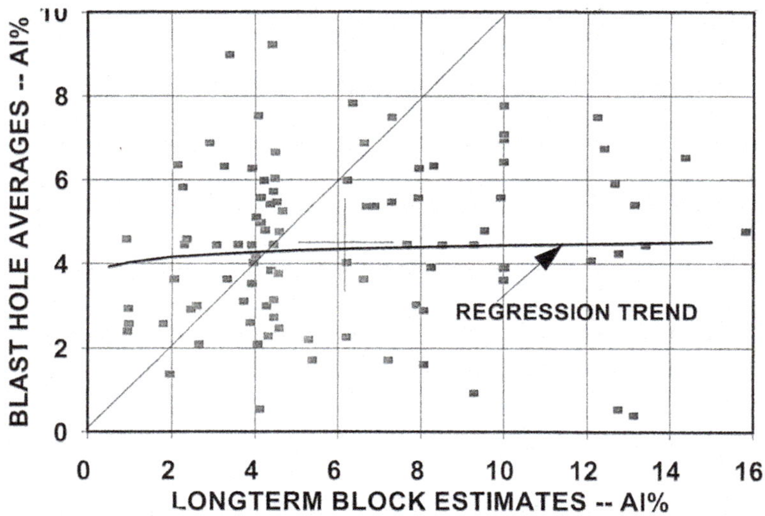

Fig. 11 Positively skewed, block estimates versus blast-hole averages

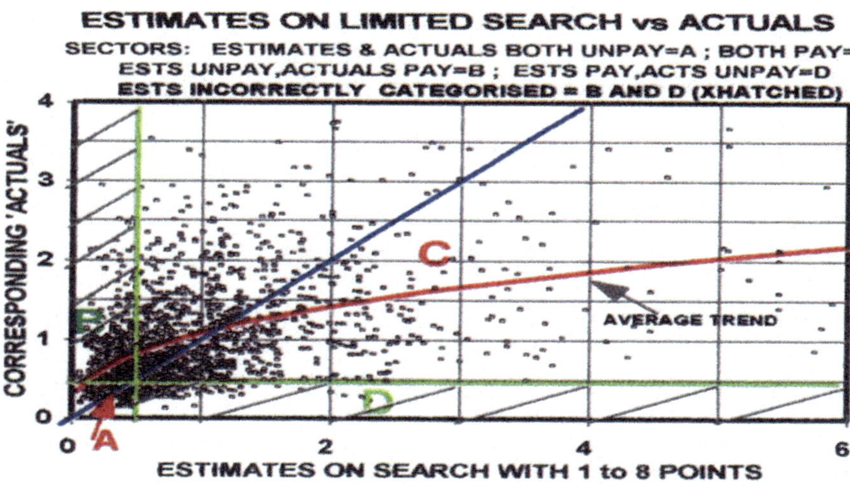

Fig. 12 Misclassification of blocks

– During the second or final stage, more data becomes available from grade control or from stope faces and auxiliary developments.

Apart from providing a basis for short- and longer-term mine planning and viability studies, evaluations are frequently required to provide resource and reserve classification figures (measured/indicated/inferred and proven/probable) and to substantiate a major capital investment and/or the raising of loans. At both stages of evaluation, the evaluation technique should ensure minimum error

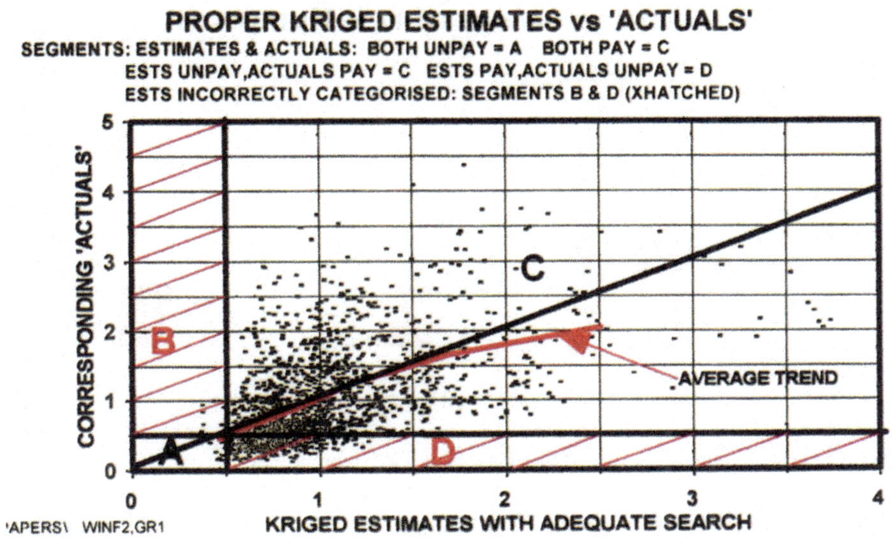

Fig. 13 Proper kriged estimates versus 'actuals'

Table 1 Geostatistical estimate with conditional biases – grades (g/t)

Recoverable LUC SMU estimates of blocks for grade categories	Corresponding average grade for actual blocks
(g/t)	(g/t)
12.0	3.0
6.0	2.3
0.3	1.7

Table 2 Geostatistical estimate with conditional biases – percentage errors

Percentage error of ore mined over respective time periods						
	6 months		1 year		3 years	
Cut-off	Tonnes	Grade	Tonnes	Grade	Tonnes	Grade
(g/t)	(%)	(%)	(%)	(%)	(%)	(%)
0.6	−15	31	−15	20	−15	16
0.7	−14	31	−16	22	−16	18
1.0	−3	27	−16	24	−17	19

variances/uncertainty. These requirements are linked closely to the expected slopes of regression of the eventual follow-up values on the original block estimates. Slopes of less than unity indicate the presence of conditional biases, with blocks in the upper-grade categories overvalued and low-grade blocks undervalued.

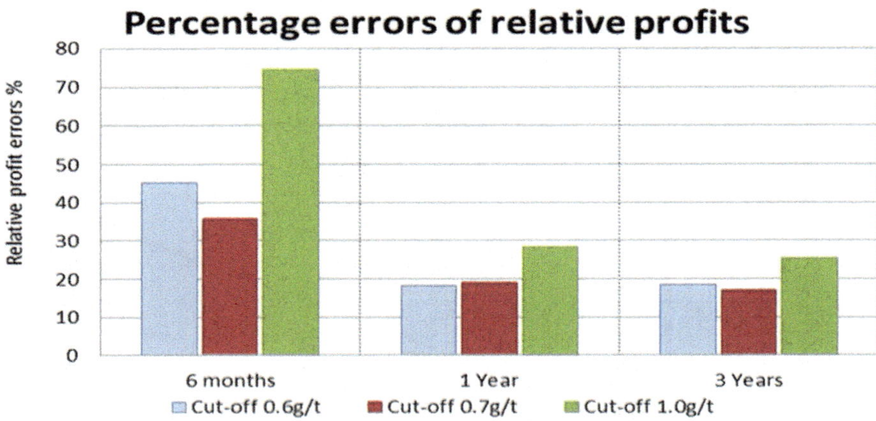

Fig. 14 Financial impact measured on the basis of relative profit

3.6.4 The Efficiency of Block Evaluations

Block evaluations subject to conditional biases have lower efficiencies. Professor Krige in 1996 proposed to define and measure the efficiency, expressed as a percentage, as follows:

$$\text{Efficiency} = (\text{BV} - \text{KV})/\text{BV} \tag{1}$$

where:

BV = block variance (i.e. the variance of actual block values, calculated from a variogram).
KV = kriging variance (i.e. the error variance of respective block estimates).

For perfect evaluations, KV = 0, the dispersion variance (DV) of the estimates (calculated from the observed kriged model) = BV, and then:

$$\text{Efficiency} = \frac{(\text{BV} - 0)}{\text{BV}} = 100\% \tag{2}$$

where only a global estimate of all blocks is practical, all blocks will be valued at the global mean, i.e.:

$$\text{DV} = 0,\ \text{KV} = \text{BV},\ \text{and Efficiency} = (\text{BV} - \text{BV})/\text{BV} = 0\,\%. \tag{3}$$

Usually blocks are valued imperfectly. With no conditional biases:

$$\text{DV} = \text{BV} - \text{KV},\ \text{and Efficiency} = (\text{BV} - \text{KV})/\text{BV} = \text{DV}/\text{BV} \tag{4}$$

However, with conditional biases present, this relationship does not hold, and then because of insufficient smoothing:

$$\mathrm{DV} > (\mathrm{BV} - \mathrm{KV}) \tag{5}$$

and:

$$\text{Efficiency} = (\mathrm{BV} - \mathrm{KV})/\mathrm{BV} < \mathrm{DV}/\mathrm{BV} \tag{6}$$

The efficiency of a block evaluation can even be negative if KV > BV. As stressed by Professor Krige, such a situation is unacceptable, and the block evaluations will be worthless; yet he encountered several such cases in practice, where the data accessed per block was inadequate.

3.6.5 Critical Control Limit Test for Kriged Block Evaluations

In order to avoid this unacceptable negative efficiency for block estimates, the following critical control limit test is proposed for the regression slope to test for conditional biases (Assibey-Bonsu and Muller 2014):

The regression slope can be written as:

$$\text{Regression slope} = (\mathrm{BV} - \mathrm{KV} + \xi \mathrm{LM})/(\mathrm{BV} - \mathrm{KV} + 2\xi \mathrm{LM}) \tag{7}$$

where: $\xi = +1$ or -1 depending on the sign of LM (ie, the kriging system set-up) and:

LM = respective Lagrange multiplier for ordinary kriging, and BV and KV are as defined above.

Where only a global estimate of all blocks is practical, all blocks will be valued at the global or subdomain mean, i.e. KV = BV and Efficiency = 0.

Substituting KV = BV into Eq. 7:

$$\text{Regression slope} = \xi \mathrm{LM}/2\xi \mathrm{LM} = 0.5 \tag{8}$$

Thus, a regression slope of less than 0.5 will always lead to a negative block efficiency estimate (i.e. worthless kriged estimates). This highlights the danger of accepting block estimates that have a slope of regression less than 0.5.

The critical regression slope limit of 0.5 should only be used to identify blocks that will result with negative kriging efficiencies. Ideal slopes of regression should be greater than 0.95, as proposed by Krige (Krige 1996).

An extensive study of some 70 cases by Professor Krige covering a wide range of spatial and data patterns used indicated a correlation between kriging efficiency and the regression slope (actuals on estimates) of 87.5 % (Krige 1996). Thus, the slope (or the extent of conditional biases present) effectively incorporates all the major factors affecting the efficiency of block evaluations.

3.6.6 Smoothing Effect of Kriging

The absence of conditional biases is unavoidably accompanied by some smoothing, and it is a fallacy to use the data search routine for block evaluation in an endeavour to reduce or eliminate it. Smoothing is inevitable and essential for conditionally unbiased estimates.

Explaining this in terms of a theoretical approach by reviewing the definition of the slope of the line of regression of actuals (Y) on estimated block values (X):

$$\text{Slope} = r\left(\sigma_y/\sigma_x\right) \tag{9}$$

where σ_y and σ_x are the respective standard deviations of actuals and estimates, respectively, and r is the correlation coefficient.

If the slope is to be unity (i.e. slope = ~1) for unbiased block estimates and (r) is less than unity, because estimates are never perfect, then:

$$\left(\sigma_y/\sigma_x\right) > 1 \tag{10}$$

i.e. the standard deviation (or variance) of the actual or real block values must be larger than that of the estimated block values. The gap between these two variances (the smoothing effect) can therefore only be reduced by increasing the correlation (r) between block estimates and actual values, i.e. by improving the efficiency of the estimation technique or by providing more data. No mathematical manoeuvring can achieve this objective.

Various post-processing techniques are available to remove smoothing effects (e.g. Assibey-Bonsu and Krige 1999; Journel et al. 2000) and should be applied only to block estimates that are conditionally unbiased.

4 Conclusion

4.1 Professor Danie Krige's Basic Points of Advice for the Practitioner

(i) Ensure data integrity.

Disastrous errors and critical risks will result by using erroneous data. Various processes that are usually set out in company or mineral resource regulatory body standards and protocols should be followed to ensure overall data integrity. The SAMREC Code, Table 1, provides good guidelines as to those aspects that should be considered and reported on in the relevant competent person reports.

(ii) Establish all the necessary geological and geostatistical models and parameters.

Geology should always be recognised as a vital element in deposit modelling. Experience has shown that geostatistical mineral resource and reserve assessment, without proper geological input, can also be disastrous and constitute a critical risk. A robust geological model is therefore a prerequisite, and if the geostatistical model does not agree with the geological one, there are grounds for serious concern. Either one or both models should be critically re-examined, so as to establish the essential correlation and validation.

(iii) Technique selection and optimal application.

Choose an appropriate geostatistical technique, and determine the search routine required to eliminate conditional biases and to ensure optimal application.

Use effective tools, including slope of regression and block efficiencies in this regard.

(iv) As blocks are mined out, do follow-up (reconciliation) studies to validate the mineral resource estimates.

This is a further important aspect, not only to ensure that estimates have the quality required and that no biases are present but also to timeously record the differences and facilitate corrective action.

(v) Research new techniques and applications, but validate new techniques properly by way of (follow-up) checks to confirm the absence of biases and the practical advantages to be gained when they are applied in practice.

4.2 *Final Thoughts*

The industry seems to be going backwards in certain areas, with a widely spread misunderstanding of the causes and consequences of conditional biases. The following are some of the possible causes:

- In certain universities, as well as training provided elsewhere in the industry, geostatistics is taught using commercially available computer programmes, with the emphasis being how to use the programmes.
- Unfortunately, this is what many mining companies expect: graduates or practitioners who are good at operating programmes ('black-box approach'). This does not give much time to teaching the fundamentals of geostatistics and the consequences of misusing the technology.
- What complicates matters is that, certain universities rarely have large databases to demonstrate the strengths and weaknesses of various methods in different environments and research is by its own nature geared towards only development of theoretical geostatistics, often based on strong stationarity assumptions.

> ...after half a century of phenomenal developments in geostatistics, conditional biases which gave birth to this subject, are still encountered in practical applications... the main concern is that this record will be tarnished by the all too ready acceptance (in certain cases) of estimates, which are still conditionally biased. For the future, I would like to see geostatistics continue to grow from strength to strength with new models, techniques and

Fig. 15 Professor Danie Gerhardus Krige

applications, but where these are all validated properly by way of (follow-up) checks to confirm the absence of biases and the practical advantages to be gained when they are applied in practice.... Professor Danie Krige.

Professor D. Krige was indeed a pioneering giant, which the South African mining industry was blessed to have had for his immense contributions. He always gave willingly and unselfishly, with the rewards not gold, platinum and diamonds but the tools for others to utilise in finding and evaluating mineral resources, so as to achieve a positive financial return whilst minimising the associated risk. He took the industry far along the road, but the journey is not over, and it now remains the responsibility of those that follow, to both adhere to his principles and indeed continue to build thereon, to ensure his legacy lives on.

Figure 15 shows Professor Danie Gerhardus Krige, who indeed made the world a better place.

Acknowledgements The author wishes to thank Gold Fields Limited for the support and time it has allowed in collating and presenting this paper.

Bibliography

Annual Report (2007) Gold fields limited 2007 annual report. p. 18. 30th June 2007

Assibey-Bonsu W, Krige DG (1999) Use of direct and indirect distributions of selective mining units for estimation of recoverable resource/reserves for new mining projects. APCOM'99 International Symposium, Colorado School of Mines, Golden, October 1999

Assibey-Bonsu W, Muller C (2014) Limitations in accepting localised conditioning recoverable resource estimates for medium term, long-term, and feasibility-stage mining projects, particularly for sections of an ore deposit. J South Afr Inst Min Metall 114:619–624

Bouw B (2011) Newmont to acquire Fronteer Gold in $2.3-billion deal. The Globe and Mail. 3rd Feb 2011

David M (1977) Geostatistical ore reserve estimation. Elsevier Scientific Publishing Co, Amsterdam

De Wijs HJ (1951/1953) Statistical ore distribution. Geologie en Mijnbouw. November 1951, January 1953

Hill L (2013) Barrick to suspend work on Pascua-Lama to conserve cash. Bloomberg. 31 Oct 2013

Jamasmie C (2014) Argentina lobbies to overturn Barrick's Pascua Lama freeze in Chile. www.mining.com. 15th Feb 2013.

Journel AG, Kyriadkidis PC, Mao S (2000) Correcting the smoothing effect of estimators: a spectral postprocessor. Math Geol 32(7):787–813, October 2000

Krige DG (1951) A statistical approach to some basic mine valuation problems on the Witwatersrand. J Chem Metall Min Soc S Afr 52:119–139

Krige DG (1952) A statistical analysis of some of the borehole values in the Orange Free State goldfield. J Chem Metall Min Soc S Afr 53:47–64

Krige DG (1955) Travaux de M.D.G.KRIGE sur l'evaluation des gisements dans les mines d'or sudafricaines. Annales des Mines. December 1955

Krige DG (1960) On the departure of ore value distributions from the lognormal model in South African gold mines. J S Afr Inst Min Metall 61:231–244

Krige DG (1966) Two-dimensional weighted moving average trend surfaces for ore valuations. Proc. Symp. on Math. Stats. and Comp. Appls. In Ore Valn. pp 13–38. S Afr Inst Min Metall

Krige DG (1978) Lognormal-de Wijsian geostatistics for ore valuation. S Afr Inst Min Metall. Monograph No. 1. 1978

Krige DG (1996) A practical analysis of the effects of spatial structure and data available and used, on conditional biases in ordinary kriging. 5th Int. Geostatistics Congress, Wollongong. 1996

Krige DG, Ueckermann HJ (1963) Value contours and improved regression techniques for ore reserve valuations. JSAfr Inst Min Metall 63:429–452

McCrae MA (2013) The worst mining deal ever: Rio Tinto buying Alcan for US$ 38.1 billion. www.mining.com. 15th Feb 2013

Minnitt RCA (2012) Thoughts from an interview Professor D. Krige gave to Professor R.C.A. Minnitt during 2012. Personal correspondence

News.com.au (2012) BHP Billiton's decision won't be covered by tax hike says SA Premier Jay Weatherill. www.news.com.au/finance/business/. 23rd Aug 2012

Ross FWJ (1950) The development of some practical applications of statistical value distribution theory for the Witwatersrand auriferous deposits. MSc (Eng.) thesis, University of the Witwatersrand, Johannesburg

Sichel HS (1947) An experimental and theoretical investigation of bias error in mine sampling with special reference to narrow gold reefs. Trans Inst Min Metall Lond 56:403–473

Sichel HS (1952) New methods in the statistical evaluation of mine sampling data. Bull Inst Min Metall Lond 61:261–288

Sichel HS, Kleingeld WJ, Assibey-Bonsu W (1992) A comparative study of the frequency distribution models for use in ore valuation. J S Afr Inst Min Metall. April. pp 91–99. 1992

Trustcott SJ (1929) The computation of the probable value of ore reserves from assay results. Trans Inst Min Metall Lond 39:482–496

Watermeyer GA (1919) Application of the theory of probability in the determination of ore reserves. J Chem Metall Min Soc S Afr 19:97–107

Part II
Theory

Functional Decomposition Kriging for Embedding Stochastic Anisotropy Simulations

J.A. Vargas-Guzmán and B. Vargas-Murillo

Abstract Functional analysis of the kriging algorithm is accomplished with *consecutive projections* of vectors in Hilbert space. The innovation unveils "functional decomposition kriging" (FDK), which can forecast fields on *spatially continuous* domains without using blocks, cells, or elements. FDK assembles the random field as a summation of *field analytic functions*, which are sample pivoted and nonstationary. Furthermore, spatially variable *uncertain anisotropy* is represented as a continuous tensor random field, which is formed from *non-orthogonal members*. FDK predicts tensor members using physical data collected at sparse sample locations. Particular interest is on *structural anisotropy tensor fields* representing curvilinear and folded patterns of *structural uncertainty*. Therefore, spatially variable eigenvector and eigenvalue *tensor fields* give continuously varying orientation and range of principal stochastic anisotropy of covariances that are used as input to *stochastic functionals*. FDK enables simulation of anisotropic properties (e.g., permeability, rock stiffness, or structural anisotropy), with *stochastic covariance parameter fields*. Integration of *field analytic functions* delivers upscaled multiresolution moments. Since FDK can be stopped, optimized, and updated without repeating computations, it is suitable for inverse, adaptive, and real-time modeling.

1 Introduction

Spatially variable and *uncertain anisotropy* can be represented as a *tensor random field* related to gradients of a property. For example, normal vectors of hydraulic potential or gradients of folded elevation can be converted into contour and

J.A. Vargas-Guzmán (✉)
Reservoir Characterization, Saudi Aramco, E-5140, 31311 Dhahran, Saudi Arabia
e-mail: vargasja@aramco.com

B. Vargas-Murillo
Reservoir Engineering Contractor, L.369, 31311 Dhahran, Saudi Arabia
e-mail: vargmur@gmail.com

J.J. Gómez-Hernández et al. (eds.), *Geostatistics Valencia 2016*, Quantitative Geology and Geostatistics 19, DOI 10.1007/978-3-319-46819-8_2

streamline vectors (i.e., complex random fields). Consequently, structural uncertainty studies involving curved patterns (e.g., sinuous sequences, meandering, braiding channels, or folded structures) must use models that match measurable local physical anisotropy data. Furthermore, spatially variable tensors can describe orthogonal, monoclinic, or triclinic anisotropy that require careful modeling.

Forecasting properties that contain data about their spatially variable directional gradients is a nontrivial problem. For example, kriging orientation tensors in embedded spaces require isotropy assumptions (van den Boogaart and Schaeben 2002). If the tensors represent correlation anisotropy, the problem demands a way to introduce variable anisotropy data into the estimation algorithm. A limitation of the kriging estimator of probability parameters for random fields is its classic formulation using constant anisotropy in a stationary and deterministic covariance model, $c(h)$. Routinely, $c(h)$ is defined on the domain of the Euclidean separation distance h, with deterministic range and variance parameters along the principal axes of orthorhombic anisotropy. However, the single anisotropy condition is not suitable for modeling random fields that should represent real patterns pertaining to spatially variable tensor anisotropy data. Hence, embedding stochastic anisotropy fields into geostatistical modeling through covariance functions with spatially variable stochastic parameters is a sought-after option for simulated realizations of random fields.

Various authors treat the curvature of spatial patterns by considering anisotropy parameters that change with spatial locality. For example, kriging and simulations with rotated anisotropic variograms for each local neighborhood were proposed (Soares 1990), following a previously simulated angular direction of local anisotropy, after conventional geostatistical facies modeling (Xu 1996). Kriging with a local variogram in an elliptic neighborhood with rotated anisotropy orientations was subsequently implemented (Horta et al. 2010). The data-versus-data covariance for the same pairs of samples entering into estimation at two adjacent locations with partially overlapping neighborhoods would conflictingly pertain to different rotated local anisotropies. Furthermore, the approach is not flexible enough in higher resolutions with sparse data because it requires all samples within the neighborhood to share the same anisotropy parameters.

Assuming the curvilinear directions of meandering streamlines are known a priori (Stroet and Snepvangers 2005), splitting a 2D simulation domain into a series of 1D independent spectral simulations along the cells connected by each streamline was proposed (Yao et al. 2007). Spectral simulation does not use neighborhoods, so it could reproduce the spatial continuity structure of simulations better than sequential simulations; however, posterior conditioning to data is ad hoc in the realizations (Gutjahr et al. 1997). Assuming independence of properties between different streamlines is unrealistic due to essential lateral continuity (e.g., depositional accretions).

Straight distances between locations can be replaced by a sum of segments along local directions (Boisvert et al. 2009). Curvilinear distances were computed between each pair of locations using locally varying anisotropy directions and produced contoured paths with guarded Newton optimization. Estimations were performed by taking the domain to a higher dimensional isotropic space, where the covariance should yield positive definite matrices, by avoiding non-Euclidean space (Curriero 2005). Although large computers may accelerate the multidimensional scaling, the approach requires the input of *deterministic* orthorhombic variogram anisotropy orientations, which must be known a priori at every location.

Anisotropy of covariances is predictable using inverse methods. The anisotropy field can be modified by correlating errors in the expected response attribute with sensitive modifications in the predicted anisotropy parameters. An example is that errors in predicted versus measured rock proportions in new drilled wells are used to enhance knowledge about anisotropy directions (Dossary and Vargas-Guzmán 2011). Reservoir models are reported to be made better by changing the orientation of geostatistical anisotropy (Caeiro et al. 2015).

Predictions of anisotropy tensors are known in computational fluid dynamics (CFD), but the results do not match field samples in detail (e.g., storm patterns, turbulence forecast, and sedimentary dynamics). Turbulent 3D flow models with physically realistic streamlines are unavailable (Ladyzhenskaya 2003). Therefore, forecasting components of non-orthogonal anisotropy tensors, with geostatistics embedding variable anisotropy at non-sampled locations, is a necessary alternative to vorticity prediction. Additionally, it is essential to embed the uncertainty of anisotropy in geostatistical predictions, considering the observed parameters for the directional covariances are spatially variable, uncertain, and usually continuous.

In this study, the spatially stationary single anisotropy constraint is circumvented by developing a truly functional analysis of the kriging algorithm. The approach is achieved via functional Hilbert space decompositions of the random field that yield continuous *field analytic functions*. Vector components of tensors with uncertain anisotropy are also predicted as non-orthogonal members. They enable stochastic parameters for input covariance functions.

2 Theory

2.1 Functional Decomposition of the Random Field Z(x)

A set of random variables $\{Z(x_i)\}$, each at a given physical location, x_i, conforms a spatial random field, $Z(x)$, which is defined in the probability space $\{\mathcal{B}, \Omega, p\}$, where p is the probability measure. The sample space Ω and Borel algebra $\mathcal{B}$ conform a measurable space. Each random variable is a vector Z_i in L_2 Hilbert space. The random variables $Z_{i=q}$ and $Z_{i=r}$ are related pairwise by inner products $Z_q \cdot Z_r$ in L_2.

A stationary and deterministic covariance function of a lag distance $c(h)$ (in general, a stationary co-cumulant function of higher-order $\kappa(h_1, \ldots h_q)$) does not exploit the possibility of spatially variable anisotropy parameters, as observed in natural phenomena (Vargas-Guzmán 2011). This is due to the variable elongation and rotation of natural correlation structures, which complicate simple mapping with stationary covariances. In order to handle these larger degrees of freedom for anisotropy in covariance functions, each location must be disassembled from the neighboring locations. The analysis follows.

Consider that the random field $Z(x)$ is the sum of regionalized addible factor fields. One option is to model each field component using a Fourier-Stieltjes representation $\{dZ_1(\omega), dZ_2(\omega)\cdots dZ_n(\omega)\}$ (Gutjahr et al. 1997) as a band in the frequency domain ω. A question arises: can components be frequency and amplitude modulated to handle spatially variable anisotropy? A powerful representation for each factor can be a "conditional component field," which removes any redundancy due to non-orthogonal components. In this representation, $Z(x)$ is a combination of conditionally independent random fields $\{R_1(x), R_2(x), \cdots, R_n(x)\}$ termed conditional components and their projections. Instead of the multiple independent spectra, "conditional spectra" is computed in the frequency domain for individual $R_j(x)|j = \{1\ldots n\}$. Each conditional power spectrum is inverse Fourier transformed into physical space to yield a valid covariance function, $c_{R_j}(h)$, for separate geostatistical modeling of each component (Vargas-Guzmán 2003).

A new contribution from our present study is to consider that the components may be *field analytic functions*, and they add to a single scalar field with nonstationary covariance that can be generalized with modulated stochastic functionals depending on physical locality. The following analysis of a random field develops such an additive model.

The standard deviation, σ_q, of a random variable $R_j(x_q)$ at location x_q, where $R_j(x_q) \in R_j(x)$, is the norm or vector modulus $|R_q|$ in L_2, and the inner product is $R_q \cdot R_r = \rho_{q,r}\sigma_q\sigma_r$, where $\rho_{q,r}$ is the linear correlation. Subsequently, a field with a stationary covariance function is represented as a sphere of vectors in L_2, with the norm supremum, $sup\|R_j(x)\|$, where $R_j(x) \in L_2$. A complete case is if $\|R_j(x_q)\| \neq \|R_j(x_r)\|$, which allows for an open ball with either smaller or larger vectors among the neighboring random variables. Thus, the sphere can be deformed with an irregular radius, but it requires prescribed smoothness and consistency.

Applying the representation theorem (Riesz 1907), a linear stochastic functional, $f_\alpha(x)$ for a single $j = k$ scalar component, is mapped in physical space from L_2 relating a vector $R_\alpha^{j=k}$ for $R_j(x_{i=\alpha})$ to all other vectors, $R_i^{j=k}$ for $R_j(x_i)$, in L_2. This isomorphic mapping is abbreviated,

$$f_\alpha(x) : R_{j=\alpha} \rightarrow R_j \tag{1}$$

Hence, a single function $\dot{f}_\alpha(x - x_\alpha)$ pertaining to $f_\alpha(x)$ does not need to be stationary in physical space, because it entails a vector $R_\alpha^{j=k}$ at a fixed physical location related to any other $R_i^{j=k}$ in a scalar random field, $R_j(x)$.

The bounded stochastic functional $f_\alpha(x)$ is defined in a finite-dimensional subspace $S_2 \subset L_2$, and can be extended to L_2 as more information becomes available

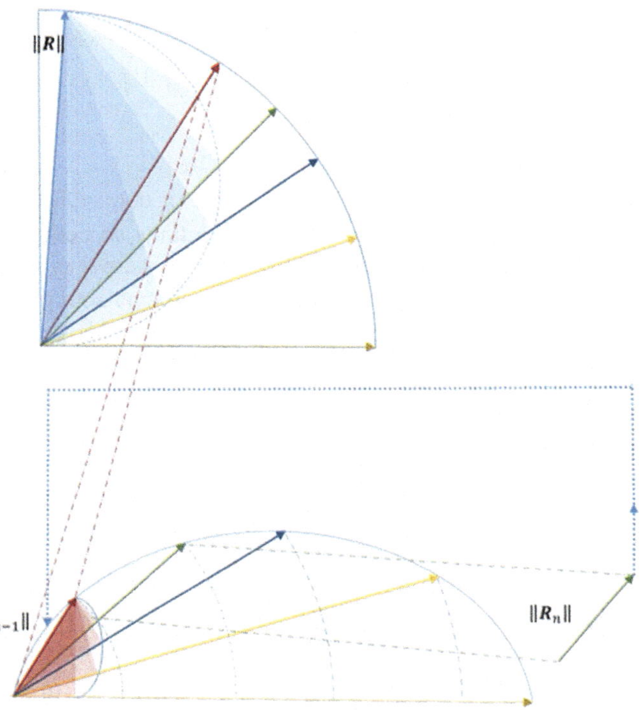

Fig. 1 Scheme of consecutive projections of norms to decompose normed space

(e.g., samples and simulated values). This is similar to the Hahn-Banach extension theorem.

Projection of a vector $R_{i=\alpha}$ onto another neighboring vector $R_{i=q}$ is an application of the orthogonal projection theorem. Furthermore, a projection in Hilbert space is applied to comfortably minimize the residual norm. For example, $\| R_{2|\alpha_1}(x_{\alpha_2}) \| = \| R_1(x_{\alpha_2}) \| - w_{\alpha_1,\alpha_2} \| R_1(x_{\alpha_1}) \|$, where $R_1(x_{\alpha_2})$ is the sampled random variable from unconditional random field $R_1(x)$ at location x_{α_2}, and w_{α_1,α_2} is the cosine between the vectors corresponding to variables $R_1(x_{\alpha_1})$ and $R_1(x_{\alpha_2})$. The projections of $R_{2|\alpha_1}(x_{\alpha_2})$ on the physical domain generate a new random field, termed $R_2(x)$, which avoids any redundancy from a prior field analytic function based on $R_1(x_{\alpha_1})$. In general, a stochastic realization of the functional $f_{\alpha_n}(x)$ must optimize $\| R_{n|\alpha_{\dots}}(x) \|$.

Remark 1 The projection theorem is generalized as a consecutive process (Fig. 1), where the component norm residual is from a domino effect, simplified in terms of uncorrelated components

$$\| R_{n|\alpha_1\ldots\alpha_{n-1}}(x_{\alpha_n}) \| = \| R_n(x_{\alpha_n}) \| - \Big(w_{\alpha_1,\alpha_n} \| R_1(x_{\alpha_1}) \| + w_{\alpha_2,\alpha_n} \| R_{2|\alpha_1}(x_{\alpha_1}) \| \\ \cdots + w_{\alpha_{n-1},\alpha_n} \| R_{n-1|\alpha_1\ldots\alpha_{n-2}}(x_{\alpha_{n-1}}) \| \Big) \tag{2}$$

Remark 2 An innovative stochastic functional $f_{\alpha_n}(x)$ is defined in the same fashion as conditional components and is extended to the infinite space L_2.

The innovative stochastic functionals are addible because they are orthogonal to each other; therefore, they can hold embedded variable anisotropy parameters for each bounded functional, provided that the total norm $\|R_j(x)\|$ for an attribute conforms a Cauchy series, at each location.

Remark 3 The posterior random field, made of variables $R_n(x)$, also has an abstraction in Hilbert space, L_2, as a new deformed ball surface (Fig. 1); regardless of that starting $R_{n-1}(x)$ variables form or not a sphere in L_2.

The space of the linear stochastic functional is the dual L_2^* of L_2; therefore, the linear model explained for the consecutive projection of the norms is applicable to the random variables as well as data. The fields $R_n(x)$ and $R_{n-1}(x)$ are generated as continuous projection of orthogonal residual data $R_{n|\alpha_1\ldots\alpha_{n-1}}(x_{\alpha_n})$ and $R_{n-1|\alpha_1\ldots\alpha_{n-2}}(x_{n-1})$. Each component follows a cosine function in Hilbert space (see Fig. 1), or a weight function that needs to be expressed in physical spatial coordinates, following Riesz representation.

The functions can also be formalized with addible spectral components, similar to modulated Fourier-Stieltjes representations. They can be generalized as complex components in frequency domain.

Remark 4 Stationarity of covariance is not a requirement of Hilbert space abstraction for the random field. The nonstationarity of innovative functionals for residual components is evident. Figure 2 is a representation of analytic functions from five successive functionals that can host stochastic parameters. A normed space is their complete representation, which embeds all the component normed subspaces.

2.2 Functional Decomposition Kriging (FDK)

In the analytical functional decomposition kriging (FDK), as created in this paper, sparse kriging weights are replaced by continuous functions of spatial coordinates that relate each orthogonal sample component to all point locations in the field. Consecutive projections on a continuous 3D domain or infinite set of point locations, $x = (x_{min}, x_{max}) \forall x \in \mathbb{R}^D$, lead to a summation of mappings. The weight functions $\vartheta_i(x) : R_{\alpha_i} \to R_i$ yield the total estimate as a spatial series of field analytic functions, which is

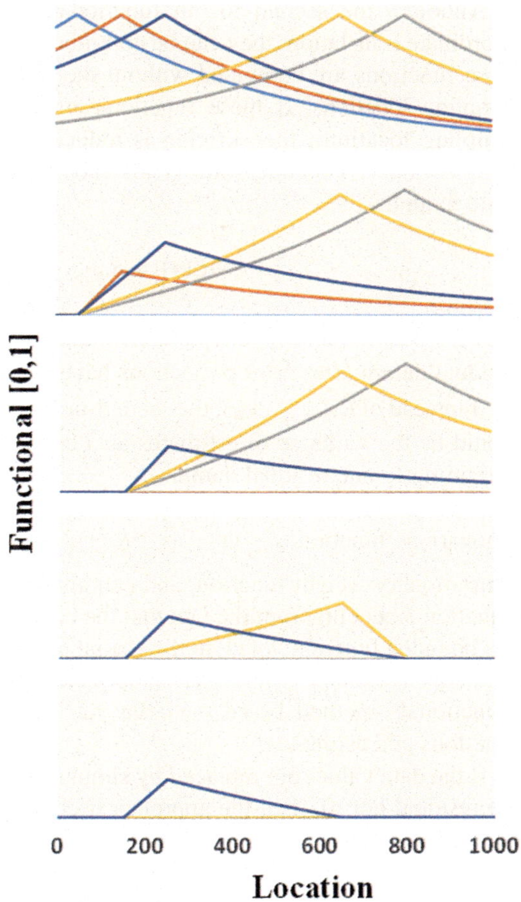

Fig. 2 Example of analytic functions for five functionals corresponding to consecutive planned sampling locations 50, 150, 800, 650, and 250

$$\hat{z}(x) = \left(\dot{R}_{\alpha_1}(x_{\alpha_1}) \cdots \dot{R}_{\alpha_n|\alpha_{n-1}\ldots}(x_{\alpha_n})\right) \begin{pmatrix} \vartheta_1(x) \\ \vdots \\ \vartheta_n(x) \end{pmatrix} \tag{3}$$

Where $\dot{R}_{\alpha_i|\alpha_{i-1}\ldots}$ represents data values corresponding to the samples of random variables with residual moduli, $\| R_i(x) \|$, in dual space, as shown in Fig. 1. The continuous weight functions, $\vartheta_1(x)$, are cosine values of the angle between pairwise vectors (i.e., equivalent to conditional correlation functions) used for estimation with each sample.

$$\vartheta_i(x) = \dot{f}_{i|\dot{\alpha}_i}(x) / \dot{f}_{i|\dot{\alpha}_i}(x_{\dot{\alpha}_i}) \tag{4}$$

Although the second to nth functionals cannot be invariant functions upon coordinate translation, they can be continuously rotated in 2D or 3D. Nevertheless, these functions are computed without the use of actual sample values, but using sampling locations. If these functions are known and anchored to all possible sampling locations, then kriging is reduced to scaled weight functions with the sample values. This operation yields each field analytic function component estimate model:

$$\hat{R}_i(x) = \dot{R}_{\alpha_i|\alpha_{i-1}\ldots}(x_{\alpha_i})\vartheta_i(x) \tag{5}$$

$\hat{R}_i(x)$ is continuous in the spatial domain x. Note that $\dot{R}_{\alpha_i|\alpha_{i-1}\ldots}(x_{\alpha_i})$ is a new sample residual datum after prior projections have been removed from the datum.

If instead of $\dot{R}_{\alpha_i|\alpha_{i-1}\ldots}(x_{\alpha_i})$, the second-order covariance is used in Eq. 5, the result would be the variance of estimates as a continuous spatial field analytic function pertaining to one residual datum.

The weights are computed from the standardization of an anchored linear innovation function, $\dot{f}_{i|\check{\alpha}_i}(x)$, as $\vartheta_i(x)$, for $i = 1\ldots n$, which represents a nonstationary weight function, and can also be interpreted as a local hat function. Equation 5 benefits from the fact that the cosine of the angle between vectors in L_2, is a bounded function, $\vartheta_i(x)$, in the spatial domain, x, which has a Fourier transform but is not stationary. This equation upgrades the estimation approach to an entirely "functional" method based on large yet simple analytic expressions for weight functions and estimates.

If the data values are replaced by simulated residuals (i.e., white noise scaled by the residual norm), then the approach becomes a functional simulation of orthogonal nonstationary parametric field analytic functions.

The FDK approach follows the consecutive projection process described in the previous section. The first functional, anchored at the first sample location, $\dot{f}_{1|\check{\alpha}_1}(x)$, is expressed on the physical space or domain x. The subindex $1|(\check{\alpha}_1)$ indicates that it is the first component f_1 functional, and the distances are offsets, e.g., $|x - x_{\check{\alpha}_j}|$. The first datum is projected onto the entire field as an analytic function. Computing a residual datum at the second sampling location removes $\widehat{R}_1(x)$ from the estimation, which yields $\dot{R}_{\alpha_2|\alpha_1}(x_{\alpha_2}) = \dot{R}_{\alpha_2}(x_{\alpha_2}) - \widehat{R}_1(x_{\alpha_2})$. Projecting the new nonredundant datum $\dot{R}_{\alpha_2|\alpha_1}(x_{\alpha_2})$ onto the entire field requires an updated weight function. When a second sample is collected, the Hilbert space L_2 would give $\dot{f}_{2|(x|\check{\alpha}_2)}(x) : R_{\alpha_2|\alpha_1} \to R_{2_j}$. A straightforward way to compute this function, as a stepwise optimization, is to anchor a function, $\dot{f}_{2|(x|\check{\alpha}_2)}(x)$. At this stage, $\dot{f}_{1|(x|\check{\alpha}_2)}(x)$ is defined as a hat function at location α_2, which may not be stationary in relation to $\dot{f}_{1|(x|\check{\alpha}_1)}(x)$, centered at the first sampled location; however a symmetry condition exists, $\dot{f}_{1|(x|\check{\alpha}_1)}(x_{\alpha_2}) = \dot{f}_{1|(x|\check{\alpha}_2)}(x_{\alpha_1})$. Consequently, following Eq. 4, the component estimate function is $\hat{R}_2(x) = \dot{R}_{\alpha_2|\alpha_1}(x_{\alpha_2})\vartheta_2(x)$. A third sample is assimilated in the same fashion. Hence, the innovative functional is generalized as

$$f_{n|(x|\dot{\alpha}_m)}(x) = f_{n-1|(x|\dot{\alpha}_m)}(x) - \vartheta_{n-1}(\alpha_m) f_{n-1|(x|\dot{\alpha}_{n-1})}(x) \tag{6}$$

Equation 6 enables all weight functions, $\vartheta_n(x)$, and field analytic functions $\hat{R}_i(x)$, estimated or simulated, to be stored in analytical form, provided that all the required priors have been previously stored. Computer saving entire analytical models is analogous to storing the algorithm parameters, but alternatively, a numerical solution can also be stored at ultrahigh resolution, and it can contain upscaled values. Nevertheless, updating models is easier if analytic weights are stored. This analytic type of FDK and functional decomposition simulation (FDS) model can be automatically updated by inputting new sample coordinates and a new realization of stochastic covariance parameters. The developer has to decide if the computer resources required are available to store all the produced field analytic functions or alternatively target numerical realizations.

A specific analytical expression for the functional requires analytic consistency. For example, using the exponential model for the covariance, the first function serves as input in all consecutive functions. It is straightforward to replace second stage component functionals to obtain third stage functionals and so forth.

FDK process is handled with a program loop in the implementation algorithm. Note that instead of inputting range parameters a_1, a_2, and a_n into the equations, the distances can be stretched or scaled in rotated coordinates, during numerical display generation. The transformed coordinates have to remove the anisotropy, as in the classic orthorhombic case, though separately for each field analytic function. The monoclinic and triclinic deformations have been handled with Cartesian equivalent coordinates.

2.3 *Kriging Data from Functions*

FDK, as introduced for the first time here, is the result of Riesz mapping on continuous space and/or time after functionals pertaining a random field in normed Hilbert space are analytically decomposed. FDK generates an explicit spatial equation for a random field in an nth dimensional domain, which can be inferred as a complete *field analytic function*. FDK enables the continuity of the field along all spatial dimensions including time, but is *quantized* by the number of discrete samples and simulated values introduced in the prediction. FDK avoids the use of matrix algebra or systems of equations; thus, the number of samples that could be potentially used is infinite, and there are no restrictions regarding their physical proximity or resolution (i.e., a continuous resolution is theoretically claimed). Practicality may require a representative elementary volume.

FDK can be applied to multivariate attributes in a vector or tensor using an unlimited number of conditional components. However, if the number of attributes is a infinitely continuous function of time in nth dimension, FDK can also predict spatial functions in multiple dimensions with sparse data from sampled functions of time or space. In this case, the model gains an additional continuum (n+1)th dimension from the curve at each location. For example, note that a 2D map or continuum random field where the point locations are random curves (i.e., a random field of function variables), one at each location, is just a 3D field. Also note that weights in FDK are continuous functions of space and not scalar numbers as in traditional kriging of sparse functions.

Applying kriging to data curves that appear as functions of space (e.g., time) is a desirable approach. For example, kriging when data values are numerical curves that are correlated in a stationary manner was proposed (Giraldo et al. 2011). Their spatial predictions are a simplification of cokriging, by replacing the large number of time domain attributes by a single integral of the covariance on time domain. Although such a simplified proposal may encounter reservations, their motivation to generate an approach to krige numerical curves is commendable. Even though the curves are linearly correlated, their aforementioned approach still requires to show a proof on why cross correlations between elements across curves at different times are eliminated. The estimates at a location appear to be made with uniform "average" weights for all variable times and follow 1D numerical curves in "discretized" blocks or cells, *not* continuous *analytic functions* in 3D space, as expected from a "function" based approach.

Note that FDK transforms cokriging into consecutive kriging solutions with conditional components. Each predicted attribute or member is made from *field analytic functions* in 3D or higher dimension. FDK leads to a model that considers stochastic covariance parameters and avoids systems of equations, the cumbersome inversion of discrete covariance matrices and ad hoc integration averaging of covariance functions prior to estimation or simulation.

2.4 Stochastic Models for Nonlinear Client Algorithms

An FDK model provides parameters for each conditional probability density function (PDF) at every spatial location in the continuum spatial domain. Functional decomposition simulation (FDS) is the continued FDK using simulated input data. FDS delivers the anisotropy tensor random fields and the scalar members of any anisotropic scalar, vector, or tensor property. The random field models are used as input to client algorithms (e.g., nonlinear transformations, flow equations, or project optimization). There are two ways to conceptualize stochastic models:

1. A stochastic model can be made of PDFs that carry conditional parameters for each location and attribute. In this case, the client algorithms use FDK predicted moments as input to deliver response moments. That is a genuine parametric

stochastic approach for the stochastic ensemble. Nonlinear stochastic client algorithms may also demand higher moments to deliver unbiased outputs (Vargas-Guzmán 2004). Higher-order moments are computed through transformations. In the case of truly non-Gaussian PDFs, the higher-order moments are not transformable from the Gaussian model parameters, and they require their own prediction, using functionals in higher-order Banach space, L_n (Vargas-Guzmán 2011).

2. FDS is no other thing than continued FDK generation of field analytic functions with countable simulated data, which can be generated infinitely. The numerical approach for nonlinear transformations is to use Monte Carlo simulations, in which case the client algorithms require numerical realizations from input parameters or geostatistical predictions. Subsequently, geostatistical simulations are performed to proceed with a continuous generation of realizations of functions with stochastic anisotropy parameters. Furthermore, the limiting factor for numerical generation is the computer storage volume and resolution. FDS advantage is that the field analytic functions can be easily stored and updated with new random values without recomputing the analytic weight functions. The visiting path is controlled with locality parameters. Numerical generation can handle unlimited number of cells and multiple resolutions in realizations, because the numbers are not stored all at once.

3 Practical Modeling Accounting for Heterogeneous Anisotropy

3.1 Geostatistics with Input Anisotropy Tensor Fields

An indicator probability field estimated with functional decomposition kriging (FDK) by using local variable anisotropy parameters is shown in Fig. 3. The model is an analytic series summation of field analytic functions from projections, corresponding to 890 samples. Only 270 samples contain measured anisotropy orientations and range ratios.

The FDK algorithm (e.g., FDIK, for functional decomposition indicator kriging) also delivers second-order moments (not shown here). Note that the FDIK model for Fig. 3 does not strictly require all the predicted anisotropy map in Fig. 5; though if needed, the anisotropy in additional simulated samples is taken from predictions to generate simulated realizations.

Each field analytic function is predicted using FDIK, with anisotropy orientation and ranges input to the functional, from measured or predicted stochastic anisotropy at sampled locations for the target attribute. The FDK algorithm requires treating each field analytic function or consecutive component as a stationary anisotropic field, as classically applied in geostatistics with a single sample. The direction cosines for anisotropy orientation conform the basis for global coordinate rotation,

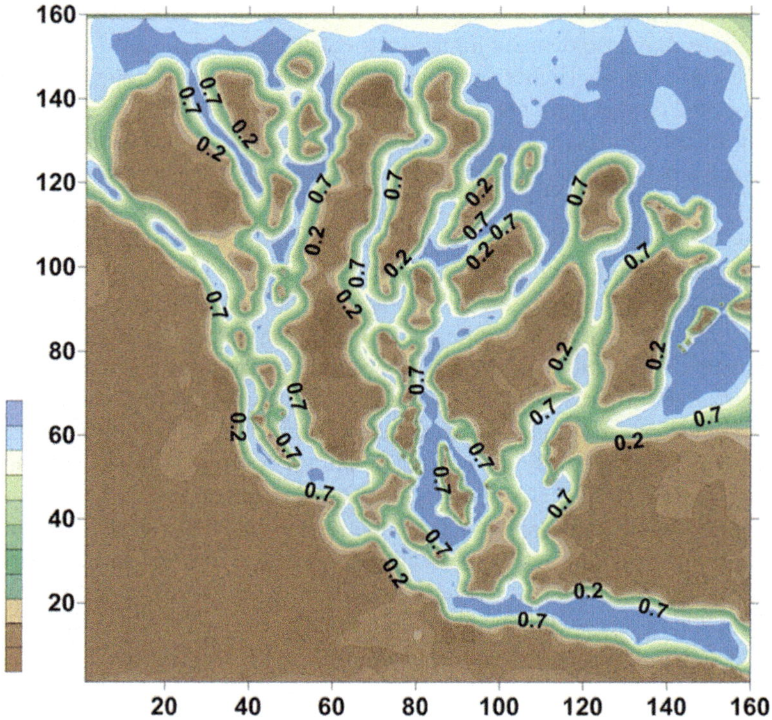

Fig. 3 Functional decomposition indicator kriging (FDIK) model for river channels in Lena Delta, based on 890 predicted field analytic functions

before updating the conditional functional and performing each Hilbert space projection. The approach must rotate and stretch the system of coordinates for each weight function (i.e., to assimilate a new sample). Consequently, the spatial model is updated with each new component. Note that due to the prior covariance between two samples is in the initial system of coordinates, it could become inconsistent, overestimated, or underestimated as the tensor principal components rotate. Then, the ranges of the covariance must be made properly consistent to match the required zero covariance with prior samples. This aspect is illustrated in Fig. 4.

3.2 *Predicted Stochastic Anisotropy Fields for Structural Uncertainty*

The idea of decomposing a random field can be intuitively extended to each of the members of the anisotropy tensor, as seen in Fig. 1. Subsequently, the tensor data representing anisotropy parameter random variables are continuously embedded

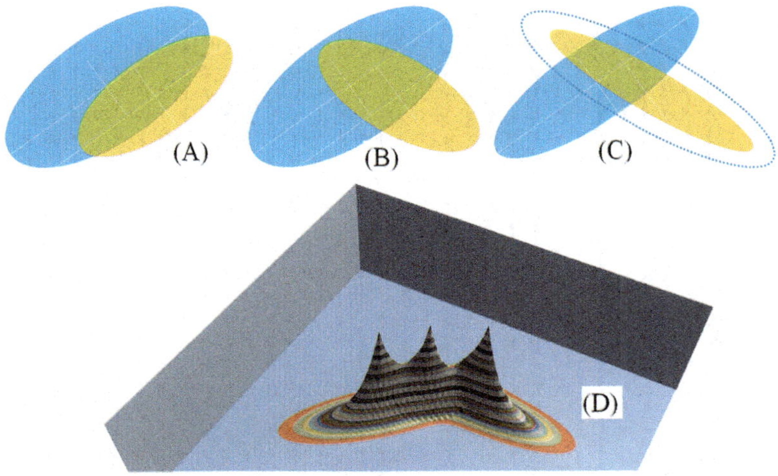

Fig. 4 Successive anisotropy. (**a**) Consistent classic. (**b**) Arbitrary. (**c**) Ranges consistent with rotation. (**d**) FDK anisotropic example for three indicator samples

into the functional for generating a random field pertaining to each attribute of anisotropy itself, at non-sampled locations. The process creates stochastic functionals that are multivariate because they contain as many parameters as delivered by the anisotropy tensor at each physical location. Furthermore, the functionals for simulated data are stochastic because they use covariance parameters from the simulated anisotropy tensors. Thus FDS is a very powerful approach because it allows for spatially variable stochastic covariance parameters simulated in the form of anisotropy tensor random fields.

Remark 5 As aforementioned, the anisotropy can be converted into vectors describing surfaces or manifolds. In the simple case of a 2D topology, the anisotropy streamlines are converted to normal vectors. These vectors represent the gradient vector of the attribute. Therefore, one can consider that a new norm becomes the sum of the norm for the attribute plus the norm for the gradient, this is, $\|Z(x)\| + \|\nabla Z(x)\|$, a Sobolev space. One finding is that FDK with stochastic anisotropy parameters in second-order covariances is a functional embedding that can be used to model a field pertaining to a Sobolev space constructed from L_2 spaces.

Case in point, uncertain anisotropy tensor members are estimated and simulated using local orientations and ratios from anisotropy data, consecutively for each vector or non-orthogonal component. Non-orthorhombic vectors are modeled in the same fashion as non-orthogonal components (Vargas-Guzmán 2003). The difference is that simulated anisotropy values are introduced into the stochastic functionals during simulation of each non-orthogonal scalar component.

Structural uncertainty is anisotropic and is represented with random variables in a tensor random field. In 3D, the three vectors in the tensor require nine cosine directors

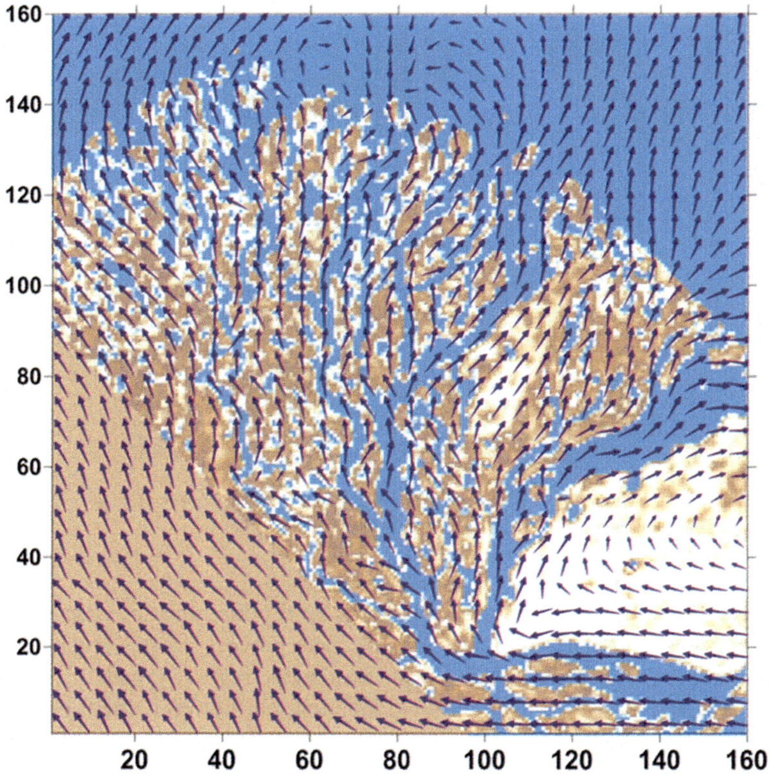

Fig. 5 Functional decomposition kriging (FDK) model of anisotropy tensors, from 270 stream-located samples from Lena Delta. Stream vector is a principal tensor component (the color *blue* in the background represents streams and water)

and three modules. In the orthorhombic case, with one vector on the vertical modeling plane, only three cosine directors and two anisotropy ratios are sufficient to handle the symmetry; the procedure is simplified further for 2D. Hence, each component can be predicted and simulated in the same way as non-orthogonal attributes with conditional components (Vargas-Guzmán 2003). However, FDK must be used to account for a circumvented embedding of anisotropy through covariance functions. For compositional reasons, it is convenient to handle cosine directors as components of unit vectors. It is not a good practice to estimate angles; to assure resulting vectors have unit modules and correct orientations, direction cosines must be estimated or simulated. A separate model of scalar anisotropy magnitudes, conditional to the orientations, is made using measured data and expert inputs or analogs. Modeling of the non-orthogonal scalar components must be performed with caution, as the anisotropy range ratios and principal components of anisotropy may conform non-orthorhombic tensor and components could hold third-order correlations.

A simplified example of FDK in 2D is presented in Fig. 5, where the vector component of the tensor of anisotropy represents averaged pseudo-streamlines. The figure depicts FDK predicted orientations of anisotropy, compared to a sedimentary image for a sector of Lena Delta. The cosine and sine components of the eigenvectors were conditionally kriged by introducing the sample anisotropy (i.e., rotation and coordinate stretching parameters) at each step, where the FDK algorithm was used with 270 selectively gathered samples at channels. The predicted anisotropy map can be easily continued with simulated residuals to produce realizations of simulated stochastic anisotropy fields.

4 Discussion and Conclusion

Functional decomposition kriging (FDK) theory offers powerful solutions for handling anisotropy in inverse, adaptive, multiresolution, and real-time methods. Because anisotropy is uncertain at non-sampled locations, FDK must predict anisotropy tensor fields before performing simulations for the target attribute.

Updating numerical model predictions with FDK geostatistics without changing the weight functions, requires simply replacing data or simulated values in the field analytic functions. Updating a specific weight function due to a change in anisotropy parameters is a swift process. If new consistent anisotropy parameters are provided for the field analytic function from a single sample location, the parametric weight function will automatically be modified without affecting previous sample locations. The conditioning algorithm modifies the target function pertaining to a single sample location, without repeating FDK computations at previous samples, but posterior residual data may need to be adjusted for posterior neighbor samples if the sequence uses the residual mode. The non-residual model is not reported here.

Modeling at multiple resolutions with arbitrary shape and size of elements requires integrating the function estimates within each element. The computational operations can be parallelized, but they do not need to be performed in the entire physical domain; instead they are applied in a selected sector of the model, as required. In theory, each field analytic function extends to the complete domain. The functional algorithm does not need to use neighborhoods, though the numerical representation for integrations and upscaling of a specific layer can be eliminated by ignoring regions pertaining to pivot locations that are significantly uncorrelated to the integrated sector model.

As a conclusion, the consecutive projection of vectors on their neighbors in functional Hilbert space leads to a powerful decomposition of a random field enabling FDK and FDS. Furthermore, an infinite number of attributes is reduced to a continuum FDK. In the context of countable attributes, vector and tensor random fields are arrays of non-orthogonal scalar member random fields. Therefore, FDK enables predictions of tensors of anisotropic attributes and structural anisotropy parameters. The resulting fields with stochastic gradients pertain to Sobolev space and may be useful for emulating turbulent flow and transport. A practical example for predicting the anisotropy and proportion of facies for curvilinear bodies (e.g., braided and meandering channels in a fluvio-deltaic environment) was illustrated. FDK anisotropy tensors fully describe truly complex structural controls, which can serve for modeling uncertain tensors of permeability of rocks and mineralization in rugose veins.

Bibliography

Boisvert J, Manchuk J, Deutsch C (2009) Kriging in the presence of locally varying anisotropy using non-euclidean distances. Math Geosci 41:585–601

Caeiro M, Soares A, Demyanov V (2015) Geostatistical modelling of complex deltaic reservoirs integrating production data through optimized history matching. Society of Petroleum Engineers, SPE 177813-MS

Curriero F (2005) On the use of non-euclidean distance measures in geostatistics. Math Geol 38:907–926

Dossary AA, Vargas-Guzmán JA (2011) Spatial modeling of complex sandstone bodies to maximize reservoir contact for wells drilled in clastic formations. GeoArabia J Middle East Pet Geosci 16:182

Giraldo R, Delicado P, Mateu J (2011) Ordinary kriging for function-valued spatial data. Environ Ecol Stat 18:411–426

Gutjahr A, Bullard B, Hatch S (1997) General joint conditional simulation using a fast fourier transform method. Math Geol 29:361–389

Horta A, Caeiro M, Nunes R, Soares A (2010) Simulation of continuous variables at meander structures: application to contaminated sediments of a lagoon. In: Atkinson P, Lloyd C (eds) GeoENV VII – Geostatistics for environmental applications. Springer, London, pp 161–172

Ladyzhenskaya O (2003) Sixth problem of the millennium: Navier–Stokes equations, existence and smoothness. Uspekhi Mat Nauk 58:251

Riesz F (1907) Sur une espèce de géométrie analytique des systèmes de fonctions sommables. C R Acad Sci 144:1409–1411

Soares A (1990) Geostatistical estimation of orebody geometry: morphology kriging. Math Geol 22:787–802

Stroet C, Snepvangers J (2005) Mapping curvilinear structures with local anisotropy kriging. Math Geol 37:635–649

van den Boogaart K, Schaeben H (2002) Kriging of regionalized directions, axes and orientations I. Directions and axes. Math Geol 34:479–503

Vargas-Guzmán JA (2003) Conditional components for simulation of vector random fields. Stoch Env Res Risk A 17:260–271

Vargas-Guzmán JA (2004) Geostatistics for power models of Gaussian fields. Math Geol 36:307–322

Vargas-Guzmán JA (2011) The kappa model of probability and higher-order rock sequences. Comput Geosci 15:661–671

Xu W (1996) Conditional curvilinear stochastic simulation using pixel-based algorithms. Math Geol 28:937–949

Yao T, Calvert C, Jones T, Foreman L, Bishop G (2007) Conditioning geological models to local continuity azimuth in spectral simulation. Math Geol 39:349–354

Can Measurement Errors Be Characterized from Replicates?

Chantal de Fouquet

Abstract Sample measurements (of grade, depth, etc.) are almost inevitably affected by errors. Several error models were studied in the literature. But the interest of replicates for selecting the error model received limited attention. If measurement errors are supposed to be additive, homoscedastic, without correlation between them, and spatially not correlated with the exact values, the variances of the measurement errors are computable from the sample, simple, and cross-variograms of replicate data sets, even if the variogram of the exact value is pepitic (Aldworth W, Spatial prediction, spatial sampling, and measurement error. Retrospective Theses and Dissertations. Paper 11842. Iowa State University Digital Repository @ Iowa State University, 1998; Faucheux et al. Characterisation of a hydrocarbon polluted soil by an intensive multi-scale sampling. Geostats 2008, proceedings of the 8th international geostatistics congress, 1–5 Dec. 2008, Santiago, Chile. Ortiz J-M, Emery X (eds) for an example, 2008). But what about the other cases? When the error is additive, its correlation with the exact value can remain undetectable. The variance of the measurement errors is thus not always computable. It's the same for an error of multiplicative type. Except in some special cases, keeping the different measurement values rather than their average improves the precision of the estimation.

1 Introduction

Measurements of grades or concentrations, depth of geologic horizon or groundwater table, etc. are inevitably spoiled by errors. Do repeated measurements at same data points (replicates) allow characterizing the amplitude of the measurement errors? How to best use replicates for the estimation?

C. de Fouquet (✉)
Centre de géosciences, Ecole des mines de Paris, Mines ParisTech, PSL. 35, rue Saint-Honoré, 77305 Fontainebleau, France
e-mail: chantal.de_fouquet@mines-paristech.fr

J.J. Gómez-Hernández et al. (eds.), *Geostatistics Valencia 2016*, Quantitative Geology and Geostatistics 19, DOI 10.1007/978-3-319-46819-8_3

Models of measurement errors are proposed in the literature and used in the estimation for a long time. In an instructive paper, Bourgault (1994) examines several models of additive or multiplicative "noise" which take heteroscedasticity (the amplitude of the error variance increases with the measured values) and conditional bias into account. A case study based on a simulated image contaminated by noise shows that the performances of factorial kriging to reproduce the spatial patterns of high signal values fall when the noise is correlated with the signal. However, the paper does not examine how to select the error model from the data. The literature remains poor on this practical matter.

In the literature, measurement errors are almost always supposed to be not correlated with the exact values. Is this usual assumption made for lack of a better option? To answer this question, three error models are revisited. Their properties are examined so as to show if the error models can be characterized from replicated data.

Kriging in the presence of replicates is then briefly examined in a practical context: how to best use all the data for the estimation? Should replicates be averaged or introduced individually in the kriging system? It is well known (Rivoirard et al. 1993) that a model for measurement errors makes the kriging matrix regular (not singular) when several data are available at the same data points. On a toy example, the precision of kriging with replicates is compared to that with their average.

In the following, the exact value is designated by "grade" and the measurement values of replicates as the "observations."

2 Three Models of Measurement Errors

Faucheux et al. (2008) showed on an example that simple and cross-variograms of replicates distributed in two sets allow detecting the presence of measurement errors. The error variances are computable, assuming that they are constant for each set and that the measurement errors are not correlated between them and spatially not correlated with the exact values. In the particular case where sample variograms are pepitic at the scale of the sampled zone, the nugget effects can be divided in two parts, the microstructure and the variance of measurement errors (Aldworth 1998).

This error model explains the case where the sample variogram of the average of the replicates is parallel to their cross-variogram and above it (Fig. 1a); another model is needed to explain Fig. 1b, where the structured components of the sample variogram of replicate averages and that of their cross-variogram appear to be proportional.

Two other models are thus examined: the case where the measurement error is additive and correlated with the exact value and the case where this error is of multiplicative type.

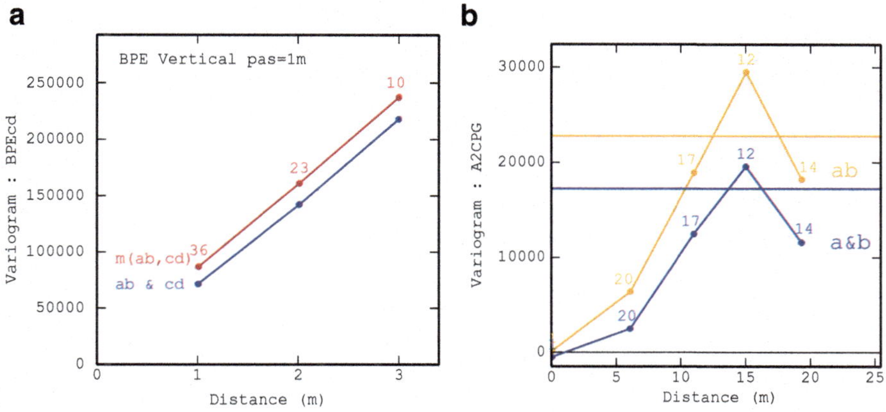

Fig. 1 Variogram of the average on two replicate sets (*upper curve*) and their cross-variogram (*lower curve*) for two measurement types and two sampling grids on the same polluted site. Hydrocarbon grades. (**a**) Vertical variogram of pyrogram measurements and (**b**) horizontal variogram of gas chromatography (Extracted from Faucheux and de Fouquet (2009))

Let us consider a site sampled by soil cores. Each homogenized core constitutes a sample, of which subsamples are extracted and analyzed. The grade at core support is modeled by a random function $Y(x)$, supposed to be stationary of order two (in order to simplify the presentation) with mean (expectation) m and covariance $C(h)$. The grade variogram is denoted $\gamma(h)$. $Z_1(x)$ and $Z_2(x)$ represent two observation sets.

2.1 Additive Error, Not Correlated with Grade

In the case of an additive measurement error, observations are written

$$Z_i(x) = Y(x) + R_i(x) \quad \text{with } R_i(x) = \mu_i + b_i T_i(x) \tag{1}$$

where the random variable $R_i(x)$ denotes the error of the measurement set i at sample point x and $T_i(x)$ the associated standardized variable. b_i is supposed to be constant for each set and positive.

In practice, the error expectation μ_i and the coefficients b_i (and thus the error variances) are unknown.

Let us first suppose that measurement errors R_1 and R_2 are mutually not correlated and spatially not correlated with grade Y. The variance of the observations of the set i is

$$Var\, Z_i(x) = C(0) + b_i^2 \tag{2}$$

and the covariance between the observations of the two sets is

$$Cov\big(\, Z_i(x),\; Z_j(x)\big) = C(0) \tag{3}$$

The variances of the measurement errors σ_i^2 are obtained from the difference between the variance of each observation set and their covariance:

$$\sigma_i^2 = Var\, Z_i(x) - Cov\,(Z_1(x), Z_2(x)) \tag{4}$$

The simple variogram of the observations per set and their cross-variogram are

$$\gamma_i(h) = b_i^2 + \gamma(h) \quad \text{and}\; \gamma_{12}(h) = \gamma(h) \tag{5}$$

The cross-variogram coincides with the grade variogram (Aldworth 1998; Faucheux et al. 2008). The simple variograms of the observations are parallel and above their cross-variogram. The variances of the measurement error can also be obtained from the difference between simple and cross-variogram:

$$\sigma_i^2 = \gamma_i(h) - \gamma_{12}(h) \tag{6}$$

When the variogram of the observations is pepitic at the scale of the sampled domain, the previous calculation allows splitting the nugget effect between the measurement error part σ_i^2 and the microstructural part, given by the cross-variogram.

In this model, the average of two observations per data point

$$Z_A(x) = \frac{1}{2}(Z_1(x) + Z_2(x)) \tag{7}$$

has the following variogram:

$$\gamma_A(h) = \frac{b_1^2 + b_2^2}{4} + \gamma(h) \tag{8}$$

which is parallel to the cross-variogram. The variogram of the difference

$$Z_D(x) = Z_1(x) - Z_2(x) \tag{9}$$

is pepitic:

$$\gamma_D(h) = b_1^2 + b_2^2 \tag{10}$$

The variogram analysis shows that this measurement error model is not suited if the simple variograms of the observations sets Z_i or of their average Z_A are not parallel to the cross-variogram $\gamma_{12}(h)$ or if the variogram of the difference is not pepitic.

Rivoirard (1983) has shown that the sample variance is equal to the average of the variogram cloud (origin included) and thus to the average of variogram points at all distances, weighted by their pair numbers. Thus, the "statistical" calculation of the measurement error variance (from sample variance and covariance, Eq. 4) is equivalent to its "geostatistical" calculation from weighted variogram and cross-variogram points (Eq. 6). In practice, these two results can slightly differ because only the first lags of the sample variograms and cross-variogram are computed.

2.2 Additive Error Correlated with Grade

The hypothesis of absence of correlation, at same point or spatially, between grade and measurement error is not always verified. For example, when the granulometry of the sample varies with grade, the loss of fine particles during the constitution of subsamples can induce a correlation between grade and measurement error at the same data point. In the case of soil coring, an insufficient cleaning of the auger between successive cores can induce a mixing between samples and thus a spatial correlation between grade and measurement errors (Bourgault 1994).

In order to describe the correlation between grade and measurement error, let us denote the error

$$R_i(x) = \mu_i + a_i(Y(x) - m) + b_i T_i(x) \tag{11}$$

with the same hypotheses on the $T_i(x)$ as previously. The link between grade and observation

$$Z_i(x) = \mu_i - a_i m + (1 + a_i)Y(x) + b_i T_i(x) \tag{12}$$

is described by the model with orthogonal residuals (Rivoirard 1994; Chilès and Delfiner 2012). The special case $a_i = 0$ corresponds to the absence of spatial correlation between measurement error and grade.

In practice, the error expectation μ_i and the coefficients a_i and b_i are unknown.

The covariance between measurement error and grade at same point is $\mathrm{Cov}\,(R_i(x), Y(x)) = a_i C(0)$. Their correlation coefficient (*sgn* denoting the sign)

$$\rho_{Y,R_i} = sgn(a_i) / \sqrt{1 + \frac{b_i^2}{a_i^2 C(0)}} \tag{13}$$

increases with a_i and $C(0)$ and decreases with b_i supposed to be positive.

The expectation and the variance of the observations of set i are, respectively, $E(Z_i(x)) = \mu_i + m$ and

$$\operatorname{Var} Z_i(x) = (1 + a_i)^2 C(0) + b_i^2 \tag{14}$$

and the covariance between the two observation sets is

$$\operatorname{Cov}(Z_1(x), Z_2(x)) = (1 + a_1)(1 + a_2)C(0) \tag{15}$$

The variance of the linear combination of observations at same data point

$$Z_L(x) = \ell_1 Z_1(x) + \ell_2 Z_2(x) \tag{16}$$

is

$$\operatorname{Var} Z_L(x) = (\ell_1(1 + a_1) + \ell_2(1 + a_2))^2 \, C(0) + \ell_1^2 b_1^2 + \ell_2^2 b_2^2 \tag{17}$$

In the special case $a_1 = a_2$ denoted a and $b_1 = b_2$ denoted b, the two measurement sets have same variance $\operatorname{Var} Z_i(x) = (1 + a)^2 C(0) + b^2$, and their covariance is $\operatorname{Cov}\,(Z_1(x), Z_2(x)) = (1 + a)^2 C(0)$, from which the coefficient b^2 can be derived. This is, for example, the case of two sets measured with the same device.

The simple variograms of observations are

$$\gamma_i(h) = b_i^2 + (1 + a_i)^2 \gamma(h) \tag{18}$$

and their cross-variogram is

$$\gamma_{12}(h) = (1 + a_1)(1 + a_2)\gamma(h) \tag{19}$$

When measurement errors and grade are correlated (with $a_i \neq -1$), the component of the variogram of observations linked to the spatial variability of grades is no longer equal to the variogram of grades but is proportional to it. In practice, the proportionality factor is unknown.

The simple variograms of observations and their cross-variogram are no longer parallel (unless $a_1 = a_2$) but are proportional, up to a nugget component on the simple variograms.

The variogram of the linear combination $Z_L(x)$ of observations is

$$\gamma_L(h) = (\ell_1(1 + a_1) + \ell_2(1 + a_2))^2 \gamma(h) + \ell_1^2 b_1^2 + \ell_2^2 b_2^2 \tag{20}$$

In particular, for the average of two replicates

$$\gamma_A(h) = \frac{b_1^2 + b_2^2}{4} + \frac{((1 + a_1) + (1 + a_2))^2}{4}\gamma(h) \tag{21}$$

and for their difference

$$\gamma_D(h) = b_1^2 + b_2^2 + (a_1 - a_2)^2\gamma(h) \tag{22}$$

When $a_1 \neq a_2$, the presence of a structured component on the variogram of the difference $Z_1(x) - Z_2(x)$ allows to detect the correlation between grade and measurement error, in accordance with the proportionality of the structured parts of the simple and cross-variograms of the observations.

When $a_1 = a_2$ (denoted a), Eqs. 18 and 19 become

$$\gamma_i(h) = b_i^2 + (1 + a)^2\gamma(h) \text{ and } \gamma_{12}(h) = (1 + a)^2\gamma(h) \tag{23}$$

that is exactly the same form as in the absence of correlation ($a = 0$) but with grade variogram $(1 + a)^2\gamma(h)$ (Eq. 5). These two models are compatible with Fig. 1a.

Remark: when the error is additive and correlated with grade, all variances, covariances, and simple and cross-variograms of the observations or of any of their linear combination (Eqs. 14, 15 and 17, 18, 19, 20, 21, 22, and 23) are identical to those obtained with the covariance $C'(h)$ and the coefficients $a_i{}'$ such that

$$C'(h) = k^2C(h)(k > 0)\text{and } 1 + a'_i = \frac{1 + a_i}{k} \tag{24}$$

Thus, the grade covariance $C(h)$ and the coefficients $1+a_i$ are determined only up to the factor k^2 and $1/k$, respectively, but they are not computable individually. As a consequence, an additional hypothesis on the grade covariance is necessary to write the kriging system for estimating the grade.

2.3 *Measurement Error of Multiplicative Type*

Let be an error of multiplicative type, possibly completed by an additive component

$$Z_i(x) = U_i(x) + (v_i + s_iT_i(x))Y(x) \tag{25}$$

where v_i denotes a deterministic parameter. The hypotheses on the standardized variables $T_i(x)$ are the same as previously. The variables $U_i(x)$ with expectation m_i^U are supposed to be not correlated and without any spatial correlation with the grade nor with the T_i.

In practice, the coefficients υ_i and s_i are unknown, as well as the expectation and the variance of the random variables $U_i(x)$ and $T_i(x)$ and those of the random function $Y(x)$.

The expectations of the observation sets are $E(Z_i(x)) = m_i^U + \nu_i m$, and their variances are

$$\text{Var } Z_i(x) = \text{Var } U_i(x) + C(0)\,\nu_i^2 + \left(C(0) + m^2\right)s_i^2 \tag{26}$$

The covariance between the two measurements at same data point is

$$\text{Cov}\,(\,Z_1(x), Z_2(x)) = C(0)\nu_1\nu_2 \tag{27}$$

When the two observation sets have same expectation and same variance, let us put $v_i = v$ and $s^2 = s_i{}^2$. With the additional hypothesis of no additive component $U_i(x)$ on the error, we obtain

$$\text{Cov}\,(\,Z_1(x),\, Z_2(x)) = C(0)\nu^2 \tag{28}$$

and

$$\left(C(0) + m^2\right)s^2 = \text{Var } Z_i(x) - \text{Cov}(\,Z_1(x), Z_2(x)) \tag{29}$$

The different parameters v, $C(0)$, and s^2 of the model cannot be separately determined.

The simple variograms of observations and their cross-variograms are

$$\gamma_i(h) = \text{Var } U_i(x) + s_i^2\left(C(0) + m^2\right) + \nu_i^2\gamma(h) \;\text{ and } \gamma_{12}(h) = \nu_1\nu_2\gamma(h) \tag{30}$$

The variogram of the linear combination $Z_L(x)$ of replicates is easily deduced from Eq. 30, by introducing the standardized variable $T(x)$ supposed to be without spatial correlation with grade nor with the additive components $U_1(x)$ and $U_2(x)$ of the measurement error:

$$Z_L(x) = \ell_1 U_1(x) + \ell_2 U_2(x) + \left(\ell_1\nu_1 + \ell_2\nu_2 + \sqrt{\ell_1^2 s_1^2 + \ell_2^2 s_2{}^2 T(x)}\right)Y(x) \tag{31}$$

These expressions are analogous to those obtained in the case of an additive error correlated with grade: Eqs. 18 and 19 have the same form as Eq. 30. These two models are compatible with Fig. 1b.

The two cases are impossible to be distinguished in practice: if present, the structured component is proportional to the simple and cross-variograms. The difference is that in Eq. 12, the coefficient of proportionality is determinist, while it is random in Eq. 25.

2.4 Contribution and Limits of the Variogram Analysis

If the two measurement sets have same mean, the simplest is to assume that the mean error of each set is zero. If the data sets have different means, additional hypotheses are needed, for writing the non-biased condition for the estimation.

From now on, the error is supposed to be additive. As previously mentioned, when the two simple variograms of the observations sets and their cross-variogram are parallel, it is not possible to determine if the error is spatially correlated with grade or not.

When the two simple variograms and the cross-variogram present a proportional (but not parallel) structured component, Eqs. 18 and 19 are a special case of the linear model of coregionalization, in which the correlation between the components of $Z_1(x)$ and $Z_2(x)$ proportional to the grade variogram is maximal (equal to 1). Indeed, if we note $(h > 0)\gamma_i(h) = C_i^0 + C_i^1\gamma(h)$ and $\gamma_{12}(h) = C_{12}^1\gamma(h)$, then $C_{12}^1 = \sqrt{C_1^1 C_2^1}$.

The previous relation can be written differently. Let us denote

$$R = \frac{1 + a_2}{1 + a_1} \tag{32}$$

the square root of the ratio of the sills of the structured component (up to the same factor k which disappears from the ratio). Equations 18 and 19 give

$$\gamma_1(h) = b_1^2 + \frac{1}{R}\gamma_{12}(h) \text{ and } \gamma_2(h) = b_2^2 + R\gamma_{12}(h) \tag{33}$$

On a practical point of view, it has been known for a long time that the absence of nugget effect on the cross-variogram of the observations indicates its absence on the grade variogram. The coefficients b_i^2 can then be derived from the nugget component of the simple variogram of the observations (Eq. 18). When the grade variogram presents a microstructural nugget component, this component is present on the cross-variogram too. The coefficient R can be computed from the structured part of the simple and cross-variograms and then applied (Eq. 33) in order to determine the microstructural part of the nugget component on the simple variograms. The difference from the nugget component of each simple variogram determines the coefficients b_i^2.

In practice, when the simple variograms of the observations are parallel and parallel to their cross-variogram, the simplest model consists in assuming the error of additive type and without correlation with grade, in order to determine the grade variogram and the pseudo-cross-variogram between grade and observations. In particular, this is the case if the simple and cross-variograms of the observations are pepitic. However, this model does not necessary describe at best the reality.

When the structured components of the simple and cross-variograms are proportional, additional hypotheses are needed in order to write the kriging system, as already mentioned.

The coefficients a_i and, up to a factor, the grade variogram being unknown, the pseudo-cross-variogram between observations and grade

$$\begin{aligned} g_i\,(h) &= \frac{1}{2}E(Z_i(x) - Y(x+h))^2 \\ &= \frac{1}{2}\left(\mu_i^2 + b_i^2 + a_i^2 C(0)\right) + (1 + a_i)\gamma(h) \end{aligned} \tag{34}$$

or their cross-variogram remains unknown too.

The properties of the additive error models remain valid if the grade is an IRF-0.

3 Kriging Within a Model of Additive Error Not Correlated with Grade

In the presence of replicates, a measurement error model makes the kriging matrix regular (i.e., nonsingular). This property is known for quite a long time (Rivoirard et al. 1993, cited in Aburto 2012; Bourgault 1994) and used, for example, in order to map a variable surveyed on profiles (geophysics, measurements from ships) on which the different values do not coincide at their intersections. An efficient model consists in assuming an acquisition error constant by profile and independent between profiles.

To precise the influence of the model of measurement error on the estimation, let us examine the elementary case of the point estimation from only two observations at the same data point. The errors are assumed to be additive, spatially not correlated with grade, with mean equal to 0, and with respective variance σ_i^2.

The (pseudo)-variogram between observations and that between observation and grade are

$$\begin{aligned} g_{12}\,(0) &= \frac{1}{2}E(Z_1(x) - Z_2(x))^2 \\ &= \frac{\sigma_1^2 + \sigma_2^2}{2} \end{aligned} \quad \text{and} \quad \begin{aligned} g_i(h) &= \frac{1}{2}E(Z_i(x) - Y(x+h))^2 \\ &= \frac{\sigma_i^2}{2} + \gamma(h) \end{aligned} \tag{35}$$

According to Eq. 5, the point kriging system of Y at $x+h$ from two replicates at point x is

$$\begin{cases} (1-\lambda)\dfrac{\sigma_1^2+\sigma_2^2}{2}+\mu=\dfrac{\sigma_1^2}{2}+\gamma(h) \\ \lambda\,\dfrac{\sigma_1^2+\sigma_2^2}{2}+\mu=\dfrac{\sigma_2^2}{2}+\gamma(h) \end{cases} \quad (36)$$

The weights of the data with respective measurement error variance σ_1^2 and σ_2^2 are $\lambda=\frac{1}{1+\frac{\sigma_1^2}{\sigma_2^2}}$ and $1-\lambda=\frac{1}{1+\frac{\sigma_2^2}{\sigma_1^2}}$: they depend on the ratio of the measurement error variances. At fixed $\sigma_1^2+\sigma_2^2$, the weight of a data decreases when its associated measurement error variance increases. When one of these variances is zero, the associated measurement is supposed to be exact, and its weight is equal to 1, whereas the weight of the other data, supposed to be affected by an error, is zero.

The Lagrange weight $\mu=\gamma(h)$ depends on the grade variogram between data and quantity to be estimated. The point kriging variance

$$\sigma_K^2(h)=2\gamma(h)+\frac{1}{\frac{1}{\sigma_1^2}+\frac{1}{\sigma_2^2}} \quad (37)$$

increases with the measurement error variances.

The kriging estimator

$$Y^K(x+h)=\frac{1}{1+\frac{\sigma_1^2}{\sigma_2^2}}Z_1(x)+\frac{1}{1+\frac{\sigma_2^2}{\sigma_1^2}}Z_2(x) \quad (38)$$

can be written

$$Y^K(x+h)=Y(x)+\frac{1}{1+\frac{\sigma_1^2}{\sigma_2^2}}R_1(x)+\frac{1}{1+\frac{\sigma_2^2}{\sigma_1^2}}R_2(x) \quad (39)$$

In this model, if the two variances of measurement errors are equal, the kriging weights are equal, assuming the variogram is the same: it is equivalent to make the estimation with all the replicates or with their average. But if the measurement error variances are different, the average of replicate values is generally not optimal. Indeed the estimation variance of *Y*(*x*+*h*) from the average of the two replicates is

$$\sigma^2(Z_A(x),Y(x+h))=\frac{\sigma_1^2+\sigma_2^2}{4}+2\gamma(h) \quad (40)$$

The precision gain of kriging in regard to the estimation from the data average is $\frac{(\sigma_1^2-\sigma_2^2)^2}{4(\sigma_1^2+\sigma_2^2)}$ which is obviously null if the measurement errors have same variance. At fixed $\sigma_1^2+\sigma_2^2$, the precision gain increases with the deviation between σ_1^2 and σ_2^2.

This result, demonstrated in a particular case, is general: if replicates (at same data point) have same measurement error variance, then it is equivalent for kriging to take their average. The demonstration is obvious: the kriging system remains unchanged by shifting the row and column associated with the replicates, assuming the variogram is the same. If the measurement error variances are different, kriging using all replicates is more precise.

4 Conclusion

Replicates are useful but they do not allow to fully characterizing the measurement error.

First of all, the usual hypothesis that the error is not spatially correlated with the grade appears to be somewhat conventional. When the two observation sets have the same variogram, this hypothesis is the simplest one, but a correlation between additive error and grade is also possible. Secondly, when the structured component appears to be proportional on the simple variograms and on the cross-variogram, the grade variogram is proportional to the cross-variogram, up to an unknown factor.

If the variance of measurement errors varies according to the set, taking the average of the replicate values is not optimal for the estimation. Realistic modeling of measurement errors allows improving the precision of the estimation, using all replicates.

In the presence of replicates, it is thus better to record all the data with the measurement characteristics (laboratory, device, date, etc.) rather than only their average per data point. Indeed, when they are numerous enough, the replicates can allow specifying a model of measurement errors, even if some indeterminations remain. Of course, if available, the precision given by the laboratory or the device characteristics should be considered for building the error model.

Acknowledgment The author thanks cordially Hélène Beucher and the reviewers for their attentive reviews and suggestions for improving the paper.

Bibliography

Aburto D (2012) Filtrage géostatistique de données géophysiques en vue de la cartographie de l'impédance d'une formation argileuse. Exemple du callovo-Oxfordien de Meuse. Thèse de doctorat ParisTech, 154 p

Aldworth W (1998) Spatial prediction, spatial sampling, and measurement error. Retrospective Theses and Dissertations. Paper 11842. Iowa State University Digital Repository @ Iowa State University

Bourgault G (1994) Robustness of noise filtering by kriging analysis. Note Math Geol 26 (6):733–752

Chilès J-P, Delfiner P (2012) Geostatistics: modeling spatial uncertainty, 2nd edn. Wiley, New York

Faucheux C, de Fouquet C (2009) Interprétation géostatistique des analyses CPG et Pollut-Eval du site 2. PRECODD LOQUAS, 9ème rapport d'avancement (D20). R2009-345CFCF. Rapport d'étude. 72p

Faucheux C, Lefebvre E, de Fouquet C, Benoit Y, Fricaudet B, Carpentier C, Gourry J-C (2008) Characterisation of a hydrocarbon polluted soil by an intensive multi-scale sampling. Geostats 2008, proceedings of the 8th international geostatistics congress, 1–5 Dec 2008, Santiago, Chile. Ortiz J-M, Emery X (eds)

Rivoirard J (1983) Remarques pratiques à propos des variances et du variogramme. *Note de cours C-75*. Ecole des mines de Paris, Fontainebleau

Rivoirard J (1994) Introduction to disjunctive kriging and non linear geostatistics. Oxford University Press, Oxford

Rivoirard J, Renard D, Léger M (1993) Quantification de l'incertitude sur les profondeurs estimées par sismique et par puits (*rapport FSH*). Ecole des mines de Paris, Fontainebleau, 14p

Modelling Asymmetrical Facies Successions Using Pluri-Gaussian Simulations

Thomas Le Blévec, Olivier Dubrule, Cédric M. John, and Gary J. Hampson

Abstract An approach to model spatial asymmetrical relations between indicators is presented in a pluri-Gaussian framework. The underlying gaussian random functions are modelled using the linear model of co-regionalization, and a spatial shift is applied to them. Analytical relationships between the two underlying gaussian variograms and the indicator covariances are developed for a truncation rule with three facies and cut-off at 0. The application of this truncation rule demonstrates that the spatial shift on the underlying gaussian functions produces asymmetries in the modelled 1D facies sequences. For a general truncation rule, the indicator covariances can be computed numerically, and a sensitivity study shows that the spatial shift and the correlation coefficient between the gaussian functions provide flexibility to model the asymmetry between facies. Finally, a case study is presented of a Triassic vertical facies succession in the Latemar carbonate platform (Dolomites, Northern Italy) composed of shallowing-upward cycles. The model is flexible enough to capture the different transition probabilities between the environments of deposition and to generate realistic facies successions.

1 Introduction

Variogram-based indicator simulation aims to distribute facies in space using first- and second-order spatial statistics as a constraint. It is widely used for modelling heterogeneous subsurface rock volumes such as hydrocarbon reservoirs and groundwater aquifers, in which data are usually sparse and deterministic methods are not appropriate. In standard oil industry practice, the facies represent regions of the reservoir where petrophysical properties such as porosity and permeability can be assumed to have statistically homogeneous distributions. Therefore, the spatial distribution of facies has a great impact on the reservoir model predictions.

T. Le Blévec (✉) • O. Dubrule • C.M. John • G.J. Hampson
Imperial College, Royal School of Mines, Prince Consort Road, London SW7 2BP, UK
e-mail: t.le-blevec15@imperial.ac.uk; o.dubrule@imperial.ac.uk; cedric.john@imperial.ac.uk; g.j.hampson@imperial.ac.uk

J.J. Gómez-Hernández et al. (eds.), *Geostatistics Valencia 2016*, Quantitative Geology and Geostatistics 19, DOI 10.1007/978-3-319-46819-8_4

While it is easy to constrain the models with the proportion and autocovariance of each facies (Alabert 1989; Armstrong et al. 2011), it is more complex to model the cross-indicator covariances between facies. For instance, SIS (sequence indicator simulation) by modelling every facies independently (Alabert 1989) does not reproduce cross-covariances between different facies, possibly resulting in non-realistic geological models.

With the aim of modelling spatial relationships between different facies, Carle and Fogg (1996) constrain cross-covariances using the parameters of a continuous-time Markov chain. An important outcome of their method is the possibility to model spatial asymmetry between the indicator variables. The probability of facies A to be on top of facies B can be different from that of facies A being under B. Such asymmetrical vertical stacking patterns of facies are common in the stratigraphic record as sedimentological processes tend to create and preserve shallowing-upward facies successions which are asymmetric (Burgess et al. 2001; Grotzinger 1986; Strasser 1988; Tucker 1985). However, the model used by Carle and Fogg (1996) is memoryless and so prevents from using a hole-effect covariance and reproducing cyclicity, which is another common feature of vertical facies successions (Burgess et al. 2001; Fischer 1964; Goldhammer et al. 1990; Grotzinger 1986; Masetti et al. 1991). Another approach uses non-parametric indicator variograms for bivariate probabilities to simulate facies with asymmetrical patterns (Allard et al. 2011; D'Or et al. 2008). The approach presented in the current paper aims to use parametric auto- and cross-covariance models that are "realizable", that is associated with valid random set models (Chilès and Delfiner 2012).

Pluri-Gaussian simulations (PGS) can handle facies interactions thanks to the use of underlying continuous gaussian variables and truncation rules defining facies ordering and geometries (Armstrong et al. 2011). Moreover, by construction, the PGS formalism leads to a general cross-covariance model between facies that is realizable (Chilès and Delfiner 2012). Developing a flexible multivariate gaussian framework allows to increase the range of facies patterns. For instance, the original linear model of co-regionalization (Wackernagel 2013), applied to the underlying gaussian functions, provides flexibility in the resulting facies thicknesses and distributions. However, the cross-correlations between the underlying gaussian functions are symmetrical and so are the facies relations.

To overcome this limitation, some authors have proposed to use spatial shifts to transform the cross-covariances between gaussian functions (Apanasovich and Genton 2010; Li and Zhang 2011; Oliver 2003). Armstrong et al. (2011) proposed to use a similar approach when defining the linear model of co-regionalization of the underlying gaussian variables. Although it is natural to expect that an asymmetrical cross-correlation between the gaussian functions should lead to asymmetrical relations between facies, this approach has not yet, to our knowledge, been fully developed and tested. Moreover, the relation between the spatial shift, the correlation and the facies asymmetry has not been studied explicitly.

In this article, we expand on the previous work described above to demonstrate that a spatial shift applied to the underlying gaussian functions can be used to create asymmetries in the vertical stacking of facies. The sensitivity of vertical facies stacking patterns to selected parameters is then investigated. Synthetic examples are produced, and the usefulness of this method is demonstrated by modelling a real facies succession from the Triassic Latemar carbonate platform (Dolomites, Northern Italy).

2 Methodology

In this section, we explain the basic principles of the pluri-Gaussian simulation (PGS) methodology and its relation to indicator functions. We then describe the shifted PGS model.

2.1 *Context and Notations*

We focus here on a simple example with three facies. The truncation rule that defines the contacts between facies and their proportion, relative to their area, can be drawn as follows (Fig. 1):

If I_1, I_2 and I_3 are the indicators of the three facies, the truncation rule defines them as follows for every location x on a vertical section:

$$I_1(x) = \begin{cases} 1, & Z_1(x) < t_1 \\ 0, & else \end{cases} \tag{1}$$

$$I_2(x) = \begin{cases} 1, & Z_1(x) > t_1, Z_2(x) > t_2 \\ 0, & else \end{cases} \tag{2}$$

$$I_3(x) = \begin{cases} 1, & Z_1(x) > t_1, Z_2(x) < t_2 \\ 0, & else \end{cases} \tag{3}$$

When the indicator of a facies equals 1, the corresponding facies is present at the location x. The marginal gaussian cumulative function G applied to each gaussian function Z_1 and Z_2 allows to have a truncation rule on which the area of a facies equals its proportion. However, if there is a correlation between the two functions, it affects the proportion as the points tend to be located along the transformation of the correlation line ρ (Fig. 1) which is plotted in the axes ($G(Z_1)$,$G(Z_2)$) and thus has for equation

$$Y = G\left[\rho^* G^{-1}[X]\right] \tag{4}$$

In the example of Fig. 1, a positive correlation increases the proportion of facies 2 over facies 3 as shown by the larger number of points generated in the domain of facies 2. With a negative correlation, it would be the opposite. A uniform truncation

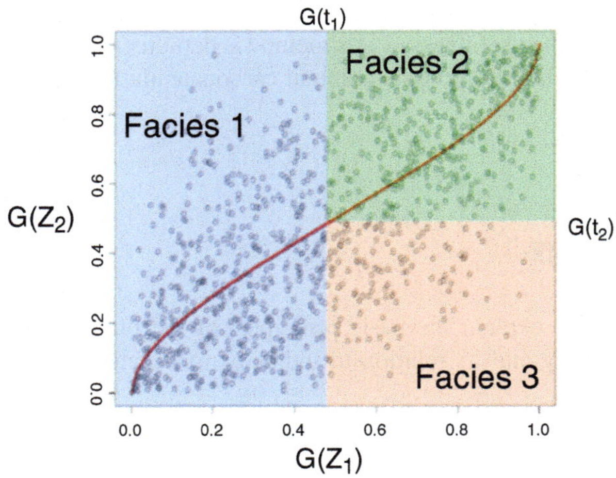

Fig. 1 Truncation rule defining three facies with two gaussian random functions Z_1 and Z_2. t_1 and t_2 are the truncations associated with each gaussian functions and G is the gaussian cumulative function. The *red curve* is defined by Eq. 4, with the correlation $\rho = 0.7$. One thousand random generations with a correlation $\rho = 0.7$ are performed thanks to the R package MASS (Venables and Ripley 2002) and displayed

rule could be obtained by applying the bi-normal gaussian cumulative function with correlation ρ on Z_1 and Z_2, but its analytical expression is not known.

The truncation rule does not contain spatial information and so does not control asymmetries. As the aim of this study is to model asymmetrical relations, the transition probability from one facies i to another j should be different in opposite directions h and $-h$:

$$t_{ij}(h) = \frac{P[I_i(x) = 1, I_j(x+h) = 1]}{P[I_i(x) = 1]} \neq t_{ij}(-h) \qquad (5)$$

Under the stationary hypothesis, the transition probability is independent of location. This transition probability results from the gaussian function parameters: correlation ρ, thresholds t_1 and t_2, gaussian correlation models ρz_1 (h) and ρz_2 (h) and the cross-correlation $\rho z_1 z_2$ (h) that can be asymmetric.

2.2 Relation Between the Indicators and Gaussian Functions

Understanding the link between the facies transition probabilities and the parameters of the underlying bi-gaussian function would help in inferring a pluri-Gaussian model resulting in the correct asymmetrical transition probabilities. Armstrong et al. (2011) show that the covariance of the facies indicator can be expressed as a multivariable integral of the underlying bi-gaussian density. For instance, the non-centred cross-covariance, between facies 2 and 3, $C_{23}(h)$, is defined as

$$C_{23}(h) = E[\, I_2(x)\, I_3(x+h)] = P[\, I_2(x) = 1, I_3(x+h) = 1] \tag{6}$$

According to Eqs. 1, 2 and 3, we have

$$C_{23}(h) = P[Z_1(x) > t_1, Z_2(x) > t_2, Z_1(x+h) > t_1, Z_2(x+h) < t_2] \tag{7}$$

This is the joint probability of four gaussian events with their dependence described by the correlation matrix:

$$\sum(\mathrm{h}) = \begin{pmatrix} 1 & \rho & \rho_{Z_1}(\boldsymbol{h}) & \rho_{Z_1 Z_2}(\boldsymbol{h}) \\ \rho & 1 & \rho_{Z_1 Z_2}(-\boldsymbol{h}) & \rho_{Z_2}(\boldsymbol{h}) \\ \rho_{Z_1}(\boldsymbol{h}) & \rho_{Z_1 Z_2}(-\boldsymbol{h}) & 1 & \rho \\ \rho_{Z_1 Z_2}(\boldsymbol{h}) & \rho_{Z_2}(\boldsymbol{h}) & \rho & 1 \end{pmatrix} \tag{8}$$

$C_{23}(h)$ can then be expressed as an integral of the quadri-variate gaussian density $g_{\Sigma(h)}(u,v,w,z)$ with the covariance matrix previously described:

$$C_{23}(h) = \int_{t_1}^{\infty}\int_{t_2}^{\infty}\int_{t_1}^{\infty}\int_{-\infty}^{t_2} g_{\sum(h)}(u,v,w,z)du\ dv\ dw\ dz \tag{9}$$

As we work with three facies (Fig. 1), the covariance between facies 1 and facies 2 is expressed by a triple integral, while a double integral defines the autocovariance of facies 1.

2.3 The Spatial Shift Applied to the Linear Model of Co-regionalization

The linear model of co-regionalization presented by Wackernagel (2013) is a flexible model for p-multivariate simulations and is chosen in this article. We also incorporate a shift on the covariance matrix C as proposed by Li and Zhang (2011). Armstrong et al. (2011) propose a way to simulate such a multivariate field from two independent gaussian functions Y_1 and Y_2 with covariances $\rho_{Y_1}(\mathrm{h})$ and $\rho_{Y_2}(\mathrm{h})$:

$$\begin{cases} Z_1(x) = Y_1(x) \\ Z_2(x) = \dfrac{\rho}{\rho_{Y_1}(a)} Y_1(x+a) + \sqrt{1 - \dfrac{\rho^2}{\rho_{Y_1}(a)^2}}\, Y_2(x) \end{cases} \tag{10}$$

The spatial shift, a, is the distance at which the correlation between the two gaussian functions Z_1 and Z_2 is maximal, and ρ is the correlation between the two simulated gaussian functions Z_1 and Z_2 at the same location. We can directly deduce from the square root term in Eq. 10 the condition of validity of the model:

$$-\rho_{Y_1}(a) < \rho < \rho_{Y_1}(a) \tag{11}$$

This condition originally results from the fact that the variance of the gaussian functions Z_1 and Z_2 is one. It is now possible to relate the covariances ρ_{Z_1} and ρ_{Z_2} of the gaussian fields Z_1 and Z_2 to the covariances of Y_1 and Y_2:

$$\begin{cases} \rho_{Z_1}(h) = \rho_{Y_1}(h) \\ \rho_{Z_2}(h) = \dfrac{\rho^2}{\rho_{Y_1}(a)^2}\rho_{Y_1}(h) + \left[1 - \dfrac{\rho^2}{\rho_{Y_1}(a)^2}\right]\rho_{Y_2}(h) \end{cases} \tag{12}$$

and the cross-correlations between Z_1 and Z_2, which are asymmetric:

$$\begin{cases} \rho_{Z_1Z_2}(h,a) = \dfrac{\rho\rho_{Z_1}(|h+a|)}{\rho_{Z_1}(a)} \\ \rho_{Z_1Z_2}(-h,a) = \dfrac{\rho\rho_{Z}(|h-a|)}{\rho_{Z_1}(a)} \end{cases} \tag{13}$$

It is interesting to see that

$$\rho_{Z_1Z_2}(-h,\, -a) = \rho_{Z_1Z_2}(h,a) \tag{14}$$

The different parameters of the model are summarized in Table 1.

Table 1 Symbols of the different parameters of the shifted pluri-Gaussian model

Signification	Parameter
First gaussian field	Z_1
Second delayed gaussian field	Z_2
Upward transition probability from facies i to facies j as a function of distance h	$t_{ij}(h)$
Proportion of facies i	P_i
Covariance function of Z_1	$\rho_{z_1}(h)$
Covariance function of Z_2	$\rho_{z_2}(h)$
Correlation coefficient between Z_1 and Z_2	ρ
Cross-correlation between Z_1 and Z_2 at distance h	$\rho_{z_1z_2}(h)$
Shift in the cross-correlation between Z_1 and Z_2	a
Range of the first gaussian function with a gaussian variogram, practical range $a_1\sqrt{3}$	a_1
Range of the second gaussian function with a gaussian variogram, practical range $a_2\sqrt{3}$	a_2

3 Results

In this section, we study the indicator transiograms derived from the shifted linear model of co-regionalization applied with PGS and with the truncation rule in Fig. 1. We first express the analytical expressions for a special case and then develop a sensitivity study in the general case thanks to numerical gaussian integrations. Gaussian variograms for the gaussian functions are used in order to have a linear behaviour at the origin on the indicator transiograms.

3.1 *Analytical Study of the Asymmetry*

We focus here on the special case where $t_1 = t_2 = 0$ as some analytical expressions can be found between the pluri-Gaussian and the transition probabilities.

3.1.1 Behaviour of the Asymmetrical Transition Probability

With the truncation rule used in Fig. 1, the transition probability between facies 1 and 2 can be written as a triple integral. Its analytical expression, developed in the appendix (Eqs. 25 and 26), is the following:

$$t_{12}(h) = -\frac{1}{4} + \frac{1}{2\pi}\left[\arccos\left(\frac{\rho\rho_{Z_1}(|h+a|)}{\rho_{Z_1}(a)}\right) + \arccos\left(\rho_{Z_1}(h)\right) + \arcsin(\rho)\right] \quad (15)$$

Therefore, the shift a and the correlation ρ must be non-zero to bring asymmetry (Fig. 2). We can also deduce the relation:

$$t_{12}(-h, -a) = t_{12}(h, a) \quad (16)$$

which means that changing the sign of the shift allows the asymmetry between the two facies to be switched.

We can see that if the correlation and shift are positive, and the transition probability tends towards a facies with low proportions, the curve has a very high concavity with a maximum before the range (Fig. 2, right). If the correlation is negative and the transition probability tends towards a facies with high proportion, the curve has an inflexion point (Fig. 2, left). In the opposite direction, the behaviour is always different, highlighting the asymmetry. If there is no shift, there is no asymmetry (Fig. 2).

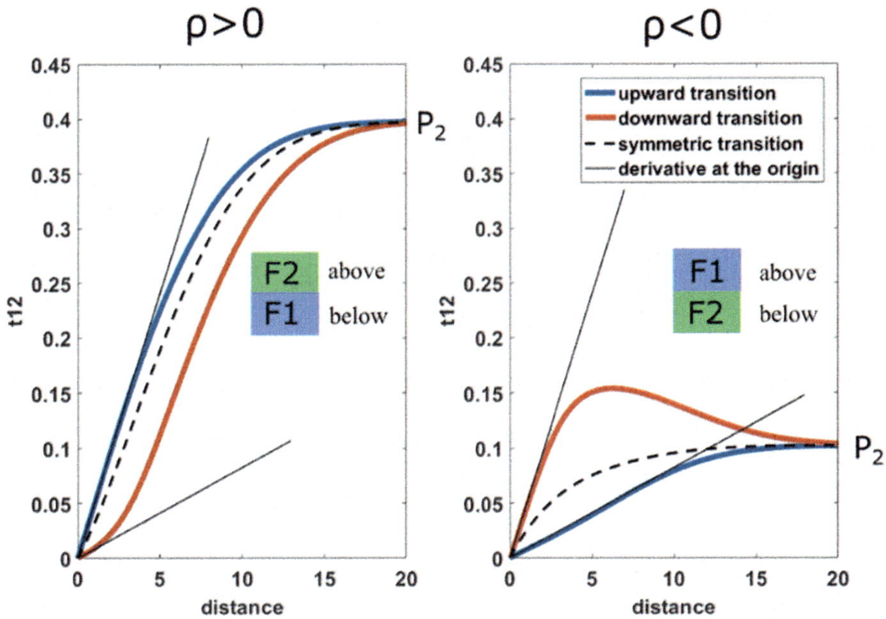

Fig. 2 Influence of a positive shift on the transition probabilities from facies 1 to facies 2 with different values of the proportion P_2 of facies 2. The coefficient ρ is either 0.8 or −0.8. The gaussian function has a gaussian variogram with range 8 (practical range = 13.85) and the shift is 3. The upward and downward transitions are deduced from Eq. 15, such as the *dotted line* obtained with a shift equal to 0, and the *black* tangents are obtained from Eqs. 17 to 18

3.1.2 Asymmetry in Facies Contacts

The frequency of contacts between two given facies can be derived from the derivative of the cross-transition probability at the origin, which is the rate of transition from one facies to the other per unit length. We can express the rate of transition upward $T_{12}{}^{+}$ and downward $T_{12}{}^{-}$ in the case of a gaussian variogram by differentiating Eq. 15:

$$T_{12}^{+}(a,\rho,a_1) = \lim_{h\to 0} t_{12}{}'(h) = \frac{1}{2\pi}\left[\frac{\sqrt{2}}{a_1} + a\frac{2\rho}{a_1^2\sqrt{1-\rho^2}}\right] \quad (17)$$

$$T_{12}^{-}(a,\rho,a_1) = \lim_{h\to 0} t_{12}{}'(-h) = \frac{1}{2\pi}\left[\frac{\sqrt{2}}{a_1} - a\frac{2\rho}{a_1^2\sqrt{1-\rho^2}}\right] \quad (18)$$

From these equations, it is clear that if the correlation ρ and the shift a are positive, the probability of having facies 2 on top of facies 1 is higher than of having facies 1 on top of facies 2. It can be interesting to see for which shift the transition rate is maximal; let's take

$$a_{\text{lim}} = a_1 \frac{\sqrt{2(1-\rho^2)}}{2\rho} \tag{19}$$

In that case, we have

$$\begin{cases} T_{12}^{+}(a_{\text{lim}}, \rho, r_1) = \dfrac{\sqrt{2}}{a_1 \pi} \\ T_{12}^{-}(a_{\text{lim}}, \rho, r_1) = 0 \end{cases} \tag{20}$$

With this shift, facies 1 cannot make a transition to facies 2 going downwards as the transition rate is 0. For the upward transition, it can be noticed that the expression of the transition rate is the inverse of the mean length of facies 1 (Lantuéjoul 2002). This implies that the upward transition rate from facies 1 to facies 3 is zero with the closing relations of the transition rate matrix $\boldsymbol{Q}$:

$$\boldsymbol{Q} = \begin{pmatrix} -1/\boldsymbol{L}_1 & 1/\boldsymbol{L}_1 & 0 \\ 0 & -1/\boldsymbol{L}_2 & 1/\boldsymbol{L}_2 \\ -1/\boldsymbol{L}_3 & 0 & -1/\boldsymbol{L}_3 \end{pmatrix} \tag{21}$$

with $\boldsymbol{L}_i$ as the mean lengths of the different facies. Therefore, this shift gives the maximum of asymmetry and allows to build perfect geologic asymmetrical sequences. However, the shift is also bounded by Eq. 11, and consequently Eqs. 19, 20 and 21 are not possible. As the transition rates increase linearly with the shift, the maximum of asymmetry is obtained for the higher shift which is the following according to Eq. 11:

$$a_{\max} = a_1 \sqrt{-\log(\rho)} \tag{22}$$

It can be noted that the expressions of a_{lim} and $a_{\max}$ converge to each other when ρ tends to one. Thus, for a correlation that tends to one, $a_{\max}$ gives upward and downward transition rates that tend, respectively, to $\boldsymbol{1/L_i}$ and $\boldsymbol{0}$, allowing to create perfect asymmetrical sequences (Eq. 21). This limit case can also be obtained by simulating only one gaussian function and use the shifted equivalent as the second gaussian function.

The expressions of the multi-gaussian integrals have allowed asymmetries for a truncation rule with cut-off at 0 to be analytically expressed. Lantuéjoul (2002) gives a solution for a general truncation rule when the correlation tends to 1. This might allow development of more general expressions with thresholds.

3.2 Sensitivity Analysis for a General Truncation Rule

The gaussian integral cannot be computed analytically in the general case with cut-offs different from 0. However, it can be computed numerically (Genz 1992)

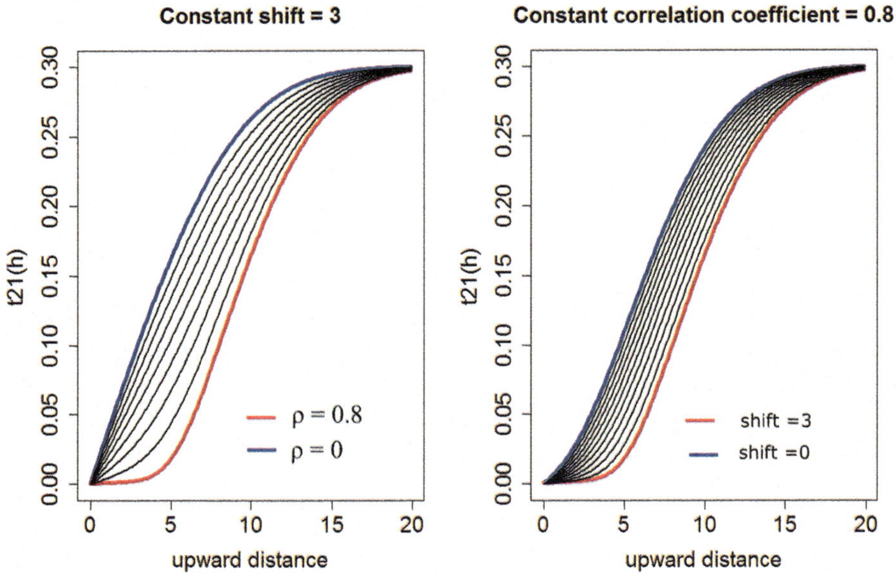

Fig. 3 Comparison of the impact of the correlation and the shift on the transition probability from facies 2 to facies 1 upwards. The step for the *black curves* is 0.1 for the correlation (*left*) and 0.3 for the shift step (*right*). The range of the first gaussian variogram is 8, the proportion of facies 1 is 0.3 and facies 2 is 0.4

using a code available on R (Genz et al. 2009; Renard et al. 2015). Consequently, we have access to all the transition probabilities, and the correlation ρ can be changed while keeping the proportions constant which is not possible analytically.

This is carried out by minimizing an objective function quantifying the difference between the targeted and simulated proportions computed with the gaussian numerical integral (Genz 1992). It can also be done with a maximum likelihood estimation of the target proportions by generating random correlated gaussian values. Understanding the impacts of the correlation and the shift at constant proportions is important for manually fitting transition probabilities (Fig. 3).

We can see in Fig. 3 that both the correlation and the shift have an impact on the tangent at the origin which provides a flexibility to match the asymmetry between facies contacts. The asymmetrical limit behaviour a_{lim} (Eq. 19) seems to have been reached with $\rho = 0.8$ and $a = 3$ as the transition rate is close to 0 for these values. The two parameters also affect differently the curvature of the transition probability increasing the flexibility of the method.

4 Case Study

This section presents a case study for illustrating the method described earlier with three facies and the truncation rule of Fig. 1. The geostatistical package RGeostats is used for the simulation (Renard et al. 2015). The transiograms are studied for two facies as the relation with the third can be automatically deduced from them.

4.1 The Latemar Data Set

Carbonate outcrops usually show significant vertical asymmetries in their facies distribution, in part explained by a gradual lateral shift in environments of depositions during sea-level highstands, followed by nondeposition during sea-level lowstands and the subsequent transgression (Catuneanu et al. 2011). For instance, the intertidal environment tends to be on top of the subtidal environment in shallowing-upward sequences (Sena and John 2013). The Latemar massif in the Dolomites of Northern Italy shows well-documented examples of asymmetrical vertical facies sequences in a carbonate platform. As reported by Egenhoff et al. (1999), a typical asymmetrical, upward-shallowing succession is bounded by a supratidal exposure surface at its top, which tends to cap intertidal-to-shallow-subtidal grainstones that overlie subtidal wackestones (Fig. 4).

4.2 Constraining the Transition Probabilities

The transition probabilities of Fig. 5 were derived from the data shown in Fig. 4. They can be fitted with the shifted linear model of co-regionalization manually through a trial-and-error process, by maximum likelihood estimation or by minimizing an objective function. In a more general context, a manual procedure is preferred as transiogram modelling is a step where geological conceptual knowledge can be incorporated. Therefore, we choose to fit manually the transition probabilities of Fig. 5.

As seen in Fig. 5, the model fitted by trial and error honours the tangent at the origin of the transition probabilities. This means that the transition rates are well constrained. Moreover, a possible hole-effect is observed in the experimental transition probabilities due to a low variance in the facies thicknesses. This effect cannot be modelled with the current model, but a hole-effect variogram on the gaussian function should be able to model it (Dubrule 2016).

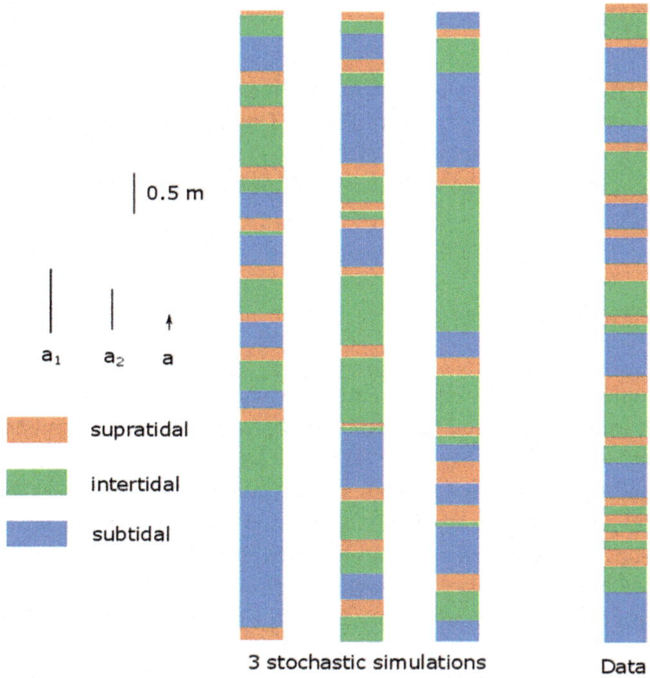

Fig. 4 Comparison between the vertical section of the Latemar section reported by Egenhoff et al. (1999) and simulations with asymmetrical pluri-Gaussian simulations. The parameters for the simulation are the same as described in Fig. 5

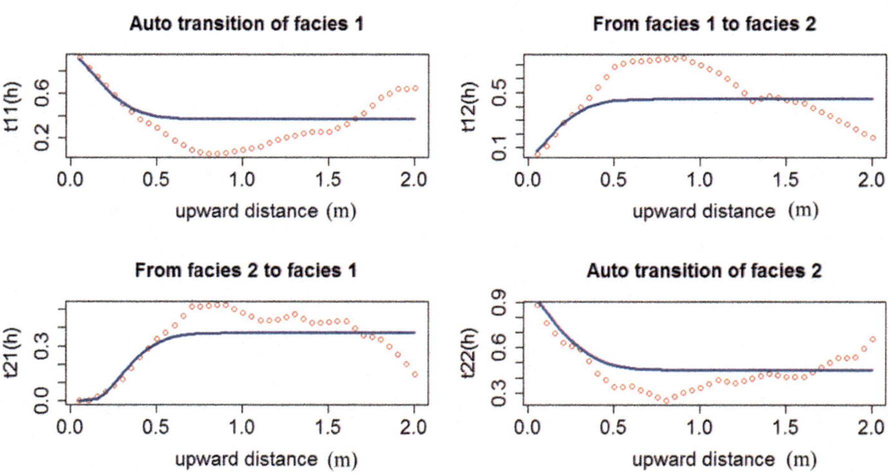

Fig. 5 Match between experimental transition probabilities (*red*) observed in Fig. 4 and the model (*blue*). Facies 1 is subtidal and facies 2 intertidal. The parameters used for the model are 0.9 for the range of the first gaussian, 0.52 for the range of the second gaussian, 0.13 for the shift and 0.8 for the correlation

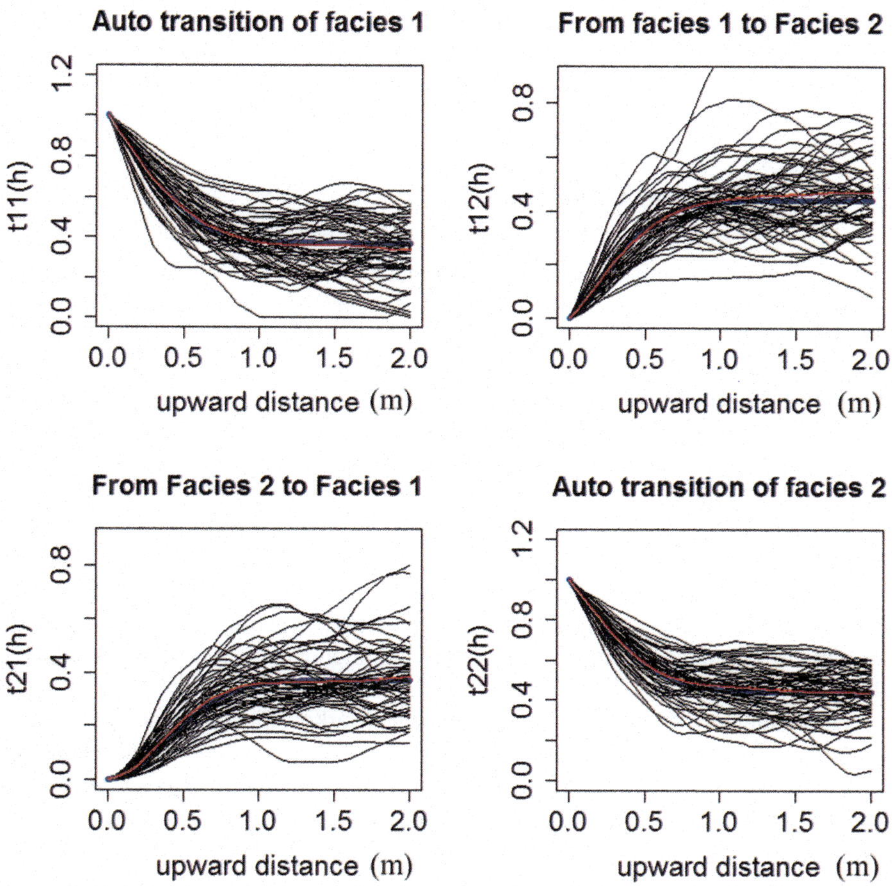

Fig. 6 Comparison between transition probabilities model (*blue*) and simulated (*grey*) and mean of the simulated (*red*) on 50 simulations of the Latemar section presented in Fig. 4. Facies 1 is subtidal, and facies 2 is intertidal

4.3 Facies Asymmetrical Simulation with Pluri-Gaussian Model

We build two gaussian fields, and then we apply the transformations described in Eq. 10 on three simulations (see Fig. 4). The asymmetry is still preserved in the simulations, with supratidal facies always on top of the intertidal facies and the intertidal facies on top of the subtidal facies. However, the limit shift a_{lim} (Eq. 19) has not been reached as the probability of having subtidal on top of supratidal is not 1, which is also observed on the data. To go further in the simulation analysis, the experimental transiograms are computed on 50 simulated sections and compared to the model variogram (Fig. 6).

This Monte Carlo study shows that the simulated transition probabilities seem to match the model well at the origin and for other distances as the mean transiogram of the simulations matches with the transiogram model (Fig. 6).

5 Discussion and Conclusion

This study has shown that the shifted linear model of co-regionalization seems well-suited to model facies transitions asymmetries using PGS. For the case of modelling three facies, the first two gaussian variograms allow to define facies mean thicknesses, while the shift and the correlation determine the asymmetrical patterns. Therefore, every transition rate of the transiogram matrix can be inferred independently making the method very flexible. Moreover, we saw analytically and numerically that the maximum rate of transitions could be reached asymptotically, which allows to build perfect asymmetrical sequences.

More precisely, the gaussian integral allows to fix the transition rates as with a Markov process (Carle and Fogg 1996). However, if the number of facies is increased, it would be more difficult to respect the different asymmetries, and manual fitting of the different transition probabilities would be more complex. Automatic procedures such as maximum likelihood estimations might address that issue.

The advantage of PGS over continuous-time Markov chains is that it provides a framework in which the resulting indicator variograms are automatically valid but also quite flexible. Beyond just transition rates, the parametrical covariances can lead to linear or fractal behaviour of the indicator variogram at the origin (Chilès and Delfiner 2012; Dubrule 2016). Other models than the linear model of co-regionalization would allow to select different behaviours for every facies. For instance, the multivariate Matern model would allow cross-transition probabilities to have different smoothness parameters for every facies (Gneiting et al. 2012), and the spatial shift could be applied to it (Li and Zhang 2011) which would also result in facies asymmetries. This is currently investigated by the authors.

Acknowledgements The authors would like to thank the Earth Science and Engineering Department of Imperial College for a PhD studentship grant for T. Le Blévec and Total for funding O. Dubrule professorship at Imperial College.

Appendix: Analytical Expression of the Triple Gaussian Integral

In a similar fashion as Kendall et al. (1994), we consider three correlated gaussian variates being in their respective intervals as a set of three dependent events. With the truncation rule displayed in Fig. 1 and thresholds that equal 0, facies 1 at location x and facies 2 at location x+h correspond to one variate being negative and two positive. The indicator covariance $C_{12}(h)$ quantifies the probability of the intersection of these three events. The correlation matrix between the three gaussian variates is the following:

$$\sum(h) = \begin{pmatrix} 1 & \rho_{Z_1}(\boldsymbol{h}) & \dfrac{\rho\rho_{Z_1}(|\boldsymbol{h}+\boldsymbol{a}|)}{\rho_{Z_1}(\boldsymbol{a})} \\ \rho_{Z_1}(\boldsymbol{h}) & 1 & \rho \\ \dfrac{\rho\rho_{Z_1}(|\boldsymbol{h}+\boldsymbol{a}|)}{\rho_{Z_1}(\boldsymbol{a})} & \rho & 1 \end{pmatrix} \tag{23}$$

The probability can be written as a triple integral of the corresponding gaussian density $g_{\Sigma(h)}(u,v,w)$:

$$C_{12}(h) = \int_{-\infty}^{0}\int_{0}^{+\infty}\int_{0}^{+\infty} g_{\Sigma(h)}(u,v,w)dudvdw \tag{24}$$

Thanks to the gaussian integral symmetry property, the probability of intersection of the events is the complementary of the probability of their union (Kendall et al. 1994). Therefore, by definition of the union, the intersection of the three events can be expressed as a sum of the corresponding single and pair events and so the triple integral as a sum of the single integrals that equal to 0.5 and double integrals with their respective correlation coefficient:

$$\begin{aligned} C_{12}(h) \\ = \frac{1}{2}\Bigg(1 - 3{*}0.5 + \int_{-\infty}^{0}\int_{0}^{+\infty} g_{\rho_{Z_1}(\boldsymbol{h})}(u,v)dudv + \int_{-\infty}^{0}\int_{0}^{+\infty} g_{\frac{\rho\rho_{Z_1}(|\boldsymbol{h}+\boldsymbol{a}|)}{\rho_{Z_1}(\boldsymbol{a})}}(u,v)dudv \\ + \int_{0}^{+\infty}\int_{0}^{+\infty} g_{\rho}(u,v)dudv\Bigg) \end{aligned} \tag{25}$$

Sheppard (1899) gives then the solution of the double integral that allows to obtain the final expression of the transition probability between facies 1 and 2 (Eq. 15):

$$\int_{0}^{+\infty}\int_{0}^{+\infty} g_{\rho}(u,v)dudv = \frac{1}{2} - \int_{0}^{+\infty}\int_{0}^{+\infty} g_{\rho}(u,v)dudv = \frac{1}{4} + \frac{1}{2\pi}\arcsin(\rho) \tag{26}$$

Bibliography

Alabert F (1989) Non-Gaussian data expansion in the earth sciences. Terra Nova 1(2):123–134

Allard D, D'Or D, Froidevaux R (2011) An efficient maximum entropy approach for categorical variable prediction. Eur J Soil Sci 62(3):381–393

Apanasovich TV, Genton MG (2010) Cross-covariance functions for multivariate random fields based on latent dimensions. Biometrika 97(1):15–30

Armstrong M, Galli A, Beucher H, Loc'h G, Renard D, Doligez B, ... Geffroy F (2011) Plurigaussian simulations in geosciences. Springer Science & Business Media, New York

Burgess P, Wright V, Emery D (2001) Numerical forward modelling of peritidal carbonate parasequence development: implications for outcrop interpretation. Basin Res 13(1):1–16

Carle SF, Fogg GE (1996) Transition probability-based indicator geostatistics. Math Geol 28 (4):453–476

Catuneanu O, Galloway WE, Kendall CGSC, Miall AD, Posamentier HW, Strasser A, Tucker ME (2011) Sequence stratigraphy: methodology and nomenclature. Newslett Stratigr 44 (3):173–245

Chilès J-P, Delfiner P (2012) Geostatistics: modeling spatial uncertainty, vol 713. Wiley, Hoboken

D'Or D, Allard D, Biver P, Froidevaux R, Walgenwitz A (2008) Simulating categorical random fields using the multinomial regression approach. Paper presented at the Geostats 2008—proceedings of the eighth international geostatistics congress

Dubrule O (2016) Indicator variogram models – do we have much choice?. Manuscript submitted for publication

Egenhoff SO, Peterhänsel A, Bechstädt T, Zühlke R, Grötsch J (1999) Facies architecture of an isolated carbonate platform: tracing the cycles of the Latemar (Middle Triassic, northern Italy). Sedimentology 46(5):893–912

Fischer AG (1964) The Lofer cyclothems of the alpine Triassic. Princeton University, Princeton

Genz A (1992) Numerical computation of multivariate normal probabilities. J Comput Graph Stat 1(2):141–149

Genz A, Bretz F, Miwa T, Mi X, Leisch F, Scheipl F, Hothorn T (2009) mvtnorm: multivariate normal and t distributions. R package version 0.9-8. URL http://CRAN.R-project.org/package=mvtnorm

Gneiting T, Kleiber W, Schlather M (2012) Matérn cross-covariance functions for multivariate random fields. J Am Stat Assoc 105(491):1167–1177

Goldhammer R, Dunn P, Hardie L (1990) Depositional cycles, composite sea-level changes, cycle stacking patterns, and the hierarchy of stratigraphic forcing: examples from Alpine Triassic platform carbonates. Geol Soc Am Bull 102(5):535–562

Grotzinger JP (1986) Cyclicity and paleoenvironmental dynamics, Rocknest platform, northwest Canada. Geol Soc Am Bull 97(10):1208–1231

Kendall M, Stuart A, Ord J (1994) Vol. 1: Distribution theory. Arnold, London

Lantuéjoul C (2002) Geostatistical simulation: models and algorithms. Springer Science & Business Media, Berlin

Li B, Zhang H (2011) An approach to modeling asymmetric multivariate spatial covariance structures. J Multivar Anal 102(10):1445–1453

Masetti D, Neri C, Bosellini A (1991) Deep-water asymmetric cycles and progradation of carbonate platforms governed by high-frequency eustatic oscillations (Triassic of the Dolomites, Italy). Geology 19(4):336–339

Oliver DS (2003) Gaussian cosimulation: modelling of the cross-covariance. Math Geol 35 (6):681–698

Renard D, Bez N, Desassis N, Beucher H, Ors F, Laporte F (2015) RGeostats: the geostatistical package (Version 11.0.1). Retrieved from http://cg.ensmp.fr/rgeostats

Sena CM, John CM (2013) Impact of dynamic sedimentation on facies heterogeneities in Lower Cretaceous peritidal deposits of central east Oman. Sedimentology 60(5):1156–1183

Sheppard W (1899) On the application of the theory of error to cases of normal distribution and normal correlation. Philos Trans R Soc London Ser A Containing Pap Math Phys Charact 192:101–531

Strasser A (1988) Shallowing-upward sequences in Purbeckian peritidal carbonates (lowermost Cretaceous, Swiss and French Jura Mountains). Sedimentology 35(3):369–383

Tucker M (1985) Shallow-marine carbonate facies and facies models. Geol Soc London Spec Publ 18(1):147–169

Venables WN, Ripley BD (2002) Modern applied statistics with S, 4th edn. Springer, New York

Wackernagel H (2013) Multivariate geostatistics: an introduction with applications. Springer Science & Business Media, Berlin

Considerations for the Use of Sequential Sampling Techniques

J. Leguijt

Abstract Sequential sampling is a well-known and efficient method to generate probabilistic realisations of models that are constrained by two-point statistics.

These two-point statistics consist of second-order moments that are defined by a variogram. The statistics describe the lateral continuity behaviour of the models. It can be shown that the sequential sampling method correctly generates samples from a probability density function (pdf), when this pdf honours only the statistics that define the lateral continuity constraints. In Bayesian statistics, this is named a prior pdf. The sequential sampling method is also used to generate models from a probability density function that is constrained by observations, similar to those that are derived from seismic data. This is known as a posterior pdf. To justify this approach, some assumptions have to be made that are not strictly valid and the result is often a significant error. The errors will be investigated using a realistic synthetic example. The probabilistic seismic inversion programme that has been developed by Shell contains a module that is able to account for lateral continuity. In this module, an alternative approach has been used to mitigate the problems with the sequential sampling method. To realise this, each location needs to be visited repeatedly.

1 Introduction

Sequential sampling is an efficient method to generate laterally continuous models based on a variogram in a probabilistic manner. When two-point statistics are used to define the lateral continuity of a model, the implicit assumption is that the underlying probability density function (pdf) is Gaussian. For this reason, this method is often called the sequential Gaussian sampling method.

The sequential sampling method is easy to implement. The present work investigated how well it performs in three different situations:

J. Leguijt (✉)
Shell International, Kessler Park 1, 2288 GS Rijswijk, The Netherlands
e-mail: jaap.leguijt@shell.com

J.J. Gómez-Hernández et al. (eds.), *Geostatistics Valencia 2016*, Quantitative Geology and Geostatistics 19, DOI 10.1007/978-3-319-46819-8_5

1. The algorithm is used to generate geostatistical models that are not constrained by measurements.
2. The algorithm is used to generate models that are only constrained by direct perfect measurements of the properties that are modelled.
3. The algorithm is used in a situation where the measurements are contaminated with noise.

First, an analysis of the sequential sampling method will be presented, using a model that only consists of two locations. Second, a more realistic example will be presented using a grid of 101 × 101 locations.

An alternative for the sequential sampling method will be presented. This is able to properly sample the posterior distribution that arises from the assimilation of measurements that are contaminated with noise.

2 Sequential Sampling Method

To understand the principle of the sequential sampling method, a model that consists of N locations is considered. The value of the modelled property at location n is denoted by v_n. Based on the available geostatistical information, a probability density function (pdf) can be defined for the values of the modelled property, at all locations (Deutsch 2002). This pdf is denoted by $p(\mathbf{v})$ and can be decomposed as follows:

$$\begin{aligned} p(\mathbf{v}) &= p(v_N, v_{N-1}, \ldots, v_3, v_2, v_1) \\ &= p(v_N, v_{N-1}, \ldots, v_3, v_2 | v_1) p(v_1) \\ &= p(v_N, v_{N-1}, \ldots, v_3 | v_2, v_1) p(v_2 | v_1) p(v_1) \\ &= p(v_N | v_{N-1}, \ldots, v_3, v_2, v_1) \ldots p(v_3 | v_2, v_1) p(v_2 | v_1) p(v_1) \end{aligned} \tag{1}$$

The variables v_N, v_{N-1}, ..., v_3, v_2, v_1 define a column vector:

$$\boldsymbol{v} = (v_N, v_{N-1}, \ldots, v_3, v_2, v_1)^T \tag{2}$$

To generate a realisation, $\boldsymbol{v}'$, from $p(\boldsymbol{v})$, first a realisation v_1', from $p(v_1)$, is generated. Next, using v_1' as a hard constraint, a realisation v_2' from $p(v_2 | v_1')$ is generated. This process is repeated until for all the coefficients of $\boldsymbol{v}' = (v_N', v_{N-1}', \ldots, v_3', v_2', v_1')^T$ a realisation has been generated. The order in which the variables are visited is often chosen randomly, although theoretically this is not necessary.

Now assume that precise, direct measurements $d_1, d_2, \ldots, d_M$ of the values v_1, v_2, ..., v_M are available, where $M<N$, and consider the following decomposition:

Fig. 1 Model with two locations

$$p(\boldsymbol{v}|d_M,\ldots,d_1) = p(v_N,\ \ldots, v_{M-1}|v_M,\ldots,v_1)p(v_M,\ \ldots, v_1|d_M,\ldots,d_1) \tag{3}$$

Sampling from $p(v_M, \ldots, v_1|d_M, \ldots, d_1)$ is straightforward. Since the measurements are precise, it is a matter of assigning the values of the measurements, $d_M, \ldots, d_1$, to the corresponding variables $v_M, \ldots, v_1$. Samples of the remaining variables, $v_N, \ldots, v_{M-1}$, can be generated using the decomposition technique that was used for the unconditional pdf.

In practice, the available measurements are not precise and are contaminated with noise. Using the same decomposition that is used to generate geostatistical models will produce incorrect results. This can be understood by examining a geostatistical problem with only two locations. The variables that are involved in this exercise are v_1 and v_2, and the corresponding measured values are d_1 and d_2 (Fig. 1).

The posterior pdf can be decomposed as follows:

$$p(v_2, v_1|d_2, d_1) = p(v_2|v_1, d_2, d_1)p(v_1|d_2, d_1) \tag{4}$$

The first term in the decompositions can be simplified to

$$p(v_2|v_1, d_2, d_1) = p(v_2|v_1, d_2) \tag{5}$$

When the value of v_1 is known, a measurement, d_1, thereof, does not add any extra information and can be skipped from the list of conditions. It would be convenient if the decomposition in Eq. 6 was valid, since then, Eq. 7 would be valid:

$$p(v_1|d_2, d_1) = p(v_1|d_1) \tag{6}$$

$$p(v_2, v_1|d_2, d_1) = p(v_2|v_1, d_2)p(v_1|d_1). \tag{7}$$

With this decomposition, it would be possible to generate a sample for v_1 and subsequently for v_2, also. This decomposition could easily be extended to any number of locations, and sequential sampling would be correct, even in a situation with imperfect measurements. This is done in many geostatistical packages, and it is based on the implicit assumption that Eq. 7 is correct.

Unfortunately, the desired decomposition is wrong. This can be understood by studying the influence diagram that is shown in Fig. 2, overleaf.

A measurement d_2 of the value of v_2 constrains the possible value of v_2 and the value of v_2 constrains the possible value of v_1. This means that a measurement d_2 of the value of v_2 also constrains the possible value of v_1. When a sequential sampling algorithm is used, based on Eq. 6, the influence of d_2 on v_1 is neglected. The assumption of Eq. 6 is only correct when v_1 and v_2 are completely independent.

Clearly, a model with two locations is primitive and expresses an oversimplified description of a geostatistical problem. However, this line of argument can be

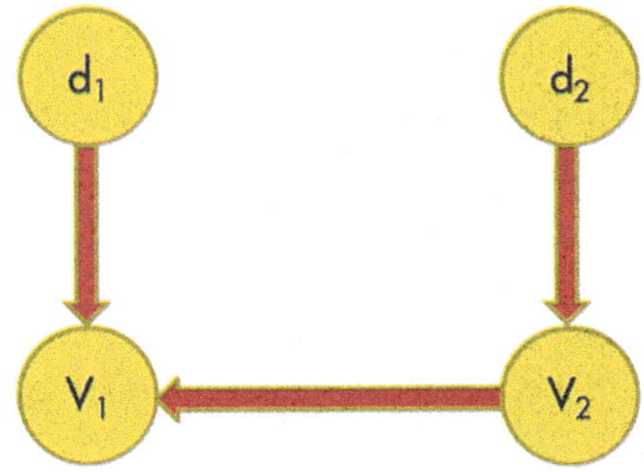

Fig. 2 Influence diagram for model with two locations

extended to models with a realistic number of locations. It demonstrates that the assumption of sequential sampling applicability is flawed. The question whether this is a problem in practice can be answered with a few numerical examples.

Example 1

This example is based on the model with two locations, shown in Fig. 1. The prior pdf that describes the relation between v_1 and v_2 is a Gaussian distribution, and its parameters are shown in Table 1.

The observed values for v_1 and v_2 are d_1 and d_2. These measurements are considered to be contaminated with noise, n_1 and n_2, such that

$$\begin{aligned} d_1 &= v_1 + n_1 \text{ and} \\ d_2 &= v_2 + n_2 \end{aligned} \tag{8}$$

It is assumed that the noise at location 1 is independent of the noise at location 2 and that the values of v_1 and v_2 are dependent on each other.

The pdfs that describe the probabilistic behaviour of the noise realisations are Gaussian distributions with an expectation value of 0.0 and a standard deviation of 2.0. The value of d_1 is -10.0 and the value of d_2 is 10.0.

The pdf that describes the probabilistic behaviour of v_1 and v_2 is a two-variable Gaussian distribution. The parameters of which are specified in Table 1.

The prior pdf is updated with the measurements using Bayes' rule, resulting in a posterior distribution. This is a Gaussian distribution from which the expectation values and the covariance matrix can be computed exactly. Samples from this posterior distribution are drawn in three different ways, as shown in Fig. 3:

1. Samples are drawn directly from the two-dimensional Gaussian distribution, using the Cholesky decomposition of the covariance matrix. This method is very efficient and is described extensively in literature. These samples are indicated with the label "Posterior 1".
2. Samples are drawn using sequential sampling, starting with location 2. These samples are indicated with the label "Posterior 2".
3. Samples are drawn using sequential sampling, starting with location 1. These samples are indicated with the label "Posterior 3".

Table 1 Parameters of Gaussian prior distribution

Statistic	Value
$<v_1>$	0.0
$<v_2>$	0.0
$\mathrm{Cov}\{v_1, v_1\}$	16.0
$\mathrm{Cov}\{v_1, v_2\}$	14.4
$\mathrm{Cov}\{v_1, v_2\}$	14.4
$\mathrm{Cov}\{v_1, v_2\}$	16.0

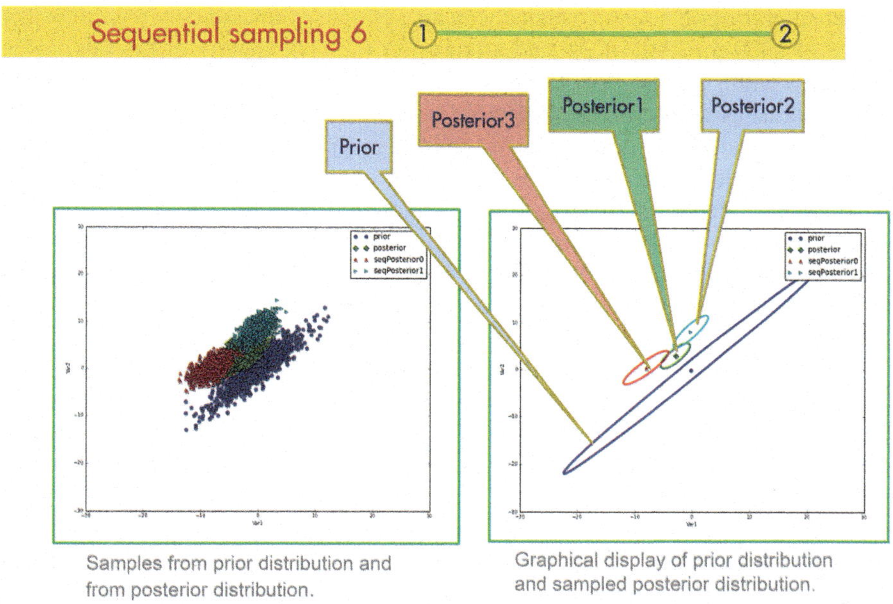

Fig. 3 Sampling results for Example 1

Samples drawn from the prior pdf are indicated with the label "Prior". The picture on the left-hand side of Fig. 3 shows the actual samples. The right-hand picture shows both the mean values of the samples that are obtained with each algorithm and the ellipsoids that are derived from the covariance matrices of each algorithm. The size of the ellipsoids is determined by the eigenvalues of the covariance matrices and the orientation by the eigenvectors. This means that the contours are one standard deviation away from the expectation value, in the direction of the eigenvectors.

Comparison of the samples from the "Prior" group with samples from the "Posterior" group clearly shows how strongly the samples are constrained by the measurements. Comparison of the samples from the "Posterior 1" group with samples from the "Posterior 2" group and the "Posterior 3" group shows the deterioration caused by the sequential sampling algorithm. The ellipsoids corresponding with the "Posterior 2" group and the "Posterior 3" group are shifted away from the ellipsoid of the "Posterior 1" group and they are also larger. This means that both the expectation values and the covariance matrices of the samples that are obtained with the sequential sampling algorithm are incorrect. This first, somewhat fictional example, has been specifically designed to show the problems with sequential sampling with a very simple model.

Example 2

The second example is a more realistic scenario. It consists of a model that is defined on a grid with 101 grid points in both x and y directions. The grid spacing between the grid points is 50 m, in both directions. The correlation between the variables at the grid nodes is defined with an isotropic exponential variogram with a range of 2000 m, a sill of 2.0 and a nugget of 0.0. The quality of the samples is inspected with χ^2, ψ^2 and φ^2 statistics, each of which is described in the Appendix. These statistics are different from those described by Leuangthong et al. (2004). The χ^2, ψ^2 and φ^2 statistics have the advantage that the statistical quality of the whole ensemble of samples can be examined.

First, the quality of the sequential sampling algorithm has been tested by sampling an ensemble of 100 realisations from the prior. Each location was constrained by a maximum of 314 nearest neighbours. This is a rather large number of neighbours, but it has been chosen such that the ensemble of samples that is generated with the sequential sampling algorithm has the correct χ^2, ψ^2 and φ^2 statistics. A map of typical realisation from this algorithm is shown in Fig. 4.

A variogram was estimated by using ten realisations, and a comparison was made with the variogram that was used to define the prior pdf, in Fig. 5. There is little discrepancy between the two variograms, which is one indication that the algorithm performs well.

From the ensemble of 100 realisations, which are generated from the prior, the average values of the diagnostic statistics, χ^2, ψ^2 and ϕ^2, are computed and compared with their expectation values. The results are listed in Table 2 and show that the average values are within one standard deviation of the expectations. This means that the sequential sampling algorithm does an excellent job when it is used to sample the prior.

When the sequential sampling algorithm is used to sample a posterior pdf, the results are not particularly encouraging. The observations that are used to constrain the geostatistical model have been generated by sampling a prior pdf that has also been defined with an isotropic, exponential variogram, with a range of 2000 m, a sill of 2.0 and a nugget of 0.0. An observation is generated for each grid location, as typical for seismic data. It is assumed that the observations are contaminated with additive noise, with a standard deviation of 1.0, and that the noise is not correlated between grid locations. These synthetic observations are shown in Fig. 6.

Realisations of the posterior distribution have been generated with the sequential sampling method and with a method that directly draws samples from the posterior. An example of a realisation from the sequential sampling algorithm is shown in Fig. 7.

The imprint of the observations is clearly visible in this realisation, but a more proper way to judge the validity of the realisations that are obtained from the sequential sampling algorithm is to compare them with realisation that is generated

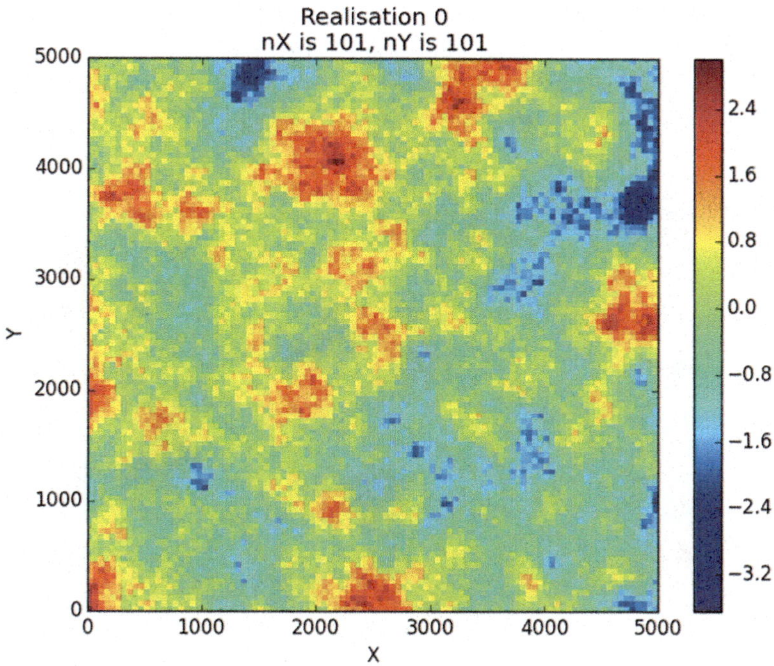

Fig. 4 Grid with prior sample for realisation 0. The x and y axes are in metres

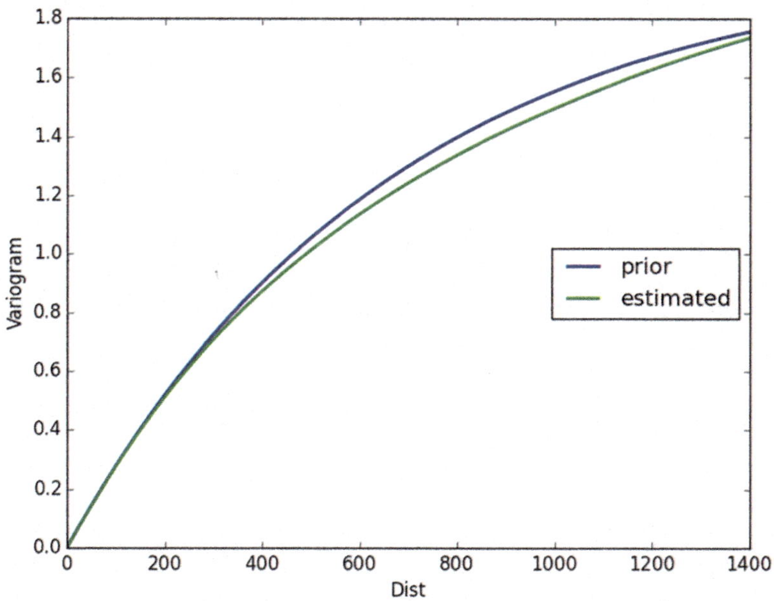

Fig. 5 Estimated variogram, prior, variogram of the prior; estimated, variogram estimated from ten realisations

Table 2 χ^2-type diagnostic statistics from sampling the prior

	Average	Expected	Difference	StdDev
χ^2	10,199.46	10,201.00	−1.54	14.28
ψ^2	10,096.87	10,098.99	−2.12	14.21
φ^2	102.60	102.01	0.59	1.44

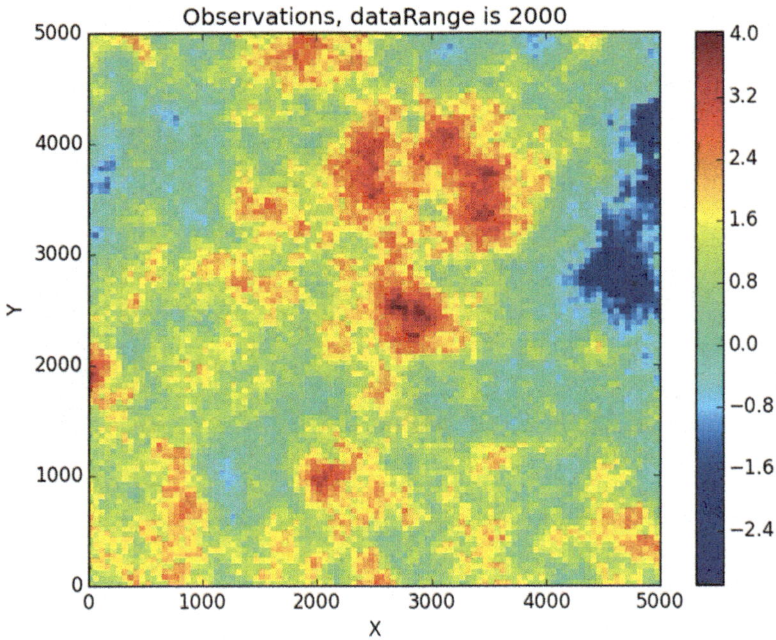

Fig. 6 Observations generated with an exponential variogram, with a range of 2000 m. *X* and *Y* axes are in metres

with an algorithm that correctly generates samples from the posterior. This algorithm is based on the Cholesky decomposition of the matrix of the true posterior distribution. This will be denoted by direct sampling and should not be confused with direct sequential sampling. With the chosen grid, the size of the covariance matrix was such that this was still possible.

First, the mean of the realisations obtained with sequential sampling method, shown in Fig. 8, is compared with the actual posterior expectation values, shown in Fig. 9.

A map of the differences between the means of the realisations obtained with the sequential sampling method and the actual posterior expectation values is shown in Fig. 10. These differences are significant when they are compared with the values of the observations and the standard deviation of the prior. The mean and the variance of all the realisations that are generated with the direct sampling method and the sequential sampling method are listed in Table 3.

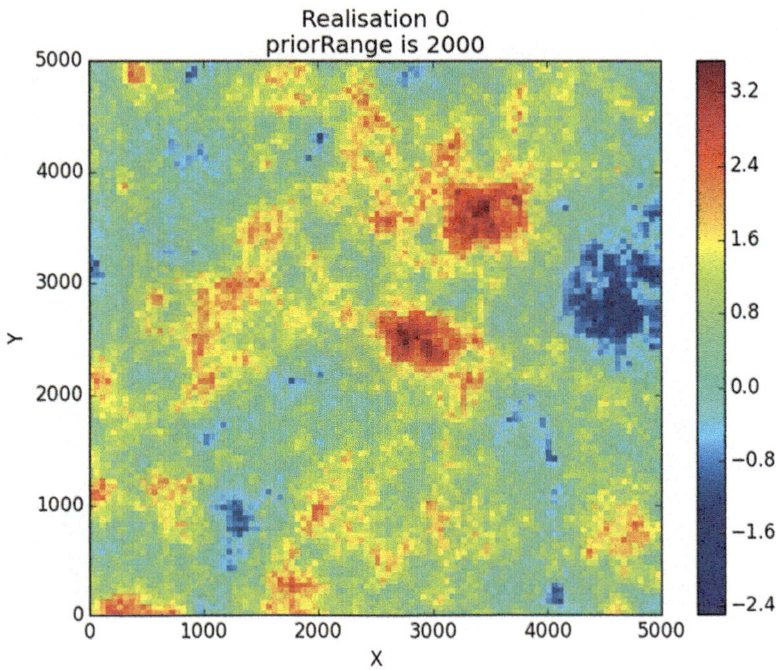

Fig. 7 Realisation from the posterior, generated with the sequential sampling method. *X* and *Y* axes are in metres

This shows that both the mean and the variance differ considerably between the two methods.

The prior variogram, the variogram estimated from the realisations that use the direct sampling method and the variogram estimated from the realisations that use the sequential sampling method are shown in Fig. 11. There is a remarkable difference between the latter two variograms, especially at larger distances. Note that the variogram that has been estimated from the realisations of the direct sampling method does not coincide with the variogram of the prior, although both the measurement and the prior are made with the same variogram. This stems from the fact that the covariance matrix of the prior and the posterior is different. A variogram that would have been estimated from the expectation value of the posterior would coincide with the prior variogram in this situation, but a realisation of the posterior also has a random component that is determined by the covariance matrix of the posterior. A more objective method to inspect the validity of a sampling method is to look at the values of the χ^2, the ψ^2 and the ϕ^2 statistics. Table 4 shows that their average values are far beyond the statistically feasible range.

The difference between the expected diagnostic statistics and the actual values of those differs by more than 100 standard deviations. This demonstrates that the sequential sampling used in practice does not properly sample the posterior pdf.

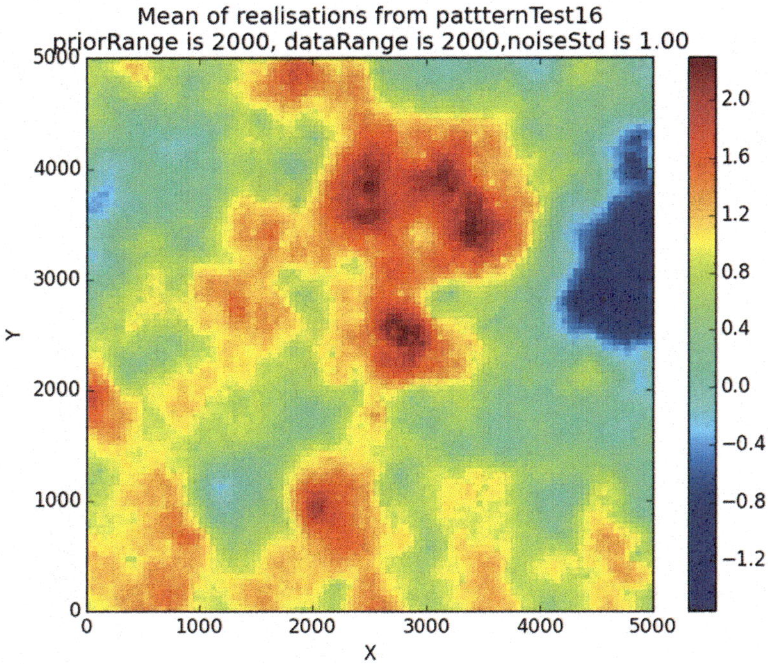

Fig. 8 Mean of realisations generated with sequential sampling

Seismically constraint stochastic models with lateral constraints suffer the same affliction. Seismic measurements are available at each location, but they are contaminated with noise. For this reason, a different method has been developed for seismically constrained stochastic modelling. In this method, each location is revisited many times, after an initial assignment of values to the model parameters at each location. The initial values can be obtained from the prior expectation values from a sample of the prior pdf.

Every time that a location is revisited, a local posterior pdf is computed using a local prior pdf that is constrained by the neighbourhood of that location. Subsequently a sample is generated from this local posterior pdf. This effectively creates a Metropolis algorithm that only creates local updates. The convergence speed of this algorithm is dependent on the type of the variogram that is used to construct the prior pdf.

3 Conclusion

In geostatistical modelling, the sequential sampling method is an efficient and reliable method to draw samples from a prior pdf. It also works well with the posterior that occurs as a result from taking accurate measurements from a few

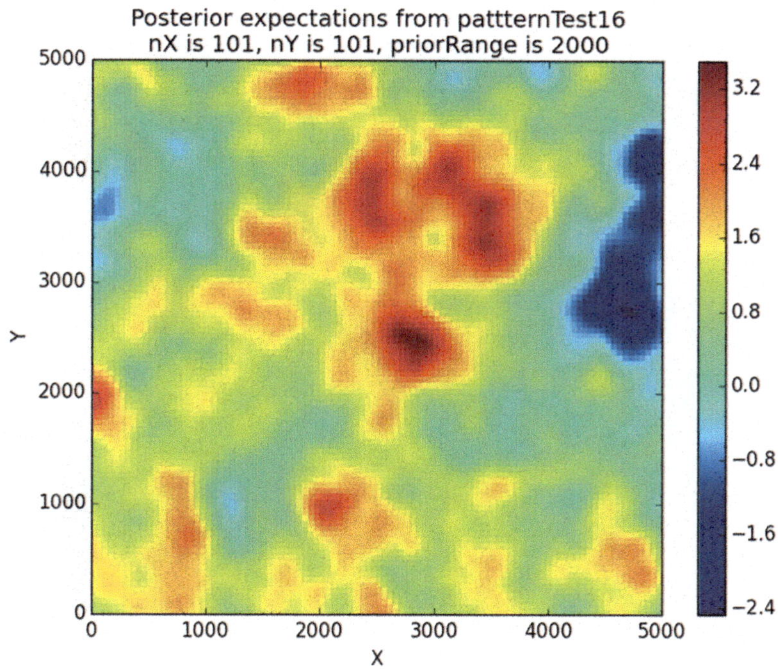

Fig. 9 Posterior expectation values

locations into account. The sequential sampling method does not correctly sample a posterior that results from taking inaccurate measurements into account from all of the locations that are considered in the model.

Appendix: Diagnostic Statistics

To judge the statistical correctness of the realisations that are generated by a geostatistical algorithm, different statistics can be used. These statistics are directly derived from these realisations. The obvious statistic is the variogram that can be estimated from the modelling results, but there are also statistics that reveal possible problems in much greater detail. A well-known statistic is the value of χ^2. This statistic is computed using all the realisations in an ensemble and it has a known expectation value and variance. This makes it suitable for such a test. It is also possible to define statistics that are similar to χ^2 but test different aspects of an ensemble. Those are called the ψ^2 and the φ^2 statistics. All these statistics are defined in next subsections.

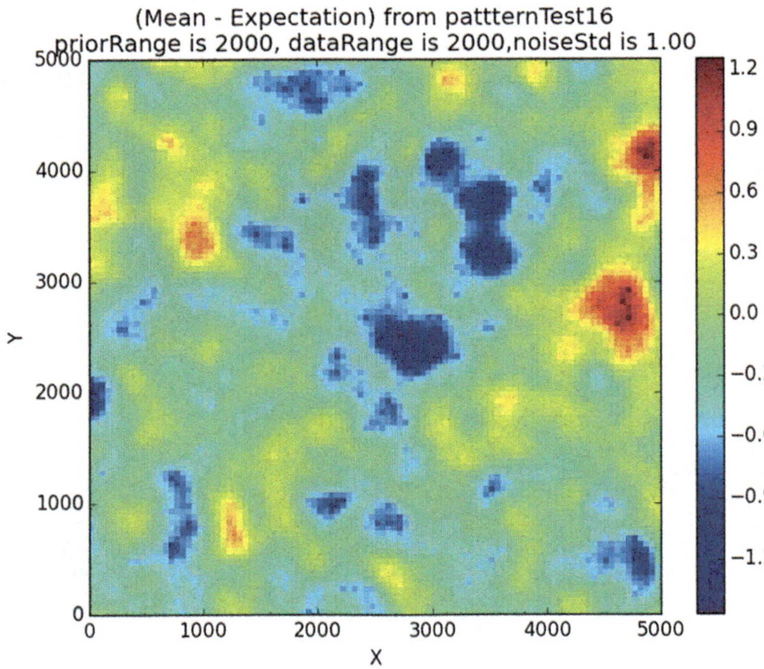

Fig. 10 Mean of realisations minus posterior expectation. *X* and *Y* axes are in metres

Table 3 Mean and variance from posterior sampling methods

	Direct	Sequential
Variance	0.875	0.634
Mean	0.766	0.978

χ^2 Statistics

The χ^2 test has been designed to test whether a realisation has been drawn from a specific normal distribution. Consider a multidimensional normal distribution $p(\boldsymbol{x})$ with an expectation value of $<\boldsymbol{x}>$ and a covariance matrix $\boldsymbol{C}_x$ as given by

$$p(\boldsymbol{x}) = \frac{1}{(2\pi)^{\frac{D}{2}}|\boldsymbol{C}_x|^{\frac{1}{2}}}\exp\left\{-\frac{1}{2}(\boldsymbol{x} - \langle\boldsymbol{x}\rangle)^T\boldsymbol{C}_x^{-1}(\boldsymbol{x} - \langle\boldsymbol{x}\rangle)\right\}, \tag{9}$$

where D is the dimension of vector $\boldsymbol{x}$. Assume that the ensemble that has to be tested consists of N realisations and is denoted by $\{x_1, x_1, \ldots, x_N\}$. The χ^2 value of realisation i is given by

$$\chi_i^2 = (\boldsymbol{x}_i - \langle\boldsymbol{x}\rangle)^T\boldsymbol{C}_x^{-1}(\boldsymbol{x}_i - \langle\boldsymbol{x}\rangle) \tag{10}$$

It can be shown that

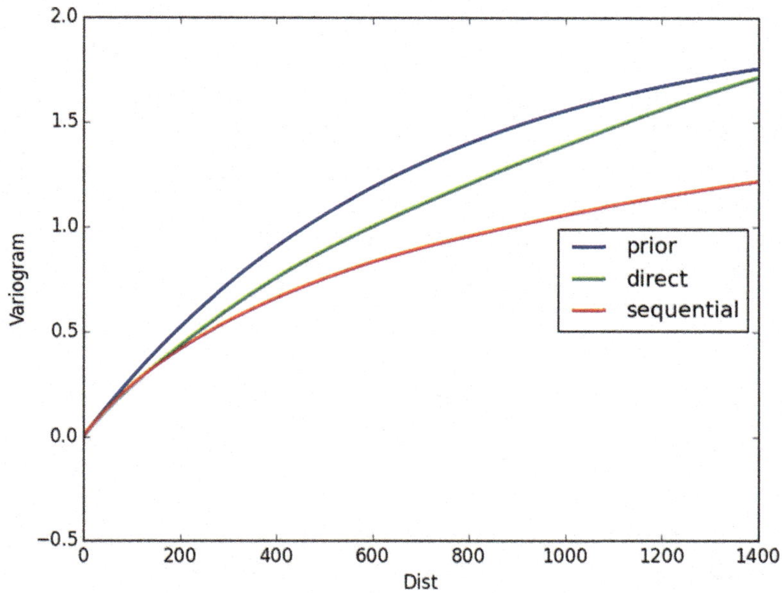

Fig. 11 Estimated variograms, prior, variogram of the prior; direct, variogram estimated from samples that are directly generated from the true posterior; sequential, variogram estimated from samples that are generated with sequential Gaussian simulation

Table 4 χ^2-type diagnostics for sequential samples of the posterior

	Average	Expected	Difference	StdDev
χ^2	14,174.93	10,201.00	3973.93	14.28
ψ^2	12,285.45	10,098.99	2186.46	14.21
ϕ^2	1889.48	102.01	1787.47	1.44

$$\begin{aligned} \langle \chi_i^2 \rangle &= D \text{ and} \\ \operatorname{Var}\{\chi_i^2\} &= 2D \end{aligned} \tag{11}$$

The average value of χ^2, taken over the whole ensemble, is

$$\overline{\chi^2} = \frac{1}{N}\sum_{i=1}^{N} \chi_i^2 \tag{12}$$

Since all the χ_i^2 values are independent, the expectation value and the variance of the average are equal to

$$\begin{aligned}\left\langle \overline{\chi^2} \right\rangle &= D \quad \text{and} \\ \operatorname{Var}\left\{ \overline{\chi^2} \right\} &= \frac{2D}{N}\end{aligned} \tag{13}$$

When the χ_i^2 values and $\overline{\chi^2}$ deviate too much from the expected value, this indicates that the vectors $\boldsymbol{x}_i$ have not been properly sampled from $p(\boldsymbol{x})$. This may have several reasons. The samples may be generated with a bias, or the "shape" of the ensemble does not match the "shape" that is described by the covariance matrix. To make a distinction between these possibilities, two other statistics, ψ^2 and φ^2, have been defined.

ψ^2 Statistics

The ψ^2 value of realisation i is given by

$$\psi_i^2 = (\boldsymbol{x}_i - \bar{\boldsymbol{x}})^T \boldsymbol{C}_x^{-1} (\boldsymbol{x}_i - \bar{\boldsymbol{x}}) \tag{14}$$

where

$$\bar{\boldsymbol{x}} = \frac{1}{N} \sum_{i=1}^{N} \boldsymbol{x}_i \tag{15}$$

The expectation value and the variance and the standard deviation of ψ_i^2 are

$$\begin{aligned}\left\langle \psi_i^2 \right\rangle &= D \frac{N-1}{N} \quad \text{and} \\ \operatorname{Var}\left\{ \psi_i^2 \right\} &= 2D \frac{N-1}{N}\end{aligned} \tag{16}$$

The ensemble average of ψ_i^2 is given by

$$\overline{\psi^2} = \frac{1}{N} \sum_{i=1}^{N} \psi_i^2 \tag{17}$$

The expectation value and the variance of $\overline{\chi}^2$ are

$$\begin{aligned}\left\langle \overline{\psi^2} \right\rangle &= D \frac{N-1}{N} \quad \text{and} \\ Var\left\{ \overline{\psi^2} \right\} &= 2D \frac{N-1}{N^2}\end{aligned} \tag{18}$$

φ^2 Statistics

While ψ^2 is a good diagnostic for the "shape" of an ensemble, there is another statistic that is suitable to diagnose a bias. This one is called φ^2 and is defined by

$$\varphi^2 = (\bar{x} - \langle \boldsymbol{x} \rangle)^T \boldsymbol{C}_{\bar{x}}^{-1} (\bar{x} - \langle \boldsymbol{x} \rangle) \tag{19}$$

The expectation value, the variance and the standard deviation of φ^2 are

$$\begin{aligned} \langle \varphi^2 \rangle &= \frac{D}{N} \text{ and} \\ \mathrm{Var}\{\varphi^2\} &= \frac{2D}{N^2} \end{aligned} \tag{20}$$

The following useful relation between $\overline{\chi^2}$, $\overline{\psi^2}$ and ϕ^2 can be derived:

$$\overline{\chi^2} = \overline{\psi^2} + \varphi^2 \tag{21}$$

Bibliography

Deutsch CV (2002) Geostatistical reservoir modeling. Oxford University Press, New York

Gelderblom P, Leguijt J (2010) Geological constraints in model-based seismic inversion, SEG Denver 2010, 2010-2825 SEG Conference Paper

Leuangthong O, McLennan JA, Deutsch CV (2004) Minimum acceptance criteria for geostatistical realizations. Nat Resour Res 13(3):131–141

A New High-Order, Nonstationary, and Transformation Invariant Spatial Simulation Approach

Amir Abbas Haji Abolhassani, Roussos Dimitrakopoulos, and Frank P. Ferrie

Abstract This paper presents a new high-order, nonstationary sequential simulation approach, aiming to deal with the typically complex, curvilinear structures and high-order spatial connectivity of the attributes of natural phenomena. Similar to multipoint methods, the proposed approach employs spatial templates and a group of training images (TI). A coarse template with a fixed number of data points and a missing value in the middle is used, where the missing value is simulated conditional to a data event found in the neighborhood of the middle point of the template, under a Markovian assumption. Sliding the template over the TI, a pattern database is extracted. The parameters of the conditional distributions needed for the sequential simulation are inferred from the pattern database considering a set of weights of contribution given for the patterns in the database. Weights are calculated based on the similarity of the high-order statistics of the data event of the hard data compared to those of the training image. The high-order similarity measure introduced herein is effectively invariant under all linear spatial transformations.

Following the sequential simulation paradigm, the template chosen is sequentially moved on a raster path until all missing points/nodes are simulated. The high-order similarity measure allows the approach to be fast as well as robust to all possible linear transformations of a training image. The approach respects the hard data and its spatial statistics, because it only considers TI replicate data events with similar high-order statistics. Results are promising.

A.A.H. Abolhassani (✉)
Electrical and Computer Engineering Department, McGill University, Montreal, QC, Canada

COSMO – Stochastic Mine Planning, McGill University, Montreal, QC, Canada
e-mail: amir@cim.mcgill.ca

R. Dimitrakopoulos
COSMO – Stochastic Mine Planning, McGill University, Montreal, QC, Canada
e-mail: roussos.dimitrakopoulos@mcgill.ca

F.P. Ferrie
Electrical and Computer Engineering Department, McGill University, Montreal, QC, Canada
e-mail: ferrie@cim.mcgill.ca

J.J. Gómez-Hernández et al. (eds.), *Geostatistics Valencia 2016*, Quantitative Geology and Geostatistics 19, DOI 10.1007/978-3-319-46819-8_6

1 Introduction

Since the early 1990s (Guardiano and Srivastava 1993), several new approaches to geostatistical simulation have been developed to move this area of research beyond the second-order or two-point methods and their limits. These methods, well developed to date, are placed under the term multipoint statistics (MPS). The basic idea of MPS approaches is that the two-point statistical tools (variogram, covariance, correlogram) of a given attribute of interest are replaced by a so-called training image (TI). The TI is then used as a source to provide multiple point statistics and spatial relations that are used along with the hard data to generate simulated realizations for the attributes of interest. The first implemented multipoint method is SNESIM (Strebelle 2002) and it is TI driven; thus, similar to all conventional MPS methods, the simulated realizations reproduce the high-order spatial relations in the TI. As a result, in applications with a dense set of hard data, the complex spatial relations in the data are overridden by those in the TI and do not assist with the application of MPS methods to applications with relatively dense datasets.

Several MPS methods are well known to date; examples are discussed next. FILTERSIM (Zhang et al. 2006) is based on the classification of both data and TIs using linear filters; it is efficient and also sensitive to the shape and size of the spatial template, and the number and form of the filters are employed. The direct sampling method (Mariethoz et al. 2010), in the other hand, does not produce a pattern database from the TIs like FIMTERSIM. Instead, in a multigrid simulation setup, first coarser grid nodes are simulated, a template is chosen about a simulation point, and the data event is extracted. This data event is then compared with the data event of a randomly chosen TI pattern, in an L2-norm distance basis. If the distance is less than a threshold, then the pattern is pasted onto the simulation grid; otherwise, another TI pattern is randomly selected and compared, and so on. The direct sampling simulation method is fast and effective in simulating based on sparse data with a given TI set and is also TI driven. Mariethoz et al. (2010) show the influence of data statistics on realizations generated from this method as a dataset increases. Other pattern-based MPS methods include the one suggested by Arpat and Caers (2007); first, a pattern database is generated by sliding a fixed template over the TIs. Then, the data event on the grid used for the simulation and at each location is compared with the data event of the pattern database, and the one with the least L2-norm distance is chosen and pasted on the grid nodes involved. Abdollahifard and Faez (2012) first cluster the pattern database generated from the TI using a Bayesian framework. Then each cluster is modeled by a set of simple linear features and the extraction of features for each incomplete pattern on the simulated grid follows. Honarkhah and Caers (2010), instead of building a raw pattern database, classify TIs using some fixed simple features and compare them, using the L2-norm, to the same features extracted from a point of the grid being simulated. The most similar pattern is pasted onto the grid nodes, until all of them are visited.

MPS simulations based on Markov methods are also available. An example in this category is the Markov mesh model by Stien and Kolbjørnsen (2011). A unilateral raster path is chosen for the data, visiting all of the points in a left-to-right and up-to-down fashion. On a chosen spatial template, a joint distribution is considered for the random field. The parameters of this distribution are then estimated from the TIs. Sampling of the local distribution function generates the realizations. These models in general are biased toward the chosen path, but are vastly used for simulating attributes of petroleum reservoirs.

In a relatively recent approach to stochastic simulation, the high-order sequential simulation extends the conventional second-order sequential simulation methods to higher orders (Mustapha and Dimitrakopoulos 2010a, 2011). The HOSIM approach first chooses a simulation point at random and considers N-nearest conditioning data as data event. A special template is built by connecting the data event to the simulation point. The conditional probability distribution function (CPDF) of the simulated grid node given the data event is then modeled by a series of weighted orthogonal functions called Legendre polynomials. The weight of each Legendre term is calculated by matching a set of particular spatial statistics, the so-called spatial cumulants that are generated from the available data. The TI is only used to complement the spatial cumulants of the available data. Note that SNESIM is similar to HOSIM with a main difference that the model used for the CPDF is much simpler and the method is TI driven.

The present manuscript presents a new patch-based high-order method, which utilizes high-order spatial statistics in the pattern's structure. Two notable differences from past approaches are that (a) it is nonstationary and that (b) it utilizes a set of TIs, rather than one, while it is data driven. Both the above address significant topics. The proposed method follows a multigrid simulation process (coarser simulation grid nodes are simulated first and become the conditioning data for finer simulation grid points. A simulation point is randomly selected from the grid to be simulated and an order $N + 1$ template is selected, based on N nearest hard data points, i.e., the data event. In addition, a N-dimensional high-order statistical feature vector, introduced herein, is calculated from the data event. By sliding the template over the TI, a pattern database is produced and then mapped into a N-dimensional high-order statistical feature space. The similarity measure of two feature vectors is defined as the weighted Euclidean distance between two vectors. The distribution of the simulated point is then estimated from the TIs using the maximum likelihood estimate (MLE) considering the similarity of each pattern. A sample is then drawn from the distribution as the realization of the simulation point. This process is continued until all grid nodes are simulated. The feature vector introduced in this paper is isotropic, that is, it is invariant to any linear transformation of the training image including rotation and transposition. This feature is also fast to calculate enabling the simulator to incorporate a large amount of TIs. The simulator is nonstationary and respects the hard data and its high-order statistics, that is, only TI patterns with similar statistics are used for simulation. As a result, the simulations are data driven.

The following sections present the proposed simulation approach; then the results of initial tests using known datasets follow. Conclusions and future work complete the presentation.

2 The Proposed Method

2.1 Overview

The goal of the proposed method is to simulate a random field $Z(\boldsymbol{x})$, in a sequential multigrid process, given a grid with nodes $\boldsymbol{x}$, a training image (TI), $z(\boldsymbol{Y}) = \{z(y_1), \ldots, z(y_M)\}$, with nodes $\boldsymbol{Y} = \{y_1, \ldots, y_M\}$ in the training image, and a sparse set of N hard data $z(x_i), i \in [1, \ldots, N]$ on a regular grid $x_i, i \in [1, \ldots, N]$. The hierarchy of the sequential simulation is illustrated in Fig. 1. The blue nodes represent the hard data (a). In the first sequence, each red node is conditioned on the four closest blue nodes and simulated. The lines represent the conditions in (b). This continues sequentially until all the nodes in the grid are simulated (Fig. 1c–e). Unicolor lines in each figure represent the spatial templates, connecting the conditioning data and a single node to be simulated. The size of the templates reduces after each sequence to maintain the same number of conditioning nodes. This is the natural representation of the multigrid approach for the simulation.

2.2 High-Order Transformation Invariant Simulation (HOSTSIM)

At each sequence, the path is chosen randomly and saved into a vector containing the indices of the visiting nodes. Each successive random variable $Z(\boldsymbol{x})$ at node $\boldsymbol{x}$ is conditioned to n-nearest neighbors, selected from the set of previously simulated nodes and the hard data $\{\boldsymbol{x}_1, \ldots, \boldsymbol{x}_n\}$ (Goovaerts 1998). A template is formed spatially by connecting each conditioning data, $\boldsymbol{x}_i$, to node $\boldsymbol{x}$, presented by a lag vector $L_x = \{\boldsymbol{h}_1, \ldots, \boldsymbol{h}_n\} = \{\boldsymbol{x}_1 - \boldsymbol{x}, \ldots, \boldsymbol{x}_n - \boldsymbol{x}\}$. Consequently, the neighborhood of $\boldsymbol{x}$ is denoted by $\boldsymbol{N}_x = \{\boldsymbol{x} + \boldsymbol{h}_1, \ldots, \boldsymbol{x} + \boldsymbol{h}_n\}$ and the data event is denoted by $\boldsymbol{d}_{\boldsymbol{N}_x} = \{z(\boldsymbol{x}_1), \ldots, z(\boldsymbol{x}_n)\}$. The goal is to estimate and draw a sample from the probability, $P(Z(\boldsymbol{x})|\boldsymbol{d}_{\boldsymbol{N}_x}, z(\boldsymbol{Y}))$, of each successive $Z(\boldsymbol{x})$ in the next visiting node on the path given its data event $\boldsymbol{d}_{\boldsymbol{N}\boldsymbol{x}}$ and the TI, $z(\boldsymbol{Y})$. This probability is intractable for continuous variables. Alternatively, a model with a set of parameters, $\boldsymbol{\theta} \in \boldsymbol{\Theta}$, may be chosen, to represent this probability independent from the data event and TI. The parameter $\boldsymbol{\theta}$ is optimized to express the data event and TI. Thus, this probability can be decomposed using the Bayes product and sum rule:

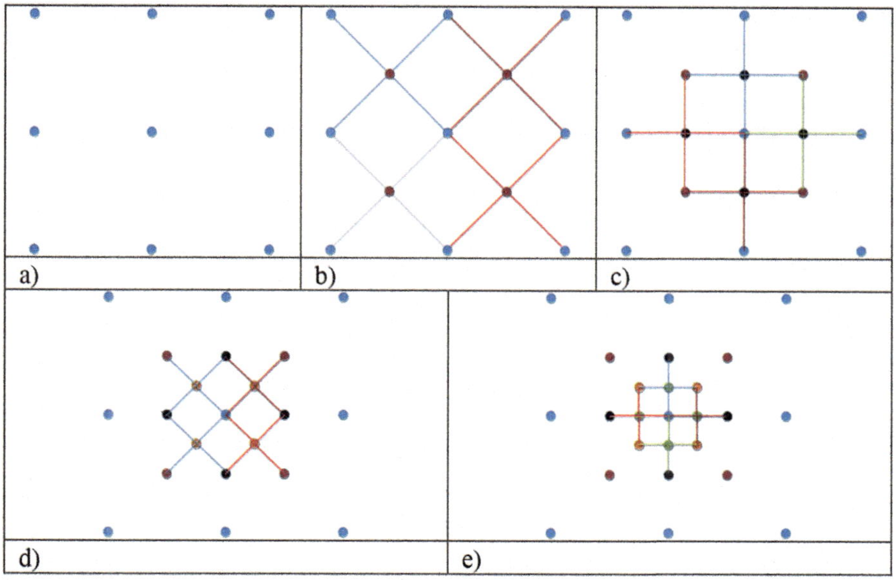

Fig. 1 The hierarchy of the sequential multigrid simulation for $n = 5$, the connecting lines present the conditioning dependencies. (**a**) The hard data. (**b**) *Red* nodes are simulated conditioned on their four nearest neighbors. (**c**) *Black* nodes are simulated conditioned on both *blue* and *red* nodes from previous sequence. (**d**) *Oranges* are simulated conditioned on previous simulated nodes and hard data. (**e**) *Greens* are simulated in the last sequence. At each sequence the resolution of the grid doubles

$$P(Z(\boldsymbol{x})|\boldsymbol{d}_{N_x}, z(\boldsymbol{Y})) = \int\limits_{\theta\in\Theta} P(Z(\boldsymbol{x})|\boldsymbol{\theta})P(\boldsymbol{\theta}|\boldsymbol{d}_{N_x}, z(\boldsymbol{Y}))d\boldsymbol{\theta}. \tag{1}$$

Estimation of the integrand further simplifies Eq. 1. Figure 2 represents the term $P(\boldsymbol{\theta}|d_{N_x}, z(\boldsymbol{Y}))$ as a function of $\boldsymbol{\theta}$. The contribution of this function is negligible except for a narrow band near an optimal value for the parameter's maximum likelihood estimate, $\boldsymbol{\theta}_{MLE}$.

Consequently, Eq. 1 could be estimated as in Eq. 2:

$$P(Z(\boldsymbol{x})|\boldsymbol{d}_{N_x}, z(\boldsymbol{Y})) \approx P(Z(\boldsymbol{x})|\boldsymbol{\theta}_{MLE})\overbrace{P(\boldsymbol{\theta}_{MLE}|\boldsymbol{d}_{N_x}, z(\boldsymbol{Y}))\Delta\boldsymbol{\theta}}^{\text{constant}}$$
$$= \frac{1}{C_{MLE}} P(Z(\boldsymbol{x})|\boldsymbol{\theta}_{MLE}). \tag{2}$$

In Eq. 2, C_{MLE} can be regarded as the normalization factor to produce a valid probability, $C_{MLE} = \int\limits_{z\in Z} P(z|\boldsymbol{\theta}_{MLE})dz$.

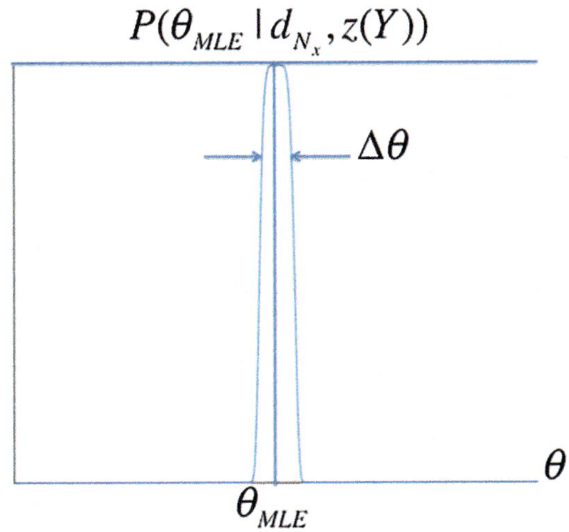

Fig. 2 Presentation of $P(\boldsymbol{\theta}|\boldsymbol{d}_{N_x}, z(Y))$ as a function of $\boldsymbol{\theta}$. In practice this probability is negligible except at a narrow $\Delta\boldsymbol{\theta}$ band near $\boldsymbol{\theta}_{MLE}$

Equation 2 implies the probability distribution function of the node $\boldsymbol{x}$ can be calculated if $\boldsymbol{\theta}_{MLE}$ is known. To estimate $\boldsymbol{\theta}_{MLE}$ based on Fig. 2, $\boldsymbol{\theta} = \boldsymbol{\theta}_{MLE}$ when $P(\boldsymbol{\theta}|\boldsymbol{d}_{N_x}, z(\boldsymbol{Y}))$ is maximum for $\boldsymbol{\theta} \in \boldsymbol{\Theta}$.

$$\boldsymbol{\theta}_{MLE} = \underset{\boldsymbol{\theta}\in\Theta}{\text{argmax}}\, P(\boldsymbol{\theta}|\boldsymbol{d}_{N_x}, z(\boldsymbol{Y})). \tag{3}$$

Using Bayes theorem:

$$\boldsymbol{\theta}_{MLE} = \underset{\boldsymbol{\theta}\in\Theta}{\text{argmax}} \frac{P(z(\boldsymbol{Y})|\boldsymbol{\theta}, \boldsymbol{d}_{N_x}) \overbrace{P(\boldsymbol{\theta})}^{\text{uniform prior}}}{\underbrace{P(z(\boldsymbol{Y}))}_{\text{independant from } \theta}}. \tag{4}$$

One assumes a uniform prior in parameters space $\boldsymbol{\Theta}$ and note that the marginal probability in TI, $P(z(\boldsymbol{Y}))$, is independent from $\boldsymbol{\theta}$. Hence, Eq. 4 becomes

$$\boldsymbol{\theta}_{MLE} = \underset{\boldsymbol{\theta}\in\Theta}{argmax}\, P(z(\boldsymbol{Y})|\boldsymbol{\theta}, \boldsymbol{d}_{N_x}). \tag{5}$$

Each node in TI is only conditioned on its neighbors and the parameters set $\boldsymbol{\theta}$. Hence, the joint distribution in Eq. 5 can be decomposed further.

$$\boldsymbol{\theta}_{MLE} = \underset{\boldsymbol{\theta}\in\Theta}{argmax} \prod_{i=1}^{M} P\left(z(\boldsymbol{y}_i)|\boldsymbol{\theta}, \boldsymbol{d}_{N_x}, \boldsymbol{d}_{N_{y_i}}\right). \tag{6}$$

A probability is maximized if the logarithm of that probability is maximized:

$$\boldsymbol{\theta}_{MLE} = \underset{\boldsymbol{\theta}\in\Theta}{argmax} \sum_{i=1}^{M} \log P\left(z(y_i)|\boldsymbol{\theta}, \boldsymbol{d}_{N_x}, \boldsymbol{d}_{N_{y_i}}\right). \tag{7}$$

At maximum the derivative with respect to $\boldsymbol{\theta}$ should be zero:

$$\frac{\partial}{\partial \boldsymbol{\theta}}\left(\sum_{i=1}^{M} \log P\left(z(y_i)|\boldsymbol{\theta}, \boldsymbol{d}_{N_x}, \boldsymbol{d}_{N_{y_i}}\right)\right) = 0. \tag{8}$$

Equation 8 should be solved for $\boldsymbol{\theta}_{MLE}$. By choosing a model with a set of parameters $\boldsymbol{\theta}$, solving Eq. 8 for $\boldsymbol{\theta}_{MLE}$ is straightforward.

2.3 Simulation Model

The exponential family is used herein to model the likelihood function in Eq. 8 with the parameter set $\boldsymbol{\theta} = \{\theta_1, \theta_2\}$.

$$P\left(z(y_i)|\boldsymbol{\theta}, \boldsymbol{d}_{N_x}, \boldsymbol{d}_{N_{y_i}}\right) = \frac{1}{c}\exp\left(-\frac{1}{2}\omega\left(\boldsymbol{d}_{N_x}, \boldsymbol{d}_{N_{y_i}}, \sigma_0^2\right)\frac{(z(y_i)-\theta_1)^2}{\theta_2}\right). \tag{9}$$

$\omega\left(\boldsymbol{d}_{N_x}, \boldsymbol{d}_{N_{y_i}}, \sigma_0^2\right)$ is introduced as the similarity measure (SM) of the data event $\boldsymbol{d}_{Nx}$ and $\boldsymbol{d}_{Nyi}$. It ensures that the TI patterns with similar data events contribute more toward building the likelihood function in Eq. 9. The SM is defined as

$$\omega\left(\boldsymbol{d}_{N_x}, \boldsymbol{d}_{N_{y_i}}, \sigma_0^2\right) = \exp\left(-\frac{\frac{1}{2}\boldsymbol{D}\left(\boldsymbol{d}_{N_x}, \boldsymbol{d}_{N_{y_i}}\right)^T \boldsymbol{D}\left(\boldsymbol{d}_{N_x}, \boldsymbol{d}_{N_{y_i}}\right)}{\boldsymbol{\Sigma}\left(\boldsymbol{d}_{N_x}, \sigma_0^2\right)}\right). \tag{10}$$

where $\boldsymbol{D}(\boldsymbol{d}_{Nx}, \boldsymbol{d}_{Nyi})$ is introduced as the high-order statistics disparity vector. $\Sigma\left(\boldsymbol{d}_{N_x}, \sigma_0^2\right)$ is the covariance matrix of the disparity vector and is calculated using the calculus of variations.

2.4 High-Order Statistics Disparity Vector

In this paper, a particular form of disparity vector is presented, which is isotropic and compares the high-order statistics of two data events. Most of the MP simulation methods choose an L2-norm for the disparity measure (Arpat and Caers 2007; Chatterjee and Dimitrakopoulos 2012; Honarkhah and Caers

2010; Mariethoz et al. 2010, and Mustapha et al. 2013). When considering two sets of data events $\boldsymbol{d}_{N_x} = \{x_1, \ldots, x_n\}, \boldsymbol{d}_{N_y} = \{y_1, \ldots, y_n\}$ of order n, to develop an isotropic L2-norm disparity measure, one must compare all possible ordering of these two data events, which results in $n \times n!$ number of operations. For $n = 5$ the number of operations is 600. This is computationally expensive and can only operate on small-size TIs. The following method is employed to reduce the computing time.

First, Vieta's formula (Funkhouser 1930) is used to calculate the coefficients of two polynomials $p_s(X)$ and $p_t(Y)$ with the roots equal to the data events $\boldsymbol{d}_{Nx}$ and $\boldsymbol{d}_{Ny}$, respectively.

$$p_s(X) = X^n + s_1 X^{n-1} + \ldots + s_n.$$
$$\begin{cases} s_1 = x_1 + x_2 + \ldots + x_n \\ s_2 = x_1(x_2 + x_3 + \ldots + x_n) + x_2(x_3 + \ldots + x_n) + x_{n-1}x_n \\ \vdots \\ s_n = x_1 x_2 \ldots x_n \end{cases} . \tag{11}$$

$$p_t(Y) = Y^n + t_1 Y^{n-1} + \ldots + s_n.$$
$$\begin{cases} t_1 = y_1 + y_2 + \ldots + y_n \\ t_2 = y_1(y_2 + y_3 + \ldots + y_n) + y_2(y_3 + \ldots + y_n) + y_{n-1}y_n \\ \vdots \\ t_n = y_1 y_2 \cdots y_n \end{cases} . \tag{12}$$

These could be regarded as two mappings $\boldsymbol{d}_{Nx} \to \mathbf{s} = \{s_1, \ldots s_n\}$ and $\boldsymbol{d}_{Ny} \to \boldsymbol{t} = \{t_1, \ldots t_n\}$. The advantage of these mappings is that they are invariant to the ordering of the domain. This invariance results from the coefficients of a polynomial being invariant to the order of the roots in a set due to Eqs. 11 and 12. $\boldsymbol{s}$ and $\boldsymbol{t}$ are in a particular form of high-order moments:

$$\begin{aligned} s_m &= \sum\nolimits_{k \in K} N(k) m^{(u)}(k), \\ m^{(u)}(k) &= \frac{1}{N(k)} \sum\nolimits_{j=1}^{N(k)} x_j \prod\nolimits_{l \in k} x_{j+l}, \end{aligned} \tag{13}$$

$$\begin{aligned} t_m &= \sum\nolimits_{k \in K} N(k) m^{(u)}(k), \\ m^{(u)}(k) &= \frac{1}{N(k)} \sum\nolimits_{j=1}^{N(k)} y_j \prod\nolimits_{l \in k} y_{j+l}, \end{aligned} \tag{14}$$

where, $\boldsymbol{k} = \{k_1, \ldots, k_{u-1}\}$ and $N(k)$ is the support for estimating the moment. The high-order statistics disparity vector is defined as follows:

$$\boldsymbol{D}\left(\boldsymbol{d}_{N_x}, \boldsymbol{d}_{N_{y_i}}\right) = \boldsymbol{s}^T - \boldsymbol{t}^T. \tag{15}$$

It is worth mentioning that the number of operations for this new disparity measure is reduced dramatically to $n \times 2^{n-1}$ per simulation node versus $n \times n!$ for the L2-norm. For $n = 5$, $op(L2 - norm) = 600$ and $op(L2 - ord) = 80$.

3 Results from HOSTSIM and Comparisons

The dataset used in this section is the Stanford V Reservoir dataset (Mao and Journel 1999). This exhaustive dataset consists of a 3D grid with porosity values. The grid consists of $130 \times 100 \times 30$ nodes, i.e., $X \times Y \times Z$. Here this dataset is cropped into a grid of $100 \times 100 \times 30$ in order to perform some linear transformation on the data, e.g., rotation. For each simulation, a layer, $\mathbf{Z} \in \{1,...,30\}$, is selected as the ground truth of the simulation, referred to as the original image. This image is then down-sampled to produce the hard dataset, containing N points. All layers except layer Z are considered as TI for each simulation. The results of the simulation produced by HOSTISIM are compared, with an order 5 template, with the ones produced by FILTERSIM, with a search grid size 11×11 and inner patch size 7×7. Note that the TIs provided for HOSTISIM are rotated 90° clockwise. The simulations are generated for layers $z = \{1,...,4\}$, from top to bottom in Figs. 3 and 4, first with $N = 625$ number of hard data points, 6.25 % of the original data, in Fig. 3, and second with $N = 169$ number of hard data points, 1.69 % of the original data, in Fig. 4. For each simulation on Figs. 3 and 4, from left to right, the original image, HOSTISIM and FILTERSIM typical realizations are all presented. For each set of results, the histogram of the original image and two simulations are also plotted. As a robust quantitative comparison, for each simulation, ten realizations are generated by HOSTISIM and FILTERSIM methods, and for each one the PSNR and SSIM scores (Wang et al. 2004) are calculated and averaged for each method over all ten realizations and provided in Figs. 3 and 4. For every single case, HOSTISIM outperforms FILTERSIM, visually by better representing the channels and low-contrast structures of the original exhaustive image, with higher PSNR and SSIM scores, and by better matching the histogram. Tables 1 and 2 are presenting the average PSNR and SSIM for HOSTISIM and FILTERSIM.

4 Comparing the Computing Times

The current implementation of the method is in Matlab, using the GPU parallel computation library. It has not been optimized nor developed in C, nor Python, as of yet. On the other hand, FILTERSIM has been optimized and developed in Python and is available in the SGEMS software platform. Despite this disparity in optimization, we ran some tests to compare them as is. The system used for the tests was a Unix OS server with eight cores Xeon CPU, 3.500 GHz with 8 MB cache size and 64 GB DDR4 memory and Nvidia Tesla k40c GPU with 12 GB DDR5 memory with 2880 cores.

For each method, two sets of tests were performed. Each set of tests consisted of generating 10 separate simulations and averaging the computing time for each simulation. The hard data used for all cases were 12×12 real-valued data on a regular grid. The goal was to generate a realization on a 100×100 SG, for each

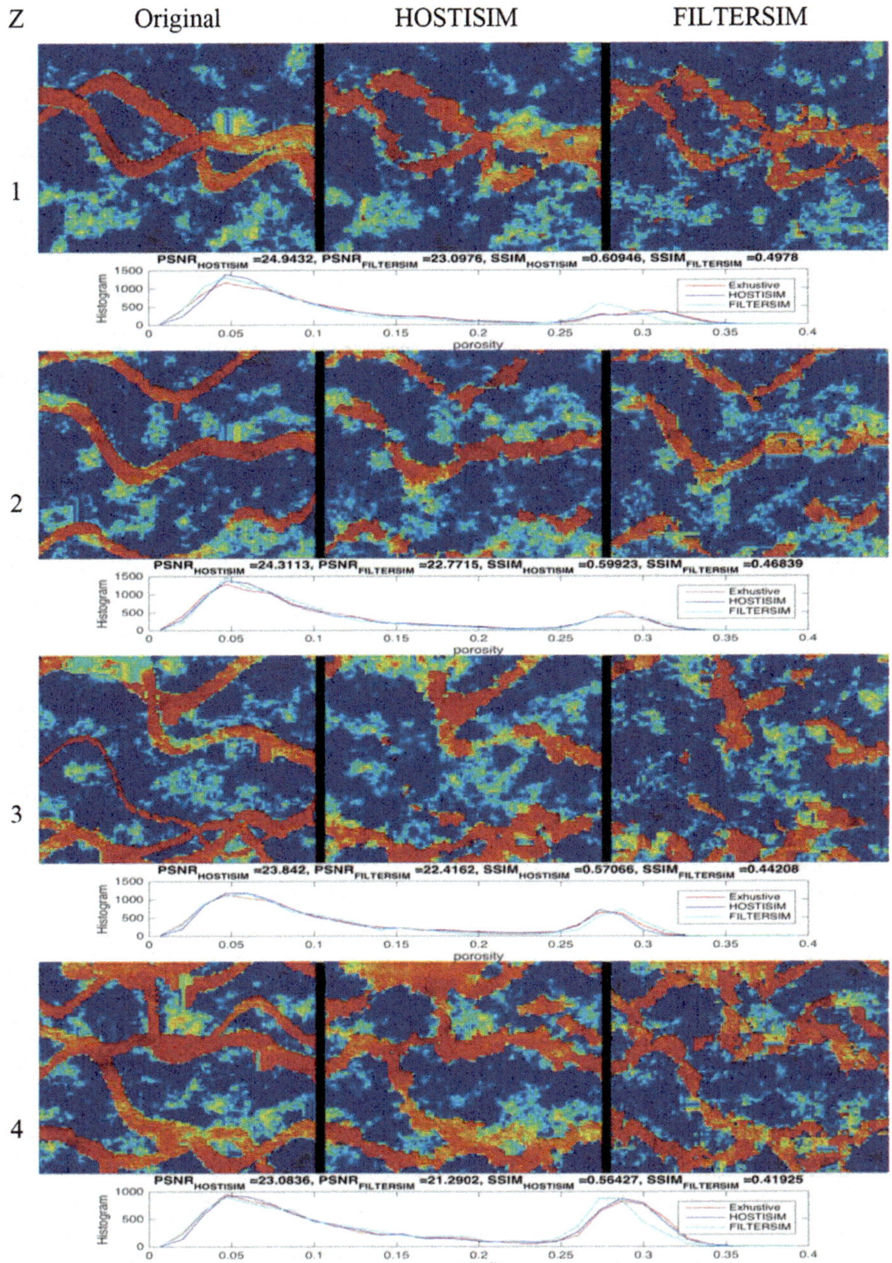

Fig. 3 The results generated given 625 hard data (6.25 %). From top to bottom, layers Z = 1,. . .4. From left to right, original image, HOSTISIM and FILTERSIM simulations. On the *bottom* the histograms of each set are presented (Red*:* Exhaustive, *dark blue*: HOSTISIM and *light blue*: FILTERSIM)

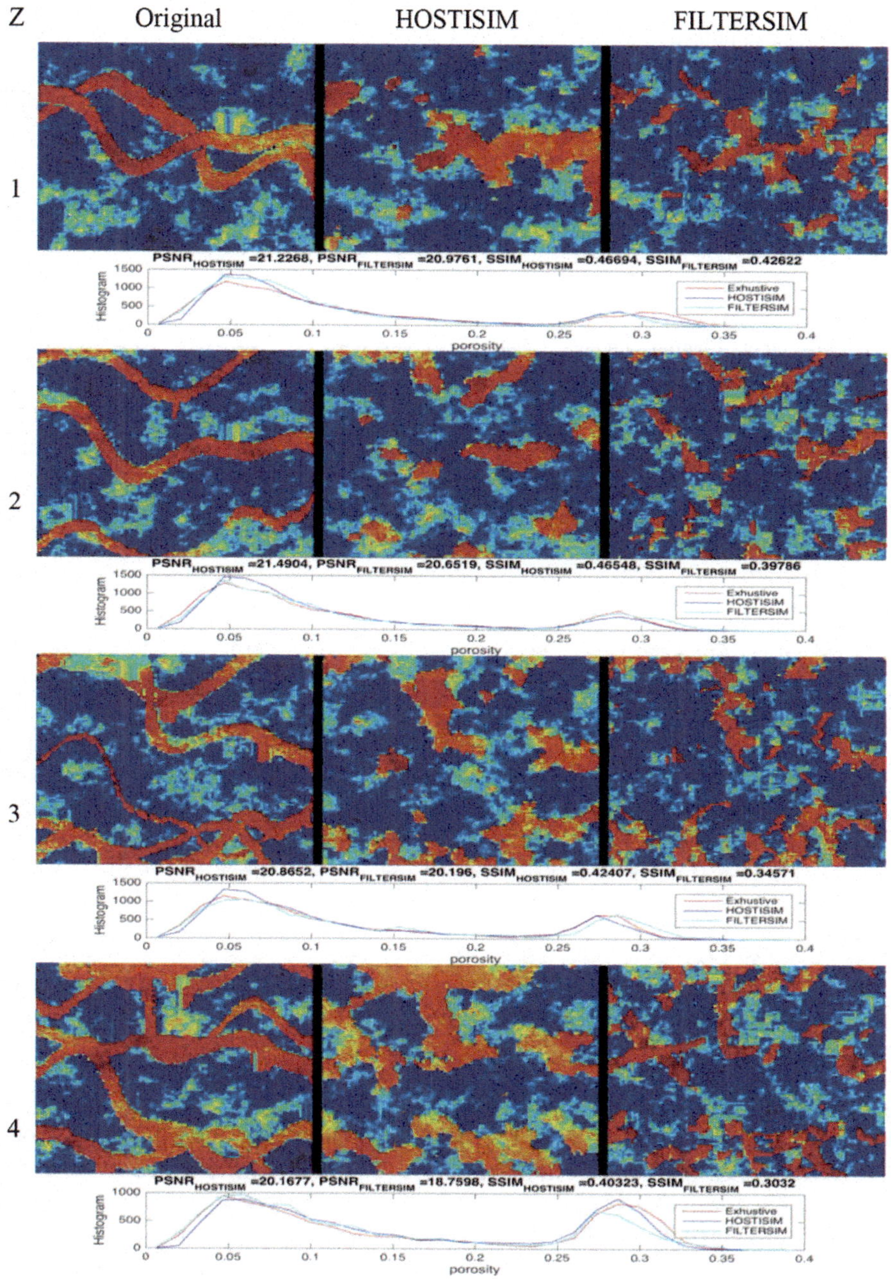

Fig. 4 The results generated given 169 hard data (1.69 %). From top to bottom, layers Z = 1,. . .4. From *left* to *right*, original image, HOSTISIM and FILTERSIM simulations. On the *bottom* the histograms of each set are presented (*Red*: Exhaustive, *dark blue*: HOSTISIM and *light blue*: FILTERSIM)

Table 1 Comparing average PSNR and SSIM for HOSTISIM and FILTERSIM methods with $N = 625$ number of hard data points (6.25 %)

	PSNR		SSIM	
Z	HOSTISIM	FILTERSIM	HOSTISIM	FILTERSIM
1	24.94	23.10	0.61	0.50
2	24.31	22.77	0.60	0.47
3	23.84	22.42	0.57	0.44
4	23.10	21.29	0.56	0.42

Table 2 Comparing average PSNR and SSIM for HOSTISIM and FILTERSIM methods with $N = 169$ number of hard data points (1.69 %)

Z	PSNR		SSIM	
Z	HOSTISIM	FILTERSIM	HOSTISIM	FILTERSIM
1	21.23	20.98	0.47	0.43
2	21.49	20.65	0.47	0.40
3	20.87	20.20	0.42	0.35
4	20.17	18.76	0.40	0.30

For more and higher-resolution results, please visit the link bellow: http://cim.mcgill.ca/~amir/HOSTISIM.html

Table 3 Comparing average computation time for HOSTISIM and FILTERSIM methods

	FILTERSIM	HOSTISIM	***
Test#1	8 s	3 s	+9 s GPU initialization in Matlab
Test#2	134 s	42 s	+12 s GPU initialization in Matlab

*** Almost constant, overhead GPU initialization time in Matlab

case. Each TI was a 100 × 100 real-valued image. In the first test, only 1 TI was used and in the second test 29 TIs were used. The average computing times are presented in Table 3.

5 Conclusions

A high-order, stochastic, and transformation invariant simulation method, HOSTISIM, is introduced in this paper. This method sequentially simulates the nodes in the grid to be simulated. At each sequence the previous simulated nodes are also considered as conditioning data. Hence, the size of the template shrinks at each new sequence. A high-order statistical disparity vector is introduced to calculate the distance between the data event of a pattern in the grid with a pattern in the TI. This disparity vector is designed to be isotropic and invariant against any linear transformation of the TI patterns. The PDF of the simulating node is then estimated using a likelihood function based on the disparity vector. This method is easy to implement and fast in performance. Since the number of operations is dramatically reduced compared to isotropic L2-norm distance measures, a large TI set can be processed. This method is nonstationary, uses a disparity measure to choose a pattern from the TI, never becomes biased from the TI, and always respects the high-order statistics of the hard data's inherent structure. This method will be expanded to accommodate irregular data locations and implemented to simulate in 3D.

Acknowledgment Funding was provided by the Natural Sciences and Engineering Research Council of Canada (NSERC) Discovery Grant 239019 and mining industry partners of the COSMO Laboratory (AngloGold Ashanti, Barrick Gold, BHP Billiton, De Beers Canada, Kinross Gold, Newmont Mining, and Vale) and the Group for Research in Decision Analysis (GERAD). Thanks are given to Prasun Lala for his assistance.

Bibliography

Abdollahifard MJ, Faez K (2012) Stochastic simulation of patterns using Bayesian pattern modeling. Comput Geosci 17(1):99–116

Arpat GB, Caers J (2007) Conditional simulation with patterns. Math Geol 39(2):177–203

Chatterjee S, Dimitrakopoulos R (2012) Multi-scale stochastic simulation with wavelet-based approach. Comput Geosci 45(3):177–189

Dimitrakopoulos R, Mustapha H, Gloaguen E (2010) High-order statistics of spatial random fields: exploring spatial cumulants for modeling complex non-Gaussian and non-linear phenomena. Math Geosci 42(1):65–99

Funkhouser HG (1930) A short account of the history of symmetric functions of roots of equations. Am Math Mon 37(7):357–365

Goovaerts P (1998) Geostatistics for natural resources evaluation. Cambridge University Press, Cambridge

Guardiano FB, Srivastava RM (1993) Multivariate geostatistics: beyond bivariate moments. Geosatistics Tróia'92, Springer, pp 133–144

Honarkhah M, Caers J (2010) Stochastic simulation of patterns using distance-based pattern modeling. Math Geosci 42(5):487–517

Mao S, Journel AG (1999) Generation of a reference petrophysical and seismic 3D data set: the Stanford V reservoir. Report—Stanford Center for Reservoir Forecasting Annual Meeting, Stanford. URL: http://ekofisk.stanford . edu/SCRF.htmlSS

Mariethoz G, Renard P, Straubhaar J (2010) The direct sampling method to perform multiple-point geostatistical simulations. Water Resour Res 46(11):W11536. doi: 10.1029/2008WR007621

Minniakhmetov I, Dimitrakopoulos R (2016a) Joint high-order simulation of spatially correlated variables using high-order spatial statistics. Math Geosci, in press

Minniakhmetov I, Dimitrakopoulos R (2016b) A high-order, data-driven framework for joint simulation of categorical variables. GEOSTAT2016, in this volume

Mustapha H, Dimitrakopoulos R (2010a) A new approach for geological pattern recognition using high-order spatial cumulants. Comput Geosci 36(3):313–334

Mustapha H, Dimitrakopoulos R (2010b) High-order stochastic simulation of complex spatially distributed natural phenomena. Math Geosci 42(5):457–485

Mustapha H, Dimitrakopoulos R (2011) HOSIM: a high-order stochastic simulation algorithm for generating three-dimensional complex geological patterns. Comput Geosci 37(9):1242–1253

Mustapha H, Chatterjee S, Dimitrakopoulos R (2013) CDFSIM: efficient stochastic simulation through decomposition of cumulative distribution functions of transformed spatial patterns. Math Geosci 46(1):95–123

Stien M, Kolbjørnsen O (2011) Facies modeling using a Markov mesh model specification. Math Geosci 43(6):611–624

Strebelle S (2002) Conditional simulation of complex geological structures using multiple-point statistics. Math Geol 34(1):1–21

Wang Z, Bovik AC, Sheikh HR, Simoncelli EP (2004) Image quality assessment: from error visibility to structural similarity. IEEE Trans Image Process 13(4):600–612

Zhang T, Switzer P, Journel A (2006) Filter-based classification of training image patterns for spatial simulation. Math Geol 38(1):63–80

A Truly Multivariate Normal Score Transform Based on Lagrangian Flow

Ute Mueller, K. Gerald van den Boogaart, and Raimon Tolosana-Delgado

Abstract In many geostatistical applications, a transformation to standard normality is a first step in order to apply standard algorithms in two-point geostatistics. However, in the case of a set of collocated variables, marginal normality of each variable does not imply multivariate normality of the set, and a joint transformation is required. In addition, current methods are not affine equivariant, as should be required for multivariate regionalized data sets without a unique, canonical representation (e.g., vector-valued random fields, compositional random fields, layer cake models). This contribution presents an affine equivariant method of Gaussian anamorphosis based on a flow deformation of the joint sample space of the variables. The method numerically solves the differential equation of a continuous flow deformation that would transform a kernel density estimate of the actual multivariate density of the data into a standard multivariate normal distribution. Properties of the flow anamorphosis are discussed for a synthetic application, and the implementation is illustrated via two data sets derived from Western Australian mining contexts.

1 Introduction

Many of the standard geostatistical simulation algorithms for continuous data are based on the assumption of Gaussianity. However, this assumption is often violated, and so prior to simulation, a transformation is required. Some notable exceptions are direct sequential simulation and multiple indicator simulation. In

U. Mueller (✉)
School of Science, Edith Cowan University, 270 Joondalup Drive, Joondalup, WA 6027, Australia
e-mail: u.mueller@ecu.edu.au

K.G. van den Boogaart • R. Tolosana-Delgado
Helmholtz Zentrum Dresden-Rossendorf, Helmholtz Institute for Resources Technology, Chemnitzerstrasse 40, 09599 Freiberg, Saxony, Germany
e-mail: boogaart@hzdr.de; r.tolosana@hzdr.de

J.J. Gómez-Hernández et al. (eds.), *Geostatistics Valencia 2016*, Quantitative Geology and Geostatistics 19, DOI 10.1007/978-3-319-46819-8_7

particular, in the multivariate setting, the requirement of multivariate Gaussianity poses a challenge. While it is straightforward to obtain marginal Gaussian distributions, for example, via individual quantile matching, their univariate Gaussianity does not imply multivariate Gaussianity. This observation is not new, for example, the stepwise conditional transformation (Leuangthong and Deutsch 2003) and the projection pursuit multivariate transform (Barnett et al. 2014) are methods designed to overcome this limitation. However, for some applications, it is important that the multivariate normal score transform is affine equivariant, that is, the anamorphosis of any data set subjected to any affine transformation is an affine transformation of the anamorphosis of the original data set. Examples of such data sets are compositional data and vectorial data where the information provided is relative. In these cases there is no canonical representation of the data in real vector space, and the introduction of a non-affine equivariant step into the analysis would break the independence of the (geo)statistical workflow from the chosen representation. However, the aforementioned methods are by construction not affine equivariant. Van den Boogaart et al. (2015) introduced an affine equivariant multivariate Gaussian anamorphosis which depends on two parameters. It has been shown in a case study that the back transformation of normal random variables reproduces the original distribution well.

The aim of this paper is to analyze the performance and choice of parameters of the anamorphosis in relation to the normality properties of the resulting multivariate distribution and spatial correlation structure. A synthetic example and two samples drawn from mining contexts are used to illustrate the behavior.

The data considered here are compositional. These are common in the mining industry, but also in geochemical surveys, where a whole suite of elements are analyzed. The information on the samples provided is in relative terms; measurements are positive and, if all constituents of the sample were analyzed, add to a constant, usually 100 %. Thus, the sample space is a D-dimensional simplex:

$$\left\{(z_1(u_\alpha),\ z_2(u_\alpha),\ \ldots, z_D(u_\alpha)) \middle| \sum_{i=1}^{D} z_i(u_\alpha) = 100,\ z_i(u_\alpha) > 0,\ u_\alpha \in \mathbf{A}\right\} \tag{1}$$

where z_i denotes the ith element that was analyzed and u_α a sample location within the study region $\mathbf{A}$. Data of this nature cannot be modeled "raw," and even before a transformation to Gaussian space is performed, a log-ratio transform ought to be performed to remove the positivity requirement as well as the constant sum constraint (Aitchison 1986). Several transformations are available for that purpose, one of which is the alr transformation, given as

$$x_i(u_\alpha) = \mathrm{alr}_D\left(z_i(u_\alpha)\right) = \ln\left(z_i(u_\alpha)/z_D(u_\alpha)\right) \tag{2}$$

for $i = 1,\ldots,D$. The choice of the divisor is arbitrary, so that there are in fact D distinct alr transformations, given a regionalized variable of D components. To ensure independence of simulation results from the choice of log-ratio transformation, the transformation to normality should be affine equivariant, that is, equivariant under full-rank linear transformations and invariant under translations. Transformations to multivariate normality that rely on quantile matching do not have this property, and so simulation results based on a multi-Gaussian transform of that type might depend on the log-ratio transform used. Therefore, a multivariate normal score transform is required that does not rely on quantile matching and that treats the vector as a whole and not as the Cartesian product of some individual quantities.

2 The Flow Anamorphosis

The normal score transformation is achieved by means of a smooth deformation of the underlying space transporting the probability mass from a kernel density estimate of the original distribution to a multivariate standard Gaussian distribution. The starting point is a set of vectors in D-dimensional space, $\{\mathbf{z}_\alpha \in \mathbb{R}^D | \alpha = 1, \ldots, n\}$, which will be regarded as centers of smoothing Gaussian kernels with spread σ_0. A mass transport is created by moving and scaling each kernel into the target distribution through an abstract time interval $t \in [0,1]$. This mass flow induces a flow field. Each point of the space is transported with this flow field from its original position in raw space to its position in transformed space. Specifically we assume that at time t, the center $\mathbf{z}_\alpha(t)$ of the αth kernel is

$$\mathbf{z}_\alpha(t) = (1 - t)\mathbf{z}_\alpha \tag{3}$$

where $t \in [0,1]$ and the spread is given by

$$\sigma(t) = (1 - t)\sigma_0 + t\sigma_1. \tag{4}$$

The spreads of all kernels evolve simultaneously according to Eq. 4. For each location, time-dependent scores are defined relative to a decomposition $\boldsymbol{M} = \boldsymbol{R}\boldsymbol{R}^T$ of the variance covariance matrix of the input data:

$$\mathbf{s}_\alpha(\mathbf{z}, t) = \frac{1}{\sigma(t)} \boldsymbol{R}^{-1}(\mathbf{z} - \mathbf{z}_\alpha(t)) \tag{5}$$

The local position X of a mass point of the αth kernel, which is at time t at location $\mathbf{z}$ as a function of time τ, is given by

$$X(\tau, \mathbf{z}, \alpha, t) = \mathbf{z}_\alpha(\tau) + \boldsymbol{R}\mathbf{s}_\alpha(\mathbf{z}, t)\sigma(\tau) \tag{6}$$

and its flow velocity is the derivative of X with respect to τ:

$$\mathbf{v}_\alpha(\mathbf{z}, t) = -\mathbf{z}_\alpha + \boldsymbol{R}\mathbf{s}_\alpha(\mathbf{z}, t)(\sigma_1 - \sigma_0). \tag{7}$$

The average velocity of the total mass at **z** is defined as the weighted average:

$$\mathbf{v}(\mathbf{z}, t) = \sum_{\alpha=1}^{n} \omega_\alpha(\mathbf{z}, t)\mathbf{v}_\alpha(\mathbf{z}, t) \tag{8}$$

where $\omega_\alpha(\mathbf{z},t)$ are weights derived from the time-dependent density of the αth kernel at this location and time. The weights are functions of the Mahalanobis distance between the vector **z** and the kernel center $\mathbf{z}_\alpha(t)$:

$$\omega_\alpha(\mathbf{z}, t) = \frac{\exp\left(-\|\mathbf{s}_\alpha(\mathbf{z}, t)\|^2\right)}{\sum_{\alpha=1}^{n} \exp\left(-\|\mathbf{s}_\alpha(\mathbf{z}, t)\|^2\right)} \tag{9}$$

The differential equation describing the motion of a parcel at time t is then

$$\frac{\partial}{\partial t} p_z(t) = \mathbf{v}(p_z(t), t), \; p_z(0) = \mathbf{z}, \tag{10}$$

where $p_z(t)$ denotes the position at time t of a vector starting at **z** at time 0.

The flow velocity **v** in Eq. 10 reflects the Eulerian formulation of the field. The equation of motion in Eq. 10 describes the deformation of the space to Gaussian space. The transformation is by construction affine equivariant (van den Boogaart et al. 2015). For the weights in Eq. 9, this follows from a property of the Mahalanobis distances established by Filzmoser and Hron (2008), and a straightforward calculation shows that if **A** is a full-rank matrix and **b** a vector, then $\mathbf{v}_\alpha(\mathbf{Az} + \mathbf{b}, t) = \mathbf{A}\mathbf{v}_\alpha(\mathbf{z}, t) - \mathbf{b}$, which in turn leads to the desired equivariance property of the transformation. So in the context of compositions, the multivariate anamorphosis constructed here provides a normal score transform which guarantees that results do not depend on the choice of log-ratio transformation.

An example of the flow anamorphosis for a bivariate data set consisting of 100 samples is shown in Fig. 1 together with the marginal distributions of the input ($\tau = 0$) and output ($\tau = 1$) data. The value of σ_0 was set to 0.02 and that of σ_1 was set to 1. The trajectories show that the motion is nonlinear. The application of the anamorphosis proceeds by first moving points toward the origin and then pushing them back out toward the periphery, resulting in a more circular scatter plot.

3 Some Properties of the Flow Anamorphosis

From the construction it is clear that the flow anamorphosis depends on the choice of suitable values for σ_0 and σ_1. The initial spread σ_0 controls the deformation of the underlying space: the greater the value, the weaker the deformation. It thus exercises the greatest influence over the shapes of the final marginal distributions and multivariate normality of the normal scores. The ranges of the marginal increase with decreasing σ_0. Multivariate normality tests performed on test data indicate that the choice of σ_0 depends both on the number of variables and the sample size. In the case of an initial distribution as shown in Fig. 1, the output distribution is bivariate normal for values of $\sigma_0 \leq 0.2$. This was ascertained for 100 replicates generated from this model (Fig. 2). Here an energy test was applied to determine multivariate normality (Szekely and Rizzo 2013).

Note that the p-value of a test under the null hypothesis follows a uniform distribution on the interval [0, 1]. Thus, the target behavior should be represented by a broad, centered boxplot covering the whole range and not a constant value of 1.0 (which would imply overfitting).

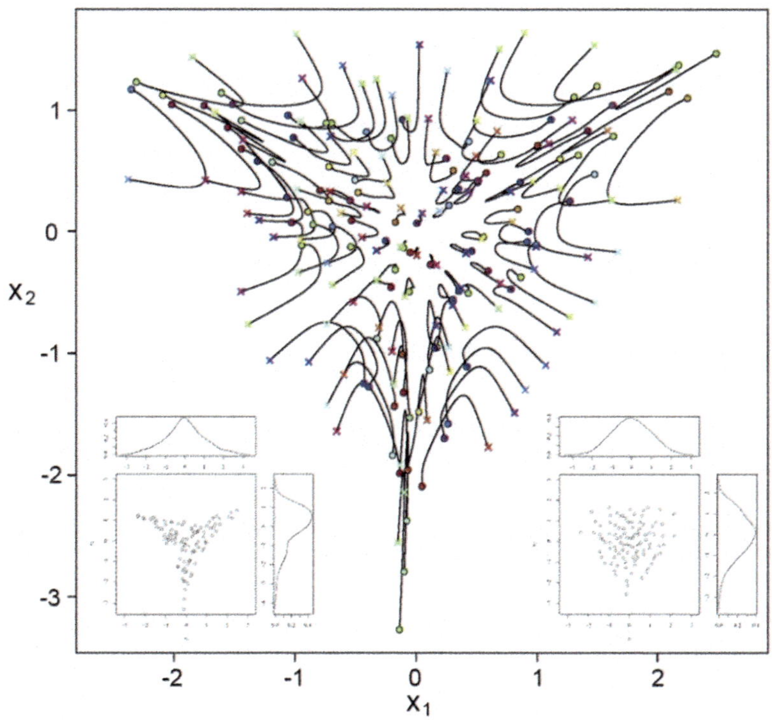

Fig. 1 Trajectories of points under flow deformation: scatter diagrams of raw data (*left*) and transformed data (*right*) with marginal distributions are superimposed in the bottom

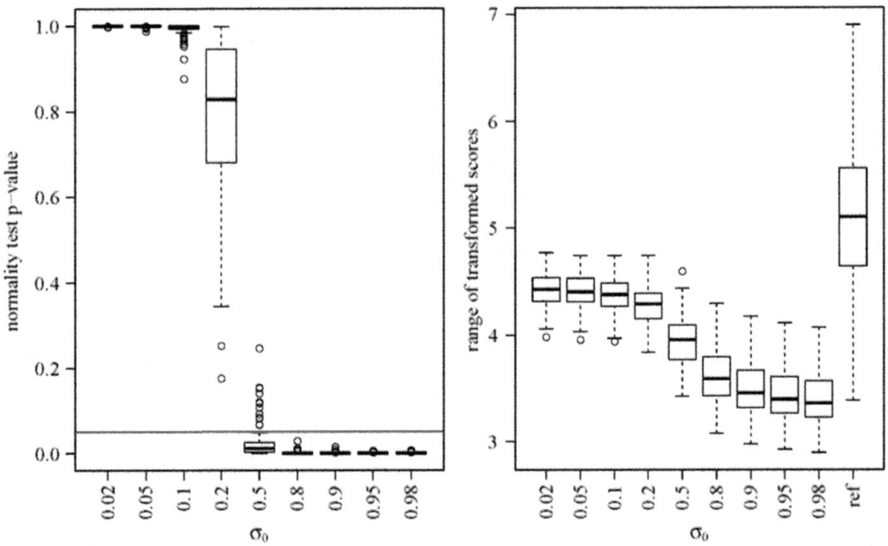

Fig. 2 Boxplots of p-values of normality tests (*left*) and ranges (*right*) based on different values of σ_0

The value of σ_1 does not impact on multivariate normality but does influence the ranges of the transformed distributions. A simplistic choice is $\sigma_1 = 1$, resulting in marginal distributions with standard deviation less than unity. The relative change in the ranges of the marginal distributions when $\sigma_1 = f(\sigma_0)$ for some suitably chosen positive function f is $f(\sigma_0) - 1$ compared to those with $\sigma_1 = 1$. These properties are illustrated in Table 1 for the case of the data shown in Fig. 1 with $\sigma_0 = 0.8$, 0.5, and 0.2 and $\sigma_1 = 1, \sigma_1 = 1 + \sigma_0, \sigma_1 = \sqrt{1 + \sigma_0^2}$, respectively.

4 Application to Spatial Data

Here the application of the flow anamorphosis is illustrated with two examples from Western Australian mining contexts. Both data come from a mining bench, so can be deemed two-dimensional. The first set is derived from an iron ore mine, the second from a manganese mine. The data configurations are shown in Fig. 3. For both sets locations were transformed to maintain confidentiality.

4.1 Iron Ore Data Set

This data set consists of 400 samples, and the variables of interest are Fe, Al_2O_3, and SiO_2. Since the data are compositional, but the three elements considered only form a sub-composition, a filler variable was introduced to achieve closure. The

Table 1 p-values for multivariate normality tests (Korkmaz et al. 2014) applied to the transforms of the data shown, ranges of the variables and relative change in range compared to $\sigma_1 = 1$ for selected values of σ_0 and σ_1

Statistic	0.80	0.50	0.20	0.80	0.50	0.20	0.80	0.50	0.20
	$\sigma_1 = 1$			$\sigma_1 = 1 + \sigma_0$			$\sigma_1 = \sqrt{1 + \sigma_0^2}$		
Mardia	0.00	0.00	0.10	0.00	0.00	0.09	0.00	0.00	0.10
Henze-Zirkler	0.00	0.02	0.90	0.00	0.02	0.89	0.00	0.02	0.90
Royston	0.00	0.03	0.64	0.00	0.03	0.64	0.00	0.03	0.64
Energy	0.00	0.01	0.90	0.00	0.01	0.90	0.00	0.02	0.92
$R(V_1)$	3.26	3.69	4.28	5.86	5.53	5.14	4.17	4.12	4.37
$R(V_2)$	3.13	3.48	3.98	5.63	5.21	4.77	4.00	3.89	4.06
Rel. change	–	–	–	0.80	0.50	0.20	0.28	0.12	0.02

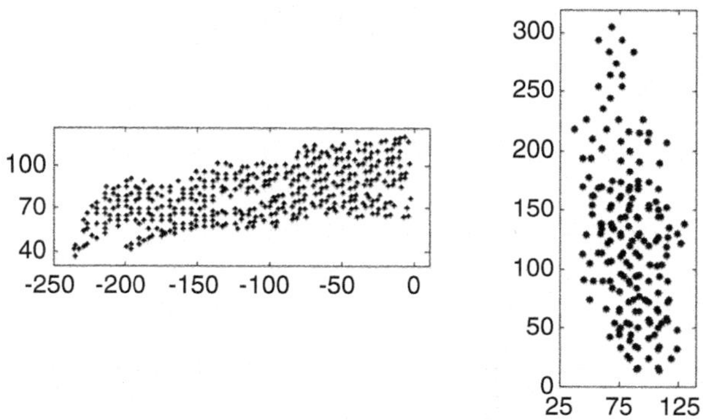

Fig. 3 Data configuration for iron ore data (*left*) and manganese data (*right*); units are in meters

data were then transformed to log-ratios relative to the filler variable via the additive logistic transform (Aitchison 1986): $\text{alr}X = \ln\left(\frac{X}{Filler}\right)$. Summary statistics for the alr-transformed variables are shown in Table 2.

The data are not aln normal, since alrAl_2O_3 fails to follow a normal distribution, and so a transformation to normal scores is required prior to any Gaussian-based simulation. The flow anamorphosis was applied to the data with $\sigma_1 = 0.3$ and $\sigma_1 = 1 + \sigma_0$, resulting in variables that are not only multivariate normal but also uncorrelated and thus statistically independent (Table 3). Several tests of multivariate normality support the conclusion: the Henze-Zirkler, Royston, and energy tests returned p-values of 0.9996, 0.4091, and 0.9530. The ranges of the transformed variables were 6.7408, 6.0568, and 6.1574, respectively (van den Boogaart et al. 2016).

Scatter diagrams of the alr-transformed input data and the output normal scores are shown in Fig. 4 and support the conclusion of independence of the output scores. Further analysis shows the flow-transformed data to be spatially decorrelated also, as shown by an average decorrelation efficiency (Tercan 1999) of 0.9563 and a mean deviation from diagonality of 0.0393. Thus, any further work including simulation can be done on the individual transformed variables.

Table 2 Summary statistics for alr-transformed iron ore data (SW denotes Shapiro-Wilk)

	$alrAl_2O_3$	alrFe	$alrSiO_2$
Min.	−5.614	0.476	−4.611
1st quartile	−4.206	0.645	−3.571
Median	−3.660	0.681	−3.080
Mean	−3.535	0.679	−2.957
3rd quartile	−2.880	0.715	−2.423
Max.	−1.233	0.884	−1.065
SW statistic	7×10^{-5}	0.9943	0.9717

Table 3 Summary statistics for data after flow anamorphosis

	FAV_1	FAV_2	FAV_3
Min.	−3.010	−2.994	−3.140
1st quartile	−0.841	−0.795	−0.874
Median	0.039	−0.026	0.028
Mean	−0.002	0.025	−0.011
3rd quartile	0.793	0.864	0.764
Max.	3.731	3.063	3.018
SW statistic	0.997	0.996	0.994

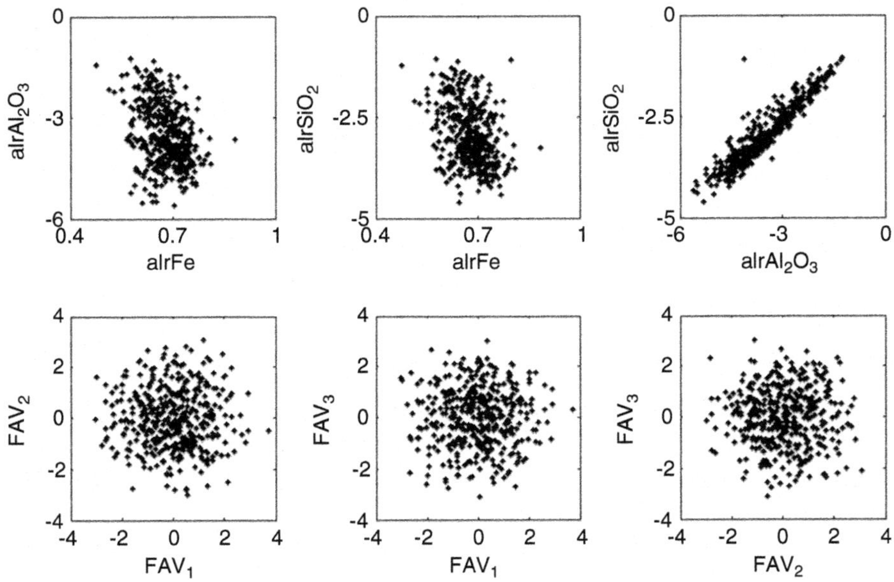

Fig. 4 Scatter diagrams of alr-transformed variables (*top*) and normal scores (*bottom*)

4.2 Manganese Data Set

This data set consists of four variables, Mn_3O_4, Fe_2O_3, Al_2O_3, and SiO_2, with Mn_3O_4 as the commodity of interest. The sample size is 176. As in the case of

Table 4 Summary statistics for alr-transformed manganese data

	$alrAl_2O_3$	$alrFe_2O_3$	$alrMn_3O_4$	$alrSiO_2$
Min.	−5.5828	−4.0621	−3.3500	−4.5475
1st quartile	−3.6665	−2.2347	1.1410	−1.0380
Median	−3.0724	−0.8765	1.6010	−0.2190
Mean	−2.8268	−0.6139	1.2560	0.06049
3rd quartile	−2.2273	0.7268	1.7200	1.2259
Max.	0.9492	4.7110	2.6000	5.8141
SW	0.9457	0.9542	0.7168	0.9795

Table 5 Summary statistics for data following two-stage flow anamorphosis

	$FA_{2s}V_1$	$FA_{2s}V_2$	$FA_{2s}V_3$	$FA_{2s}V_4$
Min.	−2.4476	−2.6904	−2.5859	−2.5850
1st quartile	−0.6873	−0.7979	−0.5852	−0.6693
Median	0.0043	0.0145	−0.0206	−0.0161
Mean	0.0014	0.0042	−0.0198	0.0098
3rd quartile	0.6421	0.7659	0.6128	0.7250
Max.	2.2276	2.5067	2.2152	2.7156
SW	0.9903	0.9937	0.9904	0.9949

the previous data, the data will be treated as compositional. This required inclusion of a filler variable followed by a transformation to log-ratios relative to that variable (Table 4).

For this data set, a much stronger deformation is required to transform the data to multivariate normality. For $\sigma_0 = 0.02$ and $\sigma_1 = 1.02$, the Henze-Zirkler and energy tests support a conclusion of multivariate normality, but the Mardia and Royston tests do not: the relevant p-values are 0.34, 0.005, 0.48, 0.003, and 0.05 for the Mardia skewness, Mardia kurtosis, Henze-Zirkler, Royston, and energy tests. This issue can be overcome by replacing a single transform by a pair of transforms, requiring weaker deformations: the data are first transformed via a flow anamorphosis with parameters $\sigma_0 = 0.2$ and $\sigma_1 = 1.2$, and then an anamorphosis with parameters $\sigma_0 = 0.2$ and $\sigma_1 = 1.2$ is applied to the output of the first transformation. With this approach, all but the Mardia tests support a conclusion of multivariate normality.

Summary statistics in Table 5 show that means are close to 0, and ranges of the transformed data lie between 85 % and 95 % of those obtained from quantile matching. Given that in the case of the flow anamorphosis, it is the underlying space that is transformed, these ranges are satisfactory. The scatter diagrams in Fig. 5 show that it is in essence the third variable of the single-step transform which leads to a rejection of multivariate normality, and from correlation tests, it follows that the two-step transformed data are statistically independent.

However, in this case, the spatial decorrelation is not as satisfactory. For the two-step anamorphosis, the mean decorrelation efficiency is 0.99, but there is evidence of remnant spatial correlation for distances of up to 15 m, which is clearly

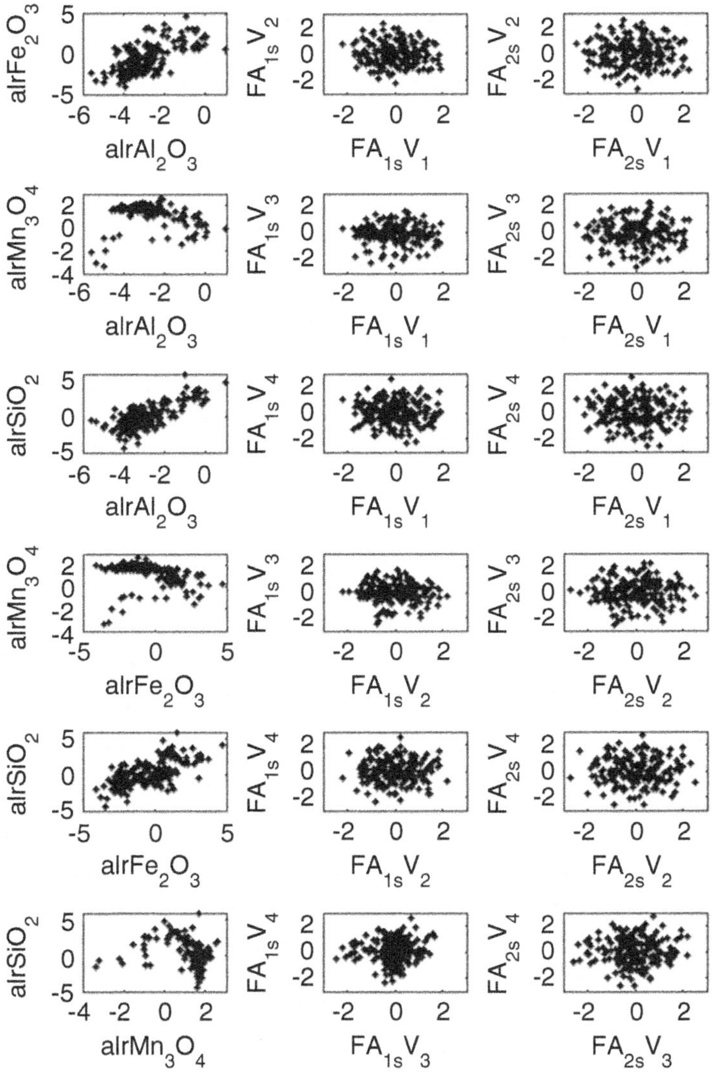

Fig. 5 Scatter diagrams of alr-transformed variables (*left* column), single-step transformed scores (*center*), and two-step transformed scores (*right* column)

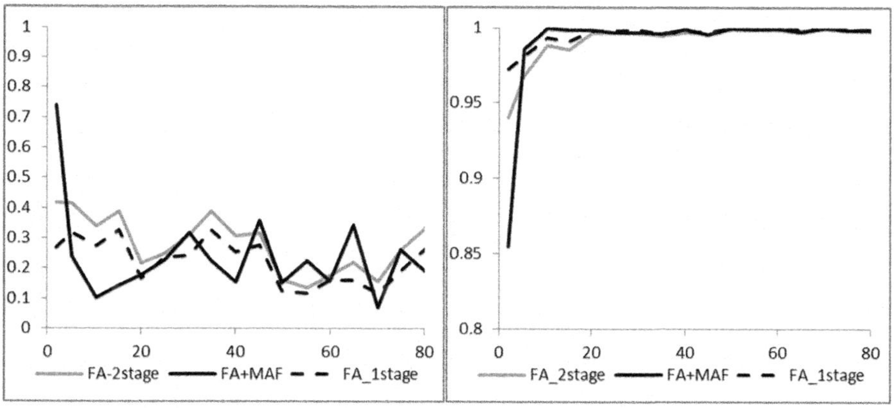

Fig. 6 Deviation from diagonality (*left*) and decorrelation efficiency (*right*) as functions of lag spacing

visible in the plot of decorrelation efficiency against lag spacing in Fig. 6. This is also evident from the mean deviation from diagonality whose value is 0.28.

Figure 6 also shows the corresponding plots for the single-stage anamorphosis and the two-step anamorphosis followed by an application of MAF decomposition (Bandarian and Mueller 2008) with lag value chosen to be 10 m. These indicate that the overall best decorrelation is achieved by the single-step flow anamorphosis although MAF achieves improvements in the mean deviation from diagonality compared to the two-step transformation.

5 Conclusion

A truly vectorial Gaussian anamorphosis should be affine equivariant, to ensure that results do not artificially depend on the particular choice of representation of those vectors. In the case of compositional data, this means that methods should be independent of the choice of log-ratio transformation, so in particular of the choice of denominator for the alr scores. The flow deformation introduced here can be used to build a bijection between the space of original vectorial observations and a Gaussian space. Our implementation is based on kernel density estimation (KDE) of the joint distribution and is affine equivariant.

The implementation has two scalar parameters, irrespective of the dimension of the input data. The first parameter describes the spread of the kernels in the KDE on the space of the observations and controls the goodness of the final fit to joint normality. The second parameter controls the spread of the final scores, but not their shape.

The examples suggest that the performance of the flow anamorphosis and its ability to satisfactorily transform the given data to multivariate normality is strongly dependent on the features of the input data. In the case of the iron ore data, the sample data appeared to be reasonably simple with relatively high correlation between the input variables, while the manganese data set showed much more complex relationships. This impacts on the strength of deformation required to successfully transform them to multivariate normality. A two-step approach might provide a feasible alternative in such a scenario. The resulting data were statistically independent, but in the case of the manganese data, a slight spatial correlation remained and a postprocessing via MAF did not remove it.

Acknowledgments The authors acknowledge financial support through the ECU CES Travel Scheme 2014, the DAAD-UA Grant CodaBlockKriging, and the Perth Convention Bureau Aspire Professional Development Grant 2016. Clint Ward, Cliffs Resources, and Matt Cobb, Consolidated Minerals, are thanked for provision of the data on which the example studies are based.

Bibliography

Aitchison J (1986) The statistical analysis of compositional data. Chapman & Hall, London

Bandarian EM, Mueller U (2008) Reformulation of MAF as a generalised eigenvalue problem. In Ortiz JM, Emery X, Geostats2008, pp 1173–1178. Santiago.

Barnett RM, Manchuk JG, Deutsch CV (2014) Projection pursuit multivariate transform. Math Geosci 46(2):337–360

Filzmoser P, Hron K (2008) Outlier detection for compositional data using robust methods. Math Geosci 40:233–248

Korkmaz S, Goksuluk D, Zararsiz G (2014) MVN: an R package for assessing multivariate normality. R J 151–162. Retrieved from http://journal.r-project.org/archive/2014-2/korkmaz-goksuluk-zararsiz.pdf

Leuangthong O, Deutsch CV (2003) Stepwise conditional transformation for simulation of multiple variables. Math Geol 35(2):155–173

Szekely GJ, Rizzo ML (2013) Energy statistics: a class of statistics based on distances. J Stat Plan Infer 143(8):1249–1272

Tercan A (1999) The importance of orthogonalization algorithm in modeling conditional distributions by orthogonal transformed indicator methods. Math Geol 31(2):155–173

van den Boogaart KG, Mueller U, Tolosana Delgado R (2015) An affine equivariant anamorphosis for compositional data. In: Schaeben H, Tolosana Delgado R, van den Boogaart KG, van den Boogaart R (eds) Proceedings of IAMG 2015, Freiberg (Saxony) Germany, September 5–13, 2015. IAMG, Freiberg, pp 1302–1311

van den Boogaart KG, Mueller U, Tolosana-Delgado R (2016) An affine equivariant multivariate normal score transform for compositional data. Math Geosci. doi:10.1007/s11004-016-9645-y

Part III
Mining Engineering

Using Samples of Unequal Length for Estimation

Marcel Antonio Arcari Bassani and João Felipe Coimbra Leite Costa

Abstract During mineral exploration, it is common to have multiple drilling campaigns. Samples from these campaigns usually have distinct sampling lengths or supports. All the available information should be incorporated when constructing a grade model. However, the variations in length among the samples must be considered during estimation. We propose to perform kriging using samples of different lengths. The kriging system is built using average covariances to account for the difference in support between the samples. The technique is applied in a mining case study and the benefits are demonstrated.

1 Introduction

In various mineral deposits, it is common to have drill hole samples obtained from different drilling campaigns. As each campaign has its own sampling protocol, samples from different campaigns are usually taken with different nominal lengths or supports. The construction of grade models would benefit if all available information is used for estimation purpose. However, the difference in length among samples must be considered during estimation.

One particular situation in which samples of different lengths occur is when the mineralization consists of a thin seam or vein and the drill hole samples cross the entire seam or vein. Each sample length corresponds to the seam thickness. In this case, practitioners usually work with the variables accumulation (product of grade and thickness) and thickness (Krige 1978; Bertoli et al. 2003; Marques et al. 2014). The grade estimates are obtained by dividing the accumulation estimates by the thickness estimates. This approach eliminates the vertical component (z), resulting in a two-dimensional (2D) model. The problem with this method is that a 2D block model may not be used directly for pit or stope optimization algorithms.

M.A.A. Bassani (✉) • J.F.C.L. Costa
Mining Engineering Department, Federal University of Rio Grande do Sul, Bento Gonçalves Avenue, sector 4, Building 75, room 101, Porto Alegre, Brazil
e-mail: marcelbassani@hotmail.com; jfelipe@ufrgs.br

J.J. Gómez-Hernández et al. (eds.), *Geostatistics Valencia 2016*, Quantitative Geology and Geostatistics 19, DOI 10.1007/978-3-319-46819-8_8

When a 3D block model is required, the common workflow to estimate grades using samples at different length involves the following steps: (1) compositing the samples and (2) kriging using the composites. However, compositing is not trivial when there are extreme variations among the lengths of the samples. In this case, compositing to a large length causes loss of information. In contrast, compositing to a short length breaks a large sample into pieces of equal grade, which is incorrect and at the same time artificially reduces the short-scale variability.

The use of average covariances in the kriging system to deal with samples of different support is well established in the literature (Deutsch et al. 1996; Yao and Journel 2000; Tran et al. 2001; Kyriakidis 2004; Pardo-Igúzquiza et al. 2006; Hansen and Mosegaard 2008; Liu and Journel 2009; Poggio and Gimona 2013). However, this approach has rarely been applied in estimation of grades in mineral deposits. Bassani et al. (2014) used average covariances in the kriging system to consider data obtained from large mined-out volumes. This paper aims at showing the applicability of kriging to estimate grades using drill hole samples of different lengths in a mineral deposit. A major bauxite deposit is used to illustrate the method. The grade model is validated by visual inspection, cross-validation and swath plots.

2 Kriging with Samples of Different Support

Consider the problem of estimating the average value of a continuous attribute z over the support V centred at location u, that is, $zv(u)$. The data consist of set of n discrete grade values defined on the supports $v_i\{z(v_i); i = 1, \ldots, n\}$. Support refers to the size (length, area or volume) in which the attribute z is measured. In our example, the support $V(u)$ refers to a selective mining unit (SMU) centred at location u, and the support vi refers to the length of the ith drill hole sample. We aim to estimate the average value of an SMU. A drill hole sample represents the average grade along the sample length. In the geostatistical literature, average values over certain support (length, area or volume) are termed block values (Journel and Huijbregts 1978; Deutsch and Journel 1998; Goovaerts 1997). Equation 1 defines the ordinary kriging estimator for the block value $z_V(u)$:

$$z_V^*(u) = \sum_{i=1}^{n} \lambda_i z(v_i) \tag{1}$$

where λ_i is the ordinary kriging weight associated to the datum $z(v_i)$. The kriging weights are the solution of the ordinary kriging system.

Equation 2 defines the ordinary kriging system accounting for the support of the data (Journel and Huijbregts 1978; Isaaks and Srivastava 1989; Goovaerts 1997):

$$\begin{cases} \sum_{j=1}^{n} \lambda_j \bar{C}(v_i, v_j) + \mu = \bar{C}(v_i, V(u)) \\ \qquad\qquad\qquad\qquad\qquad\qquad\qquad i = 1, \ldots, n \\ \sum_{j=1}^{n} \lambda_j = 1 \end{cases} \tag{2}$$

where μ is the Lagrange multiplier and $\bar{C}(v_i, v_j)$ is the covariance block to block between the block datum v_i and the block datum v_j. $\bar{C}(v_i, v_j)$ is calculated as the average of point covariances $C(u_i', u_j')$ defined between any discretizing point u_i' of the block datum v_i and any discretizing point u_j' of the block datum v_j:

$$\bar{C}(v_i, v_j) = \frac{1}{N_i N_j} \sum_{i=1}^{N_i} \sum_{j=1}^{N_j} C\left(u_i', u_j'\right) \tag{3}$$

where N_i is the number of discretizing points of the block datum v_i and N_j is the number of discretizing points of the block datum v_j. The term $\bar{C}(v_i, V(u))$ in Eq. 2 is the covariance block to block (Eq. 3) between the block datum v_i and the block to be estimated V centred at location u.

3 Case Study

3.1 Dataset Presentation

The dataset derives from a bauxite deposit located in the northern portion of the Brazilian Amazon basin. The dataset contains 686 drill holes located on a relatively regular grid of 200 × 200 m spacing along the east (X) and north (Y) directions. The original Z coordinates were transformed into stratigraphic coordinates. The variable of interest is the percentage of the total sample mass retained at the no. 14 sieve aperture (REC14). As REC14 is an additive variable, similar to grades, the methods presented here are also suitable for grades. In 343 out of the 686 drill holes, REC14 was sampled at the nominal length of 0.5 m (white points in Fig. 1). In the remaining 343 drill holes (black points in Fig. 1), there is a single sample of REC14 whose length corresponds to the ore thickness. The sample represents the average value over the total ore thickness.

Figure 2a shows the histogram of REC14 weighted by the length of the samples and summary statistics. The distribution is fairly symmetric around the mean with a low coefficient of variation. As the drill holes are approximately regularly spaced, these statistics are representative of the study area. Figure 2b shows the QQ plot between the drill hole data (length weighted) and the declustered data (obtained with a nearest neighbour estimate). The points in Fig. 2b are close to the line y = x, showing that the two distributions are similar, as expected. Figure 2c shows the

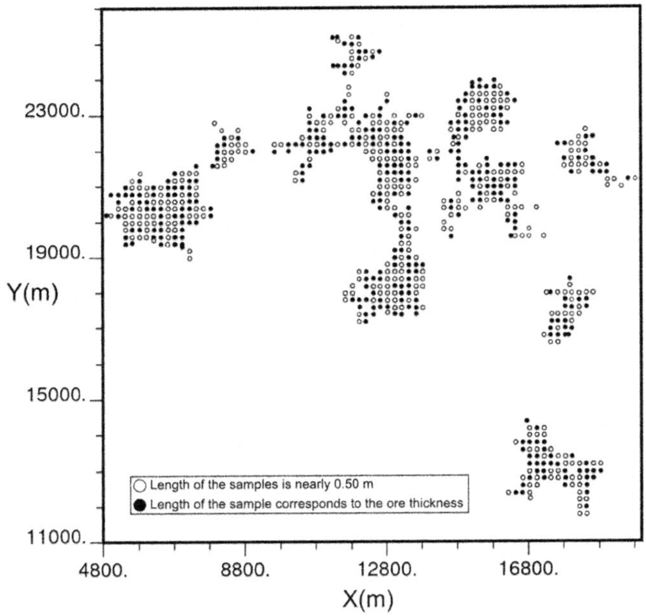

Fig. 1 Location map of the drill hole collars

histogram of the length of the samples. The length of the samples varies from 0.25 to 7.88 m. Since roughly 80 % of the samples are short, whose lengths are between 0.25 and 0.75 m, the geomodeler may feel tempted to retain only these short samples for estimation. However, keeping only the short samples results in excessive loss of information.

3.2 Variogram Analysis and Modelling

Equation 4 describes the variogram model of REC14:

$$\gamma(h) = 0.15 + 0.50 \cdot \mathrm{Sph}\left(\frac{\mathrm{NS}}{250m}, \frac{\mathrm{EW}}{250m}, \frac{\mathrm{vert}}{4.10m}\right) + 0.35\mathrm{Sph}\left(\frac{\mathrm{NS}}{4500m}, \frac{\mathrm{EW}}{4500m}, \frac{\mathrm{vert}}{4.20m}\right) \quad (4)$$

Only the samples with length between 0.25 and 0.5 m were used to calculate the experimental variogram. As kriging with samples of different support needs a variogram at a point scale to calculate the average covariances, long samples were not used to calculate the experimental variogram.

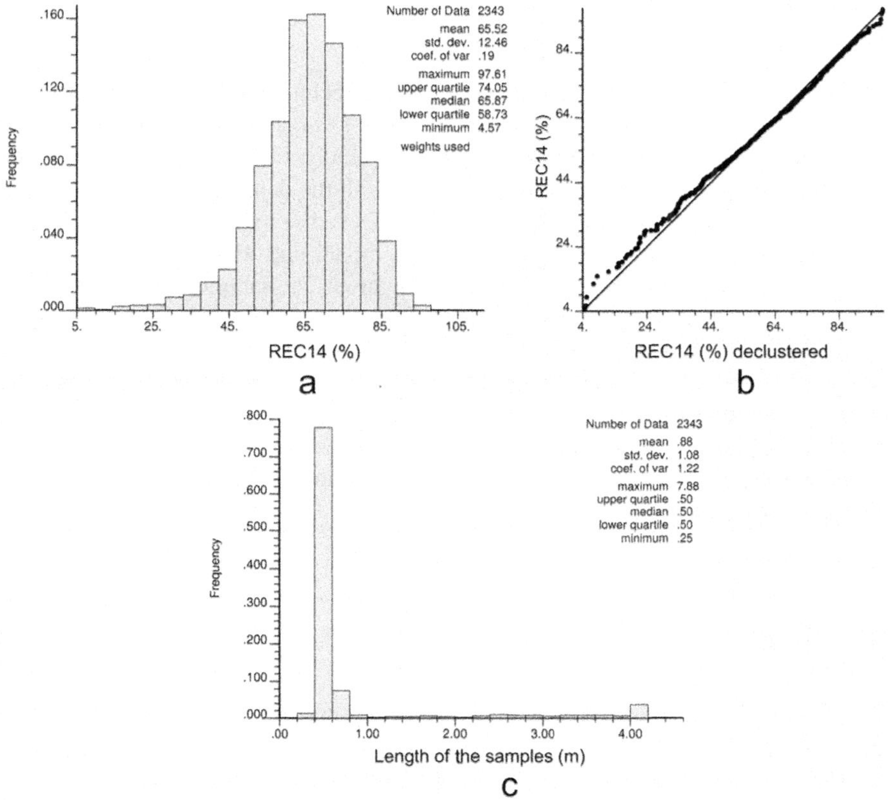

Fig. 2 Histogram of REC14 (**a**), QQ plot between drill hole and declustered data (**b**), histogram of the length of the samples (**c**)

3.3 Estimation

Ordinary kriging considering the different support in the data was used to estimate REC14. The drill hole samples were discretized along the main direction of the sample. The spacing of the discretization points corresponds to the length of the small-scale data used to calculate the experimental variogram. These discretization points of the samples were used to calculate the covariances block to block (Eq. 3) between the samples. The estimation was performed at a block model with block size of $50 \times 50 \times 0.5$ m along X, Y and Z, respectively. The block discretization was set to $5 \times 5 \times 1$. The estimates were constrained to the blocks inside the interpreted geological model.

3.4 Model Validation

The grade model was checked with the following techniques: (1) visual inspection, (2) swath plot and (3) cross-validation.

Visual inspection consists in comparing visually the grade model with the samples. The grade model must be consistent with the data.

Swath plot consists in first defining a series of swaths or slices along the X, Y and Z directions. Then, the average grade of the model and the declustered average grade of the samples within the slices are compared. The samples were declustered with a nearest neighbour estimate.

In the cross-validation, first one sample at a particular location is removed. Second, the value is estimated at that location using the remaining samples. The same estimation parameters used in the estimation of the block model were used in the cross-validation. The estimation error (difference between the estimated and true value) was calculated. Furthermore, the estimated and true values were compared using a scatter plot.

4 Results and Discussion

Figure 3 shows a plan view of the block model together with the samples. The high-grade areas of the block model are close to the high-grade samples, as expected.

The swath plots show that the block model reproduced the trend in the data along the X, Y and Z directions (see Fig. 4a–c). The local mean of the block model is similar to the local declustered mean for the three directions. In addition, the estimates neither overestimate nor underestimate systematically the local mean (see Fig. 4a–c; the local mean of the block model is not always either above or below the local declustered mean).

Figure 5a shows the histogram of estimation errors weighted by the length. The grade model is globally accurate, as the mean error is 0.14 %, which is close to zero. It represents a relative difference of 0.21 % in comparison to the mean of the data, which is 65.52 % of REC14 (see summary statistics in Fig. 2a). The scatter plot between the estimated and true values (Fig. 5b) shows that the regression line $y = ax + b$ (red line in Fig. 5b) is fairly close to the first bisector line. This diagram indicates that the grade model does not suffer from a substantial conditional bias.

The results highlight the strength of kriging to deal with samples of different support. Practitioners often overlook the fact that the kriging system handles data of different support (Journel 1986). In this paper, we used kriging considering the support of the data to estimate grades with samples of different lengths in a mineral deposit.

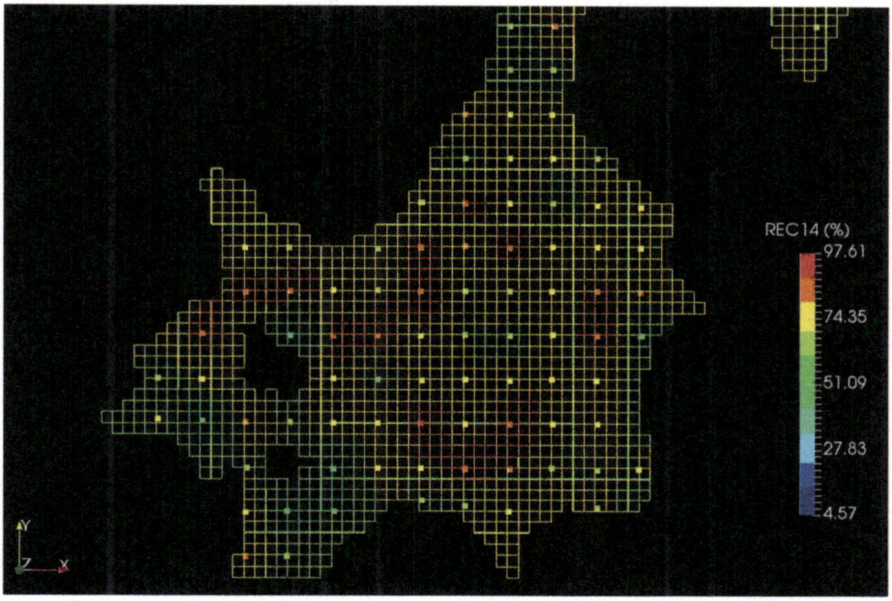

Fig. 3 Block model and samples. The *lines* represent the blocks while the points represent the samples

5 Conclusions

The study shows that kriging considering the support of the data is suitable to estimate grades in a mineral deposit using drill hole samples of different lengths. The methodology was applied to a bauxite deposit. The resultant grade model is visually consistent with the data and reproduced the trend of the data. In addition, the grade model is accurate and does not have significant conditional bias.

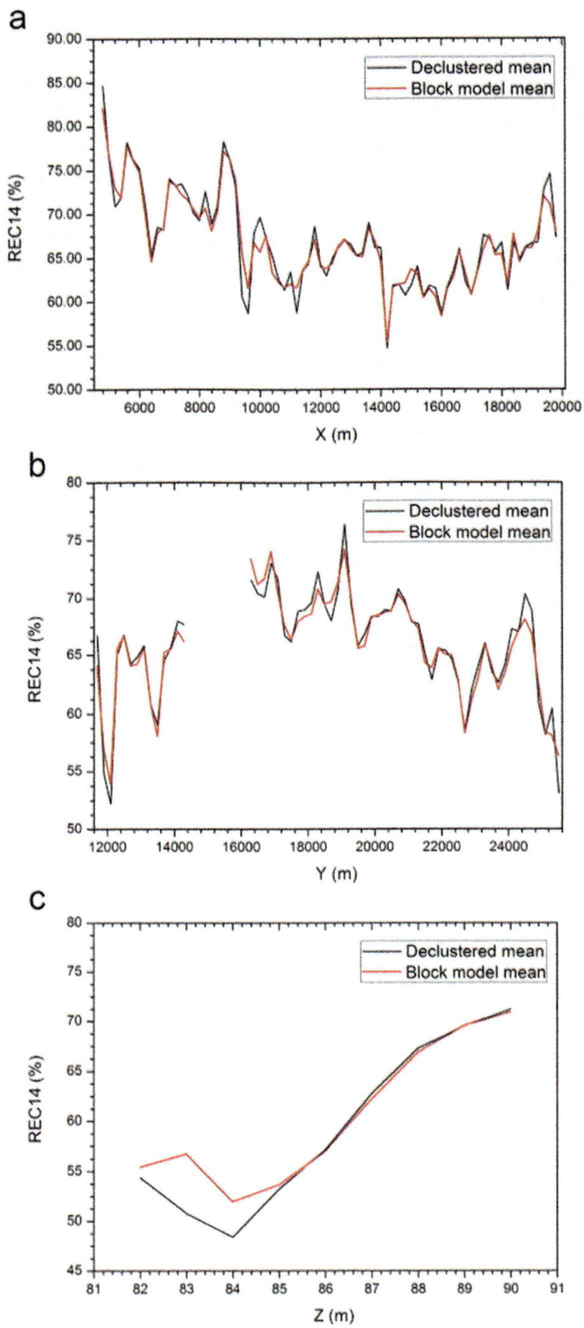

Fig. 4 Swath plots along the X (**a**), Y (**b**) and Z (**c**) directions

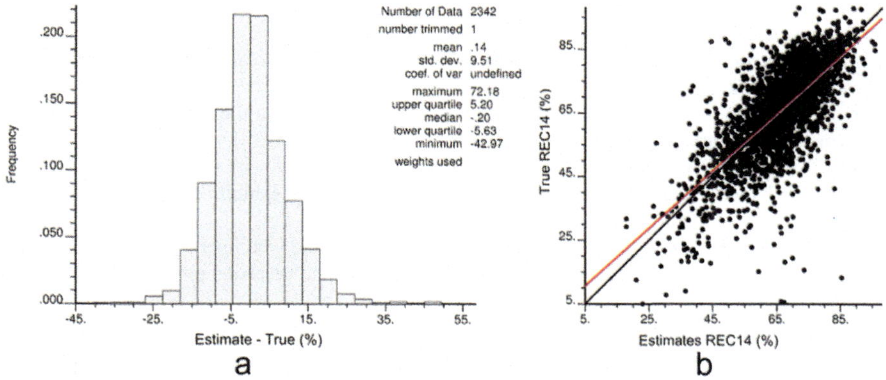

Fig. 5 Histogram of estimation errors (**a**) and scatter plot between estimated and true values (**b**)

Acknowledgements The authors would like to thank CAPES and CNPq (research agencies in Brazil) for financial support.

References

Bassani MAA, Machado PL, Costa JFCL, Rubio RJH (2014) Using production data to improve grade estimation in underground mining. Appl Earth Sci (Trans Inst Miner Metall Sect B) 122 (4):243–248

Bertoli O, Vann J, Dunham S (2003) Two-dimensional geostatistical methods—theory, practice and a case study from the 1A shoot nickel deposit, Leinster, Western Australia. In: Proceedings of the 5th international mining geology conference. The Australian Institute of Mining and Metallurgy, Melbourne, pp 189–195

Deutsch CV, Journel AG (1998) GSLIB: geostatistical software library and user's guide. Oxford University Press, New York

Deutsch CV, Srinivasan S, Mo Y (1996) Geostatistical reservoir modeling accounting for precision and scale of seismic data. SPE annual technical conference. Denver, Colorado, USA, 6–9 October 1996

Goovaerts P (1997) Geostatistics for natural resource evaluation. Oxford University Press, Oxford

Hansen TM, Mosegaard K (2008) Visim: sequential simulation for linear inverse problems. Comput Geosci 34(1):53–76

Isaaks HE, Srivastava MR (1989) An introduction to applied geostatistics. Oxford University Press, Oxford

Journel AG (1986) Geostatistics: models and tools for the earth sciences. Math Geol 18:119–140

Journel AG, Huijbregts CJ (1978) Mining geostatistics. Academic, New York

Krige DG (1978) Lognormal–de Wijsian geostatistics for ore evaluation. South African Institute of Mining and Metallurgy Monograph Series, Johannesburg, 50 pp

Kyriakidis P (2004) A geostatistical framework for area-to-point spatial interpolation. Geogr Anal 36:259–289

Liu Y, Journel AG (2009) A package for geostatistical integration of coarse and fine scale data. Comput Geosci 35(3):527–547

Marques DM, Hundelshaussen RR, Costa JFC (2014) The effect of accumulation in 2D estimates in phosphatic ore. Revista Escola de Minas 67(4):431–437

Pardo-Igúzquiza E, Chica-Olmo M, Atkinson PM (2006) Downscaling cokriging for image sharpening. Remote Sens Environ 102(1–2):86–98

Poggio L, Gimona A (2013) Modelling high resolution RS data with the aid of coarse resolution data and ancillary data. Int J Appl Earth Observ Geoinf 23:360–371

Tran TT, Deutsch CV, Xie Y (2001) Direct geostatistical simulation with multiscale well, seismic, and production data, SPE annual technical conference and exhibition. New Orleans, LA, 30 September–3 October 2001

Yao T, Journel AG (2000) Integrating seismic attribute maps and well logs for porosity modeling in a west Texas carbonate reservoir: addressing the scale and precision problem. J Pet Sci Eng 28(1–2):65–79

New Approach to Recoverable Resource Modelling: The Multivariate Case at Olympic Dam

Colin Badenhorst, Shane O'Connell, and Mario Rossi

Abstract Traditional estimation techniques significantly under-call the true monetary value of the resource on which mine plans and operations base their business. At the Olympic Dam, this is worth billions of dollars. Realising this value requires mine planning engineers to be supplied with an accurate recoverable resource model that correctly estimates the tonnes and grade for a specified support and timescale, at the time of mining.

Models estimated using linear methods and wide-spaced drilling typically fail to accurately predict recoverable resources, mainly because of incorrectly accounting for the change of support and information effect. The unavoidable smoothing property of weighted averages is also a significant obstacle. These failures are more significant in underground mining scenarios where higher cut-offs (with respect to the average grades of mineralisation) are applied. This paper discusses a different approach to recoverable resource estimation based on conditional simulation methods.

The Olympic Dam deposit is one the world's largest polymetallic deposits. The resource estimation practices at the Olympic Dam are comprised of a combination of linear and non-linear techniques to estimate 16 different grade variables critical to mine planning. Measured resources are supported by 20 m-spaced underground drilling fans where Kriged estimates perform well in terms of mine to mill reconciliation. However, this not the case for resources classified as Indicated and Inferred. Until infill drilling is undertaken, the accurate estimation of tonnes and grade to the mill is not possible with the Kriged model. This has a significant impact on life-of-mine economic valuations and ore reserve estimates of the Olympic Dam.

C. Badenhorst
Olympic Dam, BHP Billiton, 5000 Adelaide, SA, Australia
e-mail: Colin.Badenhorst@bhpbilliton.com

S. O'Connell
Manager Resource Department, Olympic Dam, BHP Billiton, 5000 Adelaide, SA, Australia
e-mail: Shane.OConnell@bhpbilliton.com

M. Rossi (✉)
GeoSystems International Inc., 33431 Boca Raton, FL, USA
e-mail: mrossi@geosysint.com

J.J. Gómez-Hernández et al. (eds.), *Geostatistics Valencia 2016*, Quantitative Geology and Geostatistics 19, DOI 10.1007/978-3-319-46819-8_9

Conditional simulation has been used to generate a recoverable resource estimate from a single realisation. This conditional simulation model takes into account both the change of support and the information effect, without the undesired smoothing effect that classic methods introduce. This paper describes the significant challenges faced in applying this approach, including issues such as which realisation to choose, data conditioning in areas with little information, ensuring that the multivariate relationships among variables are respected at a block level, software and hardware challenges and defining benchmarks for ensuring that the "correct" grade-tonnage curves are reproduced. These challenges have to be overcome whilst ensuring that the resulting estimate is a JORC compliant and is also acceptable under BHP Billiton's corporate governance standards.

1 Introduction

The Olympic Dam is Australia's largest underground sublevel open stoping mine producing around 10 Mt of ore per annum. The Olympic Dam mine has been in production since 1988 using a standard mining method of mechanised sublevel longhole open stoping (SLOS), with cemented aggregate backfill. The processing plant is a fully integrated circuit that consists of autogenous grinding mills, flotation circuits to recover copper concentrate and tailings leach circuits to recover uranium. The copper concentrate is treated in an on-site direct-to-blister-furnace (DBF) smelter, whilst an onsite refinery produces copper cathode and recovers gold and silver.

The Olympic Dam is a very large iron oxide-hosted Cu-U-Au-Ag ore deposit. The deposit is hosted entirely within the Olympic Dam Breccia Complex (ODBC) and is unconformably overlain by approximately 300 m of unmineralised, flat-lying, sedimentary rocks. The deposit was discovered in the late 1970s and the geology has been studied and described by numerous authors (Roberts and Hudson 1983; Ehrig et al. 2012). A key feature of the deposit is a central core of haematite–quartz breccia largely devoid of copper and uranium mineralisation. In general, the host breccias are more haematite-rich towards the centre of the ODBC and more granitic at the margins. Including the sulphide minerals, there are more than 100 identified ore and gangue minerals. The most common minerals include haematite, quartz, sericite, feldspar, chlorite, barite, fluorite, siderite, pyrite, chalcopyrite, bornite and chalcocite. The three primary uranium minerals account for less than 0.1 % of the total rock mass and occur as uraninite, coffinite and brannerite.

The bulk of economic mineralisation is associated with sulphide-bearing haematite-rich breccias. The majority of copper mineralisation occurs as chalcopyrite, bornite and chalcocite and dominantly manifests as binary pairings of chalcopyrite $\pm$ bornite and bornite $\pm$ chalcocite. Uranium, gold, silver and copper minerals are all correlated to a statistically significant degree.

2 Why Produce a Recoverable Resource Model?

The business case for recoverable resource modelling is simple: the true monetary value of the Olympic Dam is highly leveraged to grade. For example, depending on the specific mine plan, for every 10 % increase in grade in the early part of the plan, there is a 25 % increase in annual cash flow. Around two thirds of the life-of-mine is based on widely spaced drilling and is therefore a smoothed and under-representative view of the grade that will be realised when close-spaced drilling and high-resolution estimation is undertaken. Modelling the correct distribution of grade is worth billions of dollars to the value of the Olympic Dam.

In the past 40 or so years, there have been many attempts at producing effective recoverable resource models with varying degrees of success; some relevant discussions and case studies include (Journel and Huijbregts 1978; Journel and Kyriakidis 2004; Assibey-Bonsu and Krige 1999; Rossi and Parker 1993; Abzalov 2006; Krige and Assibey-Bonsu 1999; Rossi and Deutsch 2014 and Roth and Deraisme 2000).

In an environment of falling commodity prices, with a focus on reducing costs and improving financial metrics of cash flow, IRR, NPV and capital efficiency, an equally important consideration to costs is improving the revenue. The old adage, "grade is king", is never more true in this environment. For deposits that are relatively high grade, and where the spatial continuity of that grade is amenable to selective mining, increasing the grade for the same tonnage delivered to the mill is very effective in increasing the margin on each unit of metal produced. The mining cost per tonne of ore may increase but the cost per pound of copper can greatly decrease.

A common strategy at the Olympic Dam has been to increase metal production by increasing process plant ore tonnage throughput. Unless there is latent capacity in the plant, there is a large capital requirement in order to accommodate the increase in tonnes. Alternatively, increasing metal throughput in the operation can be achieved by raising the grade of the ore feed from the mine, which is generally much less capital intensive since the plant is large and complex.

The Olympic Dam deposit is relatively high grade with contiguous zones of 2–4 % copper grade. This continuity of high grade gives the operation an opportunity in the planning and mining processes to increase the grade delivered to the processing plant. Relatively close 20 m-spaced lines of Measured Resource drilling are required to define these contiguous high-grade ore shoots. The Olympic Dam Inferred Resource is defined by relatively wide-spaced drilling at 100–250 m notional centres. Indicated Resource is defined by both wide-spaced surface and wide-spaced underground drilling at 70–100 m centres. Comprehensive drill spacing studies (see later section) have back-tested the effect of resampling Measured Resource (20 m drill spacing) at Inferred and Indicated drill spacings in the deposit. These studies have demonstrated that a traditional Ordinary Kriging (OK) linear estimate using this wide-spaced drilling information is neither globally accurate in terms of tonnes and grade nor locally spatially accurate in representing the orientation of the high-grade zones which are to be mined (Fig. 1). This holds true for

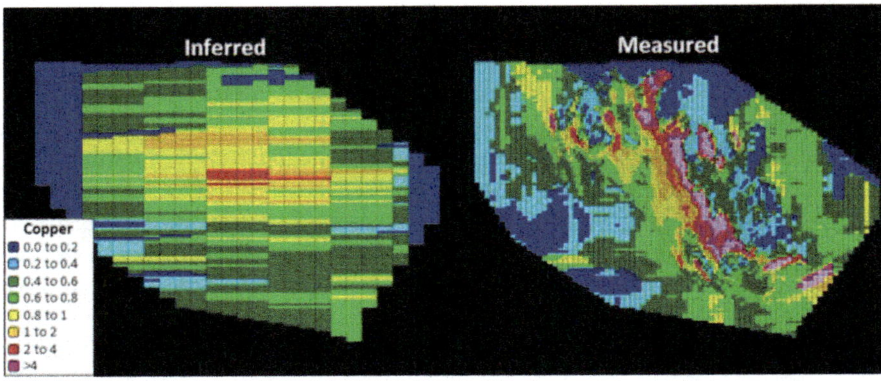

Fig. 1 Representative cross section through the Olympic Dam showing differences in *copper grade*, modelled using Inferred (250 m spaced) drill hole information and Measured (20 m spaced) drill hole information. Each model depicted above has a unique geological interpretation that is a function of the data spacing

geological interpretations and models, estimated models and simulated models, since these are a function of data spacing.

Over the longer term, how and when these higher-grade ores are exploited can also have a direct impact on the life and the consequent NPV of the operation. Since the life-of-mine has approximately two thirds of the resource at Indicated or Inferred status, there is a poor conversion of resource to reserve, which understates the overall value of the operation in the life-of-mine plans. In addition, experience at the Olympic Dam and many other operations elsewhere has shown that for the same volume of material mined, an increase in grade is realised with later infill drilling. This is due to the high-grade mineralisation controls that exist at a smaller scale, compounded by the well-documented support effect and the inherent smoothing of linear estimators such as inverse distance weighting and Kriging methods. The result is that estimates in areas with wide-spaced information under-call the true tonnage and grade that are realised once close-spaced drilling and high-resolution estimation is undertaken.

In the authors' experience, at the Olympic Dam, the most practical solution to overcoming this issue is to produce a recoverable resource model using a simulation technique. Conditional simulations which comprehensively validate against the drill hole data are the only technique that avoids smoothing by reproducing the original data variance (high granularity models), estimates the tonnes and grade at the time of mining for any drill spacing and produces a better local spatial model that can be used for practical geological, mine planning and financial valuation purposes.

Combined with traditional OK estimates of Measured Resource, estimates of Inferred and Indicated Resource produced using conditional simulation are collectively referred to as the recoverable resource model (RRM). This model is an enabler to several other mine planning value-adding initiatives which unlock the true value of the deposit. These initiatives are collectively referred to as the grade focus strategy.

3 The Grade Focus Strategy

The RRM is integral to the overall strategy to increase grade and value at the Olympic Dam. This strategy is aimed at maximising the contribution of ore grade to the value proposition for the operation. Higher ore grades have a direct impact on the revenue stream and are very effective in lowering the unit cost of metal production. This strategy is based on six key focus areas that all contribute to improving the value contribution of ore grade:

1. *Resource modelling*. By using conditional simulation to estimate grade in model areas supported by wide-spaced drilling, a higher grade from improved granularity can be realised in the mine planning process.
2. *Reserves*. By using a more sophisticated automated stope generation technique, a higher design grade can be realised through a more efficient capture of high-grade resource.
3. *Resource utilisation*. By including all potential (pre-resource) mineralisation and Inferred Resource in the mine plan, the full possible value of the underground operation can be assessed, allowing a higher cut-off to be applied, and therefore a higher grade to be realised.
4. *Cut-off grade optimisation*. Historically the operation has been valued on a fixed cut-off grade. The future plan is to implement a variable cut-off strategy, where the cut-off varies between stopes and over time. This change in stope design practice will allow the optimisation process to add value by promoting grade in time, beyond the level achieved using a fixed cut-off.
5. *Rejection of subgrade material from the ore stream*. Separate removal of below cut-off material and low-grade development to the surface raises the average ore grade and increases the proportion of high-grade stope material in the ore stream.
6. *Stope sequencing and scheduling*. Higher value mine plans can be achieved by promoting higher-grade stopes forward in the schedule and deferring lower-grade stopes in time.

Both the absolute and relative contribution of these areas to the value proposition will depend on the context of the specific mine plan that is being evaluated. The key point is that for the full value of the resource to be realised, all six areas are necessary. No modelling technique alone can realise the full value.

4 Fundamental Challenges

The development of a recoverable resource model for the Olympic Dam has been a significant challenge. Some of these challenges are purely technical in nature, but there are also fundamental challenges that inhibit the acceptance of this modelling technique.

Ultimately, if recoverable resource models are not used to generate mine plans, then they are of little practical use. One of the critical factors to success is the support of mine planning engineers in understanding the value these models hold and to use them for reserving. Without this, it is just another low-value, esoteric exercise. To extract the maximum value from the model, the cultural inertia manifests as conservatism and fear of failure and must change, or the upside in grade will never be realised.

Gaining understanding and support from decision makers in the business has been one of the single greatest challenges. Concepts that are basic to geostatisticians and resource geologists (e.g. averaging, grade above cut-off, change of support and volume variance) are obscure to decision makers and are treated with suspicion and even derision. The lack of knowledge regarding these elementary concepts in the industry, its leaders and some of the broader consultancy community to which these leaders defer, are real and significant barriers to the success of this work.

5 Technical Challenges

With regard to a deposit as large as the Olympic Dam, there are several technical hurdles to overcome in attempting to produce a recoverable resource model using conditional simulation.

The model covers an area of 6 × 3 km and extends to a depth of 2 km. It is comprised of 5 × 10 × 5 m block support, from estimates using a number of techniques for different areas, elements and minerals in the deposit. All elements and minerals in areas classified as Measured Resource are estimated using Ordinary Kriging (OK). Within areas classified as Inferred and Indicated Resource, Cu, U_3O_8, Au, Ag, S and SG are estimated using conditional simulation. The copper mineralogy is stoichiometrically calculated from the simulated Cu and S estimates. All other elements and minerals are estimated using OK.

The dimensions of the deposit present the foremost challenge. At a 2.5 m node spacing, a single model covering the entire deposit comprising a minimum of 14 data variables would require ~1.5 billion nodes and constitute a model file size in excess of 700GB. A single drill hole file used for this simulation would comprise of ~950,000 2.5 m length samples. Working with these files is impractical, and there is no simulation software capable of handling such a large model file. To deal with this challenge, the deposit is split into 16 individual sub-models and corresponding drill hole files, based on gross geological differences and also the ability of the simulation algorithm and software to perform the task in a reasonable amount of time. Once these individual models are amalgamated into one model and regularised, there are no boundary artefacts evident in the combined model. The characteristics of these individual models are listed in Table 1.

Table 1 Some characteristics of Olympic Dam simulation models

Model characteristics	Group 1 models	Group 2 models
No. of models	9	7
Model dimensions	~1 × 1 × 1 km	~2 × 1 × 1 km
Node size	2.5 × 2.5 × 2.5 m	2.5 × 2.5 × 2.5 m
No. of nodes	~64 million	~128 million
No. of simulated variables	6	6
No. of total variables	14	14
Model run time	~8 h	~16 h
Node model file size	~6.5GB	~12.4GB

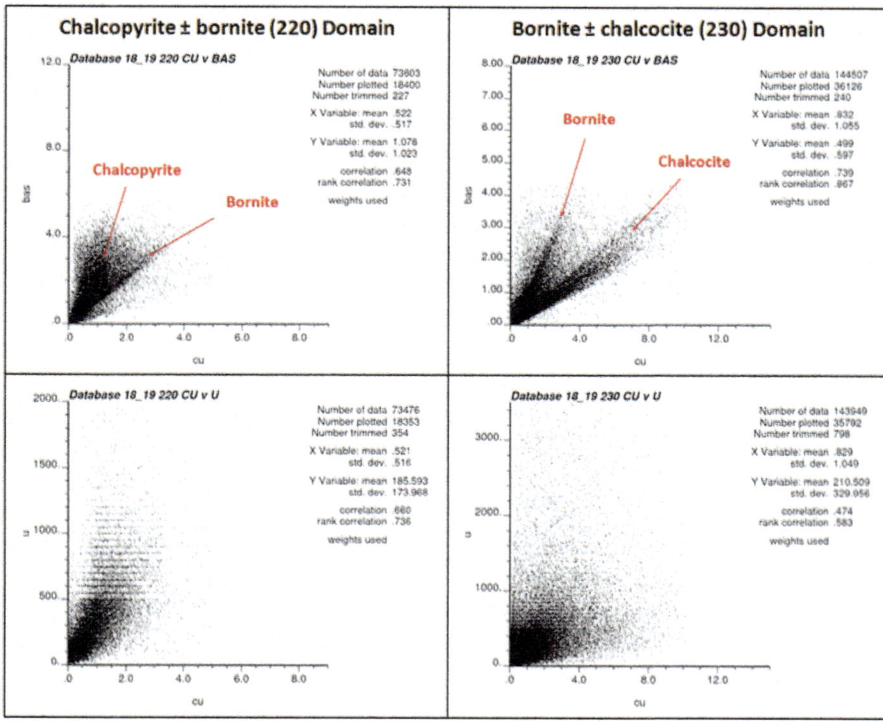

Fig. 2 *Scatter plots* of Cu v S and Cu v U_3O_8 for the chalcopyrite ± bornite (220) domain and the bornite ± chalcocite (230) domain for drilling information in Model 18_19 of the Olympic Dam

Simulation is performed separately in the chalcopyrite ± bornite (220) domain, and the bornite ± chalcocite (230) domain, to allow the unique grade relationships observed in the drilling information (Fig. 2) to be honoured. The domains are established by modelling the unique Cu to S ratios of these sulphides using a combination of deterministic modelling and probabilistic (Indicator Kriging) methods.

6 What Has Been Tested

A significant amount of work has been undertaken over the last few years to establish the foundations of the recoverable resource model at the Olympic Dam. Significant improvements in deterministic modelling of the key sulphide domains, and work on previous simulation models of the Olympic Dam, have also played a prominent role in determining the most efficient path forward. The key attributes required of techniques and software are the usability of models by mine planning, the speed and stability of the algorithm and software, the honouring of multivariate relationships between simulated variables and the ability to integrate several models from several simultaneous sources.

In these authors' experiences, conditional simulation is the preferred technique for underground mine planning work rather than other techniques such as Uniform Conditioning (UC) and Multiple Indicator Kriging (MIK), because they suffer from the same smoothing effect evident in all forms of Kriging and also produce models that are more suitable for input into open pit mine planning software.

Following extensive trials over several years, the resource team has settled on sequential Gaussian simulation (Isaaks 1990) as the most appropriate algorithm to use. Stepwise conditional transformations (Leuangthong and Deutsch 2003) are applied to account for the correlation between metals. Other algorithms, including co-simulation with Bayesian updating (Journel 1988; Rossi and Badenhorst 2010) and projection pursuit multivariate transformation (Friedman and Tukey 1974; Barnett et al. 2012), have been extensively trialled and rejected either because of poor reproducibility of input statistics, difficulty in replicating multivariate relationships observed in the drilling information, poor spatial match to input data and the inability to deal with large datasets or a combination of all of the above. SGS point simulation rather than direct block simulation (DBS) is preferred since validation against the input data of DBS shows issues caused by the proportional effect present in the original data, which has been shown to introduce biases in the final output (Leuangthong 2006).

Both commercially available and open source software have been used with varying degrees of success. Significant issues which were encountered included software errors introduced by inconsistently incorporating GSLib-based routines (Deutsch and Journel 1998), very slow operation with large datasets, stability issues and data corruption or problems with integration of models from multiple users working simultaneously on different parts of the deposit. It is clear from almost a decade of work on this topic at the Olympic Dam that it is unlikely that standard commercially available software packages can be used to produce recoverable resource estimates that meet the mine planning and the corporate governance requirements. Therefore, a modified version of the GSLib programs has been adopted to complete the work.

In order to overcome some of the known issues with SGS, including edge effects and grade blowouts between drill holes and at the edges of data, as well as to increase the program's functionality, the original GSLib FORTRAN code was modified in-house. One of the key enhancements made to the code was the implementation of multiple search passes with the ability to have different parameters for each pass. For this reason, the program has been named Dynamic Search SGS or DS-SGS. Other enhancements include the addition of domain control, not drawing from the global distribution, multiple coarse/fine grid redefined in terms of user input of x, y, z grid spacing, independent soft nodes and hard data search with assign to nodes selected and minimum number of soft nodes selection with assign to nodes option.

7 Case Study

The aim of the case study was to examine the change in grade-tonnage information for an existing well-drilled 20 m-spaced Measured Resource area (hereafter referred to as "truth"), by estimating and simulating it using wider Indicated and Inferred Resource drill spacings. The goal is to determine if the simulation parameters using the wide-spaced drilling yielded results that matched the 20 m-spaced Measured Resource truth, thus providing a mechanism to calibrate and validate simulation models in other areas of the deposit.

In order to do this, a conventional drill hole spacing-type approach was followed, but with a few differences. The starting point was to take a vast area of Measured Resource that has been drilled from underground on 20 m-spaced lines and treat this as the "truth". This area constitutes 1.3 billion tonnes of Mineral Resource and 300 Mt of Proved Ore Reserve, of which 170 Mt has been mined over the last 27 years (Fig. 3).

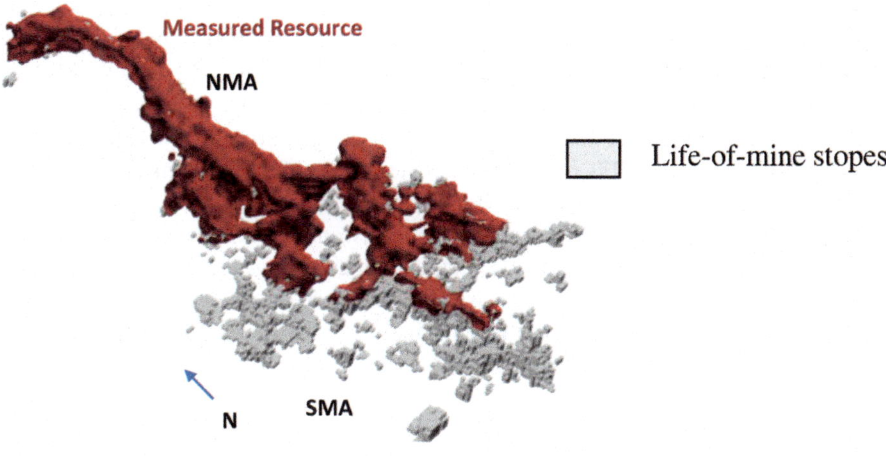

Fig. 3 Location of Measured Resource in relation to the life-of-mine stope set (Note the extensive spatial coverage of Measured Resource across the deposit)

Table 2 Resource classification and drill spacing used at the Olympic Dam

Resource classification	Drill hole spacing	Model block size
Inferred Resource	250 m	120 × 100 × 5 m
Indicated Resource	100 m	60 × 50 × 5 m
Measured Resource	20 m	5 × 10 × 5 m

This represents approximately 30 % of the expected ultimate underground reserve across the entire deposit and is representative of the material that will be mined in future years. Mine to mill reconciliation results demonstrably show that over a 3-year period (~30 Mt), the estimated tonnes and grade of Measured Resource vary by less than 1 % from that actually recovered. Thus, there is high confidence that the Measured Resource volume is a good yardstick by which to validate the simulation results.

One of the primary criteria for determining resource classification is drill hole spacing and, for convenience, will be used in this discussion. Table 2 shows the relative drill hole spacing applied at the Olympic Dam for Inferred through Measured Resource and the relationship to the estimation support.

8 Comparing Raw Drill Hole Data

A fixed volume of 20 m-spaced Measured Resource was originally drilled from the surface on 70–100 m centres and is equivalent to Indicated Resource spacing. These surface holes were resampled at 250 m-spaced centres to approximate the equivalent Inferred Resource drill spacing. This resampling was undertaken 25 times by randomly selecting holes at 250 m centres in order to capture the range of possible outcomes from variations in the drilling grid. This process approximates a drilling program that starts at 250 m spacing and is then progressively infill drilled to 100 m and 20 m spacings. The outcome is a comparison of the nearest neighbour-declustered drill hole data statistics and grade-tonnage curves of the different grid spacings of 20 m and 100 m and the 25 iterations of the 250 m spacing. The differences in results were alarming (Fig. 4).

There is considerable difference among the twenty-five 250 m iterations with 23 (92 %) of them under-calling the actual tonnes and grade by a considerable margin, whereas the 100 m-spaced drilling dataset is much closer to the 20 m-spaced dataset. In classical geostatistics, this is termed the information effect or, more precisely, in this case, the misinformation effect.

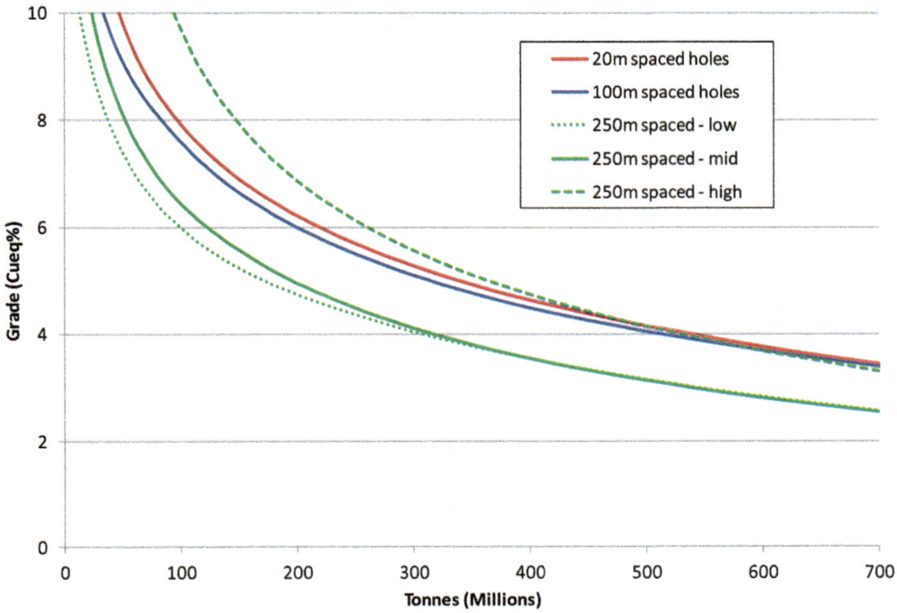

Fig. 4 Declustered sample grade-tonnage curves within the Measured Resource volume. The differences in information for the 100 m dataset and 250 m resampled datasets are clearly demonstrated. Cueq = Cu + (U_3O_8*2.44) + (Au*0.881) + (Ag*0.0048) – (S* 0.191)

9 Comparing the 100 m-Based and 25 × 250 m-Based Model Estimates

Each of the 250 m drilling datasets had its own geological model and domains and was used to create 25 separate resource models that were estimated using Ordinary Kriging into block sizes of 120 × 100 × 5m. The same process was applied using the 100 m-spaced dataset to a model with unique geological and domain characteristics and a 60 × 50 × 5m block size. The resource grade-tonnage results mirrored the differences observed in the drilling data noted previously. The 26 estimated models (25 Inferred and 1 Indicated) were then converted to reserves by the mine planning engineers running each through an automated and semiautomated stope design process.

The results from the reserve grade-tonnage curves mirrored that of the resource models and also the underlying drill hole data. That is, if the drill hole dataset was the lowest of the group on the grade-tonnage curve, then the corresponding resource and reserve models were also the lowest (Fig. 5).

A sequential Gaussian simulation was then developed using 3 of the 25 Inferred datasets by choosing a low, mid- and high iteration of drill hole data. An additional simulation model was also generated using the single 100 m-spaced Indicated dataset. The results mirrored those of the drill holes and resource and reserve

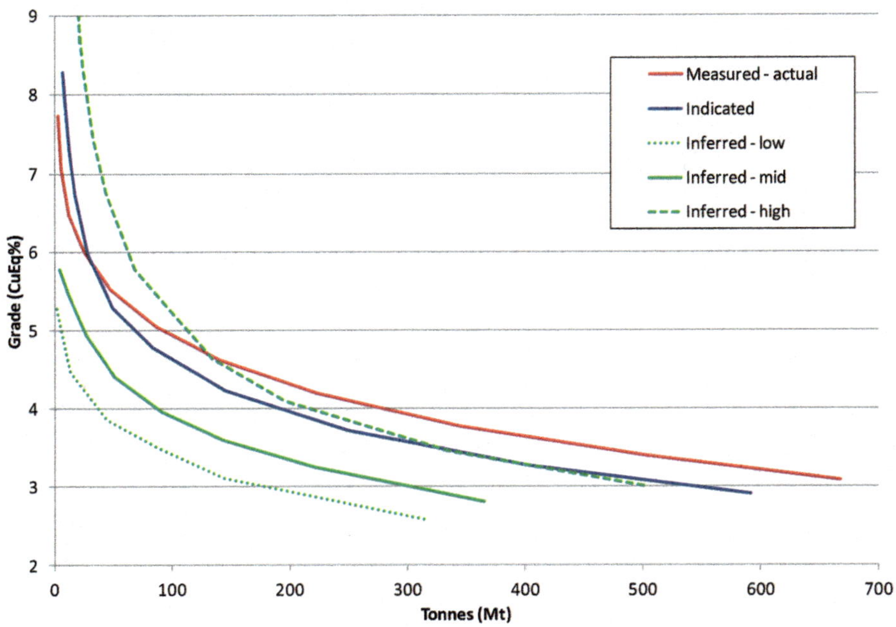

Fig. 5 Comparison between ore reserve grade-tonnage curves using Measured Resource and resampled Indicated and Inferred Resource models. $\text{Cueq} = \text{Cu} + (\text{U}_3\text{O}_8{*}2.44) + (\text{Au}{*}0.881) + (\text{Ag}{*}0.0048) - (\text{S}{*}\ 0.191)$

models discussed earlier. The results are very clear; the underlying data exerts the strongest control on whether or not the 20 m-spaced Measured Resource grade-tonnage result can be achieved from wide-spaced Inferred or Indicated Resource datasets. The conclusion is that all resource and reserve models, be they estimated or simulated, are strongly anchored to the starting drill hole data. Locally, different mine areas typically behaved differently from iteration to iteration. Most of the 25 Inferred Resource datasets were low with respect to the 20 m-spaced Measured Resource "truth", but a few were higher. This is a random feature and is attributed to chance interactions with geological influences.

10 Observations and Discussion

A critical observation is that there is a fundamental change to the drill hole information from iteration to iteration for the same spacing, as well as when the grid spacing decreases. The proportions of grades change above a cut-off and the overall volume of mineralised material increases. This happens in 92 % of the cases described; however, there were a few 250 m-spaced iterations (8 %) where the proportion and grade of drill holes were higher than the 20 m-spaced Measured Resource "truth". Locally, the pattern of uplift or downgrade can differ significantly from that observed globally.

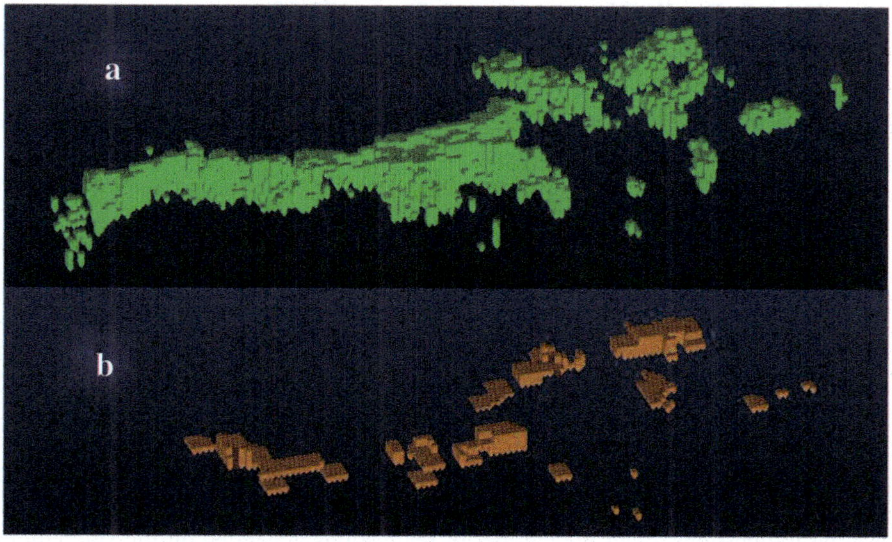

Fig. 6 Comparison between reserve stopes in the Measured Resource volume at the same cut-off, using (**a**) Measured Resource model and (**b**) one of the 25 resampled Inferred Resource models (Note how poorly the resource estimate based on Inferred Resource drill spacing estimates the mineable tonnes for the same mining area)

These differences are due to imperfect information (including sampling and estimation error), non-representativeness of data at a certain spacing and chance interactions of drill holes with complex geology and geometry. There is natural variation in the orebody which means certain drill spacing and orientation with respect to orebody geometry are not adequate to fully sample the real distribution and proportions of grades. This can lead to an over-representation of low values and under-representation of high values and vice versa, which results in vast differences in reserve stope shapes (Fig. 6). The misinformation effect is used in this case as a collective term to describe all of the aforementioned errors and effects, which is more encompassing than the traditionally used information effect. Whilst there is a relationship between the two, the term misinformation has been chosen to avoid pedanticism and confusion.

At the Olympic Dam, the misinformation effect in volumes informed by wide-spaced data leads to underestimation of the actual tonnes and grades.

Regardless of the drill spacing, the best that can be done by the practitioner is to honour the available drill hole data. This approach will not capture the range of uncertainty, and therefore a different strategy is required to account for the misinformation effect. The Olympic Dam Mine Planning Department will employ the use of modifying factors based on reconciliation to account for the global upside from the misinformation effect, allowing for local differences that may result in downside.

11 Which Realisation to Choose?

Conditional simulation is commonly used for quantifying resource uncertainty. Typically this approach involves generating numerous realisations and developing probabilities and confidence intervals from these. The multiple realisations are interpreted as values of the conditional cumulative distribution function (ccdf) of each node:

$$Prob^*\{Z(x) \leq z_c|(n)\} = F(x, z|(n)) \tag{1}$$

where "*Prob**" represents the estimated probability at location x, $Z(x)$ represents the random variable at location x, z_c represents an arbitrary cut-off, (n) represents the conditioning data used to simulate node at location x and F represents the conditional distribution function.

Unfortunately, when conditional simulation is used for recoverable resource modelling, this same "many realisations" mindset is mistakenly also applied, leading to questions about the number of realisations generated during simulation and the decision processes used to select a single realisation for further processing. What most practitioners do not realise, and most theoreticians ignore, is that the change of support variance is much more significant than the variance of the conditional distribution provided by simulation. A change in mindset is required to accept that conditional simulation can be a valuable recoverable resource technique.

As the overall volume of the simulation increases, the global difference between the realisations becomes progressively smaller. For a large model area of ~1.6 billion tonnes of resource and ~270 Mt of reserve, the difference between realisations is ~1 %. Over the entire deposit of >10 billion tonnes, the difference is <0.5 %, whereas the difference between small blocks and big blocks is >20 % for this same area.

When examining simulation realisations at the small block scale, realisations can exhibit significant local variation, leading to the erroneous conclusion that selecting a meaningful single realisation for further work is difficult or problematic. This erroneous conclusion is reached because of a failure to appreciate that during recoverable resource modelling of large areas, the difference in the change of support is an order of magnitude greater than the spread of realisations. In a sense, the realisations converge to approximate the same solution when considering large volumes. This change in variance is governed by the volume-variance relationship, which can be illustrated by examining the change in mean grade above a cut-off for varying block sizes. The tabulation below shows this effect with 100 Cu realisations for three different levels of support for a ~270 Mt parcel of material that is expected to convert to ore reserves from the North Mine Area of the Olympic Dam. The volume is a mixture of drill spacings that support Measured, Indicated and Inferred Resource.

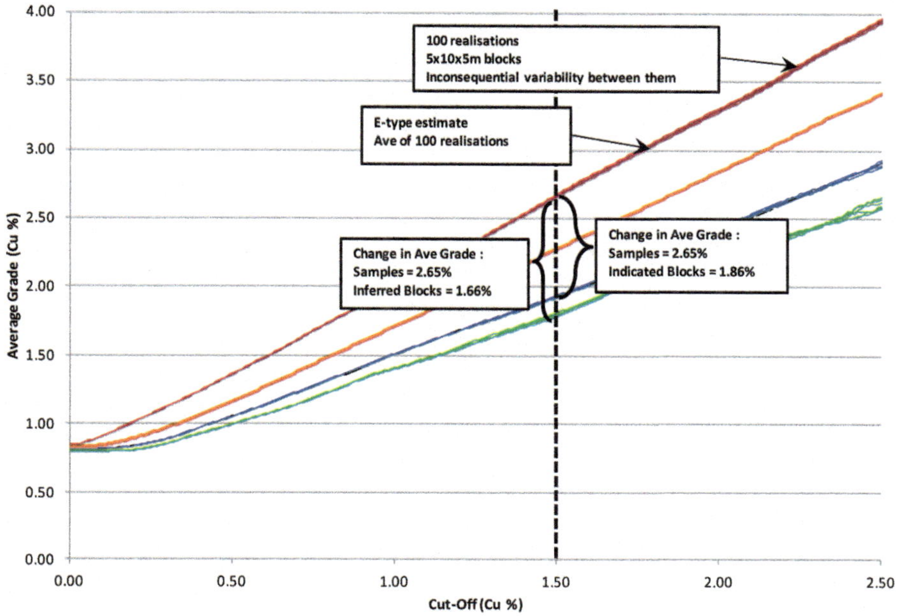

Fig. 7 Grade curve for 100 realisations of 5 × 10 × 5m (*red*), 60 × 50 × 5m (*blue*) and 120 × 100 × 5m (*green*) resource blocks showing the change in average grade for different supports. The E-type curve (*orange*) is the average of 100 realisations at the 5 × 10 × 5m support

The results clearly indicate that as the block size increases, the average grade above the cut-off substantially decreases for the same tonnage of material. This decrease in relative terms, which comes about solely from the change of support, is an order of magnitude greater than the variability between individual simulation realisations at the small block scale (Fig. 7, Table 3).

Thus, for recoverable resource modelling of this large deposit, the number of realisations in a simulation is not critical. A single realisation, chosen at random from a handful, is more than adequate to deal with the change of support problem. However, to emphasise the difference between the realisations at the local scale, a number of simulation models should be evaluated by the mine planning engineers. Moreover, as the size and area of the simulation increase, there is no global low, mid- or high realisation; this is a misnomer. One realisation that is lowest in one particular area may be the highest in another. Globally, there is no difference between realisations, so any realisation could be used to develop a mine plan as well as the basis for Mineral Resource declaration purposes.

It should be noted that this local difference between simulation realisations does not diminish the usefulness or applicability of the technique. Furthermore, the OK version of the model is just as incorrect locally; it is just that it is almost always overlooked. Inferred Resource is inferred because there is significant local uncertainty, regardless of the modelling technique used. An estimation method sensu

Table 3 Change in average Cu reserve grade above a 1.5 % Cu cut-off for 100 simulation realisations at varying support sizes

				Relative change in ave. grade		
Block size (m)	5 × 10 × 5	60 × 50 × 5	120 × 100 × 5	120 × 100 × 5 to 5 × 10 × 5	60 × 50 × 5 to 5 × 10 × 5	120 × 100 × 5 to 60 × 50 × 5
Drill spacing	20 m	70–100 m	100–250 m			
Average grade	2.65 %	1.86 %	1.66 %	60 %	43 %	12 %
Spread of realisations	~1 %	~1 %	~1 %			

stricto does not significantly alter this fact. The only way to reduce the local uncertainty is to gather more information through drilling.

The E-type estimate is often suggested as a suitable estimate for recoverable resources. The E-type estimate of $5 \times 10 \times 5$ blocks is very smooth and negates the change of support correction that is sought (see Fig. 7). The variance of the E-type under-calls the actual variance through the mere averaging process, and thus it is inappropriate as a representation of the true grade-tonnage relationship.

The purpose of the recoverable resource modelling work is to produce a block model suitable for life-of-mine underground mine planning and financial valuation, public declaration of resource and reserves and drill targeting by mine geology. The absolute accuracy of the spatial location of stopes in the RRM is not as important for underground mine planning purposes as the representation of the spatial geometry of ore and the grade architecture.

12 Conclusions and Recommendations

There is a strong and compelling business case for focussing on grade improvements at the Olympic Dam operation. Two thirds of the life-of-mine reserves are based on wide-spaced drilling, which under-calls the grade that will be realised when close-spaced drilling and high-resolution estimation is undertaken. Modelling the expected grade is worth billions of dollars to the operation.

Conditional simulation that exhaustively validate, both visually and statistically, against the drill hole data is the only technique that:

- Avoids smoothing by reproducing the original data variance (high granularity models)
- Estimates the tonnes and grade at the time of mining for any drill spacing
- Produces a better local spatial model that can be used for practical geological, mine planning and financial valuation purposes

The recoverable resource model is the main element that underpins a six-point strategy aimed at increasing the grade of the ore feed from the mine. The strategy includes improvements to mine planning and mining practices which are all required to realise the full impact of value improvement.

The development of a recoverable resource model has not been without challenges. The most significant is communicating the elementary resource concepts and principles to decision makers whom lack the specialised technical skills required to fully appreciate the importance of recoverable resource modelling. This is possibly the single largest inhibitor to a successful recoverable resource model.

Technical challenges for the Olympic Dam recoverable resource model are mostly about the size of the orebody and the inability of available software to address the large file sizes required. The only practical solution is to modify the existing GSLib SGS routine and create a fit-for-purpose algorithm. Reproduction of the correlations among multiple metals pushes the limits of application of the Stepwise Conditioning transform method employed.

The application of a single realisation for recoverable resources is outside the usual scope for conditional simulation; see, for example, Goovaerts (1997), Dimitrakopoulos (1999), Krige et al. (2004) and Van Brunt and Rossi (1999). It is therefore important to realise that in this case, a single realisation provides the additional value that the operation requires and why the usual objections to using a single realisation are not applicable.

The drill hole spacing case study demonstrably shows the impact of the misinformation effect, a term coined to describe all the unknowns in a resource estimate. All estimates, regardless of whether they are performed using traditional linear or non-linear techniques, are highly leveraged to the starting dataset. No amount of "alternate modelling", multiple simulation realisations or range analysis can fully describe the uncertainty inherent in the starting dataset.

Regardless of the drilling dataset, the best that can be done by the practitioner is to honour the available information. This approach will not capture the range of uncertainty, and thus the Olympic Dam Mine Planning Department will employ the use of modifying factors to account for the misinformation effect.

The most common applications of conditional simulation require many realisations to be evaluated. This is not the case for change of support modelling at the Olympic Dam, and perhaps on many other large deposits, since the global differences between realisations are an order of magnitude less than the difference between change of support models. However, there can be significant local differences between realisations and to quantify this impact; a handful of realisations should be given to mine planning for evaluation.

The internal company governance requirements have made it harder to produce a recoverable resource model as the basis for a Mineral Resource declaration. It is expected that the nontechnical hurdles that these requirements bring about may be overcome as management is further educated in the value of using a conditional simulation for resource estimation.

Acknowledgements The authors wish to acknowledge the contributions of geologists at the Olympic Dam, L. Voortman for his use of the original modifications of the GSLib SGS executable file and S. Khosrowshahi for valuable suggestions. E. Macmillan is thanked for kindly agreeing to review the final manuscript.

References

Abzalov MZ (2006) Localized uniform conditioning (LUC): a new approach for direct modeling of small blocks. Math J 38:393–411

Assibey-Bonsu W, Krige DG (1999) Use of direct and indirect distributions of selective mining units for estimation of recoverable resource/reserves for new mining projects. APCOM'99 Symposium. Colorado School of Mines, Golden, CO

Barnett RM, Manchuk JG, Deutsch CV (2012) Projection pursuit multivariate transform. Centre for Computational Geostatistics, Edmonton

Deutsch CV, Journel AG (1998) GSLIB: geostatistical software library and user's guide. Oxford University Press, New York

Dimitrakopoulos R (1999) Conditional simulations: tools for modeling uncertainty in open pit optimization, Optimizing with Whittle. Whittle Programming Pty Ltd, Perth, pp 31–42

Ehrig K, McPhie J, Kamenetsky V (2012) Geology and mineralogical zonation of the Olympic Dam iron oxide Cu-U-Au-Ag deposit, South Australia. Society of Economic Geologists, Special Publication 16, pp 237–267

Friedman J, Tukey J (1974) A projection pursuit algorithm for exploratory data analysis. IEEE C-23:881–890

Goovaerts P (1997) Geostatistics for natural resources evaluation. Oxford University Press, New York

Isaaks E (1990) The application of Monte Carlo methods to the analysis of spatially correlated data. Stanford University PhD thesis

Journel AG (1988) Fundamentals of geostatistics in five lessons. Stanford Center for Reservoir Forecasting, Stanford

Journel AG, Huijbregts CJ (1978) Mining geostatistics. Academic, New York

Journel AG, Kyriakidis P (2004) Evaluation of mineral reserves: a simulation approach. Oxford University Press, New York

Krige DG, Assibey-Bonsu W (1999) Practical problems in the estimation of recoverable reserves when using simulation or block kriging techniques, Optimizing with Whittle. Whittle Programming Pty Ltd., Perth

Krige DG, Assibey-Bonsu W, Tolmay L (2004) Post processing of SK estimators and simulations for assessment of recoverable resources and reserves for South African gold mines, Geostatistics Banff 2004. Springer, Banff, pp 375–386

Leuangthong O (2006) The promises and pitfalls of direct simulation. Center for Computational Geostatistics/University of Alberta, Edmonton

Leuangthong O, Deutsch CV (2003) Stepwise conditional transformation for simulation of multiple variables. Math Geol 35:155–173

Roberts DE, Hudson GR (1983) The Olympic Dam copper-uranium-gold deposit, Roxby Downs, South Australia. Bull Soc Econ Geol 78:799–882

Rossi ME, Badenhorst C (2010) Collocated co-simulation with multivariate bayesian bpdating: a case study on the Olympic Dam deposit. In: Castor R et al (eds) 4th international conference on mining innovation, MINIM. Santiago, Chile, pp 385–394

Rossi ME, Deutsch CV (2014) Mineral resource estimation. Springer, Dordrecht

Rossi ME, Parker HM (1993) Estimating recoverable reserves: is it hopeless? Geostatistics for the next century. McGill University, Montreal

Roth C, Deraisme J (2000) The information effect and estimating recoverable reserves. In: WJ Kleingeld, DG Krige (eds) Proceedings of the sixth international geostatistics congress. Cape Town, South Africa, pp 776–787

Van Brunt BH, Rossi ME (1999) Mine planning under uncertainty constraints, Optimizing with Whittle. Whittle Programming Pty Ltd., Perth

Comparison of Two Multivariate Grade Simulation Approaches on an Iron Oxide Copper-Gold Deposit

Antonio Cortes

Abstract Multivariate geostatistics takes advantage of the autocorrelation observed between variables and is a valuable tool for mining applications. This study presents a comparison of a traditional multivariate simulation technique based on a model of coregionalization against another simulation in which decorrelation is first achieved by "projection pursuit," where independent variables are simulated, then correlated back by the inverse of the decorrelation transform. The deposit used for the comparison is an iron oxide copper-gold (IOCG) ore body located in the Atacama Desert in the north of Chile. The project value is sensitive to grade, and simulation would help in assessing the grade uncertainty. The correlations observed among the different economic variables provide an opportunity to improve the simulation results. This paper starts with a brief geological description of the ore body including some of the significant geological risks. A short summary of different multivariate simulation techniques is given. The process and results of multivariate simulation of total copper, acid-soluble copper, gold, and iron within the deposit using the two methodologies are presented. The focus is on the validation and comparison of processes and results to confirm whether projection pursuit can be used as an alternative for comprehensive uncertainty analysis for resource classification.

1 Introduction

Assessing the grade uncertainty through conditional simulation provides a sound basis on which to evaluate mining projects by economic optimizations, mining designs and planning, etc.

The deposit used for this study is an iron oxide copper-gold (IOCG) ore body located in the Atacama Desert in the north of Chile. The main sector of the project is currently evaluated to pass the pre-feasibility stage.

A. Cortes (✉)
AMEC Foster Wheeler plc, av Apoquindo 3846, Santiago, Chile
e-mail: antonio.cortes@amecfw.com

J.J. Gómez-Hernández et al. (eds.), *Geostatistics Valencia 2016*, Quantitative Geology and Geostatistics 19, DOI 10.1007/978-3-319-46819-8_10

The main mineralization control is lithology and was used to define the estimation domains. Ordinary kriging was used to estimate the grades within interpreted estimation domains. The estimated variables are total copper (CuT), acid-soluble copper (CuS), iron (Fe), and gold (Au). These four variables were estimated independently without considering the correlations among them. Mineral resources were classified based on a drill hole spacing study that quantified relative confidence intervals on production increment estimates versus drill hole spacing. Reasonable prospects for economic extraction were obtained by restricting the resources to within a reasonable open-pit shell. No particular work was done to assess the geological and grade risk through conditional simulations.

This paper describes briefly both coregionalization and projection pursuit multivariate transform (PPMT) simulations and presents the results obtained from the given data set including some validation focused on the correlations among the variables. Correlations among simulated variables in the Gaussian space and raw data are compared with the correlations of the input values. Finally, conclusions and recommendations are presented.

The main objectives of the study are:

1. To describe briefly for both coregionalization and PPMT, the main practical implementation steps on a multivariate database
2. To compare the two multivariate simulation approaches in terms of reproduction of the correlations observed on the input data
3. Given the results, to discuss the usefulness of simulation in particular for this project given the information available: heterotopic database, geological modeling and parameters used for a previous resource estimation, and some conceptual information on the projected open pit

The paper does not take into account the geological uncertainty. Various assumptions were made, and a deeper analysis and improvements are recommended for future work that will contribute to the discussion that this document aims to generate. The final comparisons in terms of reproduction of the correlations (against the input data) are presented for an individual realization of both approaches. However, consistency tests were performed for a set of realizations.

2 Approaches Compared

The case study includes an initial approach using sequential Gaussian cosimulation with a linear coregionalization model.

3 Conditional Sequential Gaussian Cosimulation (SGCS)

Given a random function $\{Z(x) : x \in R^n\}$, a (nonconditional) simulation is a realization of Z(x), randomly selected from the set of all possible realizations. A model of the spatial distribution is required, and realizations reproduce the histogram of the input data. When the realization is conditioned to the data at data locations, the simulation is called conditional. There is an extensive literature which describes the theory of geostatistical simulations including:

1. Conditional simulations by turning bands algorithm (Journel 1974)
2. Simulation and coregionalization (Journel and Huijbregts 1978)
3. Sequential indicator simulation (Alabert 1987)
4. Sequential Gaussian simulation (Isaaks 1990)
5. Sequential Gaussian cosimulation (Gomez-Hernandez and Journel 1993)
6. Sequential Gaussian co-simulation (Verly 1993)

4 Projection Pursuit Multivariate Transform (PPMT)

In a multivariate case, the number and complexity of variables sometimes make it difficult to obtain multivariate Gaussian transforms and consequently to build a coregionalization model (with semi-positive definite covariance matrix). Various alternatives are used to remove complexities including principal component analysis (PCA), minimum/maximum autocorrelation factors (MAF) for removing correlation, stepwise, and PPMT (to remove nonlinearity and heteroscedasticity). PPMT facilitates multiGaussian modeling and provides an alternative way of simulating (or cosimulating) in Gaussian space and then back-transforming simulated Gaussian realizations to the original space. The main validation following this process is reproduction of the original multivariate structure as observed on the input data.

The theory and its practical application (examples and software) are covered in different documents including:

1. Guide to multivariate modeling with the PPMT, 2015 (Barnett and Deutsch 2015)
2. Projection pursuit multivariate transform (Barnett et.al. 2012)
3. Advances in the projection pursuit multivariate transform (Barnett et al. 2013)

5 Deposit Outline

The deposit is an iron oxide copper-gold (IOCG) ore body located in the Atacama Desert in the north of Chile. The deposit extends approximately 1.5 km N-S, 0.75 km E-W, and 0.45 km vertically. The CuT and Fe mineralization is mainly controlled by irregular and semi-tabular bodies of specularite breccia (BES) in which chalcopyrite (Cpy) and magnetite (Mt) are the main minerals. The database includes 58,000 m (350 drill holes) with sample values of the different variables (heterotopic). For estimation purpose, seven domains were defined, characterized by a combination of mineral zones and lithology. The projected operational scenario is an open pit with CuT as the main mined product.

6 EDA Envelope and Database

The database for the exercise includes 140 drill holes with assays values for CuT, CuS, Fe, and Au. The database is heterotopic and has the maximum number of samples for CuT and the lowest number of samples for Au. The exploratory data analysis (EDA) envelope corresponds to the main mineralized domain which corresponds to specularite breccia (BES) – usually hosted by the chalcopyrite (Cpy) mineralization. Figure 1 shows this envelope. The sample values were capped to limit the impact of outliers and then regularized (2 m composites).

7 Declustering

A nearest neighbor (NN) estimation was performed to obtain the declustering weights prior to the Gaussian transformation and subsequent steps.

The variables have positively skewed lognormal distributions with low coefficients of variation (CV) ranging from 0.6 to 0.9 except for CuS with a CV of 1.6. Table 1 gives the declustered basic statistics for the four variables considered in the study.

8 Gaussian Transform and Correlations in Gaussian Space

The declustered data were transformed to standard Gaussian values using the modeled transform function, one variable at a time. This function makes it possible to transform grades (Z) to Gaussian values (Y) and vice versa. Once the variables were transformed, basic statistics and scatterplots were produced to evaluate the correlation coefficients and the shape of the clouds of points. At first sight the scatterplots between Gaussian variables look reasonably bi-Gaussian with an elliptic shape and only some outlier points related to domaining. Figure 2 shows some examples of scatterplots of the input transformed data.

Two theoretical checks of by-Gaussianity were performed: h-scatterplots between normal pairs $[Y(x), Y(x+h)]$, and ratio of the square root of the semi-variogram and the order 1 variogram (madogram). The h-scatterplots have shapes

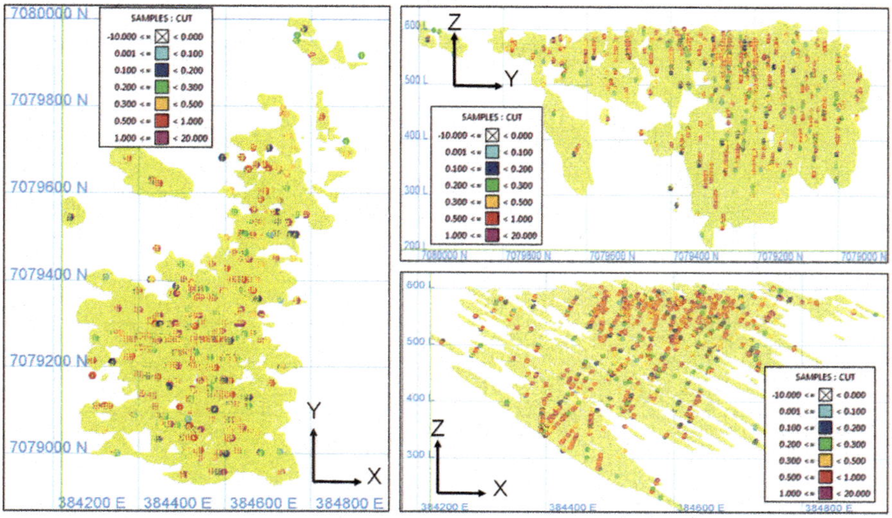

Fig. 1 Orthogonal views of EDA envelope (BES domain) and CuT samples used in the study

Table 1 Basic statistics of capped raw data in the EDA envelope

Variable	Count	Unit	Minimum	Maximum	Mean	Std. dev	CV
CuT	1386	%	0.006	5	0.77	0.74	0.96
CuS	1344	%	0.001	3	0.08	0.13	1.63
Fec	764	%	5.22	62	20.6	12.6	0.61
Au	944	g/t	0.005	1	0.11	0.10	0.90

with poor correlation (tending to circular shapes). The ratios are constant for all distances as showed in Fig. 3 for CuT and CuS as examples.

9 Gaussian Simulation

The sequential Gaussian simulation was performed over a grid restricted by the EDA domain, while search parameters for conditioning data were based on the parameters used in the resource estimation of this domain (search ellipsoid up to 200 m and a maximum number of samples of 20). A set of 100 realizations were run and then validations performed.

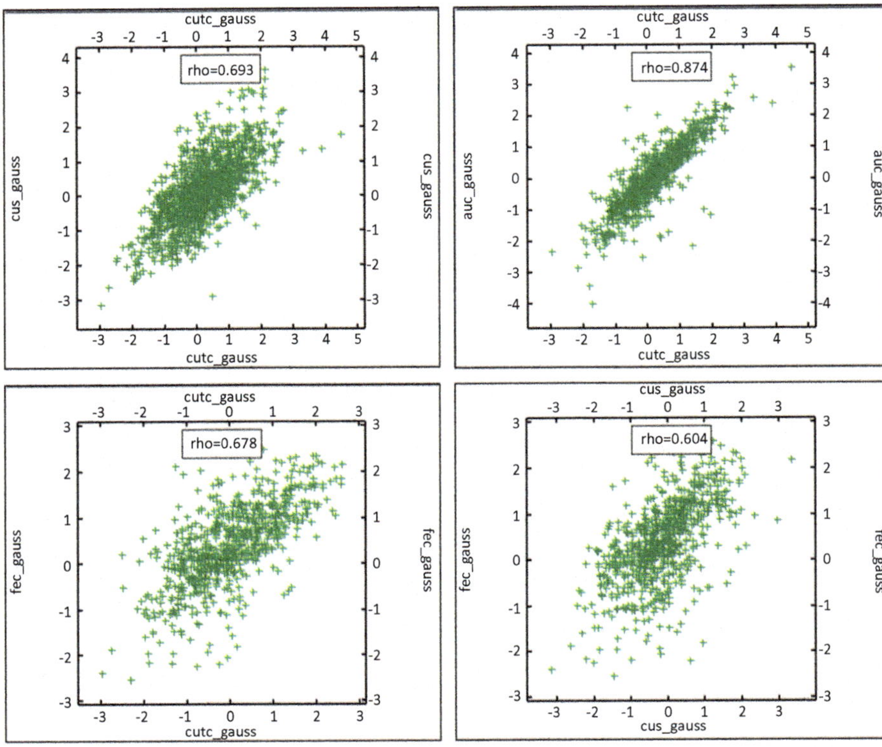

Fig. 2 Scatterplots between different variables in Gaussian space (samples)

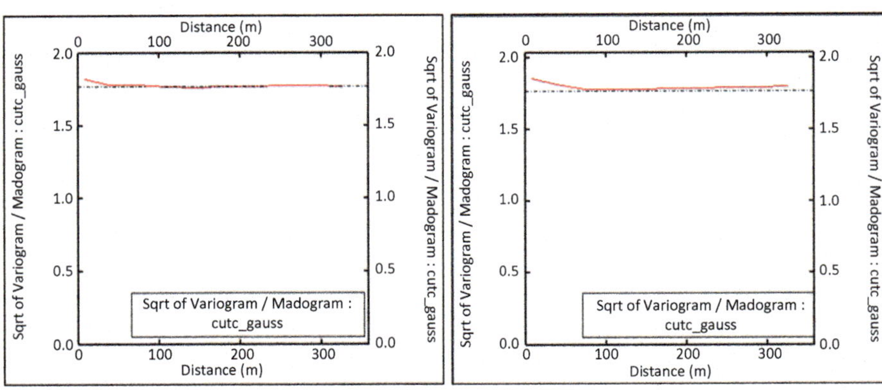

Fig. 3 Sqrt of semi-variogram/madogram for CuT (*left*) and CuS (*right*) in Gaussian space

10 Variography

The single and cross-variograms of the Gaussian values were calculated and modeled along the three main geological directions. The largest range (up to 200–250 m) and intermediate and shortest ranges were along the N-S, E-W, and vertical directions, respectively.

Figure 4 shows the single and cross variogram models for the four variables. Below the experimental curves, histograms of the number of pairs are presented.

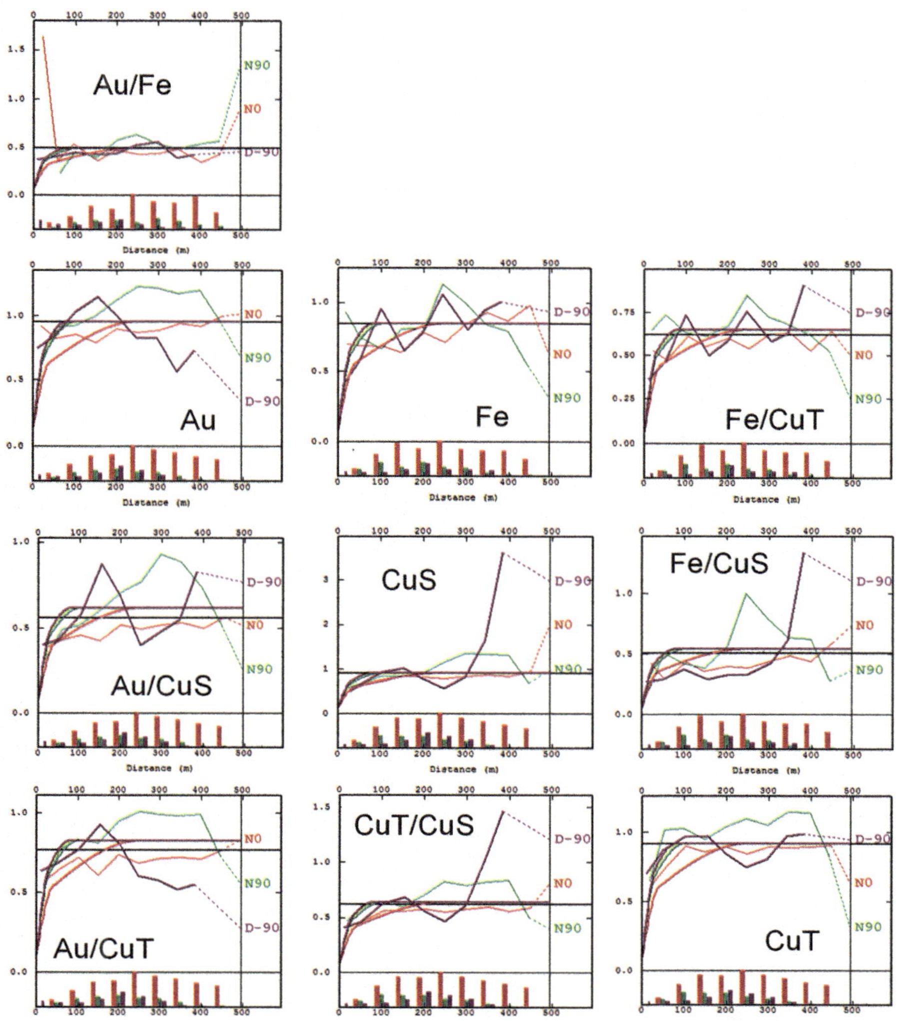

Fig. 4 Experimental and modeled variograms of CuT, CuS, Fe, and Au Gaussian values

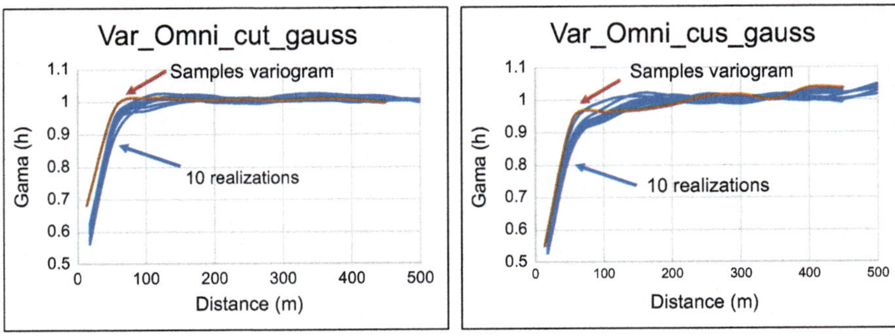

Fig. 5 Reproduction of CuT and CuS variogram models in the Gaussian space

11 Validation of Results

A set of validations in the Gaussian space were completed using the realizations. Histograms and basic statistics of the simulated Gaussian values in general show mean values very close to zero and variances slightly lower than 1.

Variograms of the simulated Gaussian values were also produced. As an example, Fig. 5 shows the CuT and CuS (Gaussian values) experimental variograms; brown denotes the variogram of the conditioning sample values, and blue denotes the variogram for ten different realizations.

Figure 6 shows the means by slices (100 m along easting, 100 m along northing, 50 m along elevation) for samples, NN model, and the E-type (from the all realizations) in the Gaussian space. The E-type has lower variance than the samples and varies locally from the declustered NN grades. The simple kriging (SK) algorithm used by the simulation could produce a portion of this due to the attraction to the mean, especially where there are less data. An example is around coordinate N7,079,700 where NN differs significantly from E-type; only 15 samples were available in this 100 m slice.

12 Correlations of Simulated Gaussian Values

To assess the obtained correlations and check them against those of the input data, scatterplots were produced for a number of realizations. Individually they show a cloud that is more homogeneous than of the input data as can be seen in the example in Fig. 7, which shows the scatterplots of the Gaussian values of the fifth realization. This cloud of points shows a good correspondence with the input data presented earlier in Fig. 2. A comparison of the simulated correlations in the Gaussian space and real data space with the projection pursuit approach is presented later.

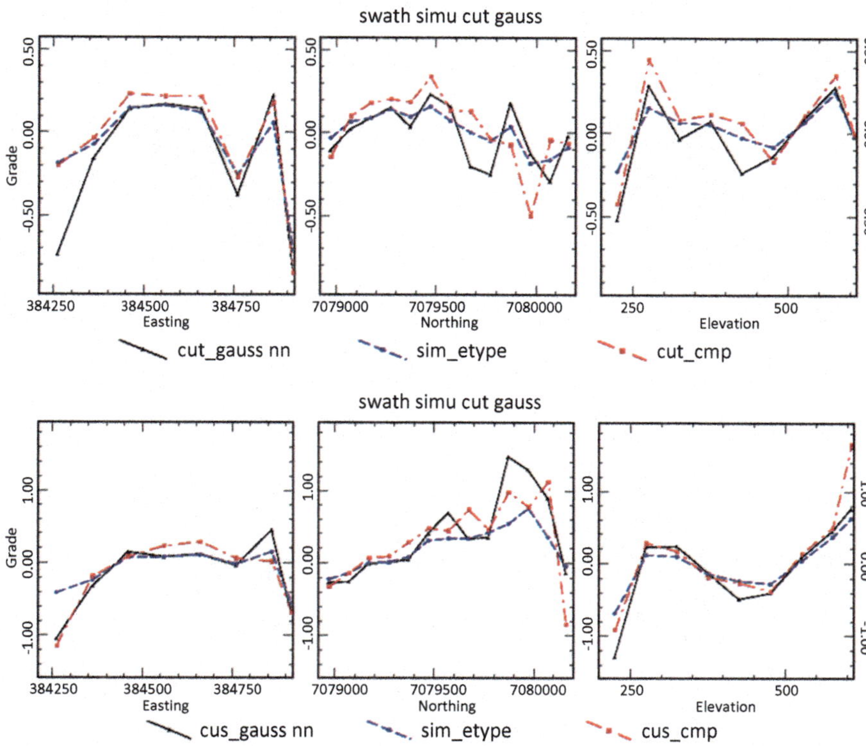

Fig. 6 Swath plots of samples, NN, and E-type on the Gaussian values

13 Statistics of Back-Transformed Values

The simulated Gaussian values were back-transformed to the original space and basic statistics calculated (particularly the mean grade and the standard deviation). Table 2 shows the main statistics from back-transformed simulated values. These are comparable with the statistics from the original data (showed previously in Table 1).

14 Simulation by Projection Pursuit

14.1 Imputation of Data

The database used for this exercise is heterotopic as showed in Table 1. This is an issue for the projection pursuit multivariate transform (PPMT), which can only be applied on homotopic databases. The exclusion of incomplete observations is not desirable because it means loss of information and an unfair comparison against the

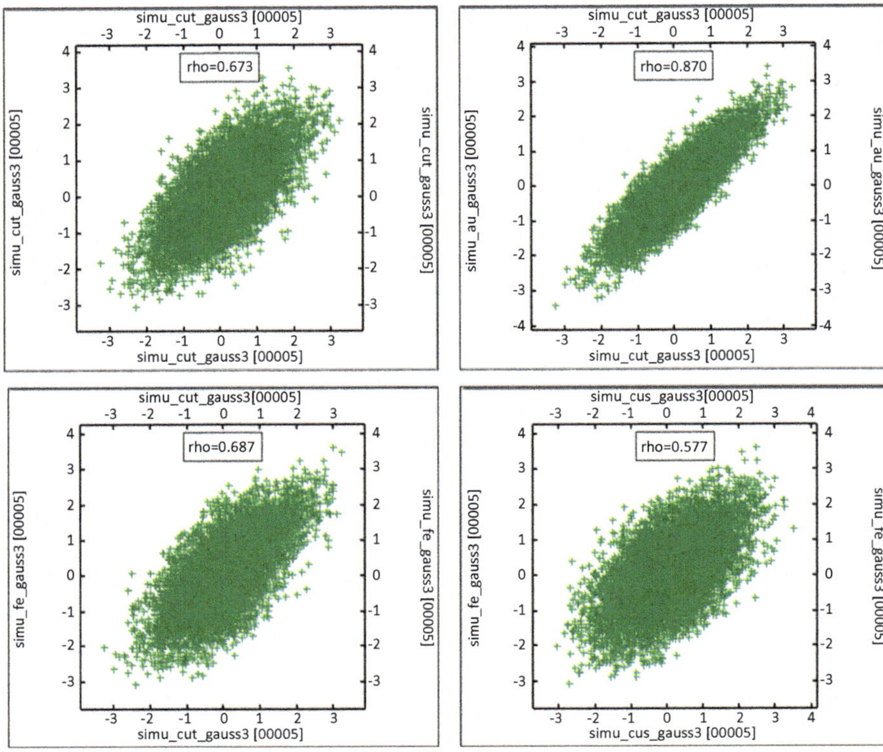

Fig. 7 Scatterplots between different variables in Gaussian space (individual realization)

Table 2 Statistics of back-transformed values

Variable	Count	Unit	Minimum	Maximum	Mean	Std. dev
CuT	10,175	%	0.006	5	0.78	0.56
CuS	10,175	%	0.001	3	0.08	0.07
Fe	10,175	%	5.22	62	21.9	9.92
Au	10,175	g/t	0.005	1	0.10	0.07

coregionalization model (which does not require a homotopic database). An imputation process is required prior to applying the PPMT.

The imputation process used in this case study is described in the document "Imputation of Geologic Data" (Barnett and Deutsch 2013). This process reproduced multivariate and spatial features of the data, by producing realizations in the normal space (the input data is a normal score heterotopic database). Once the imputation is done, some checks on reproduction of spatial variability and reproduction of basic statistics can be made.

Tests on a set of ten realizations were done to check the consistency of the imputed values with the original values. However, only one set of original plus imputed values was used in the PPMT application, i.e., one set of isotopic data is

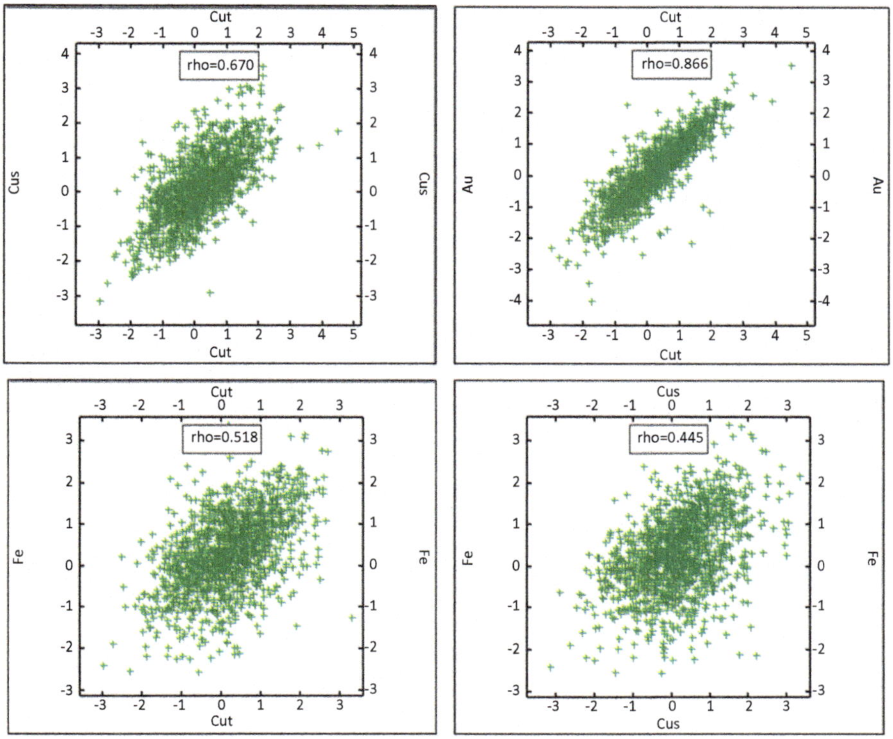

Fig. 8 Scatterplots between different variables in Gaussian space (samples + imputed data)

transformed, simulated, and then back-transformed. Then the coregionalization and PPMT results are compared, in terms of reproduction of original correlations (between variables).

Figure 8 shows the scatterplots between variables for the original data plus the imputed values (in Gaussian transformed space). The correlations observed are comparable to the correlations in Fig. 2 (original data). Q-Q plots between imputed and original values show close to 45° lines with some bias for the extreme values as can be seen in Fig. 9.

Figure 10 shows the experimental variogram of the original data (red curve), plus ten experimental variograms (in blue) of the original data and plus the imputed values. The percentages of imputed values were 3 %, 45 %, and 32 %, respectively, for CuS, Fe, and Au.

14.2 PPMT

PPMT theory and its practical application (examples and software) are given in publications referenced earlier. For this exercise, the methodology included the

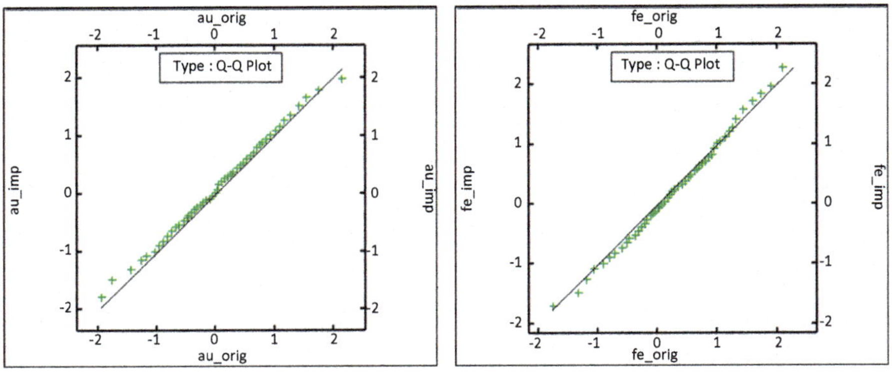

Fig. 9 Q-Q Plots for Au (*Left*) and Fe (*right*) between original (*horizontal axis*) and imputed data (*vertical axis*)

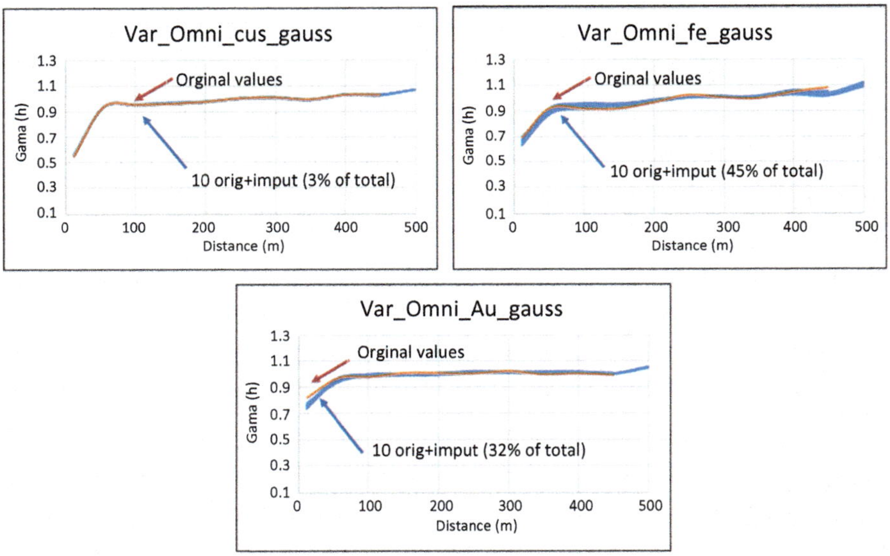

Fig. 10 Variography of original values (*red*) and original plus percentage of imputed values (*blue*)

following steps: running the PPMT algorithm (ppmt.exe) whose input corresponds to the input database (isotopic) and the output corresponds to the transformed variables and a transformation table (binary) needed later when the back-transformed algorithm is used. Transformed variables were then simulated by sgsim (sgsim.exe). Finally, these simulated values are back-transformed using the back-transformed PPMT algorithm (ppmt_b.exe).

Figure 11 shows the scatterplots between variables after 50 iterations of the PPMT algorithm. The scatterplots show uncorrelated standard Gaussian variables. Using this dataset with no correlation between the variables, one realization was

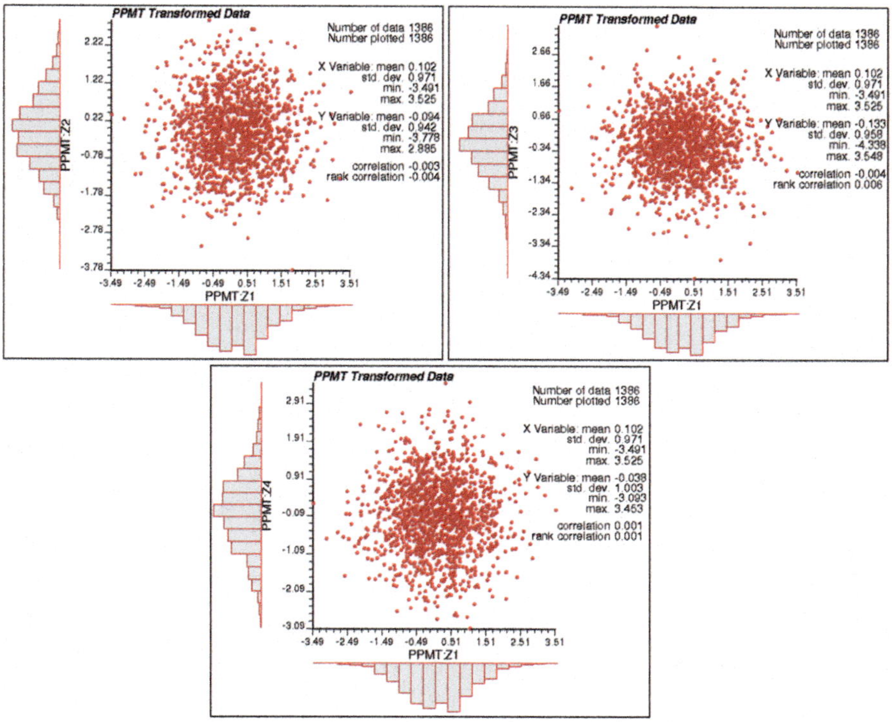

Fig. 11 Scatterplots of the PPMT transformed data (*top left*, CuT vs CuS; *top right*, CuT vs Fe; below, CuT vs Au)

performed, and then the back correlation algorithm was used to introduce the correlation between the simulated variables.

14.3 Back Transformation and Comparison

Once the back transformation PPMT_B is performed, the results (simulated values in the real space) can be checked against the original database, focusing on the examination of the correlations between variables in real space. To consider the spatial cross-correlations after back-transforming, some cross-variograms for data and simulated values were calculated. In general, both methods, linear coregionalization and PPMT, reproduce the spatial cross-correlations observed in the data.

Figures 12, 13, and 14 show scatterplots for the different approaches for CuT and CuS, CuT and Fe, and CuT and Au, respectively. It can be seen that for CuT-CuS, differences are small and both methods reproduce the primary data correlation. The main difference is in CuS grades, where there is a difference between the clouds of points around 2 % CuT: the shape of the cloud of points seems better reproduced using PPMT.

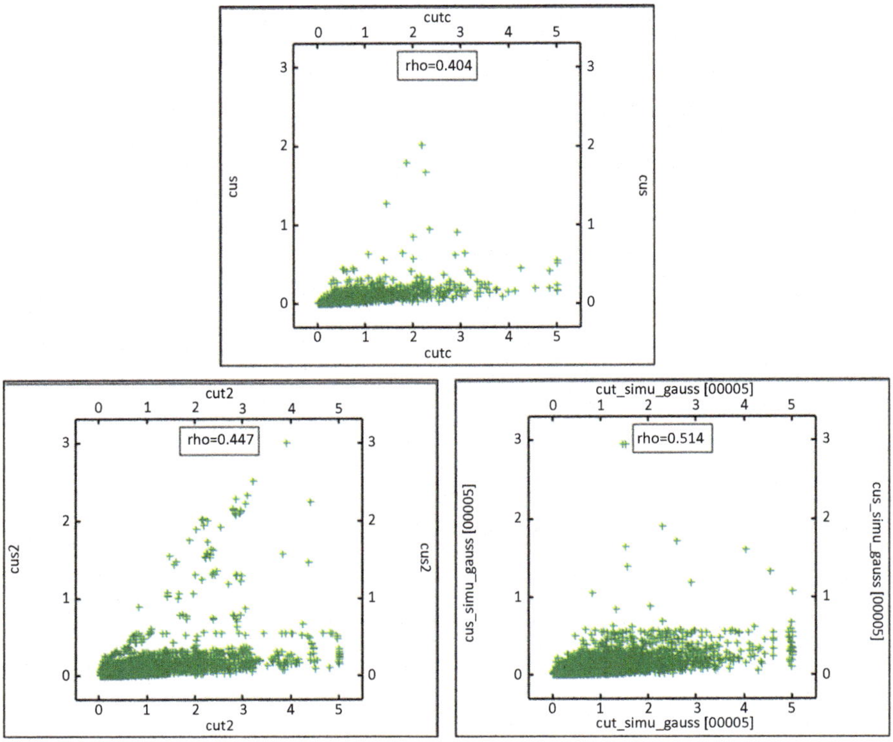

Fig. 12 CuT vs CuT scatterplot comparison (*top*, samples; below *left*, PPMT; below *right*, cosimulation)

For CuT-Fe, the cosimulation using coregionalization represents a shape closer to the original samples, especially for samples with high CuT and low Fe and with high Fe and low CuT (bottom right and top left corners, respectively).

For CuT-Au, the PPMT approach better represents the shape of the original samples, capturing well the shape of the cloud for lower Au values where CuT grades are around 2 %.

15 Conclusions

The theory and implementation of both methods is well documented; coregionalization is widely known and already included in commercial geostatistical software packages. PPMT documentation and software is also available and can be used for practitioners. The same applies for the imputation methodology required prior to applying PPMT.

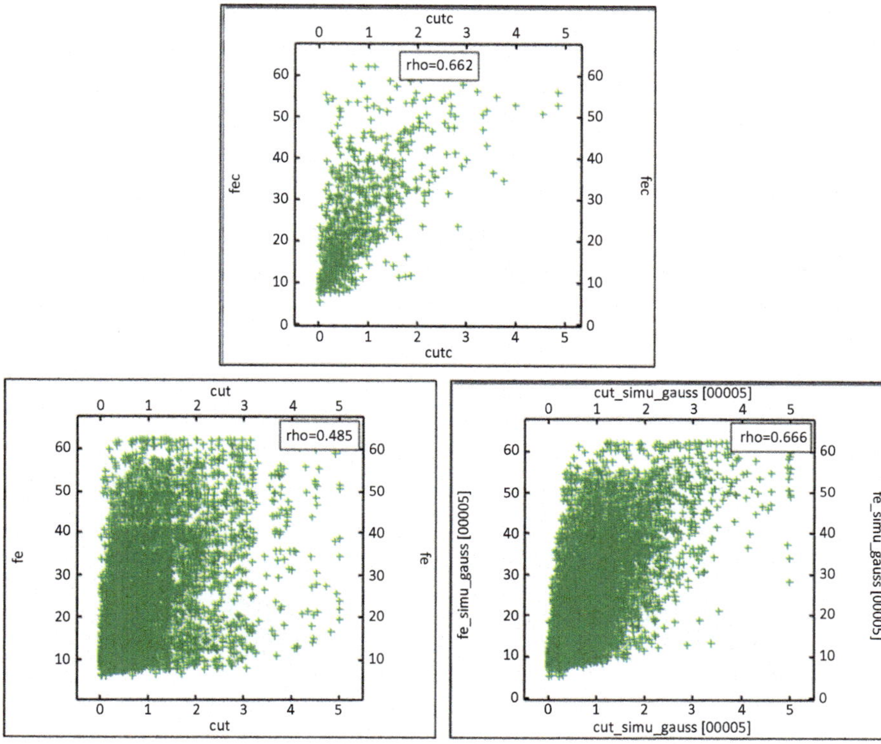

Fig. 13 CuT vs Fe scatterplot comparison between (*top*, samples; below *left*, PPMT; below *right*, cosimulation)

Both methodologies reproduce the correlations between variables based on original data and are attractive to implement on this project.

Future applications of both methodologies can be included in a comprehensive uncertainty analysis for resource classification on this project.

Acknowledgments Thanks to Georges Verly and Harry Parker for their valuable suggestions, Amec Foster Wheeler for the time for this paper, Geovariances for providing me a temporary Isatis license during the practical work, and CCG for providing the material and software to perform application exercises on the server.

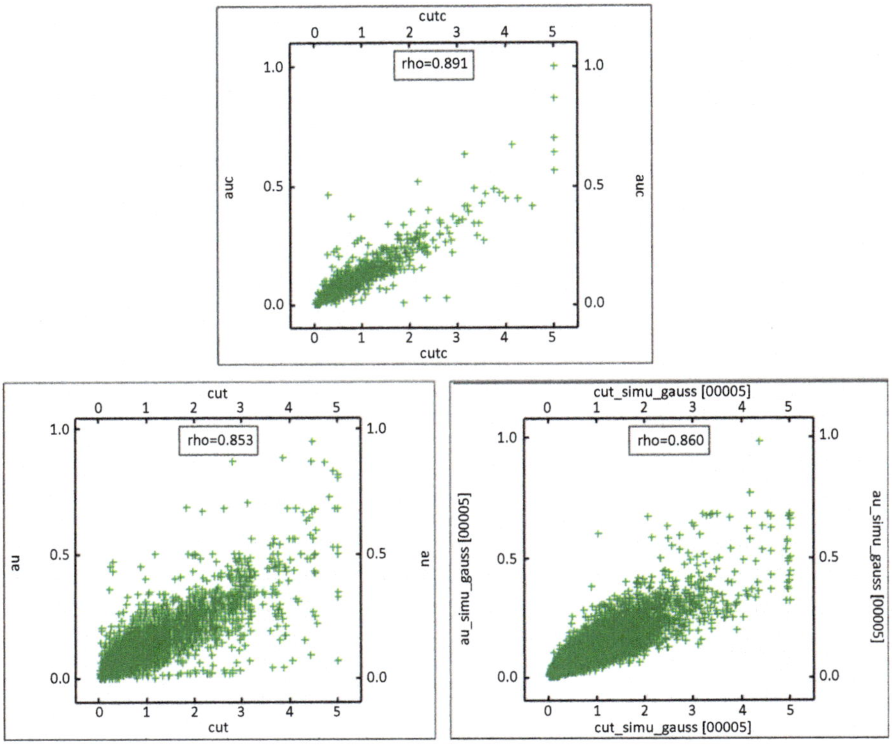

Fig. 14 CuT vs Au scatterplot comparison (*top*, samples; below *left*, PPMT; below *right*, cosimulation)

References

Alabert FG (1987) Stochastic imaging of spatial distributions using hard and soft information. M. Sc. Thesis. Stanford U., p 197

Barnett R, Deutsch C (2013) Imputation of geologic data. *Paper 102*. CCG Annual Report 15.

Barnett R, Deutsch C (2015) Guide to multivariate modeling with the PPMT. CCG, Guidebook, Series Vol 20

Barnett R, Manchuk J, Deutsch C (2012) Projection pursuit multivariate transform. *Paper 103*. CCG Annual Report 14

Barnett R, Manchuk J, Deutsch C (2013) Advances in the projection pursuit multivariate transform. *Paper 106*. CCG Annual Report 15

Gomez-Hernandez J, Journel AG (1993) Joint sequential simulation of multigaussian fields. In: Soares A (ed) Geostatistics Troia 92, part 1. Kluvert Academic Publishers, Dordrecht, pp 85–94

Isaaks E (1990) The application of Monte Carlo methods to the analysis of spatially correlated data. *Ph.D. Thesis*. Standford U., p 213

Journel AG (1974) Simulations Conditionelles de Gisements Miniers – Théorie et Pratique. Docteur Ingenieur Thesis. Nancy U., p 110

Journel AG, Huijbregts CJ (1978) Mining geostatistics. Academic, London, p 600, 2nd printing revised

Verly G (1993) Sequential gaussian co-simulation: a simulation method integrating several types of information. In: Soares A (ed) Geostatistics Troia 92, part 1. Kluver Academic Publishers, Dordrecht, pp 543–554

Complexities in the Geostatistical Estimation of Besshi-Type Mineral Deposits on the Northwest of Pinar del Río, Cuba

Abdiel Díaz-Carmona, José Quintín Cuador-Gil, Fernando Giménez-Palomares, and Juan Antonio Monosoriu-Serra

Abstract Geostatistics resources estimation of the copper (Cu) grade values was carried out in the "Hierro Mantua" mineral deposit, which is located at the northwest of Pinar del Río province, Cuba. The geologic complexities in the region of the deposit indicate the nonexistence of homogeneity in the Cu values. The structural analysis showed a high asymmetric distribution in the variable studied. The nonexistence of normality was verified by different mean and median values, a coefficient of variability greater than one, and the moving windows statistics of the mean was different. Under the previous conditions, the data was log transformed to assure the necessary stationarity in them and consequently to achieve an adequate accuracy in the resources estimation, using a rational selective mining unit (SMU). The log-transformed data revealed a homogeneous behavior in Cu grade values, demonstrated by better results in basic and moving windows statistics. Semivariograms showed defined structures with anisotropy in the 0° and 90° directions (considering 0 to the north). To apply the lognormal kriging estimation is the main objective of this work, because of the complexities of the geology in the studied area.

A. Díaz-Carmona (✉)
Department of Geology, University of Pinar del Río, Pinar del Río, Cuba
e-mail: abdiel.diazcarmona@gmail.com

J.Q. Cuador-Gil
Department of Physics, University of Pinar del Río, Pinar del Río, Cuba
e-mail: cuador@upr.edu.cu

F. Giménez-Palomares
University Institute of Pure and Applied Mathematics, Polytechnic University of Valencia, Valencia, Spain
e-mail: fgimenez@mat.upv.es

J.A. Monosoriu-Serra
Technical Superior School of the Design, Department of Physics Applied, Polytechnic University of Valencia, Valencia, Spain
e-mail: jmonsori@fis.upv.es

J.J. Gómez-Hernández et al. (eds.), *Geostatistics Valencia 2016*, Quantitative Geology and Geostatistics 19, DOI 10.1007/978-3-319-46819-8_11

1 Introduction

On the northwest of Cuba, at the province of "Pinar del Río," exists a group of mineral deposits and other small accumulations that are classified as Besshi type. The most important are "Hierro Mantua," "Unión 1," "Juan Manuel," "Unión 2," and "Flanco Este." Some of them are in the feasibility stage for future exploitation. They associate to rocks of the Esperanza Formation ($J_3^{ox} - K_1^{ne}$), which is constituted by sequences of silts, carbonate, and volcanic rocks of basic composition originated in a marginal basin. The rocks that prevail are calcareous, gritty, argillites, concordant bodies, and diabases and gabbros sub-concordant representatives of a submarine magmatism.

The mineral bodies form long lenses, which follow an NE direction, parallel to the coast. The relief altitudes are between 60 and 75 m above sea level. These bodies have many outcrops, concordant with the stockwork. Some secondary enrichment processes have taken place in some of them, as a result of a process of extreme oxidation, lixiviation, and concentration of cupper (Cu) in the primary sulfurous mineralization of volcanogenic-sedimentary origin.

Historically, classic methods were used for resources estimation in those types of mineral deposits. As it is known, these methods do not guarantee the necessary precision, mainly because of the geological complexities in the region. In order to achieve an appropriate accuracy of the resources estimation in "Hierro Mantua" mineral deposit, two objectives are established: (1) to apply a methodology for the rational SMU determination, which is used in quantifying the mineral resources in the deposit studied, and (2) to apply a nonlinear geostatistics method, lognormal kriging, for resources estimation because of the presence of Cu values asymmetric distribution. Particularly, the occurrence of a few high Cu values in "Hierro Mantua" mineral deposit makes it very difficult to handle data with lognormal distribution. This mineral deposit was taken as an example in the present research.

Regarding lognormal kriging, there have been some publications related with this estimator, classified as nonlinear. Some of the most important are Matheron (1974), Marechal (1974), Rendu (1979), Journel (1980), Krige (1981), Dowd (1982), Krige and Magri (1982), Thurston and Armstrong (1987), Armstrong and Boufassa (1988), Rivoirard (1990), Marcotte and Groleau (1997), Lee and Ellis (1997), Roth (1988), Clark (1998), Cressie (2006), Yamamoto (2007), Yamamoto and Furuie (2010), and Paul and Cressie (2011).

Some ideas included in those works are "...the sensitivity of the experimental semivariogram with regard to the few high values (Journel 1980)." High values cause the deformity of the semivariograms that prevents the use of linear kriging. According to Clark (1998), "...the potential problem in lognormal Kriging is in the "back transformation" of the logarithmic estimates to the original sample values scale, there appears to be some disagreement in the general geostatistical literature as to how this back-transformation should be carried out." In this sense, the equation for back transformation of block kriging values is not easily found in the literature. A review of lognormal estimators in an orderly way was presented in

Rivoirard (1990), but he concluded that the equations of back transformation for the block estimation are not theoretically optimal. Roth (1998) concluded that "...the practitioner must decide whether the lognormal Kriging estimator is suitable for local estimator." Cressie (2006) stated "...data that exhibit skewness may be successfully modeled as coming from a lognormal spatial process." In a practical way, back transformation is achieved by the exponential of the kriging estimate plus a non-bias term (Yamamoto 2007). Finally, in this research, Cu values are estimated following ideas presented by Rivoirard (1990), mainly the back-transformation equation, Eq. 6 on page 218. The precision of this method has been proved by some authors in different mineral deposits and other studies when skewness distributions have been found.

2 Data Description

The quality of the primary information is one of the most important elements for the resources estimation using kriging. This assumption determines the quality of the results (Armstrong and Carignan 1997). The "Hierro Mantua" deposit database used in the present research work was validated in 2008 by the Coffey Mining Company, for request of RSG Global to the "Cobre Mantua S.A. Company" (Arce-Blanco and Santana-Okamoto 2008). An initial exploration of the samples was performed, whose lengths vary between 0.05 and 4.65 m, with an average of 1.22 m. A regularization of the data support was needed; 1 m length composites inside the mineral body were calculated. Cu values from 318 wells, regularly distributed, in two 50×50 m and 100×100 m sampling networks were analyzed, creating 5,192 composites of 1 m length.

3 Exploratory Data Analysis

Table 1 shows the descriptive statistics of Cu values. The Cu mean and median values are very different and very small compared with the maximum value; the coefficient of variation is greater than 1. So, the data distribution is asymmetric, which can be seen in the histogram (Fig. 1). The Kolmogorov-Smirnov normality test value is 0.27, which is greater than 0.05. The result shows the non-normal behavior of the Cu values. For this reason, the linear geostatistics procedures cannot be applied; therefore, it is more appropriate to use nonlinear geostatistics methods (Journel and Huijbregts 1978; Isaaks and Srivastava 1989; Rivoirard 1990).

Table 1 Descriptive statistics of Cu values

Statistics	CU %
Minimum	0.005
Maximum	34.660
Mean	2.334
Median	0.979
Standard deviation	3.486
Variance	12.152
Coefficient of variation	1.493
Skewness	3.220
Kurtosis	16.289
Kolmogorov-Smirnov	0.266

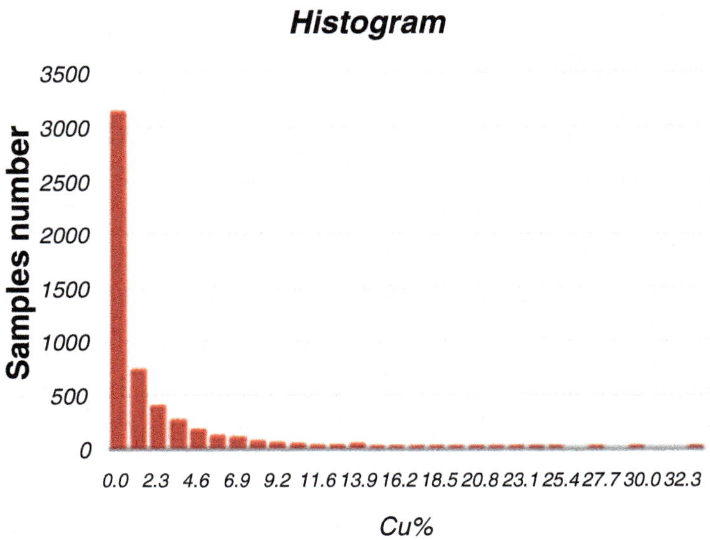

Fig. 1 Histogram of Cu values

3.1 *Lognormal Transformation Statistics*

The lognormal transformation was performed for Cu values (LnCu = Ln(Cu)); Table 2 shows the basic statistics and Fig. 2 the histogram of the transformed data. It is observed that the transformed histogram is close to normal, with very similar values for the mean and median. The Kolmogorov-Smirnov test value is 0.026, which is less than 0.05 and assures the normal behavior in the LnCu values. Previous analysis indicates the normality of the LnCu values. In addition, moving windows statistics showed stable values of the mean. The range, computed by the difference of the upper and lower limits of the mean from different moving windows, is 1/5 of the range of the LnCu values. So, the data can be assumed as

Table 2 Descriptive statistics of the LnCu values

Statistics	LnCu %
Minimum	−5.298
Maximum	3.546
Mean	−0.208
Median	−0.262
Standard deviation	1.520
Variance	2.310
Coefficient of variation	0.001
Skewness	−0.164
Kurtosis	2.865
Kolmogorov-Smirnov	0.026

stationary, and the lognormal kriging can be applied (Journel and Huijbregts 1978; Isaaks and Srivastava 1989; Cressie 1993).

3.2 Variability Analysis

The maximum separation distance between the points of the data is 650 m in deposit strike direction, 230 m in the dip direction, and 60 m in the downhole direction. The lag and directions used for the semivariogram calculation are presented in Table 3. The variability was studied approximately up to half of the maximum distance for each direction.

The experimental semivariogram and the fitted models are shown in Fig. 3. All semivariograms were computed in Gemcom software. Geometric anisotropy can be observed. The fit of the experimental semivariograms was performed by an imbricated model, composed by two spherical models with a small nugget effect. The analytical expression of the spatial variability model is as followed:

$$\gamma(h) = 0.1 + 1.21 \ \text{Sph}(26, \ 15, \ 11) + 1.11 \ \text{Sph}(114, \ 90, \ 40) \tag{1}$$

where

0.10 = nugget effect
Sph = spherical model
1.22 = sill for the first spherical structure
(25, 20, 10) = ranges for the first structure in direction: strike, dip, and downhole
1.19 = sill for the second spherical structure
(110, 90, 30) = ranges for the second structure in direction: strike, dip, and downhole

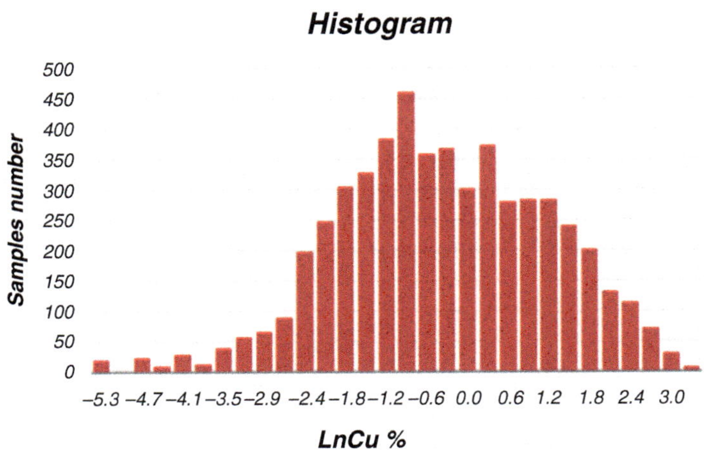

Fig. 2 Histogram of the LnCu values

Table 3 Experimental semivariogram calculation parameters

Direction	Lag (m)	Number of points	Azimuth (°)	Dip (°)	Angular tolerance (°)	Bandwidth
Strike	35	10	0	0	45	140
Dip	10	10	270	−55	45	40
Down hole	1	30	90	−42	10	10

3.3 *Validation of the Model*

The spatial variability model obtained was validated using the cross validation procedure. A high value was obtained for the correlation between estimated and real values (Fig. 4); this figure shows a high concentration of values around the slope 45°; the coefficient of Pearson is 0.92. These results show that the variability model obtained in the structural analysis describes adequately the spatial variability and correlation of the lognormal transformed data in the mineralized phenomenon under study.

Furthermore, an analysis of the estimated error was performed. 94.34 % of the real data are included in the prediction interval Zi $\pm$ 2*σ (Alfaro-Sironvalle 2007) considering σ as the standard deviation of the estimated error, and 87.11 % of the real data are included in the prediction interval considering σ as the kriging standard deviation (Szilágyi-kishné et al. 2003; De Oliveira et al. 1997).

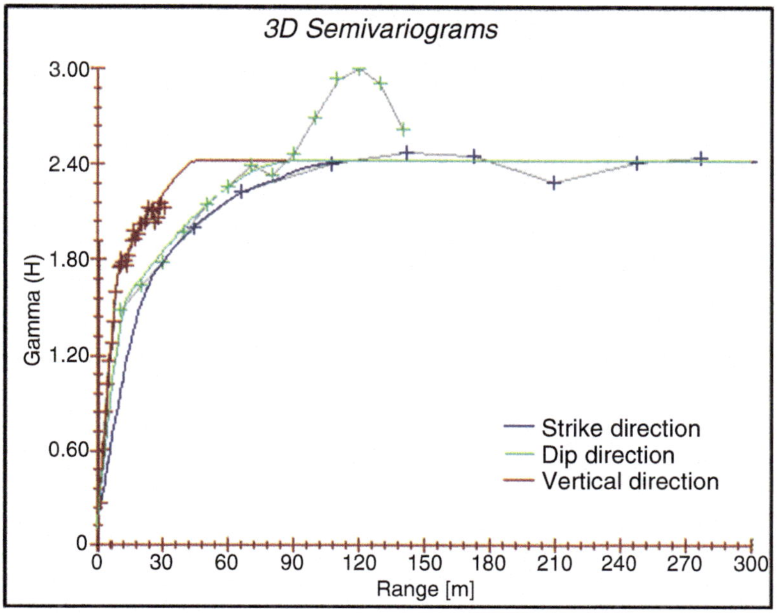

Fig. 3 Semivariograms of LnCu values

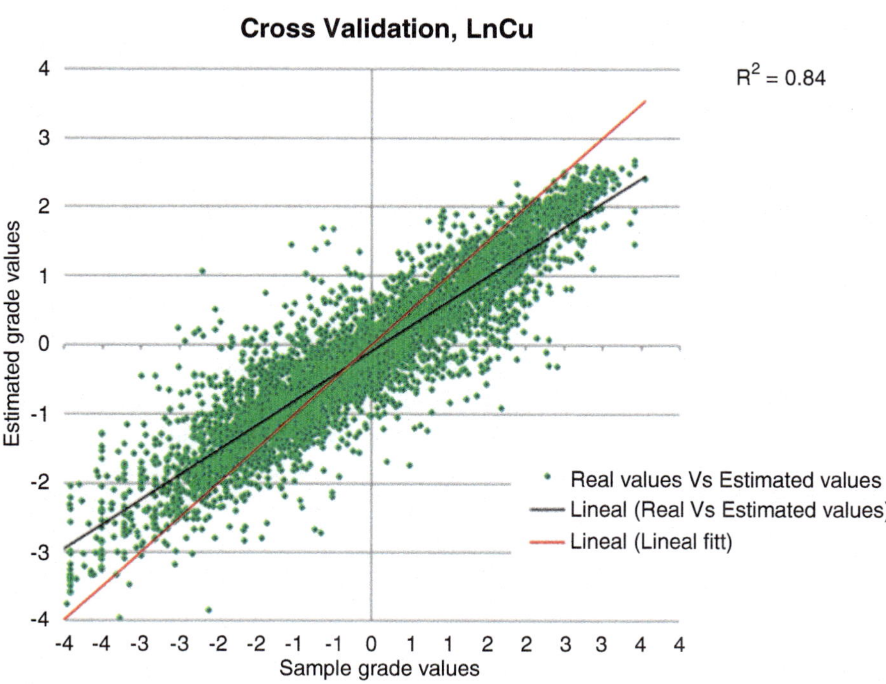

Fig. 4 Estimated value vs real values by cross validation

4 Mining Unit Size Determination

Geostatistics simulation was used to determine the rational SMU size. A realization of the regionalized variable can be obtained by geostatistics simulations. The sequential Gaussian simulation is a powerful technique (Gómez-Hernández and Cassiraga 1994) and the most used method to simulate values. In this sense, as the Cu values show extremely asymmetric distributions, the logarithmic transformed values, LnCu, were simulated in a dense network in order to reproduce possible real values. These values were computed by the average of simulated data inside of blocks with different sizes of SMU.

The procedure consists in comparing values of the possible real values with the ordinary kriging estimations for each SMU proposed. Next, the least mean squared error (MSE) was used to obtain the rational size of the blocks for estimation. This procedure guarantees a high precision for resources estimations in a local or global scale. The rational SMU size obtained was 32 × 40 m.

The procedure proposed was repeated with other realizations from sequential Gaussian simulations, and the results were similar; the 32 × 40 m SMU size was obtained as the rational SMU size. Gómez-González and Cuador-Gil (2011) made a similar research in nonmetallic deposits, based on the opinion that mining equipment must not be the factor determining the SMU size but the results of applying the right mathematics estimation methods, which take into account the natural behavior of the phenomenon.

Finally, lognormal kriging is used for resources estimation with the rational network obtained before. In future studies, multi-Gaussian kriging will be applied for more accuracy of the estimations in this ore deposit and other mineral deposits of the regions, when the logarithmic transformation does not guarantee normal distribution and consequently stationary data. The simulation was performed using "sgsim" program from GsLib (Deutsch and Journel 1998).

The sequential Gaussian simulation performed with transformed data, the LnCu, is presented in Fig. 5, using the variability model represented by Eq. 1 and a simulation network of (1 × 1 × 1 m). The simulation was verified by the basic statistics between real and simulated values that were similar, and the histograms and semivariograms between them show similar behavior (Fig. 6).

The simulation network was very dense in order to obtain possible real values inside each SMU proposed (Table 4). Estimation values were obtained by ordinary kriging for each SMU proposed. Then, errors were computed comparing the possible real values with the estimated values for each SMU, which are shown in Table 4 and Fig. 7. Mean squared error and relative errors were used for this comparison. Figure 7 shows the nomograms of the mean squared errors and relative errors vs the size of the SMU. The SMU was proposed in two forms: squared and rectangular, to take into account the anisotropy.

The influence of the SMU form was verified (Fig. 7). The least SME is present for 32 × 40 m SMU size, which is the most precise for the resources estimation. Using the rational SMU obtained before, the estimation was performed with

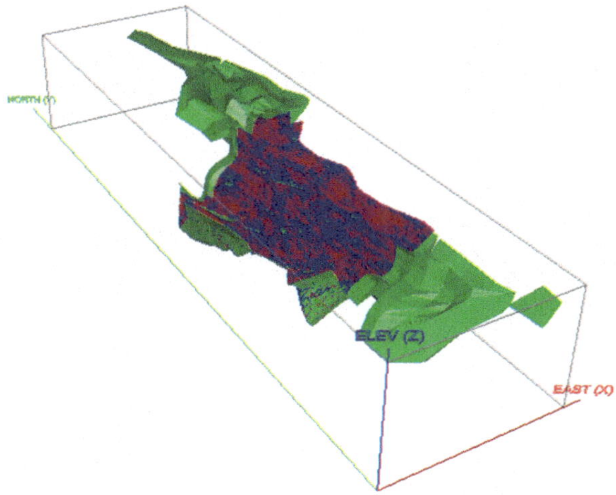

Fig. 5 Simulation of the LnCu values

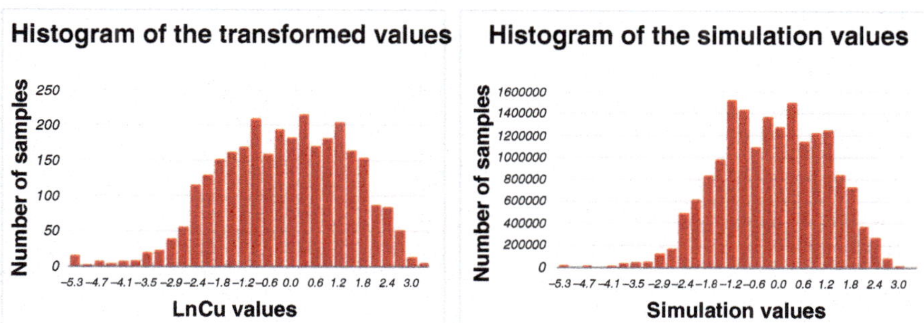

Fig. 6 Histogram of LnCu and simulated values

lognormal kriging over the mineralized body (Fig. 5). The back transformation was carried out by the following expression (Rivoirard 1990):

$$Z^*(v) = Exp\Big(L(v)^{\mathrm{KO}} + \left(\sigma^2{}_{KO} + \gamma(v,v)\right)/2 - \mu \Big) \tag{2}$$

where

$L(v)^{\mathrm{KO}}$ = estimation of the transformed values by ordinary kriging
$\sigma^2{}_{\mathrm{KO}}$ = kriging variance
$\gamma(v,v)$ = mean semivariogram block to block
μ = Lagrange multiplier

Table 5 shows the basic statistics between original and estimated grade values after back transformation. The mean values are close, and the variance has adequate values, in correspondence with the smoothen characteristics of the interpolation. No extreme values have been obtained in the estimations.

Table 4 Estimation errors for USM

USM	Mean squared error	Mean values		Differences (%)	Relative error
		Estimation	Simulation		
Squared network					
5*5	1.61	−0.26	−0.02	69 %	1.70
10*10	1.67	−0.50	−0.04	67 %	1.69
15*15	1.43	−0.39	−0.05	61 %	1.55
20*20	1.38	−0.42	−0.07	60 %	1.46
25*25	1.56	−0.64	−0.08	57 %	1.56
30*30	1.35	−0.46	−0.05	56 %	1.62
35*35	1.46	−0.58	−0.05	46 %	1.71
40*40	1.28	−0.46	−0.04	44 %	1.66
45*45	1.32	−0.55	−0.08	41 %	1.47
50*50	1.49	−0.75	−0.04	48 %	1.80
Rectangular network					
4*5	1.62	−0.25	−0.02	70 %	1.73
8*10	1.69	−0.48	−0.04	68 %	1.70
12*15	1.46	−0.37	−0.05	64 %	1.54
16*20	1.46	−0.41	−0.05	61 %	1.56
20*25	1.58	−0.65	−0.07	59 %	1.63
24*30	1.38	−0.45	−0.05	51 %	1.56
28*35	1.28	−0.52	−0.08	47 %	1.44
32*40	1.20	−0.43	−0.07	49 %	1.42
36*45	1.43	−0.55	−0.04	47 %	1.72
40*50	1.46	−0.73	−0.03	49 %	1.82
44*55	1.46	−0.59	−0.04	41 %	1.73

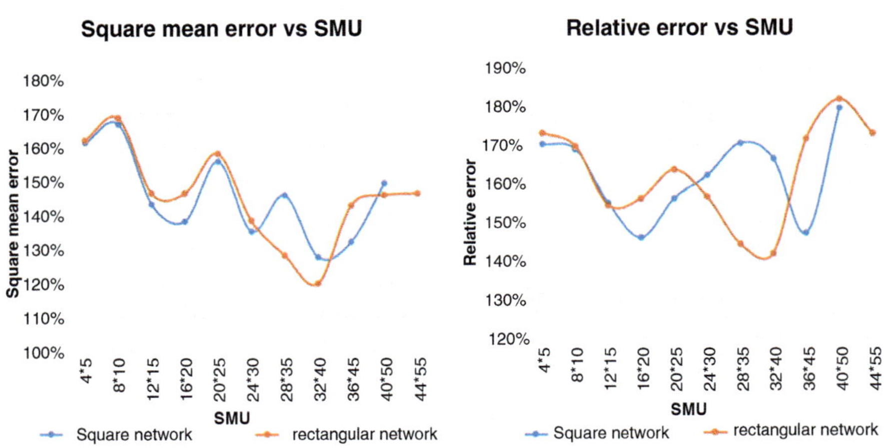

Fig. 7 Mean squared error and relative error vs SMU

Table 5 Original grade vs estimated grade values basic statistics

	N	Minimum	Maximum	Mean	Standard deviation	Variance
Sample grade values	5,192	0.005	34.660	2.227	3.583	12.838
Estimated grade values	5,192	0.015	20.368	2.156	2.854	8.146

5 Conclusion

In this article, it is shown how to obtain adequate resources estimation when datasets exhibit skewness distribution, which are very common in mining studies. The geologic complexities in the region of the deposit studied indicate that linear geostatistics cannot be applied. The non-stationarity of the Cu grade values in "Hierro Mantua" mineral deposits was proved by descriptive statistics parameters, Kolmogorov-Smirnov goodness of fit, and moving windows statistics. The lognormal kriging has been successfully applied in several mining studies and other geosciences problems.

Cu grade values were log transformed to a new variable, LnCu, in which the basic and moving windows statistics show normality and stationarity. The structural analysis of the transformed data shows geometric anisotropy in the N-S and E-W directions. The model fitted is an imbricated structure, composed by a nugget effect plus two spherical models, which was validated by cross validation.

The rational SMU size obtained showed the minimum values of mean squared error when comparing estimated and real grade values, using different sizes of SMU. This procedure was made based on the opinion that mining equipment must not be the factor determining the SMU size but the results of applying the right mathematics estimation methods, which take into account the natural behavior of the phenomenon.

The estimation performed by lognormal kriging is adequate for the resources estimation in the case study presented. Original grade vs estimated grade values basic statistics show close values of the mean and adequate values of the variance.

References

Alfaro-Sironvalle MA (2007) Estimación de Recursos Mineros, Biblioteca en línea del Centro de Geoestadística de la Escuela de Minas de París: www.cg.ensmp.fr/bibliotheque

Arce-Blanco M, Santanna-Okamoto F (2008) Memorandum database´s rebuilding. Pinar del Río, Pinar del Río, Cuba, pág. 18

Armstrong M, Boufassa A (1988) Comparing the robustness of ordinary kriging and lognormal kriging outlier resistance. Math Geol 20(4):447–457

Armstrong M, Carignan J (1997) Géostatistique Linéaire, Application au Domaine Minier. École des Mines de Paris, Fontainebleau

Clark I (1998) Geostatistical estimation and the lognormal distribution. Geocongress, Pretoria RSA

Cressie N (1993) Statistics for spatial data Rev edn. Wiley, New York, 900 p.

Cressie N (2006) Block kriging for lognormal spatial processes. Math Geol 38(4):413–443

De Oliveira V, Kedem B, Short DA (1997) Bayesian prediction of transformed Gaussian random fields. J Am Stat Assoc 92(440):1422–1433

Deutsch CV, Journel AG (1998) GSLIB: geostatistical software library and user's guide, 2nd edn. Oxford University Press, New York

Dowd P (1982) Lognormal kriging: the general case. Math Geol 14(5):474–500

Gómez-González O, Cuador-Gil JQ (2011) Determinación del tamaño racional del bloque para la estimación de recursos minerales en el yacimiento Mariel. Revista Minería y Geología 27 (4):20–39

Gómez-Hernández JJ, Cassiraga EF (1994) Theory and practice if sequential simulation. In: Armstrong M, Dowd PA (eds) Geostatistical simulation. Kluwer, Dordrecht, pp 111–124

Isaaks EH, Srivastava RM (1989) Applied geostatistics. Oxford University Press, Oxford, p 561

Journel AG (1980) The lognormal approach to predicting local distributions of selective mining unit grades. Math Geol 12(4):285–303

Journel AG, Huijbregts CJ (1978) Mining geostatistics. San Francisco Academic Press, New York, p 500

Krige DG (1981) Lognormal-De Wijsian geostatistics for ore evaluation: S. African Inst. Min. Metall, Monograph series, p. 40

Krige DG, Magri EJ (1982) Geostatistical case study of the advantage of lognormal De Wijsian Kriging with mean for a base metal mine and a gold mine. Math Geol 14:547–555

Lee YM, Ellis JH (1997) Estimation and simulation of lognormal random fields. Comput Geosci 23(I):19–31

Marcotte D, Groleau P (1997) A simple and robust lognormal estimator. Math Geol 29 (8):993–1008

Marechal A (1974) Krigeage normal et lognormal: Ecole des Mines de Paris, Centre de Morphologie Matematique, unpublished note, N376. p. 10

Matheron G (1974) Effet proportionnel et lognormalite ou: Le retour du serpent de mer. Centre de Morphologie Mathematique, Ecole des Mines de Paris, Publication N-374, p. 43

Paul R, Cressie N (2011) Lognormal block kriging for contaminated soil. Eur J Soil Sci 62:337–345

Rendu JM (1979) Normal and lognormal estimation. J Math Geol 11:407–422

Rivoirard J (1990) A review of lognormal estimators for in situ reserves. Math Geol 22 (2):213–221

Roth C (1988) Is lognormal kriging suitable for local estimation? Math Geol 30(8):999–1009

Szilágyi-kishné A, Bringmark E, Bringmark L, Alriksson A (2003) Comparison of ordinary and lognormal kriging on skewed data of total cadmium in forest soils of Sweden. Environ Monit Assess 84:243–263

Thurston M, Armstrong M (1987) A simplification of lognormal kriging to suit moderately skew data: in proceedings of APCOM; Johannesburg, South Africa, vol. 3, pp. 53–64

Yamamoto JK (2007) On unbiased backtransform of lognormal kriging estimates. Comput Geosci 1:219–234

Yamamoto JK, Furuie R (2010) A survey into estimation of lognormal data, São Paulo, UNESP, Geociências, v. 29, n. 1, p. 5–19

Definition of Operational Mining Unit (OMU) Size

Cassio Diedrich, Joao Dirk Reuwsaat, Roberto Menin, and Wellington F. De Paula

Abstract The estimation of recoverable mineral resources and reserve curves is generally based on selective mining units (SMUs) represented by a block model. The size or support of the block estimate is characterized by its volume, shape, and orientation by which it is assigned a grade that directly impacts variance and resulting confidence intervals. However, local geology variations associated with operational mining configuration and production rates are commonly disregarded when defining an SMU. This paper aims to present an operational practice for defining mining recovery curves. A methodology to achieve realistic values through an operational mining unit (OMU) that matches the expected production distribution of volumes and grades is proposed. This process computes expected actual production by considering mining sequencing and production rates, applied to a regular SMU using a common geostatistical resource estimation model. Planned dilution is calculated and implemented for a range of block sizes, and one OMU is selected that is a reasonable match to the actual production. Operational mining unit size (or sizes, if variable) will yield tonnes, grades, and metal of ore at given cutoffs (ore and waste) considering operating selectivity. Different orebody domains, mining configurations, and production rates affecting the OMU are also described. Examples are shown to illustrate the methodology using the Walker Lake public dataset and a real (Cu-Au) operational mining case.

C. Diedrich (✉)
Vale Base Metals, 337 Power Street, Door #105, Copper Cliff, Sudbury, ON P0M 1N0, Canada
e-mail: Cassio.Diedrich@vale.com

J.D. Reuwsaat • R. Menin • W.F. De Paula
Vale Base Metals, 63 Grajau Street, 1st floor, Carajás, PA 68516-000, Brazil
e-mail: joao.dirk@vale.com; roberto.menin@vale.com; wellington.paula@vale.com

J.J. Gómez-Hernández et al. (eds.), *Geostatistics Valencia 2016*, Quantitative Geology and Geostatistics 19, DOI 10.1007/978-3-319-46819-8_12

1 Introduction

Operating processes need to be considered when constructing recoverable mineral resources and reserves curves for mining projects and operations. Selective mining unit (SMU) determination is a complex engineering decision and may have an economic impact on the mining operation and will significantly affect the mine planning process. The SMU is generally defined on a block model basis and is commonly associated as the selective mining unit (SMU), the smallest volume of material on which ore/waste classification is determined. In other words, the dimensions of the SMU are supposed to be small enough to enable the mining equipment to select and mine an individual SMU at the time of mining. Resource modelers try to better understand and manage well-known problems such as change of support, the information effect, and conditional bias involved in the block model estimates. However, geostatistical estimates of mining recovery curves may be somewhat optimistic as they do not take into account factors other than block grades, commonly disregarding mining configuration (method and geometry), local geology in the production environment, and production rates.

Some specialists have studied major aspects of different block support dimensions, explained by the support effect Deraisme and Roth (2000) and simulations process (Leuangthong et al. 2003; Jara et al. 2006; Isaaks 2004, 2005). The block support is a fundamental parameter that will condition mining dilution and selectivity affecting mining costs, economics, decision-making, and production processes. The tonnes/grades of ore that the mill receives and tonnes/grades of waste scheduled in the mining plans that are associated with mining block support volumes are a result of a classification procedure with many subjective factors and are never freely and perfectly selected in any case.

The purpose of this paper is to define an operational mining unit (OMU), which is the block size(s) and orientation(s) that reasonably predicts the local and/or global tonnes of ore, waste, and head grade of recoverable reserves, considering local mining operating aspects (geology, mining configuration, and production rates). The idea is not to forecast future short-term SMU grade distributions at the time of mining and not questioning estimation methods for defining grades at SMU scales but to determine a reasonable regular and/or irregular block support for a chosen conventional SMU block model size that represents ore/waste proportions derived from a real expected mining sequencing and grade control process. In the OMU approach, there is no commitment that an individual block will be selectively mined, but it will be contributing accordingly to the "real" expected global and/or local mining recovery curves.

2 Definition of Operational Aspects

The ore control model estimates should be classified by ore type and sent to the correct destination (mill, stockpile, or dump). The problem of how to make the mineral reserve grade-tonnage curves accurately and predict the tonnes and grades above the cutoff that will be available at the time of mining still exists. Conventional SMU selection processes try to account for the impact of support and information effect on a short-term scale; however, they are insufficient to handle conditionally unbiased estimations and disregard potential external factors in the mining process. As a result, operations struggle to achieve the forecasted mining production, or, if production is achieved, the mining plan and stockpile adherence is compromised by changes in the mining configuration in order to achieve targeted values. Block SMU size should not be selected to only attain the mill targets; they also need to align with the operational process that will be implemented and to facilitate stockpiling and waste dump targets.

The methodology proposed is to incorporate the estimated ore control model, defined by mining operating practices, in the long-term mineral reserve estimations. Assuming that a verified regular unbiased global grade and tonnage estimation is used, the global proportions of planned dilution incorporated in the mining operating process can be assumed as representative in terms of block estimation. The results of applying this methodology show that commonly defined SMU size does not correspond directly with the smallest mineable volume chosen by the mining professionals. As mining selectivity is understood as the process of separating ore from waste, its global and local concept is strongly related to three functions that affect operational results. These are:

1. Geology function (Fig. 1)
2. Production rates function (Fig. 2)
3. Mining configuration function (Fig. 3)

Each of these functions can result in different local mining recovery curves. The geology function (1), at operating scale, is presented in Fig. 1a, b. For the planned grade control polygons in Fig. 1a1, a2 that have the same SMU block size and operating practices, the resulting operational ore and waste materials (b1 and b2) are different. This clearly demonstrates the impact of different local geology under real operating circumstances. The question that arises is, for identical mining configuration and production rates, should the long-term SMU size be the same where there is different local geology. The answer should be no. This scenario has been verified many times in a real mining operation where different local mining recoveries and dilution of the defined SMU have resulted in deviating from forecasted local and global mining recovery curves.

Changes in the local mining recovery curves were verified for different production rates considering the same geology and mining configuration in the operational process. The production rate function (2) at operating scale is presented in Fig. 2a, b. The planned grade control polygons in Fig. 2a1, a2, a3, a4, a5 have the same

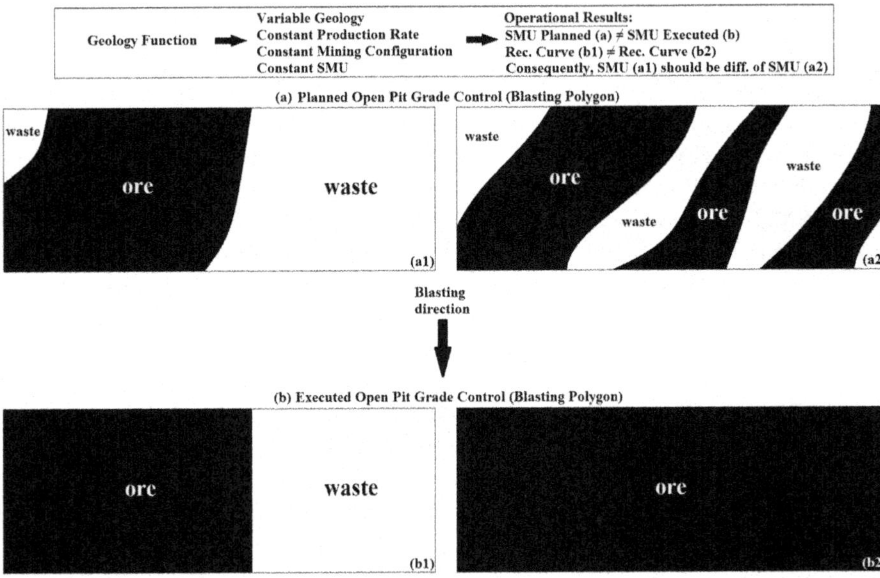

Fig. 1 Geology function operational aspect: (**a**) planned open pit grade control polygons (*a1* and *a2*) and its respective (**b**) executed open pit grade control (*b1* and *b2*)

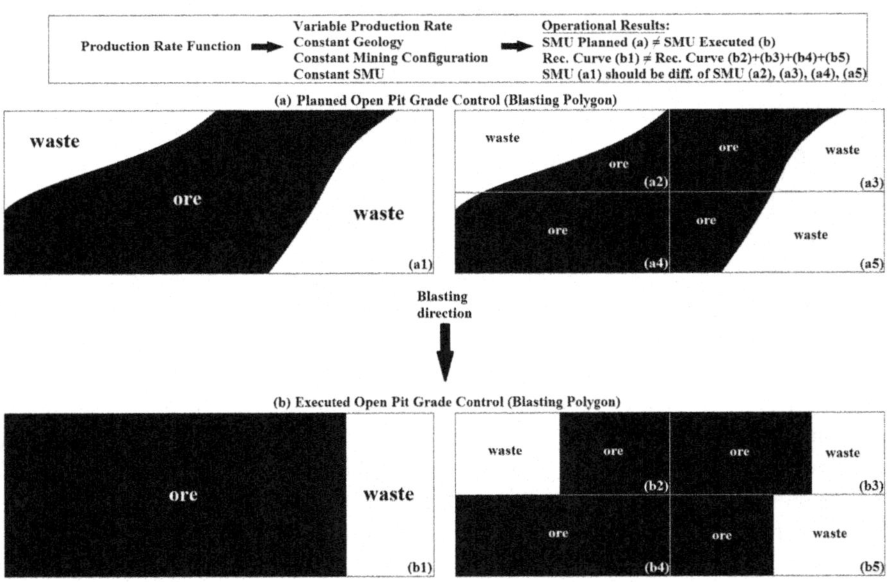

Fig. 2 Production rate function operational aspect: (**a**) planned open pit grade control polygons (*a1* and *a2*, *a3*, *a4*, *a5*) and its respective (**b**) executed open pit grade control (*b1* and *b2*, *b3*, *b4*, *b5*)

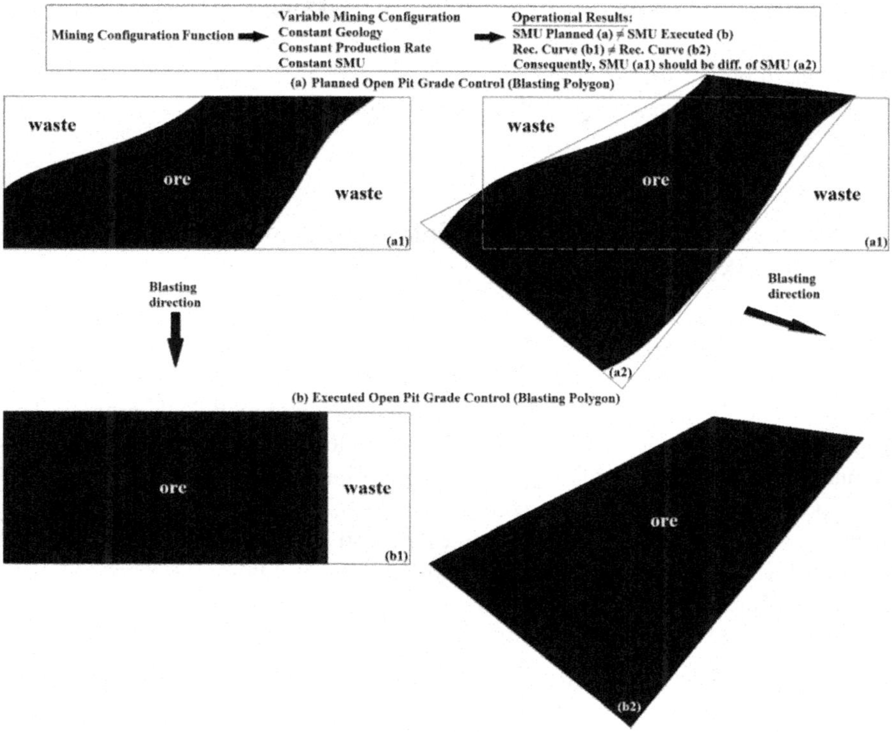

Fig. 3 Mining configuration function operational aspect: (**a**) planned open pit grade control polygons (*a1* and *a2*) and its respective (**b**) executed open pit grade control (*b1* and *b2*)

SMU block size, geology, and mining configuration, but different production rates and different resulting operational ore and waste proportions (b1 and b2, b3, b4, b5). For different production rates and related grade control processes, different mining recovery curves above cutoffs resulted. This demonstrates that SMUs used to predict ore and waste proportions should not be the same.

Different mining configurations, applied to the same local geology and using the production rates in the operational process, may change mining recovery curves as well. The mining configuration function (3), at operating scale, is presented in Fig. 3a, b. For planned grade control polygons in Fig. 3a1, a2 that have the same SMU block size, geology, and production rates, but different mining configuration, the resulting proportions of ore and waste (b1 and b2) are different. For different mining configurations and related grade control processes, different resulting mining recovery curves above cutoffs are obtained, and SMUs used to predict local values should not be the same.

Long-term mine planning takes into account that each SMU will be selectively mined as ore or waste. However, local operating particularities can result in poor reconciliation of long-term block models, and the resulting mining recovery curves will impact milling, stockpiling, and waste dump processes.

In addition to the grade control functions, external dilution factors (equipment and field operating practices) should be considered in the process by applying a factor based on operational mining experience or reconciliation. SMU should be a back-to-front exercise starting with expected operational aspects.

3 Proposed Method for Definition of Operational Mining Unit (OMU) Size

The proposed approach for OMU size(s) definition uses information from the anticipated grade control procedure based on operational mining sequencing and the use of the long-term resource model defined by a common SMU process. The idea is to determine the OMU(s) by local and/or global ore and waste proportions (tonnage and grades) calculations obtained from the resulting operational mining sequencing and proposed grade control processes. The first step is to choose a regular SMU size based on robust grade estimation practices and to define reasonably representative operating grade control process according to local geology, production rates, and mining configuration.

There are two ways of determining OMU(s) block size(s):

- *Global*: where all individual grade control polygons are used to determine global ore/waste proportions in the mining sequencing and the respective global OMU (regular/constant) size (see Fig. 9[2]) and Fig. 9[3] of case study)
- *Local*: where each individual grade control polygon is used to determine local ore/waste proportions in the mining sequencing and the respective local OMUs (regular/constant or irregular) size(s) (see Fig. 9[4] of case study)

The procedure is as follows:

1. Compute reasonable mining sequencing A (e.g., blasting polygons) based on the expected mining operating practices using a common regular SMU size.
2. Simulate the grade control practice to arrive at ore/waste dig limits. The idea is to mimic the real grade control process (e.g., Figs. 1, 2, and 3) that will be implemented in the mine for each operational mining sequencing (A).
3. Then, the ore/waste dig lines are used to calculate the expected local (A) and/or global (sum of local A) tonnes of waste (Wt) and ore (Ot), grades of waste (Wg) and ore (Og), and metal of waste (Wm) and ore (Om). The concept is to determine local and/or global OMU size(s) that matches these resulting values.
4. Choose the possible OMU size (s) by simply calculating the proportions of ore and waste and its respective targeted grade control tonnes (Wt and Ot), grades (Wg and Og), and metal (Wm and Om) results.
5. Plot the results of both grade control and OMU approach in a series of graphs containing (1) planned dilution (PD), (2) tonnes of waste and ore versus OMU size, (3) average grades of waste and ore versus OMU size, and (4) metal tonnes of waste and ore versus OMU size. In each graph, the OMU results are plotted to

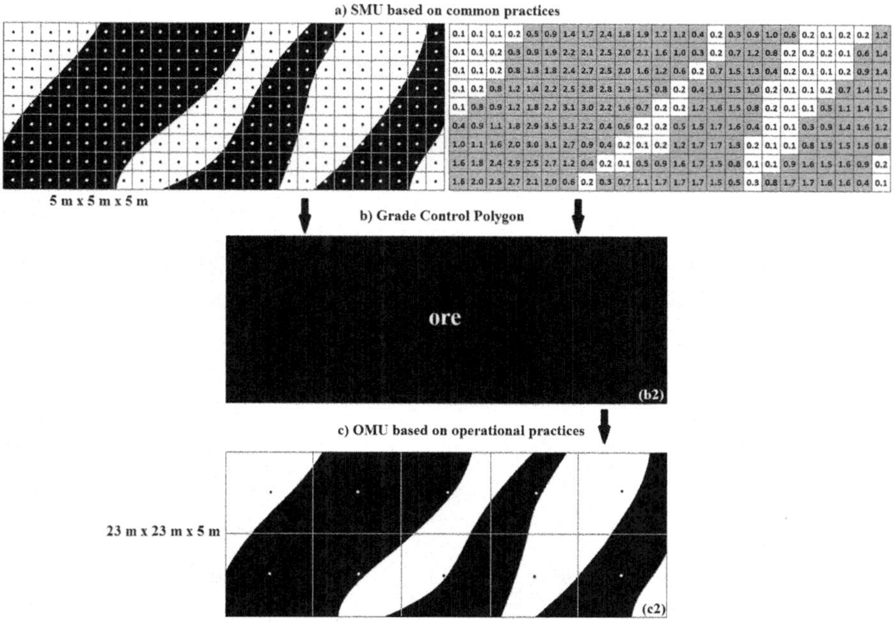

Fig. 4 (**a**) Grade control polygon with 5 × 5 × 5 m SMU block size and its respective grades (**b**) resulting operational grade control polygons based on (**a**) and (**c**) OMU block size that represents the grade control execution (target)

yield a functional relationship, while the grade control values provide a single true value that plots as a horizontal line. The OMU size is the size of which these grade control polygon (Wt, Ot, Wg, Og, Wm, and Om) lines intersect with its respective class of material (i.e., waste and ore). Figures 4, 5, 6, and 7 present a schematic illustration of the OMU process definition and resulting graphs.

6. When the resulting OMU size(s) is selected, which is (are) suitable to represent the local and/or global ore/waste proportions, the mining sequencing should be revisited through an iterative process. This is required to check the overall ore recovery and planned dilution compared to the initial mining sequencing and grade control based on a preliminary SMU size (step 1). Depending on the waste/ore proportions and the deposit characteristics, the recovery of mineralization can be significantly impacted. An adjustment of the mining sequence selectivity (see case study) may be required to maximize ore recovery or to change mining sequencing to better fit OMU block configuration. A detailed critical analysis should be made by experienced professionals.

The concept of the OMU method is to add operational factors for determining block size(s) through the calculation of planned dilution from mining and grade control process. Global and local approaches can be applied in different ways or purposes in the mining sequencing. The global approach provides ore/waste

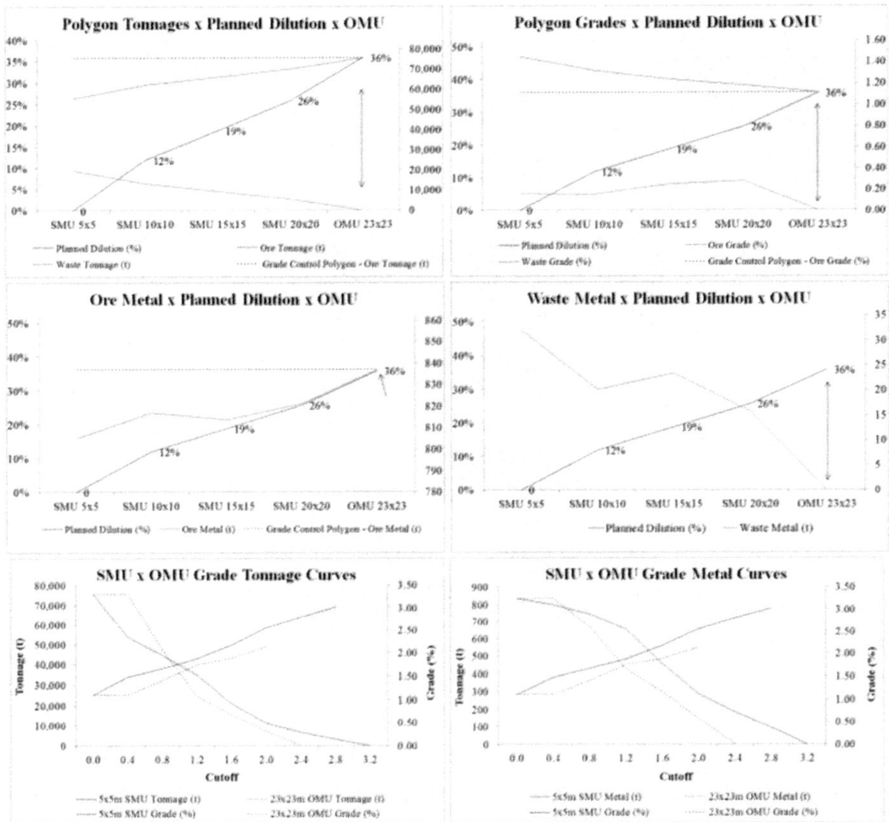

Fig. 5 Graphs of (1) tonnage and planned dilution of grade control polygons (target), SMUs, and OMU (*top left*); (2) grades and planned dilution of grade control polygons (target), SMUs, and OMU (*top right*); (3) ore metal and planned dilution of grade control polygons (target), SMUs, and OMU (*center left*); (4) waste metal and planned dilution of grade control polygons (target), SMUs, and OMU (*center right*); (5) and (6) SMU (5 × 5 m) and OMU (23 × 23 m) tonnages and grades above the cutoff grade (*bottom left* and *right*)

proportions of the overall grade control polygons applied to a regular (constant) OMU size which will reasonably honor global and local recoverable proportions of ore. The local approach provides a better cutoff distribution curve since it clearly details the grade control process by local OMU definitions that take into consideration local operating practices. So why are the local grade control polygons not used to define different OMU sizes? This could be an option to better define and achieve expected long-term recoverable proportions of ore but would result in a lack of local block discretization and mining sequencing flexibility if changes occur in real operational conditions.

A simple OMU approach application is presented in Figs. 4 and 6 for two different local grade control configurations. In Figs. 5 and 7, it is possible to determine how OMU mining recovery curves compare to regular SMU's estimation

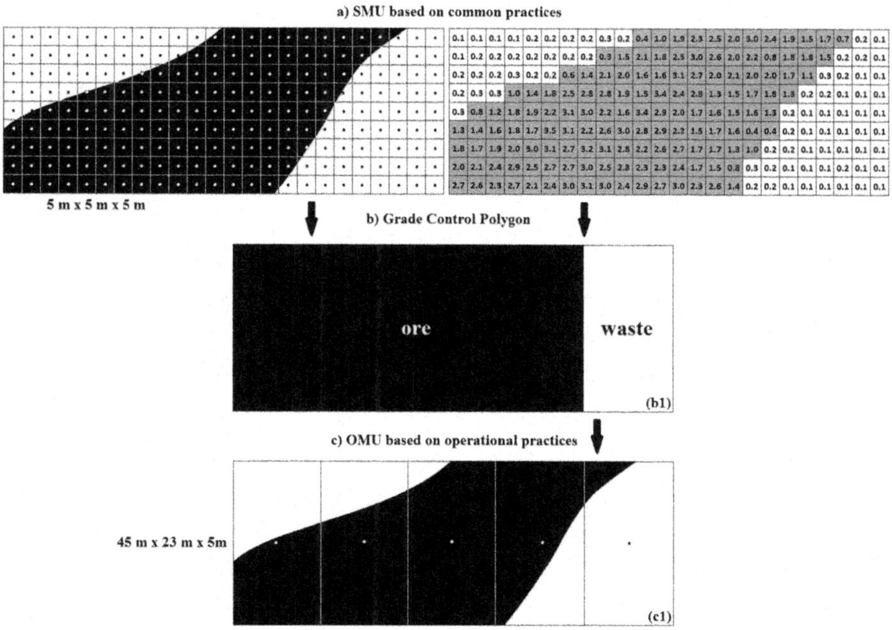

Fig. 6 (**a**) Grade control polygon with 5 ×5 ×5 m SMU block size and its respective grades (**b**) resulting operational grade control polygons based on (**a**) and (**c**) OMU block size that represents the grade control execution (target)

block support curves and how the ore and waste proportions with its respective grades are achieved. In the mining operating process, resulting external waste material incorporated as ore and ore defined as waste strongly affects the recoverable proportions of material above and below cutoffs and may change its final destination (mill, stockpile, waste dump, etc.). For these examples, an SMU size of $5 \times 5 \times 5$ m, with a generic density of 2.8 t/m^3, was chosen. Only tonnes and contained metal within the polygon boundaries were considered. Tables 1 and 2 summarize the grade control polygons (target), SMUs, and OMU resulting values. The results demonstrate how operating processes change the ore/waste proportions, how the resulting OMU block size differs from the previously defined SMU size, and how it is possible to achieve global and/or local operational ore and waste targets.

In both examples presented in Figs. 4 and 6, the operational process and local geological aspects are significant components of the ore/waste proportions in the grade control polygons. This clearly demonstrates that operating aspects are fundamental in the block size definition for achieving reasonable global and local expected mining recovery curves. Although OMU block size definition is a time-consuming process, mining professionals should think about different ways of better defining block volumes since there is no current definitive solution.

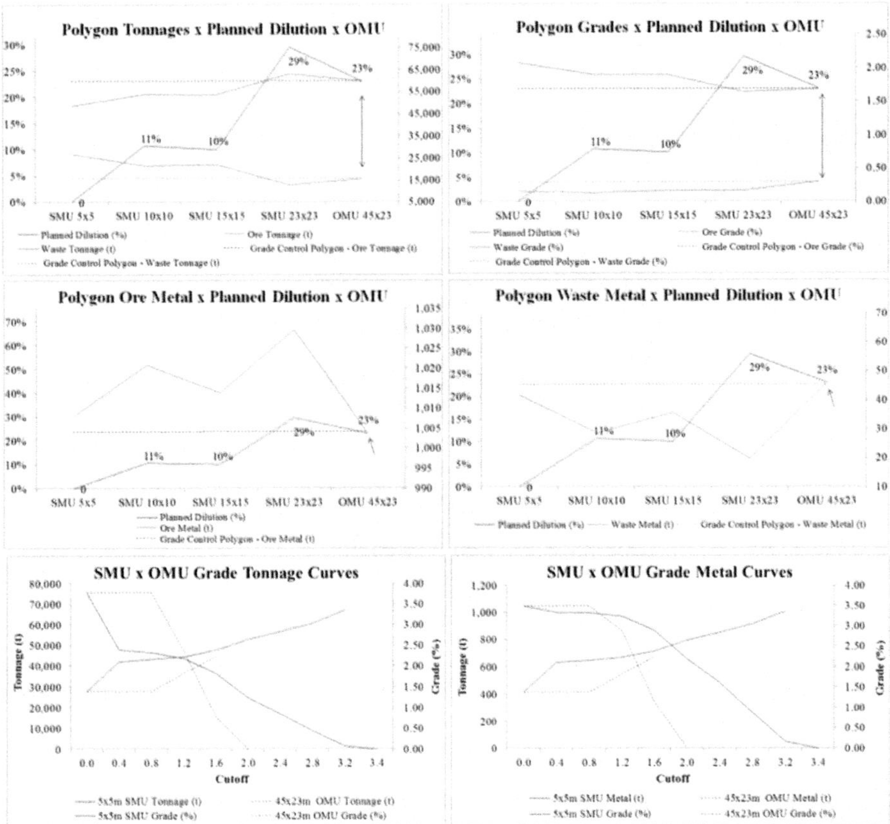

Fig. 7 Graphs of (1) tonnage and planned dilution of grade control polygons (target), SMUs, and OMU (*top left*); (2) grades and planned dilution of grade control polygons (target), SMUs, and OMU (*top right*); (3) ore metal and planned dilution of grade control polygons (target), SMUs, and OMU (*center left*); (4) waste metal and planned dilution of grade control polygons (target), SMUs, and OMU (*center right*); (5) and (6) SMU (5 × 5 m) and OMU (45 × 23 m) tonnages and grades above the cutoff grade (*bottom left* and *right*)

Table 1 Executed (target) grade control (Fig. 4 [b2]), diversified SMUs, and resulting OMU size selection (Fig. 4 [c2])

Item	Ot (t)	Og (%)	Om (t)	Wt (t)	Wg (%)	Wm (t)	PD (%)
Grade control polygon	**75,600**	**1.11**	**837**	**0**	**–**	**0**	**36 %**
SMU 5 × 5 m	55,650	1.45	805	19,950	0.16	32	0
SMU 10 × 10 m	62,300	1.31	817	13,300	0.15	20	12 %
SMU 15 × 15 m	66,150	1.23	814	9,450	0.25	23	19 %
SMU 20 × 20 m	70,000	1.17	821	5,600	0.28	16	26 %
OMU 23 × 23 m	**75,600**	**1.11**	**837**	**0**	**–**	**0**	**36 %**

Table 2 Executed (target) grade control (Fig. 6 [b1]), diversified SMUs, and resulting OMU size selection (Fig. 6 [c1])

Item	Ot (t)	Og (%)	Om (t)	Wt (t)	Wg (%)	Wm (t)	PD (%)
Grade control polygon	**59,850**	**1.68**	**1,004**	**15,750**	**0.29**	**46**	**23 %**
SMU 5 × 5 m	48,650	2.07	1,008	26,950	0.15	42	0 %
SMU 10 × 10 m	53,900	1.89	1,021	21,700	0.13	29	11 %
SMU 15 × 15 m	53,550	1.89	1,014	22,050	0.16	36	10 %
SMU 23 × 23 m	63,000	1.63	1,030	12,600	0.16	20	29 %
OMU 45 × 23 m	**59,850**	**1.68**	**1,004**	**15,750**	**0.29**	**46**	**23 %**

4 Walker Lake Example

A series of mining plans used for operational grade control, based on a block model with an SMU size of 5 × 5 × 1 m and 1.0 t/m^3 density, were compiled from the Walker Lake dataset. A series of grade-tonnage curves were analyzed at the ore definition cutoff (5 %) to determine the ore/waste proportion changes due to regular block support size and mining sequencing. Different production polygon configurations based on a block model with an SMU size of 5 × 5 × 1 m were compiled as follows:

- Production polygons ranging from 6 to 9 SMU blocks in cases 1, 4, and 7
- Production polygons ranging from 9 to 15 SMU blocks in cases 2, 5, and 8
- Production polygons ranging from 15 to 25 SMU blocks in cases 3, 6, and 9

These production polygons were then overlain (with minor adjustments) over block models with SMU sizes of 15 × 15 × 1 and 25 × 25 × 1 to develop three mining sequences and nine different mining scenarios. Figure 8 and Table 3 present the grade-tonnage curves for the nine different mining scenarios. Cases 1–3 are related to a 5 × 5 m block size, cases 4–6 are related to a 15 × 15 m block size, and cases 7–9 are related to a 25 × 25 m block size.

The mining recovery curves (Fig. 8 and Table 3) demonstrate how the ore/waste (cutoff grade of 5 %) proportions of the grade control polygons compare to the relative block sizes of the mining sequences. In mining scenarios with high production volumes, the 5 × 5 m block mining recovery curves tend to deteriorate and do not represent the waste that is incorporated to the system. When block support is increased (reblocked to 15 × 15 m or 25 × 25 m), the resulting grade control mining recovery curves tend to approximate the blocks and polygons (tonnes, grades, and metal), and ore recovery is strongly impacted by external material that is incorporated within the reblocked volumes. This may lead to significant changes to mining sequencing and proportions of ore and waste that were originally based on a 5 × 5 m block size plan. The original plan may not be realistic since the ore material could be recovered even considering the waste blocks in the grade control polygons (cases 1–3).

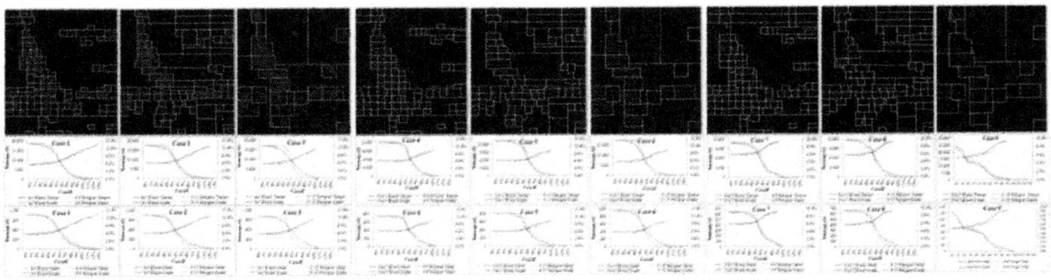

Fig. 8 Mining sequencing on 5 × 5 m (*left*), 15 × 15 m (*center*), and 25 × 25 m (*right*) block sizes, respectively

Table 3 Summary of mining sequences for different production rates (6–25 SMUs of 5 × 5 m block size) and block sizes (5 × 5 m; 15 ×15 m; 25 × 25 m) as described in the text above and its respective block and polygon values at 5 % cutoff grade

Item	Block tonnage (t)	Block grade (%)	Block metal (t)	Poly. tonnage (t)	Poly. grade (%)	Poly. metal (t)	Planned dilution (%)	% Diff. metal
Case1	12,184	6.8	827	13,389	6.4	852	9.9	3.1
Case2	12,410	6.8	841	13,991	6.3	878	12.7	4.4
Case3	10,604	7.0	740	14,003	6.0	837	32.1	13.1
Case4	10,646	6.5	692	10,682	6.4	680	0.3	−1.8
Case5	10,646	6.5	693	11,437	6.2	705	7.4	1.9
Case6	10,681	6.5	694	11,823	6.0	708	10.7	1.9
Case7	8,709	6.4	560	8,015	6.4	509	−8.0	−9.2
Case8	8,686	6.4	559	8,011	6.1	491	−7.8	−12.1
Case9	8,687	6.4	559	8,254	6.0	493	−5.0	−11.8

Case 1 (SMU 5 × 5 × 1 m on 6–9 SMU blocks production polygons) was chosen for the definition of OMU size(s) (Fig. 9—top left). Three different OMU applications were applied as follows:

1. *OMU REG* (Fig. 9 top right): operational mining unit with regular 15 × 15 × 1 m block size and regular block (centroid) position
2. *OMU RSIP* (Fig. 9 bottom left): operational mining unit with regular 15 × 15 × 1 m block size and irregular block (centroid) position
3. *OMU ISIP* (Fig. 9 bottom right): operational mining unit with irregular block size and irregular block (centroid) position

Figure 9 illustrates case 1 production polygons (6–9 SMU blocks) and SMU blocks (5 × 5 × 1 m) for the OMU definition case study. Table 4 presents recoverable values at given cutoffs for grade control polygons, SMU, and OMUs.

Figure 10 illustrates the cutoff mining recovery curves for the resulting OMU definition method. The comparison of SMU and resulting grade control polygons shows how tonnages and metal vary at 5 % and above. In fact, the operating process is less selective as forecasted by the block size.

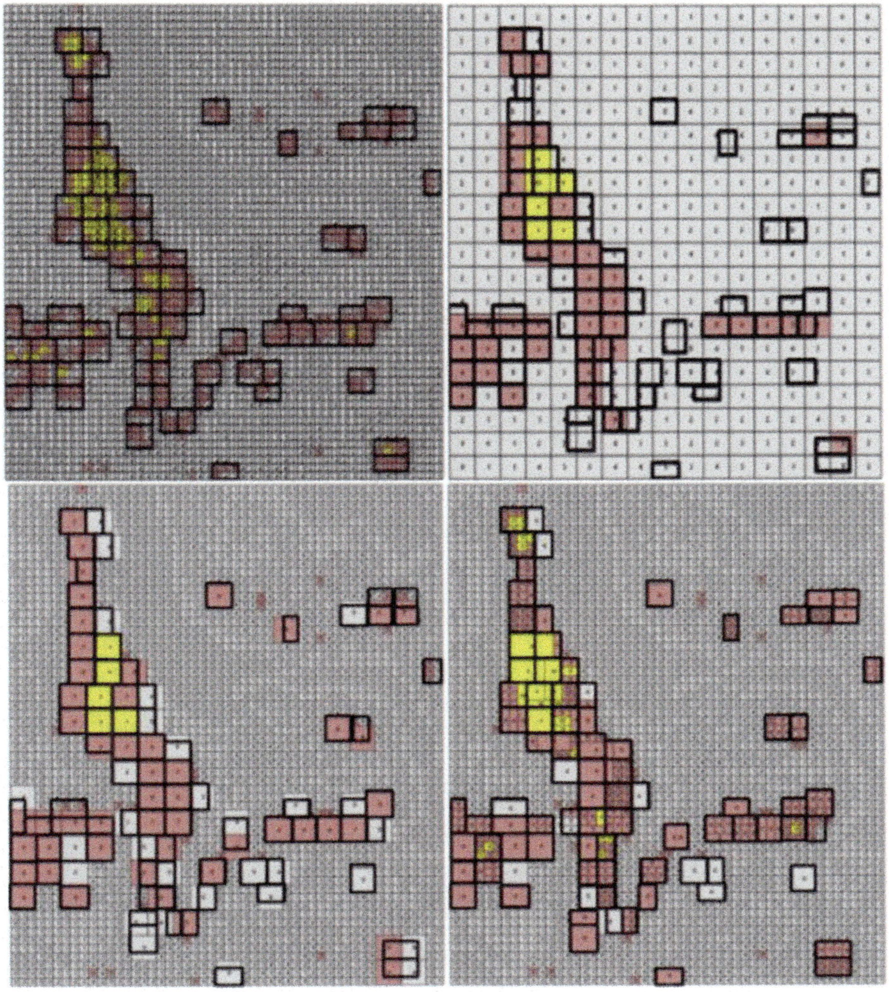

Fig. 9 (1) Case 1 production polygons (6–9 SMU blocks, *bolded lines* in all figures) and $5 \times 5 \times 1$ m (*top left*); (2) OMU REG, regular $15 \times 15 \times 1$ m block size and regular block (*centroid*) position (*top right*); (3) OMU RSIP, regular $15 \times 15 \times 1$ m block size and irregular block (*centroid*) position (*bottom left*); (4) OMU ISIP, irregular block size and irregular block (*centroid*) position (*bottom right*)

The OMU REG (regular) approach (global) shows that a 15×15 m block size fits well to the grade control curves (blue line) based on a 15×15 m regular block size. The mining recovery curves are strongly impacted by external dilution material if compared to the SMU grade control polygons, which are considered reasonable for operations to mine according the Fig. 9(1). In the RSIP approach (global), with a regular 15×15 m block size and irregular centroid position, the recovery issue is greatly improved, and its resulting block size curves provide a near-perfect match with the initial SMU grade control polygon curves at 5 % cutoff and above.

Table 4 Recoverable values at given cutoffs for grade control polygons (poly.) on case 1 (grade control target), case 1 SMU (5 × 5 × 1 m), OMU REG, OMU RSIP, and OMU ISIP

Cutoff (%)	**Poly. tonnage (t)**	**Poly. grade (%)**	**Poly. metal (t)**	**Case1 tonnage (t)**	**Case1 grade (%)**	**Case1 metal (t)**	**OMU REG tonnage (t)**	**OMU REG grade (%)**	**OMU REG metal (t)**
3	16,375	6.1	999	16,050	6.2	998	11,732	6.3	739
5	**13,850**	**6.4**	**883**	**12,500**	**6.8**	**847**	**10,646**	**6.5**	**692**
7	3,675	8.1	298	4,375	8.4	368	2,865	8.3	238
9	675	10.0	67	1,150	10.3	119	898	9.5	85
Cutoff (%)	**Poly. tonnage (t)**	**Poly. grade (%)**	**Poly. metal (t)**	**OMU RSIP tonnage (t)**	**OMU RSIP grade (%)**	**OMU RSIP metal (t)**	**OMU ISIP tonnage (t)**	**OMU ISIP grade (%)**	**OMU ISIP metal (t)**
3	16,375	6.1	999	19,125	5.7	1,087	16,100	6.1	980
5	**13,850**	**6.4**	**883**	**12,375**	**6.4**	**793**	**13,650**	**6.3**	**857**
7	3,675	8.1	298	3,375	8.2	275	3,400	8.3	282
9	675	10.0	67	675	10.0	67	675	10.3	69

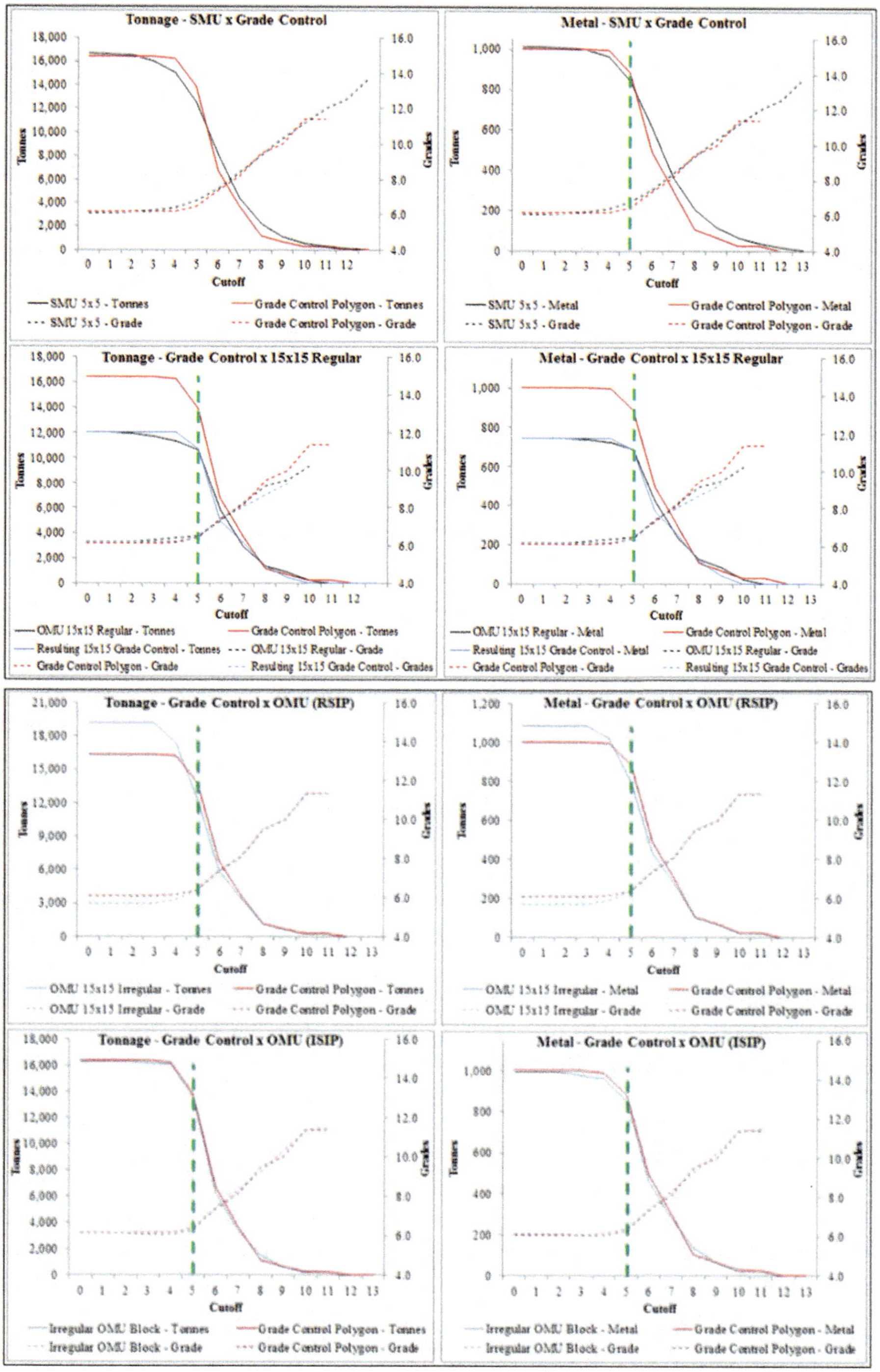

Fig. 10 Mining recovery curves of SMU, OMU REG, OMU RSIP, and OMU ISIP, respectively, from *top* to *bottom*

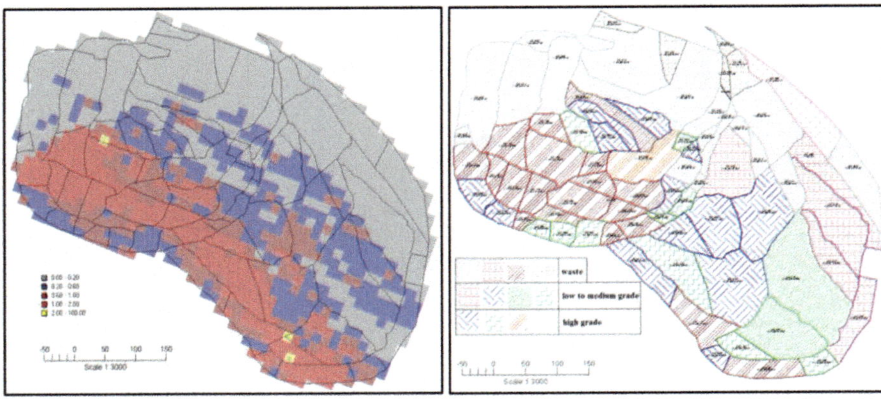

Fig. 11 SMU block size and sequencing (*left*), actual grade control polygons (*right*)

In the ISIP approach (local), considering irregular block sizes and centroid, arranged through each local sequencing polygon, a near-perfect match is achieved for the entire initial SMU grade control distribution.

5 Copper Mine Case Study (Sulfide Cu-Au) and Conclusions

A case study for a Brazilian sulfide Cu-Au mine is used to demonstrate applicability of the method by comparing to actual operational results. The grade of waste is assigned as "zero." The cutoff grade of the deposit is 0.25 % Cu. Figure 11 illustrates the 15 × 15 × 15 m SMU block estimates (left) and the actual grade control polygons (right). Table 5 lists resulting mining recovery values above cutoffs (Fig. 12).

The results demonstrate how mining recovery curves are improved using operational aspects for defining the mining unit size. For the real case study, the operational OMUs near matched the operational grade control practice. The actual values below and above cutoffs were considerably different than those from the SMU. Also, it demonstrated that a 30 × 30 m and 45 × 45 m block sizes should be applied to the material with grades below and above 1.3 % Cu, respectively.

Operational aspects should be included in the mining unit block size definition by mining professionals. Selection of the appropriate mining unit is a compromise between getting the right mining recovery estimates for tonnages of ore and waste and getting the right grade of these materials. This compromise depends on the global and/or local ore and waste proportions for a predefined operational production rates and mining configurations. The greatest impact of the support effect is where the operational aspects are not included or checked against regular SMU sizes defined by common geostatistical estimation practices. Significant differences

Table 5 Mining recovery values at given cutoffs for SMU 15 × 15 × 15 m, grade control polygons (target), OMU 30 × 30 × 15 m, and OMU 45 × 45 × 15 m, respectively

Cutoff (%)	SMU 15 × 15 Mt	SMU 15 × 15 %Cu	SMU 15 × 15,000 t Cu	Polygon Control Mt	Polygon Control %Cu	Polygon Control 000 t Cu
0	104	0.68	702	139	0.51	702
0.25	89	0.77	682	93	0.70	653
0.4	67	0.91	607	66	0.86	562
0.8	34	1.25	429	32	1.16	368
1.2	18	1.50	263	14	1.37	195
1.6	5	1.81	88	1	1.79	12
Cutoff (%)	OMU 30 × 30 Mt	OMU 30 × 30 %Cu	OMU 30 × 30,000 t Cu	OMU 45 × 45 Mt	OMU 45 ×45 %Cu	OMU 45 × 45,000 t Cu
0	120	0.58	697	133	0.52	694
0.25	90	0.73	659	92	0.70	642
0.4	66	0.87	577	65	0.85	552
0.8	32	1.18	380	30	1.14	343
1.2	13	1.45	185	10	1.42	145
1.6	3	1.77	45	1	1.74	24

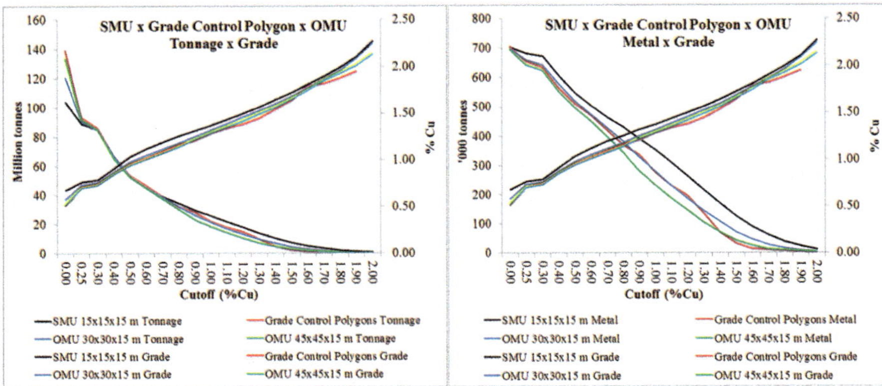

Fig. 12 Mining recovery curves of SMU 15 × 15 × 15 m, grade control polygons (target), OMU 30 × 30 × 15 m, and OMU 45 × 45 × 15 m

between grade control and operating practices to define SMU may result in poor mining, waste dumps, and stockpile plans and strongly impact financials.

Bibliography

Deraisme J, Roth C (2000). The information effect and estimating recoverable reserves. Geovariances

Isaaks E (2004, 2005) The kriging oxymoron: a conditionally unbiased and accurate predictor. Proceedings of geostatistics banff, 2nd ed. Springer, Banff, pp 363–374

Jara RM, Couble A, Emery X, Varela EM, Ortiz JM (2006) Block size selection and its impact on open-pit design and mine planning. J South African Inst Min Metall 106:205–211

Leuangthong O, Neufeld C, Deutsch CV (2003) Optimal selection of selective mining unit (SMU) size. University of Alberta, Alberta

Optimizing Infill Drilling Decisions Using Multi-armed Bandits: Application in a Long-Term, Multi-element Stockpile

Rein Dirkx and Roussos Dimitrakopoulos

Abstract Every mining operation faces a decision regarding additional drilling at some point during its lifetime. The two questions that always arise upon making this decision are whether more drilling is required and, if so, where the additional drill holes should be located. The method presented in this paper addresses both of these questions through an optimization in a multi-armed bandit (MAB) framework. The MAB optimizes for the best infill drilling pattern while taking geological uncertainty into account by using multiple conditional simulations for the deposit under consideration. MAB formulations are commonly used in many applications where decisions have to be made between different alternatives with stochastic outcomes, such as Internet advertising, clinical trials and others. The application of the proposed method to a long-term, multi-element stockpile, which is a part of a gold mining complex in Nevada, USA, demonstrates its practical aspects.

1 Introduction

The optimization of an infill drilling pattern can be linked to a more general problem present in many different applications (e.g. stock markets, research effort direction, project development, etc.). The more common term for this problem used across these various industries is the valuation of future information (under uncertainty). In terms of the economic utility theory, Schlee (1991) proves that perfect information always has a positive value. However, the requirement of a payment to retrieve this information might render the total value negative, causing a need to properly valuate this future information. In the mineral industry, Chorn and Carr (1997) assess the value of future information for decisions on additional capital

R. Dirkx (✉) • R. Dimitrakopoulos
COSMO – Stochastic Mine Planning Laboratory, Department of Mining and Materials Engineering, McGill University, FDA Building, 3450 University Street, Montreal, Quebec H3A 0E8, Canada
e-mail: rein.dirkx@mail.mcgill.ca; roussos.dimitrakopoulos@mcgill.ca

J.J. Gómez-Hernández et al. (eds.), *Geostatistics Valencia 2016*, Quantitative Geology and Geostatistics 19, DOI 10.1007/978-3-319-46819-8_13

investments and Prange et al. (2008) valuate information gathering campaigns on the sealing capacity of a fault system.

In the aforementioned work, the value of future information is a monetary value. This is very hard to quantify for a mineral deposit. Boucher et al. (2005) indirectly do this via a misclassification cost of material. This misclassification refers to ore versus waste based on a fixed cutoff grade and one processing stream. However, this is usually not the case in a mining complex. Menabde et al. (2007) show that it is optimal to use a variable cutoff grade that comes directly from the schedule optimization. This makes it impossible to evaluate misclassification errors without rescheduling the deposit. This effect is strengthened if multiple processing streams are considered in combination with the blending of material from different sources, as in the application presented below.

This paper considers additional information for a mineral deposit to be valuable if the extraction schedule is influenced. Differences in extraction schedules are often caused by a change in material types of various blocks. Therefore, value is linked to material-type changes, the main driver of schedule changes. The definition of the 'best' pattern is the one that adds the most value to the knowledge base. According to this paper's definition of value, the best infill drilling pattern is the one that causes the most material-type changes, implicitly linked to schedule changes.

Diehl and David (1982), Gershon et al. (1988) and Barnes (1989) explore infill drilling optimization by focussing on minimizing the total kriging variance or trying to locate zones of high kriging variance as prime spots for additional drilling. A flaw in using kriging variance as a measure of variability is that it only captures the geometric part of the uncertainty for a drilled deposit and does not take the variability of grades into account (Goovaerts 1997). Ravenscroft (1992) shows that working with geological simulations is the best way to represent variability in a deposit and that estimated orebody models give a flawed representation of the true variability in a deposit. Therefore, simulations are used in this approach to infill drilling optimization and the assessment of additional patterns under geological grade uncertainty. Goria et al. (2001) also propose a method based on conditional simulations to assess the value of additional drill holes. An important conclusion from this work is that additional drilling does not necessarily reduce the variability in the deposit. Boucher et al. (2005) propose a method for the optimization of infill drilling that also makes use of simulated orebody models and compares the cost of drilling additional drill holes to the misclassification cost of the material. The aforementioned works give the additional drill holes a simulated grade, drawn from simulations based on the initial exploration data. This is used as a possibly 'true' grade in the added-value calculations. A similar approach is taken herein.

The major contribution of the proposed method is the use of multi-armed bandits (MAB) for the optimization. Their benefit is that they provide an elegant solution to the requirement of testing every single simulation as possibly 'true' representation multiple times in the previous work. The MAB framework finds its origin in the world of casinos. The initial problem comes from playing a row of slot machines (commonly referred to as an *armed bandit*) in a certain sequence in order to maximize the total long-term reward from playing these machines. This provides an analogy for the well-known exploitation versus exploration trade-off present in many other applications (e.g. clinical trials, Internet advertisements, etc.). The exploration versus exploitation trade-off in infill drilling exists in the availability of many possible locations for additional drill holes with each unknown outcome.

The case study below requires simulations of a multi-element stockpile. This requires a simulation technique capable of dealing with multiple correlated variables. The simulations throughout the remainder of this work are all generated using minimum/maximum autocorrelation factors (MAF) (Switzer and Green 1984). Desbarats and Dimitrakopoulos (2000) describe the regular MAF technique in a mineral science context by applying it to the simulation of pore-size distribution. Boucher and Dimitrakopoulos (2009) apply it directly to the block support level; it is this direct block method that is used below.

The next section briefly addresses the details of the proposed method. Once the method is explained, the need for additional information for stockpiles is illustrated by the variability present in the stockpile simulations. Then an application of the infill drilling optimization for a long-term, multi-element stockpile is shown. Finally, a conclusion and recommendation for future work are given.

2 Method

The proposed method selects the best infill drilling pattern from a predefined set of patterns. The representation of each pattern for infill drilling will be called an *arm* in the MAB framework. These arms will be played. When an arm is played, the algorithm to assess the value for the pattern associated to that arm is set in motion. The steps in this value assessment are the following (Fig. 1 demonstrates the general flow of the algorithm):

- The grades in the drill holes of the selected pattern are drawn from an initial simulation, which is set to be the possibly 'true' representation of the deposit.
- The drill holes of the pattern linked to the pulled arm are used as additional data to the initial exploration data for the re-simulation of the deposit.
- After the re-simulation with additional data, the blocks of the re-simulation are classified according to a material classification guide.
- The reward or contribution from pulling that arm is defined as the percentage of blocks that have changed material classification in the re-simulation compared to the possibly 'true' deposit. Optimizing for this reward is in line with the definition of the best pattern given above.
- The Thompson sampling solution algorithm updates the current arm, and the next arm is selected.

After convergence of the algorithm, the whole procedure is repeated for other simulations of the deposit as a possibly 'true' deposit in order to test the sensitivity of the method to this choice. The best pattern is then selected over the results of all of these possibly 'true' deposits. By considering multiple simulations as possibly

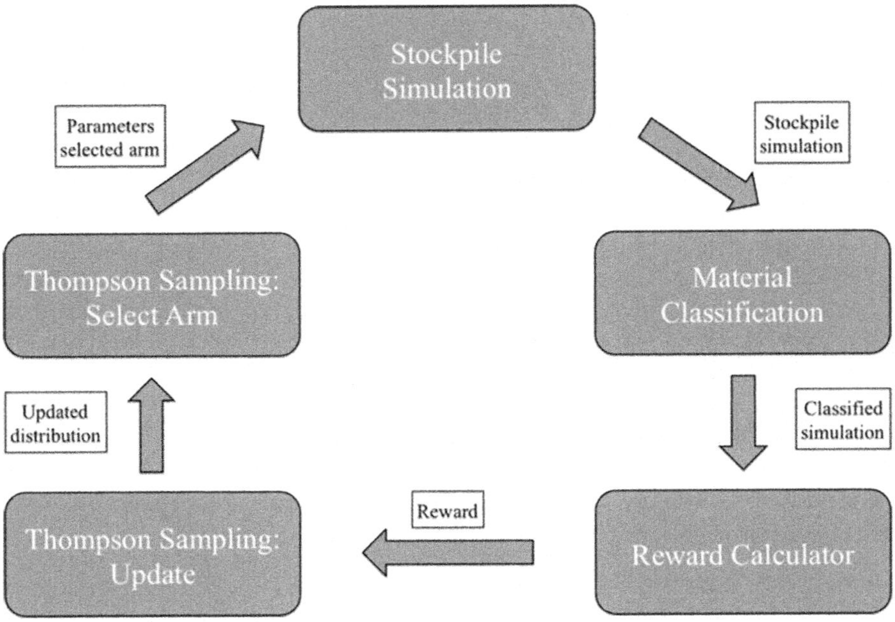

Fig. 1 High-level schematic representation of the MAB optimization algorithm

'true' deposits, the method is able to give an assessment of the performance of all patterns under geological uncertainty, which leads to the quantification of the upside potential and downside risk for each pattern.

All of the patterns in one MAB set-up are comparable in the sense that they belong to what is called the same budget class, which means that they all have the same amount of drill holes and thus represent a similar cost to be drilled. The procedure is also repeated for different budget classes of patterns where each budget class represents patterns with a different number of holes than in the other classes. A separate MAB is required to optimize within each budget class, as the rewards of patterns with more holes cannot be directly compared to the rewards of patterns with fewer holes.

3 Multi-armed Bandits and Thompson Sampling

The classic and simplest MAB problem follows the four basic rules formulated below, as in Mahajan and Teneketzis (2008). The state of an arm in a simple MAB is the internal time, which is equal to the number of times the arm has been played.

1. Only one arm is played at each time step. The reward from this play is uncontrolled.
2. Not played arms remain *frozen*, meaning they do not change state.

3. A *frozen* arm will not give a reward.
4. All arms are independent.

There are many variants of MAB problems where one or many of these rules are violated or altered, which makes them considerably more complicated to understand and solve. Therefore, the focus of the following sections will only be on these simple bandits, as these suffice to address infill drilling.

The mathematical objective function of the MAB problem, with arms, is given below (Mahajan and Teneketzis 2008)

$$J^{\gamma} = \mathbb{E}\left[\sum\nolimits_{t=0}^{\infty} \beta^t \sum\nolimits_{i=1}^{k} R_i(X_i(N_i(t)), U_i(t)) \middle| Z(0)\right] \tag{1}$$

with:

- J^{γ}, the objective function value under policy γ, defined by $U_i(t)$.
- β^t, the discount factor with time t, reflecting that earlier rewards are higher valued than later rewards, like in an NPV calculation. The time t is a measure for the number of iterations/plays.
- $R_i(X_i(N_i(t)), U_i(t))$, the reward of arm i depending on:
  - $X_i(N_i(t))$, the state of arm i at time t, which depends on:

    $N_i(t)$, the local time of arm i at time t; the number of times arm i has been played before time t.
  - $U_i(t)$, the policy decision, which determines whether or not arm i is played at time t.
- $Z(0)$, the initial state of all arms i.

The total objective function value is then found by taking the expected value over the summations over time and the number of arms conditional to the initial states of all arms. The goal is to find the optimal scheduling policy γ that determines the values of $U_i(t)$ for every arm i and time t in such a way that the total expected reward is maximized. The MAB problem can also be used to find the best performing arm, instead of a schedule of when each arm should be played. The best arm is just the one that is scheduled for continuously.

The algorithm chosen to solve the MAB is Thompson sampling (Thompson 1933), which has demonstrated excellent performance for MAB optimization (Agrawal and Goyal 2012; Scott 2010). The general idea of Thompson sampling is always play the arm with the highest likelihood of being the best arm. This gives a natural approach to the exploration versus exploitation trade-off in MAB problems. In Thompson sampling each arm is represented by a distribution linked to its reward. Usually, the prior distributions are uniform beta distributions, B (α=1,β=1), which are updated each time an arm is played, and the reward is observed. The reward percentage is used as the success probability of a Bernoulli trial. When the outcome of the Bernoulli trial is a success, the β parameter is incremented with one; otherwise, the α parameter is incremented with one.

Thompson sampling is proven to converge for the MAB problem by Agrawal and Goyal (2012) and May et al. (2012), but it has no predefined termination moment. Therefore, it is required to define a convergence criterion that works for the application in consideration. In this case a simple criterion is defined, which has shown satisfactory results. That is, if in the last 20 % of iterations one arm is selected to be played more than 90 % of the iterations, then this is the best arm.

4 Case Study: A Long-Term, Multi-element Stockpile

The case study presented below illustrates the optimization of the infill drilling decision for a long-term, multi-element stockpile of a gold mining complex in Nevada, USA. This complex consists of two open-pit mines, an underground mine and several stockpiles. The downstream processing includes an autoclave, an oxide mill and multiple heap leaches. Deleterious compounds are co-simulated with the gold grade because of strict grade requirements on them to guarantee efficient processing recoveries. Knowledge of the grade of these compounds can assure blending of ore from different sources to meet the constraints at the processing facilities. The compounds being considered are sulphide sulphur, organic carbon and carbonate, next to gold as the paying metal.

For this case study, three classes of patterns are considered, each corresponding to a different budget. The first budget class has patterns with 5 holes, the next class has patterns with 10 holes and the last class has patterns with 15 holes. Eight patterns, based on the ideas provided to us by the mine site, are considered for the optimization in each class. For every class the procedure is repeated for 20 simulations as possibly 'true' stockpiles to guarantee that the results are independent from the selected possibly 'true' stockpile. This also assesses the performance of the patterns under geological uncertainty. A validation step tests all patterns with simulations not used in the optimization procedure and shows that the results are independent of the simulations employed.

4.1 Stockpile Simulations

The exploration drilling is done on a regular grid with a 40 ft spacing. In total there are 104 drill holes with one sample each. The height of the stockpile is 20 ft. The size is ± 200,000 m^3, which corresponds to ± 400 kton. Twenty simulations are generated, each containing 204 blocks within the shape of the stockpile. The block size is 30 × 30 × 20 ft. This is the same as in the mines feeding the stockpile because the same mining selectivity is used for both the stockpiles and the mines.

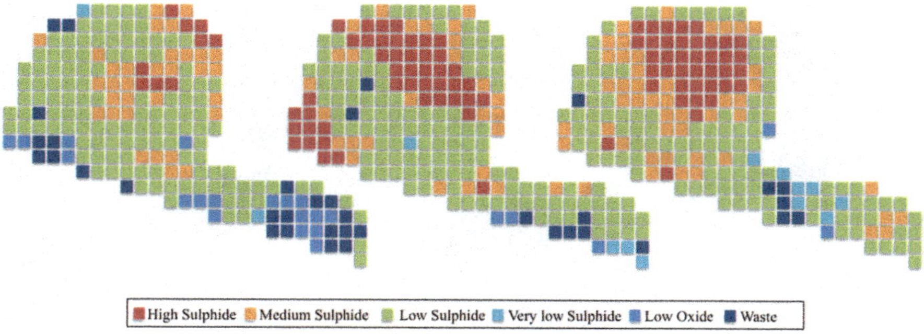

Fig. 2 Spatial material variability of the stockpile simulations at the block level

All simulations below are generated with the direct block MAF simulation method by Boucher and Dimitrakopoulos (2009).

Figure 2 shows the material variability of three stockpile simulations at the block level. The material is classified based on the material classification guide provided by the mining complex. The mining complex flags the stockpile under consideration as medium sulphide material, but Fig. 2 shows that high- and low-sulphide material usually dominates the material distribution in the stockpile. This trend is carried on in the remaining simulations.

Figure 2 also shows that there is a large amount of spatial material variability in the stockpiles, within one simulation but also from one to another simulation. This observation is the main motivation for the application of the proposed method to stockpiles. It shows that more information is required to assess the local-scale variability in stockpiles. This is especially true for the stockpile in this mining complex because, in some periods, it is the main contributor to the processed blend at the autoclave. This stream has tight constraints on the deleterious compounds like a maximum organic carbon content and an upper and lower bound on the sulphide sulphur–carbonate ratio. Figures 3, 4, 5 and 6 show the spatial material variability in two simulations for the four variables of interest. All simulations show that the same main features are respected between two simulations but that there is also local-scale variability within each simulation.

4.2 *Tested Patterns*

Figures 7, 8, 9 and 10 below show the eight patterns that are considered in the optimization. The figures always show all 15 additional drill holes for each pattern on top of the initial exploration drill holes which are represented as black crosses. However, if only the red squares ('first five holes') are considered, the patterns represent the budget class with five holes per pattern. If the red squares and the green triangles ('second five holes') are considered, the patterns represent the budget class with ten holes per pattern. The patterns have the same numbering

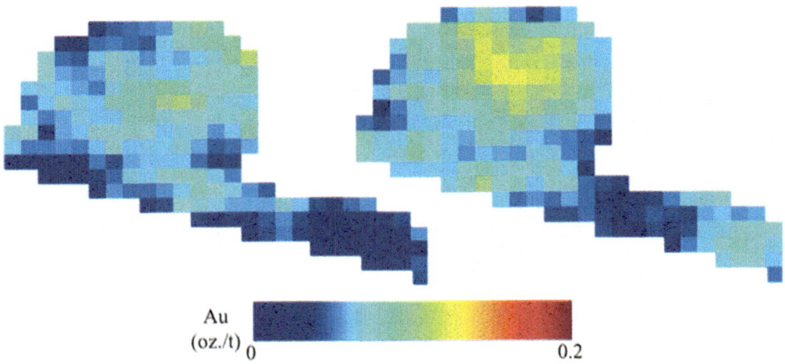

Fig. 3 Gold grade variability in two simulations at the block level

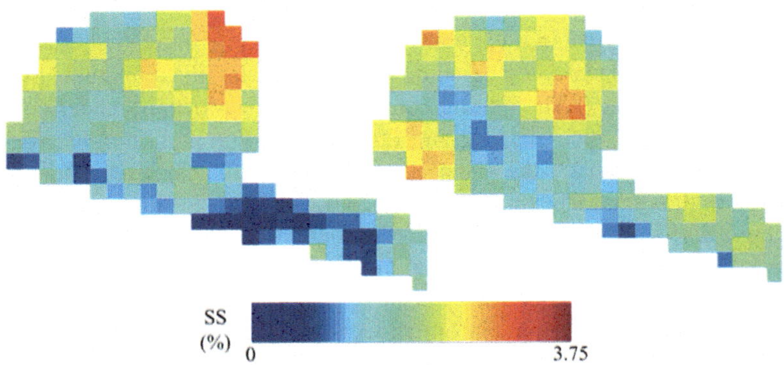

Fig. 4 Sulphide sulphur content variability in two simulations at the block level

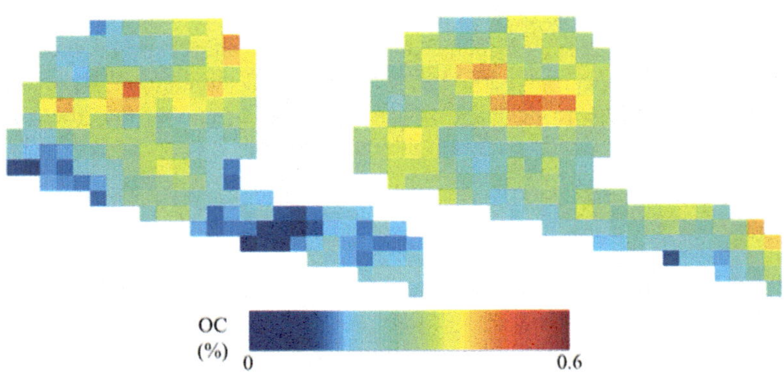

Fig. 5 Organic carbon content variability in two simulations at the block level

over the budget classes because they are always based on the same idea for the location of the holes. For example, pattern 1 is designed as being equally spread and pattern 2 to target the high-grade gold zone in the top of the stockpile.

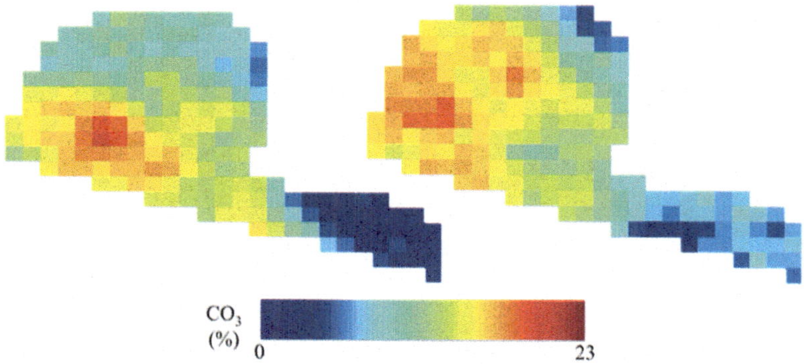

Fig. 6 Carbonate content variability in two simulations at the block level

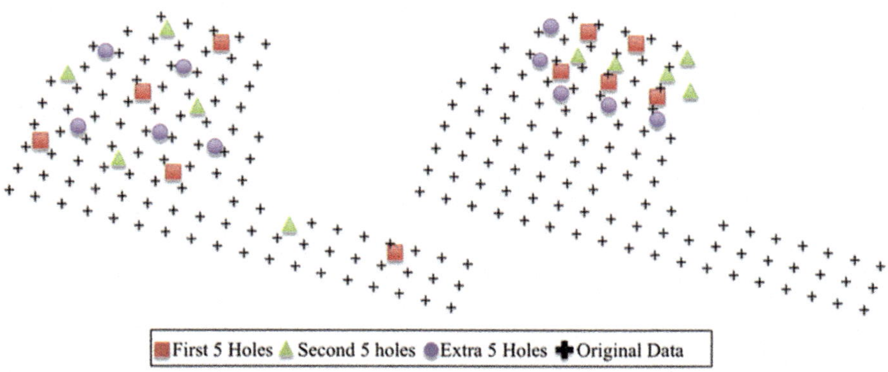

Fig. 7 Pattern 1 (*left*) and pattern 2 (*right*) with 15 additional drill holes

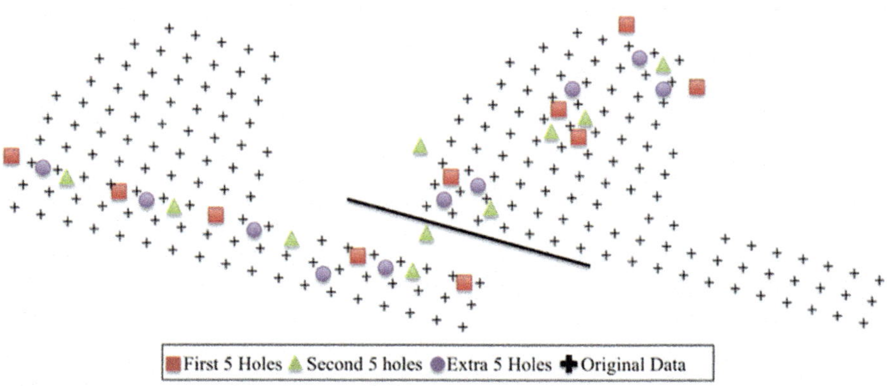

Fig. 8 Pattern 3 (*left*) and pattern 4 (*right*) with 15 additional drill holes. The *black line* indicates that the *green triangle* below it belongs to pattern 3

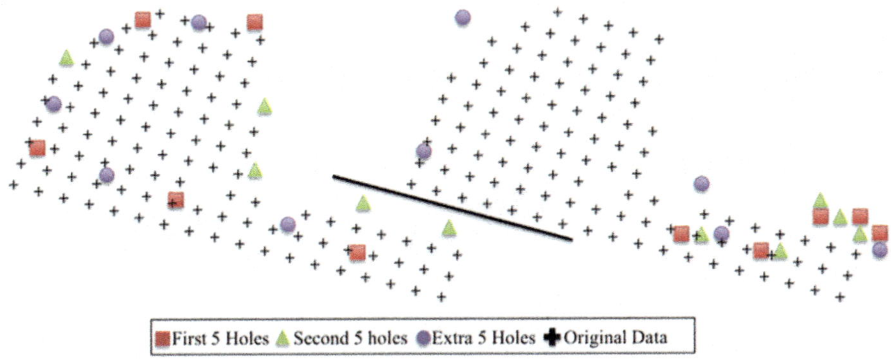

Fig. 9 Pattern 5 (*left*) and pattern 6 (*right*) with 15 additional drill holes. The *black line* indicates that the *green triangles* below it belong to pattern 5

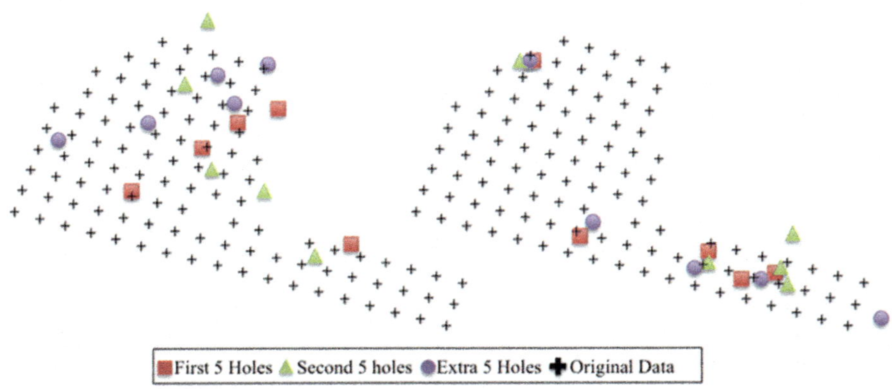

Fig. 10 Pattern 7 (*left*) and pattern 8 (*right*) with 15 additional drill holes

4.3 Case Study Results

Figure 11 shows the convergence results for the algorithm over all simulations for every pattern in each budget class. Convergence results show how often the algorithm has converged on that specific pattern in that budget class for the 20 simulations as a possibly 'true' deposit. The pattern that has been converged upon the most is called the 'winner' of that budget class. For the 10- and 15-hole budget classes, there is always a clear winner in pattern 1. However, for both budget classes, it has to be noted that pattern 7 is also a strong performing pattern, especially for the 15-hole budget class. The observation on a winner for the five-hole budget class is much less clear. Pattern 2 has 30 % convergence, but pattern 1, 4 and 5 each have 20 % too. This is only a small difference, and no strong, clear decision on the winner can be made from this.

Figure 12 shows the average reward over all 20 simulations for every pattern in each budget class. For the 10- and 15-hole budget classes, it is pattern 1 that has the

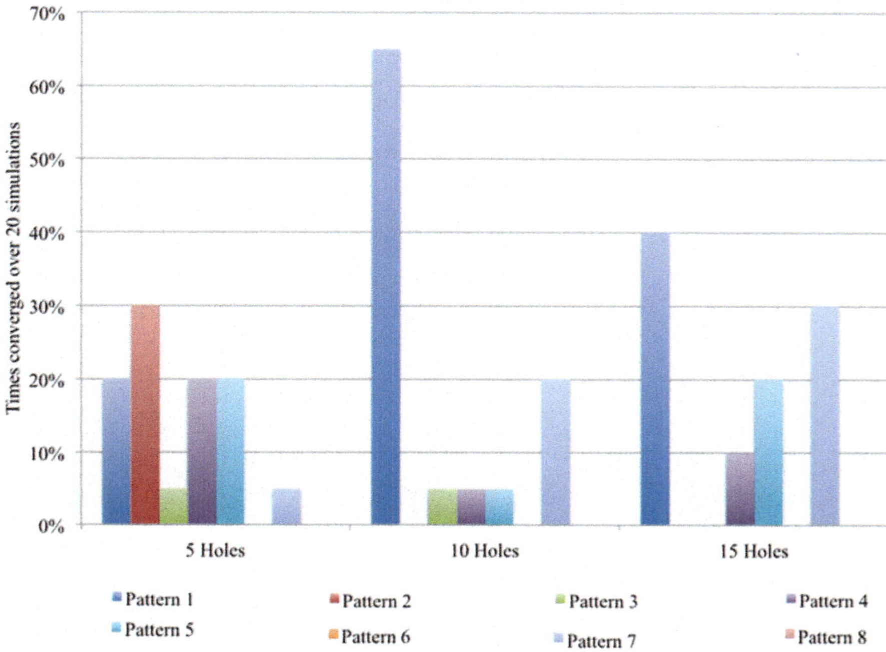

Fig. 11 Convergence over all 20 simulations for all patterns and every budget class

highest average reward, as is expected from the convergence results in Fig. 11. However, the results for the five-hole budget class are not as expected in the sense that pattern 2 does not have the highest average reward. Pattern 1 has the highest average reward for the five-hole budget class. Therefore, it is concluded that pattern 1 is more robust against geological uncertainty than pattern 2. Figure 12 strengthens the observation of no clear winner for the five-hole budget class by demonstrating that the average rewards for all patterns are very close to each other. It is hard for the MAB to distinguish between patterns with such similar rewards.

The differences between the running times observed in Table 1 can also be explained by the observations in Figs. 11 and 12 mentioned above. Because the five-hole patterns yield similar average rewards, the algorithm has a hard time finding the winner, and therefore, it requires more iterations, and thus time, to evaluate the patterns over all simulations. The number of iterations to evaluate all patterns within one budget class for all simulations is very high. In fact, it is higher than the brute-force approach, which tests every pattern 20 times for each simulation and then takes the average. This brute-force tactic would result in ($8 \times 20 \times 20=$) 3200 iterations to evaluate each budget class. The numbers higher than 3200 in Table 1 come from just a few simulations with hard convergence problems that add a lot of iterations to the total count.

Table 1 also shows the number of times the algorithm did not converge for every budget class. The number of non-converging simulations is low for every budget class. To get it to zero, it requires unnecessary large amounts of iterations because it

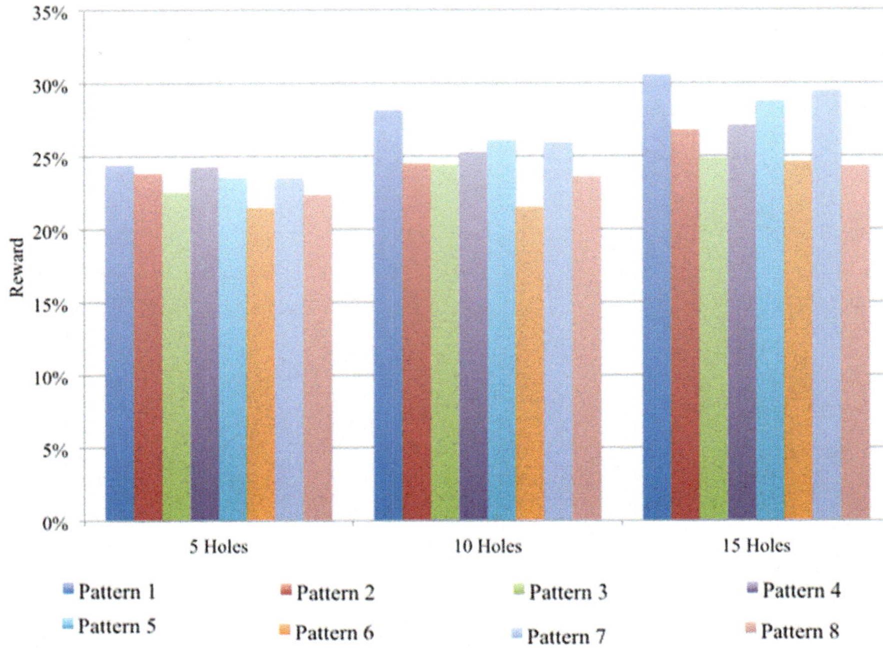

Fig. 12 Average reward over all simulations for all patterns and every budget class

Table 1 Overview of the statistics of the algorithm for each budget class

Budget class	Running time	Number of iterations	Number of times not converged
5 holes[a]	24 h 24 min 46 s	25,465	3
10 holes	13 h 9 min 31 s	13,085	4
15 holes	8 h 41 min 31 s	9378	2

[a]A different set-up is used for the five-hole pattern optimization. This set-up allows for 5000 instead of 2000 iterations as the maximum iterations without convergence. Another addition to this set-up is a forced exploration phase at the start of the algorithm forcing each pattern to be played 20 times before the algorithm starts doing its normal steps. Both measures are taken to overcome convergence issues observed in initial tests

is always possible that two patterns perform very similarly for a certain simulation. Therefore, it is not really seen as a problem because the algorithm still identifies the two patterns that perform equally well in these cases. As a tiebreaker, the pattern with the highest average reward is attributed the convergence. In most cases one of the tied patterns is pattern 1. If ties are broken in the opposite way, the results remain unchanged.

Figure 13 compares the performance of the patterns over the budget classes. This graph demonstrates the profile (P10-P50-P90) on the best pattern's reward over all simulations compared to the average of the rewards of all patterns over all simulations. The tenth percentile (P10) represents the downside risk, and the ninetieth percentile (P90) represents the upside potential of the best pattern under geological

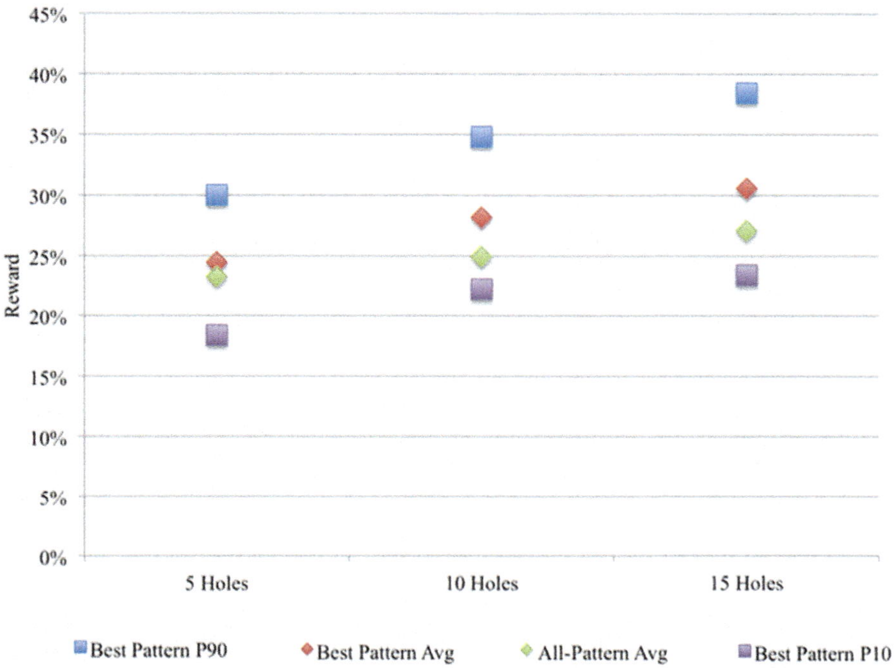

Fig. 13 Profile of the best pattern for every budget class versus the all-pattern average for every budget class

uncertainty. Figure 13 shows that the five-hole budget class performs worse than the other two budget classes. The five-hole best pattern average is lower than the all-pattern average for both other budget classes. Also, the downside risk is much higher, and the upside potential is much lower. Therefore, in combination with the reasons mentioned above, it is not recommended by the author to drill a five-hole pattern in this stockpile. Purely based on performance, pattern 1 with 15 holes is better than pattern 1 with ten holes. However, the extra five holes also cost more to drill. Another observation is that the downside risk for the 10- and 15-hole pattern 1 is almost equal. This means that they both guarantee similar results in the worst-case scenario. This is an extra argument in favour of pattern 1 with ten holes, especially if there are constraints on the budget that are more important than having the best possible knowledge.

4.4 Validation of the Result

As a validation step, the performance of the patterns is tested on an alternate set of 20 simulations, which are not used in the optimization process. Every pattern is tested 20 times for each simulation in the alternate set. One test is the same as the

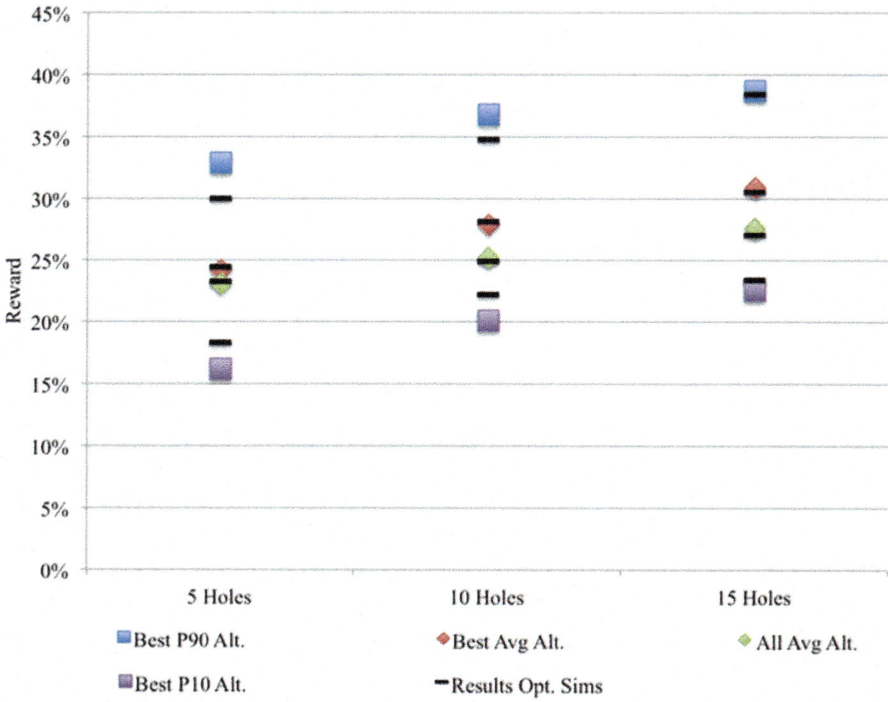

Fig. 14 Profile of the best pattern for every budget class versus the all-pattern average for every budget class based on the reward from the alternate set of simulations

evaluation procedure in the MAB described above; the stockpile is re-simulated based on the pattern data from the possibly 'true' stockpile plus the original data and the reward is calculated.

Figure 14 is the counterpart of Fig. 13. For ease of comparison, the results presented in Fig. 13 are summarized with the black lines in Fig. 14. The black lines are always very close to their counterpart from the tests on the alternate set of simulations. The results of Fig. 14 lead to the same conclusion as made for Fig. 13 above. The validation demonstrates that the proposed optimization method produces results independent of the used simulations.

5 Conclusion and Future Work

The case study presented above shows that the proposed MAB algorithm works as intended. It is able to select the best pattern within a set of patterns of the same budget and provides a decision tool to select the best pattern between sets of a different budget. Also, the case study demonstrates how the method successfully quantifies the influence of geological uncertainty on the performance of an infill drilling pattern. Additional to the performance of the proposed infill drilling optimization algorithm, it is demonstrated that stockpiles can be effectively simulated and that their local-scale variability can be much higher than expected. The method is applied to a stockpile in this case study, but its applicability is not limited to stockpiles; an infill drilling for any deposit can be optimized using the proposed method.

For future work, it would be interesting to look into updating the original simulation after every play according to the infill drilling data, instead of performing a re-simulation at every iteration. Two methods that are suitable for this are conditional simulation by successive residuals (Jewbali and Dimitrakopoulos 2011) and ensemble Kalman filters (Benndorf 2015). Another improvement to the algorithm could be to let it choose the locations of the additional drill holes independently, rather than to constrain it to a set of predefined patterns. A different focus for future research is the application of the MAB framework to other applications in the mining industry.

Acknowledgements We thank Newmont Mining Corporation for providing us with the data necessary to conduct this research and the organizations that funded this research: the Natural Sciences and Engineering Research Council of Canada (NSERC) Discovery Grant 239019 and the COSMO Mining Industry Consortium (AngloGold Ashanti, Barrick Gold, BHP Billiton, De Beers Canada, Kinross Gold, Newmont Mining and Vale) supporting the COSMO laboratory.

References

Agrawal S, Goyal N (2012) Analysis of Thompson sampling for the multi-armed bandit problem. Proceedings of the 25th Annual Conference on Learning Theory (COLT)

Barnes RJ (1989) Sample design for geological site characterization. In: Armstrong M (ed) Geostatistics 1988, vol 4. Springer Netherlands, Avignon, pp 809–822

Benndorf J (2015) Making use of online production data: sequential updating of mineral resource models. Math Geosci 47(5):547–563

Boucher A, Dimitrakopoulos R (2009) Block simulation of multiple correlated variables. Math Geosci 41(2):215–237

Boucher A, Dimitrakopoulos R, Vargas-Guzman JA (2005) Joint simulations, optimal drillhole spacing and the role of the stockpile. In: Leuangthong O, Deutsch C (eds) Geostatistics 2004, vol 14. Springer Netherlands, Banff, pp 35–44

Chorn LG, Carr PP (1997) The value of purchasing information to reduce risk in capital investments. SPE hydrocarbon economics and evaluation symposium. Society of Petroleum Engineers

Desbarats AJ, Dimitrakopoulos R (2000) Geostatistical simulation of regionalized pore-size distributions using min/max autocorrelation factors. Math Geol 32(8):919–942

Diehl P, David M (1982) Classification of ore reserves/resources based on geostatistical methods. CIM Bull 75(838):127–136

Gershon M, Allen LE, Manley G (1988) Application of a new approach for drillholes location optimization. Int J Surf Min Reclam Environ 2:27–31

Goovaerts P (1997) Geostatistics for natural resources evaluation. Oxford University Press, New York

Goria S, Armstrong M, Galli A (2001) Quantifying the impact of additional drilling on an openpit gold project. 2001 annual conference of the IAMG. Cancun

Jewbali A, Dimitrakopoulos R (2011) Implementation of conditional simulation by successive residuals. Comput Geosci 37(2):129–142

Mahajan A, Teneketzis D (2008) Multi-armed bandit problems. In: Hero A III, Castanon D, Cochran D, Kastella K (eds) Foundations and applications of sensor management. Springer, New York, pp 121–151

May BC, Korda N, Lee A, Leslie DS (2012) Optimistic Bayesian sampling in contextual-bandit problems. J Mach Learn Res 13(1):2069–2106

Menabde M, Froyland G, Stone P, Yeates G (2007) Mining schedule optimisation for conditionally simulated orebodies. In Orebody modelling and strategic mine planning. p 379–384

Prange M, Bailey WJ, Couet B, Djikpesse H, Armstrong M, Galli A, Wilkinson D (2008) Valuing future information under uncertainty using polynomial chaos. Decis Anal 5(3):140–156

Ravenscroft PJ (1992) Risk analysis of mine scheduling by conditional simulation. Trans Inst Min Metall Sect A 101:A104–A108

Schlee E (1991) The value of perfect information in nonlinear utility theory. Theory Decis 30 (2):127–131

Scott SL (2010) A modern Bayesian look at the multi-armed bandit. Appl Stoch Model Bus Ind 26 (6):639–658

Switzer P, Green AA (1984) Min/max autocorrelation factors for multivariate spatial imagery. Stanford University, Department of Statistics, Standford

Thompson WR (1933) On the likelihood that one unknown probability exceeds another in view of the evidence of two samples. Biometrika 25(3/4):285–294

Fixing Panel Artifacts in Localized Indicator Kriging (LIK) Block Models

William Hardtke and Celeste Wilson

Abstract Three types of panel artifacts are created by the localized indicator kriging (LIK) methodology. Two of these artifact types can be fixed (eliminated) by a two-step process that consists of averaging models with different panel origins and then performing a global re-localization. The first type of panel artifact, and probably the most important, is caused by change-of-support transformations, which are independent and applied on a panel-by-panel basis. The second type of artifact is the most common but is usually undetectable. This artifact is caused by the fact that there are several possible LIK models where the only difference is the panel origin, and although each of the resulting models has the same global distribution, the high-grade blocks change location. The third type of artifact is caused by using a search ellipse that is too large (nonstationary), which is more typical in areas where drill data is sparse. Unlike the first two artifact types, this type needs to be fixed by adjusting search parameters. Artifacts should be eliminated from LIK models whether they are obvious as in the case of those caused by large change-of-support transformations or are simply the added spatial noise caused by having more than one origin.

1 Introduction

The LIK methodology uses the histograms (probability distributions) from a multiple indicator kriging (MIK) estimation, which are created using a large block size (panels), and localizes those distributions into smaller SMU-sized blocks using an ordinary kriging (OK) model as a localizer (Hardtke et al. 2009). The concept of using one model to localize the distribution of a second estimation that was done at

W. Hardtke (✉)
WCH Consulting LLC, 9808 S. Rosewalk Dr., Highlands Ranch, CO 80129, USA
e-mail: william.hardtke@yahoo.ecom

C. Wilson
Newmont Mining Corp, Elko, NV, USA
e-mail: celeste.wilson@newmont.com

J.J. Gómez-Hernández et al. (eds.), *Geostatistics Valencia 2016*, Quantitative Geology and Geostatistics 19, DOI 10.1007/978-3-319-46819-8_14

a different scale was first presented by Marat Abzalov for use with uniform conditioning (Abzalov 2006). Although LIK typically yields resource models that have both the desired global target distribution and excellent location for the block values within the panels, there is always at least some degree of global imprecision in the location of the values that is caused by the panels. Any change in a block value that is due solely to the panel location can be referred to as an artifact, but there is no reason why they should be accepted as an innate part of a LIK model.

There are three types of artifacts and the most common type is undetectable and in most cases it does not have to be eliminated or even addressed. This type of artifact, which can be referred to as a change of origin artifact, is found in every LIK model and is simply the fact that a change of panel origin changes the value of every block in the model. Unless the exact value of an individual block is important as perhaps may be the case in a high-grade underground mine, this type of artifact can be ignored.

The second type of artifact is caused by change-of-support transformations and usually needs to be addressed because this type of artifact can be incredibly obvious to the point that they make the model unacceptable. This type of artifact arises because the transformations are done independently on a panel-by-panel basis. Because the transformations cause all the block values within the panel to become closer to the mean value of a panel, adjacent panels with dissimilar mean values can suddenly become quite visible. In general, the greater the transformation (the amount of variance reduction), the more pronounced the panel boundaries become.

The third type of artifact that is caused by having too large a search for the size of the panel is referred to as a non-stationarity artifact. This type of artifact is much more obvious in areas with sparse data and usually results in a pattern where each panel has its highest-grade block in the same relative position. In most cases this type of artifact is only cosmetic and unimportant because it generally occurs in areas that will be classified as inferred.

In any LIK model, there is a defined number of possible panel origins that is equal to the number of SMU-sized blocks per panel. Each of these possible panel origins will result in a unique LIK model where the block values vary locally but have a nearly identical global distribution. It is analogous to a simulation where all of realizations are equally probable and none of them is better than the others. However, because there are a finite number of panel origins for any LIK model, they can easily be combined into a single "best" answer.

Two approaches for combining LIK models to eliminate artifacts were tested during this study. In the first approach, the value for each block was randomly selected from all the possible LIK models. This resulted in a final model that eliminated all the artifacts while retaining the original global distribution, but it added a degree of spatial variability that probably is not real. This approach was abandoned early on and is not presented here.

The second approach was to average all the blocks from each of the LIK models into a single block model. This resulted in a model that had a very good spatial distribution for the gold, but the averaging of values lowered the variance of the global distribution. Then in order to correct the smoothing, the original distribution was re-localized using the averaged block model as the localizer. The final re-localized model had the best possible local distribution of gold combined with the desired global distribution. This approach gave very good results and the remainder of the paper assumes this methodology.

2 Panel Artifacts Caused by Origin

An LIK model with no change-of-support transformation typically shows little or no edge effect at the panel boundaries but does have the change of origin artifacts. Although there are no obvious grade changes at the panel boundaries, there is definitely a spatial difference in LIK models that are identical except for the origin of the panels. The LIK models in Figs. 1 and 2 have nearly identical global distributions but are quite different at the local block scale.

This small-scale variation in the LIK models is also a type of panel artifact, and although the model is pleasing to the eye, there is still some question about the true location of the high-grade blocks. In many cases a single LIK model can be arbitrarily selected and used as long as the mine design is not too sensitive to the location of individual high-grade blocks. However, untransformed models are seldom used because even a model at the SMU size is not normally mineable and needs additional smoothing.

3 Panel Artifacts Caused by Change of Support

Change-of-support transformations create panel artifacts because they are applied independently on panel-by-panel bases. The transformation changes all of the block values inside a panel, moving them toward the mean grade of that panel (Hoerger 1999). These artifacts are not always obvious if the transformation (variance reduction) is small and the grades of the adjacent panels are not too disparate. Figure 3 shows an LIK model with a moderate amount of variance reduction ($k = 0.3$), and artifacts are not very noticeable, but in Fig. 4 the variance reduction factor is increased to $k = 0.05$, which was the target distribution for this deposit causing the panel boundaries to become quite visible and the resulting model is probably unacceptable. (The k factor referred to in this paper is from the Hoerger paper and is the variance reduction factor that ranges from 1.0 to 0.0 where 1.0 is no reduction and 0.0 is the maximum reduction.)

In Fig. 4, the increased amount of transformation within the panel with the mean grade of 1.542 (upper row, second from the left) causes the low-grade (blue and cyan) values to disappear at the expense of the high-grade magenta blocks. That

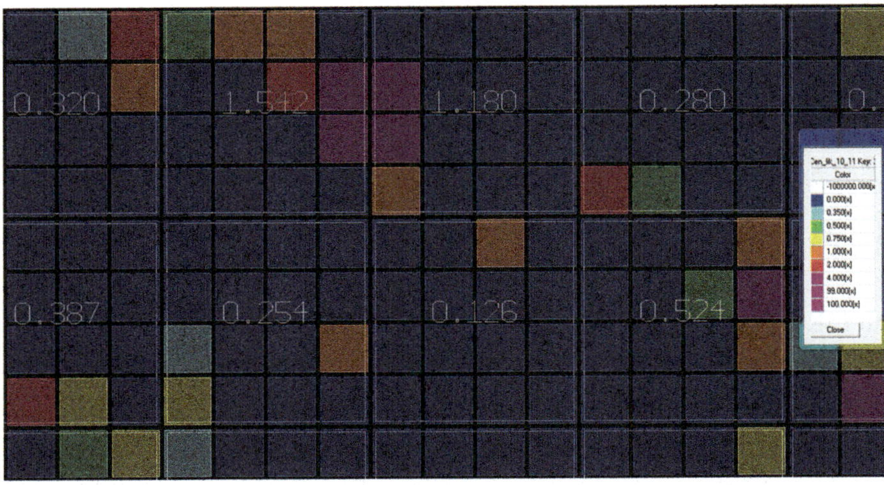

Fig. 1 Untransformed LIK model (no. 10)

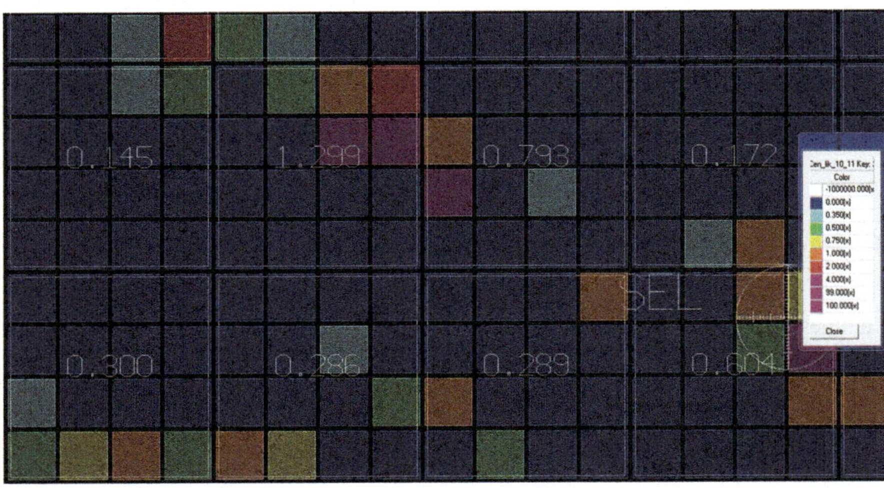

Fig. 2 Untransformed LIK model (no. 17)

change in the panel distribution results in a very obvious panel boundary where it comes into contact with the low-grade panel directly below it. In a case as obvious as this, the contact between ore and waste actually changes to the panel boundary, which would be a real problem for an engineer trying to design a mine.

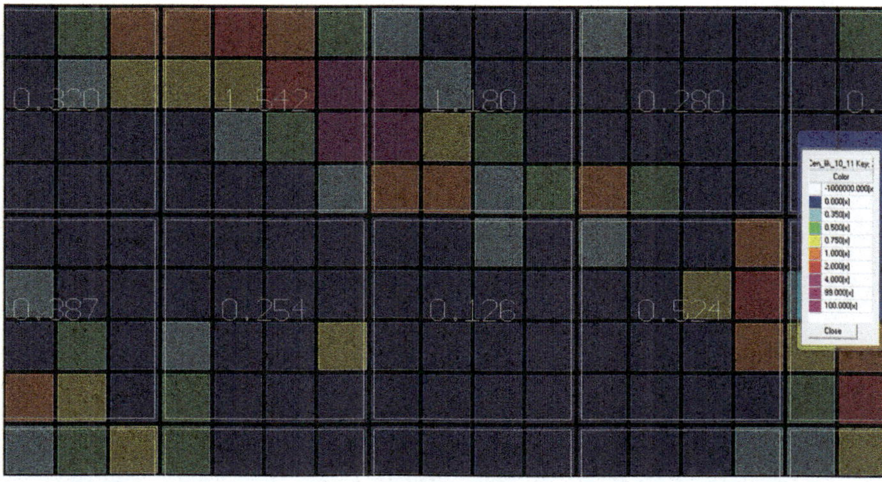

Fig. 3 LIK model (10) with a moderate amount of variance reduction

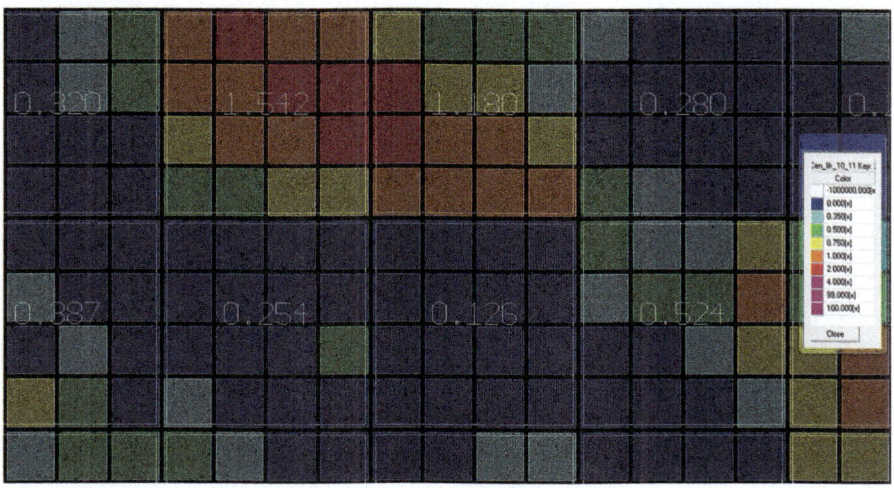

Fig. 4 LIK model (no. 10) with an extreme amount of variance reduction

4 Panel Artifacts Caused by Non-stationarity

LIK models are very sensitive to data density, and when the panel size is too small or the search distance is too large, strange-looking typically repetitive artifacts are usually created. This type of artifact is most common in low-grade areas where panels should be totally low grade, but the search ellipse is large enough to include high-grade composites that are well outside the panel being estimated.

In Fig. 5, large search distances result in high-grade blocks being estimated in an area that is obviously low grade (solid blue squares are low-grade drill intercepts). This is clearly a situation where the local search ellipse is nonstationary in nature

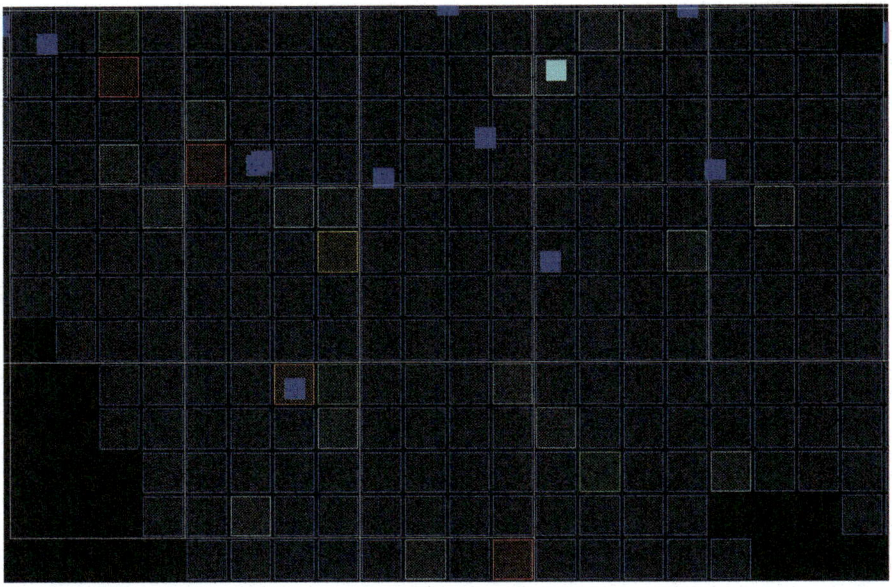

Fig. 5 LIK model showing non-stationarity artifacts

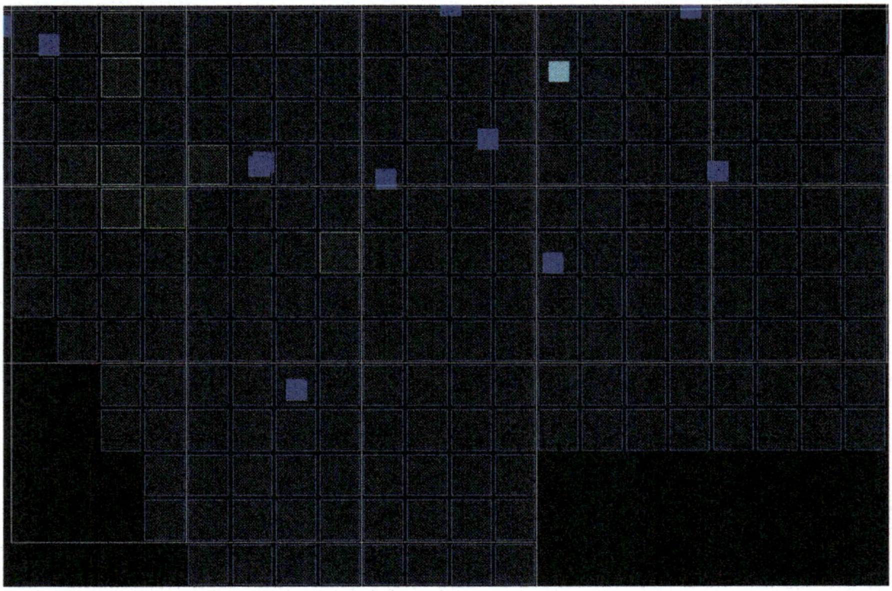

Fig. 6 The same LIK model with an appropriate search ellipse

and can easily be fixed by changing the search parameters. The same sort of thing happens all the time in OK models but is not easy to see because the effects of the high-grade composites in the search are spread more evenly over all the blocks. However, LIK partitions the misplaced high-grade values into one or two blocks per panel making them very easy to see. In this example of an area that should be all low grade (blue <0.2 g/t), the LIK model has created a few high-grade blocks (red >1.0 g/t).

This type of artifact can usually be fixed by changing the kriging parameters in the MIK model (panel size). In some cases, where the search cannot be shortened, the panel size probably needs to be increased. The methodology of averaging and making a global re-localization does not eliminate this type of artifact; it merely changes its location. Figure 6 shows the same area with a more appropriate search ellipse.

5 Example of Artifact Removal

This example of artifact removal (approach 2) is in a low-grade, high-nugget, gold deposit that requires a large change-of-support transformation. The appropriate amount of smoothing was determined by matching the LIK model to a blasthole block model (the target distribution). The blasthole model indicated that an extreme amount of grade smoothing would be needed, so a variance reduction factor of 0.05 was selected.

The two LIK models shown in Figs. 7 and 8 are identical except that they have different panel origins. Both of the models have obvious artifacts along the panel

Fig. 7 LIK model (10)

Fig. 8 LIK model (17)

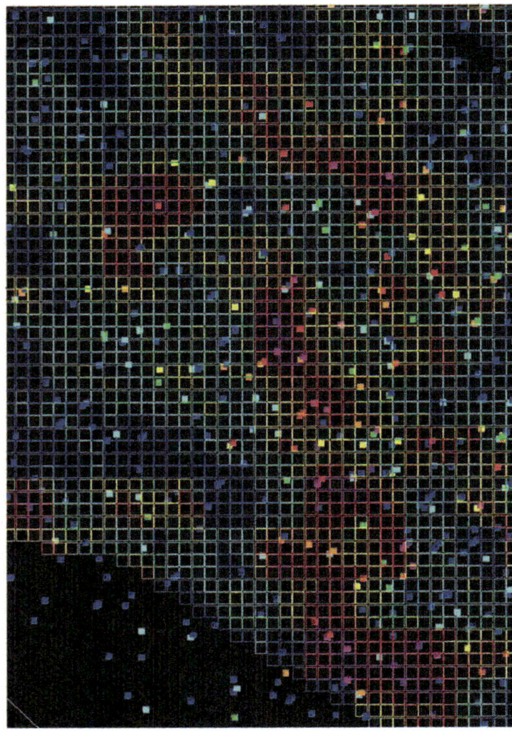

boundaries that were created by the large change-of-support transform. These artifacts take the form of the panels, which contain 16 blocks (4 by 4) and is most noticeable in extremely low-grade (blue) and high-grade (red) panels. It should also be noted that there are 14 other possible LIK models and that they all have almost exactly the same global block distribution.

It is not always clear whether artifacts need to be removed from a model or not, but in this example the answer is straightforward because the contacts between ore and waste are not be square as these model shows in many areas. Artifact removal always seems to improve the LIK models, and the only downside is the added work, so if there is any doubt as to whether it is needed, it should probably be done.

Continuing on with this example, the 16 possible LIK models were averaged together into a single block model. The idea being that the best value for any block would be the average of all the values for that block. It is evident in the averaged block model in Fig. 9 that the panel artifacts are gone and the blocks still respect the drill data, but unfortunately the averaging of the block values has increased the amount of smoothing. In short, the location of the gold is quite good but the averaged model no longer has the desired global distribution.

Therefore, the final step is to "tweak" the model, so it has the correct global distribution by doing a global re-localization using the averaged grid as the localizer. In this step all the blocks are ordered from smallest to largest, and the values for one of the original LIK distributions are substituted for the corresponding

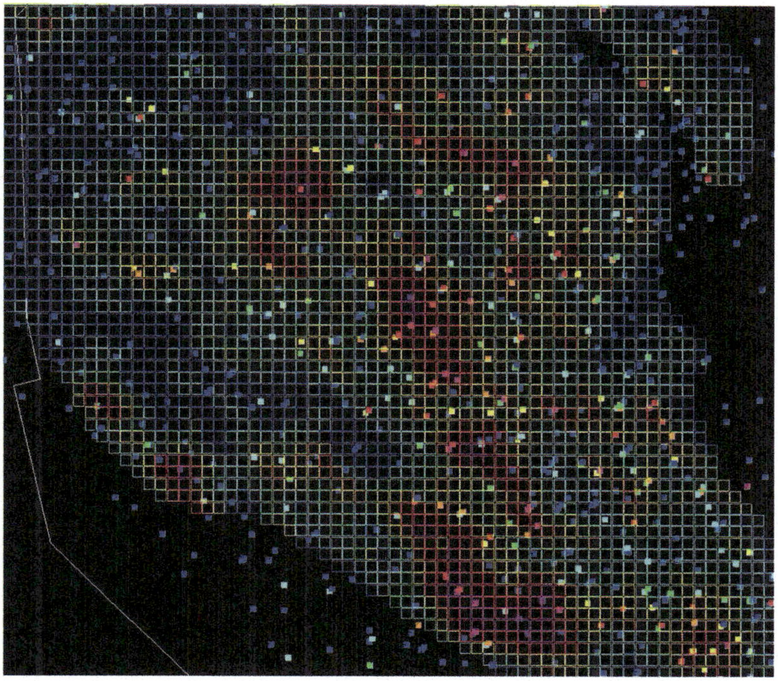

Fig. 9 Averaged model from 16 LIK models

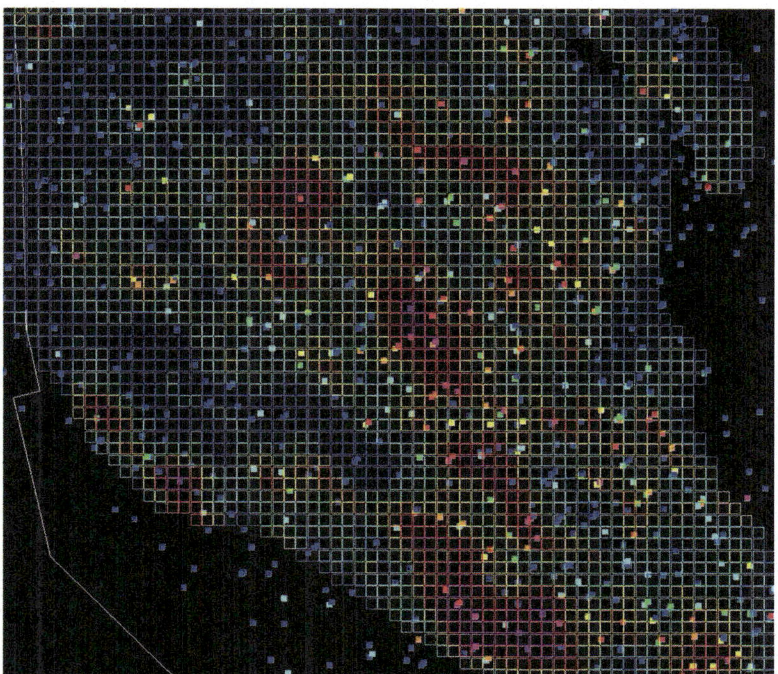

Fig. 10 The final re-localized LIK model with artifacts removed

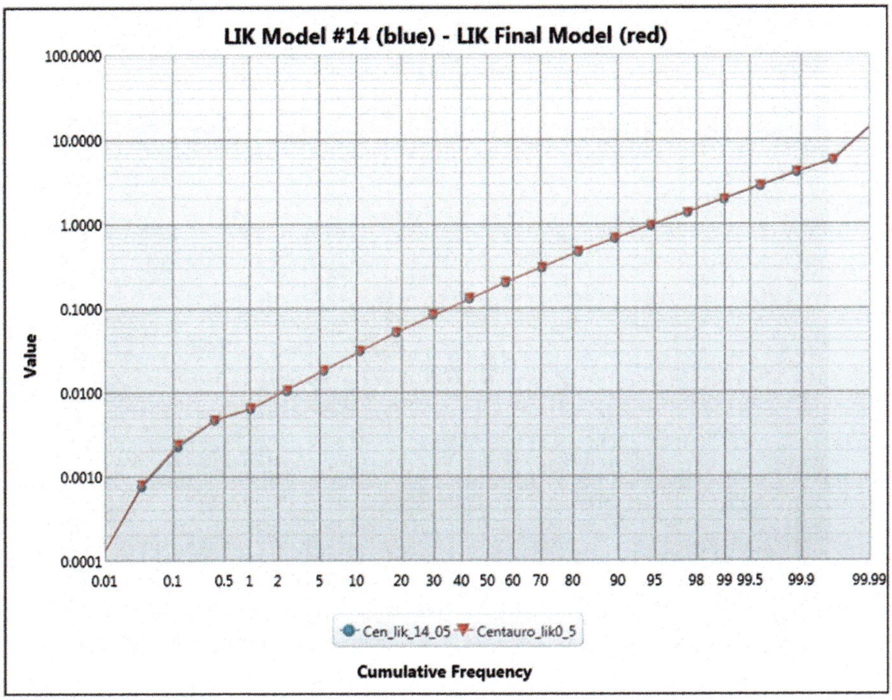

Fig. 11 Model distribution before and after fixing the artifacts

block in the averaged model. The target distribution can be any 1 of the 16 LIK models because they have the almost exactly the same distribution. Then it is a straightforward step of sorting the blocks in both models (common blocks only) and replacing the averaged values with the corresponding target value.

Figure 10 contains an example of the final model. It has no panel artifacts; it has the desired distribution, and the location of the blocks is probably as good as it can be. It looks very much like the averaged model, but after close examination it definitely has less smoothing like the target distribution.

Figure 11 shows the distribution of blocks before and after the artifact removal, and there is very little difference between the original distribution with artifacts (blue) and the final distribution after artifact removal (red).

6 Summary

All LIK models have panel artifacts but they are not always important enough to need to be removed. In most cases artifacts are just cosmetic and do not affect the quality of the model or the subsequent mine design, but when they need to be eliminated, it can be done with a two-step process. The first step is to make a new localizer by averaging all of possible LIK models into a single model. The second step is to adjust this new averaged model so it has the same global distribution as any of the LIK models.

Bibliography

Abzalov M (2006) Localised Uniform Conditioning (LUC): a new approach for direct modeling of small blocks. Math Geol 38:393–411

Hardtke W, Allen L, Douglas I (2009) Localised indicator kriging. 35th Apcom Symposium, Perth WA, pp 1–4

Hoerger S (1999) Implementation of indicator kriging at Newmont gold company.. 23rd APCOM Proceedings, pp 202–213

Implications of Algorithm and Parameter Choice: Impacts of Geological Uncertainty Simulation Methods on Project Decision Making

Arja Jewbali, Bruce Perry, Lawrence Allen, and Richard Inglis

Abstract Uncertainty in geological models and grade uncertainty are two major contributors to the total resource uncertainty of a mining project. Previous attempts at determining uncertainty in geological models using methods such as MPS (multiple-point statistics), SIS (sequential indicator simulation), and multiple applications of RBF (radial basis functions) with different parameters have shown that it is nontrivial; the uncertainty profiles are dependent on the method and the parameters selected. Most of the methods tested require additional information in the form of either local probabilities or proportions derived from the existing geological interpretation or a conceptual geological model in the form of a training image. This makes some methods amenable to use in the early stages of a project because the method allows for a more complete testing of different geological concepts. In later stage projects where there is an increased level of confidence (due to the amount of data collected) in the geologic interpretation, methods that achieve ranges of uncertainty around the interpretation likely provide a more realistic assessment of uncertainty. This paper details the continuation of research into geostatistical tools suitable for the evaluation of geological uncertainty in order to further understand the intricacies of the methods and the impact of the technique on the resulting uncertainty profile. Suggestions of which methods to use based on the amount of geological information available are provided.

A. Jewbali (✉) • B. Perry • L. Allen • R. Inglis
Newmont Mining Corporation, Denver, CO, USA
e-mail: Arja.Jewbali@newmont.com; robert.perry@newmont.com; Lawrence.Allen@newmont.com; Richard.Inglis@newmont.com

J.J. Gómez-Hernández et al. (eds.), *Geostatistics Valencia 2016*, Quantitative Geology and Geostatistics 19, DOI 10.1007/978-3-319-46819-8_15

1 Introduction

Resource uncertainty analysis incorporating both grade and geological uncertainty is done in various stages of project development to quantify the risk to the mine design and the mine schedule and the impact on the financial parameters of the project. The geology model is an important factor in determining the grade and tonnage of the resource since it provides fundamental control over the resource estimation process. For some projects uncertainty due to the geological interpretation forms a large component of the total resource uncertainty, and not taking this into account likely underestimates the risk in the resource estimate.

Trials using geostatistical methods indicate that generating alternate geological models is not trivial; uncertainty profiles depend on the method and parameters chosen. In addition, most of the methods tested require additional information in the form of rock-type proportions derived from the existing geological interpretation or a conceptual geological model in the form of a training image. Before application of the geostatistical tools, the practitioner must first decide how much confidence should be placed in the existing geological model. What is the quality of the geological model? Should the uncertainty interval for mineral content from simulated models be centered on the geology model? Answers to these questions depend on how much information has been collected, how well one understands the geologic controls of the mineralization, and how much confidence one has in the technical capability of the geologist who created the geology model. Blind application of simulation methods and parameters can lead to simulated models of geology with more or less mineralized volume when compared against the geology model. Does this mean that the resource model incorporating all existing geological knowledge is pessimistic or optimistic, or is this result a consequence of the method? These questions should be considered during uncertainty quantification since the results can have severe consequences for the project.

Moreover, other complications encountered by practitioners trying to build simulated models of geology include:

- The nature of the data collection process. Mining companies tend to preferentially collect more information in mineralized geological domains. It is not uncommon for non-mineralized geological domains to have no or very limited data.
- For methods that require a training image (multiple-point methods). A training image is necessary to derive multiple-point statistics; however, developing training images for a class of ore deposits is difficult because most ore deposits tend to have nonstationary characteristics. The use of resource models as training image is not recommended since the models are typically smooth and nonstationary and are unlikely to have the right level of connectivity and multiple-point statistics.

It will be demonstrated that some simulation methods are amenable to use in the early stages of a project because the method allows for a more complete testing of different geological concepts. In later stage projects where there is an increased level of confidence (due to the amount of data collected) in the geologic interpretation, methods that achieve ranges of uncertainty around the interpretation likely provide a more realistic assessment of uncertainty (Jewbali et al. 2014).

2 Categorical Simulation Methods

This section gives a brief description of the categorical methods applied. The techniques were used to generate simulated models of the main mineralized domain for two deposits. The geostatistical tools trialed include:

- Indicator-based methods (SIS-LVM) (Deutsch 2006).
- Methods based on distance functions (Munroe and Deutch 2008a, b; Wilde and Deutsch 2012).
- Methods based on multiple-point statistics (snesim) (Strebelle 2002; Remy et al. 2009).
- Methods based on radial basis functions (Leapfrog®) (Stewart et al. 2014) using different sets of parameters. While not a simulation method, Leapfrog® is increasingly used to quickly build alternate models of geology and to determine the impact of the geological interpretation on project financials.

2.1 Sequential Indicator Simulation with Local Varying Mean (SIS-LVM)

To determine uncertainty related to the geological interpretation, a sequential indicator simulation (SIS) approach was used. In SIS, each mutually exclusive rock type (category) is expressed as an indicator variable. These indicators are simulated, and for every location, a category is drawn according to the local conditional distribution function determined through simple kriging. Drawbacks of the SIS approach include the unstructured appearance of the simulations, the inability to impose structural control over the simulations, and its inability to handle the nonstationary nature of rock-type proportions. To partially account for these drawbacks, a deterministic categorical variable model (usually the geologic domain of interest) is filtered to calculate local varying probabilities near the boundaries of the different categories, i.e., the contacts are uncertain (Deutsch 2006). The size of the filter determines the width of the uncertainty region adjacent to the contact and is usually related to drill spacing, since the region of uncertainty should decrease as drill density increases. Next, simple kriging is used to derive the local conditional distribution.

$$I_{LVM}^{*}(u;k) - p_k(u) = \sum\nolimits_{\propto=1}^{n} \mu_{\alpha}^{sk}(u_{\alpha};k)[i(u_{\alpha};k) - p_k(u_{\alpha})] \quad (1)$$

where

$I_{\mathrm{LVM}}^{*}(u;k)$ is the simple kriged estimate at location u for category k.
$p_k(u)$ is the probability of category k at location u.
$\mu_{\alpha}^{sk}(u_{\alpha};k)$ are the simple kriging weights for data at location u_{α} for category k.
$i(u_{\alpha};k)$ are the indicators for category k at data location u_{α}.
$p_k(u_{\alpha})$ are the local probabilities for category k at location (u_{α})

The estimates for each category are performed independently which can lead to order relations deviations. Due to the noisy nature of the simulations, a post-processing step is usually applied to clean the simulations up (Deutsch 2006).

2.2 Categorical Simulation Using Distance Functions

This method is based on the interpolation of a signed distance function built using conditioning data. The distance function can be seen as the distance between a data point itself and the nearest sample belonging to another domain. Positive and negative distances are used to distinguish between inside and outside a domain. Distances depend on the orientation, geologic shape, and extension of the rock types. The methodology is as follows (Munroe and Deutch 2008a, b; Wilde and Deutsch 2012):

- Code all data points as either inside or outside the domain of interest.

$$i(\boldsymbol{u}_{\propto}) = \begin{cases} 1 \text{ if inside domain of interest at } u_{\propto} \\ 0 \text{ otherwise} \end{cases} \quad \propto = 1,,n \quad (2)$$

  where $\boldsymbol{u}$ is the location vector, $\propto$ is the sample index, and n is the number of samples.
- Next calculate the distance for each sample to the nearest sample located in another domain. If $u_{\propto}$ is inside the domain, the distance is set to negative or else the distance is positive. A factor C is added (or subtracted) to the distance function value which increases the difference between positive and negative values. The C factor is calibrated using a jackknifing approach.

$$df(\boldsymbol{u}_{\propto}) = \begin{cases} +(\boldsymbol{u}_{\propto} - \boldsymbol{u}_{\beta}) + C \ \ if \ \ i(\boldsymbol{u}_{\propto}) = 0 \\ -(\boldsymbol{u}_{\propto} - \boldsymbol{u}_{\beta}) - C \ \ if \ \ i(\boldsymbol{u}_{\propto}) = 1 \end{cases} \quad \propto \neq \beta \quad (3)$$

- Interpolate the distance function on a regular grid using a global interpolator. The boundary is located between the distance function estimates of $-C$ and C.

Simulate the boundary only for locations where the distance function estimates lie between $-C$ and C.

$$df^l(\boldsymbol{u}) = 2CG^{-1}\left(y^l(\boldsymbol{u})\right) - C \tag{4}$$

where $df^l(\boldsymbol{u})$ is the simulated distance function value, $y^l(\boldsymbol{u})$ is a unconditionally simulated standard normal value, and G^{-1} is the standard normal CDF value for $y^l(\boldsymbol{u})$.

- Determine whether a location u is inside or outside a domain, by comparing the simulated distance function value against the interpolated distance function value.

$$i(\boldsymbol{u}_\propto) = \begin{cases} \text{inside if } df^l(\boldsymbol{u}) > \text{estimate} \\ \text{outside if } df^l(\boldsymbol{u}) < \text{estimate} \end{cases} \tag{5}$$

This methodology is geared toward binary systems (in and out a particular domain) and does not handle multiple domains. The calculated distance function is nonstationary which can make variogram inference challenging. Another approach using distance functions can be found in Cáceres et al. 2011.

2.3 Multiple-Point Simulation (MPS)

A shortcoming of SIS and any two-point-based methods is that it fails to reproduce complex nonlinear geological features as seen in mineral deposits. This is due to its reliance on the variogram, which can only characterize the linear relationship between data points. Multiple-point simulation (MPS) (Guardiano and Srivastava 1993; Strebelle 2002; Remy et al. 2009) is a technique which characterizes the relationship between points with higher-order statistics. In doing so it is able to reproduce complex patterns and domain interactions (Strebelle 2002). MPS requires the use of a stationary training image (TI) to extract the higher-order statistics at various scales. However, storing and deriving the multiple-point statistics from the training image require additional RAM and CPU time. For this study, snesim (Strebelle 2002; Lui 2006) is used to generate simulated models of the different mineralized domains.

2.4 Leapfrog® Models

To build alternate models, an indicator approach utilizing radial basis functions (RBF) was used to produce a range of results. The inputs to the RBF interpolant are the categorical data in the form of a numerical indicator, the variogram, and a structural trend which is similar to locally varying anisotropy. The approach allows for locally varying directions of continuity. Volumes were created at desired values to represent probability shells. Leapfrog® software was used for this approach. The RBF interpolant is similar to the general expression of dual kriging (Stewart et al. 2014).

$$s(x) = \sum\nolimits_{i} \omega_i \varphi(|x - x_i|) + \sum\nolimits_{k}^{K} c_k q_k(x) \tag{6}$$

where

x_i are the data locations over which the interpolation is to be constrained.
ω_i are RBF coefficients (weights).
$\varphi_k(x)$ is a spatial distance function (the RBF – from which the method takes its name).
The term on the right refers to the set of K drift functions ($q_k(x)$), each having a coefficient (c_k) applied globally across all data.

3 Case Studies

This section details the use of the geostatistical tools described above to build simulated models of the main mineralized domains for the Merian deposit located in East Suriname and for the Subika deposit located in the Ahafo Region in Ghana. It starts with a description of the geologic settings of the two deposits. While both deposits are orogenic in genesis, each has different types of controls on mineralization from which to assess uncertainty.

3.1 The Merian Deposit

The Merian deposit lies within lower Proterozoic-aged rocks of the Guiana Shield in northeast Suriname, South America, approximately 100 km east of the capital Paramaribo (Fig. 1). In Suriname the Guiana Shield is composed of distinct, east-west trending belts of low-grade metamorphic rocks which are separated by large areas of granitic rocks and gneisses. Gold mineralization within the Merian deposit occurs as a vein-type Proterozoic lode-gold deposit; gold is found within and immediately adjacent to quartz veins, quartz stockworks, and irregular quartz breccia bodies. Host rocks are composed of highly folded sandstones and siltstones. Gold mineralization at Merian occurs over a strike length of approximately 3.5 km, elongate in a northwest-southeast direction, and over a width of 200–600 m.

The Merian geologic model was constructed using the following attributes collected from drill core logging:

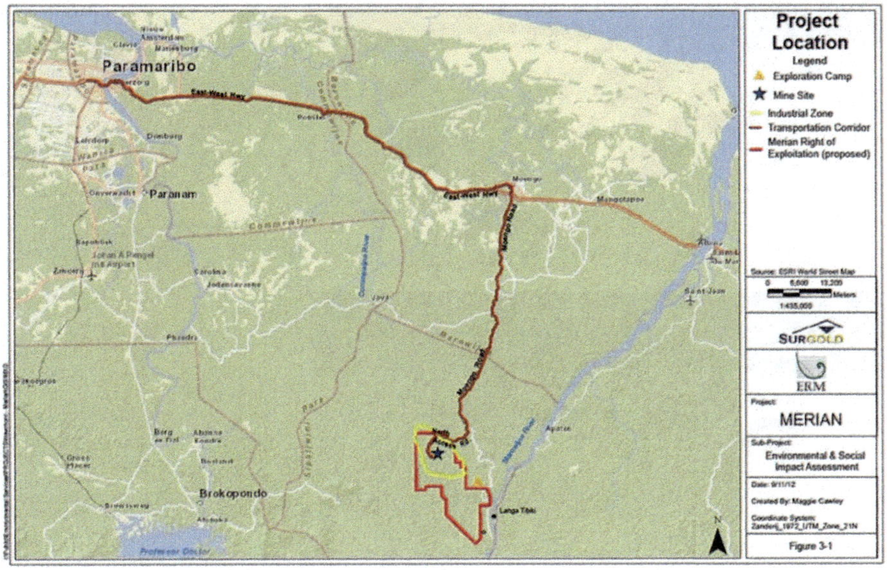

Fig. 1 Location of the Merian deposit

- Stratigraphy, faults, and folds
- Oxidation states (saprolite, fresh rock, etc.)
- Quartz vein density
- Quartz breccia

Uncertainty analysis focused on the uncertainty in the interpretation of the Quartz vein density shape which encompasses the main mineralized zone.

3.2 Simulation Methodology for Merian

Percent quartz vein content is recognized as the primary control on the geometry and grade of gold mineralization at Merian. Figure 2 displays three sections of the Merian mineralized envelope as coded in the Merian resource model. In tightly drilled areas, the drillhole spacing is approximately 25 m across and 25 m along sections.

In order to build simulated models of the quartz vein density shape and to assess the uncertainty in its contained volume, categorical simulations were generated using various techniques:

- Fifty simulations using the SIS-LVM approach with local varying probabilities were calculated using the geological interpretation and moving window sizes of 12×12 m (0.5 times the tightest drillhole spacing), 25×25 m (the tightest drillhole spacing), 50×50 m (twice the tightest drillhole spacing), and

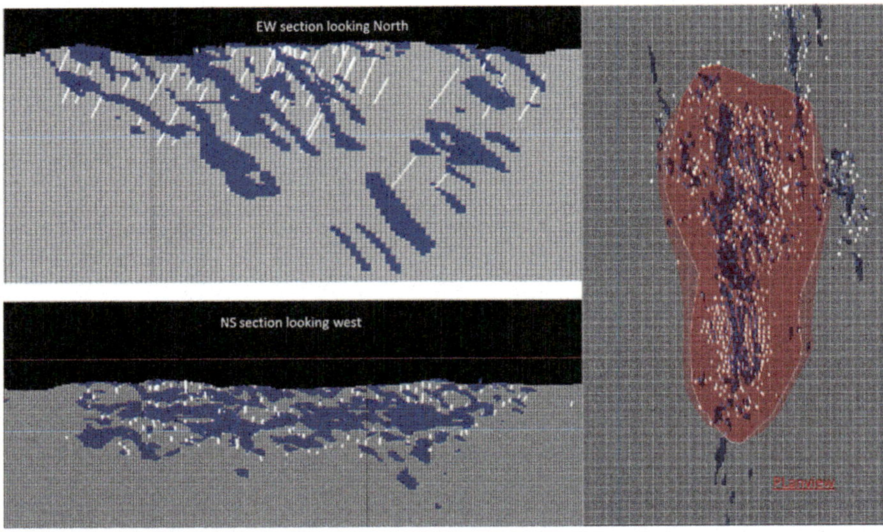

Fig. 2 Main mineralized domain for the Merian deposit with drillhole data. The area within the white boundary (*right*) is used for the study

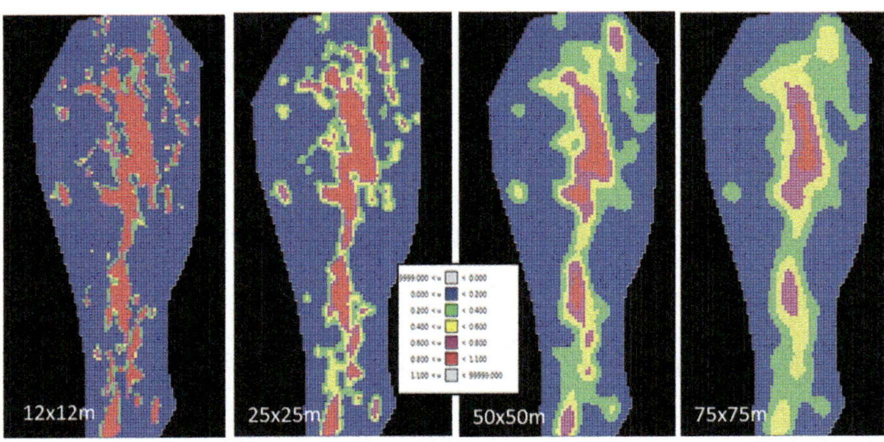

Fig. 3 Local varying probabilities to be inside the mineralized domain derived from the resource model (plan view) – high probabilities in warmer colors

75×75 m (three times the tightest drillhole spacing) (Fig. 3). It is expected that larger filter distances will yield wider bands of uncertainty. Besides the conditioning data, this method uses the geological interpretation (as per the resource model).

- Fifty simulations were generated using the distance function approach. Three C factors were used to determine the sensitivity of the uncertainty profile to the C factor ($C = 20$, approximately equal to the tightest drillhole spacing; $C = 50$, twice the tightest drillhole spacing; and $C = 100$, four times the tightest drillhole spacing) (Fig. 4). It is expected that larger C factors will result in wider

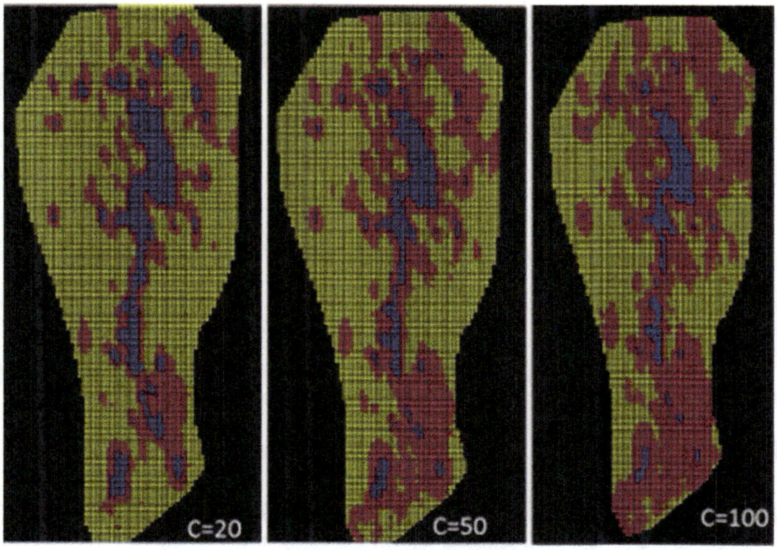

Fig. 4 Interpolated distance function for different C factors (*red* represents distances between –C and C (plan view)

uncertainty intervals. The geological interpretation was initially used to code the data as being inside or outside the mineralized domain. The logged information from the drillholes could have been used instead; however, there was some grouping of highly variable intervals into the geologic interpretation, which would not have been accounted for.

- Fifty multiple-point simulations using snesim, where the resource model in Fig. 2 was used as the training image (servo system factor of 0.5).
- Alternate models using Leapfrog®: in addition to the simulations, three volumes were produced using RBF indicator probability shells derived from conditioning data. Volumes were analyzed based on selecting volumes contoured from a range of interpolated values (P30 to P50). Indicator statistics were analyzed to determine the balanced shell (P38) which is the volume that includes as many indicator data misclassified inside the shell as it excludes indicator data misclassified outside the shell and to determine the Russian doll shell (P30) where the next larger volume incrementally includes data where the indicator mean is less than the probability of the shell. In order to produce a range of results, an arbitrary smaller (conservative) volume was selected at P46 (Inglis 2013).

3.3 Volumetric Uncertainty for Merian

Plan views of the generated simulations for the various methods are shown in Figs. 5 and 6. The figures show that the simulations generated by SIS-LVM are

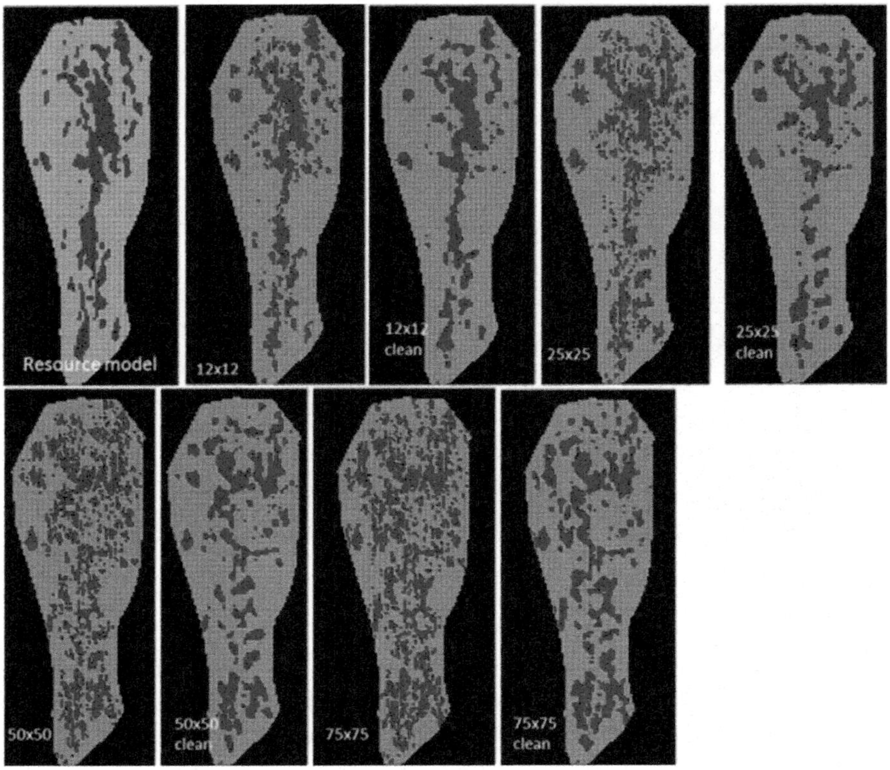

Fig. 5 Simulation results for SIS-LVM (plan view)

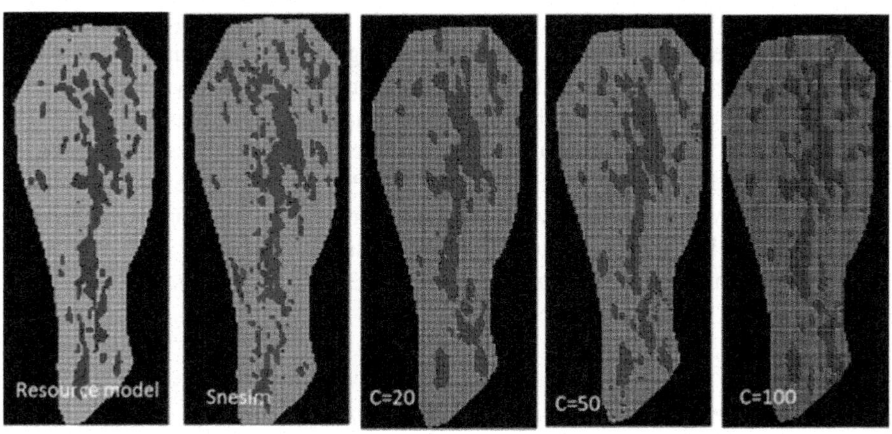

Fig. 6 Simulation results for snesim and the distance function approach (plan view)

noisier compared to the model. This is especially so, when a large window is used to generate the local probabilities. As shown in Fig. 3, this creates lower probabilities in areas where the probability to be inside the mineralized shape was previously high (with a smaller window size). This causes the simulations to have a dispersed appearance. After cleanup, the simulations for the 25 × 25, 50 × 50, and 75 × 75 still do not display the same level of continuity in the north-south direction seen in the resource model. The snesim simulations also appear noisier compared to the resource model. The distance function approach produces simulations that look similar to the resource model. Some artifacts are visible especially with large *C* factors which tend to create "donut holes" inside a mineralized volume.

Figure 7 displays the uncertainty profile for the mineralized volume. It shows the smallest, largest, and average (over all 50 simulations) volume. It also shows the width of the uncertainty interval (difference between the smallest and largest volume). In general from these figures, the following can be derived:

- For most of the methods, the volumes fluctuate above or below the volume defined by the resource model. Only three methods (SIS-LVM 75 × 75 m with post-processing clean, Leapfrog®, and distance function C50 and C100) contain the volume defined by the resource model in their uncertainty interval. Whether simulation volumes are above or below, the mineralized volume defined by the resource model depends on the chosen parameters.
- As expected, for the SIS-LVM, the wider the uncertainty window used to generate the local probabilities, the wider the uncertainty interval. The width of the uncertainty interval for SIS-LVM (75 × 75 m) is wider compared to that of SI-LVM (50 × 50 m). For this approach there appears to be a bias related to the size of the window used to create the local probabilities. For larger windows the simulations tend to generate more mineralized volume (i.e., also more dispersed).
- For the distance function approach, the models derived from larger *C* factors also tend to have wider uncertainty intervals. For smaller *C* factors, the simulated models tend to filter out small pods of mineralization (zones that one can argue are more uncertain), which is the main reason the mineralized volume of the simulated

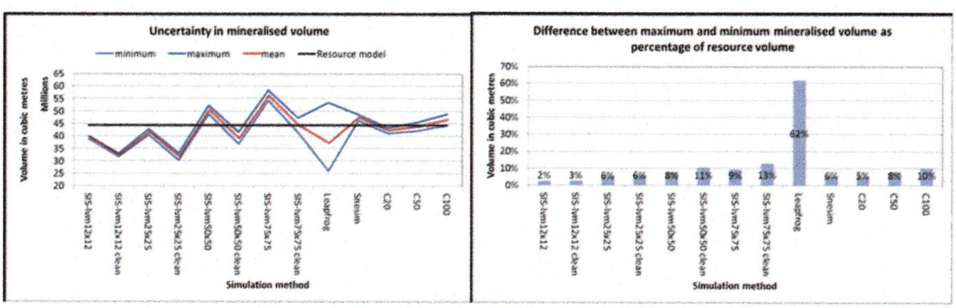

Fig. 7 Uncertainty profile for mineralized volume (*left*) and width of the uncertainty interval for mineralized volume (*right*)

models is lower compared to the mineralized volume in the resource model. For larger C factors, this effect is negated due to the large uncertainty window.

- For snesim the width of the uncertainty interval is approximately similar to SIS-LVM 25 × 25 m.
- The Leapfrog® approach delivers the widest uncertainty interval which none of the other methods are able to match.

In general after simulation of the mineralized volume, simulations of grade are generated in a hierarchical approach to derive the uncertainty interval for metal, tonnes of ore, and average grade for annual/quarterly production volumes. If, for example, the SIS-LVM 12 × 12 m was chosen for the geological simulation, one would tend to think that the resource model was too optimistic. The opposite can be said for the use of SIS-LVM 50 × 50 m, i.e., the model is too pessimistic. If a distance function approach with $C = 50$ was chosen for the geology simulations, one would possibly conclude that the resource model is reasonable in terms of contained volume.

The generation of stochastic simulations of geology that quantify uncertainty can be difficult, and the end result can be heavily influenced by the parameters and method chosen. It is therefore imperative that one understands the impact the choice of method and parameters will have on the uncertainty profile. Most of these methods (except for Leapfrog®) are based on information derived from the existing resource model whether through rock-type proportions or resource model as training image. By doing this there is an implicit assumption that there is some level of confidence in the resource model. For pre-feasibility or feasibility stage projects where there is a lot more data available, this assumption might be justified; however, for early stage projects with far less data, these methods might not be applicable.

3.4 The Subika Deposit

The Subika deposit is the southernmost of the known Ahafo deposits (Fig. 8) and is hosted entirely within the granitoid package in the hanging wall of the Kenyase Thrust. High-grade gold mineralization is focused in a dilatant fracture zone, locally referred to as the magic fracture zone (MFZ). This zone ranges from 1 to 60 m wide with a halo of lower-grade mineralization extending out to 30 m. A number of higher-grade ore shoots, which appear to be controlled by dilatant left-lateral jogs in the MFZ, are recognized and plunge steeply to the southeast.

Quartz-sericite-pyrite and iron-carbonate (QSP-Fe) alteration is the dominant alteration associated with high-grade mineralization. Alteration fluids appear to have accessed the MFZ via a network of shallow angle, brittle fractures within an overall steeply dipping shear zone. QSP alteration intensities are logged as 1, 2, or 3. The combined QSP 2/3 alteration forms the basis of the higher-grade population, while QSP 1 alteration correlates well with the lower-grade population. These two

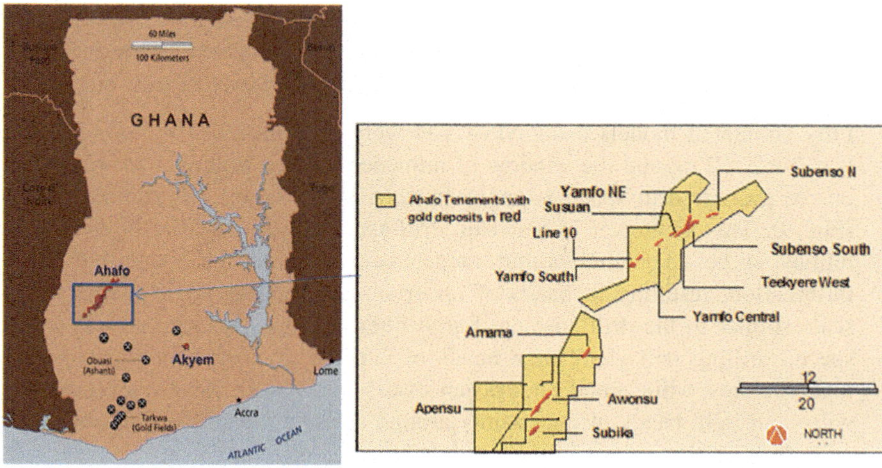

Fig. 8 Location of the Subika deposit in SouthWest Ghana

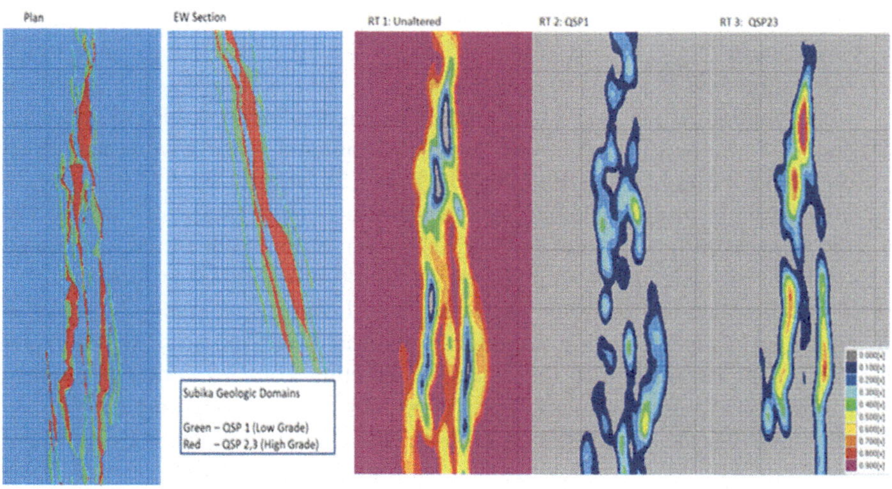

Fig. 9 Subika geology model (*left*) and block probabilities for alteration type (unaltered, QSP1 and QSP23) (*right*)

alteration categories have been used to define the geologic framework used in the resource estimates and are the main focus of this study.

3.5 Simulation Methodology for Subika

For quantification of volumetric uncertainty at Subika, four methods were used, SIS-LVM, the distance function approach, snesim, and Leapfrog®. All the methods except for Leapfrog® used the interpreted geologic model as input (Fig. 9) and attempted to reproduce the target input proportions. For these three methods, the simulated volumes fluctuated slightly above the volume of the resource model, with the exception of one MPS case. The following methods/parameters were employed:

- Fifty simulated models using SIS-LVM were constructed. The node spacing was $2 \times 2 \times 2$ m, and the window of influence applied to the LVM was based on the average drill spacing (approximately 35 m) within the simulated area (Fig. 9). This parameter is based on drill spacing which allows block probabilities to be calculated within ranges equivalent to the spacing of hard information, resulting in bands of uncertainty around geologic features at a scale similar to the drill spacing. Large filter distances associated with wide-spaced drilling will yield large bands of uncertainty around interpreted geologic features, while small filter distances associated with close-spaced drilling will yield tight bands of uncertainty around the same features. This coincides with the idea that uncertainty should decrease as drilling density increases. In addition, anisotropy derived from the variogram model was applied to the calculation of the LVM in order to preserve the preferred orientation of the structure and mineralization. The resulting SIS with the LVM used as control is shown in Fig. 10.
- Fifty simulated models were generated using the distance function approach (only applied to the QSP23 volume). The method being a binary approach cannot facilitate more than one domain. Simulated models were developed on $6 \times 12 \times 6$ m blocks with C factors of 10, 20, and 50 (Fig. 11). Figure 11 also displays the simulated models in 3D, which clearly indicate areas where the mineralization is quite thin (square box). Areas with limited data also tend to show edge effects due to extrapolation of the distance function.
- Three alternate models were generated using Leapfrog®. Three volumes were produced using RBF indicator probability shells derived from conditioning data. Volumes were analyzed based on selecting volumes contoured from a range of interpolated values (P30 to P50). Indicator statistics were analyzed to determine the balanced shell which is the volume that includes as many indicator data misclassified inside the shell as it excludes indicator data misclassified outside the shell and to determine the Russian doll shell where the next larger volume incrementally includes data where the indicator mean is less than the probability of the shell. In order to produce a range of results, an arbitrary smaller (conservative) volume was also selected (Inglis 2013).
- Fifty simulated models were generated using snesim. This case was developed on $6 \times 12 \times 6$ m blocks, using the interpreted geologic model as the training image (Fig. 12). Three different scenarios were developed, where the servo system factor was modified in each run (0.1, 0.5 and 0.9). This parameter controls how snesim reproduces the target input proportions from the TI. The higher the factor, the better the reproduction of the input target proportions. The selection of this value is somewhat subjective and should be chosen with the quality of the geologic model in mind. There are other parameters that have an impact on the result, but they were not tested during this exercise.

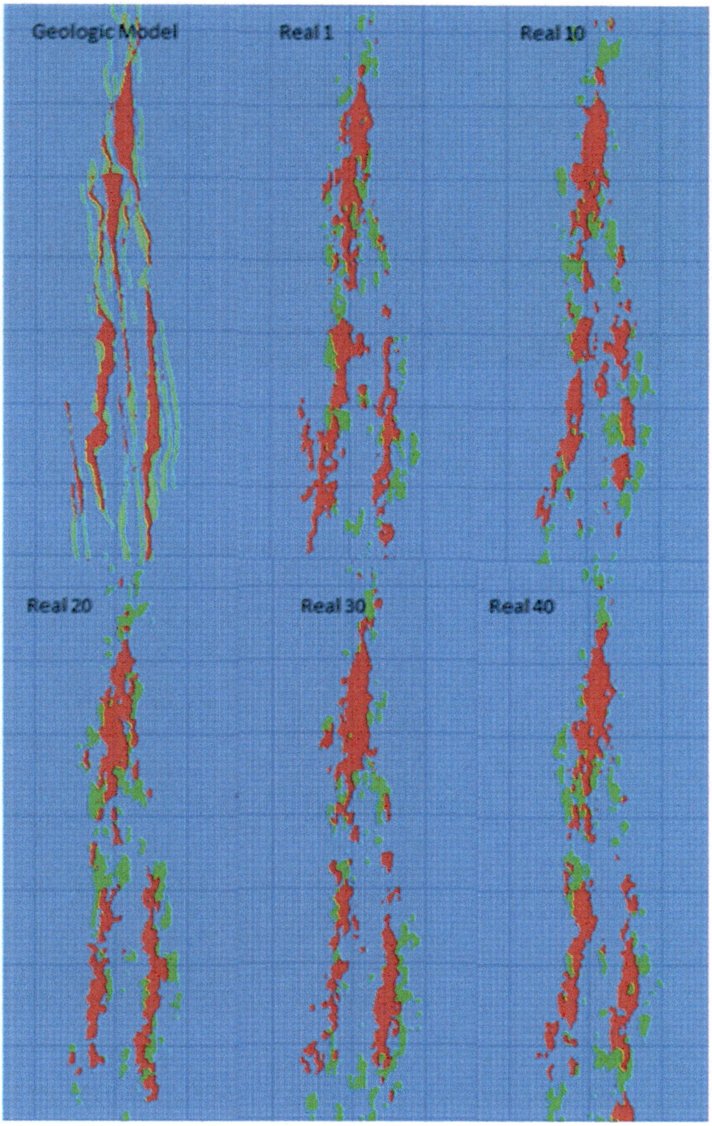

Fig. 10 Geologic model and SIS-LVM results (cleaning set to 1, mild)

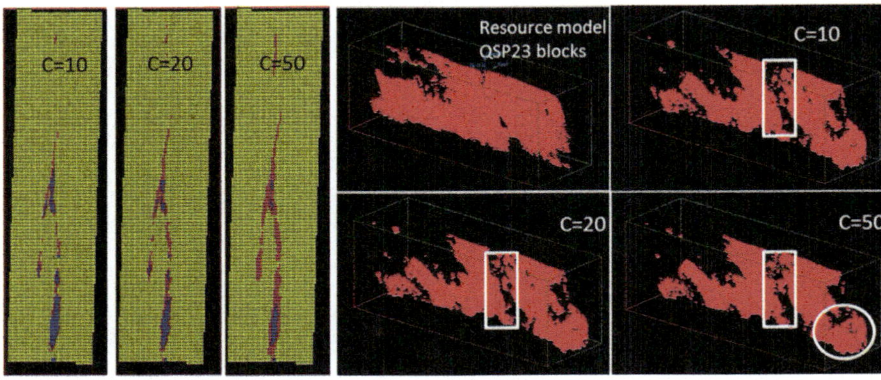

Fig. 11 Interpolated distance function for different C factors (*red*) represents distances between –C and C (plan view) and 3D view of simulated models (for QSP23) generated with the distance function approach (different C factors). Notice edge effects (*circle*) where there is limited data

Fig. 12 Geologic model and snesim (servo system factors (SSF) = 0.1, 0.5, and 0.9)

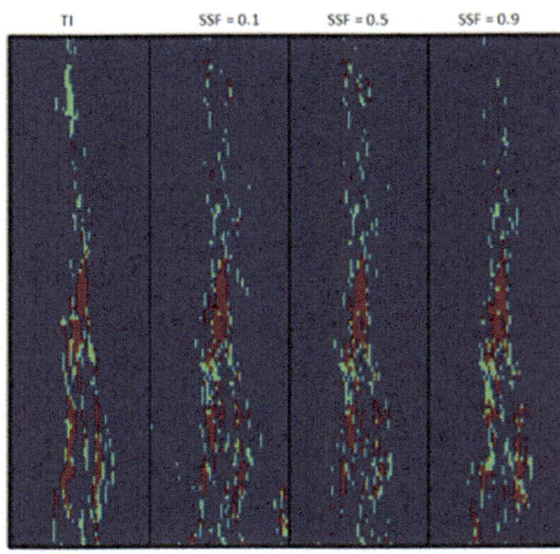

3.6 *Volumetric Uncertainty for Subika (QSP1 and QSP23)*

A summary of the uncertainty profiles for QSP1 and QSP23 is shown in Fig. 13. Note that the SIS-LVM and MPS 0.5 show similar results, while MPS 0.1 results in a wider range of uncertainty and more volume than any of the other techniques. MPS 0.9 on the other hand has a much smaller range of uncertainty and appears to be biased low on volume. It is understandable that the range of uncertainty would diminish with an increasing servo system factor because reproduction of the input target proportions is more strictly enforced; however, the reduction in volume is not so easily explained.

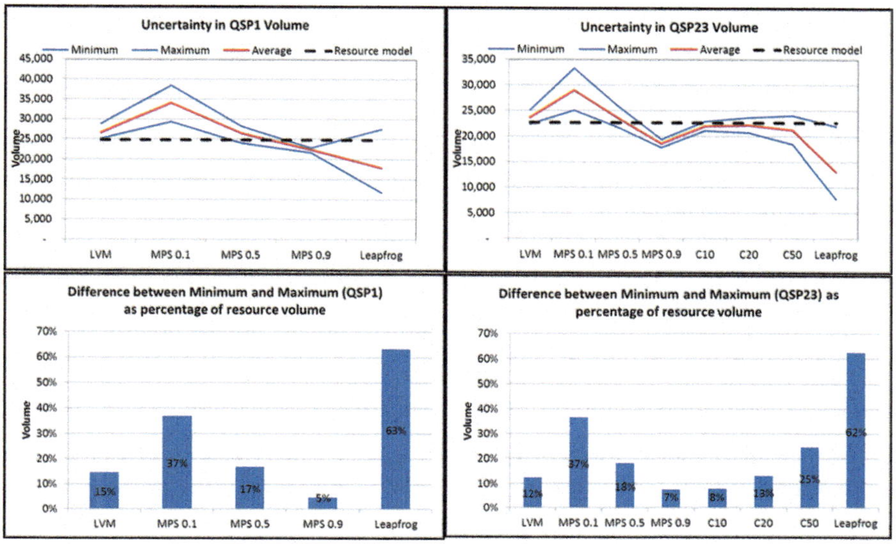

Fig. 13 Uncertainty profile for QSP1 and QSP23 (*top*) with the width of the uncertainty interval (*bottom*)

The results from the distance function approach show larger levels of uncertainty with larger *C* factors. For smaller *C* factors, the simulated models tend to filter out sections where the mineralization is thinner (zones that one can argue are more uncertain), which is the main reason the mineralized volume of the simulated models is lower compared to the mineralized volume in the resource model.

Figure 13 also illustrates the differences between the minimum and maximum volumes of each technique. This shows the variability of ranges of uncertainty depending on technique and parameter selection.

All the techniques yield promising results, and given that it is reasonable to put a high level of confidence on the geologic interpretation, it seems acceptable to try to achieve ranges of uncertainty that fluctuate around the interpretation. In earlier stages of the Subika project, this would not have been the case. Over time tonnages at Subika have shown large fluctuations caused by the wide-spaced drilling and overall complexity of the geologic framework. With drill spacing at ~35 m and a significant amount of thought and effort placed on the geologic interpretation, Newmont is to the point of applying these results to the risk associated with the mine plan.

4 Conclusions and Recommendations

The two case studies have shown that quantification of geological uncertainty for purposes of defining project risk is not trivial. Uncertainty profiles are highly dependent on method and parameter choice:

- For the SIS-LVM approach, the size of the uncertainty window has an impact on the width of the uncertainty profile, with wider profiles resulting in wider uncertainty windows (not necessarily centered on the input resource model) and more dispersed simulations.
- For MPS the servo system factor also appears to control the width of the uncertainty interval with servo system factors closer to one delivering narrower uncertainty intervals.
- For the distance function approach, the width of the uncertainty interval is a function of the *C* factors; larger *C* factors result in wider uncertainty intervals. For both deposits tested, lower *C* factors tended to filter out the thinner more discontinuous mineralized zones generally resulting in tighter uncertainty intervals and slightly less volume compared to the resource model.
- Models built using Leapfrog® showed wider bands of uncertainty.

Most of the methods (except Leapfrog®) require additional information (i.e., proportions of rock types, training image, etc.) derived from the resource model which requires some confidence in the quality of the resource model. At early stages of a project, when there is limited information, this might not be appropriate; instead one might also try to incorporate the uncertainty in the rock-type proportions/training image in the simulated models.

Leapfrog® appears to be a better tool at early stages of a project because it allows for a more complete testing of different geological concepts. In later stage projects where there is an increased level of confidence (due to the amount of data collected) in the geologic interpretation, other methods that achieve ranges of uncertainty around the interpretation likely provide a more realistic assessment of uncertainty. The case studies have also shown that in order to get ranges of uncertainty that fluctuate around the interpretation, the parameters for the various methods need to be selected carefully.

Bibliography

Cáceres A, Emery X, Aedo L, Gálvez O (2011) Stochastic geologic modelling using implicit boundary simulation. In Beniscelli J, Kuyvenhoven R, Hoal K (eds) Proceedings of the 2nd international seminar on geology for the mining industry

Deutsch C (2006) A sequential indicator simulation program for categorical variables with point and block data: BlockSIS. Comput Geosci 32:1669–1681

Guardiano F, Srivastava M (1993) Multivariate geostatistics: beyond bivariate moments. In: Soares A (ed) Geostatistics Troia 1992, vol 1. Kluwer Academic, Dordrecht, pp 133–144

Inglis R (2013) How to select a grade domain-A gold mine case study in exploration, resource and mining geology conference, Cardiff, UK,.21–22 October

Jewbali A, Perry R, Allen L, Inglis R (2014) Applicability of categorical simulation methods for assessment of mine plan risk. In: Proceedings orebody modelling and strategic mine planning symposium 2014, pp 85–98 (The Australasian Institute of Mining and Metallurgy: Melbourne). Reprinted with the permission of The Australasian Institute of Mining and Metallurgy

Lui Y (2006) Using the Snesim program for multiple point statistical simulation. Comput Geosci 32:1544–1563

Munroe MJ, Deutsch C (2008a) Full calibration of C and beta in the framework of vein type deposit tonnage uncertainty, Center for Computational Geostatistics Annual Report 10. University of Alberta, Edmonton

Munroe MJ, Deutsch C (2008b) A methodology for modeling vein-type deposit tonnage uncertainty, Centre for Computational Geostatistics Annual report 10. University of Alberta, Edmonton

Remy N, Boucher A, Wu J (2009) Applied geostatistics with SGeMS: a user's guide. Cambridge University Press, Cambridge, UK, 263p

Stewart M, de Lacey J, Hodkiewicz PF, Lane R (2014) Grade estimation from radial basis functions – how does it compare with conventional geostatistical estimation. In ninth international mining geology conference 2014, pp 129–139 (The Australian Institute of Mining and Metallurgy)

Strebelle S (2002) Conditional simulation of complex geological structures using multiple-point statistics. Math Geol 34:1–21

Wilde JB, Deutsch C (2012) Kriging and simulation in presence of stationary domains: developments in boundary modelling. In: Abrahamsen P, Hauge R, Kolbjørnsen O (eds) Geostatistics oslo 2012. Springer, Dordrech, pp 289–300

Approaching Simultaneous Local and Global Accuracy

Daniel Jasper Kentwell

Abstract When estimating block grades for mining purposes, the currently available methods allow us to maximize the accuracy of either global grade and tonnage curve prediction or local block selection but not both at once. Locally accurate block estimates provide the best result during actual selection and mining but can give highly distorted global grades and tonnages at cutoffs above zero. Globally accurate block estimates provide good prediction of grade and tonnage curves but perform badly during actual selection giving much higher misclassification rates leading to serious degradation of value of the material selected for processing. These statements hold true in varying degrees for all scales and combinations of sample spacing and block size. This paper puts forward a method that retains the properties of accurate global estimation while simultaneously approaching maximum local accuracy.

The process is a simple application of rank and replace combining two estimates, one that targets local block accuracy and one that targets actual block variability. The method is empirically demonstrated using a case study using real data. The conclusion, for this data set, is that local selection accuracy can be greatly improved (but not maximized), in comparison to existing methods, while maintaining grade and tonnage curve accuracy that results from true block variability.

Comparisons with ordinary kriging, sequential Gaussian simulation, turning bands, local uniform conditioning, and ordinary kriging with reduced sample numbers are presented.

1 Introduction

The method described in this paper is a remarkably simple one and it must surely have been used or published previously; however the author cannot find any references to anything similar. The underlying estimate is conventional ordinary kriging (OK) with an optimized search neighborhood for the specified block size designed to maximize

D.J. Kentwell (✉)
SRK Consulting, Level 9, 99 William Street, Melbourne, VIC 3000, Australia
e-mail: dkentwell@srk.com.au

J.J. Gómez-Hernández et al. (eds.), *Geostatistics Valencia 2016*, Quantitative Geology and Geostatistics 19, DOI 10.1007/978-3-319-46819-8_16

local block accuracy. The OK block grades are then ranked in grade order. A second estimate, using any method that targets true block variability (global change of support, simulation, degraded neighborhood kriging, etc.), is made and the block grades also ranked. The grades from the second estimate are then assigned to the locations defined by the OK estimate by rank order. This happens to be a direct generalization of the LUC method proposed by (Abzalov 2006). For the purposes of this paper, we will call the general method rank and replace (RR).

Some other methods of smoothing reduction and or correction that have been proposed are (Journel et al. 2000) using spectral methods and (Richmond et al. 2009) using an affine correction. Topical papers around the balance between the reproduction of global variability and the reproduction of local block accuracy required for different stages of a project are Krige (1951, 1994, 1996) and Isaaks (2005).

It will be shown in the case study that the resulting block RR estimates now have (our best estimate from available data of) the true block variability globally as well as local selectivity very close to that obtainable by OK and are superior to that obtainable by any simulation methods.

Hence, in the mining context, we now have a block model that can be used for mine planning purposes that has (our best estimate from available data of) the actual global block grade, tonnage, and conventional profit curves as well as (our best estimate from available data of) local block accuracy.

2 Case Study

2.1 The Data Set

The data set is a real gold data set from a mined-out Australian opencut mine. It consists of a widely spaced exploration drilling data set at approximately 12 m by 25 m by 1 m spacing and a corresponding close-spaced grade control data set at approximately 2.5 m by 4.0 m by 2.5 m spacing. The results presented are block models calculated from the wide-spaced exploration data with block models from the grade control data as reference.

2.2 Questions of Scale

The typical mining rule of thumb is that, for good local estimates, block sizes should approximate or be no less than half of the data spacing. This often results in blocks that are much larger than the anticipated selective mining unit (SMU) block size. Kentwell (2014) describes and quantifies some of the issues around ordinary kriging, smoothing and block size related to this and how estimates of tonnage and grade at higher cutoffs can significantly depart from the true tonnages and grades.

In order to evaluate the potential benefits of the RR method, block models at both SMU scale, 2.5 m by 4 m by 2.5 m, and data scale, 10 m by 24 m by 10 m, were compared during the study. In the interest of brevity, only the results for the SMU scale models are presented in detail here.

2.3 *Process*

For the SMU scale block model, the estimates as in Table 1 were produced. For the sake of direct comparability of methods, the variogram used for all estimates was that derived from the grade control data set with ranges and relatives nuggets/sills held constant. Both the exploration and grade control data were composited to 3 m intervals. De-clustering weights, top cutting, and restricted searches for high grades were used as appropriate to ensure all estimates validated and that average grades were within an acceptable tolerance at zero cutoff. Two RR estimates are calculated and compared, one derived from a Gaussian anamorphosis with global change of support and one derived from a single simulation.

For the RR estimate (Ana RR) derived from the global change of support, the individual ranked block grades were calculated by the following procedure:

1. Calculate the block histogram via Gaussian anamorphosis and global change of support using the exploration samples and the variogram.
2. Report the tonnage curve at 200 evenly spaced cutoffs across the full range of grades.
3. Export the cutoffs and corresponding tonnages to a curve fitting software and fit the data. In this case a Savitzky–Golay smoothing (Savitzky and Golay 1964) (a form of moving window polynomial) fit was used.

Table 1 Estimation methods

Method	Data set	Abbreviation	Comments
Ordinary kriging	Exploration	EXP 80	Optimal neighborhood to maximize regression slope and local accuracy – 80 samples
Ordinary kriging	Exploration	EXP 6	Degraded neighborhood to approximate true block variability – 6 samples
Sequential Gaussian simulation	Exploration	SGS	Conditional simulation
Turning bands	Exploration	TB	Conditional simulation – 400 bands
Local uniform conditioning	Exploration	LUC	Localized using $10 \times 24 \times 10$ m panels and 45 cutoffs localized with EXP 80
Rank and replace	Exploration	Ana RR	From global change of support and EXP 80
Rank and replace	Exploration	SGS RR	From SGS and EXP 80
Ordinary kriging	Grade control	GC 32	Optimal neighborhood to maximize regression slope and local accuracy – 32 samples
Ordinary kriging	Grade control	GC 4	Degraded neighborhood to approximate true block variability – four samples

4. Split the fitted curve into as many tonnage increments as there are blocks in the model. The resulting ascending cutoffs become the new ranked block grades.
5. Import these ranked anamorphosis grades back to the underlying OK block model using the OK rank identifier to locate each new anamorphosis grade of the corresponding rank.

This process can also be completed using the local uniform conditioning (LUC) tools in the Isatis software package but with the panel model being a single block covering the entire domain. While not detailed in this paper, the author has replicated the Ana RR results using single panel LUC in Isatis.

For the RR estimate, SGS RR derived from the sequential Gaussian simulation (SGS) the SGS grades and underlying OK grades that are both ranked and then the SGS grades that are assigned to the OK locations based in the rank identifier. The problem with deriving the grades from a set of simulations is choosing which individual simulation as they will all be a little different.

With the exception of the splitting and interpolation of the tonnage curve from the change of support anamorphosis into block size increments, all calculations for both the Ana RR and the SGS RR were performed in the Geovariances Isatis software package.

2.3.1 What Is Reality?

Even though we have a close-spaced reference data set, in this case with approximately one grade control sample within one SMU sized block and the blocks sizes and data spacing are well inside the range of the variogram, we still need to estimate the SMU model block grades. Even with this well-informed data and block configuration, significant smoothing still occurs during OK. The question is then which estimation parameters and/or method creates the closest approximation to reality to use as our reference model? We have the same problem as the one we set out to solve only at a smaller scale. At the risk of getting swamped with data, we will examine two versions of "reality." The first is the block model that results from grade control data estimated by OK using a degraded estimation neighborhood selecting only 4 (GC 4) surrounding samples and intended to reproduce global block variability. The second is the block model that results from grade control data estimated by OK with an optimum neighborhood selecting 32 (GC 32) surrounding samples and intended to insure local block accuracy.

2.4 Results

2.4.1 Prediction vs Performance

In comparing the results of the different estimation methods, we will talk about both prediction and performance. These terms are specifically defined here.

Prediction is what the models say will happen, for example, tonnages and grades at different cutoffs. This can be compared to the prediction of our reference model. Prediction is inherently global.

Performance is the interaction with our reference "reality." In other words if a particular method's block model was used to actually select/reject material at specific cutoffs, then what grades and tonnages actually result by applying that selection to the reference model? Performance is local in that it is a direct block-by-block comparison albeit that the net result is expressed as a single correlation coefficient or a cutoff curve.

In the normal case of events during planning and prior to mining, we only ever have predictions. Only after the fact and after mining and processing do we get any real information on performance.

2.4.2 Statistics

The block model statistics of the different estimation methods and correlations with the two grade control reference models are shown in Table 2. The results are listed in order of their correlation with the reference models. Note that the two RR methods produce correlations with both reference models that are very close but just less than the correlation resulting from optimal OK. As expected degraded OK,

Table 2 Statistics for SMU model estimates

Method	Mean	Variance	Correlation with GC 32	Correlation with GC 4	Min	Max	Comment
GC 4	1.72	3.75	0.88	1.00	0.00	38.56	Reference model 1
GC 32	1.70	2.17	1.00	0.88	0.00	18.25	Reference model 2
EXP 80	1.73	0.51	0.57	0.45	0.40	10.67	Best correlation with reference models 1 and 2
SGS RR	1.73	1.66	0.56	0.45	0.01	19.12	Close to best correlation with reference models 1 and 2
Ana RR	1.76	2.50	0.56	0.44	0.15	27.96	Close to best correlation with reference models 1 and 2
EXP 6	1.67	0.89	0.51	0.40	0.07	14.06	
LUC	1.74	2.13	0.41	0.33	0.07	12.61	
SGS Sim n15	1.70	1.57	0.27	0.22	0.01	19.12	n 15 (of 100) selected for best match to sample mean
TB Sim n82	1.70	2.69	0.26	0.21	0.00	23.92	

Table 3 Kriging regression slopes

Method	Kriging regression slope	Method	Kriging regression slope
GC 4	0.75	EXP 6	0.47
GC 32	1.01	EXP 80	0.75

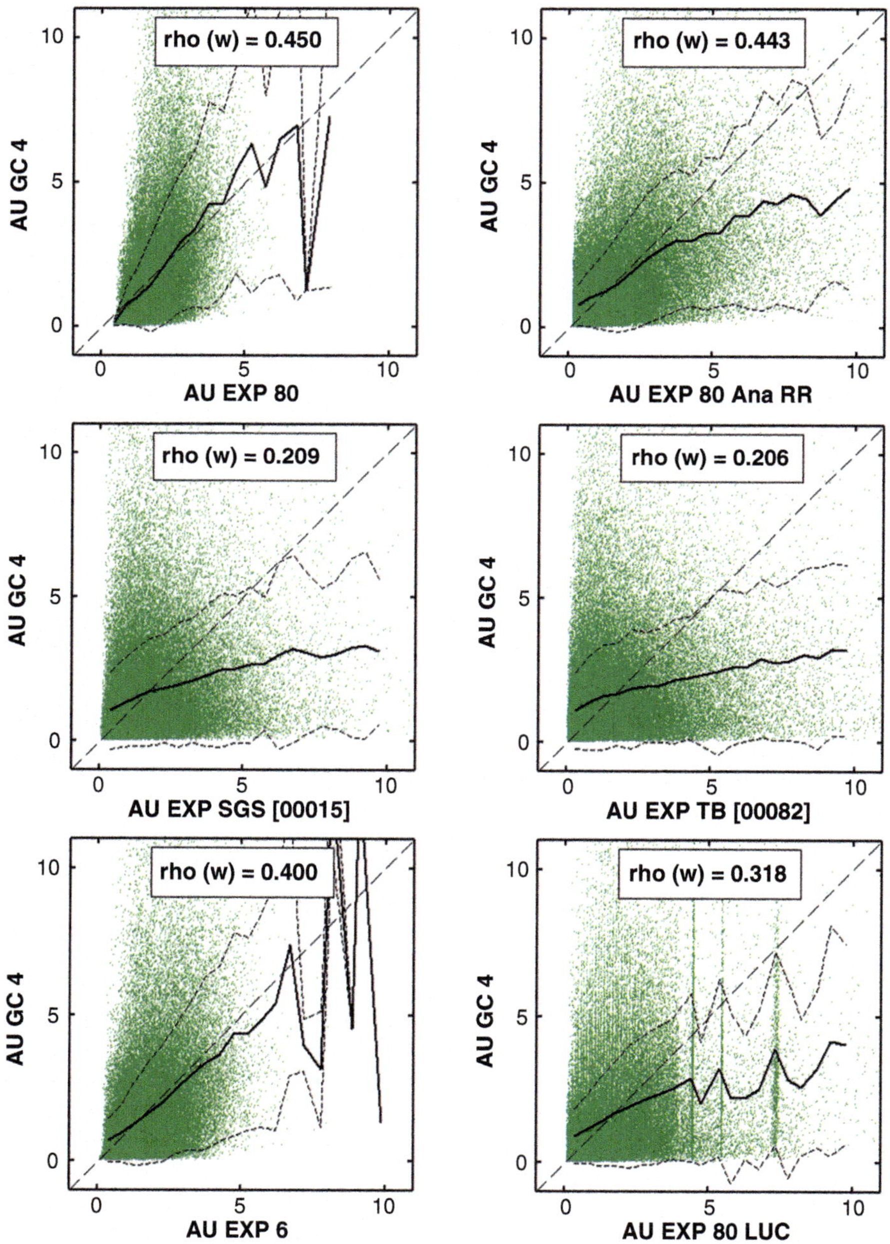

Fig. 1 Scatterplots relative to GC 4

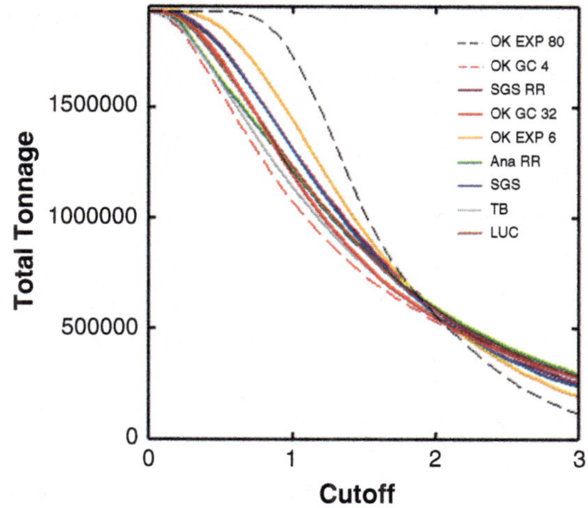

Fig. 2 Prediction tonnage curves

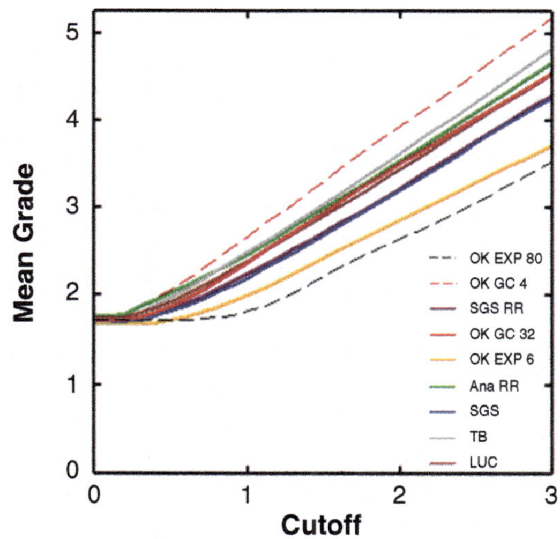

Fig. 3 Prediction grade curves

SGS, and TB produce much lower correlations. Interestingly the LUC correlation is in between optimal OK and the simulations but not as good as the RR methods. Table 3 shows the kriging regression slope averages to put the relative qualities of the kriging estimates in perspective.

Figure 1 shows the scatterplots of the correlations relative to GC 4 together with the first bisector, conditional expectation, and one standard deviation from the conditional expectation (dotted lines). Note that although the correlations for EXP 80 and Ana RR are similar, Ana RR shows a much greater spread.

2.4.3 Prediction

Although difficult to see at the scale in this document, Figs. 2, 3, 4, and 5 show the range of grade, tonnage, metal, and profit curves bounded by ordinary OK at one

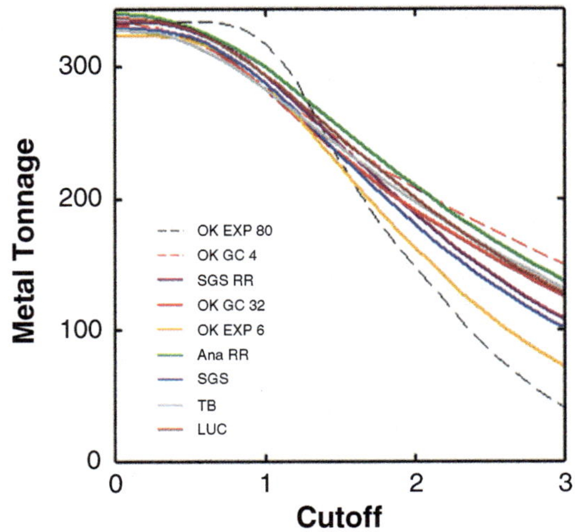

Fig. 4 Prediction metal curves

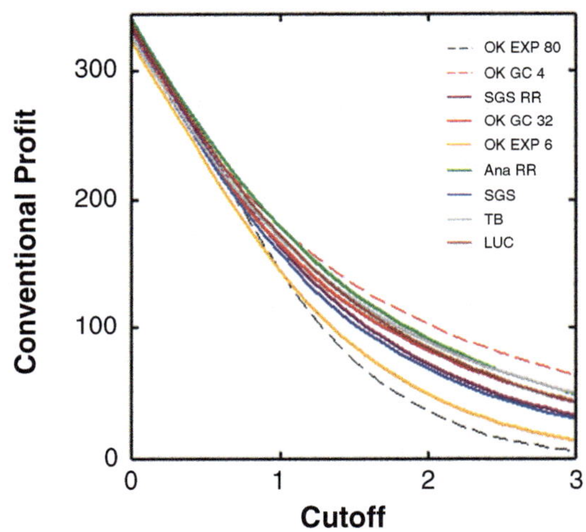

Fig. 5 Prediction profit curves

extreme and the GC 4 reference model at the other. For the grade and tonnage curves, despite some crossovers, the order, getting closer and closer to GC 4, is as follows: EXP 80, EXP 6, SGS, SGS RR, LUC, Ana RR, GC 32, and TB. The main point to take from these results is that, as expected, every other method is a better global predictor than optimal OK in that all other grade, tonnage, metal, and profit curves are closer to those of GC 4 which is our proxy for reality.

The profit curve is calculated as:

$$\text{Profit} = \text{tonnage above cut off} * (\text{grade above cut off} - \text{cut off grade}).$$

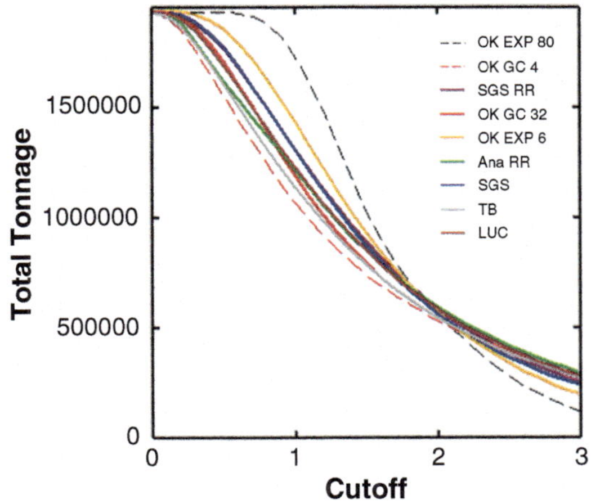

Fig. 6 Performance tonnage curves

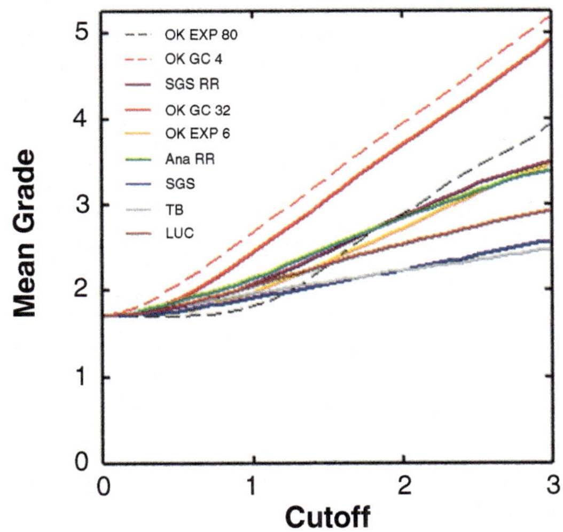

Fig. 7 Performance grade curves

3 Performance

This is where things get interesting. Performance here is with reference to GC 4 as our proxy for reality. The tonnage curves (Fig. 6) show OK EXP 80 to be the worst performer and Ana RR and SGS RR to be close to the reference curves. The grade curves (Fig. 7) show that Ana RR is the best performer up until approximately the average of the model then EXP 80 takes over as the best performer. Looking at the metal curves (Fig. 8), there is no clear best performer, but TB appears to be the worst overall. Finally the profit curves (Fig. 9) show that optimal OK 80 and Ana RR are almost exactly the same and are closest to the two reference models.

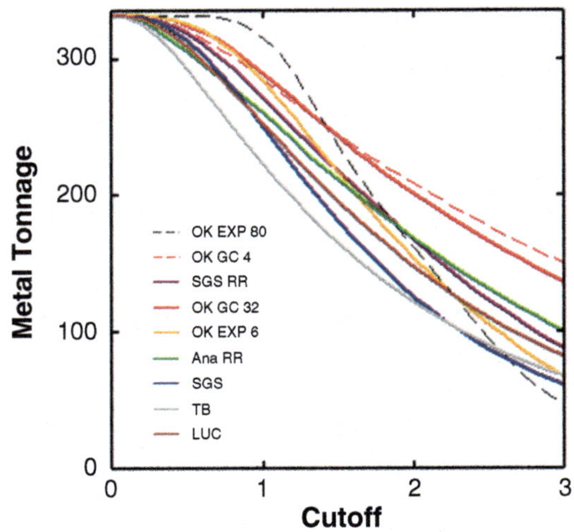

Fig. 8 Performance metal curves

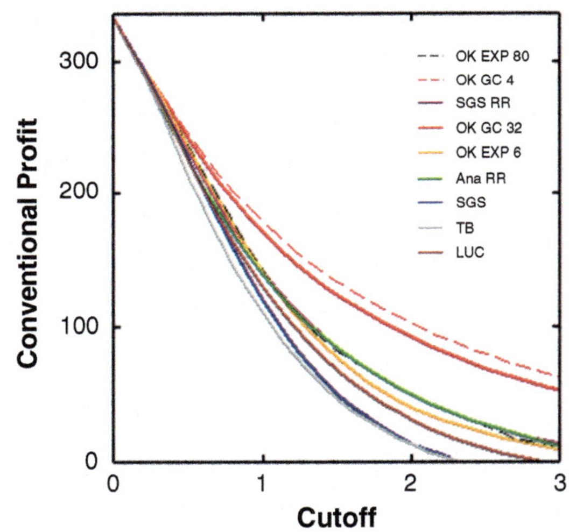

Fig. 9 Performance profit curves

4 Texture

There is another attribute of block estimates that is difficult to capture in a variogram or statistic or cutoff curve and that is the multipoint relationships or the "texture" of the grade patterns. We expect smoothing from OK estimates and we expect higher variability from simulations. Figures 10 and 11 show the textures for a long section slice through the center of the model. Note that the RR method still produces smooth textures and does not compare well with the GC 4 in terms of texture reproduction.

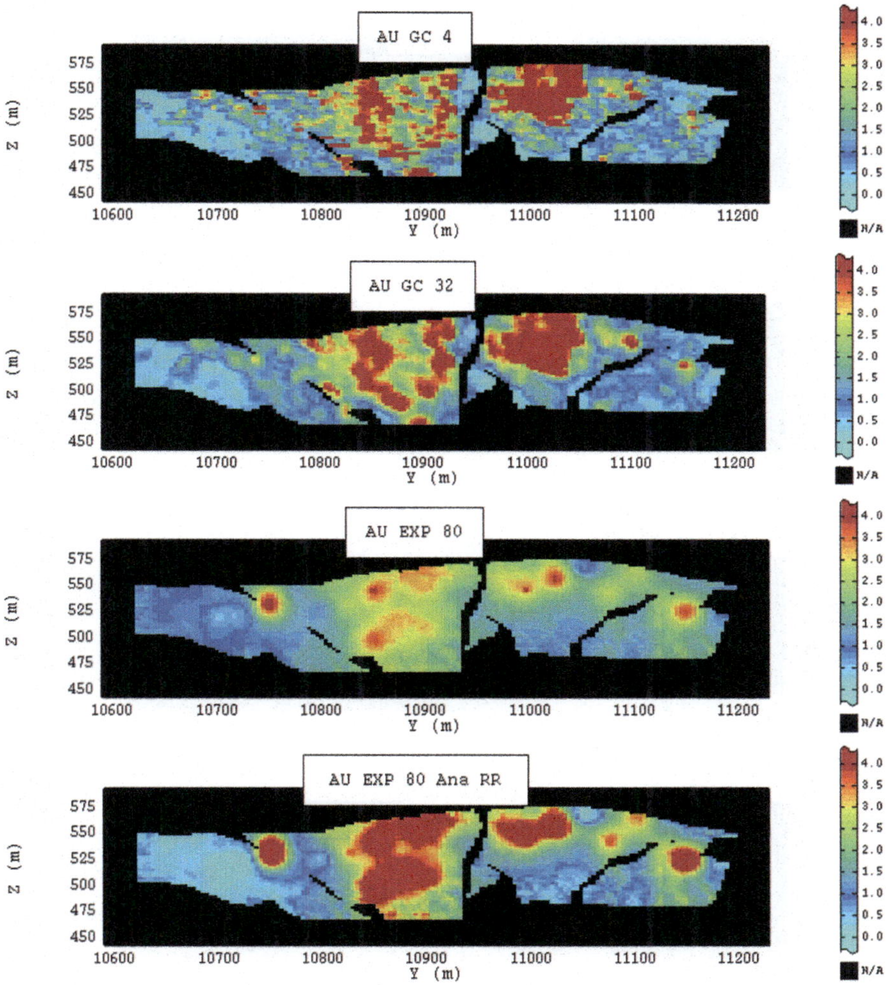

Fig. 10 Texture long sections 1

5 Conclusions

The RR method does not make small block estimation any more locally accurate than OK estimates. Small block OK and RR estimates from widely spaced data retain on average a similar level of confidence as measured by the kriging regression slope or kriging efficiency of the underlying OK estimate. A poor quality local block OK estimate is still a poor quality local block RR estimate.

The advantage of the RR method is global grade, tonnage, and profit curves are closer to reality compared to optimal OK estimates and that the estimate retains on average a level of block selectivity performance comparable to optimal OK. Also a

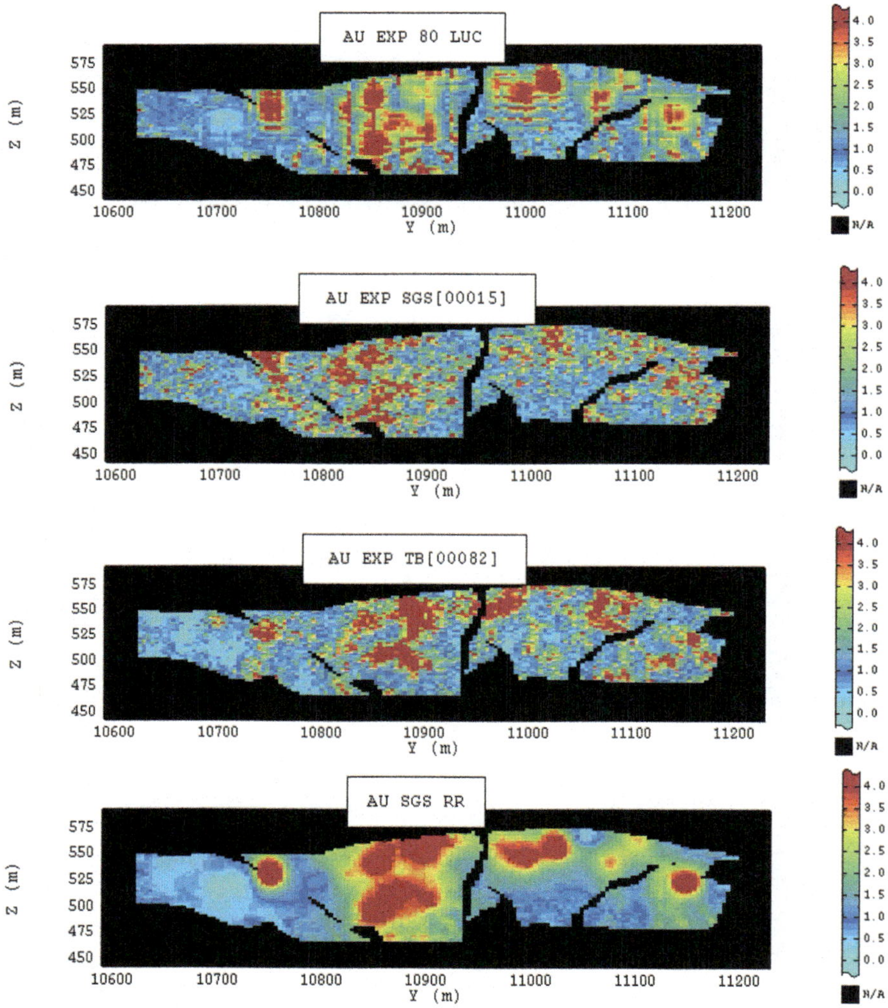

Fig. 11 Texture long sections 2

single model is available for mine planners to work with that therefore has the best available estimate both locally and globally. The RR method is better adapted to mine planning than either optimal OK or individual simulations. The RR method is effectively an alternative implementation of LUC using a single panel. It gives similar global predictions to multi-panel LUC and, for this case study, local performance which is closer to reality than multi-panel LUC.

With RR estimates the spatial "texture" of the block grades is still smooth and does not display the true texture at block scale. Approximations of true texture are better estimated by simulations (SGS or TB for example) with the inherent loss of local accuracy.

Acknowledgments The author would like to acknowledge Daniel Guibal's contribution to the ideas behind this paper. When the author originally discussed the concept for the method with Daniel, it transpired that he had already been thinking along the same lines and suggested the much more elegant process of taking the final grades directly from the global change of support anamorphosis rather than from a simulation or degraded kriging estimate. This resulted in the Ana RR methodology detailed in this paper.

Bibliography

Abzalov M (2006) Localised Uniform Conditioning (LUC): a new approach for direct modelling of small blocks. Math Geol 38(4):93–411

Isaaks E (2005) The kriging oxymoron: a conditionally unbiased and accurate predictor (2nd edn). In: Geostatistics Banff 2004. Springer, Dordrecht, pp 363–374

Journel AG, Kyriakidis PC, Mao S (2000) Correcting the smoothing effect of estimators. Math Geol 32(7):787–813

Kentwell D (2014) Aligning resource estimates with mine planning. In: Orebody modelling and strategic mine planning. The Australasian Institute of Mining and Metallurgy, Perth, 185–192

Krige DG (1951) A statistical approach to some basic mine valuation problems on the Witwatersrand. J Chem Metall Min Soc S Afr 52(6):119–139

Krige DG (1994) An analysis of some essential basic tenets of geostatistics not always practiced in ore valuations. First regional APCOM, Slovenia, pp 15–28

Krige DG (1996) A basic perspective on the roles of classical statistics, data search routines, conditional biases and information and smoothing effects in ore block valuations. Conference on mining geostatistics, Kruger National Park Sept. 1994. UNISA

Richmond A, Gaze A, Horton J (2009) Desmoothing block models for underground mine design. Seventh international mining geology conference. The Australian Institute of Mining and Metallurgy, Perth, pp 265–268

Savitzky A, Golay MJ (1964) Smoothing by differentiation of data by simplified least squares procedures. Anal Chem 38(8):1627–1639

Geostatistics for Variable Geometry Veins

Alfredo Marín Suárez

Abstract This paper presents the idea of applying a modified "moving trihedral" borrowed from the abstract Theory of the Geometry of Riemann, allowing us to model the random functions $L^2(\Omega, \sigma, P)$ of geostatistics, in a special 2-D plane within such trihedral, a feature not available in any software existing in the market. Thus, from working in the R^2 space in that trihedral plane, we obtain our results in the R^3 physical space, in which we can then derive the desired linear or nonlinear geostatistical results. The method presented here therefore can be considered akin to a kind of spatially adaptative geostatistics. And so, among other applications of this method, it becomes possible to apply geostatistics simply, without the need for a detailed, three-dimensional geological model of the mineralization, defining instead its contours in a simple plane, a feature very useful when modelling irregular veins.

1 Introduction

When it comes to performing geostatistical studies in vein-type deposits with variable widths, building the three-dimensional geological model that defines the domain for applying geostatistics is both difficult and potentially inaccurate. This is due to the limited extent of channel sampling, generally limited to some galleries, and all the more so when there are no exploration drillholes at all nor any type of exploration of the mineralized zones.

To this must be added the problem of the erratic irregularity of the veins, with abrupt changes of azimuth dip, and thickness, as observed when walking through the narrow, polymetallic veins of copper, lead, zinc and silver mineralization at

A. Marín Suárez (✉)
Docteur Ingénieur en Sciences et Techniques Minieres – Option Géostatistique École Nationale Supérieure Des Mines De Paris, Paris, France

Geostatistician Consultant, Lima, Peru

National University of Engineering, Lima, Peru
e-mail: alfredomaringeo@hotmail.com; http://www.geoestadistica.org

J.J. Gómez-Hernández et al. (eds.), *Geostatistics Valencia 2016*, Quantitative Geology and Geostatistics 19, DOI 10.1007/978-3-319-46819-8_17

Casapalca, Yauliyacu, Morococha, San Cristobal, Yauricocha and Huarón, through the filonian gold deposits of Marsa, Horizon, Powerful and Acari in Peru or through the silver deposits of Navalmedio in Almadén-Spain. Even in cases where sufficient information is available, the generally available software on the market is designed for massive deposits and calls for the definition of three-dimensional blocks hardly adaptable to the variable widths of narrow veins about 5–20 cm in lateral extent.

The geological framework of reference for this work has been that of the narrow veins system of the following deposits:

- The Peruvian polymetallic deposits, where the stratigraphic sequence of the district is made as much of sedimentary rocks as of interstratified volcanics, with ages varying from late Cretaceous to Quaternary (Bateman 1982).
- The gold vein deposits of Marsa and Poderosa in the Pataz Batholith, North-Central Peru and belonging to the late Paleozoic.
- The Navalmedio deposit in Spain, with galena mineralisation, and not much silver. The host rocks are pre-Ordovician slates of the Schist-Graywacke Complex. It is a complex vein deposit.

The general theoretical mathematical frame is shown in Annex 1 and the proposed theoretical frame in Annex 2.

2 Methodology for Practical Application of Proposed Trihedral

The previously presented version of the applied methodology, used in the veins of Centro Peru 1983 and in the vein of Navalmedio de Minas Almadén-España 1984, was programmed in Fortran and Assembler, with its respective graphic limitations. Said version was presented at Fontainebleau's Cycle of Formation Specialisée in Geostatistics as a graduate course, in the presence of Dr. Georges Matheron, who authorised its publication in 1986 (Marin 1986). The present version is based on the use of modern graphic 3-D and 2-D software, available in the market.

This methodology was applied as a consequence of irregularities in the vein thickness, strike, dip and local domains, resulting from existing geological controls in use at the aforementioned deposits. This can be seen in Fig. 1.

Let us bear in mind that vein systems are very complex (Bateman 1982) and that the construction of geological model (three-dimensional) is difficult, somewhat inaccurate and often uncertain, due to the fact that its elaboration is subjective and that it takes a lot of time, because of the unavailability of required drills, necessary to identify what is happening between one section and the other, for example, the events that take place between a geological section that corresponds to a left chimney of a given cut and another geological section that passes through the right chimney of the cut understudy.

Let us see Fig. 2.

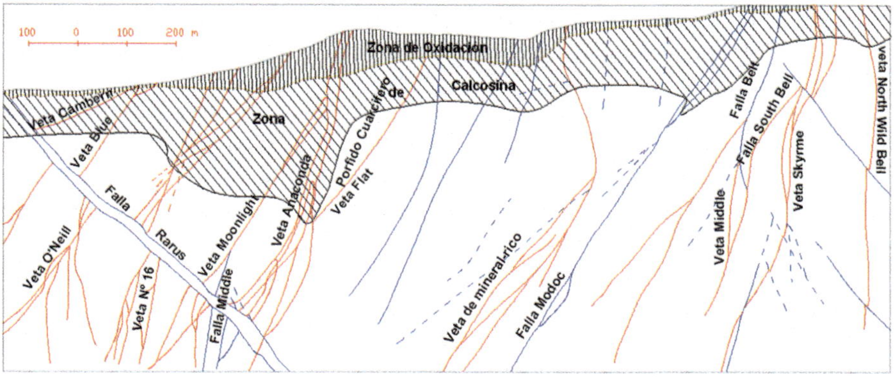

Fig. 1 A geological cross-section of a vein system with its frequent complexity, (Bateman 1982)

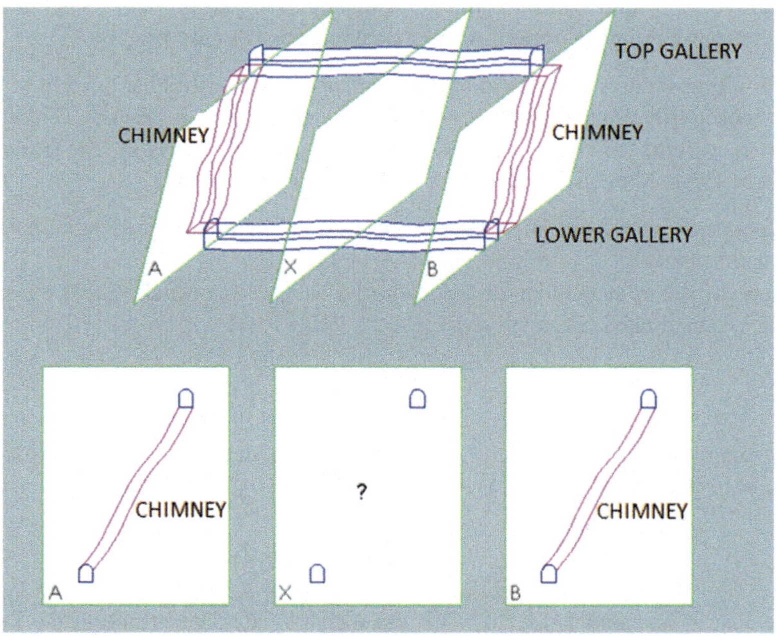

Fig. 2 Mining Stope

Sample channels and/or drill samples with its grades and real vein width, resulting from mining activity, are identified and matched with points x, y, z, corresponding to their gravity centre.

- Capping values are treated.
- "Decluster" to solve or prevent potential.
- Grouping/clustering of samples is made.

- A provisional model block is created, to include all domain sample applications. This model contains rotated and shifted coordinates in accordance with average strike and dip of vein domain.
- A geological model or auxiliary wireframe is created, in line with the initial block model, delimiting its perimeter by means of a desktop mouse and paying attention to loopholes in channel data, drill samples, lithology, geological controls and controls of structural type which must be assessed by the mine geologist in charge.
- Sample points together with their grade and width vein are projected on a single face of the predetermined geological or wireframe model
- We now have the samples on a flat surface that represent the proposed trihedral's tangent of the differentiated variable vein.
- Finally, we can work on this tangent plane with a R2 geostatistics, representing regionalized variables of accumulated grades (grade by vein width) now totally additive, also representing vein widths through random functions Z(x) in L2(Ω, σ, P) applying linear or nonlinear geostatistics as the case may be.

We will now show a practical application as defined within the tangent plane of the proposed trihedron. We want to estimate the gold grade of the ABCD block in a mineralised vein; see Figs. 3 and 4, using ordinary kriging (Marín 1986; Guibal and Remacre 1984; Maréchal et al. 1978; Chilès and Delfiner 2001).

P_i: Thickness of the vein in channel at location i is measured perpendicular to the tangent plane.

l_i: Grade of the vein in channel at location i. The average thickness in the CDEF block is estimated by:

$$P^*_{\text{CDEF}} = \lambda_1 P_1 + \lambda_2 P_2 + \lambda_3 P_3$$

The average accumulation and the whole chapter (Marin 1986) (grade-thickness product) in the CDEF block is estimated by

$$(\text{LP})^*_{\text{CDEF}} = \lambda'_1 P_1 l_1 + \lambda'_2 P_2 l_2 + \lambda'_3 P_3 l_3$$

where $\{\lambda_1, \lambda_2, \lambda_3\}$ and $\{\lambda'_1, \lambda'_2, \lambda'_3\}$ are the weights calculated through the kriging geostatistical technique.

Therefore, the estimated grade of the CDEF block (Marin 1986) is given by

$$L^* = \frac{(\text{LP})^*_{\text{ABCD}}}{P^*_{\text{ABCD}}}$$

In other words, we have estimated a block located in R3 from an estimation made in R2.

- A way of transforming the kriging variance, as a percentage measurement of the relative quality of the estimate of each unit block, is shown as follows:

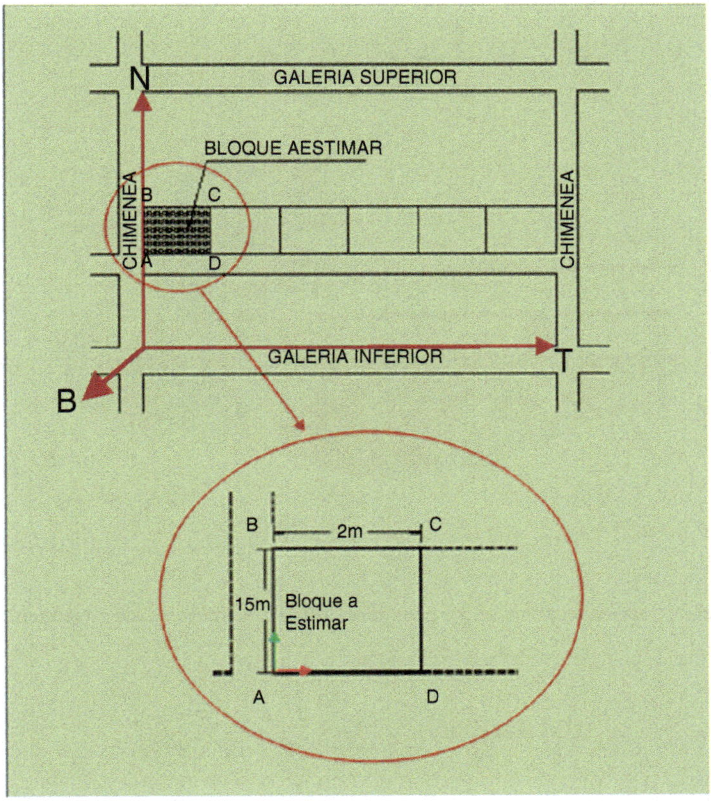

Fig. 3 ABCD, unit block in the plane

Fig. 4 CDEF, unit block in the plane, to be estimated from thickness and grade data

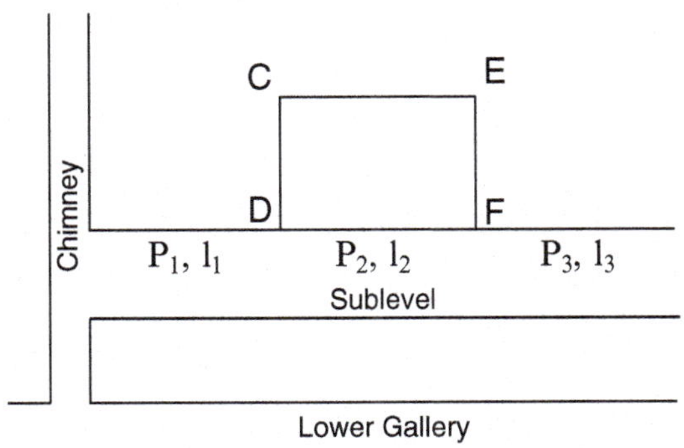

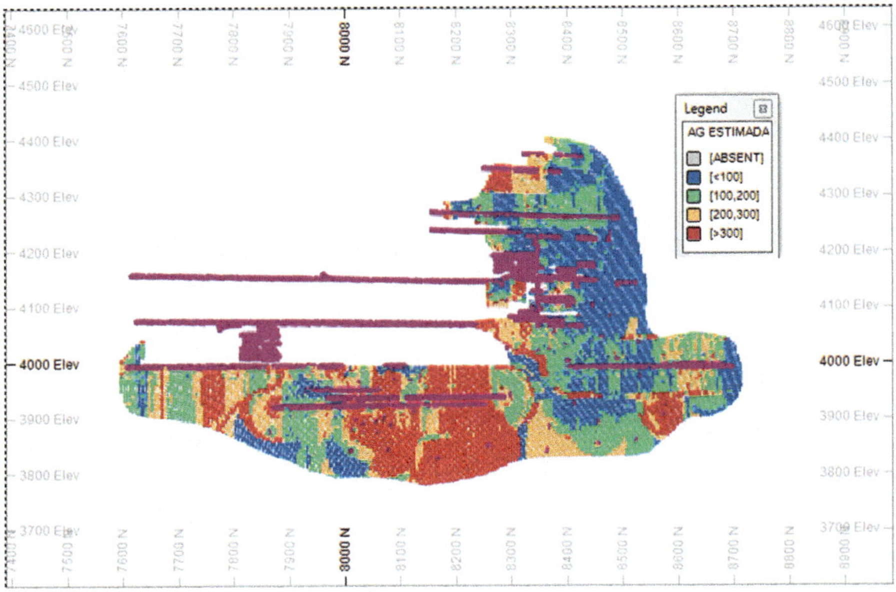

Fig. 5 Vein 11 estimated grade of Ag ppm, blocks 5 × 5, Morococha and Argentum S.A. Peru, deposits

$$\text{CM } (\%\text{quality}) = 100 \times \left(1 - \frac{\sigma^2}{\max(\sigma^2)}\right)$$

where:

- σ^2 is the estimated variance of a particular block.
- $\max(\sigma^2)$is the maximum value of the group of variances of the estimated blocks.

thus making it possible to have relative quality planes of the estimate, expressed percentage-wise: see Figs. 5 and 6.

3 Discussion and Conclusion

3.1 Practice

1. A new way of applying geostatistics is set forward, not in a fixed coordinate system, as software that is designed for massive deposits usually works (additionally forming blocks in the same manner as in massive deposits) this is, that a system of variable coordinate relative to the origin of the proposed trihedral, with its tangent plan, that is to say, trihedrals that allow a follow-up of vein inflexions, by continually changing the coordinate system – "pursuing" the veins

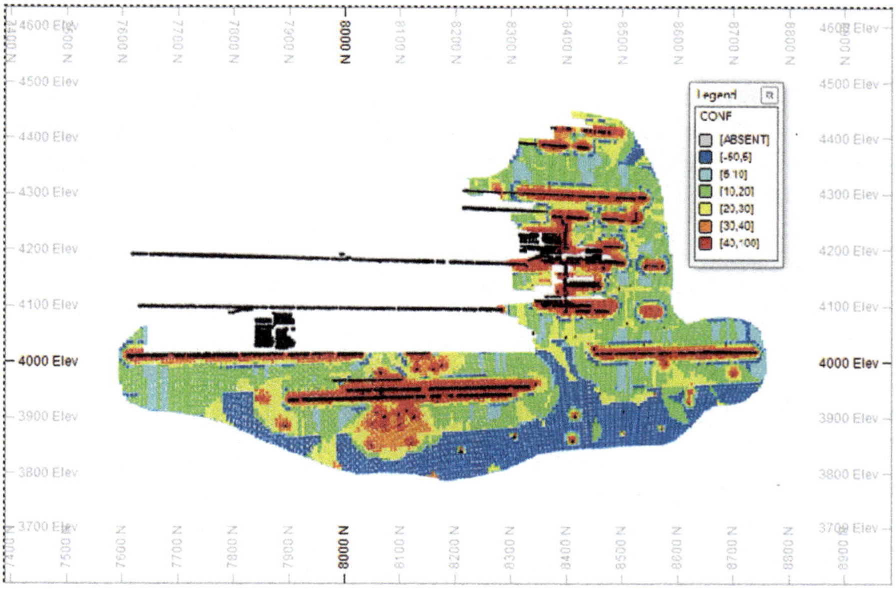

Fig. 6 Vein 11 percentage quality of the estimate of Ag ppm, blocks 5 × 5 Morococha and Argentum S.A. Peru

and their workings – with even better results when these are very thin veins as, for example, veins that average 5 cm. or so.

2. It must be pointed out that the proposed trihedral is different from Frenet and Serret standard trihedral of manifold geometry.
3. The use of the tangent plane with its proposed mobile trihedral precludes the building of the geological model as conceived in its classical form by three-dimensional software available in the market. In effect, building geological models in veins, as conceived by three-dimensional software, is difficult and inaccurate, since appropriate information for its construction is not readily available. In this case, geological modelling is limited to defining, in the tangent plane, the outline of a domain, or domains corresponding to the mineralized zone, i.e. simple lines drawn with a standard "mouse" on the tangent plane, in other words, in a simple, immediate and interactive way.
4. It must be noticed that, here, the condition is that the tangent plane contains, in the best way possible, the surface part of the vein, calculated by vectors, by the minimum square method, or with the help of block models in 3-D software available in the market.
5. This way we have the design of the methodology that will help perform paragenesis studies and estimate or simulate short- and midterm resources so that the mining engineer may proceed to calculate the exploitable reserves, in the same manner as the author of this publication, acting as a consultant, applied to mining companies. Examples of the foregoing are polymetallic veins of Cu, Ag, Zn and Pb at Morococha of the Cia Argentum S.A. and polymetallic veins of Raura of Cia Minsur S.A. deposit, both located in the Peruvian Andes.

3.2 Theory

With this work, we hope to open a field of research in theoretical and practical applications in mining, GIS and other regionalised variable domains, reframing the algorithms of current geostatistics with Riemannian geometry metrics and building Riemannian geostatistics.

Acknowledgements The author would like to acknowledge Dr. Dominique François-Bongarçon, President, AGORATEK International Consultants Inc., for his useful comments.

Annex 1: Frame Theory Riemannian Manifold

Differentiable Manifolds

A differentiable manifold is a type of manifold (In mathematics, a manifold is a topological space that locally resembles Euclidean space each point. More precisely, each point of an n-dimensional manifold has a neighbourhood that is homeomorphic to the Euclidean space of dimension n.) that is locally similar enough to a linear space to allow one to do calculus. Any manifold can be described by a collection of charts, also known as an atlas. One may then apply ideas from calculus while working within the individual charts, since each chart lies within a linear space to which the usual rules of calculus apply. If the charts are suitably compatible (namely, the transition from one chart to another is differentiable), then computations done in one chart are valid in any other differentiable chart (Warner 1983).

In other words a complex spatial feature can be represented by local, well-behaved, simple patches, possibly of lower dimensions, and under certain mathematical conditions (i.e. differentiability); they accurately describe the entirety of that feature continuously in its space of origin.

Mathematical Definition A differentiable manifold of dimension n is a set M and a family of bijective mappings

$$x_\alpha : U_\alpha \subset \mathbb{R}^n \rightarrow M$$

of open sets U_α of $\mathbb{R}^n$ into M such that

1. $\cup_\alpha x_\alpha(U_\alpha) = M$.
2. For any pair α, β with $W = x_\alpha(U_\alpha) \cap x_\beta(U_\beta) \neq \phi$, the sets $x_\alpha^{-1}(W)$ and $x_\beta^{-1}(W)$ are open in $\mathbb{R}^n$ and the mappings $x_\alpha^{-1} \text{o} x_\beta$ are differentiable (see Fig. 1).
3. The family $\{(\cup_\alpha, x_\alpha)\}$ is maximal relative to the conditions (1) and (2).

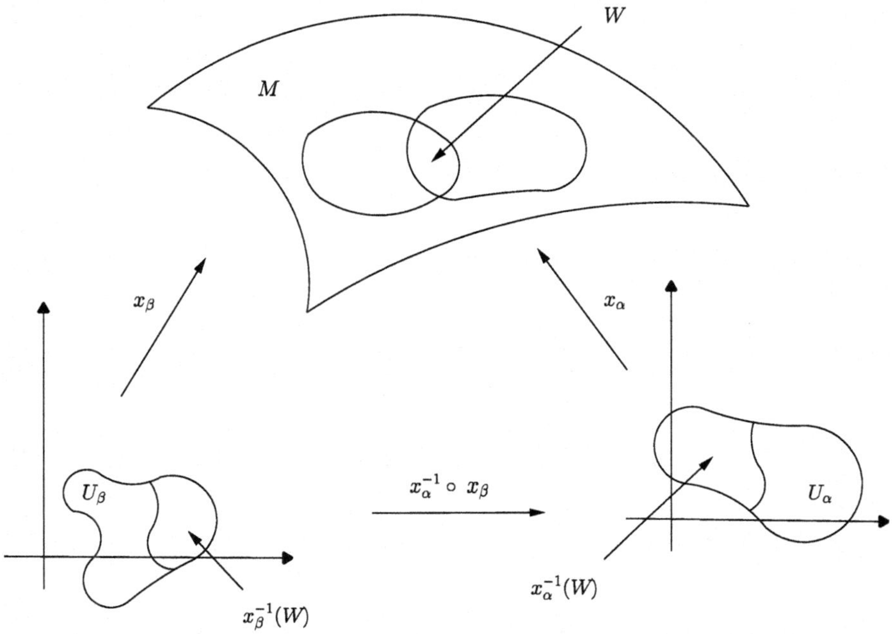

Tangent Space on Manifolds

Definition Let M be a differentiable manifold, and let p be a point of M. A linear map $v : C^\infty(M) \to \mathbb{R}$ is called a **derivation at** p if it satisfies the Leibniz rule:

$$v(\mathrm{fg}) = f(p)vg + g(p)vf; \forall f, g \in C^\infty(M)$$

The set of all derivations of $C^\infty(M)$ at p, denoted by T_pM is called a tangent space to M at p. An element of T_pM is called a tangent vector at p.

A Riemannian metric on a differentiable manifold M is a correspondence which associates to each point p of M, an inner product $\langle , \rangle_p$ on the tangent space T_pM, which varies differentiably; this metric is also called the metric tensor.

A differentiable manifold with a given Riemannian metric will be called a Riemannian manifold (Lee JM 2013; Do Carmo 1976).

Annex 2: Applying Methodology in Euclidean Space $\mathbb{R}^3$

An example of Riemannian manifold in the Euclidean space $\mathbb{R}^3$ is a surface of dimension 2 with the usual metric $\mathbb{R}^3$ and the tangent space here is a plane.

Regular Surface

A subset $S \subset \mathbb{R}^3$ is a **regular surface** if, for each $p \in S$, there exists a neighbourhood V in $\mathbb{R}^3$ and a map $x : U \to V \cap S$ of an open set $U \subset \mathbb{R}^2$ onto $V \cap S \subset \mathbb{R}^3$ such that (see Fig. 1):

1. x is differentiable.
2. x is homeomorphism.
3. $\forall q \in U, \mathrm{dx}_q : \mathbb{R}^2 \to \mathbb{R}^3$ is one-to-one.

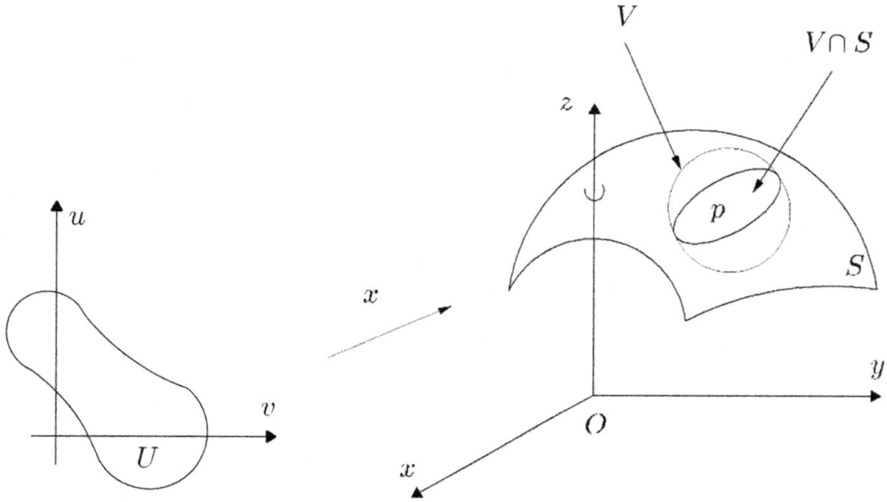

The Tangent Plane

Definition Let $v \in \mathbb{R}^3$ be a **tangent vector** to S at a point $p \in S$; it exists a differentiable parametrized curve $\alpha :]-\varepsilon, \varepsilon[\to S$ with $\alpha'(0) = v$ and $\alpha(0) = p$.

The set of all tangent vectors to S at p will be called the **tangent plane** to p and will be denoted by T_pS (Do Carmo 1976) (see Fig. 2).

$$T_pS = \left\{ v \in \mathbb{R}^3 ; v \text{ is tangent vector to } S \text{ at } p \in S \right\}$$

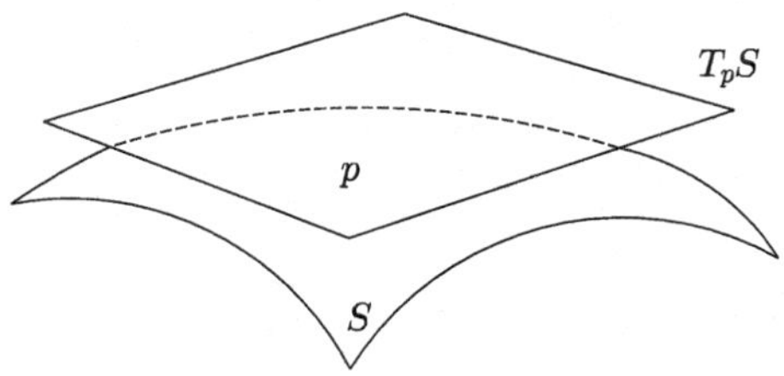

Proposed Trihedron

Now let us build a trihedron, from a point on the surface and a tangent vector to the surface at that point; this construction is a different wing of the trihedron Frenet-Serret; in very specific cases of surface geometry, these may coincide.

Let $S \subset \mathbb{R}^3$ a regular surface, parametrization (x,U) at point $p \in S$ and a regular parametrized curve $\alpha :]-\varepsilon, \varepsilon[\rightarrow \mathbb{R}^3$ content in S such that $\alpha'(0) = v$ and $\alpha(0) = p$, clearly $v \in T_pS$.

On the T_pS tangent plane, rotate the vector v angle $\pi/2$ in a sense counterclockwise forming a basis for the tangent plane T_pS (see figure). We now vector cross product $w = v \times u$ and form the proposed trihedron (see Fig. 3).

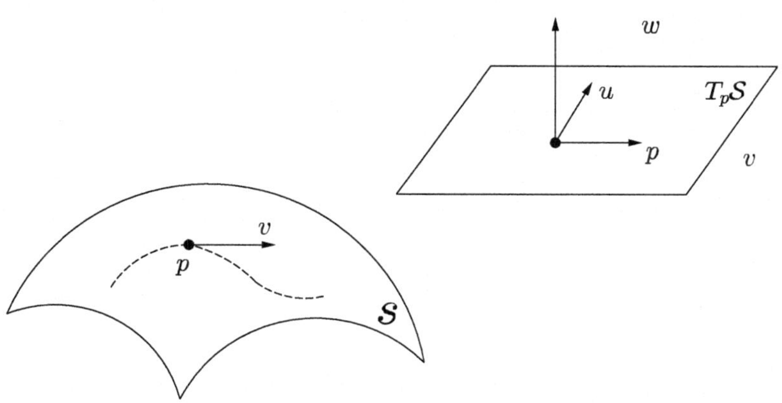

Bibliography

Bateman A (1982) Yacimientos Minerales de Rendimiento Económico. Omega S.A

Chilès J, Delfiner P (2001) Geostatistics modeling spatial uncertainty. Wiley

Do Carmo M (1976) Differential geometry of curves and surfaces. Instituto de Matematica Pura e Aplicada, Rio de Janeiro

Do Carmo M (1979) Geometria riemanniana. Instituto de Matematica Pura e Aplicada, Rio de Janeiro

Guibal D, Remacre A (1984) Local estimation of the recoverable reserves. E.N.S.M.P, France
Lee JM (2013) Introduction to smooth manifolds, 2nd edn. University of Washington, Seattle
Maréchal A, Deraisme J, Journel A, Matheron G (1978) Cours de Géostatistique non Linéaire. C-74 E.N.S.M.P. France
Marín Suárez A (1978) Méthodologie de L'estimation et Simulation Multivariable des Grands Gisements Tridimensionnels. Thèse de Docteur Ingénieur en Sciences et Techniques Minières – Option Géostatistique, présentée à L'école Nationale Supérieure des Mines de Paris
Marín Suárez A (1986) Modelo Geoestadístico de Filones de Almadén. Minas de Almadén S.A, Almadén
Marín Suárez A (2009) Diseño de una Metodología Geoestadística con Geometría de Variedades en Yacimientos Filoneanos. Facultad de Geología, Geofísica y Minas, UNSA, Arequipa
Matheron G (1962, 1963) Traité de Géostatistique Appliquée, vol 1; vol 2. Technip, Paris
Warner FW (1983) Foundations of differentiable manifolds and lie groups. Springer, New York

Drilling Grid Analysis for Defining Open-Pit and Underground Mineral Resource Classification through Production Data

Roberto Menin, Cassio Diedrich, Joao Dirk Reuwsaat, and Wellington F. De Paula

Abstract The varied types of mineral deposits and geological features around the world have led to the creation of a large number of techniques, methodologies, and definitions for mineral resource classification. The most common methods used in the mineral industry include kriging variance, drilling spacing, neighborhood restriction, and conditional simulations. These methods generally do not use reconciled production information, only long-term borehole information based on personal judgment for defining confident intervals/limits on the mineral resource classification. A drilling spacing back analysis study for defining mineral resource classification was completed considering tonnages and grades confidence intervals related to its respective production volumes, based on short-term production reconciliation of analog deposits. The definition of adequate drill holes spacing and detailed results for classifying mineral resources are demonstrated by both an open-pit and an underground project adjacent to an existing mining operation. This study has considered a Brazilian sulfide deposit (Cu-Au) operating mine as analog information.

R. Menin (✉) • J.D. Reuwsaat • W.F. De Paula
Vale Base Metals, 63 Grajau Street, 1st floor, Carajás, PA 68516-000, Brazil
e-mail: roberto.menin@vale.com; joao.dirk@vale.com; wellington.paula@vale.com

C. Diedrich
Vale Base Metals, 337 Power Street, Door #105, Copper Cliff, Sudbury, ON P0M 1N0, Canada
e-mail: Cassio.Diedrich@vale.com

J.J. Gómez-Hernández et al. (eds.), *Geostatistics Valencia 2016*, Quantitative Geology and Geostatistics 19, DOI 10.1007/978-3-319-46819-8_18

1 Introduction

New methodologies for classifying mineral resources are proposed every year with constant efforts for reducing ore deposits estimation uncertainties using new technologies. However, given different types of mineralization and its geological characteristics, it becomes impractical to define a single standardized methodology for mineral resource classification. In general the estimation classification process is based on qualified/competent person's (QP/CP) judgment. Whether or not the confidence of estimates (drilling spacing, experience, back analysis, or geostatistical methods) are based on this judgment, their validity is subject to assumptions that should be tested and validated against actual production results through reconciliation studies (Morley 2003).

Several authors have reported the major issues associated with mining operating cases that have not achieved planned production targets in its first years of operation (Burmeister 1988). The main reasons are inherent to the mineral resources and reserve classification processes, including errors associated with an inadequate sampling procedure and lack of local geological knowledge (Harquail 1991). Some studies present approaches for drill hole spacing using conditional simulation (Dimitrakopoulos et al. 2009) or estimation variance calculation (Verly et al. 2014) from long-term (hard data) drill holes information. Also a number of authors have published a series of recommendations for quantitative estimates of accuracy to classify resources and reserves (Vallée 1992; Wober 1993; Stephenson 1998). A summary used for resource classification, by compiling public disclosure of mineral projects issued by companies listed on the Toronto Stock Exchange, presents the most commonly applied resource classification practices (Silva 2013). In regard to these, none of the proposed mineral resource classification methods presents a connection to actual operating aspects or makes use of an operational reconciliation process. Therefore, mining configuration, geological knowledge, and production aspects can help in the decision for better defining mineral resources categories.

In this paper an operating mining case study (sulfide Cu-Au deposit) has been developed for defining mineral resources classification by using actual production reconciliation results. The idea is to determine drilling grid spacing that best support mineral resource categories for both open-pit and underground operating processes. Different production volumes related to its respective geology and operating practices were compared to actual exhaustive values for determining confidence levels of mineral resources categories.

Production information of a copper mine operation, based on blasting reverse circulation drilling, was used for both an underground and an open-pit projects due to its similarity in geological aspects and production scale.

2 Copper Mine Case Study and Methodology

A Brazilian large copper mine was used as analog information for both an underground and open-pit sulfide Cu-Au project. This analog deposit will be called as CMC (copper mine complex) and will be divided into two data sets, CM1 and CM2. CM1 is the one defined for the underground project (UGP). The open-pit project, referred to as OPP in the paper, took into account the CM2 data set.

The complete database of the CMC contains over 100,000 blast holes $\left(12\frac{1}{4}\text{ inch}\right)$ analysis ranging from approximately 5 to 6 m average drilling spacing. A series of different drill hole spacing grids were created for each mining method using production information. A new geological interpretation wireframe was constructed for each new grid by vertical extrusion of interpreted horizontal sections. Ordinary kriging routines were used to estimate block models of each drilling grid for both case studies.

The nearest neighbor interpolation algorithm was used for the construction of each regular drilling grid. For the selection of the drilling grid spacing (e.g., 30×30 m), a new block model was created with same expected drilling grid dimensions. After running the nearest neighbor interpolation, only the closest sample to the center of each block was selected and exported to create a new regular drilling grid with approximate 30 m spacing based on original production data.

Figure 1 illustrates how drilling grid selection was created and a sample selection was made for a regular sampling grid. The original geographic coordinates (X, Y, and Z) of each sample was kept.

According to the geological interpretation for long-term operational assumptions, drilling grid supporting each production increment (global, annual, and quarterly) was created according to long-term geological interpretation practices and interpreted on both vertical and plan sections. Vertical section traverses were created for supporting geological interpretation of plan sections without considering actual exhaustive information, only the information defined on the regular grids by the sample selection procedure.

Fifty-three vertical sections striking N-S were interpreted with distance between sections ranging from 10 to 40 m. A total of 21 plan sections, spaced at 16 m (bench height) and between -128 and 192 m sea level, were interpreted in CM1. In the same way, 21 vertical sections striking NE-SW were interpreted with distance between sections ranging from 10 to 20 m. After modeling the vertical planes and applying traverses for plan view interpretation, nine plan sections were interpreted for CM2 considering 16 m bench height and elevation varying between 64 m and 192 msl (sea level). Figure 2 presents interpreted plan sections (1a and 2b) and geological wireframes (1b and 2a) created for both CM1 and CM2 data sets.

The CMC operation long-term model ($10 \times 10 \times 16$ m block size) was used in studied cases. Figure 3 presents the 20×20 m (grid spacing) interpolation results at CM1 taking into account the geological interpreted mineralization. When different statistical domains or geological interpretations are considered, the final copper

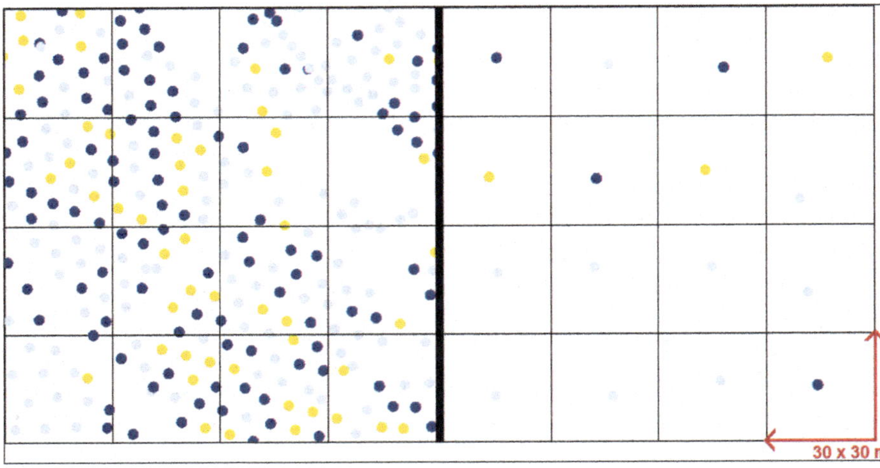

Fig. 1 Sample selection procedure for a regular drilling grid

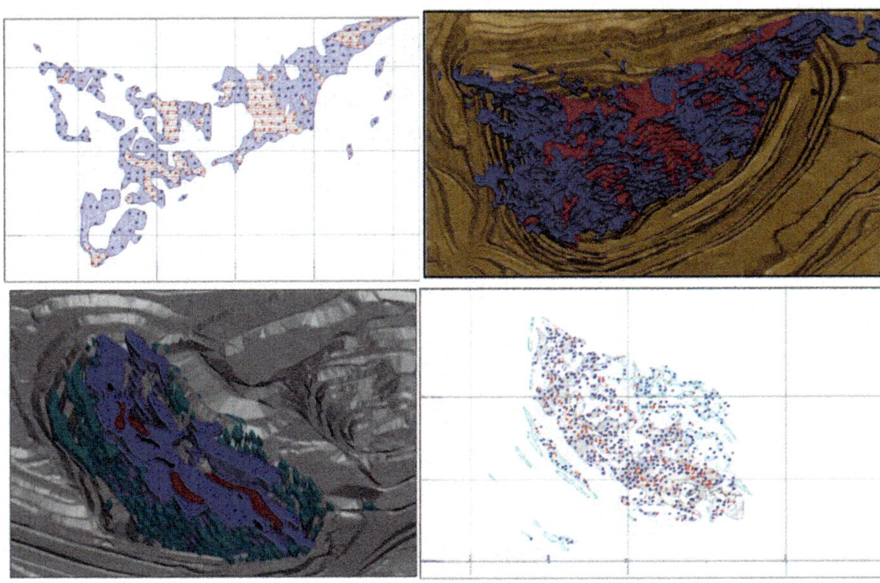

Fig. 2 Figures on *top*: (*1*) CM1 20×20 m interpreted plan section (*left*); CM1 10×10 m geological wireframes (*right*); Figures on *bottom*: (*2*) CM2 30×30 m geological wireframes (*left*); CM2 20×20 m interpreted plan section (*right*)

grades are defined by weighing the percent of each statistical domain within the block.

A surface volumetric calculation for determining differences (tonnage, grades, and metal) between drilling models in comparison with the production model was

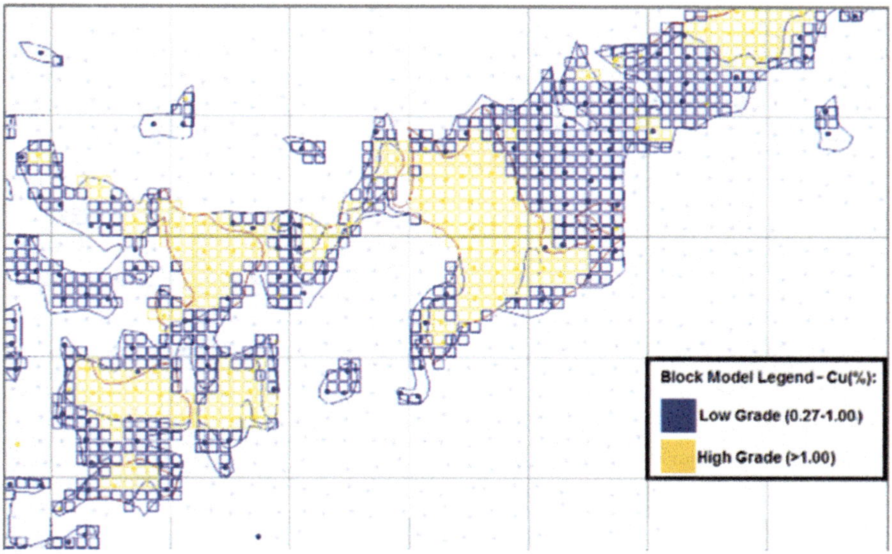

Fig. 3 Block model interpolation procedure

executed. Monthly and yearly topographic surfaces provided by operational teams were utilized for all study periods.

Figure 4 presents a 4-year period represented by each annual topographic surface.

An analysis of the OPP drilling grid spacing for mineral resource categorization was made considering seven different sampling grids based on the CM2 production data. The analysis of UGP drilling grid spacing for mineral resource categorization was made considering six different sampling grids. The difference between interpolated grids and production models CM1 and CM2 was measured for each specified period. Table 1 shows all sampling grids created for each of the case studies.

At OPP, drilling grids spaces larger than 80 × 80 m were tested. However, these provided poor results due to the nature (size and continuity) of mineralization and were not included in this paper. At UGP, spaces larger than 100 × 100 m were not tested due to lack of enough information at given drilling spacing.

CM1 data production was applied to the UGP mineral resource classification at a high-grade cutoff value (0.75 % Cu) that supports the level of selectivity of an underground mining operation.

As previously mentioned in the introduction, a number of authors have published a series of recommendations for quantitative estimates of accuracy to classify resources. Considering a systematic approach based on full operational knowledge and geological features of both case studies, the following degree of confidence limits will be applied to classifying mineral resources:

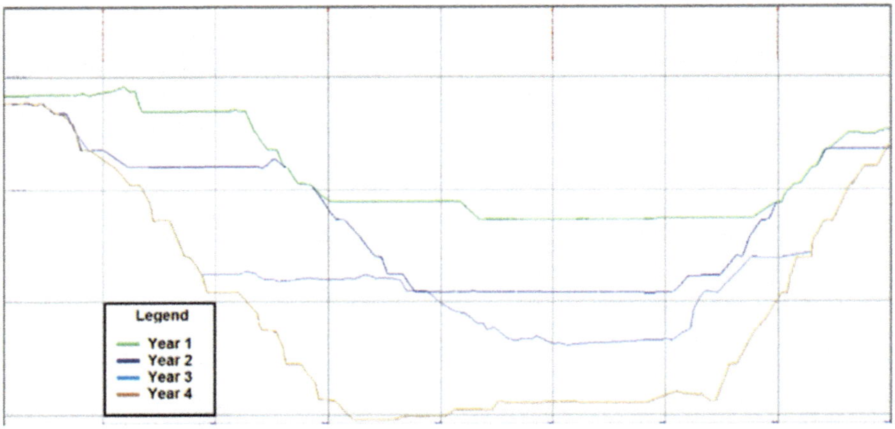

Fig. 4 Annual comparison basis

Table 1 Drilling grids

Drilling grids – CMC production data	
CM1/UGP	CM2/OPP
10 × 10 m	10 × 10 m
20 × 20 m	20 × 20 m
30 × 30 m	30 × 30 m
40 × 40 m	30 × 40 m
60 × 60 m	40 × 40 m
80 × 80 m	40 × 80 m
	50 × 50 m

- Inferred mineral resource: a level of confidence of ±15 % on the global recoverable metal content, tonnages, and grades
- Indicated mineral resource: a level of confidence of ±15 % on the recoverable metal content, tonnages, and grades over an area or volume corresponding to the footprint of 1 year of production for a given deposit type in a mine or project
- Measured mineral resource: a level of confidence of ±15 % on the recoverable metal content, tonnages, and grades over an area or volume corresponding to the footprint of one quarter of a year of production for a given deposit type in a mine or project

3 Global Production Volume (Inferred Mineral Resources)

The level of confidence for the definition of inferred resources was applied in one volume increment (global) for both CM2 and CM1. The tonnage produced in the selected comparison period was obtained through volumetric calculation between first and last year topographic surfaces. Figure 5 presents the percentage difference

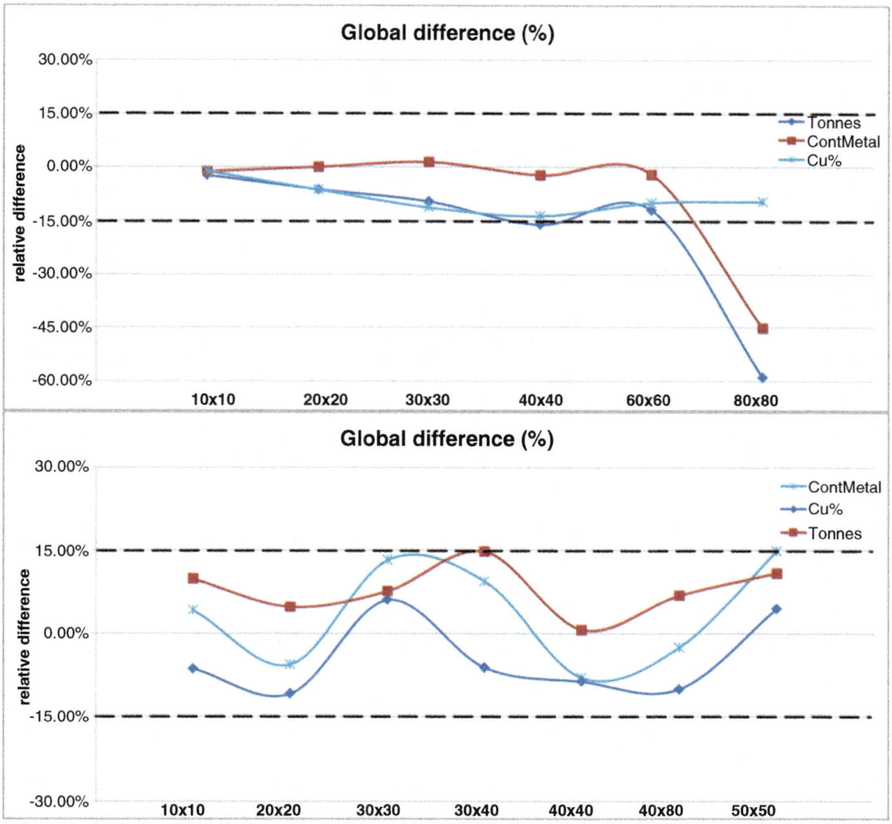

Fig. 5 (*1*) Global difference in terms of metal content (*light blue*), grade (*blue*), and tonnages (*red*) for CM1 (*top*); (2) global difference in terms of metal content (*light blue*), grade (*blue*), and tonnages (*red*) for CM2 (*bottom*)

between metal content, tonnages, and grades for all drilling grids in comparison with the CM1 and CM2 production actual models.

The resulting tonnages, grades, and contained metal obtained in the estimated models, for each drilling grid, had acceptable values for inferred resource classification in comparison with the production exhaustive models. Tables 2 and 3 present the percentage of scenarios within acceptable stipulated limits previously mentioned for both CM1 and CM2:

In the CM1 case, only the 80×80 m drilling spacing is not within acceptable limits in terms of tonnages and metal content to the established production rate. In the CM2 case, all drilling grids are within an acceptable range (percentage differences) for defining inferred mineral resources.

Table 2 Global percentage of scenarios within acceptable difference for CM1

Spacing (m)	Metal (%)	Tonnes (%)	Grade (%)
10 × 10	100	100	100
20 × 20	100	100	100
30 × 30	100	100	100
40 × 40	100	100	100
60 × 60	100	100	100
80 × 80	0	0	0

Table 3 Global percentage of scenarios within acceptable difference for CM2

Spacing (m)	Metal (%)	Tonnes (%)	Grade (%)
10 × 10	100	100	100
20 × 20	100	100	100
30 × 30	100	100	100
30 × 40	100	100	100
40 × 40	100	100	100
40 × 80	100	100	100
50 × 50	100	100	100

4 Annual Production Volume (Indicated Mineral Resources)

CM1 and CM2 global production were divided into five and four annual increments, respectively. This is in agreement with the annual production rates for both open-pit (OPP) and underground (UGP) examples. The tonnage produced in the selected comparison period was obtained through volumetric calculation between the first and the last monthly topographic surfaces in the same year (January 1st–December 31st).

The executed production rate on an annual basis ranged from 4.0–5.5 to 3.5–5.0 Mt for CM1 and CM2 open-pits, respectively. Figures 6, 7, and 8 present a comparison of the percentage differences (grades, metal, and tonnages, respectively) for each selected period for both CM1 (top) and CM2 (bottom), considering the actual production information.

As commented, a level of confidence of ±15 % on the annual recoverable contained metal, tonnage, and grades is needed to support an indicated mineral resource category. Tables 4 and 5 present the percentage of scenarios within the acceptable stipulated limits:

The high-grade portions could not be mapped for drilling grid spacing higher than 80 × 80 m in the CM1 case study. This reduced the tonnage of this material significantly where only the grades were within acceptable threshold limits for indicated mineral resources. Small differences in tonnages and grades were verified in the 60 × 60 m drill spacing. The metal content was above the stipulated threshold limits in two of the five annual increments. However, for these two scenarios, the grade and tonnages were within acceptable limits.

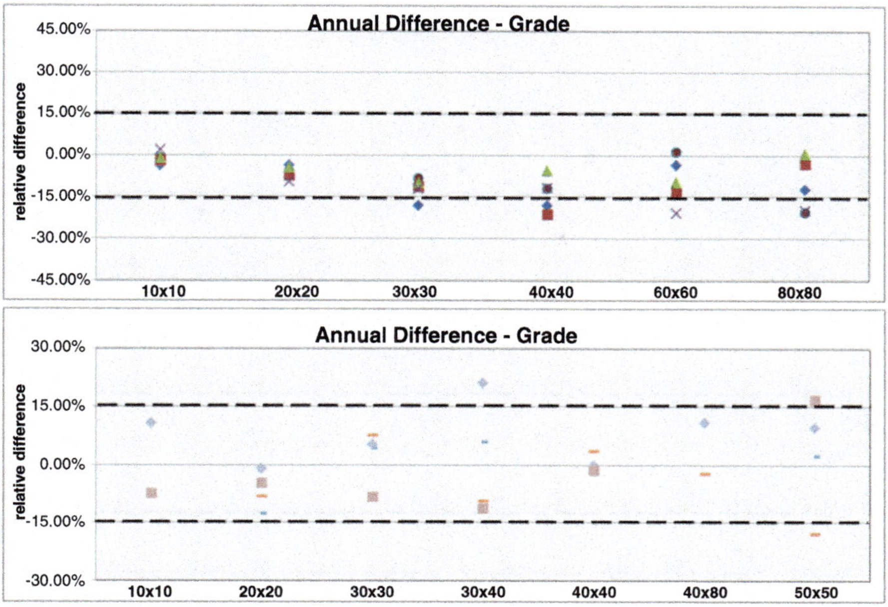

Fig. 6 (*1*) Annual difference in terms of grade for CM1 (*top*); (*2*) annual difference in terms of grade for CM2 (*bottom*)

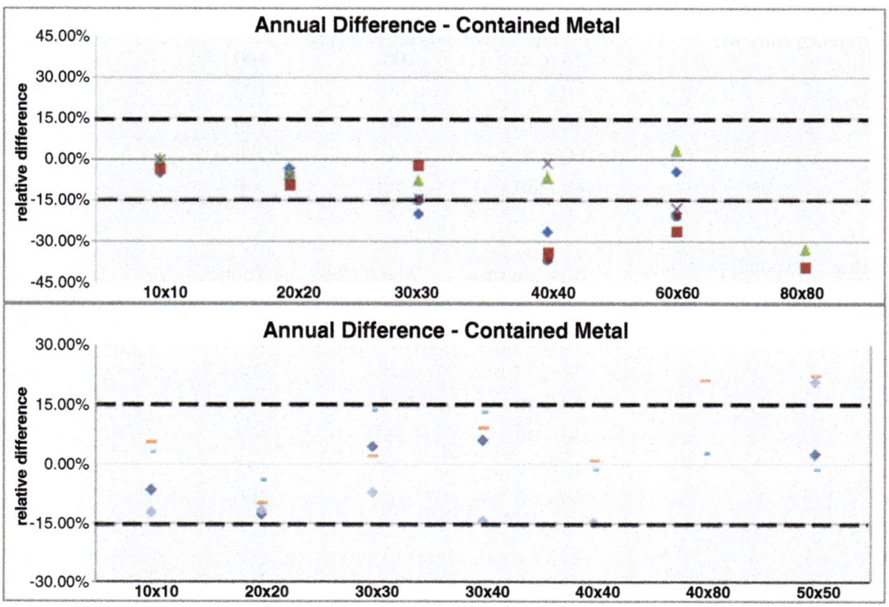

Fig. 7 (*1*) Annual difference in terms of metal content for CM1 (*top*); (*2*) annual difference in terms of metal content for CM2 (*bottom*)

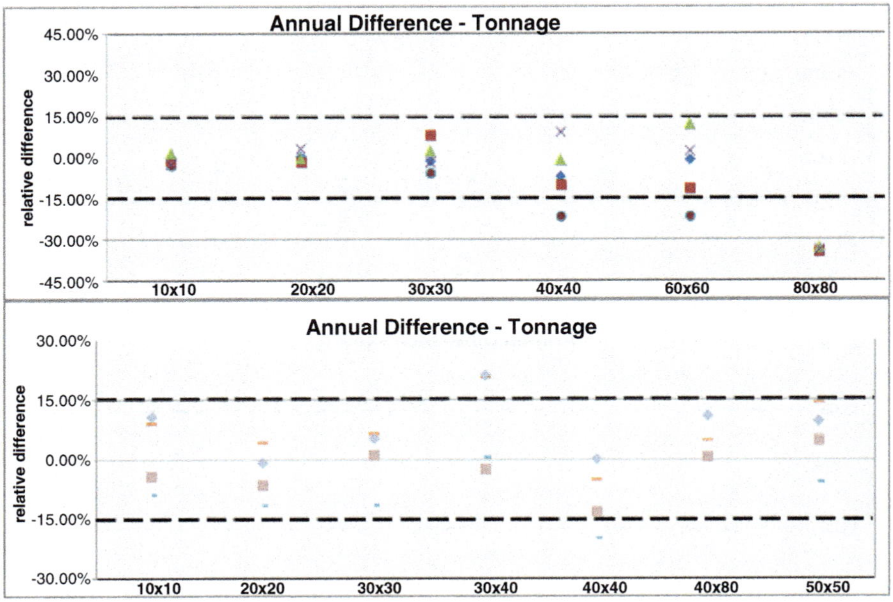

Fig. 8 (*1*) Annual difference in terms of tonnages for CM1 (*top*); (*2*) annual difference in terms of tonnages for CM2 (*bottom*)

Table 4 Annual percentage of scenarios within acceptable difference for CM1

Spacing (m)	Metal (%)	Tonnes (%)	Grade (%)
10 × 10	100	100	100
20 × 20	100	100	100
30 × 30	80	100	100
40 × 40	40	80	80
60 × 60	40	80	60
80 × 80	0	0	60

Table 5 Annual percentage of scenarios within acceptable difference for CM2

Spacing (m)	Metal (%)	Tonnes (%)	Grade (%)
10 × 10	100	100	100
20 × 20	100	100	100
30 × 30	100	100	100
30 × 40	100	50	100
40 × 40	75	75	75
40 × 80	25	75	50
50 × 50	50	100	50

In the CM2 case study, for the 40 × 40 m drilling spacing, slight differences in the grade and tonnages than those stipulated were observed. However they reflected a higher difference in the metal content compared to the values obtained by the exhaustive production model. For two of the four annual production volumes, the

30 × 40 m drilling grid scenario obtained tonnages which were slightly above the stipulated limits; however, for these two scenarios, the grade and the contained metal were within acceptable values.

5 Quarterly Production Volume (Measured Mineral Resources)

CM1 and CM2 global production were divided into quarterly increments (20 and 15, respectively) which are in agreement with the quarterly production rate executed. The tonnage produced in the selected comparison period was obtained through volumetric calculation between the first and the last monthly topographic surface of the quarter.

A level of confidence of ±15 % on the quarterly recoverable tonnages, grades, and metal content is required to support measured mineral resource categorization. The quarterly production rates of CM1 case study ranged from 0.8 to 1.4 Mt. For CM2 the ranges were from 0.75 to 1.5 Mt. The production period was divided into actual monthly basis topography (short-term CMC) for the CM1 comparison.

Figures 9, 10, and 11 present a comparison of the percentage differences (grades, metal, and tonnages, respectively) for each selected period for both CM1 (top) and CM2 (bottom), considering the actual production information.

CM1 tonnages and metal content estimation results of 60 × 60 m and 80 × 80 m drilling grid spacing are substantially different for the selected cutoff grade compared to the exhaustive production model. It is more difficult to establish spatial connectivity for the available samples when increasing drilling grid spacing, which significantly reduces the volumes of certain portions of the deposit given the lack of information for delineating mineralization. In the 30 × 30 m drilling grid spacing scenarios, only a few quarterly production intervals lied outside acceptable metal content limits in support of measured mineral resources. In the 40 × 40 m drilling grid spacing scenario, there were several quarterly production intervals outside acceptable tonnage and metal content threshold limits.

Minor differences in the tonnages and grades were observed in the CM2 case study for a 30 × 30 m drilling grid spacing. These were verified mainly in scenarios with lower amount of production information. The results demonstrated that drilling grid spacing larger than 40 × 40 m does not predict within acceptable confidence the production tonnages, grades, and metal content in a quarterly period. Tables 6 and 7 summarize the percentage of scenarios within acceptable limits defined for both cases studies.

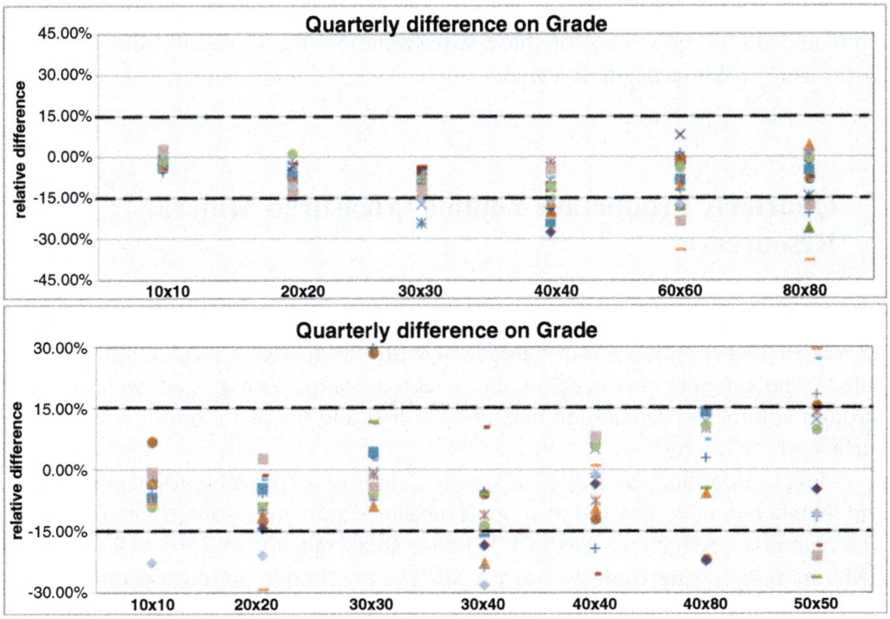

Fig. 9 (*1*) Quarterly difference in terms of grade for CM1 (*top*); (*2*) quarterly difference in terms of grade for CM2 (*bottom*)

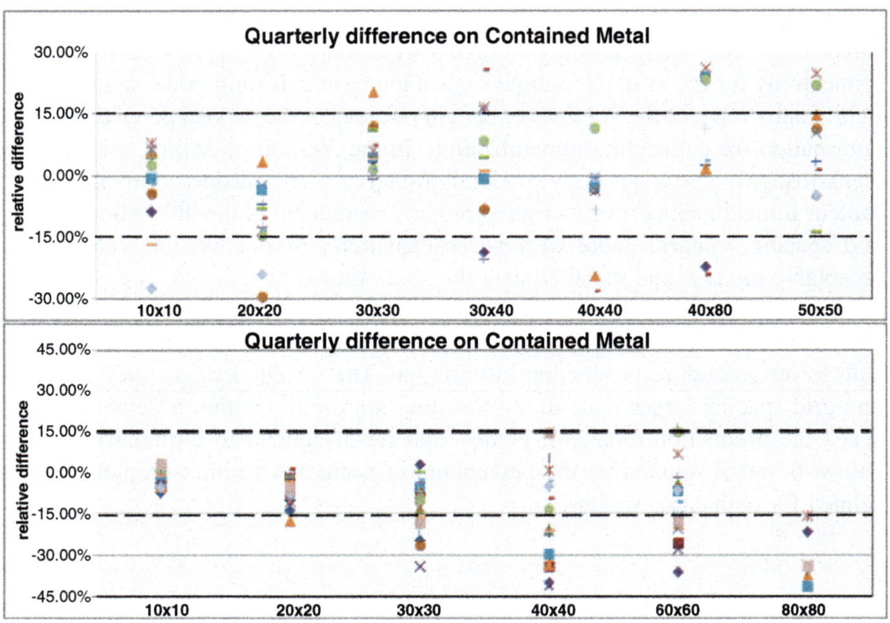

Fig. 10 (*1*) Quarterly difference in terms of metal content for CM1 (*top*); (*2*) quarterly difference in terms of metal content for CM2 (*bottom*)

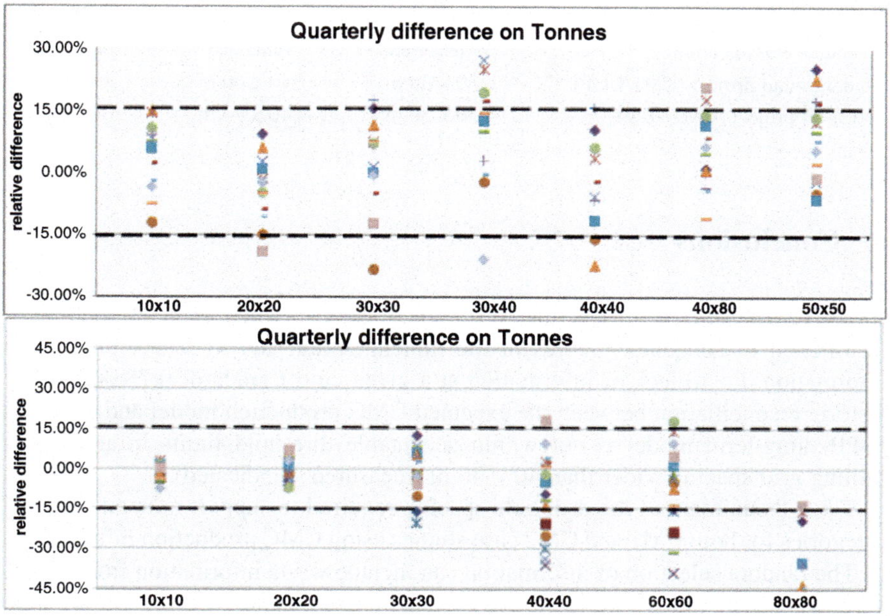

Fig. 11 (*1*) Quarterly difference in terms of tonnages for CM1 (*top*); (*2*) quarterly difference in terms of tonnages for CM2 (*bottom*)

Table 6 Quarterly percentage of scenarios within acceptable difference for CM1

Spacing (m)	Metal (%)	Tonnes (%)	Grade (%)
10 × 10	100	100	100
20 × 20	95	100	100
30 × 30	70	85	80
40 × 40	35	65	50
60 × 60	45	70	70
80 × 80	15	10	75

Table 7 Quarterly percentage of scenarios within acceptable difference for CM2

Spacing (m)	Metal (%)	Tonnes (%)	Grade (%)
10 × 10	93	100	93
20 × 20	79	100	86
30 × 30	71	86	79
30 × 40	50	43	71
40 × 40	57	71	86
40 × 80	36	79	64
50 × 50	50	79	57

Table 8 Final resource classification based on the back analysis study

Resource classification	Measured	Indicated	Inferred
Underground project (CM1/UGP)	30 x 30 m	40 x 40 m	80 x 80 m
Open-pit project (CM2/OPP)	30 x 30 m	40 x 40 m	40 x 80 m

6 Conclusions

CM1 (UGP) study indicates that drill holes spacing wider than 60×60 m is insufficient to adequately represent the mineralization and to reflect production information due to lack of information at a given cutoff grade of 0.75 % Cu. The historic reconciliation between the executed CMC production model and the CM2 (OPP) long-term model is not within acceptable threshold limits in areas with drilling grid spacing wider than 40×40 m (measured + indicated).

Table 8 summarizes the drill hole spacing required to support mineral resource categories for both CM1 and CM2 case studies using CMC production information.

The random selection of information can include waste information from neighborhoods that are mostly mineralized. This randomness in the choice of information can sometimes change the proportions of economic mineralization and waste in some local portions of the model. This is part of the "real" variability of the deposit.

The mineralization characteristics of a deposit do not define the sample spacing required for mineral resource categorization. A realistic production volume should be considered in the evaluation of drill spacing for supporting mineral resource categories. Realistic volumes are determined from analog deposits where production information is available.

As mentioned in Diedrich et al. (2016), mining selectivity is understood as the process of separating ore from waste, and its global and local concept is strongly related to three mining functions such as geology, production rates, and mining configuration that affect operational results. Operational aspects should be included in a project or operation for better defining adequate drilling spacing and for supporting mineral resources classification.

Bibliography

Burmeister B (1988) From resource to reality: a critical review of the achievements of new Australian gold mining projects during 1983–1987. Macquarie University, Sidney

Deraisme J, Roth C (2000) The information effect and estimating recoverable reserves. Geovariances

Diedrich C, Reuwsaat JD, Menin R, Paula WF (2016) Selection of operational mining unit (OMU) size, Proceedings GEOSTATS 2016. Universitat Politècnica de València, Valencia, p 16

Dimitrakopoulos R, Godoy M, Chou C (2009) Resource/reserve classification with integrated geometric and local grade variability measures. In: Proceedings orebody modelling and

strategic mine planning. The Australasian Institute of Mining and Metallurgy, Melbourne, pp 207–214
Harquail D (1991) Investing in junior mining companies. 6th Mineral Economics Symposium of CIM. Canadian Institute of Mining, Mettalurgy and Petroleum, Montreal, p 10
Isaaks E (2004, 2005) The kriging oxymoron: a conditionally unbiased and accurate predictor. In: Proceedings of geostatistics banff, 2nd edn. Springer, Banff, pp 363–374.
Jara RM, Couble A, Emery X, Varela EM, Ortiz JM (2006) Block size selection and its impact on open-pit design and mine planning. J S Afr Inst Min Metall 106:205–211
Leuangthong O, Neufeld C, Deutsch CV (2003) Optimal selection of selective mining unit (SMU) size. University of Alberta, Alberta
Morley C (2003). Beyond reconciliation: a proactive approach to using mining data. In: Large open pit mining conference. Kalgoorlie, pp 185–192
Silva DA (2013) Mineral resource classification (NI 43–101): an overview and a new evaluation technque. In: Centre for computational geostatistics. University of Alberta, Edmonton, p 307
Sinclair A, Blackwell G (2002) Applied mineral inventory estimation. Cambridge University Press, Cambridge
Stephenson P (1998) The JORC code, 1987–1997, in Geology of Australian and Papua New Guinean Mineral Deposits. In: Australasian institute of mining and metallurgy. Melbourne, pp 44–52
Vallée MA (1992) The guide to the evaluation of gold deposits: integrating deposit evaluation and reserve inventory practices. CIM Bull 85:50–61
Verly G, Postolski T, Parker H (2014) Assessing uncertainty with drill hole spacing studies – applications to mineral resources. In: Proceedings orebody modelling and strategic mine planning symposium. The Australasian Institute of Mining and Metallurgy, Perth, p 10
Wober HA (1993) Classification of ore reserves based on geostatistical and economic parameters. CIM Bull 86:73–76

A High-Order, Data-Driven Framework for Joint Simulation of Categorical Variables

Ilnur Minniakhmetov and Roussos Dimitrakopoulos

Abstract Relatively recent techniques for categorical simulations are based on multipoint statistical approaches where a training image (TI) is used to derive complex spatial relationships using patterns. In these cases, simulated realizations are driven by the TI utilized, while the spatial statistics of the hard data is not used. This paper presents a data-driven, high-order simulation approach based upon the approximation of high-order spatial indicator moments. The high-order spatial statistics are expressed as functions of spatial distances similar to variogram models for two-point methods. It is shown that the higher-order statistics are connected with lower orders via boundary conditions. Using an advanced recursive B-spline approximation algorithm, the high-order statistics are reconstructed from hard data. Finally, conditional distribution is calculated using Bayes rule and random values are simulated sequentially for all unsampled grid nodes. The main advantages of the proposed technique are its ability to simulate without a training image, which reproduces the high-order statistics of hard data, and to adopt the complexity of the model to the information available in the hard data. The approach is tested with a synthetic dataset and compared to a conventional second-order method, *sisim*, in terms of cross-correlations and high-order spatial statistics.

1 Introduction

Stochastic, or geostatistical, simulations are often required in reservoir or mineral deposit modeling and the quantification of geological uncertainty, pollutants in contaminated areas, and other spatially dependent geologic and environmental phenomena. During the past decades, Gaussian simulation techniques have been used for geostatistical simulations (Matheron 1971; David 1977, 1988; Journel and Huijbregts 1978; Cressie 1993; Kitanidis 1997; Goovaerts 1998; Webster and

I. Minniakhmetov (✉) • R. Dimitrakopoulos
COSMO, Stochastic Mine Planning Laboratory, McGill University, Montreal, QC, Canada H3A 0E8
e-mail: ilnur.minniakhmetov@mail.mcgill.ca; roussos.dimitrakopoulos@mcgill.ca

J.J. Gómez-Hernández et al. (eds.), *Geostatistics Valencia 2016*, Quantitative Geology and Geostatistics 19, DOI 10.1007/978-3-319-46819-8_19

Oliver 2007; Remy et al. 2009; Chiles and Delfiner 2012; Pyrcz and Deutsch 2014). The choice of the Gaussian distribution is driven by several factors. First of all, the Gaussian variables can be fully described by a small amount of parameters, such as the first-order statistics (i.e., average values) and the second-order statistics (i.e., covariance or variogram). Secondly, the small number of parameters allows one to run simulations on grids with several million nodes.

Natural phenomena are known to exhibit non-Gaussian distributions and have complex nonlinear spatial patterns (Guardiano and Srivastava 1993; Tjelmeland and Besag 1998; Dimitrakopoulos et al. 2010), which cannot be adequately described by second-order statistics. To overcome these limitations, multiple point spatial simulation (MPS) methods were introduced in the 1990s (Guardiano and Srivastava 1993; Journel 1993; Strebelle 2002; Journel 2003; Zhang et al. 2006; Chuginova and Hu 2008; Straubhaar et al. 2011; Toftaker and Tjelmeland 2013; Strebelle and Cavelius 2014; others). The additional information is taken into account via training images (TI), which are not conditioned on the available data but contain additional information about complex spatial relations of the attributes to be simulated. To retrieve this information from the training image, the similarity between the local neighborhood of an unsampled location and the training image is calculated in explicit or implicit form. Based on this similarity measure, the value of a node from the training image with the most similar neighborhood is assigned to the unsampled location being simulated. Generally, most of the multipoint simulation techniques are a Monte Carlo sampling of values from the TI in some form or another. No spatial models are used and, importantly, no spatial information from the hard data is retrieved. As a result, simulations of attributes reflect the TI. In cases where there are relatively large datasets, conflict between the hard data and TI's statistics is clearly observed and the resulting simulations do not reproduce the spatial statistics of the hard data (Dimitrakopoulos et al. 2010; Pyrcz and Deutsch 2014).

Several attempts have been made to incorporate more information from the hard data. Some authors suggest using replicates from the hard data in addition to TI (Mariethoz and Renard 2010); however, in practice, it is hard to find any replicates for three-point relations when data is sparse. Others (Mariethoz and Kelly 2011) apply affine transformations to better condition the hard data; however, TI is still used as the main source of information. Another approach is to construct TI based on the hard data (Yahya 2011), but the resulting simulations may be biased from the method chosen for the TI construction.

Mustapha and Dimitrakopoulos (2010a, b) proposed to use the high-order spatial cumulants as the extension of variogram models to capture complex multipoint relations during the simulation of non-Gaussian random fields. The technique estimates the third- and the fourth-order spatial statistics from hard data and complements them with higher-order statistics from the TI. However, this technique is based on the approximation of conditional distribution using Legendre polynomials, which are smooth functions and limited in generating adequate approximation of the discrete distribution of categorical variables.

The problem of describing complex multipoint relations of categorical variables was discussed in Vargas-Guzmán (2011) and Vargas-Guzmán and Qassab (2006). The author uses high-order indicator statistics to characterize spatially distributed rock bodies. In this paper, this idea is developed by introducing the connection between different orders into the mathematical model. For example, consider a third-order spatial indicator moment of a stationary random field, which is a function of two lags. When one of the lags is equal to zero, the third-order indicator moment becomes the second-order indicator moment. Besides that, instead of exponential functions, the B-spline functions are used to estimate high-order spatial indicator moments. It is known (Evans et al. 2009; Babenko 1986) that B-spines provide optimal estimation of equi-continuous functions defined on compacts. Finally, a new recursive algorithm is proposed for better approximation of high-order spatial statistics with nested boundary conditions of lower-level relations. Then, the conditional distribution for the given neighborhood is calculated from high-order indicator moments and the category is simulated. The proposed method works without a TI; however, additional information from TI can be incorporated as the secondary condition during the approximation step, for example, the conditions on derivatives or the order of continuity. In this case, the high-order spatial indicator moments are fully driven by the hard data.

The paper is organized as follows. First, high-order spatial indicator moments are introduced, as a function of distances between points for two-point and multipoint cases. Then, a mathematical model for recursive approximation of high-order spatial indicator moments is proposed. Finally, the simulation algorithm proposed is developed and tested on fully known datasets. Conclusions follow.

2 High-Order Spatial Indicator Moments

Let Ω, F, P be a probability space. Consider a stationary ergodic random vector $\mathbf{Z} = (Z_1,Z_2, \ldots Z_N)^T, \mathbf{Z} : \Omega \to S^N$, defined on a regular grid $D = \{\mathbf{x}_1, \mathbf{x}_2 \ldots \mathbf{x}_N\}$, $\mathbf{x} \in R^n, n = 2, 3$, where Ω is a space of all possible outcomes, F contains all combinations of Ω, S^N is a set of states represented by categories $S = \{s_1, s_2, \ldots s_K\}$, and P is the probability measure, or probability. For example, probability of $\mathbf{Z}_1$ being at a state s_k is defined as:

$$P(Z_1 = s_k) \equiv P\left(\left\{\omega \in \Omega : \mathbf{Z}(\omega) \in s_k \otimes S^{N-1}\right\}\right). \tag{1}$$

Without loss of generality, assume that $s_k = k$. It can be shown that the probability is equivalent to spatial indicator moment (Vargas-Guzmán 2011):

$$P(Z_{i_0} = k_0, Z_{i_1} = k_1, \ldots) = E(I_{k_0}(Z_{i_0}), I_{k_1}(Z_{i_1}), \ldots),\\ \forall i_0, i_1 \ldots = 1 \ldots N, \forall k_0, k_1, \ldots = 1 \ldots K \tag{2}$$

where E is the expected value operator and $I_k(Z_i)$ is an indicator function:

$$I_k(Z_i) = \begin{cases} 1, & Z_i = k \\ 0, & Z_i \neq k_0 \end{cases}. \tag{3}$$

From now on, indicator moments are denoted as:

$$M_{k_0 k_1,\ldots}(Z_{i_0}, Z_{i_1}, \ldots) = E(I_{k_0}(Z_{i_0}), I_{k_1}(Z_{i_1}), \ldots). \tag{4}$$

2.1 Second-Order Spatial Indicator Moment

Consider two random variables Z_{i_0} and Z_{i_1} separated by the lag $\mathbf{h}_1 = \mathbf{x}_{i_1} - \mathbf{x}_{i_0}$. Due to the stationarity assumption, their second-order spatial indicator moment for categories k_0,k_1 can be expressed as a function of the lag $\mathbf{h}_1$:

$$M_{k_0 k_1}(Z_{i_0}, Z_{i_1}) = M_{k_0 k_1}(\mathbf{h}_1). \tag{5}$$

For the sake of demonstration, consider data from the Stanford V reservoir case study (Mao and Journel 1999) on Fig. 1a and its categorization on Fig. 1b, in which the size of image is $N_x \times N_y$ pixels. Let $W_{i,j}$ be a value at pixel (i,j) of the categorized image, where $i = 1 \cdots N_x, j = 1 \cdots N_y$. If the image $W_{i,j}$ describes statistical properties of the random vector $\mathbf{Z}$, then the estimation of indicator moment $\hat{M}_{k_0 k_1}(\mathbf{h}_1)$ on the lag ($\mathbf{h}_1$=$(h,0)$ can be calculated using pairs $\{W_{i,j}, W_{i+h,j}\}$. From now on, consider that the direction of $\mathbf{h}_1$ is $\mathbf{e}_1$=(1,0) and fixed and then $M_{k_0 k_1}(\mathbf{h}_1)$ is the function of distance h.

The sections of the function $M_{k_0 k_1}(\mathbf{h})$ for fixed h equal to 0, 5, and 40 pixels which are shown on Fig. 2a–c, respectively. Figure 2d presents the sections of the function $M_{k_0 k_1}(\mathbf{h})$ for fixed values k_0,k_1. Each line corresponds to one of the 3×3 possible combinations of k_0 and k_1.

It is not hard to see that for $h = 0$ only diagonal elements are not equal to zero:

$$M_{k_0 k_1}(0) = P(Z_{i_0} = k_0, Z_{i_0} = k_I) = P(Z_{i_0} = k_0)\delta_{k_0,k_1} = M_{k_0}\delta_{k_0,k_1}, \tag{6}$$

where δ_{k_0,k_1} is Kronecker delta and M_{k_0} is the marginal distribution. Additionally, for two distant locations $h \to \infty$, the values Z_{i_0} and Z_{i_1} can be considered as independent random variables and the indicator moment can be factorized:

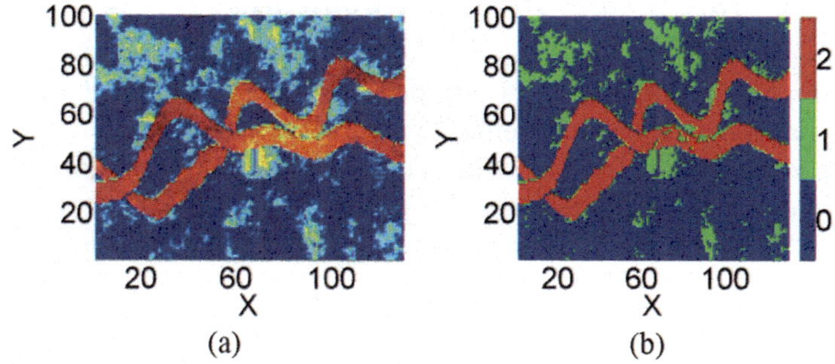

Fig. 1 Image of (**a**) continuous field and (**b**) its categorization

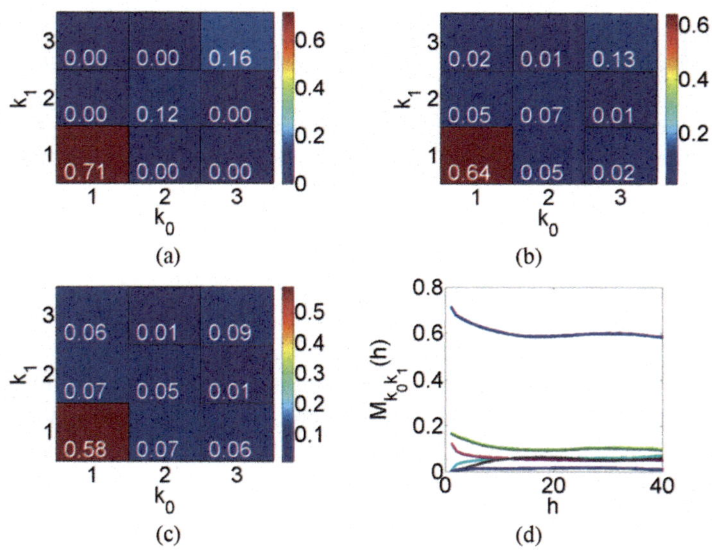

Fig. 2 The estimation of the second-order spatial indicator moment $\hat{M}_{k_0k_1}(h)$: (**a**) zero-distance $h=0$ indicator moments; (**b**) indicator moments on the lag $h=5$; (**c**) indicator moments for far separated points $h=40$; (**d**) the sections with different combinations of k_0,k_1 depends on h. Each line on (**d**) corresponds to value in one of 3×3 cells in (**a**), and its evolution across different lag separation (**b**) and (**c**)

$$M_{k_0k_1}(h \to \infty) = M_{k_0}M_{k_1}. \tag{7}$$

That can be traced in the function behavior on Fig. 2d. These functions represent two-point spatial cross-relations, similar to indicator covariances, and satisfy boundary conditions (6) and (7).

2.2 High-Order Spatial Indicator Moments

In the multipoint case, consider $n+1$ random variables Z_{i_0} and $Z_{i_l}, l = 1 \ldots n$, and then, the spatial configuration is defined by vectors $\mathbf{h}_l = \mathbf{x}_{i_l} - \mathbf{x}_{i_0}, l = 1 \ldots n$. From now on, consider that directions of $\boldsymbol{h}_1$ are defined and fixed, and then the spatial indicator moments are the functions of distances $h_l = \|\mathbf{h}_l\|, l = 1 \ldots n$. Therefore, the high-order spatial indicator moment can be expressed as follows:

$$M_{k_0 k_1 \ldots k_n}(Z_{i_0}, Z_{i_1}, \ldots Z_{i_n}) = M_{k_0 k_1 \ldots k_n}(h_1, h_2, \ldots h_n). \tag{8}$$

Hereafter, the following concise notation is used: $\mathbf{k} = k_0 \ldots k_n$ and $\mathbf{h} = h_1, \ldots, h_n$.

Similar to the case of the second-order statistics, boundary conditions can be expressed through lower order:

$$M_{\boldsymbol{k}}(h_1, \ldots, h_p = 0, \ldots, h_n) = M_{\boldsymbol{k} \setminus k_p}(\mathbf{h} \setminus h_p) \delta_{k_0, k_p}, \forall p \in 1 \ldots n, \tag{9}$$

where $\boldsymbol{h} \setminus h_p$ denotes all the lags $\boldsymbol{h}$ excluding the lag h_p. Similarly for $\boldsymbol{k} \setminus k_p$.

If the directions are quite different, then additional boundary conditions are valid:

$$\begin{aligned} M_{\boldsymbol{k}}(h_1, \ldots, h_p \to \infty, \ldots, h_n) \\ = M_{\boldsymbol{k} \setminus k_p}(\mathbf{h} \setminus h_p) M_{k_p}, \forall p \in 1 \ldots n. \end{aligned} \tag{10}$$

Thus, the high-order spatial indicator moments are bounded with lower-order moments, and this information should be taken into account during simulation.

For example, in case of three-point relations, for the image $W_{i,j}$, the sampling third-order spatial indicator moment $\hat{M}_{k_0 k_1 k_2}(h_1, h_2)$ of random variables separated by the vectors $\mathbf{h_1}$, $\mathbf{h_2}$ with directions $\mathbf{e}_1$=(1,0) and $\mathbf{e}_1$=(1,0) can be calculated using triplets $\{W_{i,j}, W_{i+h_1,j}, W_{i,j+h_2}\}$. The indicator moment $\hat{M}_{111}(h_1, h_2)$ is shown on Fig. 3. The values of the function $\hat{M}_{111}(h_1, h_2)$ on boundaries (h_1,0), (h1,50), (0,h2), and (50, h_2) correspond to two-point statistics shown on Fig. 2d.

3 Mathematical Model

In this paper, multidimensional B-spline approximation under constrains (8) and (9) is used to model the high-order spatial indicator moments $\mathbf{M}_{\boldsymbol{k}}(\mathbf{h})$. Consider the fixed categories $k_0, \cdots k_n$; then, the function $\mathbf{M}_{\boldsymbol{k}}(\mathbf{h})$ is a multidimensional function which values are known at the limited number of points $\mathbf{h}^d = (h_1^d, \ldots, h_n^d), d = 1 \ldots m$ estimated from the hard data. The calculation of sampling indicator moments $\hat{M}_k$ $(\mathbf{h}^d)$ is presented in the subsequent section. Then, the function $\boldsymbol{M}_{\boldsymbol{k}}(\boldsymbol{h})$ can be approximated using the following recursive model:

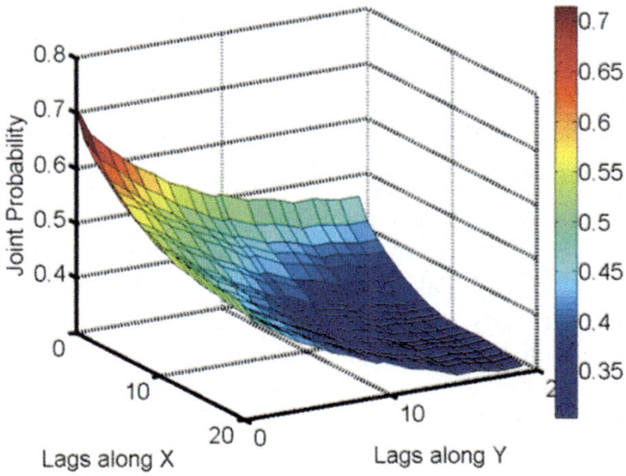

Fig. 3 The third-order spatial indicator moment $\hat{M}_{111}(h_1, h_2)$

$$M_{\boldsymbol{k}}(\mathbf{h}) = M^0_{\boldsymbol{k}}(\mathbf{h}) + \delta M_{\boldsymbol{k}}(\mathbf{h}), \tag{11}$$

$$\begin{aligned} M^0_{\boldsymbol{k}}(\mathbf{h}) = &\\ \frac{1}{\sum_{p=1}^{n} e^{-\mathrm{ah}_p} + e^{-a(1-h_p)}} &\left[\sum_{p=1}^{n} M_{k_0 \ldots k_n}(h_1, \ldots, h_p = 0 \ldots, h_n) e^{-\mathrm{ah}_p} +, \right. \\ &\left. + \sum_{p=1}^{n} M_{k_0 \ldots k_n}(h_1, \ldots, h_p \to \infty \ldots, h_n) e^{-a(1-h_p)} \right], \forall p = 1 \ldots n \end{aligned} \tag{12}$$

$$\delta M_{\boldsymbol{k}}(\mathbf{h}) = \sum_{i_l=1}^{\omega} \cdots \sum_{i_n=1}^{\omega} \alpha_{i_l, \ldots, i_n} B_{i_l, r}(h_l) \ldots B_{i_n, r}(h_n), \tag{13}$$

where user-defined parameter $\boldsymbol{a}$ determines the influence of the boundary conditions, $\alpha_{i_l, \ldots, i_n}$ are coefficients of B-spline approximation, and $B_{i,r}(t)$ is $\boldsymbol{i}$-th B-splines of order r on uniformly divided knot sequence $\{t_0, t_1, t_2, \ldots, t_p\}$, where knots are separated by step $dt = (t_p - t_0)/p$, $t_0 = 0$, and t_{p} are equal to the minimal lag distance at which the variables can be considered as independent.

The coefficients $\alpha_{i_1, \ldots, i_n}$ are found using least-square algorithm to fit points:

$$\delta M_{\boldsymbol{k}}(\mathbf{h}^d) = \hat{M}_{\boldsymbol{k}}(\mathbf{h}^d) - M^0_{\boldsymbol{k}}(\mathbf{h}^d), d = 1 \ldots m, \tag{14}$$

under zero-boundary constrains:

$$\begin{aligned} &\delta M_{\boldsymbol{k}}(h_1, \ldots, h_p = 0, \ldots, h_n) = 0, \\ &\delta M_{\boldsymbol{k}}(h_1, \ldots, h_p \to \infty, \ldots, h_n) = 0, \forall p = 1 \ldots n. \end{aligned} \tag{15}$$

In this paper, the additional regularization condition of minimum curvature (Wang et al. 2006) is used to avoid overfitting.

The high-order moments are recursively constructed by starting from the - second-order indicator moments. First, second-order indicator moments $M_{k_0 k_p}(h_p), p = 1 \ldots n$ are calculated from the basic variogram model (David 1977). Then, the trend $M^0_{k_0,k_1,k_2}(h_1, h_2)$ is calculated using Eq. 12 and relations to the lower orders (9) and (10). Further, the residuals $\delta M_{k_0,k_1,k_2}(h_1^d, h_2^d)$ can be estimated from sampling indicator moments $\hat{M}_{k_0,k_1,k_2}(h_1^d, h_2^d)$ and Eq. 14. These residuals are used as points in the B-spline approximation (13) of the multidimensional function $\delta M_{k_0,k_1,k_2}(h_1, h_2)$ under zero-boundary constrains (15). Finally, the third-order spatial indicator moments are retrieved using equation (11). The same procedure is recursively repeated for fourth, fifth, and higher orders until the desired order is achieved.

3.1 Calculation of Sampling Statistics

The octant model is used (Fig. 4) to estimate the sampling indicator moments $\hat{M}_{\boldsymbol{k}}(\mathbf{h}^d)$ from the hard data. The neighborhood area of each hard data sample is divided into $N_\phi = 8$ sectors representing N_ϕ directions. Then, each sector is divided into N_r lags and forms an $N_r \times N_\phi$ bin template. Only one point within each bin is randomly chosen to construct a replicate. Finally, the values $\hat{M}_{\boldsymbol{k}}(\mathbf{h}^d)$ are estimated from replicates using law of large numbers:

$$\hat{M}_{\boldsymbol{k}}(\mathbf{h}^d) = \frac{1}{N_{\boldsymbol{h}^\mathbf{d}}} \sum_{j=1}^{N_{\mathbf{h}^\mathbf{d}}} I_{k_0}\left(z^j_{i_0}\right) \ldots I_{k_n}\left(z^j_{i_n}\right), d = 1 \ldots m, \tag{16}$$

where the sum is taken over all $N_{\mathbf{h}^\mathbf{d}}$ replicates with the spatial configuration $\boldsymbol{h}^d$, data samples $z^j_{i_0} \ldots z^j_{i_n}$ in the replicate j are separated by lags $\boldsymbol{h}^d$, and d is the index of different spatial configurations $\boldsymbol{h}^d$. It should be noted that replicates separated by at least half the variogram range should be used for the law of the large numbers to be applicable.

The amount of information about high-order statistics that can be retrieved from data crucially depends on the number of categories K, the total number of data samples N, and the level of correlation between values. It is not hard to see that the higher order of statistics considered are the larger the number of samples available should be. In order to have an adequate number of replicates for a particular order m of statistics, the minimum number of replicates $N_{repl}(m) \leq N_{\boldsymbol{h}^d}, \forall \mathbf{h}^d$ is set up by the user. However, more advanced techniques based on an entropy or information theory should be considered (Arndt 2004). Having a minimum number of replicates $N_{repl}(m)$ for the given order of statistics m, the optimal number of lags N_r^{opt} is calculated.

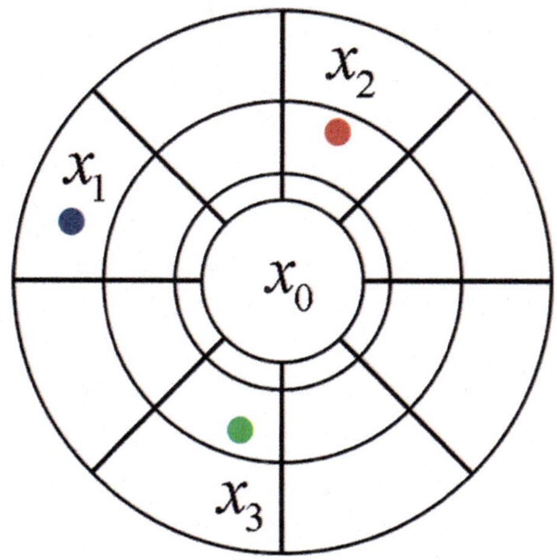

Fig. 4 Octant model for calculation of the sampling joint distribution $\hat{M}_k(\mathbf{h}^d)$

4 Simulation Algorithm

Combining all of the above, the new data-driven algorithm is formulated as follows:

Algorithm A.1

1. Starting from two-point statistics $m = 2$ until stopping criteria is reached.
   (a) Define the minimum number of replicates $N_{repl}(m)$ for given order m.
   (b) Scan the hard data using the octant model (Fig. 4) with different N_r starting with the number of the radial divisions $N_r = 2$ and find the higher N_r for which the average number of replicates is bigger than $N_{repl}(m)$.
   (c) If $N_r = 2$ and the average number of replicates is less than N_{repl}, then exit the loop.
   (d) Save all the replicates for obtained N_r and the order of statistics m.
   (e) Increase the order of statistics $m = m + 1$.
2. Define a random path visiting all the unsampled nodes.
3. For each node $\mathbf{x}_{i_0}$ in the path:
   (a) Find the closest data samples $\mathbf{x}_{i_1}, \mathbf{x}_{i_2}, \ldots \mathbf{x}_{i_n}$. The categories at these nodes are denoted by $k_1, \cdots k_n$.
   (b) For all $k_0 = 1 \cdots K$, calculate the high-order spatial indicator moments $\hat{M}_k(\mathbf{h}^d)$ using formula (15) from the replicates found in step 1 and recursive model (11–13). Note that $k_1, \cdots k_n$ are fixed. For the orders higher than maximum order $\boldsymbol{m}$, consider $\delta M_{\boldsymbol{k}}(\mathbf{h}) \equiv 0$.
   (c) Calculate the conditional distribution from joint distribution:

$$P\left[Z_{i_0} = k_0 \middle| Z_{i_1} = k_1 \ldots, Z_{i_n} = k_n\right] = A M_{\boldsymbol{k}}(\mathbf{h}), \tag{17}$$

where coefficient A is the normalization coefficient:

$$A = 1 / \sum_{k_0=1}^{K} M_k(\mathbf{h}) \tag{18}$$

(d) Draw a random value z_{i_0} from this conditional distribution (17) and assign it to the unsampled location $\mathbf{x}_{i_0}$.
(e) Add z_{i_0} to the set of sample hard data and the previously simulated values.

4. Repeat Steps 3a–e for all the points along the random path defined in Step 2.

5 Simulation Results

The proposed approach is tested on the data set from the Stanford V reservoir case study (Fig. 1a). This image is discretized on categories 0, 1, and 2 and is used as a reference image (Fig. 5a). Hard data is randomly selected from the image and shown in Fig. 5b. This represents 520 points (5 % of the image points). The results are compared with sequential indicator simulation algorithm (*sisim*; Journel and Alabert 1990; Deutsch and Journel 1998).

The simulation results for the case with 520 data samples are shown in Fig. 6. Neither the training image nor the reference image (Fig. 5a) is used during the simulation and is presented herein only for the sake of comparison. Simulations are done in two modes: using only boundary conditions (Fig. 6b) and using both conditions (14) and (15) (Fig. 6c). In the case of using just boundary conditions, the result is smooth and the width of channels is overestimated because all high-order statistics are derived from the second-order statistics. However, *sisim* simulation results (Fig. 6a) are less connected and the channels can be hardly detected. The result obtained with the account of higher-order statistics from data (Fig. 6c) reproduces the channels quite well with adequate dimensions of geometrical bodies.

On Fig. 6c, the histograms for data samples, the reference image, *sisim* simulation, and the simulation using the proposed technique are shown by blue, light blue, yellow, and red bars, respectively. The deviations from the distribution in hard data are small for both *sisim* and proposed algorithm simulations.

The second-order statistics are compared in Fig. 7a. The direction $\mathbf{e}_1$=(1,0) is used. The indicator moments $\mathbf{M}_{01}(h1)$ of simulations using *sisim* and the proposed algorithm are reproduced well. Nevertheless, the third-order statistics $\mathbf{M}_{012}(h1,h2)$ of the reference image (Fig. 7b), simulation using *sisim* algorithm (Fig. 7c), and the result of the proposed technique (Fig. 7d) are quite different. The third-order spatial indicator moments are calculated using directions $\mathbf{e}_1$=(1,0) and $\mathbf{e}_2$=(1,0). Some similarities of patterns can be traced in the bottom part of Fig. 7b, d, which correspond to statistics of the reference image and the simulation using the

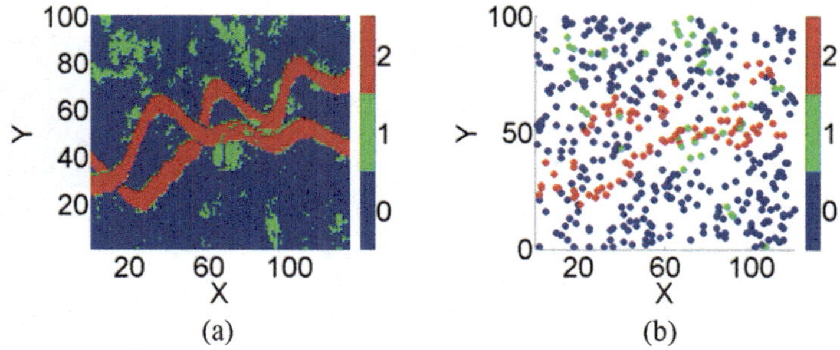

Fig. 5 Case study with 520 data samples: (**a**) the reference image, (**b**) data samples

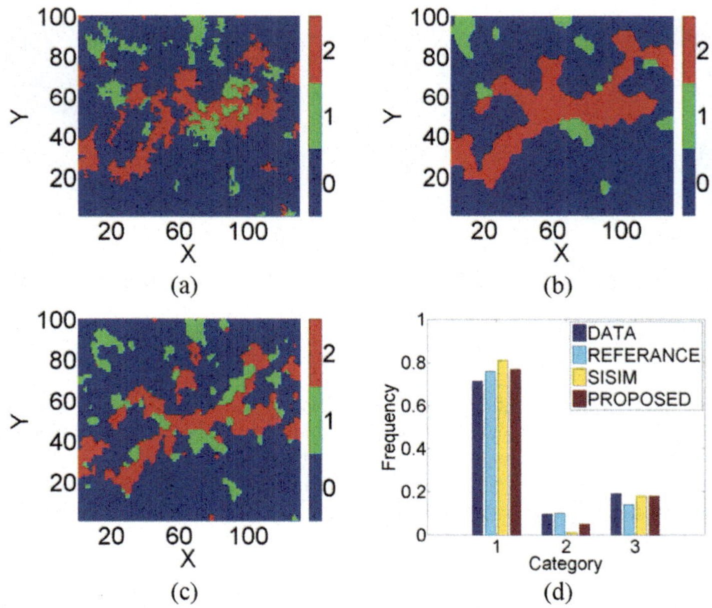

Fig. 6 Case study with 520 data samples: (**a**) sisim simulation result, (**b**) the simulation using proposed algorithm with only boundary conditions, (**c**) the simulation result using both boundary conditions and high-order statistics from data. Subfigure (**d**) shows histograms for data samples, the reference image, sisim simulation, and the simulation using the proposed technique presented by *blue*, *light blue*, *yellow*, and *red bars*, respectively

proposed algorithm. However, the point of interest is the reproduction of the high-order spatial statistics of the hard data.

The surface on Fig. 8 is a 3D view of Fig. 7d. Dots represent the statistics calculated from the hard data. Colors show the number of triplets used for the calculation of the third-order spatial indicator moment $\mathbf{M}_{012}(h1,h2)$. The higher the number of triplets, the more reliable the value of the point is. The spatial indicator

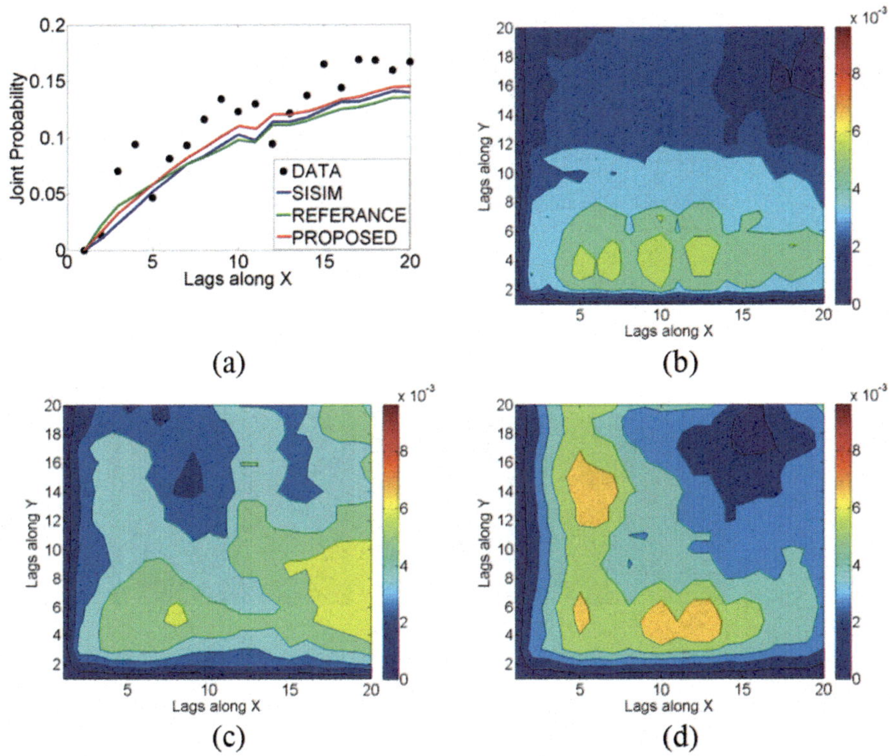

Fig. 7 Case study with 520 data samples. (**a**) Second-order spatial indicator moment $M_{01}(h_1)$ using directions $\mathbf{e}_1=(1,0)$ for data samples, the reference image, *sisim* simulation, and the simulation using proposed technique are presented by *black dots* and *green*, *blue*, *red lines*, respectively. The third-order spatial indicator moment $M_{012}(h_1,h_2)$ using directions $\mathbf{e}_1=(1,0)$ and $\mathbf{e}_1=(1,0)$ for (**b**) the reference image, (**c**) *sisim* simulation result, (**d**) the simulation using both boundary conditions and high-order statistics from data

moment $\mathbf{M}_{012}(h1,h2)$ of the simulation using the proposed algorithm tends to fit more reliable points and is consistent with the boundary conditions (Fig. 8).

6 Conclusions

This paper presented a new data-driven, high-order sequential method for the simulation of categorical random fields. The sequential algorithm is based on the B-spline approximation of high-order spatial indicator moments that are consistent with each other. The main distinction from commonly used MPS methods is that, in the proposed technique, conditional distributions are constructed using high-order spatial indicator moments as the functions of distances based on hard data. The simulations herein are generated without a TI. Note that in applications with

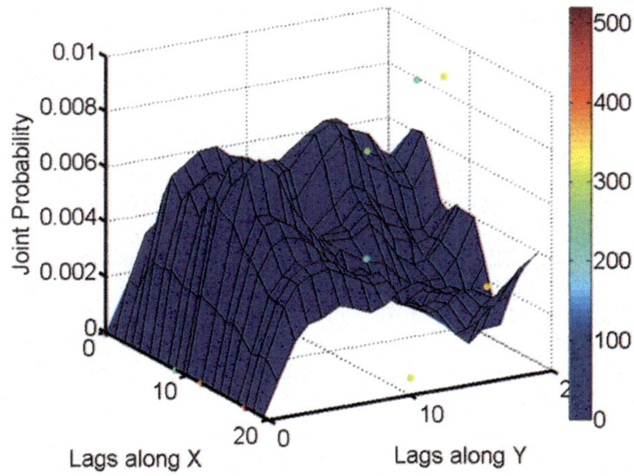

Fig. 8 The third-order spatial indicator moment $M_{012}(h_1,h_2)$ using directions $\mathbf{e}_1=(1,0)$ and $\mathbf{e}_2=(1,0)$ for the simulation using both boundary conditions and high-order statistics from data. Colors show the number of triplets found in data using template on Fig. 4

relatively large numbers of data, as in the simulation of mineral deposits, the higher-order statistics are deduced from hard data. The option of adding a TI to a dataset is available for sparser datasets.

The basic concept of the algorithm is to use recursive approximation models with enclosed boundary conditions, which are derived from the nested nature of high-order spatial indicator moments presented herein. To provide robust estimation, the regularized B-splines are used.

Another important aspect is the different amount of information that can be retrieved for different levels of relations. In the proposed method, each order of spatial statistics is approximated using the appropriate number of B-splines to provide robustness to the algorithm and to avoid overfitting. Thus, lower-order statistics are estimated with higher resolution than the higher-order statistics.

The simulation algorithm is tested on the categorized data from the Stanford V reservoir case study and compared with results of the *sisim* algorithm. No TI is used during simulations. According to the results, the proposed method reproduces the complex spatial patterns, such as channels, and preserves high-order statistics.

The proposed technique is fully data driven; however, the information from the TI can be incorporated with the suggested model as a trend to capture high-frequency features of the TI. Further research is concerned with the application to 3D models, improving the efficiency, testing for unbiasedness of the proposed approximation model, and generalization to the continuous random fields.

Acknowledgments Funding was provided by the Natural Sciences and Engineering Research Council of Canada (NSERC) Discovery Grant 239019 and mining industry partners of the COSMO Lab (AngloGold Ashanti, Barrick Gold, BHP Billiton, De Beers Canada, Kinross Gold, Newmont Mining, and Vale).

Bibliography

Arndt C (2004) Information measures: information and its description in science and engineering. Springer, Amsterdam

Babenko K (1986) Fundamentals of numerical analysis. Nauka, Moscow

Caers J (2005) Petroleum geostatistics. SPE–Pennwell Books, Houston

Chilès JP, Delfiner P (2012) Geostatistics: modeling spatial uncertainty, 2nd edn. Wiley, New York

Chugunova TL, Hu LY (2008) Multiple-point simulations constrained by continuous auxiliary data. Math Geosci 40:133–146

Cressie NA (1993) Statistics for spatial data. Wiley, New York

David M (1977) Geostatistical ore reserve estimation. Elsevier, Amsterdam

David M (1988) Handbook of applied advanced geostatistical ore reserve estimation. Elsevier, Amsterdam

Deutsch C, Journel A (1998) GSLIB: geostatistical software library and user's guide, 2nd edn. Oxford University Press, New York

Dimitrakopoulos R, Mustapha H, Gloaguen E (2010) High-order statistics of spatial random fields: exploring spatial cumulants for modeling complex non-Gaussian and non-linear phenomena. Math Geosci 42:65–99

Evans J, Bazilevs Y, Babuška I, Hughes T (2009) N-widths, sup–infs, and optimality ratios for the k-version of the isogeometric finite element method. Comput Methods Appl Mech Engrg 198:1726–1741

Goovaerts P (1998) Geostatistics for natural resources evaluation. Cambridge University Press, Cambridge

Guardiano J, Srivastava RM (1993) Multivariate geostatistics: beyond bivariate moments. Geosatistics Tróia'92:133–144

Journel AG (1993) Geostatistics: roadblocks and challenges. Stanford Center for Reservoir Forecasting

Journel AG (2003) Multiple-point Geostatistics: a State of the Art. Stanford Center for Reservoir Forecasting

Journel AG, Alabert F (1990) New method for reservoir mapping. Pet Technol 42(2):212–218

Journel AG, Huijbregts CJ (1978) Mining geostatistics. Academic, San Diego

Kitanidis PK (1997) Introduction to geostatistics—applications in hydrogeology. Cambridge Univ Press, Cambridge

Mao S, Journel A (1999) Generation of a reference petrophysical/seismic data set: the Stanford V reservoir. 12th Annual Report. Stanford Center for Reservoir Forecasting, Stanford

Mariethoz G, Kelly B (2011) Modeling complex geological structures withelementary training images and transform-invariant distances. Water Resour Res 47:1–2

Mariethoz G, Renard P (2010) Reconstruction of incomplete data sets or images using direct sampling. Math Geosci 42:245–268

Matheron G (1971) The theory of regionalized variables and its applications. Cahier du Centre de Morphologie Mathematique, No 5

Mustapha H, Dimitrakopoulos R (2010a) A new approach for geological pattern recognition using high-order spatial cumulants. Comput Geosci 36(3):313–334

Mustapha H, Dimitrakopoulos R (2010b) High-order stochastic simulations for complex non-Gaussian and non-linear geological patterns. Math Geosci 42:455–473

Pyrcz M, Deutsch C (2014) Geostatistical reservoir modeling, 2nd edn. Oxford University Press, New York

Remy N, Boucher A, Wu J (2009) Applied geostatistics with SGeMS: a user's guide. Cambridge University Press, Cambridge

Straubhaar JRP, Renard P, Mariethoz G, Froidevaux R, Besson O (2011) An improved parallel multiple-point algorithm using a list approach. Math Geosci 43:305–328

Strebelle S (2002) Conditional simulation of complex geological structures using multiple point stastics. Math Geosci 34:1–22

Strebelle S, Cavelius C (2014) Solving speed and memory issues in multiple-point statistics simulation program SNESIM. Math Geosci 46:171–186

Tikhonov AN, Arsenin VY (1977) Solution of Ill-posed problems. Winston & Sons, Washington, DC

Tjelmeland H, Besag J (1998) Markov random fields with higherorder interactions. Scand J Stat 25(3):415–433

Toftaker H, Tjelmeland H (2013) Construction of binary multi-grid Markov random field prior models from training images. Math Geosci 45:383–409

Vargas-Guzmán J (2011) The Kappa model of probability and higher-order rock sequences. Comput Geosci 15:661–671

Vargas-Guzmán J, Qassab H (2006) Spatial conditional simulation of facies objects for modeling complex clastic reservoirs. J Petrol Sci Eng 54:1–9

Wang W, Pottmann H, Liu Y (2006) Fitting B-spline curves to point clouds by curvature-based squared distance minimization. ACM Transactions on Graphics 25(2):214–238

Webster R, Oliver MA (2007) Geostatistics for environmental scientists. Wiley, New York

Yahya WJ (2011) Image reconstruction from a limited number of samples: a matrix-completion-based approach. Mater Thesis, McGill University, Montreal

Zhang T, Switzer P, Journel A (2006) Filter-based classification of training image patterns for spatial simulation. Math Geosci 38(1):63–80

Conditional Bias in Kriging: Let's Keep It

M. Nowak and O. Leuangthong

Abstract Mineral resource estimation has long been plagued with the inherent challenge of conditional bias. Estimation requires the specification of a number of parameters such as block model block size, minimum and maximum number of data used to estimate a block, and search ellipsoid radii. The choice of estimation parameters is not an objective procedure that can be followed from one deposit to the next. Several measures have been proposed to assist in the choice of kriging estimation parameters to lower the conditional bias. These include the slope of regression and kriging efficiency.

The objective of this paper is to demonstrate that both slope of regression and kriging efficiency should be viewed with caution. Lowering conditional bias may be an improper approach to estimating metal grades, especially in deposits for which high cutoff grades are required for mining. A review of slope of regression and kriging efficiency as tools for optimization of estimation parameters is presented and followed by a case study of these metrics applied to an epithermal gold deposit. The case study compares block estimated grades with uncertainty distributions of global tonnes and grade at specified cutoffs. The estimated grades are designed for different block sizes, different data sets, and different estimation parameters, i.e., those geared toward lowering the conditional bias and those designed for higher block grade variability with high conditional biases.

M. Nowak (✉)
SRK Consulting (Canada) Inc., Oceanic Plaza, 22nd Floor, 1066 West Hastings Street, Vancouver, BC V6E 3X2, Canada
e-mail: mnowak@srk.com

O. Leuangthong
SRK Consulting (Canada) Inc., Suite 1300, 151 Yonge Street, Toronto, ON M5C 2W7, Canada
e-mail: oleuangthong@srk.com

J.J. Gómez-Hernández et al. (eds.), *Geostatistics Valencia 2016*, Quantitative Geology and Geostatistics 19, DOI 10.1007/978-3-319-46819-8_20

1 Introduction

When resource modeling by kriging, a number of estimation parameters must be established such as block model block size, minimum and maximum number of data used to estimate a block, or search ellipsoid radii. Arik (1990) studied the impact of search parameters in two case studies for gold and molybdenum and considered the drill density, skewness of the grade distribution, and availability of a suitable geology model. The studies involved estimating with drill hole data and comparing against blasthole data. As Arik demonstrated and many resource modelers know, the choice of the estimation parameters is by no means an objective procedure that provides a simple recipe for all types of deposits.

This long-standing topic is not new and has been addressed by many authors. At the center of the discussion is the issue of conditional bias, wherein the expected value of the true grade conditioned on the estimates is not equal to the estimated grade (Journel and Huijbregts 1978; Olea 1991):

$$E\{Z|Z^* = z\} - z \neq 0$$

The discussion among both theoreticians and practitioners revolves around how conditional bias affects the quality of the block grade estimates (McLennan and Deutsch 2002). The first school of thought insists that the conditional bias should be as small as possible and must be dealt with, and the second school of thought believes that the conditional bias should be ignored and the variability of block estimates should be as high as the variability of underlying true block grades.

Rivoirard (1987) suggested that the size of the kriging neighborhood should consider the weight given to the mean. If the mean is given a large weight, then the neighborhood should be expanded so as to increase the slope of regression and thereby reduce conditional bias. Conversely, if the weight of the mean is low, then a localized neighborhood is adequate. Krige introduced a metric called kriging efficiency (Krige 1997) that correlates to the slope of regression. He contends that one should never accept conditional bias in an effort to reduce the smoothing effect of kriging (Krige 1997; Krige et al. 2005). Deutsch et al. (2014) further expanded on potential sub-optimality of the estimates due to large conditional biases reflected in slope of regression and in kriging efficiency measures. Deutsch proposed a new expression of kriging efficiency to aid in the assessment of quality of estimated block grades.

From a procedural perspective, Vann et al. (2003) introduced quantitative kriging neighborhood analysis (QKNA) to optimize the estimation parameters for selection of the minimum/maximum of number of samples, quadrant search, search neighborhood, and block size. The proposed criteria for evaluating quality of block grade estimates include slope of regression of true block grades on estimated block grades, weight attached to the mean in simple kriging, distribution of kriging weights, and kriging variance.

All of the above contributors have focused on metrics and efforts to minimize conditional bias. On the other end of the spectrum of this discussion, Isaaks (2005) argued that estimates cannot be both conditionally unbiased and globally accurate at the same time. The estimates may be close to conditionally unbiased but the histogram of block estimates is smoothed, which results in inaccurate predictions of the recoverable tonnes and grade above cutoff grades. He advocates that a conditionally biased estimate is necessary to obtain a globally unbiased recoverable resource above cutoff grade. There is support that during the early stages of project feasibility assessments, it is more important to accurately predict the global recoverable reserves than to produce locally accurate estimates (Journel and Kyriakidis 2004).

Despite valid points on both ends of the spectrum, it appears that the first school of thought has been winning the discussion in recent years. The authors noticed a substantial increase in application of those measures for optimization of estimation parameters. Some of the proposed measures for optimization of kriging estimates, such as slope of regression and kriging efficiency, are currently readily available in most commercial software. In some organizations, this quantitative approach has become standard in the resource estimation process irrespective of the stage of exploration and/or development of the mineral deposit.

It appears that Isaaks' sound argument for recoverable resources above an economic cutoff grade, particularly in early stage projects, appears to have been forgotten in the popularization of a quantitative approach because of software accessibility. In the wake of convenience, we seem to have lost the idea of a fit-for-purpose model, including consideration for the stage of the project.

The objective of this paper is to demonstrate that both slope of regression and kriging efficiency should be viewed with caution for optimizing estimation parameters. Lowering conditional bias may be an improper approach to estimating metal grades. In fact, it might be outright wrong especially in deposits for which high cutoff grades are required for mining.

This paper presents a summary of the two typical tools, suggested for optimization of estimation parameters, slope of regression and kriging efficiency, followed by application of these metrics applied to an epithermal gold deposit. The case study compares block estimated grades with uncertainty distributions of global tonnes and grade at specified cutoffs. The estimated grades are designed for different block sizes, different data sets, and different estimation parameters, i.e., those geared toward lowering the conditional bias and those designed for higher block grade variability with high conditional biases.

2 Two Proposed Measures for Optimization of Estimation Parameters

Block ordinary kriging is one of the most common estimation methods used for resource modeling in the mining industry (Journel and Huijbregts 1978; Sinclair and Blackwell 2002). Each block estimate can be written as:

$$Z^*(\boldsymbol{u}) = \sum_{i=1}^{n} \lambda_i Z(\boldsymbol{u}_i)$$

where $Z^*(\boldsymbol{u})$ represents the estimated block grade at location vector $\boldsymbol{u}$, and λ_i is the kriging weight assigned to sample $Z(\boldsymbol{u}_i)$. The resource model, comprised of estimated block grade at all relevant locations, forms the distribution of estimated grades that is the basis for a mineral resource statement.

Among other considerations, such as geologic confidence, grade continuity, and database quality, a cutoff grade is used to differentiate between those blocks that are reported as a mineral resource (Sinclair and Blackwell 2002). This cutoff grade is applied throughout a project, depending on the method of mineral extraction. The smoothness of this estimated grade distribution, relative to the cutoff grade, is then paramount to this discussion of accurately predicting the global mineral resource for a project.

The smoothness of the estimated grade distribution depends not only on the quantity and location of conditioning data and the modeled variogram but also on the estimation parameters such as minimum/maximum number of samples, size of search neighborhood, and type of search. While there are a number of suggested measures for assistance in the choice of optimal kriging estimation parameters, such as slope of regression, kriging efficiency, or weight of the mean from simple kriging, this paper will only focus on the first two due to their prominent use in the mineral resources sector.

2.1 Slope of Regression

When kriged estimated Z^* block grades are plotted on X axis and unknown true Z block grade are plotted on Y axis, then the regression of true values given the estimates is an indication of the conditional bias in the estimate (Journel and Huijbregts 1978) (Fig. 1). Conditional bias takes place when the expected value of true block grade Z conditional to estimated block grade Z^* is not equal to the estimated grade. The slope of the regression b is often used to summarize the conditional bias of the kriging estimate:

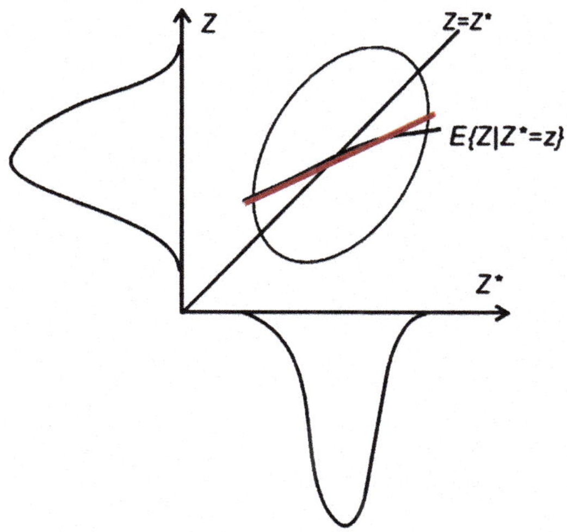

Fig. 1 Schematic illustration of conditional bias (McLennan and Deutsch 2002). The estimates Z^* are on the X axis, and the true block grades Z are on the Y axis

$$E\{Z|Z^* = z\} \approx a + \mathrm{bz} \tag{1}$$

Naturally, the block true values are unknown, but the slope b can be calculated once a variogram model is known by the following formula:

$$b = \frac{cov(Z, Z^*)}{\sigma^2_{Z^*}} \tag{2}$$

To calculate the slope from formula (2), it is enough to know kriging weights attached to samples used to estimate a block and to know covariances between samples and samples and the block. Note that actual sample grades are not taken into account in the calculation. The slope will be identical in both lower- and higher-grade areas, although potentially higher conditional bias, and by extension lower slope of regression, could be expected in the high-grade areas.

2.2 *Kriging Efficiency*

Kriging efficiency, introduced by Krige (1997), is considered a good measure of effectiveness of kriging estimates. The kriging efficiency can be calculated from two by-products of kriging procedure, block variance (σ^2_{Bl}) and kriging variance (σ^2_{kr}):

$$\mathrm{KE} = \frac{\sigma_{\mathrm{Bl}}^2 - \sigma_{\mathrm{kr}}^2}{\sigma_{\mathrm{Bl}}^2} \tag{3}$$

Kriging efficiency values can range from negative (poor estimates) to a maximum value of 1 (very good estimates). As with the slope of regression, actual assay data do not get used in the calculation. The results are purely dependent on a variogram model and on data locations used to estimate the blocks. Krige mentions that based on a number of case studies he conducted there is a correlation between the efficiency and the slope of regression (Krige 1997). For the increased slope value, there is also an increase in kriging efficiency.

3 Case Study

This case study is based on an epithermal gold deposit in British Columbia, localized along a major fault. Gold-bearing breccia, vein, and stockwork development occurs along the fault and in subsidiary dilational structures. Gold mineralization roughly parallels the fault. The deposit has been drilled out by more than 500 closely spaced holes drilled roughly on a 25×25 m grid. At one time, originally modeled from the full set of data, the lowest grade domain in the deposit had an average grade close to 0.7 g/t and the highest grade domain had an average gold grade more than 2 g/t with relatively low coefficient of variation at 1.3.

A portion of the deposit has been chosen for simulating gold grades on a dense $4 \times 4 \times 4$ m grid. The chosen portion of the deposit does not differentiate between different geological units. This simplified approach, considering closely spaced large number of drill holes, is not considered detrimental to the results of the study. A typical realization with average grades very similar to declustered assay grades from the drill holes has been chosen for this study. Maximum range of gold continuity is at 160° azimuth and is much shorter in vertical direction and at 70° azimuth (Fig. 2). The relative nugget effect of 25 % and ranges of continuity spanning from 40 to 60 m are quite typical and encountered in many gold deposits. Note that although the following analysis is based on simulated grades, considering a large number of drill holes in a relatively small area from which the simulated grades are derived, the simulated grades do represent a distribution that closely resembles actual grades in this deposit.

The simulated deposit, with an average grade of 1.26 g/t, has been "drilled" with "exploration" holes. The location of the "exploration" holes was essentially random with some limitations on allowed distances between the holes and on generally higher density of drilling in higher-grade areas. Two data sets were created: The first set (large data set) represents 281 holes drilled at 30 m spacing outside of high-grade mineralization and at down to 16 m spacing within higher-grade regions. The second set (small data set) represents more typical case with 156 drill holes drilled at 40 m distance in lower-grade areas and down to 20 m in higher-grade areas

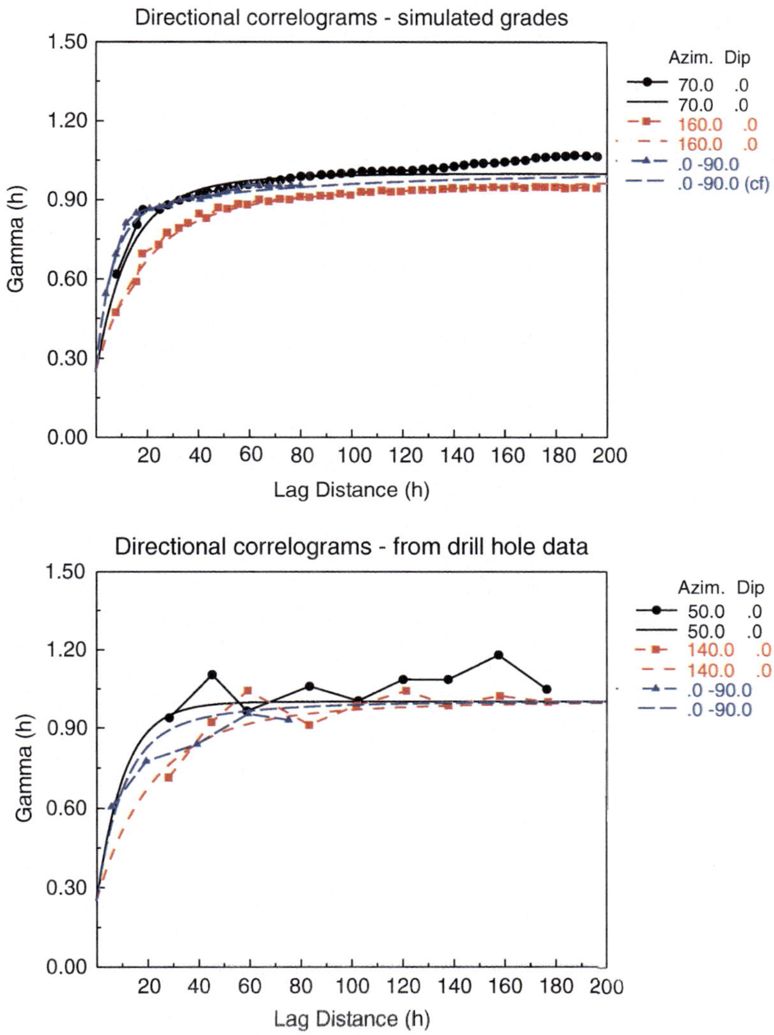

Fig. 2 Modeled correlograms from simulated grades (true data) and from exploration drilling

(Fig. 3). Drill hole spacing in the second data set is similar to what would be typically expected in a gold deposit at the early stages of exploration.

The smaller data set was used for variogram modeling. As is common in practice, the best continuity was modeled for directions slightly different than those known from simulated grades. Moreover, the modeled variogram parameters were different from those applied for the simulated reference data. Each data subset was used to estimate gold grades by ordinary kriging for different block sizes and different number of data allowed to estimate the blocks. Vertical block size was kept constant at 8 m with increased sizes along easting and northing directions.

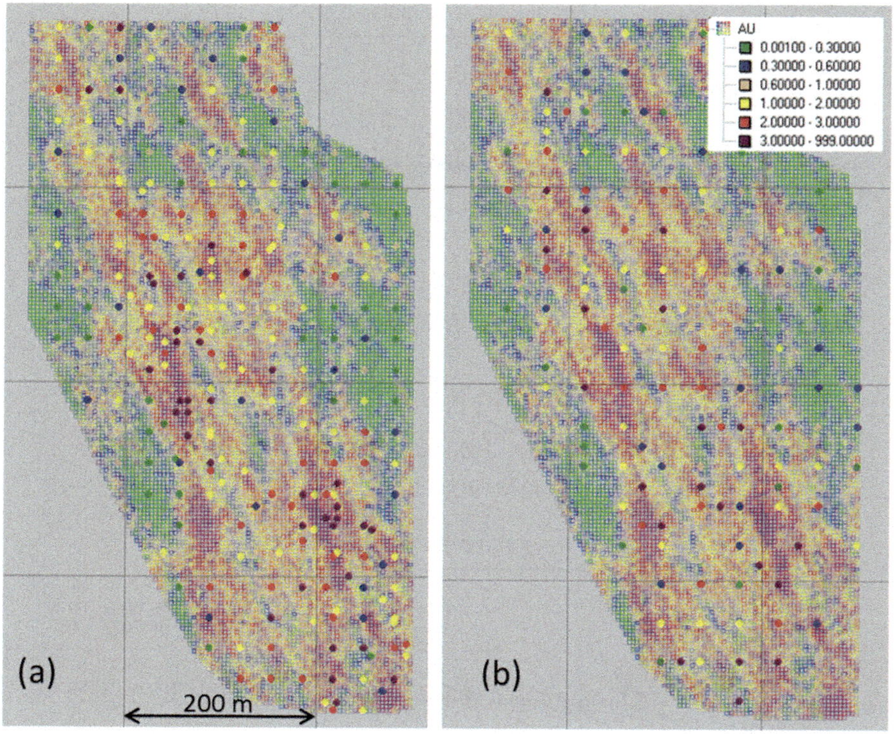

Fig. 3 Plan view of true block grades and data locations: (**a**) large data set, (**b**) small, more typical, data set

For strongly conditionally biased estimates, a small number of data was used with a minimum of 4 and a maximum of 12 composite assay grades (referred to as the "sub-optimal" design). In an attempt to lower the conditional bias, blocks were estimated with a large search ellipsoid with a minimum of 24 and maximum of 64 composite assays; this is referred to as the "optimal" design. For each estimated block grade, a slope of regression and kriging efficiency was calculated. In addition, the estimated block grades were compared to true block grades.

4 Results

Figure 4 shows how slope of regression and kriging efficiency change for different block sizes, different data sets, and estimation parameters used. As discussed, modification to the estimation parameters was limited to a number of data used starting from "poorly designed" not optimal estimation procedure with small number of data used for the estimation and ending with "well designed" optimal process with large number of data used for the estimation. As expected, the optimal

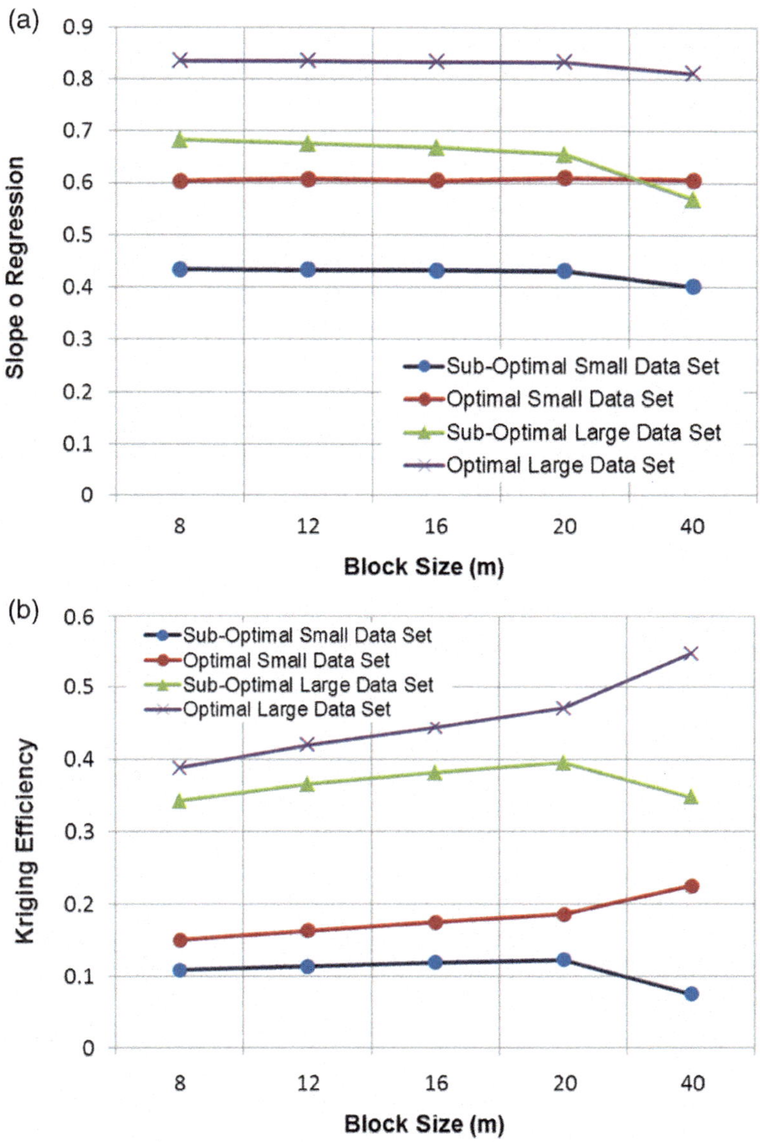

Fig. 4 Slope of regression (**a**) and kriging efficiency (**b**) for blocks estimated from two data sets for different block sizes and different estimation parameters

estimation procedure results in higher slope of regression and increases kriging efficiency.

It is interesting that for the small data set used in the estimation, the actual block size did not have any effect on the slope of regression and kriging efficiency. On the other hand, when the large data set was used there was gradual increase in kriging efficiency with the increased block size. Note that kriging efficiency is quite low regardless of the estimation type. As expected, using the large number of data (optimal case) to estimate block grades resulted in a substantial increase in the slope of regression, i.e., it resulted in substantial decrease of conditional bias. Not surprisingly, the slope of regression is quite high for the large data set and large number of data used for the estimation. These graphs clearly indicate that using a lot of data during the estimation process lowers conditional bias and increases kriging efficiency.

Now that it has been established that applying more data to the block estimates increases slope of regression and by extension decreases conditional bias, the next step involved comparisons of actual true block grades with the estimated grades for different cutoffs. Figures 5 and 6 show relative tonnage and grade differences between true and estimated block grades at the 1.0 g/t cutoff for optimal and not optimal estimates from the small data and the large data sets, respectively. Both figures show that despite high conditional bias in the sub-optimal design, the estimated tonnes and grades are closer to reference tonnage and grade in the deposit. This is also true for the estimates from the large data set, although here the differences between the not optimal and optimal models are smaller. Note that, as presented in Fig. 6, at 20 m block size estimated tonnes and grade are very similar to reference tonnes and grade. At the same time, it would be misleading to conclude that this block size produced superior estimates.

In fact, the reported tonnage and grade from different block sizes is quite similar (Fig. 7). It just happens that the estimated tonnes and grade in this specific deposit are comparable to recoverable tonnes and grade at the selective mining unit (SMU) size higher than 20 m, roughly the size of half of drill hole spacing. This is an important observation that suggests it does not matter what block size is used for estimating resources. Reported resource at 8 m or 20 m block size will be similar, but the 20 m block appears to approach the size that, if successfully applied during mining operation, would result in actually recovered tonnes and grade very similar to those estimated. As long as there is no connection made between a block size used and actual SMU considered for mining, there is nothing particularly wrong with estimates based on a small block size.

An important result of the optimization of the estimation process is high smoothing of the estimated block grades. The smoothing effect may result in large differences between estimated and actual metal content at higher cutoff values. For block estimates from the typical (smaller) data set, the estimated metal content at higher cutoffs may be as much as 70–80 % lower than the actual metal content (Fig. 8) for the "optimized" parameters, while the "sub-optimal" model yields 50–60 % less metal content relative to the reference. Similar conclusions can be made for the larger data set, with percent differences ranging from 20 to 30 % for the sub-optimal model and 30–50 % for the optimal model. Therefore, it is obvious that the lower the conditional bias the higher the smoothing effect that can be expected when estimating from sparsely spaced data. The interesting trend, however, is that for both the sparsely and densely sampled data sets, the

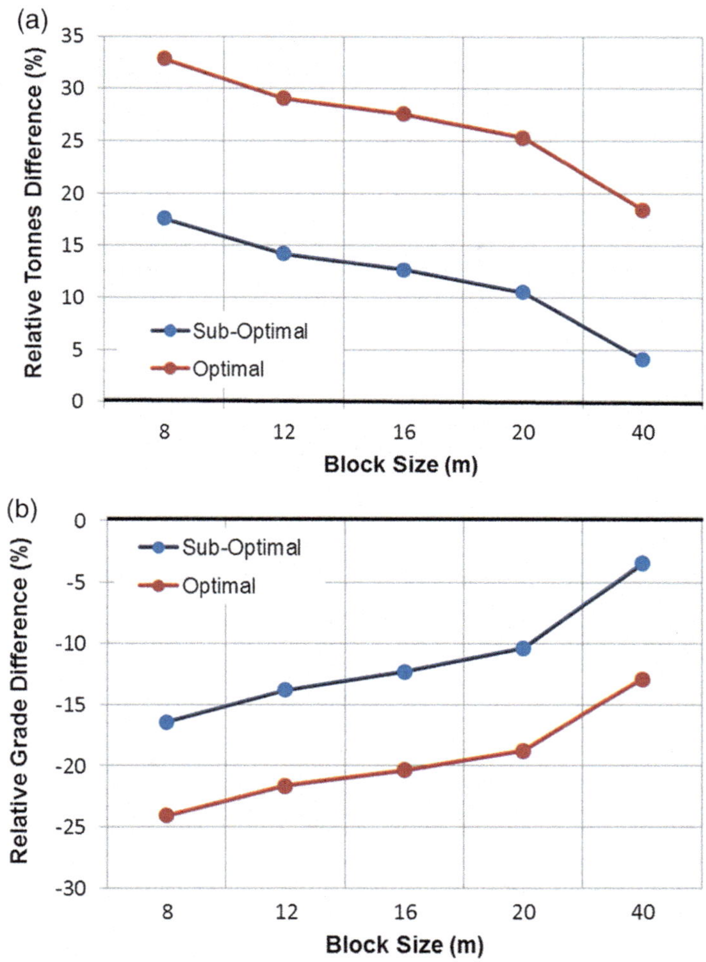

Fig. 5 Relative tonnage (**a**) and grade (**b**) differences between estimated and true block grades at 1.0 g/t cutoff. The blocks were estimated from typical, smaller data set

sub-optimal set of parameters yield the closest estimate of contained metal for cutoff grades above the 1.25 g/t mean grade.

The final task considered comparison of the estimation models with conditional simulation of the small data set. The purpose of this exercise was to compare the estimation model to a model constructed using a method that is considered to avoid conditional bias altogether (McLennan and Deutsch 2002; Journel and Kyriakidis 2004). The reference distribution was obtained via p-field simulation wherein the local distributions of uncertainty considered a local trend model. For this task, sequential Gaussian simulation was performed with no consideration for any trends, and variograms were calculated based on the small data set. As with the estimation models, the continuity directions vary slightly from those of the reference model.

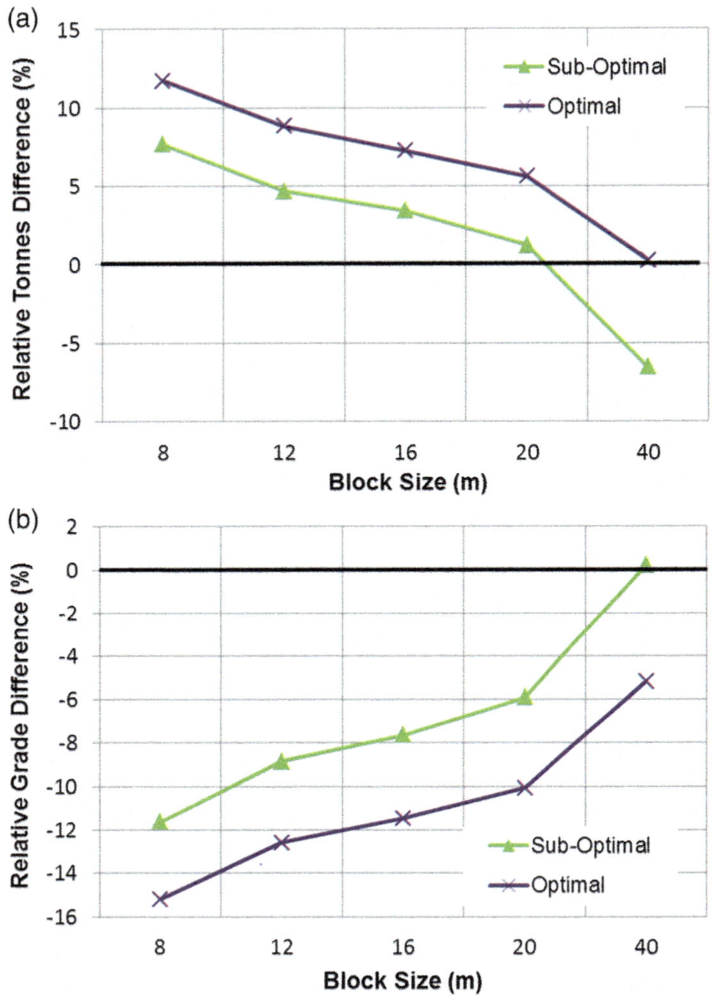

Fig. 6 Relative tonnage (**a**) and grade (**b**) differences between estimated and true block grades at 1.0 g/t cutoff. The blocks were estimated from large data set

Multiple realizations were then generated at a $2 \times 2 \times 2$ m resolution and subsequently block averaged to various block sizes: 8, 12, 16, and 20 m.

The uncertainty in grades and tonnage at a series of different cutoff grades were assessed, and the combined impact is summarized as contained metal at different cutoff grades (Fig. 9). The corresponding sensitivity curves for the sub-optimal and optimal estimation models, along with the reference model, are shown for comparison. Three interesting observations are made. Firstly, at a cutoff grade up to the mean grade of 1.25 g/t, there is no appreciable difference between the contained metal estimated using the optimal or sub-optimal parameters. Secondly, for the four block sizes considered, the sub-optimal parameters yield estimates closest to the

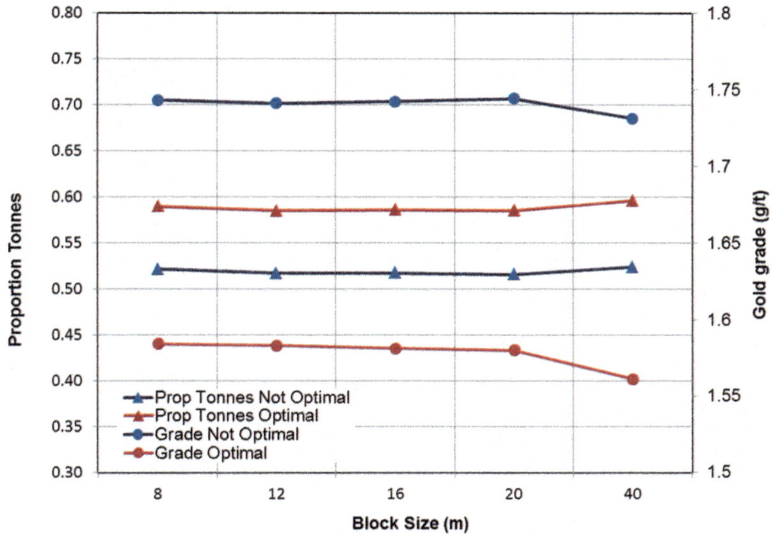

Fig. 7 Estimated from small data set grade and proportion of tonnage above 1.0 g/t cutoff for different block sizes

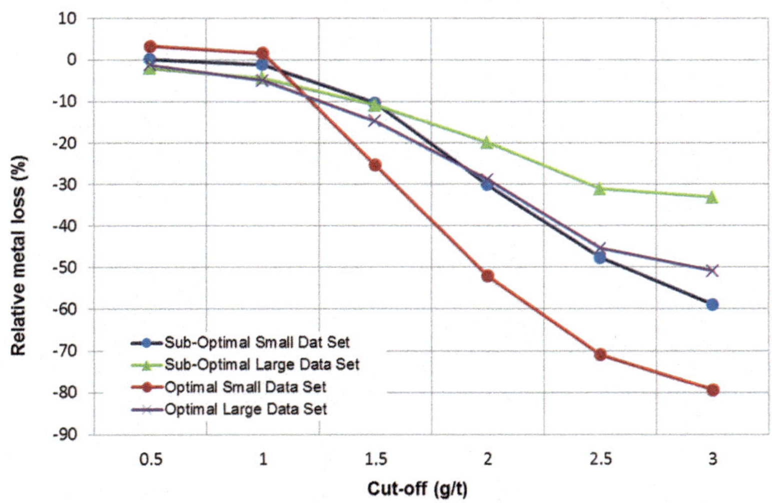

Fig. 8 Relative metal losses in estimated block grade models for different cutoffs in 16 m block models

reference model at higher cutoff grades. Thirdly, the simulation approach, which is considered to be a non-biased method, yields the distributions of uncertainty in contained metal that encompasses the reference data. This latter observation is important, particularly as a reference model is never available in practice for benchmarking purposes. This indicates that a conditional simulation approach can

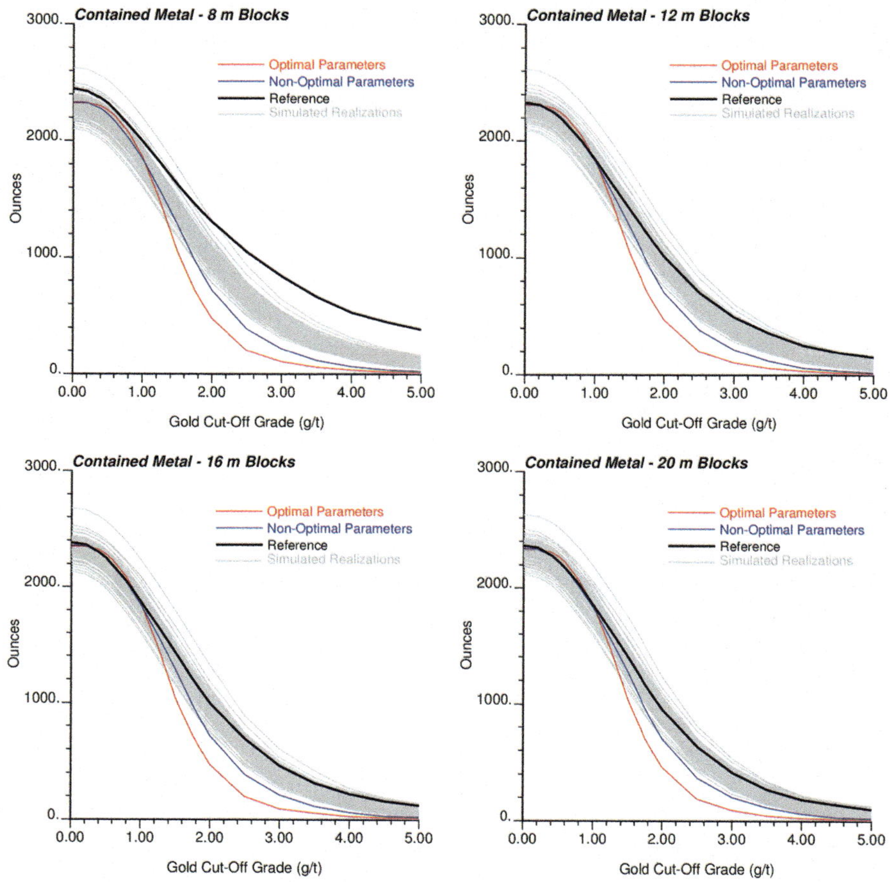

Fig. 9 Contained metal curves for estimated block models for different cutoff grades at 8 m, 12 m, 16 m, and 20 m block sizes

be used to validate the estimated tonnage, grades, and ultimately metal when determining an appropriate set of estimation parameters.

5 Discussion

Both slope of regression and kriging efficiency assume global stationarity within a specifically modeled domain. Mineral deposits are not stationary, even within a specific estimation domain. In an estimation domain, there are always small regions of high and small regions of very low grades. Slope of regression and kriging efficiency formulas do not take into account the fact that true variance of estimation errors depends on data values. In regions with higher grade or in regions with local data having more variance than in the whole domain, fluctuation of errors is larger. Disregarding these local changes in variability may lead to estimated resources that steer away from what would be expected of a typical resource estimate.

A resource block model should not be designed to produce an inventory of recoverable resource that is based on SMU size much larger than would generally be considered for mining, only because for this SMU size, when mined, the resource model will turn out to be correct. A deposit is never mined according to a resource block model. A decision on what will be mined will be based on grade control drilling, necessary both in open-pit and underground mining. A resource geologist should strive to produce a block model that predicts, reasonably well, the tonnage and grade that a mine can expect to achieve over the life of the mine, once it has sorted out its grade control procedures. In addition, the resource block models are often used for dilution calculations or blending issues. A block model designed from optimizing slope of regression, or block size, will not serve this purpose.

In short, optimizing kriged block estimates with slope of regression or kriging efficiency measures may lead to block models that do not adequately reflect true block grades. It is tempting and easy to use slope of regression and kriging efficiency for validation of block estimated grades. Those measures are commonly available in commercial software packages. Although theoretically high slope of regression, i.e., low conditional bias, is considered necessary for good quality estimates, in practice this approach may be outright harmful if the objective of the study is to predict global resource quantities above an economic cutoff grade. Both measures are a reflection of a modeled variogram and data locations and do not take into account actual assay values or their variability in the vicinity of an estimated block. Moreover, it is often quite difficult to construct a reliable variogram model, particularly in early exploration stages, and relying on its metrics to design "best" resource estimates cannot be considered best practice.

Based on the presented case study, there is strong indication that it is better to have conditionally biased block estimates for global resource quantities required for a potential investment decision, life-of-mine planning, and/or development decisions. Once a cutoff is applied to block estimated grades for reporting or further mining studies, it is better to have unsmoothed conditionally biased block estimates. In this context of achieving globally accurate predictions, it looks like the onus is back on a resource geologist to design estimation parameters that produce a realistic block model that reflects the underlying true block grades the best way possible.

Bibliography

Arik A (1990) Effects of search parameters on kriged reserve estimates. Int J Min Geol Eng 8 (4):305–318

Deutsch J, Szymanski J, Deutsch C (2014) Checks and measures of performance for kriging estimates. J South Afr Inst Min Metall 114:223–230

Isaaks E (2005) The kriging oxymoron: a conditionally unbiased and accurate predictor, 2nd edn, Geostatistics Banff 2004. Springer, Dordrecht, pp 363–374
Journel A, Huijbregts C (1978) Mining geostatistics. Academic, London
Journel A, Kyriakidis P (2004) Evaluation of mineral reserves: a simulation approach. Oxford University Press, New York
Krige D (1997) A practical analysis of the effects of spatial structure and of data available and accessed, on conditional biases in ordinary kriging, Geostatistics Wollongong '96, Fifth International Geostatistics Congress. Kluwer, Dordrecht, pp 799–810
Krige D, Assibey-Bonsu W, Tolmay L (2005) Post processing of SK estimators and simulations for assessment of recoverable resources and reserves for South African gold mines, Geostatistics Banff 2004. Springer, Dordrecht, pp 375–386
McLennan J, Deutsch C (2002) Conditional bias of geostatistical simulation for estimation of recoverable reserves, CIM Proceedings Vancouver 2002. CIM Proceedings Vancouver 2002, Vancouver
Olea R (1991) Geostatistical glossary and multilingual dictionary. Oxford University Press, New York
Rivoirard J (1987) Teacher's aide: two key parameters when choosing the kriging neighborhood. Math Geol 19:851–856
Sinclair A, Blackwell G (2002) Applied mineral inventory estimation. Cambridge University Press, Cambridge
Vann J, Jackson S, Bertoli O (2003) Quantitative kriging neighbourhood analysis for the mining geologist – a description of the method with worked case examples. 5th International Mining Geology Conference, Bendigo, pp 1–9

Operational SMU Definition of a Brazilian Copper Operation

Joao Dirk Reuwsaat, Cassio Diedrich, Roberto Menin, and Wellington F. De Paula

Abstract Achieving estimated mining recoverable grade and tonnage curves is a common challenge in mining operations. One of the most impacting parameters in the estimation process of these curves is the selective mining unit (SMU) size, defined as the smallest volume of material that can be selectively extracted as ore or waste. Such an important parameter as SMU is often defined by general practical and theoretical rules based on sampling grid without taking into account aspects such as local geology, mining configuration, and production rates. This paper presents a short-term and production reconciliation application study for defining block support size at a Brazilian copper and gold mining operation. Realistic mining recoverable curve is expected as a result of an operational SMU definition. The resulting operational SMU application demonstrates how more accurate estimates are to the actual mining recoverable curves at given cutoffs compared to the ones obtained using traditional SMU definition, commonly applied by the mining industry.

1 Introduction

Estimated mineral resources models are made to predict the deposit tonnage and grade distribution curves that will be extracted and fed into the plant (Rossi and Deutsch 2014).

Geostatistical estimates of mining recovery curves may be somewhat optimistic, as they do not take into account factors other than block grades, commonly disregarding mining configuration (method and geometry), local geology in the production environment, and production rates. Mining dilution (planned and/or

J.D. Reuwsaat (✉) • R. Menin • W.F. De Paula
Vale Base Metals, 63 Grajau Street, 1st floor, Carajás, PA 68516-000, Brazil
e-mail: joao.dirk@vale.com; roberto.menin@vale.com; wellington.paula@vale.com

C. Diedrich
Vale Base Metals, 337 Power Street, Door #105, Copper Cliff, Sudbury, ON P0M 1N0, Canada
e-mail: Cassio.Diedrich@vale.com

J.J. Gómez-Hernández et al. (eds.), *Geostatistics Valencia 2016*, Quantitative Geology and Geostatistics 19, DOI 10.1007/978-3-319-46819-8_21

unplanned) is an important driver in the operational process and generally impacts the tonnage and grade distributions, ultimately affecting milling, stockpiling, and waste dump processes. Models that try to predict these curves taking into consideration mining dilution and ore losses are called recoverable models (David 1977; Journel and Huijbregts 1978).

Operational (planned and/or unplanned) dilution is regularly accounted as global factors in the mining plan by planning engineers. However, the SMU definition should be a back-to-front exercise considering expected operational aspects, such as dilution, that will result in a different (larger in general) SMU size, depending on the level of operational selectivity.

A practical way to evaluate the operational dilution and to select the SMU size that better fits the actual operational tonnage and grade curves is presented in this paper. It is mainly based on the comparison of actual mined information (or from a simulated/mimicked mining plans based on expected operational production rates) to the commonly estimated block model provided by resource evaluators. The work is illustrated using a Brazilian copper/gold mining operation database.

2 Dilution

Dilution can be split in more than one type (Rossi 2002):

- Support dilution
- Geological contact dilution
- Mining operational dilution

Support dilution is the change in the distribution due to the size of the support being evaluated (Parker 1980). It is the type of dilution evaluated by selectivity analysis: the understanding of the support impact on the spatial data distribution to build a model that captures it.

Contact dilution is the dilution due to the mixture of materials of different qualities inside each block, which should be properly weighed.

Operational (planned and unplanned) dilution is a mixture of material due to the mining operation (grade control and equipment operation) itself. Planned dilution is directly related to the production rate and the size of the grade control polygons (ultimately defined as the operational selective size). Unplanned dilution is directly related to the equipment operation of the grade control polygons (considering already planned dilution) and its operational mining practices. This is generally more variable and difficult to predict since it is related to operational issues (failing to follow dig lines, blasting backbreak, topography errors, etc.).

3 Copper Operation Case Study

The case study was based on a copper/gold operation mine in a massive deposit with large production rate. Three block models are defined in the production process:

1. Long-term model: used in the life of mine, 5Y, and annual mining plans
2. Short-term model: used in the quarterly and monthly plans
3. Grade control model: used in the daily mining operation

The long-term model uses only diamond drill hole (DDH) information. The short-term and the grade control models use only blast holes data (approx. 6×6 m). The long-term and short-term models use a $15 \times 15 \times 15$ m SMU size. The grade control model corresponds to the estimated blasting polygons of the mining process. Figure 1 shows long-term block model along with the mining polygons on a specific level.

The tonnages and grades on this study are compared between models using all mining blasting polygons and production data existing in the mine. Therefore, all the tonnage and grade reports correspond only to the mined portion of the deposit. The long-term block model was updated considering production reconciliation information (geological contacts and short-term grade variations) for this study; thus both long- and short-term models are practically the same in this operational example. Hence, all comparisons in the study used the long-term model.

Support dilution is taken into account in the long-term model through ordinary kriging estimation plans, ensuring that block distribution is coherent with the change-of-support data analysis. Contact dilution is taken into account by tonnage weighting of the materials inside each block.

In the mining scheduling process, long-term engineers compute tonnages and grades based on the long-term model. Mining blasting polygons, however, are the main drivers of the operational process. This is ultimately considered as the grade control model and has "irregular" SMU size as a result of the local mining process and its resulting quality values. As the models (long-term compared to the grade control model) have distinct support, they have different tonnage and grade curves. Comparing these curves, the differences due to operational configuration (mining in polygons) is clear. Figure 2 shows these curves, where it shows how much is not recovered for each cutoff. This difference is a measure of the planned dilution.

Long-term mine planning take into account that each SMU will be selectively mined as ore or waste. However, mining operational process is based on blasting polygons and local operating particularities, which can result in poor reconciliation of long-term block models and the mining recovery curves, indicating lesser levels of selectivity. Hence, the importance of acknowledging planned dilution is verified.

Planning engineers usually consider planned/unplanned dilution as a constant factor. However, dilution is not constant and depends on the spatial distribution of the deposit. Homogeneous regions should have less dilution than regions with more contact variability. Therefore, if the production rate remains constant, without

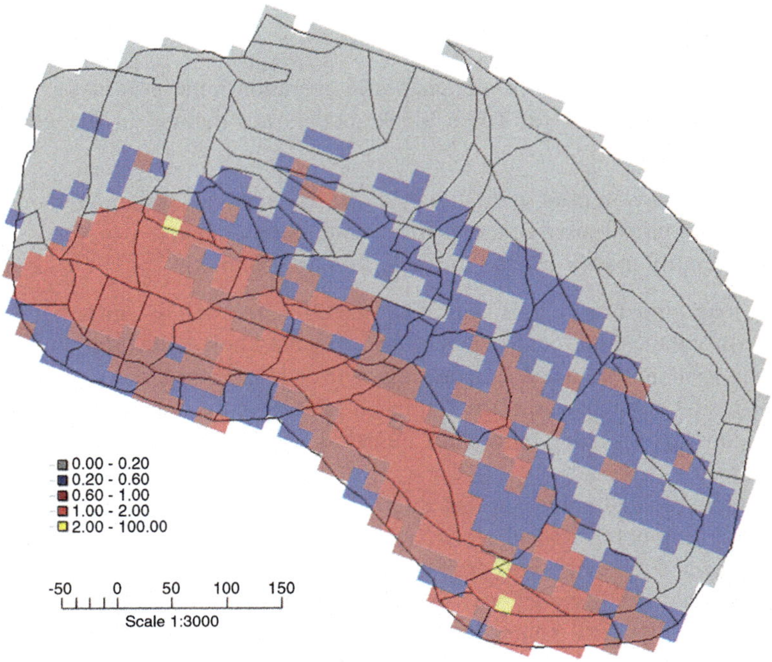

Fig. 1 Mining polygons with the long-term block model for a specific level

changing significantly the sizes of the ore polygons, it is logical to increase the SMU size to account for the operational planned dilution instead of using factors.

4 Results

The SMU size definition that will better represent actual mining recovery curves was based on an iterative long-term change-of-support process, keeping 15 m as bench height. The models were reblocked based on the official $15 \times 15 \times 15$ m long-term block model. Figure 3 presents the comparison between $15 \times 15 \times 15$ and $30 \times 30 \times 15$ m models. Density was considered as a weighting factor in the reblocking process. Figure 4 shows the tonnage and grade curves from all the models. It can be observed that the grade control model (production polygons) corresponds to a larger support size than the long-term model.

The impact of change of support is clear when computing the tonnage, grade, and metal differences between long-term and the reblocked and grade control models. Figure 5 presents the differences in tonnage, grade, and metal, respectively. The grade control curve is presented by a thick black line.

The selection of the SMU size is made by a simple visual interpretation of the metal recovery and difference curves, as it accounts for both tonnage and grade

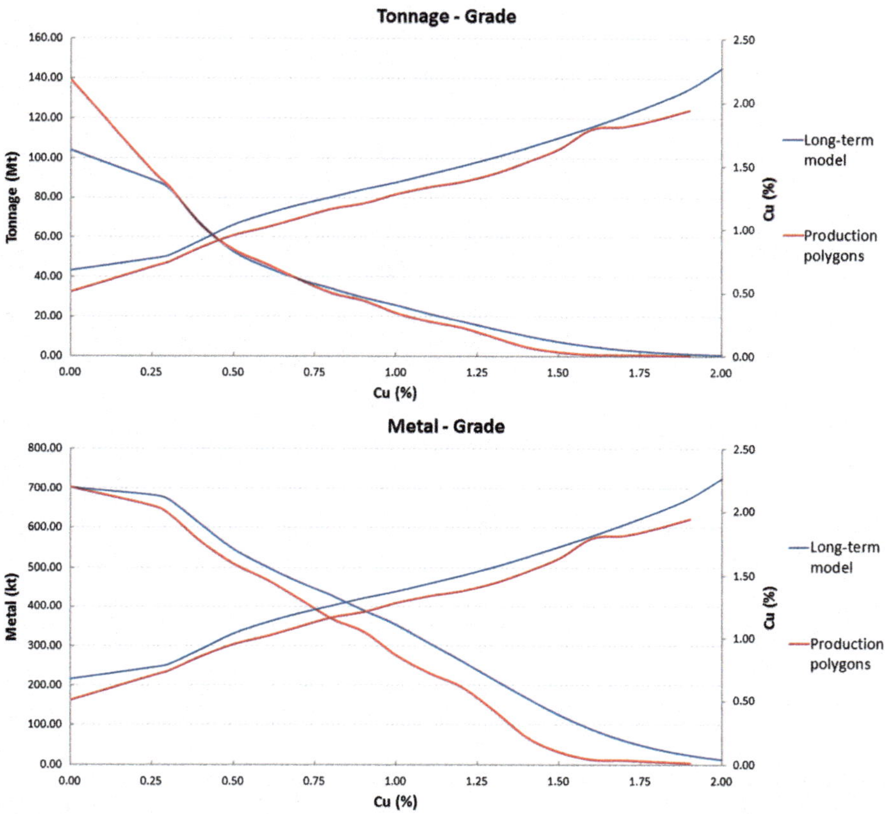

Fig. 2 Grade control and long-term model tonnage and grade curves

results. Figure 5 shows that the grade control model near matches the 30 × 30 × 15 m long-term model for most of the distribution curve (between 0.2 % Cu and 1.2 % Cu). Figure 6 shows 15 × 15 × 15 m and 30 × 30 × 15 m metal and grade mining recovery curves compared to the grade control model. The results indicated that a near-perfect match to the actual distribution curve (grade control) is unlikely, since no irregular SMU sizes are recognized and applied in the mining industry for different mining configurations, production rates, and local geology. As a consequence, impacts in mining companies' downstream processes such as milling, stockpiling, and financials are commonly verified. Mineral resource evaluators and mining planners should start to apply different mining concepts for "achieving" the expected operational mining recoverable curves at given cutoffs. Irregular local SMU sizes could be an option.

Nevertheless, in this case study, the 30 × 30 × 15 m model would better fit operational distribution curves than the 15 × 15 × 15 m as a result of the accounted actual planned dilution, avoiding the use of a gross dilution factor.

Fig. 3 Reblocking example from 15 × 15 × 15 to 30 × 30 × 15 m block size

5 Discussion

A real operational database was used for the construction of the mining recoverable curves and method definition. The method is simple and helpful in the selection of the SMU size, accounting the planned operational dilution. Unplanned dilution was not considered in this study; however it should be accounted as a full evaluation of the mining process.

In cases where no actual production information is available, planning engineers can simulate the grade control practice to arrive at ore/waste dig limits that will be implemented in the mine for operational mining sequencing. Then engineers and geostatisticians can verify the impact of the production scale on the dilution and consequently on the recoverable model. Figure 7 shows the impact of the dilution on metal distribution for several real production polygons. The difference of metal in Fig. 7 is changed to positive for better graphic visualization. The metal loss is higher for larger polygon sizes (larger production rate and mining configuration).

If the operational mining recovery curve does not match to any regular SMU support, it could be the case of using irregular operational mining unit (OMU) sizes (Diedrich et al. 2016) for defining local mining recoverable curves based on an initial mining schedule. The idea is not to forecast future short-term SMU grade distributions at the time of mining and not questioning estimation methods for defining grades at SMU scales. However it is to determine a reasonable regular and/or irregular block support for a chosen conventional SMU block model size that represents ore/waste proportions derived from a real expected mining sequencing and grade control process. In the OMU approach, there is no commitment that an

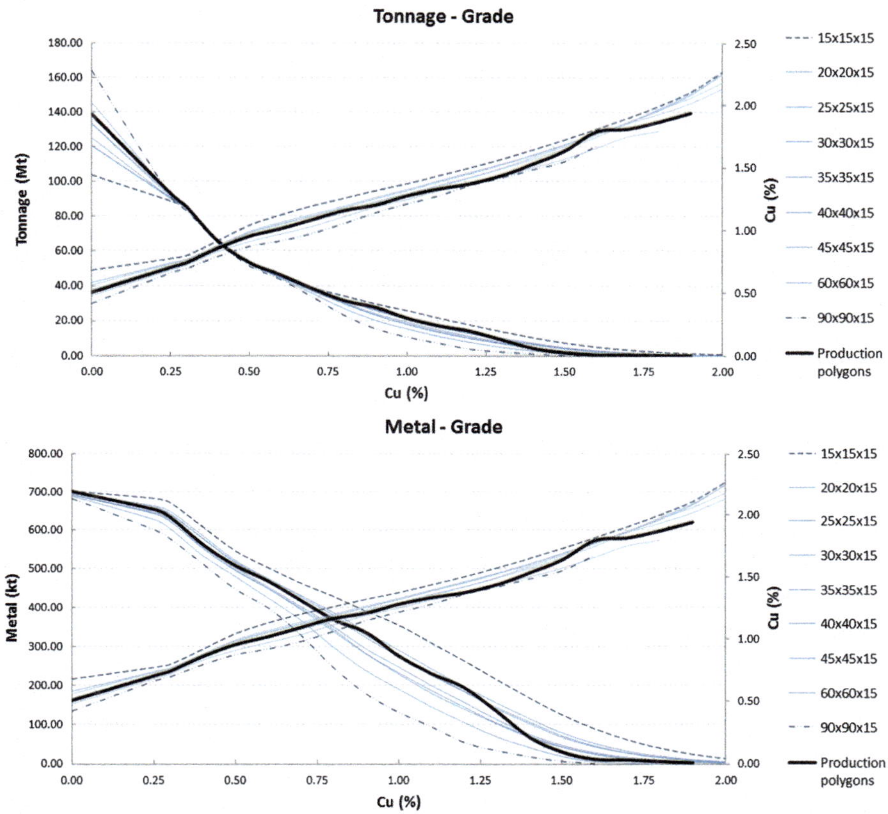

Fig. 4 Tonnage and grade curves for several models

individual block will be selectively mined, but it will be contributing accordingly to the "real" expected global and/or local mining recovery curves.

As previously mentioned, unplanned dilution could also be accounted for on the SMU selection process. However, an evaluation through reconciliation process between grade control data and operational field execution (equipment operation, processing results, etc.) is recommended. Therefore, the gains and losses can be computed and a factor can be obtained for different operating mining configurations, local geology, and production rates practices.

6 Conclusion

Recognizing the impact of operational dilution (planned and unplanned) and the fact that it is not constant over the deposit, operational aspects should be included in the mining unit block size definition and, consequently, on the resulting mining

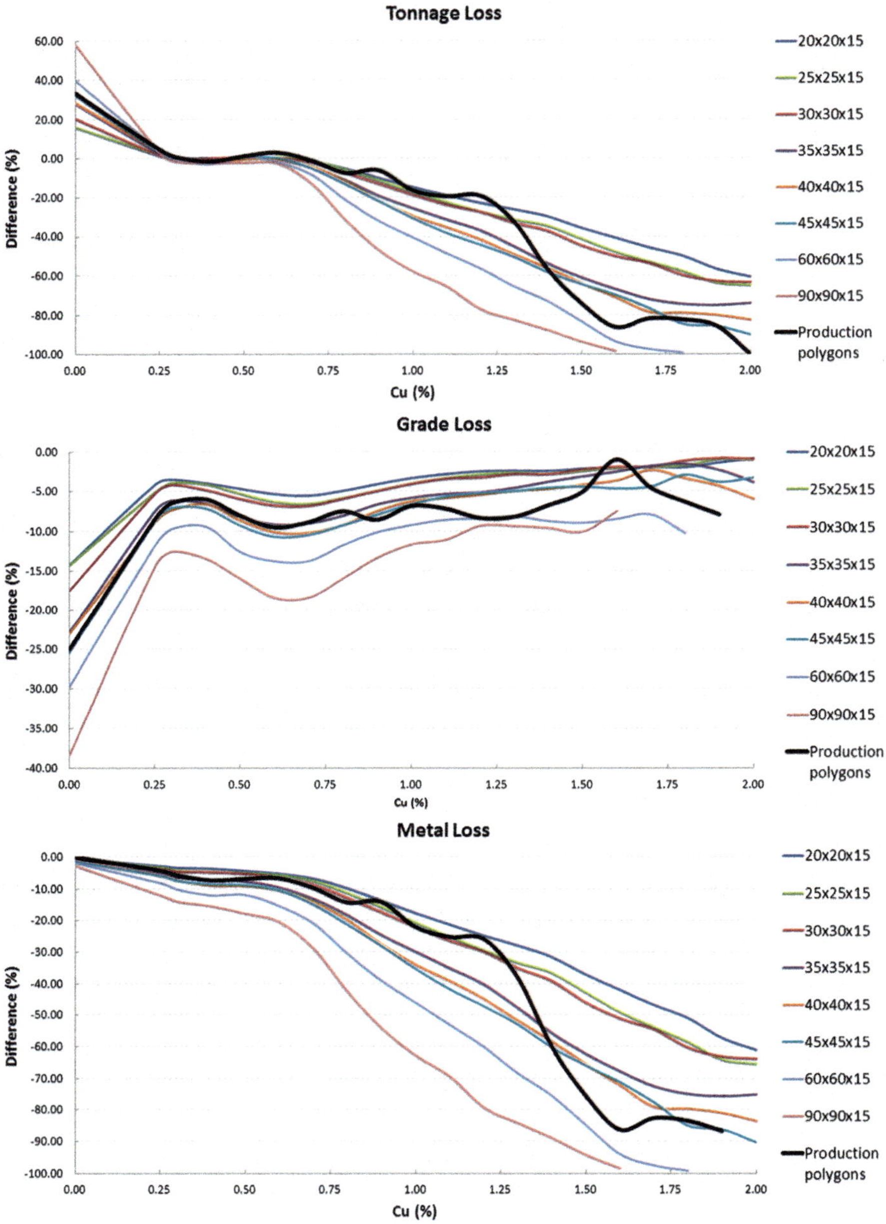

Fig. 5 Tonnage, grade, and metal differences between the reblocked and the long-term models

recoverable models, instead of using global factors. Selection of the appropriate mining unit is a compromise between getting the right mining recovery estimates for tonnages of ore and waste and getting the right grade of these materials. This

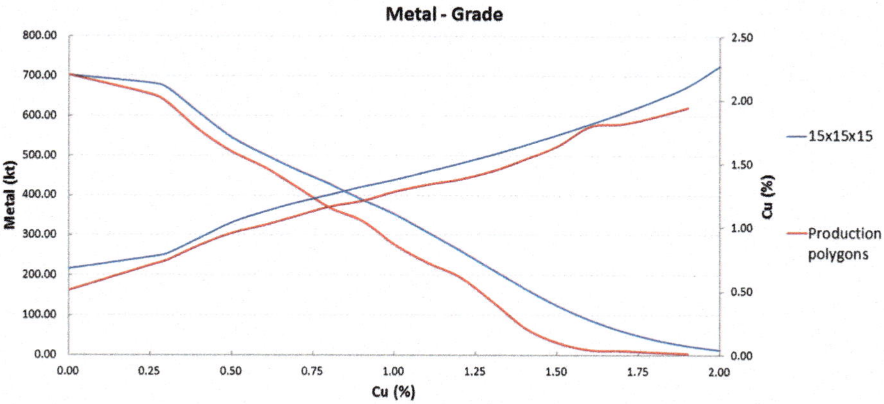

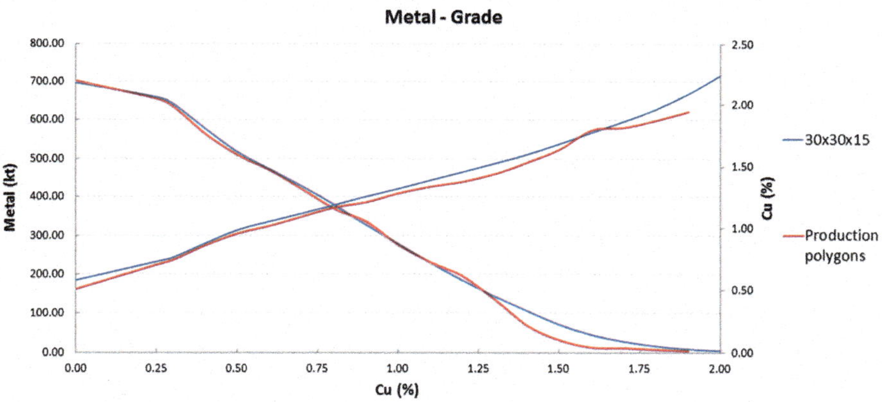

Fig. 6 Metal and grade curves of the 15 × 15 × 15 m long-term model, 30 × 30 × 15 m long-term model, and the grade control model

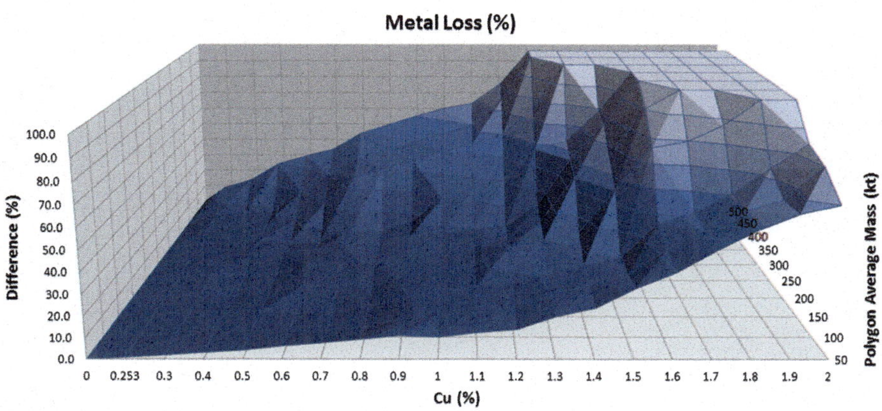

Fig. 7 Dilution sensibilities as a function of the polygon size

compromise depends on the global and/or local ore and waste proportions for predefined operational production rates and mining configurations. Significant differences between operational and long-term recovery curves may result in poor mining, waste dumps, and stockpile plans and strongly impact financials.

Bibliography

David M (1977) Geostatistical ore reserve estimation. Elsevier, Amsterdam

Diedrich C, Reuwsaat JD, Menin RG, Paula WF (2016) Selection of Operational Mining Unit (OMU) size, GEOSTATS2016. Universitat Politècnica de València, Valencia

Journel AG, Huijbregts CJ (1978) Mining geostatistics. Academic, New York

Parker HM (1980) The volume-variance relationship: a useful tool for mine planning. In: Geostatistics. McGraw-Hill, New York, pp 61–91

Rossi ME (2002) Recursos Geológicos o Reservas Mineras? Proceedings from the Sextas Jornadas Argentinas de Ingeniería de Minas. San Juan

Rossi ME, Deutsch CV (2014) Mineral resource estimation. Springer, Dordrecht

From the Spatial Sampling of a Deposit to Mineral Resources Classification

Jacques Rivoirard, Didier Renard, Felipe Celhay, David Benado, Celeste Queiroz, Leandro Jose Oliveira, and Diniz Ribeiro

Abstract In the mining industry, estimated mineral resources of a deposit are classified into inferred, indicated, or measured resources, depending upon their level of confidence. From a geostatistical point of view, this depends on the hole spacing and continuity of the mineralization in the deposit or in the different parts of the deposit. This also depends on some nominal volume on which level of confidence is sought. This corresponds typically to an annual expected production volume, not to the next week production block or to the whole deposit (unless it is small). Here we propose a geostatistical classification of mineral resources in two steps. The first step consists in measuring the spatial sampling density of the deposit (or throughout the deposit when this density varies). This is done using a specific volume, which is similar to the inverse of the classical density of sample points in space (this density being a number of samples per volume, its inverse is a volume), but which takes into account the variographic structure of the regionalized variable of interest. This first step allows comparing objectively the spatial sampling density of different deposits or parts of deposit. The second step first converts such a specific volume into a coefficient of variation on the nominal production volume resources. Then a mineral resource category is obtained by thresholding this coefficient of variation. By choosing fixed thresholds for a given commodity and type of deposit, this provides an objective classification of the resources from different deposits or parts of deposits. The proposed method is illustrated on three case studies.

J. Rivoirard (✉) • D. Renard
MINES ParisTech, Centre de géosciences, PSL Research University, 35 rue St Honoré, 77300 Fontainebleau, France
e-mail: jacques.rivoirard@mines-paristech.fr; didier.renard@mines-paristech.fr

F. Celhay • D. Benado
Codelco, El Teniente Division, Rancagua, Chile
e-mail: FCelhay@codelco.cl; DBenado@codelco.cl

C. Queiroz • L.J. Oliveira • D. Ribeiro
Vale, Belo Horizonte, Brazil
e-mail: celeste.queiroz@vale.com; leandro.jose.oliveira@vale.com; diniz.ribeiro@vale.com

J.J. Gómez-Hernández et al. (eds.), *Geostatistics Valencia 2016*, Quantitative Geology and Geostatistics 19, DOI 10.1007/978-3-319-46819-8_22

1 Introduction

Broadly speaking, the classification of the resources of a deposit depends on the geological model, the quality of samples, and, from the geostatistical point of view addressed here, on the hole spacing. The spatial sampling of a deposit is a time process. At a given time, a deposit (or part of a deposit) will be classified as inferred/indicated/measured upon the confidence of its resources (Rossi and Deutsch 2014; Jorc 2012). While the confidence can be computed for the whole domain or for each individual block from the block model, such volumes are generally not appropriate in the middle and long-term view of the management. The confidence to be sought is rather the confidence on the resources that correspond to the production volume for a given period, typically the 1-year production volume, even if this is not always precisely defined when classifying resources. An additional difficulty comes from the fact that such 1-year production volumes may not be finely delineated and could come from different places within the deposit. To summarize, the classification of a deposit, or its partition into different classes, should depend on the hole spacing as well as on expected production volumes.

Common geostatistical tools (linear geostatistics with variances, conditional simulations) are able to provide the confidence on resources corresponding to domains, but this necessitates such domains to be delineated. In the proposed approach, we will first characterize the sampling density within the deposit with appropriate tools (Rivoirard 2013; Rivoirard and Renard 2016). Then the resources of the domain or of parts of it can be categorized, given expected production volumes. In the following the methodology is presented first and then is illustrated by three applications.

2 Tools

The tools which will be used are derived from the geostatistical concept of estimation variance and have been developed to measure the sampling density, either with a regular sampling design or an irregular one. A reminder of these tools is presented here. More details are available in the above references.

Let $Z(x)$ be the regionalized variable under study, assumed to be additive. Consider a domain V, an estimation $Z(V)^*$ of $Z(V)$ by inner samples of V, and its estimation variance $\sigma_E^2(V)$. By definition the "spatial sampling density variance" (ssdv) associated to this estimation $Z(V)^*$ is (Rivoirard 2013):

$$\chi(V) = {\sigma_E}^2(V)|V|.$$

Suppose now that V is partitioned into v_i informed by samples, with consistent estimates:

$$Z(V)^* = \sum_i \frac{|v_i|}{|V|} Z(v_i)^*.$$

If the estimation errors are uncorrelated, then the estimation variances can be combined as:

$$\sigma_E^2(V) = \sum_i \frac{|v_i|^2}{|V|^2} \sigma_E^2(v_i)$$

so that χ is additive:

$$\chi(V) = \sum_i \frac{|v_i|}{|V|} \chi(v_i).$$

This is exactly the case of random stratified sampling, where each stratum v_i is sampled by samples with random locations that are uniform and independent (note that this is true for model-based geostatistics, but the same property exists in random sampling theory, Cochran (1977)). In practice the correlation between errors, if not zero, can often be neglected. Matheron (1971) has developed appropriate principles of approximation in order to compute estimation variances by combination of variances in different situations. In the particular case of a two-dimensional square regular grid and usual isotropic variogram models, V being divided into N cells with same support v, we have:

$$\sigma_E^2(V)|V| = \sigma_E^2(v)|v|,$$

that is, $\chi(V) = \chi(v)$. We can deduce that the *ssdv* of any union V of v_i is the same: the spatial sampling defined by the regular grid pattern is characterized by this *ssdv*. As a consequence the estimation variance of any set V (even not connected) made of such v_i can be derived:

$$\sigma_E^2(V) = \frac{|v|}{|V|} \sigma_E^2(v) = \frac{\chi}{|V|}.$$

Remark 1: The non-correlation between errors is not general and can be tested by simulations. Having disjoint volumes estimated by inner samples does not imply that errors are uncorrelated, even approximately (example of elongated blocks sampled by centered samples which are close to each other). On the contrary the correlation between errors often tends to vanish when considering bigger blocks.

Remark 2: Because the estimates such as $Z(V)^*$ make use of inner samples only, they are less precise than kriging and so $\sigma_E^2(V)$ that can be derived is pessimistic compared to kriging variance $\sigma_K^2(V)$. This is particularly the case when V is composed of a small number of v.

Note that $\chi(V) = \sigma_E^2(V)|V|$ is homogeneous to the square variable times the volume (or the area in two dimensions). Other expressions can be easier to manipulate (Rivoirard and Renard 2016):

- The normalized specific volume, obtained by dividing *ssdv* by the variogram sill (if any) or using the normalized variogram: $v_0 = \chi/C(0)$ (indeed this is a volume).
- The specific volume, obtained by dividing *ssdv* by the squared mean of the variable or using the relative variogram: $V_0 = \chi/M^2$. Just as *ssdv* can be used to get $\sigma_E^2(V)$, the specific volume can be used to get the nominal coefficient of variation (CoV) for the estimation of *Z*(*V*):

$$\mathrm{CoV} = \frac{\sigma_E(V)}{M} = \sqrt{\frac{V_0}{|V|}} \tag{1}$$

Such specific volumes are similar to the inverse of the classical density of points in space (number of points per unit volume). In irregular sampling pattern, when there are variations of sampling in space (not due to known and distinctly sampled areas), the *ssdv* or specific volume can be estimated and mapped just like the density of points, that is, using a moving kernel, for instance, a parallelepiped (but similarly the result will depend on this). In mining, this can be done by "superkriging," that is, kriging superblocks from inner samples only (Rivoirard and Renard 2016). These superblocks are centered at every block of the block model (usually these model blocks are small and most are not sampled).

In the following, the methodology is applied to different cases. The first case is two-dimensional with a regular sampling. The deposit is divided into big blocks which have the size of the cell and negligible errors correlation. The *ssdv* or the specific volume, which measures the density of the sampling pattern, is derived from the estimation variance of such a big block from its sample. Then they are used to compute the coefficient of variation of the resources contained in a production area of given size, which will fix the category of resources.

The two other cases are three dimensional with irregular sampling. A map of *ssdv* or specific volume, which measures the density of the sampling pattern, is derived from the kriging variance of a moving superblock from its samples. Then this map can be used to make the distinction between differently sampled areas. It also allows computing the coefficient of variation of the resources contained in a production volume of given size, which will fix the category of resources in each area.

3 Application: Regular Sampling in Two Dimensions

The case of a large deposit studied in two dimensions and sampled with regular grids is frequent and is presented here as a generic example. The two-dimensional regionalized variable of interest can be the thickness of the deposit (typically when the grade in metal presents small variations) or the metal accumulation. Corresponding resources are either volumes or abundances. In the present case, the resources will be volumes, and the variable of interest will be the thickness. It has a mean of 10 m, a sample point variance of 10, and a variogram equal to $1\,\mathrm{nug}(h) + 9\,\mathrm{sph}(h/200\ \mathrm{m})$.

The initial sampling grid is 100 × 100 m. This has been progressively centered, giving a regular grid of 71 × 71 m, and finally tightened down to a pre-exploitation grid of 50 × 50 m. The specific volumes will be computed to measure the spatial pattern corresponding to each grid. We first compute the extension variance of the grid cell centered by a sample using the variogram and divide it by the squared mean of the variable. Equivalently we can compute directly the estimation variance using the relative variogram (i.e., divided by the squared mean), which is $0.01\,\mathrm{nug}(h) + 0.09\,\mathrm{sph}(h/200\ \mathrm{m})$. The specific volume (actually a two-dimensional area) is obtained by multiplying the result by the cell area. Each grid is characterized by its specific area, which decreases when the grid is refined (Table 1). This provides an objective way to quantify the sampling efficiency corresponding to a given grid for a variable having given statistics. It enables comparison between different grids for the same deposit, as well as comparison between the sampling of different deposits of the same type.

Let us for instance consider a second deposit of the same type as the first one, but with a thickness having a mean of 8 m and a variogram equal to $6\,\mathrm{nug}(h) + 6\,\mathrm{sph}(h/200\ \mathrm{m})$. Note that it is less thick and more variable than the first deposit. Its initial sampling grid is 70 × 70 m, practically equal to the second grid of the first deposit. However, because of its variability, this grid corresponds here to a specific area of 502 m^2, that is, a spatial sampling twice less good than the 100 × 100 m grid of the first deposit (269 m^2).

To convert specific volumes or areas V_0 in resource category, we must have an idea of the exploitation rate to be considered. Let us go back to the first deposit and consider an annual exploited area of $V = 100{,}000\ \mathrm{m}^2$ (i.e., 40 blocks with size 50 × 50 m). This corresponds to a volume of ore around 1,000,000 m^3. The CoV on the average thickness, that is, on this ore volume in such an area V, depends on the grid through its specific volume V_0 and is given by Eq. 1 (Table 1). The classification itself finally depends on the thresholds based on the CoV that the company has chosen for this type of resources. Suppose that the CoV on annual resources must be less than 2.5 % for *measured* resources and less than 5 % for *indicated* resources. Then, the resources would be *inferred* where the grid is 100 × 100 m, *indicated* where it is 71 × 71 m, and *measured* where it is 50 × 50 m. But with a higher production rate, resources where the grid is 100 × 100 m could be upgraded from *inferred* to *indicated*, while with a smaller production rate, resources where the grid

Table 1 Specific area V_0 for different grids, corresponding to the relative variogram equal to $0.01\,\text{nug}(h) + 0.09\,\text{sph}(h/200\ \text{m})$. CoV represents the coefficient of variation that can be deduced for an estimation of the target variable within an area of 100,000 m^2

Grid	V_o	CoV
100×100	269 m^2	5.2 %
71×71	111 m^2	3.3 %
50×50	46 m^2	2.1 %

is 50×50 m could be downgraded from *measured* to *indicated*. Of course this is only the geostatistical point of view on resources classification, since other criteria than hole spacing should also be considered.

4 Application: Metal Grade in Three Dimensions with Irregular Sampling Pattern

This case study corresponds to the underground mine of El Teniente porphyry copper deposit, located all around its waste pipe (Fig. 1). A 520 m thick horizontal section is considered here. The model is based on $20 \times 20 \times 20$ m blocks. Several estimation units had been distinguished and mapped manually (Fig. 2). Units 1 and 2 are poor and on the edges, and we are mainly interested in the rich units 3–6. Individually these units look like spaghetti with a vertical continuity and have been used for the estimation (Fig. 3). For the purpose of classification, we will consider the set made by of all these units 3–6 and delimitated by units 1 and 2. The resources correspond to copper abundance, and the regionalized variable of interest is the copper grade. The sampling is made every 6 m along holes in various directions within sections. The spacing between sections is 50 m, and the grid is about 50×50 m although the sampling is irregular.

The sample copper grades in units 3–6 have a mean of 1.097 %, a variance of 0.236, a standard deviation of 0.486, and a coefficient of variation of 0.443. The experimental variograms are horizontally isotropic but exhibit a strong quasi-zonal anisotropy, with a high continuity observed vertically along holes (Fig. 4). The variogram has been modeled by:

$$\gamma(h) = 0.03\,\text{nug}(h) + 0.053\,\text{sph}(h/15) + 0.06\,\text{sph}(h/70) \\ + 0.08\,\text{sph}(h/220, 220, 4000)$$

where the very large vertical range is purely conventional. Because of the irregular sampling, the specific volume corresponding to the spatial sampling has been mapped using a superkriging at each block of the model. This superkriging consists in kriging with its inner samples within a moving superblock centered at each block. The superblock must be large enough to avoid irregularities, but small enough to limit the smoothing. A superblock size of $150 \times 150 \times 50$ m is chosen. At each

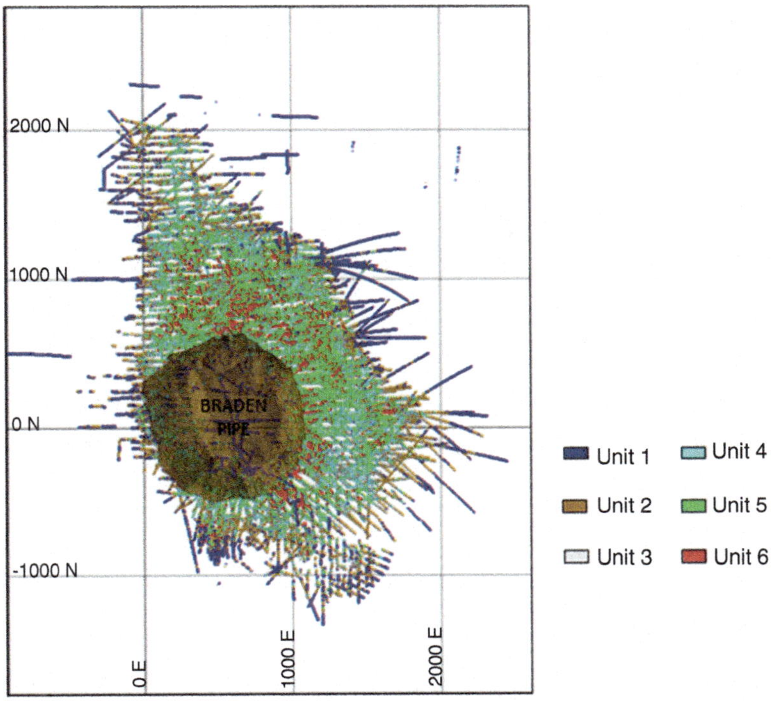

Fig. 1 Horizontal projection of drill holes, with the pipe in the center

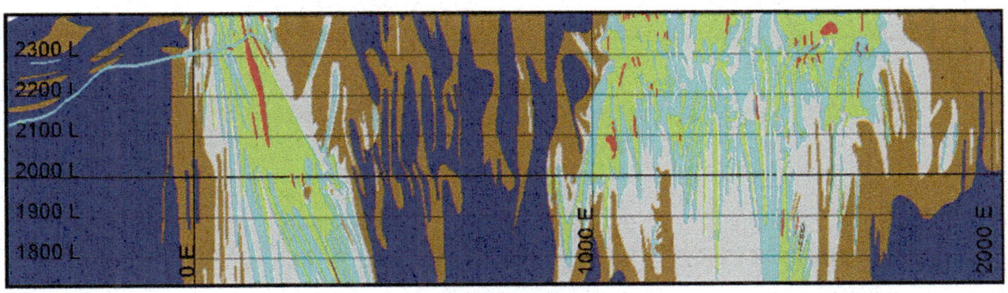

Fig. 2 E-W vertical section with the units. Same unit colors as in Fig. 1. Units 1 (*blue*) and 2 (*brown*) are poor

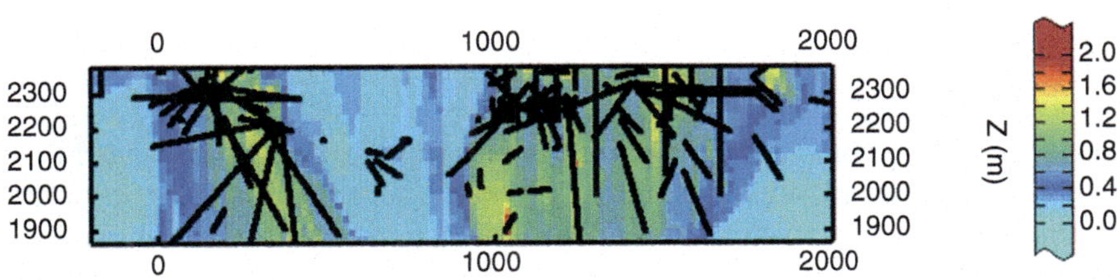

Fig. 3 E-W vertical section. Copper grade estimated by kriging. *Black lines* represent drillholes close to the section

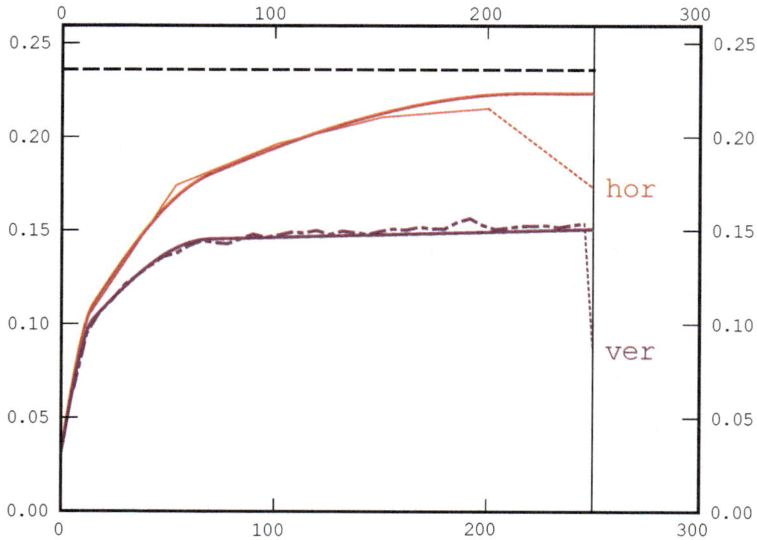

Fig. 4 Variogram of copper grade for units 3–6

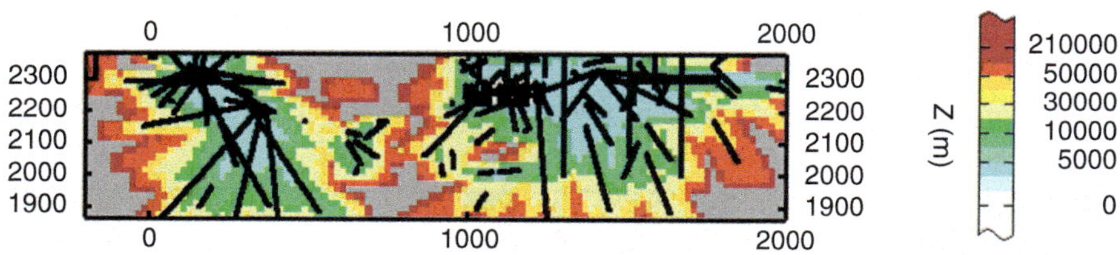

Fig. 5 E-W vertical section. Specific volume. *Green* lower than 15,000 m^3; *yellow* lower than 50,000 m^3

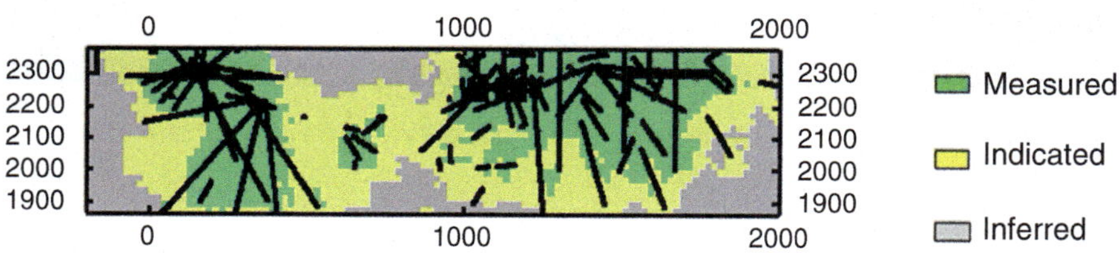

Fig. 6 E-W vertical section. Current classification based on geometric criteria

block the specific volume is obtained as the superkriging variance computed from the relative variogram, multiplied by the size of the superblock (Fig. 5).

The map of specific volume provides a delineation of the deposit in terms of sampling density. This can be compared to the current classification made on a geometric criteria based on the presence of data close to each block (Fig. 6). Based on the specific volume, *measured* resources would correspond to blocks in green:

specific volume less than 15,000 m^3 with an average specific volume of 5500 m^3. *Indicated* resources would correspond to specific volume between 15,000 and 50,000 m^3, with mean specific volume 28,000 m^3 (in yellow). Other resources (red and gray) would be *inferred*.

The annual production is about 50 Mt, that is, V~20 $Mm^3$with rock density 2.6, and the metal quantity should not deviate by more than $r = 3\,\%$. The coefficient of variation of this quantity depends on the specific volume (Eq. 1). In a Gaussian framework, the probability that a random variable with given CoV deviates by more than r from its expectation is given by $2G(-r/\mathrm{CoV})$ where G is the c.d.f. of a standard Gaussian. The CoV threshold for measured resources, corresponding to $V_0 = 15{,}000\ m^3$, is 2.7 % (probability of 1/4, i.e., 1 year out of 4, to deviate by more than 3 %), while the CoV corresponding to the average specific volume of measured resources (5500 m^3) is 1.7 % (probability of 1/10 to deviate by more than 3 %). Concerning *indicated* resources, the threshold CoV, corresponding to $V_0 = 50{,}000\ m^3$, would be 5 % (probability of 1/1.8 to deviate by more than 3 %), and the CoV corresponding to the average specific volume of indicated resources (28,000 m^3) is 3.7 % (probability of 1/2.4 to deviate by more than 3 %).

Compared to other porphyry copper deposits, the specific volumes in the best sampled areas of El Teniente are low, which means that the spatial sampling is quite good, probably because of a higher spatial continuity of grades.

5 Application: Iron Ore Proportion in Three Dimensions with Irregular Sampling Pattern

This case study is a big iron ore deposit located in Brazil. The banded iron formation from the Quadrilatero Ferrífero is called itabirite, a metasedimentary rock. The iron enrichment is given mainly by weathering which dissolves carbonate and siliceous bands, increasing the residual concentration of iron oxides, and decreasing the cohesion between grains, bands becoming friable in most part of economic ore body (Ribeiro and Carvalho 2000). A number of 15 lithotypes with distinct lithochemical characteristics, either mineralized or waste, have been distinguished. Of particular interest here will be the group of *uncontaminated facies* (denoted HC + HF + IF + IC), and the group of *ore facies* (HC + HF + IF + IC + HAR + IAR + IFR + IMN). The deposit is sampled irregularly with vertical or subvertical holes. Locally a base grid of about 70 × 70 m exists, but the sampling design is very often less dense and varies horizontally (Fig. 7) as well as vertically (Fig. 8). The block sizes used in this deposit are 50 × 50 × 13 m and 10 × 10 × 13 m. Sample data are 13 m composites, regularized by lithotype domain. Facies is available for all samples, but this is not the case for chemical variables, in particular for the Fe grade. This has a consequence on the estimation and classification. For some blocks, it is possible to estimate the facies, but not the Fe grade. Classification

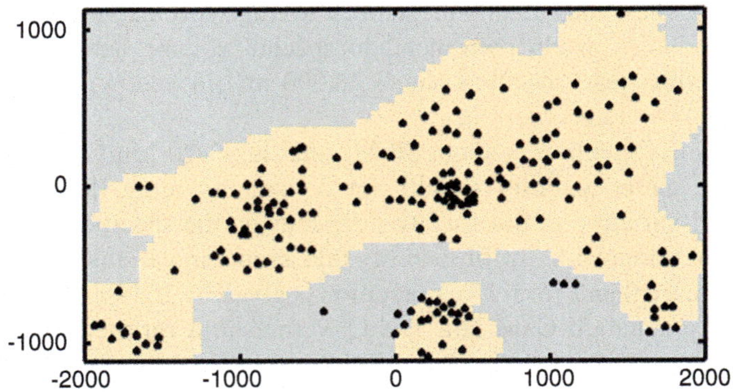

Fig. 7 Holes within a horizontal section in the middle of the deposit. In *yellow*, the area where Fe grade is estimated

Fig. 8 Samples within a 50 m thick vertical section in the middle of the deposit. Vertical exaggeration is 4

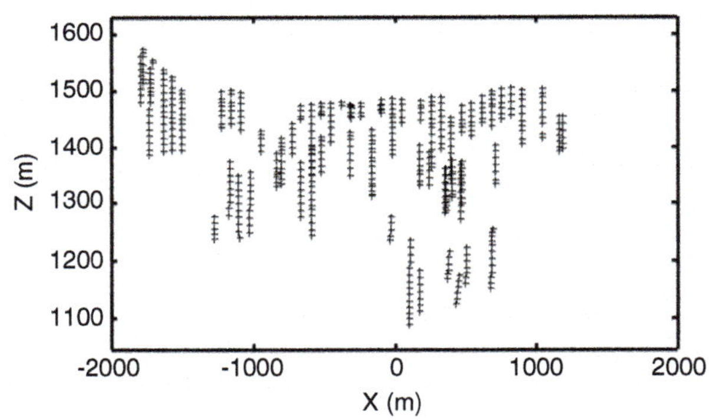

of blocks is made on facies, but only blocks where Fe grade can be estimated are considered.

The regionalized variables used to measure the spatial sampling of the deposit are the facies indicators (at samples), regularized into facies proportions (on blocks). In such deposits, the facies are numerous and many variograms are not well known, so that some simplifications are commonly made. Facies indicators are supposed to be in intrinsic correlation, that is, all facies indicators (whether individual or grouped) obey to the same variogram model (up to its vertical scale). This variogram is determined from the variogram of the indicator of uncontaminated facies, which is well known and is further used for all ore facies. The reference plane is dipping 25° to the east. The variogram is isotropic within the reference plane. The variogram model, using the reference plane, is composed of a nugget, a short isotropic 40 m component, and a large component with anisotropy coefficient of 4 (Fig. 9):

$$\gamma(h) = 0.13\,\mathrm{nug}(h) + 0.045\,\mathrm{sph}(h/40) + 0.075\,\mathrm{sph}(h/640, 640, 160).$$

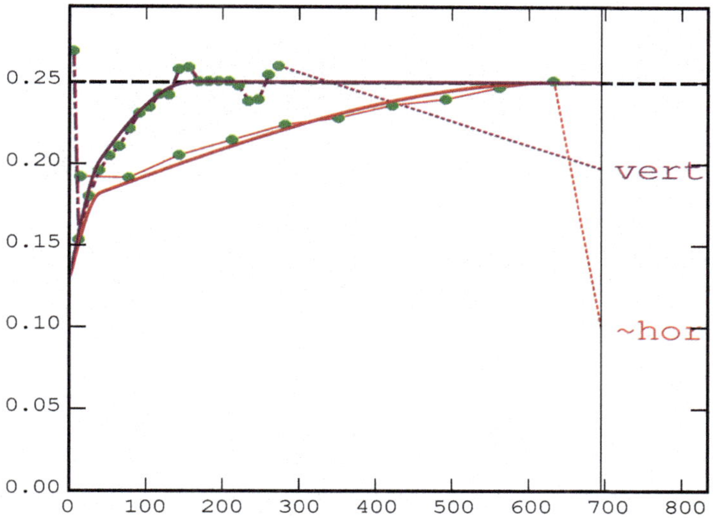

Fig. 9 Variogram of iron ore indicator of uncontaminated facies

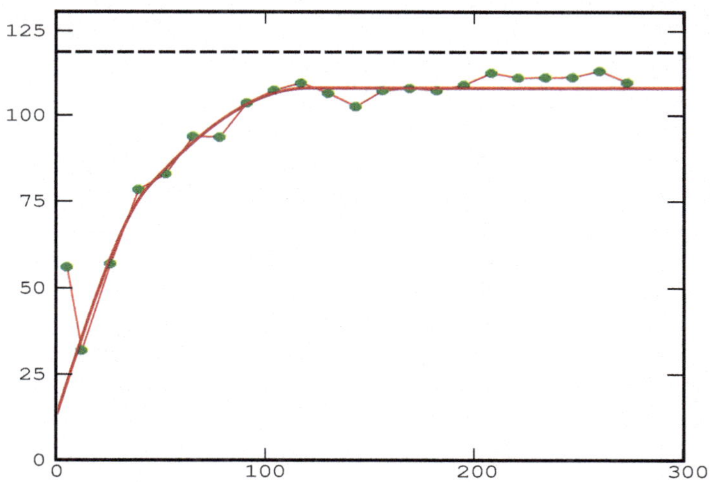

Fig. 10 Variogram of Fe grade

The mean indicator is close to 0.5 and the sill or variance is 0.25. A normalized variogram $\gamma_0(h)$ is obtained by dividing the variogram by 0.25:

$$\gamma_0(h) = 0.52\,\mathrm{nug}(h) + 0.18\,\mathrm{sph}(h/40) + 0.30\,\mathrm{sph}(h/640, 640, 160).$$

The variogram of Fe grade is fitted by an isotropic variogram model (Fig. 10):

$$\gamma(h) = 13\,\mathrm{nug}(h) + 35\,\mathrm{sph}(h/45) + 60\,\mathrm{sph}(h/120).$$

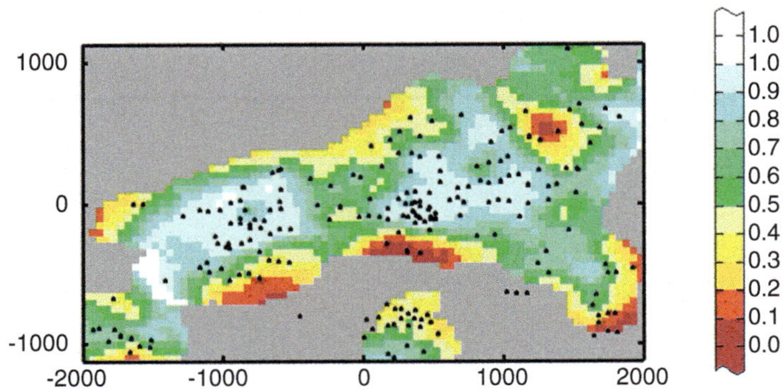

Fig. 11 Horizontal section in the middle of the deposit. Kriged iron ore proportion for $50 \times 50 \times 13$ m blocks

Resources are estimated by kriging. The normalized variogram of facies is used to krige the group of iron ore facies. Kriging is performed with an elliptic neighborhood with radius (300 m, 300 m, 100 m), and a minimum number of samples of 10. Fe grade is kriged using the same neighborhood. Blocks where Fe is not estimated contain a smaller proportion of ore than the other blocks. Kriged iron ore proportion is represented in Fig. 11 for the case of $50 \times 50 \times 13$ m blocks.

To compute the map of specific volume, a superkriging is performed at each $50 \times 50 \times 13$ m block. This consists in kriging $350 \times 350 \times 39$ m superblocks (i.e., five times the basic 70×70 m grid) from inner samples. Because the size of superblocks is smaller than the elliptical neighborhood (diameters 600 m, 600 m, 200 m) used for kriging the resources, this excludes some of the blocks which are anyway poorly estimated. As all facies or groups of facies obey to the same variogram model (up to its vertical scale), the normalized variogram is used for superkriging. Multiplied by the size of the superblock, the superkriging variance then gives the normalized specific volume at each block, valid for any facies or group of facies (Fig. 12). Thresholds of 300,000 m^3 and 1,200,000 m^3 (four times the first one) are applied on normalized specific volume to distinguish areas upon their sampling. In the best-known areas, the normalized specific volume is less than 300,000 m^3, with an average around 155,000 m^3. Where the normalized specific volume lies between 300,000 m^3 and 1,200,000 m^3, the average is 658,000 m^3, that is, a little more than four times larger (so this leads to about twice larger CoV).

Such normalized specific volumes should be multiplied by $p(1-p)/p^2 = [(1/p) - 1]$ (ratio between variance and squared mean of indicator) to give specific volumes for a facies with mean proportion p. For the uncontaminated facies ($p = 50\,\%$ whether declustered or not), this factor is 1. For the iron ore facies (declustered $p = 63\,\%$), the factor is $(1/0.63 - 1) = 0.59$. Finally, for the iron ore, the thresholds on specific volume become 177,000 m^3 and 708,000 m^3, with averages of 91,000 m^3 for specific volumes below 177,000 m^3 and of 388,000 m^3 for specific volumes between 177,000 and 708,000 m^3.

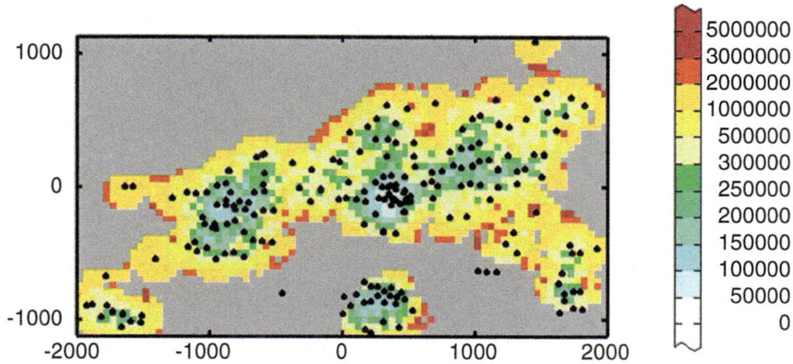

Fig. 12 Horizontal section in the middle of the deposit. Specific volume computed from $350 \times 350 \times 39$ m superblocks at $50 \times 50 \times 13$ m blocks

Consider a yearly production of 20,000,000 t, that is, 7,142,857 m^3 at rock density 2.8. With a mean ore proportion of 63 %, the total annual volume to be considered is $V = 7\ 142\ 857/0.63 = 11\ 337\ 868\ \text{m}^3$. In the best-known zones of the deposit, where the iron ore specific volume is less than 177,000 m^3 and is 91,000 m^3 on average, the CoV of the iron ore proportion in this annual volume is then $\sqrt{V_0/V} = 9.0\%$.

In such iron ore deposits, resource classification is currently done using a Risk Index, which combines resources and uncertainties as follows (Ribeiro et al. 2010):

$$\mathrm{RI} = \sqrt{\left(1 - \mathrm{Ind}^*\right)^2 + \left(\sigma^2_{\mathrm{IK}}\right)^2}$$

where Ind^* is the ore proportion of the block, obtained by kriging the iron ore indicator, and σ^2_{IK} the normalized kriging variance (i.e., using the variogram normalized by its sill). Typically, blocks are classified as *measured* when RI <0.6, *indicated* when $0.6 < \mathrm{RI} < 0.8$, and *inferred* otherwise. This RI extends the notion of confidence to risk, as a block with a proportion of ore less than 20 % corresponds to *inferred* resources, even if this proportion is perfectly known.

Another version of Risk Index can be derived using the specific volume V_0 or the normalized specific volume v_0:

$$\mathrm{New\,RI} = \left(1 - \mathrm{Ind}^*\right) + \frac{V_0}{W} = \left(1 - \mathrm{Ind}^*\right) + \left(\frac{1}{p} - 1\right)\frac{v_0}{W}$$

where p is the global ore proportion (63 %), and W is a constant. The classical RI is sensitive to the size of blocks (e.g., $10 \times 10 \times 13$ m or $50 \times 50 \times 13$ m) through the kriging variance. On the contrary both terms Ind^* and V_0 are additive, so that the new RI does not depend on the size of blocks. It can provide the same service as the classical one, with this additional advantage. By taking $W = 10{,}000{,}000\ \text{m}^3$ (practically the annual volume of reference V), the new RI (Fig. 14) was observed

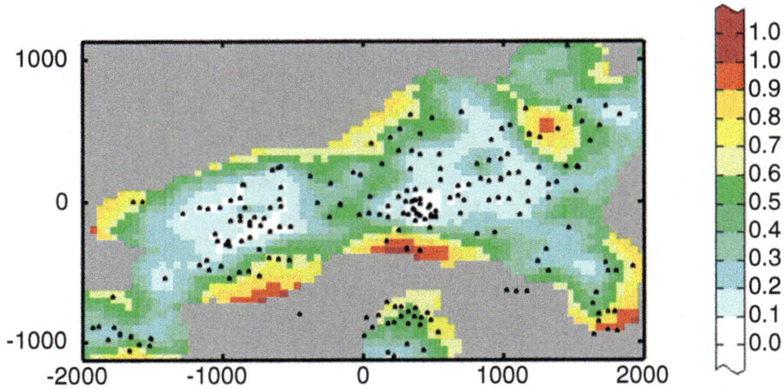

Fig. 13 Horizontal section in the middle of the deposit. RI for 50 × 50 × 13 m blocks

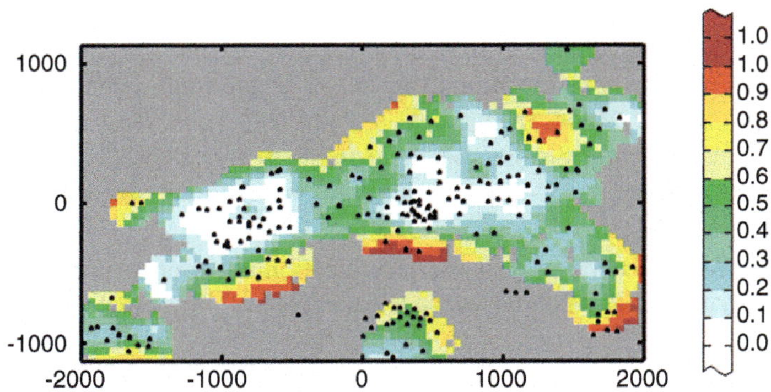

Fig. 14 Horizontal section in the middle of the deposit. New RI for 50 × 50 × 13 m blocks

to be similar to the RI for 50 × 50 × 13 m blocks (Fig. 13), less to RI for 10 × 10 × 13 m blocks.

6 Conclusion

To apply the proposed classification method, one must choose the regionalized variable corresponding to the resources. In our two-dimensional application, the thickness was used, but in other cases the metal accumulation would be more appropriate. In cases where the target variable is not additive, but is the ratio of additive variables, combining variances and specific volumes could be done using reduced differences (Journel and Huijbregts 1978). More generally, the method addresses in situ resources rather than recoverable ones. Geological cutoff grades on sample values can be applied, but mining cutoffs on selection blocks would require more sophisticated tools such as simulations.

The originality of the method is in its two steps. The first step consists in measuring (or mapping when it varies) the spatial sampling density in the deposit. The tool which is used, the specific volume, can be seen as a generalization of the density of sample points (Deustch et al. 2007), but it takes into account the variographic structure. In the second step, the specific volume is converted into coefficients of variation corresponding to large target volumes, typically the expected annual volume to be mined.

The use of coefficients of variation on production units, rather than confidence on individual blocks, is not original in resource classification, and it allows a validation by actual production data when available. It is advocated in particular by Dohm (2005). In her "logical approach," she makes the link between the coefficient of variation of "units representing likely production periods" and a "typical coefficient of variation for blocks" through a "factor production period," which is determined by conditional simulations. Our approach via a specific volume is different.

The method is simple and straightforward but it aims at giving orders of magnitude rather than precise numbers. Coefficients of variation obtained from the specific volume assume that the target volumes are estimated from their inner samples only. This may lead to pessimistic values of coefficients of variation, particularly for target volumes that are too small. The specific volume depends on the variogram and so on its knowledge. At the early stage of systematic exploration, when the hole spacing exceeds the variogram range and resources are expected to be classified as inferred, a pure nugget effect may be considered as an approximation. Moreover the target volumes need not be precisely delineated, which makes the method flexible with respect to expected production volumes. Of course the method is not meant to replace the direct estimation of the resources that are contained in delineated domains – such as the area that is to be mined, say, next year – nor their confidence. Kriging and conditional simulations are the appropriate tools for this, but are more demanding.

Acknowledgments The authors are grateful to Codelco and Vale for their support in this research. They wish to thank the participants of the workshops on mineral resource classification that were held on 26–28 November 2014 at Codelco and 1–2 December 2014 at Vale, where the present methodology was discussed and put into practice. Mapping the specific volume has been made possible, thanks to a *Superkriging* plug-in developed for Geovariances Isatis software.

Bibliography

Cochran WG (1977) Sampling techniques, 3rd edn. Wiley, New York, 428p

Deutsch CV, Leuangthong O, Ortiz JM (2007) Case for geometric criteria in resources and reserves classification. Trans-Soc Min Metall Explor Inc 322:1

Dohm C (2005) Quantifiable mineral resource classification – a logical approach. In: Leuangthong O, Deutsch CV (eds) Geostatistics banff 2004, vol 1. Kluwer Academic, Dordrecht, pp 333–342

JORC (2012) The JORC code, AusIMM, 44
Journel AG, Huijbregts CJ (1978) Mining geostatistics. Academic, London, 600p
Matheron G (1971) The theory of regionalized variables and its applications. Fontainebleau. Cahiers du Centre de Morphologie Mathématique 5. Ecole des Mines de Paris, 212p
Ribeiro D, Carvalho RM (2000) Simulation of weathered iron ore facies: integrating leaching concepts and geostatistical model. In: Armstrong et al. (ed) Geostatistics Rio 2000. Kluwer Academic, pp 111–115
Ribeiro D, Guimaraes M, Roldao D, Monteiro C (2010) An indicator geostatistical approach to support mineral resource classification. Proceedings of the 4th International Conference on Mining Innovation MININ 2010. Santiago, pp 471–478
Rivoirard J (2013) A geostatistical measure for the spatial sampling of a deposit, Proceedings of 36th APCOM Conference. Brazil, pp 209–215
Rivoirard J, Renard D (2016) A specific volume to measure the spatial sampling of deposits, Math. Geosc., accepted with minor editorial revisions https://hal-mines-paristech.archives-ouvertes.fr/hal-01305760
Rossi ME, Deutsch CV (2014) Mineral resource estimation. Springer, Dordrecht, 332p

Resource Model Dilution and Ore Loss: A Change of Support Approach

Oscar Rondon

Abstract The estimation of ore dilution and ore loss factors is critical for the evaluation of mining projects, with the proper estimation of these effects essential for Ore Reserve Estimation. Unplanned ore dilution occurs when excessive amounts of waste are mined with ore, and such dilution may result in the processing of lower than economic cut-off grade material with the ore. Unplanned ore loss occurs when material that is above the economical cut-off grade is hauled to waste stockpiles, due to poor mining practices and/or poor information regarding the local grade of the critical components.

The sources of dilution and ore loss in mining operations are many. This study focuses on the assessment of the ore dilution and loss in a Mineral Resource model, specifically, the proportion and average grade of blocks that are misclassified as ore and waste, which can be referred to as Resource Model Ore Dilution and Model Ore Loss, respectively. Unlike the well-established geostatistical conditional simulation approach for assessing model dilution and loss, in this study analytical expressions are derived that are defined under the theoretical framework of the discrete Gaussian method for change of support to quantify the expected model dilution and ore loss. Practical application of this method is demonstrated through a case study from an Iron Ore deposit in Australia.

1 Introduction

Mineral Resource estimates are reported above an economic cut-off grade on the basis of model estimated grades and not according to the real grade values. The effect of this is that all Mineral Resource models will certainly incur misclassification of mining blocks as ore or waste.

O. Rondon (✉)
CSIRO Mineral Resources, 26 Dick Perry Avenue, Kensington, WA 6152, Australia

Centre For Exploration and Targeting, University of Western Australia, Crawley, Australia
e-mail: Oscar.Rondon@csiro.au

J.J. Gómez-Hernández et al. (eds.), *Geostatistics Valencia 2016*, Quantitative Geology and Geostatistics 19, DOI 10.1007/978-3-319-46819-8_23

Misclassification results in the mine plan including tonnages of what will be found in the future to be material that is below the economic cut-off grade, which increases the planned processed or saleable tonnage compared to estimates in production forecasts. Additionally, misclassified blocks that will be found in the future to be above cut-off grade, but present as waste from the resource model, represent unplanned ore loss in the mine plan which also results in tonnages and grades discrepancies.

This study focuses on the assessment of the ore dilution and loss in a Mineral Resource model, specifically, the proportion and average grade of blocks that are misclassified as ore and waste, which can be referred to as resource model ore dilution and model ore loss, respectively.

Assessment of the Resource Model Ore Dilution and Loss calls for modelling the relationship between estimated and true block grade values. This leads to study in detail (Huijbregts 1976):

1. The distribution of true block grades
2. The distribution of estimated block grades
3. The joint distribution of true and estimated block grades

The discrete Gaussian method for change of support (Matheron 1976) provides a theoretical framework to model all these distributions (Roth and Deraisme 2000) which allows the derivation of analytical expressions to quantify the expected model dilution and ore loss as shown in this study.

Importantly, the determination of the distribution of block grades is an undetermined problem even if the point support distribution is considered known. Therefore, the choice of an appropriate change of support model is of crucial importance (Lantuejoul 1988). In this study, it is assumed that the discrete Gaussian is a plausible model for approximating the univariate and joint distributions of true and estimated block grades and as such, the results presented here are limited by the suitability of the discrete Gaussian model to the data being modelled.

2 The Discrete Gaussian Method

Let $Z(x)$ and $Z(v)$ be the grade at point x and block support v, respectively. The discrete Gaussian method (Matheron 1976) expresses $Z(x)$ and $Z(v)$ as function of two standard Gaussian variables Y and Y_v as

$$Z(x) = \phi(Y(x)) = \sum\nolimits_n \phi_n H_n(Y(x)) \tag{1}$$

where ϕ is the point support anamorphosis function derived from the point support data, $H_n\, n \geq 0$ are the normalised Hermite polynomials, $\phi_n\, n \geq 0$ are the coefficients of the expansion of ϕ in terms of Hermite polynomials and

$$Z(v) = \phi_v(Y_v) = \sum_n \phi_n r^n H_n(Y_v) \tag{2}$$

where ϕ_v is the block anamorphosis function which is derived via Cartier's relation by assuming that the two Gaussian variables $Y(x)$ and Y_v have joint Gaussian distribution with correlation $r > 0$ (Rivoirard 1994).

The coefficient r corresponds to the variance or support correction factor from point to block support and is chosen so as to respect the variance of $Z(v)$ by inverting:

$$Var(Z(v)) = \sum_n \phi_n^2 r^{2n} \tag{3}$$

An alternative method has been proposed to compute the variance correction factor (Emery 2007). The method is simpler and allows the extension of the discrete Gaussian method to the problem of modelling the local block distribution. However, the method depends only on the variogram of Y and does not depend on the anamorphosis function ϕ. Therefore, a correction factor derived with this approach will not necessarily satisfy (3) which is an important aspect of the block distribution (Chiles 2014).

After deriving the support correction factor, the distribution of real block grades $Z(v)$ can be modelled by using (2). Modelling of the distribution of the estimated block grades $Z^*(v)$ is carried out similarly (Roth and Deraisme 2000) by assuming that

$$Z^*(v) = \phi_v^*(Y_v^*) = \sum_n \phi_n s^n H_n(Y_v^*) \tag{4}$$

where Y_v^* stands for a standard Gaussian variable and ϕ_v^* is the corresponding anamorphosis function but with correction factor s chosen so as to respect the variance of $Z^*(v)$ by inverting:

$$Var(Z^*(v)) = \sum_n \phi_n^2 s^{2n} \tag{5}$$

The relation between $Z(v)$ and $Z^*(v)$ is obtained using (2) and (4) and is given by

$$Cov(Z(v), Z^*(v)) = \sum_{n,m\geq 1} \phi_n \phi_m r^n s^m E\left(H_n(Y_v) H_m(Y_v^*)\right) \tag{6}$$

A further assumption is that (Y_v, Y_v^*) has bivariate Gaussian distribution with correlation ρ, then $E(H_n(Y_v)H_m(Y_v^*)) = 0$ for $n \neq m$ and $E(H_n(Y_v)H_n(Y_v^*)) = \rho^n$ (Rivoirard 1994) which implies that

$$Cov(Z(v), Z^*(v)) = \sum_{n\geq 1} \phi_n^2 r^n s^n \rho^n \tag{7}$$

After computing the correlation ρ by inverting (7), the bivariate Gaussian distribution of (Y_v, Y_v^*) is completely specified, and the joint distribution $(Z(v), Z^*(v))$

can be modelled. This is the key property for modelling the information effect (Roth and Deraisme 2000) and for the development of the analytical expressions for the assessment of model dilution and ore loss as discussed in this study.

3 Model Dilution and Ore Loss

For the modelling of dilution and ore loss, the focus is on the true block grade $Z(v)$ conditioned by whether the estimated block grade $Z^*(v)$ is above or below an economical cut-off grade z.

The tonnage associated with model dilution is

$$P\big(Z(v) < z/Z^*(v) \geq z\big) = P\big(Y_v < y, Y_v^* \geq y^*\big)/P\big(Y_v^* \geq y^*\big) \tag{8}$$

with $y = \phi_v^{-1}(z)$ and $y^* = \phi_v^{*-1}(z)$. Since (Y_v, Y_v^*) has Gaussian distribution with correlation ρ, it is possible to show that

$$P\big(Y_v < y, Y_v^* \geq y^*\big) = 1 - F\big(y^*\big) - H_\rho\big(y, y^*\big) \tag{9}$$

where F denotes the standard Gaussian distribution function,

$$H_\rho(y, y^*) = \int_y^{+\infty}\int_{y^*}^{+\infty} g_\rho(s,t)dtds \tag{10}$$

and g_ρ is the bivariate Gaussian density with correlation ρ. Therefore, the tonnage associated with model dilution can be explicitly calculated as

$$P\big(Z(v) < z/Z^*(v) \geq z\big) = 1 - F\big(y^*\big) - H_\rho\big(y, y^*\big)/\big(1 - F\big(y^*\big)\big) \tag{11}$$

Similarly, tonnage associated with model ore loss can be obtained as

$$P\big(Z(v) \geq z/Z^*(v) < z\big) = 1 - F(y) - H_\rho\big(y, y^*\big)/F\big(y^*\big) \tag{12}$$

The expected metal associated with model dilution is

$$\begin{aligned} E\big(Z(v)1_{Z(v)<z}/Z^*(v) \geq z\big) &= \frac{1}{1 - F(y^*)}E\Big(Z(v)1_{Z(v)<z}1_{Z^*(v)\geq z}\Big) \\ &= \frac{1}{1 - F(y^*)}E\Big(\phi_v(Y_v)1_{Y_v<y}1_{Y_v^*\geq y^*}\Big) \end{aligned} \tag{13}$$

Using the development into Hermite polynomials of ϕ_v given by (2)

$$E(\phi_v(Y_v)1_{Y_v<y}1_{Y_v^*\geq y^*}) = \sum_n \phi_n r^n E(H_n(Y_v)1_{Y_v<y}1_{Y_v^*\geq y^*}) \\ = \sum_n \phi_n r^n \int_{-\infty}^{y}\int_{y^*}^{+\infty} H_n(u) g_\rho(u,t)dtdu \tag{14}$$

The function g_ρ can be expanded using the Hermite polynomials as

$$g_\rho(u,t) = \sum_k \rho^k H_k(u)H_k(t)g(u)g(t) \tag{15}$$

where g corresponds to the standard Gaussian density (Chiles and Delfiner 2008). Substituting (15) in (14) and defining

$$U_{i,j}(x) = \int_x^{+\infty} H_i(u)H_j(u)g(u)du \; i,j \geq 0 \tag{16}$$

which when applying (14) becomes

$$\sum_n \phi_n r^n \sum_k \rho^k [\delta_{nk} - U_{n,k}(y)]U_{0,k}(y^*) \tag{17}$$

with δ_{nk} the Kronecker delta function representing the orthogonality property of the Hermite polynomials

$$\delta_{nk} = \int_{-\infty}^{+\infty} H_n(u)H_k(u)g(u)du \tag{18}$$

Therefore, the analytical expression for computing the expected metal associated with model dilution is

$$E(Z(v)1_{Z(v)<z}/Z^*(v) \geq z) = \frac{1}{1-F(y^*)}\sum_n \phi_n r^n \sum_k \rho^k [\delta_{nk} - U_{n,k}(y)]U_{0,k}(y^*) \tag{19}$$

The values $U_{n,k}(\cdot)$ can be computed recursively (Rivoirard 1994), and therefore once the anamorphosis modelling is completed, the metal amount can be computed.

The expected average metal associated with model ore loss can be obtained similarly and is given by

$$E(Z(v)1_{Z(v)\geq z}/Z^*(v) < z) = \frac{1}{F(y^*)}\sum_n \phi_n r^n \sum_k \rho^k [\delta_{nk} - U_{0,k}(y^*)]U_{n,k}(y) \tag{20}$$

Using the derived tonnage and metal, the average grade $E[Z(v)/Z(v) < z, Z^*(v) \geq z]$ and $E[Z(v)/Z(v) > z, Z^*(v) \leq z]$ associated with model dilution and ore loss, respectively, can be computed.

The tonnage and metal associated with blocks correctly classified as waste or ore can also be obtained. These equations are provided in the Appendix.

The derived analytical expressions can be used in two ways: for forecasting, from sparse drilling data, the expected misclassification of blocks at the time of having close-spaced drilling data and smaller block sizes, and for assessing the expected model dilution and ore loss from a given resource model. Both applications are shown below.

3.1 Accepting Model Dilution or Ore Loss

Resource Model Dilution and Ore Loss will inevitably occur when selecting blocks on the basis of whether their estimated grades are above an economical cut-off z or not. Therefore, it is instructive to study the behaviour of both for a range of cut-off values in order to decide, for instance, whether it is preferable to accept dilution or ore loss. As a demonstration, consider the iron grades (Fe%) in an Iron Ore deposit with a nominal drill hole sample spacing of 100 m by 100 m that have been used to compute the tonnage associated with model dilution and ore loss with a block support of 50 m by 50 m by 5 m (Fig. 1). The tonnage curves reveal that, with increasing cut-off grades, the tonnage associated with model dilution increases (Jara et al. 2006) and that by accepting more dilution ore loss is minimised (Bertinshaw and Lipton 2007). The reverse effect occurs if the cut-off grade is reduced (lower dilution but higher ore loss).

Analysis of the tonnage associated with model dilution for varying block sizes and a fixed 54 Fe% cut-off (Fig. 2) shows that dilution increases with decreasing block sizes and that small blocks are subject to more model ore losses (Jara et al. 2006). Furthermore, the results shown provide an alternative way to reemphasise the dangers of estimating into small blocks from sparse drilling data (Ravenscroft and Armstrong 1985).

4 A Grade Control Model

Iron grades Fe(%) from an Iron Ore deposit in the Pilbara region in Australia are used to demonstrate the assessment of the model dilution and ore loss using the method presented above.

The drill hole data are samples from Mineral Resource definition RC drill holes within a mineralised domain with a nominal drill spacing of 25 m by 25 m which are used to carry out a kriging estimation of Fe into a grade control block model with blocks of dimension 12 m by 12 m by 6 m. Further to this, available Fe grades from grade control blastholes at a nominal drill spacing of 6 m by 6 m are kriged into the same grade control block model. These kriged grades are considered as the true block Fe grades to be the benchmark in this practical application. Since the blasthole data partially covers the domain under study, the comparison is limited to blocks that are informed by both the RC and blastholes kriged estimates. In addition, it is assumed to be no bias in either the drill hole or blasthole. The scatter

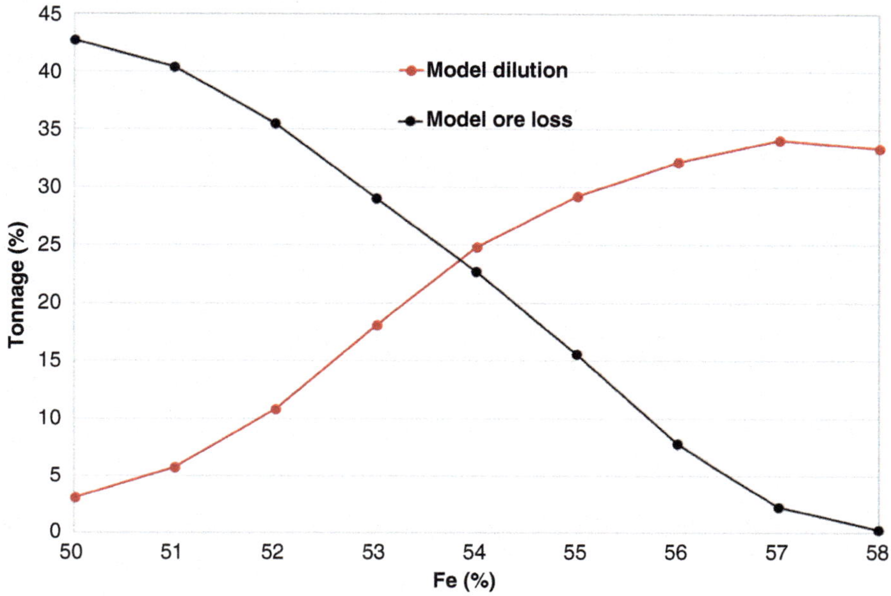

Fig. 1 Tonnage associated with model dilution (*red*) and ore loss (*black*) as function of Fe% cut-off grades

plot between both estimates (Fig. 3) indicates that the model will incur in misclassification of blocks when reporting at the cut-off of 60 Fe%, which corresponds to approximately the average Fe grade.

Using the RC drilling data, the discrete Gaussian method is used for modelling the distribution of $Z(v)$ and $Z^*(v)$. Some close-space RC data available was used to fine-tune the values of $Var(Z^*(v))$ and $Cov(Z(v), Z^*(v))$ required to derive the correction factors s and ρ for modelling the metal $Q^*(z)$, tonnage $T^*(z)$ and average grade above cut-off $M^*(z)$ of $Z^*(v)$.

The amount of metal $Q^*(z)$ above cut-off is given by

$$Q^*(z) = E\left(Z^*(v)1_{Z^*(v)\geq z}\right) = \sum\nolimits_n \phi_n s^n U_{0,n}\left(y^*\right) \tag{21}$$

and $T^*(z) = P\left(Z^*(v) \geq z\right) = 1 - F\left(y^*\right)$ which allows to compute the average grade above cut-off $M^*(z) = Q^*(z)/T^*(z)$.

The modelled tonnage $T^*(z)$ and average grade $M^*(z)$ are compared to the corresponding estimates reported from the grade control block model to assess the adequateness of the distribution of $Z^*(v)$ to represent the distribution of the kriging estimates obtained from the RC drill hole data (Fig. 4). The average grade above cut-off is closely reproduced for all cut-offs with a maximum absolute relative error of 0.75 % at 60 Fe% cut-off. Tonnage is also reasonably reproduced for all cut-offs with the exception being the tonnage at 60 Fe% where the absolute relative error is approximately 20 %. Nonetheless, the global average relative error

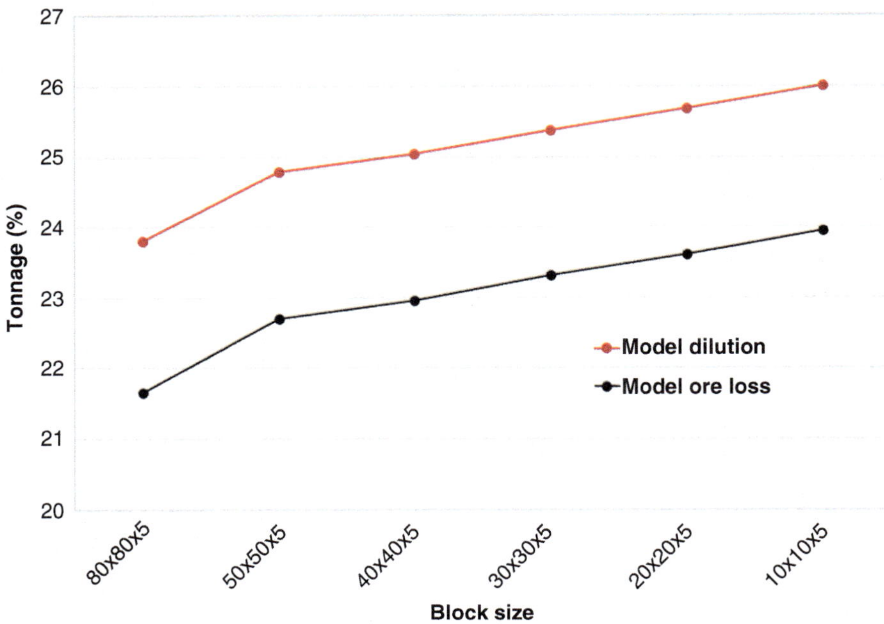

Fig. 2 Tonnage associated with model dilution (*red*) and ore loss (*black*) for different block sizes using drill hole data with a nominal drill spacing of 100 m by 100 m and a fixed 54 Fe% grade cut-off

is approximately 7 %, and therefore the distribution of $Z^*(v)$ was considered reasonably fit for purpose.

Tables 1 and 2 contain the comparison of true and estimated Resource Model Dilution and Ore Loss using a 60 Fe% and 62 Fe% cut-offs, respectively. The estimates of tonnage and average grade above cut-off are reasonably estimated by the method proposed in this study, for both type of misclassification.

The results shown are presented for only the 60 Fe% and 62 Fe% cut-offs because the model used had few blocks at high and low cut-off grades to implement the method that relies on a reasonable amount of data for modelling accuracy. Therefore, further assessment of the goodness of the proposed approach is desirable with a full set of estimated and true block grades.

5 Conclusion

This study developed analytical equations to assess the proportion and average grade of blocks misclassified as ore and waste and presented a case study with their application to a grade control model from an Iron Ore deposit in the Pilbara region of Australia. The method proposed provides only global estimates of the expected tonnage and average grade associated with these types of misclassification of

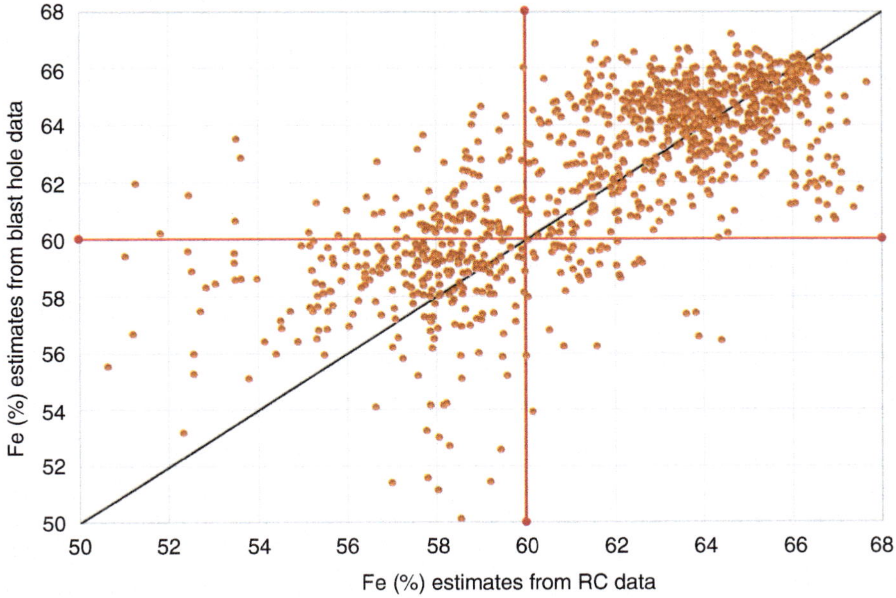

Fig. 3 Comparison of Fe kriged grades obtained from RC and blasthole data at a nominal drill spacing of 25 m by 25 m and 6 m by 6 m, respectively. *Red lines* indicate the 60 Fe% cut-off. Note that the top left sector of model ore loss has many more blocks than the lower right sector of model dilution

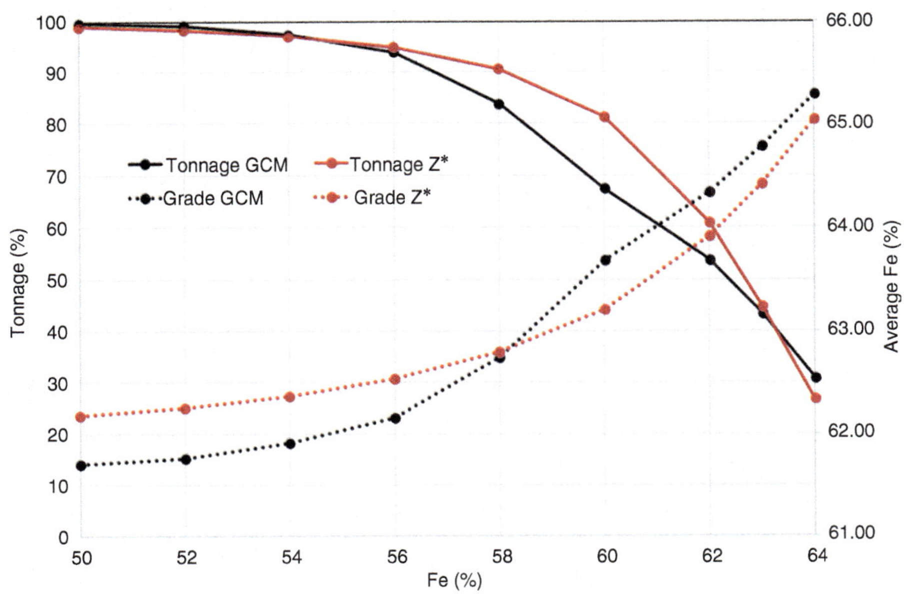

Fig. 4 Comparison of modelled tonnage and average grades above cut-off (*red*) to corresponding estimates from the grade control model (GCM) kriged grades using the RC drill hole data (*black*)

Table 1 Comparison of true and estimated model dilution and ore loss at a 60 Fe% cut-off

Misclassification type		Tonnage (%)	Grade Fe%
Model dilution	True	6.16	58.40
	Estimated	7.38	58.60
Model ore loss	True	29.85	61.46
	Estimated	28.36	61.18

Table 2 Comparison of true and estimated model dilution and ore loss at a 62 Fe% cut-off

Misclassification type		Tonnage (%)	Grade Fe%
Model dilution	True	11.15	60.77
	Estimated	12.6	61.04
Model ore loss	True	20.96	63.85
	Estimated	21.89	62.82

blocks. Therefore, it is not possible to know the spatial location of the blocks that are potentially misclassified when selecting on the basis of an estimated grade. Nonetheless, these global estimates could be used to anticipate the outcome of such selection. More advanced approaches based on conditional simulations are required to localise the blocks and classify them as ore or waste (Verly 2005) (Deutsch et al. 2000), but these approaches involve more time and effort.

The equations proposed in this paper are derived under the framework of the discrete Gaussian method for change of support and modelling of the information effect. Therefore, their practical application is largely limited by the assumptions made in both of these techniques as well as the ability of the theoretical distribution of the estimates to resemble the distribution of the Resource Model estimates.

Acknowledgements The author wishes to thank Mark Murphy for his help during the preparation of the manuscript.

Appendix

The tonnage associated with blocks correctly classified as waste is

$$P(Z(v) < z/Z^*(v) < z) = (F(y) + F(y^*) + H_\rho(y, y^*) - 1)/F(y^*)$$

and corresponding expected metal is

$$E(Z(v)1_{Z(v)<z}/Z^*(v) < z) = \frac{1}{F(y^*)}\sum_n \phi_n r^n \sum_k \rho^k [\delta_{nk} - U_{n,k}(y)][\delta_{0k} - U_{0,k}(y^*)]$$

The tonnage associated with blocks correctly classified as ore is

$$P(Z(v) \geq z / Z^*(v) \geq z) = H_\rho(y, y^*) / (1 - F(y^*))$$

and corresponding expected metal is

$$E(Z(v) 1_{Z(v) \geq z} / Z^*(v) \geq z) = \frac{1}{1 - F(y^*)} \sum_n \phi_n r^n \sum_k \rho^k U_{n,k}(y) U_{0,k}(y^*)$$

Bibliography

Bertinshaw R, Lipton I (2007) Estimating mining factors (dilution and ore loss) in open pit mines, vol 6, Sixth Large Open Pit Mining Conference. AusIMM, Perth, pp 13–17

Chiles J-P (2014) Validity range of the discrete Gaussian change of support and its variants. J South Afr Inst Min Metall 114(3):231–235

Chiles J-P, Delfiner P (2008) Geostatistics. Modeling spatial uncertainty. Wiley, New York

Deutsch C, Magri E, Norrena K (2000) Optimal grade control using geostatistics and economics: methodology and examples. Soc Min Metall Explor 308:43–52

Emery X (2007) On some consistency conditions for geostatistical change of support models. Math Geol 39(2):205–223

Huijbregts C (1976) Selection and grade-tonnage relationships. Advanced geostatistics in the mining industry, 24. D. Reidel Publishing Company, pp 113–135

Jara R, Couble A, Emery X, Magri E, Ortiz J (2006) Block size selection and its impact on open-pit design and mine planning. S Afr Inst Min Metall 106:205–211

Lantuejoul C (1988) On the importance of choosing a change of support model for global reserves estimation. Math Geol 20(8):1001–1019

Matheron G (1976) Forecasting block grade distributions: the transfer functions. In: David M, Guarasico M (eds) Advanced geostatistics in the mining industry, vol 24. D. Reidel Publishing Company, Dordrecht, pp 239–251

Ravenscroft P, Armstrong M (1985) Kriging of block models, the dangers reemphasised, vol 2. APCOM, Berlin, pp 577–587

Rivoirard J (1994) Introduction to disjuntive kriging and non-linear geostatistics. Oxford Press, New York

Roth C, Deraisme J (2000) The information effect and estimating recoverable reserves. In: Kleingeld W, Krige D (eds) Geostatistics 2000 proceedings of the sixth international geostatistics congress. Geostatistical Association of Southern Africa, Cape Town, pp 776–787

Verly G (2005) Grade control classification of ore and waste: a critical review of estimation and simulation based procedures. Math Geol 37(5):451–475

Drill Holes and Blastholes

Serge Antoine Séguret and Sebastian De La Fuente

Abstract The following is a geostatistical study of copper measurements on samples from diamond drill holes and blastholes. Both measurements are formally compared, leading to a model where a blasthole can be considered a regularization of the drill information up to a nugget effect characteristic of the blastholes. This formal link makes it possible to build a cokriging system that takes into account the different supports and leads to a block model based on blast- and drill holes. The model is tested on a realistic simulation where the true block grades, which are known, are compared to their estimate obtained by:

- Kriging using only drill holes
- Kriging using only blastholes
- Cokriging using drill and blastholes together

A preliminary conclusion is that the best estimates are obtained when only blasts or, alternatively, blast- and drill holes are used; there is no significant difference between the two, which is due to the great amount of blast information. This result justifies the usual practice of basing short-term planning on blasts only. But another conclusion may be drawn when kriging is compared to a moving average (another common practice), both based on blasts: depending on the number of data used in the neighborhood, the moving average produces a strong conditional bias. As a byproduct, we also show how it is possible to filter the blast error by kriging and to make a deconvolution to estimate point-support values using blast measurements.

S.A. Séguret (✉)
Mines ParisTech, Geosciences Center, Geostatistical Team, PSL Research University, 35 rue Saint Honoré, 77300 Fontainebleau, France
e-mail: serge.seguret@mines-paristech.fr

S. De La Fuente
Codelco, Radomiro Tomic, Santiago, Chile
e-mail: sdelafue@codelco.cl

J.J. Gómez-Hernández et al. (eds.), *Geostatistics Valencia 2016*, Quantitative Geology and Geostatistics 19, DOI 10.1007/978-3-319-46819-8_24

1 Introduction

Typically, in open-pit mines, geologists, mining engineers, and metallurgists have at their disposal two types of grade measurements: those from drill holes and those from blastholes. Because they are much more expensive, the drill holes (diamond ones in our case) are fewer than the blastholes, and sampling rates ranging from 1 over 3 to 1 over 10 or worse are frequent. Another difference concerns the way the measurements are used. Drill holes are used for medium- and long-term planning; blastholes for short-term planning without any need for geostatistics; a simple moving average is often used to estimate the block quantity of the metal. The use of two types of measurements, that are supposed to represent the same thing, raises questions about their relationship. In particular, would it not be possible to enrich the short-term estimates, now based only on blastholes, by adding the drill holes' measurement as they arrive? Another question, is it permitted to use a simple moving average based on blastholes for the block estimation?

Finally, it is often said, without real justification, that the diamond drill holes are much better than the blast ones. We ask: better in what way, better for what, and is it true?

After having modeled this relationship, we present some linear systems that make it possible to filter the nugget effect of the blastholes, removing their regularization effect and using the blast and the drill holes together in a single linear system. These are just demonstrative exercises; one can imagine other possibilities resulting from the formal link between two types of measurements known over two different supports.

2 Formal Link

The initial data (Fig. 1) are from an open-pit copper mine in Northern Chile where a subdomain was chosen for analysis because it is almost homogeneously covered by around 3000 drill hole samples (3 m long) and 13,000 blasthole samples (15 m long). In this case study, the diameters of the drill holes and blastholes are considered to be the same.

In a previous paper (Séguret 2015), the author showed that if we omit the problem of the nugget effect, both blast- and drill holes can be considered a regularization of the same phenomenon in accordance with their respective supports. But the drill holes have their own errors, independent of the blast ones, so that they do not share the nugget effect and finally, we have

$$Y_{\text{blast}}(x,y,z) = Y(x,y,z)^{*}p_{15m}(z) + R(x,y,z) \tag{1}$$

with

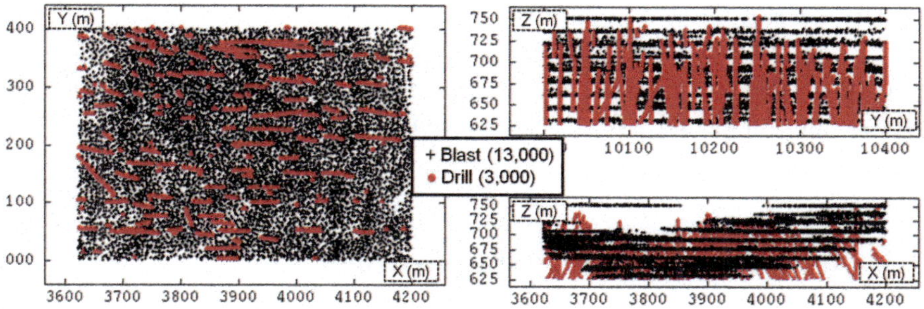

Fig. 1 Base maps of blast (*black*) and drill (*red*) measurements

Y(x,y,z), the point grade assumed to be isotropic and without any measurement error;

"*" denotes a convolution product;

$$Y(x,y,z)^*p_{15m}(z) = \int_{-\infty}^{+\infty} Y(x,y,u)p_{15m}(z-u)\mathrm{du};$$

$p_{15m}(z) = \frac{1}{15}1_{\left[0,\frac{15}{2}\right]}(|z|)$, the convolution function;

$1_{\left[0,\frac{15}{2}\right]}(|z|)$ the indicator function equal to 0 outside the interval [−7.5 m, 7.5 m] and 1 inside it;

R(x,y,z), a "white noise" residual, statistically and spatially independent of *Y(x,y,z)* and representing the blast error

The variogram of $Y_{\text{blast}}(x,y,z)$ becomes

$$\gamma_{\text{blast}}(h) = \gamma_{15m}(h) + \gamma_R(h) \tag{2}$$

with

$\gamma_R(h)$, the nugget effect due to the blast error, with the variance σ_R^2;

$\gamma_{15m}(h) = (\gamma^*P_{15m})(h) - (\gamma^*P_{15m})(0)$, regularized variogram expressed as a convolution product;

γ, the point variogram, assumed to be isotropic;

$P_{15m}(h) = (p_{15m}{}^*\dot{p}_{15m})(h) = \frac{1}{15^2}(-|h|+15)1_{[0,15]}(|h|)$, function with "*P*" (upper case) that regularizes the variogram, expressed as an auto-convolution product of p_{15m} (lower case) as previously defined by itself.

Similar equations can be established for the drill holes and the 3 m support. The model assumes that the blast and the drill samples have the same average because the independent residuals are of zero mean.

3 Resulting Linear Systems

From Eqs. (1) and (2), some linear systems can be deduced; in the following we propose three of them that are tested on a conditional simulation.

3.1 Removing the Blast Error by Kriging

One can remove the blast error by "factorial kriging" estimation (Matheron 1982), using a linear system applicable to each blast measurement and a local neighborhood of surrounding blast samples. The system is presented symbolically by matrix formalism:

$$\begin{pmatrix} \gamma_{15m} + \gamma_R & 1 \\ 1 & 0 \end{pmatrix} \begin{pmatrix} \lambda \\ \mu \end{pmatrix} = \begin{pmatrix} \gamma_{15m} + \sigma_R^2 \\ 1 \end{pmatrix} \tag{3}$$

In this system, γ_R disappears from the second member of the linear system and is replaced by σ_R^2, the value of the nugget effect. Thus, we remove, from the estimation, the part associated with the measurement error. This does not mean that there is no nugget effect in the remaining part γ_{15m}; it means that only the "natural" part remains. In our case, the complete nugget effect has to be removed because blasts and drills do not share any microstructure.

The result of the estimation is the average value of the grade over the blast support at blast sample locations with no measurement error.

3.2 Deconvolution by Kriging

It may be interesting to remove the effect on the blast of regularization by using a kriging system which estimates, for each blast measurement, a "point" value while simultaneously removing the part of the nugget effect associated with blast errors:

$$\begin{pmatrix} \gamma_{15m} + \gamma_R & 1 \\ 1 & 0 \end{pmatrix} \begin{pmatrix} \lambda \\ \mu \end{pmatrix} = \begin{pmatrix} \gamma^* p_{15m} - (\gamma^* p_{15m})(0) + \sigma_R^2 \\ 1 \end{pmatrix} \tag{4}$$

The difference with the previous system is that in the second member, $\gamma_{15m} = \gamma^* P_{15m} - (\gamma^* P_{15m})(0)$ (upper case P) is replaced by $(\gamma^* p_{15m}) - (\gamma^* p_{15m})(0)$ (lower case p). Initially developed to improve microscopy images of thin plates in the petroleum industry (Séguret1988; Le Loch 1990).

3.3 *Block Estimate by Cokriging Drill and Blast Measurements*

Finally, one can imagine renewing the mine planning block model locally by using blast and drill samples together in a cokriging system with a linked mean (same average for both measurements, Chilès and Delfiner 2012):

$$\begin{pmatrix} \gamma_{3m} & \gamma_{3m,15m} & 1 \\ \gamma_{3m,15m} & \gamma_{15m} + \gamma_R & 1 \\ 1 & 1 & 0 \end{pmatrix} \begin{pmatrix} \lambda \\ \lambda' \\ \mu \end{pmatrix} = \begin{pmatrix} \gamma^* p_{3m}{}^* p_V - (\gamma^* p_{3m}{}^* p_V)(0) \\ \gamma^* p_{15m}{}^* p_V - (\gamma^* p_{3m}{}^* p_V)(0) + \sigma_R^2 \\ 1 \end{pmatrix} \tag{5}$$

As usual, the matrix on the left in the linear system concerns only the data; here, a set of drill and blast measurements. As a consequence, it is composed on the diagonal of submatrices where the variogram of the drill holes (regularization over 3 m) and the variogram of the blastholes (regularization over 15 m) appear. The cross submatrix, which concerns the link between blast- and drill holes, is based on a regularization of the point-support variogram over both supports, which is the reason why $\gamma_{3m,15m} = \gamma^* p_{3m}{}^* p_{15m} - (\gamma^* p_{3m}{}^* p_{15m})(0)$ intervenes. Compared to a usual cokriging system that one can find in the literature (Wackernagel 2003), the linked mean constraints reduce to one line and one column, the submatrix associated with the non-bias constraints. Without this simplification, there would be two lines and two columns. In practice, the consequence is important because here the drill and blast measurements play the same role, while in a normal cokriging system, one of the measurement types would be considered auxiliary, thus reducing its relative influence on the final result. In the right-hand vector, the term p_V appears because the objective is to estimate the average grade over a v-sized block. This convolution is combined with previous regularizations.

4 Simulation

A refined simulation of point-support grades was made every meter horizontally and every 20 cm vertically. It reproduces the data set properties. Then, 100 vertical drill holes were created by averaging, every 50 m horizontally, all the values along 3 m vertically. In this way we obtained 4200 drill samples. Similarly, more than 13,000 vertical 15 m long blastholes were produced with a horizontal spacing of 12 m. The true 15^3 m^3 block value is found by averaging more than 16,000 point-support values contained in the block. The sampling ratio is approximately 1 drill hole sample to 3 blasthole samples and the domain covered by the simulation is 500 by 500 m^2 horizontally and 125 m vertically. The following figure summarizes the grids involved (Fig. 2).

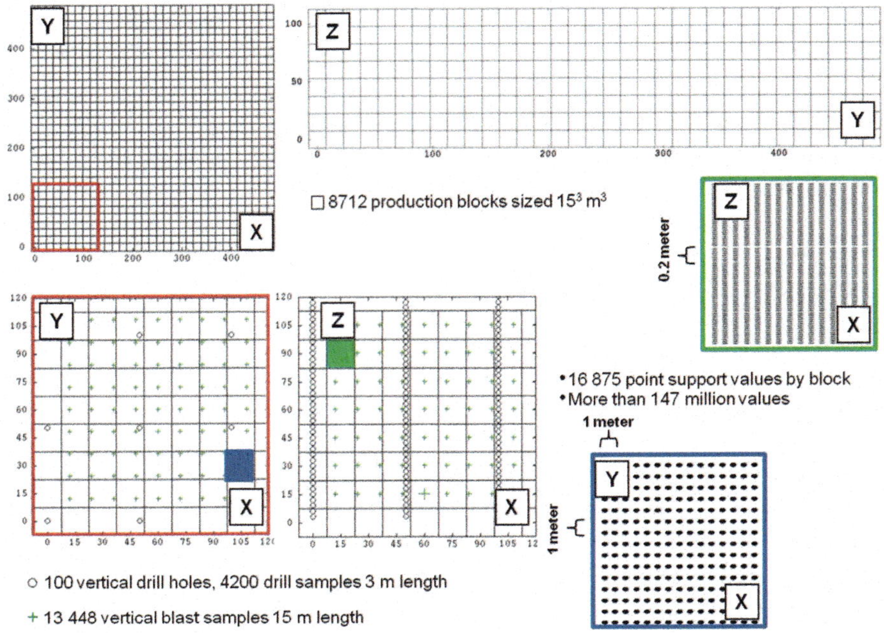

Fig. 2 Blast- and drill holes grid nodes involved in the simulation

For simplification, the drill samples have no errors. We add to the blast samples a random noise with a nugget effect of 0.02, representing the blast sampling error. The grades are realistic with a 0 minimum, 3.5 % maximum; an average of 0.63 % as in the real deposit and the distribution is correctly skewed to the right. We verify that the sills of the drill, blast, and block variograms obey the laws of regularization by following the procedure presented in Séguret (2015) which is based on the charts by Journel and Huijbregts (2003) pages 125–147. Figure 3 shows the result.

5 Removing the Blast Error by Kriging

In the first test, we propose to remove the blast error by using the linear system (3). This filter can be applied to every blast measurement, using a local neighborhood of surrounding blast samples. The neighborhood must contain the sample from which the noise is to be removed; otherwise, the filtering is not efficient.

For comparison, the estimation is made by kriging with no filtering, using the same neighborhood (but without the target sample, otherwise the kriging will obviously give back the value of the data point).

We select, among all the simulated blasts, a subset of around 1000 samples on which estimations will be conducted, using the additional samples in case of ordinary kriging and all the samples in case of nugget filtering.

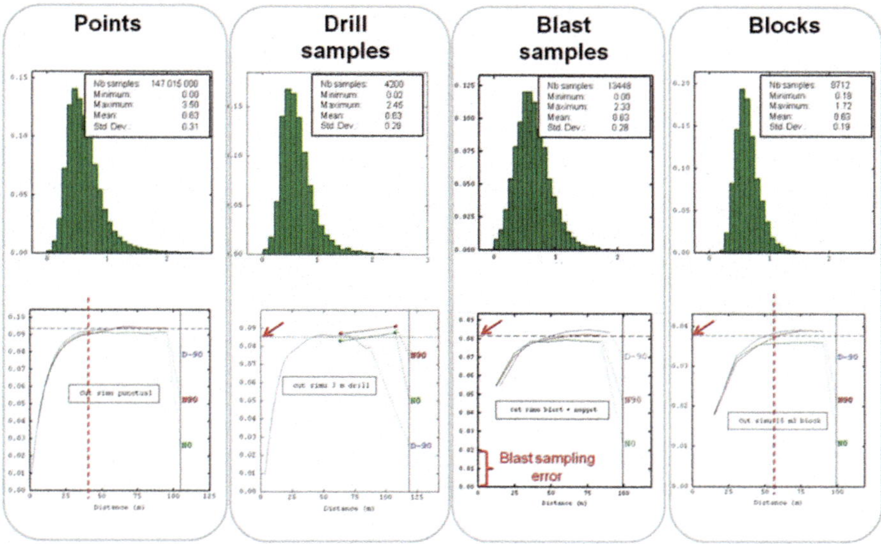

Fig. 3 From *left* to *right*, properties of point-support simulated grades, 3 m support, 15 m and blocks. *Upper* figures are experimental histograms, *bottom* variograms

The reference is the "truth," i.e., the blast without errors which we know because we work on a simulation where everything is known. Figure 4 shows the results.

On these scatter diagrams, the horizontal axis represents the true blast value without any sampling error. On the left-hand scatter diagram, the vertical axis is a usual kriging. The correlation with the truth is 0.65. On the right-hand one, when the filtering is activated, the correlation increases to 0.896. Why?

With filtering, the kriging neighborhood can incorporate the target point where the filter is applied. This point takes a high kriging weight (more than 65 %). Although noisy, this point is closer to the truth than any average based on surrounding points which explains why the filter estimate is closer to the truth.

Finally, the advantage of this linear system (3) is to enable the kriging neighborhood to incorporate the target point information.

6 Deconvolution by Kriging

Now we propose a second test: removing the effect of regularization on the blast with a kriging system that estimates a "point" value for each blast measurement, while simultaneously removing the part of the nugget effect associated with blast errors. By this procedure, we expect to restore the initial variability of the point-support value judged to be too strongly smoothed by the regularization. The linear system used is (4).

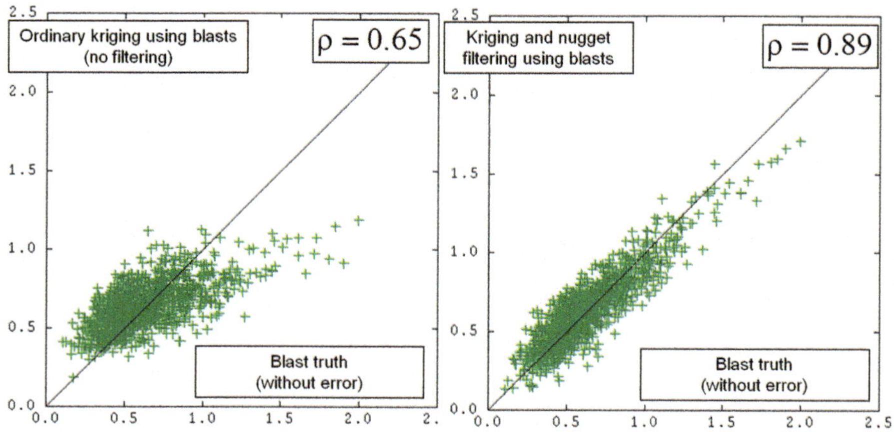

Fig. 4 Scatter diagrams between the true values (*horizontal axis*) and estimations (*vertical axis*). *Left* figure, ordinary kriging estimation; *right* figure, estimation where the blast error is removed by kriging

A comparison is made with the true point value and with previous estimates (estimating a blast with or without nugget effect). Figure 5 shows the results.

In the three scatter diagrams, the horizontal axis represents the true point-support value. The upper left-hand scatter figure presents the result when a deconvolution is made jointly with error filtering. The correlation with the true value is good at 0.85.

The upper right-hand figure presents the result of the error removal with no deconvolution. It corresponds to the previously presented system but this time as compared with the point-support value; this is the reason why the correlation is 0.78 and not 0.89 when compared to the blast values. In comparison with the left-hand figure, the deconvolution increases significantly the accuracy of the estimation.

The bottom figure shows the results when neither filtering nor deconvolution is done. The correlation is very low; it is 0.55.

As for nugget filtering, the deconvolution is efficient because the linear system authorizes the use of the target points where the filter is applied.

7 Block Estimate by Cokriging of Drill and Blast Measurements

The third test is made to renew the mine planning block model locally by using blast and drill samples together. The system used is (5), a cokriging system with linked mean because drill and blast samples have the same average, which is mandatory for carrying out all these calculations.

Our objective is to estimate the average grade at the block scale, and we compare it with two other systems: block grade estimate by kriging using only drill holes and block grade estimate by kriging using only blastholes. Figure 6 shows the results.

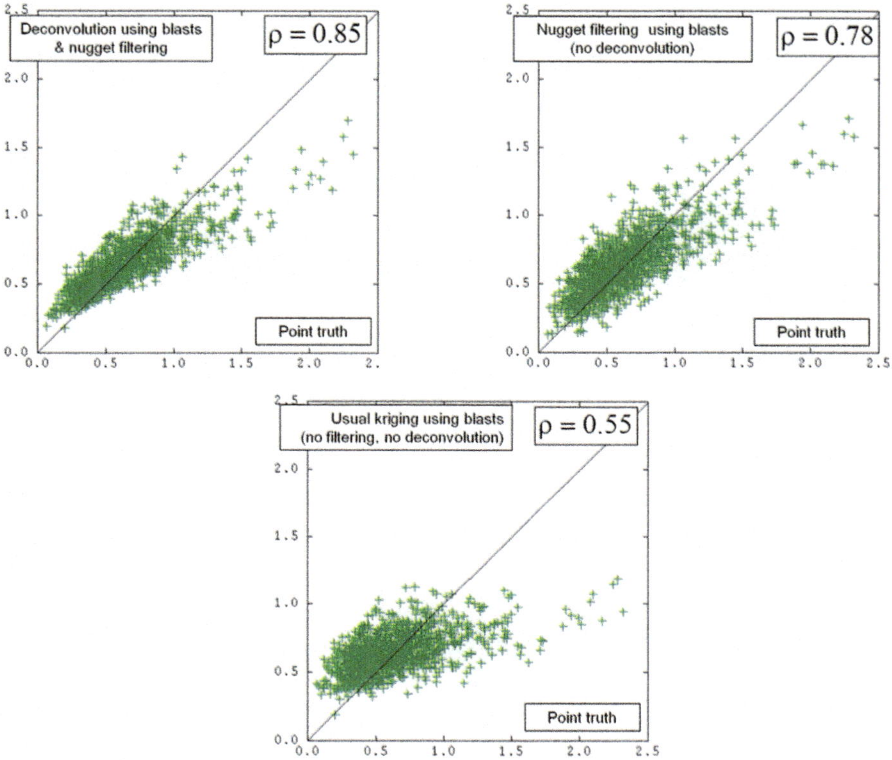

Fig. 5 Scatter diagrams between the true values (*horizontal axis*) and estimations (*vertical axis*). *Upper left* figure, deconvolution together with error filtering; *upper right* figure, error removal without deconvolution; *bottom figure*, no deconvolution, no error filtering

For the three previous scatter diagrams, the horizontal axis is the true block grade. The upper left-hand diagram is the result obtained by ordinary kriging with drill samples; the upper right-hand diagram is the result obtained with blast samples. The jump by the correlation coefficient from 0.38 (OK using drills) to 0.88 (OK using blasts) is impressive. Even if the blast samples are regularized over 15 m, the fact that they are more numerous and respect the variogram (up to a nugget effect) justifies their use when possible, in selections for mining operations, instead of the drill samples.

The bottom diagram concerns cokriging using blast and drill samples together. The performance is similar to ordinary kriging with blast samples only. In our case cokriging is not useful because the blasts are so numerous that adding a drill contribution does not improve the results. This does not mean that such a system is not helpful, for example, in short-term planning to evaluate a domain to be blasted where there are only drill holes.

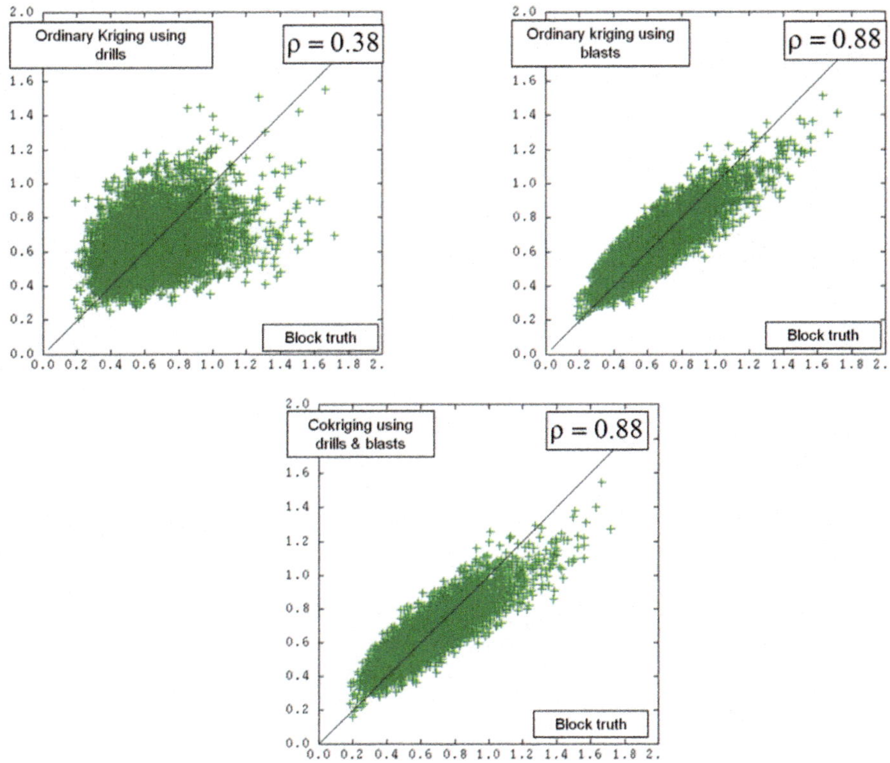

Fig. 6 Scatter diagrams between the true values (*horizontal axis*) and estimations (*vertical axis*). *Upper left* figure, estimation is kriging using drill holes; *upper right figure*, kriging using blastholes; *bottom figure*, kriging using both blast and drill samples together

8 Moving Average or Kriging?

In our case, short-term planning is based on averages using the blast samples included in the block, so the question is whether kriging can produce an improvement. We are comparing three experiments:

- Ordinary kriging with 24 surrounding blast measurements (previous work)
- Moving average with the same 24 surrounding blast measurements
- Moving average with four blast measurements at the same elevation

Figure 7 shows the results.

Replacing ordinary kriging by an average reduces the correlation with the truth from 0.88 to 0.74. This is a very large reduction which should encourage the practitioners to use kriging instead of present practices in the company.

If practitioners do not want to change their habits, one can see that with only 4 points, the result is better than when 24 points are used because the smoothing is weaker: the correlation with the truth increases from 0.74 to 0.84, a result still

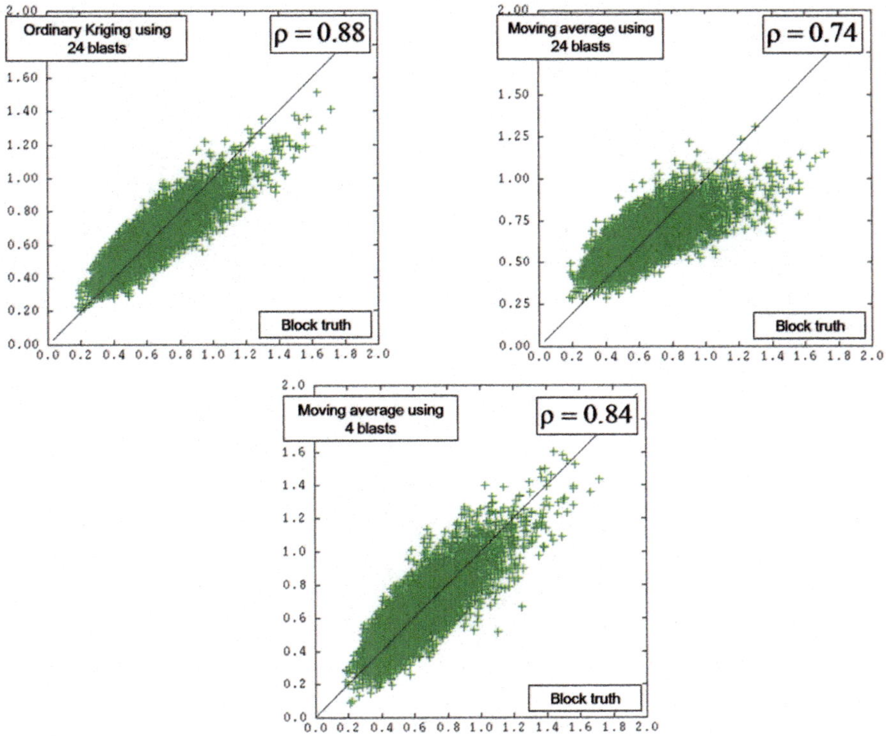

Fig. 7 Scatter diagrams between the true values (*horizontal axis*) and estimations (*vertical axis*). *Upper left*, ordinary kriging with 24 surrounding blast measurements; *upper right*, moving average with the same 24 surrounding blast measurements; *bottom*, moving average with 4 blast measurements of the same elevation

below the one obtained with kriging but very close to it. Does this mean that we recommend a moving average with only 4 points? NO! To use so few points is risky for reasons of conditional bias. To illustrate this concept, consider again the previous scatter diagram but this time, with the true block grade for the vertical axis and the estimate on the horizontal axis (Fig. 8).

In the previous figures, red continuous curves represent the mathematical expectations of the true values conditioned by different estimates. We focus on the most representative [0.3 %, 1 %] range of grades.

When kriging is done with 24 points, the conditional expectation curve is close to the first diagonal. Thus, when we select the block according to its estimation, we obtain, on average, what we expect, with perhaps a slight tendency to underestimate the high grades.

When we replace kriging by a moving average using 24 points, the red curve is still close to the diagonal, with a slight tendency to overestimate the low grades and underestimate the high ones.

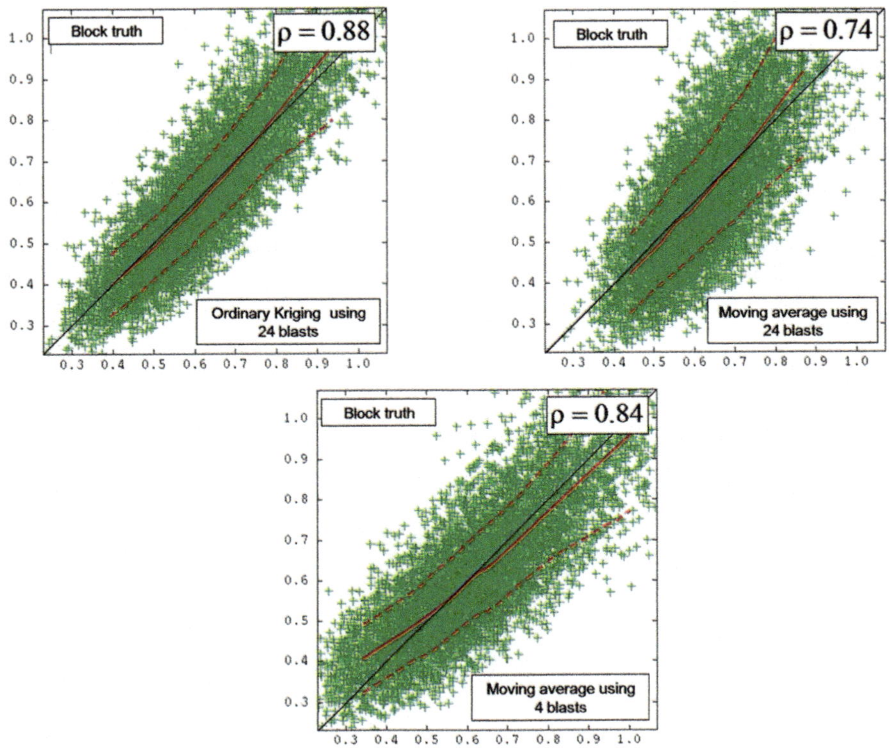

Fig. 8 Scatter diagrams between the true values (*vertical axis*) and estimations (*horizontal axis*). *Upper left*: the *horizontal axis* represents ordinary kriging with 24 surrounding blast measurements (previous work); *upper right*: the *horizontal axis* represents a moving average with the same 24 surrounding blast measurements; *bottom*: the *horizontal axis* represents a moving average with 4 blast measurements at the same elevation

When the moving average is done with only four points, the conditional bias appears clearly: in the range of the low grades, we systematically underestimate the average grade of the blocks and may decide to classify as "waste" blocks that are in reality richer than expected. Conversely, in the range of the high grades, this moving average with only four points systematically overestimates the average grade of the block so that we classify as "rich" blocks which must be considered "waste."

It is for this reason that one must use enough points in the kriging neighborhood for grade control, and reflect on the reason why kriging and geostatistics were created (Matheron 1971).

9 Conclusion

The study of a porphyry copper deposit showed a formal link between blast- and drill holes, leading to numerous linear systems able at least to filter blast errors, make blast deconvolution, or build a block model using blasts and drills together, techniques that could be used at different stages of the mining process.

Tested on a realistic simulation, these systems have proved their worth, as well as the danger of replacing kriging by a moving average, especially with few points, producing a strong conditional bias, and this is a useful reminder of the reason why kriging was created.

Overall, a formal comparison between blast- and drill holes shows that in this mine – and more generally, in this company – the quality of the blast values is as good as the quality of the drill values, contrary to conventional wisdom.

Bibliography

Chilès JP, Delfiner P (2012) Modeling spatial uncertainty. Wiley edition

Journel AG, Huijbregts CG (2003) Mining geostatistics. The Blackburn Press, New Jersey

Le Loch G (1990) Déconvolution d'images scanner, Mines ParisTech publication N-8/90/G. Center for Geostatistics, Fontainebleau

Matheron G (1971) The theory of regionalized variables and its applications, Fasc. 5. Paris School of Mines, Fontainebleau

Matheron G (1982) Pour une Analyse Krigeante des données Régionalisées, Mines ParisTech publication N-732. Center for Geostatistics, Fontainebleau

Séguret SA (1988) Pour une méthodologie de déconvolution de variogrammes, Mines ParisTech publication N-51/88/G. Center for Geostatistics, Fontainebleau

Séguret SA (2015) Geostatistical comparison between blast and drill-holes in a porphyry copper deposit. WCSB7 7th world conference on sampling and blending, Jun 2015, Bordeaux, France. 2015, doi: 10.1255/t0sf.51

Wackernagel H (2003) Multivariate geostatistics. Springer Sciences, Berlin

Building a Tonnage-Surface Function for Metal Grades and Geological Dilution: An Application to the Massive and Stockwork Zambujal Ore Deposit, Neves–Corvo Mine, Portugal

David Silva and José António Almeida

Abstract Classical metal tonnage cut-off curves are an important output of mineral deposit evaluation. However, such curves omit local morphological variations such as those found in sulphide stockwork-type deposits, in which sulphides and host rock exhibit variable proportions. This paper presents a methodology for building a stochastic model of the massive and stockwork Zambujal ore deposit with respect to both morphology and metal content. The model allows a metal tonnage cut-off surface to be constructed conditional both to metal content and to local proportions of sulphides relative to host rock. Two random variables are modelled using stochastic simulation: (i) a variable ($P(x)$) that represents the local proportion of sulphides within each mining block and (ii) an auxiliary variable ($Y(x)$) representing the relative contents of metal, that is, the metal contents recalculated for the matrix sulphides. The results are validated by comparing the global quantities obtained using ordinary kriging of the effective grades with those obtained by modelling the variables $P(x)$ and $Y(x)$.

D. Silva (✉)
Earth Sciences Department, Faculty of Sciences and Technology, Universidade NOVA de Lisboa, Campus da Caparica, 2829-516 Caparica, Portugal
e-mail: dam.silva6@gmail.com

J.A. Almeida
Earth Sciences Department and GeoBioTec, Faculty of Sciences and Technology, Universidade NOVA de Lisboa, Campus da Caparica, 2829-516 Caparica, Portugal
e-mail: ja@fct.unl.pt

J.J. Gómez-Hernández et al. (eds.), *Geostatistics Valencia 2016*, Quantitative Geology and Geostatistics 19, DOI 10.1007/978-3-319-46819-8_25

1 Introduction

The characterisation of mineral deposits combining geological information, borehole logs, and analytical measurements precedes the decision to engage in mining and usually involves the following steps (Gerst 2008; Almeida 2010; Rossi and Deutsch 2014): (a) delimiting the volume under study, for example, from digitalising sections and/or interpolating surfaces and then building a 3D representation; (b) constructing a high-resolution morphological model of the lithologies or ore types within the studied volume; (c) constructing a model of metal grades conditioned to the morphological model; and (d) parameterising the deposit in terms of the quantities of ore and metal as a function of metal grades. In these steps, geostatistical tools are usually used (Goovaerts 1997), either kriging estimation or conditional simulation, to compute quantities and map the levels of uncertainty. The particular type of mineral deposit under investigation affects the detail of the procedures used within each of the steps.

The case study examined in the present study is the Zambujal deposit of metal sulphides (Neves–Corvo Mine, Alentejo, Southern Portugal). The deposit is operated by the company Lundin Mining, who supplied the data for the study. The Zambujal is a volcanogenic copper–zinc sulphide deposit located in the southern Portuguese sector of the Iberian Pyrite Belt (Neves–Corvo mine). It encompasses two regions, massive ores (R_M) in the top part and stockwork ores (R_S) (both stringer and impregnation types) in the bottom part (Relvas et al. 2002). The stockwork ores contain sulphide metals and host rock in variable proportions, and it is important to note the extremely copper-rich ores, which reflect zone refining and late enrichment by remobilisation processes.

When quantifying the resources of an ore deposit, the cut-off grade of the primary metal is usually adopted as the unique reference. For the Zambujal deposit, and because the morphological parameter $P(x)$ (proportion of sulphides) was assessed, it was decided to test the use of this morphological variable as a second cut-off parameter; that is, a parameter that represents the resources using a combination of the metal content and the proportion of host rock. One of the reasons for conducting this particular analysis is that it is possible to consider a lower metal content in the case of ores with a significant volume of host rock, that is, a cut-off grade that is also a function of variable $P(x)$. For example, for two rock matrices with the same copper content, the copper recovery would be higher in the sample containing more host rock.

Two main issues constrain the proposed methodology. First, the stockwork ores exhibit higher recoveries during mineral processing compared with massive ores with equivalent grades. This higher recovery is because the contrast in physical properties is higher between sulphides and host rock than between different sulphide minerals. Second, copper grades in the stockwork ore sulphides are approximately double those in the massive ore sulphides. However, because of the proportion of host rock, the effective copper grades are much lower in the stockwork ores than in the massive ores, but this issue increases the relevance of the stockwork ores.

Therefore, a high-resolution morphological model was built to quantify the local proportion of sulphides in each block of the block model (the variable $P(x)$), followed by a model of relative grades (the variable $Y(x)$). By doing so, the metal contents are recalculated for the sulphide fraction rather than modelling the total metal content in the total volume. For instance, for the sample location x_i with a proportion of sulphide minerals $P(x_i)$ and effective grade of copper $Z_{\text{Cu}}(x_i)$, the relative grade of copper $Y_{\text{Cu}}(x_i)$ is expressed by:

$$Y_{\text{Cu}}(x_i) = Z_{\text{Cu}}(x_i)/P(x_i) \tag{1}$$

Both $P(x)$ and $Y_{\text{Cu}}(x)$ are regionalised variables and can be modelled by using geostatistical tools such as variograms, kriging, and stochastic simulation. The total amounts of metals and ores can be calculated directly from the estimated or simulated values of these variables. Also, with models of both these variables, parametric surfaces relating the amount of copper to both the proportion of sulphides (morphological constraint) and the grade of metal (grade constraint) are readily constructed. The ability to construct such surfaces in a straightforward way is an important objective and outcome of this paper.

The proposed methodology contains the following steps:

1. Build a two-region 3D solid model of the Zambujal ore deposit that encompasses the boundaries of the deposit and the transition between the massive ores in the top part (R_M) and the stockwork ores in the bottom part (R_S). Convert the solid model into a voxel model (mining blocks).
2. Using the densities and metal grades measured at the boreholes and knowledge about the paragenesis of the Zambujal deposit, evaluate optimal solutions for the sulphide proportion random variable $P(x)$. In the present case, ten solutions were selected for each sample.
3. Estimate local cumulative distribution functions (*cdf*s) of the random variable $P(x)$ for both region R_M and region R_S by using indicator kriging (IK). Build a high-resolution morphological model of $P(x)$ using Direct Sequential Simulation (DSS) (Soares 2001) conditional to the computed local *cdf*s and thus conditional to the low-resolution model.
4. Compute $Y_{\text{Cu}}(x)$ at the sample locations. Estimate local *cdf*s of Y_{Cu} for both region R_M and region R_S using IK. Simulate the relative grades using DSS for regions R_M and R_S simultaneously.
5. Analyse the results involving an evaluation of ore and metal tonnages, and a comparison of the tonnages with those obtained using ordinary kriging (OK). Validate the variograms of the simulated images and histograms. Map the uncertainty.
6. Build a global surface function of copper quantity conditional to the copper grades and the local proportions of sulphides relative to host rock.

It is important to note that variables $P(x)$ and copper grades (both the relative and effective grades) are not stationary within the entire deposit; rather, they are constrained to the massive and stockwork ores. For this reason, it is essential to delimit the two regions (massive and stockwork ores) and to constrain the simulations with the *cdf*s of the studied variables. A modified version of DSS in which the simulated values are resampled from the local distributions with local *cdf*s is used to simulate the entire deposit (Charifo et al. 2013; Matias et al. 2015; Roxo et al. 2016).

2 Case Study

2.1 Low-Resolution 3D Morphological Model

The upper and lower limits of the massive and stockwork regions were digitised according to the borehole data and expert information in several parallel cross sections. Then, three surfaces were constructed using linear interpolation: the top of the massive ore, the bottom of the massive ore/top of the stockwork ore, and the bottom of the stockwork ore. The final step for building the low-resolution morphological model was the conversion of the surfaces into a block model (block size $2 \times 2 \times 2$ m) of the two major ore types (Fig. 1).

2.2 Building the Sulphide Proportion Variable P(x) at Sample Locations

Before simulations were performed, the variable $P(x)$ was evaluated for all samples by combining the measured values of density $D(x)$ and metal grade $Z(x)$.

The evaluation of $P(x)$ for core samples was made using a classification procedure because this variable is not usually measured in the laboratory. To make this estimate at the sample locations, the measured levels of the most abundant chemical elements (Fe, Cu, Zn, Sn, Pb, and S) were used, as well as the densities of the sampled cores and the paragenesis of the deposit (pyrite, chalcopyrite, galena, blend, or host rock). It is important to emphasise the difference between the density of the host rock (about 2.88 t/m^3) and the densities of the minerals listed (all >4 t/m^3). A lookup table listing all admissible combinations of these minerals and the host rock was made (with a resolution of 1 % for proportions), and, for each combination, the theoretical values of the element grades (based on the chemical formulae) and the composite density were computed. For example, considering a sample comprising 30 % host rock, 30 % pyrite, and 40 % chalcopyrite ($P(x) = 0.7$), the theoretical key element grades and density should be Fe = 26.14 %, $S = 30.01$ %, Cu = 13.85 %, and $D = 4.10$ t/m^3. Then, each laboratory-measured value of grade

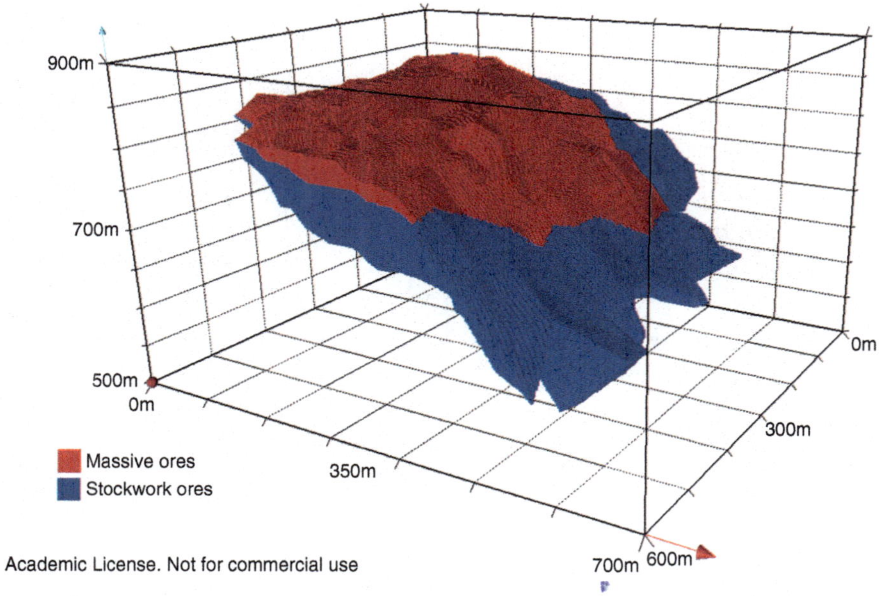

Fig. 1 Binary morphological model of the two regions R_M (massive ores) and R_S (stockwork ores) in a voxel structure

and density was compared with each item of the lookup table, and a set of the closest combination of minerals plus host rock was extracted. Ten theoretical mixtures of minerals plus host rock were considered sufficient to cover the uncertainty, and the selected measurements were in the main very close to each other. Each selected set of ten values constitutes the local *cdf* of $P(x)$ at the locations of the boreholes. These *cdfs* were then extended to the entire deposit using DSS (Soares 2001). The model of $P(x)$ is a detailed morphological model and must be used, in a second step, to condition the construction of the model of metal grades.

In summary, at this point, a *cdf* of $P(x)$ is constructed for each sample. Figure 2 shows a 3D view of a single realisation of $P(x)$ at the locations of the boreholes converted to the grid of blocks. In the next section, $P(x)$ is simulated for the entire deposit.

2.3 *High-Resolution Simulation Model for the Sulphide Proportion Variable* P(x) *and Copper Grades*

The proportion of sulphide ore minerals $P(x)$ was simulated using DSS for the two regions (R_M and R_S) simultaneously (Ruben and Almeida 2010). Instead of using a global *cdf*, the simulation of $P(x)$ was performed using a modified version of DSS that uses local *cdfs* for each region, $P(x)|R_M$ and $P(x)|R_S$ (Charifo et al. 2013). For declustering purposes, $P(x)|R_M$ and $P(x)|R_S$ were estimated using IK of 20 $P(x)$

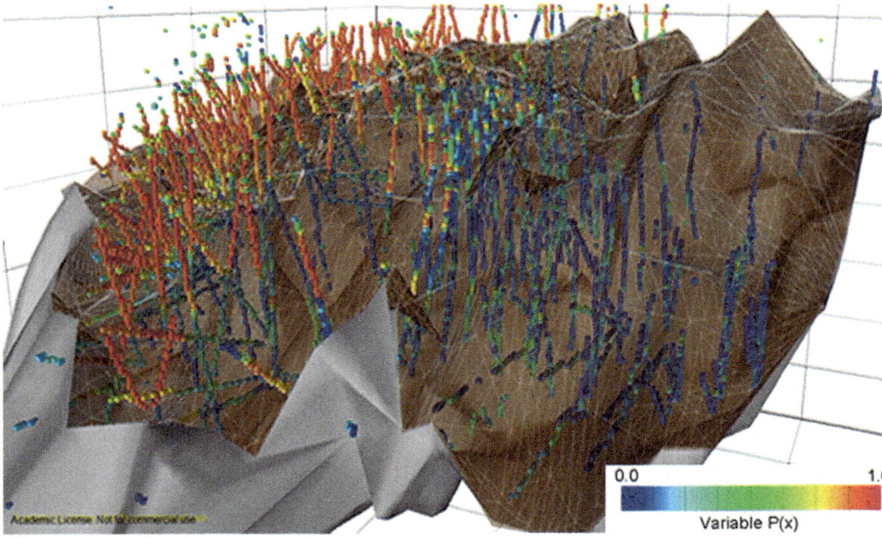

Fig. 2 3D view of a realisation of $P(x)$ at the locations of the boreholes converted to the grid of blocks. *Red* colours represent higher values of $P(x)$ (massive ores), and *blue*-based colours refer to lower values of $P(x)$ (stockwork ores)

classes. Thirty realisations of $P(x)$ were generated. Figure 3 shows the average image of $P(x)$ (top diagram) and the 10 % of blocks with lowest uncertainty (bottom diagram).

Higher values of $P(x)$ are located at the top of the structure, and lower values at the bottom (see Figs. 1 and 3), and the transition between the two regions is sharp. Histograms are well reproduced by DSS, as are the variograms of the data (Silva 2015).

For simulating metal grades, the proportion of ore minerals ($P(x)$) was combined with borehole grades to evaluate the relative metal grades ($Y(x)$). For instance, if a sample had a grade of 2 % copper and a value of $P(x) = 0.7$, then the relative grade of copper would be equal to 2 %/0.7 = 2.86 %. The simulation of $Y(x)$ follows the same approach as taken for $P(x)$, namely, the use of IK to estimate conditional *cdf* histograms of $Y(x)|R_M$ and $Y(x)|R_S$ and DSS with local histograms. Three realisations of grades were generated conditional to each of the 30 realisations of $P(x)$, giving a total of 90 realisations for $Y_{Cu}(x)$. Again, histograms are well reproduced by DSS, as are the variograms of the data (Silva 2015).

Figure 4 shows the average relative copper grades $Y_{Cu}(x)$ (top diagram) and the actual copper grades (bottom diagram).

It is important to note that the values obtained for the relative grades of copper and for the actual grades obtained through the quotient of the relative grades and $P(x)$ are theoretically valid. This theoretical validity is a good indicator that the proposed methodology can be applied; for example, a maximum of 32 % for relative copper grades is plausible. Figure 4 also shows that the relative grades in the stockwork ores are higher than those in the massive ores. The transition of

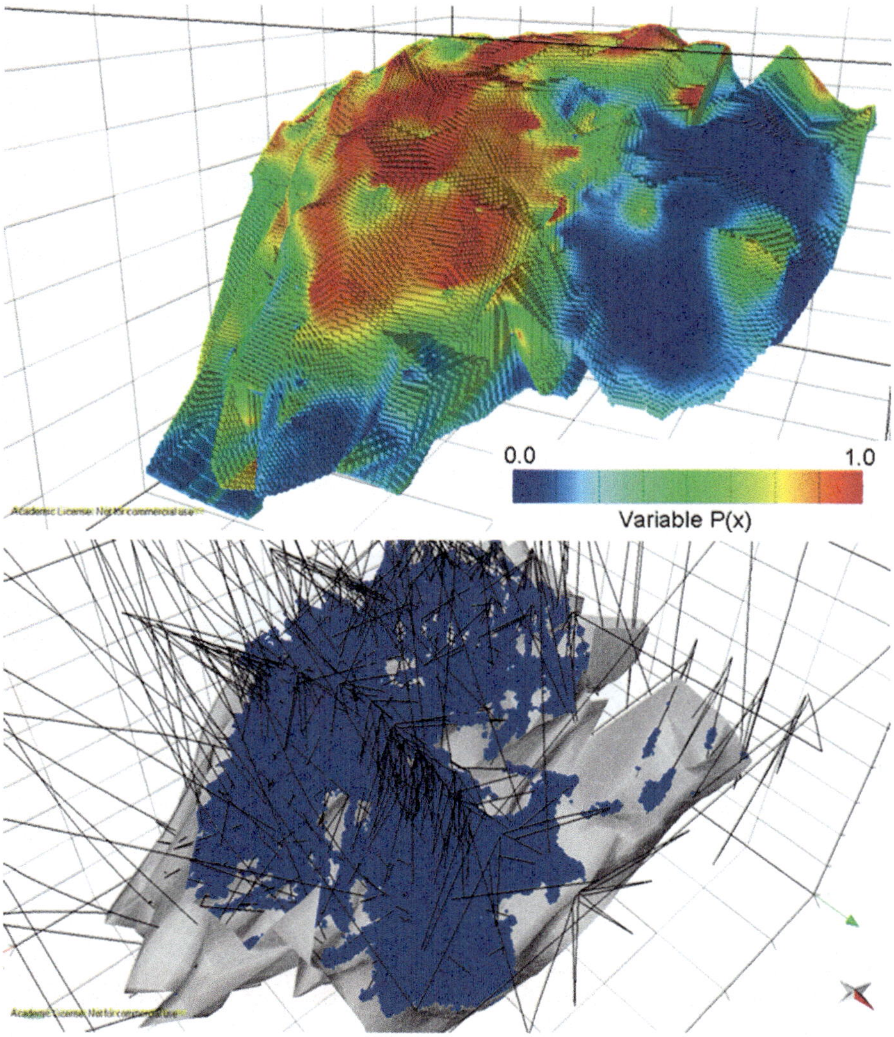

Fig. 3 3D representation of (*top*) the average values of the sulphide proportion variable $P(x)$; (*bottom*) the 10 % of mining blocks with the lowest uncertainty (in *blue*), also showing boreholes (*lines*) and the bottom surface of the stockwork ores (in *grey*)

grades between regions is smooth, which is in accordance with the variability of the grades along the boreholes.

Finally, Table 1 compares the average of grades obtained using OK and DSS for the massive ore region, the stockwork ore region, and the regions combined. Deviations are minimal, confirming that DSS with local histograms generates unbiased realisations.

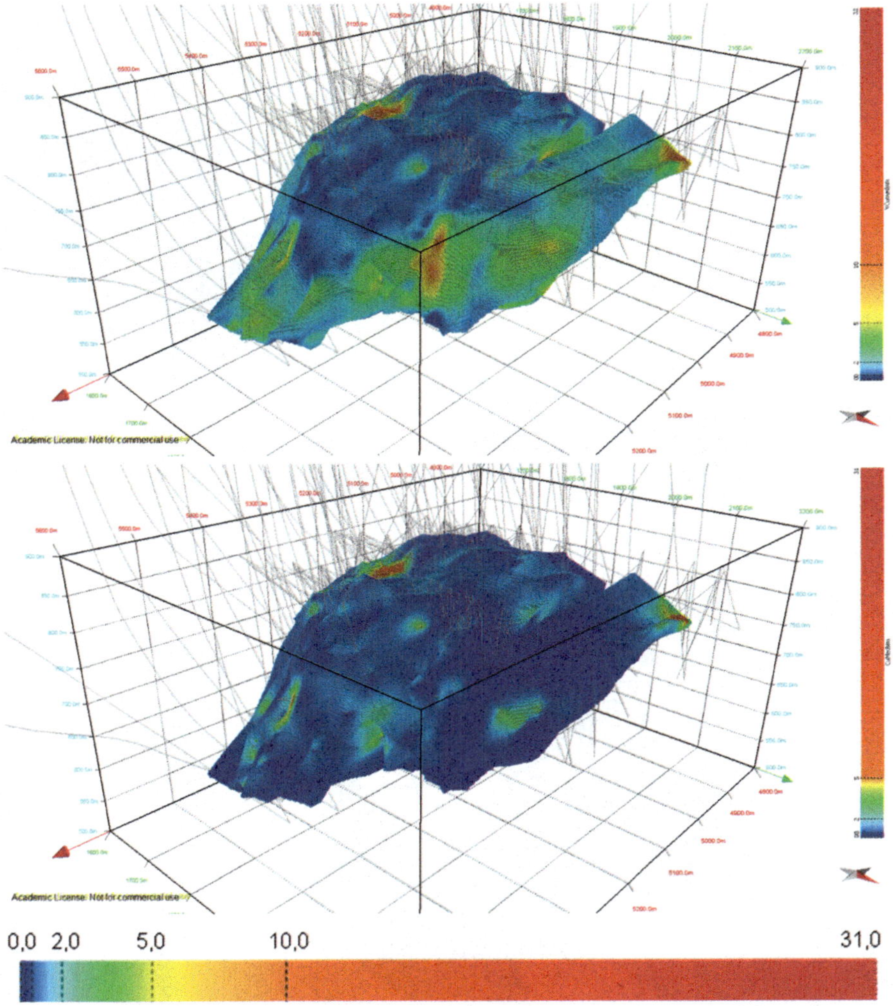

Fig. 4 (*top*) 3D representation of average relative copper grades (%). Higher relative grades are in *red*, lower values in *blue*; (*bottom*) 3D representation of average copper grades. Higher grades are in *red*, lower values in *blue*

2.4 *Construction of the Tonnage-Surface Function for Copper Grades and Geological Dilution*

Using the simulated values of the variables $P(x)$ and $Y_{\mathrm{Cu}}(x)$, average estimates for $[P(x_i)]^*$, $[Y_{\mathrm{Cu}}(x_i)]^*$ and $[Z_{\mathrm{Cu}}(x_i)]^* = [P(x_i)]^*.[Y_{\mathrm{Cu}}(x_i)]^*$ were computed for each mine block x_i. The tonnage of copper (T_{Cu}) within each mine block x_i with volume V and ore density ρ is given by

Table 1 The proportion of sulphide minerals and the relative grades of Cu for massive (RM) and stockwork (RS) ores

		R_M	R_S	$R_M + R_S$
Morphological model [$P(x)$]	OK	0.755	0.184	0.350
	DSS	0.768	0.176	0.348
	% deviation	1.6	−4.7	−0.7
Copper	$[Y_{Cu}(x)]^{OK}$	1.465	3.289	2.763
	$[Y_{Cu}(x)]^{DSS}$	1.482	3.339	2.795
	% deviation	1.1	1.5	1.1

$$T_{Cu}(x_i) = \rho.V.[P(x_i)]^*.[Y_{Cu}(x_i)]^*/100 \tag{2}$$

A surface representing the total tonnage of copper as a function of the effective grade of copper $Z_{Cu}(x)$ and the morphological variable $P(x)$ is now able to be presented (see Fig. 5).

Moving from left to right in Fig. 5, the tonnage of copper is calculated for mine blocks with increasingly higher grades; moving from bottom to top, the tonnage of copper is calculated for mine blocks with increasingly higher proportions of sulphides. The colour spectrum represents the varying tonnage of copper, indicating a decrease in the amount of metal from warm (red) colours to cool (blue) colours.

The maximum value of 528,000 tonnes in the bottom left corner (the total copper resource of the Zambujal ore deposit) corresponds to blocks with a grade of copper above 0 % and for any value of $P(x)$. The solid black reference line shown in the diagram corresponds to a threshold of 1.6 % for copper. This prototype of representation enables consideration not only of a fixed threshold but also of a variable threshold function of $P(x)$ such as those represented by dashed lines in Fig. 5.

3 Final Remarks

The proposed methodology highlights the importance of the stockwork ores in the Zambujal deposit, and as mineral concentrations are more efficient in stockwork ores, metal cut-offs should take into account the proportion of sulphides relative to host rock. Modelling the proportion variable and the relative metal grades adds detailed information to each mining block and enables metal tonnage cut-off surfaces combining metal grades and the proportion of sulphides relative to host rock (geological dilution) to be generated in a straightforward way.

Acknowledgements The authors are grateful to Lundin Mining for providing the data, information, and assistance required for the present work and to Midland Valley for providing an academic licence for Move® software. This work is a contribution to Project UID/GEO/04035/2013 funded by FCT-Fundação para a Ciência e a Tecnologia, in Portugal.

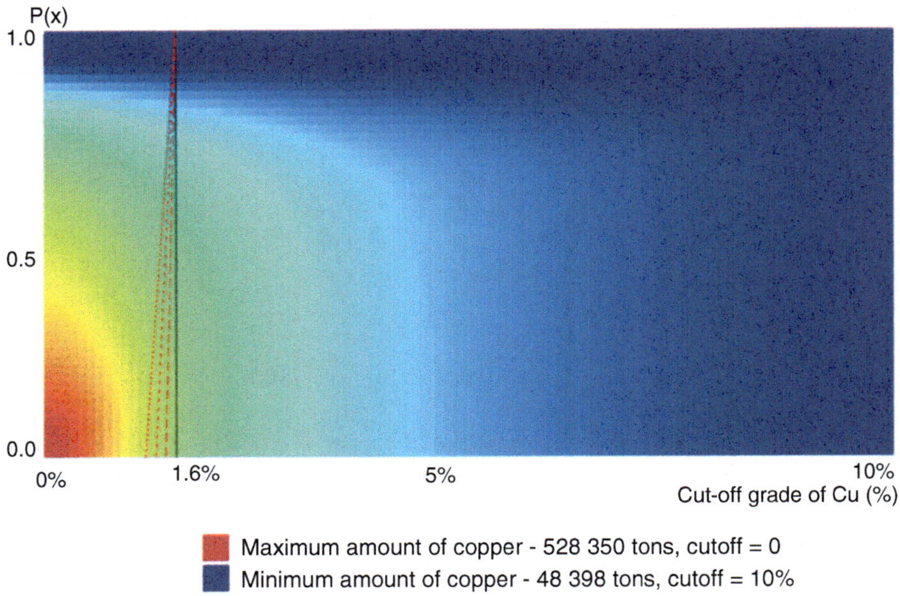

Fig. 5 Copper resources as a function of copper grade and the proportion of sulphides $P(x)$. The amount of copper increases from blue to red

Bibliography

Almeida JA (2010) Modelling of cement raw material compositional indices with direct sequential cosimulation. Eng Geol 114(1):26–33

Charifo G, Almeida JA, Ferreira A (2013) Managing borehole samples of unequal lengths to construct a high-resolution mining model of mineral grades zoned by geological units. J Geochem Explor 132:209–223

Gerst MD (2008) Revisiting the cumulative grade–tonnage relationship for major copper ore types. Econ Geol 103:615–628

Goovaerts P (1997) Geostatistics for natural resources evaluation. Oxford University Press, New York

Matias FV, Almeida JA, Chichorro M (2015) A multistep methodology for building a stochastic model of gold grades in the disseminated and complex deposit of Casas Novas in Alentejo, Southern Portugal. Resour Geol 65(4):361–374

Nunes R, Almeida JA (2010) Parallelization of sequential Gaussian, indicator and direct simulation algorithms. Comput Geosci 36(8):1042–1052

Relvas J, Jorge R, Pinto A, Matos J, Rosa C, Barriga F (2002) The neves–corvo deposit, Iberian Pyrite Belt, Portugal. In: Impacts and future, 25 years after the discovery, vol. Special Publication 9. Society of Economic Geologists, pp 155–176

Rossi ME, Deutsch CV (2014) Mineral resource estimation. Springer, Dordrecht (Vol. Science + Business Media)

Roxo S, Almeida JA, Matias FV, Mata-Lima H, Barbosa S (2016) The use of sensory perception indicators for improving the characterization and modelling of total petroleum hydrocarbon (TPH) grade in soils. Environ Monit Assess 188(3):1–19

Silva D (2015) Modelação estocástica do depósito mineral do Zambujal (Mina de Neves–Corvo): contribuição da densidade como indicador morfológico. FCT – NOVA University of Lisbon, MSc Thesis (in portuguese)
Soares A (2001) Direct sequential simulation and cosimulation. Math Geol 33(8):911–926

An Application of Direct Sequential Simulation and Co-simulation for Evaluating the Resource and Uncertainty of the Ncondezi Coal Deposit, Mozambique

Sara Ferreira Sokhin, José António Almeida, and Sofia Barbosa

Abstract The main objective of this work is to develop a model of the morphology and coal-quality variables of a sub-region of the Ncondezi coal deposit, Mozambique, with data supplied by the Ncondezi Coal Company. This coal deposit is characterized by a complex stratigraphy, with fine coal layers interleaved with layers of shales and sandstones. The morphological model consists of the spatial characterization of the random variable $P(x)$ (proportion of coal). To model the coal-quality variables and because of the large number of variables and given that some pairs of variables exhibit high correlations, an approach involving simulation of principal components as secondary variables is used. A set of selected principal components are simulated using direct sequential simulation, and the coal-quality variables are then co-simulated conditional to the previously simulated principal component values. This approach works as a global co-regionalization model. The results of the morphological model and of the model of the coal-quality variables allow global coal reserves to be calculated as well as the quantities of coal to be parameterized as a function of the coal-quality variables.

1 Introduction

Modelling coal deposits usually involves a two-step approach: first, the creation of a model of the morphology and second, the construction of a model of the properties or coal-quality variables (Cornah et al. 2013; Srivastava 2013). Coal

S.F. Sokhin (✉)
Earth Sciences Department, Faculty of Sciences and Technology, Universidade NOVA de Lisboa, Campus da Caparica, 2829-516 Caparica, Portugal
e-mail: saramatiasferreira@gmail.com

J.A. Almeida • S. Barbosa
Earth Sciences Department and GeoBioTec, Faculty of Sciences and Technology, Universidade NOVA de Lisboa, Campus da Caparica, 2829-516 Caparica, Portugal
e-mail: ja@fct.unl.pt; svtb@fct.unl.pt

J.J. Gómez-Hernández et al. (eds.), *Geostatistics Valencia 2016*, Quantitative Geology and Geostatistics 19, DOI 10.1007/978-3-319-46819-8_26

resources are commonly estimated and classified based on drill-hole data, which are often multivariate (Tercane et al. 2013). Kriging is the preferred geostatistical method in the coal industry for estimating and assessing coal resources.

Regarding the construction of the morphological model, if coal seams have considerable thicknesses, an indicator variable $I(x)$ can be adopted and modelled using geostatistical indicator kriging or indicator simulation (Almeida 2010a). Where the coal seams are considerably thinner (only a few or tens of centimetres thick), as in the present case study, it is not possible to use an indicator variable, but it is possible to use a variable that quantifies the volumetric proportion of coal in each block, $P(x)$. This variable proportion of coal can be modelled by estimation and/or simulation. However, for complex and heterogeneous deposits, kriging is insufficient and should be complemented with stochastic simulation to provide uncertainty assessment (Charifo et al. 2013; Cornah et al. 2013; Matias et al. 2015).

Of the various papers that have addressed the modelling of coal variables, Tercan and Sohrabian (2013) used direct sequential simulation (DSS) (Soares 2001) to characterize the quality of a coal deposit in Turkey in a two-step approach. In that work, principal component analysis (PCA) (Davis 1986) is applied first to the three coal-quality variables (ash, calorific value, and inherent moisture). DSS is then used to generate simulated images of the principal components, after which the simulated values of the major principal components are transformed into the values of the initial variables. Although this approach enables a multivariate set of linearly correlated variables to be simulated and the multivariate matrix of correlations to be generated, it does not impose both the histograms and the variograms of each variable. Also, anomalous combinations of results (not observed in the borehole data) can be obtained.

This paper presents a methodology to build a model of the morphology and coal-quality variables of a sub-region of the Ncondezi coal deposit, based on data from surveys conducted by the Ncondezi Coal Company. The deposit is located in the sub-basin of Moatize-Minjova, Tete Province, Mozambique (Fig. 1) (Johnson et al. 1996).

The Ncondezi deposit is a stratified coal deposit that was developed under the influence of tectonic process that played a major role in determining the stratigraphic sequence of the basin and the quality of the coal. The extensional regime experienced by the basin during its formation led to a fragmentation of blocks in a horst-graben system, which greatly influenced the lateral extent of the coal seams. This fragmentation and the resultant low levels of lateral continuity of strata would appear to be the primary reason for the difficulty of correlating coal seams between boreholes. Even between relatively close boreholes separated by distances of less than 300 m, coal-seam correlation is weak. Moreover, the tectonic instability of the basin caused variations in the depositional environment, particularly fluctuations in the groundwater level, which explain, for example, the high levels of ash in Ncondezi coal.

This coal deposit comprises a heterogeneous package of thin coal lenses, each measuring a few tens of centimetres or less in thickness, interbedded with non-coal lithologies. The total thickness of the deposit, where not eroded, is several tens of

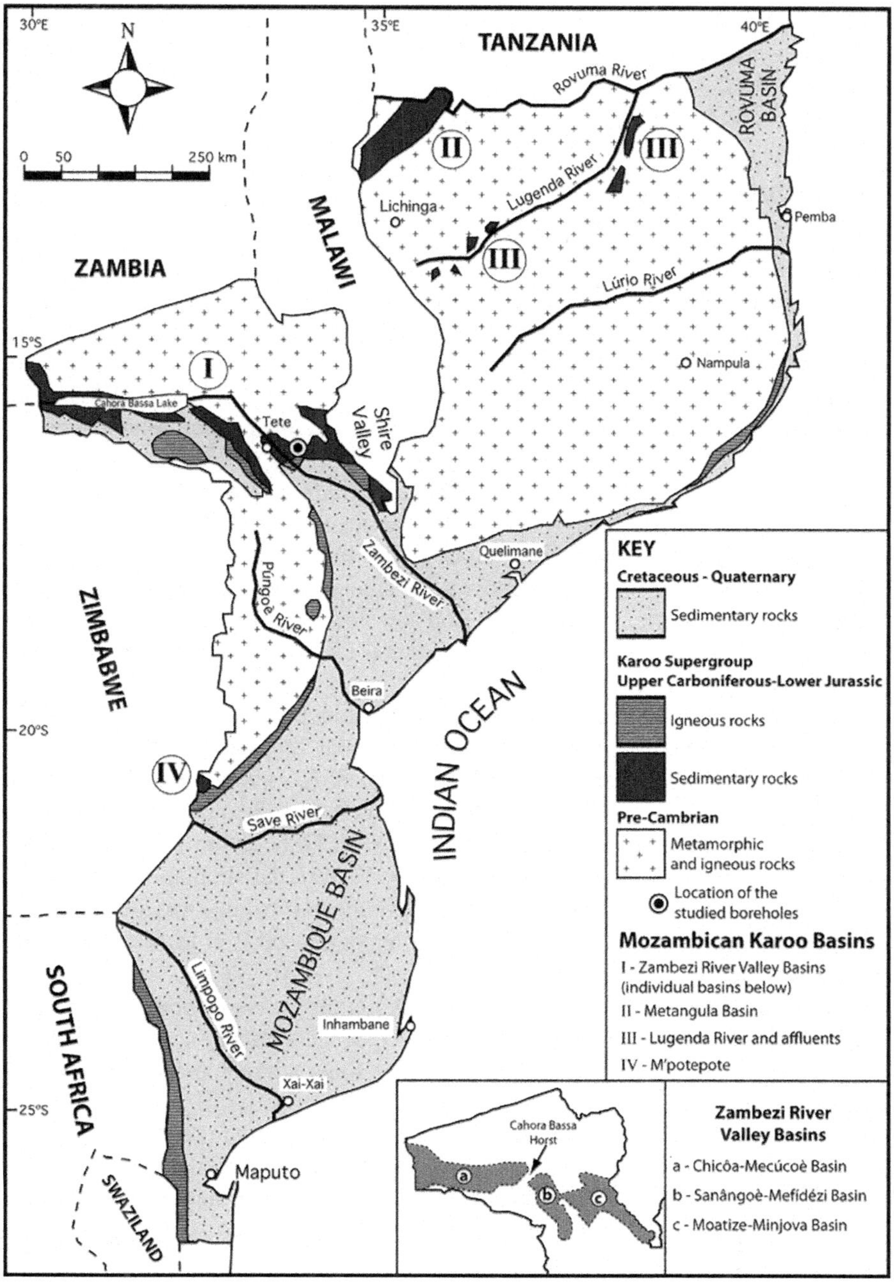

Fig. 1 Location of the three Zambezi River Valley Basins (Adapted from Fernandes et al. 2015)

metres. Lakshminarayana (2015) described this type of coal deposit and the practical implications for their extraction and referred to such deposits as "barcode" deposits.

2 Methods

Figure 2 summarizes the proposed methodology for the characterization and resource evaluation of the Ncondezi Coal deposit. The morphological model and the coal-quality data model are built independently, with the results being integrated at the end and analysed together. The coal-quality data model uses seven variables $K_i(x)$: ash yield (Ash), inherent moisture (IM), volatile matter (VM), fixed carbon (FC), calorific value (CV), total sulphur (TS), and apparent relative density (ARD).

First, a statistical study of seven coal-quality variables was performed involving univariate and bivariate analyses and multivariate PCA (Davis 1986; Isaaks and Srivastava 1989). This initial step enabled the most correlated variables to be selected in order to adopt the proposed co-regionalization simulation approach. Then, estimates of the variable $P(x)$ and the coal-quality variables $K_i(x)$ were generated for the entire set of boreholes in equal length samples. A grid of blocks was also constructed.

The modelling of the deposit used an independent two-step approach: the construction of a morphological model of the coal lithologies followed by the generation of a model of the coal-quality variables. As it is not possible to create a mesh of blocks at a centimetric scale because of grid size limitations, the morphological model consisted of simulations by DSS of the random variable proportion of coal $P(x)$ within each grid block. Before the simulation, an estimation of $P(x)$ both by ordinary kriging and by indicator kriging was performed to quantify the global resource and the global histogram of $P(x)$ to be used within the DSS.

Concerning the model of the coal-quality variables, it is essential to reproduce the multivariate correlation metrics as observed in the available coal-quality data. To avoid the calculation of cross variograms between a large number of variables and fit a full multivariate co-regionalization model, the set of linear correlated variables was modelled by using the first principal component (PC-1) as a secondary variable. Thus, PC-1 was simulated over the entire deposit using DSS, and then each coal-quality variable was co-simulated using Co-DSS using the simulated PC-1 as a secondary variable in a local co-regionalization approach via PCA loadings. This simulation approach guarantees the reproduction of the linear correlation metrics as observed in the borehole data as well as the variograms and histograms of each quality variable (Almeida 2010b).

A relevant output of the simulated block models are curves of coal quantity conditional to the coal-quality variables (Rossi and Deutsch 2014). The local

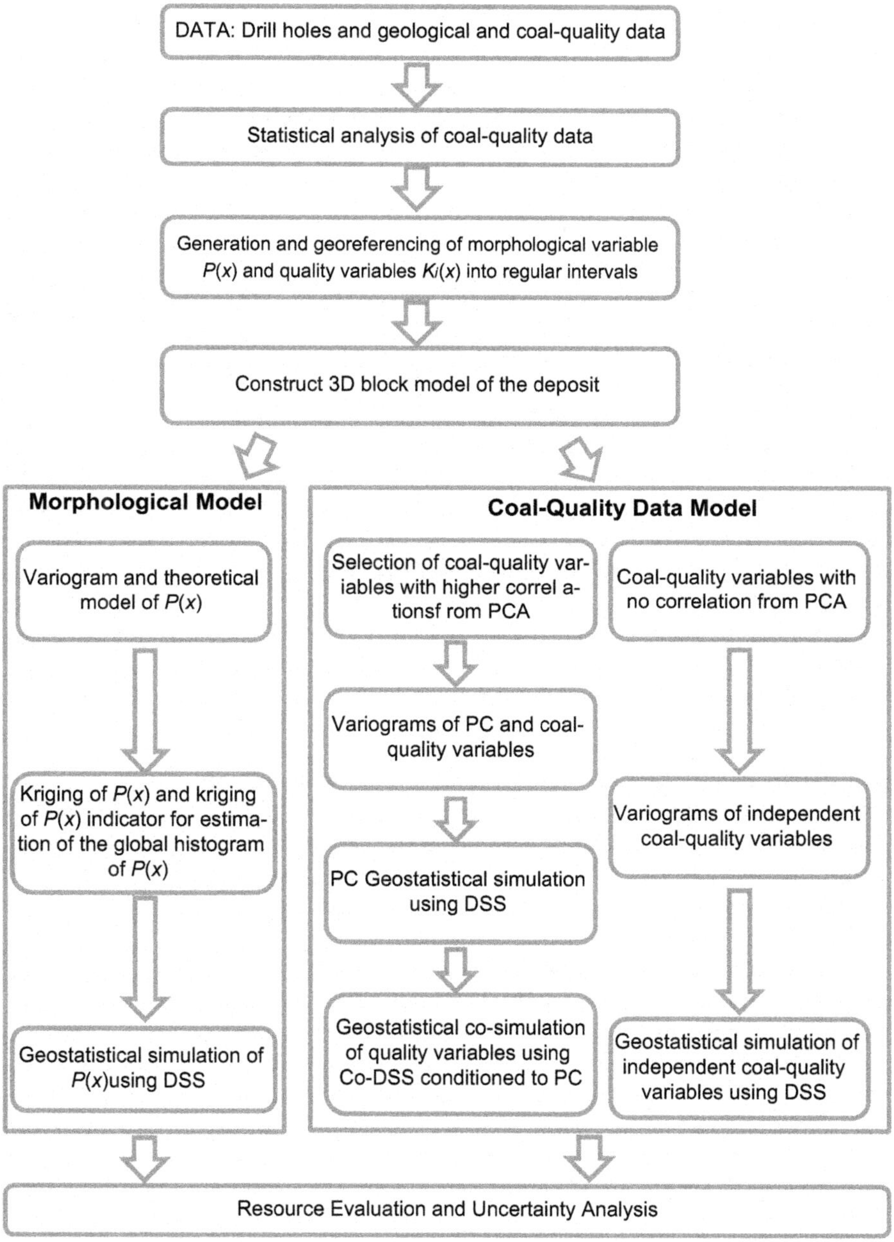

Fig. 2 Diagram of the proposed methodology used in this study

uncertainty and the probability of occurrence of extreme values are tools of prime importance for the planning and timing of resource exploitation with respect to optimizing the proportion (mixture) of raw materials coming from different mining stopes.

3 Case Study

The data consist of assays from 43 georeferenced boreholes in a study area measuring 6 km by 2.5 km and with a maximum depth of 370 m (see Fig. 7). Coal-seam intersections in the boreholes are identified by top and bottom depths, with some intersections having measurements of coal-quality variables. From a total of 6,282 m of drilling, fractions with coal account for about 12.5 % of the total length sampled.

For the present study, coal intersections were not manually linked between boreholes owing to the complexity of the deposit, the sparsity of boreholes, and the enormous number of coal lenses. The deposit is thought to be a complex interbedded horizon of carbonaceous mudstone and coal (coal zone), which in aggregate represents a large amount of potential extractable resource.

A 3D mesh of blocks was generated to develop the models, and it was decided to use a spacing of 30 m in the horizontal directions and of 1 m in the vertical direction. The ratio of the dimensions of the blocks is justified because the coal seams have a lenticular geometry and sub-horizontal orientation, and a vertical resolution of 1 m is appropriate for mining assessment and planning purposes.

3.1 *Statistical Analysis*

The results of the univariate and bivariate statistical analysis of the studied variables are presented in Tables 1 and 2, respectively. In Table 2, the highest correlations (greater than 0.6 or less than −0.6) are depicted in bold. This matrix shows that the Pearson and Spearman correlations between the variables are of similar magnitude and that the variables IM and TS are independent of the remaining variables.

The loadings of PCA and the projection of observed individual values onto the three main principal components PC-1/PC-2 and PC-1/PC-3 are presented in Fig. 3 and show the following:

Table 1 Basic statistics of the coal-quality variables

	IM	Ash	VM	FC	CV	TS	ARD
Minimum	0.4	59.2	3.1	1.6	1.0	0,05	1.44
Maximum	6.9	90.6	37.0	65.5	25.9	4.22	2.75
Mean	1.6	59.2	16.8	22.4	11.3	0.96	1.95
Median	1.4	59.0	16.9	21.9	11.4	0.78	1.95
Variance	0.47	134.51	35.19	76.62	21.15	0.406	0.029
St. deviation	0.69	11.59	5.93	8.75	4.60	0.637	0.170
Skewness	2.13	0.07	−0.02	0.61	0.06	1.636	0.339

Table 2 Correlation coefficient matrix for Pearson (bottom) and Spearman (top) of the coal-quality variables

	IM	Ash	VM	FC	CV	TS	ARD
IM		0.186	−0.275	−0.116	−0.217	−0.033	0.197
Ash	0.100		**−0.717**	**−0.878**	**−0.976**	−0.194	**0.898**
VM	−0.230	**−0.677**		0.353	**0.693**	0.074	**−0.628**
FC	−0.054	**−0.873**	0.238		**0.877**	0.261	**−0.818**
CV	−0.153	**−0.977**	**0.647**	**0.868**		0.233	**−0.909**
TS	−0.029	−0.173	0.067	0.187	0.201		−0.240
ARD	0.109	**0.883**	−0.574	**−0.789**	**−0.881**	−0.196	

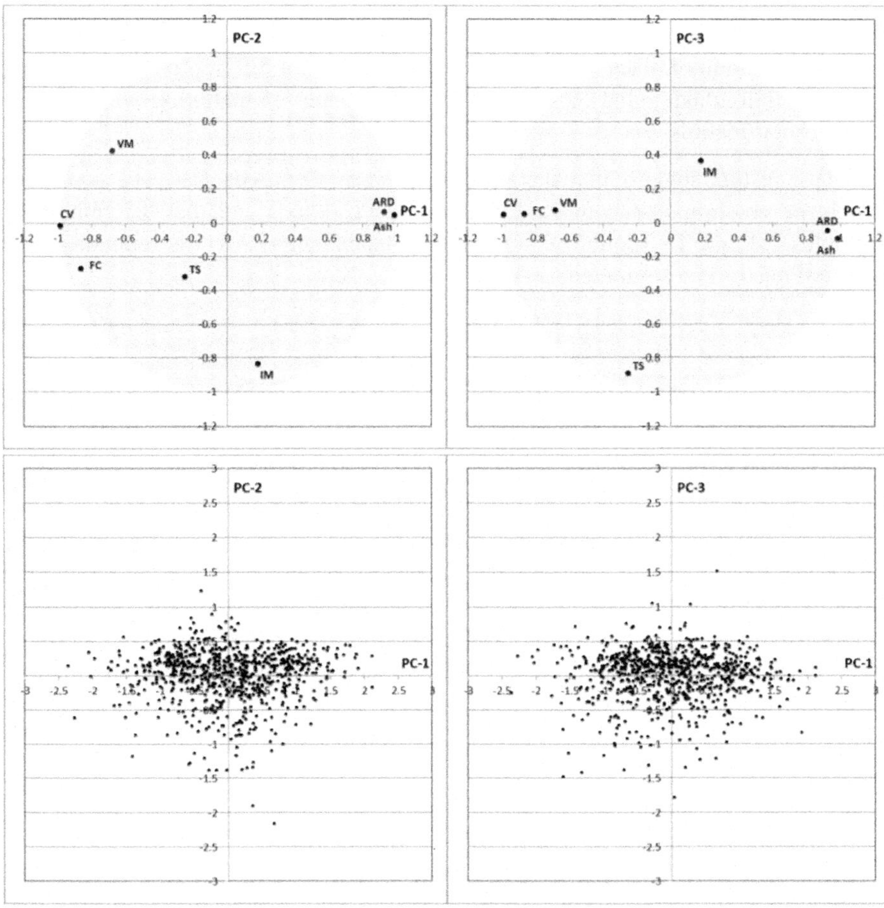

Fig. 3 (*Top*) Loadings between the PCA components PC-1/PC-2 and PC-1/PC-3 and the initial coal-quality variables. (*Bottom*) Projection of observed individual values onto the three main principal components PC-1/PC-2 and PC-1/PC-3

1. The initial seven variables can be summarized in three PCs that account for 87.5 % of the variation.
2. The variables Ash, FC, CV, and ARD are explained by PC-1, as is the variable VM but to a lesser extent. The variables Ash and ARD lie on the positive axis of PC-1 and the remaining variables on the negative axis. These results corroborate the bivariate analysis, in which these five variables are correlated with each other, some negatively.
3. The variables IM and TS are explained by PC-2 and PC-3, respectively, which means that these variables have a very high degree of independence, both with each other and with other variables. This result also confirms the results of the bivariate analysis and also reflects that these two variables have a more asymmetric distribution and are closer to lognormal type compared with the other five variables.
4. The projection of observed individual values does not differentiate populations. Some individual values are represented in dispersed patterns, which mean abnormal records.

These results suggest that the set of four variables (Ash, CV, FC, and VM) should be co-simulated with Co-DSS via the PC-1 axis and that the remaining variables IM and TS should be simulated independently. ARD is redundant and does not need to be simulated.

3.2 Variography

Before modelling by simulation, experimental standardized variograms were computed in both the horizontal and vertical directions, and theoretical models were fit for all variables involved: $P(x)$, PC-1, Ash, CV, FC, VM, IM, and TS. Figure 4 illustrates the experimental variograms and models fitted for $P(x)$, PC-1, and CV. It is important to note that the ranges for the coal-quality variables are much higher (about five times higher) than those for the $P(x)$ variable, and this is more evident in the vertical direction, meaning that the occurrence of coal is more erratic than its quality, which appears to be relatively homogeneous.

3.3 Simulation Results

Simulations were performed by implementing the presented methodology. Because of the high heterogeneity of the deposit and the lack of information, 500 realizations for each variable were generated for each variable using DSS and Co-DSS (Nunes and Almeida 2010). The average and the variance of the simulated values were generated. The results are illustrated in Figs. 5 and 6 for section OX 18/63, which intersects five boreholes. Figure 5 displays the results for $P(x)$.

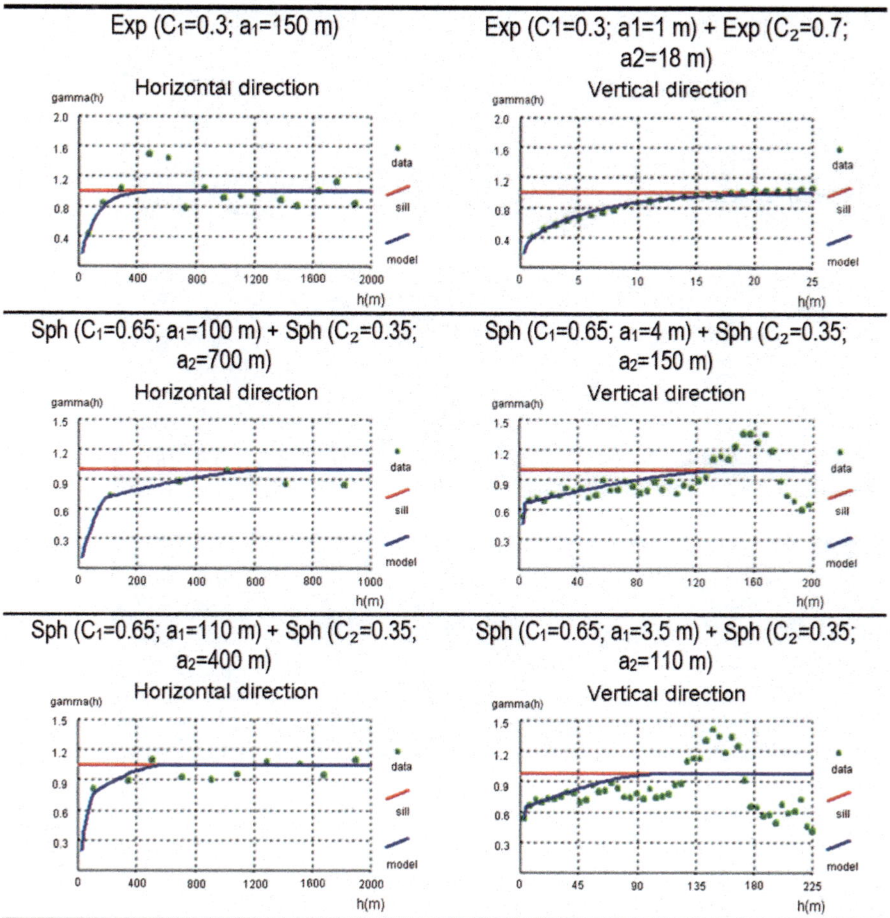

Fig. 4 Horizontal (*left*) and vertical (*right*) experimental variograms of (*top*) $P(x)$, (*middle*) PC-1, and (*bottom*) calorific values (CV), together with the respective models (*blue lines*) and sills (*red lines*)

The uncertainty is zero at the borehole locations and increases away from boreholes depending on the distance to the nearest borehole and the heterogeneity of the closest values. It should be noted that the variograms show a relatively low range (spatial continuity) in the study area, and the spacing between boreholes should be tightened to improve the quality of the estimation for the blocks with the highest uncertainties.

Figure 6 shows the results for the CV variable for section OX 18/63, and Fig. 7 shows in 3D the locations of the best coal (CV >12.5 MJ/kg) for the whole deposit.

To validate the results of the multivariate co-simulation, including the reproduction of the correlations between variables as observed in the initial data, calculations of the average Pearson correlation coefficient between homologous

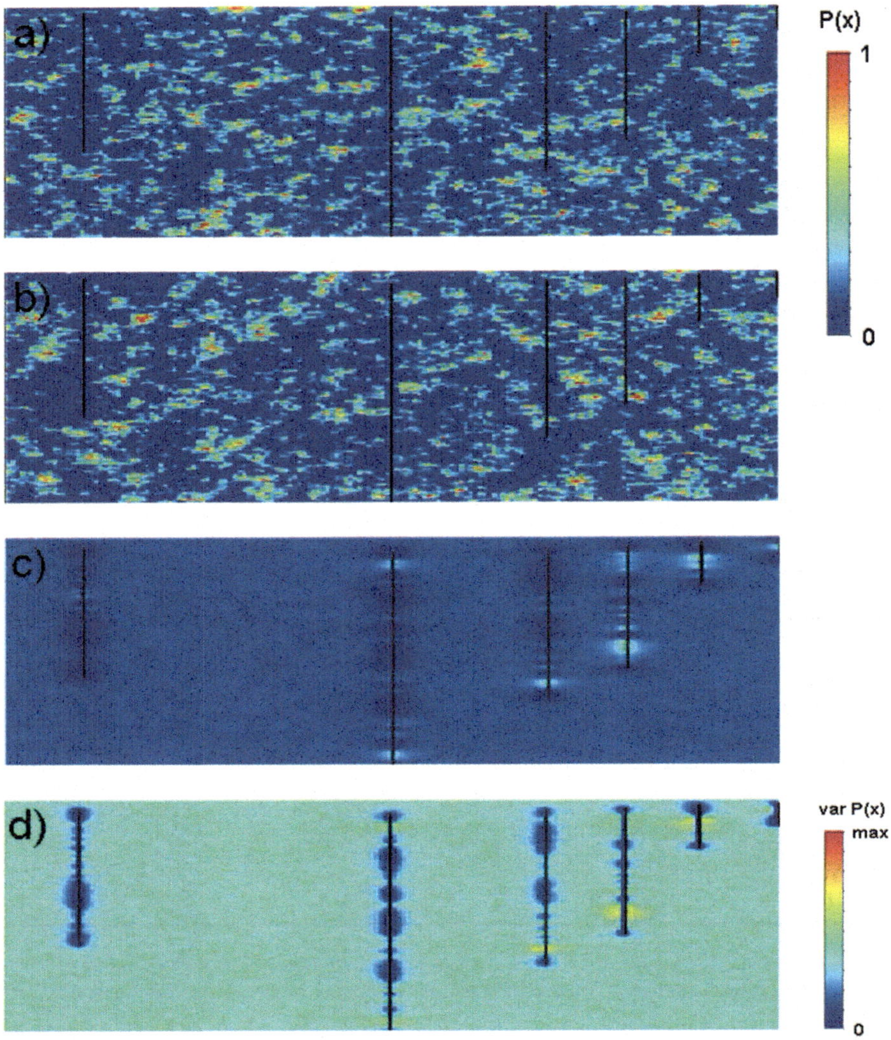

Fig. 5 Representation of the results for $P(x)$ in one section (OX 18/63): (**a**) one realization; (**b**) another realization; (**c**) the average image of the simulated set of images of $P(x)$; and (**d**) the image of the local variance constructed with the 500 simulated images of $P(x)$

images (# 1/# 1, … # 105/# 105, and so on) were achieved by sampling. In all, 2,500 pairs of simulated images were selected from the set of all variables. The results are shown in Table 3. Overall, the values of the correlation model are similar to the observed data.

The results of the simulations enable the resource to be parameterized by coal quality. For illustrative purposes, Fig. 8 presents the tonnages of coal conditional to the variable CV. A density of 1.8 t/m^3 for coal was used for the calculations.

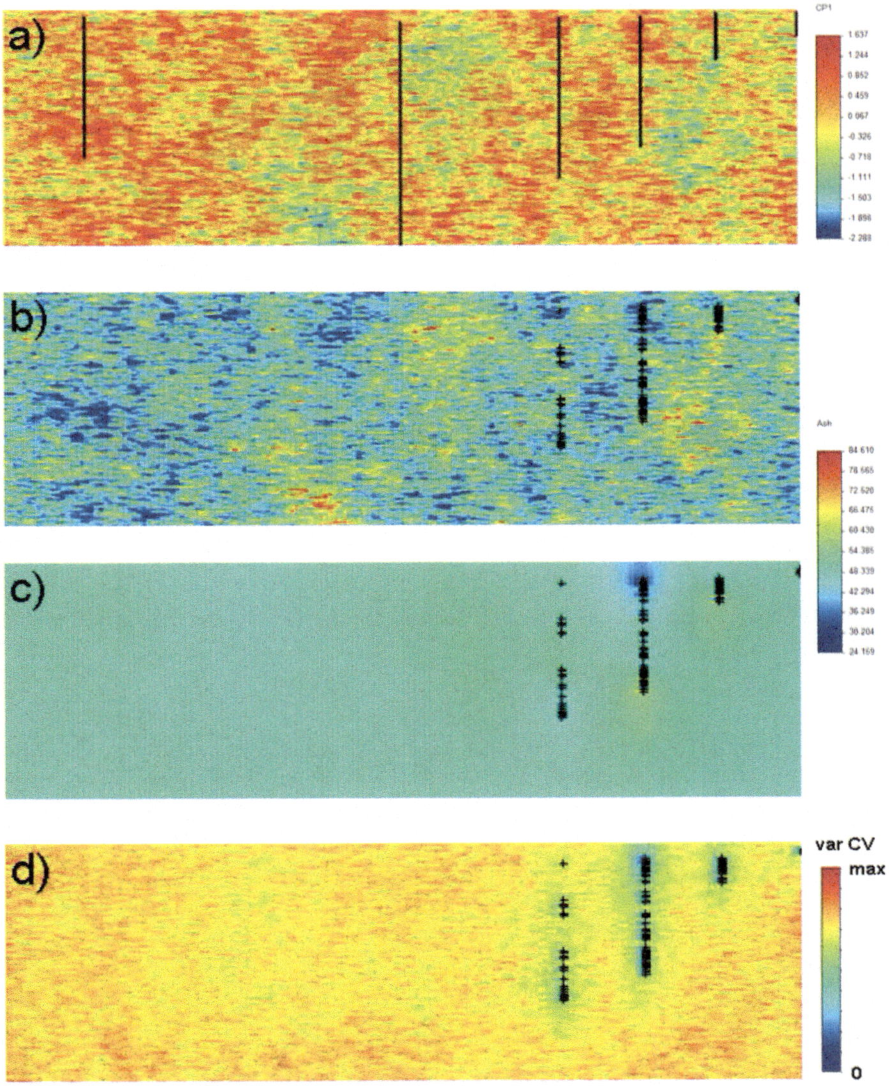

Fig. 6 Results for the coal-quality variable CV in one section (OX 18/63): (**a**) one realization of PC-1; (**b**) one realization of CV, co-simulated to the realization of PC-1; (**c**) the average image of the simulated set of images of CV; and (**d**) the image of the local variance constructed with the 500 simulated images of CV

4 Conclusions

The complex evolution of the sedimentary depositional environment associated with the Ncondezi coal deposit has not allowed individual coal seams with a reasonable thickness and lateral continuity to be formed. The morphology of the

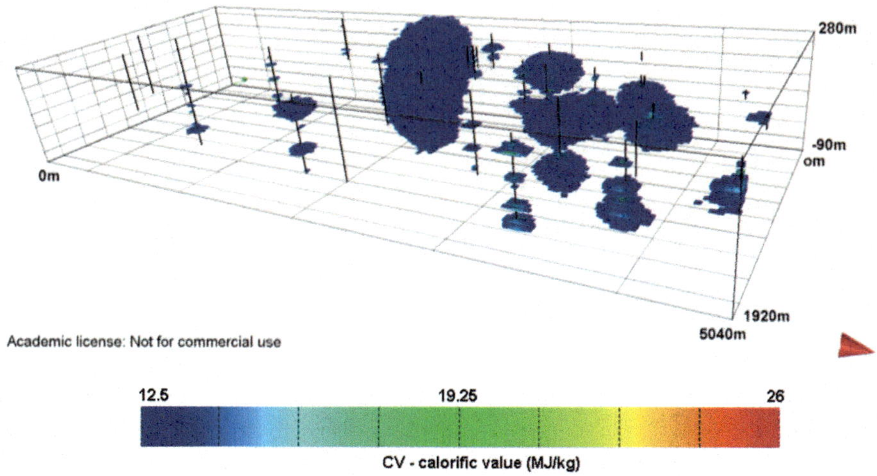

Fig. 7 3D view of the locations of coal resource with CV values higher than 12.5 MJ/kg for the whole deposit

Table 3 Correlation matrix of observed data (bottom) and simulated images (top) of coal-quality variables

	IM	Ash	VM	FC	CV	TS
IM		0.005	−0.005	−0.005	−0.005	−0.001
Ash	0.100		−0.598	−0.865	−0.972	−0.010
VM	−0.230	−0.677		0.317	0.602	0.015
FC	−0.054	−0.873	0.238		0.864	0.013
CV	−0.153	−0.977	0.647	0.868		0.010
TS	−0.029	−0.173	0.067	0.187	0.201	

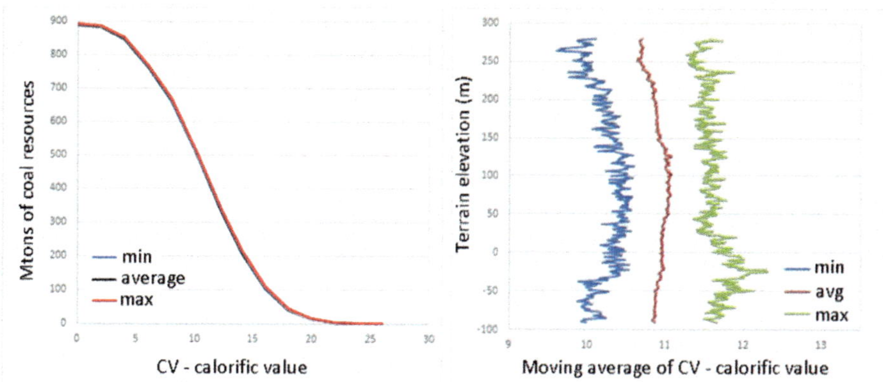

Fig. 8 (*Left*) Curves of the tonnage of coal with respect to the coal-quality variable CV. (*Right*) Moving average of CV calculated for a 1-m depth interval

coal zones is typically characterized by cyclical sequences of centimetric layers of coal, shales, and carbonaceous shales. Coal zones (stacks of thin coal lenses) range from 10 to 60 m in thickness, in which the proportion of coal to other rock/material is on average at least 50 %.

The variograms of the studied coal-quality variables of the Ncondezi deposit show high lateral variability in coal occurrence and quality, reflecting the fairly low number of surveys and samples, which are insufficient to properly characterize the deposit. For these reasons, the simulations showed a high degree of uncertainty. Also, the majority of the coal resource exhibits high ash content and low calorific value as well as a low yield of higher-quality products and would require improvement in order to be marketed. In summary, the Ncondezi is a large deposit with high tonnage, relatively low coal quality, poor spatial continuity of both morphology and coal-quality variables, and a high degree of uncertainty that could be minimized with new prospection campaigns, including new boreholes spaced more closely to form a tighter mesh.

Concerning the multivariate simulation approach tested in the present case study, it should be noted that the proposed methodology for the multivariate simulation of co-located variables imposes the individual variograms and histograms as well as the multivariate correlation metrics, providing a more straightforward methodology when compared with traditional full co-regionalization models.

Acknowledgements The authors are grateful to Ncondezi Coal Company for providing the data and information and to Midland Valley for providing an academic licence for Move® software. This work is a contribution to Project UID/GEO/04035/2013 funded by FCT-Fundação para a Ciência e a Tecnologia, in Portugal.

References

Almeida JA (2010a) Stochastic simulation methods for characterization of lithoclasses in carbonate reservoirs. Earth-Sci Rev 101(3):250–270

Almeida JA (2010b) Modelling of cement raw material compositional indices with direct sequential cosimulation. Eng Geol 114(1):26–33

Charifo G, Almeida JA, Ferreira A (2013) Managing borehole samples of unequal lengths to construct a high-resolution mining model of mineral grades zoned by geological units. J Geochem Explor 132:209–223

Cornah A, Vann J, Driver I (2013) Comparison of three geostatistical approaches to quantify the impact of drill spacing on resource confidence for a coal seam (with a case example from Moranbah North, Queensland, Australia). Int J Coal Geol 112:114–124

Davis JC (1986) Statistics and data analysis in geology, 2nd edn. Wiley, New York, 646 pp

Fernandes P, Cógne N, Chew DM, Rodrigues B, Jorge RCGS, Marques J, Jamal D, Vasconcelos L (2015) The thermal history of the Karoo Moatize-Minjova basin, Tete Province, Mozambique: an integrated vitrinite reflectance and apatite fission track thermochronology study. J Afr Earth Sci 112:55–72

Isaaks EH, Srivastava RM (1989) An introduction to applied geostatistics. Oxford University Press, New York, 561 pp

Johnson MR, Vuuren CJV, Hegenberger WF, Key R, Shoko U (1996) Stratig-raphy of the Karoo Supergroup in Southern Africa: an overview. J Afr Earth Sci 1:3–15

Lakshminarayana G (2015) Geology of barcode type coking coal seams, Mecondezi sub-basin, Moatize Coalfield, Mozambique. Int J Coal Geol 146:1–13

Matias FV, Almeida JA, Chichorro M (2015) A multistep methodology for building a stochastic model of gold grades in the disseminated and complex deposit of Casas Novas in Alentejo, Southern Portugal. Resour Geol 65(4):361–374

Nunes R, Almeida JA (2010) Parallelization of sequential Gaussian, indicator and direct simulation algorithms. Comput Geosci 36(8):1042–1052

Rossi ME, Deutsch CV (2014) Mineral resource estimation. Springer, Berlin, 332 pp

Soares A (2001) Direct sequential simulation and cosimulation. Math Geol 33(8):911–926

Srivastava RM (2013) Geostatistics: a toolkit for data analysis, spatial prediction and risk management in the coal industry. Int J Coal Geol 112:2–13

Tercan AE, Sohrabian B (2013) Multivariate geostatistical simulation of coal quality by independent components. Int J Coal Geol 112:53–66

Tercan AE, Ünver B, Hindistan MA, Ertunç G, Atalay F, Ünal S, Kıllıoğlu SY (2013) Seam modelling and resource estimation in the coalfields of Western Anatolia. Int J Coal Geol 112:94–106

Castelo de Sonhos: Geostatistical Quantification of the Potential Size of a Paleoproterozoic Conglomerate-Hosted Gold Deposit

R. Mohan Srivastava, Nicholas Appleyard, and Elton Pereira

Abstract Castelo de Sonhos, a gold deposit in Pará State, Brazil, has seen several phases of exploration since the mid-1990s. These programs have provided drill hole data, surface mapping of outcrops, geophysical surveys, geochemical surveys of soil samples, and preliminary metallurgical test work. All available data from these exploration programs have been integrated with recent advances in paleo-plate reconstructions, in modeling sedimentary depositional systems, in geostatistical simulation, and in data mining. This integration of ideas and methods from petroleum geostatistics, from classical statistics, and from plate tectonics makes it possible to predict the range of the project's potential tonnage and grade and to assess the project's upside and downside risk. This leads to an exploration target range that is probabilistically quantified, that is well grounded in data, in field observations and science, and that is testable through drilling. Not only does this quantitative risk assessment improve analysis of the project's technical and economic viability but also, importantly, it builds confidence among investors whose support is critical for advancing the project.

R.M. Srivastava (✉)
TriStar Gold Inc, Toronto, Ontario, Canada
e-mail: mosrivastava@tristargold.com

N. Appleyard
TriStar Gold Inc, Scottsdale, AZ, USA
e-mail: nappleyard@tristargold.com

E. Pereira
TriStar Mineração do Brasil Ltda, Belém, Pará, Brazil
e-mail: epereira@tristargold.com

J.J. Gómez-Hernández et al. (eds.), *Geostatistics Valencia 2016*, Quantitative Geology and Geostatistics 19, DOI 10.1007/978-3-319-46819-8_27

1 Introduction

Rising several hundred meters above the cattle-grazing lands of southern Pará State in Brazil (Fig. 1) is a plateau cut by a ravine (Fig. 2). In the 1970s and 1980s, *garimpeiros* (artisanal miners) worked the alluvial deposits on the flanks and the toe of the plateau, recovering hundreds of thousands of ounces of gold using manual methods to excavate and concentrate ore.

In 1995, Barrick Gold, reasoned that the likely source of the alluvial gold was a hard-rock deposit on top of the plateau. They launched an exploration program, initially searching for a granite-hosted gold deposit, but quickly recognizing that the host was a band of conglomerates that rims the plateau. After a small drill program, an extensive soil geochemistry survey, and a trenching program, Barrick closed the exploration program in 1996 and relinquished the mineral concessions.

Garimpeiros then moved onto the plateau, digging trenches and pits by hand near Barrick's trenches, and near the up-dip projection of the best intervals in Barrick's drill holes. Over the next decade, *garimpeiros* excavated near-surface mineralization along strike for several kilometers, to a depth of 12–15 m. In places, tunnels were dug into the face near the base of trenches, following well-mineralized bands 50–70 m into the wall (Fig. 3). The *garimpeiro* workings (*garimpos*) were not limited by lack of gold but by the difficulty of extending hand-dug slots and tunnels below the water table. The back-breaking and life-shortening work became physically impossible without access to explosives and mechanized equipment. By the late 2000s, *garimpeiro* activity had dwindled to a few aging "hobby farmers" who continued to be able to recover a few grams a day from the more prolific *garimpos*.

The past several years have been difficult for the mining industry. The price of gold, which flirted with $1,800/oz in 2012, had dropped to barely $1,000/oz by late 2015. The collapse in commodity prices made many once-promising projects unprofitable and caused funding for exploration to dry up. Mineral exploration is

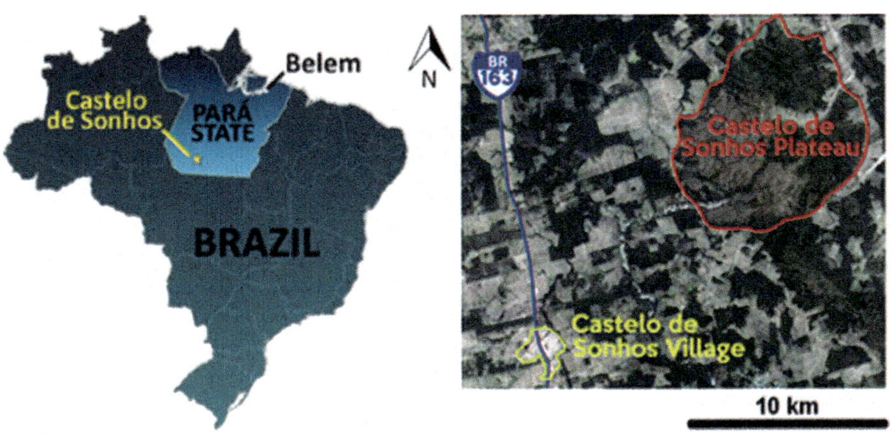

Fig. 1 Location of the Castelo de Sonhos gold project

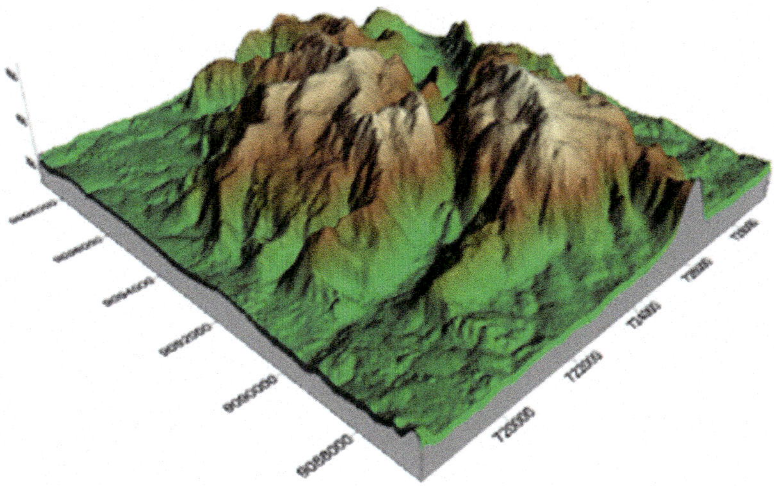

Fig. 2 Perspective view of the plateau, facing northeast, with a 10:1 vertical exaggeration

Fig. 3 Tunnels at the base of one of the *garimpeiro* trenches

a risky and costly business; with most mineral prospects never going into production, exploration can be justified only when commodity prices are so strong that the cost of the many misses is covered by the few hits.

By late 2015, the Castelo de Sonhos project was in jeopardy of being shut down a second time. More than $1,000,000 was needed just to make the final land payments that would secure TriStar 100 % ownership of the project, and much more was needed to continue exploration. Even though the drilling continued to show promising potential, lack of funding limited the areal extent of TriStar's drilling; by 2014, the last year of drilling, only 25 % of the 16 km of conglomerate outcrop and soil anomaly had been drilled (Fig. 4).

Although there were good showings of gold throughout the 250–300 m thickness of the conglomerate band, the drill holes targeted sweet spots, hunting for good news that was necessary to feed a cash-starved project. Very few holes penetrated

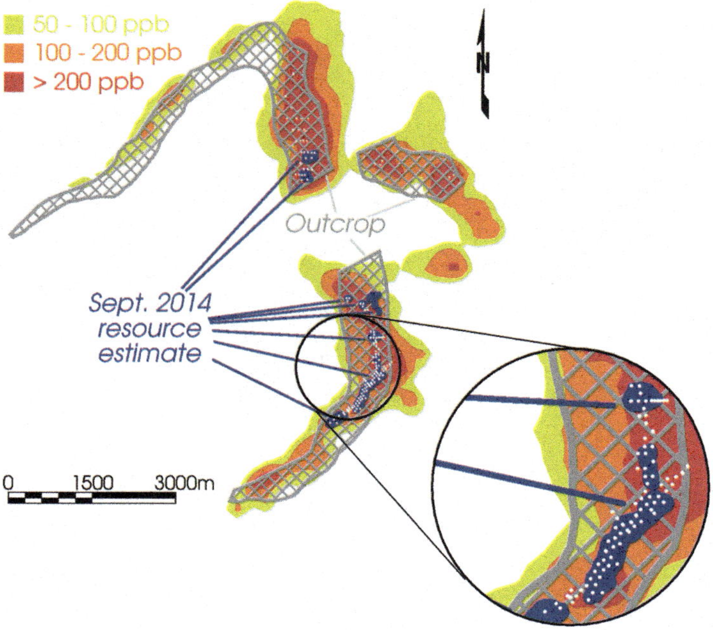

Fig. 4 Map showing the conglomerate outcrop (*hatched*), the gold anomaly in soil samples (*red-orange-yellow*), the location of drill hole collars (*white dots*), and the footprint of the 2014 resource estimate (*blue*)

the entire thickness of the conglomerate, leaving untested resource potential even in the limited areas that had been drilled.

Attracting more funding was made more difficult by a 2014 technical report that established a resource estimate of 180,000 oz of gold in indicated resources and 100,000 oz in inferred resources (Mello 2014). At this scale, the project became regarded as small, likely too small to ever become a modern industrial mine at current gold prices. It was difficult to communicate to weary and skeptical investors that this first resource estimate was "just a start" and that the project had great promise and potential.

The mining industry is littered with promises of pots of gold at the end of rainbows, with tales of the discovery of El Dorado, the fabled city of gold that the Spaniards never found. Mining promoters traffic in hyperbolic claims and wishful thinking. This is, in fact, a necessary part of sustaining investment through the high-risk phase of exploration. Many projects move forward only through sheer optimism. Exploration geologists have a tendency to fall in love with their discoveries, always maintaining faith that their project has much more potential than anyone has been able to document. Next year's drilling will prove up that potential and vindicate their faith . . . if only someone would fund the project.

Against this backdrop of boundless optimism, discussions of a project's potential are understandably met with a degree of cynicism and doubt. Arm-waving projections of blue-sky possibilities are rarely convincing to investors who have learned through experience to be suspicious of hype. This is where geostatistics makes a difference. An important difference. A project-saving difference.

This paper begins with an explanation of why mineral resources are conventionally reported as a single "best estimate" and why securities regulations limit one's ability to present resource estimates in the probabilistic context that has become familiar in other areas of applied geostatistics. It presents the concept of an "exploration target range" and shows how this provides an opportunity to introduce probabilistic analysis from geostatistical simulations, showing a space of uncertainty that spans a range of possible grades and tonnages. The controls on gold mineralization in the Castelo de Sonhos deposit are discussed, analyzing and integrating all available information and data to support the view that this is primarily an ancient placer deposit, with free gold grains being deposited in a bed of gravels and pebbles that was later buried, cemented into solid rock, then uplifted to its present position. A methodology for geostatistical simulations is presented, one that draws on tools developed for oil and gas applications but relevant to this mining problem because the gold accumulated in a sedimentary environment similar to many petroleum reservoirs. Results of the conditional simulation study are shown, leading to a P10–P90 range that provides strong technical support for the belief that the preliminary resource estimate considerably understates the project's true potential.

2 Mineral Resource Reporting: The Single-Estimate Tradition

Although the mining industry is the birthplace of applied geostatistics, mineral resource estimation makes little use of the conditional simulation tools that now dominate applied geostatistics in the many other areas where geostatistics is now applied. In practice, mineral resource estimation is still based on interpolation procedures that produce only a single estimate.

The vast majority of mining is conducted by public companies whose shares trade on stock exchanges. Governments and securities regulators, who want publicly traded companies to function with an accountability that builds public confidence in the stock market, have introduced regulations that constrain how mining companies report mineral resources. These securities regulations are intended to improve the comparability of resource estimates, so that one company's "apple" is not another company's "orange." They also aim to make available to the public the data, scientific and technical information that support companies' claims about their mineral resources.

Mining is becoming increasingly globalized. All of the world's largest mining companies operate in several countries, as do many midsize companies. By the time they reach production, many mining projects involve partners from different countries. In order to facilitate reporting of a project's mineral resources in the several jurisdictions where the partners are headquartered, and where they operate, the rules that govern reporting of mineral resources have gravitated toward a global norm. Although there are differences between the details of the securities regulations in each country, these differences are becoming smaller as many countries adopt rules that are similar to the JORC Code developed by the Australian Institute of Mining and Metallurgy and to Canada's National Instrument 43-101.

At the same time that securities regulators require adherence to specific rules and definitions, they also encourage the use of best practice guidelines developed by professional organizations. In Canada, for example, National Instrument 43-101 recommends that the best practice guidelines of the Canadian Institute of Mining and Metallurgy (CIM) be followed.

Although the geostatistics community has developed many tools for building stochastic models that span a space of uncertainty, these have not found broad use in the mining industry because securities regulations for mineral resource reporting do not embrace the possibility of different but equally likely versions of a deposit's grade and tonnage. National Instrument 43-101, for example, is unambiguous in its requirement that when multiple versions of a deposit's grade and tonnage are presented, the report must make clear which of the alternatives is the single one being presented as *the* mineral resource estimate.

The single-estimate tradition is reinforced by best practice guidelines that present advice that focuses on the use of single-estimate interpolation procedures like kriging and inverse distance weighting. With decades of effort having gone into harmonizing regulations and guidelines in different countries, it is very unlikely that the single-estimate tradition of the mining industry can be undone.

2.1 Classification

Classification is the one purpose for which conditional simulation has gained some traction in mineral resource estimation. Regulations require that mineral resource estimates be classified into three categories that reflect different levels of confidence: "measured," "indicated," and "inferred," from most confident to least.

Several mining companies and mining engineering consulting firms have developed a standard practice of classifying resources according to fluctuations in grade, tonnage, and metal content observed in multiple realizations from a conditional simulation study. But even when used for this purpose, conditional simulation is very rarely the direct basis for the reported estimate of the mineral resource; it informs only the choice of appropriate confidence categories for a resource estimate developed using a single-estimate procedure.

2.1.1 Exploration Target Range

Canadian and Australian regulations for the reporting of mineral resources explicitly permit the reporting of an "exploration target range." Canada's National Instrument 43-101 does not define what this is, a puzzling omission in a document that insists on adherence to standardized definitions and terminology. Fortunately, Australia's JORC Code does give a definition, one that captures the sense of the term as it is generally understood throughout the mining world:

> *An Exploration Target is a statement or estimation of the exploration potential of a mineral deposit in a defined geological setting where the statement or estimate, quoted as a range of tonnes and a range of grade, relates to mineralization for which there has been insufficient exploration to estimate a Mineral Resource.*

The requirement that grade and tonnage of an exploration target be reported as a range (and never as single estimates) makes conditional simulation ideal for the analysis and reporting of an exploration target.

3 Controls on Mineralization at Castelo de Sonhos

Reliable models of grade and tonnage in a mineral deposit depend on a good understanding of the controls on mineralization, the physical and chemical processes that brought gold to certain location, and not to others. This understanding informs models of spatial continuity, especially in the exploration stage, when the hard data are too few to permit robust analysis of experimental variograms. It also assists with the choice of appropriate trend models; quantification of an exploration target range is completely unconvincing if it consists simply of filling a large volume of rock with grades drawn from a homogenous distribution. Stationarity is never a comfortable assumption for mineral deposits; although it eventually becomes justifiable when the deposit is well drilled and local search neighborhoods require only an assumption of local stationarity, stationarity is a very difficult assumption to justify with sparse exploration drilling. This is largely the reason that best practice guidelines encourage mineral resource estimates to be tightly constrained to nearby drill holes. It is also the reason that the 2014 resource estimate for Castelo de Sonhos is so low. When best practice guidelines are followed, the resources (the blue areas in Fig. 4) tightly follow the available drilling. Establishing the full exploration target requires extrapolating beyond the available drilling, and this can be done well only when geologically sound trend models are used.

The existing exploration data from Castelo de Sonhos support two plausible views for the controls on gold mineralization. The first is a "syngenetic" view in which the gold arrived in the sedimentary host rock at the same time that the sediments were accumulating. The second is an "epigenetic" view in which the gold arrived long after the rocks were indurated, likely through mineralized fluids that percolated through the rock.

In the syngenetic view, free gold grains would have been moved by water and settled to the bottom when the velocity of the current was not sufficient to keep them in suspension, or to keep them as part of the bed load that rolls along the bottom. The primary direction of continuity would be stratigraphically horizontal, subparallel to bedding, and aligned with the current direction. Large-scale changes in gold grade would follow lithology, tending to be higher where large pebbles are more frequent and tending to be lower where the sediments are finer. These lithologic trends are due to the fact that the hydrodynamic conditions that cause gold grains to settle will also cause large pebbles to settle. The large-scale trends in gold grade would go from high in the proximal direction (nearer the source of the eroded gold grains) to low in the distal direction (further away from the source).

In the epigenetic view, gold would follow cracks and fissure, the high permeability pathways through which mineralized fluids could travel. Directions of maximum continuity would be aligned with these fractures; for near-surface deposits, these would be subvertical because the lack of a confining load on the top makes horizontal stress greater than vertical stress, and fractures tend to propagate perpendicular to stress. Large-scale trends in gold grade would reflect proximity to faults and fractures, with gold grades tending to be higher in brecciated rocks in fracture zones and lower as one moves into less fractured rock away from those zones.

There is geological evidence that supports both styles of mineralization. The coincidence of the conglomerate outcrop with the soil anomaly (Fig. 4) supports the syngenetic view because it suggests that the conglomerate was always the host rock and not merely a convenient stopping point where gold later came out of solution from mineralized fluids. Direct observations of thin gold films on fracture surfaces are consistent with the epigenetic view, as is anecdotal commentary on the presence of hematite alteration with high gold grades. Statistical analysis of the available data, however, confirms that the primary controls on mineralization are those of placer environment and that although gold was subsequently remobilized, it has moved only a short distance from the location where it was originally deposited among pebbles and gravels, likely in flowing water.

Figure 5 shows a schematic of the main lithologies in the Castelo de Sonhos Formation, which is composed of a conglomeratic band, 250–300 m thick, that sits between thicker units of arenite above and below. Within the main conglomerate band, there is a mixture of pebble-supported conglomerates (mC1), matrix-supported conglomerates (mC2), a micro-conglomerate (mC3), and conglomeratic arenites (mAC). The conglomeratic arenites are sandstones in which the occasional pebbles are so widely spaced that they don’t look like true conglomerates. In places where the pebbles are very widely spaced, the conglomeratic arenite looks exactly like the upper or lower arenite (mA).

Figure 6 shows side-by-side box plots of the gold grade distributions in the five lithologies, with the progression from left to right being proximal to distal: pebble conglomerates will occur closer to the source and the finer-grained arenites will occur further from the source. The gradual progression from high grade in the pebble-rich lithologies to low grade in the pebble-poor lithologies is consistent with

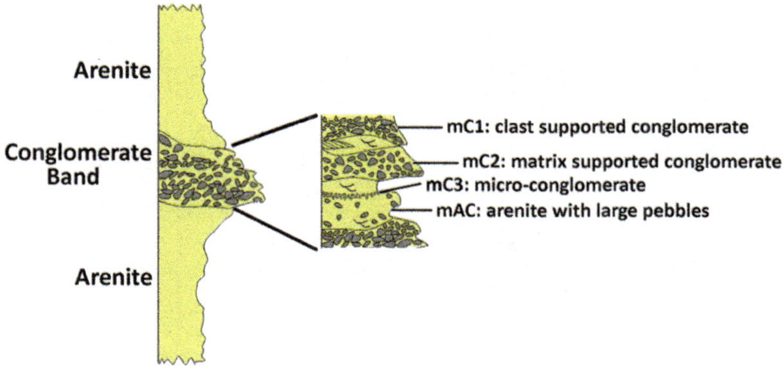

Fig. 5 Stratigraphic column of the Castelo de Sonhos Formation

the view that the hydrodynamic processes that caused pebbles to settle to the bottom also caused gold grains to settle at the same time.

Recursive partitioning (Breiman et al. 1984) provides additional statistical support for the view that the deposit is primarily an ancient placer. Figure 7 shows the recursive partitioning tree that separates high grade from low grade in the drill hole intervals coded as mC3, which is referred to as a "micro" conglomerate because it is composed entirely of small particles of gravel. The single most important factor in separating high-grade mC3 from low-grade mC3 is the thickness of the mC3 layer. Layers of micro-conglomerate less than 2 m thick have much higher gold grades (nearly 10×) than the thicker mC3 layers. This speaks to the likelihood that the micro-conglomerate acted like a natural sluice box, trapping gold that settled from the bed load above. When the same mass of gold grains is distributed in a thick mC3 layer, the gold grade ends up being lower than it does in a thin mC3 layer. For the thinner mC3 units, the next most important factor that separates higher grade from lower grade is the lithology of the overlying layer. If it is a matrix-supported conglomerate (mC2), the average grade is 3× that of the micro-conglomerates that are not overlain by a matrix-supported conglomerate. In order for bottom gravels to trap gold grains in a placer environment, the water velocity has to be sufficient to carry gold grains in the bed load, but not so high that the gold grains will be in suspension in the water column. Although the pebble-supported conglomerates likely reflect a higher water velocity, it is the matrix-supported conglomerates that would have been deposited in water that was flowing quickly, but not so quickly that there was a minimal bed load rolling at the bottom of the water column.

Other factors that support the syngenetic view include high recoveries of free gold using gravity methods in the preliminary metallurgical test, no gold in the lower arenite and very little gold in the upper arenite and only immediately above the conglomeratic band, and low variance in field measurements of the azimuths of paleo-current directions, typically from cross-bedding.

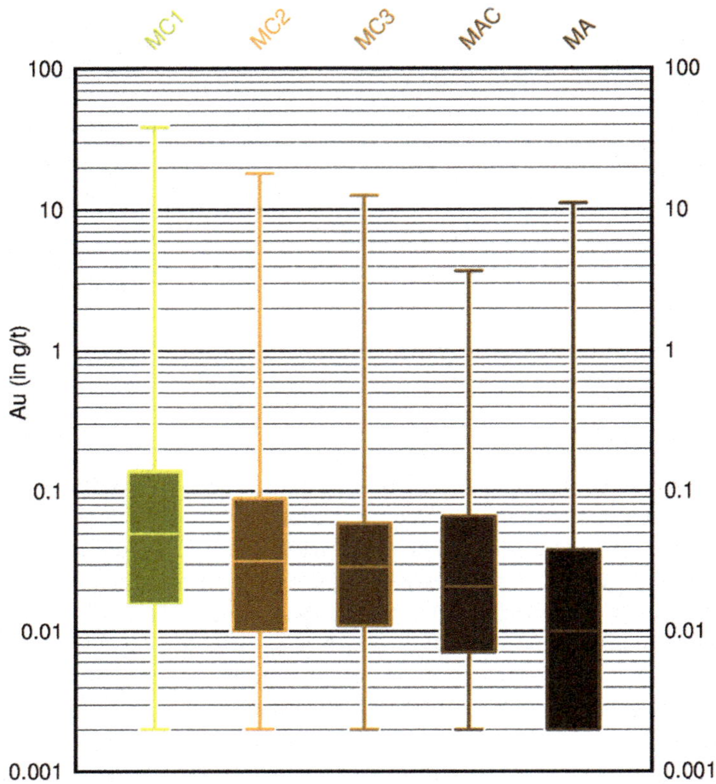

Fig. 6 Side-by-side box plots of gold grade distributions in the five lithologies

Although there are some factors that support the epigenetic view (visible gold on fracture surfaces and association of gold with hematite alteration), there are several factors that are inconsistent with an epigenetic explanation. There is no correlation between gold grades and proximity to faults or fractures, and the deposit does not contain any of the sulfide minerals, like pyrite and chalcopyrite that commonly occur with epigenetic deposits.

Placer deposits can form in many different depositional environments, including alluvial fans, braided streams, meandering rivers, fluvio-deltaic systems, and in near-shore marine environments. There is not yet enough field evidence to provide a strong opinion on the specific details of the depositional environment. Occasional hematite rims on quartz pebbles are consistent with subaerial deposition commonly seen in alluvial fans; but the low variance of paleo-current directions and the statistical evidence of subaqueous deposition (Fig. 7) point to something other than an alluvial fan. The lack of very fine-grained sediments is consistent with the separation of silts that occurs in the surface layer in a near-shore marine environment where fresh water from land meets salt water from the ocean.

Despite the lack of specific details on the depositional environment, several broad characteristics remain clear. The direction of maximum continuity will be

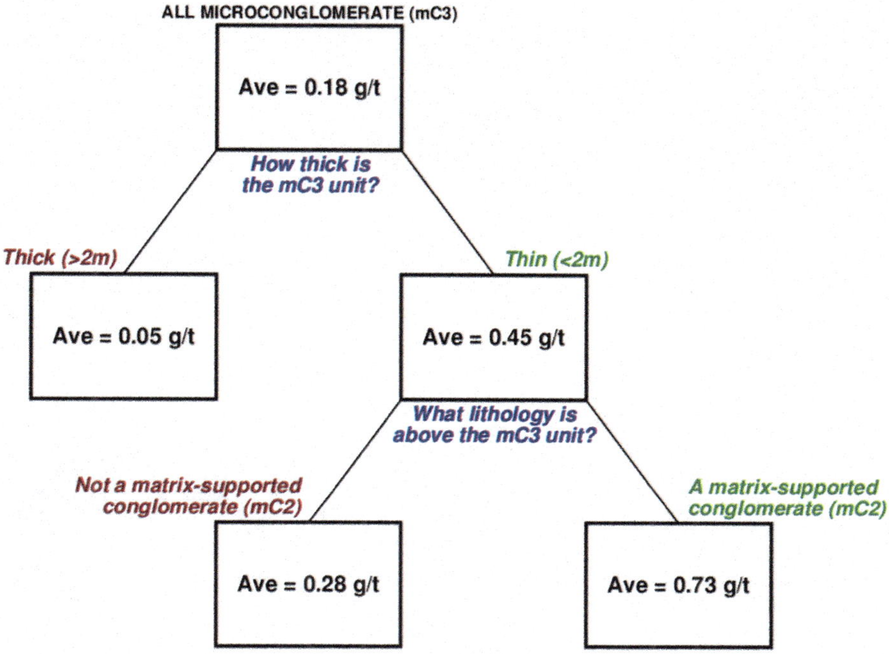

Fig. 7 Recursive partitioning tree for gold grades in drill hole intervals coded as mC3

parallel to stratigraphy, subhorizontal to bedding, and will follow sinuous channels of the type shown in Fig. 8. These types of channel geometries occur in all sedimentary environments where heavy mineral deposits form and at all scales.

The other characteristics common to all of the possible depositional environments are trends that will cause proximal-to-distal decreases in the water velocity and energy of environment, the frequency and size of pebbles, the proportions of pebble and clast-supported conglomerates, and average gold grade.

The ability to model these trends well and to capture the sinuous channel geometries in variogram models will significantly improve the realism of conditional simulations. If these characteristics are not honored, simulations will lack realism and will not inspire confidence in the attempt to quantify the exploration target range.

4 Plate Tectonics and Depositional Environment

Figure 9 shows a reconstruction of the continental plates 2.05 billion years ago (Eglington 2015), when a large supercontinent known as Nuna formed near the South Pole. In the 2 billion years since, the continental crust of Nuna has separated into four continental plates, two of which are now part of South America (the green plates on Fig. 9) and two of which are now part of Africa (the blue plates). There are

Fig. 8 Photographs showing sinuous channel geometries at several scales. From *left* to *right*: at the 10 km scale in the desert near Hamra Al Drooa in Oman (Kjell Lindgren, International Space Station, NASA); at the 100 m scale in the Thjorsa River near the Hofsjökull glacier, Iceland (Olivier Grunewald); and at the meter scale in heavy minerals in sand at Ventry Beach, Ireland (Jessica Winder)

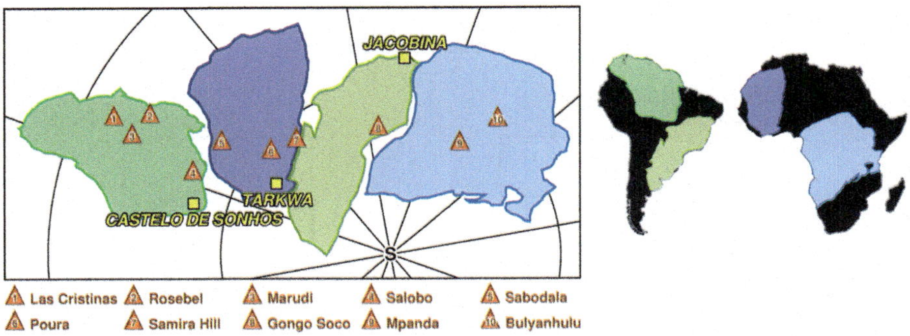

Fig. 9 Paleo-reconstruction of Nuna, approximately 2.05 billion years ago, from data provided by Bruce Eglinton (University of Saskatchewan)

several known lode gold deposits that date to 2.05 Ga or older and that can be traced back through time to their locations on Nuna (the orange triangles). These formed an arc that stretched across the continent, likely a ridge of volcanic mountains much like the modern Andes.

Zircon dates from Castelo de Sonhos place the formation of the conglomerates at 2.0–2.1 billion years ago, approximately the same age as two other paleo-placer gold deposits, Tarkwa in Ghana (2.1 Ga) and Jacobina in Brazil (2.0 Ga), both of which are now large operating mines. At the time that all three of these quartz-pebble conglomerates were forming, they sat near the shore, below lode gold deposits at higher elevations. Streams and rivers carrying eroded gold grains would have deposited their gold where the water velocity dropped, either in the alluvial plains or in the fluvio-deltaic region where rivers meet the sea.

5 Structural Geology

From its original flat-lying configuration, the band of conglomerates has been folded and faulted. The axial plane of the fold has been tilted to the northwest (in current UTM coordinates), and the hinge line has been tilted so that it dips to the southwest. The approximate horseshoe shape of the conglomerate outcrop (Fig. 4) is the result of the clipping of the folded conglomerate band against the current topography. The southern limb of the outcrop appears wider than the northern limb because of the tilt of the axial plane. The bedding has a shallow dip in the southern limb but is nearly vertical on the northern limb; so the apparent width in the southern limb is an oblique cut across the conglomerate band, while the apparent width on the northern limb is very close to the true perpendicular width.

There are two major faults that offset the conglomerate band, creating a small down-dropped block whose apparent shift to the east is due to the tilt of the fold.

A 3D model of the folding and faulting of the Castelo de Sonhos conglomerate band was created. This model allows one to wind back the clock, restoring the conglomerate band to its original depositional configuration by unfaulting, untilting, and unfolding. The original depositional configuration is the proper coordinate system for geostatistical analysis and simulation. Once conditional simulations have been built, the clock can be run forward, with the conglomerate band being folded, tilted, and faulted to bring it back to its configuration in current UTM coordinates.

6 Conditional Simulation Methodology

The simulation of gold grades at Castelo de Sonhos begins with a sequential indicator simulation (SIS) of the lithologies. This SIS procedure uses simple kriging (SK), with the local means defined by the proportion curves shown in Fig. 10.

In the vertical direction, there is a clear non-stationarity in the lithology proportions. In the oldest (deepest) layers of the conglomeratic band, there is very little of the pebble-supported conglomerate (mC1), and a lot of the conglomeratic arenite (mAC) and arenite (mA). Later during the period when the conglomerate was being deposited, the proportion of pebble-supported conglomerate increased, reaching a peak about 75 % of the way from the base of the conglomerate to the top. The vertical variation in lithology proportions is likely due to a combination of sea-level changes and changes in the sediment supply caused by uplift and erosion on the continental land mass of Nuna. Since the lithology proportions are not stationary, the conglomerate band was divided into six domains or "sequences," choosing boundaries that were designed to create stationary lithology proportions within each sequence, i.e., between each consecutive pair of boundaries.

The horizontal proportion curves shown on the right of Fig. 10 require an assumption about the direction of sediment transport. Using lithologies observed in drill holes, and maximum pebble sizes observed in outcrops, the assumption was made that the paleo-current direction was northeast to southwest, approximately aligned with the plunging direction of the fold, an assumption that is consistent with the paleo-current direction implied by the paleo-plate reconstruction (Fig. 9). As one moves down-dip (southwest) from the north-south limb of the conglomerate

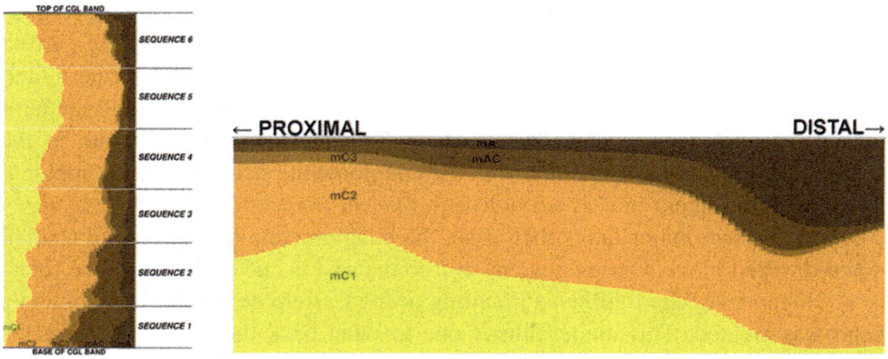

Fig. 10 Vertical and horizontal proportion curves for the five lithologies

outcrop, the proportions of the high-energy lithologies (mC1 and mC2) will tend to decrease, and the gold grades will also tend to decrease.

The SIS procedure is able to use locally varying directions of maximum continuity that are subparallel to simulated channel centerlines. These centerlines were created by borrowing a tool developed for oil and gas applications. Srivastava et al. (2013) presents a grid-less method for simulating the geometry of the centerlines of sinuous fluvio-deltaic channels, conditioned by lithology observations in well bores, with the conditioning data providing information on the proximity to sand channels. For Castelo de Sonhos, lithologies in drill holes provide information on the proximity of channels. The five lithologies were coded as integers from 1 to 5 in proximal-to-distal order: mC1 = 1, mC2 = 2, mC3 = 3, mAC = 4, and mA = 5. Within each drill hole, within each sequence, the average of these integer codes is a numerical summary of the energy of the environment. High-energy environments will have a low average, and low-energy environments will have a high average. The spatial trends in these average lithologies provide clues to the proximity to channels within each sequence.

Figure 11 shows simulated channel centerlines for the six sequences. The colored dots on this figure show the average lithology indicator for each drill hole, within the sequence boundaries. Red-to-yellow colors denote high-energy environments (low values of the average lithology integer), while green-to-blue colors denote low-energy environments (high values of the average lithology integer).

Figure 12 shows an example of the SIS lithology simulation on a stratigraphically horizontal slice at the middle of the conglomerate band, using the proportion curves (Fig. 10) for SK means, and using simulated channel centerlines (Fig. 11) to define local directions of maximum continuity.

Once lithology simulations have been created, grade simulations are done with SGS, using only the drill hole data that fall within the same lithology. Following the grade simulation, SIS is used to simulate broad zones of weak, moderate, and intense hematite alteration that slightly modify the gold grade. Figure 13 shows

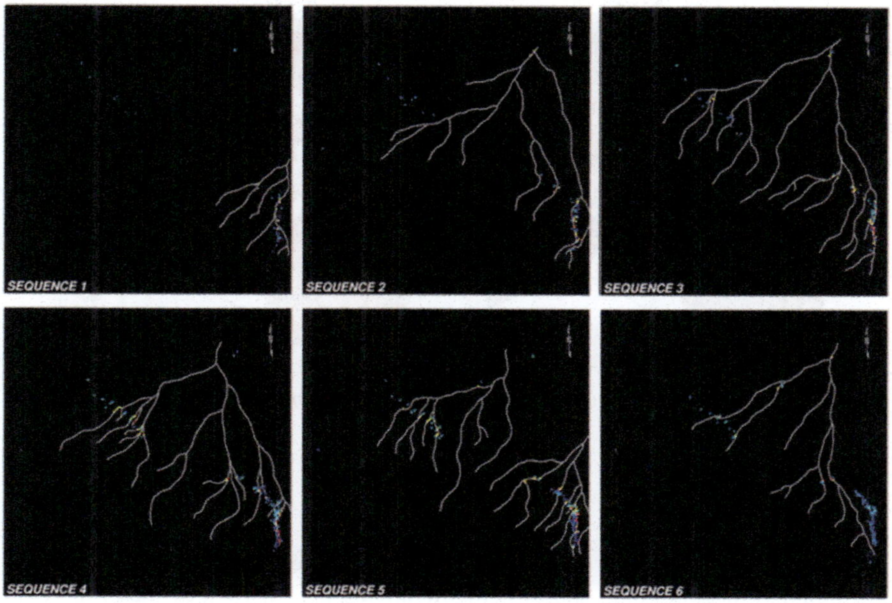

Fig. 11 Simulated centerlines of channel, conditioned by a proxy for the distance to channels: the average lithology indicator calculated from drill hole samples within each sequence

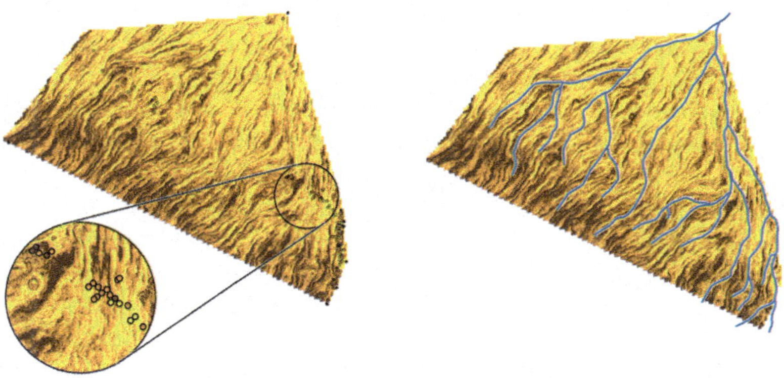

Fig. 12 An example of lithology simulation using SIS with proportion curves and locally varying directions of maximum continuity

an example of the grade simulation and the alteration simulation. The left frame in Fig. 13 shows that the gold grades decrease in the down-dip (southwest) direction. This is due to the linkage between the samples and the lithologies; when the grade within the mAC region is estimated using only the mAC samples from nearby drill holes, the fact that the gold grades are generally low in mAC (Fig. 6) entails that the simulated gold grade will also be low.

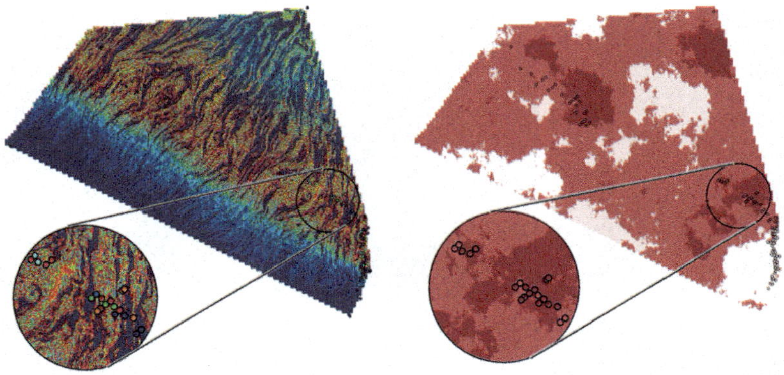

Fig. 13 Simulated gold grades (*left*) and simulated intensity of hematite alteration (*right*)

7 Results

The procedure described in the previous section was used to create 100 realizations of lithology, of gold grade and of hematite alteration. For each of these, the gold that could be extracted profitably using open-pit mining methods was calculated. Figure 14 shows the histogram of the metal content of the 100 realizations. Table 1 shows the exploration target range based on the realizations corresponding to the 10th percentile (P10) and 90th percentile (P90) of the distribution of metal content.

This conceptual exploration target range is not a mineral resource estimate. Further drilling is required to calculate a mineral resource estimate that complies with the requirements of National Instrument 43-101. If this additional drilling is done, the eventual future resource may not fall within the range expressed in Table 1.

Despite being conceptual in nature, the quantitative risk assessment (QRA) that leads to the exploration target range was an important step in advancing the project. In the months since the results of the QRA have been available, the project has been able to inspire confidence in a large and growing group of people who have invested several million dollars in the project. In the current market, when there is little funding available for mineral exploration, a project like Castelo de Sonhos could not have been advanced without the sound technical analysis of future potential that geostatistical simulation provides.

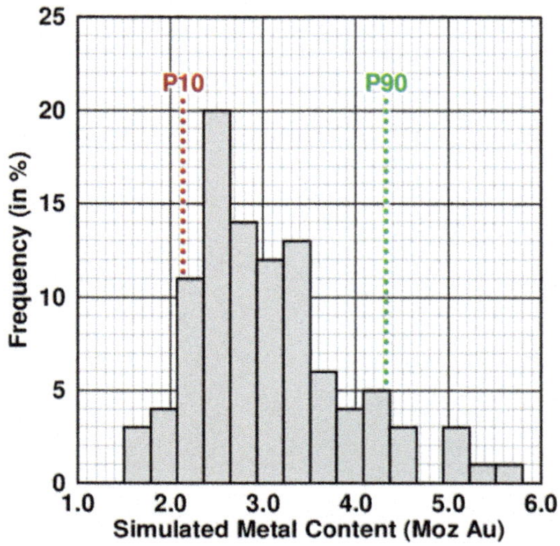

Fig. 14 Histogram of 100 outcomes of gold contained in an economically viable open-pit

Table 1 Exploration target range for Castelo de Sonhos

Tonnage:	50	–	84 Mt
Gold grade:	1.3	–	1.6 g/t
Metal content:	2.1	–	4.3 Moz

Bibliography

Breiman L, Friedman J, Stone CJ, Olshen RA (1984) Classification and regression trees. Chapman and Hall, New York

Eglington B (2015) Supercontinent cyclicity: relevant data, constraints, limitations and aspects requiring particular attention, AGU Fall Meeting. American Geophysical Union, San Francisco

Mello R (2014). Mineral resources estimation for the Castelo de Sonhos project. TriStar Gold Inc: filed on www.sedar.com

Srivastava RM, Maucec M, Yarus J (2013) Grid-less simulation of a fluvio-deltaic environment, CSPG GeoConvention. Canadian Society of Petroleum Geologists, Calgary

A Hybrid Model for Joint Simulation of High-Dimensional Continuous and Categorical Variables

Hassan Talebi, Johnny Lo, and Ute Mueller

Abstract It is a common challenge for the geosciences to jointly model the uncertainty in continuous and categorical regionalised variables and to reproduce observed spatial correlation and complex relationships in realisations. The demand for computational efficiency in the case of high-dimensional data and large simulation domains has led practitioners to utilise approaches based on decorrelation/ recorrelation and independent simulation. Among such approaches the method of min/max autocorrelation factors (MAF) has proven to be a practical technique for industrial purposes. This study presents a hybrid model for joint simulation of high-dimensional continuous and categorical variables. Continuous variables are transformed to Gaussian random functions (GRFs) via anamorphosis functions and categorical variables are obtained by truncating one or more GRFs based on the plurigaussian model. MAF factors are then derived from all GRFs. After independent simulation of MAF factors, different realisations of continuous and categorical variables are obtained via back-transformation of MAF factors followed by back-transformation for continuous and truncation for categorical variables, respectively. The proposed algorithm is illustrated through a case study.

1 Introduction

In many geoscience applications such as evaluation of mineral resources, characterisation of oil reservoirs, or hydrology of groundwater, uncertainty modelling is a common and challenging issue, as a set of multiple regionalised dependent variables from various sample spaces (e.g. continuous and categorical) need to be predicted in the target area. A further challenge is often the compositional nature of continuous data (positive values representing some part of a whole). In such data the constant sum constraint forces at least one covariance to be negative and

H. Talebi (✉) • J. Lo • U. Mueller
School of Science, Edith Cowan University, 270 Joondalup Drive, Joondalup, WA 6027, Australia
e-mail: htalebi@our.ecu.edu.au; j.lo@ecu.edu.au; u.mueller@ecu.edu.au

J.J. Gómez-Hernández et al. (eds.), *Geostatistics Valencia 2016*, Quantitative Geology and Geostatistics 19, DOI 10.1007/978-3-319-46819-8_28

induces spurious correlations. Furthermore they carry just relative information (Aitchison 1986). To transform compositional data into unbounded space and to increase mathematical tractability, different log-ratio transformations can be applied prior to using standard geostatistical techniques (Aitchison 1986; Mueller et al. 2014; Pawlowsky-Glahn and Olea 2004). Although several geostatistical models have been developed for spatial modelling of categorical or continuous variables (Chilès and Delfiner 2012), little has been done to jointly model data of different natures. As the spatial distributions of these multivariate data are often interdependent, a separate modelling of them is insufficient (Emery and Silva 2009; van den Boogaart et al. 2014).

The conventional approach to modelling multivariate data of different natures is to use a deterministic model based on categorical data and predict continuous data within each category separately. Although this approach is simple to apply, it does not consider the uncertainty in the layout of different categories (e.g. geological domains). In this approach geologists have to delineate the exact shape of each layout based on experimental data and their interpretation of earth science processes. Unfortunately, very few of these processes are understood well enough. As experimental data become sparse and geology becomes more complex, the likelihood of misclassification increases. Geostatistical simulation techniques for categorical data can be applied to define domains and quantify the uncertainty in the exact location of geological boundaries by generating multiple realisations. Subsequently continuous variables can be predicted in each simulated domain independently. This method is known as the cascade or hierarchical approach (Jones et al. 2013; Talebi et al. 2016). A substantial drawback of cascade simulation is that it does not consider the spatial dependence of continuous data across domain boundaries and potentially generates abrupt transitions in realisations of continuous variables across geological boundaries which is not always the case in practice (Ortiz and Emery 2006; Vargas-Guzmán 2008).

A more general approach is to combine multivariate Gaussian and plurigaussian simulation (Emery and Silva 2009; Maleki and Emery 2015). In this approach Gaussian data are derived by transforming continuous data to GRFs and categorical data are related to one or more GRFs by truncation. This model assumes that all GRFs are spatially cross-correlated so it can reproduce the dependencies between the categorical and continuous data and spatial correlation of continuous data across geological boundaries. An advantage of this approach is its ability to reproduce gradual transition of continuous variables across geological boundaries. By construction, this approach requires a co-simulation based on defining a linear model of coregionalisation (LMC) to jointly simulate multivariate data. Simplicity of modelling and verification of the admissibility make the LMC a popular means for defining the spatial relationships of multivariate data (Goulard and Voltz 1992). However the construction of a "satisfactory" LMC becomes harder when high-dimensional data need to be co-simulated due to restrictions on the allowed semivariogram structures. Moreover as the number of variables and simulation domains increases, co-simulation approaches based on an LMC will need considerably greater computer processing to solve large systems of equations per simulated node.

An alternative technique for joint simulation of high-dimensional continuous and categorical data is presented in this study. The plurigaussian model will be used to associate categorical data with one or several GRFs (depending on the complexity of the geological setting) through truncation, and continuous data will be transformed to GRFs via anamorphosis functions. All GRFs from the previous steps will be decomposed jointly into orthogonal factors via the MAF technique (Switzer and Green 1984; Desbarats and Dimitrakopoulos 2000) which is based on factors derived from the solution of a generalised eigenvalue problem (Bandarian and Mueller 2008). These orthogonal factors can then be simulated independently. Statistical and spatial relationships between variables are reimposed after the simulation. This joint simulation approach offers better accuracy and computational efficiency as the number of attributes being simulated increases and can be generalised to simulate several continuous and categorical variables by adding more GRFs. Although several approaches have been presented for decorrelating multiple regionalised dependent variables, data-driven MAF has proven to be a practical technique for industrial purposes.

The objective of this study is to discuss the methodology of the proposed joint simulation technique and to apply the approach to a high-dimensional real mining data set from a nickel laterite deposit in Western Australia. Finally the results will be analysed and the performance of the method will be evaluated via different statistical and geostatistical measures.

2 Methodology

2.1 Compositional Nature of Data and Log-Ratio Transformation

Compositional data are multivariate data where the components represent some part of a whole. They are measured on the same scale and are constrained by a constant sum property. A compositional data set $\{\boldsymbol{Z}(x_\alpha) = [z_1(x_\alpha), \ldots, z_D(x_\alpha)] \, | z_i(x_\alpha) \geq 0, i{=}1, \ldots, D; \alpha{=}1, \ldots, k;\}$ with D components and k observations can be represented by the equation:

$$\sum\nolimits_{i=1}^{D} z_i(x_\alpha) = 100\%, \alpha = 1, \ldots, k. \tag{1}$$

Compositional data raise some challenges for statistical and geostatistical analyses. Firstly, they are relative values and not free to vary in $(-\infty, +\infty)$. Secondly, the constant sum constraint forces at least one covariance to be negative, causing spurious correlations (Aitchison 1986). Therefore they are often transformed via a log-ratio transform (Aitchison 1986; Pawlowsky-Glahn and Olea 2004). Several transforms are available, including additive log-ratio (alr), centred log-ratio (clr), and isometric log-ratio (ilr) (Egozcue et al. 2003). In this study the alr will be used. The alr transformation is defined as

$$\zeta(x_\alpha) = \mathrm{alr}(\boldsymbol{Z}(x_\alpha)) = \left[\ln\frac{z_1(x_\alpha)}{z_D(x_\alpha)}, \ln\frac{z_2(x_\alpha)}{z_D(x_\alpha)}, \ldots, \ln\frac{z_{D-1}(x_\alpha)}{z_D(x_\alpha)}\right] \tag{2}$$

Its inverse, known as agl transformation, is able to recover $\boldsymbol{Z}(x_\alpha)$ from $\boldsymbol{\zeta} = [\zeta_1, \zeta_2, \ldots, \zeta_{D-1}]$ and is defined as

$$\mathrm{alr}^{-1}(\boldsymbol{\zeta}) = \mathrm{agl}(\boldsymbol{\zeta}) = 100\frac{[\exp(\zeta_1), \exp(\zeta_2), \ldots, \exp(\zeta_{D-1}), 1]}{\sum_{j=1}^{D-1}\exp(\zeta_j) + 1}. \tag{3}$$

2.2 *Joint Simulation Algorithm*

The proposed joint simulation algorithm provides a mechanism to simulate regionalised dependent continuous and categorical variables simultaneously. It uses the plurigaussian model (Armstrong et al. 2011; Emery 2007) to determine GRFs corresponding to the categorical variables defining the domains. These GRFs and the GRFs associated with the continuous data of interest (in the case of compositions alr-transformed data) are then subjected to a MAF transformation to derive uncorrelated factors. The latter are simulated independently via some Gaussian simulation algorithm, here turning bands simulation. The simulation is succeeded by recorrelation through the inverse MAF transform. After this back-transformation step truncation is applied to obtain categorical realisations and anamorphosis together with the alg transform in the case of compositions to obtain realisations of the continuous variables. The detailed workflow is shown in the flow chart in Fig. 1.

3 Case Study: Murrin Murrin Nickel Laterite Deposit

3.1 *Geological Description*

The Murrin Murrin nickel (Ni) laterite deposit is located in Western Australia. Laterite deposits are formed during chemical weathering of ultramafic rocks near the surface of the earth. At Murrin Murrin, the nickel laterite deposits occur as laterally extensive, undulating blankets of mineralisation with strong vertical anisotropy covering basement ultramafic rocks (Murphy 2003). From the bottom of the deposit through to the top, the following layers can normally be recognised:

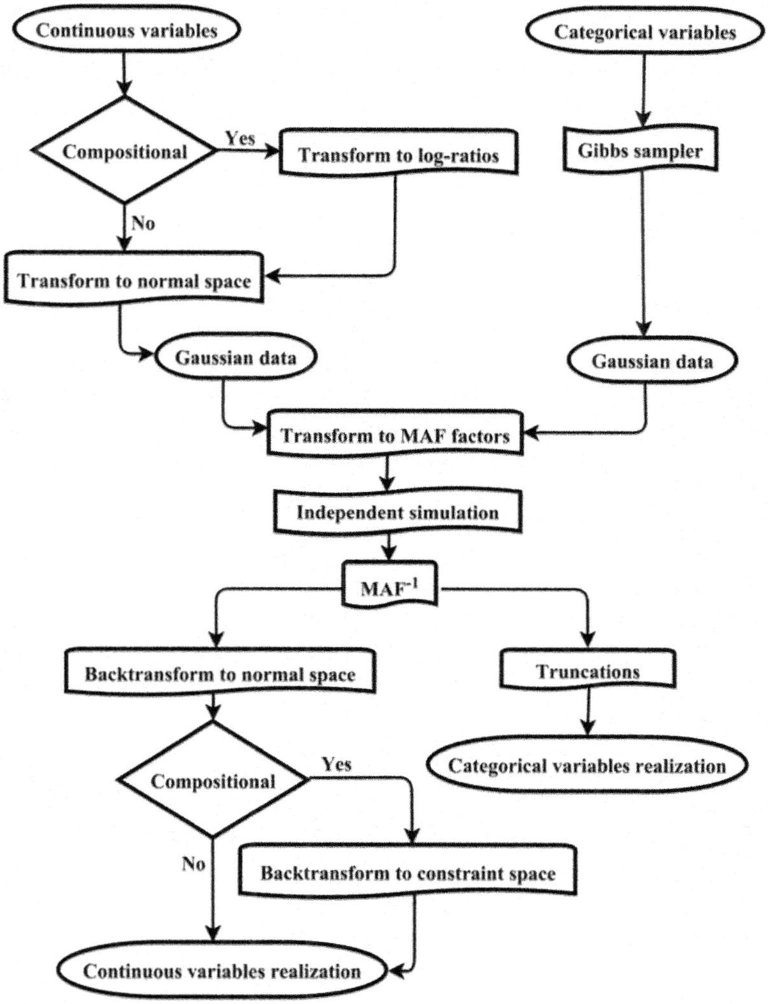

Fig. 1 Process of the joint simulation technique

ultramafic, saprolite, smectite, iron oxide, and/or clay layers with a hard ferruginous cap at the top. In this study area, there are four main geological units that control the spatial distribution of the Ni and Co. The following are short geological descriptions for each unit (Markwell 2001):

- Ferruginous zone (**FZ**): This zone is mainly composed of goethite and clay (kaolin). Although FZ tends to be less enriched in Ni, it can host significant amounts of Co.
- Smectite zone (**SM**): Consists mainly of smectite and is confined to the shoulders saprolite domes. SM is enriched in Ni and Co and depleted in Mg.

- Saprolite zone (**SA**): Consists mainly of lizardite and smectite, and high amount of Mg is present in this zone.
- Ultramafic zone (**UM**): At the base of the deposit, there is a thin layer of unweathered ultramafic rock.

3.2 Presentation of the Data Set

In total, 5145 samples, located in a volume of $1300 \times 500 \times 40$ m (Fig. 2), make up the database for this study with information on the rock types (FZ, SA, SM, and UM), five major elements (Ni, Co, Mg, Al, and Fe), and five secondary elements (Cr, Mn, Zn, Cu, and As). These ten elements plus the filler variable (i.e. continuous variables) form the compositional data, with rock type the only categorical variable. All samples were regularised to a bench height of 1 m.

Figure 3 shows a cross section of the rock types with Ni and Co distributions near boreholes for northing 180 m. It can be observed from this figure that the highly mineralised zones occur at the transitions between FZ, SA, and SM domains. This illustrates the importance of joint modelling of grade and geological units, as prior domaining may result in misclassification in the transition zones. Figure 4 and Table 1 depict the histograms and descriptive statistics of the continuous variables globally. Bimodal distributions can be recognised in Al, Fe, Filler, and Mg, while the other variables are strongly positively skewed.

3.3 Joint Simulation of Continuous and Categorical Variables

3.3.1 Categorical Data Preprocessing

According to the transition probability matrices and the geology of the deposit, UM domain can only be in contact with SA, while SA, SM, and FZ are mutually in contact. To define the truncation rule, two independent GRFs, $\{Y_1(x_\alpha), Y_2(x_\alpha)\}$, and three thresholds, $\{y_1, y_2, y_3\}$, are required (Fig. 5). Accordingly, the rock type prevailing at a given spatial location x_α is defined in the following fashion:

- Location x_α belongs to UM domain $\Leftrightarrow Y_1(x_\alpha) < y_1$
- Location x_α belongs to SA domain $\Leftrightarrow y_1 \leq Y_1(x_\alpha) < y_2$
- Location x_α belongs to SM domain $\Leftrightarrow Y_1(x_\alpha) \geq y_2$ and $Y_2(x_\alpha) \geq y_3$
- Location x_α belongs to FZ domain $\Leftrightarrow Y_1(x_\alpha) \geq y_2$ and $Y_2(x_\alpha) < y_3$

The threshold values are determined in agreement with the domain proportions calculated from the drill hole data. For variogram analysis, the two GRFs are assumed to be independent and their variograms are determined through their impact on the variograms of the domain indicators. Table 2 provides the parameters

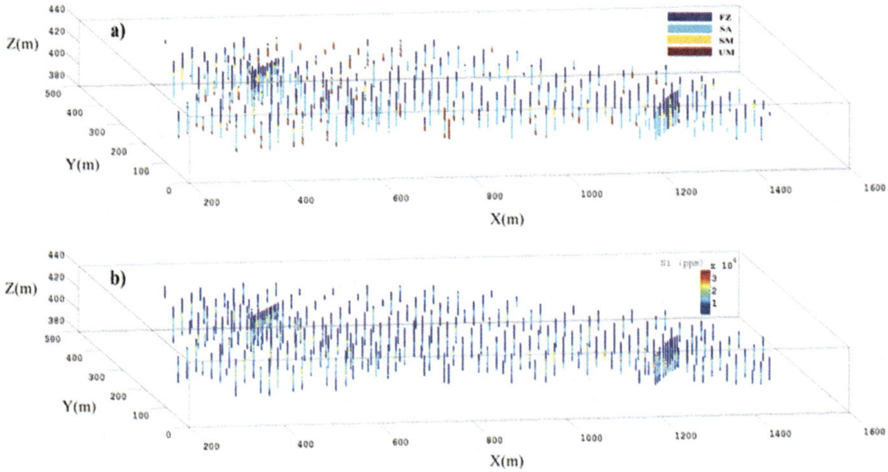

Fig. 2 Perspective view of samples showing **a** different rock types and **b** nickel grade

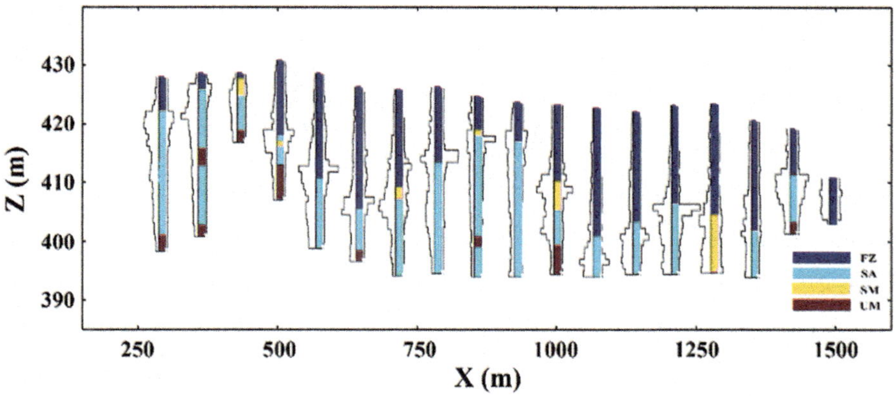

Fig. 3 Rock types (*coloured data*), nickel (*left*), and cobalt (*right*) distributions for the cross section with north coordinate 180 m

for the variogram models obtained for the two GRFs. In order to reproduce a regular boundary between different domains, cubic variogram models which are isotropic horizontally have been used for the GRFs, since they are smooth at the origin and associated with regular boundaries. Gaussian data are generated at sample locations via the Gibbs sampler algorithm as described in Armstrong et al. (2011) and plurigaussian model parameters conditional to categorical information at sample locations.

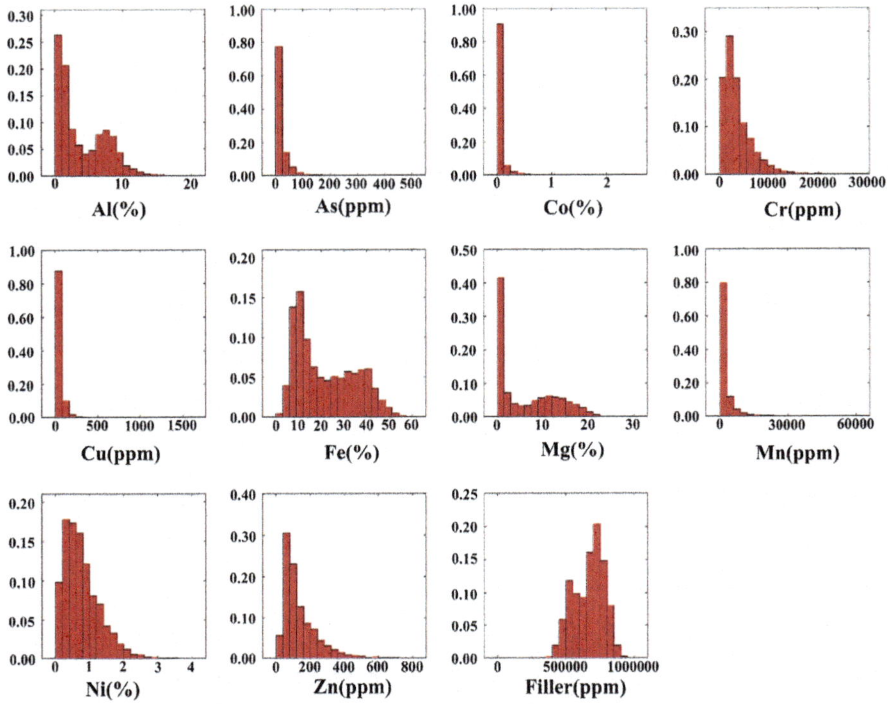

Fig. 4 Histograms of raw data

Table 1 Descriptive statistics

Variable	Min	Max	Mean	Std. dev.	Q25	Q50	Q75
Al (%)	0.05	16.90	3.92	3.42	0.90	2.50	6.90
As (ppm)	0.50	490.00	17.82	28.60	2.50	10.00	20.00
Co (%)	0.00	2.27	0.05	0.09	0.02	0.03	0.05
Cr (ppm)	40.00	27,400	3599	2849	1560	2780	4680
Cu (ppm)	2.50	1570	39.41	59.50	10.00	25.00	45.00
Fe (%)	0.40	55.90	21.59	12.92	10.10	17.90	32.70
Mg (%)	0.05	25.90	6.31	6.39	0.41	3.45	11.80
Mn (ppm)	25	59,000	2431.8	3550.20	740	1410	2530
Ni (%)	0.02	3.53	0.76	0.51	0.37	0.66	1.04
Zn (ppm)	8	720	131.49	94.38	66	100	172
Filler (%)	38.73	94.46	66.74	10.93	57.67	68.88	74.95

3.3.2 Continuous Data Preprocessing

Considering the compositional nature of the continuous data, log-ratio transformation is needed prior to any further analysis. Using Eq. 2 and the filler variable as the denominator, continuous data are transformed to additive log-ratios. As the alr-transformed data do not follow closely to a Gaussian distribution, a

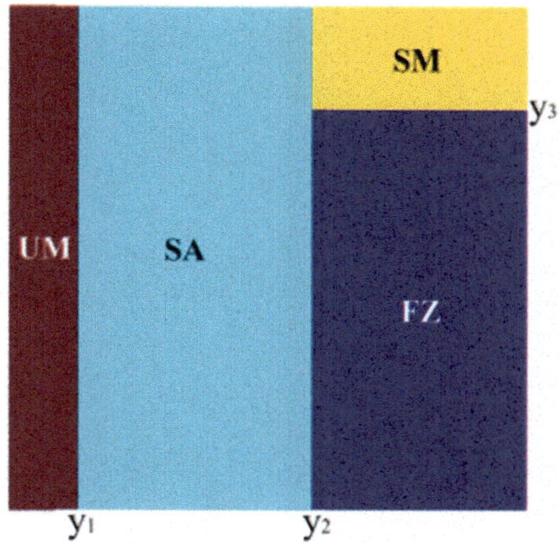

Fig. 5 Truncation rule indicating relationship between four rock types

Table 2 Parameters of variogram models of GRFs for the plurigaussian model (the anisotropy ranges are long, middle, and short range, respectively)

GRFs	Nugget	Direction	Type	Sill	Range (m)
Y_1	0	N0	Cubic	1	250,250,20
Y_2	0	N0	Cubic	1	200,200,20

transformation to normal scores is required prior to the simulation. The alr data are transformed to Gaussian space via Gaussian anamorphosis.

3.3.3 Decorrelation and Independent Simulation

The final phase of this study is to achieve realisations of the whole composition of 11 components and rock types on selective mining units (SMU) of size $10 \times 10 \times 2$ m. Based on the separation between drill holes, these SMUs are treated as point data. The Gaussian data from all previous preprocessing steps (two Gaussian variables from the Gibbs sampler algorithm and ten Gaussian variables from the alr-transformed data) are transformed to orthogonal factors via the MAF algorithm. Through testing several lag separations, a lag of 30 m has proven to be a practical choice. Experimental variograms of the 12 factors were calculated using 10 lag distance classes at a spacing of 30 m and towards several directions in horizontal plane. In the vertical direction, 25 lags at a spacing of 1 m and with an angular tolerance of 10° were used. For the first factor, which has the greatest contribution on variation, there is a presence of minor anisotropy in the horizontal plane with N-115 as the direction with greatest continuity. Table 3 shows the variogram model parameters for the MAF factors derived from the Gaussian data, while Fig. 6 shows the experimental variograms and fitted model for the first factor.

Conditional turning bands simulation was applied on the punctual grid and 100 simulations were generated for each factor.

Table 3 Variogram model parameters for the MAF factors derived from conditional Gaussian data (the anisotropy ranges are long, middle, and short range, respectively)

Factors	Nugget	Direction	Type	Sill	Range (m)	Sill	Range (m)	Sill	Range (m)
Factor 1	0.08	N115	Spherical	0.37	50,50,12	0.44	185,150,35	0.28	600,350,350
Factor 2	0.10		Exponential	0.55	60,90,15	0.40	650,300,∞	0.15	2,500,350,∞
Factor 3	0.10	N115	Exponential	0.79	90,90,20	0.19	750,350,30		
Factor 4	0.10	N120	Exponential	0.45	50,50,15	0.45	175,150,20	0.10	2,500,750,100
Factor 5	0.10	N0	Exponential	0.60	35,40,7	0.20	150,185,25	0.15	600,270,25
Factor 6	0.10		Exponential	0.72	35,35,10	0.21	500,250,13		
Factor 7	0.10	N115	Exponential	0.70	25,10,8	0.12	200,55,45	0.13	350,350,∞
Factor 8	0.08	N0	Exponential	0.64	30,30,10	0.30	85,85,17		
Factor 9	0.05	N0	Exponential	0.60	15,15,10	0.35	25,25,20		
Factor 10	0.05		Exponential	0.96	20,20,16	0.20	∞,∞,25		
Factor 11	0.05	N0	Exponential	1.00	27,16,14	0.15	∞,∞,14		
Factor 12	0.08	N0	Spherical	0.90	35,35,27	0.90	∞,∞,27		

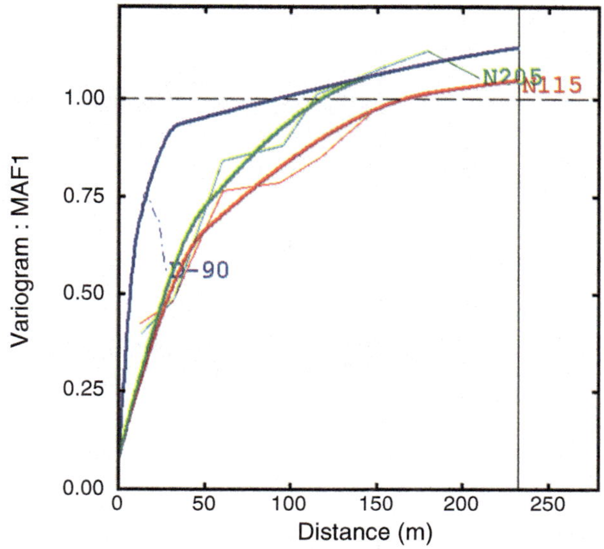

Fig. 6 Experimental variograms and fitted model for the first factor

3.3.4 Back-Transformations and Truncation

Following simulation, the MAF factor realisations were back-transformed to simulated GRFs. The simulated GRFs associated with categorical variables were then back-transformed to geological units based on the truncation rule and experimental proportions of each rock type at target nodes. To obtain the compositional realisations, the other ten simulated GRFs were first transformed to alr space and then transformed to constraint space via the agl transformation.

4 Discussion

Figure 7a, b depicts perspective views of one realisation of Ni grade and mean of the simulated Ni grade, respectively. A gradual transition from low-grade zones to high-grade zones can be seen in these figures (soft boundaries). Figure 7c, d shows one realisation of the rock types and the most probable simulated rock type, respectively. A comparison of Figs. 2 and 7 indicates a good fit with the conditioning data and geology of the deposit. Quantile-quantile plots of realisations of the major elements versus the sample data in Fig. 8 show that the global experimental distributions and simulations are reasonably well reproduced especially for nickel which is the main target economical element. However overestimation can be recognised for high Ni grades. Figure 9 shows box plots of realisation proportions for the four rock types. Compared to the proportions of the rock types in the sample data, the method has overestimated the proportions of SA and UM units and underestimated those of FZ and SM. Simulated proportions for FZ unit are close

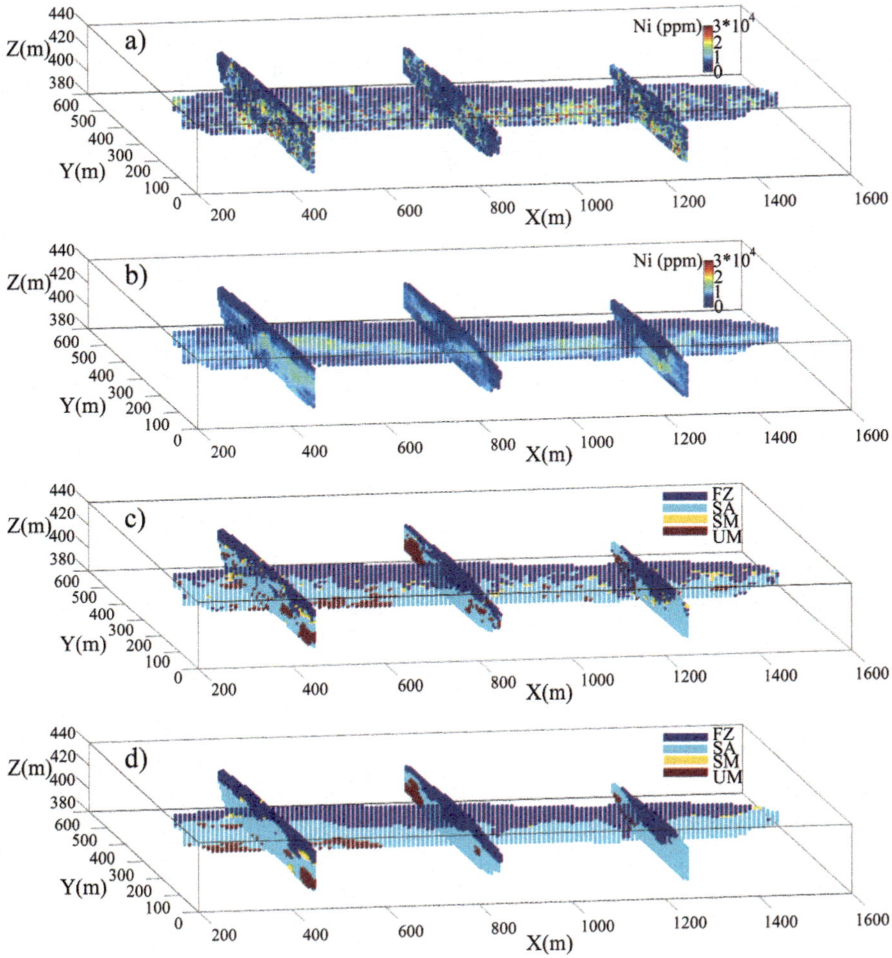

Fig. 7 Perspective view of (**a**) one realisation of Ni grade, (**b**) mean of the simulated Ni grade, (**c**) one realisation of rock types and (**d**) most probable simulated rock type

to the proportion in the exhaustive data. Figure 10 shows the experimental cross variograms between rock-type indicators and Ni grade, calculated from the sample data and the simulated model. The relationships between rock type and grade are reproduced qualitatively, with high spatial cross-correlation between SA and Ni as well as SM and Ni. For FZ and UM, there are negative correlations to the Ni grade. The underestimation of SM proportions can also be seen in the lower sill of the experimental cross variograms of the realisations in Fig. 10. Sensitivity analysis on PGS parameters might improve the performance of the joint simulation.

FZ and SA domains occur with the greatest proportions in sample data and simulated models. From Fig. 7c, d it is observed that most of the transitions between rock types in the study area occur between these two rock types. From Fig. 3 it can be seen that although SA domain is highly mineralised, high Ni grade

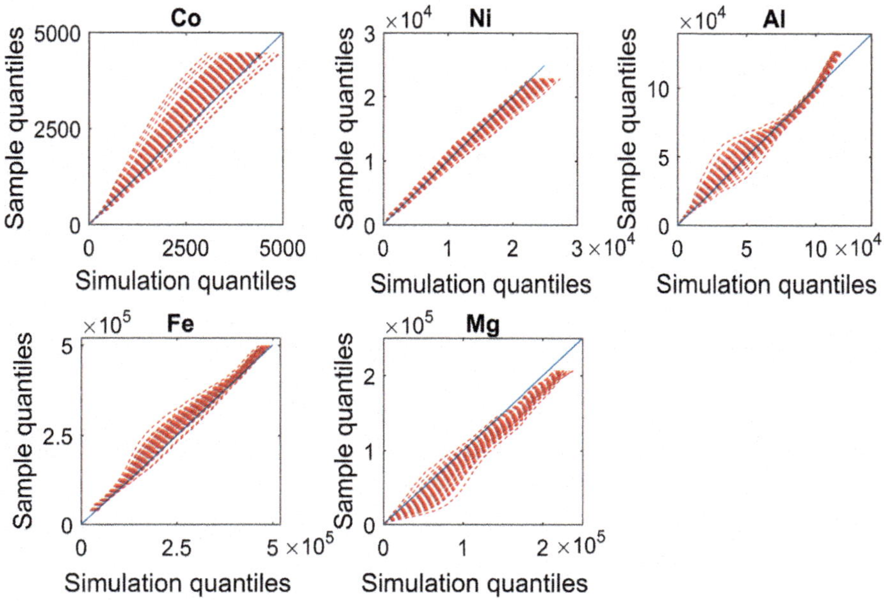

Fig. 8 Q-Q plots of realisations of the major elements against sample data

Fig. 9 Box plot of realisation proportions for the four rock types – the sample proportions are indicated by a *black horizontal line*, the exhaustive proportions by a *red horizontal line*

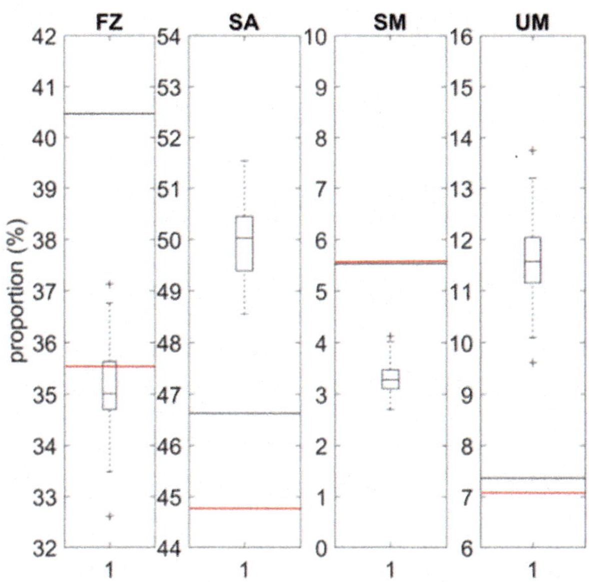

mineralisation occurs at the transition between SA and FZ. Figure 7a, b shows that the proposed joint simulation algorithm has reproduced both the highly mineralised zone at the transition area and the soft transition in realisations of Ni across geological boundaries. The ability of the method to reproduce soft transition of continuous data across geological domains is emphasised in the contact analysis

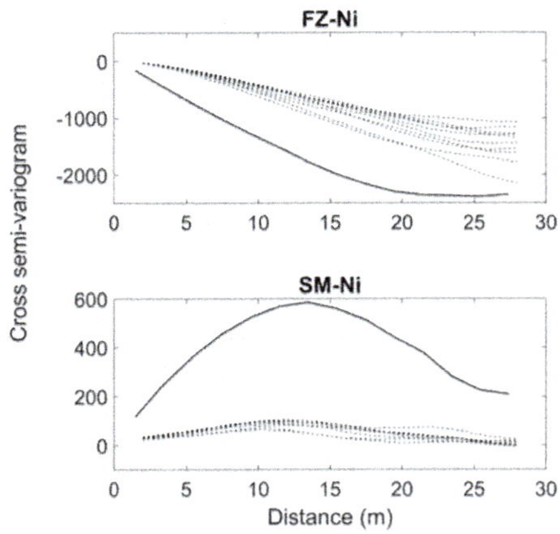

Fig. 10 Experimental cross variograms between rock-type indicators and Ni grade, for sample data (*black line*) and simulated realisations (*dashed line*)

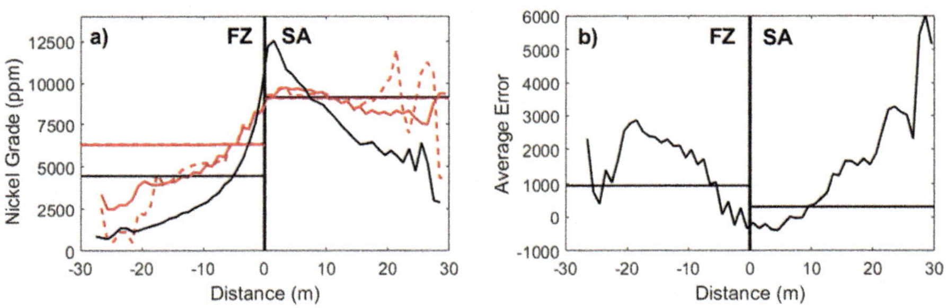

Fig. 11 (**a**) Contact analysis between FZ and SA domains for sample Ni grade (*black graph*), mean of simulated Ni grade (continuous *red graph*), and a realisation of Ni grade (*dashed red graph*). (**b**) Average prediction error for Ni grade compared with exhaustive data set

diagram of Fig. 11a. The Ni grade increases gradually from the FZ domain towards the SA domain, and the highly mineralised zone occurs at the transition zone between these two geological domains. Figure 11b depicts the average of prediction error for Ni grade compared with the exhaustive data set. Even though there is uncertainty in the exact locations of geological boundaries, the prediction errors are close to zero at the transition zone. The prediction errors are high at the boundaries of the deposit which shows that the method is sensitive to extrapolation.

5 Conclusion and Future Work

Uncertainty modelling of several dependent continuous and categorical variables is a common challenge for the geoscience modelling projects. In this study a hybrid model for joint simulation of high-dimensional continuous and categorical variables was presented based on the plurigaussian model and min/max autocorrelation factors. The proposed algorithm is able to simulate several constrained continuous (compositional data) and categorical variables simultaneously. Uncertainty measurement of the exact location of geological boundaries, incorporation of the spatial dependence between continuous and categorical variables, reproduction of spatial correlation of continuous data across geological boundaries, and the ability to simulate high-dimensional and even constrained data are some of the advantages of the proposed model. In addition it is faster and easier to apply than co-simulation algorithms based on an LMC. The performance of the method was tested on a real large mining data set producing results that are satisfactory from a practical point of view. Considering the effect of the compositional data on Gibbs sampler algorithm and sensitivity analysis on PGS parameters might further enhance accuracy of the simulated model. These issues will be addressed in future work.

Acknowledgements The first author gratefully acknowledges a travel grant by International Association for Mathematical Geosciences (IAMG) to attend and contribute to the GEOSTATS2016 conference.

Bibliography

Aitchison J (1986) The statistical analysis of compositional data. Chapman and Hall Ltd, London

Armstrong M, Galli A, Beucher H, Loc'h G, Renard D, Doligez B, . . . Geffroy F (2011) Plurigaussian simulations in geosciences. Springer, Berlin

Bandarian E, Mueller U (2008) Reformulation of MAF as a generalised eigenvalue problem. In: Ortiz J, Emery X (eds) Proceedings Eighth International Geostatistics Congress. Santiago, pp 1173–1178

Chilès JP, Delfiner P (2012) Geostatistics: modeling spatial uncertainty, 2nd edn. Wiley, New York

Desbarats AJ, Dimitrakopoulos R (2000) Geostatistical simulation of regionalized pore-size distributions using min/max autocorrelation factors. Math Geol 32(8):919–942. doi:10.1023/A:1007570402430

Egozcue JJ, Pawlowsky-Glahn V, Mateu-Figueras G, Barceló-Vidal C (2003) Isometric logratio transformations for compositional data analysis. Math Geol 35(3):279–300

Emery X (2007) Simulation of geological domains using the plurigaussian model: new developments and computer programs. Comput Geosci 33(9):1189–1201

Emery X, Silva DA (2009) Conditional co-simulation of continuous and categorical variables for geostatistical applications. Comput Geosci 35(6):1234–1246

Goulard M, Voltz M (1992) Linear coregionalization model: tools for estimation and choice of cross-variogram matrix. Math Geol 24(3):269–286. doi:10.1007/BF00893750

Jones P, Douglas I, Jewbali A (2013) Modeling combined geological and grade uncertainty: application of multiple-point simulation at the apensu gold deposit, Ghana. Math Geosci 45 (8):949–965. doi:10.1007/s11004-013-9500-3

Maleki M, Emery X (2015) Joint simulation of grade and rock type in a stratabound copper deposit. Math Geosci 47(4):471–495. doi:10.1007/s11004-014-9556-8

Markwell T (2001) Murrin Murrin Ni/Co resource estimation: MME resource modelling report. Anaconda, p 18

Mueller U, Tolosana-Delgado R, van den Boogaart KG (2014) Approaches to the simulation of compositional data – a Nickel-Laterite comparative case study. Paper presented at the Orebody Modelling and Strategic Mine Planning SMP The Australasian Institute of Mining and Metallurgy: Melbourne

Murphy M (2003) Geostatistical optimisation of sampling and estimation in a nickel laterite deposit. (MSc thesis (unpublished)), Edith Cowan University, Joondalup

Ortiz JM, Emery X (2006) Geostatistical estimation of mineral resources with soft geological boundaries: a comparative study. J South Afr Inst Min Metall 106(8):577–584

Pawlowsky-Glahn V, Olea RA (2004) Geostatistical analysis of compositional data. Oxford University Press, New York

Switzer P, Green AA (1984) Min/max autocorrelation factors for multivariate spatial imaging, Technical report No. 6. Department of Statistics, Stanford University, Stanford, p 14

Talebi H, Hosseinzadeh Sabeti E, Azadi M, Emery X (2016) Risk quantification with combined use of lithological and grade simulations: application to a porphyry copper deposit. Ore Geol Rev 75:42–51

van den Boogaart KG, Tolosana-Delgado R, Lehmann M, Mueller UA (2014) On the joint multipoint simulation of discrete and continuous geometallurgical parameters. Paper presented at the Orebody Modelling and Strategic Mine Planning Symposium, The Australasian Institute of Mining and Metallurgy, Melbourne

Vargas-Guzmán JA (2008) Transitive geostatistics for stepwise modeling across boundaries between rock regions. Math Geosci 40:861–873

Performance Analysis of Continuous Resource Model Updating in Lignite Production

Cansın Yüksel and Jörg Benndorf

Abstract Recently an efficient updating framework was proposed aiming to improve the raw material quality control and process efficiency in any type of mining operation. The concept integrates sensor data measured on the production line into the resource model and continuously provides locally more accurate resource models. A demonstration in lignite production is applied in order to identify the impurities (marine and fluvial sands) in the coal seams to lead better coal quality management. The updating algorithm applies different algorithmic parameters. This study aims to investigate the sensitivity of the performance with respect to different parameters for optimal application. Main parameters include the ensemble size, the localization and neighborhood strategies, and the sensor precision. The results should assist in future applications by determining the impact of the different parameters.

1 Introduction

One of the main challenges in lignite mining, similar to other branches of mining, is the waste intrusions in lignite seams. These marine and fluvial sand impurities can lead to high ash values (e.g., more than 15 % ash) and cannot be localized completely by exploration data and captured in the predicted deposit models.

Utilizing online sensor techniques for coal quality characterization in combination with rapid resource model updating, a faster reaction to the unexpected

C. Yüksel (✉)
Resource Engineering Section, Department of Geoscience & Engineering, Delft University of Technology, Stevinweg 1, 2628 CN Delft, Netherlands
e-mail: C.Yuksel@tudelft.nl

J. Benndorf
Institute for Mine Surveying and Geodesy, Faculty of Geosciences, Geoengineering and Mining, University of Technology Bergakademie Freiberg, Reiche Zeche, Fuchsmühlenweg 9, 09599 Freiberg, Germany
e-mail: J.Benndorf@tudelft.nl

J.J. Gómez-Hernández et al. (eds.), *Geostatistics Valencia 2016*, Quantitative Geology and Geostatistics 19, DOI 10.1007/978-3-319-46819-8_29

deviations can be implemented during operations, leading to increased production efficiency. This concept was first proposed as a closed loop framework by Benndorf et al. (2015). The developed framework is based on ensemble Kalman filter (EnKF) and basically integrates the online sensor data into the resource model as soon as they are obtained.

The first investigation (Benndorf 2015) has proven the approach to work well within a synthetic case study under a variation of several control parameters (number of excavators, precision of the sensor, update interval, measurement interval, extraction mode/production rate). The second investigation (Wambeke and Benndorf 2015) introduced an extended version of the developed framework. This extension includes a Gaussian anamorphosis of grid nodes, sensor-based measurements, and model-based predictions; to deal with suboptimal conditions, an integrated parallel updating sequence; to reduce the statistical sampling error without the need of increasing the number of realizations and a neighborhood search strategy; to constrain computation time; and to avoid the spurious correlations. Thereafter, Yüksel et al. (2016) adapted the framework to update coal quality attributes in a continuous mining environment. The applicability of the framework for a full-scale lignite production environment is validated by successful results.

To further understand the effects of used parameters during the full-scale application, to identify the sensitivity of the results and explore the performance in depth, further studies are required. For this reason, this paper aims to investigate the performance of the resource model updating framework with respect to main parameters, which are the ensemble size, the localization and neighborhood strategies, and the sensor precision. Findings of this research are expected to assist in future applications of the resource model updating concept by making it easier to achieve optimum performance.

The remainder of the article is structured as follows: First, the ensemble Kalman Filter-based approach adapted to specific application in mining is briefly reviewed. Next, application in continuous mining test case is described, and sensitivity analysis experiments are described. Findings of the study are then presented. Key findings of the study are discussed and summarized. The article concludes with a summary of the research contributions and directions for the future research.

2 A Method for Updating Coal Attributes in a Resource Model Based on Online Sensor Data

For rapid updating of the resource model, sequentially observed data have to be integrated with prediction models in an efficient way. This is done by using sequential data assimilation methods, namely, the EnKF-based methods.

With the goal of a continuously updatable coal quality attributes in a resource model, a framework based on the normal-score ensemble Kalman filter (NS-EnKF) (Zhou et al. 2011) approach was tailored for large-scaled mining applications. The

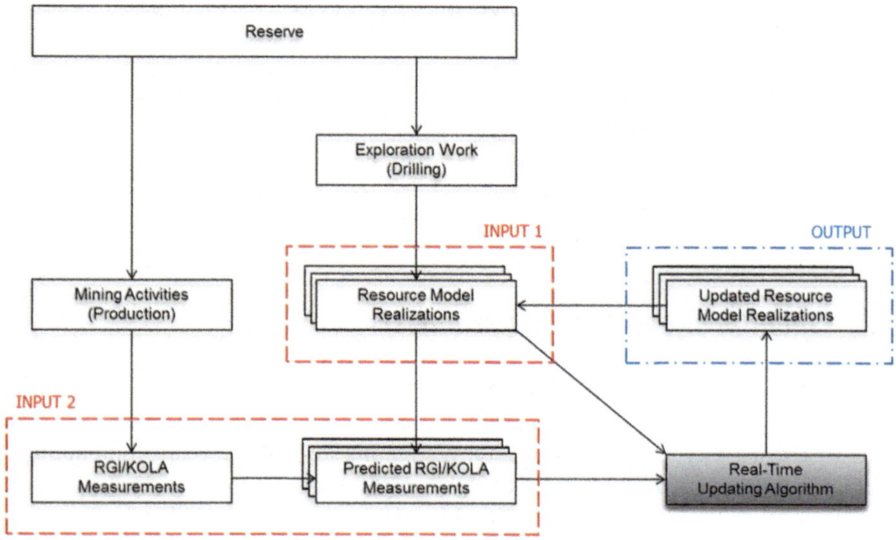

Fig. 1 Configuration of the real-time resource model updating concept (Modified from Wambeke and Benndorf 2015)

NS-EnKF is chosen to deal with the non-Gaussianity of the data by applying a normal-score transformation to each variable for all locations and all time steps, prior to performing the updating step in EnKF.

A formal description of the real-time updating algorithm is provided (Yüksel et al. 2016). Figure 1 gives general overview of the operations which are performed to apply the updating algorithm for improving the coal quality control using online data.

The concept initially starts with resource modeling by using geostatistical simulation technique, namely, sequential Gaussian simulation (SGS). This is the first required data set consisting of ensemble members to be updated. The second data set consists of a collection of actual and predicted sensor measurements. The actual online sensor measurement values are collected during the lignite production, and the predicted measurements are obtained by applying the production sequence as a forward predictor prior to resource model realizations. Once both of the input data are provided, the updated posterior resource model will be obtained. This process will continue as long as the online sensor measurement data is received.

3 Sensitivity Analysis on a Full-Scale Study

The aim of the case study presented here is to analyze the performance of the resource model updating framework method by performing sensitivity analyses on main parameters, including the ensemble size, the localization and neighborhood strategies, and the sensor precision.

3.1 *Identification of Main Parameters*

3.1.1 Number of Ensemble Members

The first sensitivity analyses focus on investigating the optimal realization number (subsequently used as ensemble size) by performing resource model updating experiments with different-sized ensembles. Defining the ensemble size that will fully represent the ore body is a very delicate problem. A lot of research in literature (Houtekamer and Mitchell 1998; Mitchell et al. 2002) focuses on the optimum ensemble size investigation and usually concludes that the analysis error decreases as the number of ensembles (realizations) increases. Contrary, the computational costs increase with the ensemble size. Therefore a sensible size of the ensemble is required.

3.1.2 Localization

The second sensitivity analyses focus on investigating the effects of localization strategies and neighborhood size on the given case. As mentioned, one of the limiting factors in EnKF-based applications is the restrained ensemble size. But having an insufficient ensemble size might cause long-range spurious correlations. In order to avoid these spurious correlations, a covariance localization technique is applied to the updating framework by Wambeke and Benndorf (2015). The spurious correlations refer to the correlations between quality attributes that are at a significant distance from one another where there is no spatial relation. Moreover, these correlations can lead to inbreeding and filter divergence. Covariance localization modifies update equations by replacing the model error covariance by its element-wise (the Schur) product with some distance-based correlation matrix (Gaspari and Cohn 1999; Horn and Johnson 1985). This replacement increases the rank of the modified covariance matrix and masks spurious correlations between distant state vector elements (Sakov and Bertino 2011).

3.1.3 Sensor Error

The final sensitivity analyses focus on testing the effect of the sensor precision. In most cases errors are involved when taking measurements, due to calibration issues of sensor technologies. For each experiment, different amounts of standard error are added to the actual measurement values. The standard error can be calculated as

$$SE_{\bar{x}} = \frac{\sigma}{\sqrt{n}} \tag{1}$$

where σ is the standard deviation of the actual measurements and n is the size (number of observations) of the actual measurements. For this study, the size of the actual measurement data set contains 700 observations, which values correspond to coal extracted from 28 mining blocks. This leads to approximately 25 actual measurement data per block. Therefore, where the added standard error is 0.1 % ash, the absolute standard deviation will be 0.5 % ash, and the variance will be 0.25 $\%^2$ ash.

Similarly, when the added standard error is 0.2 % ash, the standard deviation will be 1 % ash, and the variance will be 1 $\%^2$ ash. The variance of the actual measurements will be 6.25 $\%^2$ ash, and the standard deviation will be 2.5 % ash when the added standard error is 0.5 % ash. The variance will be 25 $\%^2$ ash when the added standard error is 1 % ash. The variance of the averaged prior model for 48 ensembles is calculated as 0.99.

To give a clear view, mentioned standard deviations are converted as the relative error of the measurements. The average measurement value is calculated as 12 % ash. This leads around 4 % ash relative error in measurement values when the added standard deviation is 0.5 % ash. Similarly, when the added standard deviation is 1 % ash, this indicates around 8 % ash relative error in measurement values. In the same way, when the added standard deviation is 2.5 % ash, this indicates around 20 % ash relative error in measurement values. Finally, when the added standard deviation is 5 % ash, this indicates around 40 % ash relative error in measurement values.

3.2 Experiment Setup

The case study is performed on a particular lignite seam in a mining operation in Germany. The seam contains multiple sand intrusions. The shape and size of these sand partings are irregular, and both characteristics are showing a large variability.

To apply the resource model updating algorithm, preparation of input data is required. First, the geological model of the defined coal seam is created on a $32 \times 32 \times 1$ m dimensioned block model based on the roof and floor information of the lignite seam. Second, a $32 \times 32 \times 1$ m dimensioned quality model capturing the wet ash content in percentages is generated with different number of simulations (24, 48, 96, 192, and 384), based on the provided drill hole data. The simulated ash

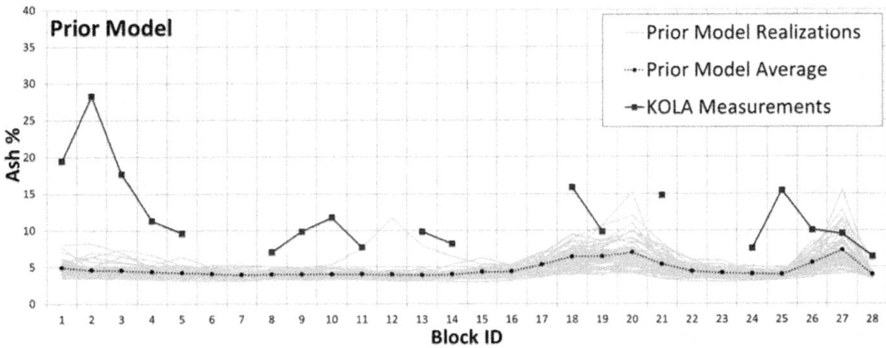

Fig. 2 Prior model and measurement data (before updating)

values are then merged with the previously defined coal seam. The block model realizations are now ready to be imported into the algorithm as the first input.

Figure 2 illustrates the prior model of 48 simulations, the averaged ash values of those simulations and related sensor measurement values, per block. A significant underestimation of the actual measurement data is observed in the prior model. This is because the prior model is created based on the drill hole data, where the local sand intrusions are not fully captured. True variability of the coal seam is captured by the online sensor measurements.

Predicted measurements are obtained by averaging the simulated ash values from each simulation set, which falls into the defined production block boundaries. The online sensor measurement data, namely, the Kohle OnLine Analytics (KOLA) data, are provided for the defined time period. KOLA system applies X-ray diffraction in order to accurately assess the components of the produced lignite. In order to determine the location of the received KOLA data, in other words to track back where the measured material comes from, the GPS data is matched with the measurement data based on the given timecodes. The located measurements in coal seam are then imported into the previously defined block model.

The second input file for the algorithm is written to a file containing the following information: the block ID, the central block location (X, Y, Z coordinates), and a series of real and predicted measurements.

A study bench is produced for a defined time period by considering all the available data (topography, RGI, GPS, and production data). Later, the study bench is divided into so-called production blocks. This was necessary to reproduce the excavated production blocks. The horizontal divisions (or production slices) are applied based on the movements of the excavator during production, provided by GPS data. The vertical divisions are based on the changes in the Z coordinates in the GPS data and capture a typical extraction sequence of bucket-wheel excavator operations. In total, the defined production bench is divided into 28 blocks and 5 slices, which gives 140 production blocks. Once the study bench is divided both in vertical and horizontal, the production blocks are now ready to be updated.

First the first block of the second slice will be updated, based on the KOLA measurements taken from that block. The series of updating experiments included seven updating experiments and continued until the ninth block (since there are no KOLA data obtained on sixth and seventh block, seven experiments are performed to update until the ninth block). In each updating experiment, only one block is updated based on the related measured KOLA value.

An empirical error measure so-called mean square difference or mean square error (MSE) is used in order to present results of the performed experiments. MSE compares the difference between estimated block value $\mathbf{Z}^*(\mathrm{x})$ and actual KOLA measurement $\mathbf{v}$ values per block, and it can be calculated as

$$\mathrm{MSE} = \frac{1}{N}\sum\nolimits_{i=1}^{N} \left(z^*(\boldsymbol{x}_i) - \boldsymbol{\nu}_i\right)^2 \tag{2}$$

where $i = 1, \ldots, N$ is the number of blocks. The mean square error graphs are calculated relative to the averaged prior model of 384 ensembles, in order to make a good comparison.

3.3 *Experiments With Respect To Main Parameters*

Table 1 provides a complete overview of the parameters used to perform the mentioned experiments. The obtained results of these experiments are provided in the next chapter. In every experiment performed for every parameter, one parameter is varied, and the others remain fixed (Table 1).

3.3.1 Number of Ensemble Members

With a view toward the real-time application of the updating resource model, the industrial case presented by Yüksel et al. (2016) focused on small- and moderate-sized ensembles (24). For the investigation of the optimum ensemble size, updating experiment series are performed with 24, 48, 96, 192, and 384 ensembles. All of the simulations are created by using SGS with same seed number and same variogram parameters.

3.3.2 Localization

The initial neighborhood size is defined as 450 m in X and Y directions and 6 m in Z direction based on the variogram of the drill hole data. For the experiments, three different neighborhood sizes (225, 450, and 900 m) are tested while the localization option was not being used. Three more experiments are performed while the localization option was being used in order to test the effect of designed

Table 1 Experimental schema

	Experiment #	Ensemble size	Localization option on/off and size (X,Y, Z) (m)	Neighborhood size (X,Y,Z) (m)	Relative sensor error (%)
Ensemble size experiments	1	24	On, 125,125,3	225,225,6	0
	2	48	On, 125,125,3	225,225,6	0
	3	96	On, 125,125,3	225,225,6	0
	4	192	On, 125,125,3	225,225,6	0
	5	384	On, 125,125,3	225,225,6	0
Localization and neighborhood strategies experiments	6	48	Off	225,225,6	0
	7	48	On, 225,225,3	450,450,6	0
	8	48	Off	450,450,6	0
	9	48	Off	900,900,6	0
	10	48	On, 450,450,3	900,900,6	0
	11	48	On, 450,450,6	900,900,6	0
Sensor error experiments	12	48	Off	450,450,6	4
	13	48	Off	450,450,6	8
	14	48	Off	450,450,6	20
	15	48	Off	450,450,6	40

localization, with varying localization and neighborhood sizes. For the experiments where the localization option was used, the localization neighborhood was assumed as half of the defined neighborhood size, except the tenth experiment. In the tenth experiment, in the X and Y direction, localization sizes were assumed as half of the defined neighborhood size. In the Z direction, the localization size remained the same. Reasons of this preference will be explained in the discussion chapter.

3.3.3 Sensor Error

For each experiment, different amounts of standard error are added to the actual KOLA measurement values. In total, five experiments are performed, where the relative measurement error varied between 4 %, 8 %, 20 %, and 40 %.

4 Results

4.1 Ensemble Size

Figures 3 and 4 present results of the updating process from the first block until the ninth block, for some of the representative ensemble sizes. For these experiment series, the localization strategies were applied; the neighborhood size was 225, 225, 6 m for X, Y, Z directions, respectively, and no sensor error is assumed.

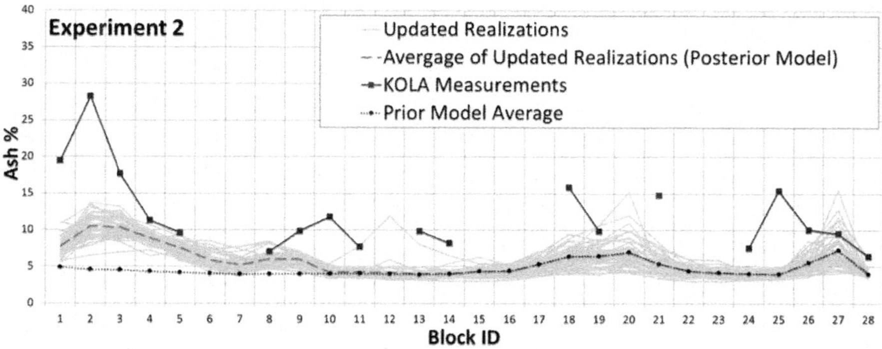

Fig. 3 Experiment 2 – ensemble size: 48

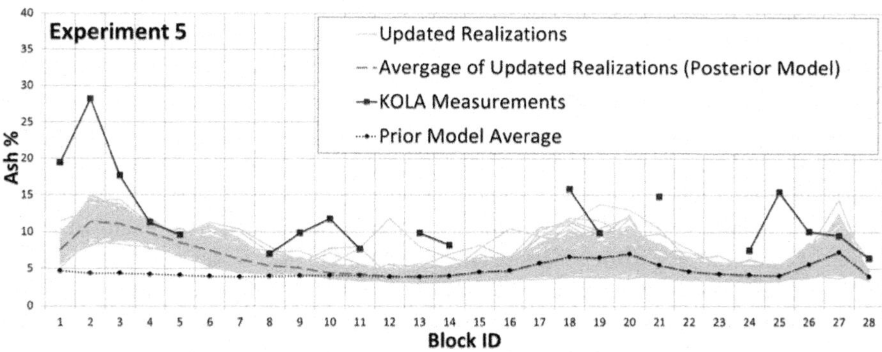

Fig. 4 Experiment 5 – ensemble size: 384

It can be seen that the average of the prior simulations substantially underestimates the actual KOLA measurements. This is caused by the data effect. The prior simulations are created based on the coal samples from drill holes spaced multiple hundred meters apart, while the KOLA measurements measure more higher ash values due to the sand intrusions in the coal seam. Integrating the KOLA measurement to the first nine blocks updates the neighborhood blocks to some relatively higher values. As expected, the update effect decreases while moving away from the last updated block, block 9.

For all different ensemble sizes, a clear improvement is observed toward the KOLA data when considering the average of the initial simulations, so-called prior model.

Figure 5 presents the relative MSE values to the prior model for each experiment performed with different ensemble sizes. The biggest reduction of the error occurs in the update of the first block. While the skewness behavior of each MSE graphs is similar, the biggest error behavior to the smallest is as follows: 48 ensembles, 96 ensembles, 192 ensembles, 384 ensembles, and 24 ensembles. Except for the results from 24 ensembles, the rest of the listing supports the literature. It is expected to observe a decrease in the MSE values while the ensemble size gets

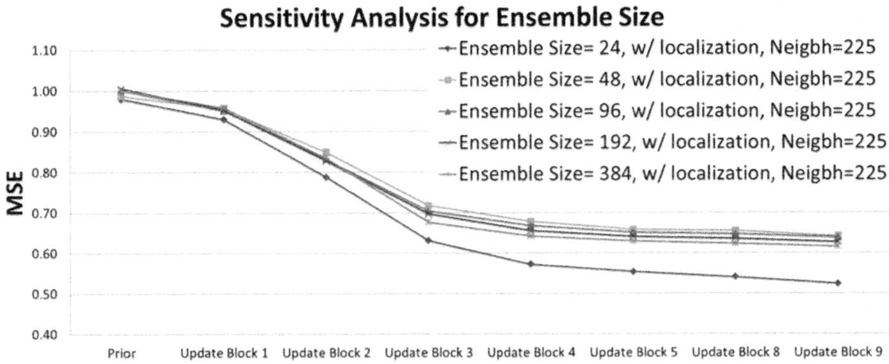

Fig. 5 Comparison graph for different ensemble-sized experiments

larger since the representativeness gets higher. However, to increase the computational efficiency and to apply the updating framework in real time during production, an economic ensemble size is required.

At first glance, higher initial variance of the 24 ensembles explains the very low MSE values. Nevertheless, a further investigation is performed in order to understand the phenomenon better. Five different sets of newly derived sets of 24 ensembles are generated with SGS, by using different random seeds for each set. New series of updating experiments are performed with the new series of 24 ensembles, and the results are compared. The comparison shows a high variety among results. MSE values obtained from the ninth block's update varied between 0.52 and 0.69. In addition, the new sets of MSE values were equal to, lower or higher than the 48 ensembles, 96 ensembles, 192 ensembles, and 384 ensembles. This big variety, which is caused by different seed numbers, shows that 24 ensembles were not sufficient to represent a statistical stable estimate of the mentioned lignite seam.

When considering the 48 ensembles, even though the 48 ensembles have the highest MSE values by comparing to the 96 ensembles, 192 ensembles, and 384 ensembles, the MSE dropped from 1.0 to 0.64. In his research, Yin et al. (2015) found that improvements while using larger ensemble sizes (after the optimum ensemble size) are relatively insignificant. Likewise, the improvements between 48, 96, 192, and 384 ensembles are obvious, yet not very significant. For this reason, this study concludes that the optimal ensemble size for this specific study is 48 ensembles.

4.2 *Localization and Neighborhood Strategies*

Figures 6, 7, 8, and 9 present results of the updating process from the first block until the ninth block, for different localization strategies and neighborhood sizes. Experiments 2 (Fig. 3) and 6 (Fig. 6), Experiments 7 (Fig. 7) and 8 (Fig. 8), and

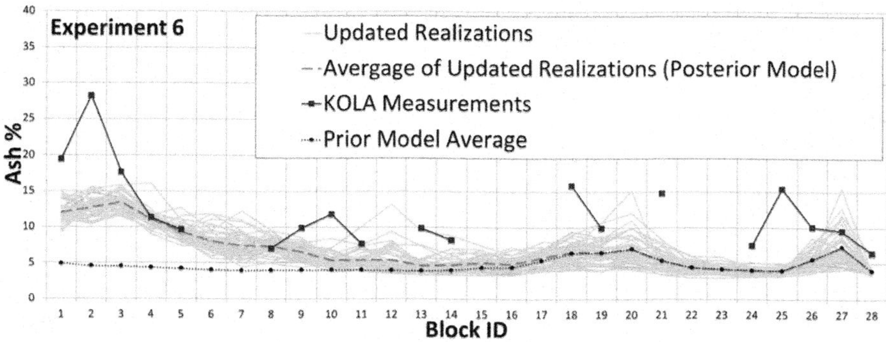

Fig. 6 Experiment 6 – localization option off, neighborhood size: 225, 225, 6 m

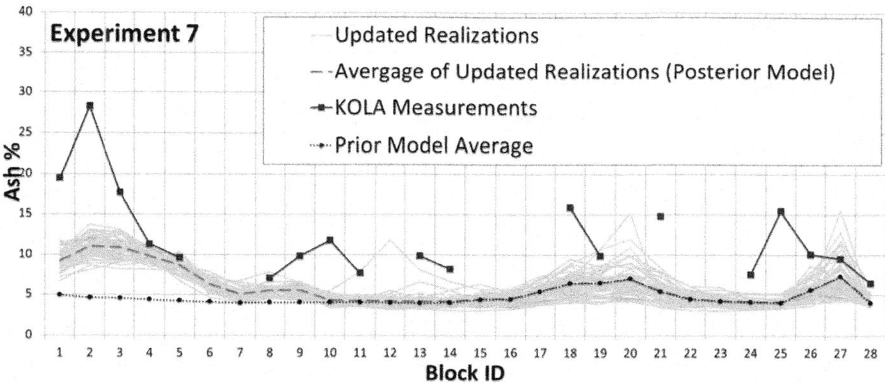

Fig. 7 Experiment 7 – localization option on (225,225,3 m), neighborhood size: 450, 450, 6 m

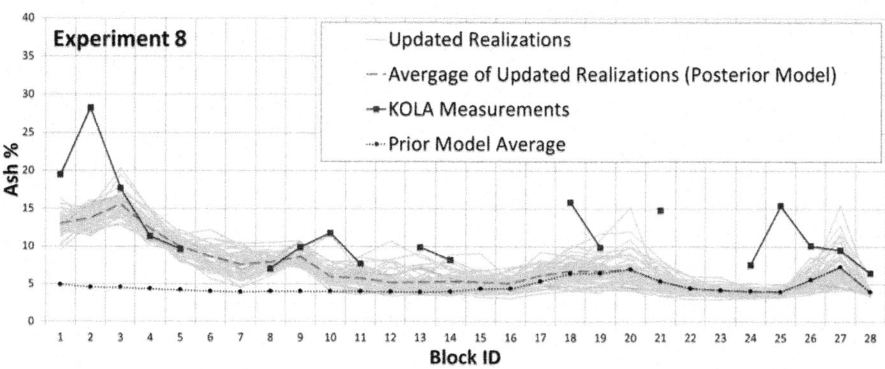

Fig. 8 Experiment 8 – localization option off, neighborhood size: 450, 450, 6 m

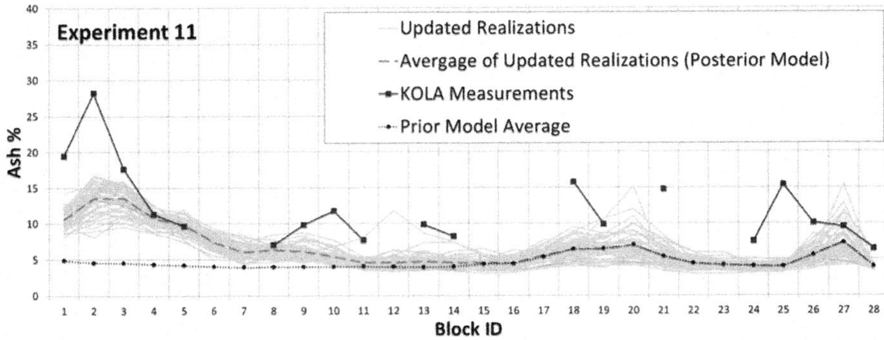

Fig. 9 Experiment 11 – localization option on (450,450,6 m), neighborhood size: 900, 900, 6 m

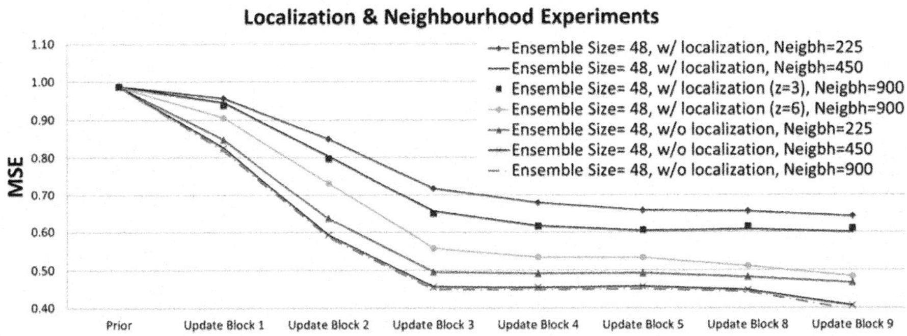

Fig. 10 Comparison graph for different localization and neighborhood strategies experiments

Experiments 9 (Fig. 9) and 10 (Fig. 10) are comparable to each other when investigating the localization option. Experiments 6 (Fig. 6), 8 (Fig. 8), and 9 are comparable to each other when investigating the neighborhood size.

Figure 10 compares all of the experiments performed in this section by plotting MSE values of each. Higher MSE values are observed when localization strategies are applied, and the neighborhood size is defined as 225, 225, 3 m. The MSE values become lower when the neighborhood size is increased and localization option is not used. This is expected because the neighborhood size was initially defined as 450, 450, 6 m based on the variogram, so performing the experiments with 225, 225, 3 m-sized neighborhood was not enough to cover the seam continuity. Minor changes are observed between the MSE values of 450, 450, 6 m neighborhood-sized experiment and 900, 900, 6 m-sized experiment due to no spatial correlation between the attributes.

The reason that applying the localization strategies did not provide any improvement in our case is due to the definition of the localization function.

Figure 14 illustrates the currently used function. Since the production block size is varying for each block, sometimes the plateau phase of the used function cannot cover a full block which is in the neighborhood. This creates un-updated values in a block and consequently the updating process of the entire block fails. For this reason, better results are obtained while the localization strategies were not in use. The future study will improve this drawback by developing the localization function in a way that it can define the block boundaries and act according to those distances.

Experiment 10 uses the localization option with the following dimensions: 450, 450, 6 m in X, Y, Z directions. The used neighborhood size was 900, 900, 6 m. As mentioned before, the initial intention was to use a localization size half the size of the neighborhood size. Yet, since the depth of a production block is 6 m, limiting the localization by 3 m decreased the expected improvements. By running the same experiment, only changing the Z localization size parameter from 6 to 3 m, the same results as found in Experiment 6 (Fig. 6) are obtained. This can be observed in Fig. 10, by comparing the related MSE values.

4.3 Sensor Precision

Figures 11, 12, and 13 present the final results of the updating process from the first block until the ninth block, for different relative sensor errors. For all the experiments performed in this section, the average prediction quality gets better in the sense that they become closer to the KOLA measurement values.

When the relative sensor error gets higher, the posterior variance appears to increase significantly. This is mainly because the KOLA measurement values are almost out of the range of the prior model (Fig. 2), and the variance of the prior model significantly underestimates the KOLA measurement values. By integrating the KOLA measurements which have lower precision (applied relative error varies between 4 and 40 % ash), the algorithm opens up the option whether the KOLA data can be right or the prior model. Subsequently, this inflates the posterior uncertainty.

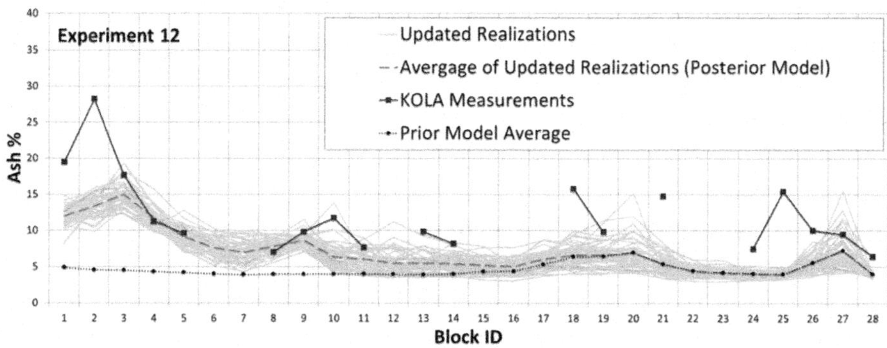

Fig. 11 Experiment 12 – relative sensor error: 4 %

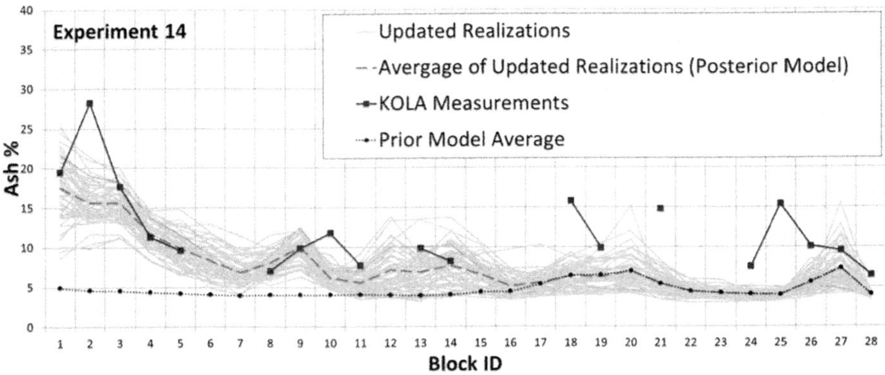

Fig. 12 Experiment 14 – relative sensor error: 20 %

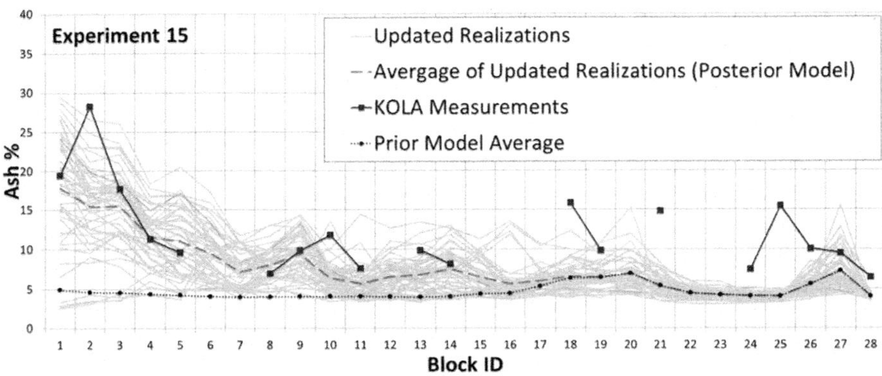

Fig. 13 Experiment 15 – relative sensor error: 40 %

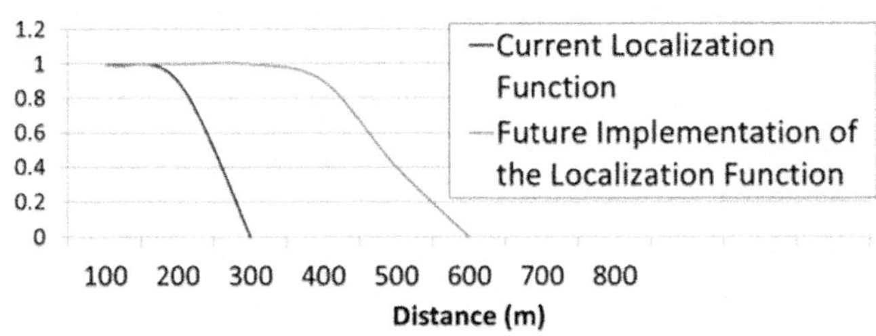

Fig. 14 Localization function illustrations

5 Conclusions and Future Work

This study analyzes the performance of the resource model updating method by performing sensitivity analyses on main parameters, including the ensemble size, the localization and neighborhood strategies, and the sensor precision in lignite production. The results should assist in future applications by determining the impact of the different parameters.

The findings of ensemble size sensitivity analysis supported the existed literature (Houtekamer and Mitchell 1998; Mitchell et al. 2002), more accurate updates are achievable by using a bigger ensemble size. Although 24 ensembles provided the best results in terms of MSE, they are not chosen as the optimum ensemble size since they were not representative enough of the lignite seam. Instead 48 ensembles were, because it was the second best and was more representative of the lignite seam.

The sensitivity analyses of the localization and neighborhood strategies concluded that the applied localization strategies need to be improved, and the neighborhood size needs to remain as 450, 450, 6 m in X, Y, Z directions, as previously defined in the variogram modeling.

Sensitivity analyses for different sensor precision showed that the lower sensor precision increases the uncertainty of the posterior model, due to the significant difference between the prior model and the actual sensor data.

In general, the KOLA data is well covered by the range of uncertainty in the updated neighborhood. It is observed that the uncertainty in the near neighborhood gets slightly smaller and more of the actual KOLA measurements are captured by this uncertainty range.

The current research was limited to a case where only one excavator is operating. Future research should apply a case study where two, three, or four excavators are operating. This will require an update to the coal quality parameters in different production benches based on one combined material measurement.

Bibliography

Benndorf J (2015) Making use of online production data: sequential updating of mineral resource models. Math Geosci 47(5):547–563

Benndorf J, Yueksel C, Shishvan MS, Rosenberg H, Thielemann T, Mittmann R, Lohsträter O, Lindig M, Minnecker C, Donner R (2015) RTRO–coal: real-time resource-reconciliation and optimization for exploitation of coal deposits. Minerals 5(3):546–569

Gaspari G, Cohn SE (1999) Construction of correlation functions in two and three dimensions. Q J Roy Meteorol Soc 125(554):723–757

Horn RA, Johnson CR (1985) Matrix analysis. Cambridge University Press, Cambridge

Houtekamer PL, Mitchell HL (1998) Data assimilation using an ensemble Kalman filter technique. Mon Weather Rev 126(3):796–811

Mitchell HL, Houtekamer PL, Pellerin G (2002) Ensemble size, balance, and model-error representation in an ensemble Kalman filter*. Mon Weather Rev 130(11):2791–2808. doi:10.1175/1520-0493(2002)130<2791:ESBAME>2.0.CO;2

Sakov P, Bertino L (2011) Relation between two common localisation methods for the EnKF. Comput Geosci 15(2):225–237

Wambeke T, Benndorf J. (2015) Data assimilation of sensor measurements to improve production forecasts in resource extraction. Paper presented at the IAMG, Freiberg (Saxony) Germany

Yin J, Zhan X, Zheng Y, Hain CR, Liu J, Fang L (2015) Optimal ensemble size of ensemble Kalman filter in sequential soil moisture data assimilation. Geophys Res Lett 42 (16):6710–6715

Yüksel C, Thielemann T, Wambeke T, Benndorf J (2016) Real-time resource model updating for improved coal quality control using online data. Int J Coal Geol 162:61–73, http://dx.doi.org/10.1016/j.coal.2016.05.014

Zhou H, Gómez-Hernández JJ, Hendricks Franssen H-J, Li L (2011) An approach to handling non-Gaussianity of parameters and state variables in ensemble Kalman filtering. Adv Water Resour 34(7):844–864

Part IV
Petroleum Geoscience and Engineering

Geostatistics on Unstructured Grids, Theoretical Background, and Applications

P. Biver, V. Zaytsev, D. Allard, and H. Wackernagel

Abstract Traditionally, geostatistical simulations are performed on regular grids, in IJK coordinates system, simulating centroids of the cells. This approach (commonly used) has severe drawbacks: the support size effect is not taken into account and some artifacts due to cells distortion may appear. On the other hand, reservoir engineers and hydrogeologists are increasingly referring to new generation of grids to perform dynamic simulation (Voronoï grids, tetrahedral grids, etc.) which require addressing the volume support effect.

In this paper, we present a theoretical framework to simulate variables directly on this new generation of grids, using a depositional coordinates system (UVT) and taking into account the support size effect.

A real field case study is subsequently presented (lithology and petrophysical modeling) to illustrate the possibilities of the new generation of simulation tools. A conclusion is provided and the remaining problems are discussed to propose some guidelines for future works.

P. Biver (✉)
Geostatistics and Uncertainties Team, Total SA, CSTJF avenue Larribau, 64000 Pau, France
e-mail: pierre.biver@total.com

V. Zaytsev
Geostatistical Team, Geosciences Centre, Mines Paris Tech. Sponsored by Total SA, CSTJF avenue Larribau, 64000 Pau, France
e-mail: victor.zaytsev@total.com

D. Allard
Biostatistics & Spatial Processes, INRA, 84914 Avignon, France
e-mail: allard@avignon.inra.fr

H. Wackernagel
Geostatistical Team, Geosciences Centre, Mines Paris Tech, 35 rue Saint Honoré, 77305 Fontainebleau, France
e-mail: Hans.Wackernagel@mines-paristech.fr

J.J. Gómez-Hernández et al. (eds.), *Geostatistics Valencia 2016*, Quantitative Geology and Geostatistics 19, DOI 10.1007/978-3-319-46819-8_30

1 Introduction

For more than 40 years, geostatistical estimations and simulations have been performed on regular so-called "sugar-box" grids. This is mainly due to historical reasons. The technology was emerging first from the mining industry, to estimate grades for open pit blocks, and it was a reasonable choice to use such support of information.

These regular grids have been kept for a long time as they allow following stratigraphy in corner point grids geometry commonly used in petroleum industry; they were also a convenient format to optimize algorithms of various kinds (sequential simulations, simulation with fast Fourier transform, multiple-point statistics simulations, etc.).

However, several new grid geometries have emerged in the last decades: tetrahedral meshes in hydrogeology and Voronoi grids with local grid refinements for petroleum industry. These grids are more convenient to solve the physical equations of flow and transport in porous media. Moreover, these grids are emerging in the geo-modeling processes with a relevant formulation of the depositional UVT coordinates system (see Mallet 2004); a dual grid approach was used to address both the dynamic simulation and the geostatistical characterization (flow simulation grid or FSG to solve the physical problem and the geological grid or GG to perform geostatistics).

The dual grid approach has several drawbacks: the geological grid resolution is driven by the smallest cells of the flow simulation grid, and, moreover, an upscaling technique is necessary to transfer information from GG to FSG; this upscaling technique needs to be general enough for the considered topologies.

As a consequence, it was necessary to adapt the geostatistical processes to use directly the flow simulation grids; but, due to the various size and geometry of elementary grid cells, it is mandatory to take into account the support size effect. We present a technique based on the formalism of the discrete Gaussian model (see Emery 2009 and Chilès and Delfiner 2012). A review of other solutions for geostatistical simulations on unstructured grids can be found in Zaytsev et al. 2016. To simulate directly on unstructured reservoir grids, an alternative method proposed by Boucher A. and Grosse H. (2015) will be also commented regarding implementation aspects.

2 Recall of the Discrete Gaussian Model

The presentation of the discrete Gaussian model (DGM) can be found in Chilès and Delfiner (2012). It can be described as follows:

- Each block of the grid v_p is attributed a parameter $r_p \in (0,1)$ which is called the change of support coefficient for this block.
- We work on the Gaussian transform Y of the variable of interest Z.

- We decompose the Gaussian anamorphosis of *Z* in a basis of normalized Hermite polynomials $\chi_i\ (Y(x))$:

$$Z(x) = \varphi(Y(x)) = \sum_{i=0}^{\infty} \varphi_i \chi_i(Y(x)) \tag{1}$$

- Using the Cartier's relation (Chiles and Delfiner 2012 p. 441), we can derive the block support distribution *Z(v)* since it can also be represented in the same polynomial basis using the decomposition (1) and the block change of support coefficient *r*:

$$Z(v) = \varphi_v(Y_v) = \sum_{i=0}^{\infty} \varphi_{i.} r^i \chi_i(Y_v) \tag{2}$$

- By double volumetric integration of the point support covariance *C(x,x′)* over the volume of interest *v*, we can derive the change of support coefficients r_p:

$$Var(Z(v_p)) = \frac{1}{|v_p|^2} \int_v \int_v C(x,x')dx\,dx' = \sum_{i=1}^{\infty} \varphi_i^2 r_p^{2i} \tag{3}$$

- Following the same principle, by double volumetric integration of the point support covariance *C(x,x′)* over two different volumes of interest v_p and v_q, we can derive the block support covariances:

$$Cov(Z(v_p),Z(v_q)) = \frac{1}{|v_p|\,|v_q|} \int_{v_p} \int_{v_q} C(x,x')dx\,dx' = \sum_{i=1}^{\infty} \varphi_i^2 r_p^i r_q^i \mathrm{cov}(Y_{v_p},Y_{v_q})^i \tag{4}$$

- Once the change of support is known for each grid cell and covariance is known between each pair of Gaussian random variables characterizing the volumes of each grid cell, we are back to a classical problem of generating a multivariate Gaussian random function with a given covariance matrix which can be solved by classical methods such as SGS.

Different formulations are available for the discrete Gaussian model; in this paper, these aspects will not be addressed; for such discussions, we refer to Zaytsev et al. (2016).

3 Practical Aspects of Applying Discrete Gaussian Model for Geostatistical Simulations on Unstructured Grids

To apply the theory of DGM to unstructured grids, key additional issues need to be addressed.

It is mandatory to integrate efficiently point support covariance (Eqs. 3 and 4) over cells that are not usual octahedrons. This issue is solved with an efficient lumping of each grid cell (see Korenblit and Shmerling 2006) followed by pseudo Monte Carlo integration using Sobol sequence of quasi-random points in the six-dimensional space of integration. This methodology is recommended for high dimension space because of the convergence speed.

Figure 1 illustrates the advantage of Monte Carlo methods for the problem of computing the variance of a block average value. In this figure, several integration methods are compared (subsequent Gauss quadrature integration, approximating the block with regularly spaced points, Monte Carlo integration). Clearly, Monte Carlo and related techniques are much more efficient.

It is also important to have an efficient procedure to navigate in the topology of the grid. The definition of searching neighborhood needs to be addressed in a general and efficient way. We propose to address this issue with a k-d tree efficient search (Bentley 1975). Expressed in the asymptotic notations for comparison of the algorithm performance (Cormen 2009), for a grid of N_b blocks, the k-d tree gives performance $O(\log N_b)$ for the neighborhood search operation, which is much faster than the naïve approach of looking through all the blocks of the grid which is at least $\Omega(N_b)$.

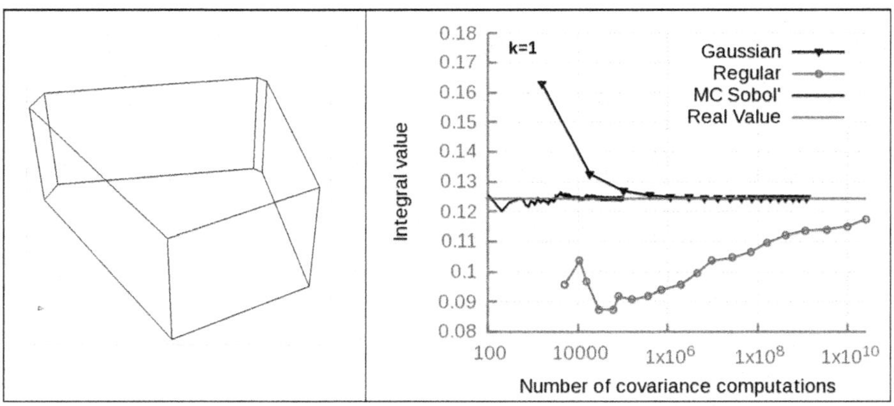

Fig. 1 Comparison of different integration methods to estimate variance of a typical grid cell (Gauss integration, regular spacing, Monte Carlo with Sobol quasi-random sequence)

4 What About Facies Modeling?

In the previous section, we addressed the problem of simulating a continuous variable on an unstructured grid.

Another important problem is to simulate categorical variables on unstructured grids. Following the same general approach, it is important to address the support size effect, at least for sub-seismic heterogeneities.

The problem can be described as follows: let us consider a categorical variable with K states $k = 1,K$. The appropriate way to handle support size effect is to simulate on the unstructured grid a proportion vector $\boldsymbol{p}=(p_1,\ldots,p_K)$ with components summing up to one.

Although using DGM for the problem of simulating categorical variables is possible, the range of applicability of the resulting model is very limited.

In our case study, the problem of facies simulation has been addressed with the method described by Gross and Boucher (2015). In their paper, the authors propose an upscaling-based approach for geostatistical simulations on unstructured grids which enables simulating the block values in a consecutive manner. Although using upscaling, this method does not require creating and storing a refined grid for the entire unstructured model, but only uses discretizations of a limited number of nodes at every step. The algorithm is based on parsimonious simulation of control points inside the cell of the unstructured mesh. The current point is simulated consistently with the other points previously simulated in the cell and the neighboring previously simulated cells; therefore, point-to-block covariances are still needed.

The key issue is to choose efficiently the number and the locations of the control points inside the cell. For this problem, sensitivity tests can be performed similarly than what has done for the integration of Eq. (3). A discretizing set of points can be considered to be good, if it enables to approximate accurately the variance of the block. Precisely, a set of points $\{x_i, i = 1 \ldots N\}$ can be used for simulating the value Z(v) if

$$Var\left(\frac{1}{N}\sum\nolimits_1^N Z(x_i)\right) \approx Var(Z(v)). \tag{5}$$

Our tests indicate that when the Sobol quasi-random sequence of discretizing points is used, a relatively small number of points (between 50 and 100; see Fig. 2) are sufficient to satisfy (5).

5 Application to a Real Field Case

The above-described methods have been applied to field X. The objective is to simulate groundwater flow on a very large grid including regional effects modeled with large grid cells and local details for a zone of interest modeled with much smaller grid cells (100–1,000 times smaller).

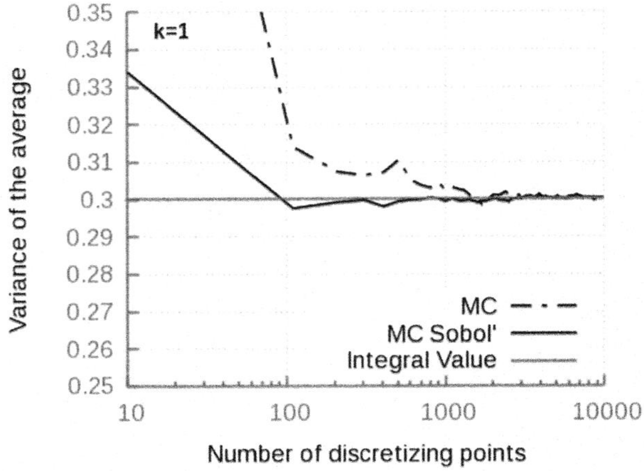

Fig. 2 Approximating the variance of blocks with the variance of a set of discretizing points. Spherical covariance in 3D, the dimensions of the blocks approximately equal to the ranges of the covariance

The area covered by the grid is 70 by 90 km^2. The previous model was built with constant values over very large domain; it was not representative of the variability of facies and porosity that can occur in this domain.

A new facies model (using the method in Gross and Boucher (2015)) was built. Three facies are modeled (shale, shaly sand, and massive sand); the target proportions over the entire grid are 20 % for shales, 45 % for shaly sands, and 35 % for massive sands; the covariance function used for the underlying Gaussian field in the truncated Gaussian simulation is a spherical model with areal ranges of 800 m by 250 m (azimuth of maximum range is 55°); the vertical range is 100 m. The coordinate system is a UVT coordinate system, built from the relevant horizons.

The facies simulation results are illustrated in Fig. 3. The simulated proportions over the grid are clearly illustrative of the support size effect (less variations in large cells, larger variations in small cells).

The important issue of modeling facies and related proportions is illustrated in Fig. 4. The dominant facies (most likely facies regarding proportions) is represented. In the area modeled with large cells, the shale facies is never dominant, and proportional modeling procedure is the only way to keep that facies into account; in the area of local grid refinement, this aspect is less important, and the traditional truncated Gaussian simulation picture is observed.

The porosity model has been built using DGM assumptions. As each facies can occur in each cell, it is therefore important to perform full field porosity model for each facies. Point-scale distributions for each facies are provided in Table 1 (we used beta distributions with p and q referring to the shape parameters); these porosity distributions are clearly different according to facies classification. Point-scale normal score variogram for porosity is a spherical model with areal ranges of 600 m by 200 m (azimuth of maximum range axis is 55°); the vertical range is 80 m.

The porosity model inside each facies is represented in Fig. 5. The support size effect is clearly visible with small variations in large cells and larger variations in smaller cells.

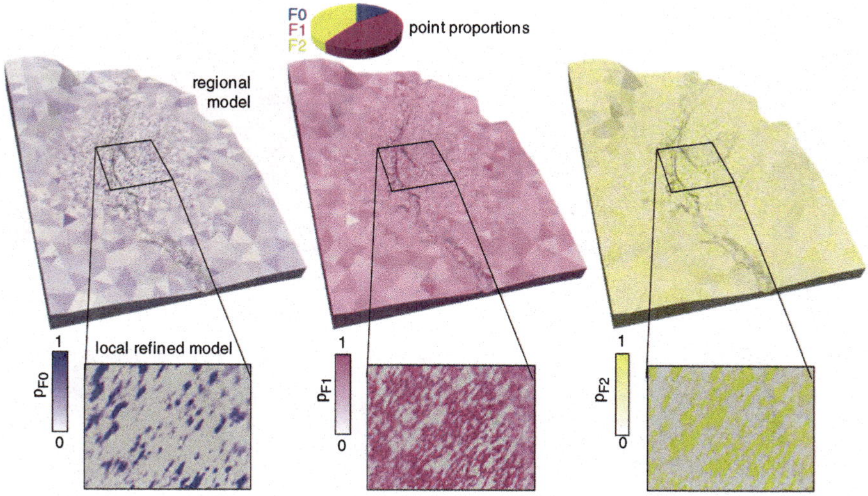

Fig. 3 Facies modeling on field X, proportions maps for silt (*blue*), silty sand (*pink*), and massive sands (*yellow*). The simulated facies proportion map is sharp in the region of the local grid refinement and becomes smoother in the regions of coarse blocks

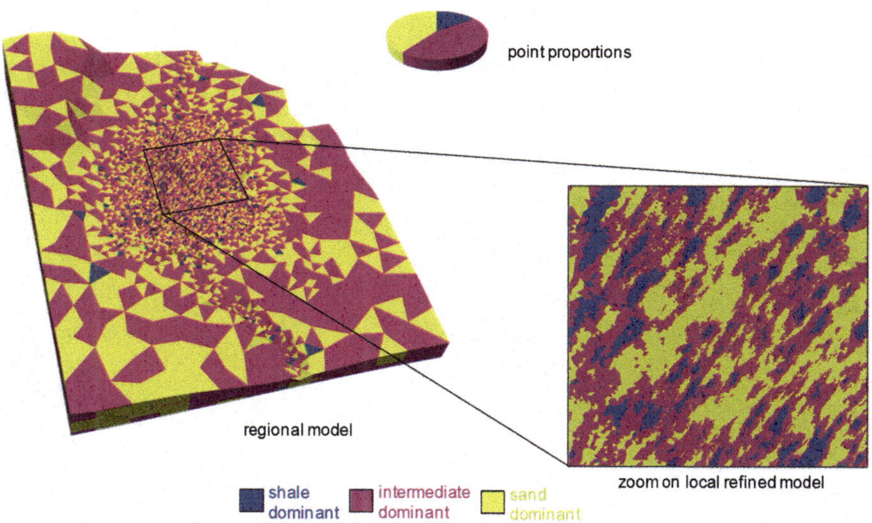

Fig. 4 Facies modeling on field X, dominant facies map

Table 1 Point-scale porosity distributions for field X

Facies	Distribution
Shales	Beta (min = 0, max = 0.4, p = 1, q = 6)
Shaly sands	Beta (min = 0, max = 0.4, p = 3, q = 3)
Sands	Beta (min = 0, max = 0.4, p = 8.5, q = 3.5)

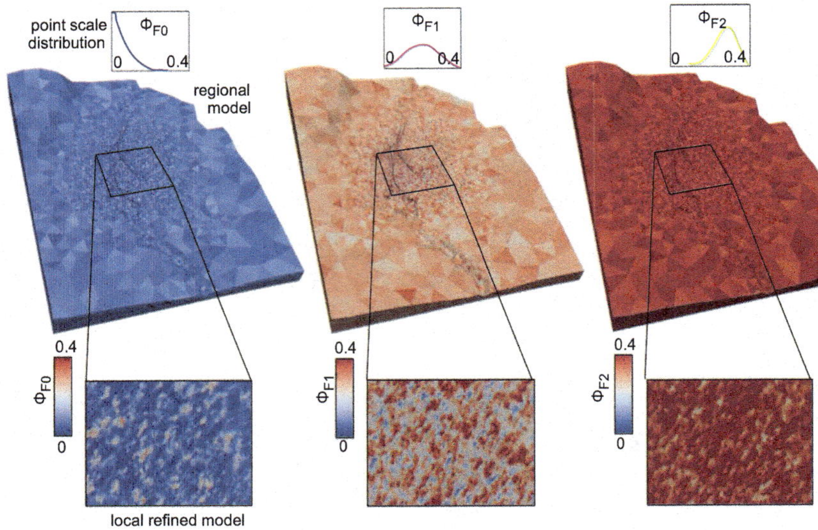

Fig. 5 Full field porosity modeling inside each facies of reservoir X

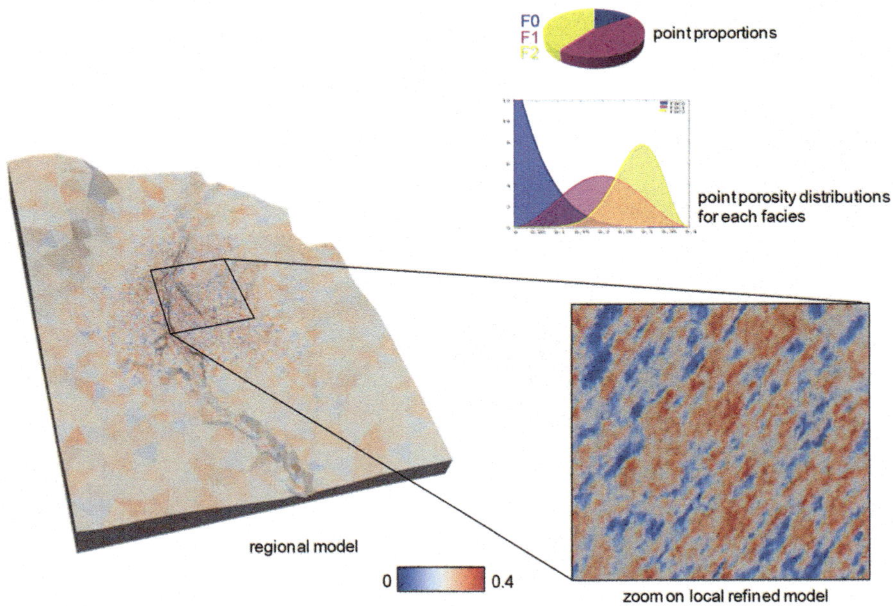

Fig. 6 Full field equivalent porosity inside reservoir X

From these facies porosities, it is possible to derive equivalent porosity for each cell simply by weighting them according to simulated proportions. This result is represented in Fig. 6; the support size effect is still visible.

6 Conclusion and Future Work

In this paper, we have presented a methodology to address geostatistical simulation on unstructured grids. Implementation issues have been discussed, and an illustrative field case example has been performed. We have shown that the world of geostatistics and the world of complex fit-for-purpose gridding can be reconciled.

The proposed workflow can be adapted to co-simulation techniques without major difficulties. A co-simulation approach with DGM on regular grids can be generalized without major modifications for unstructured grids (Emery and Ortiz 2011). In order to use the DGM for co-simulation, a linear model of co-regionalization (LMC) can be used, and conditioning kriging should be substituted with conditioning co-kriging.

However, it is important to notice than we have addressed only the domain of additive variables. An important topic is still to be treated: nonadditive variables such as permeabilities. We envisage treating these variables by using fit-for-purpose transformation like power transform as suggested in Noetinger (1996) and Deutsch (2002); but these transformations need to be calibrated by physical measures and numerical tests. This will be the next challenge of this research.

Other problems would be interesting to investigate: generalization of algorithms to nonstationary cases, addressing facies simulation techniques different from truncated Gaussian simulation.

Acknowledgments The authors would like to thank Total SA for sponsoring this research and authorizing publication.

Bibliography

Bentley JL (1975) Multidimensional binary search trees used for associative searching. Commun ACM 18:509–517

Chiles J-P, Delfiner P (2012) Geostatistics: modeling spatial uncertainty, 2nd edn. Wiley, New York

Cormen TH (2009) Introduction to algorithms. MIT Press

Deutsch CV, Tran TT, Pyrcz MJ (2002) Geostatistical assignment of reservoir properties on unstructured grids, SPE annual technical conference, SPE-77427-MS

Emery X (2009) Change-of-support models and computer programs for direct block-support simulation. Comput Geosci 35:2047–2056

Emery X, Ortiz JM (2011) Two approaches to direct block-support conditional co-simulation. C R Geosci 37:1015–1025

Gross H, Boucher AF (2015) Geostatistics on unstructured grid: coordinate system, connections and volumes, petroleum geostatistics 2015, EAGE

Korenblit M, Shmerling E (2006) Algorithm and software for integration over a convex polyhedron. In: Mathematical Software-ICMS 2006. Springer, pp 273–283

Mallet J-L (2004) Space time mathematical framework for sedimentary geology. Math Geol 36(1):1–32

Noetinger B, Hass A (1996) Permeability averaging for well tests in 3D stochastic reservoir models, SPE annual technical conference, SPE-366-53-MS

Sobol IM (1967) On the distribution of points in a cube and the approximate evaluation of integrals. Zhurnal Vychislitel'noi Matematiki i Matematicheskoi Fiziki 7:784–802

Zaytsev V, Biver P, Wackernagel H, Allard D (2016) Change-of-support models on irregular grids for geostatistical simulation. Math Geol 148(4):353–369

Using Spatial Constraints in Clustering for Electrofacies Calculation

Jean-Marc Chautru, Emilie Chautru, David Garner, R. Mohan Srivastava, and Jeffrey Yarus

Abstract Petroleum reservoir geological models are usually built in two steps. First, a 3-D model of geological bodies is computed, within which rock properties are expected to be stationary and to have low variability. Such geological domains are referred to as "facies" and are often "electrofacies" obtained by clustering petrophysical log curves and calibrating the results with core data. It can happen that log responses of different types of rock are too similar to enable satisfactory estimation of the facies. In such situations, taking into account the spatial aspect of the data might help the discriminative process. Since the clustering algorithms that are used in this context usually fail to do so, we propose a method to overcome such limitations. It consists in post-calibrating the estimated probabilities of the presence of each facies in the samples, using geological trends determined by experts. The final facies probability is estimated by a simple kriging of the initial ones. Measurement errors reflecting the confidence in the clustering algorithms are added to the model, and the target mean is taken as the aforementioned geological trend. Assets and liabilities of this approach are reviewed; in particular, theoretical and practical issues about stationarity, neighborhood choice, and possible generalizations are discussed. The estimation of the variance to be assigned to each data point is also analyzed. As the class probabilities sum up to one, the classes are not

J.-M. Chautru (✉)
Geovariances, 49bis avenue Franklin Roosevelt, 77215 Avon, France
e-mail: chautru@geovariances.com

E. Chautru
Mines ParisTech, Centre de Géosciences, 35 rue Saint Honoré, 77305 Fontainebleau, France
e-mail: emilie.chautru@mines-paristech.fr

D. Garner
TerraMod Consulting, 431 49 Avenue SW, Calgary, Alberta, Canada, T2S 1G3
e-mail: david@terra-mod.com

R.M. Srivastava
FSS Canada Consultants, 11th Floor – 120 Eglinton Ave East, Toronto, Ontario, Canada, M4P 1E2
e-mail: mosrivastava@fssconsultants.ca

J. Yarus
Halliburton, Landmark, 10200 Bellaire Blvd, Houston, TX 77072, USA
e-mail: jeffrey.yarus@halliburton.com

J.J. Gómez-Hernández et al. (eds.), *Geostatistics Valencia 2016*, Quantitative Geology and Geostatistics 19, DOI 10.1007/978-3-319-46819-8_31

independent; solutions are proposed in each context. This approach can be applied for extending class probabilities in 3-D.

1 Introduction

Petroleum reservoir geological models are usually built in two steps. First, a 3D model of geological bodies is computed, within which rock properties are expected to be stationary and have low variability. Such geological domains are referred to as "facies." In a second step, petrophysical properties are distributed within each facies.

A common issue in geological modeling is to determine such facies along each well. It is quite easy in cored wells, where it is possible to describe the rocks and make precise analyses and petrophysical measurements. Note that issues related to measurement errors, interpretation mistakes, or poorly representative sampling that will lead to an uncertainty on facies determination go beyond the scope of the present paper and are not discussed here. Defining facies is more difficult in uncored wells where only electric logs are available. The issue is usually solved by defining "electrofacies," which are classes created from clustering petrophysical log curves. Such electrofacies are then calibrated with core data to ensure their geological consistency. Electrofacies calculation results are not always of satisfactory and constant quality, which may lead to inconsistencies.

Facies are never randomly distributed in a reservoir, and accounting for the sedimentological conceptual model at the electrofacies calculation phase may improve classification performance, leading to more realistic results. Such a sedimentological model is commonly used in geological modeling for guiding facies spatial distribution simulations, through its numerical representation which is the facies proportions model calculated from local vertical proportion curves (VPC), the facies trends.

The goal of this paper is to detail several methods that take advantage of the facies proportions model at the early stage of electrofacies calculation. It is assumed here that classification is made in a supervised mode, to account for visual facies defined by the geologist; classes are defined by more than mathematical criteria. In these methods, facies proportions coming from geological analyses will be considered as the expected average probability to belong to a given facies in the neighborhood of each sample. Conditions of use of such methods are discussed, and some illustration examples are proposed.

2 The Classification Issue

Electrofacies defined during the rock-typing phase of geological modeling are always calculated using statistical clustering techniques. Very often, probabilistic classifiers are used, the output of which is not a simple set of classes determined from electric logs at each sample but a set of probabilities to belong to the different classes. Examples of such classifiers include linear discriminant analysis, artificial neural networks (with appropriate characteristics), logistic regression, Bayesian classifiers, etc. They are implemented in professional software dedicated to petrophysics and rock typing. In this context, the predicted class at a given sample (point in a well with several log measurements) is generally the one with the highest probability.

Several issues may occur, which are always due to a lack of relevant information.

- First, clustering methods do not account for the location of data samples but define rock classes from the log response only. In some cases, in particular in carbonate rocks, different facies can have similar log responses. For example, it is common in carbonates to get facies with similar porosity but different pore structure, therefore with different permeability. With the most common logs such as NPHI, RHOB, and DT, which are mainly sensitive to porosity and density or GR which is sensitive to shale content, such facies may not be properly discriminated. This will generate a lot of trouble to the reservoir engineer during the history match phase in flow simulations.
- Another common issue is due to the fact that electrofacies cannot be calculated with the same set of logs in all the wells. It is very common to have to deal with the absence of modern and accurate logs in some wells, sometimes many wells, which leads to different electrofacies calculation results in different wells. Some classes that are distinguished with the full set of logs may be merged in a single class with fewer logs.
- Varying electrofacies calculation quality is also observed when recent and old logs of same nature but different accuracy are merged. Electrofacies will be less reliable in old wells than in the recent ones.

In all the aforementioned situations, taking into account spatial information may help.

For instance, in the first case, facies with contrasted permeability are rarely homogeneously distributed in space. Very often, a given facies is dominant at a given location in the reservoir, the other ones being dominant in other areas. Therefore, a sedimentological conceptual model expressed as a 3-D model of facies proportions should provide valuable and relevant information to discriminate facies of different permeability and capillary pressure curves with similar log responses. Depending on the location in space, the same log response will be assigned to a given facies or to another one. The value at each point of the a priori 3-D model of proportions is considered here as the average probability of the presence of each facies or as the average probability to belong to a facies within a given neighborhood around the point. The neighborhood size corresponds to the size of 3-D modeling grid cells.

Such a numerical sedimentological model can be prepared by the geologist very early in the geological modeling process. It requires knowing the main facies, defined from core description, and the main sedimentological trends driving their spatial distribution, which result from the geological interpretation of the depositional environments distribution observed along the cores. The conceptual model has to be transformed into numerical data by defining local synthetic vertical proportion curves, which will be added to experimental vertical proportion curves calculated from cored wells to build a 3-D numerical model by interpolation.

For the second and third cases, the problem is not related to classes with similar log responses but to a spatial variation of electrofacies accuracy due to the heterogeneity of the logs dataset. It is a different issue, for which the potential benefit of using a 3-D model of facies proportions is not as clear. Taking into account neighbor information of high accuracy is more important, if we can assume that the spatial continuity of facies is high.

All the methods that are introduced to add spatial information to the classification process rely on the same basic principle. We propose to alter the probabilities of the presence of each facies computed from the electric logs, using in particular the geological trends provided by experts. The new predicted classes are then defined as those with highest updated probability.

3 Classification Using the Target Sample Only

Spatial information can be taken into account at each target sample without considering spatial correlation with neighboring data points, using different approaches. This particular case is relevant for improving classification results when some classes cannot be easily discriminated.

3.1 Standard Classification with Additional Variables

A first method consists in adding all the a priori probabilities to belong to a class as additional variables for a standard classification, using usual classification techniques. Calculations are then made with several logs plus as many a priori average probabilities as facies. As it is common to work with three to five logs and four to six facies, classifications with about ten variables can be expected.

It is a very simple approach, but it has some practical limitations. In particular, classifying with about ten variables will require a lot of data to get robust results. In linear discriminant analysis, for example, a multivariate density function has to be estimated from available samples. With ten variables, it requires a lot of samples to get a significant and robust estimate of the density function. In the oil and gas industry, the required number of samples will be rarely reached, especially in geological modeling for rock-typing purposes.

In addition, logs and a priori probabilities are not at the same scale and giving the same status to the two types of information in the classification process is a risky option. It may add some noise or imprecision. This approach is certainly not the most appropriate for rock typing.

3.2 *Combining Classification Results and A Priori Global Information*

3.2.1 Principle

Accounting for spatial information at a target sample location can be done in a very simple way, by calculating a weighted average of local probability to belong to a facies (from classification based on electric logs) and global probability (from a priori model built by the geologist). Consider n samples ($n > 1$), K facies ($K > 1$) and denote by $P_i = (P_{i,1},\ldots,P_{i,K})$ (resp. $Q_i = (Q_{i,1},\ldots,Q_{i,K})$) the set of probabilities calculated from the electric logs (resp. from a geological model) at sample i. The updated probability that sample i belongs to facies k is written as

$$P^*_{i,k} = \beta P_{i,k} + (1 - \beta)Q_{i,k}, \tag{1}$$

where $\beta \in [0,1]$ represents the level of confidence given to the mathematical classification relative to the geological model. Observe that taking the same β for each facies $k = 1,\ldots, K$ guarantees that the updated probabilities at a single point i are all in [0,1] and sum up to 1.

After calculating all $P^*_{i,1},\ldots,P^*_{i,K}$, one may predict the facies at sample i by choosing that with highest updated probability. The main issue here is the determination of β. For this, it is interesting to consider this weighted average as the result of a simple kriging. For a stationary process, Z observed at locations $j = 1,\ldots,$ n with known local expectation m, the simple kriging estimator Z^* of Z at a new point is the linear combination of the data that gives minimum squared error (with respect to a spatial model) and ensures no bias:

$$Z^* = \sum_j \lambda^j Z_j + \left(1 - \sum_j \lambda^j\right) m. \tag{2}$$

In the presence of measurement errors, all Z_j in Formula 2 can be replaced by $\left(Z_j + \epsilon_j\right)$, where ϵ_j is a white noise independent from Z_j with variance σ^2. In the present context, Z^* would play the role of the updated probability $P_{i,k}{}^*$ at sample i for some facies k, all $\left(Z_j + \epsilon_j\right)$ in the sum would be neighboring probabilities $P_{j,k}$ and $m = Q_{i,k}$ the local mean. In particular, if the kriging neighborhood is restricted to the target sample itself, then the simple kriging system can be developed very easily, resulting in:

$$P^*{}_{i,k} = \frac{1}{1+V_i} P_{i,k} + \frac{V_i}{1+V_i} Q_{i,k}, \qquad (3)$$

where V_i is the ratio between σ^2 and the modeled variance of the proportion $P_{i,k}$ without measurement errors. This is equivalent to Eq. 1 with $\beta = \frac{1}{1+V_i}$. Therefore, β can be viewed as a level of error in the initial classification.

More spatial information can be added if neighbor probabilities are also included in the kriging formula. However, the kriging system would need to be altered to ensure that the updated probabilities $P^*{}_{i,1},\ldots,P^*{}_{i,K}$ are in [0,1] and sum up to 1. Such extensions are discussed in the next section "Accounting for neighbor wells."

3.2.2 Application Examples

The following application examples are based on the dataset used in Garner et al. (2015).

It can be observed in Fig. 1 that the final facies is intermediate between the initial electrofacies and the facies corresponding to the highest a priori probability, as expected. The lower the uncertainty assigned to the electrofacies, the more similar to the electrofacies is the final result. When the variance characterizing this uncertainty increases, the final result becomes closer to the a priori geological trend.

It can be noted in Figs. 1 and 2 that the geological trend is not identical to the original reference facies. In particular, some shoulder effects exist at each facies border, which are due to the fact that both data are not at the same scale. The geological trend is less precise than the facies defined from core description. In some cases, like at the right side of Fig. 2, some contradictions between the original reference facies and the geological trend can be observed. Such discrepancies may occur, due to local heterogeneities or mistakes in the estimation of the sedimentological trends. In that case, combining electrofacies and geological trend cannot restore the original facies.

Figures 3 and 4 show a typical practical issue in electrofacies determination. One of the clusters in the multivariate distribution based on NPHI, RHOB, DT, and GR logs is not very efficient for discriminating the existing facies in a supervised classification, as shown in Fig. 3. The points in the cluster correspond to all of the four facies, which may lead to classification mistakes. Such an effect is shown in Fig. 4 in one well (W13) which is fully included in the selected cluster (see purple points in the figure). The initial electrofacies calculation misses the facies 2, which is actually dominant, and includes occurrences of facies 3 which is actually not present in this well. This misclassification can be partially fixed by accounting for spatial trend.

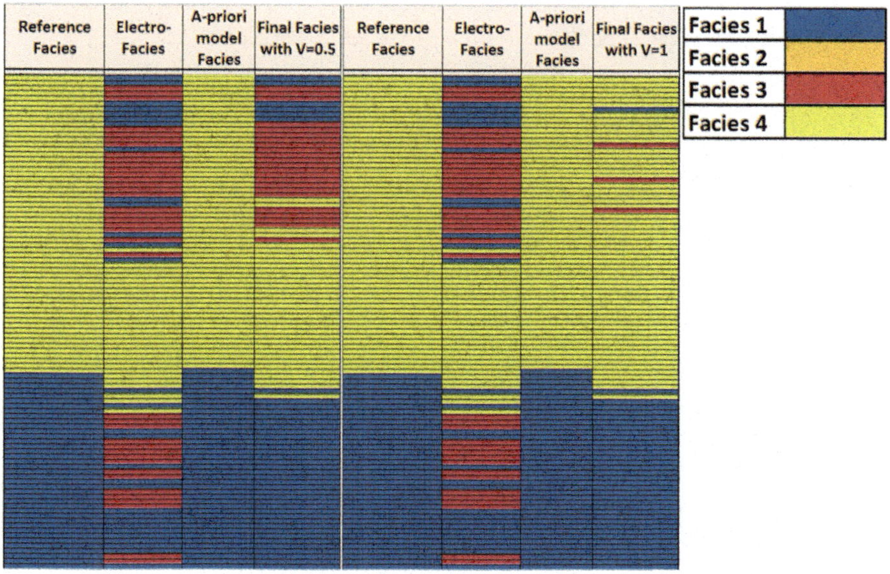

Fig. 1 Sensitivity of updated facies to "measurement error" variance values. From *left* to *right*, in both tables: the reference facies, the electrofacies, the geological trend, and the result of the approach described above

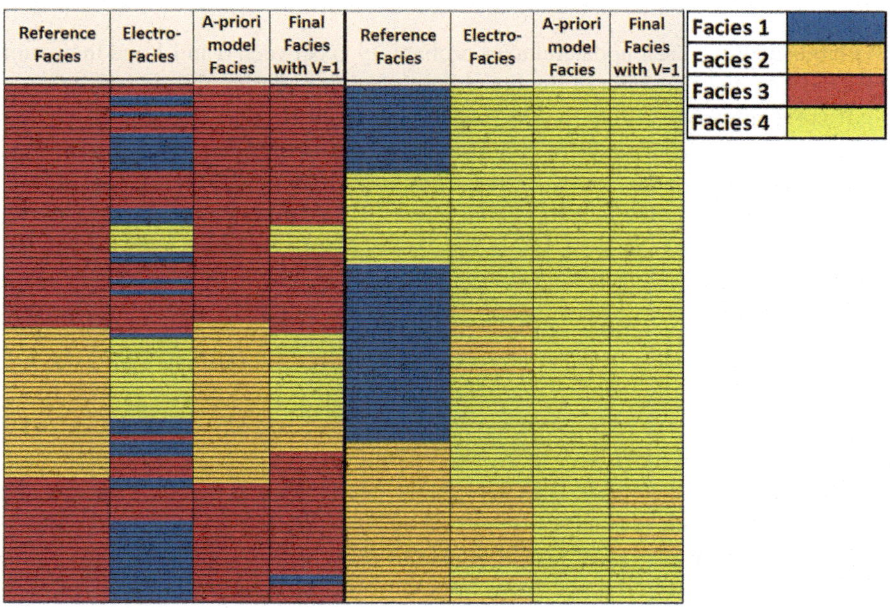

Fig. 2 Examples of updated facies in different wells. From *left* to *right*, in both tables: the reference facies, the electrofacies, the geological trend, and the result of the approach described above

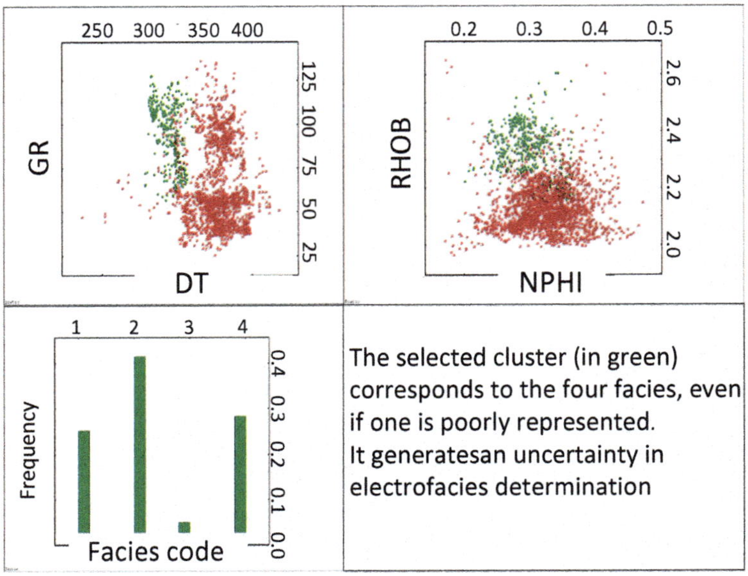

Fig. 3 Ambiguous cluster in logs leading to potential classification errors

3.2.3 Comments and Recommended Practices

With this method, each updated facies probability is between 0 and 1 and their sum is equal to 1, the two parts of the linear combination honoring this constraint.

It is recommended to apply this method on a well-by-well basis, with specific variance of "measurement error" for each well. The method is useful when several facies cannot be discriminated by the classification technique but can be distinguished according to the location in the reservoir. The main practical issue will be the determination of the most appropriate value for the variance representing the relative level of confidence on the two sources of data. It is a user-defined value, but some simple rules can be considered:

- With a null variance of "measurement error," only the classification results are considered.
- With a variance of "measurement error" equal to 1, classification results and a priori geological trend have the same weight.
- It is interesting to compare the expected facies coming from the geological trend with the initial electrofacies determined by the classification technique. In case of strong discrepancies, it is highly recommended to revisit the geological trend, to be sure of its validity, and/or to check in which cluster is the well under study. If it is located in an ambiguous cluster, unable to discriminate the facies, then electrofacies are uncertain, and it is better to put more weight on the geological trend. By the way, it can be noted that comparing electrofacies and geological trends is a way to QC the data.

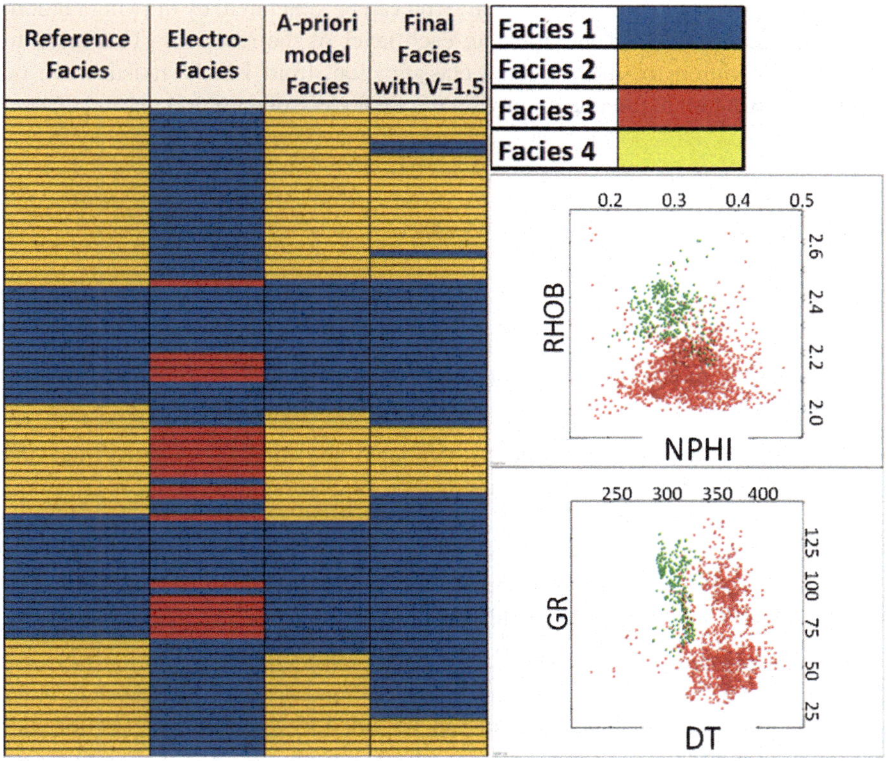

Fig. 4 An example of misclassification in a well included in the ambiguous cluster, with enhancement by accounting for spatial trends. From *left* to *right* in the table: the reference facies, the electrofacies, the geological trend, and the result of the approach described above

- In a cored well, it is easy to check the similarity of the final facies profile with the initial description. In uncored wells, this check is impossible, but it is possible to compute the proportions of each facies in the final result and compare it with the average facies proportions at the well location. It is a way to check the geological consistency of the result, assuming that the geological trend is reliable.

It is important to keep in mind that the geological trend which is used in this method results from the quantification of large-scale trends defined by the sedimentologist. Therefore, it is semi-qualitative information, which may be affected by a significant level of uncertainty. If it is defined from many cored wells, the accuracy of this trend is high, but if only few cored wells are available, the corresponding facies average probability of presence have to be used with care. Ultimately, the trend is a modeling choice, and its accuracy must be taken into account in the definition of the linear combination weights.

A method to account for geological trend uncertainty and to quantify the uncertainty associated with the final electrofacies consists in defining ranges of variation for the average probability of the presence of each facies. Several realizations of the quantified geological trend can be generated, by drawing random values in each interval of variation, in each layer of the reservoir. This leads to multiple realizations of the weighted classification, from which statistics can be calculated to estimate the uncertainty on the final electrofacies.

4 Accounting for Neighbor Wells

4.1 Principle

In many cases, electrofacies cannot be calculated in all the wells in the same conditions, due to data issues:

- The optimal set of logs may be available in some wells only, and the electrofacies in the other wells have to be calculated with fewer logs and/or with different logs;
- Due to strong differences in the date of logs acquisition, the accuracy of the data used for classification may vary a lot. Therefore, the quality of the classification results varies from one well to another.

The consequence of these data issues is that some classes which are properly discriminated in some wells are difficult or even impossible to distinguish in other wells. These issues are practical and can be aided by database preparation and thorough log normalization procedures used by petrophysicists.

A way to enhance electrofacies determination in a well where data quality is questionable is to account for the neighbor wells where the data are of high quality, as illustrated in Fig. 5. It assumes that the underlying geological phenomenon is quite continuous, at the well-spacing scale.

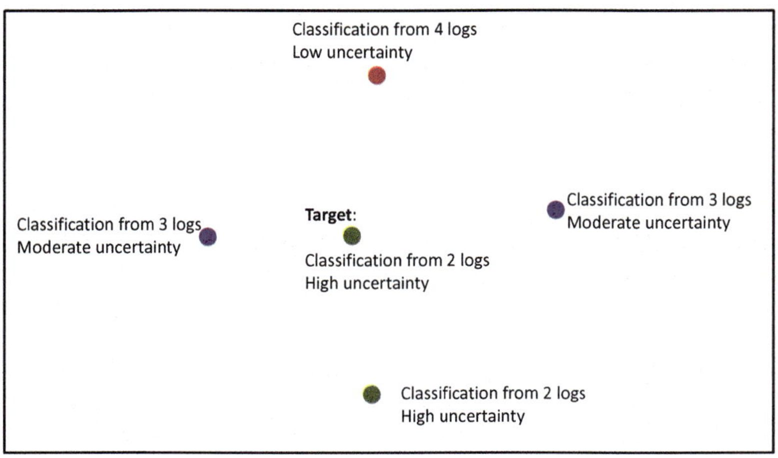

Fig. 5 Mixing wells with electrofacies of various accuracy

In practice, the method consists in estimating, at a target location, the probability to belong to a given class by kriging such probabilities from the neighborhood, using the "measurement error" option, the variance associated with each sample in the neighborhood depending on the sample accuracy. For example, classification results in the neighborhood will be considered as reliable (low or null variance) if four logs were used; it is less reliable (high variance) if only three logs were available.

It can be noted that the kriging estimation can be done:

- Only with the probabilities to belong to a class given by classification
- With the probabilities to belong to a class given by classification calculated in neighbor samples, combined with similar probabilities derived from a geological trend, at target location

The method detailed in the previous section is a limit case of the current approach, with a neighborhood reduced to the target point only.

4.2 *Constraints on Kriging*

In this method, the different variables (probabilities to belong to class) are linked together as they sum up to 1. As several samples are now included in the kriging neighborhood, this constraint is not automatically honored, and normalization procedures or specific kriging approaches have to be considered.

A first simple approach consists in performing successive independent kriging estimations for each variable, followed by a rescaling. On a numerical point of view, the constraint on the sum of probabilities will be honored. On a geological and mathematical point of view, considering as independent for calculations variables which are not independent is a strong approximation which may lead to significant inconsistencies.

An alternative approach could be to use kriging or cokriging of compositional data, where all the probabilities to belong to a facies are grouped in vectors. This approach requires the use of preliminary transformations before applying kriging or cokriging. Such transformations are generally based on logarithms, such as the additive log-ratio transformation of Aitchison (1986) detailed in Pawlowsky et al. (1994). It has been noted (Lan et al. 2006) that using log ratios in multiscale facies modeling leads to issues, mainly related to null probabilities. In our context, such occurrences are frequent, some facies having null proportions at different locations in the reservoir.

Compositional kriging as introduced by Walvoort and de Guijter (2001) can be considered instead. It extends ordinary kriging to the specific context of compositional data by simply adding the non-negativity and constant sum constraints to the minimization of the kriging mean squared error. Precisely, going back to Eq. 2, the compositional kriging weights are determined by solving the following optimization problem in vectorial form:

$$\min E\left(\|Z^* - Z\|_2^2\right) \quad \text{subject to} \quad E(Z^* - Z) = 0, Z^* \geq 0 \text{ and } \|Z^*\|_1 = 1$$

In the present context, Z^* stands for the full vector P^*_i at a sample of interest i, predicted by cokriging with neighboring vectors P_j. There, Z represents the ideal value P_i that would have been obtained if the initial classification procedure was perfect. In contrast to the additive log-ratio transformation techniques, no restriction on the values of the involved vectors is necessary; some components can be null without impeding estimation. Moreover, all components of the target vector are assessed simultaneously with this procedure.

Initially, the authors chose not to take cross correlations into account; in the present context, this means that the probability of the presence of a facies at a target point is not influenced by those of the other facies at other locations. Thus, each type of rock is compared to all others as a group instead of individually. When such an assumption is considered too restrictive, given the depositional patterns, it is possible to generalize the method to the context of cokriging. This would be the case, for instance, when two types of rock are commonly found next to each other; a sample consisting of the first would increase the probability of finding the second in a close neighborhood. Even if such types of dependence are often nonlinear, including extra facies in the cokriging formula can help capture at least part of the depositional dynamic. Like before, expert knowledge about geological trends can be included in the model through the vector of mean class probabilities at the target sample. Because the latter vary through space, so proceeding has the advantage of relaxing the stationarity hypothesis. The covariance matrix of the data can also be altered to reflect the level of confidence in the statistical clustering results; in practice, this amounts to supposing that the data contain measurement errors. The main drawback of picking cokriging over kriging is the amount of extra parameters to estimate: estimation efficiency decreases when the latter increase. It also requires stipulating a cross-correlation model, which can be difficult when samples are limited.

4.3 Application Examples

Following application examples are based on the dataset used in Garner et al. (2015).

Let us consider a well for which the initial classification delivers unsatisfactory results. The probability of each sample to belong to a given class will be estimated from the neighborhood, as shown in Fig. 6 (W7 estimated from W10, W14, W15).

A compositional cokriging was applied, using electrofacies probabilities as input data, with a priori geological trends serving as means. The level of confidence in the probabilities provided by statistical clustering techniques was included in the model by adding a variance term to the diagonal of the covariance matrix. In particular,

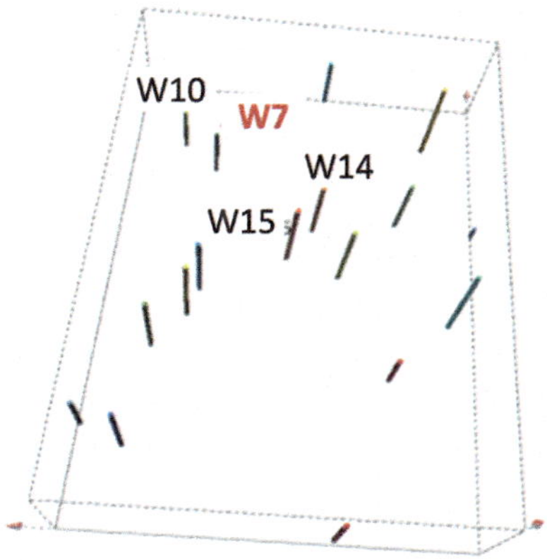

Fig. 6 Well estimated from its neighborhood

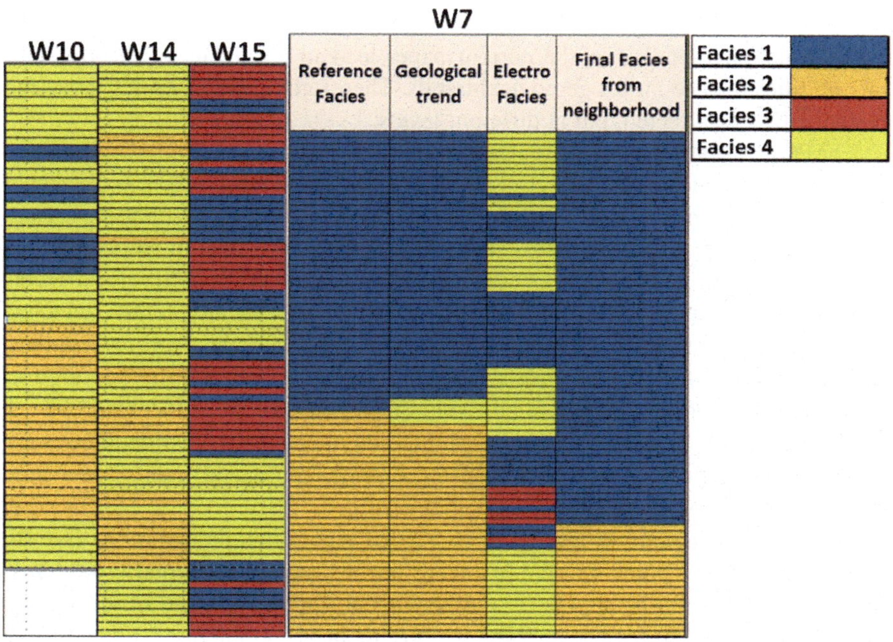

Fig. 7 Facies estimation in *W7* from electrofacies in *W10*, *W14*, and *W15*, accounting for geological trend in *W7*. Result is the last column on the *right*

entries in the covariance corresponding to well 7 were strongly penalized with a measurement error variance of 2 versus 0.95 for all other wells, which will tend to increase the impact of the geological trend. Results are summarized in Fig. 7. As expected, the identification of the types of rocks is improved as compared with electrofacies. Indeed, while only 21.43 % of electrofacies are accurate, 78.57 % of the types of rocks are correctly predicted after post-calibration. Nonetheless, let us mention that the technique is extremely sensitive to the choice of confidence parameters. They have to be picked with care, maybe using core data in a supervised approach.

5 Application to 3-D Facies Modeling

Several options for using electrofacies defined from the results of classification based on well logs are proposed in Garner et al. (2015) to build 3-D facies models:

- Plurigaussian simulations (PGS) using electrofacies as input data
- 3-D simulation of electrofacies probabilities, known at wells from classification, the final electrofacies at each point corresponding to the highest probability
- 3-D co-simulations of the well logs and application of the well-derived classification as a post-processing

The methods detailed in the current paper lead to an update of electrofacies probabilities to account for geological trends. Therefore, the three facies modeling methods will be affected, but at different levels.

For PGS, accounting for geological trends in electrofacies definition will lead to a modification of input data (electrofacies). The lithological column may change in some wells, which can lead to local adjustments of the vertical proportion curves and to variogram fitting updates. It must be noted here that in PGS, the 3-D proportion model used as soft input data is usually calculated from the electrofacies at wells. The a priori geological trend used to modify classification results, which is similar to the usual 3-D proportion model, is based on core data and on sedimentologist interpretation. Both models are actually very close, and the one used to update classification results could be used in PGS as an alternative choice.

The input data conditioning the 3-D simulation of electrofacies probabilities are modified and include a contribution of the a priori geological trend. The whole methodology shown in Fig. 8 can remain unchanged.

Details can be found in Garner et al. (2015) where this methodology is fully described. It can be noted that the first steps of the procedure (normal score transforms and MAF) can be applied to the geological trend itself. The results can be used as external drifts for constraining the turning bands simulations. It is a way to force this method to account for facies proportions in the whole 3-D model.

The first step of the third method, based on well logs co-simulations, will not be affected by the classification results update. This part of the procedure remains unchanged, but the next step, which consists in applying the classification technique to the 3-D model of logs, will be improved. Classification results can be updated, in the model, to account for the a priori geological trend. Here, the auxiliary spatial information is introduced at the last moment in the calculation.

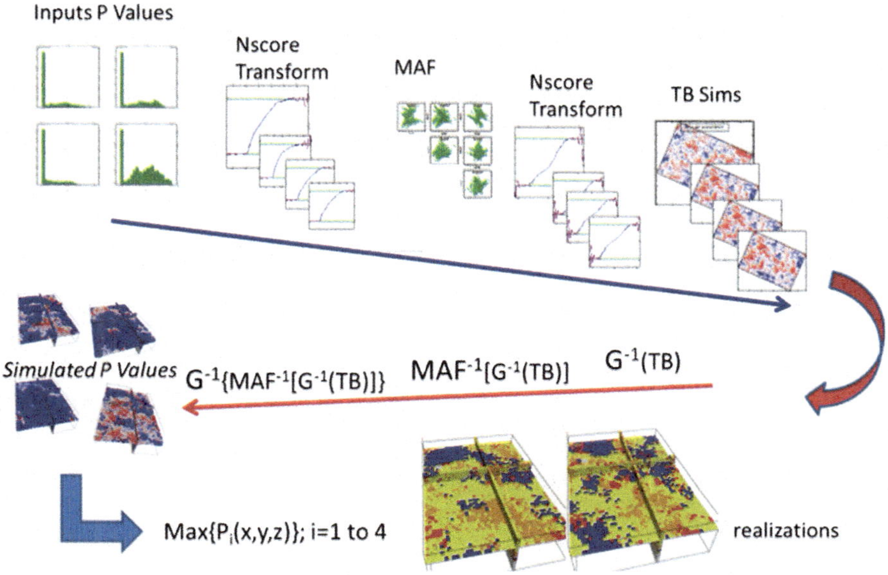

Fig. 8 3-D simulation of electrofacies probabilities

6 Conclusion

Using a priori geological trend, defined in terms of expected facies proportions which will be considered as the probability to belong to each facies, is an effective way for integrating spatial constraints in classification techniques. One may consider that it would be better to modify the classification techniques themselves, but this is a long and complex work, which has to be adjusted to each classification method. Updating classification results by means of a post-processing procedure is simpler, can be applied immediately to all the stochastic classification methods, and ensures a strong geological input.

The most common electrofacies calculation pitfalls related to available logs in the different wells, to the logs quality in the different wells, and to the logs relevance can be addressed. Depending on the technical issue and on the reservoir heterogeneity, various approaches can be used, accounting for more or less information in the target neighborhood, which gives enough flexibility to address most of the practical configurations encountered.

Bibliography

Aitchison D (1986) The statistical analysis of compositional data. Chapman & Hall, London

Garner DL, Yarus J, Srivastava M (2015) Modeling three ways from electro-facies – categorical, e-facies probabilities and petrophysics with assignment. Petroleum Geostatistics 2015, 7–11 September. Biarritz

Lan Z, Leuangthong O, Deutsch CV (2006) Why Logratios are a bad idea for multiscale facies modeling, Centre for computational geostatistics report 8. University of Alberta, Canada, p 211

Pawlowsky V, Olea RA, Davis JC (1994) Additive logratio estimation of regionalized compositional data: an application to the calculation of oil reserves. In: Dimitrakopoulos R (ed) Geostatistics for the next century. Kluwer, Dordrecht, pp 371–382

Walvoort D, de Guijter J (2001) Compositional kriging: a spatial interpolation method for compositional data. Math Geol 33(8):951–966

Pore Network Modeling from Multi-scale Imaging Using Multiple-Point Statistics

T. Chugunova and H. Berthet

Abstract Petrophysical characterization through pore network modeling has become a very active field in reservoir core analysis. The technique of X-ray microtomography allows an accurate imaging at the micrometer scale. But the engineer often faces a difficult choice – whether to resolve the heterogeneities directly influencing the petrophysical characteristics on the small volumes or to capture less resolved global tendencies on the statistically representative volumes. The idea of this work is to reconcile different scales by reconstructing the internal rock structures from the available scan images using multiple-point statistics method. This method was successfully used to reconstruct 3D pore network from 2D images. In this work we want to demonstrate on the real cases application how multiple-point statistics can be used to solve multi-scale and multi-support problem.

1 Introduction

Sophisticated imaging technologies can be used to access the tiniest internal structures of a reservoir rock sample. Modeling on the microscopic scale enables the computation of petrophysical properties such as porosity and permeability (Blunt et al. 2013) as well as enrich our understanding of multiphase flow in porous media.

The method to accurately describe a pore network at the micrometer scale is based on X-ray microtomography imaging ("3D scan"). This technique is becoming nowadays a common tool in the petroleum laboratories providing 3D high-resolution images of the pores and grains of almost any type of rock sample. The

T. Chugunova (✉)
Geostatistics and Uncertainty Service, Total SA, av. Larribau, CSTJF, 64000 Pau, France
e-mail: tatiana.chugunova@total.com

H. Berthet
Digital Rock Physics Laboratory, Total SA, av. Larribau, CSTJF, 64000 Pau, France
e-mail: helene.berthet@total.com

J.J. Gómez-Hernández et al. (eds.), *Geostatistics Valencia 2016*, Quantitative Geology and Geostatistics 19, DOI 10.1007/978-3-319-46819-8_32

trade-off between imaged volume and resolution brings however a challenge to capture heterogeneities at different scales: the smaller the imaged volume, the higher the resolution. Thus, the petrophysical characteristics (e.g., porosity, permeability) are computed with different accuracies for different scales. For example, statistically robust values are calculated from a large volume but are less resolved and have missing small structures. Otherwise, very small scales can be resolved and provide accurate porosity and permeability estimation but on a too small and hence statistically nonrepresentative volume. For instance, capillary pressure computation is very sensitive to the image resolution. When small connections in the flow pathway are badly resolved, computed capillary pressure curves diverge from experimental results (Leu et al. 2014).

Okabe and Blunt (2005) have used multiple-point statistics (MPS) method to reconstruct 3D pore network from 2D images. In their example, the internal structure was represented by a relatively simple morphology with the heterogeneities of the same scale and the same shape. A tree-based MPS algorithm (Strebelle 2002) has been used for this purpose. In this work we propose two ways to model the heterogeneities of different scales available in different supports of scan images applying the latest generation MPS algorithm. In the first application, we address the case where the high-resolution (HR) image contains the structures at several scales. In the second, the highly resolved heterogeneities from nonrepresentative sub-volumes are used to enrich a low-resolution (LR) image covering statistically representative volumes. The demonstration of the methodology and the corresponding illustrations in this paper are done for a 2D MPS model. The extension of two approaches for a 3D MPS model is straightforward.

The paper is organized as follows. We demonstrate the multi-scale MPS application on two case studies. In the first case study, we illustrate a preprocessing in order to convert a scan image to a valuable training image (TI) for MPS algorithm. Two scales of internal rock structure are present in the TI. We illustrate an example of MPS simulation using the latest generation algorithm known in the literature (Mariethoz et al. 2010) as multiple-point direct sampling (DS). The results of this simulation are captivating. Without prior knowledge it is difficult to distinguish the TI from the DS realizations. For the second case study, we introduce the problem of downscaling having the scan images of a rock sample for different scales and volumes. A low resolved image can be enriched with heterogeneities observed in a high resolved image. For this purpose, the DS simulation is applied in its nonstationary mode (Mariethoz et al. 2009) using a low-resolution image as an auxiliary variable and the fine scale highly resolved image as a main variable. The advantages and the difficulties of the proposed approaches applied to real cases are discussed.

2 Case 1

A 4-mm cylindrical sample of Castle Gate rock (outcrop sandstone, Utah, USA) was cored and placed on the sample holder of an X-ray microtomograph (Versa520XRM, Zeiss). A first image was obtained at a voxel resolution of 1.5 μm (field of view ~33 mm3). In order to access a higher level of details, we carried out a series of imaging at a voxel resolution of 503 nm (Fig. 1), now reconstructing a sub-volume of the initial image (so-called interior tomography or region of interest). Two images were acquired at two successive vertical positions and stitched together to study a larger imaged volume (~0.13 mm^3).

2.1 Training Image Construction

A successful application of MPS strongly depends on the quality of the TI standing for the heterogeneity description. Hence, a TI should be statistically representative (Emery and Lantuejoul 2013). In terms of the MPS method, it mainly means that the image should contain numerous representative patterns of the size much smaller than the TI size. The image should also be of low entropy (each pattern should have a lot of similar ones under the definition of similarity).

Nevertheless, in practice, the TI rarely satisfies all these conditions. Different MPS options were developed in the last decade to deal with the nonstationary, non-ergodic, or incomplete MPS model. For more details on the different options available for MPS algorithm, we refer the reader to Hu and Chugunova (2008) and Mariethoz and Lefebvre (2014). For example, the options of multiple grids (in tree-based MPS algorithm) or flexible template (in DS algorithm) allow better reproduction of large-scale heterogeneities but do not guarantee any more an ergodicity of the process.

The first application case we demonstrate here is a typical example of the difficulties we may face. One may observe that a raw image of a core sample (Fig. 2) is far from the required TI qualities. To transform this raw image into a more valuable TI, we proceed as follows. The raw image is opened in the common image editor (e.g., GIMP). The image is represented by a continuous property

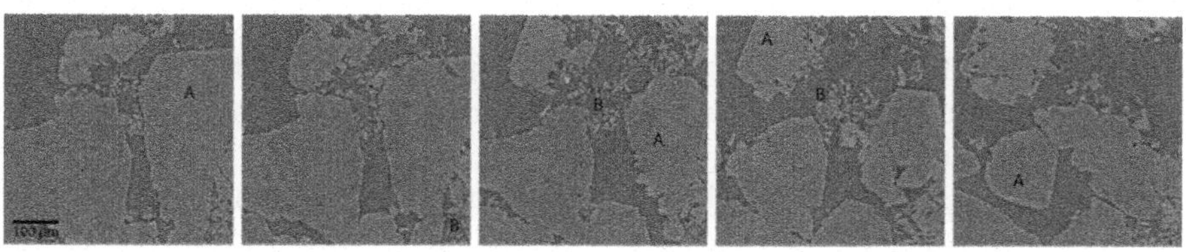

Fig. 1 2D images extracted from the 3D-reconstructed image at 503 nm voxel resolution. Large grains (**a**) and micro-porosity (**b**) are present in the structure. The edges of large grains display a phase contrast artifact due to the very high resolution of the image

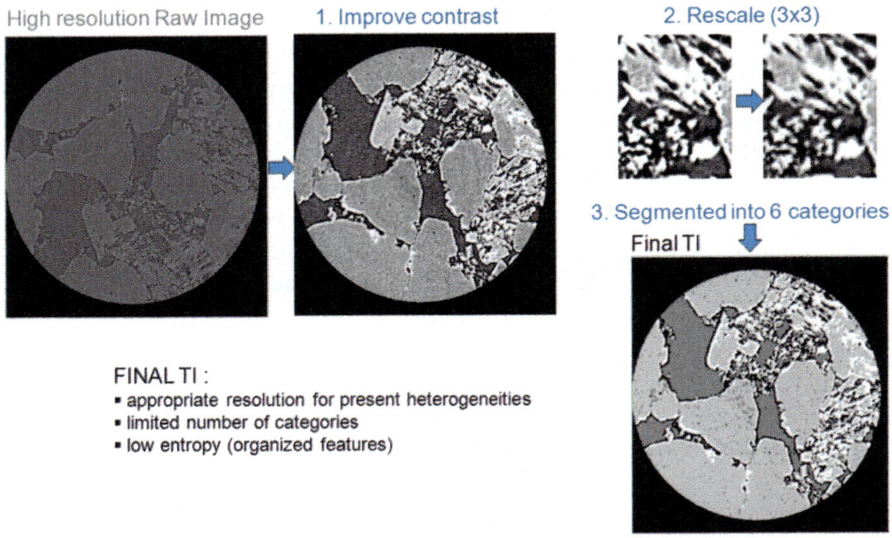

Fig. 2 Different steps of preprocessing of initial microtomography image in order to obtain a representative training image

between 0 and 1. First of all we improve the contrast of the image to better distinguish the grains and the pores. Now we clearly see that the image contains two main spatial structures corresponding to large grains and small particles. Second, we found that even the smallest particles are represented by scores of pixels in the initial image. The efficiency of MPS algorithm is not optimal when that most of the neighbors in the template are wasted to capture same values and not spatial heterogeneity. The image is rescaled (3×3 average mean upscaling) in order to optimize the representation of the given heterogeneities. Global statistics analysis confirms an absence of bias neither the loss of information after upscaling.

2.2 *DS Results with Continuous Training Image*

At this stage the preprocessed image can be used as a TI by applying DS algorithm dealing with continuous variables. The image is used as a 2D grid where each pixel is a grid cell. The areal zone around the sample is not considered during the simulation in order to exclude inappropriate patterns. Figure 3 illustrates the results of such application.

We may observe a good quality of image reproduction. We also observe that locally some simulated structures are very similar to some parts of the TI but disposed and concatenated randomly in the DS realizations. Having fixed the DS parameters such as maximal number of neighbors, threshold of similarity, and percentage of TI scan (here we refer the reader to Mariethoz et al. 2010 to get

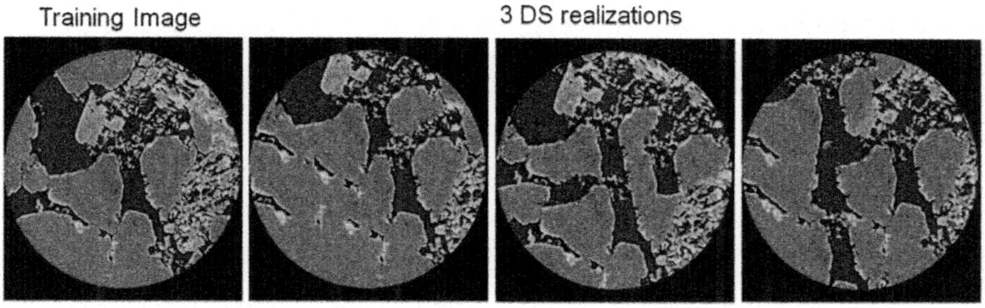

Fig. 3 Training image of continuous variable and three *DS* realizations

familiar with common DS parameterization), the DS algorithm will reproduce spatial structures up to a specific scale with low variability comparing to the TI, and the maximal variability of structures will mainly appear on the larger scale. Thus, a practitioner trades off between a copy-paste effect and maximum spatial variability. In our case the low variability scales are the scales from the smallest particle up to the large grains present in the TI. Since the large grains have a size comparable to the size of a TI, it means also that our DS model (TI + DS algorithm + DS parameters) is non-ergodic. This is also reflected by the results (Fig. 3) where the amount of small and large grains is different from one realization to another. To minimize the non-ergodicity of our DS model, in our practical application, we use several TIs coming from different parts of rock sample for 2D simulation or 3D TI (a sequence of 2D TIs) for 3D simulation.

2.3 DS Results with Boolean Training Image

One of the main steps of image processing on scan data is segmentation (binarization), leading to an estimate of porosity and local characteristics of the pore network. A threshold choice is made by a petrophysician to discriminate pore and no pore volumes. Once an appropriate level of threshold is chosen, the different levels of color of initial image are not important. One may propose to transform a TI into a binary property and to obtain the MPS results directly in the binary property ready for estimations. Here below (Fig. 4) we illustrate the results of such an experiment.

The disappointing quality of the results can be explained by comparing the spatial organization of the continuous and Boolean TIs. Indeed, the continuous TI contains the patterns where the points of low and high values are spatially organized trough the transition (from black to white color); this transition is rapid for the small particles and slow for the large grains. Thus, all patterns of the TI can be mainly separated in two groups, one for large and another for small structures. Using a binary representation of the small and large grains, there is no more transition zone of intermediate values and no more evident criteria to distinguish the MPS patterns

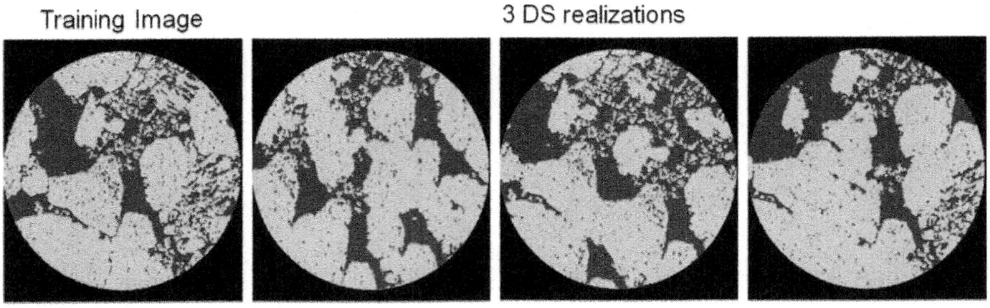

Fig. 4 Training Image of Boolean variable and three *DS* realizations

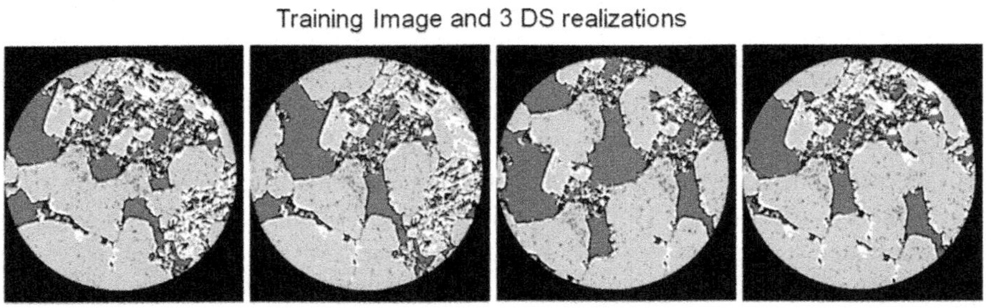

Fig. 5 Training image of categorical variable (six classes) and three *DS* realizations. TI is a second image from the *left*

from large or small structures. Practically, it means that the same combination of neighbors corresponding to a current simulation step may be compatible with both small and large structures. Consequently, the motifs started as large structures can be continued with a pattern corresponding to the small structures and inversely, which in its turn explains the quality of the results.

2.4 DS Results with Segmented Training Image

Finally, by segmenting a continuous TI to several classes (into categorical variable), we obtain the most convincing results (Fig. 5). Indeed, by segmenting the initial continuous variable in six classes, we keep different transitions for large and small particles, and so we reinforce the similarity between patterns characterizing the same structures and increase the difference between the patterns describing different scales. The threshold defining the limits of pore and no pore volumes should remain the same as chosen by a petrophysician in its pore volume calculations. The choice of intermediate thresholds' values is not important as far as it allows representing a transition effect of initial image.

As for previous tests, due to the non-ergodicity of our DS model, an amount of small and large particles varies from one realization to another so as a global porosity. It is recommended to use multiple TIs (coming from several HR images) in order to guaranty the statistically representative results.

3 Case 2

A 11-mm diameter cylindrical rock sample was cored from a reservoir sandstone petrophysics plug. The rock was chosen for its large porosity (25.4 %) and permeability (300 mD) and the presence of very small pore structures (mainly clay). A first large 3D image was obtained by successively imaging the sample at seven different vertical positions and by stitching all the images together. The result field of view is around 1.6 cm^3 at a voxel resolution of 3.1 μm (later referred to as the low resolution). A region of the rock containing large and small structures was located in the reconstructed volume and subsequently imaged at a voxel resolution of 905 nm (field of view ~5.9 mm^3, later referred to as the high-resolution image).

3.1 Downscale Problem Description

Our second case study is illustrated Fig. 6. A large low-resolution (LR) image exists for a statistically representative sample. A high-resolution (HR) image is available only for a small part of the low resolved one. This HR part contains several types of heterogeneities (grains, pores, different forms of clay). Calculated only on this small HR part, the quantity of each type of heterogeneities is not reliable, but the shapes are correctly represented. The idea is then to extrapolate different types of the heterogeneities seen on the HR image to the initial large sample. From the geostatistical point of view, it remains to apply a downscaling to the LR image.

To do so, we use a DS technique constrained by an auxiliary variable. The idea of using an auxiliary variable to characterize a TI and to constrain an MPS simulation was first proposed by Chugunova and Hu (2008) in its collocated co-simulation form in order to deal with nonstationary TI with a tree-based MPS algorithm. Then, the use of auxiliary variable was generalized to a DS algorithm with non-collocated co-simulation (Mariethoz 2009). Recently, the auxiliary variable approach was used to capture multiple-point statistics of geological heterogeneities on several scales in the reservoir modeling context (Doliges et al. 2014). In this work, we propose to use auxiliary variable technique in order to link statistically two different scales represented by HR and LR images.

The principle of this technique can be described as follows. Let's take HR TI image for a small sample and its equivalent in LR (TI and Auxiliary TI in Fig. 7). By extracting the multiple-point statistics simultaneously from both images, we may statistically estimate a relation between the patterns in the HR and the corresponding values of the LR. Then, for the simulation step, we impose the

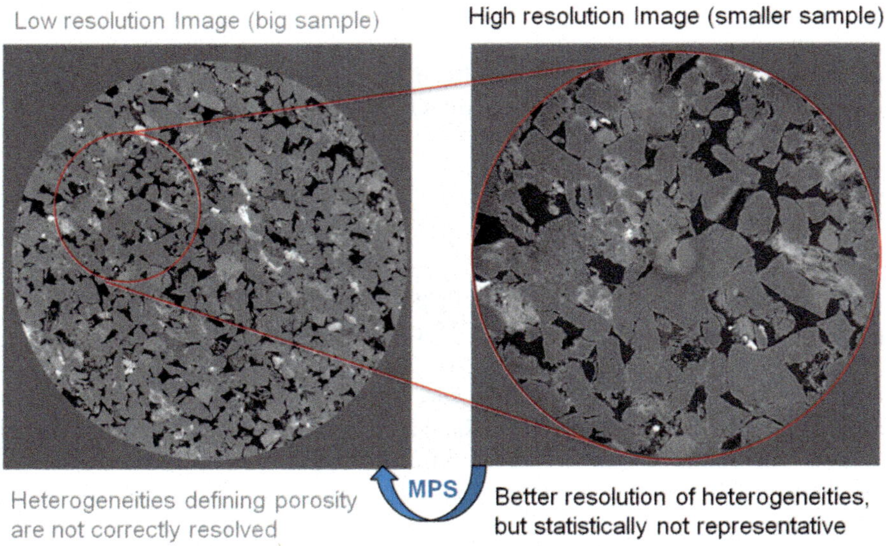

Fig. 6 LR image capturing large sample (*on the left*) and *HR* image capturing small sample (*on the right*)

available LR image of the entire sample as a constraint. For a current point to simulate, the DS algorithm will capture the neighborhood values in the simulation grid corresponding to the HR but also the neighborhood values of the LR constraint. Then, the algorithm will scan simultaneously the vector training image (HR and LR) by looking for the similar (vector) pattern. In such a way, the reproduced heterogeneities will have the structure of HR image and will be conditioned by corresponding LR values.

3.2 DS Results and Discussion

In this downscaling test, we transform the HR raw image into the TI in the similar way as in the Case 1: by improving contrasts and segmenting into several facies. Note that for the Case 2 in HR image, the smallest heterogeneities are resolved by only several pixels. The upscaling process will induce the loss of information and will reduce the accuracy of the petrophysical properties calculation. Thus, the resolution of the HR image was considered as the optimal, and no upscaling was applied.

To illustrate the results, we show one part of the target simulation grid which has approximately the same size as a HR TI. Intentionally, this illustrated zone contains a part of HR TI (Fig. 7). Thereby, we could compare the DS simulation in two zones: one covering the LR patterns with their exact replicates in the TI and another where only similar patterns exist.

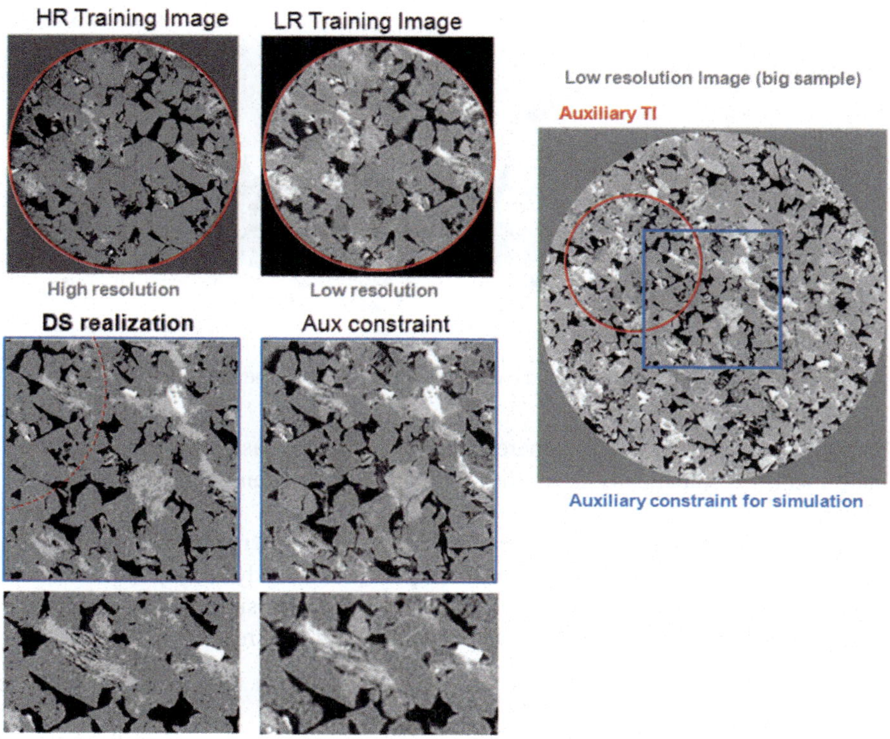

Fig. 7 Vector *TI* consists of categorical *TI* of five facies obtained by segmentation of *HR* image and auxiliary *TI* corresponding to the raw *LR* image. One *DS* realization is constrained with auxiliary constraint coming from *LR* image of large sample

The results of DS simulation are zoomed and illustrated in Fig. 8. We observe generally acceptable quality of pattern reproduction: the borders of large grains are sharp as it is a case in the TI; an evident filamentary clay feature (positioned in the overlapping zone) is also downscaled with the appropriate details. The principle of downscaling with DS constrained by the low resolved image is demonstrated.

Nevertheless, we notice that some features expected to be downscaled with clay-like structure are simulated like grains (Fig. 8). The explanation can come from the TI analysis. Coming back to the TI prerequisites, we recall here that a good TI should have a lot of repetitive similar patterns representing a same spatial structure. Whereas our TI contains not one but several types of heterogeneities rather different one to each other: grains of different sizes with sharp edges, pores, small clay patches, small clay features with lacy motifs, and small elongated clay features with filamentary motifs. Quantifying the volumes occupied by different motifs, the probability to reproduce one of the clay patterns is much smaller than the probability to reproduce grain or pore pattern. One may argue that the auxiliary constraint (LR image) should guide the DS simulation in order to distinguish grains, pores, and clays. By detailed analysis, we notice that some LR values may correspond to

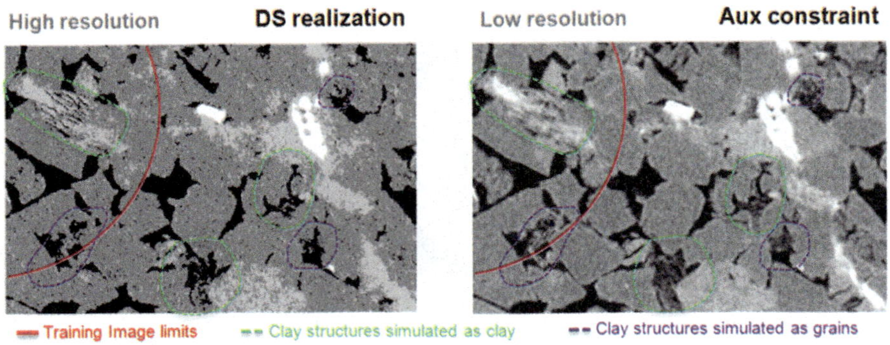

Fig. 8 Zoom on the *DS* realization and corresponding auxiliary constraint

both grain and clay structures, whereas the DS algorithm has no criteria to discriminate them. As statistically it is more probable to find grain pattern, that is what happens the most during the simulation process.

A way forward can be envisioned to improve clay heterogeneities simulation. Knowing that the regions inside grains and pores are already clearly indentified in the LR image, we may reduce our TI image as well as the simulation zone only to the heterogeneities of interest such as pore edges and clay regions.

4 Conclusions

In this work we propose two approaches to tackle the problem of multi-scale and multi-support pore network reconstruction with MPS. Two case studies applications illustrate these approaches.

In Case 1 we deal with relatively simple spatial structures but containing two scales of heterogeneities: large grains and small particles. By applying an image treatment (contrast improving, segmentation in several classes), we transform the raw X-ray image in the reliable TI. Applying DS technique with flexible template, we demonstrate a good-quality internal structure reproduction. The practitioner should nevertheless pay attention to the ergodicity aspects. Indeed, the sole TI of case 1 is non-ergodic, and it is recommended to use multiple TIs (or 3D TI) to cover a representative area of sample and to obtain statistically robust results.

In Case 2 we deal with two different supports (resolutions) but are also confronted with practical difficulties. The rock sample is characterized by two images with different resolutions and cover different volumes. The TI contains several very different spatial structures. The structures of interest are minatory comparing to the most represented patterns of pores and grains. To reconcile two different resolutions, we propose to apply the DS co-simulation with auxiliary constraint where HR image will stand for a TI and the LR image – for a constraint. We illustrate the results of DS application using the entire TI. The unresolved features like vague edges or filamentary clays are correctly resolved by this approach. But some of features instead of being simulated as clay heterogeneities are simulated as grains. This phenomenon is explained by the most frequent patterns in the TI which are the grains. As a way forward to solve these practical difficulties, we propose to reduce the TI scan area and the simulation area to the uncertain zone where the downscaling is necessary and so increase the frequency of rare clay patterns in the TI.

Having learned from these two cases, the following can be drawn. In case of heterogeneities on several scales and a spatial organization between scales (Case 1: small particles appear where the boards of large grains become broken), we recommend the first approach. In case where the spatial structure dramatically changes while changing a resolution of scan and a precision is needed for more accurate calculation (Case 2), we recommend the second approach.

Bibliography

Blunt MJ, Bijeljic B, Dong H, Gharbi O, Iglauer S, Mostaghimi P, Paluszny A, Pentland C (2013) Pore-scale imaging and modeling. Adv Water Res 51:197–216. doi:10.1016/j.advwatres.2012.03.003

Chugunova T, Hu L (2008) Multiple-point simulations constrained by continuous auxiliary data. Math Geosci 40(2):133–146

Doligez B, Ravalec M, Bouquet S, Adelinet M (2014) Multiscale reservoir geological modeling and advanced geostatistics. In: Proc. Gussow conference "Closing the Gap II", Sept. 22–24 Banff, Canada

Emery X, Lantuejoul C (2013) Can a training image be a substitute for a random field model. Math Geosci SI. doi:10.1007/s11004-013-9492-z

Hu L, Chugunova T (2008) Multiple-point geostatistics for modeling subsurface heterogeneities: a comprehensive review. Water Resour 44(11) (W11,413), doi:10.1029/2008WR006993

Leu L, Berg S, Ott H, Amstrong RT, Enzmann F, Kersten M (2014) Sensitivity study of an image processing workflow on synchrotron µ-CT images of Berea sandstone. Geophys Res Abstr 16: EGU2014–EGU10431

Mariethoz G, Lefebvre S (2014) Bridges between multiple-point geostatistics and texture synthesis: review and guidelines for future search. Comput Geosci. doi:10.1016/j.cageo.2014.01.001

Mariethoz G, Renard P, Froidevaux R. (2009) Integrating collacated auxiliary parameters in geostatistical simulations using joint probability distributions and probability aggregation. Water Resour Res 45(W08421)

Mariethoz G, Renard P, Straubhaar J (2010) The direct sampling to perform multiple-point geostatistical simulation. Water Resour Res 46(W11536), doi 10.1029/2008WR007621

Okabe H, Blunt MJ (2005) Pore space reconstruction using multiple-point statistics. J Pet Sci Eng 46-2005 p. 121–137 Elsevier

Strebelle S (2002) Conditional simulation of complex geological structures using multiple-point statistics. Math Geol 34(1):1–21. doi:10.1023/A:1014009426274

Bernstein Copula-Based Spatial Stochastic Simulation of Petrophysical Properties Using Seismic Attributes as Secondary Variable

Martín A. Díaz-Viera, Arturo Erdely, Tatiana Kerdan, Raúl del-Valle-García, and Francisco Mendoza-Torres

Abstract A novel Bernstein copula-based spatial stochastic co-simulation (BCSCS) method for petrophysical properties using seismic attributes as a secondary variable is presented. The method is fully nonparametric, and it has the advantages of not requiring linear dependence between variables. The methodology is illustrated in a case study from a marine reservoir in the Gulf of Mexico, and the results are compared with sequential Gaussian co-simulation (SGCS) method.

1 Introduction

Modeling the spatial distribution of petrophysical properties in the framework of reservoir characterization is a crucial and difficult task due to the lack of enough data and hence the degree of uncertainty associated with it. For this reason, in recent years a stochastic simulation approach for the spatial distribution of petrophysical properties has been adopted.

Seismic attributes have been extensively used as secondary variables in static reservoir modeling for petrophysical property prediction but usually assuming linear dependence and Gaussian distribution (Parra and Emery 2013).

Quite recently, copulas have become popular for being a flexible means of representing dependency relationships in the financial sector, and applications are

M.A. Díaz-Viera (✉) • T. Kerdan • R. del-Valle-García
Instituto Mexicano del Petróleo (IMP), Eje Central Lázaro Cárdenas Norte 152, CP 07730
Ciudad de México, Mexico
e-mail: mdiazv@imp.mx; tkerdan@imp.mx; rvalleg@imp.mx

A. Erdely
Facultad de Estudios Superiores Acatlán, UNAM, Mexico City, Mexico
e-mail: arturo.erdely@comunidad.unam.mx

F. Mendoza-Torres
Posgrado de Ciencias de la Tierra, UNAM, Mexico City, Mexico
e-mail: mentofran@gmail.com

J.J. Gómez-Hernández et al. (eds.), *Geostatistics Valencia 2016*, Quantitative Geology and Geostatistics 19, DOI 10.1007/978-3-319-46819-8_33

already emerging in the field of geostatistics (Díaz-Viera and Casar-González 2005; Bardossy and Li 2008; Kazianka and Pilz 2010).

A geostatistical simulation method, based on Bernstein copula approach as a tool to represent the underlying dependence structure between petrophysical properties and seismic attributes, is proposed. The procedure basically consists of applying the simulated annealing method with a joint probability distribution model estimated by a Bernstein copula in a completely nonparametric fashion (Hernández-Maldonado et al. 2012).

The method has the advantages of not requiring linear dependence or a specific type of distribution. The application of the methodology is illustrated in a case study where the results are compared with sequential Gaussian co-simulation (SGCS) method.

2 Methodology

As stated in the introduction, the main goal of this work is to show the application of a Bernstein copula-based spatial co-simulation method for petrophysical property predictions using seismic attributes as secondary variables and its comparison with the classical sequential Gaussian co-simulation method. In what follows, a brief description of both methods and a general workflow outline are presented.

2.1 Sequential Gaussian Co-simulation (SGCS)

The sequential Gaussian co-simulation (SGCS) method is very well established in the geostatistics literature, so here we will just mention the details of its application. Usually this method is applied with a linear model of coregionalization (Chiles and Delfiner 1999) which is mostly unnatural, forced, very complicated, and difficult to establish. The method assumes the existence of very strong linear dependence between primary and secondary variables, which is its main assumption and at the same time its main drawback. Here we choose to use an alternative variant, the Markov model (MM), given in Chiles and Delfiner (1999, p. 305) and implemented in SGeMS (Remy et al. 2009).

2.2 Bernstein Copula-Based Spatial Stochastic Co-simulation (BCSCS)

A Bernstein copula-based spatial stochastic co-simulation (BCSCS) method has been previously presented in a series of papers (Hernández-Maldonado et al. 2012,

2014) and has been mainly applied in one dimension for petrophysical properties at well-log scale.

The method basically consists of establishing a dependence model between a primary and a secondary variable and then uses this model in conjunction with the spatial dependence structure (variogram) of the primary variable to predict the first one using the second one as a conditioning variable. This can be done in a global optimization framework using simulated annealing method, but other methods, such as genetic algorithms, could also be applied.

The modern way to analyze dependencies is by copula approach (Joe 1997). Copula approach assumes neither a predetermined nor a priori type of dependency, but from the data one tries to establish the best model that represents the existing dependence on them.

In particular, here it is preferred to use a completely nonparametric approach to modeling the dependence by using Bernstein copulas, which gives name to the method. However, other approaches, parametric (Díaz-Viera and Casar-González 2005) and semi-parametric (Erdely and Diaz-Viera 2010), are also possible. The Bernstein copulas introduced by Sancetta and Satchell (2004) are nothing more than an approximation of the sample copula by Bernstein polynomials. Its main shortcoming is the curse of dimensionality, as it quickly becomes computationally prohibitive for more than two dimensions. Alternatives have been proposed using vine copulas (Erdely and Diaz-Viera 2016).

In summary, the algorithm consists of two stages:

1. A dependence model, using a Bernstein copula, is established from which a number of sample values are generated (see Appendix A).
2. A stochastic spatial simulation is performed using a simulated annealing method with a variogram model and a bivariate distribution function as objective functions (Deutsch and Cockerham 1994; Deutsch and Journel 1998).

Additional details about the mathematical formulation of the method and its computational implementation can be found on Hernández-Maldonado et al. (2012, 2014).

2.3 *Workflow Outline*

One of the biggest challenges in these applications is to simultaneously handle multiple scales. Here, we have two scales: a well-log scale and a seismic scale. But sometimes, due to the amount of data, an additional intermediate scale is required, since the well-log scale data is a very large dataset and upscaled well logs may have from a statistical point of view not enough data. Log data usually have a sampling interval in the range of 10–25 cm, while seismic data are in the range of several meters. So it is necessary to perform some upscaling process to make well logs compatible with seismic data. For the upscaling process, there is no single recipe because it is largely dependent on the data. Here we will use the median as upscaling procedure.

The general workflow is as follows: (1) univariate data analysis, (2) bivariate dependence analysis, (3) variography analysis, and (4) simulations.

3 Case Study

Data used in the case study are from a marine reservoir in the Gulf of Mexico. The reservoir is siliciclastic and it is formed mainly by alternating sequences of sands and shales.

3.1 Data Description

The data consist of a total porosity well-log from a single well and seismic attribute (P-impedance) obtained in a vertical (inline) section. The well log has a sample interval of 0.1 m. The section has a length of 412.5 m and covers an interval of 336.4 m in depth and was chosen so that the well was located in the middle of it (see Fig. 1). Seismic grid is made of 33 intervals of 12.5 m in X direction and 60 intervals of 5.5 m in depth direction.

At well-log scale, the P-wave impedance log is obtained from the product of P-wave velocity and density logs. At the seismic scale, a seismic inversion method was used, based on the "LP sparse spike" approach by Li (2001). Proper care was taken in incorporating low-frequency impedance trend so that the impedance from the log and the impedance from the seismic section are equivalent around the well.

The impedance in general depends on the type of rock and its petrophysical properties as well as the containing fluid types and their saturations. It is very common in reservoir geophysics to take advantage of dependency relationships between petrophysical properties (for instance, total porosity and P-impedance) at well-log scale to predict the former ones (total porosity) using seismic attributes (P-impedance) at the seismic scale.

In particular, in this work the total porosity was considered as primary variable (variable to predict) and P-wave impedance as secondary variable (conditioning variable). As mentioned in the previous section, three scales are considered: a well-log, a seismic, and an additional intermediate "one-meter" scale. Hereafter, the following notation will be used:

- PhiT_T and Ip_T for total porosity and P-impedance from original well logs (well-log scale)
- PhiT_T_1m and Ip_T_1m for total porosity and P-impedance from original well logs subsampled every meter (1-m scale)

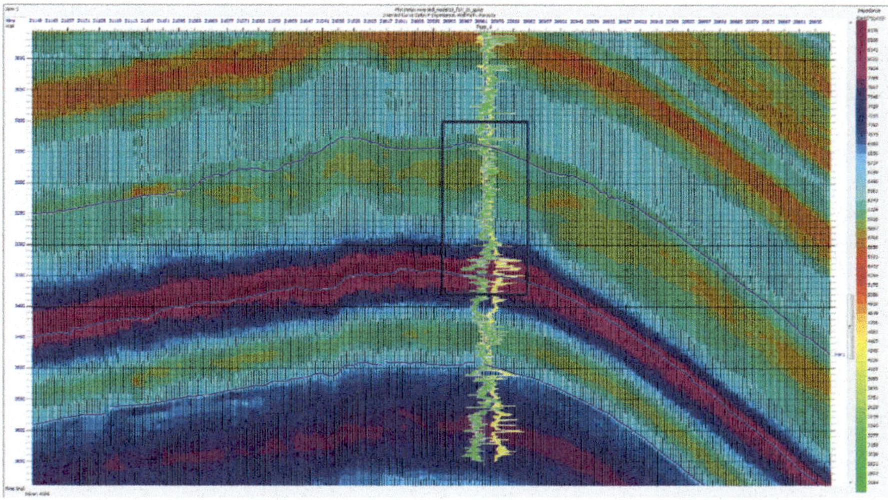

Fig. 1 *Vertical* (*inline*) section with P-impedance as a result of seismic inversion. The *color* scale represents impedance values. In the middle of the section, two logs are plotted along a well: in *yellow* P-impedance and in *green* total porosity

- PhiT_T_median and Ip_T_median for total porosity and P-impedance from original well logs upscaled using median upscaling procedure (seismic scale)
- Ip_inline for P-impedance from the vertical inline section (seismic scale)
- Ip_inline_U for P-impedance from the vertical inline section restricted to the corresponding well coordinates, i.e., only along the well trajectory (seismic scale)

3.2 Univariate Data Analysis

In Fig. 2 are shown histograms and boxplots for PhiT and Ip at the three scales, and in Tables 1 and 2 a summary of corresponding basic univariate statistics.

Note that median and mean are pretty close for 1-m scale and seismic scales, while Ip_inline_U and Ip_inline have very consistent statistics.

3.3 Bivariate Dependence Analysis

In Fig. 3 are given the scatterplots with marginal histograms and boxplots for PhiT vs. Ip (a) at well-log, (b) at 1-m, and (c) at seismic scale, respectively, while in the Fig. 3d the corresponding scatterplot for a non-conditional Ip-PhiT simulation using a Bernstein copula at 1-m scale is shown. In Fig. 4 are given pseudo-observations (sample copula) scatterplots for Ip-PhiT, (a) at well-log scale, (b) at 1-m scale, (c) at

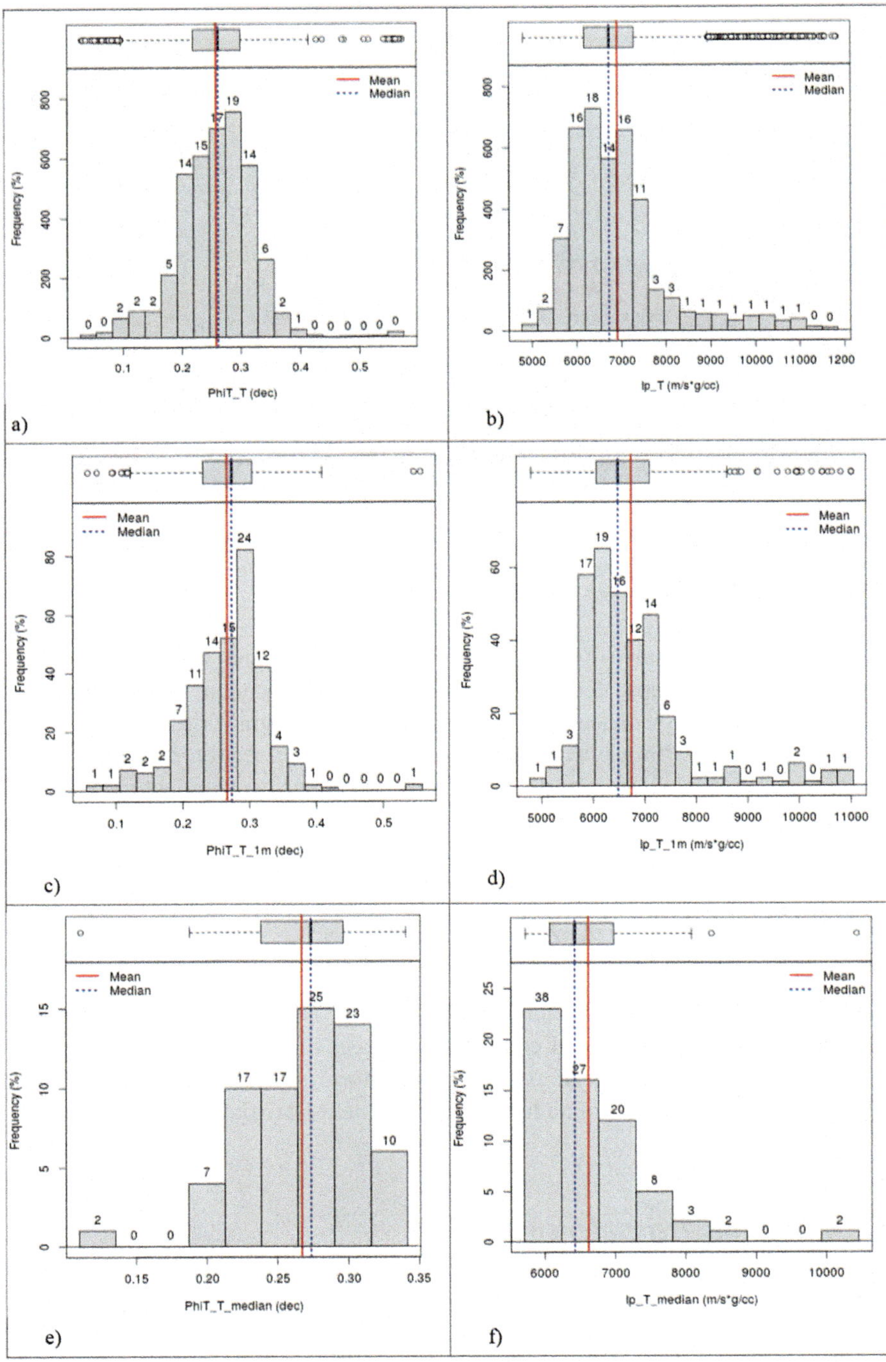

Fig. 2 Histograms and boxplots for PhiT and Ip at well-log scale (**a**, **b**), at 1-m scale (**c**, **d**), and at seismic scale (**e**, **f**), respectively

Table 1 Statistics summary of original and 1-m upscaled well logs

Statistics	PhiT_T	Ip_T	PhiT_T_1m	Ip_T_1m
n	4059	4059	337	337
Minimum	0.030	4802.22	0.057	4802.22
First quartile	0.218	6163.18	0.230	6064.66
Median	0.261	6717.99	0.273	6481.54
Mean	0.257	6906.39	0.266	6740.08
Third quartile	0.299	7270.29	0.304	7099.52
Maximum	0.571	11,812.36	0.556	11,013.43
Variance	0.004	1,264,603	0.003	1,133,016

Table 2 Statistics summary of median upscaled well logs and Ip at seismic scale

Statistics	PhiT_T_med	Ip_T_med	Ip_inline_U	Ip_inline
n	60	60	60	1980
Minimum	0.110	5730.23	5940.26	5940.16
First quartile	0.239	6074.77	6080.53	6083.72
Median	0.273	6426.24	6350.74	6340.04
Mean	0.267	6619.13	6623.50	6602.87
Third quartile	0.296	6967.11	6896.91	6893.98
Maximum	0.340	10,430.77	8726.18	8773.20
Variance	0.002	630,041	577,250	485,570

seismic scale, and (d) for a non-conditional bivariate simulation with a Bernstein copula at 1-m scale. In Table 3 a summary of corresponding correlation (Spearman and Pearson) coefficients is given. It can be observed that the dependence is weakened with the increasing of the scale.

3.4 Variography Analysis

In Fig. 5 are shown estimated variograms (a, b) and best-fit variogram models (c, d) for PhiT and Ip at seismic scale in depth direction. As is evident in Fig. 5b, the sample variogram of Ip_inline_U shows a typical behavior related with the presence of trend, which means that at least the intrinsic hypothesis is not satisfied. Then, the trend, which in this case was of second order, was estimated and removed, resulting a new variable Ip_inline_U_r2 without trend. The same previous procedure was applied to Ip_Inline and a resulting detrended variable was named Ip_inline_r2. Note, in Fig. 5d the variogram was obtained after removing trend from Ip_inline_U. While in the Fig. 6 are displayed estimated variograms and best-fit variogram models for impedance at seismic scale in (a) X and (b) depth directions, respectively, after removing trend from Ip_inline.

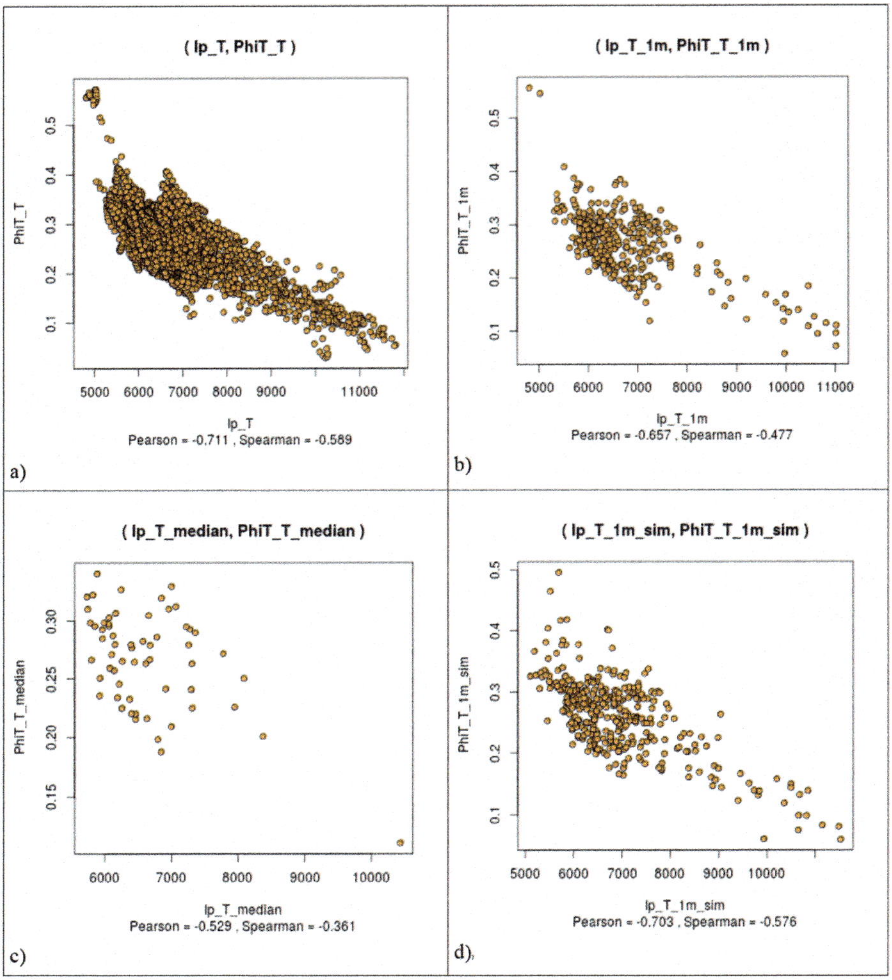

Fig. 3 Scatterplots with marginal histograms and boxplots for Ip vs. PhiT, (**a**) at well-log scale, (**b**) at 1-m scale, (**c**) at seismic scale, and (**d**) a non-conditional bivariate simulation with a Bernstein copula at 1-m scale

Because of the lack of data for total porosity in the X direction, the same variogram structure of the impedance in this direction is adopted, considering that they show almost the same structure in the depth direction (see Fig. 5). A variogram model for porosity at seismic scale is proposed so that the total variance of PhiT_T median is preserved which basically it is to consider a correlation range equal to the impedance variogram in the X direction (see Fig. 6). For both simulation methods, the following variogram model for porosity is used: model=spherical, nugget = 0.0002, and structure contribution = 0.0016; ranges, maximum = 160, medium = 50, and minimum = 1; angles, $x = 90$, $y = 0$, and $z = 0$.

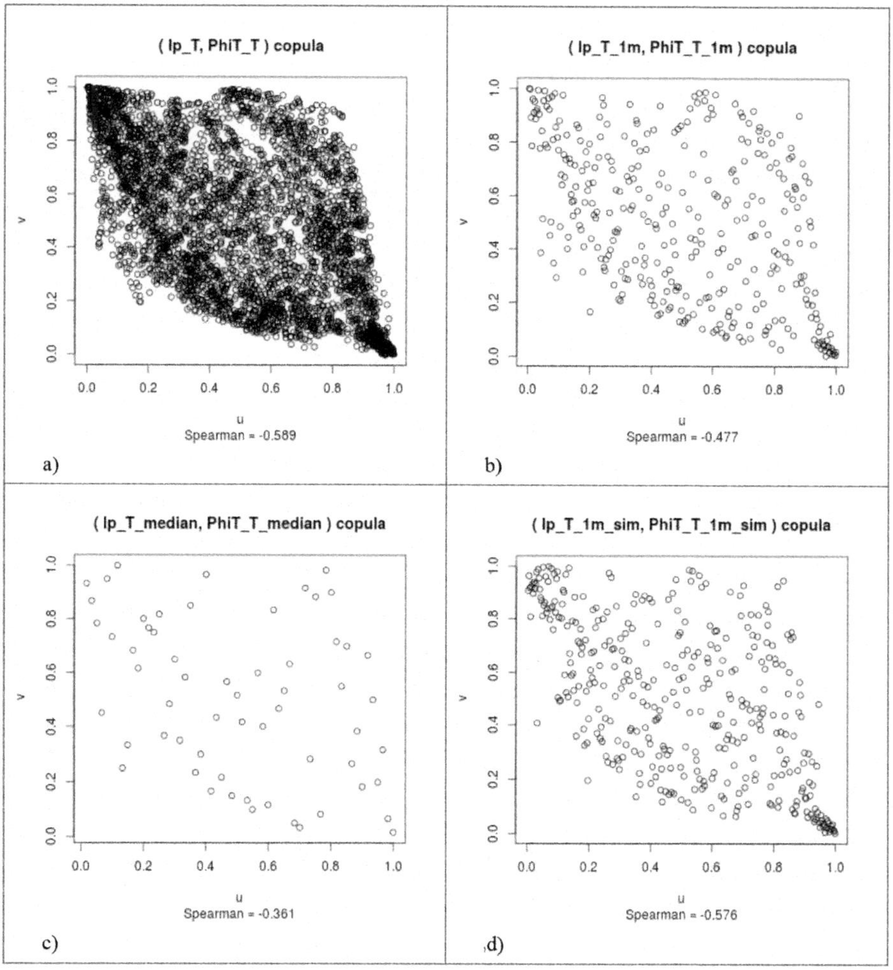

Fig. 4 Pseudo-observations (sample copulas) scatterplots of Ip-PhiT data ranks, rescaled to [0,1] (**a**) at well-log scale, (**b**) at 1-m scale, (**c**) at seismic scale, and (**d**) a non-conditional bivariate simulation with a Bernstein copula at 1-m scale

Table 3 Summary of correlation coefficients for Ip vs. PhiT at well-log, 1-m, and seismic scales and for a non-conditional bivariate Ip-PhiT simulation with a Bernstein copula at 1-m scale (BCS_1m)

Correlation coefficients	Well-log scale	One-meter scale	Seismic scale	BCS_1m
Spearman	−0.589	−0.477	−0.361	−0.576
Pearson	−0.711	−0.657	−0.529	−0.703

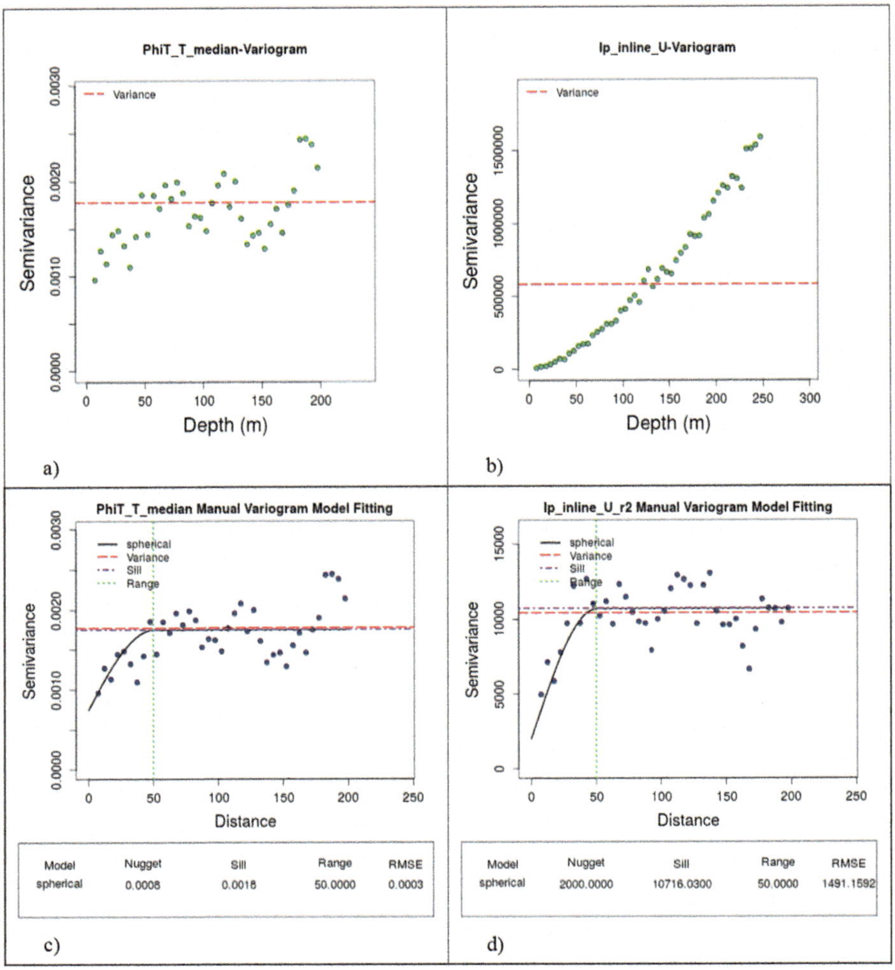

Fig. 5 PhiT and Ip estimated variograms (**a**, **b**) and best-fit variogram models (**c**, **d**) at seismic scale in depth direction. Note, in **d** variogram after removing trend from Ip

3.5 SGCS Simulations

A sequential Gaussian co-simulation (SGCS) with Markov model variant (MM1), implemented in SGeMS (Remy et al. 2009), was performed with the following parameters: primary variable, PhiT_T_median; secondary variable, Ip_Inline_r2; grid, $33 \times 60 \times 1$ (the same as Ip_Inline_r2); Kriging type, simple kriging (SK); max conditioning data, 12; correlation coefficient, -0.657; search ellipsoid, 160, 50, and 1; variogram model of primary variable=spherical, nugget = 0.0002, and structure contribution = 0.0016; ranges, maximum = 160, medium = 50, and minimum = 1; angles, $x = 90$, $y = 0$, and $z = 0$.

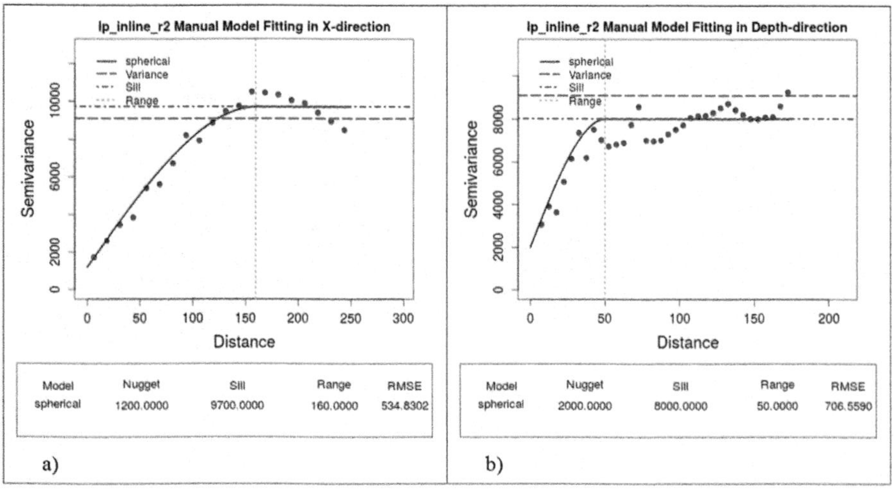

Fig. 6 Estimated variograms and best-fit variogram models for Ip at seismic scale after removing trend in (**a**) X and (**b**) depth directions, respectively

The resulting simulation is named PhiT_SGC and its map in the vertical (inline) section is given in the Fig. 8a.

3.6 BCSCS Simulations

A Bernstein copula-based spatial stochastic co-simulation (BCSCS) was performed using the procedure explained before. First, a dependence model, using a Bernstein copula at 1-m scale (see Fig. 4b), was obtained from which 40,000 conditional bivariate simulations (BCsim40000_cond) conditioning by secondary variable were generated (see Fig. 7). Then, as simulated annealing program was used a modified version of SASIM from GSLIB (Deutsch and Journel 1998) with the following parameters: primary variable, PhiT_T_median; secondary variable, Ip_Inline; grid, $33 \times 60 \times 1$ (the same as Ip_Inline); objective function, variogram and bivariate distribution function; paired data, 40,000 conditional bivariate simulations using a Bernstein copula (BCsim40000_cond); number of primary thresholds = 10, number of secondary thresholds = 10, number of variogram lags = 40, variogram model of primary variable = spherical, nugge t= 0.0002, and structure contribution = 0.0016; ranges, maximum = 160, medium = 50, and minimum = 1; angles, $x = 90$, $y = 0$, and $z = 0$. A map of the resulting simulation is given in the Fig. 8b.

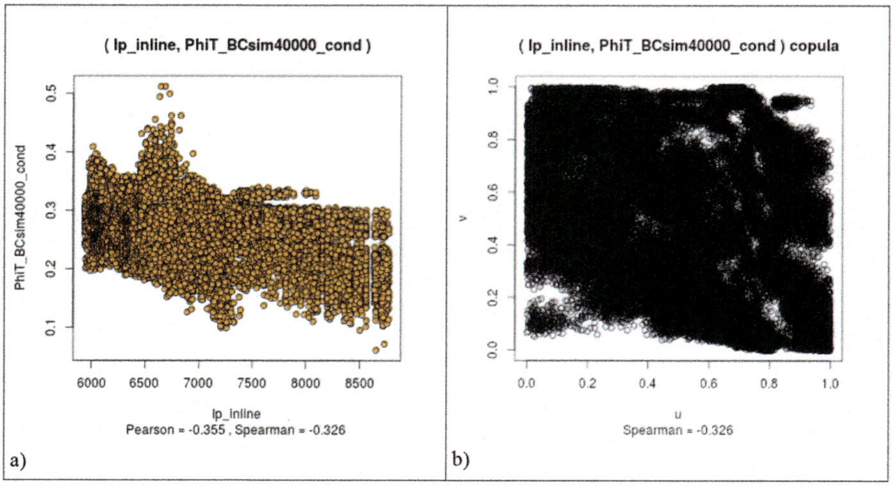

Fig. 7 (**a**) Scatterplot with marginal histograms and boxplots and (**b**) pseudo-observations (sample copulas) scatterplot for 40,000 conditional bivariate simulations using a Bernstein copula at 1-m scale

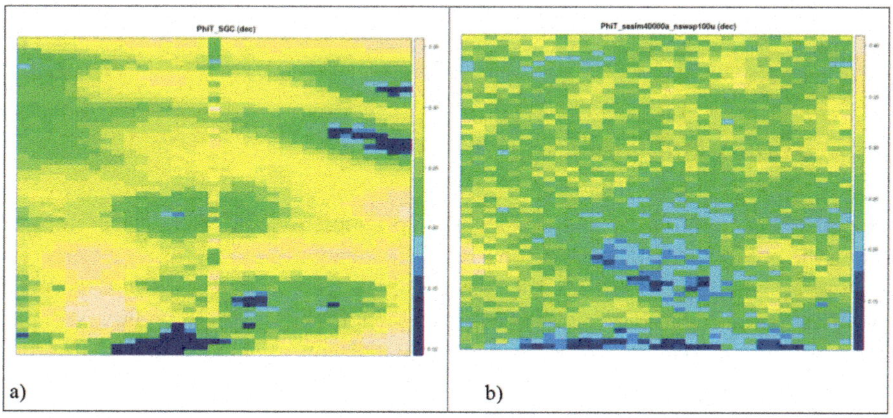

Fig. 8 Maps for (**a**) a PhiT sequential Gaussian co-simulation, and (**b**) a PhiT Bernstein copula-based co-simulation in the vertical (inline) section

3.7 *SGCS vs. BCSCS: A Comparative Summary*

In comparison with the SGCS method, the BCSCS method better reproduces the statistics in terms of variance and extreme values (see Fig. 9), and both methods reproduce quite well the spatial structure (see Fig. 10), but the sequential Gaussian co-simulation shows spurious correlation dependence, which does not exist in the original data, highlighted in red color in Fig. 11. This is the main reason of the difference between Fig. 8a and 8b.

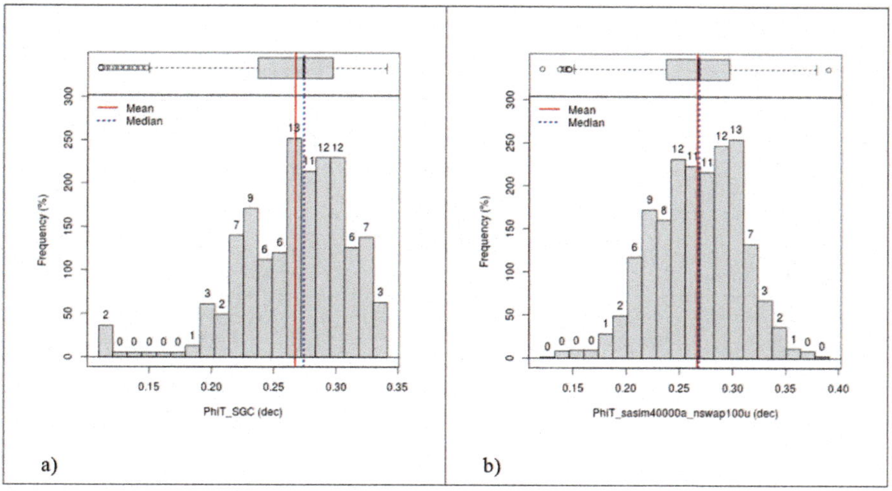

Fig. 9 PhiT histograms and boxplots for (**a**) a sequential Gaussian co-simulation and (**b**) a Bernstein copula-based co-simulation, respectively

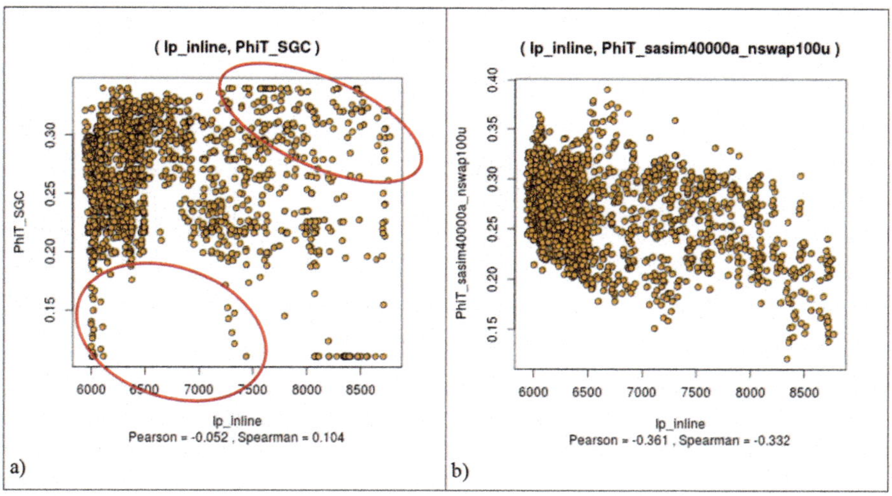

Fig. 10 Ip vs. PhiT scatterplots with marginal histograms and boxplots for (**a**) a sequential Gaussian co-simulation and (**b**) a Bernstein copula-based co-simulation, respectively. Simulated values with spurious dependence, which does not exist in the original data, are highlighted in *red* color

4 Final Remarks and Future Work

A Bernstein copula-based spatial stochastic co-simulation (BCSCS) method presented in this paper possesses several advantages over the classical sequential Gaussian co-simulation method (SGCS). Firstly of all, it does not require of a strong

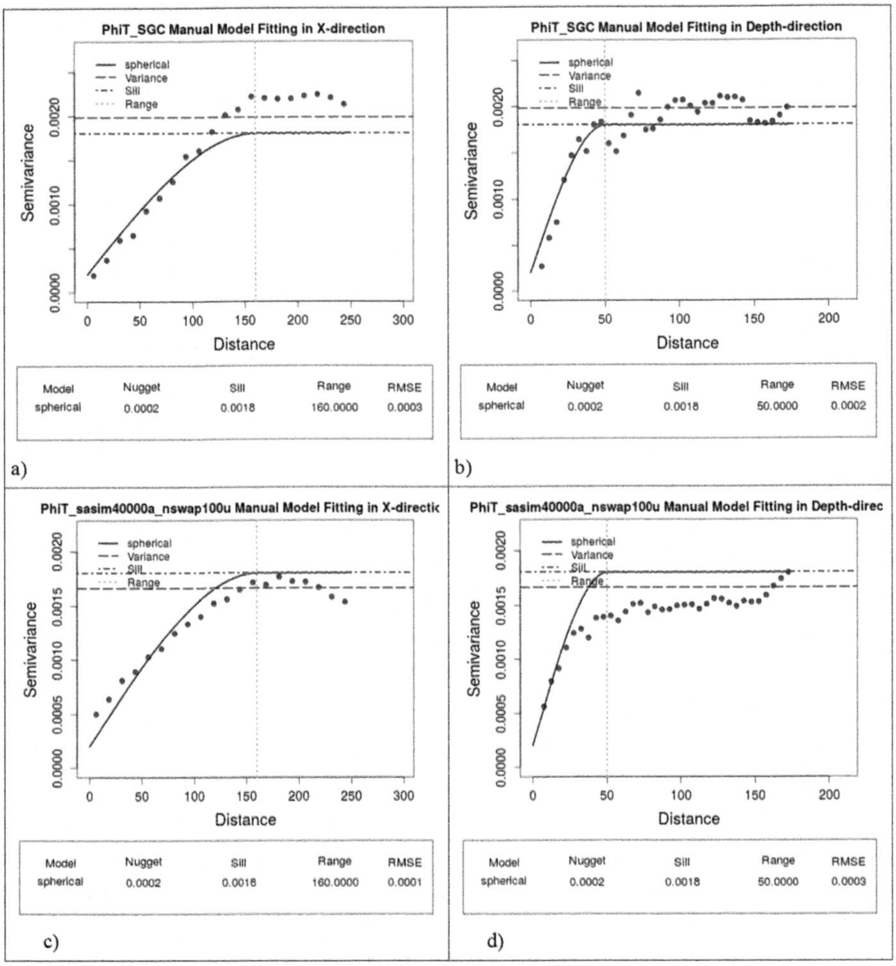

Fig. 11 Estimated variograms and best-fit variogram models in X and depth directions for a sequential Gaussian co-simulation (**a**, **b**) and a Bernstein copula-based co-simulation (**c**, **d**), respectively

linear dependence between variables; on the contrary, it can capture and reproduce the existing dependence between them. The method is nonparametric which means that it does not need a specific distribution function. Since the Bernstein copula is based on the sample distribution function, it may reproduce the variability and the extreme values. Another advantage is that there is no need to make back transformations, which are potentially biased, since copulas are invariant under strictly increasing transformations.

Instead of using a single seismic attribute, it could be used the best combination of them depending on the primary (explanatory) variable by applying standard multivariate statistical procedures such as principal component and factorial analysis. Another option would be using a multivariate copula with three or more variables to directly exploit their dependencies.

This work can be easily extended to 3D problems but it depends on the computing power available. Although in this work the aim was to show the performance of the simulation method, a simpler and efficient alternative, perhaps more convenient for 3D large problems, could be the median regression approach already shown in previous works (Erdely and Diaz-Viera 2010; Hernández-Maldonado et al. 2012) in 1D.

Acknowledgments The present work was supported by the IMP project D.61037 "Interpretación Sísmica Cuantitativa Guiada por Litofacies para la Caracterización de Yacimientos."

Appendix A: Copula-Based Approach for Dependence Modeling

A theorem by Sklar (1959) proved that there exists a functional relationship between the joint probability distribution function of a random vector and its univariate marginal distribution functions. In the bivariate case, for example, if (X, Y) is a random vector with joint probability distribution $F_{XY}(x, y) = P(X \leq x, Y \leq y)$, then the marginal distribution functions of X and Y are $F_X(x) = P(X \leq x) = F_{XY}(x, \infty)$ and $F_Y(y) = P(Y \leq y) = F_{XY}(\infty, y)$, respectively, but in the marginalization of F_{XY}, some information is lost since the only knowledge of the marginal distributions F_X and F_Y is not generally possible to specify F_{XY} because the marginals only explain the probabilistic individual behavior of the random variables they represent. *Sklar's theorem* proves that there exists a function $C_{XY} : [0, 1]^2 \to [0, 1]$ such that

$$F_{XY}(x, y) = C_{XY}(F_X(x), F_Y(y))$$

C_{XY} is called *copula function* associated to (X, Y) and contains information about the dependence relationship between X and Y, independently from their marginal probabilistic behavior. C_{XY} is uniquely determined on $\operatorname{Ran} F_X \times \operatorname{Ran} F_Y$, and therefore, if F_X and F_Y are continuous, then C_{XY} is unique on $[0, 1]^2$. Among several properties of copula functions, see Nelsen (2006), we have the following:

- $C(u, 0) = 0 = C(0, v)$
- $C(u, 1) = u, C(1, v) = v$
- $C(u_2, v_2) - C(u_2, v_1) - C(u_1, v_2) + C(u_1, v_1) \geq 0$ if $u_1 \leq u_2, v_1 \leq v_2$
- C is uniformly continuous on its domain $[0, 1]^2$.
- The horizontal, vertical, and diagonal sections of a copula C are all nondecreasing and uniformly continuous on $[0, 1]$.

- $W(u,v) \leq C(u,v) \leq M(u,v)$ where $W(u,v) = \max(u+v-1,\ 0)$ and $M(u,v) = \min(u,v)$ are also copulas known as the lower and upper Fréchet-Hoeffding bounds.
- A convex linear combination of copula functions is also a copula function.
- If X and Y are continuous random variables with copula C_{XY}, and if α and β are strictly increasing functions on *Ran X* and *Ran Y*, respectively, then $C_{\alpha(X)\beta(Y)} = C_{XY}$. Thus, C_{XY} is invariant under strictly increasing transformations of X and Y.

Copula functions are a useful tool to build joint probability models in a more flexible way since we may choose separately the univariate models for the random variables of interest and the copula function that better represents the dependence among them, in each case in a parametric or nonparametric way. In the case of a multivariate normal model, for example, all the marginal distributions have to be normally distributed, with no tail dependence at all and with finite second moments for the correlations to be well defined. In fact, the multivariate normal model is a particular case when the underlying copula is Gaussian and all the univariate marginals are normally distributed.

In case F_X and F_Y are continuous, by elementary probability we know that $U = F_X(X)$ and $V = F_Y(Y)$ are continuous Uniform(0, 1) random variables and the underlying copula C for the random vector (U, V) is the same copula corresponding to (X, Y), and by Sklar's theorem we have that the joint probability distribution function for (U, V) is equal to $F_{UV}(u,v) = C(F_U(u), F_V(v)) = C(u,v)$. Therefore, in case F_X and F_Y are known and F_{XY} is unknown, if $\{(x_1,y_1),\ldots,(x_n,y_n)\}$ is an observed random sample from (X, Y), the set $\{(u_k,v_k) = (F_X(x_k), F_Y(y_k)) : k = 1, \ldots, n\}$ would be an observed random sample from (U, V) with the same underlying copula C as (X, Y), and since $C = F_{UV}$ we may use the (u_k, v_k) values (called copula observations) to estimate C as a joint empirical distribution:

$$\hat{C}(u,v) = \frac{1}{n}\sum_{k=1}^{n} 1_{\{u_k \leq u\ ,\ \ v_k \leq v\}}$$

Strictly speaking, the estimation $\hat{C}$ is not a copula since it is discontinuous and copulas are always continuous. If F_X, F_Y, and F_{XY} are all unknown (the usual case), we estimate F_X and F_Y by univariate empirical distribution functions:

$$\hat{F}_X(x) = \frac{1}{n}\sum_{k=1}^{n} 1_{\{x_k \leq x\}} \quad \hat{F}_Y(y) = \frac{1}{n}\sum_{k=1}^{n} 1_{\{y_k \leq y\}}$$

Now the set of pairs $\{(u_k, v_k) = \left(\hat{F}_X(x_k), \hat{F}_Y(y_k)\right) : k = 1, \ldots, n\}$ is referred to as *copula pseudo-observations*. It is straightforward to verify that $\hat{F}_X(x_k) = \frac{1}{n}\mathrm{rank}(x_k)$ and $\hat{F}_Y(y_k) = \frac{1}{n}\mathrm{rank}(y_k)$. In this case the concept of *empirical copula*, see Nelsen

(2006), is defined as the following function $C_n : I_n^2 \rightarrow [0, 1]$, where $I_n = \left\{\frac{i}{n} : i = 0, \ldots, n\right\}$, given by:

$$C_n\left(\frac{i}{n}, \frac{j}{n}\right) = \frac{1}{n}\sum_{k=1}^{n} 1_{\{\text{rank}(x_k) \leq i,\ \text{rank}(y_k) \leq j\}}$$

Again, C_n is not a copula, but it is an estimation of the underlying copula C on the grid I_n^2 that may be extended to a copula on $[0, 1]^2$ by means of, for example, Bernstein polynomials, as proposed and studied in Sancetta and Satchell (2004), which leads to what is known as a *Bernstein copula* nonparametric estimation $\widetilde{C}$: $[0, 1]^2 \rightarrow [0, 1]$ given by:

$$\widetilde{C}(u, v) = \sum_{i=0}^{n}\sum_{j=0}^{n} C_n\left(\frac{i}{n}, \frac{j}{n}\right)\binom{n}{i} u^i (1-u)^{n-i} \binom{n}{j} v^j (1-v)^{n-j}$$

As summarized in Erdely and Diaz-Viera (2010) in order to simulate replications from the random vector (X, Y) with the dependence structure inferred from the observed data $\{(x_1, y_1), \ldots, (x_n, y_n)\}$, we have the following:

Algorithm 1

1. Generate two independent and continuous Uniform(0, 1) random variates u and t.
2. Set $v = c_u^{-1}(t)$ where $c_u(v) = \frac{\partial \widetilde{C}(u,v)}{\partial u}$.
3. The desired pair is $(x, y) = (\widetilde{Q}_n(u), \widetilde{R}_n(v))$ where $\widetilde{Q}_n$ and $\widetilde{R}_n$ are empirical quantile functions for X and Y, respectively.

For a value x in the range of the random variable X and a given $0 < \alpha < 1$, let $y = \varphi_\alpha(x)$ denote the solution to the equation $P(Y \leq y | X = x) = \alpha$. Then the graph of $y = \varphi_\alpha(x)$ is the *α-quantile regression curve* of Y conditional on $X = x$. In Nelsen (2006), it is proven that:$P(Y \leq y | X = x) = c_u(v)\big|_{u=F_X(x),\, v=F_Y(y)}$

This result leads to the following algorithm to obtain the *α-quantile regression curve* of Y conditional on $X = x$:

Algorithm 2

1. Set $c_u(v) = \alpha$.
2. Solve for v the regression curve, say $v = g_\alpha(u)$.
3. Replace u by $\widetilde{Q}_n^{-1}(x)$ and v by $\widetilde{R}_n^{-1}(y)$.
4. Solve for y the regression curve, say $y = \varphi_\alpha(x)$.

Bibiography

Bardossy A, Li J (2008) Geostatistical interpolation using copulas. Water Resour Res 44:1–15

Chiles JP, Delfiner P (1999) Geostatistics: modeling spatial uncertainty. Wiley series in probability and statistics. Applied probability and statistics section. Wiley, New York

Deutsch CV, Cockerham PW (1994) Geostatistical modeling of permeability with annealing cosimulation (ACS). SPE 69th Annual Technical Conference and Exhibition. Society of Petroleum Engineers, New Orleans, p 523–532

Deutsch CV, Journel AG (1998) GSLIB: geostatistical software library and user's guide, 2nd edn. Oxford University Press, New York

Díaz-Viera M, Casar-González R (2005) Stochastic simulation of complex dependency patterns of petrophysical properties using t-copulas. Proc IAMG'05: GIS Spat Anal 2:749–755

Erdely A, Diaz-Viera M (2010) Nonparametric and semiparametric bivariate modeling of petrophysical porosity-permeability dependence from well-log data. In: Jaworski P, Durante F, Härdle W, Rychlik T (eds) Copula theory and its applications. Lecture notes in statistics 198. Springer, Berlin, pp 267–278

Erdely A, Diaz-Viera M (2016) A vine and gluing copula model for permeability stochastic simulation. In: Bozeman J R, Oliveira T, Skiadas CH (eds) Stochastic and data analysis methods and applications in statistics and demography, pp 199–207

Hernández-Maldonado V, Díaz-Viera M, Erdely A (2012) A joint stochastic simulation method using the Bernstein copula as a flexible tool for modeling nonlinear dependence structures between petrophysical properties. J Petrol Sci Eng 92–93:112–123

Hernández-Maldonado V, Díaz-Viera M, Erdely A (2014) A multivariate Bernstein copula model for permeability stochastic simulation. Geofis Int 53(2):163–181

Joe H (1997) Multivariate models and dependence concepts. Chapman & Hall, London

Kazianka H, Pilz J (2010) Copula-based geostatistical modeling of continuous and discrete data including covariates. Stoch Environ Res Risk Assess 24(5):661–673

Li Q (2001).LP sparse spike impedance inversion. Hampson-Russell Software Services Ltd. – CSEG

Nelsen RB (2006) An introduction to copulas (Springer series in statistics), 2nd edn. Springer, New York

Parra J, Emery X (2013) Geostatistics applied to cross-well reflection seismic for imaging carbonate aquifers. J Appl Geophys 92:68–75

Remy N, Boucher A, Wu J (2009) Applied geostatistics with SGeMS: a user's guide. Cambridge University Press, New York

Sancetta A, Satchell S (2004) The Bernstein copula and its applications to modeling and approximations of multivariate distributions. Economet Theor 20(3):535–562

Sklar A (1959) Fonctions the repartition à n dimensions et leurs marges. Publ Inst Stat 8:229–331

Robust MPS-Based Modeling via Spectral Analysis

Morteza Elahi Naraghi and Sanjay Srinivasan

Abstract Spatially distributed phenomena typically do not exhibit Gaussian behavior, and consequently methods constrained to traditional two-point covariance statistics cannot correctly represent the spatial connectivity of such phenomena. This necessitates the development of multiple-point statistics (MPS)-based algorithms. However, due to the sparse data available to infer these MPS statistics, one needs to have a training image (TI) to accurately model higher-order statistics. Training images are usually inferred from outcrops and/or conceptual models and are subject to uncertainty that have to be accounted for in MPS algorithms.

In this study, we propose a new method for ranking different sets of TIs corresponding to different geological scenarios. These set of TIs represent the associated uncertainty and contain features of different shapes (channels, ellipses, fracture, etc.) as well as different sizes and orientation. We analyze the polyspectra of the different TIs (power spectrum and bispectrum) to distinguish between different TIs. We show that object size and orientation can be inferred from the power spectrum, while the object shapes can be inferred from the bispectrum. Therefore, the combination of power spectrum and bispectrum can be used as an identifier for each TI.

We then infer the power spectrum and bispectrum from the available conditioning data. Since the data is scattered and sparse, we use a nonuniform fast Fourier transform (NUFFT) method based on a basis pursuit algorithm to estimate the Fourier transform of the scattered data. We then use the Fourier transform to calculate the power spectrum and the bispectrum. Then, the identifier features are calculated from the higher-order spectra. Finally, we choose the TI with the closest identifier to that of conditioning data as the representative TI.

We implement the proposed algorithm on different geologic systems such as channelized reservoir, fractured reservoirs, and models with elliptic objects with different sizes and orientations and examine its performance. We study the sensitivity of the algorithm to the available conditioning data. We show that the algorithm performs well with very sparse conditioning data. This algorithm can address one of the main issues pertaining to MPS algorithms, which is ensuring the

M. Elahi Naraghi (✉) • S. Srinivasan
Petroleum and Geosystems Engineering Department, The University of Texas at Austin, Austin, TX, USA
e-mail: morteza.naraghi@utexas.edu

J.J. Gómez-Hernández et al. (eds.), *Geostatistics Valencia 2016*, Quantitative Geology and Geostatistics 19, DOI 10.1007/978-3-319-46819-8_34

consistency between the training image and conditioning data in order to develop robust models that have improved predictive ability. In case MPS-based simulations are performed accounting for the uncertainty in TI, the method can be used to rank the prior TIs so as to yield robust estimates for uncertainty.

1 Introduction

One of the most important issues in reservoir modeling is to generate maps of flow and geologic attributes that play a great role in hydrocarbon production. Mostly, the primary source of information is borehole measurements that are only available at sparse locations. Integrating the available data along wells within robust models for spatial continuity, multiple reservoir models can be generated that can be used to quantify the uncertainty associated with reservoir predictions. Investigation of new methods to integrate information from multiple sources that are at different resolution and precision accuracy is still an active area of research.

Different algorithms have been developed to stochastically simulate reservoir properties using sparse measured data. All these methods aim to draw realizations of the random function $Z(\mathbf{u})$ based on the joint distribution function describing the random function. Some popular methods are sequential simulation (Journel 1983; Isaaks and Srivastava 1989; Goovaerts 1997; Chiles and Delfiner 1999), iterative approaches (Maksimov et al. 1993), etc. that have become the core tools in many current geostatistical applications.

These simulation methods are based on variogram functions – a measure of spatial variability based on a two-point spatial template. It has been pointed out that in order to ensure legitimacy of the kriged/simulated values, the first step should be to model the inferred variograms in such a way that the covariance matrix is guaranteed to be positive definite. To achieve this, the variogram functions need to be modeled using some well-known positive definite functions. Alternatively, Yao (1998) proposed that the covariance function could be modeled non-parametrically in Fourier domain to satisfy the positive definiteness assumption. This overcomes the limitation of parametric modeling of the covariance function, where one is limited to a few families of positive definite functions for covariance modeling, which limits our ability to model complex heterogeneities.

The major drawback of this traditional variogram-based geostatistical modeling is that they are not able to reproduce complex spatial patterns such as fluvial channels. The variogram is inadequate to capture complex curvilinear features. To reproduce the curvilinear structures and pattern continuity, the anisotropy direction of variogram must be changed locally (Deutsch and Lewis 1992; Xu 1996; Boisvert and Deutsch 2011). One could also correct for additional connectivity of the geological patterns by modifying the variogram ranges (Gringarten and Deutsch 2001).

All the previous algorithms are based on two-point connectivity functions only and cannot reproduce complex curvilinear objects such as fluvial channels. Initially Srivastava (1992) proposed to go beyond bivariate moments by introducing multiple-point geostatistics that considers variability at more than two locations taken jointly. Caers and Journel (1998) proposed the idea of borrowing conditional probabilities directly from a training image (TI), allowing the use of higher-order or multiple-point statistics to reproduce geological structures and patterns.

Strebelle (2000) developed the first structured multiple-point statistics algorithm of "single normal (extended) simulation" (called *snesim*) for simulating categorical variables. The approach of *snesim* is based on inferring the probability of various outcomes at the central node of a spatial template based on the pattern of outcomes on the remaining nodes of the template from the training image. In order to reduce the computational complexity of computing the conditional probabilities, instead of scanning the training image for each conditioning data template, it stores all probabilities in a search tree by a one-time scanning of the training image. The search tree data structure allows a fast retrieval of all required conditional probabilities during the simulation. One of the shortcomings of snesim, however, is that the training image needs to be categorical, and it does not work with continuous variables.

Several other algorithms for mps simulation have been proposed. *GrowthSim* (Eskandari and Srinivasan 2007; Huang and Srinivasan 2012) introduced the notion of simulating multiple-point simulation event conditioned to multiple-point data event in the vicinity of the simulation node. This method is in contrast to traditional multiple-point statistics algorithms where the simulation progresses one node at a time. Huang and Srinivasan (2012) demonstrated the GrowthSim algorithm for developing the reservoir model for a deepwater turbidite system. They also showed the capability of GrowthSim algorithm to represent nonstationary features.

Cumulants are combinations of statistical moments, such as mean or variance that allow the characterization of non-Gaussian random variables (Rosenblatt 1985). Higher-order spatial cumulants can capture the complex geological features and geometrical shapes of the physical phenomena. Dimitrakopoulos et al. (2010) first used the concept of cumulants in the spatial context to characterize nonlinear stationary and ergodic spatial random fields.

They showed that higher-order cumulants are related to the orientation of the spatial template. Each geological process requires its own choice of cumulants for optimal pattern analysis. They also showed that cumulants up to and including fifth order are sufficient for efficiently characterizing complex spatial geometries observed in training images.

Mustapha and Dimitrakopoulos (2010b) provided a computer code for calculating higher-order spatial cumulants. Furthermore, spatial cumulants were used as the basis for the simulation of complex geological phenomena by Mustapha and Dimitrakopoulos (2010b). The simulation takes advantage of spatial cumulants in the high-dimensional space of Legendre polynomials in a sequential framework, called *hosim*. It proceeds by randomly choosing a spatial node $\mathbf{u}$, estimating the conditional probability of the random variable given the neighboring data and previously simulated nodes and, finally, drawing a value for that node from the distribution. The process repeats until all the nodes of the grid have been visited. The process is similar to sequential simulation; the only difference is the method of deriving the analytical expressions for the local probability density functions. In *hosim*, Legendre coefficients are inferred using multiple-point spatial templates and used to derive the expressions for multivariate conditional distributions. The proposed cumulant-based method is assumed to be less dependent on the training image statistics than *snesim* method but more data-driven; in the sense that it first tries to infer the multiple-point statistics from the data, and only if not enough replicates could be found, the training image will be used for inference.

Many pattern-based algorithms such as Filtersim (Zhang 2006), direct sampling (Mariethoz and Renard 2010), Wavesim (Chatterjee et al. 2012), dispat (Honarkhah and Caers 2010), CCSIM (Tahmasebi et al. 2012), and Bunch-DS (Rezaee et al. 2013) have been proposed to improve the computational efficiency of the simulation algorithm over previously introduced ones.

However, all the proposed algorithms rely heavily on training images and the patterns depicted in the TI. We discuss the development of an MPS algorithm that can be applied even if an exhaustive training image is not available. We discuss the inference of higher-order moments from available data.

This paper mainly focuses on selecting the best training image from a set of multiple scenarios, which is the problem addressed in (Pérez et al. 2014) with another point of view. This paper is organized as follows: we first explain what higher-order cumulants and polyspectra are and describe their features. We present how they can be calculated and what features can be inferred from them. We subsequently discuss the inference of polyspectra from sparse conditioning data. We demonstrate the application of the proposed algorithm on different reservoir models such as channelized reservoir, fractured reservoirs, and models with elliptic objects with different sizes and orientations. We study the sensitivity of the algorithm to the available conditioning data. We show that the algorithm performs well even when the conditioning data is very sparse.

2 Method

The modeling algorithm presented in this paper is based on two steps: (1) calculating the polyspectra from training images and conditioning data and (2) feature extraction from polyspectra (size, orientation, and shape of the objects). These steps are explained in the next sections.

2.1 Higher-Order Cumulants and Polyspectra

As mentioned earlier, the first step of most of geostatistical simulations is to infer the multiple-point connectivity functions. In this section, we discuss higher-order statistics and present the definition of polyspectra of random function and show a fast algorithm for inferring it even when the training image is not available.

2.1.1 Higher-Order Statistics of Random Functions

Given a random variable, Z, its moment-generating function is defined as (Rosenblatt 1985):

$$M[\omega] = E[e^{\omega z}] = \int_{-\infty}^{\infty} e^{\omega z} f_Z(z) dz \tag{1}$$

The rth ($r \geq 0$) moment of Z is $\mathrm{Mom}^r[Z] = E[z^r] = \int_{-\infty}^{\infty} z^r f_Z(z) dz$. Provided that the moment-generating function M has a Taylor expansion about the origin:

$$M[\omega] = E[e^{\omega z}] = E\left[\sum\nolimits_{r=0}^{\infty} \frac{\omega^r z^r}{r!}\right] = \sum\nolimits_{r=0}^{\infty} \frac{\omega^r \mathrm{Mom}[Z]}{r!} \tag{2}$$

Then the rth moment of Z is the rth derivative of M at the origin.

For a k-th order stationary random function $Z(\boldsymbol{u})$, the k-th order moment of this process, denoted as $M_{k,z}(\boldsymbol{h}_1, \boldsymbol{h}_2, \ldots, \boldsymbol{h}_{k-1})$, is defined as the joint k-th order moment of the random variables $Z(\boldsymbol{u})$, $Z(\boldsymbol{u}+\boldsymbol{h}_1)$, …, $Z(\boldsymbol{u}+\boldsymbol{h}_{k-1})$, i.e.,

$$M_{k,z}(\boldsymbol{h}_1, \boldsymbol{h}_2, \ldots, \boldsymbol{h}_{k-1}) = E\{Z(\boldsymbol{u}), Z(\boldsymbol{u}+\boldsymbol{h}_1), \ldots, Z(\boldsymbol{u}+\boldsymbol{h}_{k-1})\} \tag{3}$$

The k-th order polyspectrum is defined as the $(k-1)$-dimensional discrete Fourier transform of the k-th order moment ($M_{k,z}(\boldsymbol{h}_1, \boldsymbol{h}_2, \ldots, \boldsymbol{h}_{k-1})$), i.e.,

$$S_{k,z}(\omega_1, \omega_2, \ldots, \omega_{k-1}) = \sum\nolimits_{\boldsymbol{h}_1=-\infty}^{\infty} \cdots \sum\nolimits_{\boldsymbol{h}_{k-1}=-\infty}^{\infty} M_{k,z}(\boldsymbol{h}_1, \boldsymbol{h}_2, \ldots, \boldsymbol{h}_{k-1}) \times \exp\left[-j\sum\nolimits_{i=1}^{k-1} \omega_i \boldsymbol{h}_i\right] \tag{4}$$

The $\omega_1 - \omega_2 - \cdots - \omega_{k-1}$ frequency space is the domain of support for $S_{k,z}(\omega_1, \omega_2, \ldots, \omega_{k-1})$. $S_{2,\,z}(\boldsymbol{w})$ (which is the Fourier transform of covariance function) is known as the power spectrum (in many papers the notation $P(\boldsymbol{w})$is used). $S_{3,z}(\omega_1, \omega_2)$ and $S_{4,z}(\omega_1, \omega_2, \omega_3)$ are known as bispectrum (also noted as $B(\omega_1, \omega_2)$) and trispectrum (also noted as $T(\omega_1, \omega_2, \omega_3)$) of the random process and have been widely used for many applications in signal processing such as system identification, image reconstruction, etc. (Chandran and Elgar 1993; Hall and

Giannakis 1995; Giannakis and Mendel 1989). Many symmetries exist in arguments of $M_{k,z}(\boldsymbol{h}_1, \boldsymbol{h}_2, \ldots, \boldsymbol{h}_{k-1})$ making their calculation manageable.

By plugging Eq. 3 in Eq. 4, the power spectrum, bispectrum, and trispectrum can be calculated by

$$S_{2,z}(\boldsymbol{\omega}) = |F(\boldsymbol{\omega})|^2 \tag{5}$$

$$S_{3,z}(\boldsymbol{\omega}_1, \boldsymbol{\omega}_2) = F(\boldsymbol{\omega}_1)F(\boldsymbol{\omega}_2)F^*(\boldsymbol{\omega}_1 + \boldsymbol{\omega}_2) \tag{6}$$

$$S_{4,z}(\boldsymbol{\omega}_1, \boldsymbol{\omega}_2, \boldsymbol{\omega}_3) = F(\boldsymbol{\omega}_1)F(\boldsymbol{\omega}_2)Z(\boldsymbol{\omega}_3)F^*(\boldsymbol{\omega}_1 + \boldsymbol{\omega}_2 + \boldsymbol{\omega}_3) \tag{7}$$

More generally, it can be proved that (Brillinger 1965)

$$S_{k,z}(\boldsymbol{\omega}_1, \boldsymbol{\omega}_2, \ldots, \boldsymbol{\omega}_{k-1}) = F(\boldsymbol{\omega}_1) \ldots F(\boldsymbol{\omega}_{k-1})F^*(\boldsymbol{\omega}_1 + \boldsymbol{\omega}_2 + \cdots + \boldsymbol{\omega}_{k-1}) \tag{8}$$

where $\boldsymbol{F}\ (\boldsymbol{w})$ is the Fourier transform of $\boldsymbol{Z}\ (\boldsymbol{u})$ and F^* is its conjugate transform. As can be seen in Eqs. 5 and 6, the power spectrum is related to the amplitude of the Fourier transform, but bispectrum is related to both amplitude and the phase of the Fourier transform.

If one aims to infer the moments from a training image, the Fourier transform of the training image can be used in Eq. 8 to calculate the polyspectra and use inverse Fourier transform to calculate the moment as in Eq. 4. This would be computationally more efficient than the case that the pairs with prescribed lag size are searched in the training image to calculate the moments. For example, for a simple 1-D training image with N data points, calculating covariance by calculating power spectrum density will be done in O (N log N) operations instead of O(N^2) (by searching the pairs in the training image and calculating covariance function directly). In addition, this allows us to model the connectivity function without constraining to a few parametric functions. For example, it will be shown in the results section that using the covariance inferred from power spectrum instead of using covariance models based on positive definite functions improves the reproduction of connectivity in the simulated image.

However, when the training image is not available, calculating the Fourier transform from scatter data points (conditioning data) and inferring the polyspectrum will be challenging. This in fact is the main reason that when the training image is not available, one is only limited to variogram-based (two-point statistics) simulation techniques. In order to solve this issue, we propose a computationally efficient and robust method for calculating higher-order spectra of a random function from scattered data points using a nonuniform fast Fourier transform (NUFFT) (Fessler and Sutton 2003; Keiner et al. 2009; Khalighi et al. 2015; Drach et al. 2015).

2.1.2 Estimation of Higher-Order Spectra from Scattered Data Points

Estimation of higher-order moments by calculating higher-order polyspectra improves the computational efficiency of the process due to the fast performance of FFT. However, FFT requires an exhaustive dataset or image (such as a training image) and not amenable to calculating higher-order moments from scattered data. In this section, we present a new methodology for fast calculation of Fourier transform from scattered data based on compressed sensing theory (Donoho 2006). This will be done by assuming that the Fourier transformation of geologic variable is nearly sparse in frequency domain (have only a few nonzero coefficient). This assumption has been used in many image reconstruction studies (Gan 2007; Lustig et al. 2008; Wang et al. 2008). We propose to infer Fourier transform of reservoir variables from the available scattered data by imposing sparsity constraint on Fourier transform. A brief discussion of compressed sensing theory follows.

Compressed sensing has been recently proposed for reconstruction of sparse signals from partial observations in a complementary "incoherent" domain using convex optimization (Candès et al. 2006; Donoho 2006; Candes and Tao 2006). Assume a sparse signal x_N and its transformation coefficients y_N when subject to transformation matrix $A_{N\times N}$:

$$y_{N\times 1} = A_{N\times N} x_{N\times 1} \tag{9}$$

In the notation of Eq. 9, x is the Fourier transform coefficient, A is the inverse Fourier transform matrix, and y is the actual image. Because the image exhibits spatial correlation, its Fourier transformation is sparse.

When performing conditional geostatistical simulations, construction of an image or a map knowing only $(K(K \ll N)$ observations is desired. These sparse observations $y_{K\times 1}$ can be reconstituted as:

$$y_{K\times 1} = A_{K\times N} x_{N\times 1} \tag{10}$$

This problem is ill-posed and has multiple solutions. However, a unique solution can be found by imposing a sparsity constraint on x, by solving the following l_1 norm optimization problem:

$$\min_{x\in R^N} \|x\|_1 \text{ subject to } y_{K\times 1} = A_{K\times N} x_{N\times 1} \tag{11}$$

This is the basis pursuit problem in the field of signal processing (Chen and Donoho 1994). It should be noted that because of the spatial correlation of the images, x (the Fourier transform) has a sparse representation. Another approach for solving this problem is to change the problem to a regularization problem and solve the following (Zhdanov 2002):

$$\min_{x \in R^N} \|y - Ax\|_p + \lambda \|x\|_q \tag{12}$$

Usually three alternatives for the objective function norms lq and lp in the above equation are considered: the linear least squares (LLS) [p=q = 2], the least absolute deviation (LAD) [p=q = 1], and the least mixed norm (LMN) [p = 2, q = 1].

As mentioned before, the Fourier transform coefficient of images is i sparse; therefore, in our case, x is the Fourier transform coefficient of the geologic map, A is the inverse Fourier transformation matrix, and y is the scattered measured geology variables. By solving the optimization problem, Fourier transform can be calculated and used to estimate the higher-order moments.

This idea has also been used for geostatistical simulation (Jafarpour et al. 2009) where the Eq. 11 has been used to estimate a Fourier-based transform (DCT in that paper) and the reservoir properties are calculated by inverse transform. However, the proposed algorithm can only generate one realization and uncertainty cannot be quantified.

2.2 *Feature Extraction from Polyspectra*

In this section, we present how the polyspectra can be used to extract information about size, shape, and orientation of features in images. We focus on the properties of power spectrum and bispectrum.

2.2.1 Image Orientation

One of the properties of Fourier transform is that if an image is rotated, its Fourier transform also rotates (Fig. 1) (Reddy and Chatterji 1996). Therefore, by applying a rotational transform in the Fourier transform space, one can account for changes in the azimuth direction of continuity in the spatial domain. Since the amplitude of the Fourier transform is related to power spectrum as in Eq. 4, this allows us to find the orientation of the objects from its power spectrum. For instance, Fig. 2 shows three different maps of channelized reservoir with different orientations, and their power spectrum is shown in Fig. 3. As can be seen in Fig. 3, the angle subtended by the main axis of the power spectrum represents the orientation of the objects. Figures 4 and 5 make the same point for different type of reservoirs with ellipsoid objects.

2.2.2 Size Identification from Power Spectrum

Another useful property of Fourier transform is that if the Fourier transform F{$g(t)$}) is denoted as $G(f)$, i.e.,

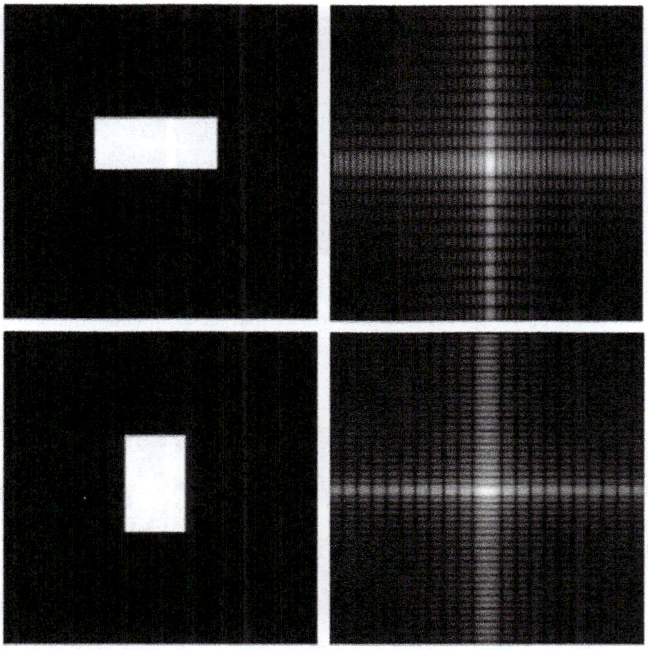

Fig. 1 The effect of rotation in the spatial domain on the frequency characteristics: The *top row images* are the original image (*left*) and its power spectral density, and the *two bottom images* are the rotated image and its power spectral density. It can be seen that the Fourier transform has been also rotated (http://www.cse.iitd.ac.in/~parag/projects/DIP/asign1/transrotfft.shtml)

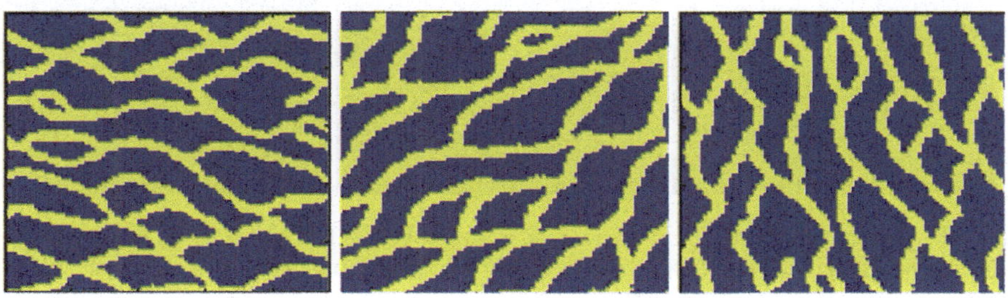

Fig. 2 Three different maps of facies type for channelized reservoirs with different orientation of channels

$$F\{g(t)\} = G(f), \text{ then } F\{g(ct)\} = \frac{G\left(\frac{f}{c}\right)}{|c|}$$

where c is a constant. This allows us to relate geobody dimension to the bandwidth of the power spectrum ratio. For instance, Fig. 6 shows two different maps of channelized reservoir with different sizes, and their power spectrum is shown in Fig. 7. As can be seen in Fig. 7, the bandwidth of the power spectrum represents the

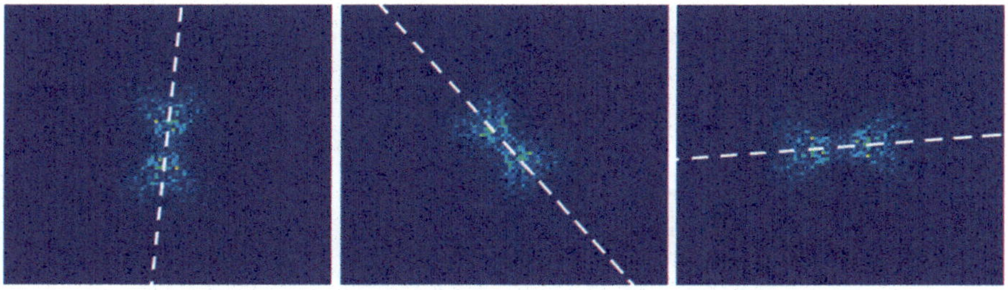

Fig. 3 The power spectrum of the facies map shown in Fig. 2. The *white dashed line* shows the main orientation of the power spectrum. It can be seen that the azimuthal orientation of continuity shown by the spatial objects is the same as that shown by the power spectrum

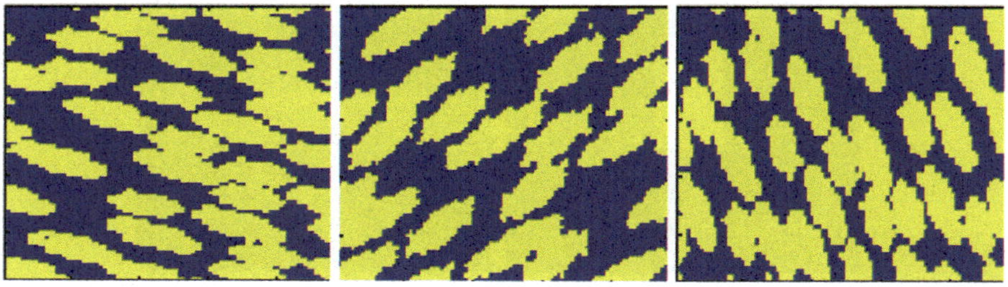

Fig. 4 Three different facies models with ellipsoidal objects in different orientations

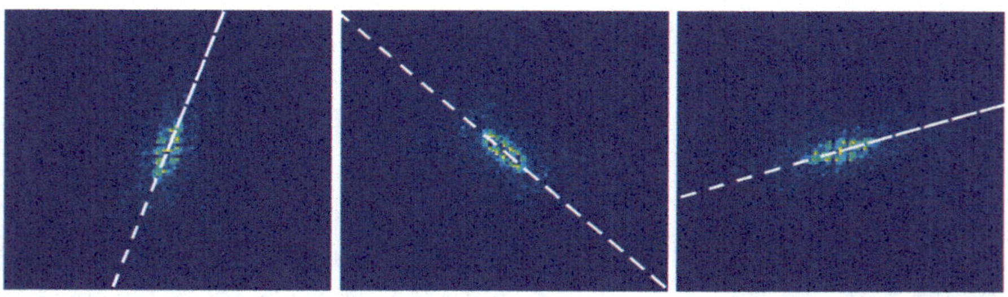

Fig. 5 The power spectrum of the facies map shown in Fig. 4. The *white dashed line* shows the main orientation of the power spectrum. It can be seen that the rotation of the objects shows the same rotation as the power spectrum

size of the objects. Figures 8 and 9 make the same point for different reservoir models with ellipsoidal objects.

2.2.3 Feature Extraction from Bispectrum

As shown previously, the size and orientation of the objects can be inferred from the bandwidth and orientation of the power spectrum. However, the power spectrum

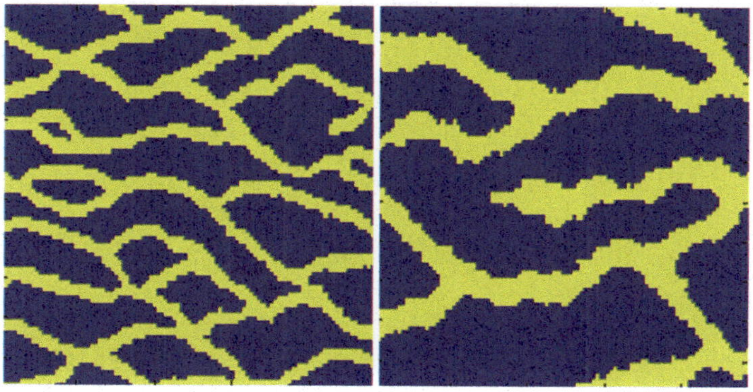

Fig. 6 Two different models for channelized reservoirs with different sizes of channels

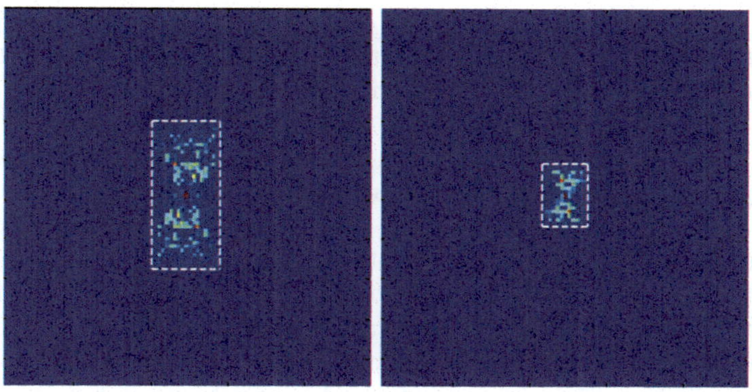

Fig. 7 The power spectrum of the facies map shown in Fig. 6. The *white dashed line* shows the bandwidth of the power spectrum. It can be seen that the size of the objects corresponds to the bandwidth of the power spectrum

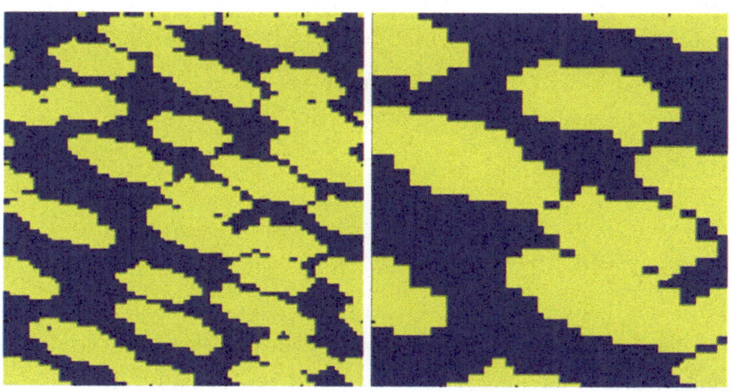

Fig. 8 Two different reservoir models with ellipsoidal objects of different dimensions

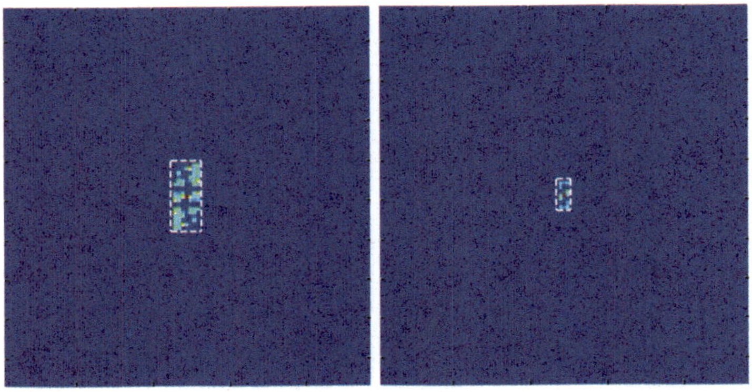

Fig. 9 The power spectrum of the facies models shown in Fig. 8. The *white dashed line* shows the bandwidth of the power spectrum. It can be seen that the size of the objects corresponds to the bandwidth of the power spectrum

cannot help with distinction of the object type. In this section, we present how to distinguish between different objects using the bispectrum.

The bispectrum is a triple product of Fourier coefficients, $S_{3,z}(\omega_1, \omega_2) = Z(\omega_1)Z(\omega_2)Z^*(\omega_1 + \omega_2)$, and is a complex-valued function of two frequencies, unlike the power spectrum, which is a second-order product of Fourier coefficients and a function of only one frequency. Unlike the power spectrum, the bispectrum retains information about the phase of the Fourier transform of the sequence. The phase of the Fourier transform is a nonlinear function of frequency, and this nonlinearity is extracted by the biphase (the phase of the bispectrum). For example, left and right asymmetric sequences will have opposite signs for the biphase. These properties form a basis for the use of the bispectrum for extracting features from patterns. Additionally, parameters can be defined from the bispectrum that are invariant to translation, scaling, and amplification. In particular, the phase of the integrated bispectrum along a radial line of slope (see Fig. 10) satisfies these properties (Chandran and Elgar 1993; Chandran et al. 1997). Parameters

$$P(a) = \arctan\left(\frac{I_i(a)}{I_r(a)}\right) \tag{13}$$

where

$$I(a) = I_r(a) + i\, I_i(a) = \int_{f_1=0}^{1/(1+a)} B(f_1, af_1)df_1 \tag{14}$$

for $0<a<1$ can serve as features for patterns. The variables I_r and I_i refer to the real and imaginary parts of the integrated bispectrum, respectively, and $i = \sqrt{-1}$. It has been shown in Chandran and Elgar (1993) that these parameters satisfy the desired

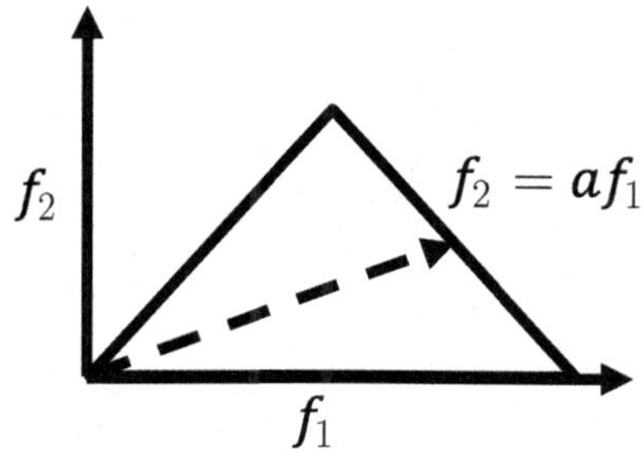

Fig. 10 Region of computation of the bispectrum. Features are obtained by integrating the complex bispectrum along a radial line with slope = a (*dashed line*). The phase of this integral is translation and scale invariant

invariance properties. It means that scaling and rotating will not change this identifier. For example, scaling the original sequence results in an expansion or contraction of the Fourier transform that is identical along the $f1$ and f_2 directions. Thus, the bispectral values along a radial line in bifrequency space map back onto the same line upon scaling. The real and imaginary parts of the integrated bispectrum along a radial line are multiplied by identical real-valued constants upon scaling, and therefore, the phase of the integrated bispectrum is unchanged (Nikias and Raghuveer 1987; Chandran and Elgar 1993; Chandran et al. 1997).

Since this identifier is not sensitive to rotation and scaling, it can be used as an identifier to distinguish between the shapes of the objects within the image.

To demonstrate the feasibility of this application, we run an example on 4 1-D signals. These signals are shown in Fig. 11. We also generate different signals with different scaling factors as shown in Fig. 12. To show the sensitivity of the identifier to noise, we also added Gaussian noise to each signal. We then calculated $P(a)$ for each signal as a function of a. Figure 13 shows the scatter plot of $P(1)$ versus $P(1/16)$ for each signal. As shown in Fig. 13, given $P(1/16)$ and $P(1)$, the two similar looking bolts clearly can be distinguished.

It should be noted that this example is just to demonstrate the feasibility of the bispectrum identifier for shape classification. This method can be easily applied to 2-D images. The only difference is that the scalar a becomes a vector of size 2. Later in the results section, we will show the applicability of the algorithm for 2-D training images in the results section.

3 Results

In this section, we present the results of the explained methodology. First, we show the results of calculating higher-order spectra from both TIs and conditioning data. Then, we present how the best set of TIs in terms of shapes of the objects can be detected using the phases of the bispectrum. Finally, we show how the size and orientation of the objects can be corrected using the features from power spectrum.

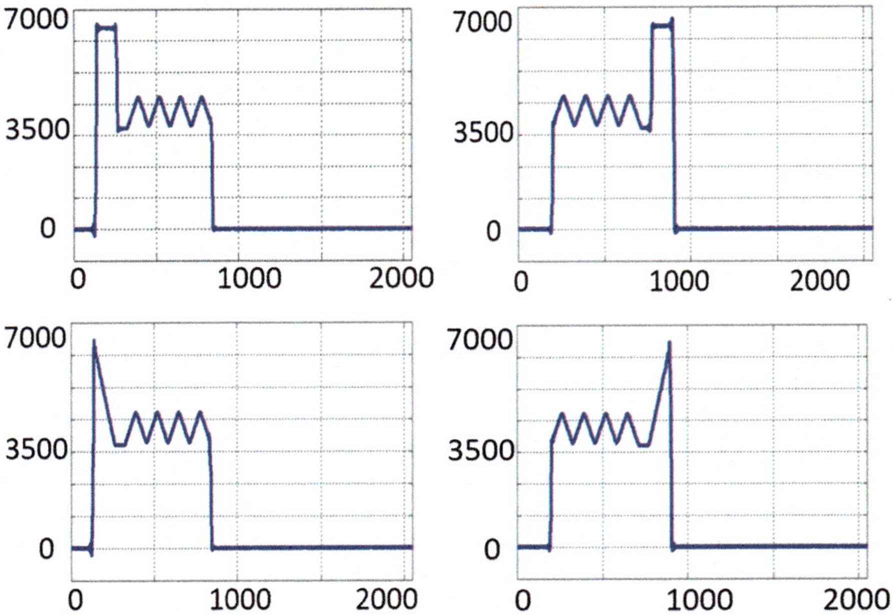

Fig. 11 Profiles of two types of bolts with *left* and *right* orientation

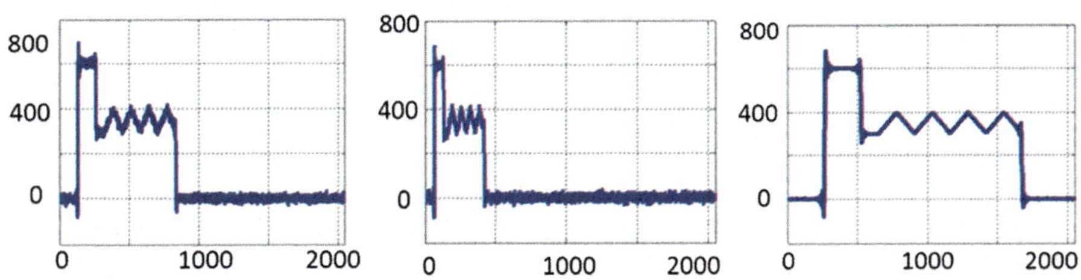

Fig. 12 Examples of the 1-D signal with different scaling factor (1, 0.5, 2 from *left* to *right*)

Fig. 13 The values of a feature vector consisting of {P(a), a = 1/16 and a = 1) for bolts 1 and 2 with *left* and *right* orientations, for scale variations between 0.5 and 2.0

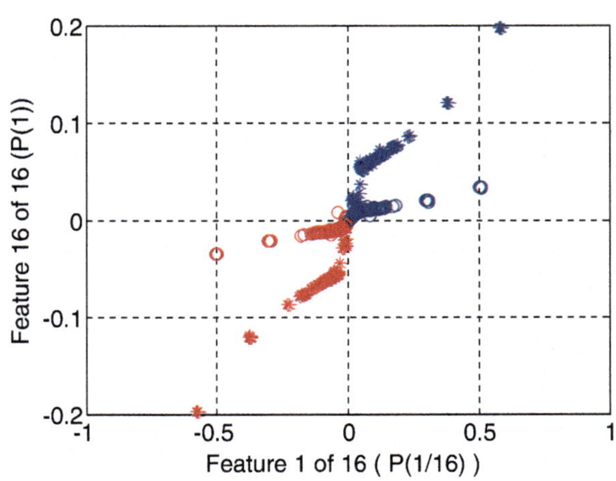

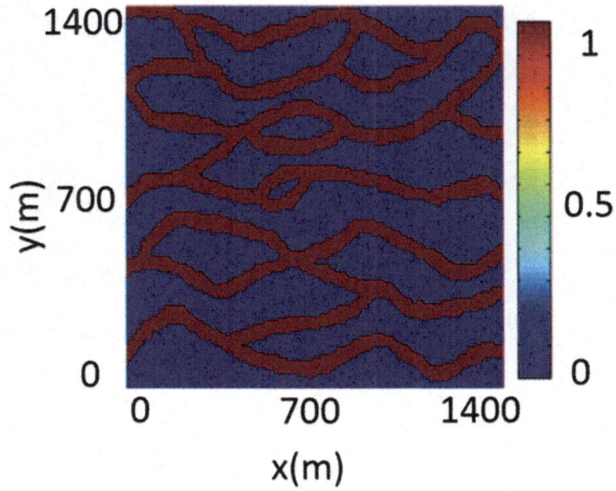

Fig. 14 Training image used for this study

3.1 Calculating Higher-Order Spectra Training Image and Conditioning Data

As mentioned above, the moments of random function $z(\boldsymbol{u})$ can be calculated by using FFT of the training image. Figure 14 shows the training image in this study, and Fig. 15 shows the covariance map of the training image using direct searching of pairs in the training image and covariance calculated using FFT. As can be seen in Fig. 15, the covariance map can be correctly calculated by means of FFT, which is much faster than directly calculating by searching the pairs in the training image.

Then, we selected 500 data points randomly from the training image shown in Fig. 14 and calculated the covariance and third-order moment using the methodology discussed for computing the polyspectra using sparse data. The scattered conditioning data are shown in Fig. 18. Figures 19 and 20 show the calculated covariance map and three-point connectivity functions with different template configurations, respectively. This process is fast, and it can be seen that the connectivity maps calculated using the scattered data point resemble the polyspectra computed using the exhaustive image as shown in Figs. 15 and 17.

3.2 Selecting Best Training Image Given Bispectrum

In this section, we present the results for feature extraction from bispectrum for 2-D training images. We selected three different training images for different geologic scenarios as shown in Fig. 21. Then we generated several images with different orientation and geobody size. Figure 22 shows four examples of these images. We extracted four sets of scattered data from four random images of the data set and

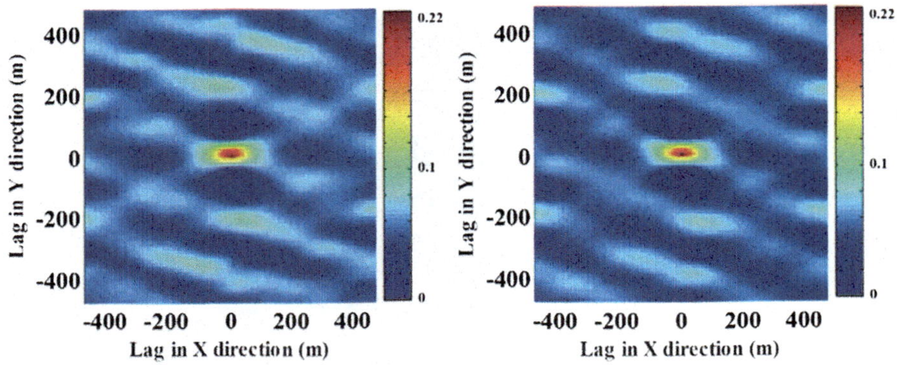

Fig. 15 Covariance map calculated by directly searching training image (*left*) and by using FFT (*right*)

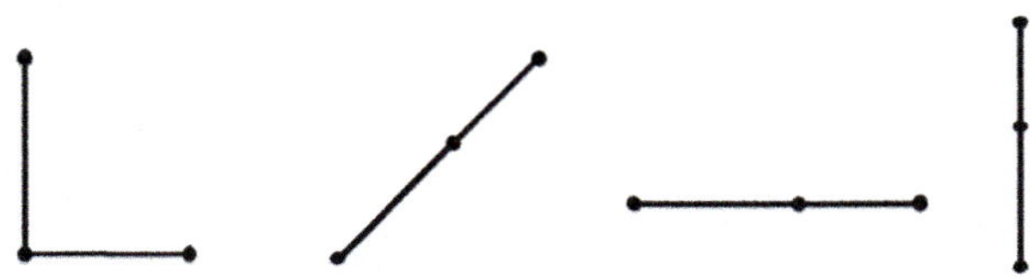

Fig. 16 Three-point connectivity inferred using different spatial templates (from *left* to *right*: L shape, 45 degree, *x*-*x* and *y*-*y* connectivity)

calculated the biphases of the bispectrum of all the images and the scattered data sets. The biphases computed only using the sparse conditioning data are superimposed (diamond points) on the scatter observed on the basis of the exhaustive images. As can be seen, the three different scenarios can be successfully distinguished using these features from both scattered data sets and training images. Therefore, we can find the correct training image (in terms of the object shape) by selecting the image with the closest feature value to the one of the scattered data set (Fig. 23).

3.3 Size and Orientation Detection

In the previous section, we presented how the bispectrum can be used to distinguish among different training images with different objects. However, the orientation and scale of the objects cannot be detected from bispectrum. In this section, we present how the power spectrum can be used to distinguish the orientation and the scale of the objects. We generated the rotated realization from a TI (45°) and used it as the correct map of the reservoir. We then sampled some conditioning data from the correct map as shown in Fig. 24. Then, we calculated the power spectrum of both available TI and the scattered data set. Figure 25 shows the covariance map of

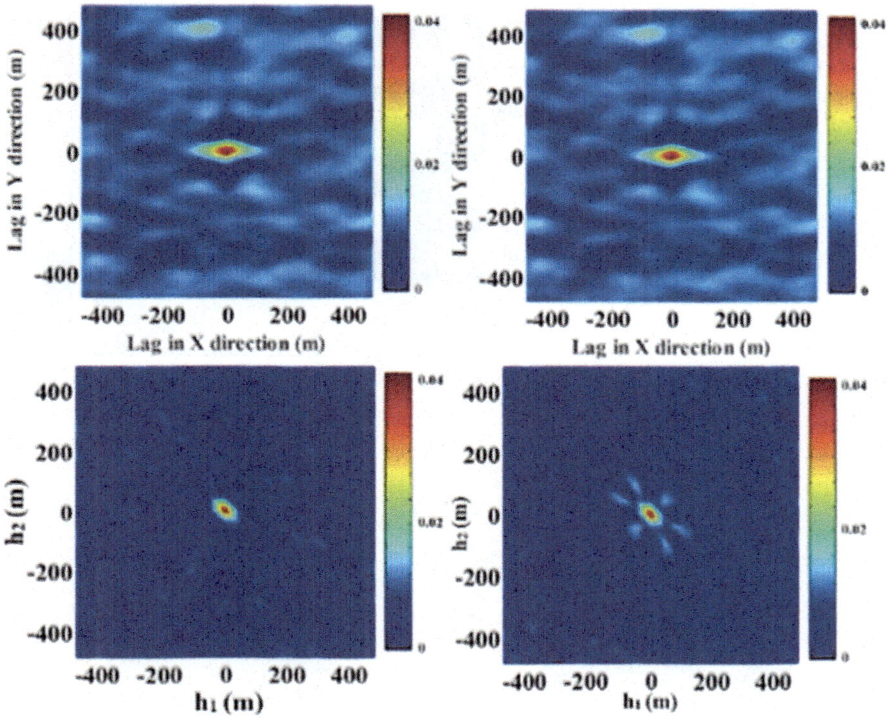

Fig. 17 The third-order moment of the image shown in Fig. 14. The *left* column is calculated by calculating bispectrum, and the *right* column is calculated by direct search of pairs in the image

Fig. 18 Scattered conditioning data used for calculating covariance. The number of data points is 500 (~2.3 % of total data points)

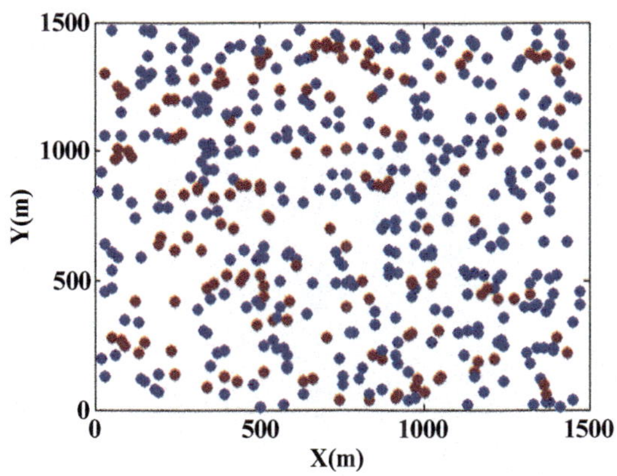

the TI and conditioning data as well as the corresponding power spectrum. As shown in Fig. 25, the orientation direction of the two power spectrums is different (40°). Thus, we rotated the available TI to account for the mismatch of the orientation of the channels. To show the importance of the orientation of TI in simulations, we used both original and the rotated TIs to perform simulation using

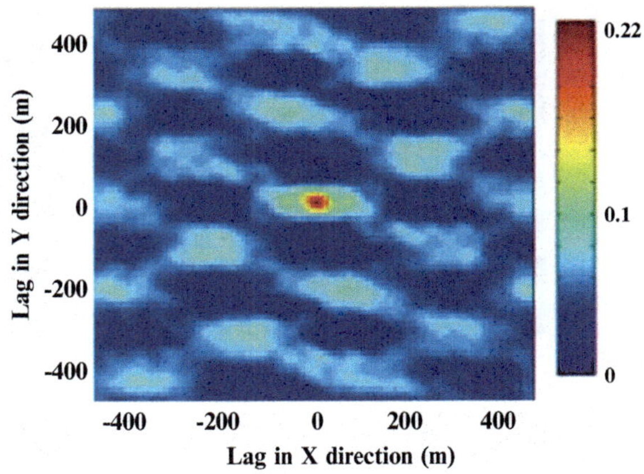

Fig. 19 Covariance map calculated by Fourier transform of scattered data by sparsity constraints

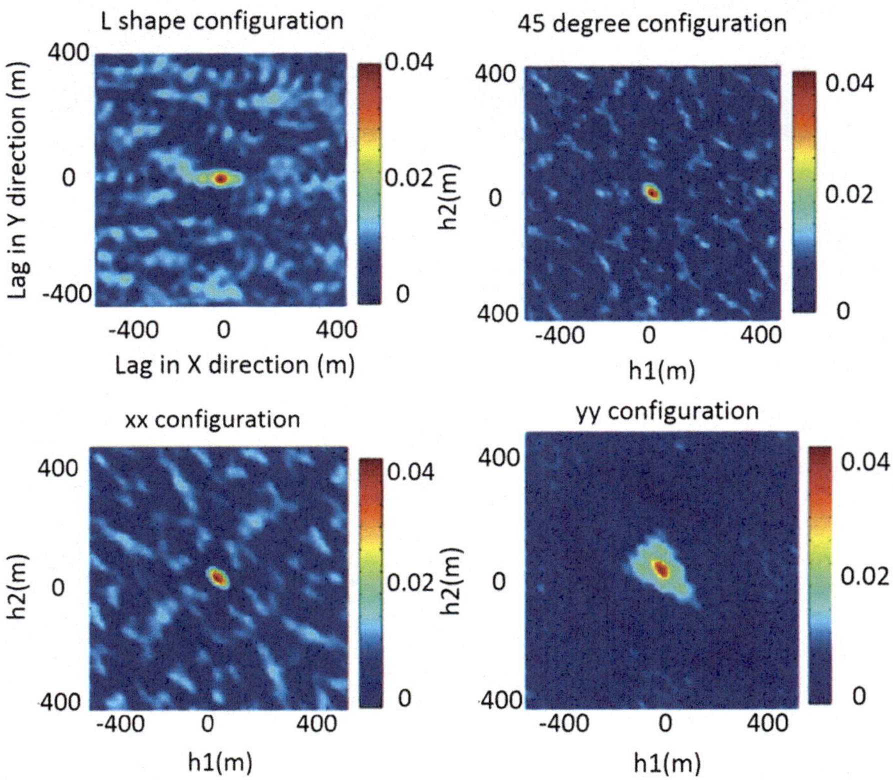

Fig. 20 Calculated three-point connectivity with different configurations using scattered data in Fig. 18 (the templates are, from left to right, L shape, 45 degree, xx configuration, and yy configuration, as shown in Fig. 16)

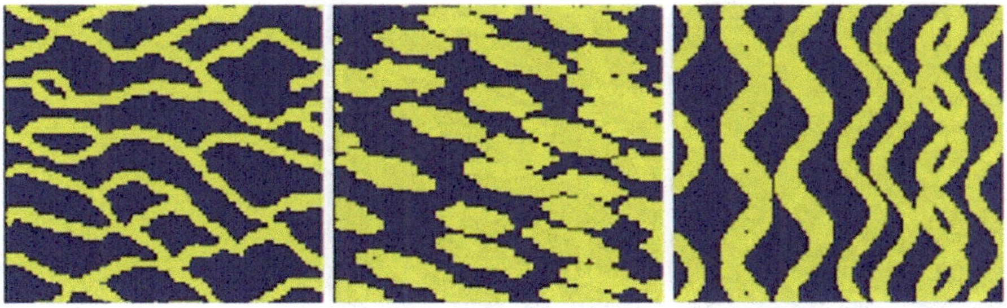

Fig. 21 The training images used in this study with different objects and geologic scenarios

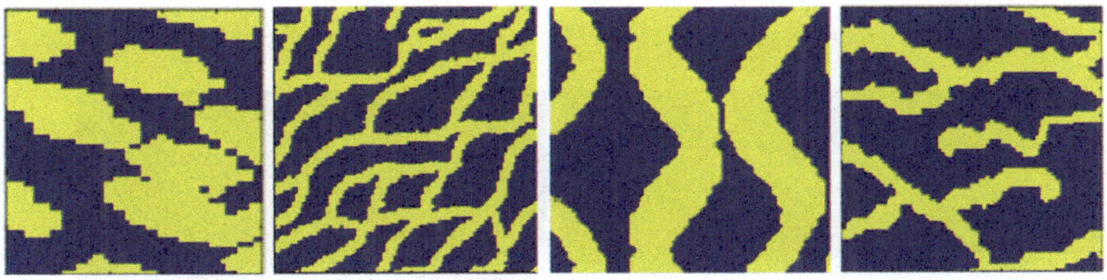

Fig. 22 Different rotated and scaled images from the reference training images in Fig. 21

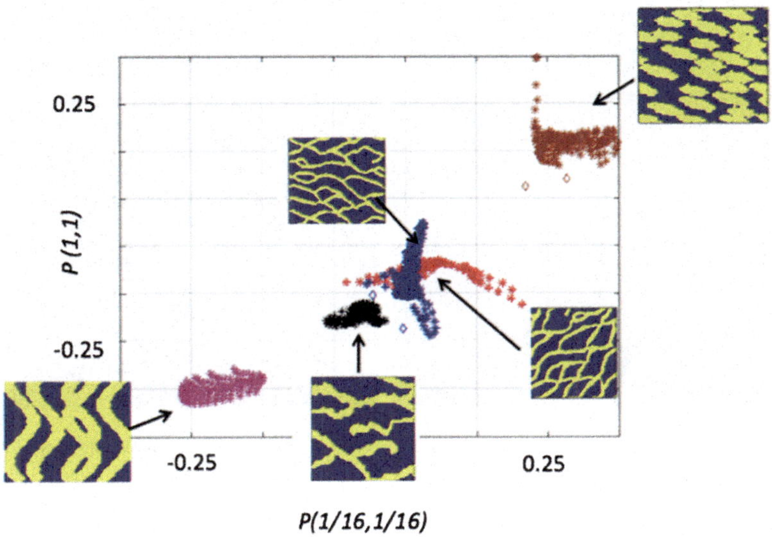

Fig. 23 The values of a feature vector consisting of {P(1,1), P(1/16,1/16)} for all of the different images with different scale and orientations

direct sampling (Mariethoz and Renard 2010). As can be seen in Fig. 26, the results of the simulation using the rotated TI is much closer to the correct map indicating the importance of the orientation of objects within TI in simulation.

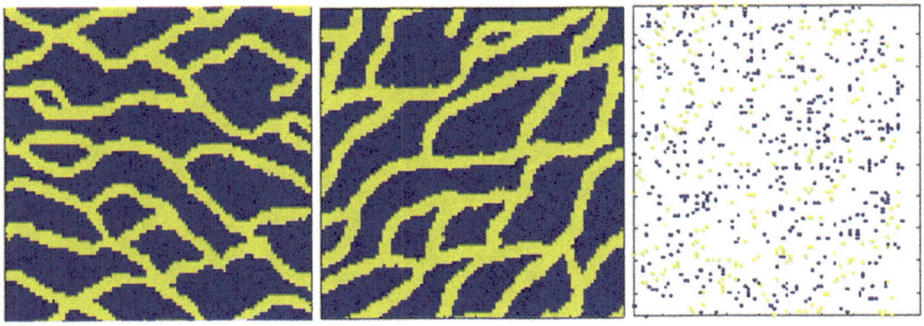

Fig. 24 The available TI (in the *left*), the corrected map with different orientation (in the *middle*), and the scattered data sampled from the correct map (in the *right*)

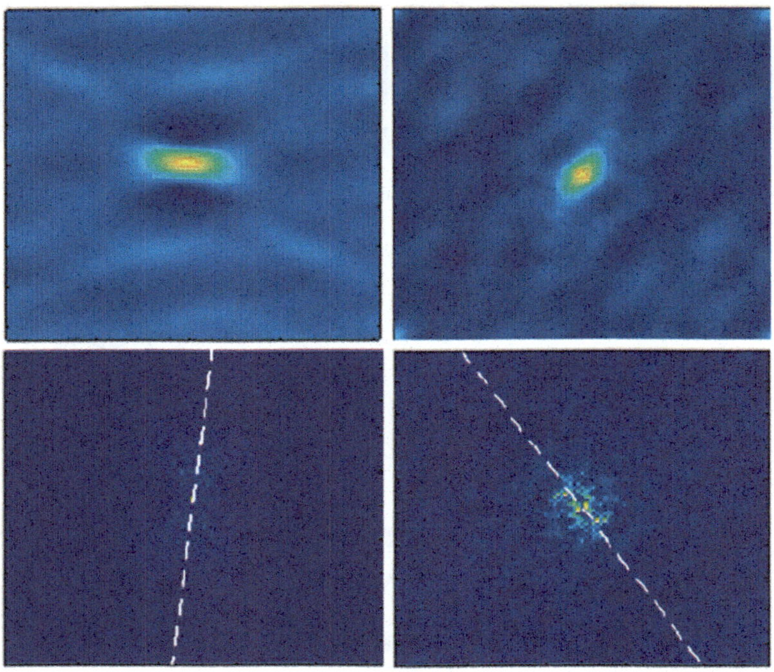

Fig. 25 The covariance map of the available TI (*top left*), the covariance map of the scattered data (*top right*), the power spectrum of the available TI (*bottom left*), and the power spectrum of the scattered data (*bottom right*). *Dashed line* represents the main orientation of the power spectrum

We also generated another realization form a TI with different geobody size, and used it as the correct map of the reservoir. We sampled some conditioning data from the correct map as shown in Fig. 27. Then, we calculated the power spectrum of both available TI and the scattered data set. Figure 28 shows the covariance map of the TI and conditioning data as well as the corresponding power spectrum. As shown in Fig. 28, the bandwidth of the two power spectrums are different. Thus, we

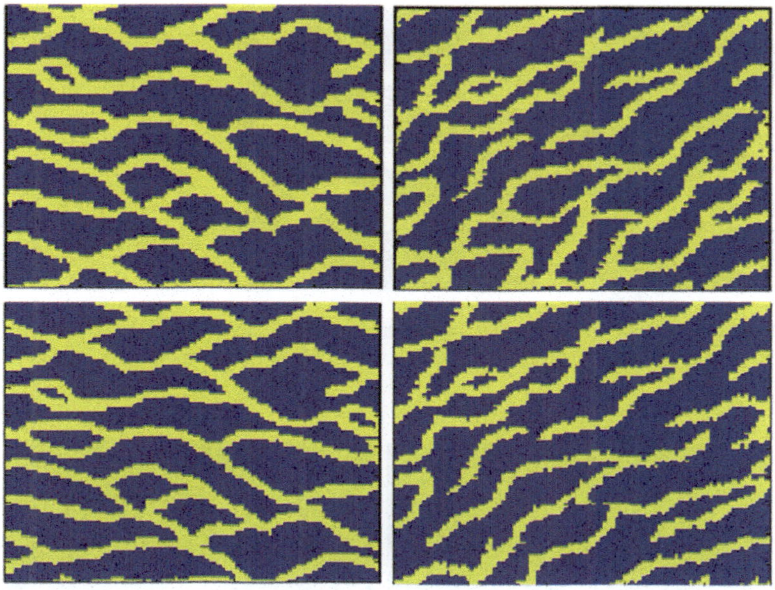

Fig. 26 The available TI (*top left*), the corrected TI (*top right*), the simulation using the available TI (*bottom left*), and the simulation using the corrected TI (*bottom right*)

generated another TI to account for the mismatch of the size of the channels. To show the importance of the geobody size of TI in simulations, we used both original and the corrected TIs to perform simulation using direct sampling. As can be seen in Fig. 29, the results of the simulation using the corrected TI are much closer to the correct map indicating the importance of the size of objects within TI in simulation.

4 Summary and Conclusion

In this paper, we presented a new method for modeling higher-order connectivity functions. The method is based on calculating polyspectra. We showed the method is faster than direct searching within the images due to the use of FFT. Also, we proposed a new method for modeling polyspectra from scattered data. We showed that the method has high accuracy for both training images and scattered data. We then proposed a new method for ranking different sets of TIs corresponding to different geological scenarios. These set of TIs represent the associated uncertainty and contain features of different shapes (channels, ellipses, fracture, etc.) as well as different sizes and orientation. We analyzed the spectra of the different TIs (power spectrum and bispectrum) to distinguish between different TIs. We showed that object size and orientation can be inferred from the power spectrum, while the object shapes can be inferred from the bispectrum. Therefore, the combination of power spectrum and bispectrum can be used as an identifier for each TI.

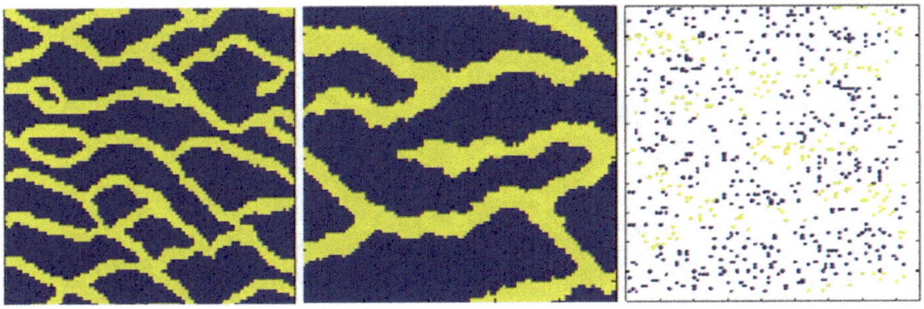

Fig. 27 The available TI (*in the left*), the corrected map with different geobody size (*in the middle*), and the scattered data sampled from the correct map (*in the right*)

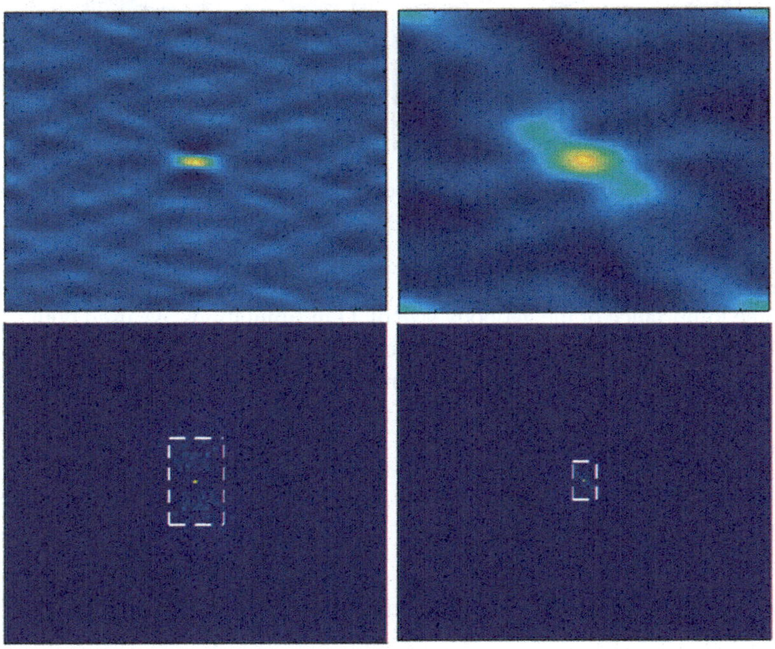

Fig. 28 The covariance map of the available TI (*top left*), the covariance map of the scattered data (*top right*), the power spectrum of the available TI (*bottom left*), and the power spectrum of the scattered data (*bottom right*). *Dashed line* represents the bandwidth of the power spectrum

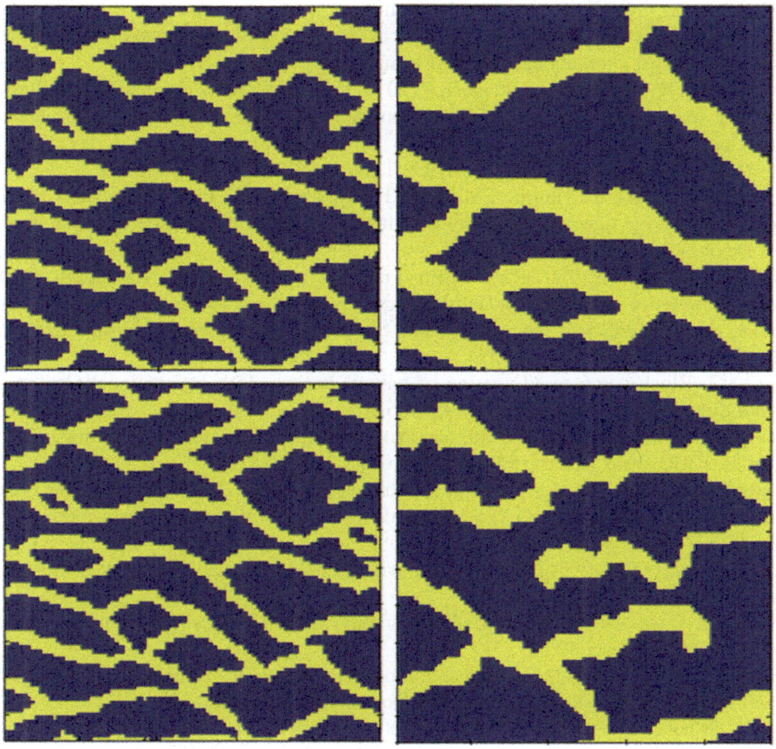

Fig. 29 The available TI (*top left*), the corrected TI (*top right*), the simulation using the available TI (*bottom left*), and the simulation using the corrected TI (*bottom right*)

Bibliography

Brillinger DR (1965) An introduction to polyspectra. Ann Math Stat 36:1351–1374

Boisvert JB, Deutsch CV (2011) Programs for kriging and sequential Gaussian simulation with locally varying anisotropy using non-Euclidean distances. Comput Geosci 37:495–510

Caers J, Journel AG (1998) Stochastic reservoir simulation using neural networks trained on outcrop data. SPE annual technical conference

Candes EJ, Tao T (2006) Near-optimal signal recovery from random projections: Universal encoding strategies? IEEE Trans Inf Theory 52:5406–5425

Candès EJ, Romberg J, Tao T (2006) Robust uncertainty principles: exact signal reconstruction from highly incomplete frequency information. IEEE Trans Inf Theory 61:489–509

Chandran V, Elgar S (1993) Pattern recognition using invariants defined from higher order spectra-one-dimensional inputs. IEEE Trans Sig Process 41:205–212

Chandran V, Carswell B, Boashash B, Elgar S (1997) Pattern recognition using invariants defined from higher order spectra: 2-D image inputs. IEEE Trans Image Process 6:703–712

Chatterjee S, Dimitrakopoulos R, Mustapha H (2012) Dimensional reduction of pattern-based simulation using wavelet analysis. Math Geosci 44:343–374

Chen S, Donoho D (1994) Basis pursuit. Conference record of the twenty-eighth asilomar conference on signals, systems and computers IEEE, pp 41–44

Chiles JP, Delfiner P (1999) Geostatistics modeling spatial uncertainty. Wiley, New York
Deutsch CV, Lewis RW (1992) Advances in the practical implementation of indicator geostatistics. 23rd international APCOM symposium
Dimitrakopoulos R, Mustapha H, Gloaguen E (2010) High-order statistics of spatial random fields: exploring spatial cumulants for modeling complex non-Gaussian and non-linear phenomena. Math Geosci 42:65–99
Donoho DL (2006) Compressed sensing. IEEE Trans Inf Theory 52:1289–1306
Drach A, Khalighi AH, ter Huurne FM, Lee CH, Bloodworth C, Pierce EL, Sacks MS (2015) Population-averaged geometric model of mitral valve from patient-specific imaging data. J Med Devices 9:030952
Eskandari K, Srinivasan S (2007) Growthsim–a multiple point framework for pattern simulation. EAGE Conference on Petroleum Geostatistics
Fessler JA, Sutton BP (2003) Nonuniform fast fourier transforms using min-max interpolation. IEEE Trans Sig Process 51:560–574
Gan L (2007) Block compressed sensing of natural images. 15th International Conference on Digital Signal Processing. IEEE, pp 403–406
Giannakis GB, Mendel JM (1989) Identification of nonminimum phase systems using higher order statistics. IEEE Trans Acoust, Speech Sig Process 37:360–377
Goovaerts P (1997) Geostatistics for natural resources evaluation. Oxford Press, New York
Gringarten E, Deutsch CV (2001) Teacher's aide variogram interpretation and modeling. Math Geol 33:507–534
Hall TE, Giannakis GB (1995) Bispectral analysis and model validation of texture images. IEEE Trans Image Process 4:996–1009
Honarkhah M, Caers J (2010) Stochastic simulation of patterns using distance-based pattern modeling. Math Geosci 42:487–517
Huang YC, Srinivasan S (2012) Efficient conditional simulation of spatial patterns using a pattern-growth algorithm. Geostat Oslo. Springer, Dordrecht, pp 209–220
Isaaks EH, Srivastava RM (1989) An introduction to applied geostatistics. Oxford University Press, New York, p 561
Jafarpour B, Goyal VK, McLaughlin DB, Freeman WT (2009) Transform-domain sparsity regularization for inverse problems in geosciences. Geophysics 74:R69–R83
Journel A (1983) Nonparametric estimation of spatial distributions. Math Geol 15:445–468
Keiner J, Kunis S, Potts D (2009) Using NFFT 3–a software library for various nonequispaced fast Fourier transforms. ACM Trans Math Softw (TOMS) 36:19
Khalighi AH, Drach A, ter Huurne FM, Lee CH, Bloodworth C, Pierce EL, … & Sacks MS (2015, June) A comprehensive framework for the characterization of the complete mitral valve geometry for the development of a population-averaged model. In International Conference on Functional Imaging and Modeling of the Heart (pp. 164–171). Springer International Publishing
Lustig M, Donoho DL, Santos JM, Pauly JM (2008) Compressed sensing MRI. Sig Process Mag, IEEE 25:72–82
Maksimov MM, Galushko VV, Kutcherov AB, Surguchev LM, Bratvold RB (1993) Parallel nested factorization algorithms. SPE Symposium on Reservoir Simulation
Mariethoz G, Renard P (2010) Reconstruction of incomplete data sets or images using direct sampling. Math Geosci 42:245–268
Mustapha H, Dimitrakopoulos R (2010a) High-order stochastic simulation of complex spatially distributed natural phenomena. Math Geosci 42:457–485
Mustapha H, Dimitrakopoulos R (2010b) A new approach for geological pattern recognition using high-order spatial cumulants. Comput Geosci 36:313–334
Nikias CL, Raghuveer MR (1987) Bispectrum estimation: a digital signal processing framework. Proc IEEE 72:869–891
Pérez C, Mariethoz G, Ortiz JM (2014) Verifying the high-order consistency of training images with data for multiple-point geostatistics. Comput Geosci 70:190–205

Reddy BS, Chatterji BN (1996) An FFT-based technique for translation, rotation, and scale-invariant image registration. IEEE Trans Image Process 5:1266–1271

Rezaee H, Mariethoz G, Koneshloo M, Asghari O (2013) Multiple-point geostatistical simulation using the bunch-pasting direct sampling method. Comput Geosci 54:293–308

Rosenblatt M (1985) Stationary sequences and random fields. Birkhuser, Boston

Srivastava RM (1992) Reservoir characterization with probability field simulation. SPE Form Eval 7:927–937

Strebelle S (2000). Sequential simulation drawing structures from training images. Ph. D. Thesis, Stanford University

Tahmasebi P, Hezarkhani A, Sahimi M (2012) Multiple-point geostatistical modeling based on the cross-correlation functions. Comput Geosci 16:779–797

Wang Y, Yang J, Yin W, Zhang Y (2008) A new alternating minimization algorithm for total variation image reconstruction. SIAM J Imaging Sci 1:248–272

Xu W (1996) Conditional curvilinear stochastic simulation using pixel-based algorithms. Math Geol 28:937–949

Yao T (1998) SPECSIM: a Fortran-77 program for conditional spectral simulation in 3D. Comput Geosci 24:911–921

Zhang T (2006) Filter-based training pattern classification for spatial pattern simulation. Ph. D. Thesis, Stanford University

Zhdanov MS (2002) Geophysical inverse theory and regularization problems. Elsevier, Amsterdam

Efficient Uncertainty Quantification and History Matching of Large-Scale Fields Through Model Reduction

Jianlin Fu, Xian-Huan Wen, and Song Du

Abstract Uncertainty quantification (UQ) and history matching (HM) have become a regular routine for reservoir management and decision-making in petroleum industry. The zonation method was widely used to reparametrize correlated fields with one lumped constant or multiplier specified for each zone such that the dimensionality of problems can be reduced and the HM problem can be efficiently solved. However, this ad hoc method faces a challenge to find the optimal zones. Moreover, it may fail to honor the geological (or geostatistical) features after the lumped constants or multipliers are applied, resulting in patches. In this work, we present several PCA-based techniques to address this problem by reducing the dimensionality of problem but not subject to the limitations of the zonation method.

1 Introduction

Uncertainty quantification (UQ) and history matching (HM) have become a regular routine for efficient reservoir management and decision-making in petroleum industry. There are many types of uncertain parameters (e.g., permeability, porosity, geometry, aquifer, oil-water contact, relative permeability, PVT parameters, etc.) that have been found to have evident impacts on reservoir performance. Among them, spatially correlated reservoir geo-models (e.g., permeability) have been a

J. Fu (✉)
Institute for Water and Environmental Engineering (Currently with Chevron), Universitat Politècnica de València, Valencia, Spain
e-mail: fu_jianlin_ac@yahoo.com

X.-H. Wen
Chevron ETC, 1500 Louisiana St., Houston, TX 77002, USA
e-mail: xwen@chevron.com

S. Du
Petroleum Engineering Department (Currently with Chevron), Texas A&M University, College Station, TX, USA
e-mail: dusong83@gmail.com

J.J. Gómez-Hernández et al. (eds.), *Geostatistics Valencia 2016*, Quantitative Geology and Geostatistics 19, DOI 10.1007/978-3-319-46819-8_35

challenge for UQ and HM when combining with other parameters. These geo-models are usually created by geostatistical tools with a high dimensionality. Traditionally, this type of correlated parameters is reparametrized using a zonation method with one lumped constant or multiplier specified for each zone such that the dimensionality of problems can be reduced and the relative significance of these parameters for reservoir performance prediction can be studied in comparison with other types of parameters. However, this ad hoc method faces a challenge to find the optimal zones such that the real impact can be accurately quantified. Moreover, it may fail to honor the geological (or geostatistical) features after the lumped constants or multipliers are applied.

Several techniques are available to address this problem. For low-dimensional problems (e.g., thousands of parameters), principal component analysis (PCA) is used to efficiently reduce the dimensionality of geostatistical models for accurate UQ and HM. For high-dimensional problems (e.g., large-scale correlated permeability fields), the distance-based kernel PCA is used to reparametrize the field for UQ, and geostatistical features can be preserved during HM. These approaches can reduce the dimension of parameters yet preserve the key spatial geostatistical features in the parameter fields. These methods have been implemented in our in-house software platform. In this paper, we present a simple synthetic example to demonstrate the effectiveness of this methodology.

2 Methodology

Principal component analysis (PCA) method can be used to re-parameterize the correlated fields for UQ and HM by reducing the large-scale fields to several independent parameters according to a threshold of the energy level (in terms of eigenvalues) such that the spatial heterogeneity can be accounted for. PCA is a statistical procedure that converts a set of correlated parameters into a set of linearly uncorrelated parameters (i.e., principal components) using an orthogonal transformation. This transformation is constructed in a way that the first principal component has the largest possible variance, and each succeeding component in turn has the highest variance possible under a constraint that it is orthogonal to the preceding components.

However, there remain two challenges for the PCA-based method when dealing with more practical problems in petroleum engineering. The first one is the curse of dimensionality. The existing PCA methods cannot efficiently handle large-scale cases due to the inherent difficulty of numerical algorithm for high-dimensional eigen-decomposition. A snapshot-based PCA (SPCA) method (e.g., Sirovich 1987; Cardoso et al. 2009; Vo and Durlofsky 2014) was proposed to treat an arbitrarily large-scale reservoir model, e.g., with a million cells. The main idea of SPCA is that, instead of direct eigen-decomposition of the large-scale covariance matrix (e.g., of million by million), one may generate an ensemble of stochastic realizations (e.g., several hundred models) using fast geostatistical tools to represent the heterogeneity of field. Then, one can eigen-decompose the small-scale, so-called "kernel" matrix (e.g., Scholkopf et al. 1998, 1999) with a smaller dimension (e.g., hundreds by hundreds), such that the main pattern of eigenvectors and eigenvalues of the original large-scale covariance matrix can be recovered from these small-scale kernel matrices and the associated snapshots. These eigenvectors and eigenvalues can be used to reconstruct the images of correlated random fields. Unfortunately, the eigenvectors of the covariance matrix cannot be accurately recovered in terms of magnitude with this method. Another challenge is how to generate random models that are consistent with the existing information for HM.

To address these challenges, we implement a distance-based kernel PCA (DKPCA) for large-scale problems (Scheidt and Caers 2009; Park 2011). The re-parameterization procedure of DKPCA is the same as SPCA except the way to reconstruct new models. The concept "distance" between random realizations is used such that the new models are generated through an optimization process that minimizes the HM error through the linear combination of the ensemble realizations under distance space (Kwok and Tsang 2004; Park 2011). By doing so, the stochastic models generated always share the same geostatistical features as the given ensemble and thus are "geologically" consistent. With the DKPCA, moreover, the conditioning data can be always honored as long as all of the individual realizations respect the hard data, which is straightforward for many existing geostatistical tools.

3 An Illustrative Example

3.1 Experimental Setup

The model presents an oil-water two-phase black-oil reservoir and has 1681 ($41 \times 41 \times 1$) active cells with each one $50 \times 50 \times 50$ [ft^3]. The porosity is assumed to be homogeneous and is equal to 0.2. The isotropic permeability (e.g., xperm) is heterogeneous with a multi-Gaussian model specified for ln$k$. The reservoir is initially produced with five wells: four producers (PROD1, PROD2, PROD3, and PROD4) located in the four corners and one injector (INJW1) in the center. Figure 1 shows the reservoir and well configuration.

All of the four wells are put into operation at the same time. The injector is operated by a specified bottom-hole pressure, while the producers are constrained by a constant bottom-hole pressure subject to a maximum oil rate. The constants are fixed for a certain period but may vary over time. The simulation time spans 2000 days. The data from the first 1000 days are used for history matching, while results from the second 1000 days will be used for checking the quality of predictions of the HM models. The production data include both water production rate (WPR) and oil production rate (OPR). We are going to generate stochastic realizations for xperm by matching the historical production data at four producers.

Two scenarios of synthetic experiments are performed for HM. The first one is a so-called zonation method, i.e., to divide the field into several regions (patches or zones) such that the field can be reduced to small number of parameters for HM. The second one is the DKPCA-based method for preserving geological consistency in the models. In the two scenarios, we will use the same number (6) of parameters for

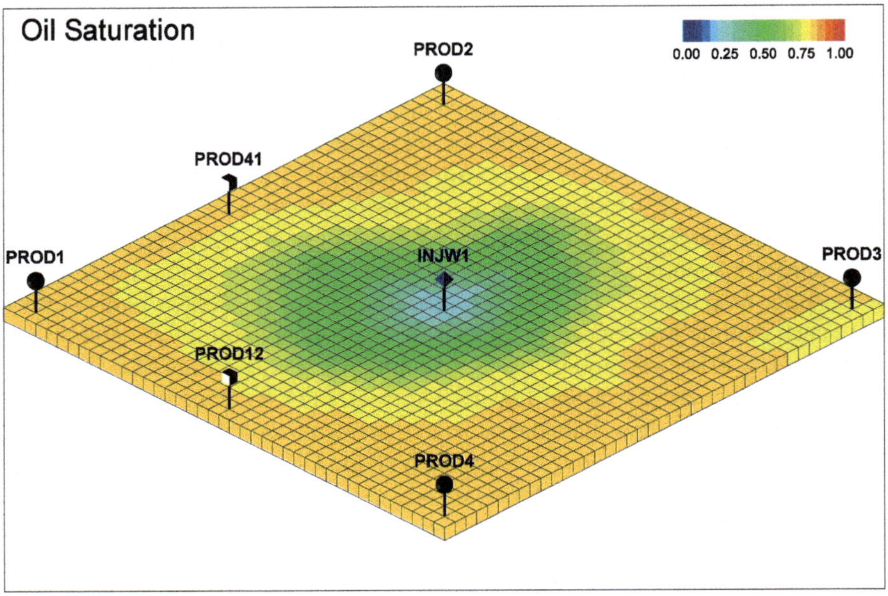

Fig. 1 Reservoir model and well configuration

comparison purpose. Now that the large-scale HM problem (with 1681 parameters) is reduced to a very small-scale one (with six parameters), a variety of optimization methods can be used to create multiple history-matched models. The particle swarm optimization (PSO) is employed in this work as search engine to minimize the HM error (Isebor 2013).

3.2 *Scenario 1: History Matching and Prediction with Existing Wells*

The results of oil production rate (OPR) before and after HM with the first 1000-day data are compared in this section. The predictions of the second 1000 days with the existing wells are also plotted and compared.

Figure 2 plots the OPR profile of PROD1 by comparing the individual initial guesses (the gray lines) generated by experimental design with the history-matched results (the color lines) using the traditional zonation method. Obviously, the results from initial guesses significantly deviate from the historical data, while the historical data of the first 1000 days (the magenta dots) can be reasonably reproduced by history-matched models. Although not shown in this case, it was also observed in other cases that an improper zonation may lead to a poor HM or even fail to do the job. It also shows that the models could also reasonably predict OPR for the second 1000 days for well PROD1.

Figure 3 plots the OPR result for PROD1 from the models generated by DKPCA-based method. Obviously, the final-matched models are able to match

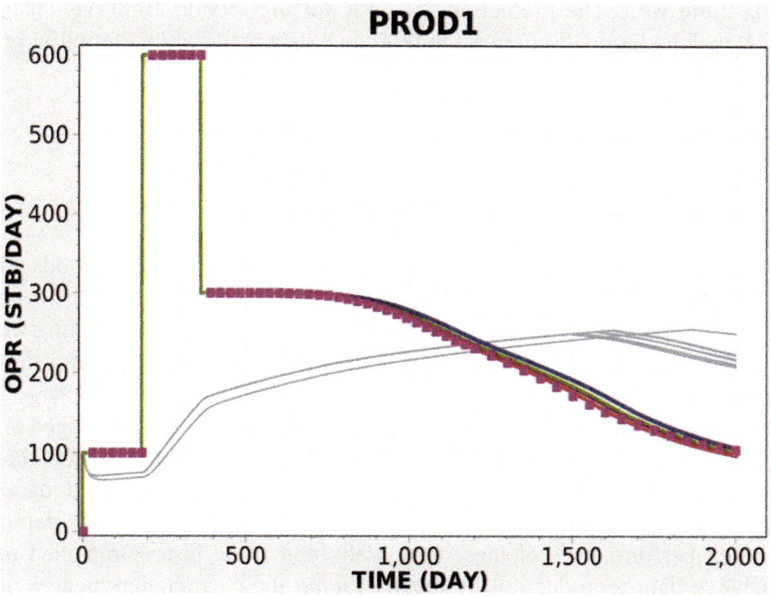

Fig. 2 Oil production rate of PROD1 from the zonation method: *color lines* (the final matched), *gray lines* (the initial guess), and *magenta dots* (historical and reference data, only the first 1000-day data are used for history matching)

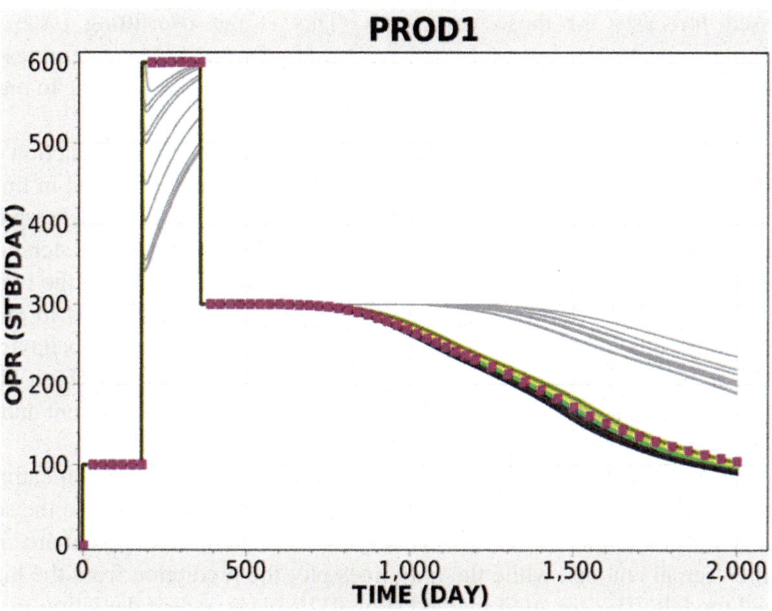

Fig. 3 Oil production rate of four producers from the DKPCA method: *color lines* (the final matched), *gray lines* (the initial guess), and *magenta dots* (historical and reference data, only the first 1000-day data are used for history matching)

the data quite well. The predictions of OPR for the second 1000 days using the DKPCA-matched models capture the reference data well, while the initial guesses fail to do so.

3.3 Scenario 2: Prediction with New Infill Wells

Results from Scenario 1 show that both the zonation and DKPCA methods are able to history match the first 1000-day's data equally well and are able to provide reasonable predictions of the second 1000-day production with the existing wells. A more interesting test is to check the predictability of the HM models for new well locations where no production data are used during the HM process.

The drilling of infill wells represents a very common practice for staged oil field development. Prediction of performances for these new wells under changing production conditions is challenging. In some cases, the historical data from existing wells may contain useful information to identify reservoir parameters that impact performances of these new wells, but some history-matched models may achieve data reproduction by locally tuning their parameters near well-bore only, leading to lack of precision in predictions. This is the underfitting issue. In other cases, the reservoir properties near new wells remain largely unsampled, and the historical data contain very limited information to infer them. However, some history-matched models may bring false information from optimization, resulting in biased forecasts for those new wells. This is the overfitting issue. Both underfitting and overfitting models are not robust for forecast of long-term reservoir performances. Therefore, prediction of new wells may serve as a metric to measure the robustness of reservoir simulation models.

Specifically in this case, a campaign with two new infill production wells (PROD12 and PROD41 in Fig. 1) is run, and the results are compared in this part so as to further check the prediction capability of history-matched models generated before subject to new production conditions. Note that the history-matching scenario is the same as the previous part, but in the prediction scenario, the two new wells (PROD12 and PROD41) replace PROD1, PROD2, and PROD4 to produce the reservoir. Although the well-level performances have the similar behavior, the metric that we are looking at in this scenario is the field's OPR, which plays a critical role in finance and operations for efficient reservoir management and asset development.

Figures 4 and 5 compare the OPR of field with the new production campaign. Note that the first 1000-day's data were used in history matching while the second 1000-day for prediction. The magenta dots show the reference data that are created from the "actual" model, while the gray lines plot the prediction from the history-matched models. The zonation method (Fig. 4) displays a clear deviation from the reference data. The DKPCA-based method (Fig. 5) obviously is able to obtain a better result: the reference data well reside in the envelope of predictions from the DKPCA models.

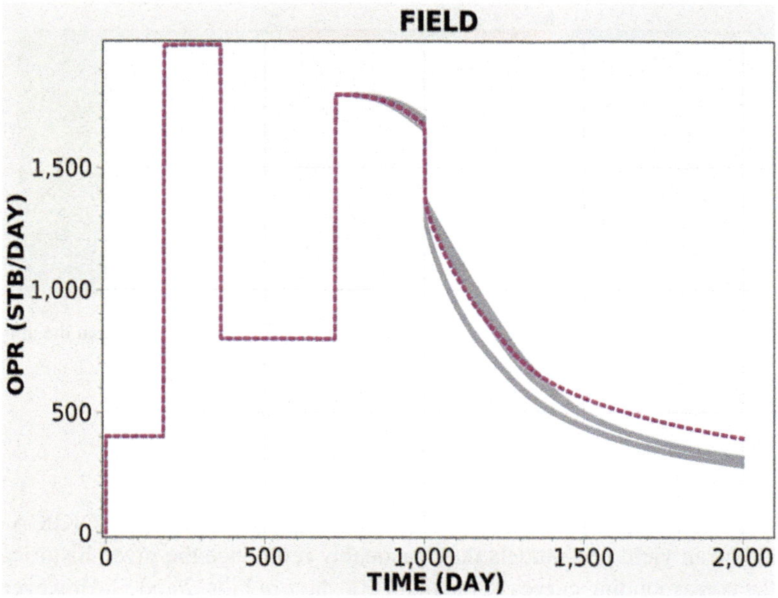

Fig. 4 OPR prediction of the field from the zonation method: The first 1000-day data are used for history matching, while the second 1000-day for prediction. Note that the *gray lines* represent the results from history-matched models and the *magenta dots* are historical (the first 1000-day) or reference (the second 1000-day) data

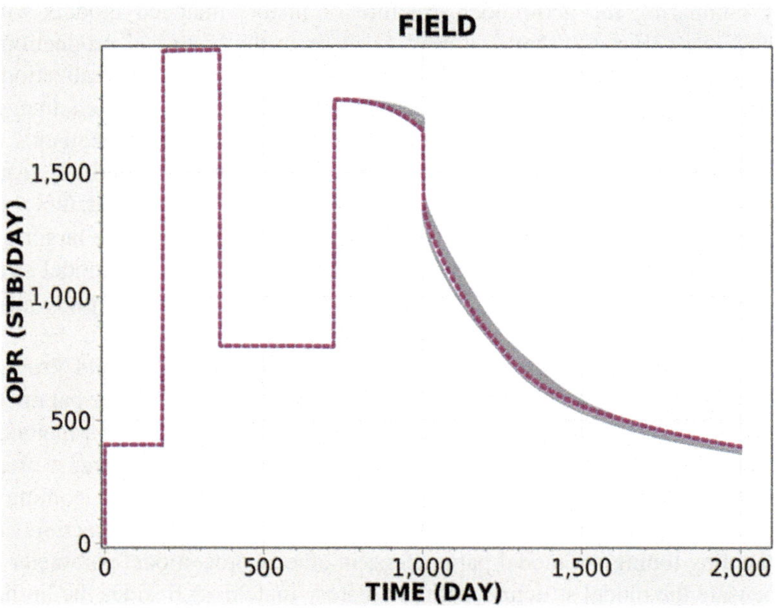

Fig. 5 OPR prediction of the field from the DKPCA method: The first 1000-day data are used for history matching, while the second 1000-day for prediction. Note that the *gray lines* represent the results from history-matched models, and the *magenta dots* are historical (the first 1000-day) or reference (the second 1000-day) data

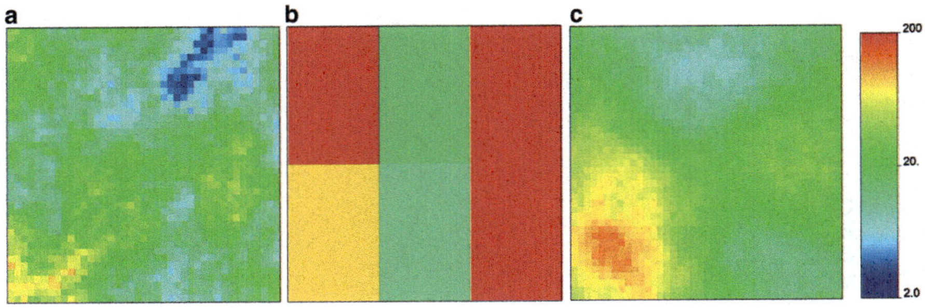

Fig. 6 Typical ln*k* images: (**a**) reference field, (**b**) zonation, (**c**) DKPCA. Note that the unit of *k* in the *color bar* is in [md]

4 Discussion

Two history-matching methods (i.e., the zonation-based and the DKPCA-based methods) can yield geo-models that reasonably reproduce the given historical data (see the corresponding curves of the first 1000 days in Figs. 2 and 3). However, their prediction abilities are quite different (see the second 1000 days in Figs. 4 and 5). The model complexity (i.e., the dimensionality) is not a root cause for such inaccuracy since all of the models are history matched with the same parameter size (6) and with an equal precision in terms of HM residuals.

By comparing the geo-model structure of history-matched models with the reference case (Fig. 6a) that was used to generate the historical production data, we found that the zonation method (see Fig. 6b for a representative realization) fails to describe the pattern of spatial heterogeneity. The DKPCA-based image (see Fig. 6c for a representative realization) correctly reproduces the statistics of the original model, and the spatial pattern is also reasonably identified from the historical data. The two scenarios of model predictions show that the zonation method has the worse structure and, as a consequence, produces the larger prediction bias; the DKPCA-based method accurately captures the geo-model structure and remains consistent during HM and, as a result, offers the more accurate predictions.

As showed in several studies (e.g., Refsgaard et al. 2006; Clark and Vrugt 2006; Fu 2008; Doherty and Christensen 2011; White et al. 2014), when the parameters to be estimated are assigned a compensating role (e.g., by the zonation method), the potential of prediction errors may actually increase, rather than decrease after HM (see Figs. 4 and 5). By compensating, we mean that the error coming from inadequate representation of the subsurface reality in model structure can be mitigated by tuning the model parameters; in other words, model parameter errors compensate the model structure errors in history matching. Besides the limitations of methodology (e.g., unable to sample the entire model space of the posterior distribution (Fu and Gómez-Hernández 2009a, b), the improper representative of actual geo-model structure is another important factor that is mainly responsible for this paradox. This calibration-induced bias tends to result in a wrong conclusion and thus is particularly dangerous in practice for decision-making.

In summary, accounting for the spatial variability pattern is important to build robust simulation models for reliable forecasts. The "robustness" was traditionally referred to an HM method that can generate models to match the data no matter what prior was specified (whether correctly or wrongly). This is valuable when the useful information from the subsurface is very limited for structure identification. However, when there are sufficient data (including geological, geophysical, and engineering data) to inform the heterogeneity structure of subsurface reservoir, the separation of structure uncertainty and parameter uncertainty is valuable since it precludes the clutter of their complications, leading to a better understanding of reservoir behavior and robust predictions. In other words, when the geo-model structure is adequate, the compensating role due to model calibration can be excluded from the process of history matching or inverse modeling for a solution with improved robustness (Taguchi and Clausing 1990) in terms of predictions.

5 Conclusion

This paper compares a PCA-based method with a traditional zonation method for history matching. We have observed that the proposed method is efficient for history matching large-scale heterogeneous fields. We also demonstrate the value of the proposed method in building simulation models for improved forecasts owing to a full consideration of the spatial heterogeneity pattern, a dominant factor for reservoir management and asset development in reservoir engineering.

Bibliography

Cardoso MA, Durlofsky LJ, Sarma P (2009) Development and application of reduced-order modeling procedures for subsurface flow simulation. Int J Numer Methods Eng 77(9):1322–1350

Clark MP, Vrugt JA (2006) Unraveling uncertainties in hydrologic model calibration: assessing the problem of compensatory parameters. Geophys Res Lett 33:L06406

Doherty JE, Christensen S (2011) Use of paired simple and complex models to reduce predictive bias and quantify uncertainty. Water Resour Res 47:W12534

Fu J (2008) A Markov Chain Monte Carlo method for inverse stochastic modeling and uncertainty assessment, Unpublished Ph. D. thesis, Universidad Politecnica de Valencia, Valencia, Spain, p 140

Fu J, Gómez-Hernández JJ (2008) Preserving spatial structure for inverse stochastic simulation using blocking Markov chain Monte Carlo method. Inverse Probl Sci Eng 16(7):865–884

Fu J, Gómez-Hernández JJ (2009a) A blocking Markov chain Monte Carlo method for inverse stochastic hydrogeological modeling. Math Geosci 41(2):105–128

Fu J, Gómez-Hernández JJ (2009b) Uncertainty assessment and data worth in groundwater flow and mass transport modeling using a blocking Markov chain Monte Carlo method. J Hydrol 364:328–341

Isebor OJ (2013) Derivative-free optimization for generalized oil field development, Ph. D. thesis, Stanford University, p 169
Kwok JT-Y, Tsang IW-H (2004) The pre-image problem in kernel methods. IEEE Trans Neural Netw 15(6):1517–1525
Park K (2011) Modeling uncertainty in metric space, Ph. D. thesis, Stanford University, p 229
Refsgaard JC, Van der Sluijs JP, Brown J, Van der Keur P (2006) A framework for dealing with uncertainty due to model structure error. Adv Water Resour 29(11):1586–1597
Scheidt C, Caers J (2009) Representing spatial uncertainty using distances and kernels. Math Geosci 41(4):397–419
Scholkopf B, Smola A, Muller KR (1998) Nonlinear component analysis as a kernel eigenvalue problem. Neural Comput 10:1299–1319
Scholkopf B, Mika S, Burges CJC, Knirsch P, Muller KR, Ratsch G, Smola A (1999) Input space versus feature space in kernel-based methods. IEEE Trans Neural Netw 10(5):1000–1017
Sirovich L (1987) Turbulence and the dynamics of coherent structures, Part I: Coherent structures. Q Appl Math XLV(3):561–571
Taguchi G, Clausing D (1990) Robust quality. Havard Bus Rev 1990:56–75
Vo HX, Durlofsky LJ (2014) A new differentiable parameterization based on principal component analysis for the low-dimensional representation of complex geological models. Math Geosci 46(7):775–813
White JT, Doherty JE, Hughes JD (2014) Quantifying the predictive consequences of model error with linear subspace analysis. Water Resour Res 50:1152–1173

Revealing Multiple Geological Scenarios Through Unsupervised Clustering of Posterior Realizations from Reflection Seismic Inversion

Mats Lundh Gulbrandsen, Knud Skou Cordua, Thomas Mejer Hansen, and Klaus Mosegaard

Abstract In this study, we analyze 26,000 posterior realizations obtained through Monte Carlo sampling from the posterior distribution of a reflection seismic inverse problem and show that the posterior realizations cluster around multimodal peaks. This problem is based on a seismic trace recorded in the southern part of Jutland, Denmark. Prior information is based on observations of lithology sequences of the geology in the area, and the multimodal modes in the posterior realizations will hence represent different geological scenarios. In order to uncover the multimodal nature of the posterior distribution, grouping of posterior realizations is done using an unsupervised clustering technique, namely, the K-means clustering algorithm. In order to quantify the choice on the number of clusters in the realizations, the gap statistic method is used. The clustering method is applied on both categorical model parameters representing lithological units (LU) and on the continuous parameters representing the acoustic impedance (AI). These techniques allow quantifying the probability of the different possible geological scenarios that are consistent with the seismic and geological observations. Results demonstrate that the cluster characteristics are significantly dependent on the types of parameters considered. If the goal of the inversion is to identify different geological scenarios, using a parameterization based on lithological units is more informative than a parameterization based on acoustic impedance.

M.L. Gulbrandsen (✉) • K.S. Cordua • T.M. Hansen • K. Mosegaard
Solid Earth Physics, University of Copenhagen, Juliane Maries Vej 30, 2100 Copenhagen, Denmark
e-mail: mats.lundh@nbi.ku.dk

J.J. Gómez-Hernández et al. (eds.), *Geostatistics Valencia 2016*, Quantitative Geology and Geostatistics 19, DOI 10.1007/978-3-319-46819-8_36

1 Introduction

A widely used approach for inversion of reflection seismic data is based on a linear forward model and Gaussian assumptions about the prior distribution of the elastic parameters and the noise model (Buland and Omre 2003; Jullum and Kolbjørnsen 2016; Hampson et al. 2005). The solution to such inverse problems is Gaussian distributed, which is simply a single-modal distribution.

In reality, the Gaussian prior assumption and the linear forward model may not adequately describe the information available. In order to handle this, these problems can instead be formulated probabilistically, which leads to a solution described by a typically unknown and non-Gaussian posterior distribution, (Larsen et al. 2006; Ulvemoen and Omre 2010; Zunino et al. 2014; Bosch et al. 2007). In this way, the solution to the inverse problem can be characterized by a sample from the posterior distribution, which will represent a set of realizations (i.e., possible solutions) that are all consistent with observed data and prior information (Mosegaard and Tarantola 1995). The posterior distribution of such non-Gaussian inverse problems may be highly multimodal: i.e., realizations from the posterior probability distribution may be located at isolated clusters in the high dimensional space spanning all possible earth models. If such clusters exist, each cluster will correspond to a group of realizations with similar appearance, and each cluster may represent one geological scenario. For example, one cluster may represent a commercially viable reservoir, while another cluster may represent a non-commercially viable reservoir. Using a probabilistic approach to inverse problem theory, the relative probability of each local scenario can easily be quantified. This may be hugely beneficial in characterizing reservoir models of all kinds. It should be noted that this study does not intend to *analyze* the different geological scenarios but demonstrates a technique to reveal their nature, which would be helpful in this respect.

In the following we will analyze the result of a probabilistically formulated inversion of a normal incidence seismic data set from the southern part of Jutland, Denmark. Initially we will demonstrate that the posterior sample contains many quite different realizations. Then we use a clustering algorithm to locate multimodal clusters (i.e., similar posterior realizations). This analysis will be done both with respect to the realizations representing lithological units and with respect to the realization's corresponding acoustic impedance parameterization. Finally the analysis obtained from the two different ways of parameterizing data will be compared.

1.1 The Inverse Problem

Figure 1 shows 26,000 realizations resulting from sampling of the posterior distribution for a probabilistically formulated inversion of a reflection seismic data set

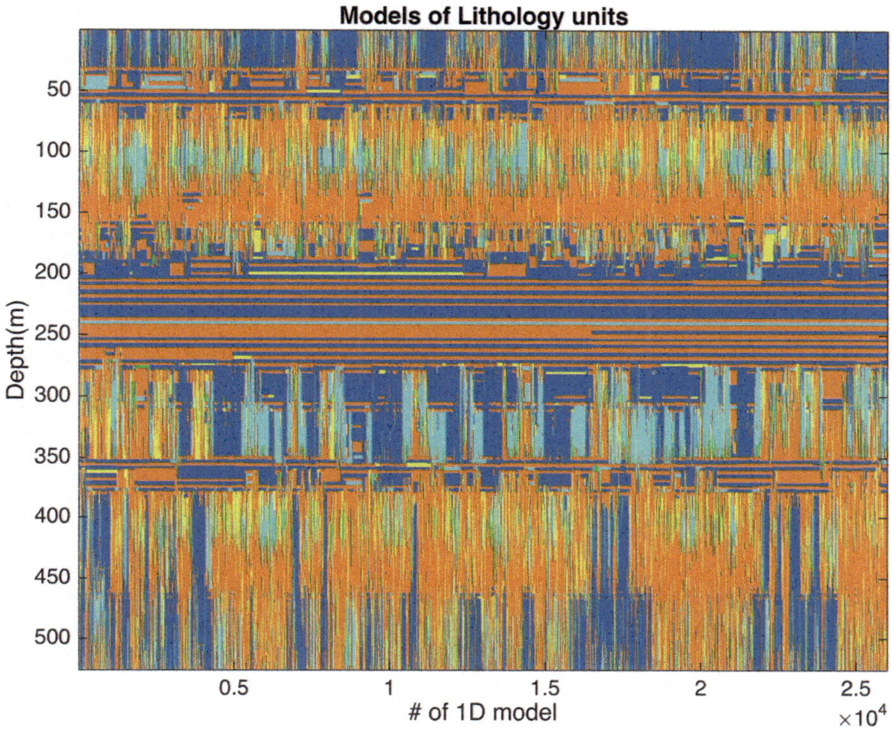

Fig. 1 26,000 1D lithology profiles resulting from an inversion of reflection seismic data. The colors *blue*, *turquoise*, *green*, *yellow*, and *orange* represent the lithology units halite, limestone, lamination, 50/50 anhydrite and limestone, and anhydrite, respectively

from of the Zechstein unit in the southern part of Jutland, Denmark (Cordua et al. submitted). This distribution is based on a prior distribution that is statistically consistent with geological scenes observed in borehole logs. Hence, the posterior realizations are all consistent with both the observed geological scenes and the observed seismic data (within the expected data uncertainty). A presentation of the inversion method used to obtain the posterior realizations is outside the scope of this paper. However, it can be assumed that the method in fact samples the true posterior distribution. Later we actually show that the method in fact revisits the same modes of the posterior distributions several times during the sampling period.

In Fig. 1 it can be seen that this highly underdetermined seismic inverse problem has several possible solutions of different nature. Figure 1 further suggests some clusters of similar posterior realizations. In this study, we try to quantify this apparent clustering to get a better understanding of how the realizations actually are distributed. This is done using the unsupervised clustering method called K-means. In addition to provide a better visual representation of the results, the clustering algorithm allows quantifying the probability of the possible geological scenarios.

Table 1 The different lithological units and the corresponding AI values as used in (Cordua et al. submitted)

Lithological units (color Fig. 1)	AI values [kg/m^2 s]
Halite (blue)	$1.0*10^7$
Dolomitic limestone (turquoise)	$1.2*10^7$
Lamination (green)	$1.4*10^7$
50/50 limestone/anhydrite (yellow)	$1.6*10^7$
Anhydrite (orange)	$1.8*10^7$

Figure 1 represents one way of displaying the posterior realizations. Each color represents a lithological unit. Another way could be to parameterize these units with their elastic property, acoustic impedance (AI). Table 1 shows the relation between the lithological units and their corresponding AI values, as used in the inversion algorithm (Cordua et al. submitted).

2 Locating Cluster-Specific Geological Scenarios

The task is now to group all the individual posterior realizations into different clusters. A general description of a *cluster* is difficult, and no unique definition exists. However, it can be valuable to think of clusters as groups of which all members are closer together with the other members from the same group than members from other groups, even though this not always has to be a mathematical fact. There are several ways of defining these mutual distances and several different clustering algorithms exist. There is no *correct* algorithm, since the different methods represent different ways of analyzing the data and the different methods should be decided based on the problem. In this study, the K-means clustering algorithm will be used to group the posterior realizations both with respect to the lithological units and with respect to the acoustic impedance values.

2.1 K-Means Clustering

Assuming the number of clusters, K, is given, the aim is to group the D-dimensional data $\boldsymbol{x} = \{\boldsymbol{x}_1, \boldsymbol{x}_2, \ \ldots \boldsymbol{x}_N\}$ into the K clusters. The clusters are represented by the D-dimensional vectors $\boldsymbol{\mu}_k$, where $k = 1 \ldots K$, and each vector $\boldsymbol{\mu}_k$ is a *prototype* vector for the kth cluster (Bishop 2006). More specifically, the prototype vectors represent the centers of each cluster, and hence, the K-means clustering model is a centroid model. The objective goal of the method is to minimize the sum of the square distances between all points and their closest prototype vector. Mathematically this can be described as minimizing the objective function J (Bishop 2006):

$$J = \sum_{n=1}^{N} \sum_{k=1}^{K} r_{nk} \|\boldsymbol{x_n} - \boldsymbol{\mu_k}\|^2, \tag{1}$$

where r_{nk} is a set of binary indicator variables, $r_{nk} \in \{0, 1\}$, each associated with one data point $\mathbf{x_n}$, and stating which cluster each point is assigned to. In order to find the values for r_{nk} and $\boldsymbol{\mu_k}$ that minimizes Eq. 1, an iterative procedure is performed. Every iteration consists of two steps. The first step optimizes on the indicator variables and the next step on the cluster centers. Prior to the simulation, K prototype vectors are randomly chosen among the D-dimensional points $\boldsymbol{x}$ (note that there are several other ways to choose the starting points as well). In the first step of the iterations, each point is assigned a cluster based on the prototype vector, i.e., cluster center, they are closest to. This is in fact optimizing J with respect to r_{nk}, since each data point is independent and taking the smallest distance between $\boldsymbol{x_n}$ and $\boldsymbol{\mu_k}$ will therefore minimize J. For the realizations of continuous variables, i.e., acoustic impedance, the Euclidean distance is used, and for categorical variables, i.e., the lithological units, the Hamming distance[1] is used. The next step is to optimize on $\boldsymbol{\mu_k}$, and since J is a quadratic function of $\boldsymbol{\mu_k}$, the optimal $\boldsymbol{\mu_k}$ is found by setting the derivative of the objective function with respect to $\boldsymbol{\mu_k}$ to zero:

$$2\sum_{n=1}^{N} r_{nk}(\boldsymbol{x_n} - \boldsymbol{\mu_k}) = 0 \tag{2}$$

Solving Eq. 2 with respect to $\boldsymbol{\mu_k}$ gives

$$\boldsymbol{\mu_k} = \frac{\sum_n \boldsymbol{r_{nk} x_n}}{\boldsymbol{r_{nk}}} \tag{3}$$

From Eq. 3 it is seen that $\boldsymbol{\mu_k}$ represents the mean of all points within each cluster. The second step of each iteration is hence to compute the mean of all points assigned to each cluster and define that new mean as the optimized cluster prototype. This procedure is repeated until the values converge. It should be noted that depending on the data set, two different simulations (i.e., two different starting positions) might not converge to the same local minimum. You can never be sure if you reach the global minimum of the objective function, so it might be a good idea to run the simulations more than once.

2.2 Gap Statistic

The K-means algorithm requires a predefined number of clusters. However, not knowing the complete nature of the data set, this choice can be tricky. One way of

---

[1] The hamming distance is the number of position (coordinates/elements) that differs between two vectors of the same length.

approaching this issue could simply be to try different values and visually examine the results. Another way could be to use a statistical approach to get an idea of which number of clusters that best represents the sample. In this study an analysis of the gap statistic (Tibshirani et al. 2001) is done. The idea of the gap statistic is to find the number of clusters which standardize the comparison of log (W_k) with a distribution of no obvious clustering, when W_k is defined as the within-cluster dissimilarity:

$$W_k = \sum\nolimits_{r=1}^{K} \frac{1}{2N_r} D_r, \tag{4}$$

where N_r is the number of points within clusters and D_r is the distance between all points within cluster r. In this analysis the Euclidean distance is used for D_r when the AI data is analyzed, and the Hamming distance is used for the lithological units. The distribution with no obvious clustering is computed by taking the average of 20 simulations of uniformly distributed data within the range of the sample. Computing the within-cluster dissimilarity of the uniformly distributed data, U_k, the gap statistic is defined as

$$G = \log(U_k) - \log(W_k). \tag{5}$$

The optimal number of clusters is then the smallest number of k, which fulfill the criteria $G(k) \geq G(k+1) - s_{\{k+1\}}$, where $s_{\{k+1\}}$ is $\mathrm{std}(\log(U_k))\sqrt{1 + \left(\frac{1}{20}\right)}$ and "std" denotes the standard deviation. Figure 2 shows the logarithm of the within-cluster dissimilarity for both the sample (blue) and for the simulated data (red). This simulation is done with respect to the lithological units. As expected the blue curve decreases with an increasing number of clusters, while the red curve is more or less flat. In Fig. 3 the gap curve is plotted. Based on this plot, it is seen that the optimal choice of clusters for this simulation is 15 (this is the first k where $G(k+1) - 1\mathrm{std}$ is less than $G(k)$).

This method is based on simulations, and the optimal number of clusters may therefore vary a little for different simulations. It should also be mentioned that the method is totally general and independent on any clustering method, so this analysis can be done with different clustering techniques.

3 Results

Figures 2 and 3 show the results from an analysis of the gap statistic of the posterior realizations in Fig. 1 with respect to the lithological units. The gap statistic simulations are however ambiguous, and the information from one simulation is hence not enough to base our choice of the optimal number of clusters. Figure 4 shows the histograms of ten simulations of the gap statistic using the K-means

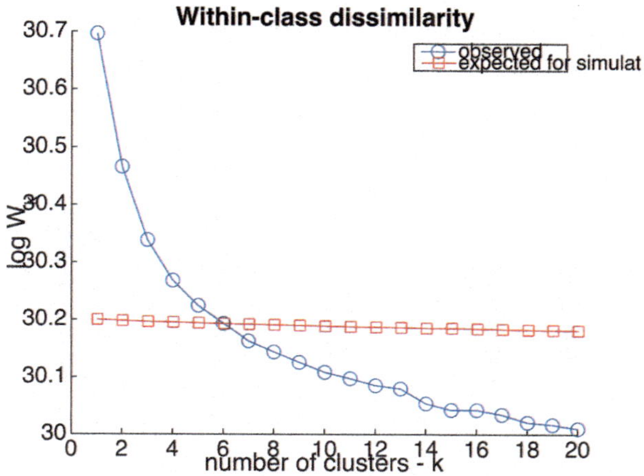

Fig. 2 The logarithmic within-cluster dissimilarity of the observations (*blue*) and the simulated distribution (*red*) as functions of the number of clusters

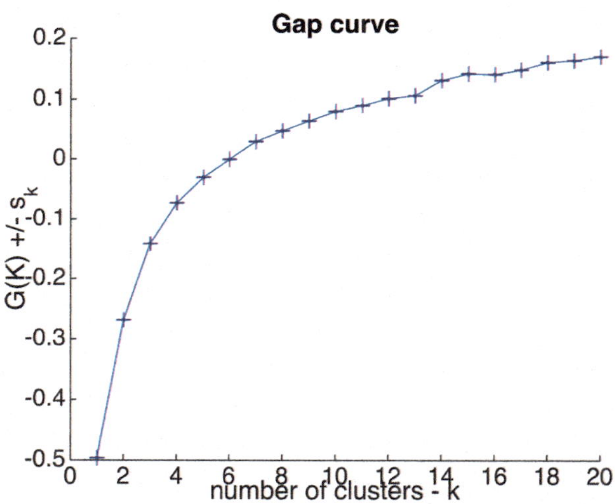

Fig. 3 The gap statistics with its corresponding 1std error bars as a function of number of clusters

algorithm on both data sets. It is seen that the "optimal number of clusters" (K_{opt}) varies between 6 and 20. It should be noted that the upper boundary of the simulation was set to 20 clusters; hence, a result of $K_{opt} = 20$ indicates that no optimal number of cluster was found within the test range (i.e., the optimal number of cluster may exist outside of this range). (Tibshirani et al. 2001) states however that it can be important to examine the whole gap curve and not only the optimal choice. Figure 5 shows an example of a gap curve representing a simulation where no optimal cluster numbers were found within the test range. It can however be seen that the criteria for concluding on an optimal number of clusters almost are met, both at $k = 10$ and $k = 16$.

All the gap curves for the ten simulations for both data sets have been analyzed. This analysis combined with a visual examination of the clusters has resulted in the choice of running the K-means clustering algorithm with respect of grouping

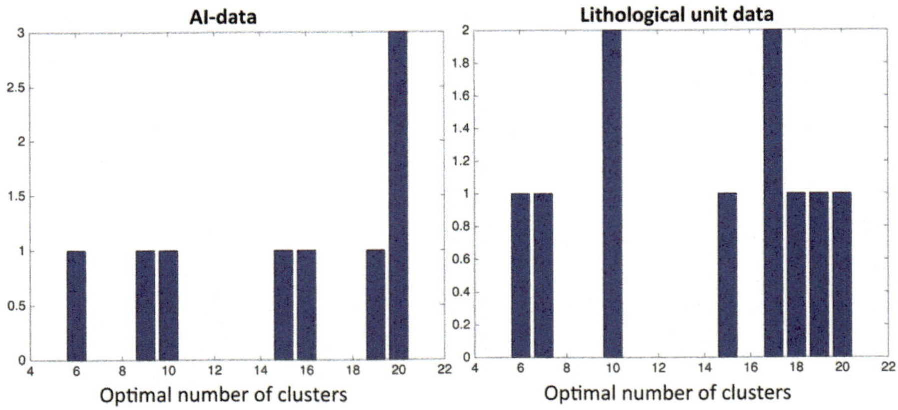

Fig. 4 The histograms of optimal number of clusters from 10 runs of the gap statistic method using the K-means clustering algorithm on the AI data (*left plot*) and the lithological unit data (*right plot*), respectively, with respect to the optimum choice of K (x-axis)

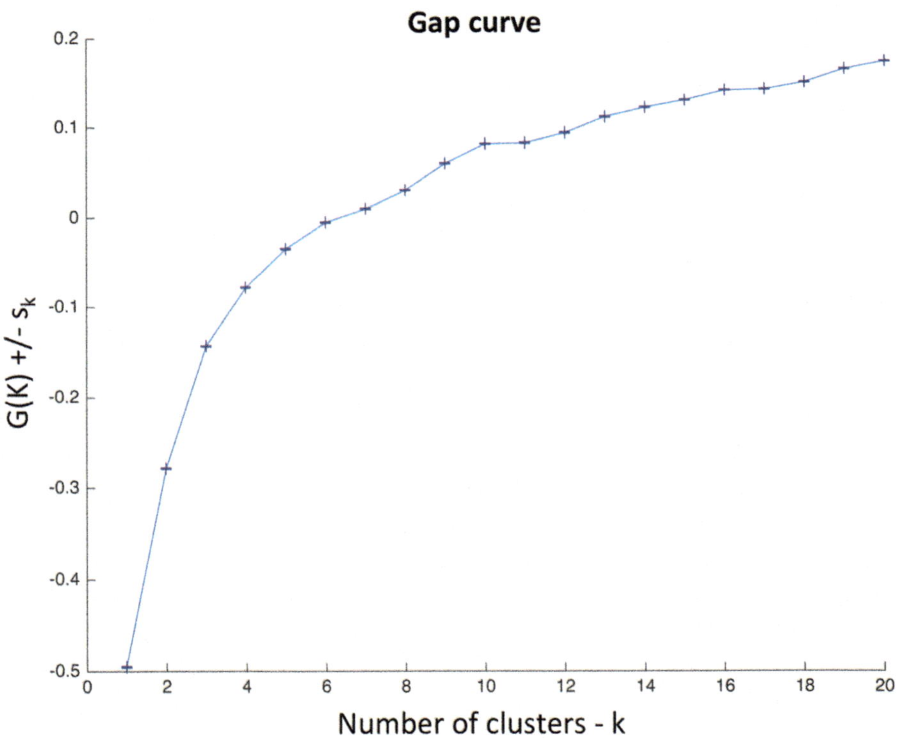

Fig. 5 The gap curve from one simulation of the gap statistic analysis using the K-means algorithm with respect to the AI data

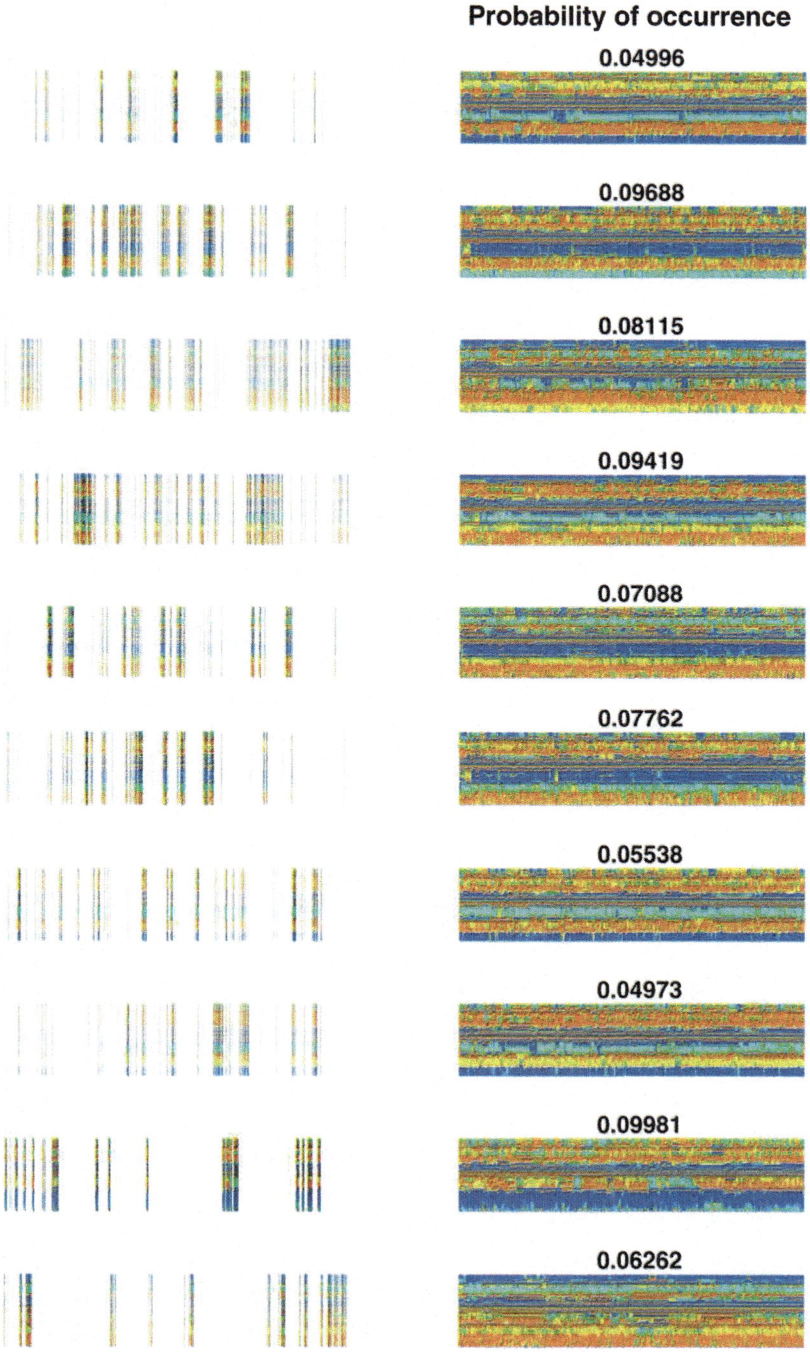

Fig. 6 The 15 clusters resulting from the K-means clustering of the lithology profiles in Fig. 1. The *left* panel shows which models among the realizations belong to each cluster and the *right* panel shows these models put together. The "probability of occurrence" represents the number of models in each cluster relative to the whole sample. The lithology units are represented with the same colors as in Fig. 1

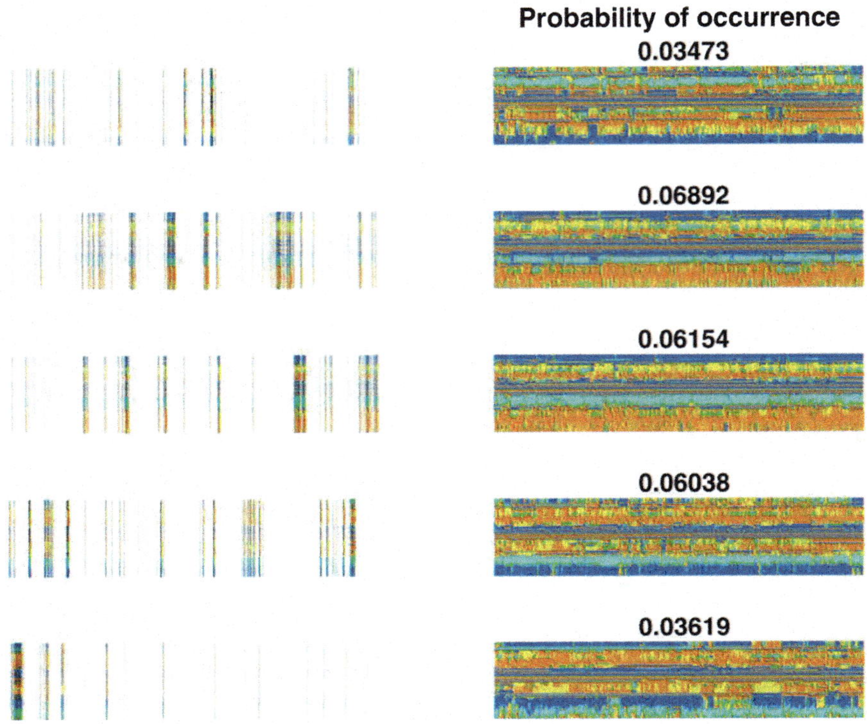

Fig. 6 (continued)

15 clusters for both parameterizations. These results are presented in Figs. 6 and 7. The left panels of the two figures show which realizations from Fig. 1 that is grouped in each cluster, and the right panels show all these realizations put together to see the overall structure of each cluster. Note that the size of each plot in the right panel of the two figures does not represent the actual size of the clusters. The relative sizes are printed above each plot and represent the percentage of models belonging to those specific clusters.

Analyzing the plots in the left panels of Figs. 6 and 7, we can see that the realizations constituting the different clusters more or less are evenly distributed throughout the set of all realizations. Even though the inversion algorithm itself is outside the scope of this study, it can be mentioned as a curio that the distributions of models (seen in Figs. 6 and 7) from the same clusters indicate that the sampling algorithm actually visits the same modes of the posterior distributions several times during the sampling period. This is an underlying assumption of sampling the posterior distribution using the Metropolis algorithm. However, Figs. 6 and 7 demonstrate that this actually is the case.

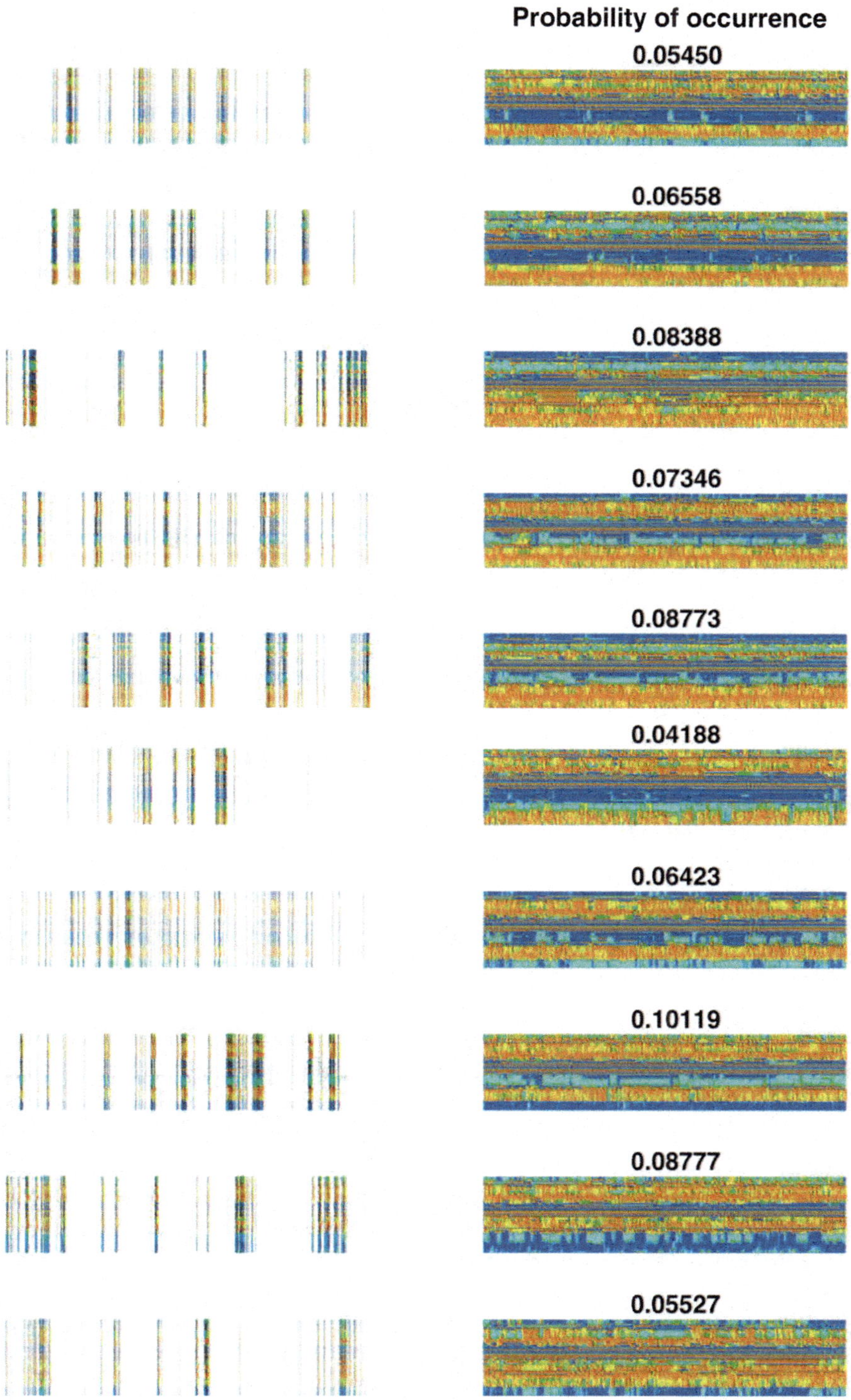

Fig. 7 The 15 clusters resulting from the K-means clustering of the AI realizations converted from the realizations shown in Fig. 1, using the relation in Table 1. The *left* panel shows which models among the realizations belong to each cluster and the *right* panel shows these models put together. The "probability of occurrence" represents the number of models in each cluster relative to the whole sample. The lithology units are represented with the same colors as in Fig. 1

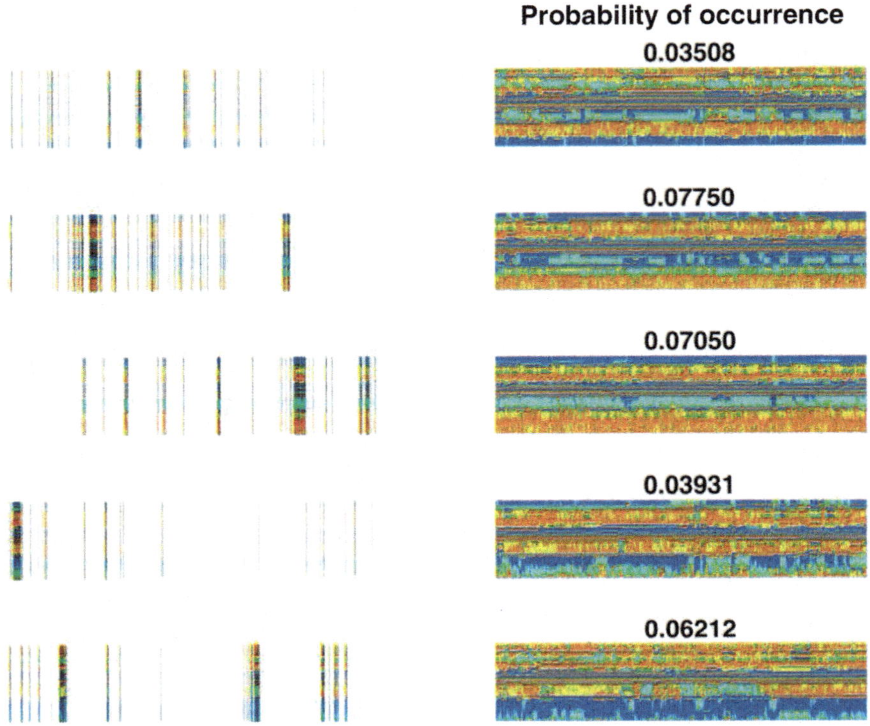

Fig. 7 (continued)

4 Discussion

4.1 Other Clustering Algorithms

As stated earlier there is no such thing as a correct clustering algorithm. The different algorithms and methods can be thought of as looking at your data with different classes. In this specific study a few different *agglomerative* hierarchical clustering techniques also have been tried out. Hierarchical clustering does not, unlike the K-means clustering, need a predefined number of clusters. Instead the hierarchical clustering techniques can be divided into two main groups, namely, the *agglomerative* and *divisive* group (Everitt et al. 2001). The agglomerative techniques all consist of a series of fusion of the *N* individual data points into groups, where the divisive techniques start out with all points belonging to the same cluster and successively separate all points into finer and finer groups. The merging or splitting of clusters is done with respect to different similarity/dissimilarity or distance measures between different clusters. All the different ways of measuring distances give rise to the different clustering methods. For the data set in this study, only techniques representing the agglomerative group have been tested, namely, the *single linkage*, *complete linkage*, *average linkage*, and *centroid linkage*. None of the results perform as well as the K-means method for this study, which is a statement purely based on visual inspection of the clusters. None of these results

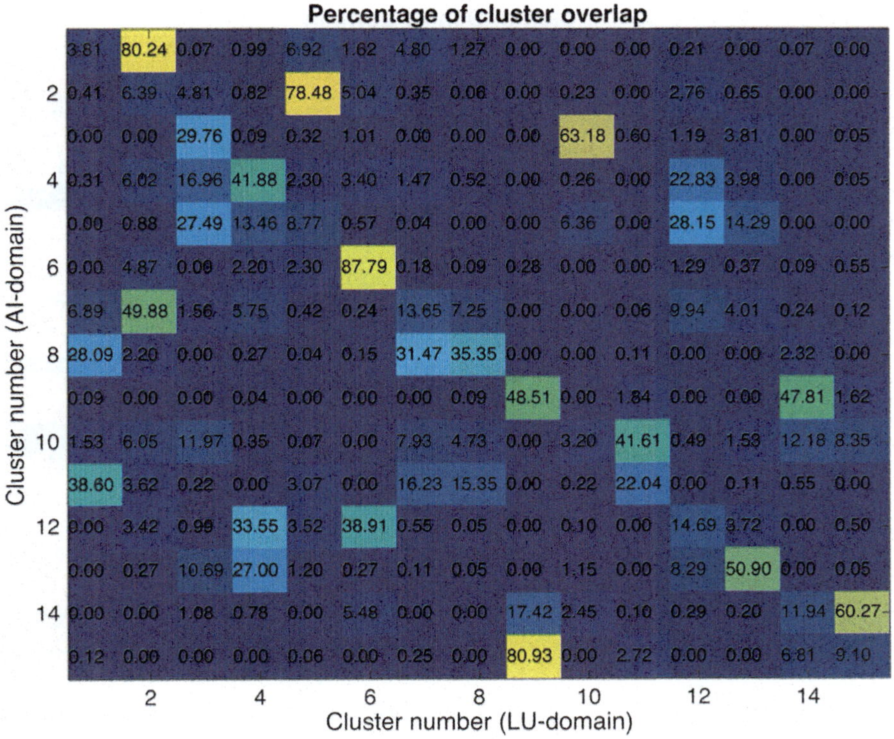

Fig. 8 The percentage of models overlapping between the 15 clusters of the AI realizations (y-axis) and the 15 clusters representing the lithological units (x-axis)

will be presented here since the comparison of the different methods is outside the scope of this paper. Distribution models such as the expectation maximization (assuming data can be described as a Gaussian mixture model) are not considered in this study since they are unsuited for clustering categorical variables.

4.2 Comparison of Clustering Results

The importance of being able to cluster the data in the lithological unit (LU) domain is illustrated in Fig. 8. The figure displays the percentage overlap between realizations belonging to the different clusters arising from applying the K-means clustering algorithm on the AI data (vertical axis) and the lithological units (horizontal axis), respectively. As an example, it is seen from Fig. 8 that cluster 15 in the AI domain is pretty well represented in the lithological unit (LU) domain. 80.9 % of all realizations in cluster 6 from the AI domain are grouped together in one cluster, namely, number 9, in the LU domain. The rest of the realizations are mainly distributed between cluster 14 and 15, with 6,8 % and 9,1 %, respectively. If we, however, look at cluster 9 in the AI domain, we can see that the models belonging to this cluster are split between two different clusters in the LU domain, namely,

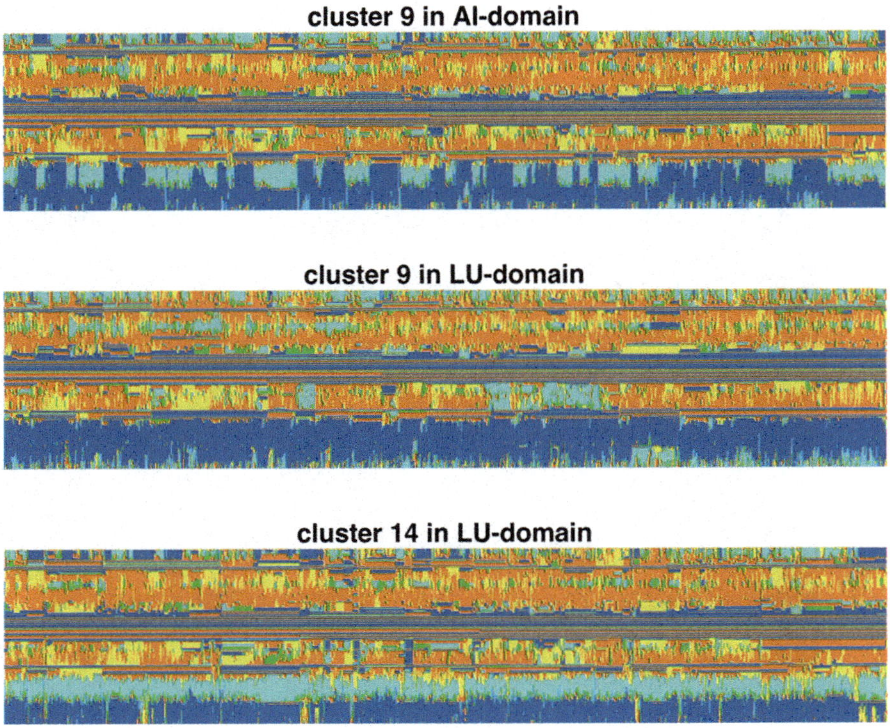

Fig. 9 The *upper* plot shows the models representing cluster 9 in the AI domain. The *middle* and *lower* plot shows cluster number 9 and 14 from the LU domain

cluster number 9 and 14 with 48,5 % and 47,8 %, respectively. These clusters are plotted in Fig. 9. From the deeper part of the realizations displayed in Fig. 9, it is clear that the clustering algorithm distinguishes between the blue (halite) and the turquoise (limestone) when applied to the LU domain but cluster realizations with both these sections together in the AI domain. This is because these models are much closer together in the continuous parameter space, than in the discrete parameter space, where the distance is the same between all the categorical variables. It should be noted that the clustering simulations for the two data sets have the same starting points, i.e., the simulations start in the same random realizations. This is important to emphasize, since it suggests that the differences illustrated in Fig. 8 are actual differences of the two domains and not differences due to two different local minima of the objective function (Eq. 1).

5 Conclusion

We have shown that K-means clustering can be used to get a better understanding of how the multimodal landscape representing the solution space of a 1D reflection seismic inversion is distributed. By clustering the posterior realizations, we get a much clearer picture of the potential geological scenarios and the probabilistic distribution between them. The combination of probabilistic inverse problems and cluster analysis thus allows us to perform scenario-based inversion of reflection seismic data.

It is however important to emphasize that the posterior probability distributions change for the different parameterizations, which in turn can result in very different multimodal landscapes. Clustering the same realizations with different parameterization will hence result in different clusters, and it is important to know what is analyzed. In this study we show that one cluster in the AI domain actually is represented in two different geological scenarios, and it is hence important to analyze the domain representing the parameterization of interest.

Bibliography

Bishop CM (2006) Pattern recognition and machine learning. Springer, Heidelberg

Bosch M, Rodrigues J, Navarro A, Díaz M (2007) A Monte Carlo approach to the joint estimation of reservoir and elastic parameters from seismic amplitudes. Geophysics 72(6):029–039

Buland A, Omre H (2003) Bayesian linearized AVO inversion. Geophysics 68(1):185–198

Cordua KS, Gulbrandsen ML, Hansen TM, Mosegaard K (submitted) Constructing scene-consistent prior distributions using a Markov-frequency-matching model: applied to seismic inversion. GJI

Everitt BS, Landau S, Leese M (2001) Cluster analysis. Arnold, London

Hampson DP, Russell BH, Bankhead B (2005) Simultaneous inversion of pre-stack seismic data. SEG Annual Meeting. Houston: Society of Exploration Geophysicists

Jullum M, Kolbjørnsen O (2016) A Gaussian-based framework for local Bayesian inversion of geophysical data to rock properties. Geophysics 81(3):R75–R87

Larsen AL, Ulvmoen M, Omre H, Buland A (2006) Bayesian lithology/fluid prediction and simulation on the basis of a Markov-chain prior model. Geophysics 71:R69–R78

Mosegaard K, Tarantola A (1995) Monte Carlo sampling of solutions to inverse problems. J Geophys Res 100(B7):12,431–12,447

Tibshirani R, Walter G, Hastie T (2001) Estimating the number of clusters in a data set via the gap statistic. J Royal Stat Soc 63(part 2):411–423

Ulvemoen M, Omre H (2010) Improved resolution in Bayesian lithology/fluid inversion from prestack seismic data and well observations: part 1—methodology. Geophysics 75(2): R21–R35

Zunino A, Mosegaard K, Lange K, Melnikova Y, Hansen TM (2014) Monte Carlo reservoir analysis combining seismic reflection data and informed priors. Geophysics 80:R31–R41

Object-Based Modeling with Dense Well Data

Ragnar Hauge, Maria Vigsnes, Bjørn Fjellvoll, Markus Lund Vevle, and Arne Skorstad

Abstract Although object models are popular with geologists due to their ability to control the geometries that are produced, they tend to have convergence issues when conditioning on complex well patterns. In this paper, we present a new well conditioning algorithm that utilizes more local data when generating channels. We show that this algorithm performs better than the currently commercially available state-of-the-art object model and thus makes object models viable in modern mature field well settings.

1 Introduction

Object models are one of the earliest geostatistical approaches to facies modeling, originating with Bridge and Leeder (1979). This model lends itself nicely to both geometrical descriptions of geology and Bayesian statistical modeling. A mathematically rigorous description can be found in Holden et al. (1998). Other approaches can be found in Deutsch and Wang (1996) and Viseur et al. (1998).

In the unconditional case, object models are simple to handle and can create very realistic geology. They are the primary choice for generating training images for multipoint methods such as Strebelle (2002). Low well densities are also easy to handle both stringently and ad hoc, but as the wells get dense compared to the object size, these models tend to have conditioning problems. This is seen either as overshooting of target volume fractions or by lack of conditioning in some well observations.

As the well density increases, the first noticeable problem is generally that each object is not able to cover enough observations. Thus, more objects are needed to

R. Hauge (✉) • M. Vigsnes • B. Fjellvoll
Norwegian Computing Center, 114, Blindern, 0314 Oslo, Norway
e-mail: Ragnar.Hauge@nr.no; Maria.Vigsnes@nr.no; Bjorn.Fjellvoll@nr.no

M.L. Vevle • A. Skorstad
Roxar Software Solutions AS, Lysaker Torg 45, 1366 Lysaker, Norway
e-mail: MarkusLund.Vevle@emerson.com; arne.Skorstad@emerson.com

J.J. Gómez-Hernández et al. (eds.), *Geostatistics Valencia 2016*, Quantitative Geology and Geostatistics 19, DOI 10.1007/978-3-319-46819-8_37

cover all observations, leading to a concentration of objects around wells. Typically, objects also get smaller, to maintain volume fraction.

An even further increase of well density leads to outright failures. The object model no longer achieves correct volume fractions, as its algorithm is not able to place enough objects between wells. There may also be cases of unconditioned object observations, again because the algorithm is not able to place objects there without colliding with other wells.

Another popular approach to facies modeling is multipoint methods, as described in Strebelle (2002). These always guarantee well conditioning and volume fractions but may give poor geometries. A natural idea is then to hybridize, such as in Henrion et al. (2010). They combine an object model with a truncated Gaussian post-process to ensure well conditioning.

Another example of hybridization can be found in Vargas-Guzman and Al-Quassab (2006), where they combine object models with indicator kriging. In this paper, the entire algorithm is hybridized, but ultimately, it relies on the flexibility of the indicator kriging to ensure well conditioning. The problem for all hybridization schemes is that it is very difficult to control how the hybridization modifies the objects and thus preserves object geometry.

There are three possible reasons for why the object model fails with real data:

1. The object geometries are not flexible enough to describe true reservoir geometries.
2. The parameters that are set for the object model are too restrictive and do not allow description of the true reservoir geometries.
3. The algorithm fails to sample low-density areas of the prior, but it is here the match with complex data is found.

It is interesting to note that failures in object models are more pronounced in real-world cases. As long as synthetic data sampled from unconditional realizations are used, object models can perform really well, as seen in Hauge et al. (2007). Failures with real data could be due to mismatch between parameters and the truth but also indirectly through the mapping between the stratigraphic and modeling grid, which can make well observations incompatible.

We argue that a major part of the problem lies in the third point and will show this by presenting a well conditioning algorithm that significantly improves object model performance with respect to dense wells.

In this paper, we use the Bayesian framework of Holden et al. (1998). We believe that this is the best way to get a reasonable uncertainty description for a given geological scenario. Our ideas for well conditioning algorithms are based on those in Skorstad et al. (1999) and Hauge et al. (2007). These are implemented in commercial software and considered state of the art of what is commercially available.

There are other approaches to well conditioning. In Deutsch and Wang (1996), they essentially generate unconditional channels and see if they fit the wells. This works in models with few wells but quickly fails in more complex settings. A more sophisticated approach is described in Viseur et al. (1998), where they force objects to fit well observations. However, this non-iterative approach is not easy to fit into a Bayesian framework.

A recent interesting approach, which also argues that the algorithms are the problem, is that of Boisvert and Pyrcz (2014). Here, they look at the well conditioning problem as an optimization problem and get good results. However, establishing a sampling algorithm from an optimization is generally a difficult problem, and we do not see how this can be generalized for any explicitly defined model.

The next section describes the model we use, followed by a section with the sampling algorithm for this model. The sampling algorithm is the new contribution in this paper. Finally, we present some examples, followed by concluding remarks.

2 Object Model

We will here introduce our object model and its key parts. We also take a brief look at some key differences between an object model and a multipoint model.

2.1 *Mathematical Model*

In this paper we use the same model as in Holden et al. (1998), but we will focus only on the parts that are important for well conditioning. For a deeper understanding of the full model, we refer to Holden et al. (1998). Choosing this point of view means that we consider channel objects in a shale background; however, the methodology is easily generalized.

The probability density for a realization r given well data d and volume fraction target v is given by

$$f(r|d, v) \propto f_g(r) I_d(r) I_v(r) f_i(r). \quad (1)$$

The first term here, f_g, is the geological prior, controlling the shape of objects. I_d and I_v are indicators showing that all wells are conditioned and the volume fraction is correct. Finally, f_i is an interaction term, ensuring that objects do not get too close to each other.

For a completely determined model, f_g should consist of an intensity term, giving the probability of the number of objects n in r, and the product of geometry priors for each of the objects. We have found that in practice the distribution for n is very troublesome; hence, we use

$$f_g(r) = \prod_{i=1}^{n} f_o(o_i),$$

where f_o is the geometry distribution for a single object, i.e., the distribution for size and shape. The reason for ignoring the intensity term is that we do not know the distribution for the number of objects in a realization. This distribution must be estimated from the volume fraction. However, complexities in shape, interactions, and erosion patterns can lead to inconsistencies in this estimate, giving a model where the volume fraction term is very difficult to satisfy given the distribution for the number of objects. Hence, we rely only on the volume fraction to control the number of objects. This works since the interaction keeps the objects from hiding inside each other.

The well conditioning indicator, I_d, is 1 if all well observations are satisfied and 0 otherwise. This means that only realizations which honor the well data have positive likelihood.

The volume fraction indicator, I_v, is 1 if all volume fraction criteria are fulfilled. In this paper, we will only consider a global volume fraction, but any volume fraction criteria that can be given by an indicator are valid. When working with large objects, it is important to have enough tolerance on the volume fraction, since adding or removing one channel can change the volume significantly.

We will not discuss the interaction term, f_i, here since it is included only to create a bound on the number of objects.

2.2 Geometrical Model

Each channel is parameterized along a line, which is denoted as the channel line. The direction of the channel line is determined by the azimuth and dip, which are typically given a normal prior distribution. The location within the simulation area is most often uniform.

Let us for simplicity consider the case with no dip, where the lateral extent of the object is determined in the xy-plane. The channel line defines a local x-axis where $y = 0$. We let the channel line and the local coordinate system of the channel start at the point where the normal to the line in the xy-plane is tangential to the simulation area, setting $x = 0$ here. Similarly, the channel line ends at the other end of the simulation area, at $x = L$, where L is the length of the line. See Fig. 1.

We then add a 1D Gaussian field, m_h, to this line, to define the local center of the channel laterally. Another 1D Gaussian field, w, is defined, independent of m_h. This field is the local width of the channel.

Thus, the left and right channel edges, $l(x)$ and $r(x)$, at local x-coordinate x are independent and are given by

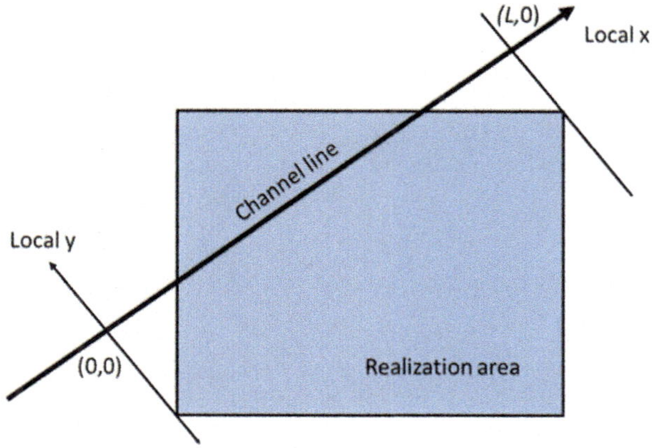

Fig. 1 Channel line relative to realization area and the definition of the local coordinate system for a channel

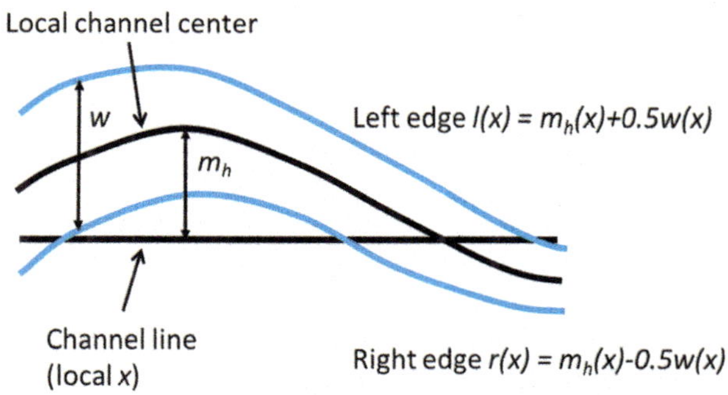

Fig. 2 *Left* and *right* edge of channel seen in the xy-plane. The edges are defined by the channel line and the 1D Gaussian fields m_h and w

$$l(x) = m_h(x) + 0.5w(x),$$
$$r(x) = m_h(x) - 0.5w(x).$$

This is shown in Fig. 2.

The variogram for the Gaussian fields, m_h and w, are set equal. The range, expected width, its local standard deviation, and the standard deviation for m_h are drawn from a prior distribution and will thus vary from channel to channel. By using this parameterization, we can ensure that the channel has positive width by truncating only one Gaussian field. We treat a dip of an object as a vertical translation. That is, given a dip of φ, we translate everything at local coordinate x vertically with $x\sin(\varphi)$. This gives the advantage of not having to do any rotations to compensate for dip and gives almost the same results as long as the dip is small, which it should be in a channel setting.

To create the full 3D representation of the channel, we add two 2D Gaussian fields, m_v and h. The field m_v (similar to m_h) defines the vertical center of the channel and h defines the height. These fields are defined on a coordinate system with the same x-axis as the local coordinate system but with $y_{2D} = 0$ on the right edge and $y_{2D} = \mathrm{E}(w)$ on the left edge. In this coordinate system, the channel

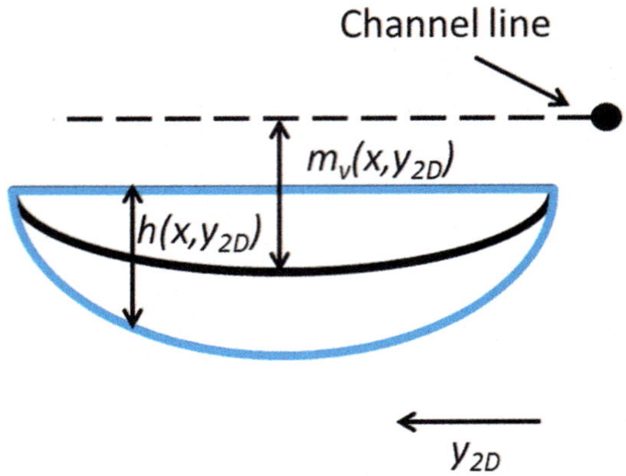

Fig. 3 Cross section (in yz-plane) of channel in plane normal to local x-axis. The expected *top* and *base* of the object is shown, as defined by *channel line*, m_v and h

becomes a rectangle, and we can generate 2D Gaussian fields on it. The top and base of the channel in local position (x,y_{2D}) are given by

$$t(x, y_{2D}) = m_v(x, y_{2D}) - 0.5h(x, y_{2D}),$$
$$b(x, y_{2D}) = m_v(x, y_{2D}) + 0.5h(x, y_{2D}).$$

This is shown in Fig. 3, using only the expected fields.

Fixing the y-coordinate at both edges leads to a stretching and compression of the Gaussian fields in this direction as the channel width varies but gives a nice lateral correlation along both edges. The expectation of m_v and h can be given as trends, defining the channel shape.

The uncertainty in these 2D fields is important to handle well conditioning. If there were no uncertainties in these, an object with trends as shown in Fig. 3 would have a completely flat top. This object could condition multiple well observations only if the observations started at the same depth. Similarly, the thickness of the observation would uniquely define how far from channel center it is located. By adding uncertainty to these fields, we get an uncertainty on the vertical undulation of the channel. This gives us the flexibility needed in the well conditioning and can be thought of as a compensation for non-perfect translation from stratigraphy to the modeling grid.

2.3 Object Models Versus Grid-Based Models

A notable difference between object models and grid-based models such as indicator simulation or multipoint is that for some combinations of input parameters and well data, object models explicitly fail, either to condition all well data or to achieve the correct volume fraction. The reason is that object models are implemented in a way that preserves a minimum of geometrical integrity. All objects generated will have positive probability density under the model. This also implies that some facies realizations will have zero probability density, as they cannot be generated by the given objects, so there is a true null space.

Pixel-based models will never assign zero probability to a realization. Standard multipoint implementation, as in Strebelle (2002), shows the pragmatic stance taken here: Initially, probabilities are assigned to the patterns seen in the training image, based on how often they are seen. No assumptions are made for patterns that are not seen; however, we know that some patterns should be strictly forbidden, whereas others should have a positive probability. The next step is to fill in facies so that the patterns that have assigned probabilities are preserved locally. Inevitably, this creates conflicts elsewhere, and we run into unseen patterns. This is then solved by assigning probabilities to the patterns with similarities to the ones that got an initial probability. This entire process is ad hoc and simulation dependent; however, it ensures that the simulation never stops. A result is always achieved, with correct volume fraction and well conditioning. When a multipoint model meets a complicated well set, it will thus sacrifice geometry, to a degree that object models never will.

The challenge for object models is whether the null space is too large for them to be useful in densely drilled reservoirs. However, what we argue here is that there is still quite a bit to gain with better algorithms. As well data gets denser and more complicated, we must be able to explore less likely realizations, as this is where the data will take us.

3 Algorithm for Generating Channel Realizations

When we want to sample from distributions such as Eq. 1, the Metropolis-Hastings (MH) algorithm (Hastings 1970) is a good tool. The MH algorithm is very robust but may have convergence issues.

3.1 The Metropolis-Hastings Algorithm

Again, we refer to Holden et al. (1998) for details of how to implement the MH algorithm in this setting. The most important aspect of the MH algorithm is that it generates a Markov chain that converges to the desired distribution. All we need to provide is a Markov chain sampling algorithm that covers the state space. We must be able to compute the sampling densities and the true density of the realization up to a scaling constant.

This allows us to manipulate the drawing algorithm any way we want. A drawback is inherent in the fact that we need to use this algorithm: We are not able to draw directly from the true distribution but must rely on the iterative convergence of the MH algorithm. This means that object models based on this algorithm will be slow compared to single pass grid-based algorithms.

There are two extremes for the choice of the sampling algorithm. One is to go for an algorithm that is very easy to compute so that proposals can be generated very fast. The idea here is that a low acceptance rate does not matter as long as each iteration is very fast. This is more or less the approach in Deutsch and Wang (1996), where they do not look at well data at all but generate unconditional channels. The idea is that discretized on a grid, there is a positive probability of getting the correct facies in all wells. This has extremely slow convergence in complex well situations. The other extreme is going for very computer-intensive proposals with high acceptance rates. The algorithms described in Hauge et al. (2007) and Skorstad et al. (1999) move in that direction. In this paper we take that approach even further.

3.2 Generating Conditioning Points for the Edges

We first draw the channel line for the object and transform the relevant well data to local coordinates. Each object is parameterized at regular intervals along the channel line, called sections. Lateral conditioning points are channel edge points drawn in areas where the channel is close to wells to ensure consistency. These are put in the sections on both sides of the observation, to ensure conditioning. See Fig. 4 for an illustration. Sections without conditioning points are later filled through simulation and kriging. Initially, the algorithm is a 2D version of the one in Skorstad et al. (1999).

We divide the facies observations into four categories: Negative observations are shale observations, and we must avoid them. Positive observations are sand observations that we will condition with this channel. We often build a channel around an observation, and the user may also have specified other observations belonging to the same object. Possible observations are other channel observations that may or may not be conditioned by this object, and irrelevant observations are channel observations conditioned by other channels or missing observations.

4 Deciding Whether a Possible Observation Is Positive or Irrelevant

This is done based on the probability of having the channel edges encompass both the possible observation and existing positives, as well as the likelihood of the top and base fields matching the end points of the possible observation. There is also a stochastic element in this choice, to span the realization space. So an observation that fits nicely into the expected channel shape will have a high probability of being

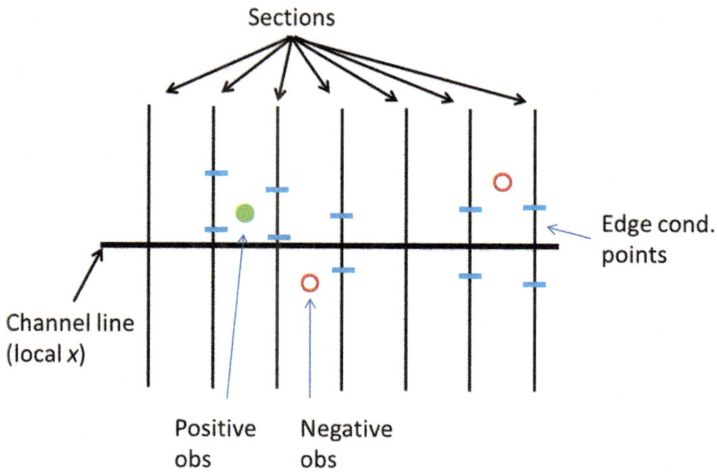

Fig. 4 Three *vertical* wells in an xy cross section and the corresponding conditioning point for edges. Given these, *positive* observations are inside the channel, and *negative* are outside

turned positive, whereas observations with wrong thickness, height, or lateral location have high probabilities of being classified as irrelevant.

5 Draw Conditioning Points in a Section

When drawing conditioning points in a section, we know the distribution for the lateral edges, the vertical center displacement, and the vertical thickness. We have chosen to condition these on the three nearest previously drawn conditioning points, if such exists.

First, we find the expected vertical position of the object center. If we have positive observations, we condition on these. Otherwise, this is done from the prior. Given the vertical location, we are ready to map out the lateral possibilities.

We then reduce the well data to what we need to consider for this particular object. Since we are now generating the lateral conditioning, we remove all well data that is too far above or below the current vertical location. This distance depends on the vertical standard deviation for the center but is also a tuning parameter. So far, we have chosen to only take a narrow interval close to the center and assume that the top and base fields can handle the rest. The wells are divided into facies observations and projected down to the xy-plane of the object.

At this stage, all observations are already categorized as positive (object must condition), negative (object must avoid), or irrelevant (object may pass through). In a well, a positive observation is typically adjacent to negative observations, and these are set to be irrelevant here, since we must partially cover them laterally.

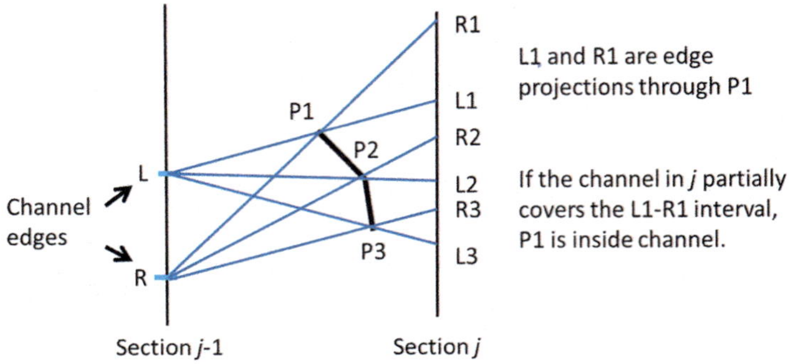

Fig. 5 Cross section in the xy-plane showing two sections. The edges are known in the left section and projected through the well observation points

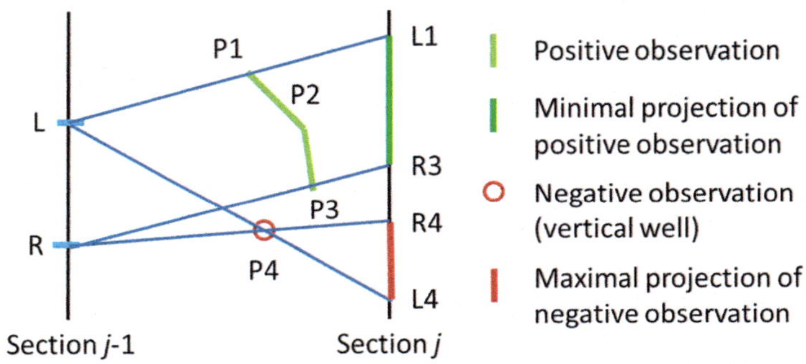

Fig. 6 Cross section in the xy-plane showing two sections and two well observations at relevant depth. From the projections, we see that the channel edges in section *j* must lie so that the L1–R3 interval is completely covered, while none of the interval R4–L4 is covered

At first, a starting point along the channel is selected, typically where we have one or more positive observations between section *j*-1 and *j*. Assume that the channel edges in section *j*-1 are known. We then create a map of legal locations for the edges in section *j* by drawing straight lines from the edges in section *j*-1 through all the well observations and onto section *j*; see Fig. 5. Given the two edge projections through a point, the interpretation is simple: If any part of the interval these generate in section *j* lies between the channel edges in section *j*, the point is inside the channel laterally.

This means that all edge projections together create a simple map for section *j*, from where the channel edges must be drawn based on the following criteria: We must avoid the intervals generated by negative points and cover (at least parts of) the intervals generated by positive points. This is illustrated in Fig. 6.

So far, the algorithm is a slight simplification of the one described in Skorstad et al. (1999). However, we now extend the algorithm by also creating another map, looking at the transition from section j to $j+1$.

We do this by scanning through a discretized space of potential center locations of section j, using the expected width for the channel in section j. This width is updated to condition positive observations between sections j-1 and j for the given center, if necessary. For each center location, we check that it is legal under the first map. Given a legal center location and appurtenant edges, we use the same rules as above to create a map in section $j+1$ based on these edges.

We then compute the probability of drawing a valid channel from j to $j+1$ given this map, if we disregarded the well observations when drawing. This probability is the channel center goodness, and the channel center should be drawn where this is high. We thus draw the edges in section j based on what is legal from the first map and what gives high channel center goodness in the second. If the second map gives no positive probabilities, we try again with a smaller width, if possible. Otherwise, the channel has failed.

The idea of this second map is shown in Fig. 7. Here, we see how the second map helps us to avoid negative observations that are just beyond the horizon of our first map, thus greatly increasing the chance of generating a valid channel in a dense well environment. The cost is that we have to do the scan through possible center locations.

If the channel edges in section j-1 are unknown, e.g., at the start of the algorithm or after a jump between sections, we make the map from j to $j+1$ and draw the edges in j based on this. This works best if there are no observations between j-1 and j.

As long as we have positive observations in the next section, we keep going sequentially from section to section. When this is no longer true, we activate a slightly more complex algorithm to find where to condition next. This algorithm is beyond the scope of this paper, and it will often work to just keep processing all sections in one direction and then continue with the other.

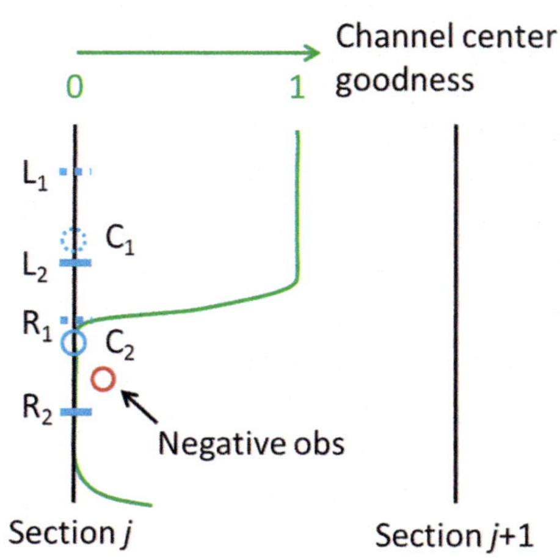

Fig. 7 Cross section in the xy-plane of sections j and $j+1$, with the channel center goodness curve projected onto section j. A channel centered in *C1* will guarantee that we avoid the negative observation, while a center in *C2* will fail

6 Generating the Channel Edges

Given all the conditioning points, which are symmetrically located on the right and left edge, it is trivial to simulate and krige the channel edges.

6.1 Generating Vertical Conditioning Points

The vertical conditioning is simpler than the lateral and is why we try to do much of the conditioning here. First of all, we can disregard all well observations that are outside the channel edges. We then identify the transition points at the ends of positive observations. We also identify points where negative observations are close to or inside the expected object and points where positive observations are close to or outside the top or base of the expected object.

The set of transition points are possible exact observations of the top or base of the object. We assume that a well going downward passes in through the top of an object and out through the base and opposite for a well moving upward. This is a slight approximation but greatly simplifies the problem. If an observation goes up through the top into another channel, it is considered internal, and not an edge, since the channel above has eroded into this channel.

Given the observations of top and base, we draw additional conditioning points for top and base that ensures correct conditioning where we have high probability of errors. At first we draw at the most crucial locations, that is, where the probability of well conflict is largest. Multiple points close to each other are grouped to draw conditioning points that satisfies them all. Figure 8 illustrates vertical conditioning points. Based on the vertical conditioning points, the top and base of the object are simulated and kriged to fit the observations.

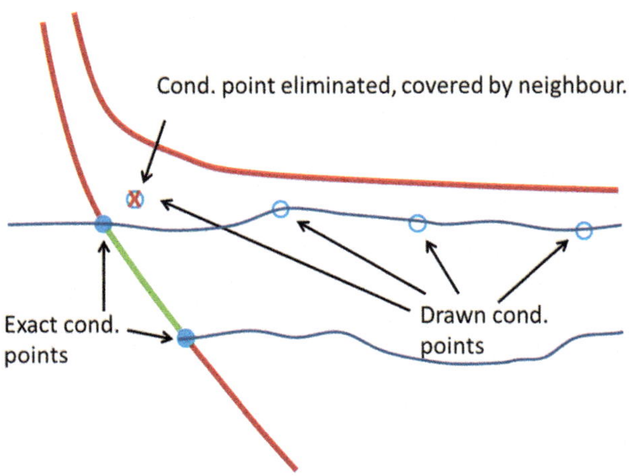

Fig. 8 Cross section in the xz-plane, showing vertical conditioning points and *top* and *base* of conditional channel

7 Examples

We present three examples here. In two of them, we show classic failure cases for object models and how our new algorithm handles them, whereas the third case shows the improved trade-off between prior model and data in our new algorithm.

7.1 *Dense Vertical Wells*

In this example, we have a densely drilled field. The dimensions of the field are 1,000 m by 1,000 m, and the well spacing is 50 m in both directions, as shown at the left in Fig. 9. We have generated an unconditional realization with two channels and sampled these into the wells, as facies observations. Trying to run a commercial state-of-the-art object model on this case fails, meaning that we could not obtain a single realization even after hours of run time. With our new algorithm, we are able to generate a realization in less than a minute on an ordinary PC. Since both use the same model, the improvement is due to the algorithm. A realization generated by the new algorithm is shown to the right in Fig. 9.

Not only do we generate realizations, but by monitoring our proposals, we see that we also generate almost 100 % valid proposals. That is, we almost never have to abandon a channel proposal due to well issues not handled by our algorithm. This is also an important speedup, since time wasted on generating invalid objects either leads to more or slower iterations. On the downside, there seems to be some kriging overshoot in the upper channel, with the spike coming out. This is due to densely drawn points not following the variogram completely, here a function of us

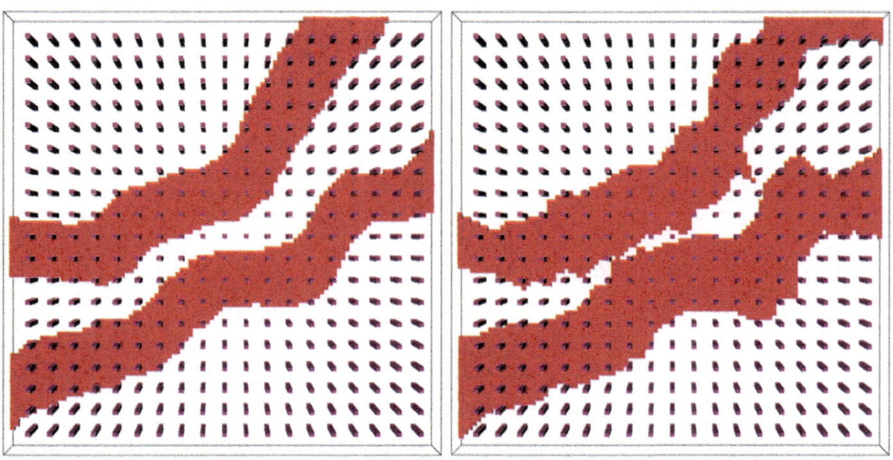

Fig. 9 At the *left* is the dense vertical well pattern with the observed channels, and at the *right* a simulation of two channels conditioned to these well observations

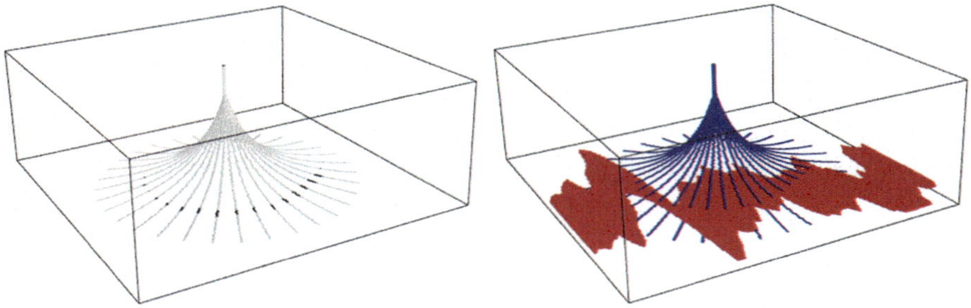

Fig. 10 At the *left* is the well pattern from a drilling pad, where we see one channel in *black*. At the *right* is a channel conditioned to these observations with our new algorithm

conditioning only on the three nearest points. Kriged points may then generate extremes as here.

7.2 Wells from a Drilling Pad

Although there are fewer wells in this example, the general pattern is more complex, as all wells are deviating from a common center. We only have a single channel in our unconditional simulation; again, this is sampled into the wells as facies observations. The well pattern and a realization generated from by the new algorithm are shown in Fig. 10.

The results are the same as before. Our new algorithm is able to generate valid channels in almost all proposals, although we note that the issue with overshooting in the kriging of the edge fields is more severe here. We believe this is a tuning issue that can be solved by choice of variogram and distance between conditioning points. The commercially available algorithm fails to condition this pattern, meaning that not even one valid proposal was generated when run for hours.

7.3 Biased Synthetic Case

The two previous cases have been very clinical and only looked at the improvement in complex cases where all parameters were set correctly. But the first indication of failure in an object model is that it starts to sacrifice its prior distribution. We believe this happens too fast in existing models and has created a test example to explore this.

We have taken a set of vertical wells and sampled a realization into these, as shown in Fig. 11. This is our data set, and so far, it is similar to the ones above.

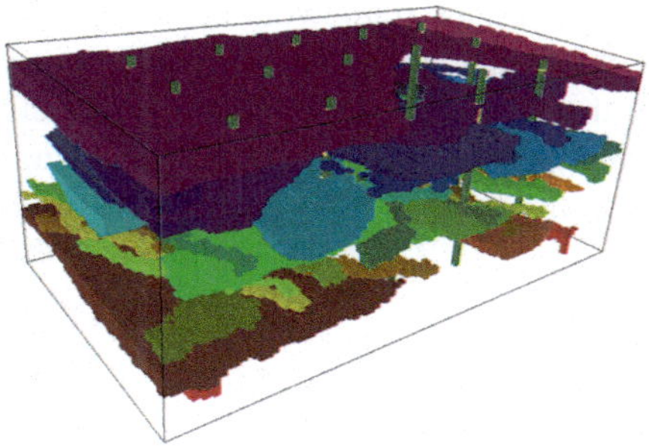

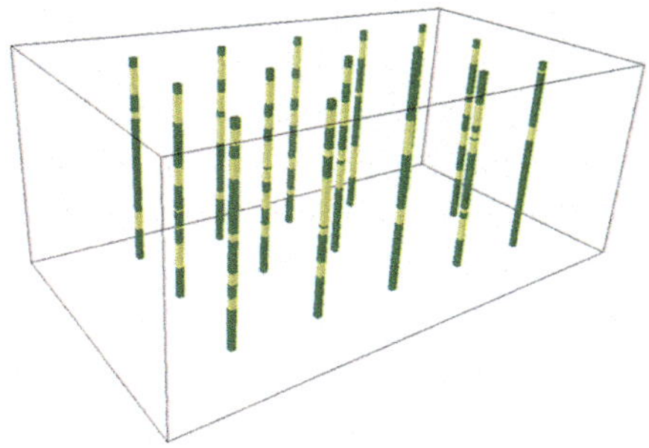

Fig. 11 Original reservoir and well pattern

Table 1 The number of channels originally, and the mean and standard deviation (in parentheses) for the number of channels with 10 runs of our new and the commercial algorithm

	Original	New algorithm	Commercial
# of channels	40	25.9 (0.53)	29.4 (0.79)

However, when generating conditional simulations, we have doubled the expected width of the objects. This is a particularly effective test, since the width of the channel is the parameter best describing how difficult it is to place. A very narrow channel is easy to thread between wells, but a wide one requires more finesse.

The object modeling algorithm is Bayesian and thus has a prior distribution for the width of channels. What we would expect now is a trade-off between the channel width we used to generate the data and the new prior model. Due to edge effects, we use the number of channels as a proxy for actual width; otherwise, wide channels that only graze the modeling box may confuse the issue. The results are summarized in Table 1.

What we see here is that the new algorithm is much better at keeping the number of channels low, as it is much better at navigating wide channels through the well pattern without generating conflicts in the wells. This means that even when the parameters are off compared to the truth, we are able to generate realizations that stay truer to the prior model. The new algorithm is thus better able to explore the relevant parts of the prior distribution.

8 Concluding Remarks

Given the space available, we were not able to present the full model, which also includes levees and crevasses, with full conditioning of the entire hierarchy. Even with the focus on well conditioning, what is presented here is only the key idea in a large and complex framework. As indicated by one of the examples, the tuning is still not optimized. But already, we see significant advances in the complexity of what can be handled. This means that object models can play an important role in fields where the well data amounts previously was not handled by such models.

Although we only present the basic channel and shale framework here, it should be clear how this can be generalized to any object that can be straightened out in a local coordinate system, since this is where the well conditioning occurs. Objects built around piecewise linear backbones would thus fit right into this conditioning, whether they are object- or rule-based.

Bibliography

Boisvert J, Pyrcz M (2014) Conditioning 3D object based models to a large number of wells: a channel example. Mathematics of planet earth: proceedings of the 15th annual conference of IAMG. Springer, p 575–579

Bridge J, Leeder M (1979) A simulation model of alluvial stratigraphy. Sedimentology 26:617–644

Deutsch CV, Wang L (1996) Hierarchical object-based stochastic modeling of fluvial reservoirs. Math Geol 28:857–880

Hastings WK (1970) Monte Carlo sampling methods using Markov chains and their applications. Biomterika 57:97–109

Hauge R, Holden L, Syversveen A (2007) Well conditioning in object models. Math Geol 39:383–398

Henrion V, Caumon G, Cherpeau N (2010) ODSIM: an object-distance simulation method for conditioning complex natural structures. Math Geosci 42:911–924

Holden L, Hauge R, Skare Ø, Skorstad A (1998) Modeling of fluvial reservoirs with object models. Math Geol 30:473–496

Skorstad A, Hauge R, Holden L (1999) Well conditioning in a fluvial reservoir model. Math Geol 31:857–872

Strebelle S (2002) Conditional simulation of complex geological structures using multiple-point statistics. Math Geol 34:1–21

Vargas-Guzman J, Al-Quassab H (2006) Spatial conditional simulation for facies objects for modelling complex clastic reservoirs. J Petrol Sci Eng 54:1–9

Viseur S, Shtuka A, Mallet J (1998) New fast, stochastic, boolean simulation of fluvial deposits. Proceedings, 1998 SPE ATCE, New Orleans, USA. SPE 49281

Machine Learning Methods for Sweet Spot Detection: A Case Study

Vera Louise Hauge and Gudmund Horn Hermansen

Abstract In the geosciences, sweet spots are defined as areas of a reservoir that represent best production potential. From the outset, it is not always obvious which reservoir characteristics that best determine the location, and influence the likelihood, of a sweet spot. Here, we will view detection of sweet spots as a supervised learning problem and use tools and methodology from machine learning to build data-driven sweet spot classifiers. We will discuss some popular machine learning methods for classification including logistic regression, k-nearest neighbors, support vector machine, and random forest. We will highlight strengths and shortcomings of each method. In particular, we will draw attention to a complex setting and focus on a smaller real data study with limited evidence for sweet spots, where most of these methods struggle. We will illustrate a simple solution where we aim at increasing the performance of these by optimizing for precision. In conclusion, we observe that all methods considered need some sort of preprocessing or additional tuning to attain practical utility. While the application of support vector machine and random forest shows a fair degree of promise, we still stress the need for caution in naive use of machine learning methodology in the geosciences.

1 Introduction

In petroleum geoscience, sweet spots are defined as areas of oil or gas reservoirs that represent best production potential. In particular, the term has emerged in unconventional reservoirs where the reserves are not restricted to traps or structures, but may exist across large geographical areas. In unconventional reservoirs the sweet spots are typically combinations of certain key rock properties. Total organic carbon (TOC), brittleness, and fractures are some of the properties influencing possible production. In identifying these sweet spots, the operators face the

V.L. Hauge (✉) • G.H. Hermansen
Norwegian Computing Center, 114 Blindern, 0314 Oslo, Norway
e-mail: Vera.Louise.Hauge@nr.no; Gudmund.Hermansen@nr.no

J.J. Gómez-Hernández et al. (eds.), *Geostatistics Valencia 2016*, Quantitative Geology and Geostatistics 19, DOI 10.1007/978-3-319-46819-8_38

challenge of working with large amounts of data from horizontal wells and modeling the complex relationships between reservoir properties and production.

In general, a more data-driven approach for sweet spot detection allows for a more direct use of less costly reservoir data, such as seismic attributes. Moreover, such an approach may potentially avoid parts of the expensive reservoir modeling. In particular, the time-consuming computations needed to build a full reservoir model can be avoided. Fast and reliable classification of the sweet spots is of high significance, as it allows for focusing efforts toward the most productive areas of a reservoir. This makes machine learning algorithms desirable, since these are typically fast to train, often easy to regularize, and have the ability to adapt and learn complex relationships.

The use of machine learning methodology for predicting and detecting potential areas of interest is gaining attention and is not new to the geosciences. A multidisciplinary workflow in order to predict sweet spot locations is presented in Vonnet and Hermansen (2015). An example of support vector machine application on well data for prediction purposes is given in Li (2005). In Wohlberg et al. (2006), the support vector machine is demonstrated as a tool for facies delineation, and in Al-Anazi and Gates (2010), the method is applied for predicting permeability distributions.

In this paper we continue this exploration and view sweet spot detection in a machine learning setting, framed as a traditional supervised learning problem, i.e., classification. These are data-driven algorithms that aim to learn relationships between the reservoir properties and sweet spots from labeled well-log training data. We illustrate different popular machine learning algorithms through a case study, considering a real and challenging data set with a weak signal for sweet spots. The algorithms we consider and compare are logistic regression, *k*-nearest neighbor (kNN), support vector machines (SVMs), and random forest.

We will emphasize a more moderate and cautious approach to uncritical use of machine learning for classification, wherein the awareness of what we can learn is of significance for interpreting the results. The main challenge here is related to the low data quality and the limited evidence for sweet spots (see Sect. 2). In such cases, the focus should be on the confidence of evidence of sweet spots, despite a potentially low discovery rate. There is usually a high cost associated to exploration and development of a field. It is therefore generally better to sacrifice some sweet spots (i.e., detection rate) in order to gain accuracy and precision. This is our main focus and we compare the ability of these machine learning algorithms to learn from a weak signal. We show how a simple modification can be used to improve such methods and how this improves recovering of the potential and providing sufficiently confident evidence of sweet spots. We also discuss the inadequacy of simple summary statistics for model validation and show that generally a more detailed investigation is needed in order to assess the actual performance.

In Sect. 2 we describe our real data set and set the sweet spot detection in a binary classification setting. Next, in Sect. 3, we discuss the machine learning algorithms used in this case study. Section 4 outlines the setup for training and validating the machine learning methods, before the numeric results are presented and discussed. Lastly, Sect. 5 concludes the case study.

The training and validation of machine learning methods and the predictions and numeric comparisons are carried out in R, using the package `e1071`, `class`, and `randomForest`.

2 Data and the Problem

The case study consists of labeled observations from four vertical blocked wells in a reservoir, providing a total of 315 observation points. For each observation point, there are six reservoir properties available for training, henceforth referred to as features. These are the seismic attributes P-wave velocity (Vp), S-wave velocity (Vs), density, acoustic impedance (AI), 4D residual of pre-stack time migration and average magnitude of reflectivity. In addition, total organic carbon (TOC) and gamma ray (GR) are provided in the wells, which are used to set the labels, i.e., sweet spots. See Fig. 1 and Table 1 for details regarding the number of observations and fraction of sweet spots to non-sweet locations in the wells.

Note that the first four features (Vp, Vs, density, and AI) have been corrected for a depth trend. Thus, the a priori background model for the parameters has been removed, since this introduced a systematic bias in the predictions.

To illustrate the complexity of this data set and the weak relationship in the underlying relationship between sweet spots and reservoir properties, we plot four cross plots of a selection of pairwise combinations of the six features in Fig. 2. These plots indicate a quite strong linear correlation between Vp and density and also between density and AI. Moreover, there is no clear relationship between AI and 4D residual and 4D residual and the average magnitude of reflectivity. This seems to be caused by the high level of noise in the measured 4D residuals. In all four cross plots, there is no trace of geometric delimitation of the sweet spots. Indicating that the well data is not easily linearly separable in the feature space, hence we expect a complex, or limited, relationships.

The reservoir model used for predicting sweet spots has dimensions 280 by 350 by 100 cells. Figure 3 shows the top layer and a vertical slice of the acoustic impedance. Note that the upper left and lower right corners of the lateral view do not contain defined values. Six wells are marked with a circle in the lateral plot. Wells 1, 2, 5, and 6 have given the features defining the sweet spots. The two additional wells in the reservoir, Wells 3 and 4, lack values for total organic carbon and gamma ray. They cannot be used to define the sweet spots and are therefore excluded from further analysis.

The S-wave velocity in the labeled data set appears to be artificially constructed from P-wave velocity, as the estimated correlation between the two is above 0.99, which is also verified by plotting. Traditionally, we would be inclined to exclude one of these variables in the statistical analysis, e.g., to avoid collinearity. However,

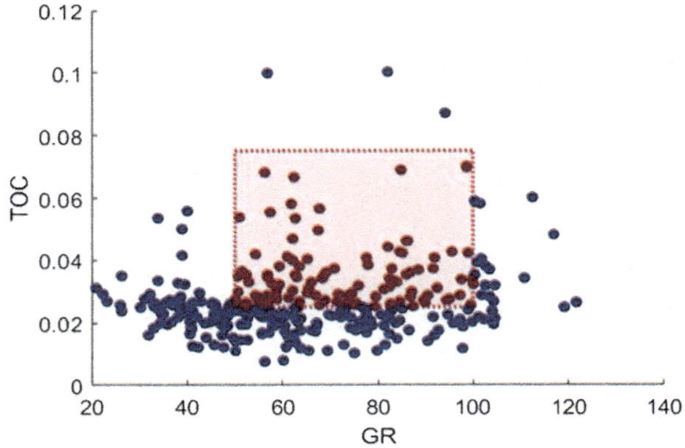

Fig. 1 The ranges of TOC and GR used to define the sweet spots. The observations in the *red region* are defined (by the geologist) as sweet spots

Table 1 Number of sweet spots and non-sweet spots in the four wells

	Sweet spots: non-sweet spots
Well 1	40: 63
Well 2	9: 40
Well 5	38: 63
Well 6	19: 43

we will keep both features in our training data to test and illustrate the robustness of the (probabilistic) model-free machine leaning methods.

The sweet spot classification is a binary classification problem, where we identify the two classes: sweet spots and non-sweet spots. In a binary classification there are four possible outcomes summarized below:

True negative (TN)	*False positive (FP)*
Correctly classified true non-sweet spots	Wrongly classified true non-sweet spots as sweet spots
False negative (FN)	*True positive (TP)*
Wrongly classified true sweet spots as non-sweet spots	Correctly classified true sweet spots

A perfect performance of a classification is identifying all true non-sweet spots as non-sweet spots and all true sweet spots as sweet spots. To measure the performance of a classification, several accuracy and error measures are available. In sweet spot detection, our primary goal is to obtain precise knowledge on the locations of possible sweet spots and the corresponding accuracy and precision of the predictions. We therefore focus on the performance measures True Detection Rate (TDR) and True Prediction Rate (TPR) defined as

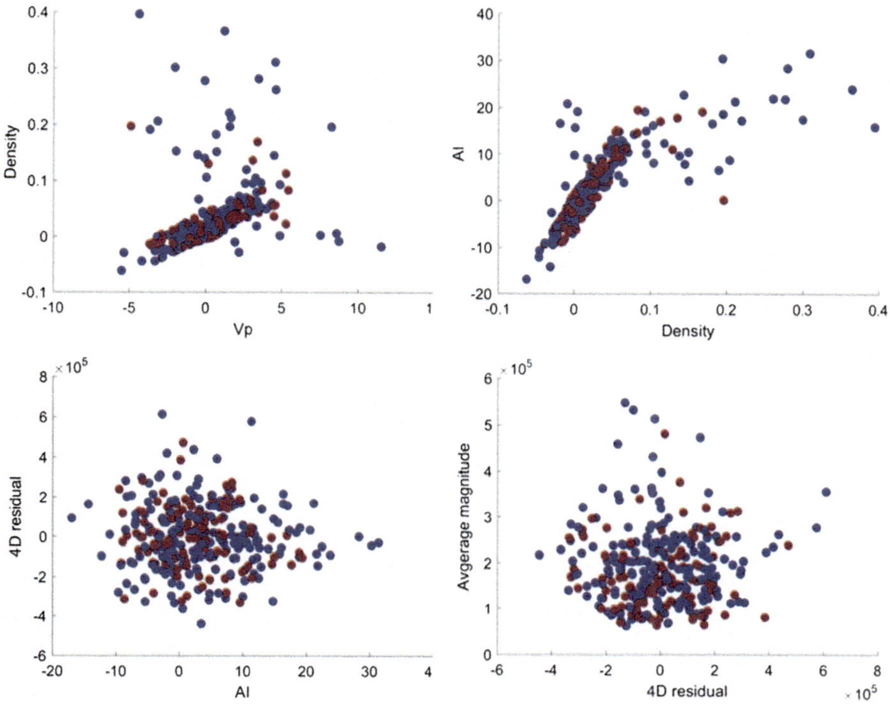

Fig. 2 Cross plots of a selection of pairwise combinations of the six features. All features are plotted in the normalized domain, hence no units along the axis. Again, *red color* marks observations defined as sweet spots; *blue color* marks non-sweet spots

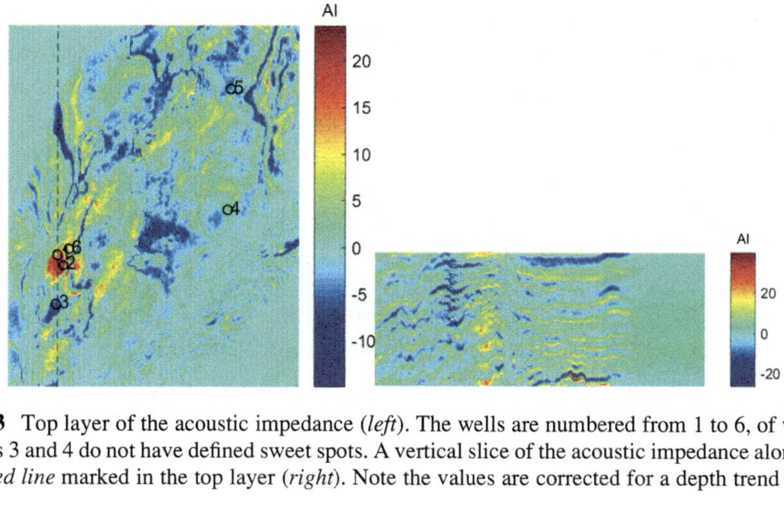

Fig. 3 Top layer of the acoustic impedance (*left*). The wells are numbered from 1 to 6, of which Wells 3 and 4 do not have defined sweet spots. A vertical slice of the acoustic impedance along the *dashed line* marked in the top layer (*right*). Note the values are corrected for a depth trend

$$\text{TDR} = \frac{\text{TP}}{\text{TP} + \text{FN}} = \frac{\text{number of correctly predicted sweet spots}}{\text{number of true sweet spots}} \tag{1}$$

$$\text{TPR} = \frac{\text{TP}}{\text{TP} + \text{FP}} = \frac{\text{number of correctly predicted sweet spots}}{\text{number of predicted sweet spots}} \tag{2}$$

The TDR is a measure of the recall (or sensitivity) of the classification and describes how well the classification method correctly detects the sweet spots that actually are sweet spots. The TPR is a measure of precision and gives the proportion of predicted sweet spots that are actual sweet spots. To combine the measure of recall and precision, we will use the Fβ-score defined as the weighted harmonic mean of recall and precision:

$$\text{F}\beta\text{score} = \frac{\left(1+\beta^2\right) \cdot \text{TPR} \cdot \text{TDR}}{\left(\beta^2 \cdot \text{TPR}\right) + \text{TDR}} = \frac{\left(1+\beta^2\right) \cdot \text{TP}}{\left(1+\beta^2\right) \cdot \text{TP} + \beta^2 \cdot \text{FP} + \text{FN}} \tag{3}$$

In the following, we will use the balanced weighting with $\beta = 1$, denoted F1 score. The more general Fβ score will become of value for tuning the algorithms. Especially for the SVMs this score will be used as a mean of favorable balancing of TPR and TDR to avoid overfitting and collapsing the model to the uninteresting solution of predicting all locations as either sweet or non-sweet spots.

In the sweet spot setting, we argue that TPR is of most importance, as an assurance of correct sweet spot predictions. On the other hand, a carefully balanced focus on the TDR will ensure that more sweet spots are found, at the cost of including misclassified sweet spots. Again, care is needed when tuning methods against these measures.

Moreover, we expect that there is an overrepresentation of sweet spots in the data. This seems obvious, since the initial or any wells are not placed randomly into the field, but they are placed exactly where the developers expect they have the greatest potential for success, i.e., in the sweet spots. This suggests that there is most likely a confounding, or omitted, variable not observed. The information and process underlying the positioning of wells can be thought of as an unobserved (and highly complex) variable influencing both the response and the explanatory variables. This may in turn result in an unbalanced data problem (too many sweet spots) and introduce potentially complex correlations among the explanatory variables and the response; see, among others, He and Garcia (2009) and King and Xeng (2001) for additional discussion. It is generally hard, or even impossible, to correct for such; see Li et al. (2011) for an attempt to correct the support vector machine. The logistic regression model is particularly sensitive; see also Mood (2010). As a final remark, if we consider the overall reservoir from which well logs are collected, we will expect a minority of the sweet spots, causing an additional imbalance, this time in the opposite direction. Proper treatment of such effects and possible extensions are outside the scope of this paper.

3 Machine Learning Methods

In general, machine learning refers to algorithms and statistical methods for data analysis. Here, we will focus on machine learning methodology for prediction of binary class labels, i.e., two class problems. It should be pointed out that all methods discussed can easily be generalized to multiclass problems. We will consider four common and popular supervised learning algorithms, which are the logistic regression, random forest, *k*-nearest neighbor (kNN), and support vector machine (SVM).

3.1 Logistic Regression

Logistic regression is a classical and popular model-based classification algorithm. We refer the reader to, e.g., Hastie et al. (2009) or any introductory textbook in statistics for a general introduction. The logistic regression model provides estimates for the probability of a binary response as a function of one or more explanatory variables. Since it is model based, it is possible to obtain proper and valid statistical inference, e.g., for statistical tests for feature selection. In addition, compared to some machine learning algorithms, e.g., kNN, SVM, or tree-based models, the outputs of a fitted logistic regression model can be interpreted as actual class probabilities under the model conditions. Most (model-free) machine learning algorithm only output class labels, and the probabilistic proxies are obtained and tuned from the raw outputs to mimic an output from a probabilistic model; see, for instance, Platt (1999) for an algorithm for obtaining class probabilities for SVMs.

The logistic regression model has certain well-known challenges. Firstly, compared to simple machine learning algorithms, like the kNN and SVMs, fitting a logistic regression model requires some form of semi-complex and iterative optimization algorithm (like gradient decent). On the other hand, since it is based on a low-dimensional parametric model (the number of parameters is essentially number of features + 1), the fitted model is very efficient for predicting in large grids. Another challenge is that the logistic regression is sensitive to collinearity and confounding; it is not particularly robust against outliers and may become hard to tune automatically (i.e., select the appropriate number of features to use); see, among others, Menard (2002) for details on applied use of logistic regression.

3.2 Random Forest

Random forest is the ensemble of multiple decision or classification trees; see, e.g., Hastie et al. (2009). A decision tree is a greedy approach that recursively partitions the feature space. A single decision tree will easily overfit the training data to the test data and has potentially a large bias. In particular, with noisy data, the generalization of a single decision tree is poor. To avoid overfitting, the ensemble of decision trees, i.e., random forest, averages multiple decision trees based on different resampling of training data. Each of the trees in the ensemble has potentially a high variance, and the averaging of the ensemble reduced this variance. In general, random forest is computationally efficient and is easily interpreted. For more details, we refer the reader to Breiman (2001).

3.3 k-Nearest Neighbor (kNN)

The k-nearest neighbor (kNN) algorithm is one of the simpler and more robust supervised learning algorithms. An introduction can be found in any introductory textbook in machine learning. The algorithm classifies a new observation, or location, by comparing it with the k-nearest observations in the training set and classifies the new observation according to the dominant class. This algorithm is completely model-free and nonparametric. However, each new prediction needs a unique nearest neighbor search. This makes the algorithm less efficient for large data sets and prediction grids. The best choice of the number of neighbors, k, depends upon the data. In our case we perform a cross validation to find this parameter. In general, small values of k may result in noisier results. Larger values of k reduce the effect of noise, but make boundaries between classes less distinct. This algorithm will always improve with more data, and the method is known to work well in simpler classification problems; see also Beyer et al. (1999).

3.4 Support Vector Machine (SVM)

Lastly, support vector machine (SVM) classifies data by finding a hyperplane that separates the data. In the case of linearly separable data in two dimensions, the separating hyperplane is a separating line. Figure 4a shows an illustration of a linearly separable case with the separating line marked as the black line and the data points (support vectors) defining the line marked with circles. The dashed lines mark the margins, i.e., the distance from the separating line to the nearest data points.

For data sets that are not completely separable, the concept of soft margin is introduced to allow some data to be within the margin. SVM now attempts to find a hyperplane that separates the data as cleanly as possible, however, not strictly enforcing that there are no data in the margin (hence the term soft margin). The soft margin is controlled through a regularization parameter, often referred to as C. A large value for this regularization parameter aims for a smaller soft margin and fewer misclassified points. On the other hand, a small value for the regularization parameter aims for a larger soft margin, allowing more points to be misclassified

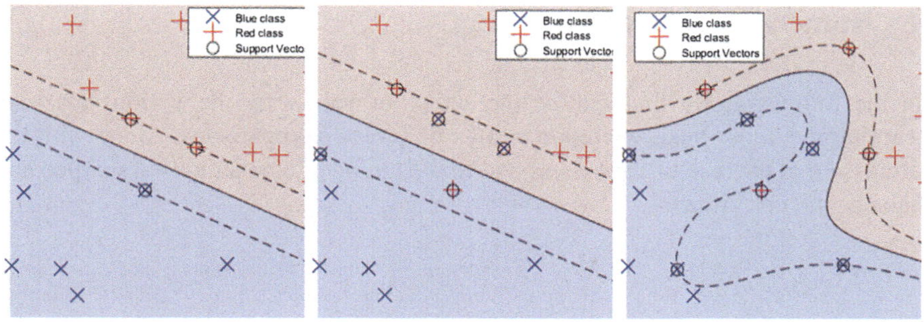

a) Separating line. b) Separating line with data points within the margins. c) Non-linear separating boundary

Fig. 4 Illustration of SVM for (**a**) a linearly separable data set, (**b**) a non-separable data set with soft margins, and (**c**) a nonlinear separating hyperplane

and yielding a smoother decision boundary. Figure 4b has interchanged three points between the blue and red classes, making the data set linearly inseparable. This figure shows the separating plane as the black line, support vectors again marked with circles, and we observe that some points are allowed to appear within the margins (dashed lines).

The SVMs handle nonlinear classification by applying the so-called kernel trick, which allows for nonlinear decision boundaries, while the algorithm for the linear SVM still can be applied for determination of the hyperplane. The kernel trick can be thought of as mapping the observation points into some higher-dimensional space, in which an optimal separating hyperplane is found. Projecting the hyperplane back to the original space yields a nonlinear decision boundary. A typical choice for the kernel function applied is the radial basis function; see, e.g., Hastie et al. (2009). The radial basis function kernel is a scaled version of the Gaussian kernel, in which the squared Euclidean distance between two features is scaled by a free parameter. In the following, we will denote this kernel parameter γ. Adjusting these parameters allows the decision boundary to go from finely detailed decision boundary to a coarser distinction between the classes. Figure 4c shows a nonlinear separating boundary.

The use of SVMs is of great interest as a sweet spot classifier, as it is known to perform well in classification problems where the decision regions of the feature space are of a smooth geometric nature, as we expect to be the case in several applications in the geosciences. The SVM is often referred to be the "out-of-the-box" classifier and is known to be of high accuracy and has the ability to deal with high-dimensional data, i.e., usually no preselection of features is needed.

For a more extensive introduction to SVMs, the reader is referred to Bishop (2006) and Cortes and Vapnik (1995).

4 Numeric Comparisons

In the following, we first outline the setup for validating the various machine learning methods. Next, we report results of several comparisons. Along with the discussion of the results, we present additional tuning of the methods to sharpen and balance the performances.

4.1 *Training, Testing, and Validation of Methods*

To evaluate the machine learning methods, we use the labeled data and carry out a fitting (training and testing) and validation setup. In 100 rounds of validation, we assign 30–70 % of the labeled data set (randomly) for validating. The rest is left for fitting. In the validation, the fitted methods are applied on the validation data set, and F1 score, True Prediction Rate (TPR), and True Detection Rate (TDR) are recorded. For fitting of the methods (training and testing), again 30–70 % is assigned (randomly) for testing the methods, leaving the rest of the data set for training. In both training and testing, cross validation is used to obtain optimal parameters for the algorithms. Here we have focused on maximizing mainly the TPR value, but also various Fβ-scores. After the 100 rounds of training, testing, and validating, we average the obtained performance measures.

Note that when validating the methods, we randomly choose the observations from all of the four wells. We also consider a more real-case predicting study, where we sequentially hold out one well, fitting the methods on the remaining three wells, and investigate performance on the held-out well.

The optimal parameters, found by cross validation, refer to the parameters yielding, e.g., the largest TPR score. For the kNN we find the optimal number of nearest neighbors $0 < k < 30$. For the SVM we cross validate for the regularization parameter $2^{-5} < C < 2^{10}$ and kernel parameter $2^{-10} < \gamma < 2^{5}$. For the random forest algorithm, ensembles of up to a couple of 1000 trees were tested. Interestingly we saw no significant change in performance for ensembles of more than 100 trees.

4.2 *Results*

Table 2 summarizes the performances of random forest, kNN, and SVM applied as described above. We report the obtained performance measures from the fitting, as well as on the validation set (in bold). The last column in Table 2 reports performance of a random classifier, which randomly (with equal probability) assign predictions as sweet or non-sweet spots. For the logistic regression, we were not able to obtain any results better than TPR of 0.10. The failure of the logistic

Table 2 Summary of the performance of random forest, kNN (tuned k), and SVM (tuned C and γ)

	Random forest		kNN		SVM		Random
F1	0.24	**0.23**	0.51	**0.50**	0.50	**0.49**	0.40
TPR	0.33	**0.32**	0.35	**0.34**	0.36	**0.34**	0.33
TDR	0.21	**0.19**	1.00	**1.00**	0.88	**0.89**	0.50

All tuning is optimized for TPR. The first column for each method is the measures obtained on the testing sets. The second column, marked with bold, is the measures obtained on the validation sets. Last column is the performance of a random classifier

regression is explained by the weak signal (correlation between the explanatory variables and the sweet spots are all less than 0.1) together with nonlinear separation in the feature space, as previously described. Inspection of residuals and several corrections, such as feature and subset selections, and change of threshold on the probabilistic output were tested, however, with no success. Other possible extensions to logistic regression, for instance, by introduction of hidden layers (neural nets) (see, e.g., (Bishop 2006)), are beyond the scope of this paper, and the logistic regression is left out of the following discussion.

From Table 2 we observe that all methods perform more or less equally with the random classifier, with quite low detection rate. Note that both kNN and SVM seem to perform well with the high detection rates. These rates, however, are a consequence of several cases of classifying all predictions as sweet spots, hence finding all, at the cost of significant misclassification rates. This suggests that additional fine-tuning, or preprocessing, is needed to improve the potential.

In further tuning of the methods, we were able to obtain better measures for all reported methods. Specifically, the tuning of random forest consists of excluding the features 4D residual and average magnitude of reflectivity. These features have a negative variable importance measure score; see Liaw and Wiener (2002). It is interesting to note that the kNN algorithm was essentially (with unchanged scores) insensitive to this preprocessing. Furthermore, the SVM actually did worse on the reduced data set, suggesting that SVM is able to make the feature selection on its own.

Therefore, to further improve performance, we included an additional fine-tuning parameter with the aim of obtaining a higher level of precision, i.e., TPR score, by increasing the threshold used by each algorithm to classify observations into the respective classes. This makes it harder, by requiring more evidence, to classify locations as sweet spots.

For kNN this tuning was on the threshold for the majority vote in the neighborhood and for SVM, tuning the threshold, or cutoff, on the decision function score. As alluded to above, the random forest algorithm was not sensitive to additional fine-tuning. For both kNN and SVM, the threshold-tuning comes in addition to tuning of the model parameters. By changing this cutoff, the hope is that we are able to modify the fraction of sweet spots detected. Table 3 thus summarizes the best achieved rates as obtained by using the above described tuning.

Table 3 Summary of the performance for random forest (excluding the features 4D residual and average magnitude of reflectivity), kNN (also threshold tuned), and SVM (also threshold tuned)

	Random forest		kNN		SVM	
F1	0.32	**0.34**	0.41	**0.38**	0.21	**0.27**
TPR	0.38	**0.40**	0.38	**0.35**	0.49	**0.44**
TDR	0.30	**0.33**	0.61	**0.58**	0.20	**0.26**

All tuning is optimized for TPR. Columns are for test and validation sets as given for Table 2

Comparing reported results in Table 3 with Table 2, we see that (prior) feature selection for random forest increases both precision and detection and seems to be the winner among the three. Additional tuning provided no significant improvements to the kNN algorithm, suggesting that sophisticated versions of kNN are required, e.g., the popular (Friedman 1994) or the more involved (Goldberger et al. 2005). The SVM algorithm received a considerable increase in the TPR score, indicating a good potential for additional fine-tuning of the SVM toward the most important properties (e.g., a predefined balance between TPR and TDR).

To evaluate and to get a better understanding of how the obtained performance measures will transfer to the real field, we now fit the models by sequentially holding out one of the wells. Firstly, Fig. 5a shows predictions in all four wells using random forest with four features as specified in Table 3. Here we get a visual impression of how well sweet spots are predicted. We note several missing sweet spots in the predictions, as well as sweet spots detected where the labeled data show non-sweet spots. We accompany the plots of predictions with Table 4, reporting obtained performance measures in the wells.

Next, Fig. 5b shows predictions in all four wells using kNN, with only tuning of the number of neighbors, k (as specified for Table 2). This poor performance is included to illustrate how "good" performance measures indeed transfer to real field. Although we might be led to believe in the predicting power of kNN from Table 2, here kNN is either useless (as in Wells 1 and 6) or yields quite noisy predictions (as in Wells 2 and 5). Also, note that for Well 5 the performance measures in Table 4 are indeed the same as one would expect from a random classifier.

Acknowledging the need for additional balancing of TPR and TDR for a best trade-off, we introduced for SVM additional tuning of the weighting of TPR and TDR. This is done by optimizing the parameters, by cross validation, against the Fβ-score, Eq. 3, for different values of β. The developments of the three performance measures TPR, TDR, and Fβ score, as a function of the weight β, for the four wells are shown in Fig. 6. Note that a TDR of 1.0 corresponds with predicting all points as sweet spots, hence detecting all, at the cost of a large amount of misclassifications.

Selecting an appropriate weight β for each well yields predictions in the wells as displayed in Fig. 5c. By appropriate we here refer to the weights that best balance TPR and TDR, typically at the point where TPR and TDR cross in Fig. 6. Here, the "optimal" balance point is determined by inspecting of Fig. 6. Table 4 reports the

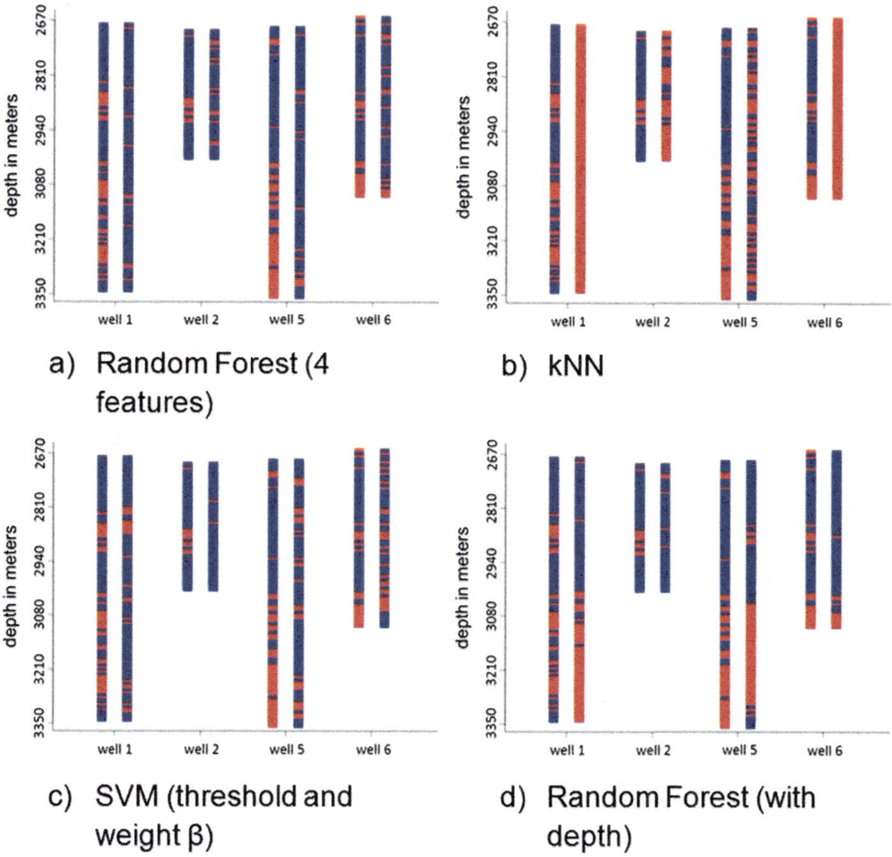

Fig. 5 Prediction of sweet spots in the four wells using (**a**) random forest with four features and (**b**) kNN, (**c**) fine-tuned SVM, and finally (**d**) random forest with four features and without corrections of the depth trend. *Leftmost* well column for each well is the labeled data, while the *rightmost* well column for each well is the prediction

Table 4 Obtained performance measures when sequentially holding out one well at a time

	Random forest		SVM			kNN	
	TPR	TDR	TPR	TDR	β	TPR	TDR
Well 1	0.55	0.15	0.60	0.30	0.30	0.39	1.00
Well 2	0.25	0.33	0.00	0.00	0.40	0.19	0.78
Well 5	0.54	0.18	0.50	0.37	0.30	0.38	0.55
Well 6	0.53	0.42	0.38	0.53	0.45	0.31	1.00

For the random forest and kNN, the two columns are TPR and TDR. For the SVM, we report, in addition to TPR and TDR, the weight β used in the optimization for parameters

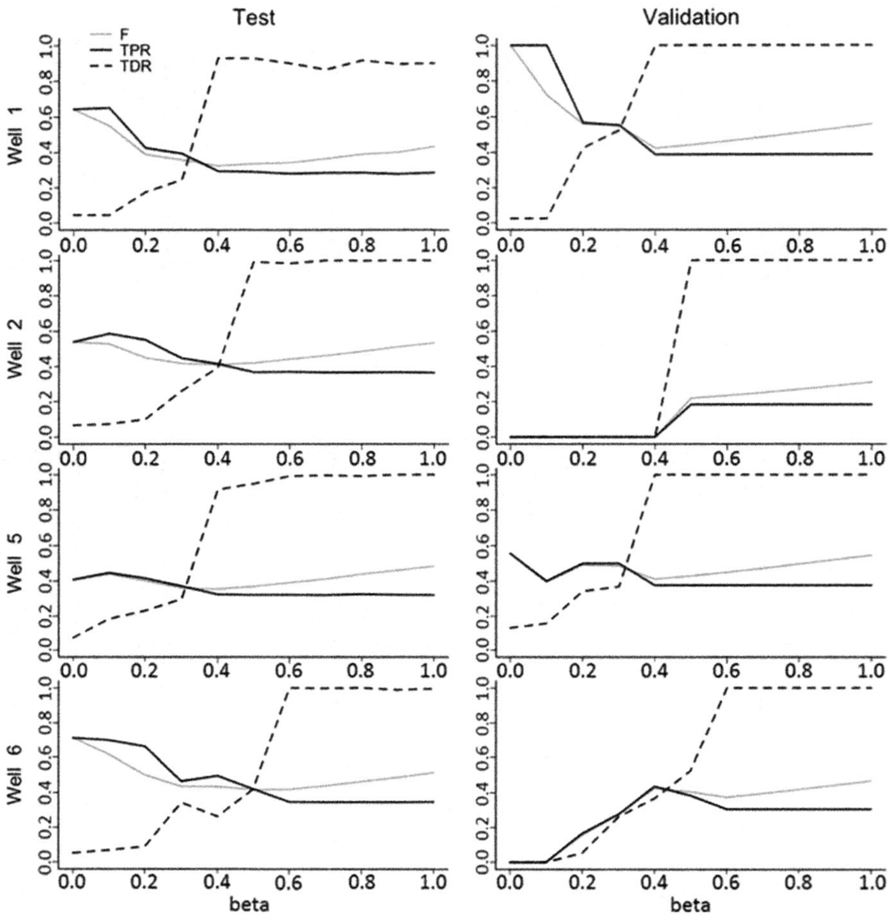

Fig. 6 Development of the three performance measures TPR (*solid*), TDR (*dashed*), and Fβ score (*gray*) as a function of the weight β for the four wells. *Left* column shows the development on the testing set, while the *right* column shows the validation set

weight β used for each of the wells. In general, we now observe in Fig. 5c that the detection has increased compared to random forest, as well as the precision is kept at an acceptable level, indicating good generalization potential.

Note again that predictions in Well 2 fail – more or less – for all methods. Extracting wells as validation sets introduces a grouping of the observations. There is reason to be skeptical regarding the results for Well 2. Figure 7 shows the pairwise cross plots of some of the features, distinguished by color on the four different wells. We observe that for Well 2 (red color) the features do not coincide with the three other wells. Therefore, this well can be interpreted as significantly noisier or to be representing something different. It is, of course, generally hard for a predictor to predict something it has never seen before. On the other hand, a

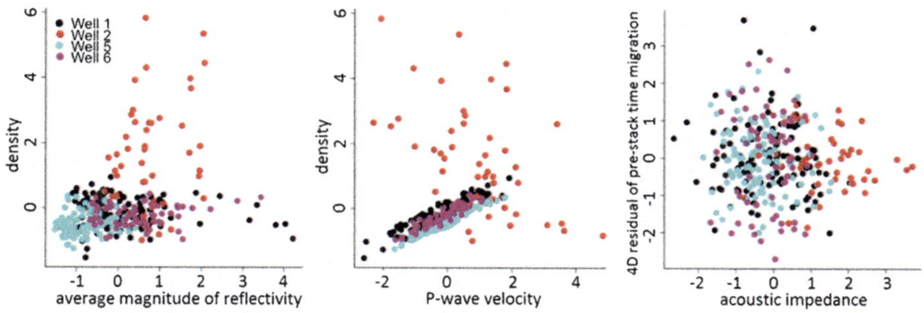

Fig. 7 Cross plots of three pairs of features, distinguished on *color* for the four wells. *Red color* is for Well 2. Values are in the normalized domain, hence no units on the axis

simple linear classifier may still provide reasonable results, depending on the structure of the underlying problem.

As pointed out earlier, four of the features in the data set have been corrected for a depth trend. Figure 5d displays the obtained predictions applying random forest by including depth as an independent feature. We observe a seemingly good match, indicating possible spurious relationship. In the well data in our case study, the majority of the defined sweet spots are indeed located toward the bottom of the reservoir. However, none of the other methods performed acceptably with the depth trend; results were indeed worse.

5 Conclusion

In this paper we have illustrated the application of machine learning methods to a small, but challenging, real field case study of sweet spot detection. The data set has weak evidence of sweet spots, and validation of the methods supports the difficulty of detection. To increase the performance of the methods, we illustrate and discuss a simple solution. As a concluding summary, random forest, given proper preprocessing and feature selection, seems a safe and simple choice, at least for the described data set. Next, SVM shows flexibility and a good potential by responding well to tuning of parameters. SVM is able to obtain acceptable rates and proves transferrable to field predictions. However, unguided use of SVM easily leads to poor performance. The simple kNN with the described tuning does not seem to yield trustworthy results, and logistic regression failed already in the onset of these analyses and did not recover. In general, machine learning algorithms should be used with caution and proper preprocessing, and guided tuning seems to be needed for obtaining reasonable performance.

Acknowledgment We thank Arne Skorstad and Markus Lund Vevle, both at Emerson Process Management Roxar AS, for the data set and for answering questions related to it.

Bibliography

Al-Anazi A, Gates I (2010) A support vector machine algorithm to classify lithofacies and model permeability in heterogeneous reservoirs. Eng Geol 114(3–4):267–277

Beyer K, Goldstein J, Ramakrishnan R, Shaft U (1999) When is "nearest neighbor" meaningful? In: Database theory — ICDT'99, vol 1540. Springer, Berlin, pp 217–235

Bishop CM (2006) Pattern recognition and machine learning (Information science and statistics). Springer, New York

Breiman L (2001) Random forest. Mach Learn 45(1):5–32

Cortes C, Vapnik V (1995) Support-vector networks. Mach Learn 20(3):273–297

Friedman J (1994) Flexible metric nearest neighbor classification. Stanford University

Goldberger J, Roweis S, Hinton G, Salakhutdinov R (2005) Neighborhood components analysis. Adv Neural Inf Process Syst 17:513–520

Hastie TJ, Tibshirani R, Friedman JH (2009) The elements of statistical learning: data mining, inference, and prediction. Springer, New York

He H, Garcia E (2009) Learning from imbalanced data. IEEE Trans Knowl Data Eng 21 (9):1263–1284

King G, Xeng L (2001) Logistic regression in rare events data. Polit Anal 2:137–163

Li J (2005) Multiattributes pattern recognition for reservoir prediction. CSEG Natl Conv 2005:205–208

Li L, Rakitsch B, Borgwardt K (2011) ccSVM: correcting support vector machines for confounding factors in biological data classification. Bioinformatics 27(13):i342–i348

Liaw A, Wiener M (2002) Classification and regression by randomForest. R News 2(3):18–22

Menard S (2002) Applied logistic regression analysis. Sage, Thousand Oaks

Mood C (2010) Logistic regression: why we cannot do what we think we can do, and what we can do about it. Eur Sociol Rev 26(1):67–82

Platt JC (1999) Probabilistic outputs for support vector machines and comparisons to regularized likelihood methods. In: Advances in large margin classifiers. MIT Press, Cambridge, pp 61–74

Vonnet J, Hermansen G (2015) Using predictive analytics to unlock unconventional plays. First Break 33(2):87–92

Wohlberg B, Tartakovsky D, Guadagnini A (2006) Subsurface characterization with support vector machines. IEEE Trans Geosci Remote Sens 44(1):47–57

Theoretical Generalization of Markov Chain Random Field in Reservoir Lithofacies Stochastic Simulation

Xiang Huang, Zhizhong Wang, and Jianhua Guo

Abstract This paper mainly focuses on the theoretical generalization of Markov chain random field (MCRF) model and discusses its application in reservoir lithofacies stochastic simulation. We first introduce the fully independent and conditional independent assumptions of multidimensional Markov chain models. The Equivalence of Markov property and conditional independence is derived explicitly based on the Bayes' theorem, which completes the theoretical foundation of MCRF. The MCRF model is then applied to the lithofacies identification of a region in China, and the results are compared with those by fully independent assumption. Analyses show that conditional independent-based MCRF model performs better in maintaining the percentage composition of each lithofacies and reproducing the geological continuity of lithofacies distribution.

1 Introduction

Lithofacies identification is an important procedure in petroleum exploitation industry. At present, determinative modeling and stochastic modeling techniques in oil-gas reservoir engineering are available to characterize reservoir heterogeneity (Liu 2008; Zhang et al. 2010). Determinative modeling requires detailed underground characterizations, but it is especially difficult to acquire these information. Stochastic modeling is mainly based on variograms and transition probabilities (Carle and Fogg 1996; Weissmann and Fogg 1999). Variograms are symmetric, but the actual reservoir formation has directional property, so that the interclass relationship between the associated position of reservoir categorical attribute is asym-

X. Huang (✉) • Z. Wang
Department of Statistics, Central South University, Changsha, China
e-mail: huangxiang@csu.edu.cn; wzz8713761@163.com

J. Guo
School of Geosciences and Info-Physics, Central South University, Changsha, China
e-mail: gjh796@mail.csu.edu.cn

J.J. Gómez-Hernández et al. (eds.), *Geostatistics Valencia 2016*, Quantitative Geology and Geostatistics 19, DOI 10.1007/978-3-319-46819-8_39

metric. Transition probabilities can closely link reservoir categorical variables and show the spatial (asymmetric) correlation relationships. As a measure for characterizing spatial variability (Huang et al. 2016b), transition probabilities can be used as fundamental tools for the spatial Markov chain theory.

Markov chain model was introduced by Markov in 1906 and applied to stratigraphy in the 1940s. One-dimensional (1-D) Markov chains (or transition probabilities) have long been used in geosciences; see Krumbein and Dacey (1969) and Carle and Fogg (1997) for some reviews in geology and soil science. Multidimensional (M-D) Markov chain models can be traced to Lin and Harbaugh (1984) in geology for modeling lithological (or sedimentological) structures. A spatial hidden Markov chain was introduced by Huang et al. (2016a) for estimation of petroleum reservoir categorical variables in M-D space. The M-D Markov chain model proposed by Elfeki and Dekking (2001) and Li et al. (2012) is considered nearest known neighbors in cardinal directions with the fully independent assumption. This M-D Markov chain is composed of multiple fully independent 1-D Markov chains, and they are forced to move to the same location with equal states. The fully independent assumption causes the small-class underestimation problem. Li (2007b) solved this problem based on Markov chain random field (MCRF) theory and proposed a single spatial Markov chain (SMC) that moves in an M-D space, with its transition probabilities at each given location entirely depending on its nearest neighbors in different directions under the conditional independence assumption (Li 2007a, b; Huang et al. 2016c). Although the conditional independence assumption has been widely used to construct M-D models, the rationality to use this assumption has not been well discussed. The MCRF theory was not sufficiently described mathematically in previous publications (Li and Zhang 2013). The problem is discussed and solved in this paper on the basis of the equivalence of Markovianity (Markov property) and conditional independence.

We first put forward the background information of M-D spatial Markov chain models in Sect. 2. The fully independent assumption and conditional independence assumption are described in Sects. 2.1 and 2.2, respectively. The Equivalence of Markovianity and conditional independence is then derived explicitly based on the Bayes' theorem (Sect. 3), which gives a solid logical proof for MCRF theory. In Sect. 4, we use MATLAB programming to give a simple reservoir lithofacies simulation example to compare the performance of conditional independence-based MCRF method and fully independent-based model. Finally, Sect. 5 concludes.

2 Spatial Markov Chain Models

Oil-gas reservoir field can be divided into discrete gridding units, and the corresponding states (lithofacies) can be regarded as a sequence of random category variables $F_1, F_2, \ldots, F_n$ defined on the n ordered spatial site set $S = \{1, 2, \ldots, n\}$, in which each random variable F_s takes a state value f_s in the m state set

$\Omega = \{1, 2, \ldots, m\}$. The sequence $F_1, F_2, \ldots, F_n$ satisfying Markovianity is defined as an SMC or spatial Markov process if

$$\Pr(f_s | f_1, f_2, \cdots, f_n) = \Pr(f_s | f_{s_1}, f_{s_2}, \cdots, f_{s_l}) \tag{1}$$

where $s_1, s_2, \ldots, s_l$ is its nearest known locations of current unknown site s in different directions. In order to study the complex distribution of subsurface reservoir categorical variables, the SMC models, based on the fully independent and conditional independence assumptions, respectively, are proposed and systematically discussed in this paper.

2.1 Spatial Markov Chain Models with Fully Independent Assumption

The SMC can be constructed by coupling l one-dimensional Markov chains together, but these one-dimensional chains are forced to move to the same location with equal states under the fully independent assumption. The fully independent is defined as

$$\Pr(f_s | f_{s_1}, f_{s_2}, \cdots, f_{s_l}) = C \cdot \Pr(f_s | f_{s_1}) \Pr(f_s | f_{s_2}) \cdots \Pr(f_s | f_{s_l}) \tag{2}$$

where

$$C = \left[\sum_{f_s=1}^{m} \Pr(f_s | f_{s_1}) \Pr(f_s | f_{s_2}) \cdots \Pr(f_s | f_{s_l}) \right]^{-1}, \tag{3}$$

in order to ensure that $\sum_{f_s=1}^{m} \Pr(f_s | f_{s_1}, f_{s_2}, \cdots, f_{s_l}) = 1$.

If two-point conditional probabilities are replaced by transition probabilities $p^r_{f_{s_r} f_s}$, Eq. 2 can be expressed as

$$\Pr(f_s | f_{s_1}, f_{s_2}, \cdots, f_{s_l}) = \frac{p^1_{f_{s_1} f_s} p^2_{f_{s_2} f_s} \cdots p^l_{f_{s_l} f_s}}{\sum_{f_s=1}^{m} p^1_{f_{s_1} f_s} p^2_{f_{s_2} f_s} \cdots p^l_{f_{s_l} f_s}} \tag{4}$$

where $p^r_{f_{s_r} f_s}$ denotes a transition probability in the rth direction from state f_{s_r} to state f_s.

From Eq. 4, the conditional probabilities of two- and three-dimensional Markov chain models are derived as follows:

$$\Pr\left(Z_{i+1,j+1} = k \middle| Z_{i,j+1} = u, Z_{i+1,j} = t\right) = \frac{p_{\text{uk}}^{h} p_{\text{tk}}^{v}}{\sum_{k=1}^{m} p_{\text{uk}}^{h} p_{\text{tk}}^{v}} \tag{5}$$

$$\Pr\left(Z_{i+1,j+1,r+1} = k \middle| Z_{i,j+1,r+1} = u, Z_{i+1,j,r+1} = t, Z_{i+1,j+1,r} = q\right) = \frac{p_{\text{uk}}^{h} p_{\text{tk}}^{v} p_{\text{qk}}^{y}}{\sum_{k=1}^{m} p_{\text{uk}}^{h} p_{\text{tk}}^{v} p_{\text{qk}}^{y}} \tag{6}$$

where $Z_{i,j}$ and $Z_{i,j,r}$ are the coupled Markov chain (CMC) and triplex Markov chain (TMC), respectively, all defined in the state space $\Omega = \{1, 2, \ldots, m\}$, including k,t, q, and u. p_{tk}^{v}, p_{uk}^{h}, and p_{qk}^{y} represent three transition probabilities in the vertical z, horizontal x, and y directions. Equations 5 and 6 can be found in Elfeki and Dekking (2001) and Li et al. (2012), which have been applied in reservoir modeling. However, the transition in Eq. 2 requires moving to the same location with equal states, which causes big-class-overestimation or small-class-underestimation problem (Li 2007b).

2.2 Spatial Markov Chain Models with Conditional Independence Assumption

The acquisition mode and calculation of high dimensional transition probabilities (conditional probabilities) is more complex than that of low dimensional transition probabilities in SMC models. If the higher dimensional transition probabilities are represented by the lower dimensional transition probabilities, it is a great aid and advantage in simplifying the acquisition mode and reducing the calculation work. Taking $s \in S$, its neighboring set is $\eta_s = \{s_1, s_2, \ldots, s_t\}$. By the Markovianity, we have that

$$\Pr\left(f_s \middle| f_{\eta_s}, f_{\overline{\eta}_s}\right) = \Pr\left(f_s \middle| f_{s_1}, f_{s_2}, \ldots, f_{s_t}\right) \tag{7}$$

where $\overline{\eta}_s$ denotes the non-neighboring set of site s. The categorical variables F_{s_1}, $F_{s_2}, \ldots, F_{s_t}$ corresponding to the neighboring sites $s_1, s_2, \ldots, s_t$ may be conditionally dependent or conditionally independent. At most q sites in $s_1, s_2, \ldots, s_t$ are non-neighbors, i.e., for $\forall\ s_i$, $i \in 1, 2, \ldots, t$, the other sites s_j satisfy $s_j \notin \eta_{s_i}, (j \in 1, 2, \ldots, t) \cap (j \neq i)$. Assuming that the q sites are $s_1, s_2, \ldots, s_q$ and corresponding categorical variables are $F_{s_1}, F_{s_2}, \ldots, F_{s_q}$. The conditional independence assumption is given as

$$\Pr\left(f_{s_1},f_{s_2},\ldots,f_{s_q}|f_s\right)=\Pr(f_{s_1}|f_s)\Pr(f_{s_2}|f_s)\cdots\Pr\left(f_{s_q}|f_s\right). \tag{8}$$

Thus, given a categorical variable at site s, the categorical variables at their non-neighboring locations in the neighboring set of site s are conditional independent.

For example, if we choose the first-order neighborhood system in a regular two-dimensional rectangular lattice, given a site s, there are four nearest locations in cardinal directions, and the corresponding categorical variables are conditionally independent. In a regular three-dimensional rectangular lattice, the corresponding categorical variables at the six nearest locations from cardinal directions are conditionally independent. This seems to be basically consistent with the conditional independence assumption proposed by Li (2007b), i.e., given a site, its nearest neighbors are conditionally independent. For the second-order neighborhood system in a two-dimensional rectangular lattice (horizontal plane), at most four diagonal directions are conditional independent. In a three-dimensional rectangular lattice, at most eight diagonal directions are conditional independent. This shows that the conditional independence depends on the order choice of a neighborhood system and no pixels in any directions are conditionally independent; thus conditional independence assumption should be checked whenever possible (Journel 2002).

For the SMC, we assume that at most $l+1$ categorical variable $F_{s-1},F_{s_1},F_{s_2},\ldots,F_{s_l}$ are conditional independent for its nearest unknown variable F_s, where $s-1$ is the last-stay location of the SMC (Li 2007b). Thus, the conditional independence of a single spatial Markov chain is defined as

$$\Pr(f_{s-1},f_{s_1},\ldots,f_{s_l}|f_s)=\Pr(f_{s-1}|f_s)\Pr(f_{s_1}|f_s)\cdots\Pr(f_{s_l}|f_s) \tag{9}$$

Although a large number of applications for the conditional independence assumption can be found in Bayesian network, indicator geostatistics, image processing, and multiple-point statistics (Huang et al. 2013), some of these applications are subjective. We can check the rationality of the use of conditional independent assumption according to the method mentioned above.

3 Theoretical Generalization of MCRF

Let A, B, and C be three mutually disjoint subsets of the spatial site set $S=\{1,2,\ldots,n\}$, suppose A and B are non-neighbors, i.e., for $\forall s\in A$, there does not exist $i\in B$ and $i\in\eta_s$. F_A, F_B, and F_C are three mutually disjoint subsets of categorical random variables in F and are defined on the sets A, B, and C, respectively. We have that $\Pr(f_A f_B|f_C)=\Pr(f_A|f_C)\Pr(f_B|f_A f_C)$.

By Markovianity, we obtain

$$\Pr(f_B|f_Af_C) = \Pr(f_B|f_C).$$

Therefore,

$$\Pr(f_Af_B|f_C) = \Pr(f_A|f_C)\Pr(f_B|f_C). \tag{10}$$

We find that F_A and F_B are conditionally independent given F_C. The conditional independence in the random field F implies that if A, B, and C are any three disjoint subsets of the spatial site set S, A and B are non-neighbors; we have that F_A and F_B are conditionally independent given F_C.

We have proven that if the Markovianity holds, then the conditional independence holds. Now we assume that the conditional independence holds, it needs to be proven that the Markovianity holds. Take an arbitrary $s \in S$, and let $A = \{s\}, C \subset \eta_s, B \subset \overline{\eta_s}$. Obviously, A, B, and C are three mutually disjoint subsets of S; in addition, A and B are non-neighbors. Because the conditional independence in the random field F holds, we have that F_A and F_B are conditionally independent given F_C, i.e.,

$$\Pr(f_Af_B|f_C) = \Pr(f_A|f_C)\Pr(f_B|f_C).$$

With some transformations, we can obtain

$$\Pr(f_A|f_Bf_C) = \frac{\Pr(f_C)\Pr(f_Af_B|f_C)}{\Pr(f_Bf_C)} = \Pr(f_A|f_C), \tag{11}$$

hence the Markovianity holds. Then we have the equivalence between the Markovianity and conditional independence. By using the definition of conditional probability, the general expression of the conditional probability at any location s in a MCRF is expressed as

$$\Pr\left(f_s|f_{s-1},f_{s_1},f_{s_2},\ldots,f_{s_l}\right) = \frac{\Pr(f_{s-1})}{\Pr\left(f_{s-1},f_{s_1},f_{s_2},\ldots,f_{s_l}\right)} p_{f_{s-1}f_s} p^1_{f_sf_{s_1}} p^2_{f_sf_{s_2}} \cdots p^l_{f_sf_{s_l}} \tag{12}$$

where $p^r_{f_sf_{sr}}$ denotes a transition probability in the rth direction from state f_s to f_{s_r}, and $p_{f_{s-1}f_s}$ denotes a transition probability along moving direction of the spatial Markov chain from state f_{s-1} to f_s.

Li (2007b) proposed the general expression of the SMC in an MCRF theory, which is given as follows:

$$\Pr(f_s|f_{s-1},f_{s1},f_{s2}\cdots,f_{s1}) = \frac{p_{f_{s-1}f_s}p^1_{f_sf_{s1}}p^2_{f_sf_{s2}}\cdots p^l_{f_sf_{sl}}}{\sum_{f_s=1}^{m} p_{f_{s-1}f_s}p^1_{f_sf_{s1}}p^2_{f_sf_{s2}}\cdots p^l_{f_sf_{sl}}} \quad (13)$$

Note that Eqs. 12 and 13 are essentially equivalent. We derive Eq. 12 by using the equivalence of Markovianity and conditional independence. The general solution given by Li (2007b) was using the Bayes' theorem (or the definition of conditional probability).

Some special SMC models are provided and can be found in Li (2007a). For example, the conditional probabilities of two- and three-dimensional Markov chain models are provided as follows:

$$\Pr\left(Z_{i+1,j+1} = k \middle| Z_{i,j+1} = u, Z_{i+1,j} = t\right) = \frac{p^h_{\mathrm{uk}}p^v_{\mathrm{kt}}}{\sum_{k=1}^{m} p^h_{\mathrm{uk}}p^v_{\mathrm{kt}}} \quad (14)$$

$$\Pr\left(Z_{i+1,j+1,r+1} = k \middle| Z_{i,j+1,r+1} = u, Z_{i+1,j,r+1} = t, Z_{i+1,j+1,r} = q\right) = \frac{p^h_{\mathrm{uk}}p^v_{\mathrm{kt}}p^y_{\mathrm{kq}}}{\sum_{k=1}^{m} p^h_{\mathrm{uk}}p^v_{\mathrm{kt}}p^y_{\mathrm{kq}}} \quad (15)$$

These models are simple and useful in modeling categorical variables without the overestimation or underestimation problem.

4 Case Study

4.1 Data Sets

The data we used for our research are gathered from Tahe area of the Tarim Basin in Xinjiang Uyghur Autonomous Region, China. There are three major lithofacies in this work area: mudstone, sandstone, and conglomerate. The conglomerate is relatively low in content. We have got four wells' lithologic data in the three-dimensional space (Fig. 1). Three wells are located in the corners of this work area; another well is located inside. The distance in east-west direction of the two wells is 6000 m and 8000 m in south-north direction, the simulated space is split into a $60 \times 80 \times 100$ grid system, and each cell is a $100 \times 100 \times 1$ m cuboid. Note that the coordinate values represent the number of grids.

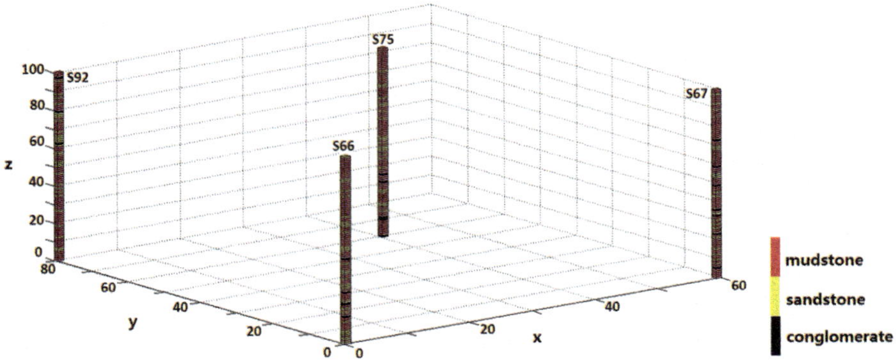

Fig. 1 The 3-D work area with four wells (S66, S67, S75, S92), x axis and y axis indicate east-west direction and south-north direction, respectively. z axis indicates vertical direction. Note that the coordinate values represent the number of grid

4.2 Simulation Results and Analysis

According to the four wells and step size we choose, the unknown grids in this work area are being simulated by using MCRF algorithm, which has been discussed in Sect. 3. The four realizations of lithologic stochastic simulation are shown in Fig. 2. These realizations generally reflect the basic rule of sand body development in this area. The sandstone thickness is mostly in 1–5 m and presents a thin layer of output. It is continuous in horizontal direction, which can extend to several kilometers. Mudstone is the background lithofacies, and it is widely distributed in this area. Conglomerate is not well developed in this work area, which is not continuous neither in vertical direction nor horizontal direction. For one thing, we can safely draw a conclusion from Fig. 2a–d that the conglomerate can only extend 200–300 m in horizontal direction; the average thickness of this type of lithofacies is no more than 3 m. For another, the conglomerate is more likely to appear in sandstone distribution area, which means that the transition probability between conglomerate and sandstone is higher than that of mudstone. Mudstone and sandstone are more continuous than conglomerate in horizontal direction, and the occurrence frequency of mudstone is significantly greater than the other lithofacies in vertical direction.

We also use the method based on fully independent assumption for comparison. The simulation results has been shown in Fig. 3. Compared with Fig. 2, the sandstone layer is thinner in vertical direction and less continuous in horizontal direction. We also find that conglomerate is rarer distributed, which means that this fully independent assumption may underestimate the small class proportion. This conjecture is verified in Table 1. In our analysis, the well data (Table 1) are used for validation. It is obvious that, compared with model based on fully independent assumption, MCRF, a method based on conditional independence assumption, is relatively better at maintaining the percentage composition of each lithofacies.

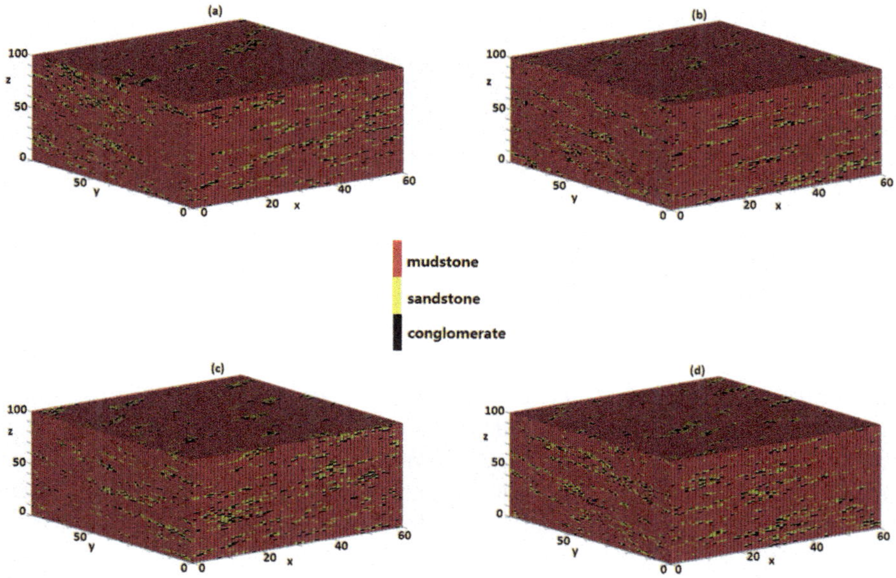

Fig. 2 (**a**) the first simulation result; (**b**) the second simulation result; (**c**) the third simulation result; (**d**) the fourth simulation result

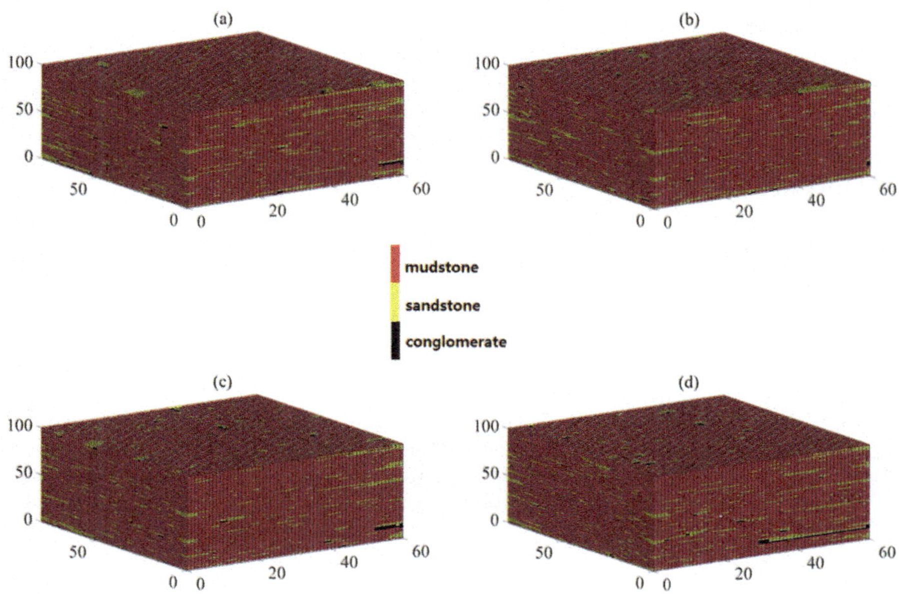

Fig. 3 (**a**) the first simulation result; (**b**) the second simulation result; (**c**) the third simulation result; (**d**) the fourth simulation result

Table 1 Lithofacies proportions in well data and averaged from four simulated realizations

Lithofacies	Well data	Simulation method	
		MCRF	Fully independent assumption
Mudstone	78.70 %	79.81 %	83.14 %
Sandstone	15.75 %	16.02 %	14.83 %
Conglomerate	5.55 %	4.17 %	2.03 %

5 Conclusions

The focus of this article is the theoretical generalization of MCRF model. In addition, a simple reservoir lithofacies simulation example is also given to illustrate the superiority of the conditional independence assumption than the fully independent counterparts. MCRF model uses conditional independence assumption for stochastic simulation. The rationality of this assumption, however, has not been well discussed in previous studies. We solve this by proving the equivalence of Markovianity and conditional independence. Compared with fully independent-based method, MCRF simulation has more advantage in reproducing the geological continuity of lithofacies distribution and performs better in maintaining the percentage composition of each lithofacies.

Acknowledgments This study is funded by the *Fundamental Research Funds for the Central Universities of Central South University* (*No. 2016zzts011*) and the *National Science and Technology Major Project of China* (*No. 2011ZX05002-005-006*). We thank Dr. Dongdong Chen for the helpful discussion regarding the three-dimensional stochastic simulation.

References

Carle SF, Fogg GE (1996) Transition probability-based indicator geostatistics. Math Geol 28 (4):453–476

Carle SF, Fogg GE (1997) Modeling spatial variability with one and multi-dimensional continuous-lag Markov chains. Math Geol 29(7):891–917

Elfeki A, Dekking MA (2001) A Markov chain model for subsurface characterization: theory and applications. Math Geol 33(5):568–589

Huang T, Lu D, Li X, Wang L (2013) GPU-based SNESIM implementation for multiple-point statistical simulation. Comput Geosci 54(4):75–87

Huang X, Li J, Liang Y, Wang Z, Guo J, Jiao P (2016a) Spatial hidden Markov chain models for estimation of petroleum reservoir categorical variables. J Petrol Explor Prod Technol. doi:10.1007/s13202-016-0251-9

Huang X, Wang Z, Guo J (2016b) Prediction of categorical spatial data via Bayesian updating. Int J Geogr Inf Sci 30(7):1426–1449

Huang X, Wang Z, Guo J (2016c) Theoretical generalization of Markov chain random field from potential function perspective. J Cent South Univ 23(1):189–200

Journel AG (2002) Combining knowledge from diverse sources: an alternative to traditional data independence hypotheses. Math Geol 34(34):573–596

Krumbein WC, Dacey MF (1969) Markov chains and embedded Markov chains in geology. Math Geol 1(1):79–96
Li W (2007a) A fixed-path Markov chain algorithm for conditional simulation of discrete spatial variables. Math Geol 39(2):159–176
Li W (2007b) Markov chain random fields for estimation of categorical variables. Math Geol 39 (3):321–335
Li W, Zhang C (2013) Some further clarification on Markov chain random fields and transiograms. Int J Geogr Inf Sci 27(3):423–430
Li J, Yang X, Zhang X, Xiong L (2012) Lithologic stochastic simulation based on the three-dimensional Markov chain model. Acta Pet Sin 33(5):846–853 (in Chinese)
Lin C, Harbaugh JW (1984) Graphic display of two and three dimensional Markov computer models in geology. Wiley, New York
Liu W (2008) Geological modeling technique for reservoir constrained by seismic data. Acta Pet Sin 29(1):64–68 (in Chinese)
Weissmann GS, Fogg GE (1999) Multi-scale alluvial fan heterogeneity modeled with transition probability geostatistics in a sequence stratigraphic framework. J Hydrol 226(1):48–65
Zhang R, Zhang S, Chen Y, Chen B, Hou Y, Huang J (2010) Stochastic simulations of clastic reservoirs in East China. Acta Pet Sin 31(5):787–790 (in Chinese)

Deepwater Reservoir Connectivity Reproduction from MPS and Process-Mimicking Geostatistical Methods

Rhonika Kaplan, Michael J. Pyrcz, and Sebastien Strebelle

Abstract Deepwater unconfined lobe depositional systems are important reservoir targets. High drilling cost and subsalt imaging greatly limit the quantity and quality of data. In the absence of sufficient data, analog architectural studies have identified a variety of potentially important reservoir quality related geometries and trends resulting from the well-understood depositional processes. Internal lobe trends (proximal, dominated by amalgamated sands, to distal, dominated by non-amalgamated sands) impact horizontal connectivity and coupled with compensational lobe stacking impact vertical connectivity (alternating proximal and distal lobe components are superimposed locally).

Current geostatistical algorithms, pixel based or object based, using semivariograms, training images, or geometric parameters, enable the reproduction of spatial statistics inferred from available conditioning data and analogues but rarely integrate information related to depositional processes. Indeed, because conventional geostatistical models are constructed without any concept of time or depositional sequence, their ability to incorporate sedimentological rules, which explain facies geobodies interactions and intra-body porosity/permeability heterogeneity, is quite limited.

Process-mimicking methods provide an improved ability to honor these flow unit stacking patterns and trends, but trade-off precise conditioning to [moderate to dense] well data and detailed seismic informed trend models. To guide geostatistical reservoir modeling practice, a study assesses the incremental impact of process-mimicking relative to a common multiple-point statistics (MPS) approach with respect to reservoir flow response.

A surface-based (a variant of process-mimicking) method coupled with hierarchical trends efficiently reproduces realistic deepwater lobe geometry, stacking

R. Kaplan (✉)
Reservoir Management Unit, Chevron Energy Technology Company, 1500 Lousiana St, Houston, USA
e-mail: rhonika@chevron.com

M.J. Pyrcz • S. Strebelle
Strategic Research Unit, Chevron Energy Technology Company, 1500 Louisiana St, Houston, USA
e-mail: mpyrcz@chevron.com; stsb@chevron.com

J.J. Gómez-Hernández et al. (eds.), *Geostatistics Valencia 2016*, Quantitative Geology and Geostatistics 19, DOI 10.1007/978-3-319-46819-8_40

patterns, and internal lobe heterogeneity. A spectrum of deepwater lobe reservoir models, using archetypal well log data and flow diagnostics and benchmarked with the current MPS approach, quantifies incremental flow significance of these reservoir features. This quantification supports guidance and best practice for geostatistical reservoir modeling workflow design for unconfined deepwater lobes reservoirs.

1 Introduction

Current geostatistical algorithms, pixel based or object based, using semivariograms, training images, or geometric parameters, enable the reproduction of spatial statistics inferred from available conditioning data and analogues but rarely integrate information related to depositional processes. Indeed, because conventional geostatistical models are constructed without any concept of time or depositional sequence, their ability to incorporate sedimentological rules, which explain facies geobodies interactions and intra-body porosity/permeability heterogeneity, is quite limited (Pyrcz et al. 2012).

While we do acknowledge the value and practical success of the more traditional geostatistical models mentioned above, there still remains an opportunity for developing models that incorporate stratigraphic rules that relate to the underlying geologic processes and hence offer an improved representation of depositional heterogeneity. Process-mimicking facies models attempt to increase the level of integration of the geological conceptual model by integrating rules, based on the geological process. The rules constrain the sequential construction of reservoir architecture represented by object or, in our experiment, surfaces (Pyrcz and Deutsch 2014).

Comparisons of modeling approaches are risky and may be misleading. The purpose of this paper is not to determine that one modeling approach is better than another. The authors are motivated by their practical experience in mentoring and directing reservoir modeling on deepwater assets worldwide. The subsurface reservoir modeling teams have consistently asked, “Will more geological process information and realism impact the connectivity of the reservoir model?” This paper demonstrates that in some cases the impact is significant, and with connectivity quantification suggests that there are cases where one technique might be more appropriate than the other, given the reservoir modeling goals aligned with business need.

In our study we use a surface-based method to generate our geologically realistic lobe model (see Fig. 1). Surface-based methods are a modified version of object-based methods that produces and track surfaces that delineate objects. Within a process-mimicking framework, surface-based methods incorporate aggradation and erosion of surfaces based on geometric templates. For greater details on this method

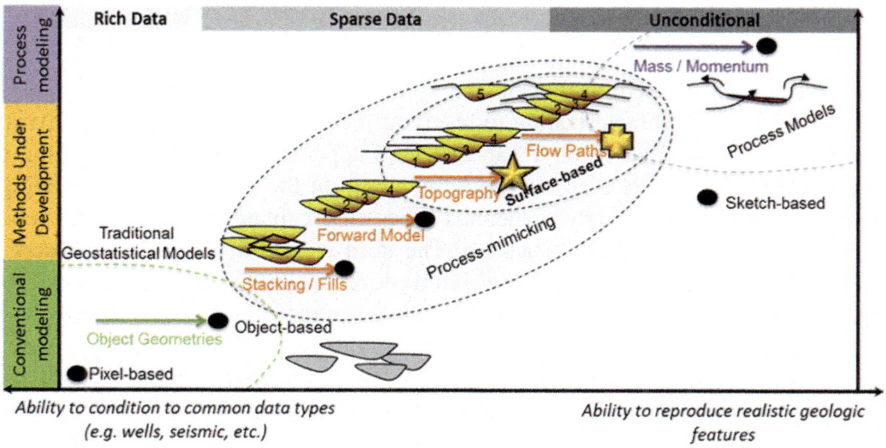

Fig. 1 Continuum of process-mimicking approaches that allow us opportunities to produce more geologically realistic reservoir models

and other process-mimicking approaches, the reader can refer to various publications including Pyrcz and Deutsch (2005), Wen (2005), Miller et al. (2008), Michael et al. (2010), Sylvester et al. (2010), and Pyrcz et al. (2015).

This technique allows for inclusion of realistic lobe stacking patterns and internal lobe heterogeneity, which is not typically captured by conventional modeling approaches. Stacking is a fundamental characteristic of lobes, as well as the internal architecture (proximal, dominated by amalgamated sands, to distal, dominated by non-amalgamated sands).

This study endeavors to quantify the added value of utilizing a process-mimicking approach compared to conventional modeling approaches. A simple multiple-point statistics (MPS) model is utilized as a benchmark, because it represents the common modeling approach in this setting. To achieve our objective we propose to generate typical deepwater lobe reservoir models using archetypal well log data and perform flow diagnostics to estimate the impact recovery. Previous studies have shown the importance of reservoir connectivity and heterogeneity on reservoir performance (e.g., Larue and Hovadik 2006). Our results quantify the incremental impact of connectivity (lobe stacking) and heterogeneity (within-lobe trends) for different net-to-gross scenarios modeling with process-mimicking models benchmarked with standard MPS workflow. This is useful to justify the additional effort to adopt emerging process-mimicking methods rather than utilizing widely available MPS workflows.

2 Methodology

The experiment is based on the application of three parallel modeling workflows from stratigraphic characterization to reservoir modeling to connectivity analysis. The first workflow is based on a standard MPS approach with associated training image construction. The second workflow, based on the surface-based modeling process described previously, generates lobes that capture the large-scale lobe geometry and compensational stacking. The third workflow includes the features of the second workflow with the added within-lobe reservoir property (porosity and permeability) trends that include fining (decrease in reservoir quality) distally and laterally, as well as sedimentary cycles vertically. Each workflow from 1 to 3 represents increasing integration of the stratigraphic characterization. The quantification of impact is based on a flow diagnostic tool that calculates the dynamic Lorenz coefficient, a global measure of the degree of connectivity complexity. The experiment includes 20 stochastic realizations from each workflow for 40, 60, and 80 % NTG (ratio of lobe related to overbank facies).

2.1 Geologic Characterization

Deepwater lobe characterization is based on a hierarchical architecture. Individual lobe elements are lenticular in cross section and lobate in map view, with high aspect ratios on the order of 1,000:1 (Beaubouef et al. 1999; Sullivan et al. 2004; Prélat et al. 2009, 2010). They represent the unconfined deposition of genetically related sediment gravity flows. Lobe elements that are genetically related show similar grain size and facies distribution and similar architectural styles and stack in a compensational manner to form a lobe complex.

Lobe elements are characterized by individual depositional packages known as stories. These include general trends in depositional grain size with fining from proximal to distal and inner to outer lobe and vertical cycles (Fig. 2).

2.2 Geostatistical Modeling

Lithofacies modeling, for Workflow 1 geostatistical modeling, utilizes MPS. The MPS variant is the standard SNESIM approach (Strebelle 2002). The training image is composed of sand facies lobes in an overbank background (see Fig. 3). It was built using a simple object-modeling tool that generates lobular shapes with dimensions 60,000 ft in length, 40,000 ft in width, and 40 ft in thickness. The global proportions are constrained by firstly designing training images with the target facies proportions and updating conditional distributions from the training images based on mismatch with target global proportions during simulation.

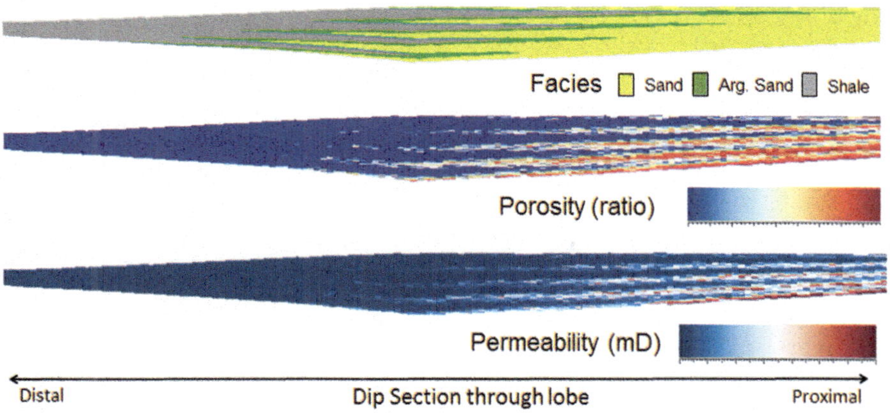

Fig. 2 Dip section of a simple conceptual model of within-lobe heterogeneity for a simple lobe geometry

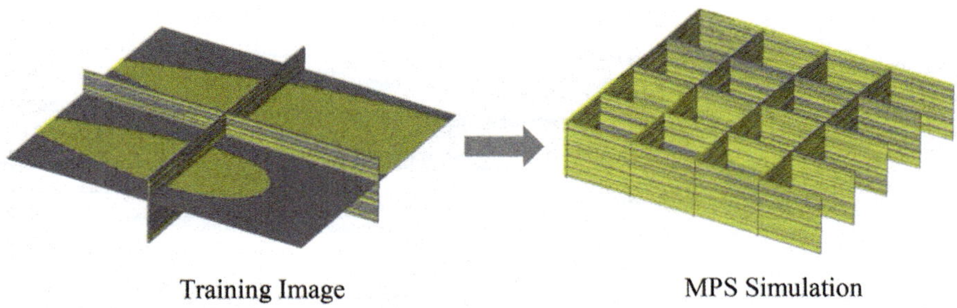

Fig. 3 Training image used in MPS simulation at a sand proportion of 60 %

For workflows 2 and 3 our surface-based method was applied to simulate compensationally stacked lobes (Pyrcz et al. 2015). The same dimensions mentioned above were used in the construction of these lobes. Our tool also generates an azimuth field within the lobe objects, as well as longitudinal, transverse, and vertical trends which are combined to produce our hierarchical trend. More information on the hierarchical trend approach is found in Pyrcz et al. (2005). The hierarchical trend is crucial because it is used to guide the petrophysical modeling in Workflow 3 (Fig. 4).

Sequential Gaussian simulation is applied to simulate porosity within the reservoir facies using a histogram of values ranging from 0.04 to 0.26, and a variogram range of 2,000 ft in the longitudinal direction and 1,300 ft in the transverse direction is assumed. For workflows 1 and 2, no trends are applied for the porosity simulation. However, for workflow 3 within-lobe trends, which mimic the hierarchical architecture recognized in deepwater lobes, are added. These trends are integrated as a secondary variable with a correlation coefficient of 0.8.

Fig. 4 Surface-based lobe model with sand and argillaceous sand proportion of 60 %

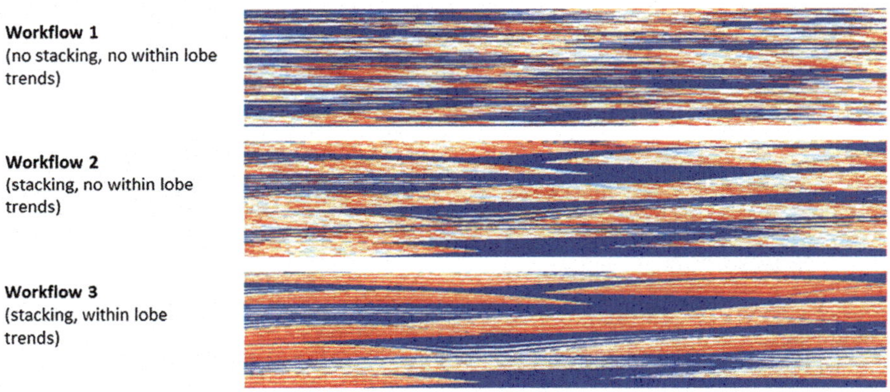

Fig. 5 Strike sections of a porosity realization from each of the workflows

Cloud transform is applied to simulate permeability conditional to the simulated porosity realizations using the same variogram mentioned previously. Permeability values range from 5 to 30 mD within the reservoir facies. Examples of strike sections for realizations from each workflow are shown in Figs. 5 and 6.

2.3 Flow Diagnostic Assessment

Flow diagnostic assessment was used to derive quantitative information on reservoir connectivity. These methods utilize standard reservoir property models (porosity and directional permeability) along with efficient, simplified flow simulation to provide immediate information on the impact of modeling decisions on flow heterogeneity (Shook and Mitchell 2009; Shahvali et al. 2012; Møyner et al. 2015).

In this case study a steady-state pressure field is calculated that induces a displacement or flux from injectors to producers across the reservoir. Importantly,

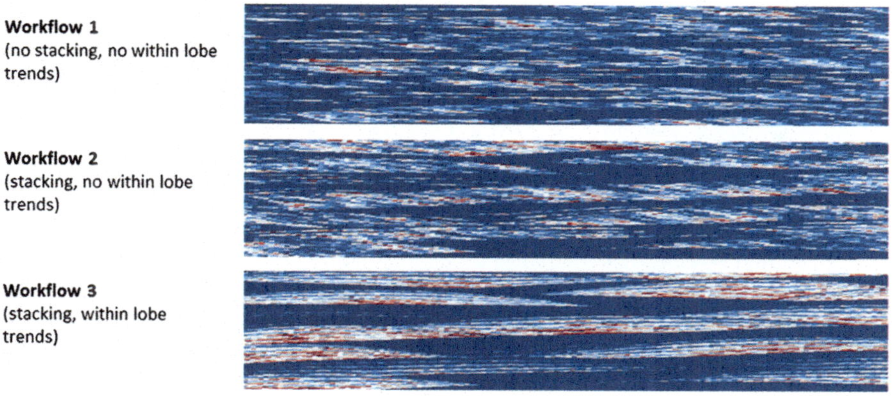

Fig. 6 Strike sections of a permeability realization from each of the workflows

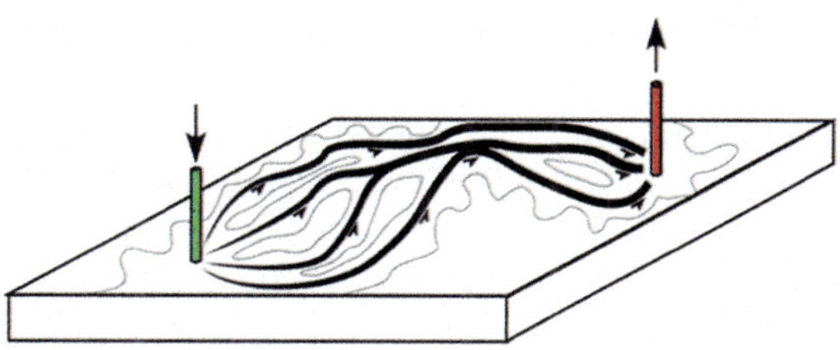

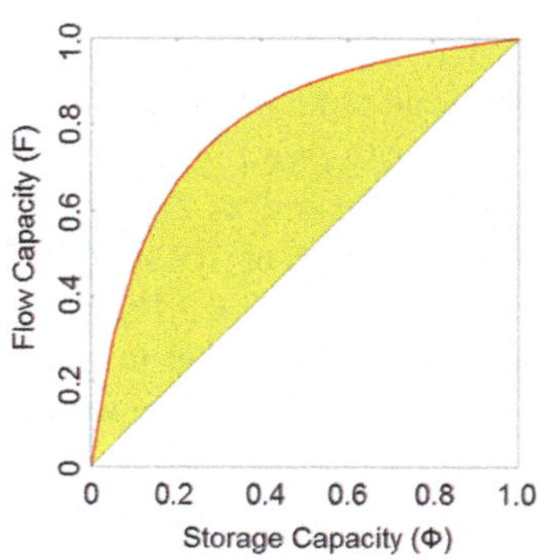

Fig. 7 Dynamic Lorenz coefficient to quantify reservoir connectivity heterogeneity as a summarization of the storage capacity subtracted from the flow capacity based on a set of injectors and producers

the computed flux does not account for the full physics of the flow between two wells as the calculation of flux within the model makes several assumptions. Firstly, the flow is assumed to be steady state and single phase; therefore, the phase of the flow will not change due to buoyancy, well controls, or fluid mobility changes. Secondly, a main assumption is that the flow is incompressible and well-driven flow; any independent reservoir compartment or fault block in the model must have at least one producer and one injector well to initiate some flow.

This approach yields a number of outputs including dynamic Lorenz coefficient (DLC) and time of flight information. Once the flux across the volume and time of flight are calculated, the F-Φ relationships can be calculated. When plotted this generates a curve representing the relationship between normalized, cumulative flow capacity (F) against storage capacity (Φ). Specifically, the curve explains the ratio between the volume of injected and swept reservoir pore volume. Twice the area under this curve and above the 1:1 line is the DLC and can be used to rank the heterogeneity of multiple models (see Fig. 7). This is a good quantification of flow complexity (reservoir heterogeneity) as it describes whether flow in the reservoir represents a simple piston displacement or highly impacted by high permeability streaks and barriers (indicated by high DLC).

3 Results

The flow diagnostic assessment quantified by DLC (a good indicator of recovery factor) suggests that the level of reservoir connectivity and heterogeneity is impacted by the additional features such as lobe stacking pattern and within-lobe trends that are available with process-mimicking model. Visually, we see this behavior in Fig. 8 where the volume that is being swept shrinks and is less uniform as we go from the simplistic model from workflow 1 to a more complex model from workflow 3.

The expectation over several model realizations, for each workflow, was calculated and plotted. It demonstrates increasing flow complexity with increasing model complexity (see Fig. 9).

We assessed at each net-to-gross scenario which features, stacking patterns (connectivity) or within-lobe architecture, contributed more significantly to the change in the DLC. The proportion increase of DLC, as we move from a purely geostatistical method to a geological method that incorporated stacking patterns and within-lobe architecture, was calculated. At lower net-to-gross regimes, we see

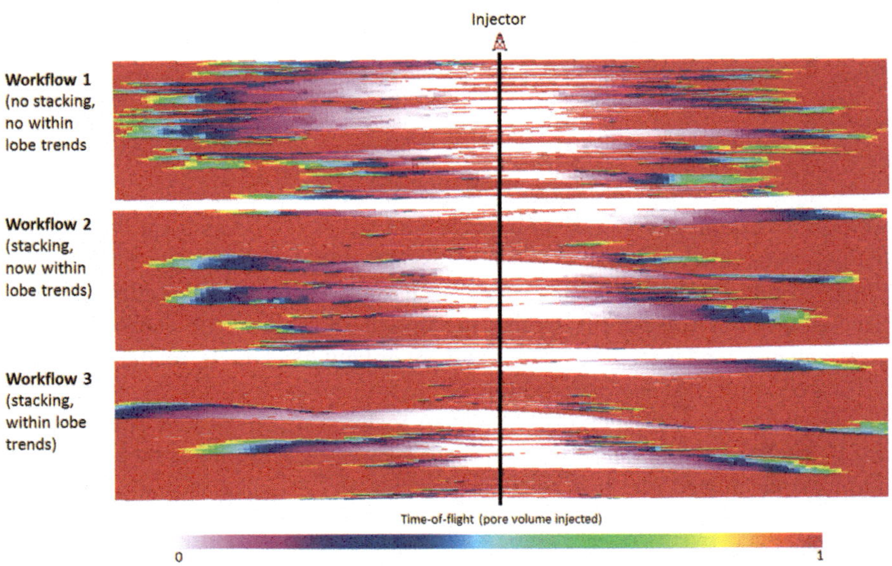

Fig. 8 Time-of-flight represents the time it takes (in pore volumes injected) for injected fluid to travel from an injector to a given point in the reservoir

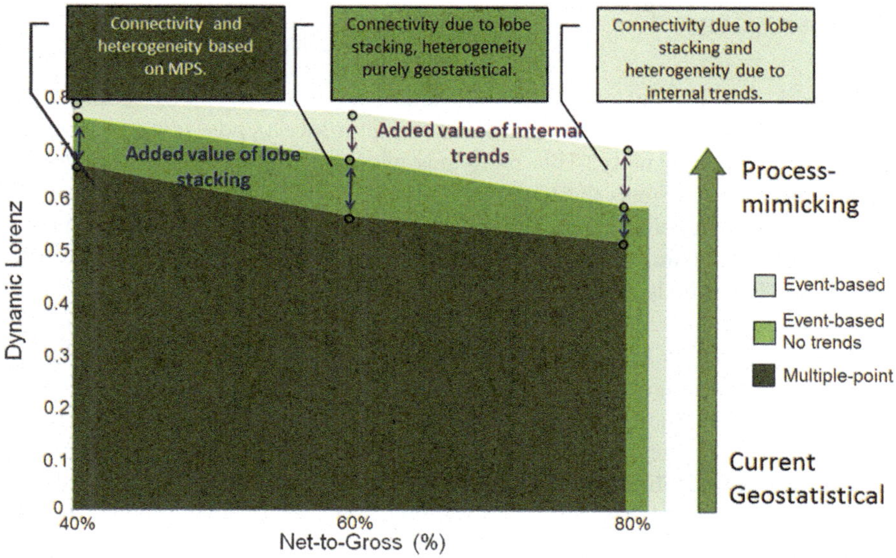

Fig. 9 Graph showing the DLC at 40 %, 60 %, and 80 % net-to-gross. It is apparent that as we add compensational stacking and within-lobe trends, we see an increase in the DLC

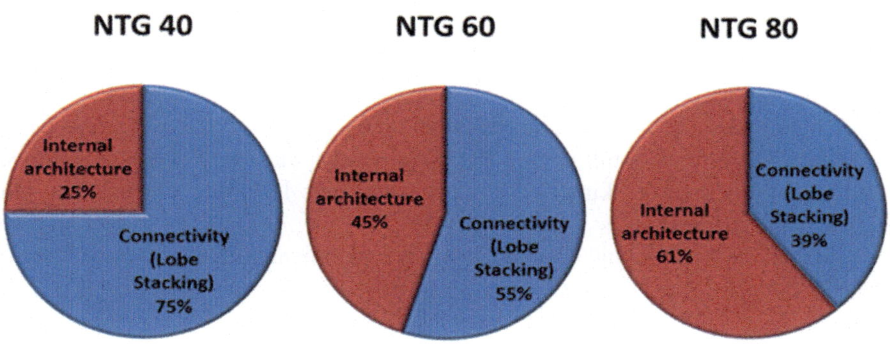

Fig. 10 Graphs showing the proportion of DLC increase due to lobe stacking pattern and within-lobe internal architecture

that connectivity has more of an effect on the DLC; hence more emphasis should be placed on modeling shale placement (e.g., drapes, baffles, etc.) within the reservoir. As the net-to-gross increases, the modeling of internal trends (lobe architecture) becomes more critical (see Fig. 10).

4 Discussion

This work has demonstrated that the additional reservoir model complexity, namely, lobe stacking patterns and detailed within-lobe trends, impacts reservoir connectivity heterogeneity. This motivates adoption of new process-mimicking technology that integrates these geologic features, depending on modeling objective. For example, this approach will be more appropriate for evaluating the detailed flow behavior of a deepwater lobe reservoir in the presence of mud drapes and within-lobe trends.

Conversely, this work suggests that current widely available reservoir modeling approaches (such as MPS) may underestimate reservoir fluid flow complexity, without the inclusion of addition constraints to directly capture these heterogeneities. In our experience, this may translate into overestimation of recovery factor and time to water breakthrough. Generating more realistic models may assist in managing and mitigating risk that may arise with conventional modeling methods.

Future work includes performing a full flow simulation which provides us with measures of heterogeneity other than the DLC.

5 Conclusions

Deepwater lobes are important reservoir targets. New insights from geological characterization suggest a hierarchy of heterogeneities including compensationally stacked lobes and within-lobe reservoir property trends. The incremental impact of these heterogeneities is determined through three workflows: (1) a traditional MPS workflow, (2) a process-mimicking workflow that captures lobe stacking patterns, and (3) a process-mimicking workflow that captures lobe stacking patterns and within-lobe trends.

Each of these additional complexities has a significant impact on flow heterogeneity as represented by DLC.

Acknowledgments The authors express appreciation to Chevron Energy Technology Company for support of this work and permission to publish and to Brad Mallison and four anonymous reviewers for the helpful critical comments.

Bibliography

Abrahamsen P, Fjellvoll B, Hauge R, Howell J, Aas T (2008) Process based on stochastic modeling of deep marine reservoirs. Petroleum Geostatistics 2007 Proceedings, European Association of Geoscientists & Engineers

Beaubouef RT, Rossen C, Zelt F, Sullivan MD, Mohrig D, Jennette DC (1999) Deep-water sandstones, brushy canyon formation. In: Field guide for AAPG Hedberg Field Research Conference, number 40. West Texas. AAPG Continuing Education Course. pp 1–48

Larue DK, Hovadik J (2006) Connectivity of channelized reservoirs: a modeling approach. Pet Geosci 12(4):291–308

Michael HA, Li H, Boucher A, Sun T, Caers J, Gorelick SM (2010) Combining geologic-process models and geostatistics for conditional simulation of 3-D subsurface heterogeneity. Water Resour Res 46(5):1–20

Miller J, Sun T, Li H, Stewart J, Genty C, Li D, Lyttle C (2008) Direct modeling of reservoirs through forward process-based models: can we get there. International Petroleum Technology Conference proceedings. pp 259–270

Møyner O, Krogstad S, Lie K-A (2015) The application of flow diagnostics for reservoir management. Soc Pet Eng J 20(2):306–323

Prélat A, Hodgson DM, Flint SS (2009) Evolution, architecture and hierarchy of distributary deep-water deposits: a high-resolution outcrop investigation from the Permian Karoo Basin, South Africa. Sedimentology 56(7):2132–2154

Prélat A, Covault JA, Hodgson DM, Fildani A, Flint SS (2010) Intrinsic controls on the range of volumes, morphologies, and dimensions of submarine lobes. Sediment Geol 232(1–2):66–76

Pyrcz MJ, Deutsch CV (2005) Conditional event-based simulation. Quant Geol Geostat 135–144

Pyrcz MJ, Deutsch CV (2014) Geostatistical reservoir modeling, 2nd edn. Oxford University Press, New York, p 448

Pyrcz MJ, Leuangthong O, Deutsch CV (2005) Hierarchical trend modeling for improved reservoir characterization. International Association of Mathematical Geology Annual Conference, Toronto

Pyrcz MJ, McHargue T, Clark J, Sullivan M, Strebelle S (2012) Event-based geostatistical modeling: description and applications. In: Abrahamsen P et al (eds) Geostatistics Oslo 2012

Pyrcz MJ, Sech R, Covault JA, Willis BJ, Sylvester Z, Sun T (2015) Stratigraphic rule-based reservoir modeling. Bull Can Pet Geol

Shahvali M, Mallison B, Wei K, Gross H (2012) An alternative to streamlines for flow diagnostics on structured and unstructured grids. SPE J 17(3):768–778

Shook GM, Mitchell KM (2009) A robust measure of heterogeneity for ranking earth models. The F PHI Curve and Dynamic Lorenz Coefficient: Society of Petroleum Engineers

Strebelle S (2002) Conditional simulation of complex geological structures using multiple-point statistics. Math Geol 34(1):1–21

Sullivan MD, Foreman JL, Jennette DC, Stern D, Jensen GN, Goulding FJ (2004) An integrated approach to characterization and modeling of deep-water reservoirs, Diana Field, Western Gulf of Mexico. AAPG Mem 80:215–234

Sylvester Z, Pirmez C, Cantelli A (2010) A model of submarine channel-levee evolution based on channel trajectories: implications for stratigraphic architecture. Mar Pet Geol 28(3):716–727

Wen R (2005) SBED studio: an integrated workflow solution for multi-scale geo modelling. European Association of Geoscientists and Engineers conference proceedings

Modeling of Depositional Environments: Shoreline Trajectory – The Link Between Sequence Stratigraphy and Truncated Gaussian Fields

Lars Edward Rygg Kjellesvik, Erling Igor Heintz Siggerud, and Arne Skorstad

Abstract The key to understanding the reservoir and fluid properties of any hydrocarbon system, clastic or carbonate, is to understand the depositional environment. The distribution of sediments, reservoir quality, as well as source and seal is controlled by well-understood geological processes that can be interpreted and described. By understanding these processes, it is possible to make qualified predictions of the consequential distribution of sediments by proactive geostatistical modeling. This paper directly links the main components of the depositional systems, in a sequence stratigraphic context, to the controlling parameters of the Truncated Gaussian Fields algorithm and uses this to predict sediment distribution in time and space. The approach is demonstrated at several scales, all the way from seismic-scale basin models to reservoir models, ensuring a systematic geological modeling approach to the entire value chain, from exploration, through appraisal, field development, and into production.

L.E.R. Kjellesvik (✉)
Digital Geology AS, Johs. Nores Vei 4, 1384 Asker, Norway
e-mail: lars.edward@digitalgeology.no

E.I.H. Siggerud
Digital Geology AS, Fagrabrekka 14, 7056 Ranheim, Norway
e-mail: erling@digitalgeology.no

A. Skorstad
Roxar Software Solutions AS, Lysaker Torg 45, 1366 Lysaker, Norway
e-mail: arne.skorstad@emerson.com

J.J. Gómez-Hernández et al. (eds.), *Geostatistics Valencia 2016*, Quantitative Geology and Geostatistics 19, DOI 10.1007/978-3-319-46819-8_41

1 Introduction

Up until the late 1970s, the understanding of sediment distribution was confined to recognition of lithologically restricted packages of sediments (litho-stratigraphy), where interpretation was based on correlating the same lithology between data points. Sequence stratigraphy (Mitchum et al. 1977) developed this by stating that logical predictions of sediment distribution could be made by correlating unconformities. This fundamentally changed the understanding of reservoir architecture. Being highly qualitative it did however not easily lend it selves to quantitative predictions of sediment distribution.

This happened in parallel with the introduction of geostatistical methods to populate reservoir models (Matheron et al. 1987). The main objective of a geostatistical reservoir model was to reproduce a static image of the reservoir architecture based on conceptual geological models and statistical measures. There was little focus on replicating the result of the depositional processes. As a result, the geostatistical models were able to introduce heterogeneity into reservoir models, but not to predict the results of the sedimentary process. In other words, if the conceptual model was wrong, the geostatistical model would not reflect this. Furthermore, the geostatistical model could not predict changes in the stratigraphy over larger areas.

This paper aims to demonstrate how Truncated Gaussian Fields can be used to model the results of sedimentary processes. It will be demonstrated how this methodology directly can be used to logically predict sediment distribution away from points of "hard" information such as an outcrop or well at any scale (Siggerud et al. 2015).

2 The Shoreline Trajectory – Link to Truncated Gaussian Fields

An important milestone in the ability to predict sediment distribution within the dynamic sedimentary system, introduced with Sequence Stratigraphy, was reached with the introduction of the shoreline trajectory (Helland-Hansen and Gjelberg 1994). The shoreline trajectory is a theoretical "line" drawn through all the nick points of clinoform successions making up a sedimentary system (cf. Fig. 1).

Although highly theoretical (Helland-Hansen and Gjelberg 1994) proposed that changes in the angle of the trajectory would result in systematic differences in sediment distribution. The recognition of the angle would therefore enable logical prediction of sediment distribution (Patruno et al. 2015). While working on Eocene outcrops in NE Spain, (Siggerud and Steel 1999) demonstrated the usefulness of shoreline trajectory in the field, where the consequences of the change in angle could be observed in Fig. 2. This realization is especially important because it

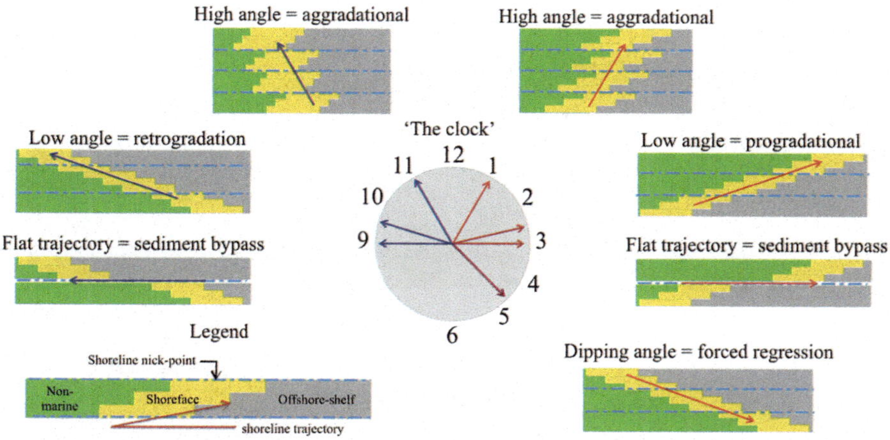

Fig. 1 The conceptual model elucidating the relationship between changes in shoreline trajectory and the extension of the facies belts and subsequent reservoir architecture. Note the use of a clock to denote the angle; a common mistake is in trying to depict the exact angle while in fact it is the relative change (from, e.g., 10 to 12 o'clock) that is important, modified after (Helland-Hansen and Gjelberg 1994)

essentially allows you to describe the sedimentary systems by a simple parametrization of the shoreline trajectory, paleo-coastline, and variability of the boundaries, which is paralleled to the parametrization of truncated Gaussian fields.

Truncated Gaussian Fields are fully specified probability distributions defined by a trend, truncation levels, and a variogram (Beucher, Galli, Le Loc'h, and Ravenne) (Wenlong and Journel 1993). In this setting, the trend is defined by the depositional direction and the shoreline trajectory, the truncation levels by the paleo-coastline, and the variability of the boundaries is reflected by the variogram as illustrated in Fig. 3. In a Bayesian context, the parametrization of the sedimentary system then describes a fully defined prior distribution and conditional simulations of the truncated Gaussian field that reflect the posterior distribution.

Modeling the depositional system with truncated Gaussian fields will therefore give an unbiased prediction of the distribution of sediments in time and space, honoring the geological understanding and constrained by the available data.

This is critical, not only to represent the sediment distribution, but to proactively test the consequences of the sedimentological understanding for any sedimentary system (silisiclastic or carbonate) (Siggerud 2008). This is important since the sequential development does not have an endless number of possible scenarios, but rather an uncertainty within the angle of trajectory, which can be captured using a Truncated Gaussian Field as seen in Fig. 4. In other words, the variability in the extension of the sedimentary facies belts for any given setting can be logically tested with respect to the extension and variability, illustrating the uncertainty for each succession.

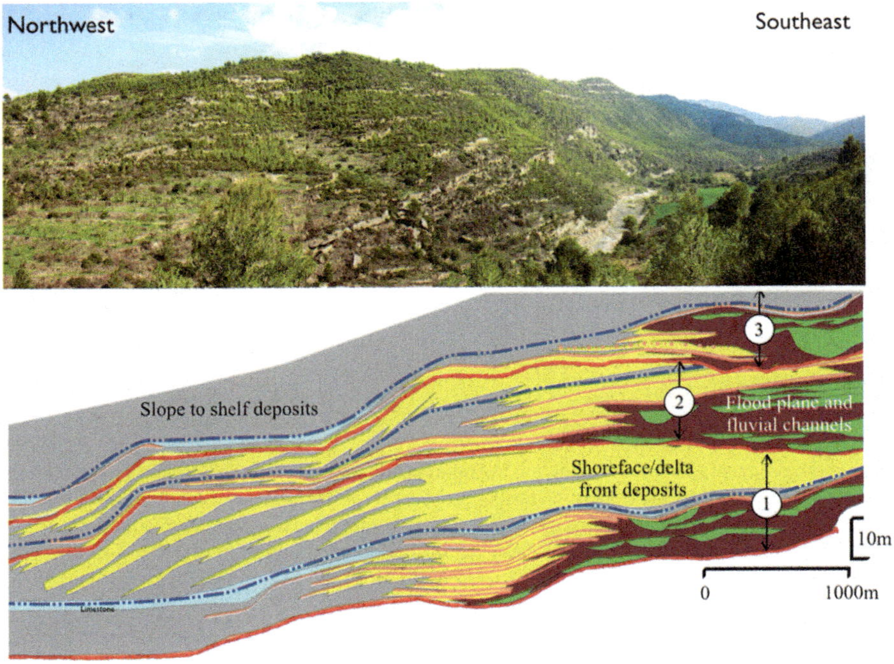

Fig. 2 Eocene outcrop example in NE Spain (*above top*) where the dipping foresets of the low-angle shallow shelf clinoforms can be observed as shown schematically in the figure (*yellow color*); note also the stacking light-colored sands and red beds in the skyline reflecting the aggradation (high-angle trajectory) of the transgressive systems tract (shown in *dark red* and *green*)

3 Practical Application, Basin-Wide to Reservoir Scale

While most common use of geostatistical geomodeling has been limited to fields and reservoir scale (few kilometers), this paper demonstrates the extended use of the ideas outlined above. While the ideas of shoreline trajectory were originally related to shallow marine clinoform developments (sequence scale, which is hundreds of meters; Burgess et al. 2008), it is equally applicable on megasequence or basinal scale (tens to hundreds of kilometers: Siggerud 2012).

Drawing on extensive knowledge from several decades of fieldwork in the Arctic islands of Svalbard and work in the Norwegian Barents Sea, a megasequence subdivision of the post-Permian succession was established, using the facies stacking pattern to establish an understanding of the depositional trajectory for each megasequence. In addition, an interpretation of lithology, sedimentary facies, and facies associations was undertaken for 255 outcrops and wells, serving as a basis for the sedimentological sequence stratigraphic model. A megasequence is here defined as a seismic-scale sequence, bounded by regional peak transgressions, ranging from a few hundreds to more than 1,000 m in thickness and spanning several million years of time (Steel 1993). In this case the near 6000 m thick

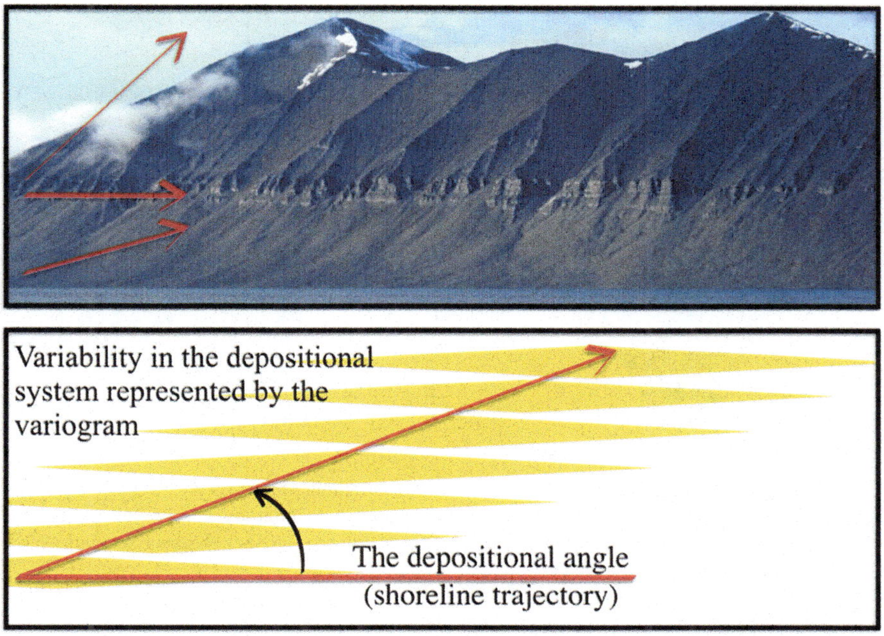

Fig. 3 Outcrop example from Svalbard demonstrating changes in shoreline trajectory from intermediate to very low- and high-angle trajectory (*above top*). The *lower image* illustrates the variability around the shoreline captured by variograms

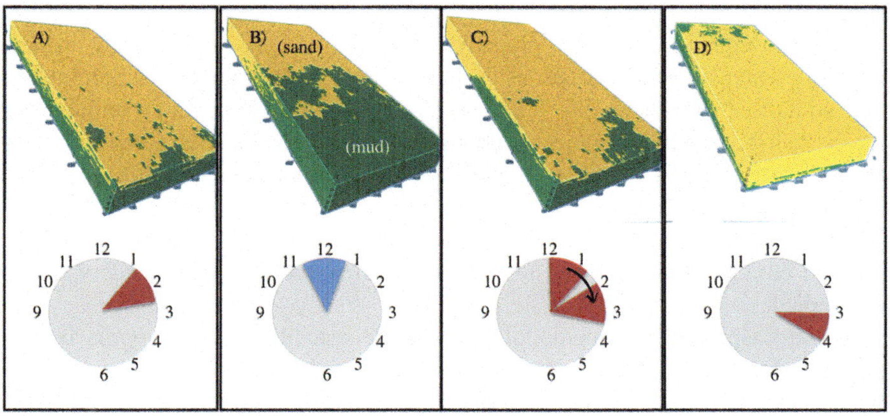

Fig. 4 The concept of shoreline trajectory as shown in Fig. 1 here visualized for a low angle (**a**), high angle (**b**), low to high angle (**c**), and a "negative" angle (forced regression) following a relative sea level fall (**d**). In the latter case note how the system begins to detach landward, as can be observed for the quaternary deposits along the Norwegian coastline

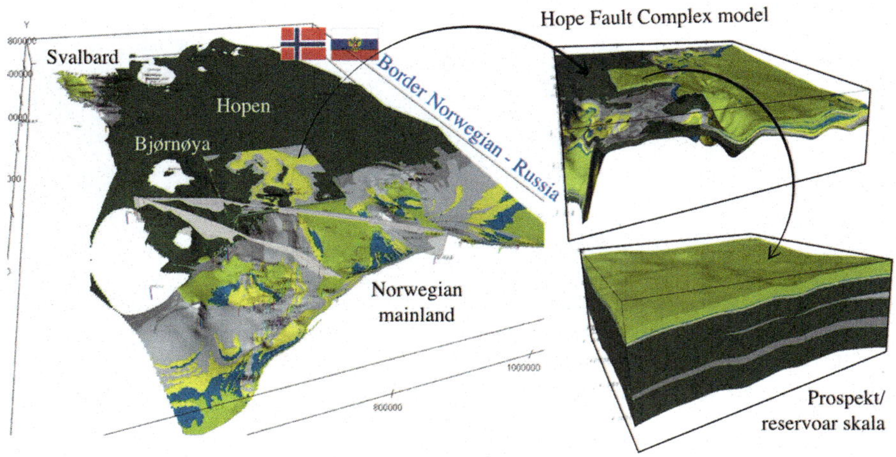

Fig. 5 Bird's-eye view of the 650,000 km^2 basin-wide geomodel based on the 12 megasequences. Inserted is the sequence-scale Hoop Fault Complex model, maintaining the overall trend while elucidating the details on semi-regional scale, while even more detailed prospect-scale model

sedimentary succession was split into 12 megasequence that was identified and correlated across the entire Barents Sea, covering some 650,000 km^2, probably making it the largest 3D geomodel ever built (Siggerud et al. 2015) (Fig. 5).

Un-eroded thickness maps were created using the stratigraphy to establish the sedimentary sequences at the time of deposition in order to clarify the logical orientation of the sedimentary system for each megasequence (the trend in the Truncated Gaussian Field) (Fig. 6). This, combined with the shoreline trajectory for each megasequence, enabled proactive modeling of the sedimentary system depicting extension and limitations of all lithologies across the entire Barents Sea (including Svalbard shelf) Fig. 7. Secondly, while not only illustrating sedimentary facies distribution and volumes of deposited sediments, it also enabled proactive modeling of source rock presence and improved migration probabilities using the sand probability distribution derived from the geomodel (Throndsen et al. 2016). As such the sedimentary distribution within the Norwegian Barents Sea is no longer random but can be logically tested regardless of sedimentological understanding and model.

A third aspect of this methodology is in the scalability of the approach. It is possible to zoom in on an area of particular interest within the original megasequence model, while carrying the regional information. This is illustrated with the prolific Hoop Fault Complex within the central Barents Sea, cf. Fig. 5. A detailed geomodel for this area was constructed while carrying all the main trends and architecture from the main megasequence model. In the past this has proven to be a challenge, because of the lack of a logical regional distribution of the sedimentary system. Most other paleoreconstructions depict a somewhat arbitrary range of shoreline locations for the Middle Triassic within the Hoop area (Riis et al.

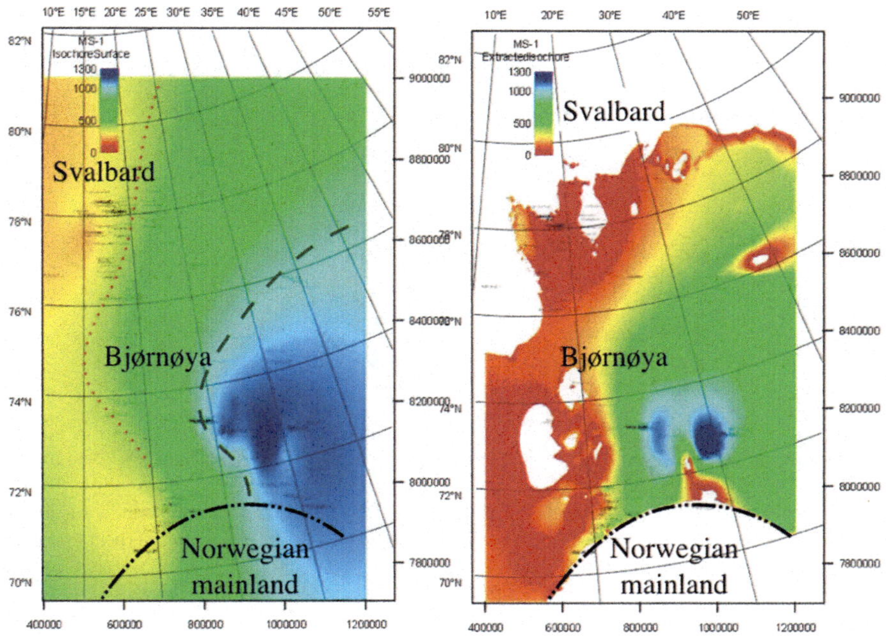

Fig. 6 Example on the construction of un-eroded and current day (eroded) thickness maps for megasequence two (Middle Triassic). The *dashed green line* on the left picture depicts the position of the basin floor, while the *orange dotted line* elucidates the shoreline transition, thereby illustrating the wide shallow shelf in the north as evident from seismic data

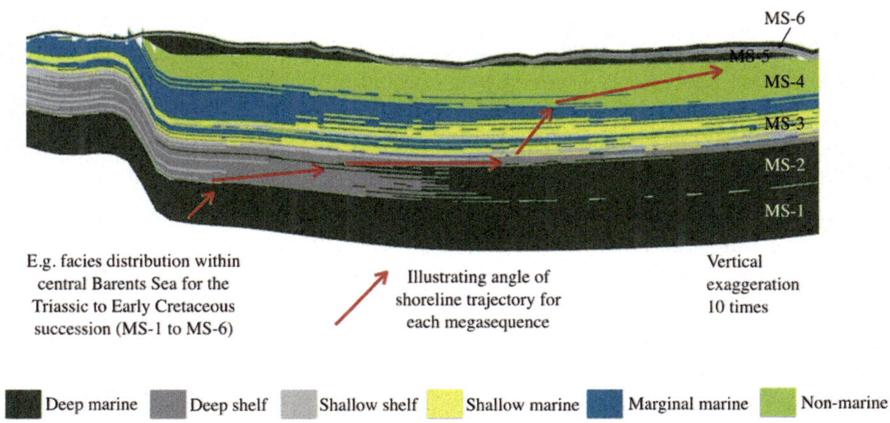

Fig. 7 Example of a vertical section from the Barents Sea Geomodel. Colors reflect the main depositional environment as shown in the legend, and the shoreline trajectory is illustrated schematically. Note the variability in the facies belts and that there will be higher-order variations within the megasequences which reflect variations on sequence scale

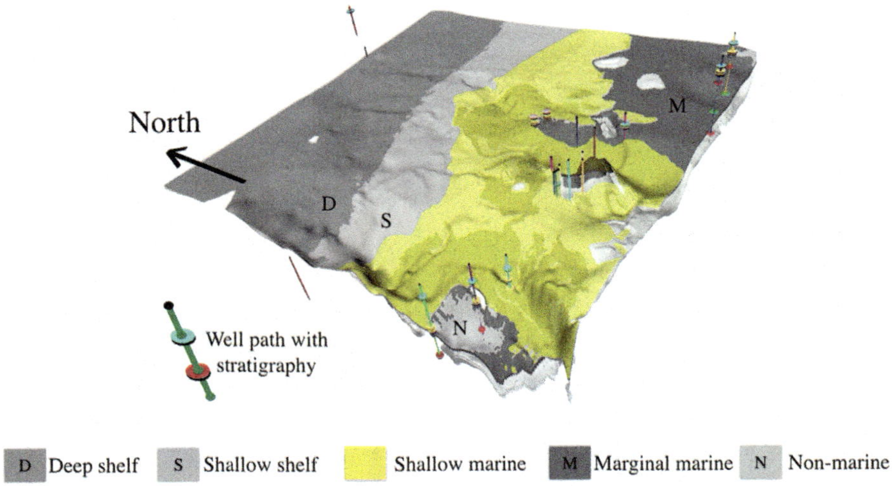

Fig. 8 Detailed time section of one layer of the of the Hoop Fault Complex geomodel depicting the main facies associations (depositional environments) where the shallow marine (best reservoir potential) is shown highlighted in *yellow*

2008; Matheron et al. 1987) (Beucher, Galli, Le Loc'h, and Ravenne). Our model, using the understanding of the shoreline trajectory while honoring the well data and sedimentary understanding gleaned from the seismic data, gives the logical distribution of the Middle Triassic depositional system as elucidated in Fig. 8.

A fourth point is that alternative scenarios from different sedimentological models can be tested by coherently altering the angle of the trajectory. This gives a true handle on the uncertainty of the interpretation and provides highly valuable feedback to the geologist as to the physical probability of the proposed depositional model. What this means is that while in a correlation panel any "irregularity" in the sedimentological interpretation can be "accounted" for, the proactive modeling will provide "bulls-eyes" where the sedimentary system proposed is not physically (and mathematically) possible.

A fifth point to be remembered is that the proactive use of truncated Gaussian field modeling is not dependent on an enormous database of well points and/or seismic. Remembering the relationship as outlined above between the truncated Gaussian field and the sequence stratigraphy and shoreline in particular, any depositional system can readily be proactively modeled, whereby (again) the logical and quantitative consequences can be elucidated. An example is work undertaken by the authors on basin scale across the Arctic islands in Svalbard. Here the lack of extensive outcrop data and basic structural mapping in the past had not yielded a satisfactory understanding of the reservoir distribution. The shoreline trajectory enabled the construction of different detailed sequence stratigraphical models, as shown in reduced uncertainty and enabled systematic testing of alternative hypothesis as seen in Fig. 9.

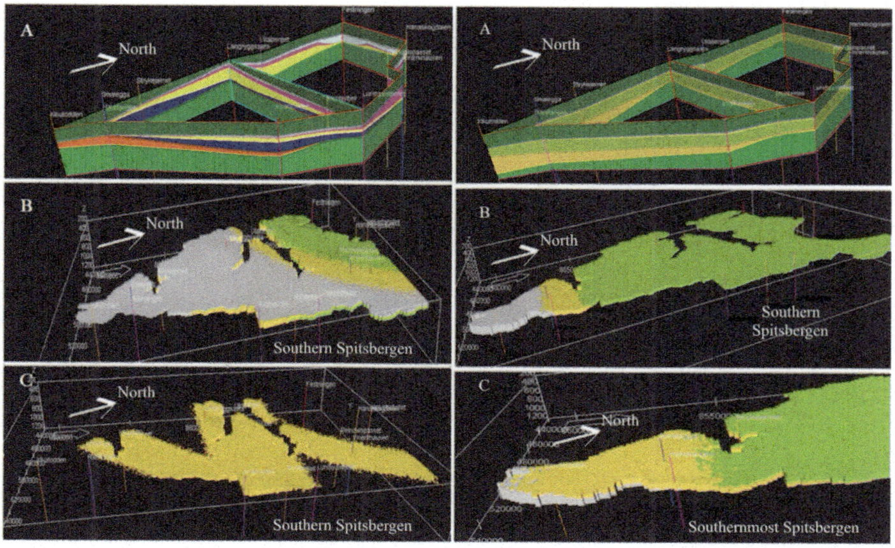

Fig. 9 Two depositional models as they appear in the model (**a**) thickness fence diagram (**b**) facies distribution: *gray* = offshore, *yellow* = shallow marine, and *light green* = nonmarine (**c**) show only the shallow marine deposits in the left and an alternative hypothesis to the right

Similarly, systematic use of outcrop data providing the physical relationships between height and width of numerous incised valleys that can be observed in the field on Svalbard and mappable across the north-central Barents Sea was used to propagate the Hoop Fault Complex with fluvial channel deposits where again the regional and semi-regional sediment distribution were derived from the larger geomodels (Fig. 10).

4 Summary and Conclusions

As mentioned in the introduction, the aim of this paper has been to elucidate how Truncated Gaussian Fields are suited to proactively model the consequences of our geological understanding of any depositional environment or the sediment distribution in time and space. The direct link between the parametrization of the truncated Gaussian field and the parameters controlling and steering the sedimentary succession is demonstrated with the principle of the shoreline trajectory. Consequently, this allows an unbiased prediction of the distribution of sediments and uncertainty, constrained by the available data.

Armed with this understanding, one can logically transfer from basin to field scale maintaining the overall sedimentary system while capturing the details on reservoir scale. While being systematic and highly predictive, it also is quantitative,

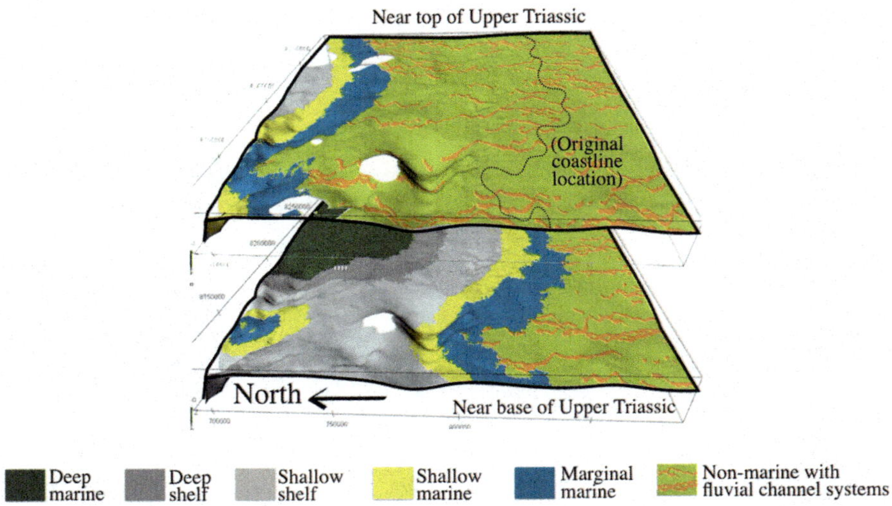

Fig. 10 Bird's-eye overview of two time slices depicting sediment distribution for the prospect area within the Hoop semi-regional geomodel (see Fig. 5). The images show the occurrence of fluvial channels correctly honoring the width, orientation, interconnectedness, and frequency of the channels in addition to the shoreline location as evident from the regional model

thereby opening for reservoir characterization and maturation modeling. We strongly believe that systematic use of sequence stratigraphy and shoreline trajectory with truncated Gaussian fields is the way forward in not only understanding sediment distribution but logically predicting it on all scales.

Bibliography

Armstrong M, Galli M, Beucher H, Loc'h G, Renard D, Doligez B, … Geffroyj F (2011) Plurigaussian simulations in geosciences. Springer

Beucher H, Galli A, Le Loc'h G, Ravenne C (n.d.) Including a regional trend in reservoir modelling using the truncated Gaussian method. Geostat Troia. pp 555–566. Troia

Burgess PM, Steel RJ, Granjon D (2008) Stratigraphic forward modeling of basin-margin clinoform systems: implications for controls on topset and shelf width and timing of formations of shelf-edge deltas. In: Hampson GS (ed) Recent advances in models of siliciclastic shallow-marine stratigraphy. SEPM (Society for Sedimentary Geology), Tulsa

Helland-Hansen W, Gjelberg JG (1994) Conceptual basis and variability in sequence stratigraphy; a different perspective. Sediment Geol 92(1–2):31–52

Matheron G, Beucher H, de Fouquet C, Galli A, Guerillot D, Ravenne C (1987) SPE16753 conditional simulation of the geometry of fluvio-deltaic reservoirs, SPE Annual Technical Conference and Exhibition. SPE, Dallas

Mitchum RM, Vail PR, Thompson S et al (1977) Seismic stratigraphy and global changes in sea level, part 2: the depositional sequence as the basic unit for stratigraphic analysis. In: Payton C (ed) Seismic stratigraphy: applications to hydrocarbon exploration, vol 26. AAPG, Tulsa

Patruno S, Hapson GJ, Jackson CA-L (2015) Quantitative characterization of deltaic and subaqueous clinoforms. Earth-Sci Rev 142:79–119

Riis F, Lundschien BA, Høy T, Mørk A, Mørk MB (2008) Evolution of the Triassic shelf in the northern Barents Sea region. Polar Res 27:318–338

Siggerud EI (2008) Application of shoreline trajectory in geomodelling, cretaceous helvetiafjellet formation, Arctic Norway, AAPG Extended Abstract, AAPG International Conference and Exebition. AAPG, Cape Town

Siggerud EI (2012) In: Siggerud EIH, SGS Petroleum Geological Excursions (eds) Helvetiafjellet formation – a discussion, vol 2. SGS Publications, Trondheim

Siggerud EI, Steel RJ (1999) Architecture and trace fossil characteristics of a 10.000–20.000 year, fluvial-marine sequence, SE Ebro Basin, Spain. J Sediment Res 69:365–387

Siggerud EI, Ballardini L, Kjellesvik LE (2015) A quantitative tool for predicting sediment distribution, exemplified from the post-permian succession Svalbard & Barents Sea. Recent Advances in Exploration Technology, NCS-Conference and prospect-fair. Fornebu

Steel RJ (1993) Triassic-Jurassic megasequence stratigraphy in the Northern North Sea: rift to post-rift evolution. In: Parker J (ed) Petroleum geology of NW Europe; Proceedings of the 4th Conference. Geological Society, London, pp 299–335

Throndsen T, Kjellesvik LE, Siggerud EI (2016) 4D facies modelling in petroleum system analysis of the Barents Sea TFO area. In: Carstens H (ed) Hydrocarbon habitats. GeoNova, Oslo, pp 6–7

Wenlong X, Journel AG (1993) GTSIM: Gaussian truncated simulations of reservoir units in a West Texas carbonate field. SPE

Facies Inversion with Plurigaussian Lithotype Rules

Lewis Li, Siyao Xu, and Paul Gelderblom

Abstract Accurate incorporation of geological concepts such as lithological facies distributions is an important aspect of building reservoir models. Consequently, accounting for facies in seismic inversion generates models conditioned to geological concepts and plays an important role in decision-making. Shell's proprietary probabilistic model-based seismic inversion engine *Promise* is generally applied to invert for continuous variables, such as NTG, saturation, and layer thickness from seismic data. In some depositional environments, the spatial variability of reservoir properties is characterized at fine geological scale by facies and the corresponding petrophysical properties; hence, an implementation of facies in seismic inversion is desirable. In this study, we propose a novel methodology for lithological facies inversion utilizing Plurigaussian rock-type rules. Direct inversion of facies may result in unrealistic facies contacts; therefore, the proposed technique instead inverts for a pair of "guide" variables using Promise. The guide variables are then classified into facies using a methodology inspired by Plurigaussian simulations, where a defined lithofacies rule map is used to constrain facies proportions and contacts. The required inputs for the workflow are the lithofacies rules and variogram estimates of the guide variables. Both of these can be derived from a prior estimate of the facies distribution and can also take into account geological constraints from a human expert. We demonstrate the workflow for a three facies case with a synthetic wedge model and seismic data of a marine survey.

L. Li (✉)
Department of Energy Resources Engineering, Stanford University, Stanford, CA 94305, USA
e-mail: lewisli@stanford.edu

S. Xu
Shell International Exploration and Production Inc, Houston, TX 77082, USA

P. Gelderblom
Shell Global Solutions International BV, Rijswijk, The Netherlands

J.J. Gómez-Hernández et al. (eds.), *Geostatistics Valencia 2016*, Quantitative Geology and Geostatistics 19, DOI 10.1007/978-3-319-46819-8_42

1 Introduction

Seismic inversion is the process by which acoustic impedance and other elastic properties are derived from the seismic traces. The results of seismic inversion should honor geological knowledge as well as observed data, such as well logs. One of the aims of seismic inversion is to characterize the spatial variability of the reservoir, by extracting subsurface information from the seismic data and other data sources. *Promise* (Leguijt 2001, 2009) is the Shell proprietary model-based seismic inversion tool. It can perform trace-by-trace inversion and has been designed to naturally honor the prior models. Moreover, geological constraints have been deployed aiming to improve the geological realism of inversion results (Gelderblom and Leguijt 2010). In the trace-by-trace inversion, the algorithm generates an ensemble of local model realizations at each trace location, taking the seismic data and the well data into account. If a geologically constrained inversion is performed, an ensemble of reservoir-size model realizations will be generated taking also the lateral continuity of properties into account. Applications of the algorithm have been performed on continuous variables, such as NTG, porosity, and layer thickness. However, a reservoir is normally characterized by lithological facies. A lithological facies prediction algorithm based on small extensions to the current algorithm is presented in this paper.

In our methodology, we invert for a pair of continuous variables and apply a classification technique inspired by Plurigaussian simulations (Armstrong et al. 2011) to obtain the inverted facies. Seismic inversion with the application of geostatistics has been studied and published before (Doyen 2007; Larsen et al. 2006; Ulvmoen and Omre 2010; Grana and Della Rossa 2010; Rimstad and Omre 2010; Gelderblom and Leguijt 2010; Gunning et al. 2014). Our work is different in the sense that this method is a combination of model-based inversion in depth domain, vertical and lateral continuity constraints, and facies modeling, which did not appear simultaneously in any previous publication. In comparison to the popular multistep inversion methods (Dubois et al. 2007), the facies classification in stochastic inversion methods is performed inside the inversion process, as opposed to as a post-processing procedure. In multistep methods, acoustic impedance values are first determined before facies probabilities and properties are known. Next, the facies are chosen on the basis of the acoustic impedance values. Since seismic inversion is generally performed at the acoustic scale, which corresponds to the vertical resolution of the seismic dataset, a downscaling process is needed to generate a fine-scale static model for dynamic reservoir simulation (Hesthammer et al. 2001; Doyen et al. 1997). In our algorithm, seismic inversion can directly produce fine-scale model realizations, in which the subseismic scale properties are constrained by geological and lithological knowledge. The specifics of our inversion algorithm will be described in the next section, followed by a description of a methodology to derive necessary input parameters. Finally, the workflow will be demonstrated on two cases, a synthetic wedge model and a field case from a deepwater offshore reservoir.

2 Plurigaussian Inversion

The proposed method, as shown in Fig. 1, uses a pair of underlying continuous variables (herein termed guide variables) to represent the facies. Individually, these variables lack physical meaning; rather they serve as random fields that express the likelihood of a facies being present in that location. A separate categorization step needs to be applied to convert this continuous variable into a discrete facies representation. A rock model is then used to obtain reservoir properties from the resulting facies map. A synthetic seismogram is then computed and compared with the observed data. The inversion loop continues by adjusting the underlying guide variable according to a given variogram using the Iterative Metropolis Simulation (Gelderblom and Leguijt 2010). This is repeated until the observed seismic data is matched to a certain tolerance. The key to this process is accurately and correctly inferring the discrete facies map from the continuous guide variable. It is also important that this categorization must also incorporate prior geological information such as facies ordering and proportions.

To this end, we propose a technique based on Plurigaussian simulations (Armstrong et al. 2011). The procedure encodes the prior geological information as a lithofacies rule map (LRM), shown in Fig. 2. The X-axis represents the CDF of the first guide variable from 0 to 1, and the Y-axis is the second. In addition to the LRM, an estimate of the variograms describing the spatial continuity of the guide variables is required for Plurigaussian inversion.

The Plurigaussian simulation has increased flexibility in the sense that it can produce realizations in which the occurrence frequency of specific facies transitions is determined by the user-specified lithofacies rules. In the next section, we will discuss obtaining the required inputs from a prior model.

3 Deriving Input Parameters

The specific inputs to the Plurigaussian inversion process are:

1. Lithotype Rule: This is used to convert the guide variable from a continuous variable into facies (a categorical variable).
2. Prior Facies Probabilities: As a Bayesian method, the proposed algorithm requires prior facies probabilities for each facies.
3. Lateral and Vertical Variogram: To ensure lateral continuity, the inversion process needs an estimate of the variograms of both guide variables in both the vertical and lateral directions.
4. Rock Physics Model: This is then used to convert the facies into reservoir properties, on which forward modeling is performed to yield the synthetic seismic.

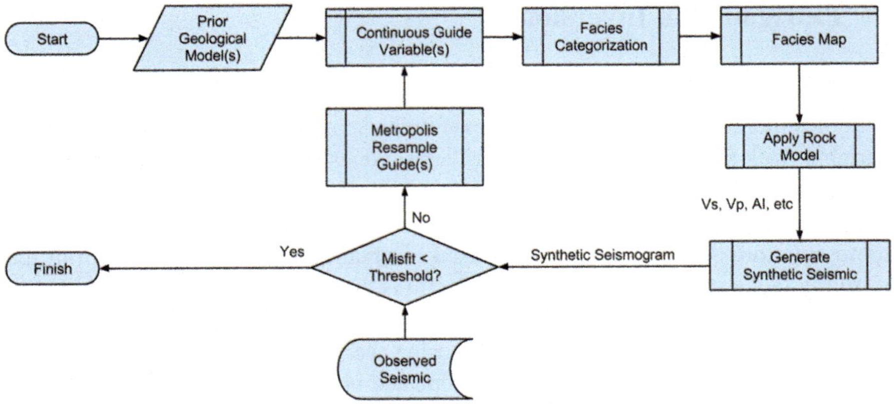

Fig. 1 Overall workflow for fine-scale facies inversion. The inversion process solves for a pair of continuous guide variables that will be classified into facies. A prior geological and reservoir property model is the required input for the process and is used for constructing the initial guess for the guide variables and their respective spatial continuities

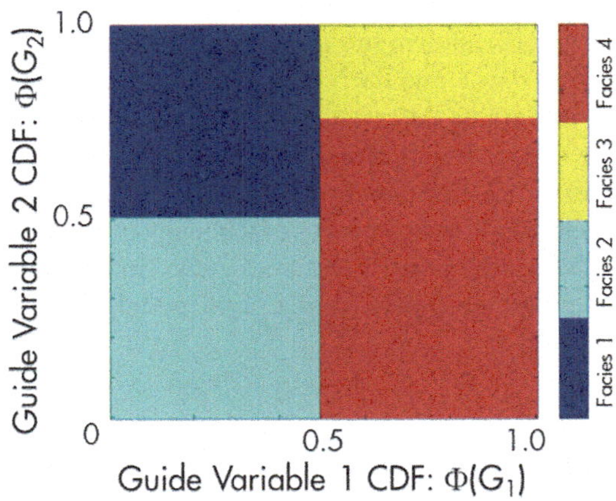

Fig. 2 Illustration of Plurigaussian simulation. Two coregionalized guide variables are jointly thresholded to obtain a facies classification. This can be viewed as a lithofacies rule map, where each axis represents the CDF of a guide variable. For a location x in the geological model, this lithofacies rule map (LRM) can be used to "look up" the facies type for that location: if g_1 and g_2 are the realizations of the "guide" variable at x, then the color of the point $(\Phi(G_1), \Phi(G_2))$ in the LRM defines the facies ($\Phi(.)$ is the standard normal CDF)

We will assume that prior facies probability maps and rock physics models are provided, and a preliminary static geological model is available to derive other inputs. The following subsections will discuss how we can derive the lithotype rule and variograms.

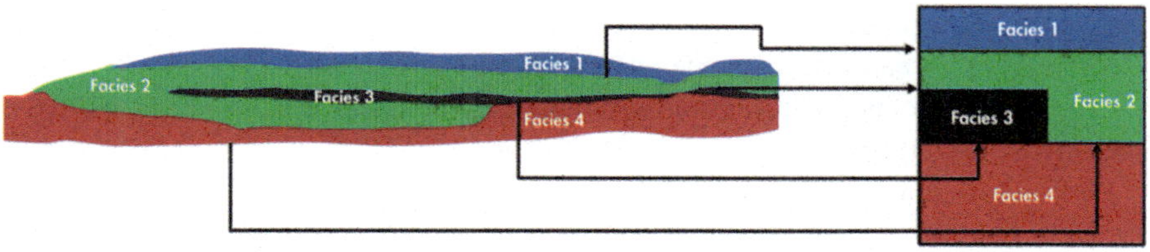

Fig. 3 Illustration of how a geological model can be used to construct a lithofacies rule. The global proportions of each facies are used to determine the area of the rectangles in the rule map, while the frequency of contacts is used for placement of the areas

3.1 Lithofacies Map

The central tenet of the Plurigaussian simulation is an explicit delineation of its lithotype rule. This rule map is what enables us to convert a continuous random field into a facies map. These rules also allow the modeler to constrain the facies maps to certain geological knowledge, namely, proportions and contacts between the different facies types. Obviously, the lithotype will need to be derived from some prior geological model. For illustration, consider the scenario in Fig. 3. In this schematic stratigraphic sequence, there are four facies, and the lithotype rules for this sequence should reflect the following propositions:

Proposition 1 The area of a given facies F_i on the lithotype rule map will be equal to the proportion of the facies in the inversion result.

Proposition 2 The lack of adjacency between any two facies on the lithotype rule will prohibit contact between two facies in the inverted result.

3.2 Variogram Analysis

Inversion with lateral continuity necessitates the specification of variograms for both guide variables. But the guide variables are artificial variables, introduced to make this form of simulation possible. They do not represent a real-world quantity on which a variogram could be empirically estimated. The only variograms we can establish empirically are those of the indicator variable for a facies type.

For N facies, N-1 experimental indicator variograms (for a given facies F_i) can be computed as in Eq. 1:

$$\begin{aligned}\gamma_{Fi}(x,x+h) &= \frac{1}{2}Var[\mathbb{1}_{F_i}(x) - \mathbb{1}_{F_i}(x+h)] \\ &= \frac{1}{2}\Big\{E\Big[(\mathbb{1}_{F_i}(x) - \mathbb{1}_{F_i}(x+h))^2\Big] - (E[\mathbb{1}_{F_i}(x) - \mathbb{1}_{F_i}(x+h)])^2\Big\}\end{aligned} \tag{1}$$

where the location and offset vectors are $x, h \in \mathbb{R}^3$ and $E(\cdot)$, $Var(\cdot)$, $\mathbb{1}_{F_i}(\cdot)$ are the expectation operator, the variance operator, and the indicator operator.

Assuming second-order stationarity, the second term reduces to 0, and the expression simplifies to Eq. 2:

$$\gamma_{F_i}(x, x+h) = \frac{1}{2}\left\{E\left(\left[\mathbb{1}_{F_i}(x) - \mathbb{1}_{F_i}(x+h)\right]^2\right)\right\} \tag{2}$$

Empirically, this can be evaluated as

$$\gamma^{\star}{}_{F_i} = \frac{1}{2N} \sum_{|x-x_0|=h} \left[\mathbb{1}_{F_i}(x) - \mathbb{1}_{F_i}(x_0)\right]^2 \tag{3}$$

The variogram model parameters (type, nugget, range) of the guide variables must be chosen such that when realizations of the guide variables are converted to facies realizations as described above, those facies realizations have indicator variograms consistent with the observed empirical indicator variograms (Eq. 3).

The indicator variogram is defined as

$$\begin{aligned} \gamma_{F_i}(x, x+h) = \frac{1}{2}\Big\{ & P[\mathbb{1}_{F_i}(x) = 1] + P[\mathbb{1}_{F_i}(x+h) = 1] \\ & -2P[(\mathbb{1}_{F_i}(x) = 1) \bigcap (\mathbb{1}_{F_i}(x+h) = 1)]\Big\} \end{aligned} \tag{4}$$

where P is the probability by counting indicator frequencies. For each facies i, denote the region on the lithofacies rule where it is defined as R_i. Therefore, for guide variables G_1 and G_2,

$$\mathbb{1}_{F_i} = 1 \equiv [G_1(x), G_2(x)] \in R_i \tag{5}$$

In comparison to the LRM, where $\Phi(G_1)$, $\Phi(G_2) \in [0,1]$, the guide variables $G_1(x)$, $G_2(x) \in [-\infty, \infty]$ in this section. This means Eq. 4 can be reexpressed in terms of facies regions as

$$\begin{aligned} \gamma_{F_i}(x, x+h) = \frac{1}{2}\Big\{ & P[G_1(x), G_2(x) \in R_i] + P[G_1(x+h), G_2(x+h) \in R_i] \\ & - 2P[(G_1(x), G_2(x) \in R_i) \cap (G_1(x+h), G_2(x+h) \in R_i)]\Big\} \end{aligned} \tag{6}$$

The first two terms on the right side of Eq. 6 reduce to

$$\frac{1}{2}\{P[G_1(x), G_2(x) \in R_i] + P[G_1(x+h), G_2(x+h) \in R_i]\} = A_i \tag{7}$$

A_i is the area of region R_i or equivalently the global proportion of facies i. Under the assumption that the guide variables are Gaussian, the final term of Eq. 6 can be evaluated by taking the integral of a quadvariate Gaussian distribution centered at μ and covariance Σ. As described in Armstrong et al. (2011), the quadvariate Gaussian represents the joint pdf of $G_1(x)$, $G_2(x)$, $G_1(x+h)$, and $G_2(x+h)$ falling on a given facies F. This allows us to rewrite Eq. 6 as

$$\gamma_{F_i}(x, x+h) = A_i - \iint_{A_i} \iint_{A_i} f(x_1, y_1, x_2, y_2) dx_1 dx_2 dy_1 dy_2 \tag{8}$$

The guide variables are assumed to have $\mu = 0$. Assuming that the guide variables are independent and have covariance functions $\rho_1(h)$ and $\rho_2(h)$, respectively, the covariance matrix Σ can be expressed as

$$\Sigma = \begin{bmatrix} 1 & 0 & \rho_1(h) & 0 \\ 0 & 1 & 0 & \rho_2(h) \\ \rho_1(h) & 0 & 1 & 0 \\ 0 & \rho_2(h) & 0 & 1 \end{bmatrix} \tag{9}$$

Using Eq. 8, it is possible to evaluate the corresponding expected indicator variogram of a facies, given the variograms of two guide variables and the accompanying lithofacies rule. A typical procedure for tuning the variogram of the guide variables would be:

1. Measure, estimate, or postulate (based on prior geological knowledge) the indicator variograms γ_{Fi} for all facies F_i.
2. Choose variograms for each guide variables γ_{G1} and γ_{G2}. Furthermore, decide upon a coregionalization model to evaluate the covariance matrix.
3. Transform guide variable variograms into expected indicator variogram using Eq. 8.
4. Compare to empirical variogram from Step 1. Go back to Step 2 and repeat until appropriate fit.

One of the difficulties with this methodology is that all N empirical indicator variograms should be fitted with the same set of variograms for the guide variables. In practice, this may be a daunting task especially as the number of facies increases. An approximate approach is to take an average of the empirical indicator variograms and attempt only to fit to this mean variogram.

In this Bayesian scheme, the variogram model specified by the user is used in the prior distribution of the "guide" variable. The parameters of the prior distribution are not updated during the inversion. An empirical variogram of the guide variables of the posterior ensemble will have a different range and structure from the prior variogram model, which is the consequence of conditioning to the data. However, the prior fitting process described in this paper is based on the prior facies proportions. If the facies proportion has been updated significantly by the inversion, the variogram must be updated as well. This problem requires further investigation in future works.

Table 1 Variogram parameters used to generate synthetic ground-truth guide variables

Name	Long range	Short range	Vertical range	Azimuth	Nugget	Type
Guide 1	600	400	20	90	0.0001	Exponential
Guide 2	400	200	20	0	0.0001	Exponential

4 Case Study: Synthetic Wedge Model

4.1 Synthetic Wedge

For demonstrating this process, a study was conducted on a synthetic wedge model with maximum thickness of 90 m and three facies (clean sand, shale, and sand-shale mix). The two guide variables were simulated using Sequential Gaussian Simulation and the variogram parameters outlined in Table 1.

The grid thickness was chosen to be 2 m to allow for the generation of thin beds of sand as seen in Fig. 4. The lithofacies rule used for categorizing the facies is depicted in Fig. 5. The thresholds were chosen to be 0.4 for thresholds of both guide variables. A rock model for laminated sand and shale (Allen 1984) was chosen. In this rock model, the three lithological facies can be defined by net-to-gross ratio (NTG) of 0 %, 50 %, and 100 %. The thresholding procedure yielded a global proportion of 40 % shale, 36 % sand, and 24 % sand-shale mix.

4.2 Seismic Scale Inversion

To compare with subseismic scale performance of our algorithm, a seismic scale inversion was first performed. In this depositional environment, it is more appropriate to characterize reservoir rocks using continuous properties. Hence, the seismic scale inversion was directly performed on NTG. The grid cells were set to have a thickness of 20 m, and the prior NTG was set to be completely shale. The sand fraction prior distribution was set to a truncated Gaussian distribution with mean 0.5 and variance of 0.5, truncated at 0 and 1. An example of an inverted realization is given in Fig. 6. The sand fraction exhibits the general trend of the true underlying facies map but lacks the granularity to resolve the thin beds. Furthermore, the seismic scale inversion has produced a far larger estimate of the global average net-to-gross ratio (48.5 %) than the ground truth (42.6 %). The inversion result lies between the prior of 50 % and the true model of 42.6 %, which is the expected result of a Bayesian method. In this test, the variograms generating the ground-truth wedge model (Table 1) were used for the Plurigaussian facies inversion, and an alternative variogram similar to the indicator variogram of the truth model is used for the seismic scale inversion.

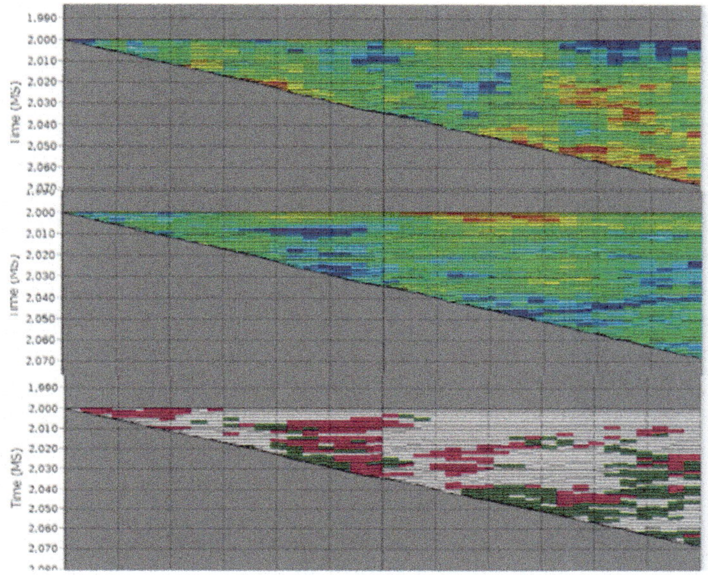

Fig. 4 2D profile view of a medial slice of the ground truth. The *top two images* indicate the guide variables, while the *bottom* image is the generated facies map

Fig. 5 LRM used to generate ground-truth facies in the synthetic wedge model

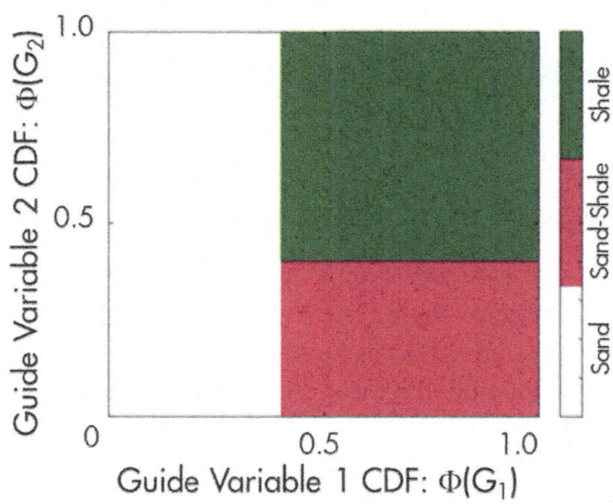

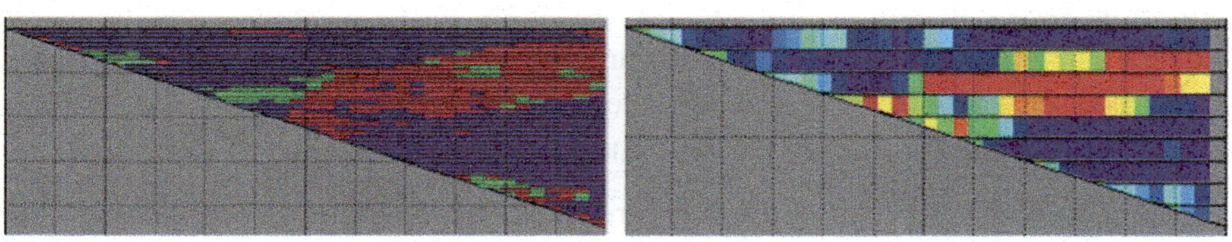

Fig. 6 2D profile view of seismic scale inversion for NTG in comparison with true NTG (slice 14)

4.3 Plurigaussian Inversion

The fine-scale case assumes that both horizontal and vertical variograms can be estimated accurately, and the prior facies probabilities are assumed to be equal throughout the model grid. In this example of three facies, the prior probability for each facies is 33 %. The inverted result is shown in Fig. 7. The inverted result yields an overall sand fraction of 41.4 %, which is very close to the ground-truth value of 42.6 %. This illustrates a reduction in the error misfit when compared to the seismic scale case. The reduction of misfit may be attributed to the finer scale on which the inversion was performed. Further studies are required to understand the impact of model scales on the quality of *Promise* inversion results. Moreover, it should be noted that the Plurigaussian inversion was able to generate realizations with multiple contacts between facies (purple is in contact with both red and green facies).

5 Case Study: Deepwater Lobate System

The algorithm was also tested with a marine survey on a deepwater channel-lobe system. Internal petrophysical studies indicated that the reservoir consists of three facies (clean sand, shale, and a mixture of the two), and prior facies probability maps have been postulated for each facies (Fig. 8).

5.1 Preparing Inversion Inputs

Lithofacies Rule The lithofacies rule can be generated given the expected facies proportions and the facies contact proportions. In this study, an initial geological concept model of three facies has been built for the reservoir. The facies proportions were obtained from the initial model, while the contacts between facies were estimated by computing the facies transition probabilities (Table 2) from the initial model.

The nine facies transition probabilities constitute nine constraints, but for three rectangular facies, only two thresholds are available for regular rectangular facies map, resulting in an overdetermined system. Therefore, a least squares approach is used to fit the thresholds and results in the LRM shown in Fig. 9.

Variogram Fitting The final step before inversion is obtaining estimates for the variograms. A prior most likely facies map is obtained by identifying the facies with

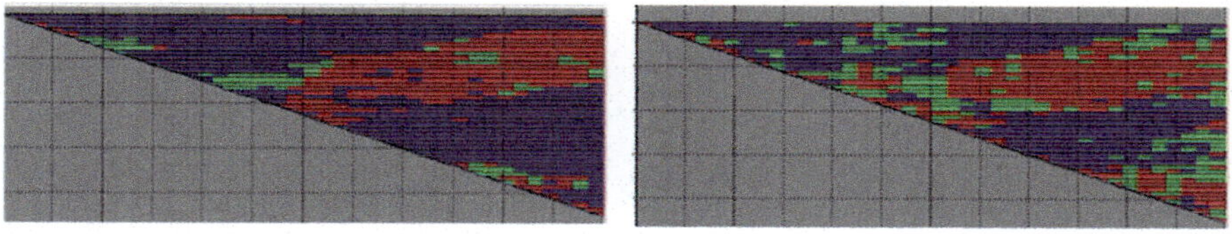

Fig. 7 2D profile view of slice 14 of the ground truth (*left*) compared to the inverted result (*right*) of 3D inversion with correct horizontal and vertical variograms

Fig. 8 Postulated prior facies probability maps for an offshore reservoir in deepwater channel-lobe system

Table 2 Facies transition probabilities

Facies type	Shale	Sand-shale mix	Sand
Shale	–	40 %	60 %
Sand-shale mix	56 %	–	43 %
Sand	27 %	73 %	–

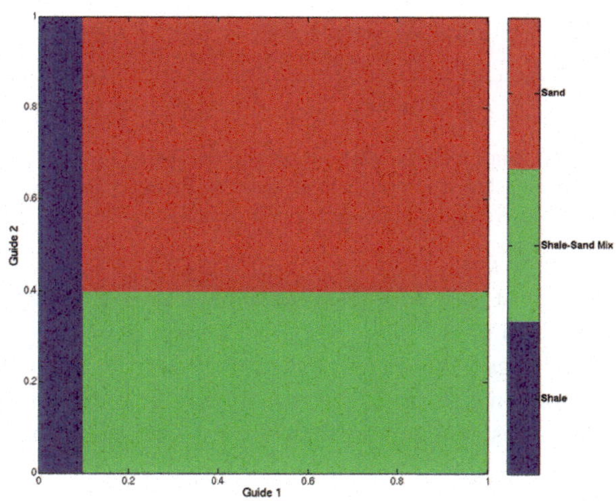

Fig. 9 Estimated LRM for offshore reservoir

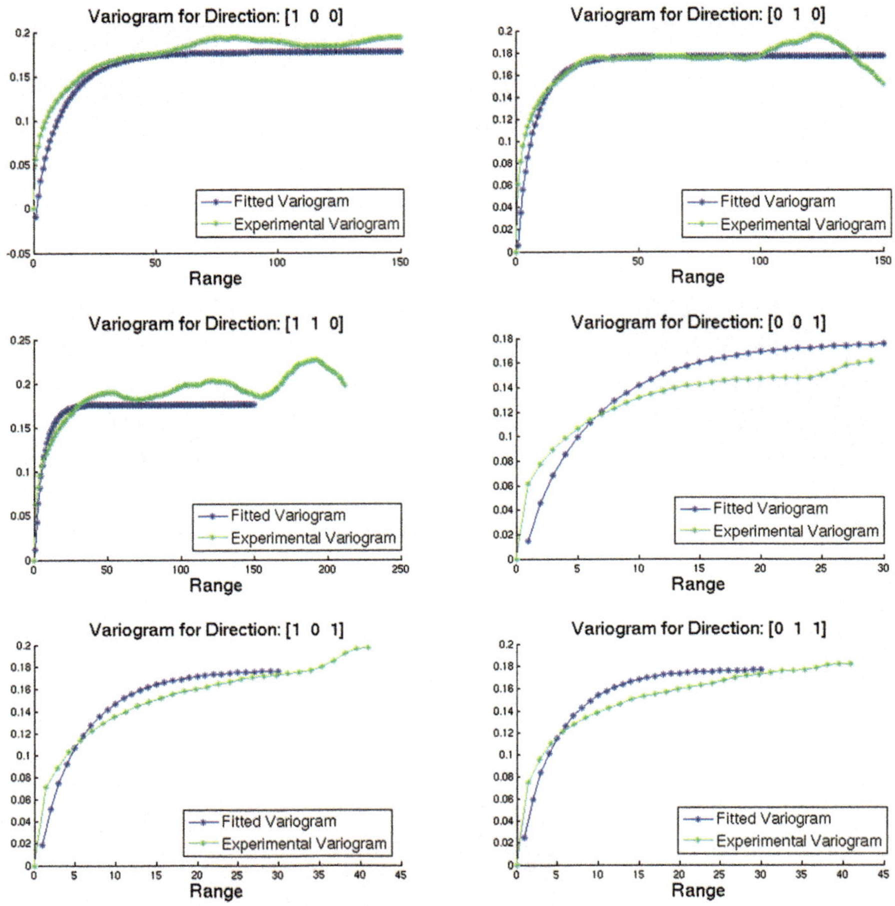

Fig. 10 Empirical indicator variograms computed for the sand-shale mix on the most probable facies map shown in *green* along with the corresponding indicator variograms by modeling two continuous variograms for each guide variable

the highest probability from the input probability maps. By computing the indicator variogram on the most likely facies map, and computing the integral in Eq. 8 over the averaged thresholds, one continuous variogram can be fitted for each guide variable. In a 3D grid, the variograms must be fitted to six directions in the space for each facies. Variograms fitted for the sand-shale mix facies are shown in Fig. 10.

The fitted variogram parameters are shown in Table 3. It should be noted that these parameters are not necessarily unique. Ideally, a single set of parameters

Table 3 Variogram parameters estimated for the testing data

Name	Long range	Short range	Vertical range	Azimuth	Nugget	Type
Guide 1	1620	1250	5	45	0.0001	Exponential
Guide 2	1500	750	20	90	0.0001	Exponential

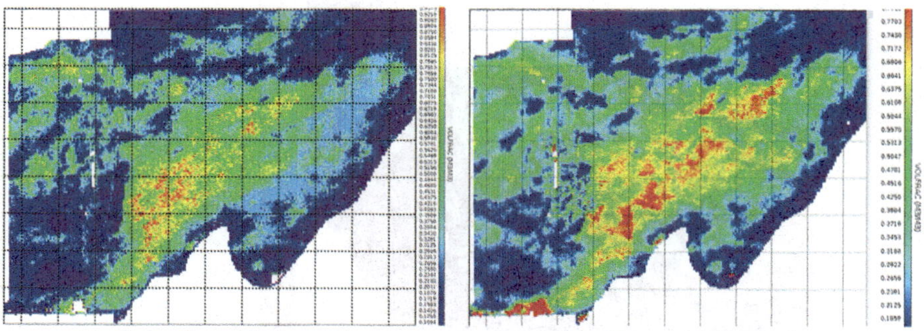

(a): Seismic scale inversion (b): Fine scale inversion

Fig. 11 Arithmetic average map of NTG over the P50 of the posterior realizations of the inversion

should be used for each facies; however, in practice this is a difficult task. Consequently, the averaged variogram over all three facies was used.

5.2 *Inversion Results*

For comparison, a seismic scale inversion (30 m cells) was performed with the resulting average NTG map of the P50 realization of the inversion results shown in Fig. 11a. Using the estimated LRM and variograms, fine-scale inversion (2 m) was performed, with the resulting maps shown in Fig. 11b. The fine-scale inversion indicates isolated regions of higher NTG than the coarse-scale result. The overall trends for both approaches are similar, but the fine facies inversion provides additional insight on areas of higher NTG which would be useful for drilling decisions. The top view map and a vertical cross section shown in Fig. 12 demonstrate results of the P50 realization of the fine-scale facies inversion. Lenticular shape sand bodies of channels and lobes are identifiable in the vertical cross section. In the top view, the transition between sand and shale is observed in the top view map, which demonstrates the effect of the Plurigaussian lithofacies rule.

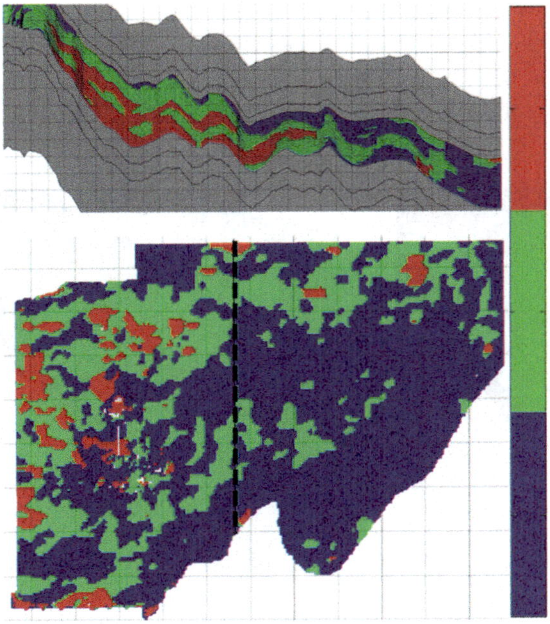

Fig. 12 A vertical cross section and a map of the top view of the P50 posterior realization

6 Conclusions

In this paper, a stochastic facies inversion algorithm was demonstrated using a Plurigaussian-inspired technique. As part of the algorithm, a method to estimate variograms for the underlying guide variables based on their mathematical relationship with the resulting indicator variable is presented. In comparison to conventional multistep inversion, this method directly generated lithological facies model and provides subseismic scale constraints based on geological and lithological knowledge. Using a synthetic case as a benchmark, it was found performing the new algorithm with accurate input parameters resulted in realizations that closely resembled the ground truth. The algorithm was also demonstrated with real seismic data from a marine survey on a deepwater channel-lobe system. The inversion process itself produces inversion results that are similar to the seismic scale inversion but indicates specific pockets of higher NTG that could not be seen on the seismic scale inversion. This could play an important role in drilling decisions.

Bibliography

Allen D (1984) Laminated sand analysis. SPWLA 25th Annual Logging Symposium. New Orleans: Society of Petrophysicists and Well-Log Analysts, 20

Armstrong M et al (2011) Plurigaussian simulations in geosciences. Springer Science & Business Media, New York

Coléou T, Manuel P, Azbel K (2003) Unsupervised seismic facies classification: a review and comparison of techniques and implementation. Lead Edge 22(10):942–953

Doyen P (2007) Seismic reservoir characterization: an earth modelling perspective. EAGE publications, Houten

Doyen PM, Psaila DE, Den Boer LD, Jans D (1997) Reconciling data at seismic and well log scales in 3-D earth modelling, SPE Annual Technical Conference and Exhibition. Society of Petroleum Engineers, San Antonio, p 10

Dubois M, Bohling G, Chakrabarti S (2007) Comparison of four approaches to a rock facies classification problem. Comput Geosci 33(5):599–617

Gelderblom P, Leguijt J (2010) Geological constraints in model-based seismic inversion, SEG International Exposition and 84th Annual Meeting. Society of Exploration Geophysicists, Denver, pp 2825–2829

Gómez-Hernández JJ, Rodrigo-Ilarri J, Cassiraga E (2016) Extended abstract example for GEOSTATS2016, Instructions for GEOSTATS2016 participants. Universitat Politècnica de València, Valencia, pp 1–4

Grana D, Della Rossa E (2010) Probabilistic petrophysical-properties estimation integrating statistical rock physics with seismic inversion. Geophysics 75(3):21–37

Gunning JS, Kemper M, Pelham A (2014) Obstacles, challenges and strategies for facies estimation in AVO seismic inversion. 76th EAGE Conference and Exhibition 2014

Hesthammer J, Landrø M, Fossen H (2001) Use and abuse of seismic data in reservoir characterisation. Mar Pet Geol 18(5):635–655

Larsen AL, Ulvmoen M, Omre H, Buland A (2006) Bayesian lithology/fluid prediction and simulation on the basis of a Markov-chain prior model. Geophysics 71(5):69–78

Leguijt J (2001) A promising approach to subsurface information integration, 63rd Conference Extended Abstracts. European Association of Geoscientists & Engineers, Amsterdam, p 4

Leguijt J (2009) Seismically constrained probabilistic reservoir modeling. Lead Edge (Soc Explor Geophys) 28(12):1478–1484

Rimstad K, Omre H (2010) Impact of rock-physics depth trends and Markov random fields on hierarchical Bayesian lithology/fluid prediction. Geophysics 75(4):93–108

Ulvmoen M, Omre H (2010) Improved resolution in Bayesian lithology/fluid inversion from prestack seismic data and well observations: part 1—methodology. Geophysics 75(2):21–35

Combined Use of Object-Based Models, Multipoint Statistics and Direct Sequential Simulation for Generation of the Morphology, Porosity and Permeability of Turbidite Channel Systems

Inês Marques, José Almeida, Mariana Quininha, and Paulo Legoinha

Abstract This work presents a new methodology for simulating the morphology and petrophysical properties of hydrocarbon reservoirs in turbidite channel systems. The simulation of the morphology uses an object-based algorithm that imposes multipoint statistics of azimuth angle classes and *cdf*s of width and thickness as measured from training images of channels. A facies is then assigned to each block of the reservoir grid according to a conceptual model of facies both laterally and vertically within the channels. Following this, as each facies has a specific *cdf* for both porosity and permeability, simulated images of these petrophysical properties are generated using direct sequential simulation (DSS) with local histograms. For illustrative purposes, a case study of a reservoir in the Lower Congo Basin is presented.

1 Introduction

Turbidite channel systems are one of the most common types of siliciclastic reservoirs in deep-water settings. Such systems by some estimates account for between 1200 and 1300 oil and gas fields worldwide (Stow and Mayall 2000). Geologically, these

I. Marques (✉) • P. Legoinha
Departamento de Ciências da Terra and GeoBioTec, FCT Universidade Nova de Lisboa, Campus da Caparica, 2829-516 Caparica, Portugal
e-mail: cmarques.ia@gmail.com; pal@fct.unl.pt

J. Almeida
Earth Sciences Department and GeoBioTec, Faculty of Sciences and Technology, Universidade NOVA de Lisboa, Campus da Caparica, 2829-516 Caparica, Portugal
e-mail: ja@fct.unl.pt

M. Quininha
Departamento de Ciências da Terra, FCT Universidade Nova de Lisboa, Campus da Caparica, 2829-516 Caparica, Portugal
e-mail: mquininha@gmail.com

J.J. Gómez-Hernández et al. (eds.), *Geostatistics Valencia 2016*, Quantitative Geology and Geostatistics 19, DOI 10.1007/978-3-319-46819-8_43

systems are composed of sets of channels with similar architecture, referred to as "channel complexes". The considerable size and detritic properties of these complexes mean that they form good hydrocarbon reservoirs, which has prompted great interest in the petroleum industry to better characterize these structures. However, modelling these complexes is a challenge because of their curvilinear geometries, the connectivity between channels and the internal distribution of facies or rock types (Babonneau et al. 2010; McHargue et al. 2011; Hashemi et al. 2014).

According to the literature, three classes of algorithms are used to model channel reservoirs: simulation with two-point statistics (Luis and Almeida 1997), simulation with multipoint statistics (Strebelle 2002; Liu 2006) and object-based simulation (Deutsch and Wang 1996). Each method has advantages and drawbacks regarding the modelling process: simulations based on two-point statistics are straightforward in terms of calculations but do not fully capture the curvilinear behaviour of the channels; simulations based on multipoint statistics can capture the shape of the channels with large templates but are computationally very intensive; and object-based modelling requires a simplified geometry for each class of objects, and the conditioning to data, porosity and permeability can be difficult as the output is typically in the form of polylines. Thus, there is great interest in the development of new and more effective stochastic algorithms to characterize these curvilinear structures, namely, by mixing the above approaches (Kuznetsova 2012; Kuznetsova et al. 2014; Quininha 2015; Quininha et al. 2015; Marques 2015).

In this work, an integrated methodology is presented for simulating both the morphology and the petrophysical properties (porosity and permeability) of turbidite channel complex systems. The simulation of the morphology involves two main stages: first, the complex pathline is generated, and second, the channels are generated conditional to the complex pathline. Both the pathline and the channels are simulated using an innovative algorithm that combines multipoint statistics of classes of azimuth angles embedded within a stochastic object-based modelling approach. The algorithm computes the multipoint statistics of the azimuth classes from training images and stores the probabilities of their occurrence in a dynamic data structure termed a "search tree". For assessing porosity and permeability, the simulated channels are intersected as individual objects by a high-resolution grid of blocks, and a facies is assigned to each small block according to a conceptual facies model with both lateral and depth variations in the channel section. Then, as each facies of the conceptual model has a prior probability distribution function for porosity and another for permeability, simulated images of these petrophysical properties are generated using direct sequential simulation (DSS) (Soares 2001) with local histograms. Finally, the high-resolution grid of blocks is upscaled to a larger grid to be used in flow simulators.

2 Methodology

The methodology comprises five main stages: (i) data preparation, (ii) generation of the morphology of the complex and the channels through multipoint statistics of training images, (iii) conversion of the vector morphological model to a grid of blocks with the assignment of facies to each block, (iv) simulation of the porosity and the permeability conditional to the facies model, and (v) upscaling of the grid of blocks in order to be used in a flow simulator.

2.1 Data Preparation

To apply the proposed methodology, it is first necessary to prepare the data and information as well as to establish the dimensions of the reservoir and the size of each reservoir block. Representative training images of the complex pathline and of the turbidite channel centreline are selected. A set of control points within the reservoir volume where a channel complex occurrence is confirmed (from seismic information or well data) are now defined. Each control point has an associated maximum distance, and the complex pathline should approximate each control point according to a random distance drawn between zero and the maximum distance. Two 1D list values (for thickness and width dimensions, respectively) to assign to the simulated channels are now simulated by DSS and Co-DSS (Soares 2001) imposing a correlation metric between the two variables. A conceptual facies model is designed that accounts for the lateral and vertical distributions of the porosity and permeability of each facies.

2.2 Simulation of Morphology and Assignment of Facies

As stated above, a channel complex is composed of sets of two or more channels, and therefore it is necessary to simulate the complex pathline first, which will influence the simulation of the individual channel centrelines. The following steps are used to generate the complex pathline:

1. Scan the training image of the complex pathline as a polygonal line and adjust the line for line segments of equal length.
2. Determine the azimuth angle of each line segment, by convention measured from the north direction clockwise. The angles are a continuous, circular variable ranging from 0° to 360°.
3. Convert the azimuth angles into classes of angles (build a categorical variable). For example, if the classes have a range of 20°, there will be 18 classes.
4. Calculate the residual of each azimuth angle (the difference between the azimuth angle itself and the azimuth angle that corresponds to the centre of the class). At this point, the training image of the complex pathline is defined by two lists of 1D values: the first contains the azimuth classes and the second contains the residual values. These two lists allow the precise geometry of the training image to be recreated.
5. Calculate and adjust a variogram for the residual values and simulate new lists of 1D values using DSS.

6. Evaluate the multipoint statistics of the azimuth classes from the training image using a 1D template with a maximum predefined dimension, and create the search tree (Marques 2015). The training image is scanned by the template, and all the azimuth class sequences found in the image are stored in the search tree along with their conditional probabilities. The structure of the search tree histogram is quite simple: first, the proportions of each single class are stored, and then the conditional probability of sequences of two classes is calculated and stored, followed by the conditional probability of sequences of three classes, and so on, until the maximum size of the template is reached. Because the probabilities are stored as relative values, the sum of the proportions of all the sequences composed of the same number of classes is always equal to 1.
7. Simulate the complex pathline through polyline object-based simulation conditional to the multipoint statistics of the azimuth class sequences stored in the search tree. A starting point is selected within the volume of the reservoir, and several line segments with the same length are added to the point. The orientation angles of these segments are computed, first by Monte Carlo simulation of an azimuth class conditional to the multipoint statistics stored in the search tree and then by adding a simulated residual value to the central angle of the class previously simulated.
8. Adjust the complex pathline according to the control points previously defined. Each segment of the polyline is rotated locally until it reaches the drawn generated distance for each control point.

The simulated pathline of the complex will condition the simulation of the centrelines of the individual channels, which are also represented as polygonal lines. The simulation of the individual channels is constrained to the reservoir area by respecting a defined tolerance distance between the pathline of the channel complex and the channel centrelines. The tolerance distance at the top of the reservoir should be greater than that at the bottom of the reservoir to give the complex the shape of a large channel. The following steps are used to generate the morphology of the turbidite channels:

9. Using the training image of the individual channel, compute turbidite channel centrelines at different depths within the reservoir volume, following the same procedure as that used to generate the complex pathline between steps 1 and 7 above.
10. Assign thickness and width dimensions previously simulated during the data preparation phase to each vertex of the centrelines. At this stage, each channel is represented by four polygonal lines: the top of the channel, two lateral margins and the base of the channel.

Having generated the morphology model, a conceptual facies model of the section of a channel is applied to the simulated pathlines in order to obtain a global facies model for the reservoir. In the present study, a model with five different regions was considered (Fig. 1): Region I, coarse-grained sand; Region II, medium-grained sand; Region III, fine-grained sand; Region IV, fine-grained sand and silt; and Region V, clay. Region V corresponds to the outer channel region.

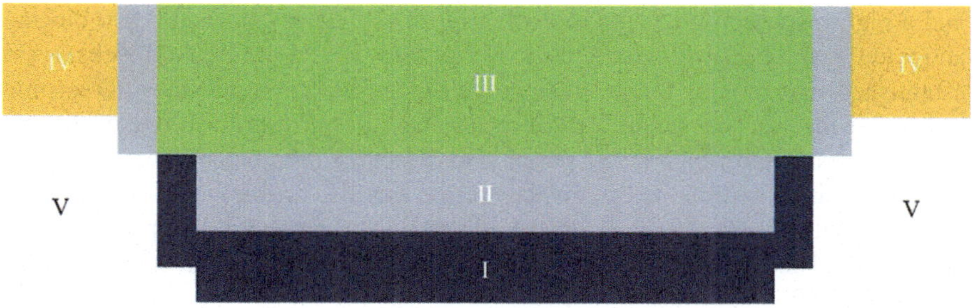

Fig. 1 Proposed conceptual facies model defining five regions within a cross-section of a channel

The vector model is converted to a high-resolution grid of blocks, and then, considering the position of the central point of each block relative to the centreline of the turbidite channel, a facies from the conceptual model is assigned.

The vertical aggradation and migration of the channels can influence the facies distribution because more recent channels can cut and fill older ones. For this reason, the facies attribution is generated from the oldest channel to the most recent one.

2.3 Porosity and Permeability Simulation and Upscaling

At this stage, each high-resolution reservoir block has a facies assigned according to the conceptual model of facies and the simulated pathlines of the channels. The simulation of porosity and permeability conditional to the facies images is now performed using an adapted version of DSS (proposed by Soares 2001, with the algorithm parallelization proposed by Nunes and Almeida 2010) that uses local histograms of the variables (here, the porosity and permeability conditional to each facies) instead of a global histogram (Roxo et al. 2016). The stochastic simulation process allows several equally probable scenarios of both properties to be generated for the same morphological model.

The upscaling of both petrophysical properties allows the images to be used in a dynamic flow simulator. The upscaling of porosity is calculated by the arithmetic mean of the porosity values of the high-resolution blocks; for permeability, the upscaling follows the classical approach of combining the arithmetic and harmonic means of the small blocks, generating a tensor of permeability for each upscaled block.

3 Case Study

For illustrative purposes, we applied the proposed methodology to a region of the turbidite channel system in the Lower Congo Basin, offshore Angola, located on the West African passive margin. The reservoir has dimensions of $2000 \times 6200 \times 50$ m, and a high-resolution grid with a block size of 10 × 10 × 1 m (6,200,000 blocks) and a low-resolution grid with a block size of 50 × 50 × 5 m (49,600 blocks) were established. The training image of the complex was drawn by analysing 3D seismic images of the reservoir area. This analysis also allowed the locations of control points to be defined where it was possible to detect the presence of the complex. A training image of a turbidite channel centreline was also drawn with a very sinuous path to evaluate the capacity of the algorithm to represent complex circular geometries. The training images of the complex pathline and the channel centreline are shown in Fig. 2.

A uniform distribution law [13 m, 25 m] was used to generate several 1D lists with 1000 values of thickness, and another uniform law [50 m, 150 m] was used to generate more 1D lists with 1000 values of width, to be attributed to the channels. A linear correlation of 0.7 was considered between the two dimensions.

As data for porosity and permeability were not available for this particular reservoir, Gaussian distribution laws for porosity and permeability were adopted

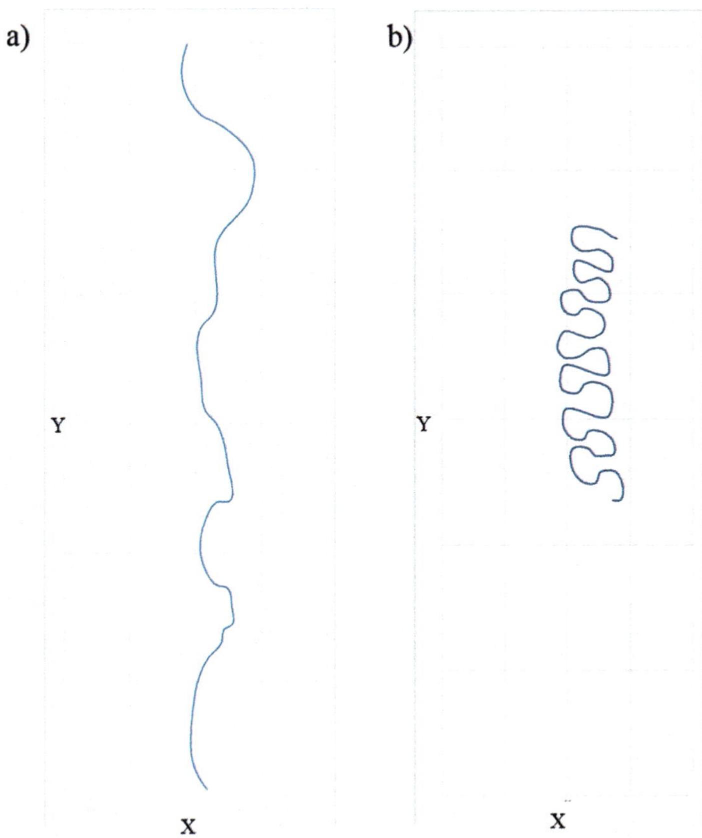

Fig. 2 (**a**) Training image of the complex pathline produced by analysing 3D seismic images of the reservoir. (**b**) Training image of the channel centreline

for each facies of the conceptual model, with values of mean and standard deviation as appropriate for the different lithologies. The box plots of porosity and permeability for each facies of the conceptual model are shown in Fig. 3.

For the case study, a template with a maximum dimension of 10 cells and 18 classes of 20° for the complex pathline and 36 classes of 10° for the channel centrelines was used. The complex pathline was adjusted for line segments of 100 m, and the channel

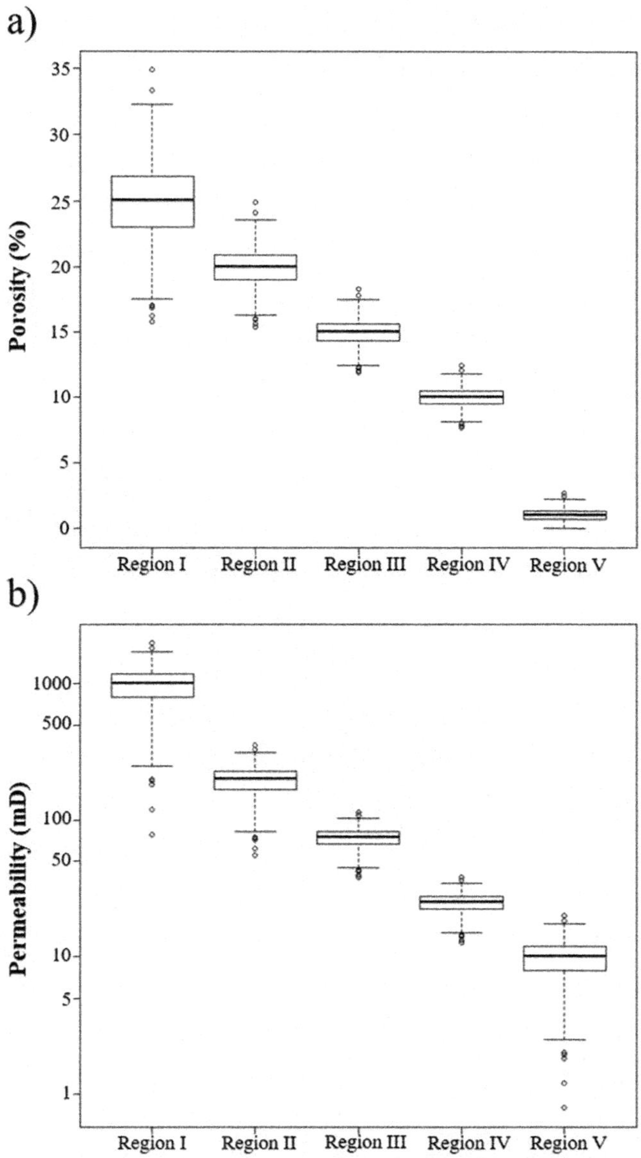

Fig. 3 (**a**) Box plots of porosity for each region of the conceptual facies model. (**b**) Box plots of permeability for each region of the conceptual facies model

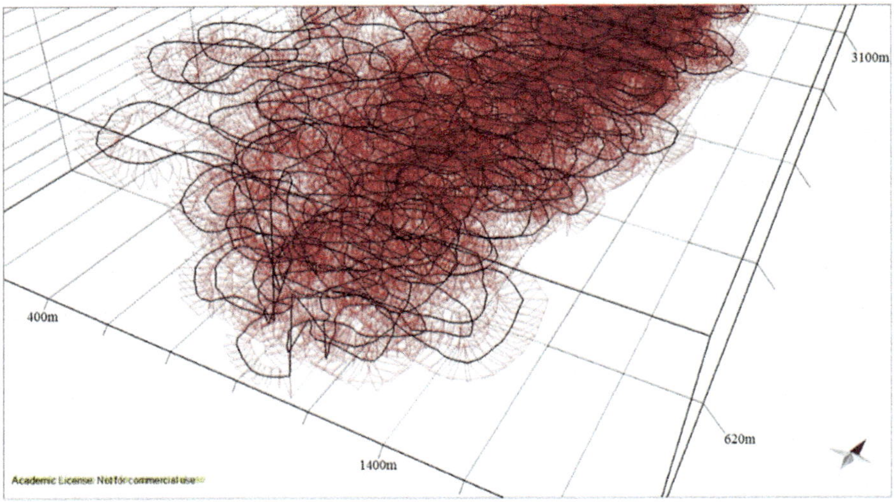

Fig. 4 A portion of a 3D morphology model of a simulated complex with 25 channels, showing the channel centrelines and channel sections

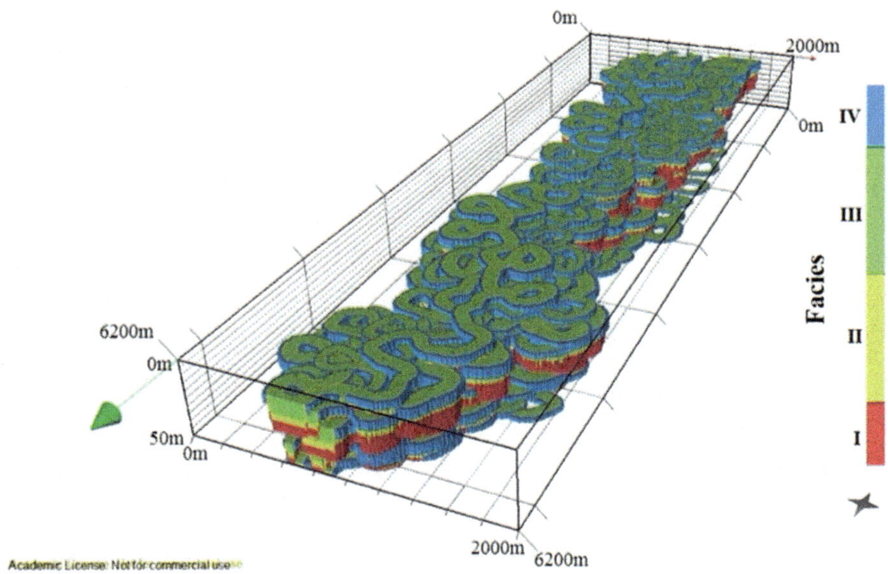

Fig. 5 3D representation of a simulated channel complex with 25 channels

centrelines were adjusted for line segments of 25 m. The tolerance radius between the complex pathline and the control points was defined as 250 m.

Figure 4 presents a portion of a 3D vector model of a complex comprising 25 channels, in which it is possible to see the centrelines of the channels and the sections of the channels separated by a distance of 25 m. Figure 5 presents an image

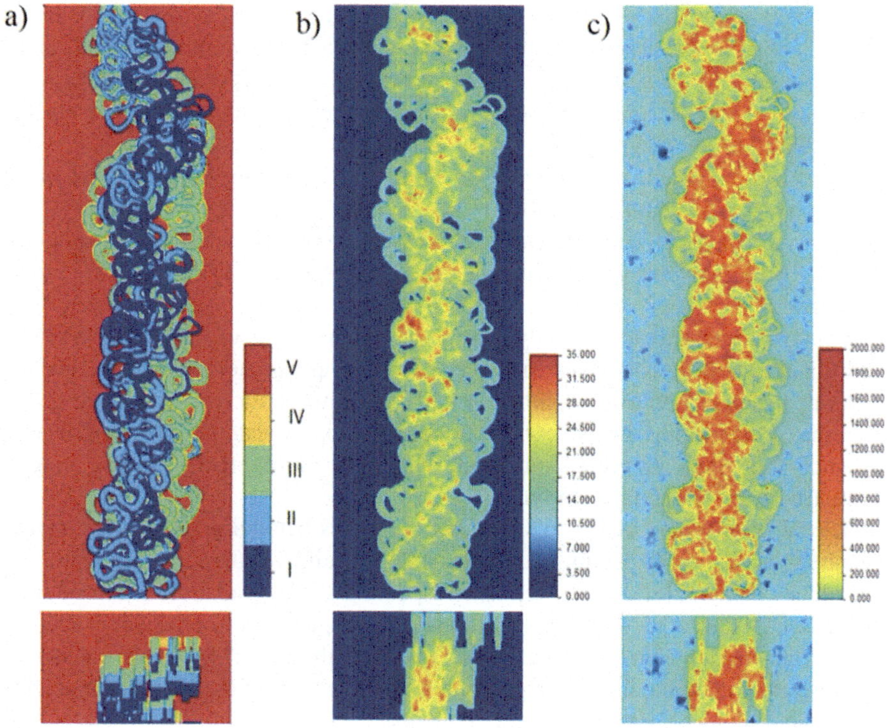

Fig. 6 (**a**) Map view ($Z = 30$ m) and cross-section view of the facies distribution of a simulated channel complex. (**b**) Map view ($Z = 30$ m) and cross-section view of the porosity (%) of a simulated channel complex. (**c**) Map view ($Z = 30$ m) and cross-section view

of a complex comprising 25 channels, in which it is possible to observe the morphology model and the facies distribution (Region V was excluded to allow a better view of the model). Figure 6a–c, respectively, presents map views and cross-section views of the facies distribution, porosity and horizontal permeability of a complex generated with 25 channels at a depth of 30 m. The variograms of the petrophysical properties have a spherical model with a range equal to 100 m in the north–south direction and a range equal to 20 m in the east–west direction.

4 Discussion

Figures 4, 5 and 6 show that the algorithm used is very effective when representing the complex curvilinear geometries that characterize the channels, even though the training image of the channel is very sinuous. Also, it can be seen that the set of generated channels follows a pathline (the complex pathline).

The thickness and width values of the channels were generated based on uniform distribution laws, carrying a high level of uncertainty. However, these values should be re-evaluated when there is information available on the dimensions of the channels (e.g. from 3D seismic images). It is important to remark that the proposed methodology and the software have been developed to incorporate available information about the thickness and width of the channels and about the porosity and permeability of the facies in the form of any *cdf*s.

The mean and standard deviation of the distribution laws of the petrophysical properties (porosity and permeability) decrease from Region I to Region V, and therefore the range of values of Region I is larger compared with the other regions. The applied histograms are plausible, but they must be altered when the real values are known.

This work presents results only for template dimensions of ten cells. However, additional tests were conducted to evaluate how this parameter would influence the generated images. A lower number of cells were not sufficient to reproduce the patterns of the training images. On the other hand, a template with many cells (for instance, more than 50) gives simulations that are too similar to the training images, detracting from the simulation process.

The 2D map views of Fig. 6 show that the facies belonging to the most recent channels (at shallower depths in the reservoir) prevail over the facies belonging to the older channels (at greater depths), respecting the aggradation and migration processes. Also, the views demonstrate that the images of petrophysical properties are in concordance with the image of the facies distribution. The warmer colours on the porosity and permeability images match the zones on the facies image that have the higher values of permeability and porosity.

In this case study, well data were not available and therefore not used. However, the proposed methodology is able to impose well data as control points to locally rectify the pathline of the channel complex. By analysing the facies data, it would be possible to evaluate the location of the well within the conceptual facies model of the channel and establish an appropriate tolerance radius between the well and the channel centreline. Also, if porosity and permeability measurements are available, their values would condition the DSS with local histograms of the petrophysical properties as conditioning data.

5 Concluding Remarks

This work has presented an innovative and effective methodology to simulate channel complex systems. Three fundamental aspects should be highlighted. The first aspect is that the morphological simulation is developed in two phases: firstly, the generation of the complex pathline and, secondly, the generation of the channels conditional to the complex. The second aspect is that the facies assignment to the blocks of the grid follows a type of hierarchy, being assigned from the oldest channels to the most recent. This respects what happens in reality when a new channel cuts and fills existing channels. The third aspect is that the categorical models of the facies images influence the generation of the petrophysical properties.

The algorithm converts a continuous variable (azimuth angle) into a categorical one (azimuth class), avoiding the problem of the azimuth being a circular variable. Despite the fact that multipoint statistics may be considered a somewhat inefficient method, even with all the recent computational optimisation, the proposed multipoint algorithm runs at 1D, using azimuth classes from training images, and stores the multipoint statistics in a dynamic data structure. This allows the use of templates with large dimensions but with low computer processing demands.

The parameters involved in the algorithm are input as *cdf*s. Therefore, the parameters can be used in all case studies in which the distribution functions are known or can be assumed.

The present case study may be considered only partially real, as a more detailed analysis of seismic information and well data would ideally be needed, which lies beyond the scope of the study. However, the study demonstrates that the methodology and associated algorithm can be applied to real sizes of reservoirs, and the achieved results respect the complex geometries of such reservoirs.

Acknowledgements The authors are grateful to Partex Oil and Gas Group for providing information and to Midland Valley for providing an academic licence for Move® software. This work is a contribution to Project UID/GEO/04035/2013 funded by FCT-Fundação para a Ciência e a Tecnologia, in Portugal.

Bibliography

Babonneau N, Savoye B, Cremer M, Bez M (2010) Sedimentary architecture in meanders of a submarine channel: detailed study of the present Congo turbidite channel (ZaiAngo Project). J Sediment Res 80:852–866

Deutsch CV, Wang L (1996) Hierarchical object-based stochastic modeling of fluvial reservoirs. Math Geol 28:857–880

Hashemi S, Javaherian A, Ataee-pour M, Tahmasebi P, Khoshdel H (2014) Channel characterization using multiple-point geostatistics, neural network, and modern analogy: a case study from a carbonate reservoir, southwest Iran. J Appl Geophys 16:47–58

Kuznetsova A (2012) Stochastic simulation of the morphology of fluvial sand channels reservoirs. Master thesis, FCT Universidade Nova de Lisboa

Kuznetsova A, Almeida JA, Legoinha P (2014) Improved realism of channel morphology in object modelling with analogue data constrains. ECMOR XIV – 14th European conference on the Mathematics of Oil Recovery

Liu Y (2006) Using the SNESIM program for multiple-point statistical simulation. Comput Geosci 32(10):1544–1563

Luis J, Almeida JA (1997) Stochastic characterization of fluvial sand channels. In: Baafi EY, Schofield NA (eds) Geostatistics Wollongong'96, vol 1. Kluwer Academic Publ, Amsterdam, pp 477–488

Marques I (2015) Reservatórios turbidíticos – simulação da morfologia por estatísticas multiponto e avaliação das propriedades petrofísicas. Master thesis, FCT Universidade Nova de Lisboa (in Portuguese)

McHargue T, Pyrczb MJ, Sullivanb MD, Clarka JD, Fildania A, Romansa BW, Covaulta JA, Levya M, Posamentierb HW, Drinkwaterc NJ (2011) Architecture of turbidite channel systems on the continental slope: patterns and predictions. Mar Pet Geol 28:728–743

Nunes R, Almeida JA (2010) Parallelization of sequential Gaussian, indicator and direct simulation algorithms. Comput Geosci 36(8):1042–1052

Quininha M (2015) Simulação de estruturas meandriformes por objetos e estatísticas multiponto e avaliação da porosidade – aplicação a reservatórios siliciclásticos. Master thesis, FCT Universidade Nova de Lisboa (in Portuguese)

Quininha M, Almeida JA, Legoinha P (2015) Multipoint statistics of azimuth angle classes: an application to the simulation of channel structures and the evaluation of porosity in siliciclastic reservoirs. 17th annual conference of the International Association for Mathematical Geosciences

Roxo S, Almeida JA, Matias FV, Mata-Lima H, Barbosa S (2016) The use of sensory perception indicators for improving the characterization and modelling of total petroleum hydrocarbon (TPH) grade in soils. Environ Monit Assess 188(3):1–19

Soares A (2001) Direct sequential simulation and cosimulation. Math Geol 33(8):911–926

Stow D, Mayall M (2000) Deep-water sedimentary systems: new models for the 21st century. Mar Pet Geol 17:125–135

Strebelle S (2002) Conditional simulation of complex geological structures using multiple-point geostatistics. Math Geol 34:1–22

Recent Advancements to Nonparametric Modeling of Interactions Between Reservoir Parameters

Håvard Goodwin Olsen and Gudmund Horn Hermansen

Abstract We demonstrate recent advances in nonparametric density estimation and illustrate their potential in the petroleum industry. Here, traditional parametric models and standard kernel methodology may often prove too limited. This is especially the case for data possessing certain complex structures, such as pinch-outs, nonlinearity, and heteroscedasticity. In this paper, we will focus on the Cloud Transform (CT) with directional smoothing and Local Gaussian Density Estimator (LGDE). These are flexible nonparametric methods for density (and conditional distribution) estimation that are well suited for data types commonly encountered in reservoir modeling. Both methods are illustrated with real and synthetic data sets.

1 Introduction

Proper understanding and modeling of relationships between various reservoir properties is an important part in several applications in geostatistics. This is indeed the case for prediction of seismic and geological attributes, in the imputation of reservoir characteristics in areas with few observations and conditioning reservoir properties on inverted seismic parameters (Kolbjørnsen and Abrahamsen 2005). One important application of the estimated density is to use it for trans-Gaussian Kriging, where the data are transformed to the Gaussian domain before the Kriging interpolation is applied (Cressie 1993). Therefore, obtaining a reliable joint model is a critical component related to several aspects of reservoir modeling. Here, we will take a probabilistic standpoint and investigate methods for assessing such relationships by nonparametric estimation of the corresponding joint (probability) density.

The straightforward approach is to estimate the joint density, or distribution, using either parametric or simple nonparametric methods, where standard methods like the binned estimator or Kernel Smoothing (KS) (see Sect. 3 and Fig. 2) are

H.G. Olsen (✉) • G.H. Hermansen
Norwegian Computing Center, 114 Blindern, 0314 Oslo, Norway
e-mail: Havard.Olsen@nr.no; Gudmund.Hermansen@nr.no

J.J. Gómez-Hernández et al. (eds.), *Geostatistics Valencia 2016*, Quantitative Geology and Geostatistics 19, DOI 10.1007/978-3-319-46819-8_44

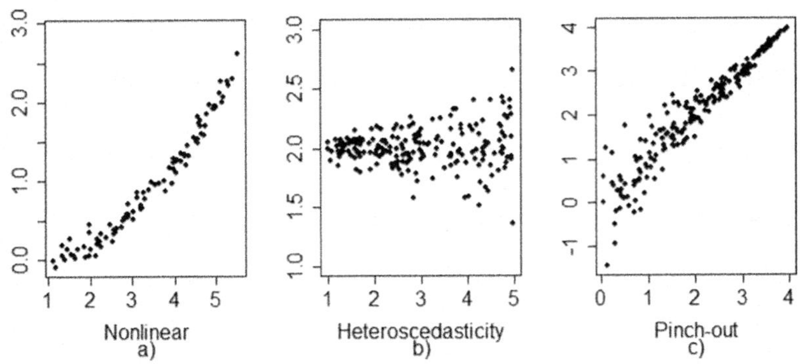

Fig. 1 Examples of three different data features where parametric methods often perform poorly, nonlinear (**a**), heteroscedasticity, (**b**) and pinch-out (**c**)

among the more popular. The parametric methods may run into problems for data with certain special or distinct features like pinch-outs, nonlinearity, and heteroscedasticity (see Fig. 1 above for an illustration). The simple nonparametric methods also have limitations and may perform poorly if the data either has a high number of variables or if there are few data points. They typically struggle to adapt pattern that is easily seen visually; see also (Leuangthong and Deutsch 2003) for additional discussion.

Therefore, in order to construct a reliable joint model, that can handle data commonly encountered in reservoir modeling, it is necessary to rely on more advanced methods.

Here, we will focus on two recent advances in the field of nonparametric density (and conditional distribution) estimation. The first is a variation of the Cloud Transform with directional smoothing (CT, all mentions of CT will hereafter mean with directional smoothing, unless otherwise stated) (Hermansen et al. 2016), and the other is a nonparametric density estimation called Local Gaussian Density Estimator (LGDE) (Otneim and Tjøstheim 2016a). Both are good and robust estimators for lower-dimensional problems, with CT having a slight edge over LGDE; however, the latter is more stable in problems with a high number of variables where the curse of dimensionality comes into effect. This curse refers to the problem of the exponential increase in volume when dimensions increase, together with limited data support (Sammut and Webb 2010). The KS requires certain amount of observation within a boundary to make a stable and reliable estimation, but in a high-dimensional volume data tend to become very sparse. Parametric models do not have the high-dimensional problems, and we will later see how nonparametric models can use some parametric structures to overcome this issue.

In Sect. 2 we will illustrate one limitation of the standard kernel methods on a synthetic data and also outline the basic mechanics underlying the improvements of CT. The two main methods, CT and LGDE, are properly introduced in Sects. 4 and 5, respectively; both will be tested with synthetic and real data. The results are summed up in Sect. 6.

2 The Problem and the Solution

In most applications, kernel density estimators improve the simplistic binned estimator, as illustrated in (Kolbjørnsen and Abrahamsen 2005) and Fig. 2. However, estimators based directly on a standard KS often become sensitive and unstable in areas with few observations. Therefore, direct and careless application may result in quite undesired and introduce biased behavior when, e.g., used for interpolation and extrapolation. Also, kernel-based methods may struggle when applied on high-dimensional data, as the curse of dimensionality (described in previous section) comes into effect; depending on the number, distribution, and sparsity of the observations, such methods may even become unreliable in low-dimensional problems.

To illustrate the problem (and foreshadowing the solution), we consider the synthetic data case shown in Fig. 2, which illustrates a pinch-out effect of the data. Data that pinch out is a common feature for highly correlated data, see, e.g., (Hermansen et al. 2016) for real data cases with this effect. The figure also shows the conditional cumulative distribution function (cdf) of permeability (log) given porosity for CT (b), binned estimator (c), and standard KS (d). The CT method shown here is one of the main methods investigated in this paper and will be described in detail in Sect. 4.

Somewhat simplified, the conditional cdf for permeability is traditionally obtained by first estimating the joint density between both variables, here permeability and porosity, and the corresponding conditional density (and distribution) is obtained from the joint density by (numerical) integration. The conditional cdf is an important part in simulations and predictions of permeability values, where we can condition on additional info like certain values/levels of porosity. The binned estimator (c) divides the data into equal proportions and estimates a one-dimensional distribution in each bin, which will create artificial discontinuities. Standard KS, which avoids this artifact, is easily seen as a better solution; see (Kolbjørnsen and Abrahamsen 2005) for a more thorough comparison.

The estimated distributions in Fig. 2 clearly illustrate one improvement that directional smoothing (b) has over standard kernel methods (d). While both methods work well in areas with much data, the difference is easily seen at the boundaries. Here, the KS (sort of) flattens out too early, while CT continues along and follows the main "signal" in the data cloud. This structure is easily observed visually; however, traditionally it is very hard to model this mechanically. If such structures exist, the KS is almost always bound to commit several errors at the boundaries. CT on the other hand adapts, more or less automatically, to such pinch-out effect by utilizing certain transformation and normalization of the data (see

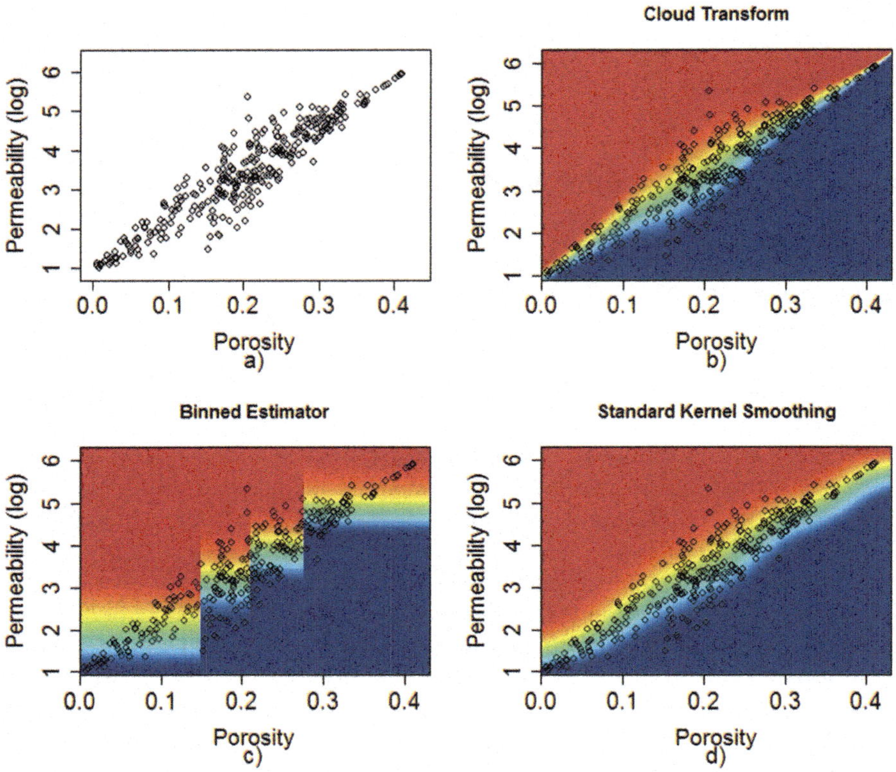

Fig. 2 Plot (**a**) shows the scatter plot of two synthetic data, and (**b**) is the corresponding conditional cumulative distribution for Cloud Transform, (**c**) binned estimator, and (**d**) standard Kernel Smoother. The colors represent values from 0 (*blue*) to 1 (*red*). These 2D cdf plots illustrate the conditional cdf of permeability (log) given porosity

Sect. 4 for details). Additional illustrations with the traditional KS are shown in Fig. 5.

For a specific porosity value in Fig. 2, the corresponding vertical slice will give the one-dimensional cdf. In Fig. 3, we have illustrated an example of the difference between KS and CT for permeability (log) given porosity value equal to 0.1, illustrating the potential of including porosity in the model.

3 Kernel Density Estimation

To understand the basic mechanics of the CT method, and also exactly how it improves on the standard methods, some basic knowledge of the traditional methodology is needed. Kernel smoothers refer to a general class of nonparametric methods, used for function estimation, such as density and regression. In short, such

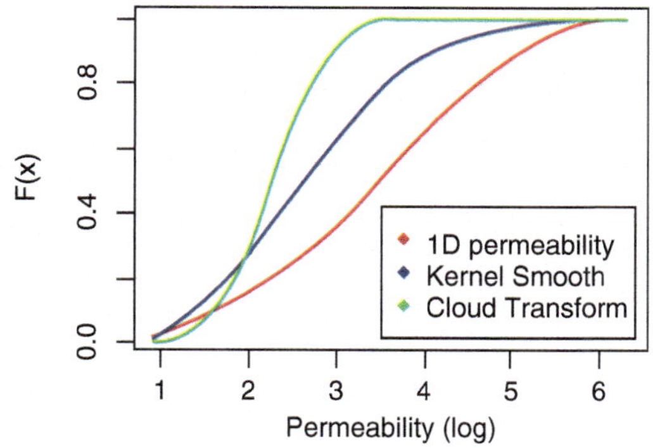

Fig. 3 Three 1D cdf of permeability taken from Fig. 2. *Red curve* is the 1D cdf of permeability (log) data in Fig. 2a. The *blue* and *green curves* are the 1D cdf of permeability (log) given porosity obtained by setting porosity equal to 0.1 in Fig. 2 for KS (d) and CT (b), respectively. Note that the level of porosity seen in Fig. 2 alters the estimated cdf

methods use nearby locations (like a moving window) to borrow strength for the estimation, e.g., density, distribution, or function, at every location; see Eq. 1 and Fig. 4 below. Given a k-dimensional data set $\boldsymbol{x}_1, \ldots, \boldsymbol{x}_k$ from an unknown distribution f, the kernel density estimator is defined as

$$\hat{f}_H(\boldsymbol{x}) = \frac{1}{n}\sum\nolimits_{i=1}^{n} K_H(\boldsymbol{x} - \boldsymbol{x}_i), \tag{1}$$

with $K_{\boldsymbol{H}}(\boldsymbol{z}) = |\boldsymbol{H}|^{-\frac{1}{2}}K\left(\boldsymbol{H}^{-1}\boldsymbol{z}\right)$. Here, $K(\cdot)$ is the kernel function, $\boldsymbol{H}$ is a k-dimensional symmetric and positive definite matrix of bandwidths, and $\boldsymbol{x}$ is the value where the density is estimated. A standard kernel choice is the Gaussian kernel; with the bandwidth matrix $\boldsymbol{H}$, the kernel function is given by $K_{\boldsymbol{H}}(\boldsymbol{z}) = \frac{1}{2\pi}|\boldsymbol{H}|^{-\frac{1}{2}}\exp\left(-\frac{1}{2}\boldsymbol{z}^T\boldsymbol{H}^{-1}\boldsymbol{z}\right)$ (Wand and Jones 1993).

Figure 4 shows the porosity plotted against permeability (log) with an ellipse indicating the bandwidth in x and y direction, indicated by h_1 and h_2, respectively. This is $K_{\boldsymbol{H}}$ from Eq. 1 for a 2D data set. In this illustration the density is estimated in the center of the ellipse at point (x_i, x_j). Only the points inside the ellipse contribute significantly to the density in this point, and the choice of the kernel function $K_{\boldsymbol{H}}$ decides how each point contributes.

In addition to choosing the kernel function $K(\cdot)$, an important part of kernel density estimation is to determine the bandwidths in the matrix:

$$\boldsymbol{H} = \begin{bmatrix} h_1 & h_{12} \\ h_{12} & h_2 \end{bmatrix}, \tag{2}$$

where the elements indicate the smoothing along the axes. Different setups of $\boldsymbol{H}$ will result in different shapes of the ellipse shown in Fig. 4; $h_1 = h_2$ and $h_{12} = 0$ gives a circle, $h_1 > h_2$ and $h_{12} = 0$ gives an ellipse without orientation (it follows the

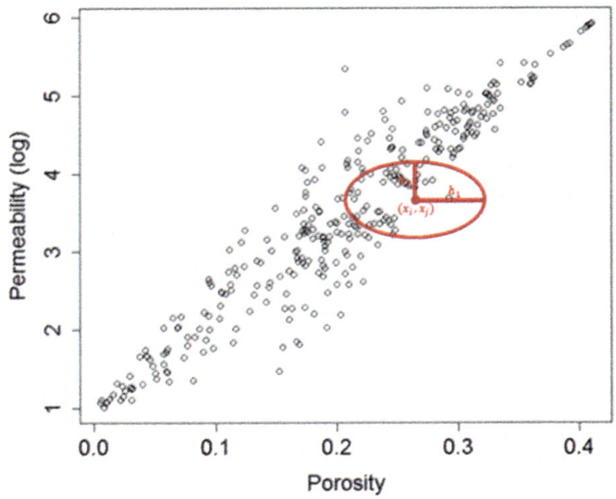

Fig. 4 Porosity vs permeability (log) with an illustration of two bandwidths in a Gaussian kernel

axis like the example in Fig. 4), and finally $h_1 > h_2$ and $h_{12} \neq 0$ gives an ellipse with an orientation. From Figs. 2 and 4, it seems clear that the last option provides the best use of the information (direction) observed in the data.

The choice of bandwidths is important, e.g., too small may result in overfitting, and very large bandwidths make it impossible to detect the finer details. Choosing the optimal bandwidths is a nontrivial task. Under reasonable restrictive assumptions (Gaussian kernel and the underlying data is Gaussian), there exist "optimal" bandwidth (rates) recommendations, and for the one-dimensional case, this is $h = 1.06\sigma n^{-0.2}$, where σ is the standard deviation and n is the number of data points; see Silverman (1998). Still, most problems usually require some manual tuning, which becomes hard for high-dimensional problems.

In two- or higher-dimensional problems, it is more difficult to determine and tune the bandwidths and one must typically rely more on estimation from iterative methods. A popular choice is cross-validation, where sections of the data will be left out for testing and the remaining data will be used for estimation. The bandwidths are then chosen based on the performance of a predetermined criterion, but choosing a good criterion is a problem on its own and relies on a clear understanding of the purpose of the model. One advantage of CT (discussed in Sect. 4) is that it solves some of these difficulties and will often give good results (as seen in Fig. 2) when used out of the box with almost no tuning required. The issue with choosing bandwidths is illustrated in Fig. 5, which shows the same synthetic data set as in Fig. 2, but estimated with two different bandwidths using the standard Gaussian KS. In Fig. 2, we used the previously mentioned optimal 1D bandwidth for each dimension, whereas in Fig. 5, this is scaled trying to obtain a better fit.

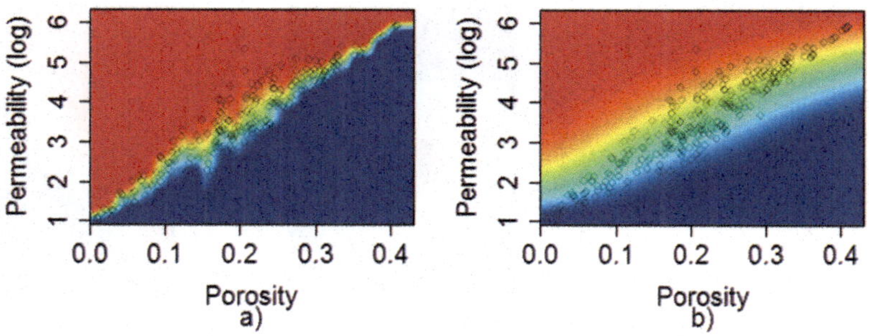

Fig. 5 Densities estimated with KS and two different bandwidths. Both have used optimal bandwidth based on the standard deviation of the data, but in (**a**) we have scaled this by a factor 0.25 and in (**b**) we have scaled this by a factor of 2

4 Cloud Transform (CT) with Directional Smoothing

We will now present the main components of the CT method. This was developed to solve some of the challenges of traditional kernel smoothers for density estimation, like bandwidth estimation, and adapting to the main structure of the data. The CT method presented here is an extension of the one presented in (Kolbjørnsen and Abrahamsen 2005) and builds on the work described in (Hermansen et al. 2016).

The main idea is to perform the kernel smoothing in a transformed domain (see Fig. 6b) which makes the method more robust against outliers and better behaved in areas where there are few observations. The kernel smoother borrows information of nearby observations, which is problematic in areas with few observations. Therefore, it performs better in the transformed domain, for most bandwidth choices, as each point will have more evenly support.

The main focus will be on the implementation of two-dimensional data sources, but we will also illustrate how the method can be extended to higher dimensions and show an example of a three-dimensional case.

Let (y_i, x_i), $i = 1, \ldots, n$ be independent observations of (Y,X) from an unknown joint distribution $f(y,x)$. Here, we will construct the nonparametric estimate for the conditional cdf, $\hat{F}(y|X = x)$, based on these observations. In short, the CT method estimates this conditional cdf by the following four main steps; see (Hermansen et al. 2016) for more details:

1. Normalize the response variable $\hat{y}_i = \left(y_i - \hat{E}(Y|X = x_i)\right)/\sqrt{\widehat{Var}(Y|X = x_i)}$, where the estimated expectation and variance are obtained by a local linear regression and a weighted sample variance.
2. Transform x_1 to $T_i = \hat{F}(x_i)$ where $\hat{F}(\cdot)$ is the estimated cdf.
3. Estimate the joint density and the conditional cumulative distribution in the transformed domain using the pairs $(\tilde{y}_i, T_i)$ and a standard Gaussian kernel.
4. Transform data back to the original domain by inverting step 1 and 2.

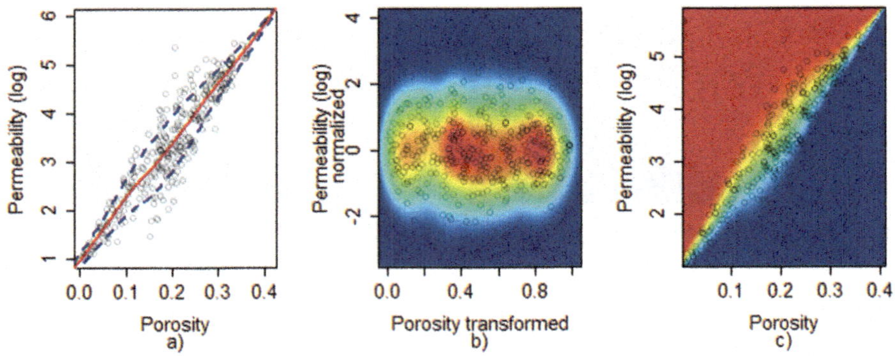

Fig. 6 The process of CT. Plot (**a**) is the original data with estimated mean (*red solid*) and one standard deviation (*blue stippled*). The normalized data is shown in (**b**) with the corresponding density estimated with a standard Kernel Smoother. Plot (**c**) is the backtransformed conditional cdf in the original domain

Figure 6 shows the process of CT. Plot (a) shows the original pinched-out data with estimated mean (red line) and one standard deviation (blue lines); these are used to transform the data to the domain seen in plot (b). Here, the density is estimated with a KS before transforming back to the original domain, shown in plot (c). An advantage of step 1 and 2 above is that the transformed data will be somewhat similar across different data sets and makes it easier to set more general bandwidths.

We have further applied the method on a real data case. The variables available in this case were P-wave velocity (Vp, ms/ft), density (Rho), gamma ray (GR), and total organic content (TOC, weight percent). This data set is of a conventional reservoir and consists of four blocked wells with a total of 315 observations. Figure 7 shows two 2D examples where we have used GR conditioned on Vp and GR conditioned on Rho.

The previous examples have dealt with two variables and specifically exemplified by estimating permeability (log) conditioned on porosity. In this section, we will outline how CT can be extended to higher-dimensional problems, $(Y_1, X_1, \ldots, X_k)$ with k fixed (Hermansen et al. 2016):

1. (a) Normalize each variable sequentially. First, let $\widetilde{\boldsymbol{x}}_2$ be the normalized $\boldsymbol{x}_2$ conditioned on $\boldsymbol{x}_1$, and then let $\widetilde{\boldsymbol{x}}_3$ be normalized $\boldsymbol{x}_3$ conditioned on $\widetilde{\boldsymbol{x}}_2$ and $\boldsymbol{x}_1$, and so on for all variables $\boldsymbol{x}_\mathrm{j}$. The normalization is done by computing the corresponding conditional means and standard deviations.
   (b) Finally, normalize $\boldsymbol{y}$ by computing the conditional mean and sd given $\widetilde{\boldsymbol{x}}_k, \ldots, \widetilde{\boldsymbol{x}}_2, \boldsymbol{x}_1$.

2. Transform all the normalized x variables from step 1 (a) sequentially. First, let $T_{1i} = \hat{F}_1(x_{1i})$, where $\hat{F}_1(\cdot)$ is the marginal cdf of X_1, and then let $T_{2i} = \hat{F}_{2|1}(\widetilde{x}_{2i}|x_{1i})$, where $\hat{F}_{2|1}(\cdot)$ is the conditional cdf of normalized X_2 given X_1. This is continued for all variables.

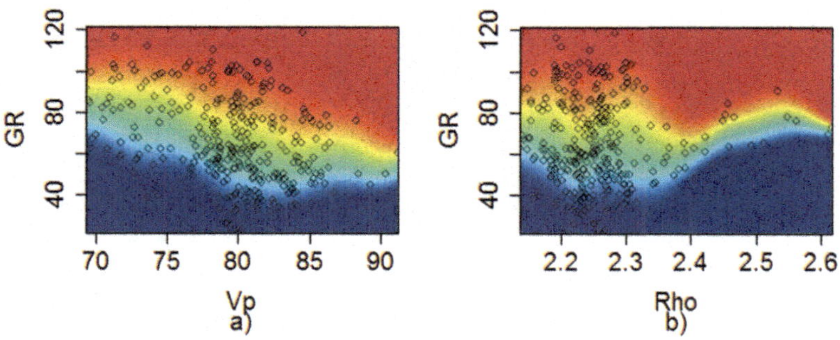

Fig. 7 Estimated conditional cdf with CT on a real data case, overlain by the observations, GR conditioned on Vp (**a**) and GR conditioned on Rho (**b**)

3. The n-dimensional joint density and conditional cdf can be computed in the transformed domain by a standard kernel density method.
4. Finally, the density can be transformed back by using inverse operations from step 1 and 2.

There are some general problems with the generalized version of CT. Firstly, the normalization and transformation steps become quite involved for data that goes beyond three dimensions, as each transformation and normalization step is conditioned on all the previous normalization and transformations.

Moreover, we are still vulnerable to the curse of dimensionality since there is a traditional kernel density smoother at the core of the algorithm after the normalization and transformation. However, CT will generally be more robust, compared to standard KS, since the kernel estimation takes place in a more well-behaved space, but it will eventually run into the same high-dimensional challenges; see (Hermansen et al. 2016) for a more detailed discussion.

We will now show CT applied in three-dimensional real data case. In this model we will condition TOC on Vp and Rho; see Fig. 8 for a plot of the data. There are 15 high TOC values in this data set (orange to red dots). According to Roxar ASA, who provided the data, these points may be erroneous, so we have chosen to exclude them. In addition, we have excluded the low Vp value (under 60) as we suspect that this may also be an error.

The result of 3D cdf estimated with CT is shown in Fig. 9. Plot (a) shows the data and (b) shows the corresponding 2D cdf of TOC given Rho (Vp values are ignored). The bottom row ((c) and (d)) shows two slices of the 3D cdf cube. It illustrates the effect of the third parameter on the estimated density. From this we can see the distribution of TOC given Rho is different for the two chosen Vp values (71.1 and 89.2). The lower Vp values give a wider cdf (larger variance). The one-dimensional cdfs corresponding to Rho equal 2.2 is plotted in Fig. 10, where it is easier to see the differences. From this we can conclude that to include Vp variable will give additional information to the estimated density. The performance of CT, in relation to LGDE, is discussed more in Sect. 6.

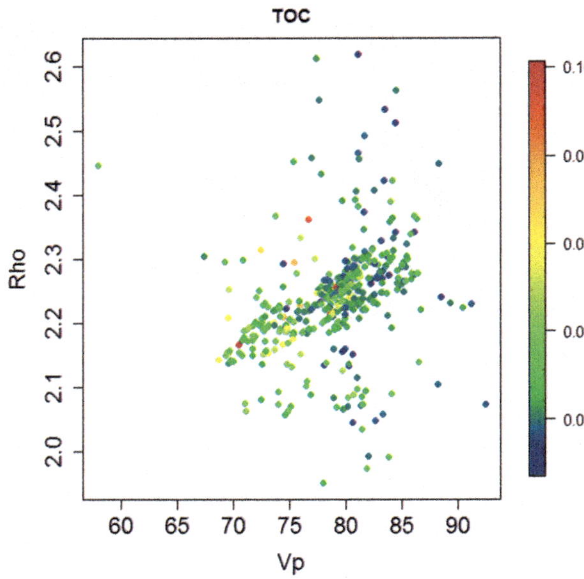

Fig. 8 Scatter plot of Vp vs Rho with TOC values given by contour colorings (*right bar*). These are the data from the real data case

We have previously outlined how CT can be estimated in higher dimensions. However, the curse of dimensionality, described in Sect. 1, may still break down the KS in high dimensions due to the limited data and the increasing volume. In the next section, we will therefore introduce an alternative nonparametric estimator, which is generally more suited for high-dimensional problems.

5 Local Gaussian Density Estimator

Here, we present an alternative method called Local Gaussian Density Estimator (LGDE) developed in the papers (Otneim and Tjøstheim 2016a, b). Here, we see this as a response to the shortcomings of CT in high-dimensional problems. It shares many of the same strengths; however, our experience is that LGDE is not as flexible and is not able to achieve the same level of performance in low-dimensional problems.

The LGDE method consists of three main steps for estimating the joint and conditional density, $f(\boldsymbol{x}, \boldsymbol{y})$ and $f(\boldsymbol{y}|\boldsymbol{x})$, respectively. The first step is to estimate the marginal cumulative distributions for each dimension, $\hat{F}(x_i)$, and transform each marginal to standard normal. We have used a one-dimensional KS with a Gaussian kernel in our testing.

The second step is to locally maximize the likelihood function

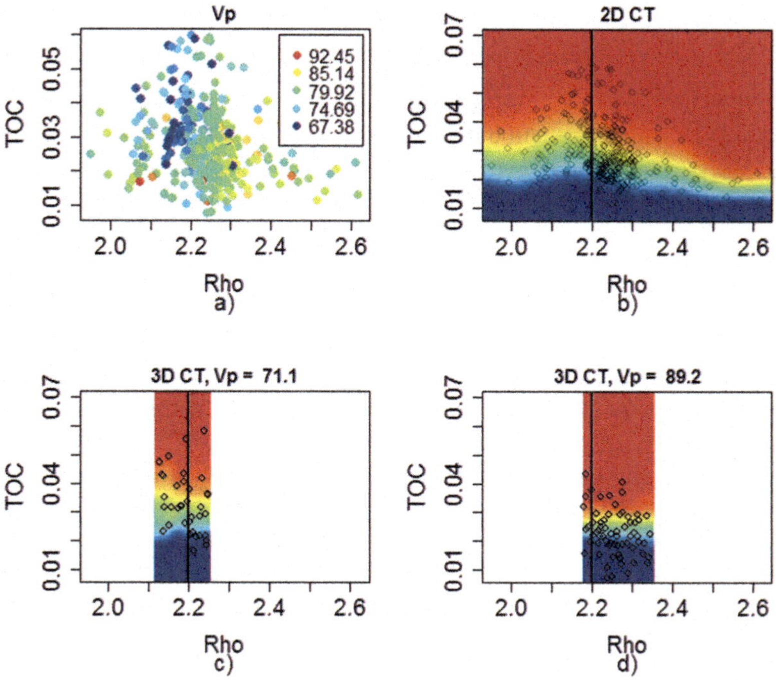

Fig. 9 Result of the 3D conditional cdf estimated with CT, here TOC is conditioned on Vp and Rho. Figure (**a**) shows the 3D data set and (**b**) is the 2D CT of TOC given Rho (Vp is ignored). The *bottom row* shows two slices of the 3D density, for Vp equal to 71.1 in (**c**) and Vp equal to 89.2 in (**d**). Only sections of the 3D cube where there is ample data support are shown, and only closest 20 % data to the chosen Rho value is plotted. The *black lines* indicate the corresponding 1D cdf in Fig. 10

Fig. 10 Three 1D cdf extracted from Fig. 9 of Rho equal to 2.2 (*black lines*). *Red line* is from the 2D cdf (Vp ignored), and *blue* and *green* are from the two slices of the 3D cdf cube, a low and high Vp value, respectively

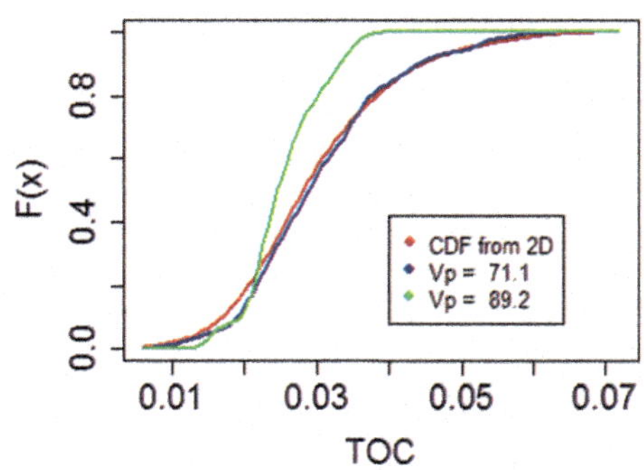

$$L(\theta, \boldsymbol{x}) = \frac{1}{n}\sum\nolimits_{i=1}^{n} K_{\boldsymbol{H}}(\boldsymbol{X}_i - \boldsymbol{x})\log\psi(\boldsymbol{X}_i, \theta) - \int K_{\boldsymbol{H}}(v - \boldsymbol{x})\psi(v, \theta)\mathrm{d}v \qquad (3)$$

at the locations $\boldsymbol{x}$ from the sample $\boldsymbol{X}_1, \ldots, \boldsymbol{X}_n$ (Otneim and Tjøstheim 2016a). This estimates the unknown density by fitting a parametric family of densities $\psi(\cdot, \theta)$

locally, where $\hat{\theta}(\boldsymbol{x})$ maximizes the likelihood. The unknown density $\hat{f}_X(\boldsymbol{x})$ is then equal to $\psi(x, \hat{\theta}(\boldsymbol{x}))$. Here, ψ is chosen to be a standardized multivariate normal distribution, that is, $\psi(z, \theta) = \psi(z, \boldsymbol{R}) = (2\pi)^{-\frac{p}{2}}|\boldsymbol{R}|^{-\frac{1}{2}}\exp\left(-\frac{1}{2}z'\boldsymbol{R}^{-1}z\right)$, where $\boldsymbol{R} = \boldsymbol{R}(z) = \{\rho_{ij}(z)\}$ is the correlation matrix. $K_H(\cdot)$ is chosen to be Gaussian kernel.

The last step is to combine to a final density and transform the data back to the original domain. The density given by (Otneim and Tjøstheim 2016a) is

$$f(\boldsymbol{x}) = f_Z\left(\Phi^{-1}(F_1(x_1)), \ \ldots, \Phi^{-n}\left(F_p(x_p)\right)\right) \prod_{i=1}^{p} \frac{f_i(x_i)}{\phi\left(\Phi^{-1}(F_i(x_i))\right)}, \tag{4}$$

where f_i and F_i are the marginal distributions, and $\Phi^{-1}(F_i(x_i))$ is the transformation of the original vector x_i to the normal domain (by first the marginal F and then the inverse standard normal Φ^{-1}).

We have been investigating the conditional density, which previously has been obtained by using the relation $f(y|x) = f(x, y)/f(x)$. This can however be unstable due to the division of small values of the marginal $f(x)$. The LGDE method makes use of the Gaussian framework to make a stable estimation of the conditional density directly. For a Gaussian distribution, the conditional distribution has an explicit formula; if $\boldsymbol{x} = [\boldsymbol{x}_1; \boldsymbol{x}_2]$ is multinormal distributed with $\boldsymbol{\mu} = [\boldsymbol{\mu}_1; \boldsymbol{\mu}_2]$ and $\Sigma = [\Sigma_{11}\Sigma_{12}; \Sigma_{21}\Sigma_{22}]$, then $\boldsymbol{x}_1|\boldsymbol{x}_2 = \boldsymbol{a}$ will be multinormal with $\overline{\boldsymbol{\mu}} = \boldsymbol{\mu}_1 + \Sigma_{12}\Sigma_{22}^{-1}(\boldsymbol{a} - \boldsymbol{\mu}_2)$ and $\overline{\boldsymbol{\Sigma}} = \Sigma_{11} - \Sigma_{12}\Sigma_{22}^{-1}\Sigma_{21}$. With this (Otneim and Tjøstheim 2016b) obtains an estimator similar to Eq. 4 for $f(\boldsymbol{y}|\boldsymbol{x})$.

The main advantage of the LGDE method is for high-dimensional problems where it is shown to overcome some of the problems related to the curse of dimensionality. This is done by a simplification which is based on letting the local correlation matrix at each point, $\hat{\rho}_{ij}$, only depend on their own pair of variables, that is, $\hat{\rho}_{ij}(x_1, \ldots, x_p) = \hat{\rho}_{ij}(x_i, x_j)$. This simplification reduces the estimation of $f(\boldsymbol{x})$ to a series of bivariate problems. Otneim and Tjøstheim (2016a) have tested the LGDE method for several data cases and compared it to standard kernel density. They have generated data from known distributions, varying between two and ten dimensions and evaluate the results through a relative square error (IRSE) $\int \left(\hat{f}(\boldsymbol{x}) - f(\boldsymbol{x})\right)^2 / f(\boldsymbol{x})\mathbf{dx}$. The result is that LGDE outperforms the KS for most cases and dimensions, chi-squared marginals with Gaussian copula, lognormal marginals with *t*-copula, *t*-marginals with Clayton copula, uniform marginals with Clayton copula, and multivariate t-distribution. The case where LGDE struggled against KS was a mixture of two Gaussian models; here, KS performed better for the lower dimensions (less than 4). See (Otneim and Tjøstheim 2016a) for more discussions.

While the main purpose of the LGDE method is for the high-dimensional data sets, we demonstrate the use of the method in two dimensions. The results are shown in Fig. 11a for the synthetic data and Fig. 11b, c for the real data, which is the

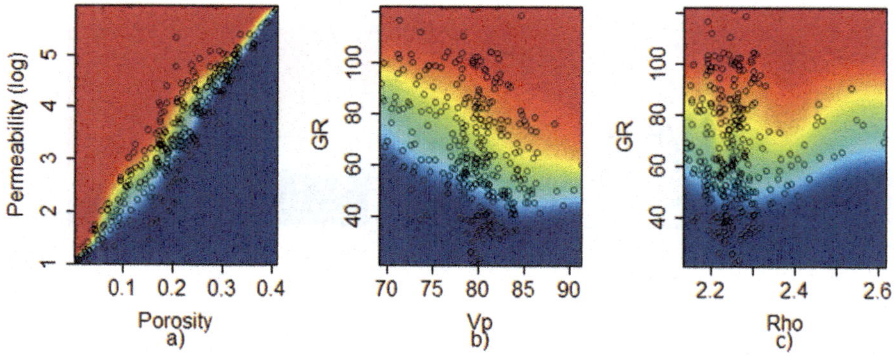

Fig. 11 Result of the LGDE method used on the same data as in Fig. 2 shown in (**a**). Plots (**b**), (**c**) show the estimated cdf on the real data case, GR conditioned on Vp (**b**) and GR conditioned on Rho (**c**)

same data previously used for CT in Figs. 2 and 7, respectively. The LGDE in Fig. 11a is seen as an improvement of the KS in Fig. 2 as it is able to achieve some pinch-out effects. Figure 12 shows the LGDE method applied on the real data set for three dimensions; this is similar to the example in Fig. 9 for CT. We compare CT and LGDE on this data set in Sect. 6; see Fig. 13 and Table 1. While the comparisons show some small favor to CT, we see that LGDE also works well.

6 Discussions

Our experience using these methods is that both CT and LGDE work well for two- and three-dimensional problems, but that CT has the edge both in performance and stability. To compare the performances of the methods on the real data sets (Figs. 9 and 12), we backtransform using the estimated cdf and look at the residuals in the normal transformed domain. This means to first transform each data point to the uniform domain with the estimated cdf and then transform them to the standard normal domain by using the inverse cdf of the standard normal distribution. A perfect transformation, where all dependencies to the underlying parameters (which we condition on) are removed, will give normal distributed residuals with mean and standard deviation scattered around zero and one, respectively. The reason we base our test on transforming to standard normal is that this is of particular interest in geostatistics where methods like Kriging often uses a workflow of first transforming the data toward Gaussianity; this is called trans-Gaussian Kriging (Cressie 1993). See (Hermansen et al. 2016) for a more thorough explanation on this transformation test, where they used it on CT in two and three dimensions.

The sliding average results of the backtransformations can be seen in Fig. 13 where the mean should be around zero. The summed absolute deviations are listed in Table 1, where we have also included results for standard deviations and a model

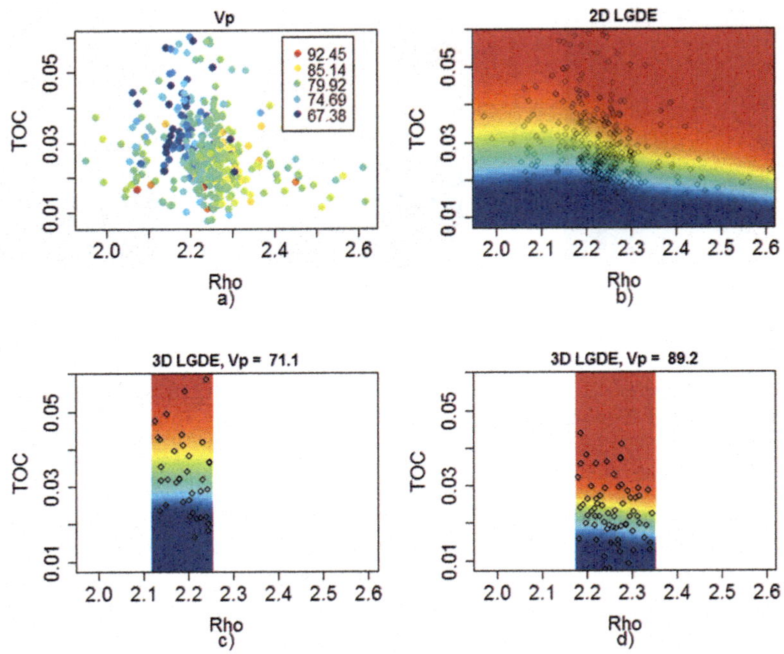

Fig. 12 Result of the 3D conditional cdf estimated with LGDE, here TOC is conditioned on Vp and Rho. Plot (**a**) shows the 3D data set and (**b**) is the 2D LGDE of TOC given Rho (Vp is ignored). The bottom row shows two slices of the 3D density, for Vp equal to 71.1 in (**c**) and Vp equal to 89.2 in (**d**). Only sections of the 3D cube where there is ample data support are shown, and only closest 20 % data to the chosen Rho value is plotted

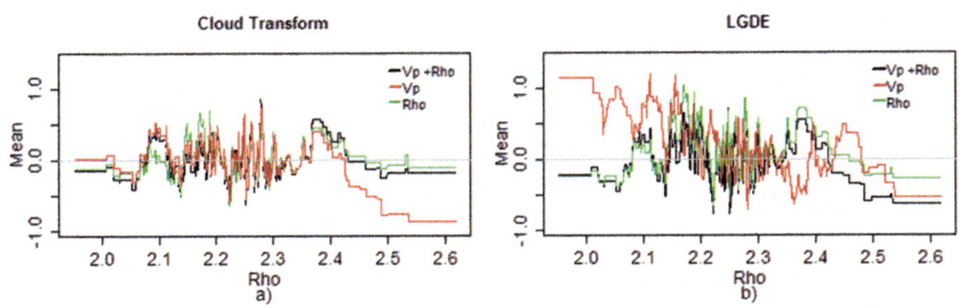

Fig. 13 Backtransformed data using the estimated cdfs (2D and 3D) for CT (**a**) and LGDE (**b**). For the 2D estimations, only Rho is conditioned on; Vp is ignored. The *red* and *green lines* are result of backtransforming the Vp and Rho data, respectively. The *black lines* are from the 3D model, where both Vp and Rho are used as conditioned on. The goal of the plot is to see if the 2D estimations have systematic bias in the direction of the variable not used in the model, and we see that the Vp variable trails off. This indicates that Rho should also be included when modeling the response TOC

with only using the Rho variable. From this table, and this criterion, we see that CT performs better than the LGDE (lower values) for the 3D models (Vp + Rho). We also see that in the 2D models that backtransforming the ignored variable gets high

Table 1 Summary of the backtransformed variables, by summing the absolute deviation in Fig. 13

		VP		Rho	
		Mean	SD	Mean	SD
CT	VP + Rho	209.1	158.8	204.4	235.0
	Vp	198.3	179.7	328.5	266.3
	Rho	387.4	238.0	169.2	208.0
LGDE	VP + Rho	253.3	220.5	321.4	233.0
	Vp	201.8	224.6	475.5	230.4
	Rho	392.2	245.6	267.0	237.3

For the standard deviations, we have subtracted 1 from the values before summing, so the smaller number indicates the best transformation

values; see line 2 and 5 in Table 1 and Fig. 13 where Vp was the only variable conditioned on (Rho is ignored). This indicates that 2D model does not capture all the underlying signals and that this variable should be included in the model.

We emphasize that one should not rely too much on the results in Table 1 for a comparison of the methods, as this is only for one data case and this criterion. To create test criteria for comparisons, one must consider what the estimated densities will be used for. Another evaluation of the methods could be to draw data from several known distributions, and then repeatedly measure the difference between the true and estimated distributions, like was done for LGDE by (Otneim and Tjøstheim 2016a).

6.1 *Adaptive Bandwidths*

The kernel density estimation in Eq. 1 was set up with a constant bandwidth matrix. One disadvantage with constant bandwidths is that the number of data inside the kernel varies, depending on the clustering of the data. An improvement to this is letting the bandwidth vary with the data, e.g., use a small bandwidth in areas with many observations, and increase the bandwidth in areas with few observations. This fits within the kernel density estimator described in Sect. 3, but with replacing the bandwidth $\boldsymbol{H}$ with $\boldsymbol{H}(\boldsymbol{x})$, one that varies along x in each direction. See (Sain 1994) for a comparison of adaptive kernel estimation methods.

An example of estimating a density in one dimension with a varying bandwidth is shown in Fig. 14. Both are estimated with a Gaussian Kernel, but the pdf in (a) is estimated with a constant h and the pdf in (b) with an adaptive h. From this figure it is clearly seen that the pdf in (b) is better suited to the underlying data.

As all three methods discussed in this paper use kernel smoothing at one point, this extension could apply to all methods. However, CT and LGDE utilize transformations of the data before applying the KS, which typically helps in regions with sparse data.

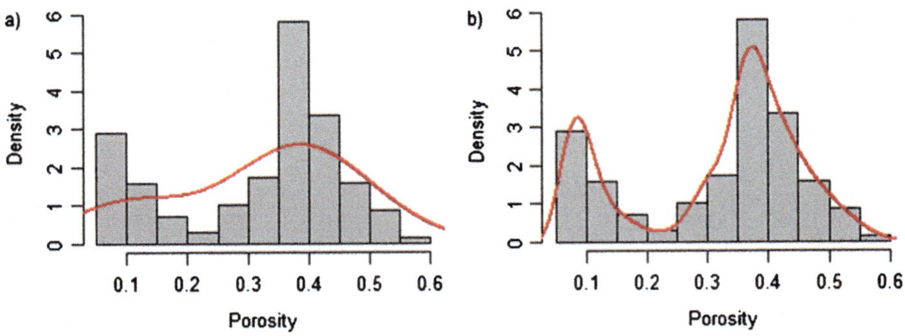

Fig. 14 Porosity data with pdf (*red line*) estimated with constant bandwidth (**a**) and with an adaptive bandwidth (**b**)

7 Conclusion

We have investigated two alternatives to standard kernel density estimators, Cloud Transform (CT) with directional smoothing, and Local Gaussian Density Estimator (LGDE). While the standard kernel smoother potentially has some problems on its own, as seen in Fig. 2, the two proposed methods utilize certain improvements, e.g., both transform the data before estimating the densities.

We have seen that both CT and LGDE work well for lower-dimensional problems. However, a major advantage for LGDE is that it easily scales to high-dimensional problems without introducing the increasing complexities between the variables that occur for CT. As discussed in Sect. 4, generalizing CT to higher dimensions adds in new layers of normalization and transformation that depends on the other normalized and transformed variables. This will increase the complexity of the method for each dimension, in contrast to LGDE which is already laid out for n dimensions.

This complexity issue of the normalization and transformation step of CT is mainly implementation wise, but the kernel method in CT may eventually break down to the curse of dimensionality. LGDE uses a simplification to overcome this curse, based on reducing the high-dimensional estimation to a set of bivariate problems. An improvement to CT could be to replace the standard kernel method with the local Gaussian approximation.

The LGDE method is a semiparametric density estimator, part nonparametric and part parametric, with Gaussian distribution chosen as the parametric family in Eq. 3. This will make the LGDE estimate unreliable if the data are far from Gaussian, as the Gaussian assumption will be broken. As previously mentioned the LGDE did struggle in comparison to KS when it was tested on a bimodal distribution for lower dimensions; see (Otneim and Tjøstheim 2016a). Other parametric families may be also be considered. In addition, in the tail of the distribution, where the data are sparse, LGDE will fit a Gaussian tail, which will be erroneous if the general structure of the data is far from Gaussian.

Acknowledgment We thank Arne Skorstad at Emerson Process Management – Roxar AS for providing the real data case. We also thank Dag Tjøstheim and Håkon Otneim for an early draft of the LGDE papers and for providing R code.

Bibliography

Cressie N (1993) Statistics for spatial sata. Wiley series in probability and mathematical statistics

Hermansen G, Kolbjørnsen O, Abrahamsen P (2016) Extensions and applications of the cloud transform with directional smoothing. Norwegian Computing Center

Kolbjørnsen O, Abrahamsen P (2005) Theory of the cloud transform for applications. Springer

Leuangthong O, Deutsch CV (2003) Stepwise conditional transformation for simulation of multiple variables. Int Assoc Math Geol 35:155–173

Otneim H, Tjøstheim D (2016a) The local Gaussian density estimatorfor multivariate data. Stat Comput

Otneim H, Tjøstheim D (2016b) Non-parametric estimation of conditional densities. A new method. Chapter 3.3 in the PhD thesis of Håkon Otneim, Department of Mathematics, University of Bergen, Norway

Sain SA (1994) Adaptive kernel density estimation. PhD thesis

Sammut C, Webb GI (2010) Encyclopedia of machine learning. In: Encyclopedia of machine learning. Springer US, pp 257–258

Silverman B (1998) Density estimation for statistics and data analysis. Chapman & Hall/CRC, London

Wand M, Jones MC (1993) Comparison of smoothing parameterizations in bivariate kernel density estimation. Am Stat Assoc 88:520–528

Geostatistical Methods for Unconventional Reservoir Uncertainty Assessments

Michael J. Pyrcz, Peter Janele, Doug Weaver, and Sebastien Strebelle

Abstract New methods are required to support unconventional reservoir uncertainty modeling. Unconventional plays add additional complexity with greater uncertainty in direct reservoir measures (e.g., unreliable permeability measures in low-permeability rock) and weakened relationships between currently measurable reservoir properties and production results (production mechanisms may not be well understood). As a result, unconventional plays are often referred to as "statistical plays," suggesting the reliance on statistical characterization of production distributions as a function of well counts. The application of the techniques described herein can be utilized to integrate all available information to determine appropriate levels of drilling activity to reduce uncertainty to an acceptable level.

Geostatistical approaches provide opportunities to improve the rigor in the dealing with statistical plays. Rigor is introduced through integration of methods that account for representative statistics, spatial continuity, volume-variance relations, and parameter uncertainty.

Analog production data from US shale gas plays are utilized for demonstration. These datasets, after debiasing, are sources for analog production rate distributions and spatial continuity. Given these statistics along with a decision of stationarity, geostatistical workflows provide repeatable uncertainty models that may be summarized over a spectrum of model parameters, drilling strategy, and well counts.

These geostatistical methods do not replace the need for expert judgment, but they improve the rigor of statistical-based approaches that are essential in statistical plays.

M.J. Pyrcz (✉) • S. Strebelle
Strategic Research Unit, Chevron Energy Technology Company, 1500 Louisiana St, Houston, USA
e-mail: mpyrcz@chevron.com; stsb@chevron.com

P. Janele
Exploration Review Team, Chevron, 1500 Louisiana St, Houston, USA
e-mail: ptja@chevron.com

D. Weaver
Exploration and New Ventures, Chevron, 1500 Louisiana St, Houston, USA
e-mail: weavdr@chevron.com

J.J. Gómez-Hernández et al. (eds.), *Geostatistics Valencia 2016*, Quantitative Geology and Geostatistics 19, DOI 10.1007/978-3-319-46819-8_45

1 Introduction

Modeling of unconventional reservoirs uncertainty presents new challenges relative to conventional reservoirs. There may be limited sample coverage and highly uncertain measures of reservoir properties due to extremely low-permeability rock, and reservoir properties may be weakly related to production due to measurement imprecision and complicated production mechanisms. Olea et al. (2011) suggest that conventional reservoir modeling workflow assessments with pore volume modeling through conventional porosity measures and mapping are unfeasible for resource estimation. This has motivated (1) a direct modeling of well estimated ultimate recovery (EUR) (or initial production, IP, as a proxy) as a regionalized variable (Olea et al. 2011) and (2) a strong reliance on statistical methods such as bootstrap to evaluate uncertainty (SPEE 2010).

This precludes common geostatistical reservoir modeling workflows that sequentially model regions, facies, porosity, permeability, and saturations for volumetrics and production forecasts (Pyrcz and Deutsch 2014). Nevertheless, there are opportunities to employ geostatistical theory to address technical limitations and to improve reservoir assessments and uncertainty modeling in unconventional reservoirs. We first discuss the statistical play concept and currently applied bootstrap methods for well aggregation-based uncertainty and then introduce new workflows based on geostatistical concepts. We acknowledge the geostatistical simulation of "production density" (cell-based EUR) workflow that has been discussed by Olea et al. (2011) as an alternative to the workflows presented. The demonstrations in the paper utilize custom code and GSLIB software (Deutsch and Journel 1998).

1.1 Statistical Play

Unconventional plays are often considered statistical plays. With a traditional play, the local data measures are reasonably accurate and in combination with local geologic factors allow for the mapping of reservoir properties that are suitable for forecasting with the application of flow simulation. In contrast, for a statistical play, well measures are highly uncertain, and the relationships between the measured values production is weak (see Fig. 1). In the face of this difficulty, Olea et al. (2011) suggests directly modeling well production measures such as EUR (although we utilize IP as a production proxy due to availability). This may first require a standardization to account for well parameters such as completed length. Well production is then treated as the regionalized variable of interest and summarized for reservoir uncertainty modeling. Note, while geologic factors naturally enter the traditional workflows through hierarchal depo- and lithofacies and porosity, permeability, and saturation distributions and relations, the challenge is to integrate geologic factors in this statistical play workflow. Careful sub-setting and selection

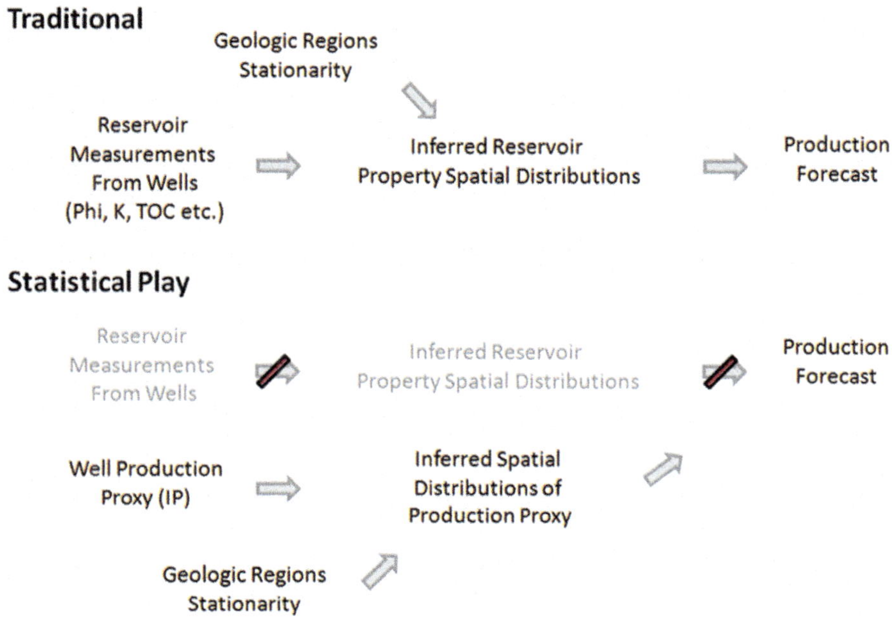

Fig. 1 Workflows for traditional and statistical play assessment. In a statistical play, the well-based petrophysical measures are highly uncertain and poorly related to production. This motivates direct modeling of production as a regionalized variable (Figure from Pyrcz and Deutsch 2012)

of analogs and matching and sub-setting modeling area of interest with consideration of geologic setting and stationarity are essential.

1.2 *Bootstrap-Based Workflow*

The statistical play concept has been formalized within the SPEE#3 Monograph that outlines methods to model uncertainty for undeveloped reserves based on the bootstrap method (SPEE 2010). The bootstrap is a statistical resampling technique that permits the quantification of uncertainty in statistics by resampling (with replacement) from the original data (Efron 1982). Importantly, this method does not account for neither spatial context such as location of samples nor spatial continuity.

The SPEE#3 Monograph approach proceeds by identifying analogous wells, modeling a distribution of EUR, determining the number of wells in the well plan, and repeated bootstrap sampling to evaluate the uncertainty in production results over a well aggregate measure (e.g., the average or sum of production). This approach benefits from simplicity and ease of use.

2 Methodology

There are opportunities to improve unconventional uncertainty assessments with the addition of concepts widely known within geostatistical reservoir practice. The fundamental geostatistical concepts of representative statistics, spatial continuity, and volume-variance relations are demonstrated with examples based on domestic US shale plays. This is followed by demonstration of additional workflows that utilize these geostatistical concepts for well aggregate and unconventional block uncertainty.

2.1 Representative Statistics

Representative statistics are a prerequisite for uncertainty modeling. Any uncorrected sampling bias will bias the uncertainty model. In the proposed workflows, the production distribution should be representative of the area of interest. This step requires the pooling of all available information and correction of the data for sampling representativity (spatial clustering and biased sampling). These corrections may include declustering to correct for sample clustering and soft data debiasing to correct for un- or under-sampled subsets of the property distribution (Deutsch and Journel 1998; Pyrcz et al. 2006).

Cell-based declustering is demonstrated with three US domestic shale plays (Haynesville, Barnett, and Fayetteville) based on IP data (see Fig. 2). The wells were filtered to pool similar vintage (recently drilled) and types of wells (horizontal lengths and stages). Even with large domestic US IP datasets (100's of wells), data representativity is a major concern. This example indicates 4–8 % reduction in play mean IP due to accounting for sample clustering. While declustering does generally improve results, it is not guaranteed to do so in all circumstances, yet these results provide evidence for preferential sampling in these shale gas plays that must be treated when formulating representative statistics for uncertainty modeling.

2.2 Spatial Continuity

A semivariogram provides a mathematically consistent model of spatial continuity for a variable of interest. This model indicates the expected dissimilarity (and under the assumption of stationarity, the correlation) for all offset distances and directions. An experimental semivariogram is calculated directly from the data as half the expectation of the square difference of data offset by a lag vector, **h**. Expectation is calculated by pooling all data pairs with the same **h** offsets (within a decision of lag angle and distance tolerance).

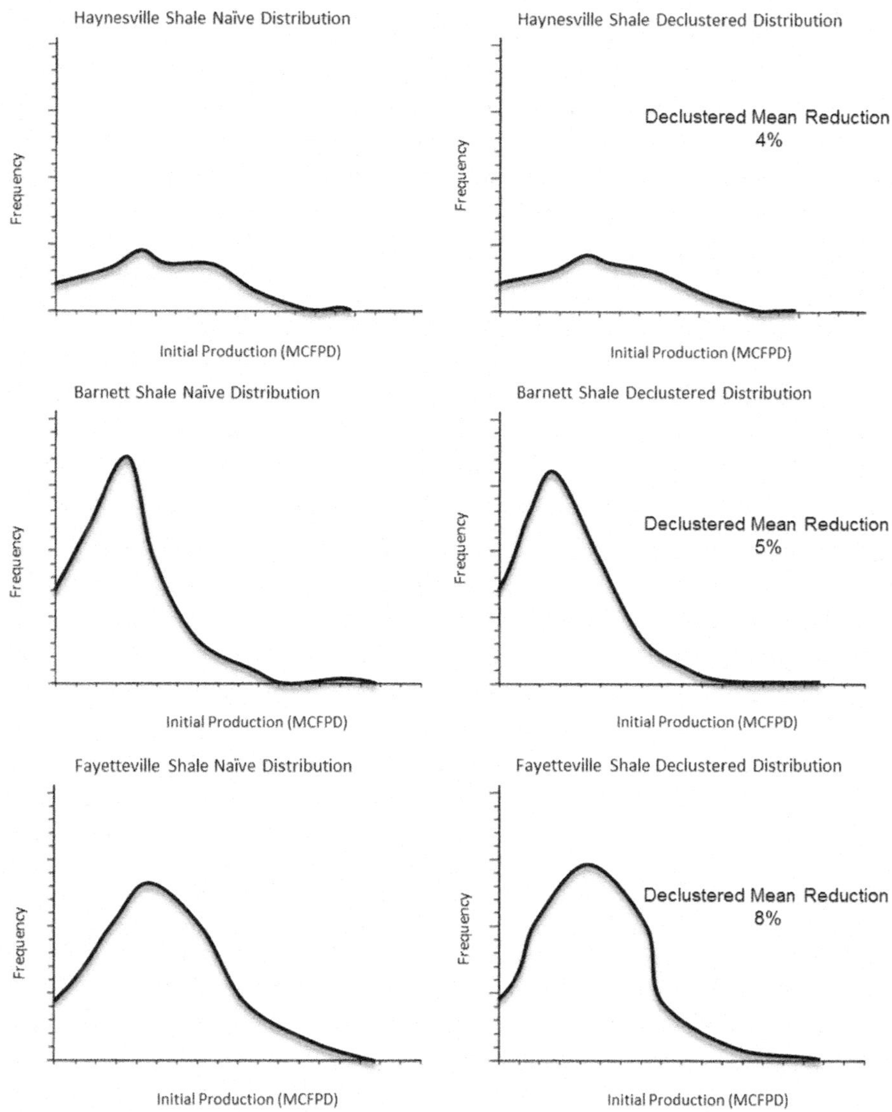

Fig. 2 Naïve and cell-based declustered IP distributions for three US domestic shale plays. Display includes distribution shape change and change in mean

Modeling and interpretation of production rate semivariograms offer insights into the spatial behavior of production rates and the ability to predict away from wells. For a single well, this is easy to visualize and calculate. At the well location, there is no uncertainty and perfect correlation. At some small distance, below the minimum sample spacing, the correlation decreases by the relative nugget effect. In general correlation decreases with lag distance and at the range, no correlation exists any longer; the well provides no information. Beyond the range there may be some small negative correlation indicative of spatial trends.

The semivariogram is a fundamental input for geostatistical simulation and the workflows presented later, but there are benefits in directly utilizing insights from the semivariograms in unconventional reservoirs. A simple application of the semivariogram is to determine the maximum extent of information from a well in map view based on the semivariogram range. In general, wells that have a separation of twice the range are not redundant and carry the maximum information content. Also, well location maps with range circles (or ellipses in the case of anisotropy) provide visualization of data coverage and may be applied to assist in well planning. The circle size may be reduced to reflect an acceptable level of spatial uncertainty.

Experimental isotropic semivariograms are calculated for the normal score transform of well IP (positioned at the well collars) with the filtered IP data for three US domestic shale plays (see Fig. 3). Each of experimental semivariogram indicates (1) 30–40 % relative nugget effect, (2) long correlation ranges, and (3) indication of long-range trends. These shale gas plays indicate a high degree of variability between adjacent wells, but also some degree of correlation or information over long distances. Large relative nugget effects are not expected to be due to allocation issues, but may represent strong initial production sensitivity to artificial fracture sets.

2.3 Volume-Variance Relations

Dispersion variance is a generalized form of variance that accounts for the scales of the samples, scale of the area of interest, and spatial continuity. In geostatistics, given the spatial and scale context, all variances are dispersion variances. Volume-variance relations refer to the general relationship between the variance and scale (Pyrcz and Deutsch 2014).

The dispersion variance may be calculated from volume integration of the semivariogram values (known as gamma bar), $\overline{\gamma}$:

$$D^2(\text{block}, \text{play}) = \overline{\gamma}(\text{play}, \text{play}) - \overline{\gamma}(\text{block}, \text{block}) \quad (1)$$

Since semivariograms tend to monotonically increase, as the scale of the estimate (the lease block in the example above) increases, the dispersion variance is reduced; therefore, the estimation uncertainty decreases under the assumption of stationarity.

For a variety of unconventional problems, we are concerned with how dispersion variance scales as we move from well scale to exploration block scale. Krige's relation indicates the additivity of variance or in other words the partitioning of variance over scales:

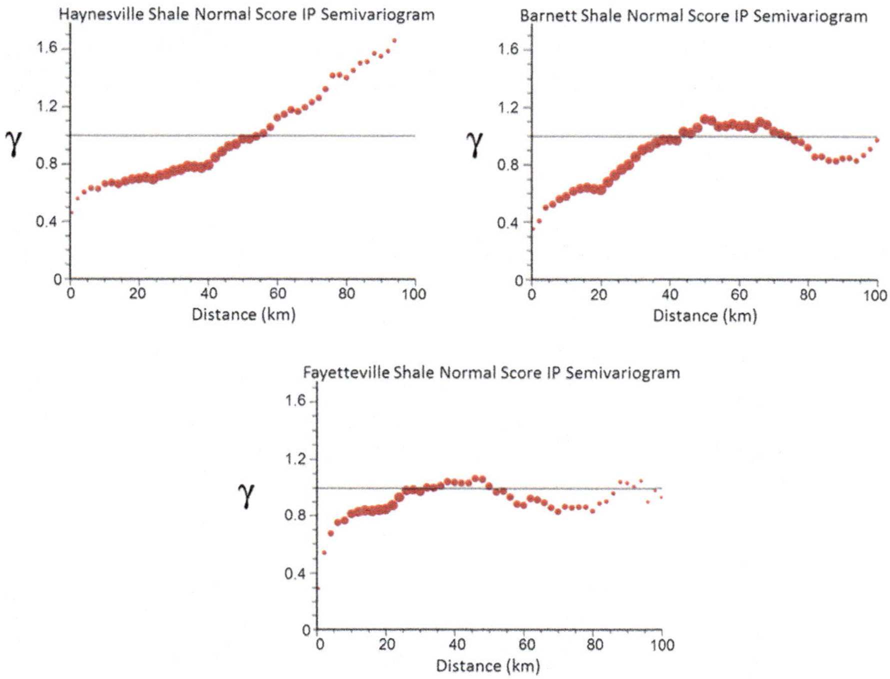

Fig. 3 Semivariograms of normal score transformed IP data for three US domestic shale plays. The sizes of the experimental points are scaled by number of pairs available

$$D^2(\text{well}, \text{play}) = D^2(\text{well}, \text{block}) + D^2(\text{block}, \text{play}) \tag{2}$$

It follows that selecting a larger lease block will reduce dispersion variance between lease blocks within the play, D^2 (block, play), but will retain a larger dispersion variance within the lease block, D^2 (well, block). While estimates of larger lease block scale are easier, individual well results within the larger lease block will have more variability.

Volume-variance relations based on gamma bar values are formulated under the assumption that for the variable of interest, the variogram is stationary over the domain, averaging is performed with nonoverlapping volumes, and variable averages linearly (Journel and Huijbregts 1978; Frykman and Deutsch 2002). In the proposed workflow based on global kriging, well EUR (IP when EUR is not available) is the variable of interest. This is reasonable if it is assumed that:

1. The EUR variogram is stationary over the block, $\mathbf{V}$.
2. Well EUR is represented as the total volume of production at the scale of the effective drainage of a nominal well, $\text{EUR}_v(\boldsymbol{u})$.
3. $\text{EUR}_v(\boldsymbol{u}_\alpha)$, $\forall \boldsymbol{u}_\alpha \in \mathbf{V}$ is an exhaustive set nonoverlapping data support production volumes of the block, $\mathbf{V}$.
4. There is no interaction in EUR between adjacent $\text{EUR}_v(\boldsymbol{u})$.

Under these simplifying assumptions, block production volume, EUR_V, is additive and is calculated as the summation of the well support subsets' production volumes, $\mathrm{EUR}_V = \sum_{\boldsymbol{u}_\alpha \in V} \mathrm{EUR}_v(\boldsymbol{u}_\alpha)$. By extension, IP is treated as a subset of the EUR volume.

3 New Workflows and Results

Given the geostatistical reservoir modeling concepts of representative statistics, spatial continuity, and volume-variance relations, workflows are proposed to support unconventional uncertainty modeling. The first is a direct method to assess lease block uncertainty with global kriging. The second is a method to assess well aggregate uncertainty accounting for spatial concepts.

3.1 Block Uncertainty Modeling with Geostatistical Methods

Kriging is general and may directly integrate scale information. For example, kriging may be applied to calculate models of data scale uncertainty represented by a mesh of the kriging variances over the lease block. Also, kriging may be applied with volume-integrated semivariogram models to directly estimate and assess uncertainty of a production property over other scales, such as lease block scale, through the kriging estimate and variance, respectively. Deutsch and Deutsch (2010) propose the use of global kriging to directly estimate block uncertainty (and declustered block mean).

The kriging variance equation derived in standard geostatistics textbooks is composed of global dispersion variance, closeness, and redundancy terms (Pyrcz and Deutsch 2014). (1) The global dispersion variance (the variance of the property of interest at the correct scale) is the maximum uncertainty possible at the specified volume support. This accounts for the size of the block being estimated. (2) Closeness accounts for the correlation of the data to the volume being estimated. This component accounts for the respective coverage of the data within this volume being estimated. For example, if for a specific dataset, the block expands further away from the data, overall closeness decreases and the kriging variance increases. (3) Redundancy accounts for the correlation between the data. For example, if the spatial continuity range relative to well spacing is large, then the well data redundancy is large and the kriging variance increases.

Closeness and redundancy components interact resulting in interesting behavior in block uncertainty. A simple example is provided with representative histogram (Haynesville declustered example) and a stationary 36×36 km lease block. The well data count was varied from 0 to 100 wells (assuming uniform well spacing) and the spherical variogram range was varied from 0 to 50 km. The resulting

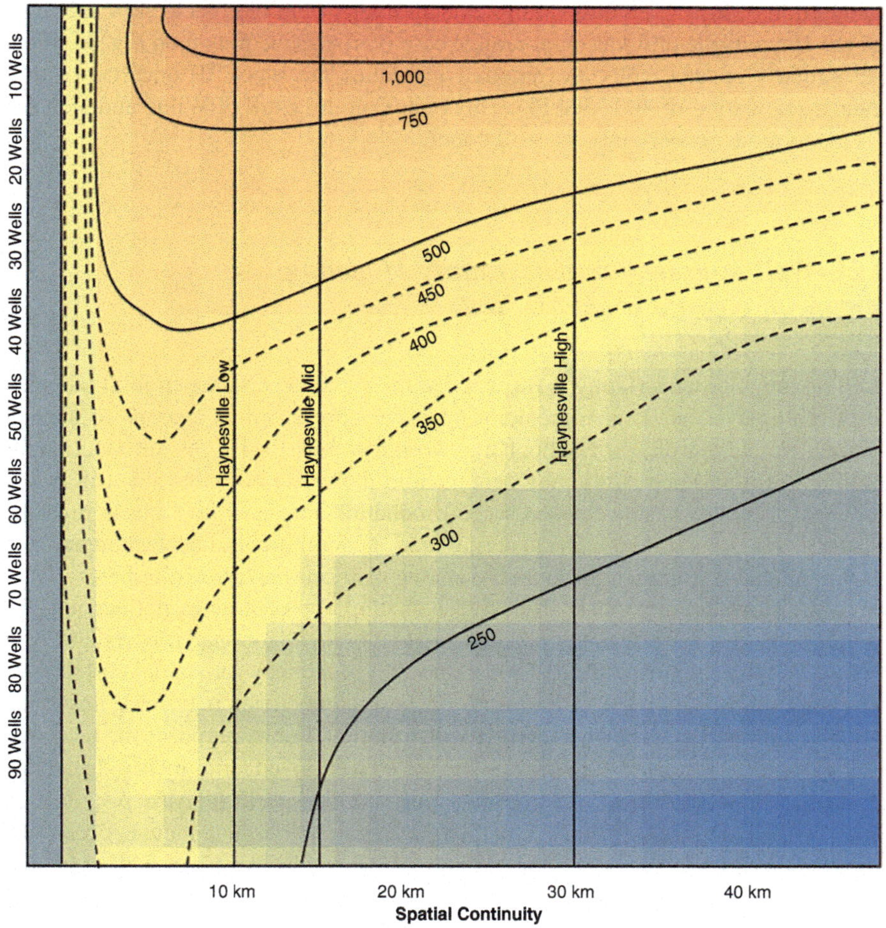

Fig. 4 Global kriging uncertainty vs. number of wells and spatial continuity. Response surface is standard deviation of IP (MCFPD)

response surface is displayed as block scale average IP uncertainty represented by standard deviation (MCFPD) (see Fig. 4).

The block uncertainty response surface exhibits interesting features. (1) Reduction in spatial continuity (i.e., variogram range) decreases the maximum uncertainty possible, because short-range features average out quickly as we scale up resulting in a lower dispersion variance. (2) Increase in spatial continuity increases the redundancy between the well data and, therefore, increases block uncertainty. (3) Increase in spatial continuity increases the closeness between well data and the entire block and, therefore, decreases block uncertainty. (4) Increase in number of wells decreases uncertainty if the wells are spatially correlated with the block.

This response surface may be applied directly to assess the number of wells required to reduce lease block uncertainty to an acceptable level (with the assumption of histogram and variogram stationarity over the block). For example, for the Haynesville mid-variogram range case (vertical line drawn on Fig. 4), about 50 regularly spaced wells are required to reduce the block IP uncertainty to a standard deviation of 400 MCFPD. This builds on the work of Wilde and Deutsch (2013) to model uncertainty vs. well count.

3.2 Well Aggregation Uncertainty Modeling with Geostatistical Methods

Well aggregation uncertainty is the calculation of the uncertainty in joint outcome from multiple wells. This aggregation may be represented as average or sum of production over a well set. The bootstrap workflow described in the introduction for well aggregation uncertainty may benefit from integration of geostatistical spatial concepts. As shown above, even large production rate analog datasets must be checked for spatial bias. Declustering and general debiasing techniques are recommended to ensure representative analog distributions are applied.

In addition, the previously discussed bootstrap approach does not account for spatial context of production as related to (1) closeness (or spatial correlation) of new wells with previously drilled wells and (2) redundancy (or spatial correlation) between new wells (SPEE 2010). Both of these components may have important impact on the well aggregate uncertainty distribution. For example, as the proposed wells are more correlated with previously drilled well(s), uncertainty is reduced due to the spatial constraints from the previously drilled well(s). Also pad drilling clusters the new wells resulting in a high degree of redundancy between the proposed wells that increases uncertainty, while widely scattered wells (drilling for information) result in low to no redundancy that decrease the well aggregate uncertainty.

Spatial bootstrap is a variant of bootstrap that accounts for the spatial context of the samples (Journel 1994). With bootstrap samples randomly selected from the reference distribution, but with spatial bootstrap, the sample locations are retained and the samples are simulated at their respective locations accounting for spatial continuity models that impose correlation between samples and with available primary and secondary data. A simple method for calculating a single spatial bootstrap sample set is to calculate a complete simulated realization over the area of interest conditional to all available data and then to extract the local simulated values at the proposed well locations and then to summarize these local values for a well aggregate realization. This process is repeated to calculate the required number of joint well realizations to build the aggregate distribution to model uncertainty of the well aggregate.

The use of spatial bootstrap for well aggregation is demonstrated with a synthetic 24×35 km lease block. In Fig. 5 the available 20 well data locations are superimposed on a map of local IP uncertainty (standard deviation of local

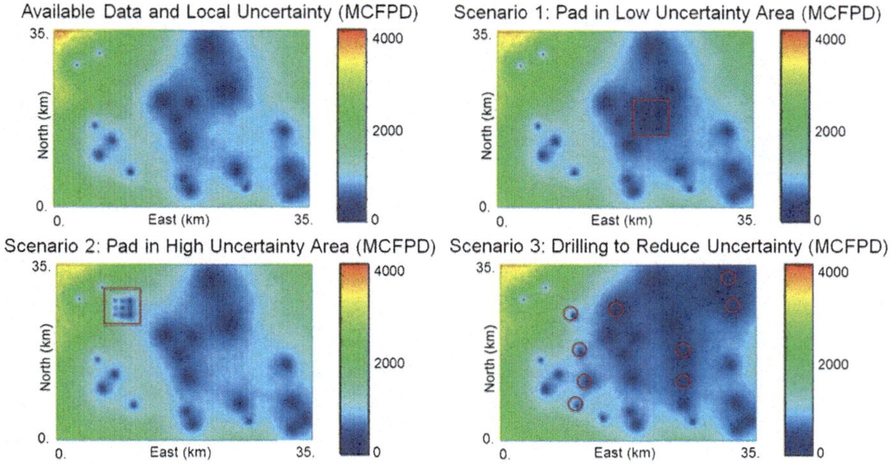

Fig. 5 Original well data and local uncertainty in IP and three scenarios for proposed wells, including 3 × 3 pads in low and high uncertainty locations and drilling for information. The proposed well locations are outlined in *red*

production data uncertainty in MCFPD from the kriging variance). Three proposed nine well scenarios are shown, including 3 × 3 pads in low and high uncertainty locations and drilling for information with their updated local uncertainty maps. Global IP distribution and spatial continuity model is based on a declustered IP dataset from a modeled analog dataset.

The resulting well aggregate IP distributions are shown in Fig. 6. The spatial context significantly impacts the resulting well aggregate uncertainty model. The mean is influenced by previously drilled well IP within the range of spatial continuity. The uncertainty for the well aggregate result is significantly higher for the pad in the second scenario due to the limited correlation with previously drilled wells along with highly redundant proposed wells, and the third scenario has the lowest uncertainty due to limited correlation between proposed wells resulting in averaging out of variability in the aggregate summary.

4 Discussion

This work has demonstrated the application of geostatistical methods to support block and well aggregate uncertainty modeling for unconventional plays. Concepts such as representative statistics, spatial continuity, and volume-variance relations are essential for accuracy of uncertainty models. Without consideration for representative statistics, uncertainty models are likely biased by preferential well sampling. Without spatial continuity and location context, the well data are assumed to be independent. In the presence of well redundancy and/or poor data coverage of

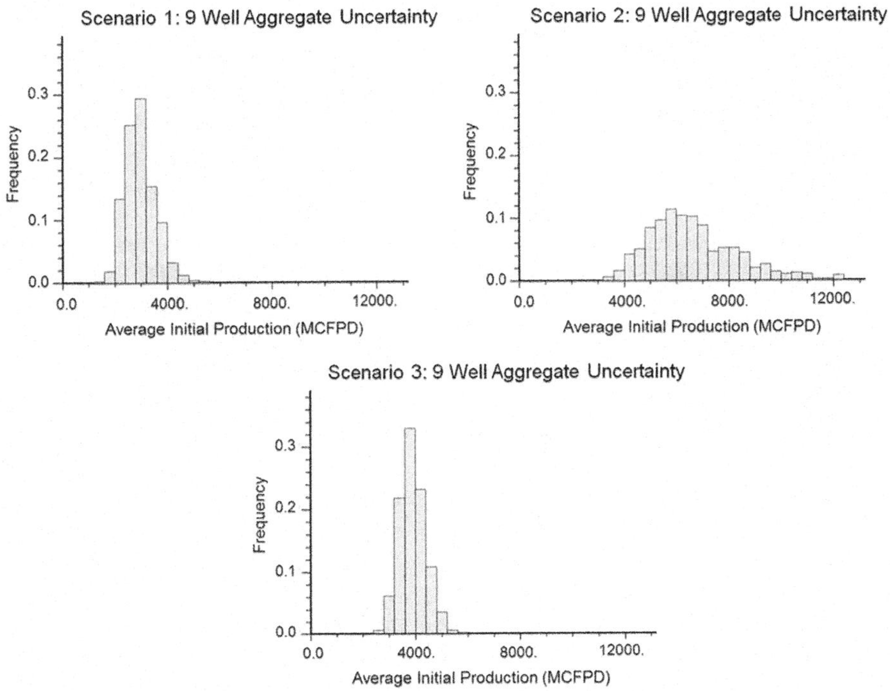

Fig. 6 Well aggregate uncertainty distributions (average IP) for three scenarios of nine proposed wells based on representative statistics, spatial continuity, and spatial bootstrap

the block, this may result in underestimation of uncertainty. Without the concept of volume-variance relations, the scale of the block is assumed to be the scale of the data, resulting in overestimation of block uncertainty.

5 Conclusions

Unconventional plays may be treated as statistical plays. Yet, in statistical plays, there are opportunities to improve rigor with integration of geostatistical concepts such as data representativity, spatial continuity, and volume-variance relations. The simple demonstrations in this paper have shown (1) the need for declustering with dense shale play IP datasets and (2) the spatial continuity features of shale play IP and (3) response surface for block uncertainty to support well count choices and a (4) workflow to assess well aggregate uncertainty with the integration of spatial concepts.

Acknowledgments The authors are appreciative to Chevron Energy Technology Company for support of this work and for allowing this publication. Also, the constructive reviews of two anonymous reviewers are appreciated.

Bibliography

Deutsch JL, Deutsch CV (2010) Some geostatistical software implementation details. CCG Paper 2010-412, Centre for Computational Geostatistics, University of Alberta, Edmonton

Deutsch CV, Journel AG (1998) GSLIB: geostatistical software library and user's guide, 2nd edn. Oxford University Press, New York

Efron B (1982) The jackknife, the bootstrap, and other resampling plans. Society for Industrial and Applied Math, Philadelphia

Frykman P, Deutsch CV (2002) Practical application of geostatistical scaling laws for data integration. Petrophysics 43(3):153–171

Journel AG (1994) Resampling from stochastic simulations. Environ Ecol Stat 1:63–84

Journel AG, Huijbregts C (1978) Mining geostatistics. Academic, New York City, 600 p

Olea RA, Houseknecht DW, Garrity CP, Cook TA (2011) Formulation of a correlated variables methodology for assessment of continuous gas resources with an application to the Woodford play, Arkoma Basin, eastern Oklahoma. Boletín Geológico y Minero, 122 (4), pp 483–496. ISSN: 0366-0176

Pyrcz MJ, Deutsch CV (2012) Uncertainty models for exploration and appraisal of shale gas. CCG Paper 2012-131, Centre for Computational Geo-statistics, University of Alberta, Edmonton

Pyrcz MJ, Deutsch CV (2014) Geostatistical reservoir modeling, 2nd edn. Oxford University Press, New York, p 448

Pyrcz MJ, Gringarten E, Frykman P, Deutsch CV (2006) Representative input parameters for geostatistical simulation. In: Coburn TC, Yarus RJ, Chambers RL (eds) Stochastic modeling and geostatistics: principles, methods and case studies, vol II and 5. AAPG Computer Applications in Geology, pp 123–137

Society of Petroleum Evaluation Engineers (2010) Guidelines for the practical evaluation of undeveloped reserves in resource plays. Monograph 3

Wilde BJ, Deutsch CV (2013) A methodology for quantifying uncertainty versus data spacing applied to the oil sands. CIM J

Productivity Prediction Using Alternating Conditional Expectations

Emmanuel T. Schnetzler

Abstract We present an approach to predict spatial distribution of a variable from a set of geophysical and interpreted grids using Alternating Conditional Expectations (ACE). This technique is based on nonparametric transformations of the predictor and response variables in order to maximize the linear correlation of the transformed predictors with the transformed response. ACE provides a powerful method to detect underlying relationships between the variables and use them in a regression framework to predict the response variable. A case study is presented illustrating the approach using a set of grids derived from geophysical attributes (gravity, magnetic, electromagnetic) and interpreted grids (isopach, total organic carbon, etc.) as predictor variables to estimate early hydrocarbon production.

1 Introduction

Data-driven approaches are increasingly being applied to geological settings. This is particularly the case in a multivariate setting when multiple predictor variables have complex and often nonlinear relationships with the response variables. Machine learning algorithms can be difficult to interpret. ACE provides more insight into the relationships extracted from the data.

2 Multiple Linear Regression

In the classical multiple linear regression approach, the response variable y is modeled as a linear combination of the predictor variables $x_1, x_2, \ldots, x_p$:

E.T. Schnetzler (✉)
NEOS, 6210 Stoneridge Mall Rd, Suite 450, Pleasanton, CA 94588, USA
e-mail: eschnetzler@neosgeo.com

J.J. Gómez-Hernández et al. (eds.), *Geostatistics Valencia 2016*, Quantitative Geology and Geostatistics 19, DOI 10.1007/978-3-319-46819-8_46

$$y = \beta_0 + \beta_1 x_1 + \beta_2 x_2 + \cdots + \beta_p x_p \tag{1}$$

This imposes a strong assumption of linearity that is not appropriate when complex relationships exist.

In some cases, nonlinear relationships are present and known, in which case parametric transformations of the variables can be applied to build a model. This is the case in the Box-Cox family of transformations for continuous variables (Box and Cox 1964). However, this also imposes a strong preconceived model on the data that is not valid in complex cases.

3 Alternating Conditional Expectations

3.1 Theory

A number of nonparametric regression techniques have been applied successfully when complex unknown relationships are present between predictor variables and a response variable.

In this family of techniques, the relationships between predictor and response variables are built from the data via optimal transformations. In some, only the predictor variables are transformed (Generalized Additive Models). In others, both predictor and response variables are transformed. This is the case of the Alternating Conditional Expectations technique (Breiman and Friedman 1985). The equation takes the form

$$\theta(y) = \varphi_1(x_1) + \varphi_2(x_2) + \cdots + \varphi_p(x_p) = \sum_{i=1}^{p} \varphi_i(x_i) \tag{2}$$

Each variable (predictor and response) is transformed with optimal transformations. The transformations are derived from the data using an iterative process aimed at maximizing the linear correlation between transformed predictor variables and transformed response variable. An iterative approach is applied that works in an alternating fashion to minimize Eq. 3 with respect to one function while keeping the other functions constant:

$$e^2 = E\left\{\left[\theta(y) - \sum_{i=1}^{p} \varphi_i(x_i)\right]^2\right\} \tag{3}$$

Some constraints can be put on the shape of the transformations, such as monotone or linear. The approach also allows the use of categorical variables as predictors.

ACE has been shown to be able to model complex nonlinear relationships as illustrated in *Estimating Optimal Transformations for Multiple Regression Using the ACE Algorithm* (Wang and Murphy 2004).

Ryan Barnett and Clayton Deutsch looked at an application to a geometallurgical example in a nickel laterite deposit where Ni was predicted from five predictor variables: Fe, SiO_2, MgO, Co, and Al_2O_3 (Barnett and Deutsch 2013). The application shows good results from the approach at finding underlying relationships although the spatial distribution was not presented.

3.2 Application to Spatial Data

Regression techniques can be applied in a spatial setting: down a borehole in 1D, over a 2D area, or in a full 3D setting. In a 2D case, one can consider predicting a response variable away from data locations where the variable is known (typically wells), using a set of exhaustive datasets covering the area of interest.

The general workflow illustrated on Fig. 1 follows the steps:

- Select the response variable to be predicted and gather a consistent dataset (training points).
- Select a set of relevant exhaustive predictor variables on a 2D grid (data layers) and extract values of the predictor variables at the response variable locations (wells).
- Build optimal transformations for response and predictor variables from values at the training locations.
- Apply the transformations to the full 2D grid to predict the response variable over the area.

4 Application to the Denver-Julesburg Basin

4.1 Overview

ACE is applied to a real case in an area of 3000 m^2 located in the western part of the Denver-Julesburg Basin in Colorado. Major oil-producing areas shown in Fig. 2 in the Niobrara Formation include the Hereford in the northwest, the East Pony in the northeast, and the Wattenberg Field in the center.

4.2 Input Data

The available data include production information and a set of geophysical and geological layers.

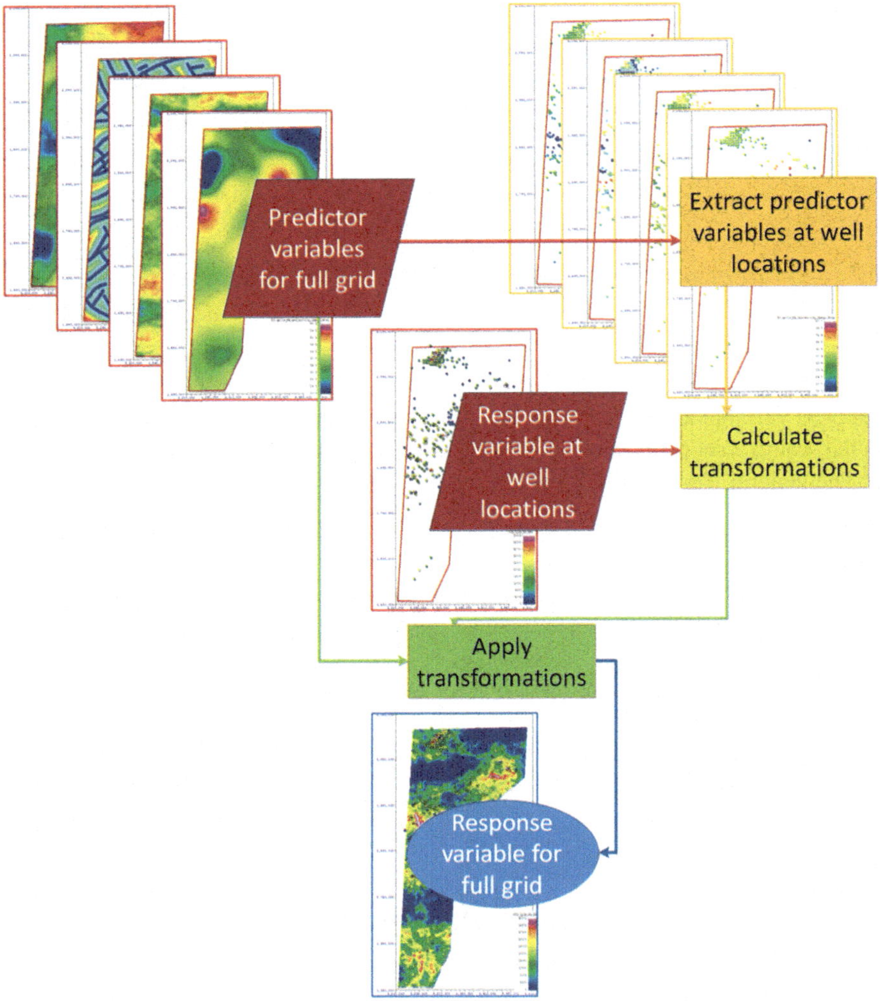

Fig. 1 Workflow for ACE application to spatial data

5 Response Variable

The response variable selected is the early cumulative oil production, calculated as the sum of the production of the best 3 months in the first year of production. Because historical production includes wells drilled over several decades using different technologies, it is necessary to subset the data to a consistent set of wells; in this case, only horizontal wells were used. This controls for part of the variability between wells, although some further filtering could control for additional variables, in particular engineering factors. Exploratory data analysis shows that in this case the subset of

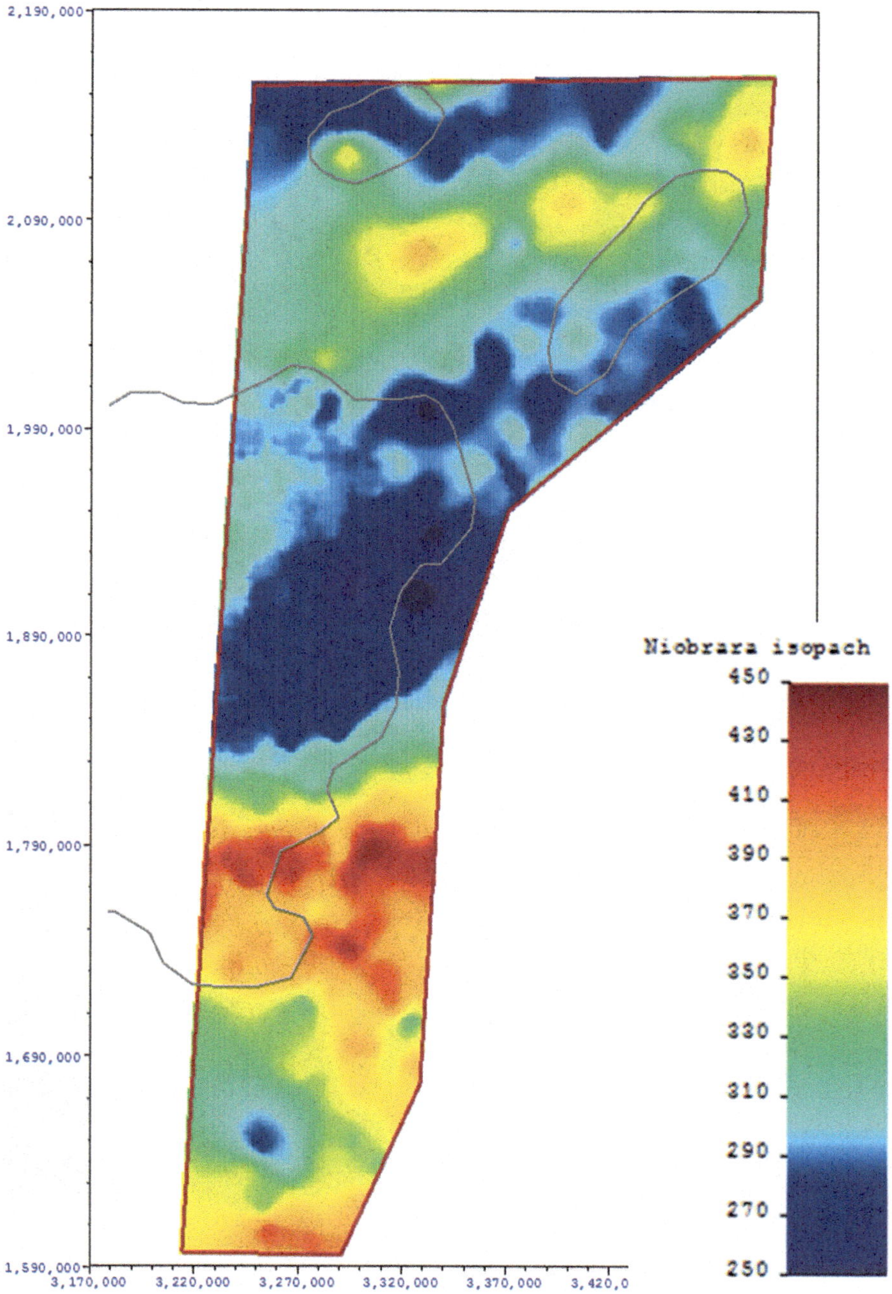

Fig. 2 Overview of the study area and Niobrara isopach

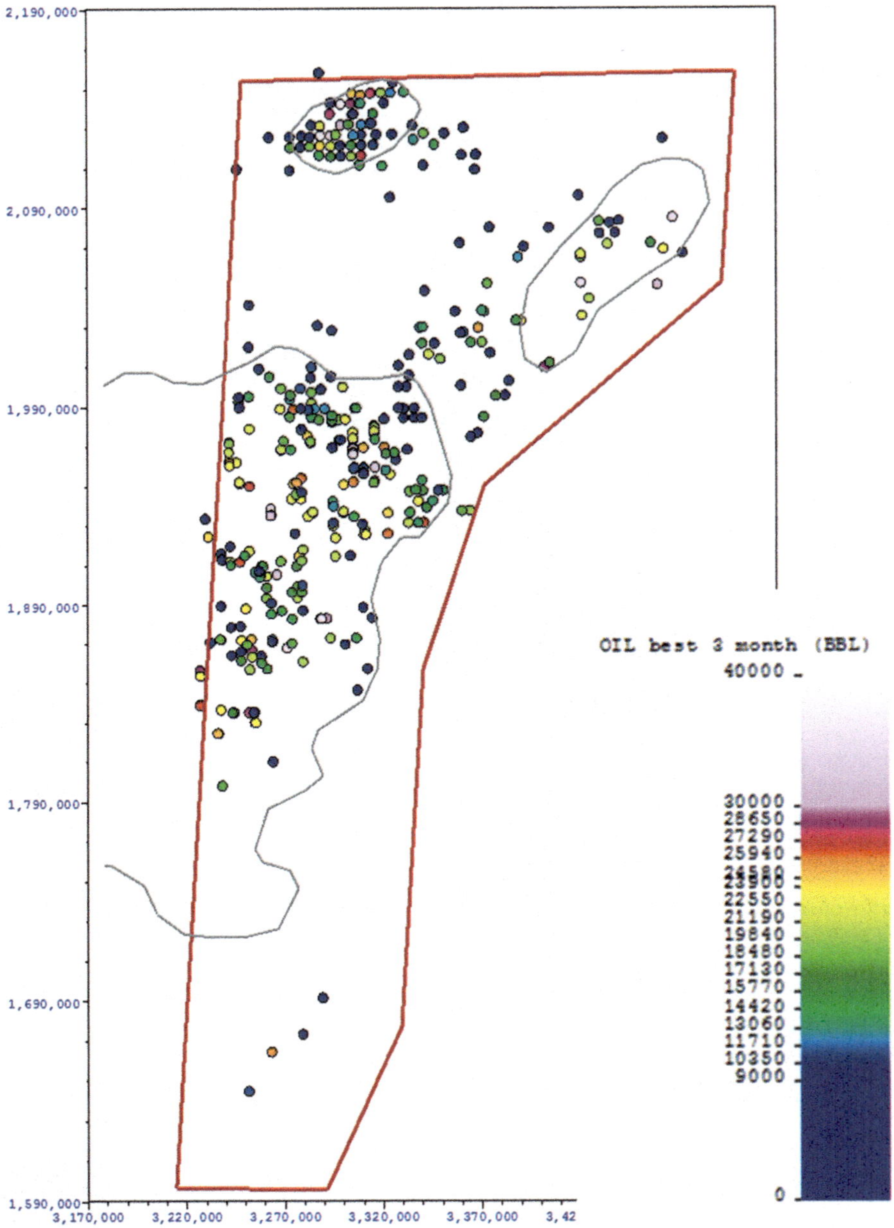

Fig. 3 Spatial distribution of early oil production (BBL)

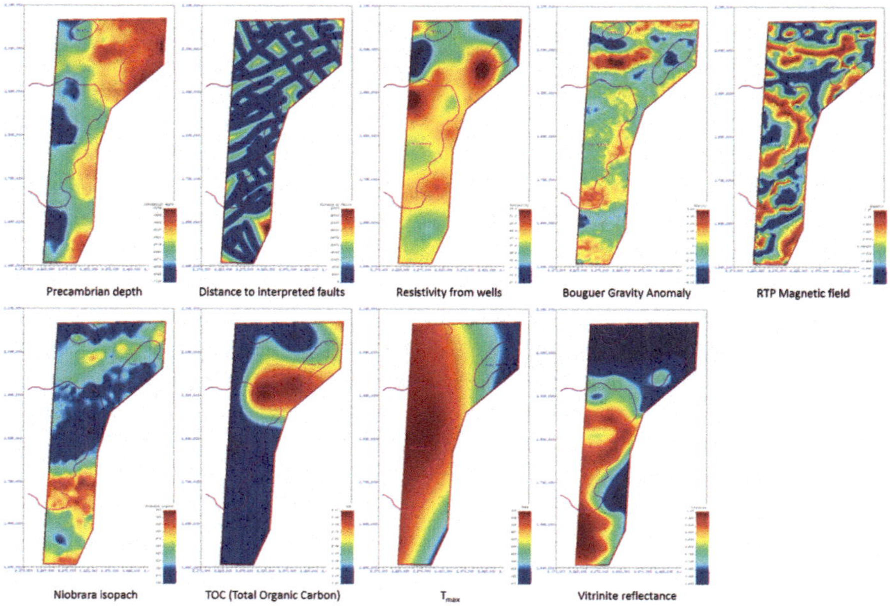

Fig. 4 Data layers input to the ACE run as listed in the text

horizontal wells is sufficiently homogeneous. Figure 3 illustrates the spatial distributions of horizontal wells with corresponding early production in BBL.

6 Predictor Variables

A set of nine predictor variables available for the full area of interest and deemed related to productivity is compiled. Figure 4 shows an overview of the data layers listed below.

Top row, left to right:

- Precambrian depth
- Distance to interpreted faults
- Resistivity from wells
- Bouguer gravity anomaly
- RTP (reduced to pole) magnetic field

Bottom row, left to right:

- Niobrara isopach
- TOC (total organic carbon)
- T_{max} (temperature at which the maximum rate of hydrocarbon generation occurs in a kerogen sample during pyrolysis analysis)
- Vitrinite reflectance

7 ACE Result

The result of the prediction (early oil prediction in BBL) from the Alternating Conditional Expectations is shown in Fig. 5. The three major oil-producing areas are highlighted in the prediction, in part driven by the data. Additional areas with limited data control are predicted as high-producing locations, in particular in the northeast, south of the East Pony Field, and in the south where only a few training data points are available.

Correlation between actual and predicted production is 0.38 for multiple linear regression compared to 0.74 for ACE (Fig. 6).

Predictor variables can be ranked according to their influence on the response variable, as shown in Table 1, with increasing influence from top to bottom.

The optimal transformation can be plotted and analyzed to understand the effect of each variable on the response variable. Figure 7 shows the optimal transformation of the response variable (early oil production, in BBL).

8 Conclusions

Many regression techniques can be used to predict a variable from a set of predictor variables. When relationships are complex and cannot be modeled parametrically either through linear model or more complex functions, a nonparametric approach is more appropriate.

Machine learning algorithms (Support Vector Machines, Neural Networks, etc.) are powerful but tend to be difficult to interpret. Nonlinear, nonparametric approaches based on optimal transformations of the variables such as Alternating Conditional Expectations and some modifications designed to address some limitations, AVAS, for example (Tibshirani 1988), provide the opportunity to inspect the optimal transformations and put them in a physical context for validation.

A number of potential pitfalls need to be kept in mind when applying any regression technique. Overfitting the training data is a common problem in predictive modeling that should be closely watched through cross-validation. The predictor variables should be chosen with care, and it should be possible to formulate a reason why each is likely to be related to the response variable, even if the link cannot be expressly defined. The method relies on colocated relationships and does not take into account relations to neighbors such as is the case in co-kriging or co-simulation.

The example presented shows how ACE can be applied in a spatial setting to predict a grid of early oil production. The case study illustrates the improvement in correlation between transformed predictors and transformed response. Other variables such as petrophysical properties (e.g., porosity) could be considered for this approach, and while the case study is 2D, it is applicable in 3D.

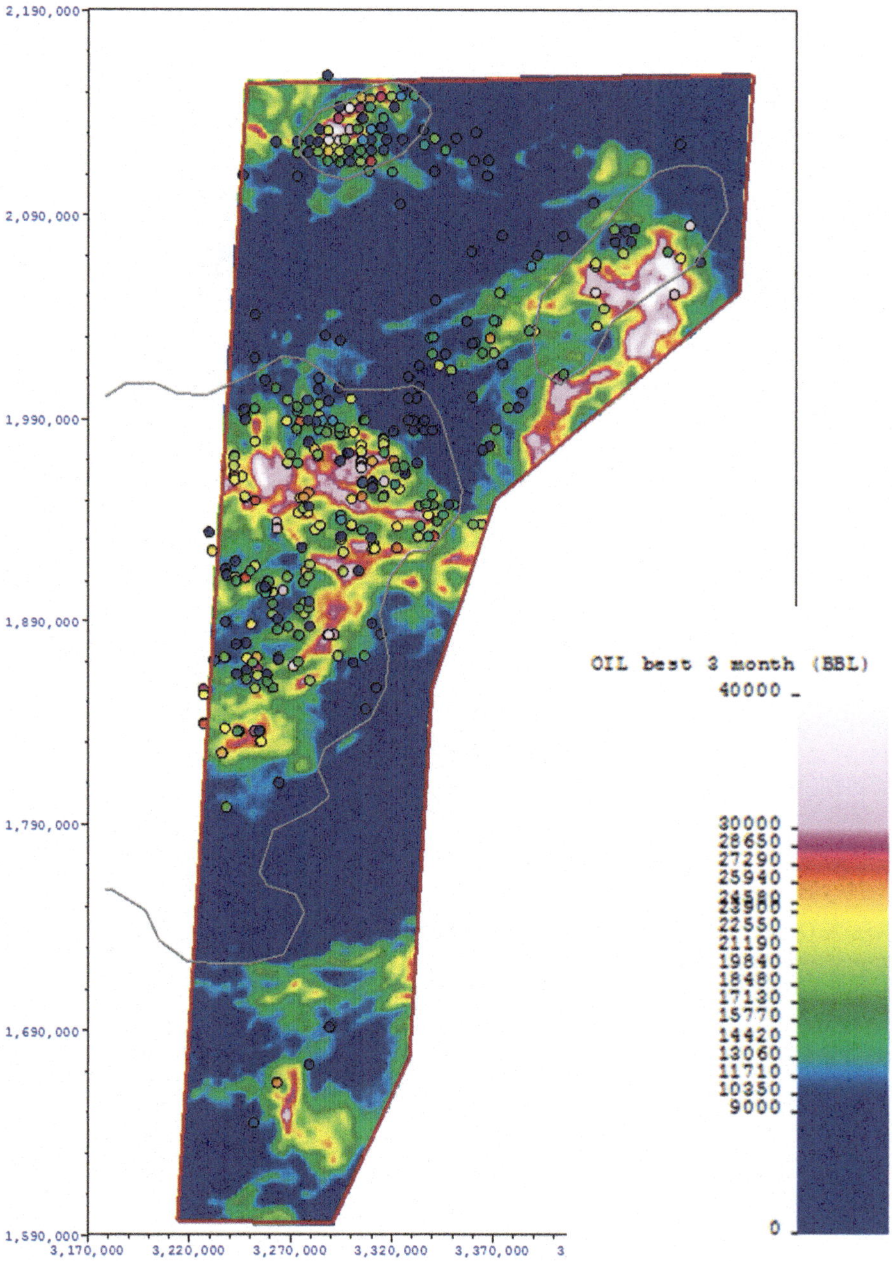

Fig. 5 Result of the ACE run

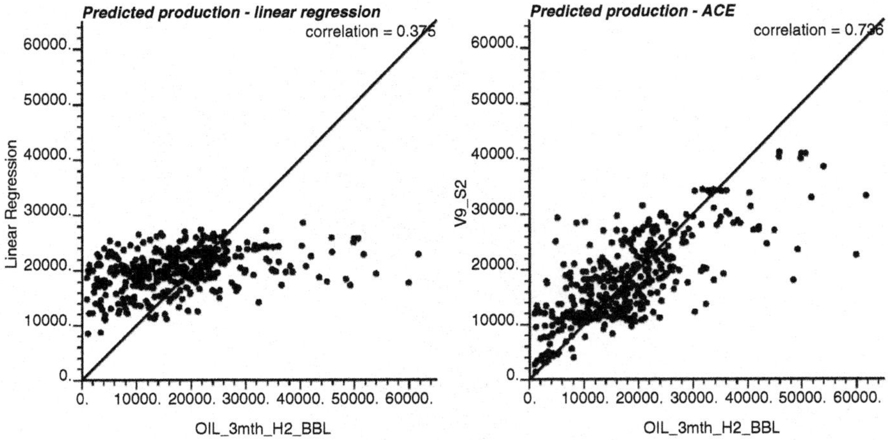

Fig. 6 Correlation between true and predicted production for linear regression and ACE

Table 1 Influence of each predictor variable on the response variable

Variable	Influence
Magnetic	0.156
Gravity	0.165
Distance to faults	0.179
Vitrinite	0.193
Niobrara isopach	0.194
Precambrian depth	0.223
Resistivity	0.234
T_{max}	0.254
TOC	0.267

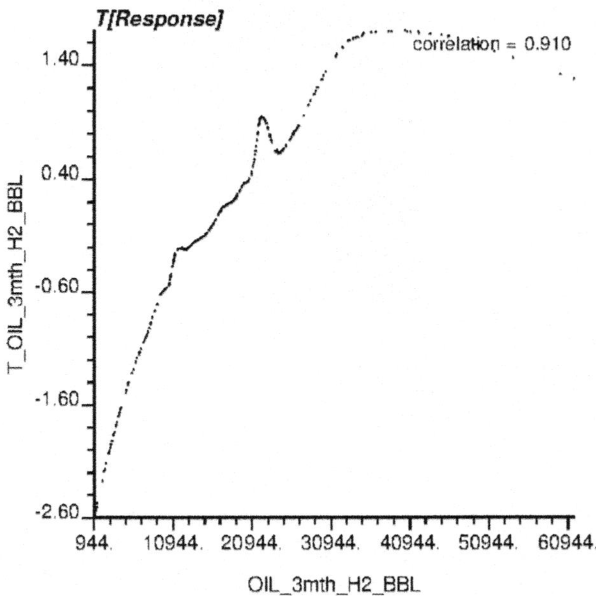

Fig. 7 Optimal transformation of the response variable

Bibliography

Barnett RM, Deutsch CV (2013) Tutorial and tools for ACE regression and transformation. Centre for Computational Geostatistics (CCG) Annual Report 15, 401, University of Alberta, Edmonton

Box EG, Cox RD (1964) An analysis of transformations. J R Stat Soc Ser B 211–252

Breiman L, Friedman JH (1985) Estimating optimal transformations for multiple. Regression and correlation. J Am Stat Assoc 80:580–598

Kuhn M, Johnson K (2013) Applied predictive modeling. Springer, New York

Tibshirani R (1988) Estimating transformations for regression via additivity and variance stabilization. J Am Stat Assoc 83:394–405

Wang D, Murphy M (2004) Estimating optimal transformations for multiple regression using the ACE algorithm. J Data Sci 2:329–346

The Adaptive Plurigaussian Simulation (APS) Model Versus the Truncated Plurigaussian Simulation (TPS) Model Used in the Presence of Hard Data

Bogdan Sebacher, Remus Hanea, and Andreas Stordal

Abstract In this study, we present a comparison between the traditional truncated plurigaussian simulation (TPS) model conditioned to hard data (facies observations) and the adaptive plurigaussian simulation (APS) model in the presence of facies probability fields that incorporate hard data. In practice, the prior probability fields of the facies are developed by a group of experts (geologists, geophysicists, geo-modelers) in the early phase of the reservoir exploration, using various data, such as core information, seismic data, well log data, etc. Here we create ourselves a set of probability fields of the facies occurrence that incorporate the hard data and condition the plurigaussian simulation to it. We show that the APS is able to better quantify the prior uncertainty than the traditional TPS conditioned to hard data.

1 Introduction

The truncated plurigaussian simulation (TPS) model is a powerful technique that creates realistic and complicated distributions of the geological formations using variogram information (two-point geostatistics). It was first introduced in the literature by Galli et al. (1994) and Loc'h et al. (1994) as a natural generalization of the truncated Gaussian simulation model and was further developed by Le Loch and Galli (1997) to work for nonstationary case (TPS conditioned to nonstationary

B. Sebacher (✉)
TU Delft, Delft, The Netherlands
e-mail: bogdansebacher@yahoo.com

R. Hanea
Statoil ASA, Stavanger, Norway
e-mail: rhane@statoil.com

A. Stordal
IRIS, Stavanger, Norway
e-mail: Andreas.S.Stordal@iris.no

J.J. Gómez-Hernández et al. (eds.), *Geostatistics Valencia 2016*, Quantitative Geology and Geostatistics 19, DOI 10.1007/978-3-319-46819-8_47

proportion maps). Its easy implementation within an inverse modeling process (in reservoir engineering assisted history matching, AHM) has made this methodology to be used either in synthetic fields (Le Ravalec-Dupin et al. 2004) or even in real fields (Deraisme and Farrow 2005). One of the challenges of the methodology was the conditioning of the facies fields to the hard data (facies observations) collected at the exploration wells (Lantuejoul 2002). In addition, the facies observations at the well locations must be kept during AHM process. In the papers of Liu and Oliver (2005), Agbalaka and Oliver (2008, 2009), Astrakova and Oliver (2014), Sebacher et al. (2013), and Zhao et al. (2008), the authors proposed various methodologies to condition the facies fields to facies observations and also to keep them during the history matching process. A common characteristic of all methodology was that the prior seems to be over-constrained by the geostatistical properties of the Gaussian fields (Agbalaka and Oliver 2008; Sebacher et al. 2013).

In this paper, we present a solution to prevent the drawback of the previous methodologies. The solution consists of defining of prior probability fields of the facies occurrence and modifies the TPS methodology in order to condition the simulation to the prior probability fields. The methodology is called the adaptive plurigaussian simulation (APS) model (Sebacher 2014) and can be simply characterized as the TPS conditioned to soft data (facies probability fields). Consequently, in the APS, the truncation map is changing with location and the facies observations are always honored irrespective of the Gaussian field values. This is always happening when the prior probability fields of the facies incorporate the hard data (in probabilistic terms). The traditional TPS uses a single truncation map for the entire domain, and the Gaussian fields are generated such that in combination with the truncation map yield facies fields conditioned to facies observations. Consequently, the Gaussian fields used in the TPS (conditioned to hard data) are not stationary, while in the APS the Gaussian fields are typically stationary. We show in an example that the facies fields obtained with the TPS conditioned to hard data exhibit a bias in their spatial distribution. We conclude that an aspect analyzing the spatial distribution of the probability fields of the facies is calculated from an ensemble of realizations. The geostatistical properties of the Gaussian fields used for the facies field simulations with TPS drastically influence the facies fields, introducing a bias in the spatial distribution of the facies. This means that the TPS constrains the prior too much, especially when many facies observations are available and when some of the observations are inside the correlation range of the Gaussian fields. In contrast, when the APS is used, the probability fields of the facies types calculated from an ensemble of realizations reproduce very well the prior probability fields and the geostatistical properties of the Gaussian fields have little influence. Moreover, the prior probability fields incorporate the information about the expected facies proportions because the mean of the probability fields is the global proportion of the facies types. The APS is able to condition the facies simulation to this indicator because the APS uses stationary Gaussian fields and each grid cell has its own truncation map. The TPS uses a single truncation map of which parameters are defined based on the prior information about expected facies proportions but for the stationary case (using stationary Gaussian fields). In order to generate facies fields conditioned to hard data, the TPS truncates Gaussian fields that are not stationary. Consequently, the simulated facies fields with TPS are no longer conditioned to expected facies proportions.

The experiment is performed using a synthetic example with three facies types of which facies fields exhibit a particular topology, characteristic of a real field from the North Sea.

2 The Adaptive Plurigaussian Simulation (APS) Model

In order to give an easy description of the APS method, we describe the methodology for a reservoir model where three facies types are present (denoted F_1, F_2, and F_3). In each field, the probability is 1 at the well locations (or any other location where we have a facies observation) where the associated facies type occurs and 0 where it does not occur. This means that the facies observations are already incorporated into the probability fields. Important prior information is regarding the transition (contact) between any two facies type and any topological information. Suppose that any two facies types can intersect each other. We denote by p_k the probability field associated to facies type k (where $k = 1, 2, 3$).

The APS algorithm (for three facies types) for simulating a facies field is summarized as follows:

1. Given the prior probability fields p_1, p_2, p_3 and information of facies connections, create a layout of the simulation map (or truncation map[1]). The simulation map consists of a reliable decomposition of the square $(0, 1^2)$ in three sub-domains, each having assigned a facies type. The area of each sub-domain is equal to the probability of the associated facies type. In Fig. 1 is shown a layout of the simulation map used in the APS procedure in the next section. The construction of the simulation map must take into account the information about the number of the facies, the facies connection, and the topological characterization of the reservoir geology.
2. Generate samples from two (stationary) Gaussian random fields, Y_1 and Y_2, with predefined dependence structure. The geostatistical properties of the Gaussian fields are correlated with the prior information of the spatial distribution of the facies types (indicator variogram).
3. Transform the Gaussian random fields to uniform random fields, α_1 and α_2, using the integral transform. For each grid cell j of the reservoir domain, $\alpha_1(j) = \Phi_1(Y_1(j))$ and $\alpha_2(j) = \Phi_2(Y_2(j))$, where Φ_1 and Φ_2 are the marginal Gaussian cumulative distribution functions (CDF) of Y_1 and Y_2.

[1] We prefer to call it the simulation map instead of truncation map because we consider that in the APS we simulate a facies type from a discrete distribution rather than truncate some real values.

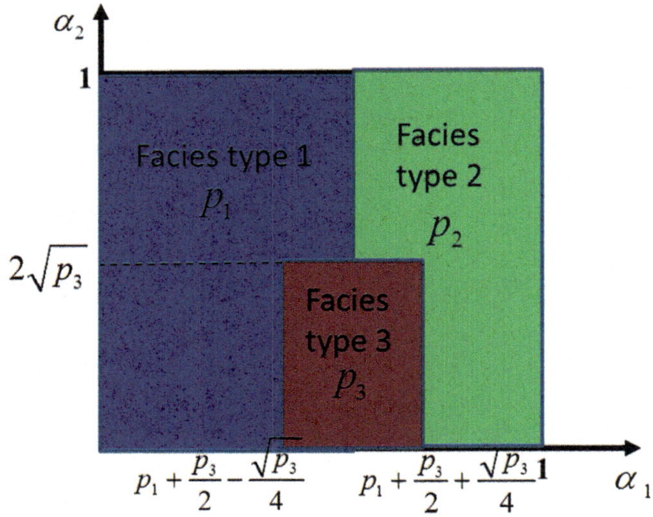

Fig. 1 The layout of the facies simulation map

4. For each grid cell j of the reservoir domain, build its simulation map from the layout (using the marginal probabilities p_1^j, p_2^j, p_3^j collected at the grid cell j) and set the facies type in grid cell j to F_k if the point $(\propto_1(j), \propto_2(j))$ is situated in the sub-domain assigned to facies type k (in the simulation map assigned to grid cell j).

Consequently, each grid cell has its own simulation map that depends on the set of probabilities collected at that location. At the location where we have facies observation, its simulation map consists of the square $(0, 1)^2$, occupied by the facies type observed there. This means that irrespective of the Gaussian field values, the observed facies will be always simulated so the hard data is preserved. With this method, we always simulate facies fields honoring the marginal facies probabilities and, consequently, the expected facies proportions.

3 APS vs TPS

The traditional plurigaussian simulation (TPS) model has two main ingredients:

- The Gaussian random fields defined on a region of interest. The region of interest in reservoir engineering is the reservoir domain, but with a discrete structure (grid cells).
- A truncation map defined on a multidimensional real space. The truncation map is defined by the intersection of some curves that divide the space into regions, each having assigned a facies type.

The number of the Gaussian fields is equal to the dimension of the space where the truncation map is designed. If the dimension of the space is two, a pair of Gaussian random fields, Y_1 and Y_2, is simulated, and, at each grid cell j, a facies type is assigned depending on where the point of simulated values (Y_1^j, Y_2^j) is situated in the truncation map. Consequently, in the TPS methodology, the truncation map tailors the facies fields according to the Gaussian field values at each grid cell. This methodology can be easily rewritten in a similar manner as we have presented the APS before. Let us consider for simplicity that the Gaussian fields Y_1 and Y_2 follow at each location a normal distribution of which CDF's are Φ_1 and Φ_2. Then, the function $F : R^2 \to (0,1)^2, F(x_1,x_2) = (\Phi_1(x_1), \Phi_2(x_2))$ is a bijection and has increasing components. Consequently, it defines a bijection between the truncation maps designed in the space $\mathbf{R}^2$ and the ones designed in the space $(0, 1)^2$. Thus, assigning the facies type based on the projection of (Y_1^j, Y_2^j) in a truncation map from $\mathbf{R}^2$ is equivalent with assigning a facies type based on projection of $(\Phi_1(Y_1^j), \Phi_2(Y_2^j)$ in a truncation map designed in $(0, 1)^2$. Consequently, the APS can be viewed as the TPS model conditioned to probability fields of the facies type (soft data) or is the methodology that incorporates the probability fields in the TPS model. In addition, the APS with uniform probability fields is the TPS.

In the traditional TPS, when facies observations are available at some locations, the Gaussian field values are generated such that the simulated facies fields satisfy the observations (interval conditioning). This implies a change in the mean and variance functions of the Gaussian fields compared with the case where the facies observations are not present, i.e., the Gaussian fields are no longer stationary. We will show that this modification produces a bias for the probability field of each facies type calculated from an ensemble of realizations.

Let us consider a rectangular domain with 50*100 grid cells with three facies types present. One facies type exhibits a long correlation from west to east (denoted facies type 2) and a small correlation from north to south. Another facies type (denoted facies type 3) is characterized by bodies of rocks that occur most likely on the edge of the other facies, and the last facies type is the medium where all other facies types are propagated (denoted facies type 1). This topology is characteristic to a real field from the North Sea (Hanea et al. 2014). We apply the TPS methodology for creating realistic facies distributions, using two Gaussian fields. Initially, we set both Gaussian fields stationary with its marginal having a standard normal distribution and having a Gaussian covariance type. The first Gaussian field is anisotropic, modeled with a long correlation range of 50 grid cells and a small correlation range of 5 grid cells and with the principal direction being the horizontal direction. The second Gaussian field is isotropic with the correlation range of five grid cells.

The truncation map used for generation of the facies fields is shown in Fig. 2, where the parameters of its lines are calculated based on the expected facies proportions $p_1 = 0.425$, $p_2 = 0.425$, and $p_3 = 0.15$. This truncation map is obtained applying the inverse of function F to the simulation map from Fig. 1. Consequently,

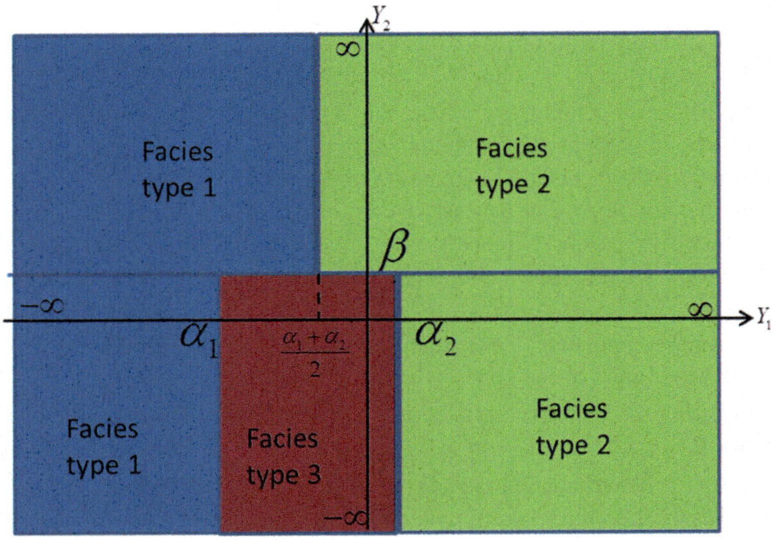

Fig. 2 The truncation map from R^2

the thresholds α_1, α_2, and β (see Fig. 2) are calculated based on the inverse of function F:

$$\alpha_1 = \Phi_1^{-1}\left(p_1 + \frac{p_3}{2} - \frac{\sqrt{p_3}}{4}\right),\ \alpha_2 = \Phi_1^{-1}\left(p_1 + \frac{p_3}{2} + \frac{\sqrt{p_3}}{4}\right), \beta = \Phi_2^{-1}\left(2\sqrt{p_3}\right) \quad (1)$$

This truncation map, in combination with the Gaussian fields presented before, generates facies fields with realistic topological and geometrical structure (Hanea et al. 2014), but is not necessarily conditioned to facies observations. We consider 13 locations from where we observe the facies types that occur at those positions. The positions and the facies observations are presented in Table 1 and in Fig. 3.

In order to generate conditional facies fields, we follow a conditional simulation approach (Armstrong et al. 2003). Firstly, at the observation grid cells, we generate pairs of Gaussian values such that those pairs yield correct facies observations in accordance with the truncation map. Secondly, the simulated Gaussian values are used in a conditional sequential Gaussian simulation process in order to populate with values the remaining grid cells. With this method, we generate an ensemble of 120 samples of pairs of Gaussian fields. These samples generate by truncation of 120 different facies fields with correct facies observations. We note that the procedure of conditioning of the Gaussian fields destroys their stationarity. This means that the mean function of the Gaussian fields is not 0 at each location, and consequently, the simulated facies fields are no longer conditioned to expected facies proportions. This raises the question if the truncation map is still valid (i.e., produces facies fields conditioned to expected facies proportions).

Table 1 The facies observations

x coordinate	5	5	5	25	25	50	50	50	75	75	95	95	95
y coordinate	5	25	45	15	30	5	25	45	15	30	5	25	45
Facies observation	2	2	1	3	2	2	3	1	1	2	2	1	3

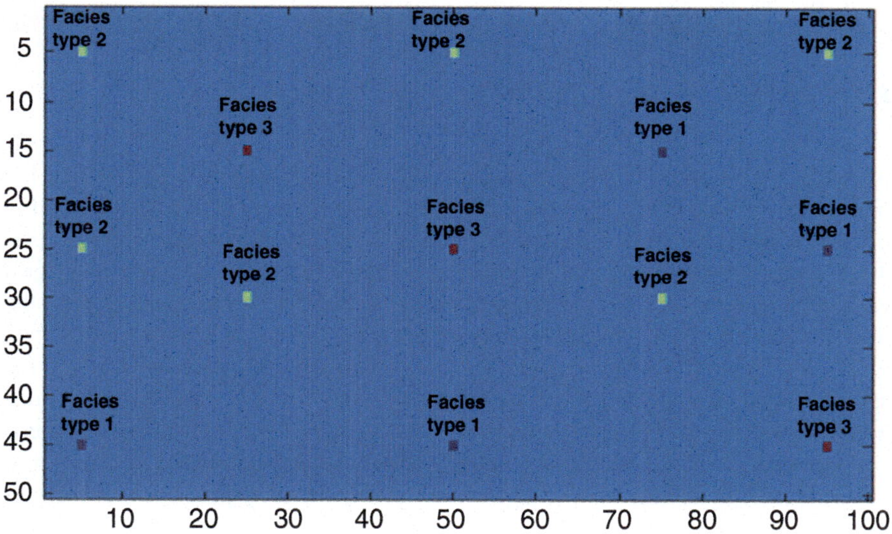

Fig. 3 The facies observations into the domain

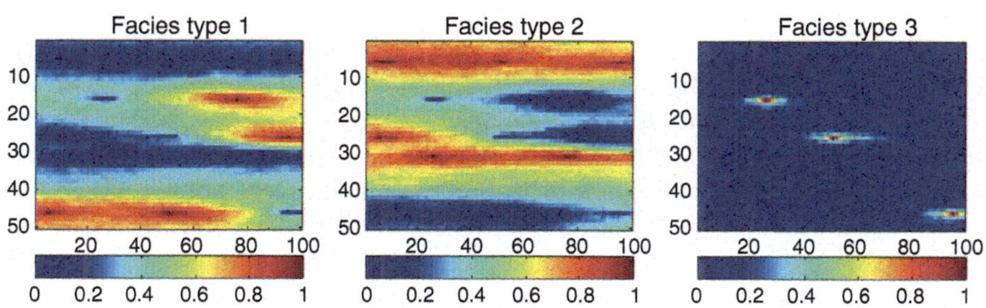

Fig. 4 The probability fields of the facies types in the TPS

In Fig. 4 is shown the probability fields of the facies types obtained from the ensemble of facies fields generated with the TPS method and conditioned to the facies observations presented in Table 1.

One may observe that these probability fields are biased in the sense that the geostatistical properties of the Gaussian fields drastically influence their spatial distribution. For instance, if we look at the top of probability field of the facies type

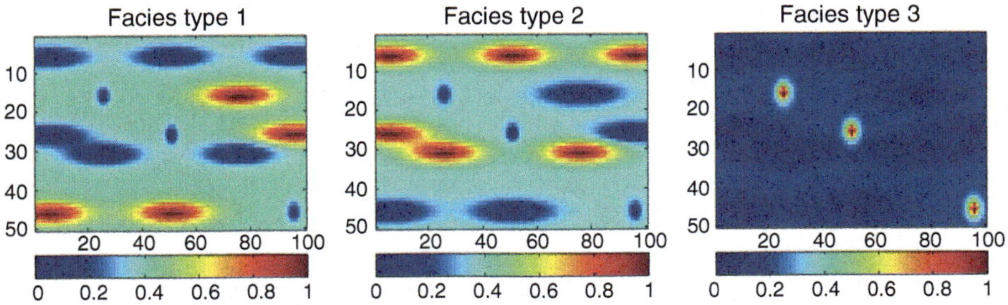

Fig. 5 The "prior" probability fields of the facies types

two, one may observe a long correlation in the horizontal direction due to the placement of three facies observations of the facies type two within the correlation influence of the first Gaussian field. The distance between the observation grids is 40 grid cells, whereas the long correlation range of the Gaussian fields is 50 grid cells. This causes the bias for the probability fields. The bias increases with the number of the facies observations situated within the correlation influence.

We also apply the APS methodology using stationary Gaussian fields having the same geostatistical setup as before. In order to do that, we need to have prior probability fields of the facies occurrence that incorporate the facies observations.

In practice the prior probability fields of the facies are developed by a group of experts (geologists, geophysicists, geo-modelers) in the early phase of the reservoir exploration, using various data, such as core information, seismic data, well log data, etc. In the papers of Beucher et al. (1999) and Doligez et al. (2002), the authors present methodologies to incorporate facies probability maps that comes out from seismic inversion, in the geostatistical simulation of the facies distribution. Here using a synthetic example, we have to create ourselves these probability fields, but we want to avoid the strong bias observed in the probability fields from the ensemble obtained with traditional TPS. Using an empirical approach, we generate a set of "prior" probability fields of the facies that incorporate the facies observations (Fig. 5). We generate these fields with a small correlation around the observation locations and, in addition, with the property that the mean of each probability field is equal with the expected facies proportions of the associated facies type. This last property ensures that the simulated facies fields are conditioned to expected facies proportions. The layout of the simulation map used in the APS is presented in Fig. 1. We generate an ensemble of 120 samples of pairs of stationary Gaussian fields, and, using the APS, the simulation map layout, and the probability fields, we generate the ensemble of facies fields.

The probability fields for each facies type, calculated from the ensemble of realizations with APS, are presented in Fig. 6. One clearly observes that the prior probability fields are very good preserved and the geostatistical properties of the Gaussian fields have little influence. This was expected because the APS samples correctly at each grid cell from the prior marginal distribution of the facies types.

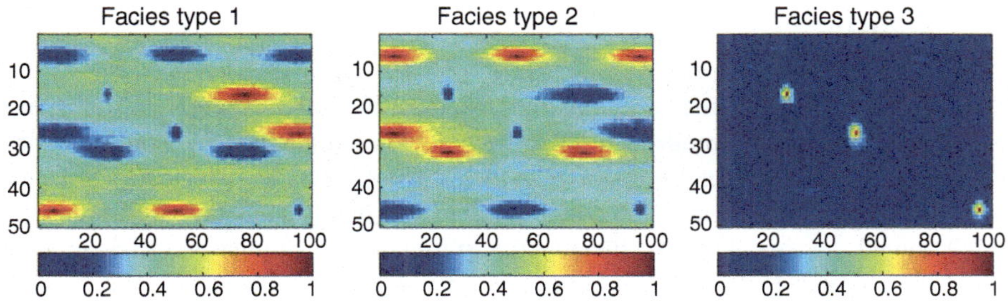

Fig. 6 The probability fields of the facies types in the APS

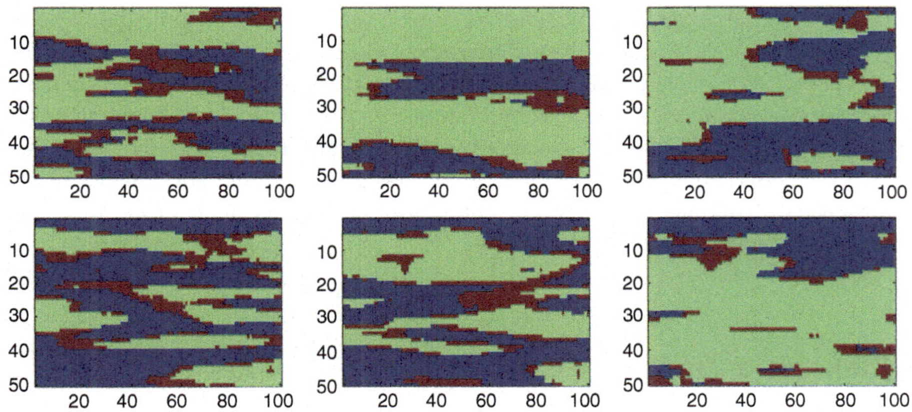

Fig. 7 Facies fields generated with TPS (*top*) and APS (*bottom*)

The ensembles of facies fields approximate the prior *pdf* of the facies field conditioned to the facies observations and the probability fields from Figs. 4 and 6 approximate of the prior marginal distribution of the facies fields. In order to completely quantify the quality of the simulation, we still have to look at the facies fields from the topological and geometrical perspective, to see if they represent the same geological concept. In Fig. 7 we present three ensemble members of facies fields in both simulations; at the top is shown the simulation with the TPS conditioned to facies observations and at the bottom the facies fields obtained with the APS (facies type 1 is in blue, facies type 2 is in green, and facies type 3 is in red). One can observe that there is no difference regarding topology and geometry of the facies among these fields. Consequently, we have two ensembles of facies fields with consistent facies observations at the well locations, both representing the same geological concept, but the ensemble obtained with APS quantifies better the prior uncertainty of the facies distribution.

4 Conclusions

In this study, we presented a comparison between two methodologies of the same geological simulation model, i.e., the plurigaussian simulation. We compared the truncated plurigaussian simulation model conditioned to hard data and the plurigaussian simulation model conditioned to soft data (facies probability fields). We named the second model the adaptive plurigaussian simulation (APS) model. We have shown that the ensemble generated with the APS offers a better uncertainty quantification of the facies distribution than the ensemble generated with the traditional approach of conditioning TPS on facies observations. In other words, the TPS conditions the Gaussian fields to facies observations, which may significantly change the probability fields based on the geostatistical parameters of the Gaussian fields. The APS algorithm simulates facies fields directly from the probability fields and thereby does not change the underlying Gaussian fields. It is independent of the geostatistical parameters, and it can preserve the initial probability fields, which are already conditioned to the facies observations. A consequence is that with the APS method we are able to better characterize the prior uncertainty than the traditional TPS method. Consequently, the use of the plurigaussian simulation model in the presence of hard data gave better results if prior probability fields of the facies are firstly developed.

Bibliography

Agbalaka C, Oliver D (2008) Application of the enkf and localization to automatic history matching of facies distribution and production data. Math Geosci 40(4):353–374

Agbalaka CC, Oliver DS (2009) Automatic history matching of production and facies. SPE reservoir simulation symposium. Society of Petroleum Engineers

Armstrong M, Galli A, Loc'h G (2003) Plurigaussian simulations in geosciences. Springer, New York

Astrakova A, Oliver DS (2014) Conditioning truncated pluri-Gaussian models to facies observations in ensemble-kalman-based data assimilation. Math Geosci 47(3):345–367

Beucher H, Fournier F, Doligez B, Rozanski J (1999) Using 3D seismic-derived information in lithofacies simulations. A case study.SPE 56736. SPE – annual conference of the Soc. of Petrol. Eng. SPE, Houston, pp 581–592

Deraisme J, Farrow D (2005) Geostatistical simulation techniques applied to kimberlite orebodies and risk assessment of sampling strategies. Geostat Ban: 429–438

Doligez B, Fournier F, Jolivet G, Gançarski S, Beucher H (2002) Seismic facies map integration in geostatistical geological model: a field case. EAGE – 64th conf. and exhibition of the European Assoc. of Geoscientists & Engineers. EAGE, Florence, pp 215–218

Galli A, Beucher H, Le Loc'h G, Doligez B (1994) The pros and cons of the truncated Gaussian method. In: Geostatistical simulations. Kluwer, Dordrecht, pp 217–233

Hanea R, Ek T, Sebacher B, Saetrom J, Sollien D (2014) Geologically realistic facies updates for a north sea field. 76th EAGE conference and exhibition. Amsterdam

Lantuejoul C (2002) Geostatistical simulation: models and algorithms. Springer, Berlin

Le Loch G, Galli A (1997) Truncated plurigaussian method: theoretical and practical points of view. Geostat Wollongong 96(1):211–222

Le Ravalec-Dupin M, Roggero F, Froideveaux F (2004). Conditioning truncated Gaussian realizations to static and dynamic data. SPE J 9(4)

Liu N, Oliver D (2005) Ensemble Kalman for automatic hystory matching of geologic facies. J Pet Sci Eng 47(3–4):147–161

Loc'h L, Beucher G, Galli H, Doligez A (1994) Improvement in the truncated Gaussian method: combining several Gaussian functions. ECMOR IV, Fourth European Conf. on the Math. of Oil Recovery

Sebacher B (2014) Data assimilation under geological constraints, Phd thesis. Gildeprint

Sebacher B, Hanea R, Heemink A (2013) A probabilistic parametrization for geological uncertainty estimation using the ensemble kalman filter (EnKF). Comput Geosci 17(5):813–832

Zhao Y, Reynolds A, Li G (2008) Generating facies maps by assimilating production data with enkf. Proceedings of SPE/DOE Symposium on Improvement Oil Recovery (pp. 20–23). SPE113990

An MPS Algorithm Based on Pattern Scale-Down Cluster

Yu Siyu, Li Shaohua, He Youbin, Tao Jinyu, and Dai Weiyan

Abstract A key evaluation indicator of multiple-point geostatistics modeling algorithm is ensuring model quality as well as harmonizing the modeling calculation of time-consuming and space-consuming RAM. Due to the inherent flaws of SIMPAT, poor efficiency of similarity match computation between the data event and the whole pattern of train image led to the impracticability of SIMPAT many years after it was proposed. Some improvement following methods based on SIMPAT, such as Filtersim and DisPat, still did not resolve the problem. After studying key points of SIMPAT, this paper proposes PSCSIM algorithm based on a pattern scale-down clustering strategy which uses an interval sampling technique. Unlike SIMPAT, PSCSIM replaces the one-step similarity computation with the two-step similarity computation: firstly, comparing the representative patterns of the pattern cluster to the data event to find the most related pattern cluster and, secondly, matching the similarity of whole patterns in a pattern cluster with the data event to search the target pattern. With the same condition, this paper made a comparison of modeling in two dimensions and three dimensions among with PSCSIM, SIMPAT, Snesim, Filtersim, and DisPat in the end. As a result, PSCSIM greatly improves modeling efficiency on the premise of quality assurance.

1 Introduction

Most oil fields in China have entered into the middle and later development stage. In order to effectively explore the remaining oil, it is necessary to finely characterize the reservoir. For example, the grid of the reservoir model should be refined to 10×10 m in a plane and 0.2 m vertically. For some old oil fields, the nodes of the reservoir model can reach up to ten million, and the greater the volume of the model, the longer time the simulation needed. Reservoir stochastic modeling method usually requires the establishment of multiple equal probability models. These models are used to predict the risk development and increase the simulation

Y. Siyu (✉) • L. Shaohua • H. Youbin • T. Jinyu • D. Weiyan
College of Geosciences, Yangtze University, Wuhan 430100, China
e-mail: 573315294@qq.com

J.J. Gómez-Hernández et al. (eds.), *Geostatistics Valencia 2016*, Quantitative Geology and Geostatistics 19, DOI 10.1007/978-3-319-46819-8_48

time. Therefore, the computational efficiency of the reservoir modeling algorithm became the current focus of the research. Apart from the proposed SIMPAT which is a multipoint geostatistics modeling algorithm based on pattern (Arpat 2005) in the Stanford Reservoir prediction center session in 2003, unlike the object-based algorithm and traditional multipoint geostatistics algorithm Snesim (Strebelle 2002; Ortiz and Emery 2005; Jef 2001; Mariethoz et al. 2010), SIMPAT firstly applies the idea of image rebuilding to the reservoir modeling field. The core idea of SIMPAT is finding out the data pattern which has the greatest similarity with the estimated "data event" in the study area from the priori geological model, through the comparison of the similarity between data patterns and data events, using the most similar data pattern cover and freezing the correspondent node until all nodes in the work area are simulated. The bottleneck of computing the speed of SIMPAT is the similarity calculation between the data event and all of the data patterns in the database. Many improved algorithms are proposed based on SIMPAT (Julien and Philippe 2011; Honarkhah and Caers 2012; Comunian et al. 2012; Abdollahifard and Faez 2013; Tao et al. 2013; Jeff 2007; Straubhaar et al. 2013; Mariethoz et al. 2010; Yanshu et al. 2008), such as Filtersim (Zhang 2006), DisPat (Honarkhah 2011), and Filtersim using filter technology to cluster data pattern. Although the simulation speed is faster than SIMPAT, it is still slow in the actual field reservoir modeling. DisPat employs multidimensional scaling analysis and K-means clustering to process the data, the calculation efficiency of DisPat is higher than SIMPAT and Filtersim; however, the similarity matrix of the data pattern established by multidimensional scaling analysis requires a great deal of computer memory. This paper presents a clustering multipoint geostatistics modeling algorithm PSCSIM which reduces dimension and cluster base on the pattern (pattern scale-down cluster simulation), effectively improving the efficiency of modeling and controlling the memory footprint.

2 The Basic Principle of SIMPAT Algorithm

Data patterns (pattern, called Pat) are based on the style of multipoint geostatistics modeling algorithm in computing the minimum unit constituting a priori geological concept model "of training image (TrainImage, referred as TI)" local space-related multipoint structure. The data pattern is a data model (template, abbreviated T) which is a framework for the training template scan image acquisition. First, define the data model as follows:

$$T = \left\{h_m, m = i + j \times I + k \times I \times J \middle| \, i \in [0, I-1], j \in [0, J-1],\ k \in [0, K-1]\right\}$$

I, *J*, and *K* are the length and breadth dimensions of the data template T. Figure 1 is a two-dimensional example of data pattern at position u and is obtained through scanning the training image:

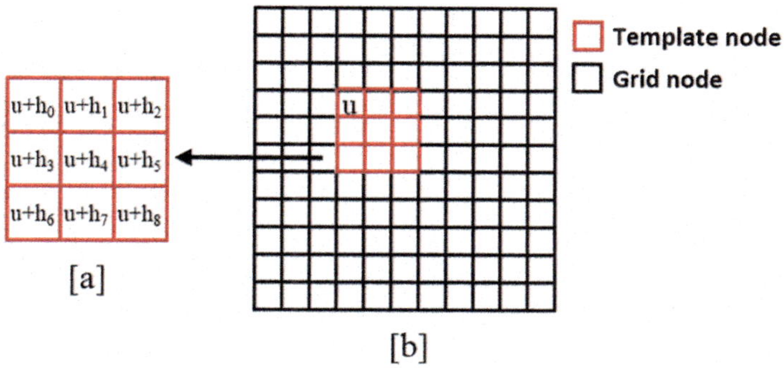

Fig. 1 Sketch of pattern and realization. (**a**) Pattern based on template. (**b**) Realization used to scan by template

$$\mathrm{Pat}_T(u) = \{\mathrm{TI}(u+h) \big| h \in T\}$$

The similarity is the index which can evaluate the degree of similarity between two objects, and it is widely used in pattern recognition, computer vision, and other fields. Common similarity functions are Minkowski function (L_k norm), Hsim function, Pearson coefficient, and Jaccard similarity coefficient. Similarity function takes two objects as input variables to determine a nonnegative real number which can represent similarity. SIMPAT using Manhattan distance (L_1 norm) calculates the degree of similarity between data patterns or between the data pattern and data event. It is defined as

$$L_1(X, Y) = \sum_{i=1}^{d} |x_i - y_i|$$

Distance calculations between data patterns X and Y based on Manhattan distance are equal to the absolute value of the difference between two points (Fig. 2f). The distance between X and Y plus one is the inverse of similarity, and the greater the distance, the smaller the similarity, where the value of similarity ranges from 0 to 1. The similarity is calculated as

$$s(X, Y) = 1/(d(X, Y) + 1)$$

Figure 2a–e is the Manhattan distance of five data patterns. The Manhattan distance of data patterns in Fig. 2a is equal to 1. X and Y are the most similar data pattern. The similarity of X and Y in Fig. 2c is minimum.

The core idea of the pattern-based multipoint geostatistics algorithm SIMPAT is that it adopts the degree of similarity between data patterns and data events in the training image, getting all the priori geological model through by scanning the training images and storing it in the computer memory to avoid repeated scanning of training images. Calculating the similarity between data patterns and data events,

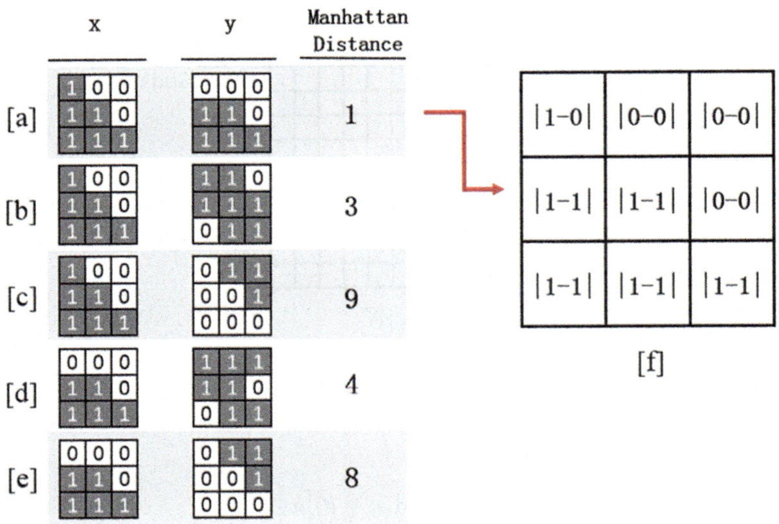

Fig. 2 Principle of calculating similarity of patterns by Manhattan distance (Arpat 2005)

matching the most similar data events and data pattern in the database, and covering and freezing data event until the simulation are completed. The time complexity of SIMPAT is proportional to the size and scale of pattern database, and the size of the pattern database is closely related to the size of the training image and data pattern. The larger the size of the training image and data pattern, the more information the reservoir geological model library contains and the longer the simulation time. When SIMPAT is applied to an actual oil field, in which grids reach one million to ten million, the simulation time can be up to several hours or even days. By analyzing the modeling features of SIMPAT and considering the memory footprint drawbacks of previously improved methods, the new PSCSIM algorithm is proposed to balance the contradiction between time-consuming and memory footprint.

3 The Core Idea of PSCSIM Algorithm

3.1 Pattern Scale-Down Cluster Method Based on the Adjacent Equally Spaced Sampling

The core concept of the new algorithm is the thumbnail pattern. The commonly used methods for the dimension reduction of data pattern are adjacent equally spaced sampling method, bilinear interpolation and cubic convolution interpolation method, etc. In this paper, the basic idea of adjacent equally spaced sampling method is calculating the coordinate mapping relationship between the thumbnail pattern and data pattern by the reverse transform of coordinates. If the coordinates mapped to the data pattern are the floating point, then it changes to an integer. The formula calculates nodes of thumbnail pattern based on the adjacent interval sampling method follows:

$$\mathrm{ThumPat}(i,j,k) = \mathrm{Pat}\left(i \times r_i, j \times r_j, k \times r_k\right)$$

$r_i = I_{\mathrm{Pat}}/I_{\mathrm{ThumPat}} r_j = J_{\mathrm{Pat}}/J_{\mathrm{ThumPat}}, r_k = K_{\mathrm{Pat}}/K_{\mathrm{ThumPat}}, I_{\mathrm{Pat}}$, and I_{ThumPat} is the dimension of data patterns and thumbnail pattern in the direction of I.

Figure 3 is the principle diagram of calculating the thumbnail pattern based on the adjacent equally spaced sampling method, where the dimension of the data pattern is 7×7 as shown in Fig. 3a and the dimension of thumbnail pattern is 4×4 as shown in Fig. 3c. The nodes in the thumbnail pattern are arranged with equal spaces in the data pattern. The probability of similar patterns classified as the same kind of data pattern after dimension reduction processing is much larger than the dissimilar data patterns.

The data pattern corresponding to the same thumbnail pattern was deemed as the same type of pattern (pattern cluster, referred as PatClusters) as it was able to cluster the pattern database after reducing the dimension of the data pattern. Figure 4a shows the PatClusters which contain six data patterns, and Fig. 4b is the thumbnail pattern corresponding to data pattern. Pattern database is clustered into a lot of PatClusters after the dimension reduction processing, and they are collectively called pattern clusters.

3.2 Represent Pattern Based on the E-Type (Ensemble Average)

Only when data pattern in PatClusters has a link between data event and PatCluster can it participate in the actual modeling, and that link is called represent pattern. The represent pattern has the same dimensions as the data pattern. This paper computed represent pattern based on the ensemble average method. The formula calculates nodes of represent pattern as follows:

$$\mathrm{RepPat}(i,j,k) = \sum_{1}^{n} \left(\mathrm{Pat}_{\mathrm{in\ PatCluster}}(i,j,k)\right)/n$$

where n is the number of data patterns in pattern cluster, where the value of nodes in represent pattern is equal to the average value of nodes in the entire data pattern. Figure 4 shows the represent pattern of pattern cluster (4c). Each pattern cluster has a unique represent pattern. The difference between represent pattern and data pattern is that the value of represent pattern is not binary but continuous.

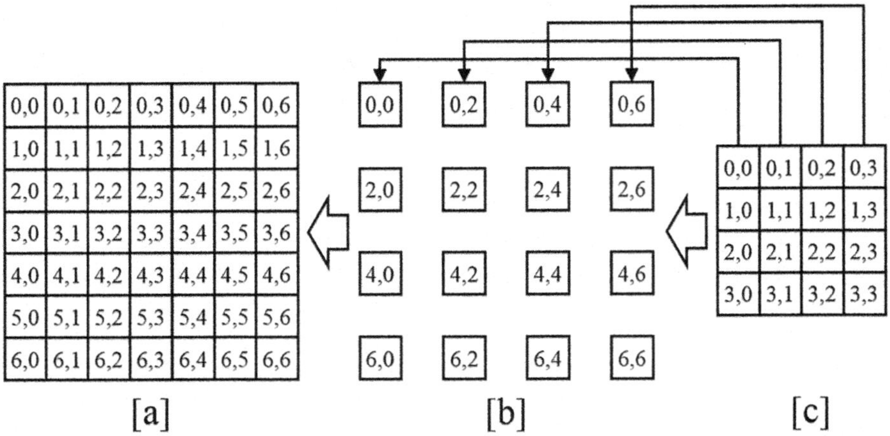

Fig. 3 Principle sketch of the procedure for scaling down pattern by space sampling method. (**a**) Pattern based on template, (**b**) process of reverse transformation, and (**c**) thumbnail pattern of pattern

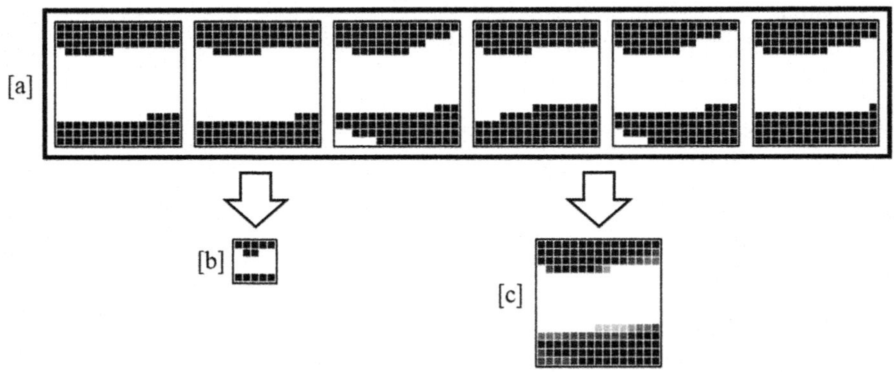

Fig. 4 The relationship of pattern cluster, thumbnail pattern, and represent pattern. (**a**) Pattern cluster, (**b**) thumbnail pattern, and (**c**) represent pattern

4 The Implementation Steps of PSCSIM Algorithm

This paper proposed a pattern-based multipoint geostatistics algorithm PSCSIM based on thumbnail pattern and represent pattern. PSCSIM is a kind of algorithm based on the similarity comparison, the specific process of accessing all nodes on the path successively, comparing the similarity of nodes, and constructing the model. Its core principle is to cluster the similar data pattern in the pattern database, converting the huge pattern database to relatively small pattern clusters. In performing the simulation, compare the similarity of represent pattern and data events (data event, referred as Dev) first. Then, find the most similar pattern cluster

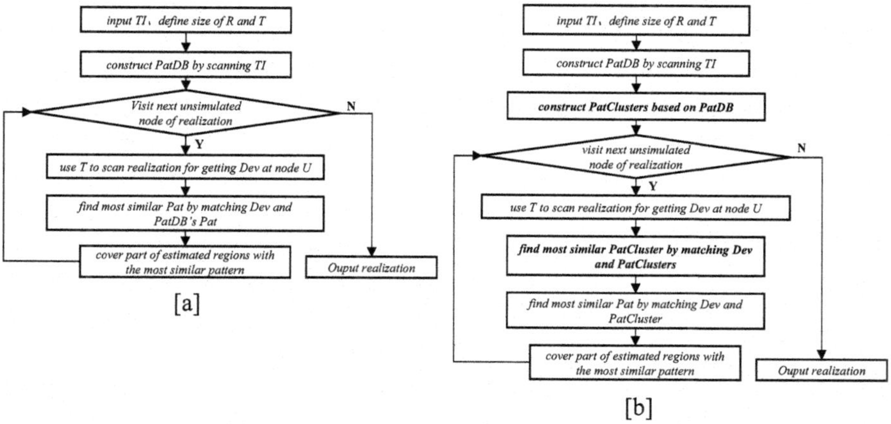

Fig. 5 Flow charts of SIMPAT and PSCSIM. (**a**) Flowchart of SIMPAT and (**b**) flowchart of PSCSIM

from the pattern clusters and also find the data pattern which is most similar to the data event from the most similar pattern cluster.

The difference between PSCSIM and SIMPAT is the introduction of pattern clusters and the calculation of similarity based on pattern clusters. The flowchart of SIMPAT and PSCSIM is shown in Fig. 5.

Process of PSCSIM

1. In inputting the training image, define the size of simulate realizations and data patterns.
2. Using data pattern, scan the training image and construct pattern database.
3. Clustering all data pattern in the database based on the pattern scale-down cluster strategy, the data pattern corresponding to the same thumbnail pattern is deemed as the same type of pattern, and construct pattern clusters.
4. Build a simulated realization-based random path.
5. Mesh nodes randomly access the nodes on the path. If there are no simulation nodes, go to step 6; otherwise, proceed to step 10.
6. Take the data template as a unit, scanning the simulate realizations to obtain the data event at the current simulation node.
7. The first similarity comparison: compare the similarity of represent pattern in data events and PatClusters and identify data patterns most similar with the data events.
8. The second similarity comparison: compare the similarity of data event and all data patterns in pattern clusters and identify the data pattern most similar to the data event.
9. Cover part of the estimated regions with the most similar data pattern and freeze these areas. Then, return to step 5.
10. Complete the simulation. Output the realizations.

PSCSIM proposed the strategy of "pattern scale-down cluster," where the concept of pattern clusters is proposed based on SIMPAT and the similarity matching process by the single-step comparison based on pattern database expands to the two-step comparison based on pattern clusters.

5 Case Study

5.1 Comparison of the Time Consumed in Calculation

Taking two binary images as the training image, compare the efficiency of PSCSIM, SIMPAT, Snesim, Filtersim, DisPat, and other mainstream multipoint geostatistics algorithms. The first training image is a two-dimensional fluvial digital model (Fig. 6a) with a 250×250 dimension and a 1×1 m grid cell. The second training image is a three-dimensional fluvial digital model (Honarkhah 2011) (Fig. 6d) with a dimension of $69 \times 69 \times 39$ and a $1 \times 1 \times 1$ m grid cell. Using SIMPAT and PSCSIM in conducting the non-conditional simulation, the simulation results are shown in Fig. 6b–c (black is the channel phase, whereas white is the background phase) and Fig. 6e–f (red is the channel phase, whereas blue is the hollow background phase). PSCSIM showed better internal priori geological structure characteristics of training image through the comparison of SIMPAT and PSCSIM.

The CPU and memory of the testing hardware are 2.0 GHz and 8 GB, respectively. The time consumed by PSCSIM, SIMPAT, Snesim, and Filtersim algorithm generated 100 realizations as follows:

1. Taking two-dimensional model as training image, grid dimension of the simulated realizations is 250×250, dimension of data template is 9×9, and dimension of thumbnail pattern is 5×5 using three-grid simulation. Computation time of SIMPAT is 5245 s, and PSCSIM algorithm is 204 s. The computational efficiency of PSCSIM algorithm improved by 25 times compared to SIMPAT algorithm.
2. Taking three-dimensional model as training images, the grid dimensions of the simulated realizations are $69 \times 69 \times 39$, dimensions of data template are $11 \times 11 \times 5$, and dimensions of thumbnail pattern are $5 \times 5 \times 3$ using the two-grid simulation. The calculation time of SIMPAT is 96,588 s and PSCSIM algorithm is 1498 s. The computational efficiency of PSCSIM algorithm increased by 64-folds compared to SIMPAT algorithm (Table 1).

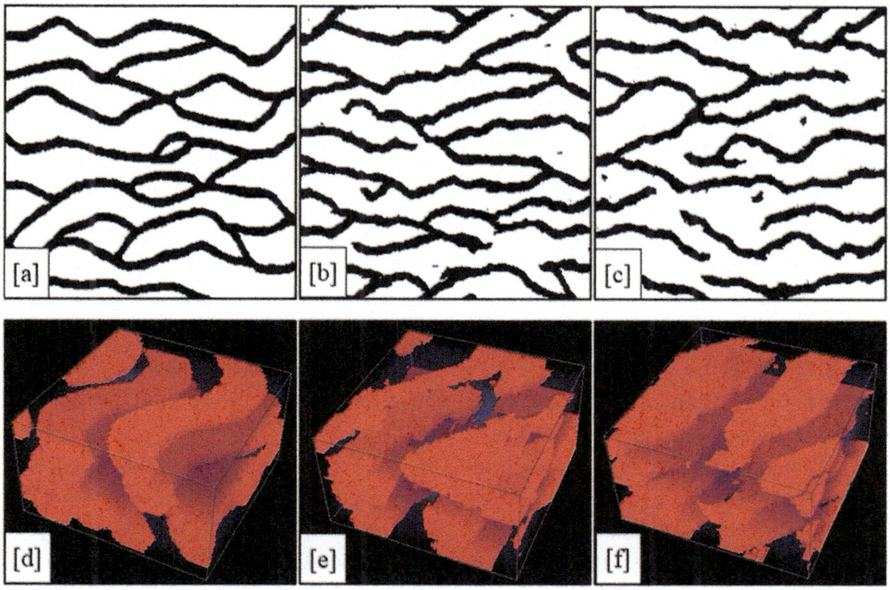

Fig. 6 Training images and realizations in two dimensions and three dimensions. (**a**) Largetrain, (**b**) realization of PSCSIM for Largetrain, (**c**) realization of SIMPAT for Largetrain, (**d**) fluvsim, (**e**) realization of PSCSIM for fluvsim, and (**f**) realization of SIMPAT for fluvsim

Table 1 Consumed time of computing 100 realizations by MPS algorithms

	Computation time (s)	
Algorithms	Largetrain (250 × 250)	Fluvsim (69 × 69 × 39)
SIMPAT	5245	96,588
Filtersim	1790	50,938
Snesim	597	5170
PSCSIM	204	1498
DisPat	120	900

5.2 *Memory Footprint Comparison*

Multipoint geostatistics modeling algorithm stored statistics of multipoints in a computer memory; therefore, how much memory the algorithm occupies is one of the important indicators to evaluate in multipoint geostatistics modeling algorithm. To further verify the practicality of PSCSIM, take two-dimensional models and three-dimensional models as training image, analyzing the computer memory demand of PSCSIM and other mainstream algorithms:

1. Taking two-dimensional model Largetrain as training images, the dimension of data pattern is 11 × 11 and the dimension of data pattern is 5 × 5 after reducing

Table 2 Consumed RAM of PSCSIM and other mainstream MPS algorithms

Algorithms	Memory footprint (MB)	
	Largetrain (250 × 250)	Fluvsim (69 × 69 × 39)
SIMPAT	35	329
Filtersim	60	143
Snesim	50	64
PSCSIM	36	440
DisPat	6799	20,684

its dimensionality using one-grid simulation. PSCSIM, SIMPAT, Filtersim, and Snesim use the same order of memory (<100 MB), the distance matrix that DisPat algorithm established for the multidimensional scaling analysis demand for memory reaching 6799 MB.

2. Taking three-dimensional model fluvsim as the training image, the dimension of data pattern is $11 \times 11 \times 5$, and the dimension of data pattern is 5×5 after reducing its dimensionality using one-grid simulation. PSCSIM occupied 440 MB of memory, and the memory DisPat occupied reached up to 20.6 GB completely beyond the ordinary amount of memory the computer can provide (Table 2).

Comprehensive analysis of both consumed time and memory shows that though the traditional multipoint algorithms Snesim and SIMPAT take less memory, they are low in efficiency. DisPat improves the computational efficiency; however, the memory footprint is too large. PSCSIM better balances the contradiction between the time consumed and the memory footprint than other multipoint geostatistics modeling algorithm because it has a great reference for the utility of multipoint geostatistics modeling algorithm.

6 Conclusions

1. This paper found the efficiency bottleneck of SIMPAT through the analysis of basic principles of multipoint geostatistics modeling algorithm, which includes the process of similarity calculation by data event and data pattern, putting forward the concept of data pattern dimension reduction and introducing the adjacent equally spaced sampling method to geostatistics. Using adjacent equally spaced resampling method to reduce the dimension of data patterns reserved multipoint statistical information of space while effectively reducing the data dimension.
2. A huge number of data patterns are clustered to establish a database library based on the data pattern of which dimension is reduced. This method used the ensemble average method of an E-type to calculate the representative data pattern. With the proposed PSCSIM algorithm, the algorithm improved one

similarity comparison of SIMPAT to twice the similarity comparison by first comparing the similarity between the data event and the representative data pattern. Then, find the most similar pattern classes, compare the data event with the data pattern in pattern classes individually, and find out the most similar data pattern.
3. Taking the 2D and 3D model as an example, compare the calculation time and memory footprint of PSCSIM and the mainstream multipoint geostatistics modeling algorithm such as SIMPAT, Snesim, and Filtersim under the same conditions. The test results validate that PSCSIM can guarantee the quality of modeling results and greatly improve the computational efficiency and effectively balance the computational efficiency and memory footprint.
4. Some aspects of PSCSIM need further study. In addition to the adjacent equally spaced resampling methods mentioned herein, using bilinear interpolation and cubic convolution interpolation method to reduce the dimension may cause less information loss and further improve the quality of data pattern cluster. The size of lower-dimensional data pattern plays a key role on the clustering effect of the new algorithm. The data library contains many pattern clusters after reducing and clustering the dimension. The number of data patterns in the pattern clusters is different. For example, there may be only one data pattern in the cluster; thus, the occurrence probability of such pattern cluster in the global geological model is very small, or we can consider removing it.

References

Abdollahifard MJ, Faez K (2013) Fast direct sampling for multiple-point stochastic simulation [J]. Arab J Geosci

Arpat GB (2005) Sequential simulation with patterns. Stanford University

Boisvert JB, Pyrcz MJ, Deutsch CV (2007) Multiple-point statistics for training image selection. Nat Resour Res 16(4):313–321

Caers J (2001) Geostatistical reservoir modelling using statistical pattern recognition. J Pet Sci Eng 29(3-4):177–188

Comunian A, Renard P, Straubhaar J (2012) 3D multiple-point statistics simulation using 2D training images. Comput Geosci 40:49–65

Honarkhah M (2011) Stochastic simulation of patterns using distance-based pattern modeling. Stanford University

Honarkhah M, Caers J (2012) Direct pattern-based simulation of non-stationary geostatistical models. Math Geosci 44(6):651–672

Huang T, Li X, Zhang T, Lu D-T (2013) GPU-accelerated direct sampling method for multiple-point statistical simulation. Comput Geosci 57:13–23

Mariethoz G, Renord P, Straubhaar J (2010) The direct sampling method to perform multiple-point geostatistical simulation. Water Resour Res 46(11)

Ortiz JM, Emery X (2005) Integrating multiple-point statistics into sequential simulation algorithms. Quant Geol Geostat 14:969–978

Straubhaar J, Renard P (2011) An improved parallel multiple-point algorithm using a list approach. Math Geosci 43(3):305–328

Straubhaar J, Walgenwit A, Renard P (2013) Parallel multiple-point statistics algorithm based on list and tree structures. Math Geosci 45(2):131–147

Strebelle S (2002) Conditional simulation of complex geological structures using multiple-point statistics. Math Geol 34(1):1–21

Yanshu Y, Changmin Z, Shaohua L, Guowei H (2008) A post-processing method based on information degree for stochastic modeling of reservoir. Acta Petrolei Sin 29(6):889–893

Zhang T (2006) Filter-based classification of training image patterns for spatial simulation. Stanford University

Integrating New Data in Reservoir Forecasting Without Building New Models

Sebastien Strebelle, Sarah Vitel, and Michael J. Pyrcz

Abstract In most hydrocarbon reservoir development projects, geological models are fully rebuilt on a regular basis to integrate new data, in particular observations from new wells, for up-to-date forecasts. Not only this common practice is very time consuming as rebuilding models can take weeks or even months, but it also leads to major, hard-to-justify, fluctuations in reservoir volume or flow performance forecasts, especially when the modeling staff changes or a new modeling technology, workflow, or software is adopted. Rationalizing the geological model updating process is required to provide stable and reliable forecasting and make timely, well-informed, reservoir management decisions. This paper presents an innovative methodology to quickly reassess model forecasts, such as reservoir oil-in-place or oil recovery, without rebuilding any geological models provided that the new data observations are reasonably consistent with the current models. The proposed methodology uses a Bayesian framework whereby the multivariate probability joint distribution of new data predictions and forecast variables needs to be modeled. Assuming that this joint distribution is multi-Gaussian, the first step consists in computing proxies, e.g., response surfaces using experimental design, to estimate from the set of current geological models the distribution (mean and variance) of new data predictions and forecast variables as a function of the input modeling parameters (e.g., property variograms or training images, trends, histograms). Because the model stochasticity (i.e., spatial uncertainty away from wells) typically entails significant uncertainty in the prediction of new local data observations, computing the previous proxies requires generating multiple stochastic realizations for each combination of input modeling parameters. Then, using those proxies and Monte Carlo simulation, the full multivariate probability joint distribution of new data predictions and forecast variables is estimated. Plugging the actual new data values into that joint distribution finally provides new updated

S. Strebelle (✉) • M.J. Pyrcz
Strategic Research Unit, Chevron Energy Technology Company, 1500 Louisiana St., Houston, USA
e-mail: stsb@chevron.com; mpyrcz@chevron.com

S. Vitel
Earth Sciences Department, Chevron Energy Technology Company, 1500 Louisiana St., Houston, TX, USA
e-mail: sarah.vitel@chevron.com

J.J. Gómez-Hernández et al. (eds.), *Geostatistics Valencia 2016*, Quantitative Geology and Geostatistics 19, DOI 10.1007/978-3-319-46819-8_49

probabilistic distributions of the forecast variables. This new methodology is illustrated on a synthetic case study. In addition to quickly reassess reservoir volume and flow performance predictions, this new approach can be used to select new data observation types and impact maps to assess potential well locations that would optimally reduce forecasting uncertainties.

1 Introduction

Most reservoir modeling projects involve updating current models with new information (e.g., logs from new wells, new seismic processing, or early production data). Models are typically rebuilt from scratch, with very little quantitative effort to check the consistency of the new data with the current models and to estimate the impact of those same new data on the project forecasts (e.g., oil-in-place or ultimate recovery). Yet, if the new data are consistent with the current models, i.e., if the new data values could have been predicted by some of the current models, there may be no need to rebuild models; a relationship between new data measurements predicted from current models and corresponding forecasts could be developed and used to update current forecasts with the actual new data measurements. This would not only save a considerable amount of time, allowing rapid reservoir management decisions in response to new information, but it would also reduce the risk of irrational fluctuations of the model forecast uncertainty range due to successive subjective reinterpretations of the data, arbitrary changes in modeling decisions, and/or introduction of new modeling technologies.

The proposed approach identified as "direct forecast updating" is quite similar to "direct forecasting," a new reservoir modeling methodology that aims at making reservoir forecasts by integrating data without performing any complex conditioning or inversion (Scheidt et al. 2015a; Satija and Caers 2015); the main difference is that prior models, which are fully unconstrained in direct forecasting, are replaced with reservoir models constrained by previously collected data. One particular focus of this paper is the direct updating of reservoir forecasts using new well data or, to be more specific, using statistical measures computed from those new well data, for example, well net-to-gross or well hydrocarbon pore column. When building models to make global forecasts, modelers only assess and model global uncertainties, such as reservoir facies proportions or porosity and permeability distributions. However, to be able to build a relationship between new well data predictions and global forecasts, local variability at the new well locations also needs to be captured in the current models. That local variability is derived from both geostatistical simulation stochasticity (seed number) and local input modeling parameter uncertainties, for example, local petrophysical property trends. In this paper a new methodology is proposed to account for such local variability when updating current forecasts directly with new well data.

2 Methodology

To explain the methodology proposed in this paper, the following simple case study is considered: the original oil-in-place (OOIP) forecasts of a reservoir need to be updated after a new well was drilled and an average net-to-gross value $\text{NTG} = \text{NTG}_m$ was estimated from the logs at the new well location. Statistically speaking, we want to compute $P\{\text{OOIP}|\text{NTG} = \text{NTG}_m\}$, which can be rewritten using Bayes' formulation as:

$$P\{\text{OOIP}|\text{NTG} = \text{NTG}_m\} = \frac{P\{\text{OOIP and NTG} = \text{NTG}_m\}}{P\{\text{NTG} = \text{NTG}_m\}} \tag{1}$$

Let $\boldsymbol{\theta}$ be the set of input modeling parameters representing the major global geological uncertainties identified in the reservoir, for example, the global reservoir rock volume or the reservoir porosity distribution.

The numerator and denominator of Eq. 1 can be rewritten using integrals over the whole input modeling parameter uncertainty space:

$$P\{\text{OOIP}|\text{NTG} = \text{NTG}_m\} = \frac{\int P\{\text{OOIP and NTG} = \text{NTG}_m|\,\boldsymbol{\theta}\}d\boldsymbol{\theta}}{\int P\{\text{NTG} = \text{NTG}_m|\,\boldsymbol{\theta}\}d\boldsymbol{\theta}} \tag{2}$$

The input modeling parameter uncertainty space can be sampled by drawing n equiprobable combinations $\boldsymbol{\theta}_i(i = 1 \ldots n)$ of global input modeling parameters:

$$P\{\text{OOIP}|\text{NTG} = \text{NTG}_m\} = \frac{\frac{1}{n}\sum P\{\text{OOIP and NTG} = \text{NTG}_m|\,\boldsymbol{\theta}_i\}}{\frac{1}{n}\sum P\{\text{NTG} = \text{NTG}_m|\,\boldsymbol{\theta}_i\}} \tag{3}$$

For each combination $\boldsymbol{\theta}_i$ of global input modeling parameters, multiple realizations can be generated to capture local uncertainties. Very often, generating multiple stochastic realizations by changing the random seed numbers of the geostatistical simulations is sufficient to capture local variability. However, in more complex cases, additional local variability such as local property trends may need to be accounted for.

One solution to compute $P\{\text{OOIP and NTG} = \text{NTG}_m|\,\boldsymbol{\theta}_i\}$ for any combination $\boldsymbol{\theta}_i$ of input parameters is to use a multi-Gaussian model, in which case only the means and standard deviations of the NTG and OOIP, as well as the correlation coefficient between NTG and OOIP, need to be estimated as a function of $\boldsymbol{\theta}_i$. The multi-Gaussian assumption can be tested using various methods (Mecklin and Mundfrom 2005) by generating a large number of realizations for some representative combinations $\boldsymbol{\theta}_i$ of input parameters. Provided that the multi-Gaussian assumption is not rejected, the NTG and OOIP means and standard deviations, as well as the correlation coefficient between NTG and OOIP, can be modeled using a design of experiments built for the global input modeling parameters space $\boldsymbol{\theta}$:

1. For each of the N experimental design runs, which correspond to a specific combination $\boldsymbol{\theta}_i(i = 1 \ldots N)$ of global input modeling parameters, generate L stochastic realizations.
2. For each of the L realizations, compute the OOIP and NTG value at the new well location.
3. Calculate the means and standard deviations of the L values of NTG and OOIP, as well as the correlation coefficient between the NTG and OOIP values.
4. Using the N experimental design runs, model a response surface for the NTG and OOIP means and standard deviations, as well as the correlation coefficient between NTG and OOIP.

Using the previous response surfaces, under the multi-Gaussian assumption, $P\{\text{OOIP and NTG} = \text{NTG}_m|\,\boldsymbol{\theta}_i\}$ and $P\{\text{NTG} = \text{NTG}_m|\,\boldsymbol{\theta}_i\}$ can be computed for a very large number n of combinations $\boldsymbol{\theta}_i$ randomly drawn from Monte Carlo simulation; this provides a new updated OOIP probability distribution according to Eq. 3.

The proposed approach has several advantages:

- First, the multi-Gaussian assumption, combined with the use of response surfaces to estimate the parameters of the multi-Gaussian model for any combination $\boldsymbol{\theta}_i$ of input parameters, allows fully determining the bivariate distribution P {OOIP and NTG}; there is no need to use any arbitrary interpolation technique such as the traditional Kernel smoothing (Park et al. 2013; Scheidt et al. 2015b).
- Then, the exact NTG_m value can be directly plugged into the multi-Gaussian function $P\{\text{NTG} = \text{NTG}_m|\,\boldsymbol{\theta}_i\}$ for any combination $\boldsymbol{\theta}_i$ of input parameters; there is no need to determine a quite arbitrary bandwidth around the new data measurements (Scheidt et al. 2015a).
- $P\{\text{NTG} = \text{NTG}_m|\,\boldsymbol{\theta}_i\}$ provides the probability that the NTG value measured at the new well location will be observed for the specific combination $\boldsymbol{\theta}_i$ of global input modeling parameters. Typically, one would expect the correlation between NTG and OOIP over multiple stochastic realizations to be quite low. If this is indeed the case, i.e., if NTG and OOIP are conditionally independent, which can be tested, Eq. 3 can be rewritten as:

$$P\{\text{OOIP}|\text{NTG} = \text{NTG}_m\} = \frac{\frac{1}{n}\sum P\{\text{OOIP}|\,\theta_i\}P\{\text{NTG} = \text{NTG}_m|\,\theta_i\}}{\frac{1}{n}\sum P\{\text{NTG} = \text{NTG}_m|\,\theta_i\}} \qquad (4)$$

In that new Eq. 4, the updated OOIP forecasts can be interpreted as the linear combination of the OOIP forecasts corresponding to each possible combination $\boldsymbol{\theta}_i$ of input parameters weighted by the probabilities that the NTG value at the new well location be observed in the stochastic model realizations generated for $\boldsymbol{\theta}_i$.

Extending the previous methodology to multiple new measurements, e.g., the NTG values from several new wells, is straightforward, provided that the multi-Gaussian assumption holds.

3 Illustrative Case Study

As an illustrative example, the following synthetic data set mimicking an actual Chevron reservoir is used with a reference model corresponding to a tidal dominated reservoir with 25 % sandbars. The synthetic example was generated using the multiple-point statistical simulation program snesim (Strebelle 2002). There is no horizontal facies proportion trend, but a significant vertical trend; see the three horizontal sections of the model displayed in Fig. 1.

The tidal sandbar porosity distribution is approximatively normal, with a 19 % mean, and the permeability distribution is lognormal with a 50 md mean. Porosity was simulated using SGS, while permeability was simulated from porosity using SGS with collocated cokriging and a 0.9 correlation coefficient. The background shale porosity and permeability are assumed to be close to 0. The reference OOIP is about 80 M bbl.

Table 1 provides the list of global input modeling parameters and their corresponding uncertainties.

The reference model is consistent with the uncertainty ranges defined for the different global input modeling parameters. A D-optimal design of experiments (Atkinson et al. 2007) was used to generate 99 models, and a quadratic response surface was computed to estimate the initial OOIP probabilistic distribution displayed in Fig. 2. The initial P50 value (93 M bbl.) significantly overestimates the OOIP from the reference model, while the uncertainty range is relatively broad with a P10 value of 61 M bbl. and a P90 value of 128 M bbl.

Three wells were used to condition all the initial models. The objective of this case study is to update the initial OOIP forecasts using the NTG values measured at two alternative new well locations. The first location was randomly selected and has a relatively low NTG of 4 %, whereas the second location is very close to an existing well and has a NTG value of 20 % similar to the NTG of that well (see Fig. 3).

The methodology presented in the previous section was applied by generating 10 stochastic realizations for each of the 99 runs of the D-optimal experimental design. For each alternative new well location, the NTG and OOIP means and standard deviations, as well as the correlation coefficient between NTG and OOIP, were computed for the 99 experimental design runs, and quadratic response surfaces were computed as a function of the input modeling parameters.

Then, 10,000 combinations $\boldsymbol{\theta}_i$ of input modeling parameters were generated from Monte Carlo simulation. Figure 4 shows the histograms of predicted NTG values for each new well location. The actual NTG value observed at the first new well location corresponds to the 9th percentile of the prediction distribution, while

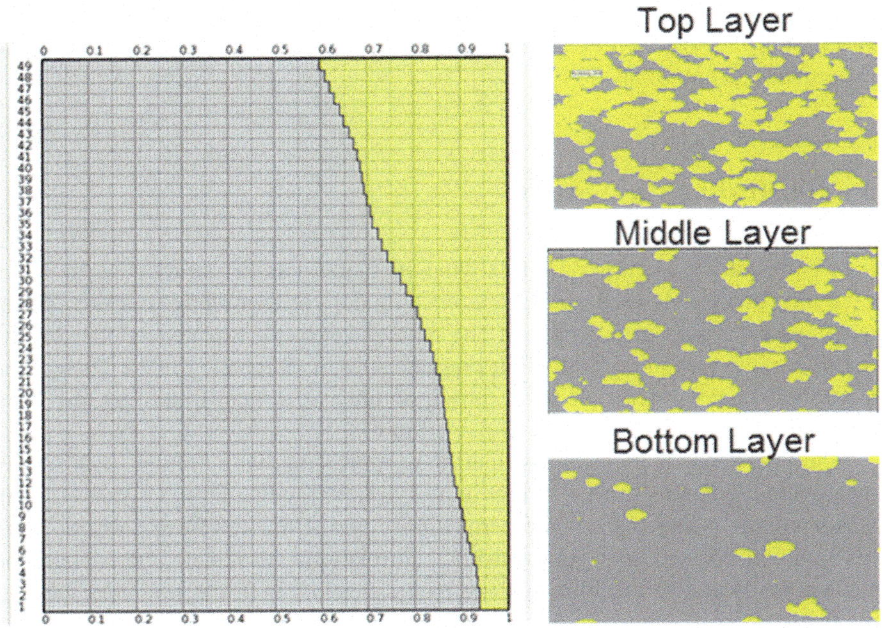

Fig. 1 Facies proportion curve and three horizontal sections of the reference model

Table 1 Global input modeling parameters, uncertainty ranges, and reference model

Input modeling parameters	P10	P50	P90	Reference
Reservoir sand proportion	20 %	30 %	40 %	25 %
Sand geobody shapes	Ellipses	Variogram-based	Channels	Ellipses
Vertical trend	None	Medium	High	Medium
Horizontal trend	None	Medium	High	None
Porosity average	15 %	17 %	19 %	19 %
Permeability average	10 md	50 md	250 md	50 md
Poro/perm correlation	0.5	0.7	0.9	0.9

the actual NTG value observed at the second new well location corresponds to the 64th percentile. Therefore, in both cases, the new NTG measurements can be considered as consistent with the existing models, and the previously described forecast updating process can be applied.

Figure 5 provides a bubble graph displaying the bivariate distribution $P\{\text{OOIP and NTG}\}$ resulting from Eq. 3 for the first new well location.

Table 2 provides the P10, P50, and P90 OOIP values for the two alternative new well locations.

For the first well, the P50 value of the updated forecasts is closer to the reference OOIP value (80 M bbl.), while the uncertainty forecast range significantly decreased: the new P10-P90 difference is 46 M bbl. vs. 67 M bbl. initially. In contrast, as expected, because the second well is very close to an existing well, its

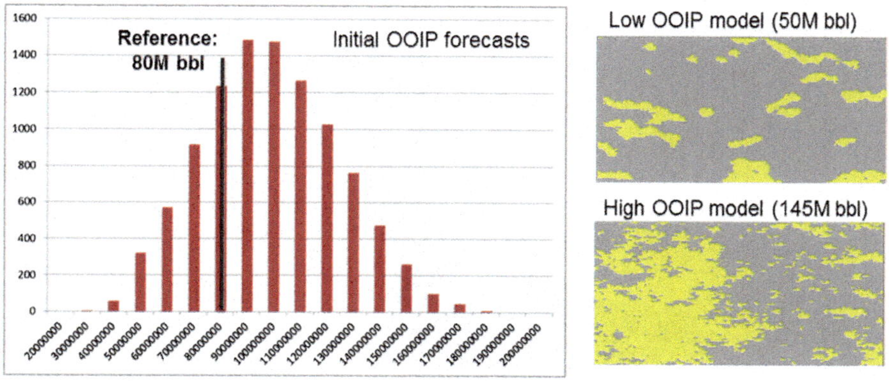

Fig. 2 Initial forecasts and horizontal sections of two models generated using the D-optimal experimental design

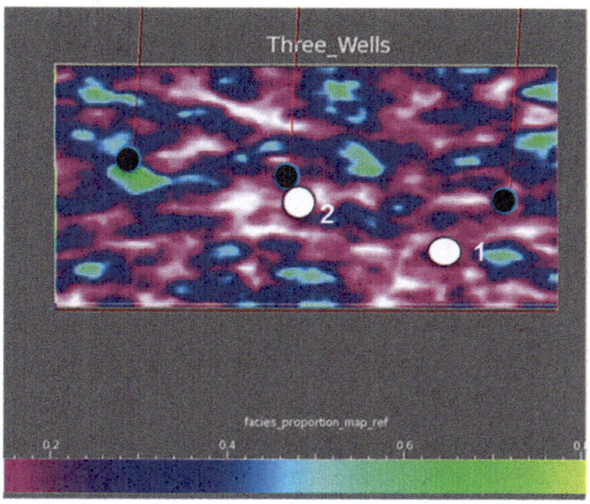

New well location	NTG
1 (random)	4%
2 (close to existing well)	20%

Fig. 3 NTG map from reference model, with locations of the three existing wells (*black dots*) and two alternative new wells (*white dots*)

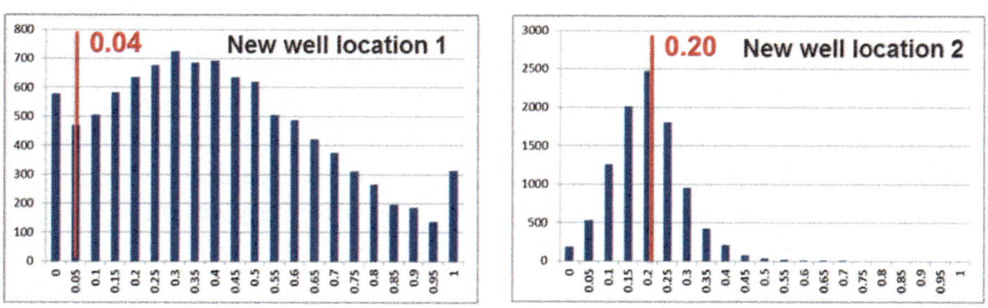

Fig. 4 Histograms of the NTG predictions for both new well locations. The *red line* corresponds to the actual observed NTG value

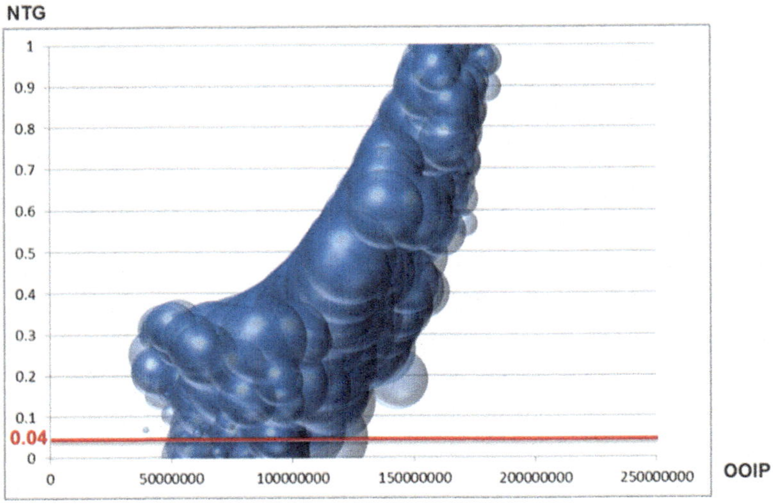

Fig. 5 Bivariate distribution $P\{\text{OOIP and NTG}\}$ for the first new well location. Each bubble corresponds to a particular combination $\boldsymbol{\theta}_i$ of global input modeling parameters; it is centered at the estimated NTG and OOIP mean values, and its size is proportional to the estimated NTG standard deviation (only 200 bubbles are displayed). The *red line* corresponds to the actual NTG value observed at that first new well location

Table 2 Updated forecasts for the two alternative new well locations

New well location	P10	P50	P90
1 (random)	57	79	103
2 (close to existing well)	66	97	125
Initial forecasts	61	93	128

impact on OOIP forecasts is extremely limited; thus the P10, P50, and P90 values of the updated forecasts (66, 97, and 125 M bbl.) are very close to the initial forecasts (61, 93, and 128 M bbl.).

The same methodology based on Eq. 3 can be applied to the case where both wells 1 and 2 are drilled. This requires the computation of an additional response surface: the correlation between NTG values at wells 1 and 2 for any combination $\boldsymbol{\theta}_i$ of global input parameters. Figure 6 provides the scatterplot of predicted NTG values at well location 1 versus predicted NTG values at well location 2 for 10,000 combinations $\boldsymbol{\theta}_i$ of input modeling parameters. The actual NTG values (0.04 for the first well and 0.2 for the second well) are in the predicted ranges. Thus the combination of the two new NTG measurements can be considered as consistent with the existing models, which confirms that the previously described forecast updating process can be applied. Note that several methods exist to quantitatively check that consistency between new observed values and predictions, in particular the Mahalanobis distance (Mahalanobis 1936).

When both new well locations are used, the P10, P50, and P90 values of the updated forecasts are 65, 86, and 105 M bbl. The new P10-P90 difference is 40 M

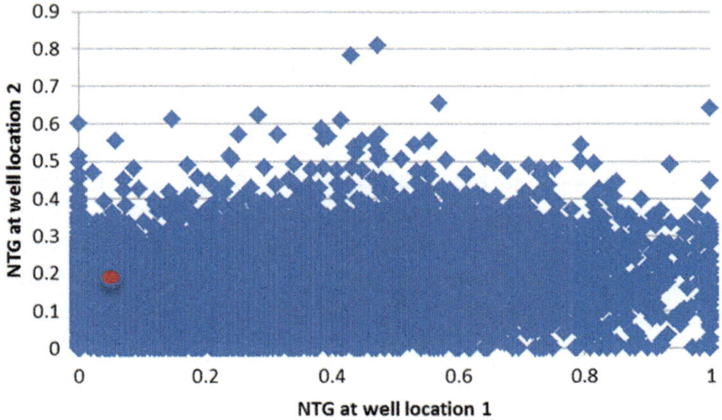

Fig. 6 Scatterplot of predicted NTG values at well location 1 versus predicted NTG values at well location 2 for 10,000 combinations $\boldsymbol{\theta}_i$ of input modeling parameters. The *red dot* corresponds to the actual NTG values observed at the new well location (0.04 for well location 1 and 0.2 for well location 2)

bbl., which is, as expected but not guaranteed, smaller than the P10-P90 difference obtained for each well considered individually.

4 Discussion

The results obtained in the case study above show that reservoir forecasts can be directly updated in the presence of new information without rebuilding any models provided that the new information is consistent with the existing reservoir models. It should be noted that there is no guarantee for the P50 value of the updated forecasts to be closer to the true reservoir value or for the updated uncertainty range to systematically decrease; it all depends on the new data measured value. However, getting more accurate and precise forecasts is expected on average as the number of additional new wells increases.

In most cases, the previous methodology can be simplified by replacing some response surface with constant values or straightforward functions. For example, in the previous case study, it can be observed that, as expected, OOIP varies very little across multiple stochastic realizations for any particular combination $\boldsymbol{\theta}_i$ of global input parameters. On average over the 99 experimental design runs, the coefficient of variation (ratio between standard deviation and mean) is only 0.004. This means that only the response surface for the OOIP mean need be modeled; the OOIP standard deviation could be directly estimated by multiplying the OOIP mean by 0.004. This simplification provides updated P10, P50, and P90 OOIP values very close (less than 0.5 % relative difference) to the updated forecasts obtained using a full response surface for the OOIP standard deviation. Ignoring completely the

OOIP standard deviation, i.e., setting it to a constant 0, still provides updated forecasts with less than 1 % relative difference compared to the initial methodology. In contrast, the coefficient of variation for the NTG is 0.803 for the first new well location and 0.243 for the second well location (close to an existing well) on average over the same 99 experimental design runs, which demonstrates the importance of capturing all the local variability at new well locations, a quite challenging exercise. Finally, the correlation coefficient between NTG and OOIP is systematically low for the 99 experimental design runs, 0.03 on average. Thus Eq. 4, which assumes the conditional independence of NTG and OOIP for any combination $\boldsymbol{\theta}_i$ of global input modeling parameters, could have been used. This simplification would have again yielded very similar updated P10, P50, and P90 OOIP values (less than 0.1 % difference).

Expanding the proposed methodology to more than two new measurements and/or forecast variables is straightforward from a theoretical point of view, but will need to be tested in future work. In particular, when a large amount of new data is available, our methodology may require the identification of a limited number of physical metrics representing or summarizing the new data. For example, in the illustrative case study presented in this paper, the average NTG at the new well location was used instead of the whole facies log. If summary physical metrics are difficult to identify or compute, brute-force dimensionality reduction techniques such as nonlinear PCA (Scheidt et al. 2015a) could be applied.

Also, the proposed approach calls for the use of experimental design and the construction of response surfaces, which limits the application to continuous and ordinal input modeling parameters. However, other direct forecasting methodologies could reuse the main idea of this paper: explicitly account for local variability at new measurement locations by modeling new measurement predictions and global forecasts as the sum of an average value over multiple simulated realizations and a residual. The average value captures the impact of the global modeling uncertainty parameters, while the residual captures local variability, especially geostatistical simulation stochasticity.

5 Conclusions

A simple methodology using a Bayesian framework with a traditional multi-Gaussian assumption is presented in this paper to update forecasts after acquiring new well log data, which allows making rapid reservoir management decisions in response to new information. One main advantage of this methodology is that it fully accounts for local uncertainties, in particular model stochasticity, when estimating the impact of new local data on reservoir forecasts.

The proposed methodology was successfully applied to a simple synthetic case study, but it needs to be further tested on more complex synthetic data sets and actual reservoir modeling projects. Another next step is to use that methodology to select what new data should be collected and where it should be collected to optimally decrease reservoir forecasts uncertainty similar to the impact map approach of Zagayevskiy and Deutsch (2013).

Bibliography

Atkinson A, Donev A, Tobias R (2007) Optimum experimental designs, with SAS. UOP Oxford, Oxford

Mahalanobis P (1936) On the generalised distance in statistics. Proc Natl Inst Sci India 2(1):49–55

Mecklin C, Mundfrom D (2005) A monte carlo comparison of the type I and type II error rates of tests of multivariate normality. J Stat Comput Simul 75(2):93–107

Park H, Scheidt C, Fenwick DH et al (2013) History matching and uncertainty quantification of facies models with multiple geological interpretations. Comput Geosci 17:609–621

Satija A, Caers J (2015) Direct forecasting of subsurface flow response from non-linear dynamic data by linear least-squares in canonical functional principal component space. Advances in Water Resources 77:69–81

Scheidt C, Renard P, Caers J (2015a) Prediction-focused subsurface modeling: investigating the need for accuracy in flow-based inverse modeling. Math Geosci 47(2):173–191

Scheidt C, Tahmasebi P, Pontiggia M et al (2015b) Updating joint uncertainty in trend and depositional scenario for reservoir exploration and early appraisal. Comput Geosci 19:805–820

Strebelle S (2002) Conditional simulation of complex geological structures using multiple-point statistics. Math Geol 34(1):1–21

Zagayevskiy Y, Deutsch CV (2013) Impact map for assessment of new delineation well locations. JCPT 2013:441–462

Statistical Scale-Up of Dispersive Transport in Heterogeneous Reservoir

Vikrant Vishal and Juliana Y. Leung

Abstract Numerical methods are often used to simulate and analyze flow and transport in heterogeneous reservoirs. However, they are limited by computational restrictions including small time steps and fine grid size to avoid numerical dispersion. The ability to perform efficient coarse-scale simulations that capture the uncertainties in reservoir attributes and transport parameters introduced by scale-up remains challenging. A novel method is formulated to properly represent sub-grid variability in coarse-scale models. First, multiple sub-grid realizations depicting detailed fine-scale heterogeneities and of the same physical sizes as the transport modeling grid block are subjected to random walk particle tracking (RWPT) simulation, which is not prone to numerical dispersion. To capture additional unresolved heterogeneities occurring below even the fine scale, the transition time is sampled stochastically in a fashion similar to the continuous time random walk (CTRW) formulation. Coarse-scale effective dispersivities and transition time are estimated by matching the corresponding effluent history for each realization with an equivalent medium consisting of averaged homogeneous rock properties. Probability distributions of scale-up effective parameters conditional to particular averaged rock properties are established by aggregating results from all realizations. Next, to scale-up porosity and permeability, volume variance at the transport modeling scale is computed corresponding to a given spatial correlation model; numerous sets of "conditioning data" are sampled from probability distributions whose mean is the block average of the actual measured values and the variance is the variance of block mean. Multiple realizations at the transport modeling scale are subsequently constructed via stochastic simulations. The method is applied to model the tracer injection process. Results obtained from coarse-scale models where properties are populated with the proposed approach are in good agreement with those obtained from detailed fine-scale models. With the advances in

V. Vishal (✉)
School of Mining and Petroleum Engineering, University of Alberta, 7-207 Donadeo Innovation Centre for Engineering, Edmonton, AB T6G 1H9, Canada
e-mail: vvishal@ualberta.ca

J.Y. Leung
School of Mining and Petroleum Engineering, University of Alberta, 6-285 Donadeo Innovation Centre for Engineering, Edmonton, AB T6G 1H9, Canada
e-mail: juliana2@ualberta.ca

J.J. Gómez-Hernández et al. (eds.), *Geostatistics Valencia 2016*, Quantitative Geology and Geostatistics 19, DOI 10.1007/978-3-319-46819-8_50

nanoparticle technology and its increasing application in unconventional reservoirs, the method presented in this study has significant potential in analyzing tracer tests for characterization of complex reservoirs and reliable assessment of fluid distribution. The approach can also be employed to study scale-dependent dispersivity and its impacts in miscible displacement processes.

1 Introduction

Accurate predictions of flow and transport in natural porous media are crucial in the management of valuable subsurface resources including water aquifers and hydrocarbons reservoirs. These predictions are usually assessed with uncertainty due to (1) underlying heterogeneity or spatial variation in rock and transport properties at multiple scales and (2) conditioning data. This uncertainty, though can be reduced by improved geophysical (e.g., seismic), core, well log, pressure test, and tracer test data, cannot be entirely eliminated. Reservoir models are typically constructed and subjected to flow and transport simulation to capture the aforementioned uncertainties. Although fine-scale models could capture detailed description of the heterogeneity, simulation with these models can be computationally demanding. A commonly adopted alternative is to replace these fine-scale models with lower-resolution coarse-scale (scale-up) models. During this process of coarsening, a number of transport properties (e.g., dispersivity), along with reservoir properties (e.g., porosity and absolute permeability), must be scaled up accordingly. Properly scale-up models should not only honor the conditioning data, but they should also account for the associated subscale heterogeneities.

Mass transfer mechanisms in single-phase flow involve: (1) advection or convection, (2) diffusion, and (3) mechanical dispersion. These phenomena are generally described by the parabolic advection-dispersion equation (ADE). Common solution techniques, such as finite volume and finite element, suffer various computational restrictions including numerical dispersion; poor predictions are observed if flow is advection-dominated (high Péclet number). An alternative solution framework is the classical random walk particle tracking (RWPT), which is free from numerical dispersion and imposes no restriction on grid size.

Another conceptual issue associated with the ADE is its inability to capture non-Fickian transport behavior. ADE considers dispersion as a sum of diffusion and convective spreading, ignoring any additional mixing introduced by the interaction of these two mechanisms. It neglects heterogeneity that is not resolved completely at the volume support on which the ADE and its associated parameters are defined. It is true that if heterogeneities at all scales are modeled explicitly, the ADE reflects non-Fickian behavior of transport (Salamon et al. 2007; John 2008; Rhodes et al. 2009; Li et al. 2011).

To model the non-Fickian behavior due to unresolved heterogeneities, the continuous time random walk (CTRW) has been developed as a viable option. As opposed to the RWPT (which models the dispersion of space walks and reproduces the ADE), CTRW takes into account dispersion of the particles' times of flight by sampling the transition time from a probability distribution function (pdf).

In this paper, a statistical framework is implemented to facilitate the representation of sub-grid variability in coarse-scale models. To capture additional unresolved heterogeneities occurring below even the fine scale, the transition time is sampled stochastically in a fashion similar to the CTRW formulation. Coarse-scale conditional probability distributions of effective dispersivities and transition time are established.

2 Governing Equations of Flow and Transport

Solute (or particle) transport in porous media is traditionally described by the (Fickian-based) advection-dispersion equation (ADE):

$$\frac{\partial c(\boldsymbol{X},t)}{\partial t} = \nabla \cdot (\boldsymbol{D}\nabla c(\boldsymbol{X},t)) - \nabla \cdot (\boldsymbol{V}c(\boldsymbol{X},t)). \tag{1}$$

$c(\boldsymbol{X}, t)$ is the volumetric solute concentration, t is the time, $\boldsymbol{X}$ is the spatial coordinates, $\boldsymbol{V}$ is the macroscopic flow velocity, and $\boldsymbol{D}$ is the dispersion coefficient tensor. The flow velocity is estimated by solving the continuity equation for incompressible flow, Eq. 2, and the Darcy Eq. 3:

$$\nabla \cdot \boldsymbol{V} = 0\,. \tag{2}$$

$$\boldsymbol{V} = -\frac{\boldsymbol{k}}{\mu}\nabla p. \tag{3}$$

$\boldsymbol{k}$ is the permeability tensor; p and μ are the pressure and viscosity of the fluid, respectively. Equation 1 can be solved using the particle-tracking method, which simulates mass transport by discretizing injected mass concentration into a number of particles (walkers) of equal mass; over a given time step, the movement of each particle is controlled by advection and dispersion/diffusion. The particle tracking scheme is summarized by Eq. 4 (Kinzelbach 1986; Tompson and Gelhar 1990; LaBolle et al. 1996; Delay et al. 2005; Salamon et al. 2006):

$$\begin{aligned} &\boldsymbol{X}_p(t+\Delta t) = \boldsymbol{X}_p(t) + \boldsymbol{A}\left[\boldsymbol{X}_p(t)\right]\Delta t + \boldsymbol{B}\left[\boldsymbol{X}_p(t)\right].\boldsymbol{\xi}(t)\sqrt{\Delta t}. \\ &\boldsymbol{A} = \boldsymbol{V} + \nabla \boldsymbol{D}, \boldsymbol{B}\boldsymbol{B}^T = 2\boldsymbol{D}. \end{aligned} \tag{4}$$

$\boldsymbol{X}_p(t)$ is the position of a particle at time t; Δt is the time step; $\boldsymbol{\xi}$ is a vector of independent, normally distributed random numbers with zero mean and unit variance. If the transition time is considered to be deterministic and constant for all particles, the above formulation reduces to the classical RWPT framework and converges to the ADE. To capture non-Fickian features due to unresolved heterogeneities occurring below the fine scale, Srinivasan et al. (2010) proposed a different particle-tracking model, in which the transition time is considered to be stochastic and vary among particles:

$$\begin{aligned} \boldsymbol{X}_p(\tau+\Delta\tau) &= \boldsymbol{X}_p(\tau) + \boldsymbol{A}\left[\boldsymbol{X}_p(\tau)\right]\Delta\tau + \boldsymbol{B}\left[\boldsymbol{X}_p(\tau)\right]\boldsymbol{\xi}(\tau)\sqrt{\Delta\tau}. \\ t(\tau+\Delta\tau) &= t(\tau) + \eta(\tau,\Delta\tau). \end{aligned} \tag{5}$$

($\boldsymbol{\xi}$, η) is a set of random series that is drawn from the joint probability density function of transition length ξ and time η. In this formulation, time steps are incorporated implicitly and sampled from time distribution density function.

This formulation of the RWPT with stochastic transition time is equivalent to the commonly adopted CTRW framework for modeling non-Fickian transport. The governing equation in CTRW is the Fokker-Planck with memory equation, Eq. 6, which is derived from the equivalence of the classical generalized master equation and CTRW (Berkowitz et al. 2002). Solute particle migration is described as a series of jumps over different transition times. The jump position ($\boldsymbol{X}$) and transition times (t) follow a joint probability density function $\psi(\boldsymbol{X}, t)$ that describes the jump at position and time t:

$$s\widetilde{c}(\boldsymbol{X},s) - c_o(\boldsymbol{X}) = -\widetilde{M}(s)(\boldsymbol{V}\cdot\nabla\widetilde{c}(\boldsymbol{X},s) - \nabla\cdot(\boldsymbol{D}\nabla\widetilde{c}(\boldsymbol{X},s))). \tag{6}$$

where $\widetilde{c}$ denotes the Laplace transform of c and $c_o(\boldsymbol{X})$ is the initial condition. The particle transport velocity $\boldsymbol{V}$ and the dispersion tensor $\boldsymbol{D}$ are defined by the first and second moments of the pdf of the transition length, $p(\boldsymbol{X})$:

$$\begin{aligned} v_i &= \frac{1}{t_1}\int_{\Re} p(\boldsymbol{X})X_i d\boldsymbol{X} = \frac{1}{t_1}\sum_{\boldsymbol{X}} X_i p(\boldsymbol{X}). \\ D_{ij} &= \frac{1}{2t_1}\int_{\Re} p(\boldsymbol{X})X_i X_j d\boldsymbol{X} = \frac{1}{2t_1}\sum_{\boldsymbol{X}} X_i X_j p(\boldsymbol{X}). \end{aligned} \tag{7}$$

t_1 is the median transition time or lower cutoff time. The memory function $\widetilde{M}(s)$ that accounts for the unknown heterogeneities below the model resolution is defined as

$$\widetilde{M}(s) = \frac{st_1\widetilde{\varphi}(s)}{1-\widetilde{\varphi}(s)}. \tag{8}$$

Equation 6 is obtained by decoupling the pdf $\widetilde{\psi}(\boldsymbol{X},s)$ as

$$\widetilde{\psi}(\boldsymbol{X},s) = p(\boldsymbol{X})\widetilde{\varphi}(s). \tag{9}$$

$\widetilde{\psi}(\boldsymbol{X}, s)$ is the Laplace transform of the $\psi(\boldsymbol{X}, t)$, and $\widetilde{\varphi}(s)$ is the pdf of the transition time in Laplace space. The Gaussian function is typically considered for $p(\boldsymbol{X})$, while the truncated power-law, TPL, (Eq. 10) or exponential (Eq. 11) functions are considered for $\varphi\ (t)$ in the case of non-Fickian and Fickian transport, respectively (Margolin et al. 2003; Dentz et al. 2004; Berkowitz et al. 2006; Gao et al. 2009):

$$\varphi(t) = \frac{(1 + t/t_1)^{-1-\beta}}{t_1 r^{\beta}\Gamma(-\beta, r)} e^{\left(-\frac{t_1+t}{t_2}\right)}, r = \frac{t_1}{t_2}, t_1 < t_2, 0 \leq \beta \leq 2. \tag{10}$$

$$\varphi(t) = \frac{1}{t_1} e^{\left(-\frac{t}{t_1}\right)} \tag{11}$$

Γ is the incomplete gamma function, β is an exponent controlling the nature of transport behavior, and t_2 is the upper cutoff time (Dentz et al. 2004; Cortis et al. 2004).

To illustrate that the implemented particle-tracking model follows the CTRW framework, single-phase tracer transport through a one-dimensional homogeneous porous medium, with the following initial and boundary conditions, is considered (Kreft and Zuber 1978):

Boundary conditions:

$$c(X_1 = 0, t) = c_o \text{ and } c(X_1 = L, t) = 0. \tag{12}$$

Initial conditions:

$$c(X_1, t = 0) = 0. \tag{13}$$

The corresponding solution in the Laplace space for the flux-average concentration (c_f) is determined by

$$\widetilde{c}_f(X_1, s) = c_o \exp\left[-\frac{vX_1}{2D_L}\left\{\sqrt{1 + 4\frac{sD_L}{\widetilde{M}(s)v^2}} - 1\right\}\right]. \tag{14}$$

D_L is the longitudinal dispersion and it is equal to $\alpha_L v$, where α_L is the longitudinal dispersivity. Equation 14 is computed using the CTRW toolbox that is readily available in the public domain (Cortis and Berkowitz 2005). The solutions are compared against those obtained with the particle-tracking model used in this study for two scenarios, where transition time is sampled from Eqs. 10 and 11, respectively. A total of 10^4 particles are employed. Fluid velocity $(v) = 0.0342$ km/year; $\alpha_L = 0.5$ km; $t_1 = 2.84 \times 10^{-2}$ year, $t_2 = 4.44 \times 10^4$ year; $\beta = 1.35$. The ensuing effluent histories obtained from the two models are compared in Fig. 1. To demonstrate the capability of the numerical model for two or three dimensions, a couple of classical cases, such as transport through layered media when flux is perpendicular and parallel to the layers, are tested, and the results are shown in Fig. 2.

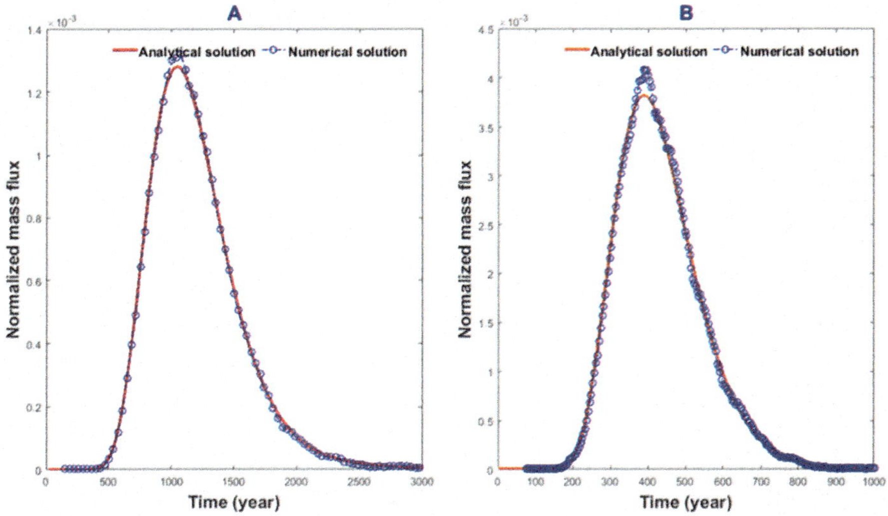

Fig. 1 Normalized mass flux at $X_1 = L = 15.2$ km corresponding to $c_o = 1$ for: (**A**) non-Fickian model and (**B**) Fickian model

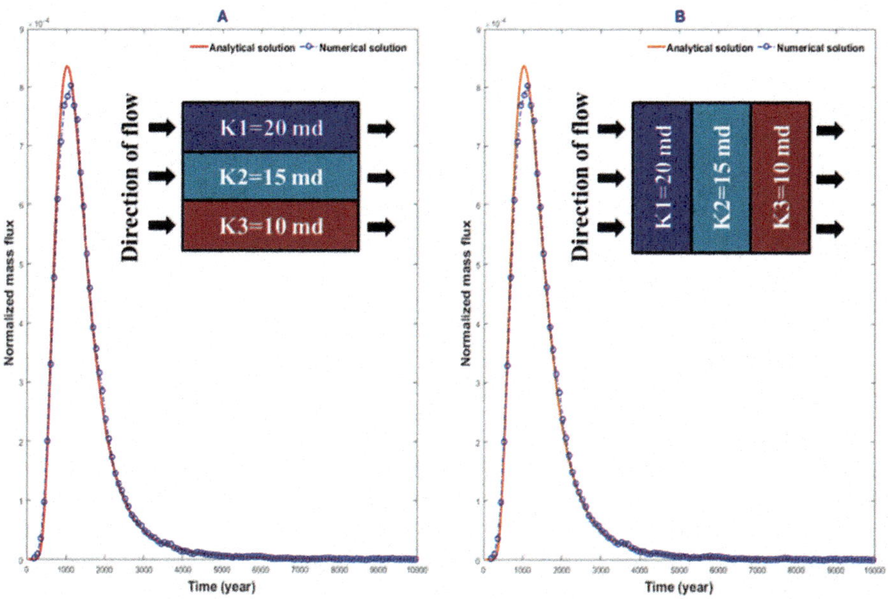

Fig. 2 Normalized mass flux at $X_1 = L = 16.0$ km corresponding to $c_o = 1$ for a non-Fickian model when flux is: (**A**) parallel and (**B**) perpendicular to the layered porous media

3 Method

3.1 Scale-Up of Reservoir Properties

The change in volume support introduces additional uncertainty in the coarse-scale models due to the averaging of subscale heterogeneity. This uncertainty is referred to as subscale variability. The variance of mean of the attribute is a measure of the variability associated with the spatial average at a particular spatial volume V, and it decreases with V. For a multivariate Gaussian random variable (Z) with a variance (σ^2) and a autocorrelation model ρ_{corr}, its spatial mean ($\bar{Z}$) is obtained via linear averaging; therefore, the corresponding variance of mean or $\text{Var}(\bar{Z})$ can be computed by integrating over all possible lag distance η over a volume V (Lake and Srinivasan 2004):

$$\text{Var}(\bar{Z}) = \frac{2\sigma^2}{V^2}\left(\int_{v}\int_{\eta}\rho_{\text{corr}}(\eta)\ d\eta\ d\xi\right). \tag{15}$$

Apart from the volume variance, the spatial correlation length also changes with volume support. Averaged semi-variogram $\bar{\gamma}$ is estimated numerically by averaging the point-scale variogram γ on two volume supports V and V' (Journel and Huijbregts 1978):

$$\bar{\gamma}\left(V, V'\right) = \frac{1}{VV'}\int_{V}\int_{V'}\gamma\left(v, v'\right)dvdv' \approx \frac{1}{nn''}\sum_{i=1}^{n}\sum_{j=1}^{n'}\gamma\left(\boldsymbol{h}_{\text{ij}}\right). \tag{16}$$

A procedure, which was described in Leung and Srinivasan (2011) and was also implemented in Vishal and Leung (2015), is adopted to construct coarse-scale models of reservoir properties that capture the subscale variability:

1. Compute the variance of mean ($\text{Var}(\bar{Z})$) according to Eq. 15.
2. Estimate averaged variogram $\bar{\gamma}$ according to Eq. 16.
3. Draw multiple sets of conditioning data via parametric bootstrapping of a Gaussian likelihood function, whose mean and variance are the block average of the fine-scale conditioning data and $\text{Var}(\bar{Z})$, respectively.
4. Perform conditional simulation at the coarse scale (i.e., transport modeling scale) using histogram (mean = fine-scale global mean; variance = $\text{Var}(\bar{Z})$) and $\bar{\gamma}$ (which accounts for the spatial correlation) for a particular set of conditioning data from step #3.
5. Repeat step #4 for the remaining sets of conditioning data from step #3.

In this work, porosity (ϕ) is modeled according to above mentioned steps, while absolute permeability (k) is assumed to be correlated with porosity as $k = a \times \phi^b$, where a and b are empirical constants. This assumption would imply that linearly averaging is also valid in the transformed space of $(k/a)^{1/b}$; therefore $\bar{k}$ is estimated from $\overline{\phi}$ using the same empirical relation, assuming the transform relationship is invariant with scale (Leung and Srinivasan 2011). In principle, permeability can be scaled up in the same fashion as for the transport parameters, as explained in the next section, particularly if a simple correlation with porosity is unattainable.

3.2 Scale-Up of Solute Transport Parameters

As discussed in the introduction, subscale variability also introduces uncertainty in transport-related attributes at the coarse scale and contributes to non-Fickian behavior. A particle-tracking model is employed for transport modeling. The transition time is sampled from the TPL distribution, which is characterized by β, t_1, and t_2. A workflow, which is modified from the one presented in Vishal and Leung (2015), is implemented to construct probability distributions of the coarse-scale TPL parameters. In this work, only probability distribution of β is estimated, since β primarily controls the nature of particle migration and, thus, functionally characterizes the dispersion behavior. Results obtained from the case study (next section) also suggest that proper scale-up of β alone would have reasonably captured the detailed fine-scale response. The steps are described as follows:

1. Assign n_b bins to the histogram of $\overline{\phi}$.
2. For a given bin, perform unconditional sequential Gaussian simulation to generate n_s sub-grid realizations of $\overline{\phi} \sim N(\overline{\phi}_b, \sigma^2)$ using the fine-scale variogram γ. The corresponding permeability value at each location is assigned according to $k = a \times \phi^b$.
3. Construct an equivalent homogeneous model with averaged properties for each n_s sub-grid model from step #2.
4. Compute steady-state velocity fields (incompressible flow) and solute transport for both n_s heterogeneous models and n_s homogeneous models constructed in steps #2–3, respectively.
5. Estimate an effective value of β (i.e., β^*) by minimizing the root mean square error in the breakthrough concentration profile between each heterogeneous model and its equivalent homogeneous model. β^* corresponds to the value of β at the coarse scale.
6. Repeat steps #2–5 for other bins (step #1). Aggregate results over all realizations and construct the probability distribution of $P\{\beta^* | \overline{\phi}_i\}$ for $i = 1,\ldots,n_b$.
7. To construct the coarse-scale models, β^* is assigned at each location by sampling from $P\{\beta^* | \overline{\phi}\}$.

4 Case Study

A synthetic two-dimensional model of size 500 × 500 m ($\Delta x = \Delta y = 1$ m) is considered. The true fine-scale porosity (ϕ) model is shown in Fig. 3**A**. It follows a normal distribution $N(0.25, 0.07)$ and spherical variogram (correlation lengths along the X- and Y-direction are 200 m and 50 m, respectively). The absolute permeability is modeled as k (md) = 25,000 × ϕ^2. The value of k ranges from 81.0 to 3058.4 md (mean = 1566.1 md), while ϕ ranges from 0.05 to 0.35 (mean = 0.25). In addition, α_L and α_T at the 1-m^2 volume support is equal to 0.5 m and 0.05 m, respectively. The following TPL parameters are used: $t_1 = 1.25$ year, $t_2 = 10{,}000$ year, and $\beta = 1.25$.

The objective is to construct a set of coarse-scale 50 × 50 models ($\Delta x = \Delta y = 10$ m), which are subsequently subjected to transport modeling. A normalized variance of mean of 0.8 is obtained based on Eq. 2, and scale-up variogram (correlation length in X-direction is 215 m and Y-direction is 65 m) is calculated based on Eq. 16 corresponding to a volume support 10 × 10 m. This relatively high value suggests that a substantial amount of subscale variability is present.

A total of 100 realizations of porosity and permeability at the coarse scale are constructed following the method for scale-up of reservoir properties described in the previous section. For each realization, the corresponding TBL parameters (β, t_1, and t_2) at the coarse scale are constructed following the method for scale-up of solute transport parameters described in the previous section. An example of a randomly selected realization is shown in Fig. 3**B**, which exhibits some dissimilarities when compared to Fig. 3**A**: (1) there are fewer low values in the coarse-scale model, since the $Var(\overline{\phi}) < Var(\phi)$; in addition, values are correlated over a longer distance in the coarse-scale model since the ranges associated with $\overline{\gamma}$ are higher. Each realization is then subjected to particle-tracking modeling. The breakthrough

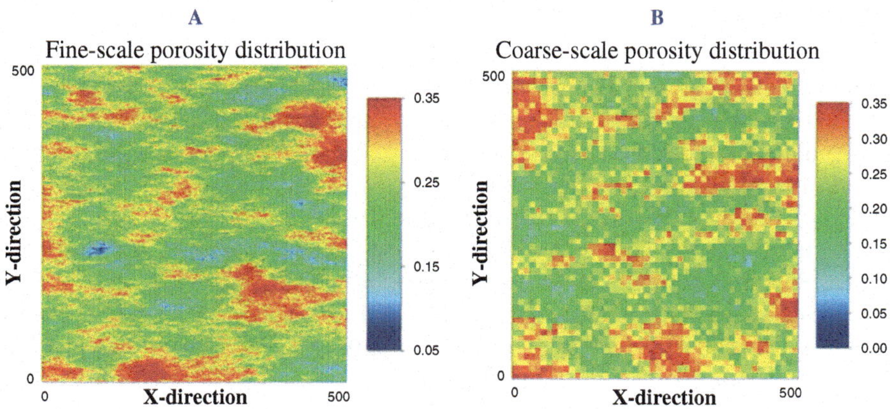

Fig. 3 Distribution of porosity: (**A**) true fine-scale model and (**B**) a randomly selected realization of the coarse-scale model

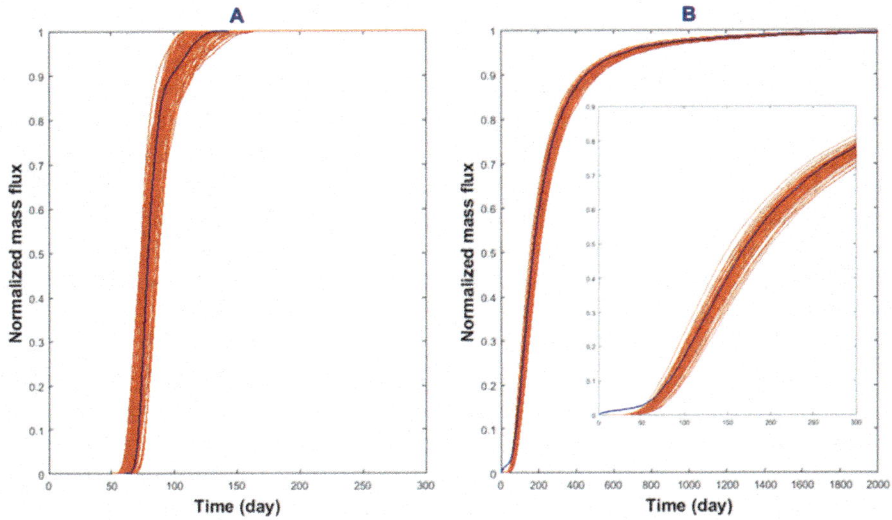

Fig. 4 Normalized cumulative mass flux profile for: (**A**) Fickian model and (**B**) non-Fickian model

profiles for all 100 realizations are compared to the true fine-scale model in Fig. 4**B**. Also shown in Fig. 4**A** is a case where there is no unresolved heterogeneity occurring below the fine scale; constant transition time is assumed; therefore, α_L^* and α_T^*, instead of β^*, are modeled (Vishal and Leung 2015). It is clear that significant non-Fickian behavior is observed when there is unresolved heterogeneity. In both cases, the true fine-scale response is captured by the variability exhibited among all 100 coarse-scale realizations.

5 Conclusions and Future Works

1. A particle-based non-Fickian transport model is adopted, which is capable of modeling unresolved heterogeneities occurring below the fine scale.
2. A multi-scale workflow is proposed to scale-up reservoir and transport parameters (e.g., β in the transition time distribution), such that subscale variability is properly represented in the coarse-scale models.
3. Results from the case study show that coarse-scale transport responses are capable of capturing the fine-scale transport response.
4. The case study illustrates the applicability of the method in 2D models. Previous works have already demonstrated the feasibility of the workflow for scale-up of reservoir properties in three dimensions. Future work should focus on the scale-up of solute transport parameters in 3D models. It is hypothesized that the workflow can be readily extended to multiple dimensions by performing the sub-grid calculations (i.e., steps #4–5) over a 3D volume.

Bibliography

Berkowitz B, Klafter J, Metzler R, Scher H (2002) Physical pictures of transport in heterogeneous media: advection-dispersion, random-walk, and fractional derivative formulations. Water Resour Res 38(10):W1191

Berkowitz B, Cortis A, Dentz M, Scher H (2006) Modeling non-Fickian transport in geological formations as a continuous time random walk. Rev Geophys 44(2)

Cortis A, Berkowitz B (2005) Computing "anomalous" contaminant transport in porous media: the CTRW MATLAB toolbox. Ground Water 43(6):947–950

Cortis A, Gallo C, Scher H, Berkowitz B (2004) Numerical simulation of non-Fickian transport in geological formations with multiple-scale heterogeneities. Water Resour Res 40(4)

Delay F, Ackerer P, Danquigny C (2005) Simulating solute transport in porous or fractured formations using random walk particle tracking. Vadose Zone J 4(2):360–379

Dentz M, Cortis A, Scher H, Berkowitz B (2004) Time behavior of solute transport in heterogeneous media: transition from anomalous to normal transport. Adv Water Resour 27 (2):155–173

Gao G, Zhan H, Feng S, Huang G, Mao X (2009) Comparison of alternative models for simulating anomalous solute transport in a large heterogeneous soil column. J Hydrol 377(3):391–404

John AK (2008) Dispersion in large scale permeable media (Dissertation)

Journel AG, Huijbregts CJ (1978) Mining geostatistics. Academic, London

Kinzelbach W (1986) Groundwater modelling: an introduction with sample programs in BASIC, vol 25. Elsevier, Amsterdam

Kreft A, Zuber A (1978) On the physical meaning of the dispersion equation and its solutions for different initial and boundary conditions. Chem Eng Sci 33(11):1471–1480

LaBolle EM, Fogg GE, Tompson AF (1996) Random-walk simulation of transport in heterogeneous porous media: local mass-conservation problem and implementation methods. Water Resour Res 32(3):583–593

Lake LW, Srinivasan S (2004) Statistical scale-up of reservoir properties: concepts and applications. J Pet Sci Eng 1-2:27–39

Leung JY, Srinivasan S (2011) Analysis of uncertainty introduced by scaleup of reservoir attributes and flow response in heterogeneous reservoirs. SPE J 16(3):713–724

Li L, Zhou H, Gómez-Hernández JJ (2011) A comparative study of three-dimensional hydraulic conductivity upscaling at the macro-dispersion experiment (MADE) site, Columbus Air Force Base, Mississippi (USA). J Hydrol 404(3):278–293

Margolin G, Dentz M, Berkowitz B (2003) Continuous time random walk and multirate mass transfer modeling of sorption. Chem Phys 295(1):71–80

Rhodes M, Bijeljic B, Blunt MJ (2009) A rigorous pore-to-field-scale simulation method for single-phase flow based on continuous-time random walks. SPE J 14(01):88–94

Salamon P, Fernàndez-Garcia D, Gómez-Hernández JJ (2006) A review and numerical assessment of the random walk particle tracking method. A review and numerical assessment of the random walk particle tracking method. J Contam Hydrol 87(3):277–305

Salamon P, Fernandez-Garcia D, Gómez-Hernández JJ (2007) Modeling tracer transport at the MADE site: the importance of heterogeneity. Water Resour Res 43(8)

Srinivasan G, Tartakovsky DM, Dentz M, Viswanathan H, Berkowitz B, Robinson BA (2010) Random walk particle tracking simulations of non-Fickian transport in heterogeneous media. J Comput Phys 229(11):4304–4314

Tompson AF, Gelhar LW (1990) Numerical simulation of solute transport in three-dimensional, randomly heterogeneous porous media. Water Resour Res 26(10):2541–2562

Vishal V, Leung JY (2015) Modeling impacts of subscale heterogeneities on dispersive solute transport in subsurface systems. J Contam Hydrol 182:63–77

A Comparative Analysis of Geostatistical Methods for a Field with a Large Number of Wells

Maria Volkova, Mikhail Perepechkin, and Evgeniy Kovalevskiy

Abstract This paper discusses methods for interpolating the well data that describes a hydrocarbon reservoir. The main difficulty lies in the requirement that the interpolation result should correspond not only to borehole data but also to some additional criteria, such as variograms and histograms. In a broader sense, it should reflect our knowledge of the geological environment. The authors are interested in what capabilities the different stochastic methods, sequential Gaussian simulation (SGS), multiple-point statistics (MPS) and fuzzy model simulation, can provide.

1 Introduction

Let us have a look at the problem in more detail. An interpolation of well data requires taking into account the following:

1. A geological environment is of a categorical nature, and it is thus impossible for the quantitative properties to be interpolated among different categories. The quality of property interpolation in such an environment is controlled by comparing the histograms of initial well data with the histograms computed for values in the interwell space.
2. The interpolated values should represent the actual variability of a geological environment, where the term "actual" means meeting a certain criterion. But in all cases, the variability of interpolated values should not depend on the spacing of well data points. This requirement can be achieved only by means of multiple stochastic realizations. Where a variogram is used as a mentioned criterion, the quality of interpolation is controlled by comparing the variograms of initial well data with the variograms computed for values in the interwell space.
3. It is geostatistical techniques that are mostly applied to compute stochastic realizations. All of these methods are based on the assumption of stationarity,

M. Volkova • M. Perepechkin • E. Kovalevskiy (✉)
GridPoint Dynamics (GPD JSC), Narodnogo Opolcheniya St, 34/1, Moscow, Russia
e-mail: m.volkova@gpd.email; m.perepechkin@gpd.email; e.kovalevskiy@gpd.email

J.J. Gómez-Hernández et al. (eds.), *Geostatistics Valencia 2016*, Quantitative Geology and Geostatistics 19, DOI 10.1007/978-3-319-46819-8_51

i.e. it is assumed that the geological environment is statistically uniform. Is a real geological environment statistically uniform? Certainly it is not. So, in stochastic methods, the key to a successful interpolation is dividing the interpolated parameter into deterministic (non-random) and random components. The problem is that such division is always ambiguous and difficult to achieve.

4. The deterministic features of initial well data (trends, categories, anomalous zones, etc.) are not obvious and, if not specially marked, will be erased in the realizations. Stochastic realizations are "sterile" with respect to the deterministic features and for this reason are not preferred by geologists. This is especially so with the realizations calculated based on variograms.
5. In order to overcome this "sterility", a number of "nonclassical" geostatistical methods have been developed. All of them are of a heuristic nature. The object modelling technique generates realizations which include sand bodies of a particular shape. Its heuristics consists in that the shape is defined by a geologist. Being able to manage the shape of individual bodies, the geologist however cannot manage the configuration of the set thereof.
6. To enable managing such combinations, multiple-point geostatistics (MPS) has been introduced and employed. Its heuristics consists in using a training image. A new version of MPS, known as "direct sampling" (DS), allows computing realizations of both categorical and quantitative parameters (e.g. for the porosity value).

This paper is aimed to investigate the capabilities the three stochastic methods (sequential Gaussian simulation, multiple-point statistics and the fuzzy geological model simulation) can provide. For this, a real field consisting of nearly 200 wells has been taken as an example. Used as initial data for interpolation are ASP (normalized SP) well log values that are closely linked with the porosity parameter and are presented along the entire section of each well.

2 Trial Field and Initial Data

A part of an existing large field in Western Siberia is shown below in Fig. 1. The shown part was used as a trial field.

Figure 1 shows not only a trial field but also a three-dimensional model calculated based on the well data. It includes a geometrical framework (as a pile of stratigraphic surfaces) resulting from the correlation of the well log data and the subsequent interpolation of the stratigraphic marks. The frame is used as a base for constructing a stratigraphic grid of $100 \times 100 \times 170$ (170 cells on $\boldsymbol{Z}$). The model also includes an ASP cube, which is the result of a deterministic quasi 3D (layer-by-layer) interpolation of the ASP well-logging curves calculated with a $\boldsymbol{1/R^2}$ weighting. The stratigraphic surfaces and the ASP cube are shown in the paleo-reconstruction where the surfaces are transformed to the horizontal planes. The initial ASP curves are presented on the paths of the wells in the window for the

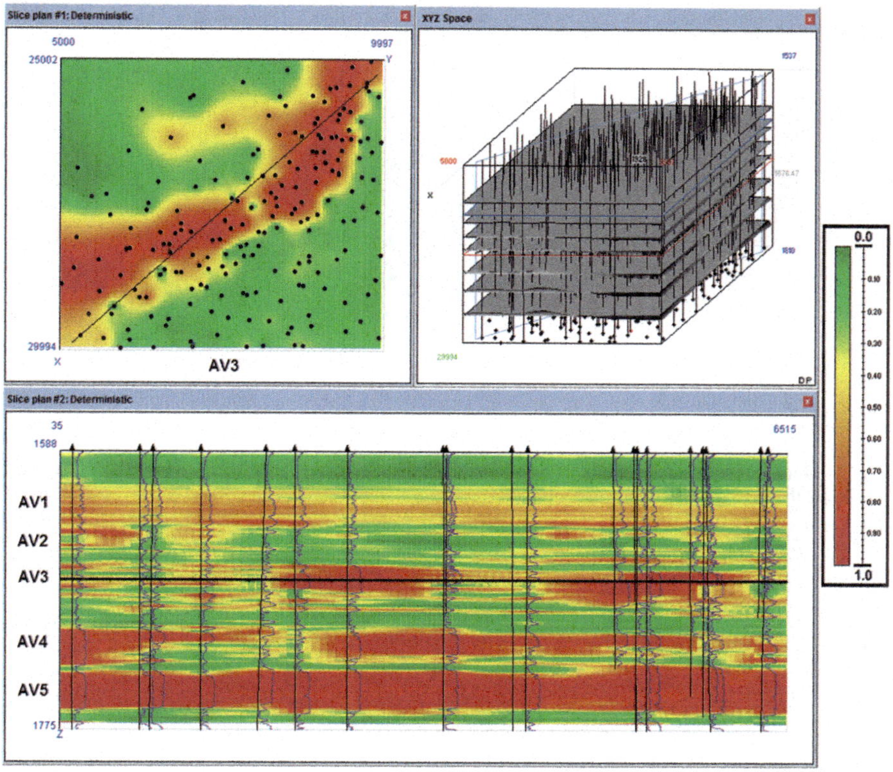

Fig. 1 Deterministic interpolation of the ASP well data. The *red* and *green* colours approximately demonstrate sands and clays, respectively. On the *top right* is the stratigraphic frame of the model. In the figures presented below, the colour legend is the same

vertical slice. Both (vertical and horizontal) slices show the intersection line. Hereinafter, the stratigraphic grid of the model will remain unchanged. What will change are the ASP cube calculation methods.

The ASP cube, as illustrated in Fig. 1, is considered as a starting point of our investigation. Why only as a starting point? Because the quality of such an interpolation is very low. First, it does not take into account the categorical nature of the environment, being therefore incapable of reproducing a well data histogram. The biggest error is seen for the AV4 horizon (see Fig. 2, right).

Second, this interpolation type incorrectly displays the variability of the environment. If the range of the horizontal variogram for borehole data is about 1,000 m, its range for the interpolated values is about 2,000 m (see Fig. 3, left). The horizontal variability of the deterministic ASP cube depends on the well spacing. At the same time, the vertical variogram is displayed well (see Fig. 3, right). The latter is the result of a layered interpolation. The other differences in the variograms are not of great importance. The difference in the sills of the variograms for the borehole and cube data follows from the difference in their histograms and

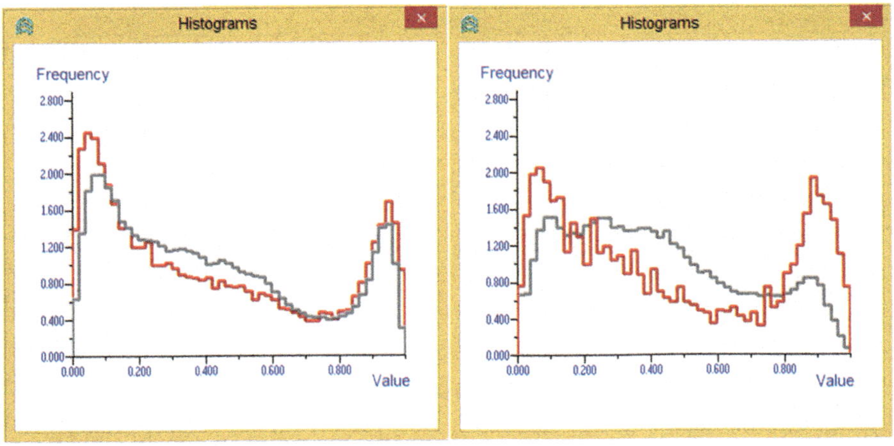

Fig. 2 Deterministic interpolation, histograms of the ASP values (with the *red curves* calculated based on the initial well data; and with the *grey curves*, on the interpolated values; and with the data for the AV1–AV5 horizons shown on the *left* and separately for the AV4 horizon on the *right*)

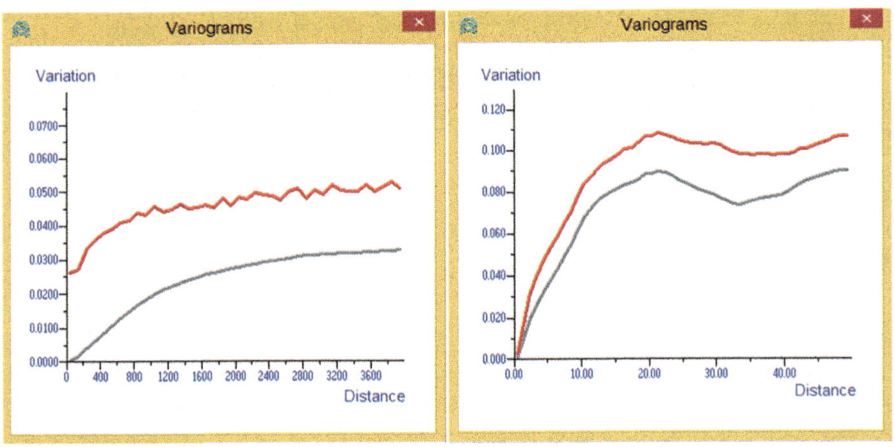

Fig. 3 Deterministic interpolation, variograms for the ASP values (the *red* and *grey curves* representing the calculated initial well data and the interpolated values, respectively). Shown on the *left* are the horizontal variograms, and on the *right* are the vertical variograms

the difference in the sills of the horizontal and vertical variograms, from a greater variability of the environment in the vertical direction.

Based on Figs. 2 and 3, we can conclude that the deterministic interpolation (see Fig. 1) gives a false view of the environment. However, the cube shown in Fig. 1 has one advantage: it excellently displays the deterministic features contained in the initial well data. It is immediately seen that the AV1 horizon is a heterogeneous reservoir of a poor quality, that there is a massive buried channel in the AV3 horizon and that the AV5 horizon is a solid reservoir.

3 Stochastic Modelling by Sequential Gaussian Simulation with Normal Score Transformation

Here we will try, within an interpolation, to reproduce a histogram and variogram of the initial well data. It is obvious that the reproduction of a horizontal variogram is possible only by means of stochastic realizations. The most often used method therefore is sequential Gaussian simulation (SGS). Its application, however, requires subdividing the environment into statistically uniform categories (the criterion for statistical uniformity being the Gaussian histogram of values). The SGS method is applied separately for interpolation within each category. The histogram of initial well data would otherwise be roughly distorted.

At the same time, the normal score (NS) transformation method exists, allowing the application of SGS when the histogram of initial data is not Gaussian (Deutsch and Journel 1998). NS enables the exact reproduction of a non-Gaussian histogram of initial data. The result of the interpolation of our ASP well data by SGS with NS transformation is illustrated in Fig. 4, and the histograms and variograms are shown in Fig. 5, with a very good fit.

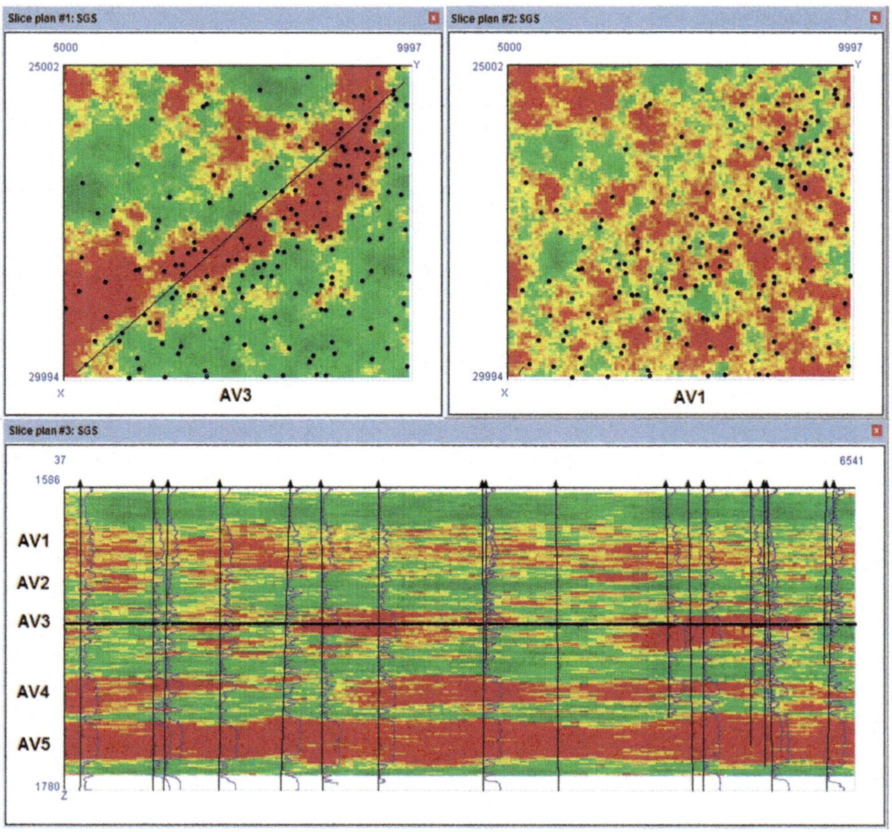

Fig. 4 Interpolation of the ASP well data by the SGS method, using the normal score (NS) transformation (one realization). A horizontal section within the AV1 horizon is shown on the *top right*

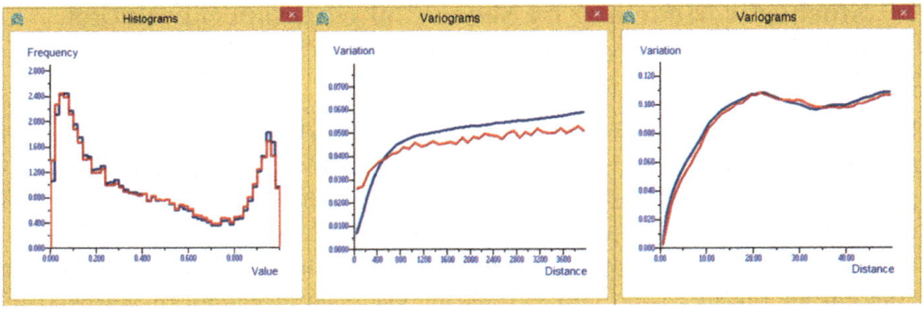

Fig. 5 Interpolation of the ASP well data by the SGS method with NS transformation. Histograms and variograms of the ASP values (with the *red curves* indicating values calculated based on the initial well data and the *blue curves* on the interpolated values). In the centre are the horizontal variograms, and the vertical variograms are on the *right*

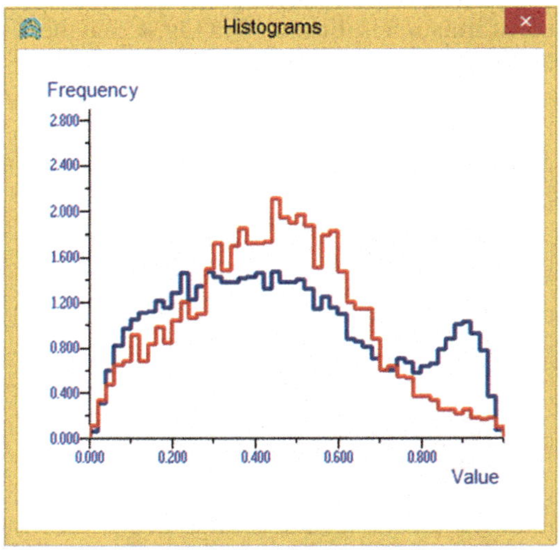

Fig. 6 Interpolation of the well ASP data by the SGS method with NS transformation. Histograms of the ASP values within the AV1 horizon, with the *red line* indicating values calculated based on the initial well data and the *blue line* on the interpolated values

However, it should be kept in mind that the NS method allows fitting of the histograms of initial and interpolated values only for the entire cube. To prove that this method can lead to errors, it is quite enough to look at the "local" results it can deliver. Figure 6 displays the histograms for the well data and for the interpolated values for the AV1 horizon. What we see is that they have essential differences. Figure 4 (top right) shows the horizontal section of the ASP cube within the AV1 horizon. It is noteworthy that all the red and green "spots" (high and low ASP values) are located exactly between the wells. The AV1 horizon well data has no such values.

Of course, the above problem of the vertical nonstationarity has been well addressed. Each layer with different statistics (e.g. AV1) should be calculated separately as well as each zone with different statistics on the XY plane. Let's have a look at the AV3 horizon once more (Fig. 4 top left). The presence of the reservoir zones outside the channel predicted by NS is very unlikely. These problems were mentioned in the Introduction section.

4 Stochastic Modelling by Multiple-Point Statistics

The second of the three techniques used to represent the actual variability of the environment is multiple-point statistics (Strebelle 2002). Appropriate tools have been developed and tested for this use (Volkova 2015). We have made such a calculation as well. We used a fragment of the deterministic model shown above in Fig. 1 as a training image (TI). It is a small cube of 13 % of the total field with the highest density of wells. The quality of the TI can be argued, but it is the best TI we can obtain. The calculation was carried out using the direct sampling (DS) algorithm (Mariethoz et al. 2010). Parameters of the DS algorithm are as follows: the number of neighbours, 40; their weights, $\boldsymbol{1/R^2}$, where $\boldsymbol{R}$ is the distance in the 3D space measured in the cells; the threshold value, 0.10; and the maximum fraction for TI scanning, 1 %. The result of the interpolation (as well as the contour of the fragment taken as the TI) is illustrated in Fig. 7, and the histograms and variograms are presented in Fig. 8.

The direct sampling algorithm distinguishes areas having distinct statistical properties. For example, the AV1 horizon is modelled almost correctly (see AV1 in Figs. 9 and 10). At the same time, the method leads to some errors. The well data does not indicate that the AV5 horizon reservoir has a gap (see AV5 in Fig. 9).

5 Stochastic Modelling Based on the Fuzzy Model

The third of the above-named methods employed to represent the actual variability of the environment is the fuzzy model (Kovalevskiy 2015). This method does not rely on statistical generalizations. Computations are performed on a 3D stratigraphic grid. In the most recent edition (aimed at simplification), the calculation algorithm (for one realization) consists of the following steps:

1. We randomly choose an empty cell (as the forecast cell). Then we select an $\boldsymbol{n}$ number of the non-empty cells nearest to that empty cell. (Those cells initially contain borehole data.) For example, $\boldsymbol{n}$ can be equal to 6. The selection should provide azimuthal control to ensure selection of a single nearest cell from each azimuth sector (see Fig. 11, left). The selection of cells from the upper or lower layer of the grid can be performed as an option.

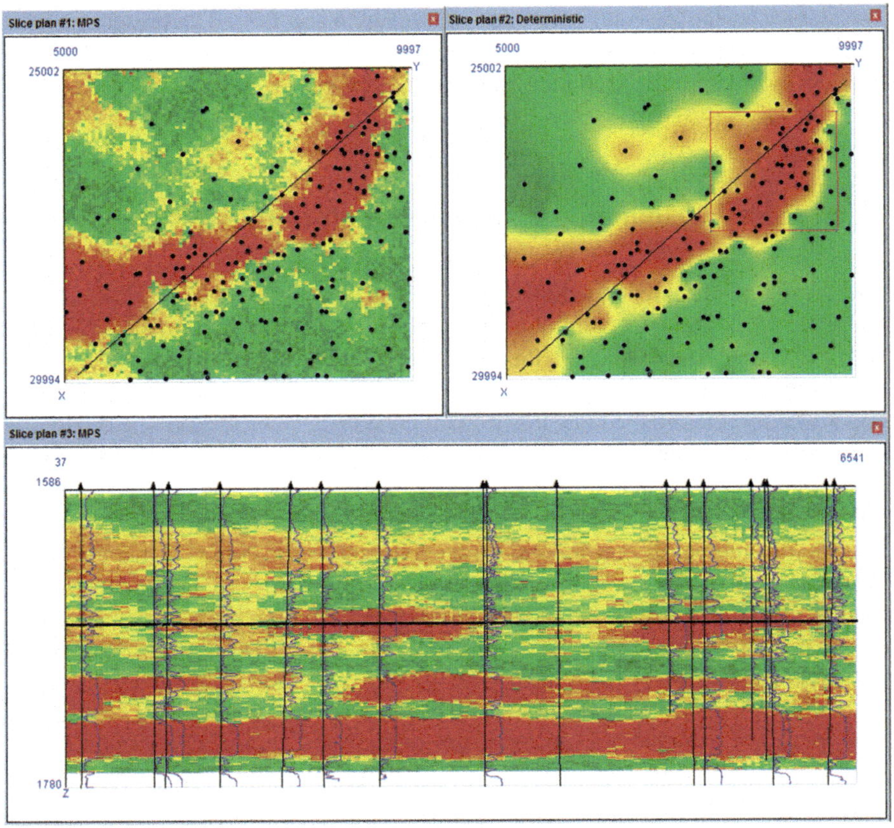

Fig. 7 Interpolation of the ASP well data by the MPS method, using the direct sampling algorithm (one realization). A horizontal section of the deterministic cube is added on the *top right*, with an indication of the fragment taken as a training image

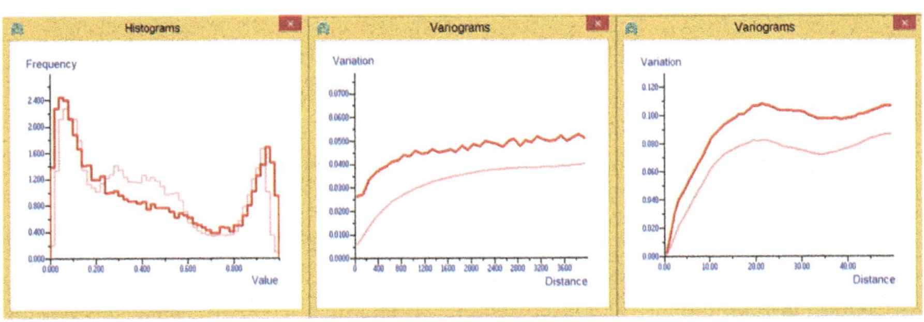

Fig. 8 Interpolation of the ASP well data by the MPS method, using the direct sampling algorithm. Histograms and variograms of the ASP values (with the *red curves* calculated based on the initial well data and the *pink curves* on the interpolated values). In the centre are the horizontal variograms, and on the *right* are the vertical variograms

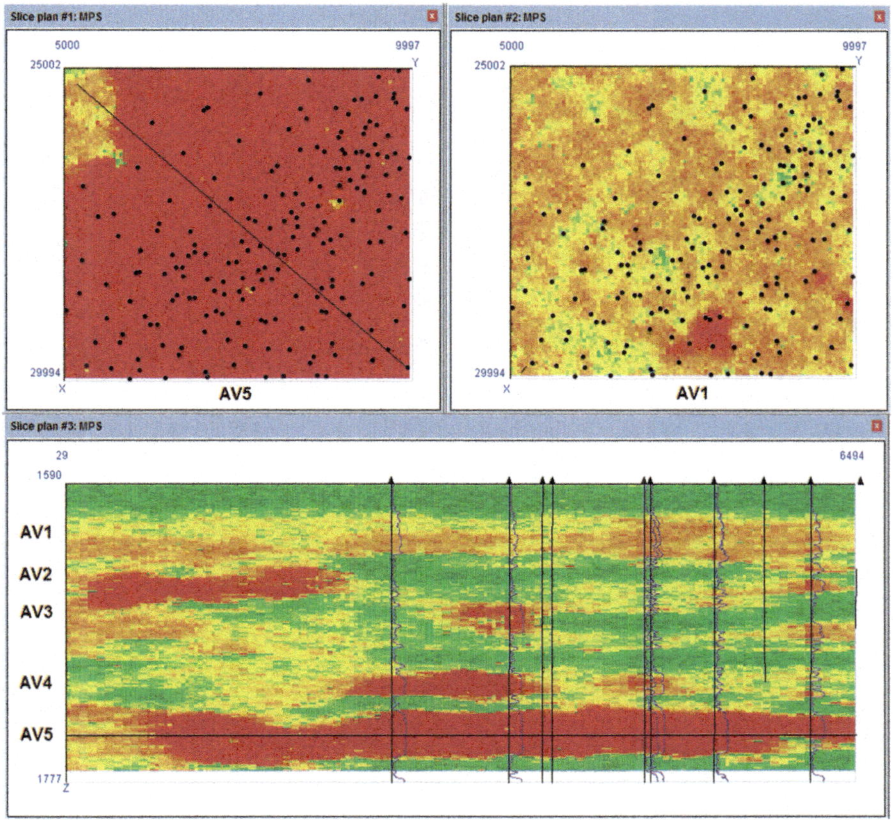

Fig. 9 Interpolation of the ASP well data by the MPS method, using the direct sampling algorithm. On the *top right* is the horizontal section of the AV1 horizon. On the *top left* is the horizontal section of the AV5 horizon

Fig. 10 Interpolation of the ASP well data by the MPS method, using the direct sampling algorithm. Histograms of the ASP values within the AV1 horizon (with the *red line* calculated based on the initial well data and the *pink line* on the interpolated values)

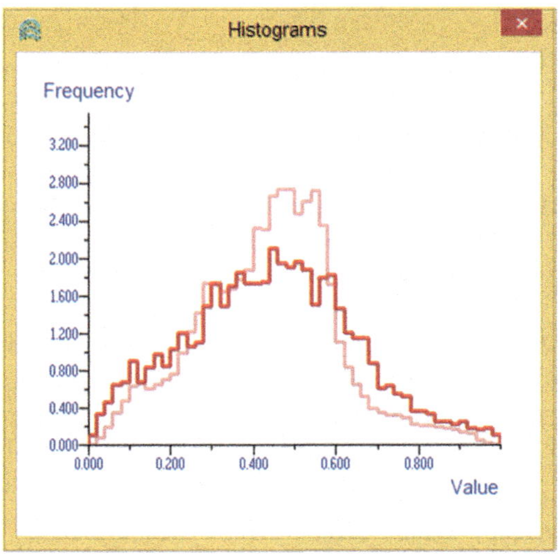

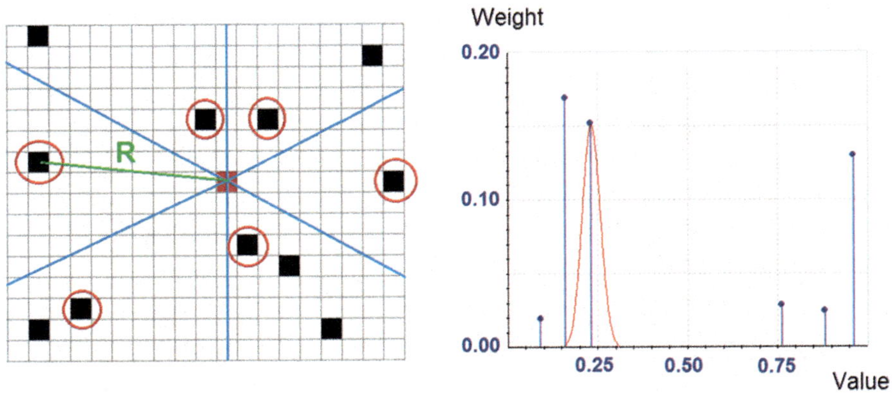

Fig. 11 Selection of the data affecting the forecast cell (on the *left*). On the *right* is the weight of values calculated as $1/R^2$. The one cell's value is presented with its estimated accuracy

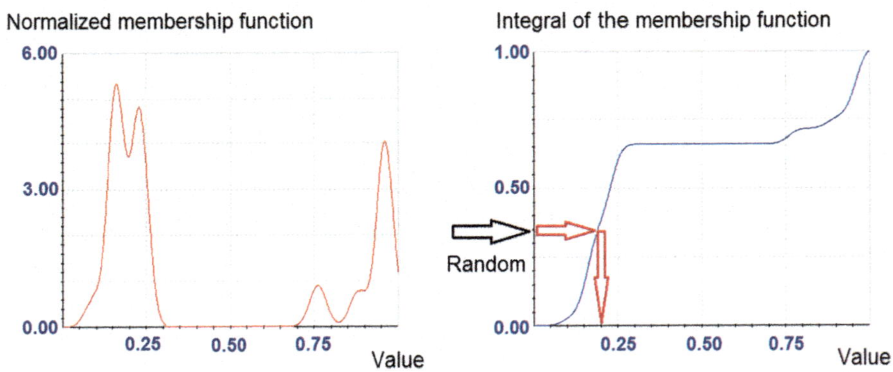

Fig. 12 On the *left* is the fuzzy presentation of the parameter value in the forecast cell. On the *right* is the random selection from the integral distribution

2. Each selected cell is represented by two numbers: the value in the cell and its "weight" (see Fig. 11, right). The weight is calculated as $1/R^2$, where R is the distance from the cell with data to the forecast cell. Each cell value is presented with its estimated accuracy (set by expert evaluation).
3. On the basis of the selected values and taking into account the accuracy of each value, we compute the fuzzy value in the forecast cell (see Fig. 12, left). We do it by a simple summation of the inputs. After that, we normalize the membership function so that its integral is equal to 1.
4. Next, we compute the integral distribution of the membership function (which is an analogue of integral probability distribution – see Fig. 12, right). After that, we select a random number from the uniform distribution on the interval [0, 1]. Using the integral distribution of the membership function, we transform the selected random number to another, also random number, but one distributed

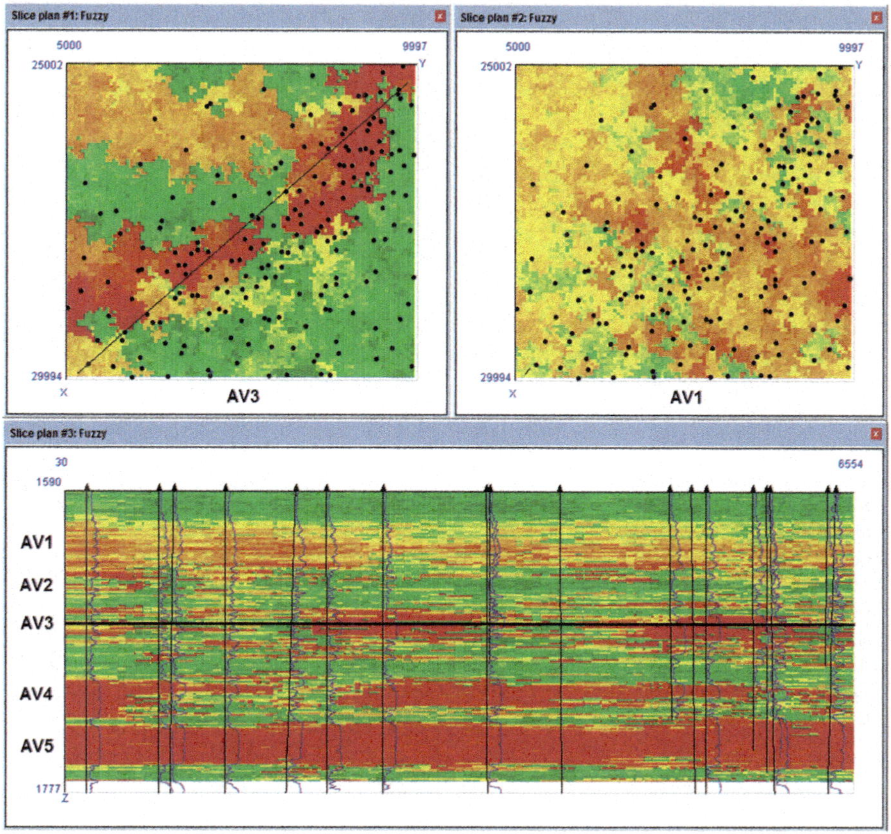

Fig. 13 Interpolation of the ASP well data as a realization of the fuzzy model. A horizontal section of the AV1 horizon is added on the *top right*

according to the membership function. And it is this second random number that we insert into the forecast cell and that is assigned the status of a data point. Then we select the next forecast cell and repeat the same process. Thus, the described algorithm is very similar to a standard sequential simulation.

The so calculated stochastic realization of the fuzzy model is illustrated in Fig. 13 and the histograms and variograms in Figs. 14 and 15.

As regards selection of non-empty cells from the upper or lower layer of the grid, we have not selected such cells in this calculation. However, because the initial well data was presented as the vertical (or almost vertical) columns of cells, each of the simulated predictions is also made for the entire vertical column, with individual random numbers used for each forecast cell.

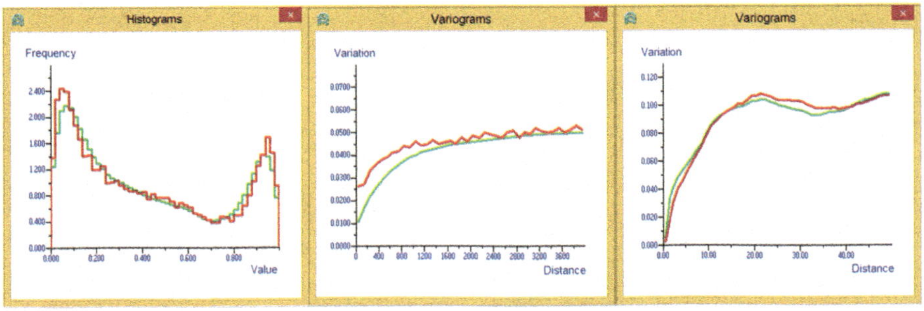

Fig. 14 Interpolation of the ASP well data as a realization of the fuzzy model. Histograms and variograms of the ASP values (with the *red curves* calculated based on the initial well data and the *green curves* on the interpolated values). In the centre are the horizontal variograms, and on the *right* are the vertical variograms

Fig. 15 Interpolation of the ASP well data as a realization of the fuzzy model. Histograms of the ASP values within the AV1 horizon (with the *red line* calculated based on the initial well data and the *green line* on the interpolated values)

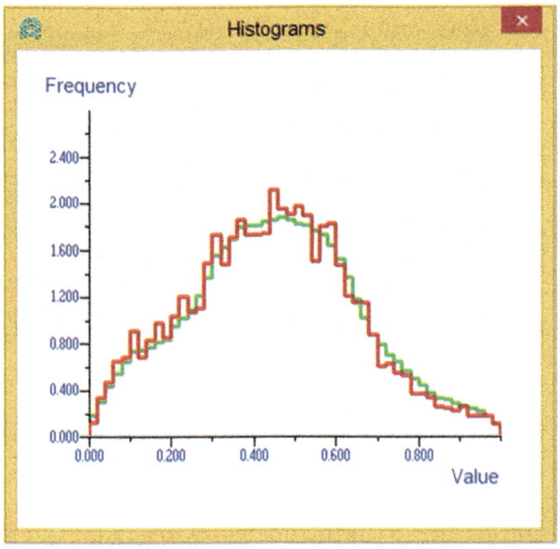

6 Conclusions

In conclusion, it is worth paying attention to the following. As was said above, the deterministic interpolation technique (reproducing neither a histogram nor a variogram) has an important advantage: it excellently displays the deterministic features of the data. Indeed, it is possible to see in the AV3 horizon very specific meandering facies (see Fig. 16(1)). This means that the well data indicates that meandering facies are present.

On the other hand, no meandering facies can be seen as the result of the interpolation by the SGS method (see Fig. 16(2)), while, just to repeat, the well data indicates their presence. The interpolated values obtained using this technique

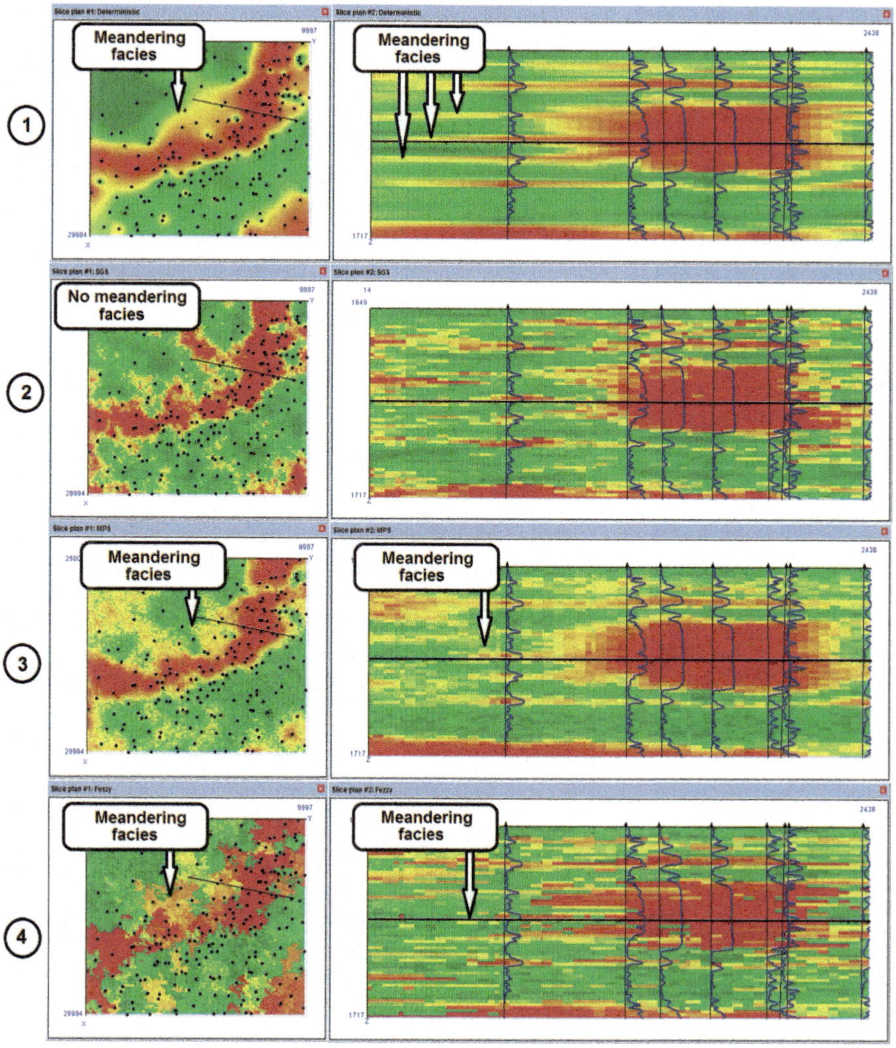

Fig. 16 *1* deterministic interpolation, *2* realization of SGS, *3* realization of MPS, *4* realization of the fuzzy model

is "sterile" with respect to those deterministic features, which have not been specially marked.

This is that very "sterility" which was mentioned in the Introduction section. The histograms and variograms of the initial data are displayed well, with the caveat that SGS can lead to errors where the local histograms differ from the overall histogram.

The result of interpolation by the MPS method (see Fig. 16(3)) displays the meandering facies well. On the whole, it can be said that it works, with the variograms being displayed well and the histograms being displayed satisfactorily.

It is important that MPS allows dealing with nonstationary data and a nonstationary training image. It can be assumed that the result may be improved by improving the training image.

The fuzzy model realization as regards showing the meandering facies is also good (see Fig. 16(4)). It clearly shows that the facies (meandering facies, channel facies, etc.) cannot be considered as statically uniform. What we can see is a gradual transition from one into the other. The well data histograms and variograms are displayed well, even locally. So it can be stated that this approach seems to have great potential.

Bibliography

Deutsch CV, Journel AG (1998) GSLIB: geostatistical software library and user's guide. Oxford University Press, Oxford

Kovalevskiy EV (2015) Fuzzy geological model: stochastic realizations preserving deterministic features of data. EAGE Conference "Petroleum Geostatistics 2015"

Mariethoz G, Renard Ph, Straubhaar J (2010) The direct sampling method to perform multiple-point geostatistical simulations. Water Resour Res 46(W11536) doi:10.1029/2008WR007621

Strebelle S (2002) Conditional simulation of complex geological structures using multiple-point statistics. Math Geol 34:1–22

Volkova M (2015) Stochastic modeling on the base of multiple-point statistics with seismic attribute maps as training images. Geophysics 6:68–73 [in Russian]

Part V
Hydro(geo)logy

Building Piezometric Maps: Contribution of Geostatistical Tools

B. Bourgine, M. Saltel, N. Pedron, and E. Lavie

Abstract This paper shows practical applications of two kriging techniques: kriging with boundary conditions and kriging with external drift, which are used in order to introduce hydrogeological constraints when producing piezometric maps. These techniques are not new to geostatisticians, but not very well known by many hydrogeologists. Kriging with boundary conditions is implemented here with a discrete formulation that allows the use of non-differentiable or of anisotropic covariance models. Kriging with external drift is applied here in the case of an unconfined aquifer, the external drift being a smoothed version of the topography. A method is proposed to select the appropriate smoothing radius. Results show a clear improvement compared to standard kriging.

1 Introduction

Building piezometric maps is a rather common task for many hydrogeologists. Indeed, piezometric maps are essential to characterize the aquifer behavior (recharge and outlet area, flow path) or to quantify and monitor the aquifer temporal evolution due to groundwater mining. They are frequently built using standard kriging.

However, standard kriging does not take into account physical laws that govern hydraulic head distribution. This sometimes leads to physically inconsistent results. In addition, geostatistical methods that enable to take into account hydrogeological

B. Bourgine (✉)
BRGM (French Geological Survey), 36009, 45060 Orléans Cedex 2, France
e-mail: b.bourgine@brgm.fr

M. Saltel • N. Pedron
BRGM, Direction Régionale Aquitaine, 33600 Pessac, France
e-mail: m.saltel@brgm.fr; n.pedron@brgm.fr

E. Lavie
Conseil Régional Aquitaine, 33077 Bordeaux, France
e-mail: e.lavie@aquitaine.fr

J.J. Gómez-Hernández et al. (eds.), *Geostatistics Valencia 2016*, Quantitative Geology and Geostatistics 19, DOI 10.1007/978-3-319-46819-8_52

constraints in the interpolation are not always used neither known by hydrogeologists. This is why some of them still prefer to draw these maps by hand. In this context, BRGM and Aquitaine Region signed an agreement entitled "Groundwater management in Aquitaine Region" for the 2008–2013 period, financially supported by the Adour Garonne Water Agency. The aim was to identify tools and procedures in order to facilitate the production of piezometric maps of the various aquifers of the Aquitanian Basin, using appropriate interpolation techniques. The objective was also to propose a standardized procedure in order to facilitate the work, from data control to the final map.

This paper presents two applications of geostatistics in hydrogeology: (1) kriging with boundary conditions and (2) kriging with a smoothed topography taken as an external drift, where we propose a procedure to select the degree of smoothing. The results are general and applicable to other examples as those illustrated here.

Before detailing these applications, we present a short review of previous work.

2 Brief Review of Previous Work

Kriging hydraulic head has been used for a long time to build reference maps either to characterize high or low water levels or to calibrate the parameters of flow simulations (Renard and Jeannée 2008). As usually hydraulic heads are not stationary, kriging with a drift should be used. An example can be found in (Aboufirassi and Marino 1983), where the drift depends on the coordinates. In unconfined aquifers, the hydraulic head is generally well correlated with topography. Thus hydraulic heads can be interpolated by co-kriging with topography (Hoeksema 1989) or by using a variable derived from topography as an external drift (Desbarats et al. 2002). In this last case the authors compare two external drift variables: the topography itself and another variable "TOPMODEL" which combines several parameters such as the slope and the upslope area draining to a given point. This second method is supposed to take into account the fact that the water table elevation is much smoother than topography (Wolock and Price 1994) but authors conclude that the results "are not always physically plausible." Renard and Jeannée (2008) use a smoothed topography as external drift and obtain smoothing by a standard moving average algorithm. The smoothing radius is set empirically.

Boundary conditions are important constraints to take into account. The two main ones are the Dirichlet condition (known and constant head) and the Neumann condition (known and constant flow). The first one is easily satisfied by adding data points with given hydraulic head along the boundary (e.g., on the banks of a lake). The second one is generally used to take into account no flow boundary conditions. A first application proposed by Delhomme (1979) was described by Chilès and Delfiner (1999). This work was continued by Le Cointe (2006) during a joint work with J.P. Delhomme as tutor. In this work, a co-kriging of heads and their derivatives is implemented using a finite difference approach. At last, Kuhlman and Pardo

Igúzquiza (2010) proposed the exact formulation using mathematical derivation of covariances.

In this paper, we will investigate the use of kriging with boundary conditions, using the finite difference approach, and compare different types of kriging. We will also present a practical application of kriging with external drift based on smoothed topography and propose a method to select the smoothing radius.

3 Kriging with Boundary Conditions: Finite Increments Approach

3.1 Methodology

We start here from a preliminary internal unpublished work initiated in 2003–2004 at the BRGM with the cooperation of J.P. Chilès (2004), which corresponds partly to the work presented later by Le Cointe (2006). The aim was to interpolate hydraulic heads taking into account:

- Measured hydraulic heads from piezometer data or constant heads from Dirichlet boundary conditions (e.g., river elevation).
- No flow boundary conditions, which are observed, for example, in case of an impervious boundary, of a groundwater ridge, or at the watershed boundary. In this case the gradient of hydraulic head is equal to zero in the direction perpendicular to the no flow boundary. It is also possible to take into account a nonzero value for the gradient in a given direction if such a value is known.

The hydraulic head is considered as a non-stationary random variable:

$$Z(x) = m(x) + Y(x) \tag{1}$$

The deterministic drift $m(x)$ is expressed as usually as:

$$E[Z(x)] = m(x) = \sum\nolimits_{l=0}^{L} a_l f^l(x) \tag{2}$$

where $f^l(x)$ are known drift functions that can depend on the coordinates or can be also external drift and where a_l are unknown coefficients.

The residual $Y(x)$ can be non-stationary, with a covariance $\sigma(x, x')$. In case of a *IRF-k*, the polynomial components with degree lower or equal to k are part of $m(x)$.

Data is denoted Z_a and can belong to two groups:

- The first group represents known hydraulic heads at point x_α:

$$Z_\alpha = Z(x_\alpha) \tag{3}$$

- The second group corresponds to the value of the component of the gradient at point x_α along a unit vector u_α and is used to represent non-flow boundary conditions or fixed gradients:

$$Z_\alpha = \frac{\partial Z}{\partial u_\alpha}(x_\alpha) \tag{4}$$

In this paper, we use a finite increment formulation. The value of the component of the gradient at point x_α is represented by two *dummy points* $x_{\alpha 1}$ and $x_{\alpha 2}$ separated by a distance $2b_\alpha$ along the unit vector u_α (the macroscopic gradient is calculated on length $2b_\alpha$). Thus a data of the second group has the following form:

$$Z_\alpha = \frac{Z(x_\alpha + b_\alpha u_\alpha) - Z(x_\alpha - b_\alpha u_\alpha)}{2b_\alpha} \tag{5}$$

With $x_{\alpha 1} = x_\alpha - b_\alpha x_\alpha$ $x_{\alpha 2} = x_\alpha + b_\alpha x_\alpha$ $w_\alpha = \frac{1}{\|x_{\alpha 1} - x_{\alpha 2}\|}$, Eq. 5 can be rewritten as

$$Z_\alpha = w_\alpha(Z(x_{\alpha 2}) - Z(x_{\alpha 1})) \tag{6}$$

Kriging can be expressed in the classical way. The estimator at point x_0 is:

$$Z^*(x_0) = \sum_\alpha \lambda_\alpha Z_\alpha \tag{7}$$

Note that the Z_α are of one of the two forms shown in (3) and (4). In the particular case of no flow boundary condition, the Z_α of the second group are equal to zero (partial derivative equal to zero perpendicularly to the boundary) and don't contribute directly to the estimator in Eq. 7. But in fact they modify the weights and consequently contribute indirectly to the estimation.

The kriging system can be written in a matrix form (Chilès and Delfiner 1999):

$$\begin{bmatrix} \Sigma & F \\ F' & 0 \end{bmatrix} \begin{bmatrix} \lambda \\ \mu \end{bmatrix} = \begin{bmatrix} \sigma_0 \\ f_0 \end{bmatrix} \tag{8}$$

where Σ is the matrix of the covariances $\sigma_{\alpha\beta}$ between data points, F the matrix of drift functions $f^l(x_\alpha)$ at data points x_α, λ the vector of weights λ_α, μ the vector of Lagrange parameters μ_α, σ_0 the vector of covariances $\sigma_{\alpha 0}$ between data points and estimated point, and f_0 the vector of drift functions at estimated point.

The covariances $\sigma_{\alpha\beta}$ have different expressions according to the type of data. If both α and β are indices of points of type "hydraulic heads," the covariance is simply $\sigma_{\alpha\beta} = \sigma(x_\alpha, x_\beta)$ where σ is the covariance of $Y(x)$. If both α and β are indices of points of type "component of a gradient," the covariance $\sigma_{\alpha\beta}$ can be written as (using the finite increment formulation)

$$\sigma_{\alpha\beta} = w_\alpha w_\beta \left(\sigma\left(x_{\alpha 2}, x_{\beta 2}\right) - \sigma\left(x_{\alpha 1}, x_{\beta 2}\right) - \sigma\left(x_{\alpha 2}, x_{\beta 1}\right) + \sigma\left(x_{\alpha 1}, x_{\beta 1}\right)\right) \quad (9)$$

At last, if α is a point of type "hydraulic heads" and β a point of type "component of a gradient," the covariance $\sigma_{\alpha\beta}$ is $\sigma_{\alpha\beta} = w_\beta \left(\sigma\left(x_\alpha, x_{\beta 2}\right) - \sigma\left(x_\alpha, x_{\beta 1}\right)\right)$.

The drift functions take also different forms according to the type of data.

If α is a hydraulic head, $f^l(x_\alpha)$ is of the classic form. For example: $f^0(x_\alpha) = 1$; $f^1(x_\alpha) = Xcoord(x_\alpha)$; $f^2(x_\alpha) = Ycoord(x_\alpha)$ (*Xcoord* and *Ycoord* being the X and Y coordinates of the data point).

If α is component of a gradient, $f^l(x_\alpha)$ is obtained by the following equation: $f^l(x_\alpha) = w_\alpha\left(f^l(x_{\alpha 2}) - f^l(x_{\alpha 1})\right)$.

3.2 Applications of Kriging with Boundary Conditions

The expressions above were introduced in GDM software (Geological Data Management, http://gdm.brgm.fr) and were applied in a large aquifer of the Aquitaine Region (France): the Middle Eocene aquifer. This aquifer shows many lateral facies variations, from continental fluviatile deposits in the eastern part to marine deposit to the west. It covers a 17,290 km^2 area around Bordeaux city (Fig. 1) and is intensively pumped. First measurements of hydraulic heads started in 1873.

The oldest available piezometric map represents the initial state of hydraulic head corresponding to the 1870–1900 period. The aquifer is now monitored by a network of piezometers that has been set up since 1959.

Nowadays a reference piezometric map of the average level of the last quarter of the year is built every year. For the year 2007, that will be illustrated here, 119 piezometers are available. The aquifer is connected to the river network in the northern and eastern parts of the studied area. Therefore points along the river can be added as constant heads, the hydraulic head being equal to the surface topography at these points. The aquifer is confined in almost the whole area, except in the few square kilometers near the outcrops where it is unconfined.

The variogram of the hydraulic head (computed from piezometers only) is non-stationary (Fig. 2) and can be fitted, for example, by a power model. Several kriging options were tested and are presented in (Bourgine and Pédron 2011), including co-kriging using data from different years and kriging with an external drift based on a hydrodynamic flow model, but we will focus here on kriging with boundary conditions.

Since 1993, a hydrodynamic flow model has been developed in order to improve groundwater management and to find solutions to reduce impact due to pumping (Saltel et al. 2012). The hydraulic head obtained from this model for year 2007 is displayed in Fig. 3. The red ellipse indicates an area where the aquifer becomes impervious, along its south boundary, just to the north of *Villeneuve sur Lot* City. This boundary condition was integrated in the flow model, and consequently contour lines are perpendicular to this boundary.

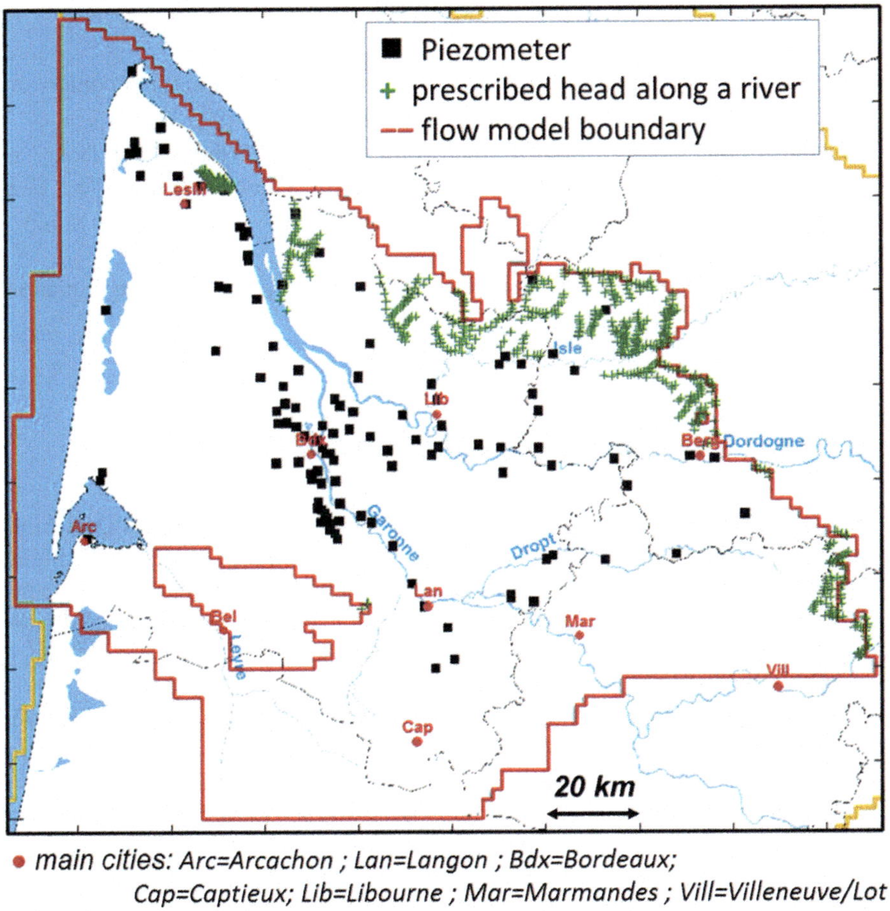

Fig. 1 Available data for the Middle Eocene aquifer around Bordeaux City (France)

Another piezometric map was computed independently by ordinary kriging using piezometer and river data (Fig. 4). Obtained contour lines are not perpendicular to the south boundary and are not very stable. Moreover, the estimated contour line near *Villeneuve sur Lot* is 60 m, whereas the flow model gives a value of 50 m.

To improve this map, no flow boundary conditions were applied along the south boundary. They are represented by vertical red segments on Fig. 5: the red dots represent the point of application of the constraint (the x_α) and the red segment the segment joining the two *dummy points* $x_{\alpha 1}$ and $x_{\alpha 2}$ of Eq. 6. The length of the segments is 4 km and the x_α are separated also by approximately 4 km. The result obtained by kriging under these boundary conditions is displayed Fig. 5 and is much more consistent with the hydrogeological hypothesis.

Boundary conditions can also be used to account for groundwater ridge. Figure 6 shows the example of the *Médoc* groundwater ridge, located in the western part of this aquifer, near the Atlantic Ocean. The upper left map (map a) shows the

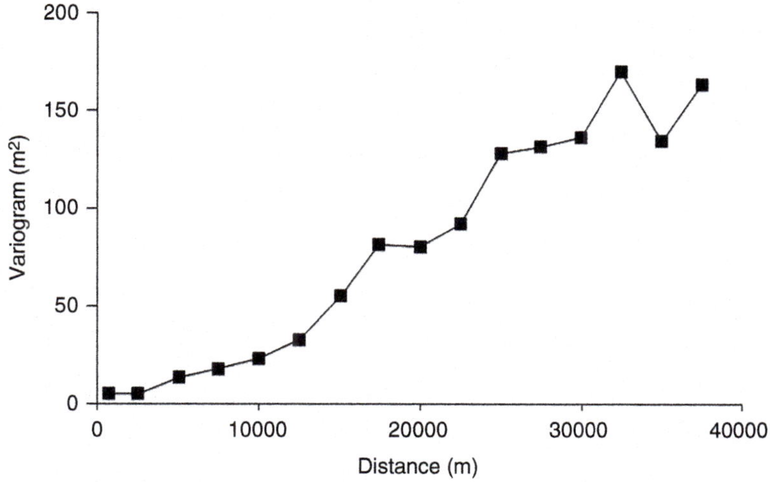

Fig. 2 Variogram of hydraulic head for year 2007 (piezometers data only)

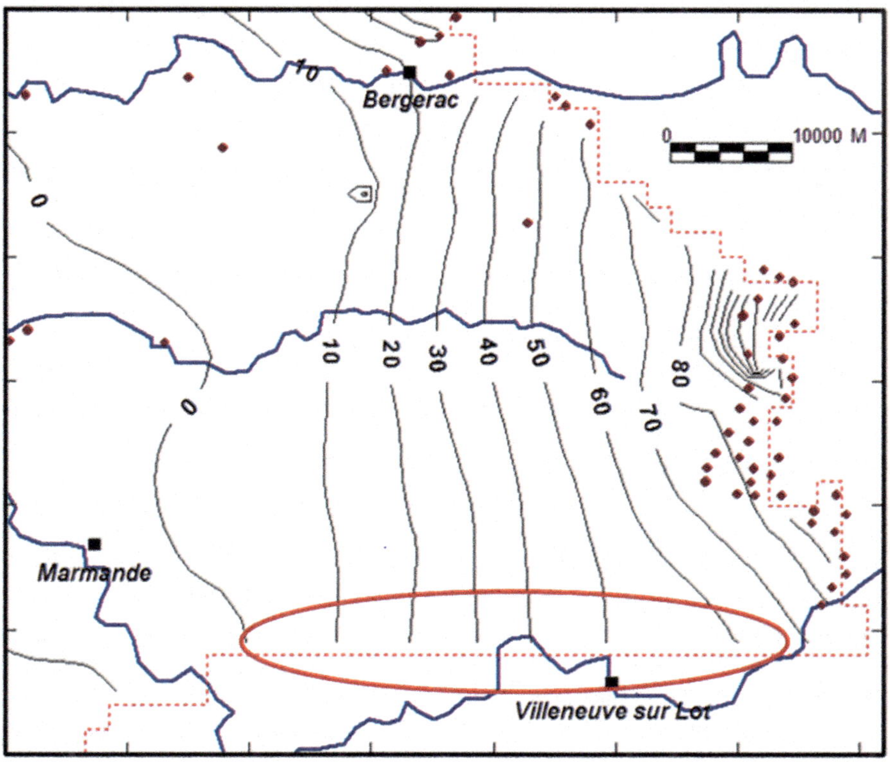

Fig. 3 Hydraulic head computed from the flow model

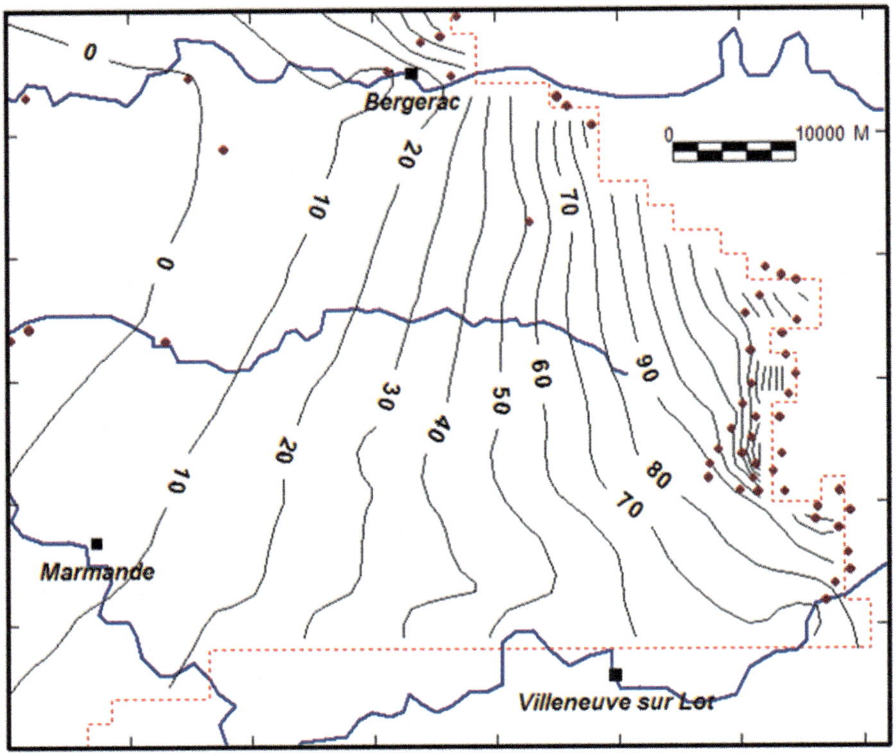

Fig. 4 Hydraulic head computed by ordinary kriging

hydraulic head derived from the hydrodynamic flow model for year 2007. Note that additional constant heads have been added in the ocean (green point along the western boundary) in order to calibrate the flow model.

The same data are used again with ordinary kriging (map b): the obtained groundwater ridge is less continuous and the map is unstable in this area. Boundary conditions are then introduced as no flow data perpendicularly to the groundwater ridge. The new computed map (Fig. 6 – (map c)) is now much more continuous and satisfactory. The comparison of the southwestern part of map (a) with the same part of maps (b) or (c) shows that the high gradient area seen on map (a) near the southern aquifer boundary is not reproduced on maps (b) and (c). This difference is due to a supplementary point (named 08512X0001), not measured in 2007, that was introduced in the flow model to calibrate it. After adding it to the 2007 piezometer data with the same hydraulic head value, we get the map (d) that is now very satisfactory. This shows that even if introducing boundary conditions in the kriging improves the results, a sufficient number of measurements of hydraulic head is essential to calibrate the correct value.

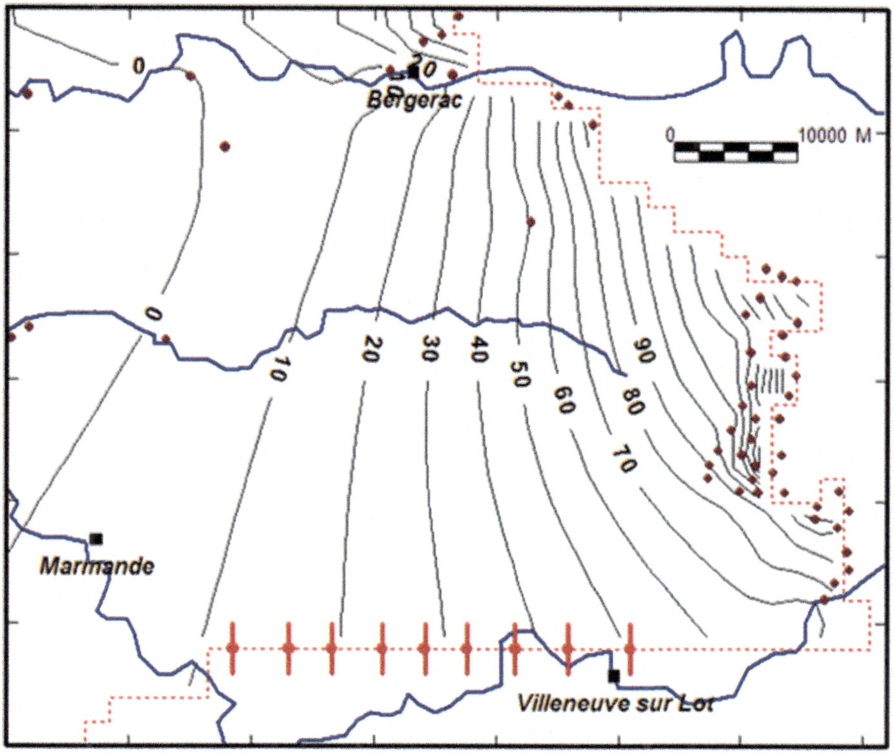

Fig. 5 Hydraulic head computed using no flow boundary conditions

4 Kriging Heads with External Drift Based on Smoothed Topography

4.1 Proposal of a Methodology to Smooth the Topography

As recalled in the review of previous work, kriging hydraulic heads with external drift based on smoothed topography can be used in unconfined aquifers, but the remaining question lies in the way to smooth the topography. We propose to smooth the topography using a standard moving average and to set the smoothing radius using the following method:

- Several smoothing radius are tested. In each cell of the initial topography grid, we compute a new field equal to the initial topography averaged within the specified radius.
- For each smoothing radius, we compute the variogram of the residual:

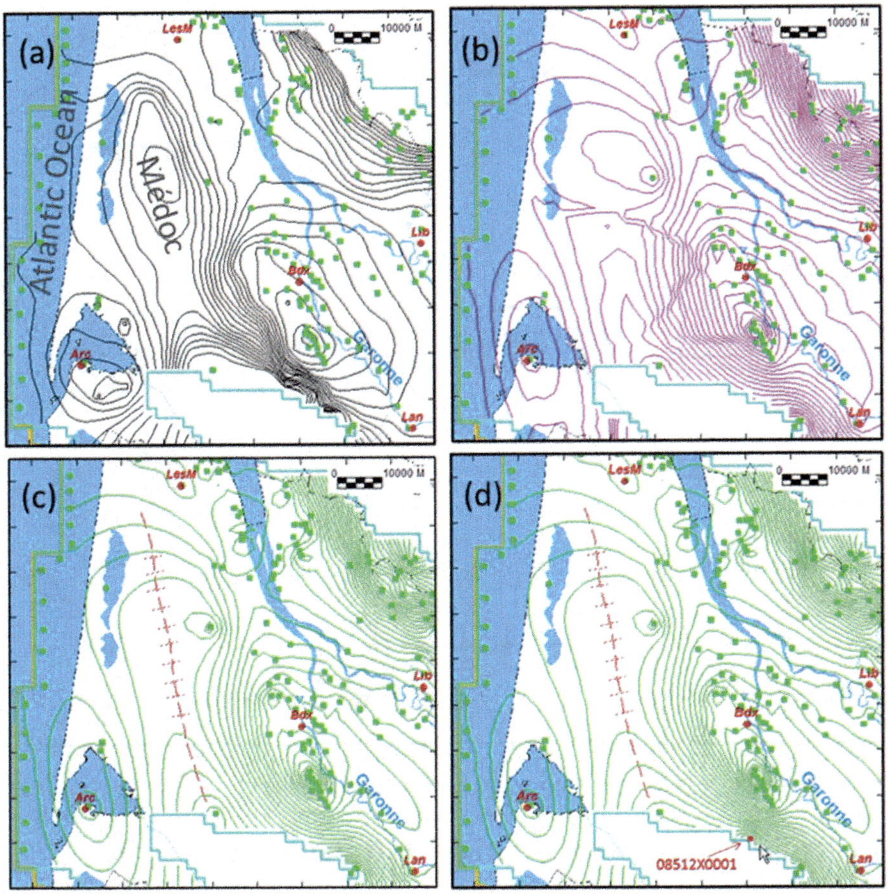

Fig. 6 Use of boundary conditions given by a groundwater ridge; (**a**) hydraulic head from the flow model; (**b**) from ordinary kriging; (**c**) from kriging with boundary conditions; (**d**) idem **c** with a supplementary data

$$R(x) = HH(x) - ZS(x) \tag{1}$$

where $R(x)$ is the residual at point x, $HH(x)$ the measured hydraulic head, and $ZS(x)$ the smoothed topography at this point.

- The scatter diagram of both variables and the linear regression of hydraulic head against the smoothed topography are also computed.

The smoothing radius is then chosen close to the value that minimizes the variance of the regression error and that gives the most correlated variogram as well as a nearly linear scatter diagram.

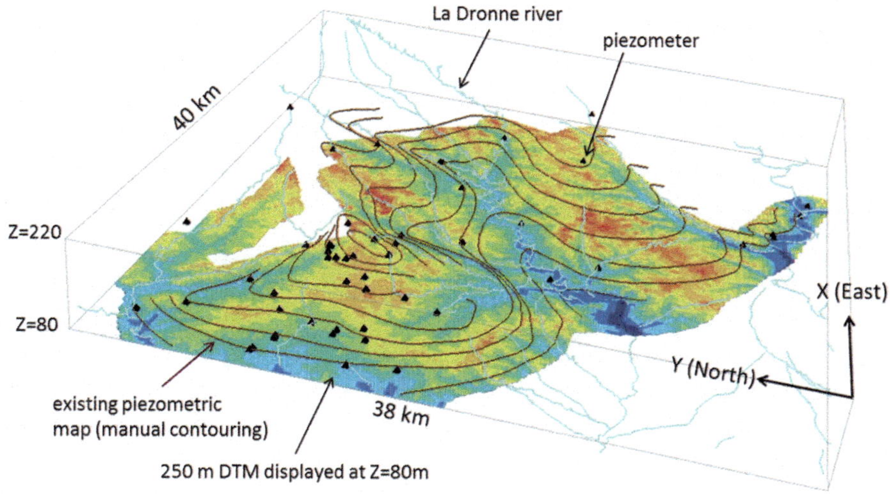

Fig. 7 Available data for the Turonian aquifer

4.2 *Application*

This method has been applied successfully for several unconfined aquifers (Saltel and Bourgine 2015). We report here the example of an aquifer located in the Turonian chalk located near Perigueux City (Dordogne department, southwest of France) (Bourgine and Pédron 2011). In this aquifer, 67 piezometers are monitored in an area of 38 × 40 km. A manual contouring is available for the high waters of year 2002 (Fig. 7). This figure shows that the manual piezometric map is much smoother than topography but follows more or less the topography. As the groundwaters are converging toward the *La Dronne* river, the contour lines were drawn perpendicularly to the river axis.

To quantify the correlation between hydraulic head and topography (initial or smoothed), the topography was smoothed using a moving average. The radius of smoothing was increased from 0 (no averaging compared to initial 50 m resolution digital elevation model) to 3,000 m. For each window size the standard deviation of the error of the linear regression of the measured hydraulic head as a function of smoothed topography was computed, as well as the variogram of the residual $R(x)$ (Fig. 8).

Figure 8 (left) shows that the standard deviation of the regression error reaches a minimum for a window radius of 1,000 m. This radius is chosen as the optimal one and the variogram of the residual $R(x)$ is computed (Fig. 8 – right). It can be fitted with an exponential model with a practical range of 5,300 m.

Figure 9 shows the linear regression between the measured hydraulic head and (left) the initial topography (50 m resolution) or (right) the topography smoothed with a radius of 1,000 m. The correlation is much better in the second case. In this case it is clear that the window radius of 1,000 m gives the best results. We have

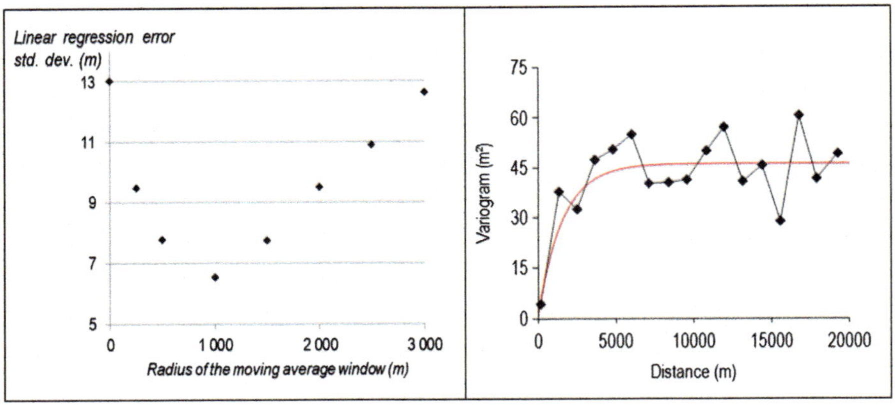

Fig. 8 Residual standard deviation versus radius of the moving average (*left*) and variogram of residual for the optimum size (*right*)

encountered other situations where the minimum in Fig. 8 can be chosen between two or three window radius. In this case we compare the variograms of the residuals and chose the radius that gives the best variogram continuity. It is also possible to compare the linear regressions obtained for various radii and select the radius that gives the best results. At last it is also possible to perform a cross validation for each window radius and select the radius that gives the best results in the key areas of the aquifer to be studied.

At last kriging with the smoothed topography as external drift was performed in the whole aquifer, and results were compared to other kriging options: ordinary kriging and universal kriging with polynomial drift, as well as with manual contouring. Kriging was performed on a 500 m cell-size grid.

Table 1 shows a comparison between manual contouring and three types of kriging. The second column gives the number of meshes in the estimated grid for which the estimated hydraulic head is above topography. Considering that the aquifer is not confined, this should not be observed. Results show that kriging with external drift is far ahead from other methods and gives the best results. In addition, we indicate in the table the mean and maximum value of the difference between interpolated value and topography, when interpolated value is above topography. Ordinary kriging and universal kriging give results of equivalent quality and are not significantly better than manual contouring, whereas kriging with external drift gives clearly the best results.

Another way to quantify the quality of the three kriging is to use a subset of data as validation data set. For this we selected an area south of the *Mareuil* City where the water table draws a dome (Fig. 10). The 19 piezometers located in the dome (red points on the figure) were discarded, and only the other piezometers located outside this dome (blue triangles) were used as input data for kriging.

Results of this validation test are shown Fig. 11. The upper part of the figure shows the interpolations obtained using all data (blue triangles + red points). All the interpolations are consistent and close to manual contouring. The lower part of the

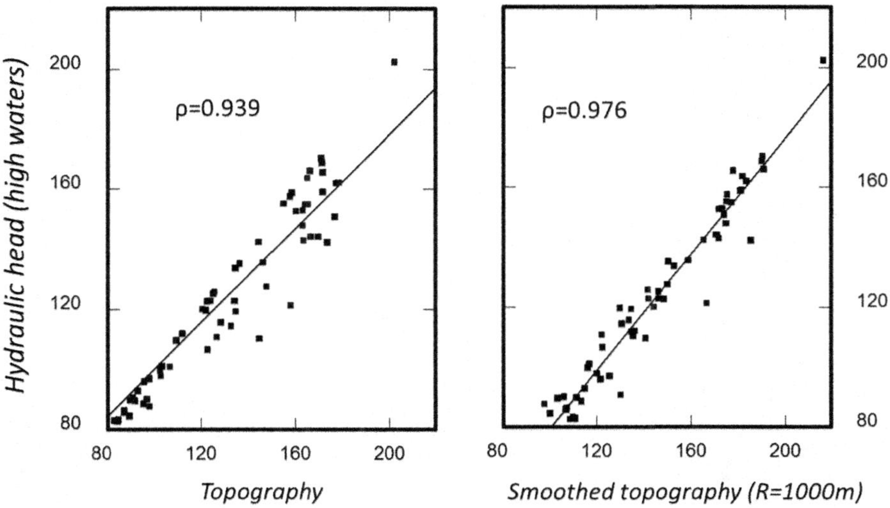

Fig. 9 Linear regression between hydraulic head and topography (initial and smoothed)

Table 1 Comparison of three kriging options with manual contouring

Calculation option	Nb of meshes with computed head > topography	Mean crossing (m)	Max. crossing (m)
Manual contouring	120	5.9	32
Ordinary kriging	111	7.8	25
Universal kriging	116	4.7	15
Kriging with exter-nal drift	14	2.1	7.5

figure shows the result obtained when only the points represented by blue triangles on Fig. 10 are used. As expected, ordinary kriging and kriging with polynomial drift are not able to reproduce the dome at the center of the figure. Conversely kriging with external drift gives very satisfactory results, as it tends to mimic the shape of the smoothed topography when interpolation is performed far from the data.

4.3 Discussion

Similar tests were performed in other areas of the aquifer, and similar results were obtained.

As a conclusion of this example, kriging with external drift based on smoothed topography gives consistent (and the best) results in this unconfined aquifer. The influence of external drift becomes dominant far (compared to variogram range) from the data, which is one of the well-known properties of this method. In areas

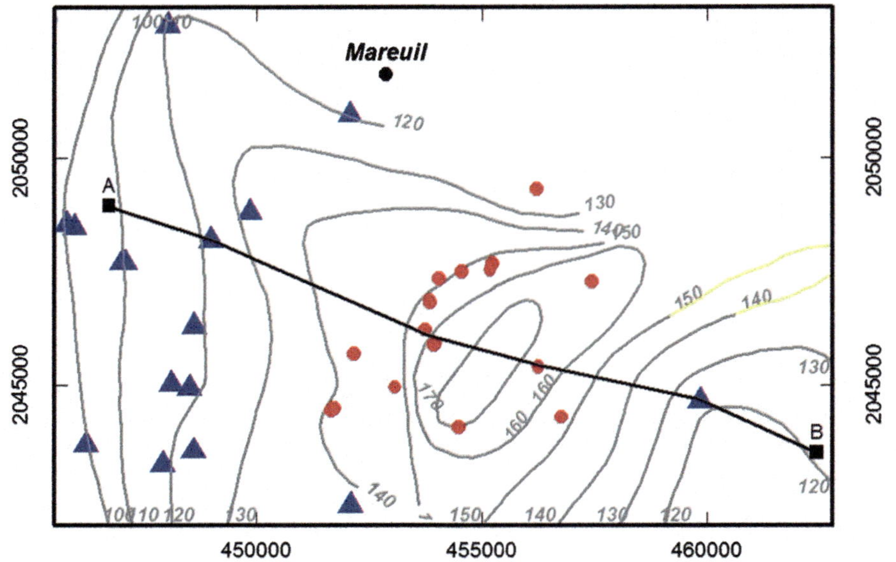

Fig. 10 Test zone in the "Mareuil dome" area. *Blue triangles* are points kept as input for interpolation, and *red points* are data used to validate the interpolation result. *AB* cross section shown next figure. Coordinates in Lambert II projection, meters

with sparse data, it can be a drawback if the correlation is not linear or weaker, but that was not observed here.

One of the limitations of this method is that the window size used to smooth the topography is constant over the whole studied area. Moreover it does not take into account hydrodynamic properties of the aquifer. It is likely that results could be improved by letting the size of the smoothing window vary, as a function of rock hydrodynamic properties, or according to other topographical of hydrogeological information (draining valley, dry valley, plateau, terrain slope or change of slope, etc.), but that was not investigated in this work.

5 Conclusions

This work illustrates the use of two types of kriging in order to take into account hydrogeological constraints when building piezometric maps. These two types of kriging (kriging with external drift based on smoothed topography and kriging with boundary conditions) are not new but not very well known by hydrogeologists. For both methods we obtain much better results than with ordinary kriging or universal kriging that are often proposed "by default" in many software. For kriging with external drift, we propose a method that facilitates the choice of the smoothing radius when smoothing the topography. Both kriging methods were implemented in

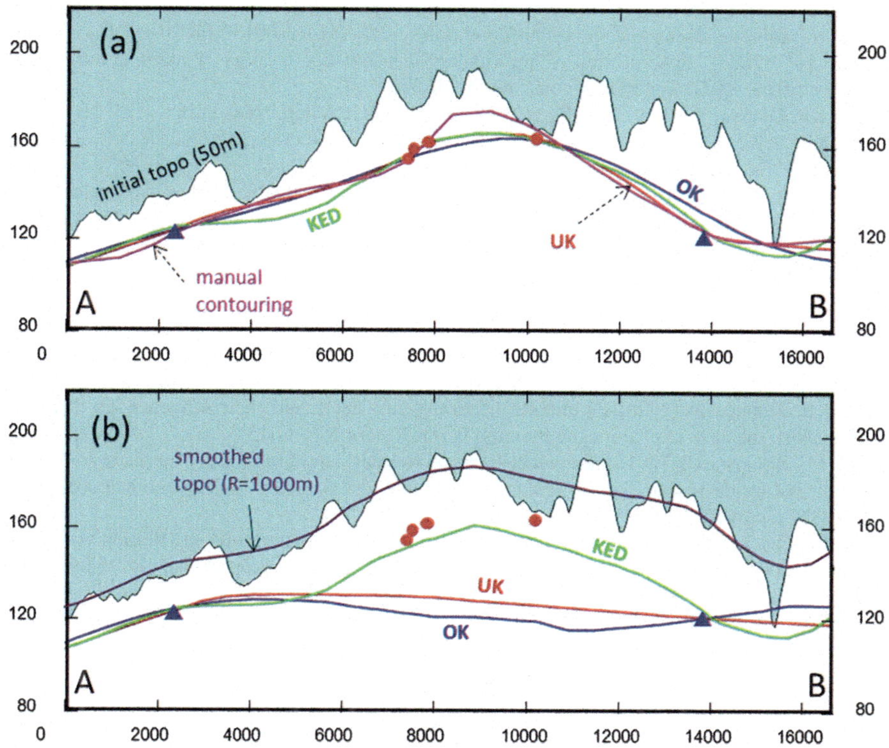

Fig. 11 Comparison of kriging results on a cross section (**a**) kriging from all data, (**b**) kriging from data represented by the *blue triangles*. *OK* ordinary kriging, *UK* universal kriging, *KED* kriging with external drift

GDM software. This software, developed by the BRGM, is used for many applications (geological modeling, geochemistry, environment, hydrogeology, geostatistical analysis). A report was also written and training organized, in order to describe the methodology to be followed by hydrogeologists. In this way a better standardization of procedures is obtained that facilitates the work.

Of course the expertise of the hydrogeologist remains fundamental in order to define, for example, the appropriate boundary conditions that should be taken into account to assess the quality of the resulting map and to introduce additional constraints when necessary.

Bibliography

Aboufirassi M, Marino M (1983) Kriging of water levels in the Souss Aquifer, Morocco. Math Geol 15:537–551

Bourgine B, Pédron N (2011) Piézométrie des systèmes aquifères : outils de cartographie et optimisation de réseaux de surveillance – Année 2. BRGM/RP-60153-FR
Chilès J-P (2004) Spécification d'algorithmes géostatistiques pour l'interpolation de la piézométrie. BRGM internal report – not published
Chilès J-P, Delfiner P (1999) Modelling spatial uncertainty. Wiley, New York
Delhomme J-P (1979) Kriging under boundary conditions. American Geophysical Union Falls Meeting. San Francisco
Desbarats A-J, Logan C-E, Hinton M-K, Sharpe D-R (2002) On the kriging of water table elevations using collateral information from a digital elevation model. J Hydrol 255:25–38
Hoeksema R (1989) Cokriging model for estimation of water table elevation. Water Resour Res 25:429–438
Kuhlman K, Pardo Igúzquiza E (2010) Kriging. J Hydrol 384:14–25
Le Cointe P (2006) Kriging with partial differential equations in hydrogeology. Master II report, Université Pierre et Marie Curie, www.sisyphe.upmc.fr/~m2hh/arch/memoires2006/Lecointe2006.pdf
Renard F, Jeannée N (2008) Estimating transmissivity fields and their influence on flow and transport: the case of champagne mounts. Water Resour Res 44:157–182
Saltel M, Bourgine B (2015) Piézométrie des systèmes aquifères: outils de cartographie et optimisation de réseaux de surveillance – Module 2 – Année 4 – Convention 2008–2013. BRGM report RP-63575-FR
Saltel M, Faucher A, Bourgine B (2012) Gestion des Eaux souterraines en Région Aqutaine. Développements et maintenance du Modèle Nord Aquitain de gestion des nappes – Module 1 – Annéee 3. BRGM RP-61614-FR
Wolock D, Price C (1994) Effects of digital elevation model map scale and data resolution on a topography-based watershed model. Water Resour Res 30(11):3041–3052

A Gradient-Based Blocking Markov Chain Monte Carlo Method for Stochastic Inverse Modeling

Jianlin Fu, J. Jaime Gómez-Hernández, and Song Du

Abstract Inverse modeling for subsurface flow and transport in porous media is expected to improve the reliability of predictions in that the realizations generated are consistent with the observations of states. A gradient-based blocking Markov chain Monte Carlo (McMC) method is presented for stochastic inverse modeling. The method proposed effectively takes advantage of gradient information for tuning each realization to create a new "candidate" proposal, and hence it is capable of improving the performance of McMC. The gradients are efficiently computed by an adjoint method. The proposal mechanism is based on the optimization of a random seed field (or probability field), and thus it is able to preserve the prior model statistics. The method proposed has better performances than the single-component McMC and also avoids directly solving a difficult large-scale ill-conditioned optimization problem simply by turning it into a sampling procedure plus a sequence of well-conditioned optimization subproblems. A synthetic example demonstrates the method proposed.

1 Introduction

Inverse modeling for flow in heterogeneous porous media refers to a process of producing geological models based on some measurements of model responses that are related to model parameters according to known physical laws. Two typical

J. Fu (✉)
Institute for Water and Environmental Engineering (Currently with Chevron), Universitat Politècnica de València, Valencia, Spain
e-mail: fu_jianlin_ac@yahoo.com

J.J. Gómez-Hernández
Institute for Water and Environmental Engineering, Universitat Politècnica de València, Valencia, Spain
e-mail: jgomez@upv.es

S. Du
Petroleum Engineering Department (Currently with Chevron), Texas A&M University, College Station, TX, USA
e-mail: dusong83@gmail.com

J.J. Gómez-Hernández et al. (eds.), *Geostatistics Valencia 2016*, Quantitative Geology and Geostatistics 19, DOI 10.1007/978-3-319-46819-8_53

examples are history matching in petroleum engineering and model calibration in hydrogeology. Many inverse methods have been developed in the last few decades. The Markov chain Monte Carlo (McMC) method stands itself from other competitors probably because of its several merits. First, McMC can, in theory, sample the entire space of the posterior probability density function, allowing extensive evaluation of model uncertainty and prediction uncertainty. Second, McMC is very flexible; it does not limit to any prior assumption about the type of spatial variability and can generate non-Gaussian fields, e.g., channels (Alcolea and Renard 2010). Third, the ill-posed inverse problem is cast as a sampling problem (by construction) without any explicit objective function optimization involved and thus becomes well-posed. Fourth, McMC is relatively simple to implement, and the complex forward simulator can be called in a noninvasive black-box way. On the other hand, McMC has also some drawbacks, e.g., slow rate of convergence and high computational demand.

To improve the convergence rate of McMC, Fu and Gómez-Hernández (2009a, b) proposed a so-called blocking McMC (BMcMC) scheme to accelerate the generation of inverse-conditional realizations and demonstrated its capability and efficiency. It is well known that a block scheme can accelerate the convergence of McMC. Besides the efficient scaling, another possible reason for the success of BMcMC is that it accounts, through a geostatistical approach, for the strong correlation between the parameters within one coarse block; beyond the block, the parameters are less dependent of or decoupled from others as displayed in natural geological phenomena. Unlike the truncated Karhunen-Loeve expansion, BMcMC treats the block as the updating unit and can be viewed as an implicit dimension reduction technique (the relative dimension is reduced because of the use of blocks) but without damaging the structure of the models.

The main contribution of this work is to develop a new proposal mechanism by combining gradient information with BMcMC, such that optimal (based on the current state) realizations can be efficiently generated due to the local optimization property of the gradient-based method and, at the same time, the model space can be sufficiently explored owing to the global optimization property of BMcMC. Besides the specific differences in the algorithm from other gradient-based methods (e.g., Duane et al. 1987; Casey et al. 2008; Girolami and Calderhead 2011; Hanson 2002; Martin et al. 2012; Qi and Minka 2002; Zhang and Sutton 2011), one of the striking features of the present method is the preservation of the model structure for stochastic inverse modeling.

A 2D transient single-phase flow inverse modeling problem is used to illustrate the proposed method. Results show that the method proposed further improves the performances of BMcMC simply because of the efficient coupling of gradients with the generation of geostatistical models. The gradient information uncovers the most sensitive places that significantly influence the objective function and, thus, can be used to guide the search of the updating places such that the McMC computation can be sped up. The gradients cannot guarantee a global minimum but can cause a fast locating of possible regions of the several minima. The BMcMC can guarantee the final finding of the global minimum in theory, but the efficiency might be problematic. The method proposed effectively combines both strengths of the traditional gradient method and BMcMC.

2 Blocking McMC

Consider a random function (RF) discretized at n grid nodes and assume that there are m hard data (i.e., direct observations) and k linear or nonlinear state data where the term "nonlinear" means that the dependent state data are a nonlinear function (e.g., through flow and transport partial differential equations) of model parameters. Specifically, let $\boldsymbol{x}$ denote the RF, $\boldsymbol{x}_1 = \boldsymbol{x}_{\text{obs}}$ denote the m hard data, and $\boldsymbol{y} = \boldsymbol{y}_{\text{obs}}$ denote the dependent state data. The objective of a stochastic modeling is to generate realizations of $\boldsymbol{x}$ conditioned to $\boldsymbol{x}_1$ and inverse conditioned to $\boldsymbol{y}$, i.e., $\boldsymbol{x}|\boldsymbol{x}_1,\boldsymbol{y}$. If we adopt a multi-Gaussian distribution to characterize the RF, then $\tilde{\boldsymbol{x}} \sim N(\mu, \boldsymbol{C}_x)$, where $\boldsymbol{\mu}$ is the prior mean of the RF and $\boldsymbol{C}_x$ is the two-point covariance. After conditioning to the hard data, the distribution remains multi-Gaussian, $\boldsymbol{x}|\boldsymbol{x}_1 \sim N(\mu_{x|x_1}, C_{x|x_1})$, where $\boldsymbol{\mu}_{x|x1}$ is the conditional mean of the RF and $\boldsymbol{C}_{x|x1}$ is the conditional covariance, which, respectively, correspond to the simple kriging (SK) estimate and the SK covariance commonly used in geostatistics.

The joint prior probability density function (pdf) of the multi-Gaussian random field $\boldsymbol{x}$ conditioned to the hard data is

$$\pi(\boldsymbol{x}|\boldsymbol{x}_1,\theta) \propto \exp\left\{-\frac{1}{2}(\boldsymbol{x} - \mu_{x|x_1})^T \boldsymbol{C}_{x|x_1}^{-1} (\boldsymbol{x} - \mu_{x|x_1})\right\}, \tag{1}$$

where the hyperparameter $\boldsymbol{\theta}$ represents the prior information about the RF given by the prior mean $\boldsymbol{\mu}$ and the prior covariance $\boldsymbol{C}_x$, which are necessary to compute the SK estimate and the SK covariance. These prior values have to be either adopted as a priori subjective estimates or modeled after some measurements. Essentially, this pdf (1) measures the dissimilarity (or distance) of a realization from the prior; the closer to the prior, the higher the probability. Usually the covariance reflects our limited knowledge on the subsurface reality. The realizations generated from this distribution by Monte Carlo are all considered as "equally likely."

We assume a multi-Gaussian error for the discrepancy between the observed state $\boldsymbol{y}$ and the state resulting from the approximate solution of the state equations $y_{\text{sim}} = g(\boldsymbol{x})$, $y_{\text{sim}}|\tilde{\boldsymbol{x}} \sim N\left(g(\boldsymbol{x}), C_y\right)$, where $\boldsymbol{C}_y$ describes the degree of discrepancy between the transfer function $g(\boldsymbol{x})$ and the true but error-prone observation $\boldsymbol{y}$.

The joint pdf of $\boldsymbol{y}$ for a realization of the parameters $\boldsymbol{x}$ is given by

$$\pi(y|\boldsymbol{x}) \propto \exp\left\{-\frac{1}{2}(y - g(\boldsymbol{x}))^T C_y^{-1}(y - g(\boldsymbol{x}))\right\}. \tag{2}$$

The error covariance matrix $\boldsymbol{C}_y$ is generally assumed diagonal. For the deterministic optimization problem, the objective function J_{ML} is often defined as $-\ln\pi(\boldsymbol{y}|\boldsymbol{x})$,

$$J_{\mathrm{ML}} = \frac{1}{2}(\boldsymbol{y} - g(\boldsymbol{x}))^T \boldsymbol{C}_y^{-1}(\boldsymbol{y} - g(\boldsymbol{x})), \tag{3}$$

the optimal solution of which corresponds to the maximum likelihood (ML) estimate. Similarly, this pdf (2) measures the possibility (or simply the distance in (3)) of the simulated results of a realization to reproduce the data by recognizing the uncertainties of both data and simulations.

Following Bayes' theorem, we can derive the posterior distribution of $\boldsymbol{x}$ given the observations $\boldsymbol{x}_1$ and $\boldsymbol{y}$, and the prior model $\boldsymbol{\theta}$, and write the posterior pdf as

$$\begin{aligned}&\pi(\boldsymbol{x}|\boldsymbol{x}_1, \boldsymbol{y}, \theta)\\ &\propto \exp\left\{-\frac{1}{2}(\boldsymbol{x} - \mu_{x|x_1})^T \boldsymbol{C}_{x|x_1}^{-1}(\boldsymbol{x} - \mu_{x|x_1}) - \frac{1}{2}(\boldsymbol{y} - g(\boldsymbol{x}))^T \boldsymbol{C}_y^{-1}(\boldsymbol{y} - g(\boldsymbol{x}))\right\}.\end{aligned} \tag{4}$$

For the deterministic optimization problem, the regularized objective function J_{MAP} is often defined as -ln$\pi(\boldsymbol{x}|\boldsymbol{x}_1,\boldsymbol{y},\boldsymbol{\theta})$ by dropping the constant, i.e.,

$$J_{\mathrm{MAP}} = \frac{1}{2}(\boldsymbol{x} - \mu_{x|x_1})^T \boldsymbol{C}_{x|x_1}^{-1}(\boldsymbol{x} - \mu_{x|x_1}) + \frac{1}{2}(\boldsymbol{y} - g(\boldsymbol{x}))^T \boldsymbol{C}_y^{-1}(\boldsymbol{y} - g(\boldsymbol{x})), \tag{5}$$

the solution of which corresponds to the maximum a posteriori (MAP) estimate.

Now the task for stochastic inverse modeling is how to draw independent identically distributed (*i.i.d.*) multivariate samples (not just an estimate) from this posterior distribution. In the sequel, the explicit dependency of $\boldsymbol{x}$ on the prior model $\boldsymbol{\theta}$ and on the conditioning data $\boldsymbol{x}_1$ is dropped out to lighten the notation, thus $\pi(\boldsymbol{x}) \equiv \pi(\boldsymbol{x}|\boldsymbol{x}_1, \theta)$ and $\pi(\boldsymbol{x}|\boldsymbol{y}) \equiv \pi(\boldsymbol{x}|\boldsymbol{x}_1, \boldsymbol{y}, \theta)$. Plus, $\boldsymbol{x}$ is normalized by removing the mean. A blocking McMC was proposed by Fu and Gómez-Hernández (2009a) to draw *i.i.d.* realizations from (5).

The main idea of BMcMC is that the proposed member $\boldsymbol{x}^*$ is built from the previous member $\boldsymbol{x}$ by modifying an entire block of grid nodes. The mechanism of how the block updating works was shown in Fig. 1 of Fu and Gómez-Hernández (2009a): $\boldsymbol{x}$ denotes the entire field; $\dot{\boldsymbol{x}}$ denotes a subset of the entire field centered at the block to be updated for the purpose of approximating the prior multivariate pdf, $\pi(\dot{\boldsymbol{x}}) \approx \pi(\boldsymbol{x})$; $\hat{\boldsymbol{x}}$ is the block that will be modified in the updating from $\boldsymbol{x}$ to $\boldsymbol{x}^*$; $\breve{\boldsymbol{x}}$ is a skin of grid cells around the block; and $\ddot{\boldsymbol{x}}$ represents the "extended block" made up of the updating block $\hat{\boldsymbol{x}}$ plus the skin $\breve{\boldsymbol{x}}$. More details can be found in Fu and Gómez-Herndez (2009a).

3 Gradient-Based BMcMC

The central idea of the gradient-based BMcMC method (gBMcMC) is to propose a new candidate $\boldsymbol{x}^*|\boldsymbol{x}$ using gradient information based on the block updating of McMC (by "maximizing" the likelihood $\pi(\boldsymbol{y}|\boldsymbol{x}^*)$), such that the convergence rate and the acceptance ratio can be increased and thus the computational efficiency can be enhanced. Another possible way is to find an optimal $\boldsymbol{x}^*|\boldsymbol{x}$ by maximizing the a posteriori density $\pi(\boldsymbol{x}^*)\pi(\boldsymbol{y}|\boldsymbol{x}^*)$ using the gradient-based MAP method. A probability perturbation gBMcMC method is presented here for this purpose.

Instead of focusing on $\boldsymbol{x}^*|\boldsymbol{x}$ itself for the proposal function $q(\boldsymbol{x}^*|\boldsymbol{x})$, we introduce a probability perturbation gBMcMC method that works on the random deviate $\boldsymbol{z}^*|\boldsymbol{z}$ used for the generation of the block conditional realization:

$$\hat{x} = \boldsymbol{L}_{21}\boldsymbol{L}_{11}^{-1}\breve{x} + \boldsymbol{L}_{22}\hat{z} \tag{6}$$

where $\boldsymbol{L}_{11}$, $\boldsymbol{L}_{21}$, and $\boldsymbol{L}_{22}$ can be computed from

$$\ddot{\boldsymbol{C}} = \begin{bmatrix} \ddot{\boldsymbol{C}}_{11} & \ddot{\boldsymbol{C}}_{12} \\ \ddot{\boldsymbol{C}}_{21} & \ddot{\boldsymbol{C}}_{22} \end{bmatrix} = \boldsymbol{LU} = \begin{bmatrix} \boldsymbol{L}_{11}\boldsymbol{U}_{11} & \boldsymbol{L}_{11}\boldsymbol{U}_{12} \\ \boldsymbol{L}_{21}\boldsymbol{U}_{11} & \boldsymbol{L}_{21}\boldsymbol{U}_{12} + \boldsymbol{L}_{22}\boldsymbol{U}_{22} \end{bmatrix}, \tag{7}$$

in which $\ddot{\boldsymbol{C}}$ is the prior covariance matrix for the block $\ddot{\boldsymbol{x}}$. Working on the random deviate allows the method to maintain the structure consistency for all the stochastic realizations even though the optimization method itself does not. Although the gradient can be computed for the entire field, we only utilize part of this information that is closely related to the updating block such that we can locally work on that block. The advantages of using only a block's information are the following: (1) McMC is more efficient based on block updating, (2) the dimensionality of the local optimization problem used to tune the field can be limited to a relative small size (i.e., of $\hat{\boldsymbol{x}}$) and thus the optimization problem can be efficiently solved, and (3) if the multiscale adjoint method is used to compute the gradient, only the local gradient is required and can be reconstructed selectively such that the computational efficiency can be greatly improved (Fu et al. 2010, 2011).

Again, we only need to tune a local block $\hat{\boldsymbol{x}}$ in proposing $\boldsymbol{x}^*|\boldsymbol{x}$. The derivative of the objective function $J = J_{\text{ML}}$ in (3) with respect to the random deviate $\hat{\boldsymbol{z}}$ (or equivalently the probability field) is computed by

$$\frac{\partial J}{\partial \hat{z}} = \frac{\partial J}{\partial x}\frac{\partial x}{\partial \hat{z}} \approx \frac{\partial J}{\partial \hat{x}}\frac{\partial \hat{x}}{\partial \hat{z}}, \tag{8}$$

where $\frac{\partial J}{\partial \hat{x}}$ is simply the computed gradient of the objective function with respect to model parameters using the adjoint method (Fu et al. 2010, 2011). Note that $\frac{\partial \hat{x}}{\partial \hat{z}}$ is simply the lower triangle matrix $\boldsymbol{L}_{22}$ in (6).

In this way, the structure information contained in the prior covariance is explicitly applied to the realization generated, hence ensuring model consistence.

More details can be found in Fu and Gómez-Hernández (2008). The accuracy of the approximation in (8) largely depends on the correlation length of the field and the flow scenario. However, the quality of the approximation does not damage the quality of the solution because it only serves to find a candidate for the construction of the Markov chains.

Sensitivity analysis shows that the size of the computational template associated with each updating block is generally not necessarily very big (Fu and Gómez-Hernández 2009a); a common multidimensional gradient-based optimizer will work fine in finding the "optimal" $\hat{z}_{opt}$ for an updating block. The BFGS method (Broyden-Fletcher-Goldfarb-Shanno; also called the quasi-Newton method) is one of the most efficient gradient-based optimization methods and is applied here for the local optimization problem. Its key feature is to iteratively build up a good approximation to the inverse Hessian matrix (Press et al. 1992). Within this method, the deviate at the iteration l is updated by

$$\hat{\mathbf{z}}_l = \hat{\mathbf{z}}_{l-1} + \hat{\boldsymbol{H}}_l^{-1}(\hat{\boldsymbol{g}}_l - \hat{\boldsymbol{g}}_{l-1}), \tag{9}$$

where $\hat{\boldsymbol{H}}_l^{-1}$ is the inverse Hessian matrix.

Several additional comments are pertinent:

1. Although one can start gBMcMC for stochastic inverse modeling with a history matched model, which can be obtained by some deterministic approaches (Fu et al. 2010, 2011), we assume that this large-scale model is not available in this work but one can initialize gBMcMC with an unconditional guess model that was generated by some geostatistical tool (Gómez-Hernández and Journel 1993).
2. Because updating one block may not necessarily reduce the objective function too much (especially if the block is located at an insensitive zone), a general recommendation is to perform only a couple of iterations for solving the local optimization problem.
3. Only the relatively sensitive locations (e.g., above a certain sensitivity threshold) should be subject to the gradient-based model update, while the insensitive locations enter the standard McMC update without prior gradient optimization. By doing so, many local optimization problems can be avoided.
4. In the case that the convergence is problematic for the local optimization, an escape mechanism is required to ensure the continuous evolution of the chain without break, for example, the chain can be advanced by the best result at hand even though the convergent result is not available, or the chain keeps the current state and finds another place to tune, i.e., the so-called "run-and-hit" scheme; note that a "hit-and-run" scheme may not work fine with a gradient-based proposal since if an "optimal" proposal is rejected, it is usually hopeless to tune the same block at the current state to match the data.

An adaptive version of the probability perturbation gBMcMC can be formulated when the gradient-based update is always located at the most sensitive places, i.e., the block with the highest gradients. Again, a "run-and-hit" scheme may be invoked if the objective function does not decrease too much after several iterations by tuning the most sensitive places.

4 An Illustrative Example

4.1 *Experimental Setup*

Consider a 2D transient single-phase flow problem on a confined aquifer. All parameters have consistent length and time units, the absolute magnitude of which is irrelevant for the purpose of demonstrating the efficiency of the proposed method, and only relative magnitudes are mentioned when necessary. The aquifer is discretized in 100×100 cells as shown in Fig. 1. The reference lnK field is generated using GCOSIM3D (Gómez-Hernández and Journel 1993) with a prior distribution lnK ~ $N(0,1)$ and an exponential isotropic variogram. The correlation range is set to 25 cells.

The four boundaries are set to be non-flow. The initial pressure head field is assumed to be zero everywhere in the aquifer. The time discretization for flow simulations employs the so-called time multiplier scheme which assumes that the time increment for each step is multiplied by a constant time-step coefficient. The simulation time of total 500 time units is discretized into 100 steps. The advantage of this scheme is that it allows for an adequate time discretization at the early stage

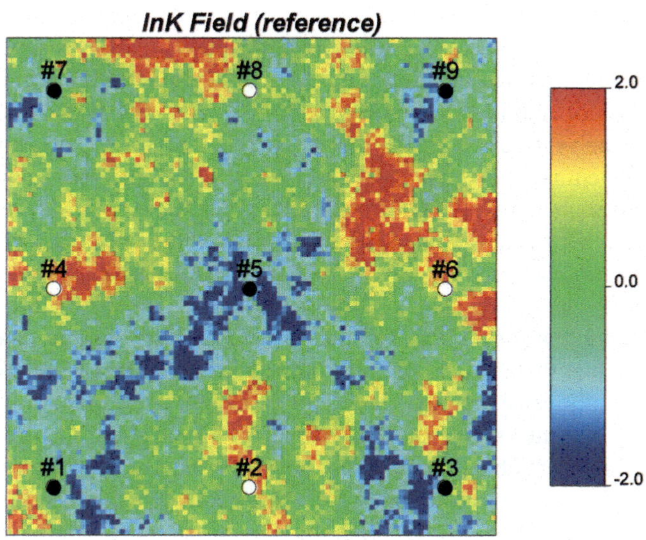

Fig. 1 The reference lnK field and the well configuration: Four empty *circles* denote injection wells and five *solid dots* are extraction wells

of simulation such that the simulated transient pressure head distribution is to the least degree influenced by the time discretization. Nine wells are drilled throughout this confined aquifer (Fig. 1): four of them are injection wells with a constant flow rate and the other five are production wells with a constant pressure. As the state, the hydraulic head at the four injection wells is continuously collected for the first 50 time steps (approximately until after 40 time units).

The stochastic inverse modeling problem is to infer the hydraulic conductivity field according to the observed hydraulic head data at the four injection well bores. As a consequence of the method used for the generation of the conductivity fields, lnK will preserve the prior distribution statistics, i.e., lnK ~ $N(0,1)$, and the exponential covariance with the range equal to 25. Measurement errors of state (i.e., the pressure head) are assumed to have a normalized error variance of 5 % at all locations, and the error covariance matrix is diagonal.

5 Results

We first want to compare the performance of the BMcMC algorithms, i.e., how the convergence rates and acceptance ratios of various algorithms look like. Figure 2 plots the traces of three BMcMC algorithms with different block sizes. The x axis shows the chain index, scaled in log; the y axis shows the log normalized mismatch, that is, the mismatch (i.e., the objective function) is first normalized to one by dividing by the original initial value and then a log value of the normalized mismatch is taken as the y value. An obvious observation is that, for the relatively large block size (i.e., of 25×25), the two gradient-based algorithms (i.e., the gBMcMC and adaptive gBMcMC) do improve the convergence rate of BMcMC (Fig. 2b); moreover, as expected, the adaptive scheme that is able to locate the most sensitive places for model updating (i.e., the adaptive gBMcMC) seemingly has a faster convergence rate than the others (both gBMcMC and BMcMC). However, another observation is that for a small block size (i.e., of 8×8), such improvement is not very evident (Fig. 2a). This is mainly because the objective function can be brought down much more quickly by the gradient-based methods if the block size is larger; on the other hand, the difficulty of the underlying optimization problem is also increased and may become very challenging for a large-scale problem. In other words, updating the entire field may have the fastest convergence rate, but the optimization problem for finding the updates is too challenging. In addition, from Fig. 2, one can find that the adaptive algorithm has the highest rate of convergence, yet the quality of inverse solution seemingly is the worst. By quality, we mean the objective function value to measure the mismatch. Note that the adaptive algorithm has the largest average objective function; yet, since the y axis is taken as log of the normalized mismatch, the real difference of the objective functions between them is quite small in reality. These observations are supported by several parallel chains.

Next, we want to see the quality of the history matching, i.e., how the historical data are reproduced and how reliable the predictions are. Figure 3 plots the histories

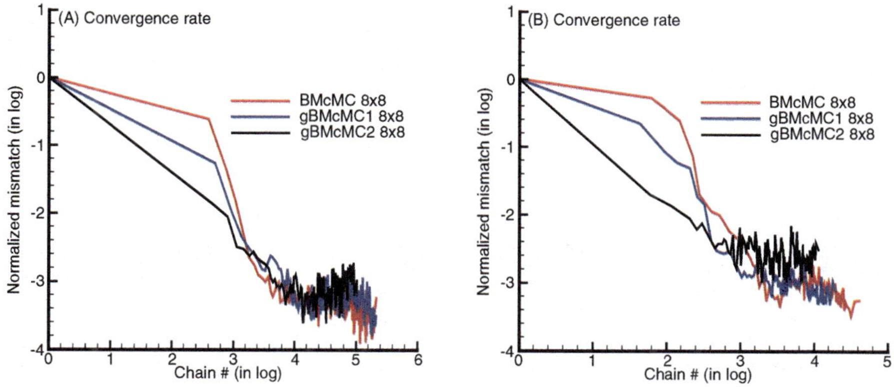

Fig. 2 Convergence rate of McMC inverse modelings: gBMcMC1 shows the gBMcMC method and gBMcMC2 represents the adaptive gBMcMC method

of the pressure head at the four injection wells from the three inverse modeling algorithms. For comparison, the unconditional cases that were generated by GCOSIM3D are also plotted. The conditioning state data are plotted in circles, while the gray lines plotted 200 independent realizations from different methods. Note that the inverse modelings only constrain the models to the first 40-day's state data, the rest of states is treated as the prediction. Clearly, the unconditional cases cannot reproduce the historical data but stochastic inverse-conditional realizations match well the given data; also, stochastic inverse modelings are able to offer much better predictions than the unconditional cases. Moreover, the data are efficiently enveloped by the inverse-conditional realizations, meaning that the realizations generated by the stochastic inverse modeling algorithms (BMcMC, gBMCMC, and adaptive gBMcMC) well represent a model of uncertainty consistent with the data. The failure of unconditional realizations in representing the uncertainty is probably because the well performances are very sensitive to the heterogeneity of lnK around the well bores. In addition, the adaptative gBMcMC has a relatively larger uncertainty than the others simply because it has larger objective function values (Fig. 2 also shows the mismatches).

Finally, we point out that the main spatial pattern of the reference field has been reasonably reproduced by the proposed method using various block sizes. The variograms follow the prior specifications and are consistent with the reference field (Fig. 4). Note that the prior model represents the models before history matching; BMcMC plots history matched models using BMcMC method; while gBMcMC represents the models after history matching using the method proposed. Therefore, the method proposed can effectively preserve the spatial statistics and structure for stochastic inverse modeling.

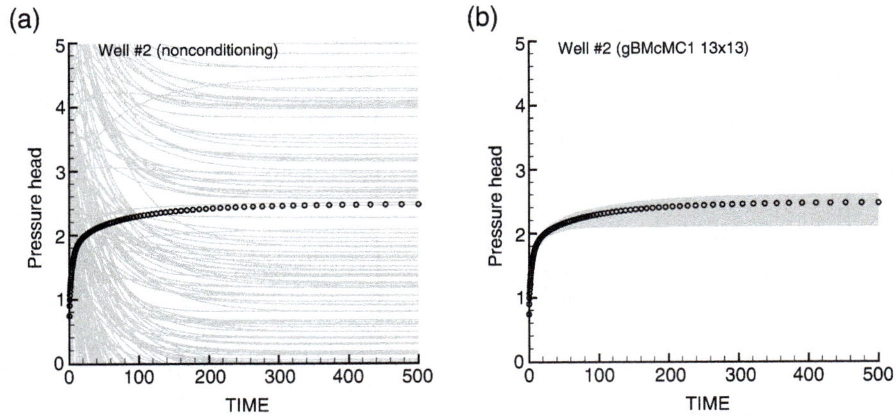

Fig. 3 Reproduction of the historical state data at well #2: (**a**) before history matching and (**b**) after history matching

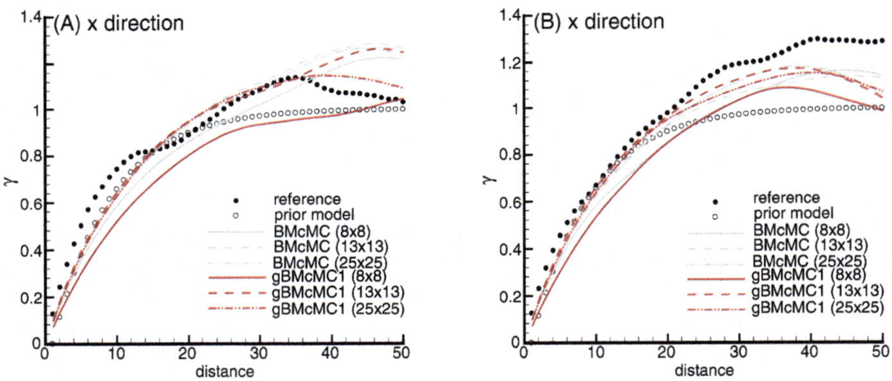

Fig. 4 Ensemble mean of semivariogram of MCMC inverse modeling: (**a**) x direction and (**b**) y direction

6 Discussion and Conclusions

We present a gradient-based probability perturbation BMcMC method for the stochastic inverse modeling and uncertainty quantification of flow in porous media. The method proposed further improves the performances of BMcMC simply because of the efficient coupling of gradients with the generation of geostatistical models. The gradient information uncovers the most sensitive places that significantly influence the objective function and, thus, can be used to guide the search of the updating places such that the McMC computation can be sped up. The gradients cannot guarantee a global minimum but can cause a fast locating of possible regions of the several minima. The BMcMC can guarantee the final finding of the global minimum in theory, but the efficiency might be problematic. The method proposed effectively combines both strengths of the traditional gradient method and BMcMC.

In the algorithms that we present, we only consider updating one block with a fixed size. As we observed before for BMcMC (Fu and Gómez-Hernández 2009a), a varying block scheme is also expected to improve the performance of gBMcMC similarly. As observed, a larger updating block usually has a faster rate of convergence but a moderately small block yields solutions with a better quality. One may start McMC with a large block until convergence is reached and then switch to a scheme with a small block to output realizations. Or simply pushing to an extreme situation, one may initialize the chain with a history matched model that is obtained from a deterministic method (Fu et al. 2011) and then use gBMcMC to generate the stochastic realizations. A "multiblock" scheme may also be considered; this is achieved by simultaneously updating the field with several blocks that are located at the most sensitive positions as discovered by the gradients. In this way, the dimensionality of problem can be reduced compared to the original large-scale gradient-based optimization problem, and thus the computational challenge is largely relaxed, but at the same time, the accuracy of the final results does not deteriorate since the most sensitive places are always updating. This is somehow similar to the localization scheme: the most sensitive places are often found in the very near vicinity around the observation points (Fu et al. 2011).

Since the high-resolution gradient computation is also very expensive, just like the forward simulation, a multiscale strategy may further apply and improve the computational efficiency of BMcMC. There are two possible options to this goal. One is the so-called two-stage McMC method proposed by Dostert et al. (2006). Another more efficient way is to use the so-called multiscale gradient method (Fu et al. 2010, 2011). The gradients are computed only at the coarse scale, but the fine-scale gradients can be selectively reconstructed for the target block using the basis functions and coarse-scale gradients. The MH test is only applied once to the fine scale and avoids the two-stage procedure. Hence, this multiscale method is a good candidate for gradient-based BMcMC.

The use of a gradient "mode" by accounting for model uncertainty may also be considered for improving the performance of gradient-based stochastic inverse modeling. A possible way is to use an ensemble gradient that is computed from an ensemble of realizations such that the mode of gradients can be used for locating the position of the targeted posterior distribution. For BMcMC, the ensemble gradient will gradually progress toward such mode along the evolution of McMC as more individual gradients of the models generated are computed and added into the database for assembling the gradient mode. Note that even though the ensemble mean of stochastic realizations is homogeneous, it does not necessarily mean that the ensemble gradient is homogeneous. This is because it is the heterogeneity (i.e., the spatial variability), rather than the mean values of fields, that controls the gradient.

Bibliography

Alcolea A, Renard P (2010) Blocking moving window algorithm: conditioning multiple-point simulation to hydrogeological data. Water Resour Res 46:W08511

Casey FP, Waterfall JJ, Gutenkunst RN, Myers CR, Sethna JP (2008) Variational method for estimating the rate of convergence of Markov-chain Monte Carlo algorithms. Phys Rev E 78 (4):046704

Dostert P, Efendiev Y, Mohanty B (2009) Efficient uncertainty quantification techniques in inverse problems for Richards' equation using coarse-scale simulation models. Adv Water Resour 32:329–339

Duane S, Kennedy AD, Pendleton BJ, Roweth D (1987) Hybrid Monte Carlo. Phys Lett B 195:216–222

Fu J, Gómez-Hernández JJ (2008) Preserving spatial structure for inverse stochastic simulation using blocking Markov chain Monte Carlo method. Inverse Probl Sci Eng 16(7):865–884

Fu J, Gómez-Hernández JJ (2009a) A blocking Markov chain Monte Carlo method for inverse stochastic hydrogeological modeling. Math Geosci 41(2):105–128

Fu J, Gómez-Hernández JJ (2009b) Uncertainty assessment and data worth in groundwater flow and mass transport modeling using a blocking Markov chain Monte Carlo method. J Hydrol 364:328–341

Fu J, Tchelepi HA, Caers J (2010) A multiscale adjoint method to compute sensitivity coefficients for flow in heterogeneous porous media. Adv Water Resour 33(6):698–709

Fu J, Caers J, Tchelepi HA (2011) A multiscale method for subsurface inverse modeling: single-phase transient flow. Adv Water Resour 34(8):967–979

Girolami M, Calderhead B (2011) Riemann manifold Langevin and Hamiltonian Monte Carlo methods. J R Stat Soc B 73(2):123–214

Gómez-Hernández JJ, Journel AG (1993) Joint simulation of multiGaussian random variables. In: Soares A (ed) Geostatistics Troia '92, vol 1. Kluwer, Dordrecht, pp 85–94

Gómez-Hernández JJ, Sahuquillo A, Capilla JE (1997) Stochastic simulation of transmissivity fields conditional to both transmissivity and piezometric data: I. Theory. J Hydrol 203:162–174

HansonTM (2002) Use of probability gradients in hybrid MCMC and a new convergence test, Los Alamos Report LA-UR-02-4105, pp 1–11

Martin J, Wilcox LC, Burstedde C, Ghattas O (2012) A stochastic Newton MCMC method for large-scale statistical inverse problems with application to seismic inversion. SIAM J Sci Comput 34(3):A1460–A1487

Press WH, Teukolsky SA, Vetterling WT, Flannery BP (1992) Numerical recipes in C: the art of scientific computing, vol 2. Cambridge University Press, Cambridge, p 994

Qi Y, Minka TP (2002) Hessian-based Markov chain Monte Carlo algorithms, First Cape Cod Workshop on Monte Carlo Methods, Cape Cod, Massachusetts, September, 2002, pp 1–13

Zhang Y, Sutton C (2011) Quasi-Newton methods for Markov chain Monte Carlo. In: Shawe-Taylor J, Zemel RS, Bartlett Pl , Pereira FCN, Weinberger KQ (eds) Advances in Neural Information Processing Systems (NIPS), vol. 24, pp 1–9

Geostatistical Modelling and Simulation Scenarios as Optimizing Tools for Curtain Grouting Design and Construction at a Dam Foundation

V. Gavinhos and J. Carvalho

Abstract Foundation treatment is a crucial stage in dam construction. An intense construction stage dedicated to ground improvement and quality control is often implemented. Foundation permeability is the parameter to be monitored and optimized during the construction stage. This work describes the application of geostatistics to permeability data obtained by the Lugeon test during curtain grouting at a dam foundation. The usual construction methodology consists in applying a sequentially phased drilling strategy materialized by drilling and grouting a first line of primary boreholes followed by a secondary collinear line of in-between boreholes. The studied site is located at the Sabor river mouth, northeastern part of Portugal. Groundwater flows in a fractured medium. The main objective of this study is assessing the advantage of a geostatistical modelling approach in identifying the zones in which the project design criteria may demand for a first- and a secondary-stage grouting treatment defined by a threshold permeability of 1 Lugeon unit, Lu. After a preliminary statistical, spatial continuity characterization and error study, estimation and probability maps are compared based on ordinary and indicator kriging and indicator simulation. The obtained models are compared with the information of the acquired construction phase data. Obtained models are finally compared with the initial geological–geotechnical information indicating that the described approach can be used as an optimizing tool for curtain grouting design at a dam foundation.

V. Gavinhos (✉)
Faculty of Engineering, University of Beira Interior, DECA/GEOBIOTEC, Covilhã, Portugal
e-mail: vasco.gavinhos@gmail.com

J. Carvalho
Faculty of Engineering, University of Porto, DEM/CERENA, Porto, Portugal
e-mail: jorcarv@fe.up.pt

J.J. Gómez-Hernández et al. (eds.), *Geostatistics Valencia 2016*, Quantitative Geology and Geostatistics 19, DOI 10.1007/978-3-319-46819-8_54

1 Introduction

Dams are among the most complex and sensitive manmade structures both in terms of project design and construction. To prevent water passing under the main structure and particularly to avoid the effects of under pressure and foundation erosion, safety standards are observed very closely. This makes the foundation treatment design and construction a crucial phase in any dam project. With the increasingly overall quality and safety awareness regarding the foundation conditions, most designers adopt an intensive construction stage dedicated to ground improvement and quality control. The overall criteria and general strategies are established prior to construction stage, but the real amount of work and the adaptations to real conditions are made with close monitoring of some geological–geotechnical parameters (like fracture density, weathering), permeability, and grout intake. A very common parameter used in decision making for depth and density of drilling is permeability. In this case study the project design team (EDP 2011) established the foundation permeability as the parameter to be monitored, used in decision making, and, finally, optimized to values under 1 Lu, all during the construction stage.

The foundation treatment consists basically of a consolidation stage and a sealing stage where a deep curtain of grouted drill holes is built. Curtain grouting is a series of parallel and vertical drill holes along the dam foundation which, after grouting and sealing the surrounding fractured rock, work as an underground barrier to water percolation. The usual construction methodology consists in applying a sequentially phased drilling strategy in which drill holes are grouted by stages (upward or downward according to the quality of the rock). The curtain is materialized by successively drilling and grouting a first line of primary boreholes followed by a secondary collinear line of in-between boreholes. This implies that the permeability of the secondary drilling locations may be affected by the first phase grouting.

The general approach is to drill all primary-stage grout positions by sectors to a target depth and evaluate the permeability of every 5 m in each drill hole and immediately decide to go deeper and/or prepare for second-stage treatment if permeability above 1 Lu is found.

Geostatistical models are used to identify the areas to be targeted with further treatment at an early stage of construction development and enable decision making for technical specification of further treatment and, in addition, allow production management and planning to be optimized.

2 Objective and Methodology

The main objective of this study is assessing the advantage of a geostatistical modelling approach in identifying the zones in which the project design criteria may demand for an extended first phase and a secondary phase grouting treatment defined by a threshold permeability of 1 Lugeon unit, Lu.

Once initial data sets are prepared, quality controlled and loaded into the geostatistical software (Surfer12 and Gslib), a typical work flow is followed with exploratory statistical analysis, spatial continuity analysis, variogram model cross-validation, estimation with ordinary kriging, indicator kriging and qualitative model evaluation based on the geological model comparison, and a first approach to simulations using sequential indicator kriging (sism).

3 Geological Setting

The studied site is located at Sabor river mouth, northeastern part of Portugal. This river crosses two main geological settings along its course: schist and granitic complexes. At the site the dam foundation rock mass is described as a medium to good-quality schist alternating with quartz-rich wakes with a permeable fracture network developed in a very compact rock matrix. Regionally there is an important tectonic active fault – Vilariça Fault – crossing the center of the valley graben located to the west side of the site (Fig. 1). The river flows from northwest to southeast.

Groundwater flow in this fractured medium is governed by three main planar orientations, with various fillings and close aperture. Due to fracture network and foliation/schistosity pattern, the dam foundation has low permeability although some occurrences of small faults and altered veins work as preferred pathways.

Regional foliation/schistosity and some important fracture systems occur with dips of about 40°–50° along average bearings of N 70°E (EDP 2007). A cross section of the rock mass shows the interpreted model from the early stages of the project, Fig. 2.

4 Technical Features

The Feiticeiro Dam is the first of two dams built at the end of the Sabor river basin. Along with the Laranjeiras Dam (4 km upriver), it comprehends the hydroelectric complex of Baixo Sabor project from Energias de Portugal (EDP). Specifically the Feiticeiro Dam is a conventional gravity concrete dam with 22 blocks up to 45 m high with a 290 m development along a straight axis approximately perpendicular to the river current.

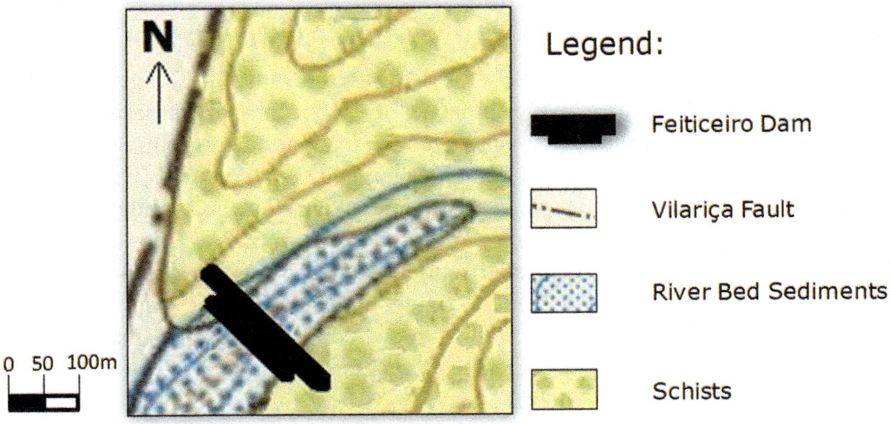

Fig. 1 Geological setting of the Feiticeiro Dam (Adapted from IGM 1989)

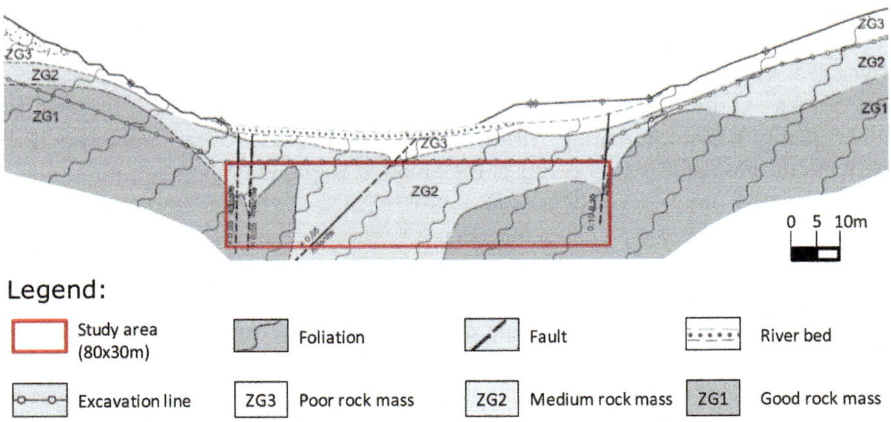

Fig. 2 Cross section of the study area. Geological–geotechnical profile from early studies of the project. View from upstream to downstream (Adapted from EDP 2007)

Construction began at level 95 m (asl) up to level 140 m. Full storage level is located at 138 m (asl). It is equipped with two reversible pump/turbines with independent circuits with a total power of 34 MW, and it works as a storage reservoir for the main production units from the Laranjeiras Dam (123 m high).

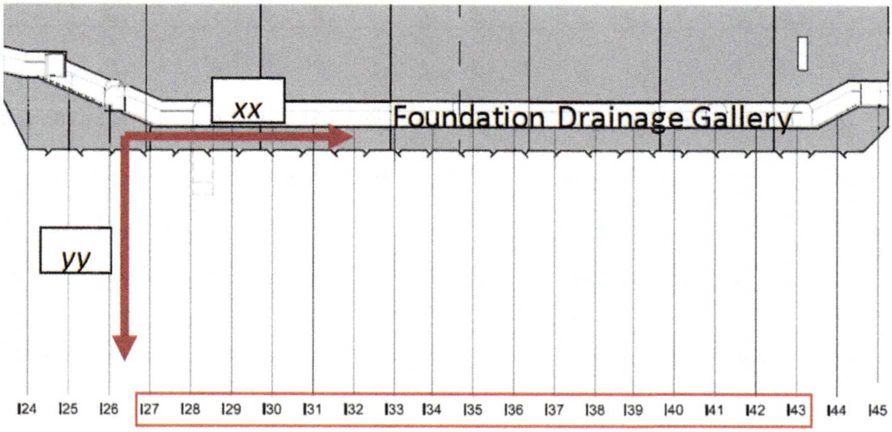

Fig. 3 Location of the sampled area: primary drilling campaign, spaced 5 m. Axis convention for data georeferencing (Adapted from EDP 2007)

5 Data

The data chosen for this procedure was acquired at the construction site at a location of production where there was enough information for a probabilistic approach. The first sector to be drilled and monitored for permeability was the center of the river valley (riverbed), from inside the already built foundation drainage gallery (FDG) of the Feiticeiro Dam (Fig. 3).

The sampling occurs within the cycle of *wireline* core drilling where a pressure packer test (Lugeon test) is performed every 5 m in descending mode until at least 30 m depth. Then, for the 17 primary drill holes inside the bottom rectangle in Fig. 3, 102 Lugeon values were collected. The interpretation was made according to the (Houlsby 1976) guidelines, assuming the values to represent the point at the bottom of each 5 m chamber. Also these drill holes are separated by 5 m from each other along the FDG. From this description the resulting sampling grid is regular (5×5 m^2) from 0 to 80 m along X and from 5 to 30 m along Y, Fig. 4.

This 2D grid was obtained by shifting the depths up toward the positive Y axis by a vertical translation of 30 m. So the value 30 in Fig. 4 actually corresponds to a depth of 5 m and the value 10 corresponds to the depth of 20 m.

6 Descriptive Statistics

The random variable to be characterized is the permeability of the first drilling campaign – P1, Table 1.

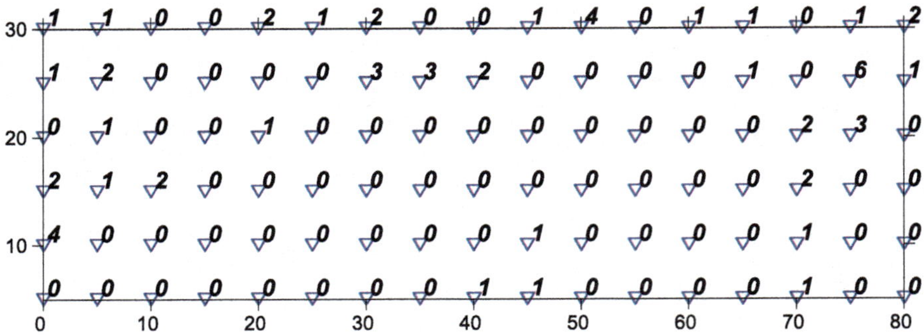

Fig. 4 Spatial representation of the data grid (Surfer12), with measured (interpreted) Lugeon values. Assumed punctual at the *bottom* of each 5 m chamber

Table 1 P1 permeability data ordering and classification

Unit. classes of P1	Nr. of samples	Cumulative
0	69	69
1	18	87
2	9	96
3	3	99
4	2	101
5	0	101
6	1	102

There is a total of seven permeability classes. The predominant classes are low permeability (0 Lu and 1 Lu) with some scarce high values (4 Lu and 6 Lu). Table 2 summarizes the descriptive statistics of the permeability 102 values sample set.

As expected, data distribution is very positively skewed. High skewness (*2,43*) and kurtosis (*6,94*) indicate asymmetry and departure from normality. The high variation coefficient indicates influence from high values which can become problematic in the estimation process.

Next charts on Figs. 5 and 6 graphically complement the information on Table 2 concerning the statistical characteristics of the primary permeability data set. As previously observed the data distribution follows a clear lognormal tendency. Data stationarity was evaluated using moving windows statistics for the mean and variance and polynomial regression analysis using the function *polytool* (Matlab 13, Mathworks) (Fig. 6).

No significant trend was identified along horizontal and vertical directions. This is an expected behavior for permeability in the context of dam foundation rock masses. Stationarity and homoscedasticity were assumed and ordinary kriging was used in estimation.

Table 2 Summary statistics of P1 permeability data

Data	**102**	Average	**0,6** Lu	Max.	**6,0** Lu
		Median	0,0 Lu	Variance	1,14 Lu^2
		Mode	0,0 Lu	Standard deviation	1,07 Lu^2
		Min.	0,0 Lu	Skewness	2,43
		1st Q.	0,0 Lu	Kurtosis	6,94
		3rd Q	1,0 Lu	Variation coef.	1,84

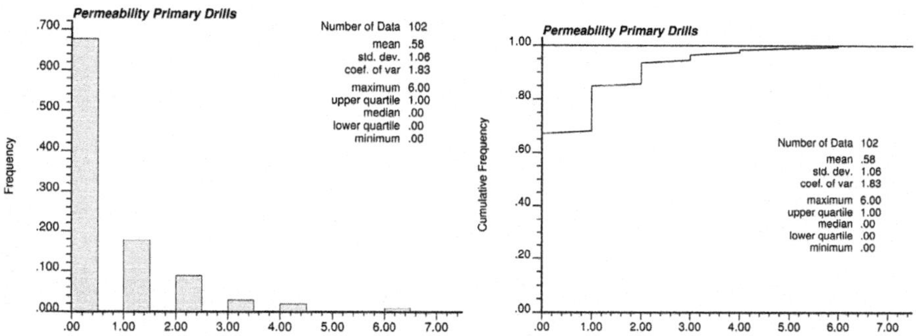

Fig. 5 Permeability P1 data frequency and cumulative histograms

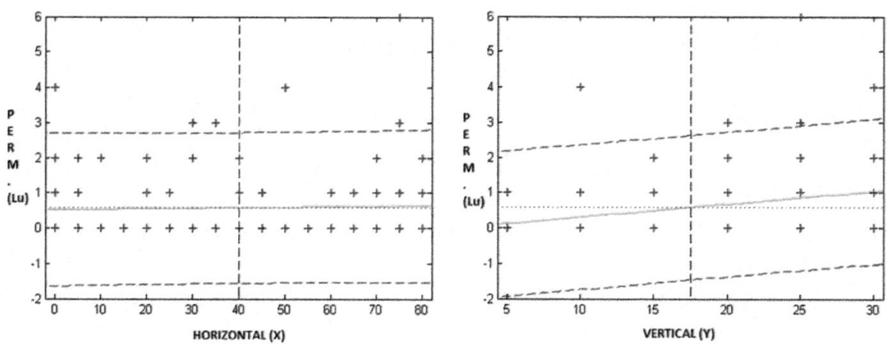

Fig. 6 Data regression models along horizontal and vertical directions (*polytool*-Matlab13)

7 Analysis of Continuity

Variogram analysis was performed using Surfer 12 (Golden Software). Figure 7 shows the initial omnidirectional experimental variogram fitted by a spherical model.

To investigate a probable anisotropic behavior, four major directions – 0°, 45°, 90°, and 135° – were tested with different angular tolerances (from 22,5° up to 45°) (Fig. 8).

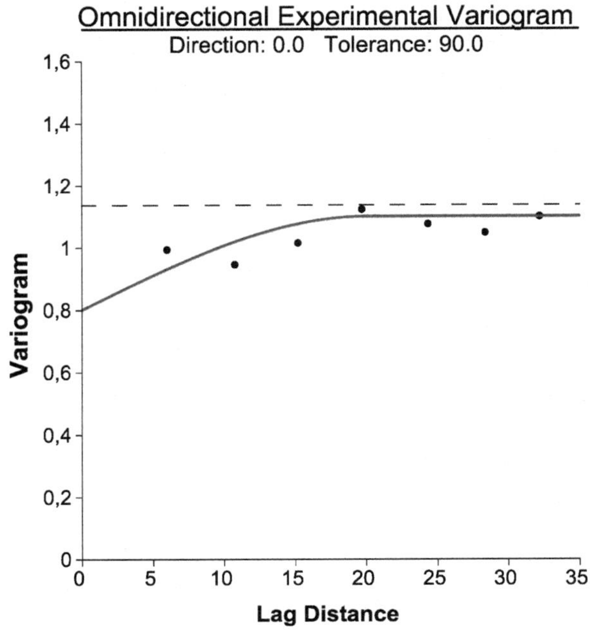

Fig. 7 Omnidirectional variogram with spherical fitted model

Direction 45° presented the major range. The 3D model of the variogram surface (Fig. 9) confirms the higher continuity along direction 45°.

This evidence is corroborated by the real geological setting where there are important structures oriented according to the identified ranges of anisotropic behavior.

The process of variogram model adjustment was affected by an overall randomness tendency of the data. Nevertheless cross-validation was systematically used to check error variance during variogram parameter adjustment. The adopted values are summarized in Table 3.

From the refinement of the adjustments, the nugget effect was reduced to 0,75, and a final geometric spherical variogram model was assumed, modeled with an anisotropic ratio of 1,7 along direction 45°.

8 Estimation

A contour map of the ordinary kriging estimation, based on the referred variogram model, is shown in Fig. 10. The underlying regular grid was defined as 100 × 32 nodes with a spacing of about 0,8 m covering all data points in the sampled area of the foundation (Figs. 3 and 4).

The error study by cross-validation (Fig. 11) and residual analysis showed an error variance of 0,6 and a maximum error of 3,6.

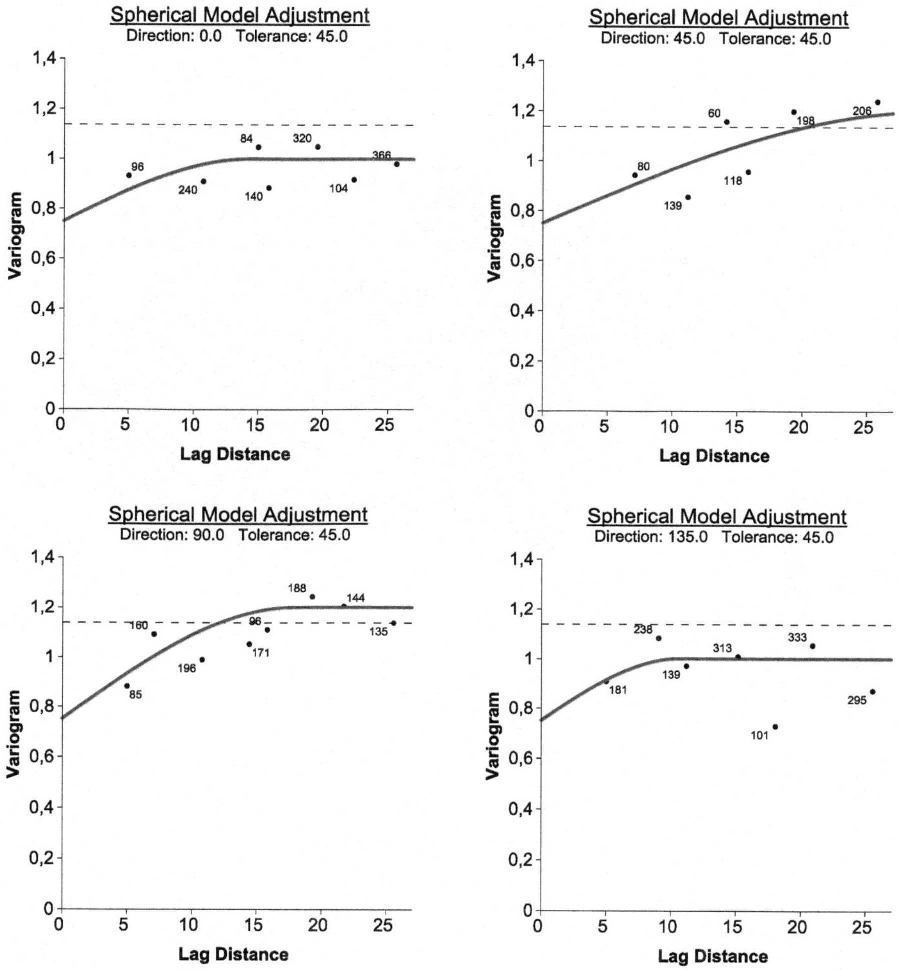

Fig. 8 Directional variograms for the four selected directions: 0°, 45°, 90°, 135°, and fitted spherical models

This map shows that error is concentrated in the areas that most likely will be treated with secondary drilling and grouting.

The observation of the geological–geotechnical profile of the foundation rock mass (Fig. 2), when compared with the estimated permeability map obtained by ordinary kriging (Fig. 10), shows clear correlation in particular in which concerns the referred regional foliation/schistosity pattern as well as some important fracture systems occurring with dips of about 40°–50° along average bearings of N 70°E (EDP 2007). Although the good correlation between the geological models with the estimation map (Fig. 12) is a positive sign of the adequacy of the analysis and of the reliability of the permeability spatial variability model, the main objective is to

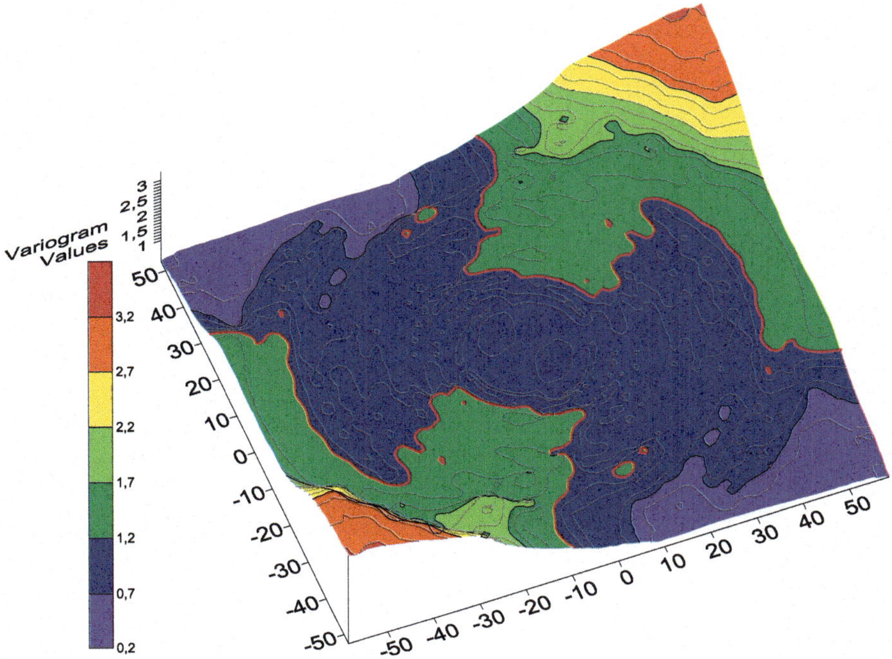

Fig. 9 3D variogram surface model (Surfer12)

Table 3 Summary of spherical model adjustment parameters for selected directions

Parameter	0°	45°	90°	135°
Nugget effect	0,75	0,75	0,75	0,75
c (var. contribution)	0,25	0,45	0,45	0,25
a (range)	20	30	25	18

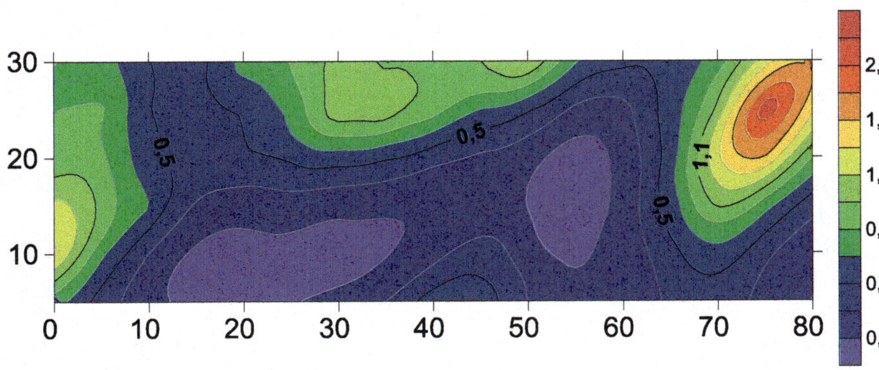

Fig. 10 Contour map–ordinary kriging estimation (Surfer12)

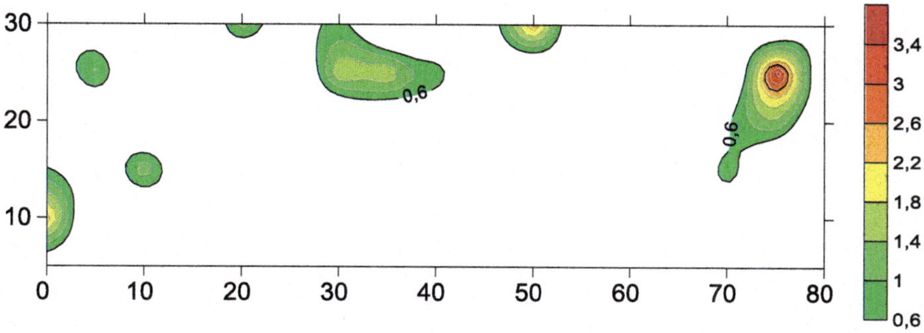

Fig. 11 Error map (residuals) from the kriging estimation. Values above error variance (0,6) (Surfer12)

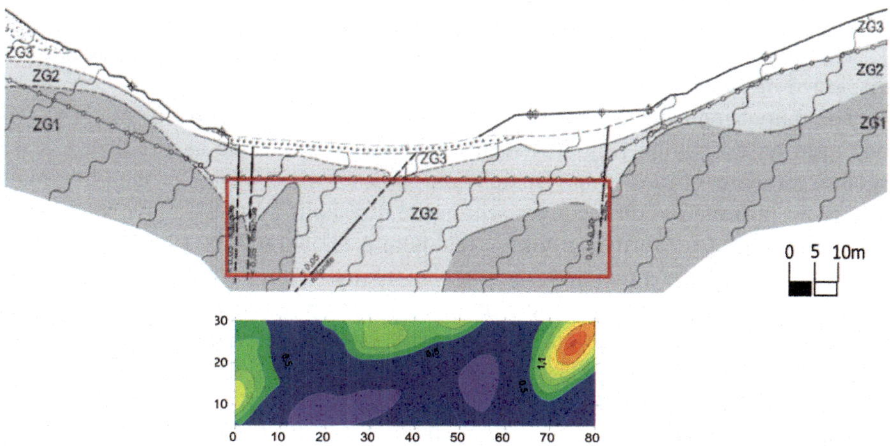

Fig. 12 Geological–geotechnical model compared with OK estimation map (From Figs. 2 and 10)

obtain probability maps to provide assistance in decision-making processes like project review, production planning, and strategies of execution.

Indicator kriging was the first approach to build comprehensive probability maps of the variable, above and below the cutoff of 1 Lu, in order to identify the areas of the foundation to be treated with further grouting. The choice of indicator kriging relies on four fundamental features:

- It is nonparametric and does not depend on prior assumptions about the shape of the distributions, unlike parametric methods.
- Being a nonlinear interpolator, it is useful for highly skewed data like the set used in this investigation.
- It is less prone to over-smoothing of the variable than ordinary kriging and other linear interpolation methods.
- It gives a direct probabilistic estimation of the areas to be further treated at the foundation.

For this purpose, in addition to Surfer 12, GsLib software package was used.

The binary indicator transform of the data consisted in defining the cutoff permeability (for P1 data), following (Soares 2006), as

$$I(1,p) = \begin{cases} 1, \text{ if } P1 \leq 1 \\ 0, \text{ if } P1 \ > 1 \end{cases} \tag{1}$$

Given the short amount of data available, only two threshold data sets were considered by grouping the classes [0–1] and [2–6].

Although the P1 data set gave fairly adjustable experimental variograms, the second phase, P2, data set showed randomness, and it was more difficult to model also due to the number of pairs being even smaller. For indicator kriging the main anisotropy angle, 45°, identified previously in the variogram study of the P1 data, was also consistent with class [0–1] but for class [2–6] an omnidirectional exponential model was adopted (Fig. 13 and Table 4).

The major geometric anisotropy range is 45°. Zonal anisotropy is residual and discarded of the analysis.

The output indicator probability values were mapped on a grid identical to the one used on estimation (Fig. 10) with 3,200 nodes, providing at this time a least square estimate of the *ccdf* (Deutch and Journel 1998). The probability map providing information on the areas within the threshold 1 Lu, e.g., low permeability, with no need for further grouting treatment is shown in Fig. 14.

From Fig. 14 the predicted areas to be treated with further grouting of the foundation become clear. Also the areas of very low permeability become very well defined. It is observed that the areas that were more affected by error in the estimation by ordinary kriging are the ones that bear more certainty to be treated with a secondary stage of grouting.

For a general risk analysis, it is possible to use the E-type estimate. This is simply the average permeability derived by weighting the permeability in each cutoff class by the probability for that class, e.g., it is an optimal estimation of the *ccdf* value at each grid node location (Goovaerts 1997).

When assessing the permeability in terms of the E-type estimates, it is assumed that each grid cell has a single value. In this case the task is simple in terms of computation and visualization because we only have two classes. So, the map in Fig. 15 is consistent with the cutoff map in Fig. 14 because high mean values are associated with high original data values, class [2–6], and complementary low mean values are associated with class [0–1].

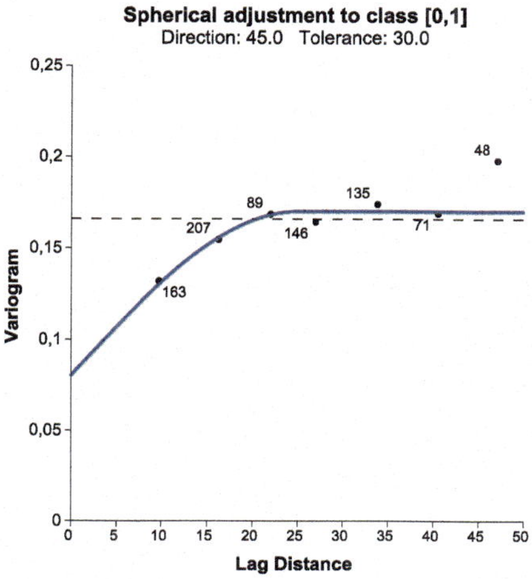

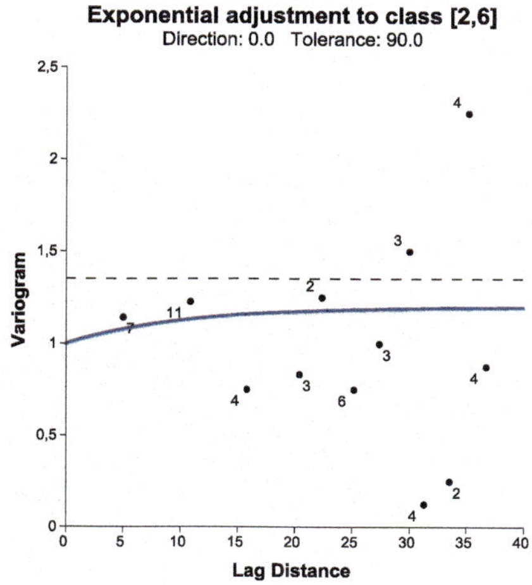

Fig. 13 Indicator variograms of the threshold data sets: class [0–1] – example of the 45° fit; class [2–6] omnidirectional fit

Table 4 Spherical model parameters for class [0–1] variogram

Parameter	0°	45°	90°	135°
Nugget effect	0,08	0,08	0,08	0,08
c (var. contribution)	0,08	0,09	0,12	0,09
a (range)	5	25	13	19

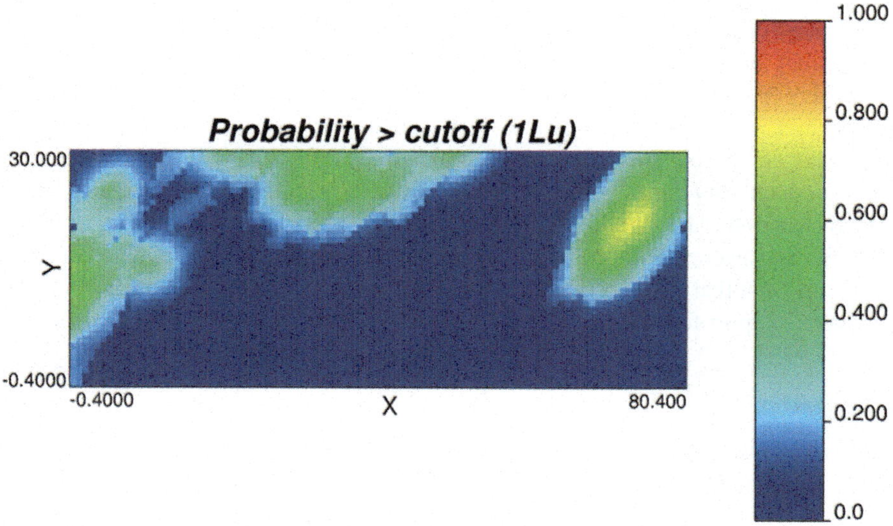

Fig. 14 Probability map of the permeability being above 1 Lu

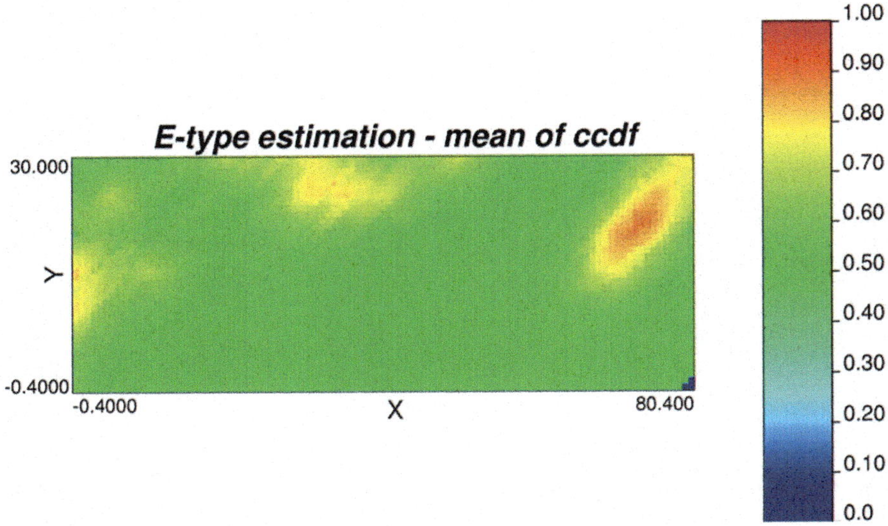

Fig. 15 E-type estimation of the mean of the conditional cumulative distribution function (*ccdf*)

9 Exploratory Approach to Simulation

To consolidate this exploratory approach on using geostatistics as an optimization tool for decision support in dam foundation curtain grouting, the use of simulation was tested based on the indicator kriging approach performed earlier, using the sequential indicator algorithm *sisim* (from GsLib).

Figure 16 shows an example of a random realization obtained by indicator sequential simulation. It shows a very satisfactory consistency with previous analysis which concerns the clear identification of the distinctive main low- and high-permeability zones with the known additional significant advantage of keeping the original data variability.

This consistency with the previous analysis and with the original data corroborates the idea that the geostatistical approach can be a useful and valuable tool in decision support for optimization of project design and production management.

10 Conclusion

The main objective of this study is assessing the advantage of a geostatistical estimation and simulation modelling approach in identifying the zones in a dam foundation in which the project design criteria may demand for an extended first-stage and a secondary-stage grouting treatment. This exploratory approach showing

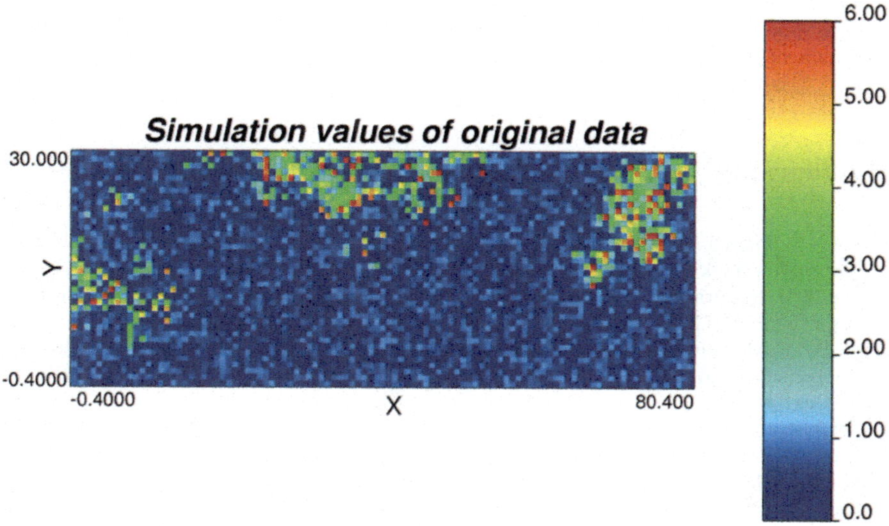

Fig. 16 Indicator sequential simulation realization map obtained with *sisim*, from GsLib

initial very promising results is intended to be developed and applied in the project of a new dam in river Tâmega, northern part of Portugal.

After a preliminary statistical, spatial continuity and error analysis study, the obtained estimation, probability, and simulation maps were compared. All estimation and simulation results appear to be very consistent with each other and with the known geology and geotechnical information of the site. Ordinary kriging estimation, indicator kriging, and sequential indicator simulation allowed defining probabilistically quantified foundation zones needing treatment. Indicator kriging, being a nonlinear interpolator, was useful for highly positively skewed data like the data set used in this investigation.

Obtained models were successfully correlated with the initial information of the project base study, namely, the geological–geotechnical model, indicating that the described approach has an interesting potential as an optimizing tool for curtain grouting design at the dam foundation during the construction stage.

A crucial further step in this investigation will be reducing model uncertainty by including more data and to use secondary variables to complement information. Nevertheless any accepted and validated method of predicting the permeability in unsampled areas will be very useful in construction stages that last up to 3 years long, which represent a significant percentage of cost and planning. In terms of project and production management, a simple exercise of discarding secondary drilling and grouting labor from the identified impermeable zones would give a clear improvement in planning, cost control, and also environment and safety gains, indicating this methodology as having an interesting potential for support in efficiency decisions.

Acknowledgments The authors thank the support of GEOBIOTEC and the contributions of EDP. This work is financed by national funds through FCT-Fundação para a Ciência e Tecnologia, I. P., within the project GEOBIOTEC – UID/GEO/04035/2013.

Bibliography

Deutch CV, Journel AG (1998) Geostatistical software library and user's guide. Oxford University Press, New York

EDP (2007) Relatório Geológico Geotécnico Escalão Jusante: Empreitada geral de construção do Aproveitamento Hidroeléctrico do Baixo Sabor

EDP (2011) Memória de Projecto de Execução: Tratamento de Fundação do Escalão Jusante. Empreitada geral de construção do Aproveitamento Hidroeléctrico do Baixo Sabor

Goovaerts P (1997) Geostatistics for natural resources evaluation. Oxford University Press, Oxford

Houlsby AC (1976) Routine iinterpretation of the Lugeon water test. Q J Eng 9:303–313

IGM (1989) Geological chart 11C – Torre de Moncorvo. Serviços Geológicos de Portugal, Lisboa

Soares A (2006) Geoestatística para as Ciências da Terra e do Ambiente. IST – Instituto Superior Técnico, Lisboa

Inverse Modeling Aided by the Classification and Regression Tree (CART) Algorithm

Julio César Gutiérrez-Esparza and J. Jaime Gómez-Hernández

Abstract Inverse modeling in hydrogeology is a powerful tool to improve the characterization of hydraulic conductivity and porosity. In the last few years, the use of data assimilation techniques, such as the ensemble Kalman filter, has proven very effective in this field. However, in some cases, the parameter updates by the filtering process may create artificial heterogeneity in certain zones in order to reduce the estimation error. This may happen when observations are scarce in time or space, but also when the parameters being updated are not the only responsible of the behavior of the aquifer (for instance, when high piezometric heads are due to an undetected recharge event, and the filter keeps reducing the conductivity to increase the gradients around high piezometric head observations). This study pretends to avoid those artifacts by the use of classification and regression trees. The decision and regression trees will be implemented using the CART algorithm with the aim of discriminating whether an updated parameter field is acceptable, and in case it is not acceptable how to proceed. When the algorithm marks as unacceptable a parameter field, it is swapped with another parameter field. The method is demonstrated for a contamination event in a synthetic aquifer based on real data. A numerical model has been created to reproduce flow and transport as observed in the real aquifer. The model has a rectangular-shaped area of 3000 m long by 500 m wide. For the inverse modeling process, two ensembles of fields are used, one for hydraulic conductivity and one for porosity; if needed, recharge can be modified smoothly using spline interpolation. The ensemble Kalman filter is used to update porosities and conductivities, and, if the decision algorithm requests it, the recharge is also modified.

J.C. Gutiérrez-Esparza (✉)
Institute for Water and Environmental Engineering, Universitat Politècnica de València, Camino de Vera s/n, 46022 Valencia, Spain
e-mail: enggtz@gmail.com

J.J. Gómez-Hernández
Institute for Water and Environmental Engineering, Universitat Politècnica de València, Valencia, Spain
e-mail: jgomez@upv.es

J.J. Gómez-Hernández et al. (eds.), *Geostatistics Valencia 2016*, Quantitative Geology and Geostatistics 19, DOI 10.1007/978-3-319-46819-8_55

1 Introduction

In the last decade, ensemble Kalman filtering (EnKF) has been used to characterize aquifer heterogeneity by assimilating observations such as piezometric heads and concentrations. The filter has shown very good results when reproducing piezometric heads (Zhou et al. 2011), but not as good when reproducing concentrations (Xu et al. 2013; Liu et al. 2008). It has been shown that matching concentrations is easier when assimilating both piezometric heads and solute concentrations than if only piezometric heads are used (Li et al. 2012), because piezometric heads do not contain enough information about the transport process. Furthermore, for the purpose of matching concentrations, results are improved if both porosity and hydraulic conductivity are updated simultaneously.

One problem we can face is that when the aquifer is large and the density of observation small, the filter update may be restricted to small areas around the observations, producing nonrealistic anomalies in the spatial distribution of the parameters.

Another problem we have found is related to observations that are too close in space and that are much more different between them than what could be attributed to measurement errors. In these cases, some observations must be treated as outliers and removed from the dataset, otherwise, the filter will introduce artifacts trying to match these dissimilar observations.

When the filtering is not able to provide good results in a specific area of the aquifer, it could be a sign that the conditions imposed for the simulation are not correct, but this is hard to know a priori. A typical problem is the treatment of recharge as deterministic, since there is always large uncertainty in its estimation.

The main objective of this manuscript is to address these issues by using a decision tree so the parameter or parameters that are causing troubles in the filtering process can be modified when needed. To do so, we use the classification and regression tree (CART) algorithm (Breiman et al. 1984); this algorithm creates a decision tree by a binary recursive partitioning procedure, up until a previously imposed stopping rule is met; at this point the tree is "pruned" to obtain a clearer and more understandable decision tree.

In this context, we present the CART-aided EnKF as an alternative to the standard EnKF to avoid the problems mentioned above. The CART algorithm discriminates which of the updated parameter fields should be replaced by another field in order to avoid convergence issues or if the parameter field is incongruent with the case being analyzed. Section 2 explains the EnKF methodology as well as the CART algorithm. Section 3 explains the synthetic example used throughout this paper. Results and discussions are found in Sect. 4. Finally, conclusions are in Sect. 5.

2 Methodology

2.1 The EnKF

Kalman filtering methods have been used for many years to estimate both system states and parameters in an optimal manner (Haykin 2001). The EnKF differs from the original Kalman filter in the use of an ensemble of realizations in order to obtain a better estimation of the augmented state covariance matrix $\boldsymbol{P}$, necessary for optimal updating of parameters and system state variables.

In the EnKF, the augmented state vector $\boldsymbol{x}$ is a collection of parameters and state variables; in our case $\mathbf{x}$ consists of logarithm of hydraulic conductivity (*LnK*) and porosity (Φ)—the parameters—, and hydraulic heads (h) and concentrations (c)—the state variables. Vector $\mathbf{x}$ is updated with the aim of reducing the difference between model predictions and observations, through the assimilation of state observations in time.

For a number of realizations, N_{re} in the ensemble $\boldsymbol{x} = (x_1, x_2, \ldots, x_{N_{re}})$, any realization i, x_i includes parameters and state variables:

$$x_i = \begin{pmatrix} \boldsymbol{A} \\ \boldsymbol{B} \end{pmatrix}_i = \begin{pmatrix} \left(a_1, a_2, \ldots, a_{N_e \cdot N_p}\right)^T \\ \left(b_1, b_2, \ldots, b_{N_n \cdot N_s}\right)^T \end{pmatrix}_i \tag{1}$$

where vector $\mathbf{A}$ is the vector of parameters and vector $\mathbf{B}$ is the vector of system state variables, N_e is the number of elements in which the model is discretized (in our example over a finite element mesh), N_p is the number of parameters (in our case, 2, loghydraulic conductivity and porosity), N_n is the number of nodes in which heads and concentrations are calculated (in our example, at the vertices of the finite element mesh), and N_s is the number of state variables (in our case, 2, heads and concentrations). For the parameters and state variables considered in our example, Eq. (1) becomes

$$x_i = \begin{pmatrix} \left(LnK_1, \Phi_1, LnK_2, \Phi_2, \ldots, LnK_{N_e}, \Phi_{N_e}\right)^T \\ \left(h_1, c_1, h_2, c_2, \ldots, h_{N_n}, h_{N_n}\right)^T \end{pmatrix}_i \tag{2}$$

The filtering process then consists of two steps; the first one is called the forecast step and the second one the update step. In the forecast step, an initial ensemble of parameters is used as input for the transient groundwater flow and solute transport model:

$$x_t = F(x_{t-1}) \tag{3}$$

where $F(\cdot)$ is the transfer function (numerical model) used to calculate state variables, and it would correspond to the solution of Eq. (8) for heads and of Eq. (9) for concentrations shown below. The transfer function only modifies the state variables, leaving parameters unchanged.

After the forecast is done, the update step starts by collecting observations at observation locations. The discrepancy between observations and predictions is used to update the augmented state vector according to the following equation:

$$x_t^u = x_t^f + G_t\left(z_t + \varepsilon - Hx_t^f\right) \tag{4}$$

where x_t^u is the updated augmented vector; x_t^f is the forecasted vector obtained with the flow and transport model; z_t is the observation vector at time t; ε is the observation error following a normal distribution of zero mean and diagonal covariance R; H is the observation matrix, it is composed of 1s and 0s when observations are located at mesh nodes; and G_t is known as the Kalman gain and is given by the following expression:

$$G_t = P_t^f H^T\left(HP_t^f H^T + R_t\right)^{-1} \tag{5}$$

where P_t^f is the covariance matrix of the augmented state, computed from the ensemble of realizations.

In order to reduce spurious correlations, we have used covariance localization as a way to constrain correlation up to a certain range. Using the distance-dependent localization function used by Xu (Xu et al. 2013):

$$\rho_{\widetilde{X}Y}(d) = \rho_{YY}(d) = \begin{cases} -\frac{1}{4}\left(\frac{d}{a}\right)^5 + \frac{1}{2}\left(\frac{d}{a}\right)^4 + \frac{5}{8}\left(\frac{d}{a}\right)^3 - \frac{5}{3}\left(\frac{d}{a}\right)^2 + 1, & 0 \le \mathrm{d} \le \mathrm{a}; \\ \frac{1}{12}\left(\frac{d}{a}\right)^5 - \frac{1}{2}\left(\frac{d}{a}\right)^4 + \frac{5}{8}\left(\frac{d}{a}\right)^3 + \frac{5}{3}\left(\frac{d}{a}\right)^2 - 5\left(\frac{d}{a}\right) + 4 - \frac{2}{3}\left(\frac{d}{a}\right)^{-1}, & \mathrm{a} \le \mathrm{d} \le 2\mathrm{a}; \\ 0 & \mathrm{d} > 2\mathrm{a}. \end{cases} \tag{6}$$

where d is the Euclidean distance and a is a length parameter controlling the influence distance of the covariance (200 m in our case for both $\rho_{\widetilde{X}Y}$ and ρ_{YY}). Then, Eq. (5) can be rewritten as follows:

$$G_t = \rho_{\widetilde{X}Y(d)} \circ P_t^f H^T\left(\rho_{YY(d)} \circ HP_t^f H^T + R_t\right)^{-1} \tag{7}$$

In Eq. (7) ° denotes the Schur product between correlation coefficients and covariance terms.

2.2 *Flow and Transport Equations*

The groundwater flow equation in saturated porous media can be written in the form:

$$\nabla \cdot (K\nabla h) + w = S_s \frac{\partial h}{\partial t} \tag{8}$$

where ∇ denotes the gradient operator, $\nabla\cdot$ is the divergence operator, K is the hydraulic conductivity $[LT^{-1}]$, h is the hydraulic head $[L]$, w represents the source/sink terms $[LT^{-1}]$, S_s is the specific storage coefficient, and, finally, $\frac{\partial h}{\partial t}$ is the partial derivative of h with respect to time t $[T]$.

The transport equation for a nonreactive solute can be expressed as

$$\Phi \frac{\partial c}{\partial t} = -\nabla \cdot (qc) + \nabla \cdot (\Phi D \nabla C) \tag{9}$$

where Φ denotes porosity, $\frac{\partial c}{\partial t}$ is the partial derivative of concentration c $[ML^{-3}]$ with respect to time t $[T]$, q is the Darcy velocity $[LT^{-1}]$, and D is the dispersion coefficient tensor $[L^2T^{-1}]$.

3 CART Algorithm

The CART decision tree is a binary recursive partitioning procedure capable of processing continuous and nominal attributes both as targets and predictors (Springer-Verlag 2007); data can be treated in its raw form and it is not sensitive about abnormal data or outliers.

First, the tree is grown to its maximum size, and then, it is pruned back to reduce its complexity, where the nodes to be pruned are the ones that contribute the least to the overall performance of the tree when using training datasets. Then, performance is measured via cross-validation or with independent data.

To make a decision tree, an initial node is created; nodes will split the information based on splitting rules in order to best comprehend the studied subject. When growing the tree, the splitting rules compare an attribute with some threshold or condition, for example:

IF X CONDITION then go left, otherwise go right.

where CONDITION can be related to a threshold or could be related to an attribute value, e.g.,

$$\text{if } X_j >= X_j^R. \tag{10}$$

Here X_j is one of the possible values of variable j in our training dataset, while X_j^R is the threshold or attribute for variable j that will discriminate the data. This is repeated, until all the dataset is analyzed and terminal nodes contain information only for one class. For every split, the information is homogenous, so that the left split and the right split separate the information in an optimum way; to do so the threshold X^R is searched within all the available values of attributes in the training dataset with the CART algorithm. For variables that are continuous, as in our case, the CART algorithm then solves the next minimization problem:

$$\text{argmin}_{X_j \leq X_j^R\ j=1,\ldots,M}[P_1\ \text{Var}(Y_1) - P_r\ \text{Var}(Y_r)] \tag{11}$$

where M is the number of variables or attributes in the training dataset, Y is the vector of responses for our training dataset, $\text{Var}(Y_1)$ is the variance of other responses that will go to the left node, and P_1 and P_r are the probability of that variable to go left or right (in our case both probabilities are equal). An impurity measure is used to obtain an estimation of the homogeneity of the right and left nodes after a split, using the impurity function $i(t)$:

$$i(t) = \sum\nolimits_{k=1}^{N} (y_k - \mu_k)^2 \tag{12}$$

where $i(t)$ is the impurity measure for node t, y_k is the response for training set k, and μ_k is the mean of responses in the node.

After all nodes in the tree have been created, pruning is the next step, and one way to do it is to set a threshold for the reduction in the impurity measure, below which no split will be made. A preferred approach is to grow a very large tree and then prune the tree back to an optimal size (Moisen 2008). For our case we used the cost-complexity function used by Timofeev (2004) with no parameters:

$$C(T) = R(T) + (\bar{T}) \rightarrow \min T \tag{13}$$

where $R(T)$ is the misclassification error observed in a cross validation dataset and $\bar{T}$ is a complexity measure of the tree $\bar{T}$ which is equal to the sum of terminal nodes in the tree.

The flowchart in Fig. 1 shows how the three parts are coupled. First, we start with the flow and transport model and the initial ensemble of parameters for *LnK* and Φ, where part of the ensemble is used as training dataset for the regression tree, and the other part is used in the CART algorithm to prune a tree. Finally, the EnKF is used to update the fields.

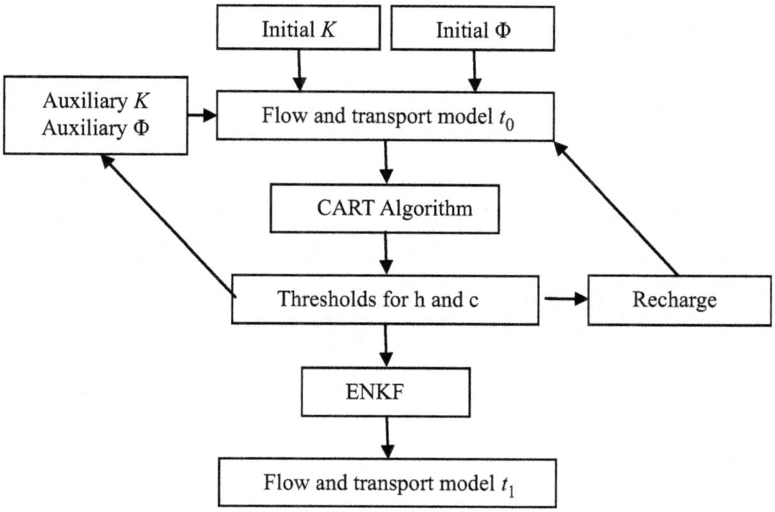

Fig. 1 Flowchart for the proposed methodology

4 Synthetic Example

We created a synthetic aquifer based on a true one, in which a contamination event has occurred. The model has a rectangular-shaped area of 3000 m long by 500 m wide. It has been discretized in 5 m by 5 m cells for the purpose of generating the heterogeneous fields. (Later an irregular triangular mesh will be used for the simulation of flow and transport.) For the parameters of the aquifer, four ensembles (two ensembles for *LnK* and two for porosity) were created using SGeMS (Remi 2004) by sequential Gaussian simulation. Table 1 shows the parameters for the geostatistical simulations.

Figure 2 shows four random hydraulic conductivity realizations to give an idea of the range of variability in *LnK*.

For flow and transport an irregular triangular mesh is used. This mesh has 5,489 elements and it is superimposed on the heterogeneous fields of *LnK* and Φ in order to compute values for each element. When a cell falls over several nodes of the heterogeneous realization, the arithmetic average of values within the cell is assigned to it, and when the cell is smaller than 5 m by 5 m, the value of the heterogeneous field is directly assigned to the cell. This averaging procedure is applied for both parameters.

Figure 3 shows the mesh used for solving the flow and transport equations. The red line marks the model border and the blue line the contaminant source area.

Dirichlet boundary conditions are imposed in all four boundaries of the model, the values of which are derived from the real aquifer that serves as reference. Recharge is also taken from the real aquifer. Similarly, the information about the contamination source is based on the information on the real aquifer.

Table 1 Parameters used in geostatistical simulation

Field	Variogram type	λ_x	λ_x	Sill
LnK	Spherical	75	165	9
Φ	Spherical	75	165	0.012

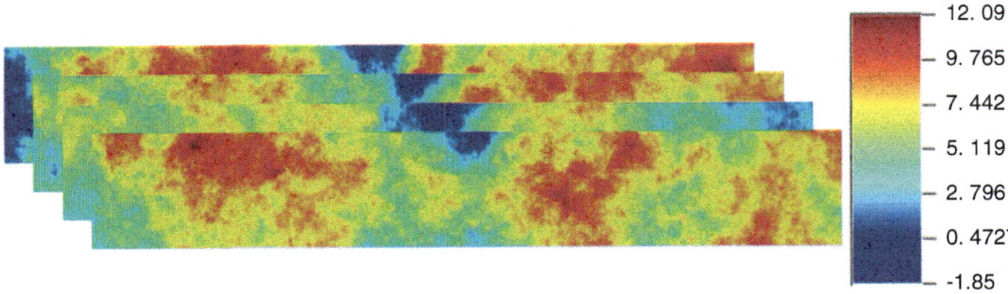

Fig. 2 *LnK* fields for realization 3, 15, 26, and 64 in ensemble 1

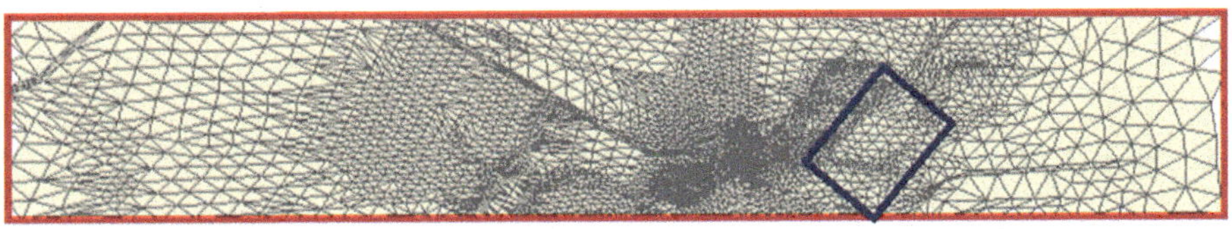

Fig. 3 Finite element mesh for flow and transport calculations

5 Results and Discussion

After 50 time steps and using ensembles of 100 realizations per parameter (together with a reserve of 100 auxiliary realizations for each parameter), some fields of hydraulic conductivity and porosity after updating are shown in the next set of figures. Figures 4, 5, 6, 7, 8, and 9 show *LnK* fields before and after the first time step updating. In general, conductivities increase after the first update because the initial hydraulic head field was not very consistent with the observations at that time step.

Porosity fields are also shown from Figs. 10, 11, 12, and 13; here four fields are shown at the initial time step prior to any modification and after 50 time steps.

LnK after 50 time steps is shown for the same three fields in Figs. 14, 15, and 16. It can be seen that an increase in *LnK* has occurred in the western-central part of the aquifer. That is due to a drastic difference between the recharge needed in order to obtain a good estimation and the one actually imposed to the model.

Hydraulic head is shown at four observation points in Figs. 17, 18, 19, and 20. In general, the adjustment of hydraulic heads is not as good as we expected, and we attribute it to the uncertainty on the recharge. At this moment we are working on introducing modifications in the recharge field to improve the reproduction of the hydraulic heads with realistic hydraulic conductivity patterns. To show the

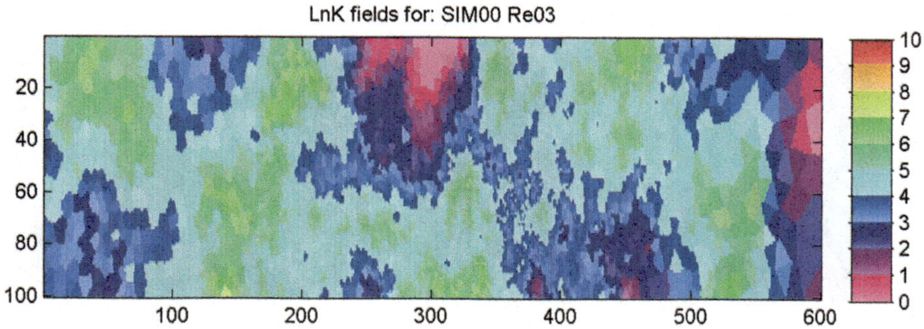

Fig. 4 *LnK* field for realization 3 before the first time step updating

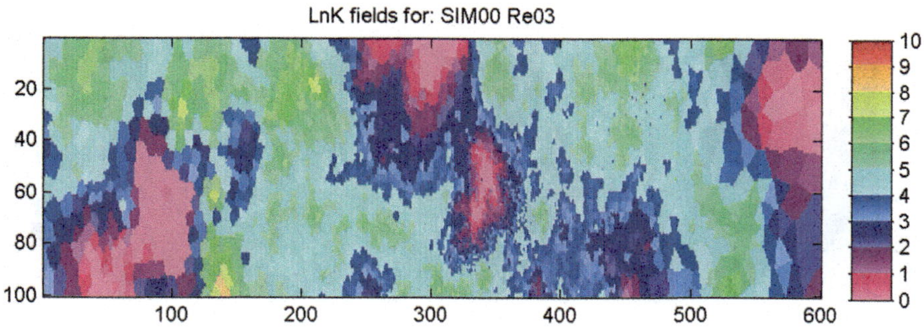

Fig. 5 *LnK* field for realization 3 after the first time step updating

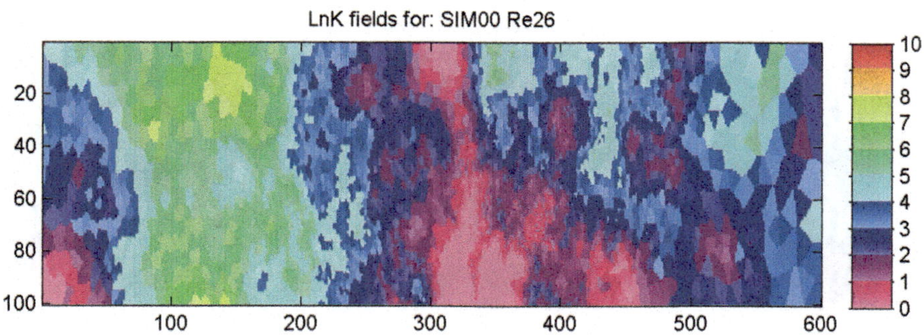

Fig. 6 *LnK* field for realization 26 before the first time step updating

importance of recharge in the matching of the model, we have run a test without modifying the recharge patterns, and the filter produces unrealistic conductivity fields such as the one in Fig. 21.

Concentrations at observation points are shown in Figs. 22 and 23; the large spread is similar as in the hydraulic head calculations, since the recharge is being modified without any optimization criteria. Posterior optimization of recharge

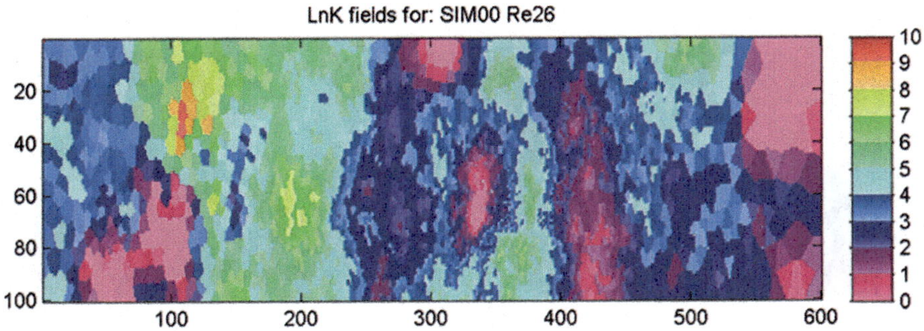

Fig. 7 *LnK* field for realization 26 after the first time step updating

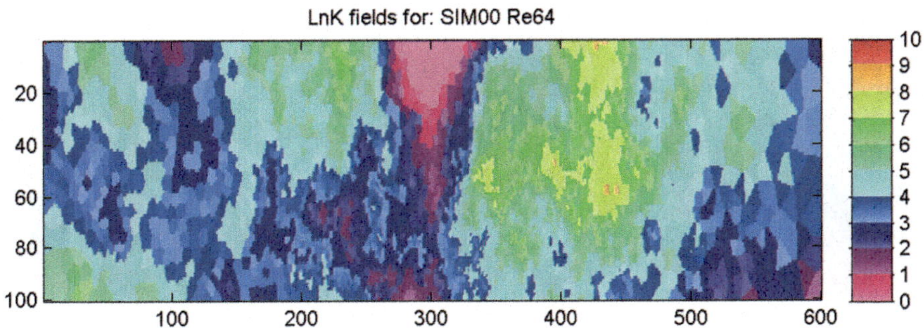

Fig. 8 *LnK* field for realization 64 before the first time step updating

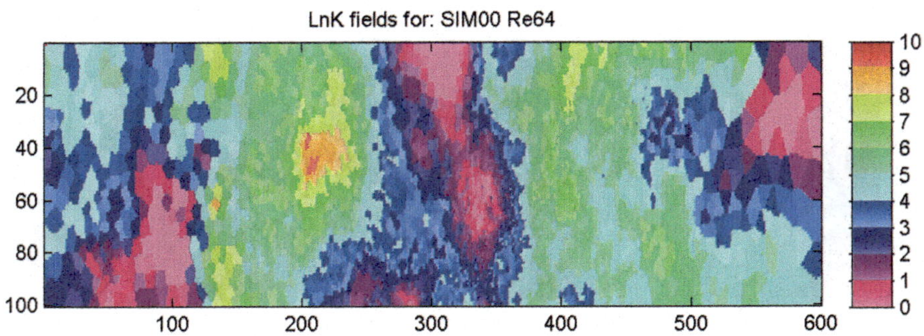

Fig. 9 *LnK* field for realization 64 after the first time step updating

could lead to an improvement in the concentrations and hydraulic head calculations.

For the updating of the final steps, the CART algorithm thresholds, that is, the error above which recharge had to be touched, were 0.275 m for h and 150 ppb for c. Recharge had to be modified in all time steps since it was impossible to get predictions with errors below the set thresholds.

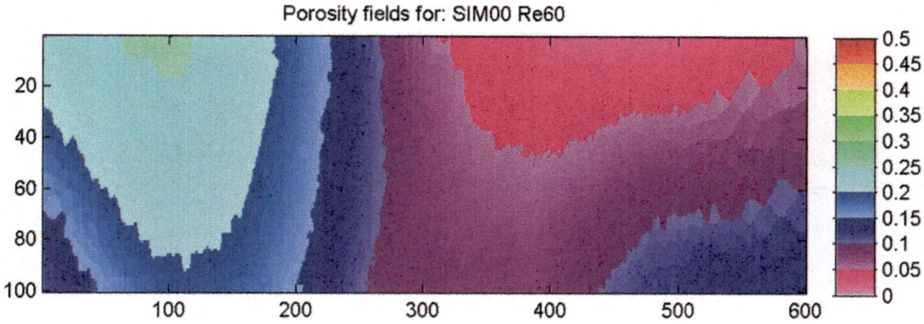

Fig. 10 Porosity field at initial time for realization 60

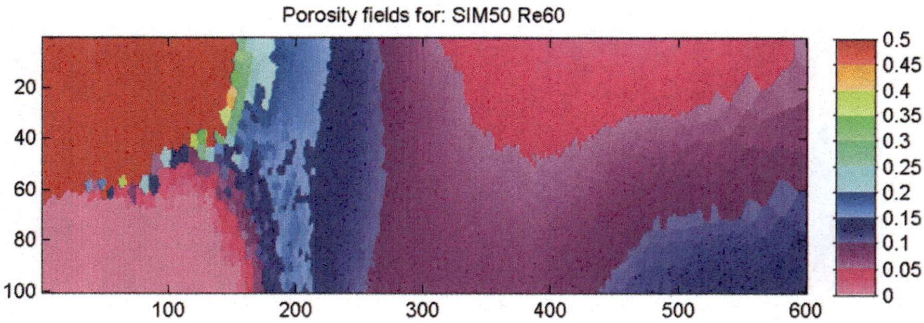

Fig. 11 Porosity field after 50 time steps for realization 60

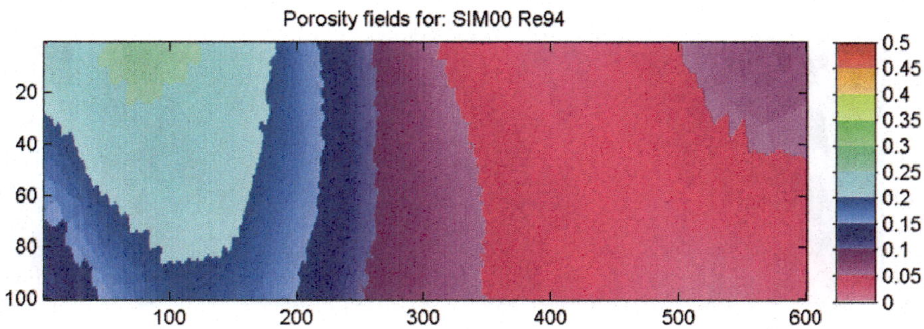

Fig. 12 Porosity field at initial time for realization 94

6 Conclusions

The application of data assimilation algorithms to a realistic case faces the problem that when some of the components defining the conceptual model is improperly characterized, the filtering algorithm may result in very unrealistic heterogeneity patters; patterns that are built to compensate the conceptual model errors. As it is the case of our example with the initial estimation of the recharge, the hydraulic

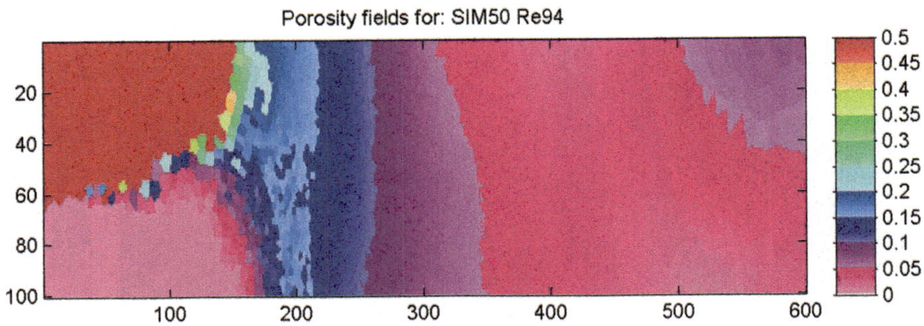

Fig. 13 Porosity field after 50 time steps for realization 94

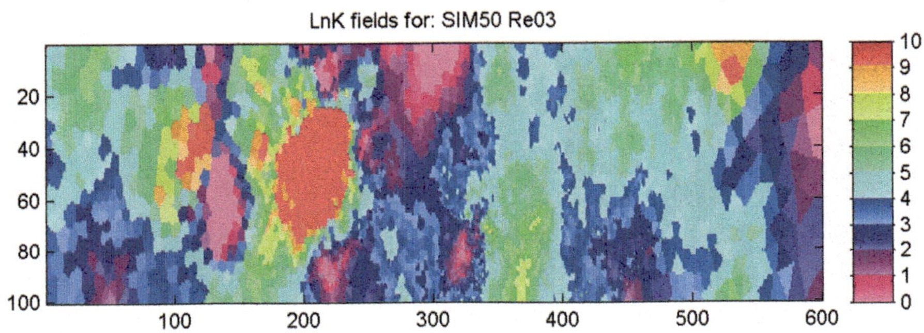

Fig. 14 Final *LnK* field for realization 3

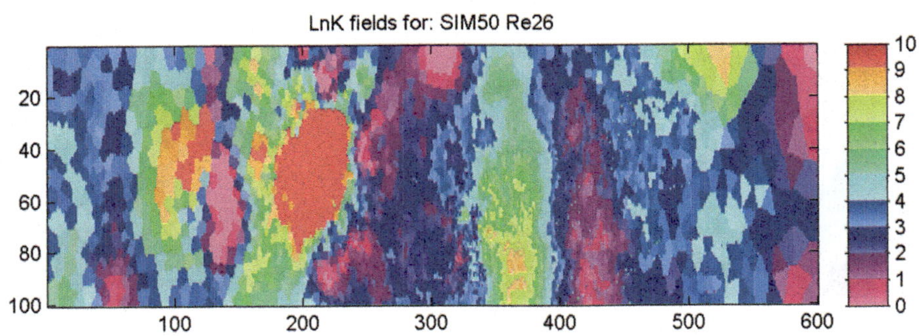

Fig. 15 Final *LnK* field for realization 26

head predictions do not match the observations, so the EnKF modifies drastically the parameters. Since it was clear that modifying conductivities alone was not enough to match observations, we have implemented a decision tree to force modifications in the normal procedure of the filtering algorithm. For instance, when an updated conductivity field is not plausible, the CART algorithm enters into play proposing a modification in the workflow, either by replacing the

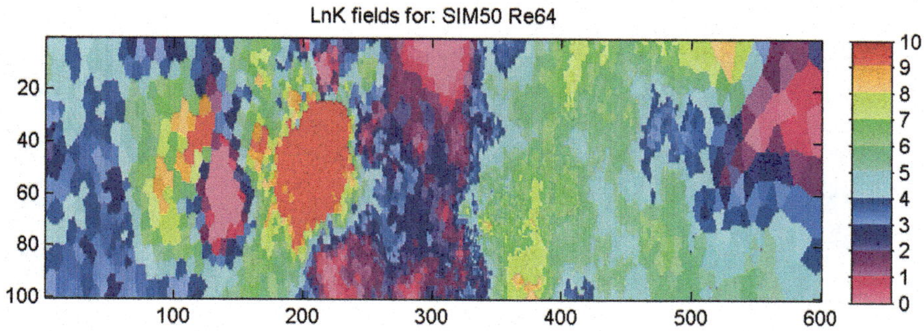

Fig. 16 Final *LnK* field for realization 64

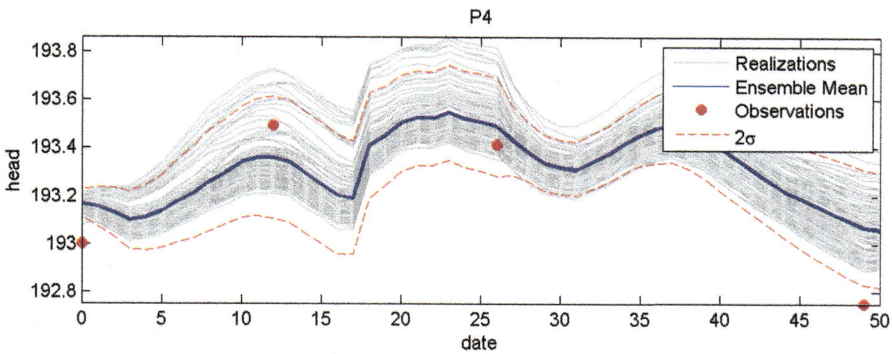

Fig. 17 Hydraulic head for observation point P4

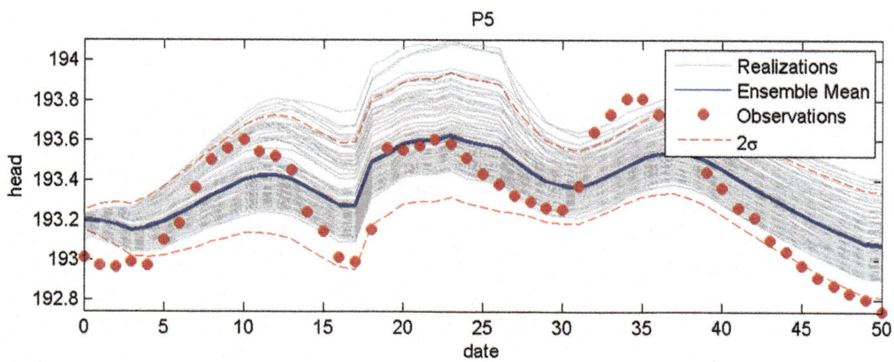

Fig. 18 Hydraulic head for observation point P5

conductivity field or by activating the updating of a different parameter, in order to re-steer the filter toward plausible results.

For future work it seems necessary to include the recharge as another parameter to update by the filter.

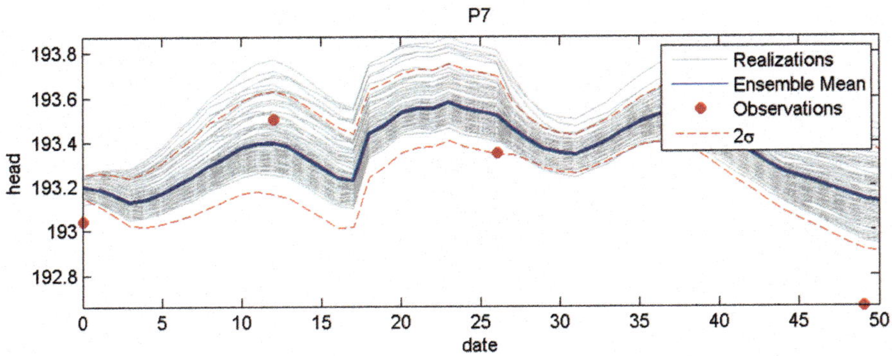

Fig. 19 Hydraulic head for observation point P7

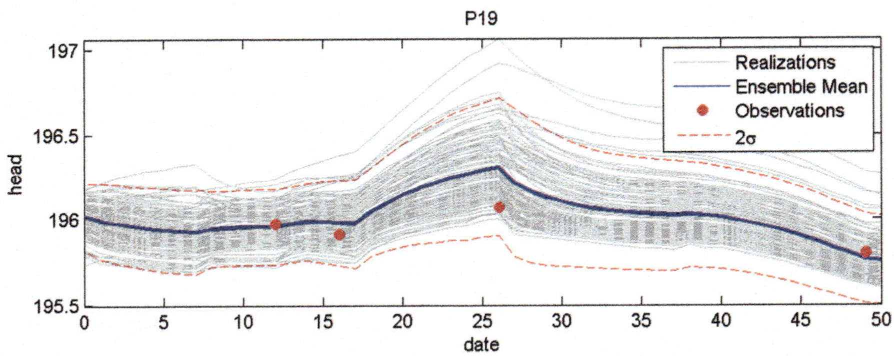

Fig. 20 Hydraulic head for observation point P19

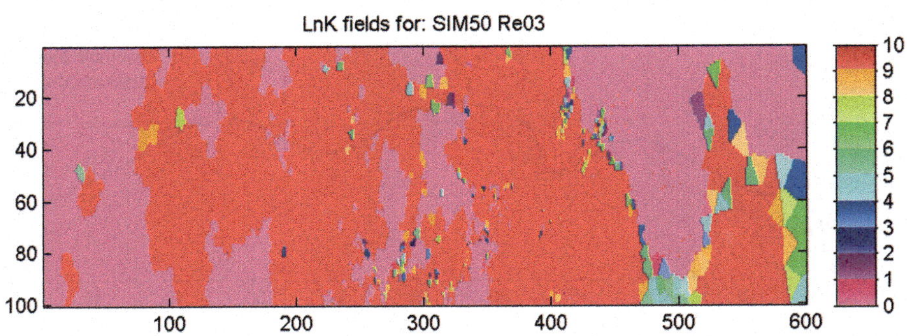

Fig. 21 *LnK* field for realization 3 after 50 time steps if no recharge modification is implemented

Acknowledgment Financial support to carry out this work was received from the Spanish Ministry of Economy and Competitiveness through project CGL2014-59841-P and through the Ministry of Culture, Education, and Sports FPU program.

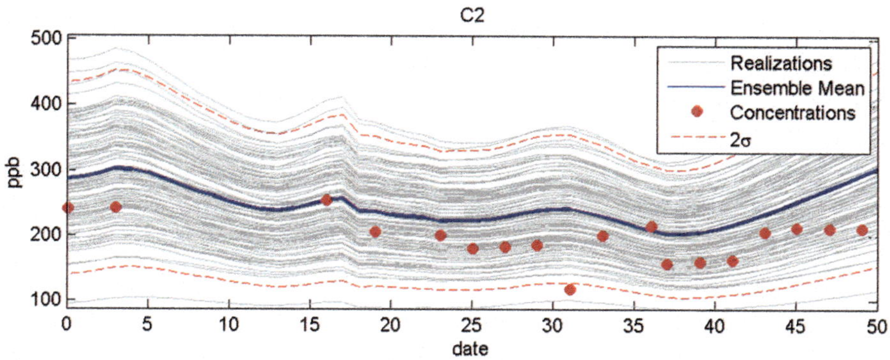

Fig. 22 Concentrations at measurement point C2 after 50 time steps

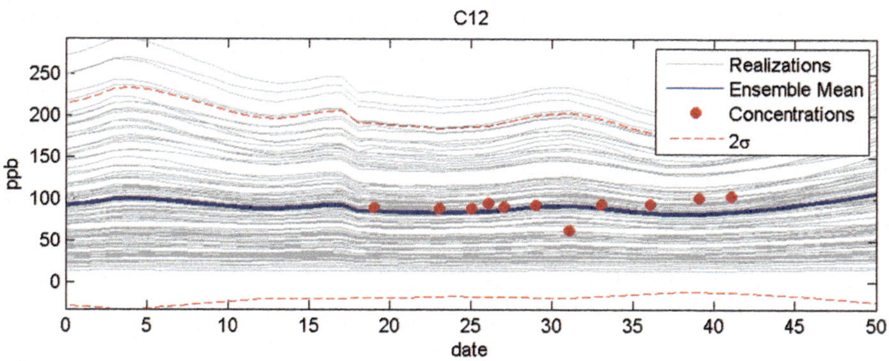

Fig. 23 Concentrations at measurement point C12 after 50 time steps

Bibliography

Breiman L, Friedman J, Oshen R, Stone CJ (1984) Classification and regression trees. Taylor & Francis, Oxon

Haykin S (2001) Kalman filtering and neural networks. Wiley, New York

Li L, Zhou H, Gómez-Hernández JJ, Hendricks Franssen H-J (2012) Jointly mapping hydraulic conductivity and porosity by assimilating concentration data via ensemble Kalman filter. J Hydrol 428:152–169

Liu G, Chen Y, Zhang D (2008) Investigation of flow and transport processes at the MADE site using ensemble Kalman filter. Adv Water Resour 975:986

Moisen GG (2008) Classification and regression trees. Elsevier, Oxford

Remi N (2004) Geostatistical earth modelling software: user's manual. Blackwell, Palo Alto

Springer-Verlag (2007) Top 10 algorithms in data mining. Knowl Inf Syst 14(1):1–37

Timofeev R (2004) Classification and Regression Trees (CART) Theory and Application Msc. Thesis. Humboldt University, Berlin

Xu T, Gómez-Hernández JJ, Zhou H, Li L (2013) The power of transient piezometric head data in inverse modeling: an application of the localized normal-score EnKF with covariance inflation in a heterogenous bimodal hydraulic conductivity field. Adv Water Resour 54:100–118

Zhou H, Gómez-Hernández JJ, Hendricks Franssen H-J, Li L (2011) An approach to handling non-Gaussianity of parameters and state variables in ensemble Kalman filtering. Adv Water Resour 54:844–864

Numerical Simulation of Solute Transport in Groundwater Flow System Using Random Walk Method

Nilkanth H. Kulkarni and Rajesh Gupta

Abstract This paper presents the new random walk solute transport model (RWSOLUTE) for solute transport simulation in groundwater flow system. This model is novel in using an efficient particle tracking algorithm. The proposed model is validated against analytical and other reported numerical solutions for chosen test case. The accuracy and stability of the RWSOLUTE model solutions are verified through mass balance error checks and Courant stability criteria. Further the sensitivity of the model solutions is analyzed for varying values of time step size and particle mass.

1 Introduction

Groundwater quality degradation in many parts of the world due to poorly planned municipal, agricultural, and industrial waste disposal practices has continuously drawn the attention of research worker to modify methods of predicting and analyzing the impact of the migration of the dissolved solutes reliably in the aquifers. The pollution of this vital water resource resulted into a serious environmental problem which may damage human health and destroy the ecosystem. Thus it has become essential to assess the severity of groundwater pollution and chalk out the strategies of aquifer remediation, which are made possible by the use of the statistical numerical models.

N.H. Kulkarni (✉)
Department of Civil Engineering, Shri Guru Gobind Singhji Institute of Engineering & Technology, Vishnupuri, Nanded 431 606, India
e-mail: kulkarninh@yahoo.com

R. Gupta
Department of Civil Engineering, Visvesvaraya National Institute of Technology Nagpur, Nagpur, India
e-mail: rajeshguptavnit@hotmail.com

J.J. Gómez-Hernández et al. (eds.), *Geostatistics Valencia 2016*, Quantitative Geology and Geostatistics 19, DOI 10.1007/978-3-319-46819-8_56

The groundwater pollution problem becomes severe because of the migration of solutes by advection and hydrodynamic dispersion from the point of its introduction in aquifers (Freeze and Cherry 1979). Numerical models of groundwater flow and solute transport are properly conceptualized versions of a complex aquifer system which approximate the flow and transport phenomena through the set of assumptions pertaining to the geometry of the domain, heterogeneities of the aquifer, and type of flow regime.

Prickett et al. (1981) presented the transport model on the basis of the random walk method. The complex aquifer system is treated as a continuum, which implies that the fluid and solid matrix variables are continuously defined at every point in the aquifer domain (Wang and Anderson 1995). The equations that describe solute transport in highly heterogeneous porous media are among the most difficult equations to solve because of their hyperbolic nature due to sharp concentration fronts so the random walk method is found to be suitable for solute transport simulation in heterogeneous porous media. Neuman (1984) proposed an adaptive Eulerian–Lagrangian formulation which solves the advective component of steep concentration front by single-step reverse particle tracking of moving particles clustered around each front and away from such fronts, and the dispersive transport is solved by Lagrangian formulation on a fixed grid. It is found that the use of particle clusters at steep concentration fronts only caused numerical dispersion.

A mixed Eulerian–Lagrangian method is proposed by Sorek (1988) which decomposes the transport equation into pure advection along the characteristic path lines and residual dispersion as the propagation of the solute mass residue at a fixed grid point. This method worked well for coarse grid and high Peclet numbers. Illangasekare and Döll (1989) developed a discrete kernel approach for solving the linear governing partial differential equation of groundwater flow to obtain the flow velocities resulting from different pumping and injection schemes. Further the two-dimensional transient solute transport in water table aquifers is simulated by a method of characteristic model. However, the simulation of sharp concentration fronts at injection wells and divergent velocity fields at pumping wells caused numerical smearing of the solutions. Bentley and Pinder (1992) discussed the Eulerian and Lagrangian methods (ELM) to reduce the smearing and oscillations in the solutions of the advection–dispersion equation. It is noticed that the accuracy of ELM is governed by the choice of time step size in second-order Runge–Kutta method.

Zhang et al. (1993) proposed a computationally efficient Eulerian–Lagrangian method for solving advection–dispersion equation in both steady and transient velocity field. The method uses single-step reverse particle tracking technique for steep concentration fronts, and separate weighting factors which relate to grid Peclet and Courant numbers are used for upstream and downstream region of the advection front. In transient velocity field, the model determines the weighting factors automatically based on the mass balance errors. This model is found to be more suitable for the solution of the advection-dominated problems. Bellin et al. (1994) presented a new Eulerian–Lagrangian method for modeling flow and transport of conservative solutes in heterogeneous porous formations. They used geostatistical model to generate physically plausible random velocity fields. In this model the solute transport problem is solved by particle tracking approach with suitable grid refinements to correctly handle the velocity fluctuations. Their model is particularly found to be useful for large field scale problems where conventional numerical methods are not computationally efficient. Lu (1994) discussed a semianalytical method of particle tracking analysis under transient flow conditions. In this method the particle velocity is interpolated linearly in both space and time coordinates within each finite difference cell. This method uses analytical integration instead of numerical integration to obtain the trajectory of a particle within each cell. Uffink (1988) presented a random walk solute transport model that is different than the conventional model. The random walk is used as a mathematical concept to create an analogue process that obeys the classical advection–dispersion equation. The paper also discusses the conditions required for a complete equivalency between random walk and advection–dispersion equation. It proposes the use of proper coefficients in Fokker–Plank equation to eliminate the occurrence of unphysical high solute concentrations due to trapping of particles near stagnation points or in the parts of the aquifer with a much lower permeability than the surrounding area.

Wen and Gomez-Hernandez (1996) proposed a constant displacement scheme of particle tracking for simulation of advective transport. This scheme automatically adjusts the time step for each particle according to local pore velocity so that each particle always travels a constant distance. The application of this scheme for two-dimensional solute transport in groundwater flow system is found to be four times computationally faster than constant time step scheme which is conventionally used in particle tracking in random walk model simulations. Banton et al. (1997) developed a time domain random walk method which is different than the classical random walk method in the sense that it directly calculates the arrival time of a particle cloud at a given location providing the direct solute concentration breakthrough curve. The major advantage of this method lies in the fact that it is quite flexible in case of the choice of space and time step sizes. The method is found to be reliable from the comparison of its results with those of analytical and classical random walk solutions. Unlike the classical random walk method, this method does not generate high numerical dispersion.

Dentz et al. (2004) investigated the time behavior of solute transport in a heterogeneous medium using a spatially biased continuous time random walk method. This method is governed by the joint probability density function for an event–displacement with an event–time. The proposed model calculates the solute concentration distribution by using a generalized advection–dispersion equation with Laplace transformation. It gives an equivalent advection–dispersion equation which is solved to analyze the time dependence of the resident and flux concentrations. It is demonstrated that the model results are in close agreement with efficient random walk simulations based on the same joint probability density function. Delay et al. (2005) reported that the random walk technique of solute transport is especially well suited to the transport phenomena which are governed by strong variations in fluid velocity over relatively short distances. It is found that this technique is quite capable of simulating non-Fickian solute transport so it is more preferred than the Eulerian techniques which are potentially unstable or hampered by numerical diffusions.

The proposed model overcomes the numerical dispersion and oscillations which are prevalent in existing random walk models. The motivation behind the development of the present model is to dampen the numerical dispersion which masks the actual physical dispersion making particle tracking difficult. The new model employs an efficient particle tracking algorithm to track the positions of particles at new time steps.

The objectives of this study are (1) to develop the RWSOLUTE model, (2) to validate the developed numerical model, (3) to investigate the performance of the model for damping of numerical oscillation and numerical dispersion along with the average mass balance error, (4) to ensure the stability of the model solutions based on the criteria of Courant numbers, and (5) to examine the sensitivity of the model solutions to time step size and particle mass.

2 Mathematical Model

2.1 *Advection–Dispersion Equation*

The governing equation of solute transport in two-dimensional transient unconfined groundwater flow system that includes point and distributed sources/sinks of solute due to injection/extraction wells and distributed recharge/discharge can be described by the following equation:

$$\begin{aligned} R\frac{\partial c}{\partial t} &= \frac{\partial}{\partial x}\left(D_{xx}\frac{\partial c}{\partial x}\right) + \frac{\partial}{\partial x}\left(D_{xy}\frac{\partial c}{\partial y}\right) + \frac{\partial}{\partial y}\left(D_{yx}\frac{\partial c}{\partial x}\right) \\ &\quad + \frac{\partial}{\partial y}\left(D_{yy}\frac{\partial c}{\partial y}\right) - V_x\frac{\partial c}{\partial x} - V_y\frac{\partial c}{\partial y} + \sum\nolimits_{i=1}^{n_w}\frac{(c-c_i')}{\theta\, b}Q_i\,\delta\,(x_o - x_i,\, y_o - y_i) \\ &\quad + \sum\nolimits_{j=1}^{n_p}\frac{q_j}{\theta}(c - c_j') + \frac{cS_y}{\theta\, b}\frac{\partial h}{\partial t} \end{aligned} \tag{1}$$

where R is the retardation factor [dimensionless], c is the solute concentration $[M/L^3]$; D_{xx}, D_{xy}, D_{yx}, and D_{yy} are hydrodynamic dispersion coefficients $[L^2/T]$; V_x and V_y are the components of average linear groundwater velocity $[L/T]$; c_i' is the solute concentration of the injected water at ith injection well $[M/L^3]$; n_w is the number of injection wells in the domain; θ is the effective porosity of the aquifer [percent]; b is the saturated thickness of the aquifer $[L]$; c_j' is the solute concentration of the recharge water at jth node with distributed recharge $[M/L^3]$; S_y is the specific yield [dimensionless]; h is the hydraulic head averaged over vertical [L]; t is the time [T]; x and y are spatial coordinates [L]; Q_i is the pumping rate when ($Q_i < 0$) and injection rate when ($Q_i > 0$) at ith pumping and/or injection well [L^3/T]; n_w is the number of pumping and/or injection wells in the domain; n_p is the number of nodes in the domain with distributed discharge and/or recharge; $\delta(x_o - x_i, y_o - y_i)$ is the Dirac delta function; x_o and y_o are the Cartesian coordinates of the origin [L]; x_i and y_i are the coordinates of ith pumping and/or injection well, [L]; and q_j is the distributed discharge rate when ($q_j < 0$) and recharge rate when ($q_j > 0$) at jth nodes [L/T].

The hydrodynamic dispersion coefficients in the tensor form can be given as

$$\begin{bmatrix} D_{xx} & D_{xy} \\ D_{yx} & D_{yy} \end{bmatrix} = \frac{\alpha_L}{|\bar{v}|}\begin{vmatrix} v_x^2 & v_x v_y \\ v_x v_y & v_y^2 \end{vmatrix} + \frac{\alpha_T}{|\bar{v}|}\begin{vmatrix} v_y^2 & -v_x v_y \\ -v_x v_y & v_x^2 \end{vmatrix} \tag{2}$$

where α_L and α_T are longitudinal and transverse dispersivities [L] and $|\bar{v}|$ is the magnitude of the average linear groundwater velocity [L/T].

An initial concentration of the solute is prescribed in the entire aquifer domain Ω by

$$c(x, y, 0) = c_0(x, y) \qquad (x, y) \in \Omega \tag{3}$$

where c_0 is the initial solute concentration [M/L^3].

Equation (1) is subject to the Dirichlet boundary condition which is given as

$$c(x, y, t) = c_1 \qquad (x, y) \in \Gamma_1; t \geq 0 \tag{4}$$

where c_1 is the prescribed solute concentration over aquifer domain boundary $\gamma\varkappa$ Γ_1, [M/L^3].

Neumann boundary conditions are considered as

$$[v \times c(x, y, t) - [\boldsymbol{D}] \quad \nabla c(x, y, t)]'.\{\boldsymbol{n}\} = v \times c'(x, y) \in \Gamma_2; t \geq 0 \tag{5}$$

where vc' is the specified advective solute flux across the boundary Γ_2 [$M/L^3/T$] and [$\boldsymbol{D}$] ∇c is the dispersive solute flux across the boundary Γ_2 [$M/L^3/T$].

Although only numerical solutions of the Eq. (1) are sought, it is very important to get head distribution from the following mass balance equation for the groundwater flow which is given as

$$\nabla.v = S_y \frac{\partial h}{\partial t} + q \tag{6}$$

where v is the average linear groundwater velocity having two components in x- and y-directions and is obtained from a momentum equation by using Darcy's law which is given as

$$v_x = -T_{xx}\nabla h \, , v_y = -T_{yy}\nabla h \tag{7}$$

where T_{xx} and T_{yy} are components of the transmissivity tensor $[L^2/T]$ which are approximated as $T_{xx} \approx K_{xx}\, h$ and $T_{yy} \approx K_{yy}\, h$, provided the change in the head in unconfined aquifer is negligible as compared to its saturated thickness (Illangasekare and Döll 1989); K_{xx} and K_{yy} are components of the hydraulic conductivity tensor $[L/T]$.

2.2 Random Walk Solute Transport Model (RWSOLUTE)

The Random Walk Solute Transport Model (RWSOLUTE) (Kulkarni 2008) uses real particles generated on the basis of total solute mass in the aquifer. Unlike conventional random walk models, this model tracks real particles instead of imaginary particles, which minimizes book keeping efforts. It also employs an efficient particle tracking algorithm by taking weighted average of particle velocities computed at four time instances in a given time step.

This model does not directly solve the governing advection–dispersion equation. It decouples that equation into advection and dispersion parts. The advective and dispersive transport of the solute mass is simulated with the help of particles. Unlike the other Eulerian–Lagrangian methods, only tracer particles are moved in the flow field. Each particle represents the fraction of the total solute mass in the aquifer system. The number of particles in a computational cell is worked out as

$$NP^t_{i,j} = \frac{c^t_{i,j}(\theta \times \Delta x \times \Delta y \times b) + c^{'}_{i,j}\left(\left(Q_{i,j}\delta\left(x_0 - x_i, y_0 - y_j\right)\right) + q_{i,j}\right)(\theta \times \Delta x \times \Delta y \times b)}{PM} \tag{8}$$

where $NP^t_{i,j}$ is the total number of the particles to be generated in the computational cell (i, j) at a given point of time t and PM is the fixed particle mass, $[M]$.

The advective transport is solved by tracking the movement of the particles along flow lines. The advective displacement components of the particle in x- and y-direction during a given time step due to average linear groundwater flow velocity can be given as

$$\begin{aligned} dx^a_p &= V_{x_p} \times \Delta t + \propto_L \frac{{V_{x_p}}^2}{|V|}\Delta t \\ dy^a_p &= V_{y_p} \times \Delta t + \propto_T \frac{{V_{y_p}}^2}{|V|}\Delta t \end{aligned} \tag{9}$$

where dx_p^a, dy_p^a are the advective displacement components of p the particle during a time step Δt, [L]; $dl_p^a = \sqrt{\left(dx_p^a\right)^2 + \left(dy_p^a\right)^2}$ is a resultant displacement of the particle, [L]; and V_{x_p} and V_{y_p} are the particle velocities, [L/T].

The particle velocity components in Eq. 9 are obtained from the nodal velocity components using bilinear interpolation method. The time weighted value of the particle velocity is calculated by using fourth-order Runge–Kutta method. In the bilinear interpolation method, the computational cell in which the particle has moved after advection during a given time step is identified. The groundwater velocity components at the representative node of the identified cell along with the velocities at the three closest nodes that are obtained from the velocity field are used to interpolate the particle velocity. For the location of the particle in the first quadrant, the x-component of the particle velocity is interpolated using the following equations:

$$
\begin{aligned}
V_{xp}^1 &= \left(1 - \left(\frac{x_p^t - x_{i,j}}{\Delta x}\right)\right) V_{x_{i,j}} + \left(\frac{x_p^t - x_{i,j}}{\Delta x}\right) V_{x_{i+1,j}} \\
V_{xp}^2 &= \left(1 - \left(\frac{x_p^t - x_{i,j}}{\Delta x}\right)\right) V_{x_{i,j+1}} + \left(\frac{x_p^t - x_{i,j}}{\Delta x}\right) V_{x_{i+1,j+1}} \\
V_{x_p} &= \left(\frac{y_p^t - y_{i,j}}{\Delta y}\right) V_{xp}^1 + \left(1 - \left(\frac{y_p^t - y_{i,j}}{\Delta y}\right)\right) V_{xp}^2
\end{aligned}
\tag{10}
$$

Similarly the y-component of the particle velocity is interpolated using the following equations:

$$
\begin{aligned}
V_{yp}^1 &= \left(1 - \left(\frac{y_p^t - y_{i,j}}{\Delta y}\right)\right) V_{y_{i,j}} + \left(\frac{y_p^t - y_{i,j}}{\Delta x}\right) V_{y_{i,j+1}} \\
V_{yp}^2 &= \left(1 - \left(\frac{y_p^t - y_{i,j}}{\Delta y}\right)\right) V_{y_{i+1,j}} + \left(\frac{y_p^t - y_{i,j}}{\Delta y}\right) V_{y_{i+1,j+1}} \\
V_{y_p} &= \left(\frac{x_p^t - x_{i,j}}{\Delta x}\right) V_{yp}^1 + \left(1 - \left(\frac{x_p^t - x_{i,j}}{\Delta x}\right)\right) V_{yp}^2
\end{aligned}
\tag{11}
$$

where Δx and Δy are the dimensions of the computational cell, [L], and (i, j), $(i+1, j)$, $(i, j+1)$, and $(i+1, j+1)$ are the indices of the four closest nodes to the particle.

The above components of the particle velocity are weighted for taking into account the variation of the velocity during a time step by fourth-order Runge–Kutta method. The particle velocity components are obtained by the following equations:

$$\begin{aligned} V_{xp}^{t+\Delta t} &= V_{xp}^{t} + \frac{1}{6}\left(\left({}^{1}V_{xp}\right) + 2\left({}^{2}V_{xp}\right) + 2\left({}^{3}V_{xp}\right) + \left({}^{4}V_{xp}\right)\right)V_{yp}^{t+\Delta t} \\ &= V_{yp}^{t} + \frac{1}{6}\left(\left({}^{1}V_{yp}\right) + 2\left({}^{2}V_{yp}\right) + 2\left({}^{3}V_{yp}\right) + \left({}^{4}V_{yp}\right)\right) \end{aligned} \tag{12}$$

where the superscripts 1, 2, 3, and 4 stand for the velocity components at t, $t + \Delta t/2$, $t + \Delta t/2$, and $t + \Delta t$ time levels. There are two velocity components (corresponding to superscripts 2 and 3) at time instant $t + \Delta t/2$.

The dispersive transport of the particle due to hydrodynamic dispersion is calculated from the random movement of particles following the Gaussian normal distribution. The dispersive transport of a particle in x-and y-direction during a given time step can be given as

$$\begin{aligned} dx_p^d &= Z_{L_P} \sin\varphi + Z_{T_P} \cos\varphi = Z_{L_P}\frac{dx_p^a}{dl_l^a} + Z_{T_P}\frac{dy_p^a}{dl_l^a} \\ dy_p^d &= Z_{L_P} \cos\varphi - Z_{T_P} \sin\varphi = Z_{L_P}\frac{dy_p^a}{dl_l^a} - Z_{T_P}\frac{dx_p^a}{dl_l^a} \end{aligned} \tag{13}$$

where dx_p^d, dy_p^d are the dispersive transport components of pth particle during Δt, $[L]$; $Z_{L_p} = N(0,\sigma_L^2)\&Z_{T_p} = N(0,\sigma_T^2)$ are random numbers; $N(0,\sigma_L^2)$ & $N(0,\sigma_T^2)$ are normally distributed random numbers with zero mean and one standard deviation and which range from -6 to $+6$; φ is the angle between the direction of advective transport dl_p^a and the y axis; and $\sigma_{L_p} = \sqrt{2\alpha_L dl_p^a}$ and $\sigma_{T_p} = \sqrt{2\alpha_T dl_p^a}$ are standard deviations. Thus, the particle under consideration moves during a given time step first by advective transport in proportion to the velocity of groundwater flow, and its location after this movement is determined. Then the change in location during the same time step due to dispersive movement is superimposed over this location. Thus, the total displacement of a particle in x- and y-direction during a given time step can be given as

$$\begin{aligned} x_p^{t+\Delta t} &= x_p^t + dx_p^a + dx_p^d \\ y_p^{t+\Delta t} &= y_p^t + dy_p^a + dy_p^d \end{aligned} \tag{14}$$

where $x_p^{t+\Delta t}, y_p^{t+\Delta t}$ is the location of the pth particle at $t + \Delta t$ time level and x_p^t, y_p^t is the initial location of pth particle. This procedure is repeated for all the particles generated in the cell. The change in nodal solute concentration during a time step is given as

$$c_{i,j}^{t+\Delta t} = c_{i,j}^t + \left(NP_{i,j}^{t+\Delta t}\right)PM \tag{15}$$

where $NP_{i,j}^{t+\Delta t}$ is the updated number of the particles residing in the computational cell after both advective and dispersive transport. The model takes care of prescribed concentration boundary condition by adding the particles at the corresponding boundary cells to maintain the prescribed level of solute concentration throughout the simulation. In case of zero solute flux boundary, the dispersive solute flux is often assumed as zero, and thus the boundary solute flux consists advective flux only. The stability of the model solutions is ensured through the proper selection of time step size such that a particle should not cover more than one fifth of the grid size in one advection step; this will further avoid the oscillations in particle density (Prickett et al. 1981). But this model is likely to suffer from low sensitivity of model solutions to the variation of certain transport parameters which may be masked by statistical variation of solute mass.

2.3 Numerical Stability and Accuracy

The stability of the transport model solutions can be ensured by selecting proper time step size which will satisfy the Courant number criteria that can be given as

$$\begin{aligned} C_x &= \frac{V_x \times \Delta t}{\Delta x} \leq 1 \\ C_y &= \frac{V_y \times \Delta t}{\Delta y} \leq 1 \end{aligned} \tag{16}$$

where C_x and C_y are the Courant numbers calculated for the maximum magnitudes of the x- and y-components of the velocity vectors, respectively. It relates the distance a particle travels during one time step to the spatial increment. The Courant number controls the oscillations of the numerical solution arising from the temporal discretization.

The accuracy of the transport model is verified with respect to the average mass balance error in the solutions of Eq. (1) which is given as

$$E_c = \frac{100\left(\Delta\, Mf_c^T - \Delta\, Ms_c^T\right)}{M_c^0} \tag{17}$$

where E_c is the average mass balance error in the solute transport solutions [percent], ΔMf_c^T is the net solute flux in the aquifer system during the simulation period [ppm], ΔMs_c^T is the change in solute mass in the aquifer system [ppm], and M_c^0 is the initial solute mass in the aquifer system [ppm].

3 Results and Discussion

3.1 Validation of RWSOLUTE Model

Figure 1 shows the schematic of aquifer modeled in test case (Sun 1996), which is aimed at validating the RWSOLUTE model. The aquifer is 80 m long and 70 m wide. The left and right sides of the aquifer are subject to no-flow boundary conditions, while the top and bottom sides are subject to prescribed head boundary conditions with the constant head of 100 m. The groundwater is flowing in from the bottom to the top side at the uniform velocity of 1.0 m/day.

The boundary conditions for solute transport simulation involve zero solute flux boundaries across the left and right sides and a concentration of 1 ppm on the bottom side and zero concentration at the top side of the aquifer. Starting from an initial state of zero solute concentration, the aquifer is gradually contaminated due to transport of solute from a line source of pollution at the bottom side due to both advection and dispersion.

For solute transport simulations by the proposed model, for advective transport simulation, the aquifer is discretized by using a mesh-centered finite difference grid with uniform nodal spacing in x- and y-direction as 5 m resulting into total 255 nodes involving 60 boundary nodes. For dispersive solute transport simulation,

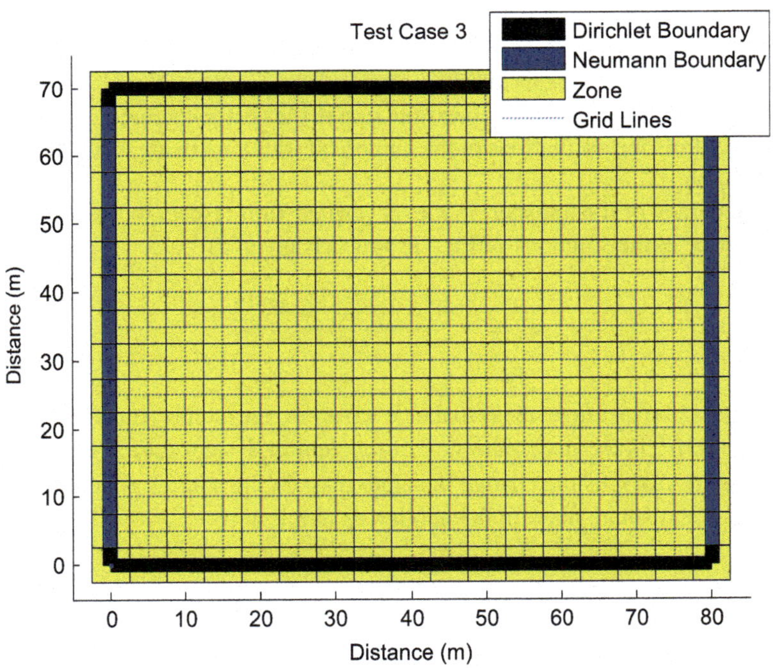

Fig. 1 Schematic of aquifer modeled in test case for the simulation of solute transport in uniform one-dimensional groundwater flow in unconfined aquifer with line source of pollution

a triangular finite element mesh is used with isosceles triangular elements of size 5 m resulting into 448 finite elements. The aquifer parameters used in this test simulation are aquifer thickness ($b = 30$ m), effective porosity ($\theta = 0.30$), transmissivity ($T = 10$ m^2/d), specific yield ($S_y = 0.10$), and longitudinal dispersivity ($\alpha_L = 10$ m).

Figures 2 and 3 show comparison of transverse concentration profiles obtained at the bottom side from the proposed model with the reported analytical solutions (Sun 1996) and FESOLUTE and RWSOLUTE solutions (Kulkarni 2008). It is found from the comparison of the concentration profiles obtained from solute transport simulation by proposed model that the initial error in numerical solutions dampens out as the solution progresses through the time. The RWSOLUTE concentration profile shows numerical overshoots at early times and deviates leftward from the analytical concentration profile at later times. At distance of 10 m from the line source of pollution, the percentages of the deviation between MMOCSOLUTE and RWSOLUTE and analytical solutions are 5 % and 30 %, respectively. However, RWSOLUTE model is justified as it simulates the randomness in actual solute transport behavior.

3.2 Comparison of Mass Balance Error in RWSOLUTE and Other Reported Models

The rise in mass balance curve is very sharp in the early stages of the simulation up to 3 days. Thereafter the mass balance error curves increase gradually till the end of the simulation period of 30 days. It is found from the results that the mass balance error curve of RWSOLUTE simulations deviates from other reported model simulations by an order of 34 %. The RWSOLUTE simulations are more erroneous because of the randomness in generation of the particles to simulate the solute transport process. As the simulation progresses, the numerical oscillations get dampened, and consequently the errors in the later stages of the simulation are comparatively less than the error in early stages of the simulation.

3.3 Effect of Time Step Size on RWSOLUTE Solutions

It is seen from Fig. 4 that the RWSOLUTE simulated concentration profiles for the 3.75-day time step size are severely affected by numerical oscillations and dispersions.

The solutions are found to be reasonably accurate for the time step size of 1.25 days. Even for the choice of time step size of 0.25 day, an overshoot in the numerical solutions of around 30 % is observed at 10 m downstream from the line source of pollution. The unacceptable overshoot of 100 % is noticed at the

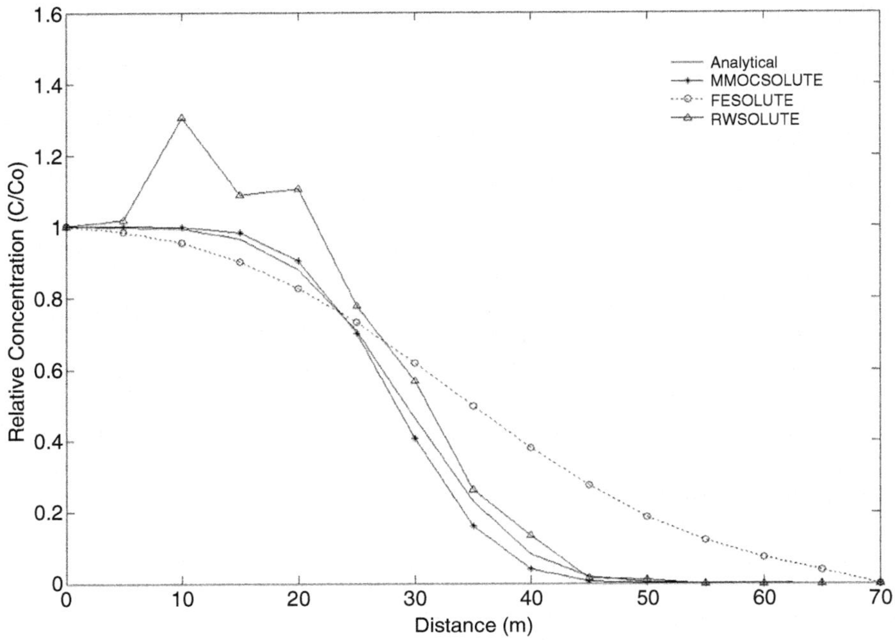

Fig. 2 Validation of RWSOLUTE model with analytical solutions and reported model solutions

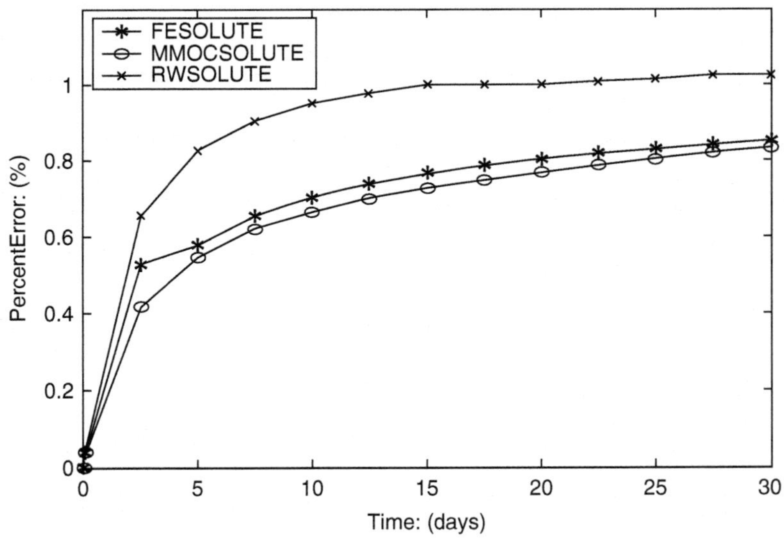

Fig. 3 Comparison of mass balance errors in RWSOLUTE and other reported models

distance of 25 m downstream of the source for 3.75-day time step simulation. Thus, the RWSOLUTE simulations are found to be severely restricted by the choice of time step size.

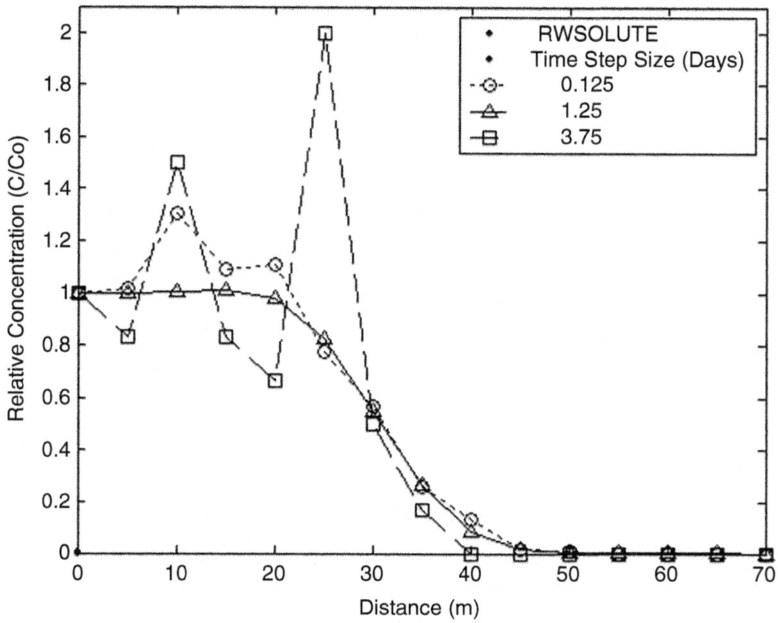

Fig. 4 Effect of time step size on model RWSOLUTE solutions

3.4 Effect of Particle Mass on RWSOLUTE Solutions

Figure 5 shows the effect of particle mass variation on the shape of the concentration profiles simulated by RWSOLUTE model. The fixed particle mass signifies the fraction of the total solute mass residing in the unit volume of groundwater at a given point of time. The particle mass is a fixed quantity while the number of particles is only varying due to the transport of the solute mass in a given time step. The particle mass is varied from 1.0 to 10.0 g. The numerical solutions experienced oscillations up to the distance of 20 m from the source of the pollution for all the chosen values of the particle mass. The maximum overshoot is observed in the concentration profile simulated for the 10.0-g particle mass to the order of 50 %, and the maximum undershoot is found in the concentration profile simulated for 5.0-g particle mass which is of the order of 30 %.

4 Conclusion

1. Validation of RWSOLUTE model for chosen test case shows that the numerical dispersion in solute concentration profiles is 24 %.
2. The investigations pertaining to the effect of time step size (Courant number) on proposed transport models reveal that the RWSOLUTE solutions suffer from

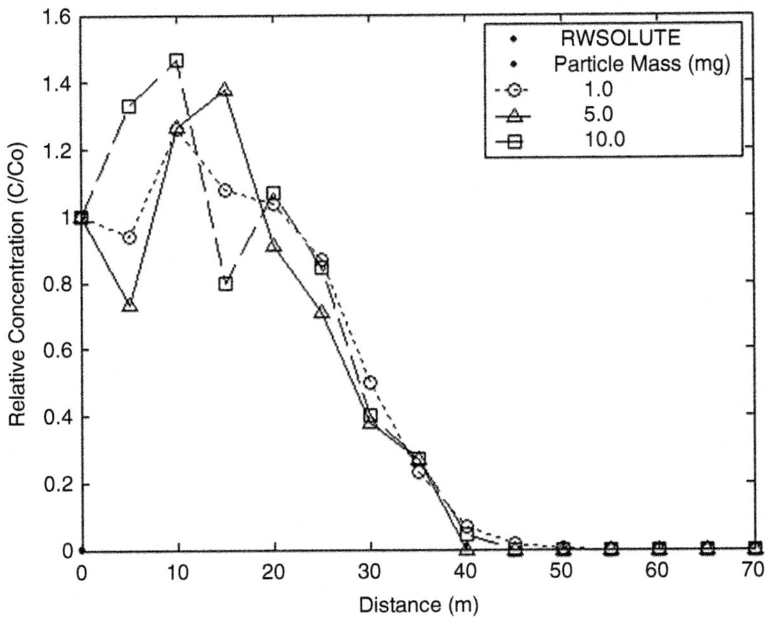

Fig. 5 Effect of particle mass on RWSOLUTE model solutions

high overshoots in the solutions to the order of 28 % higher than analytical solutions even for smaller time step of 0.25 day.

3. RWSOLUTE model produces 20 % higher mass balance error compared to the other two models. The results of RWSOLUTE simulations show that the numerical undershoot and overshoot in solute concentration profiles are of two orders magnitude greater than other two reported model solutions because of approximation in dispersion term.
4. It is also found that the RWSOLUTE solutions are sensitive to the choice of particle mass and in turn to the number of particles generated to represent the solute mass in aquifer system. Study noted that almost all model solutions develop numerical oscillations with the increase in time step size. It is found from the RWSOLUTE simulations that large mass balance errors up to 25 % may occur when less number of particles is used in the solute transport simulation.

Bibliography

Banton O, Delay F, Porel G (1997) A new time domain random walk method for solute transport in 1 – D heterogeneous media. Ground Water 35(6):1008–1013

Bellin A, Rubin Y, Rinaldo A (1994) Eulerian-Lagrangian approach for modeling of flow and transport in heterogeneous geological formations. Water Resour Res 30(11):2913–2924

Bentley LR, Pinder GF (1992) Eulerian-Lagrangian solution of the vertically averaged groundwater transport equation. Water Resour Res 28(11):3011–3020

Delay F, Ackerer P, Danquigny C (2005) Simulating solute transport in porous or fractured formations using random walk particle tracking. Vadose Zone J 4(2):360–379

Dentz M, Cortis A, Scher H, Berkowitz B (2004) Time behavior of solute transport in heterogeneous media: transition from anomalous to normal transport. Adv Water Resour 27 (2):155–173

Freeze RA, Cherry JA (1979) Groundwater. PrenticeHall Inc, Englewood Cliffs

Illangasekare TH, Döll P (1989) A discrete kernel method of characteristics model of solute transport in water table aquifers. Water Resour Res 25(5):857–867

Kulkarni NH (2008) Numerical experiments on the solute transport in groundwater flow systems. IIT Bombay, Mumbai

Lu N (1994) A semianalytical method of path line computation for transient finite-difference groundwater flow models. Water Resour Res 30(8):2449–2459. http://doi.org/10.1029/94WR01219

Neuman SP (1984) Adaptive Eulerian – Lagrangian finite element method for advection – dispersion. Int J Numer Methods Eng 20(2):321–337

Prickett TA, Naymik TG, Lonnquist CG (1981) A "random-walk" solute transport model for selected groundwater quality evaluations, vol 65. Illinois State Water Survey Champaign, Champaign

Sorek S (1988) Eulerian-Lagrangian method for solving transport in aquifers. In: Groundwater flow and quality modelling. Springer, Dordrecht, pp 201–214

Sun N-Z (1996) Mathematical modelling of groundwater pollution. Springer, New York

Uffink GJM (1988) Modeling of solute transport with the random walk method. In: Groundwater flow and quality modelling. Springer, Dordrecht, pp 247–265

Wang HF, Anderson MP (1995) Introduction to groundwater modeling: finite difference and finite element methods. Academic, San Diego

Wen X-H, Gomez-Hernandez JJ (1996) The constant displacement scheme for tracking particles in heterogeneous aquifers. Ground Water 34(1):135

Zhang R, Huang K, van Genuchten MT (1993) An efficient Eulerian-Lagrangian method for solving solute transport problems in steady and transient flow fields. Water Resour Res 29 (12):4131–4138

A Comparison of EnKF and EnPAT Inverse Methods: Non-Gaussianity

Liangping Li, Haiyan Zhou, J. Jaime Gómez-Hernández, and Sanjay Srinivasan

Abstract The EnKF has been extensively used for real-time data assimilation in areas such as reservoir/groundwater modeling. One of the big challenges of the EnKF is how to handle the non-Gaussianity of aquifer properties, particularly for channelized aquifers where preferred flow conduits are encountered. EnPAT is a pattern-based inverse method and was developed to deal with the non-Gaussianity of model updating. In this work, we compared the EnKF and EnPAT on a benchmark example. The results show that EnPAT can better reproduce the curvilinear geological features and thus has a better transport prediction.

1 Introduction

For developing and managing production of oil fields, it is important to predict the locations of hydrocarbon-bearing areas accurately and thus determine optimum production well locations. This is critical because of the significant cost of drilling wells. However, one of most challenging tasks is to characterize the geological structure and identify reservoir properties such as permeability and porosity, which define the locations of hydrocarbons. The challenge mainly stems from inherent strong heterogeneity of the deposits and from scarcity of data.

L. Li (✉) • H. Zhou
Geology and Geological Engineering, South Dakota School of Mines and Technology, 501 E. Saint Joseph St., Rapid City, SD 57701, USA
e-mail: Liangping.Li@sdsmt.edu

J.J. Gómez-Hernández
Institute for Water and Environmental Engineering, Universitat Politècnica de València, Valencia, Spain
e-mail: jgomez@upv.es

S. Srinivasan
Petroleum and Natural Gas Engineering, PennState University, 110 Hosler Building|University Park, State College, PA 16802, USA
e-mail: szs27@psu.edu

J.J. Gómez-Hernández et al. (eds.), *Geostatistics Valencia 2016*, Quantitative Geology and Geostatistics 19, DOI 10.1007/978-3-319-46819-8_57

Geostatistical approaches are usually employed to represent the spatial heterogeneity of reservoir properties, conditioned on static data such as measured permeability. Besides the static data, dynamic data such as oil production rate also can be used to condition the models and improve the characterization. The procedure of constructing reservoir models conditioned to dynamic data is termed inverse method, where the objective is to identify the parameter values at un-sampled locations by integrating those dynamic data into the model, thus improving the prediction of oil production in the future.

The EnKF has been often used to update models by integrating the dynamic data. Its advantages include real-time data integrating and computational efficiency (Evensen 2003). One remarkable disadvantage is that it can't preserve the high-order statistics which is a reflection of the geological structures. The EnPAT is developed to deal with this problem and is based on multiple-point geostatistics. It can honor not only the dynamic data but also the geology in accordance with the geological field conditions (Zhou et al. 2011). A benchmark comparison between the EnKF and EnPAT is conducted in this work. It clearly shows that a better transport prediction is achieved using the EnPAT.

2 Example

A single-phase transient groundwater flow example is presented to compare the effectiveness of the EnKF and the EnPAT for dynamic data integration when considering non-multi-Gaussian conductivity fields. The aquifer has $50 \times 50 \times 1$ cells of size $1\ \text{m} \times 1\ \text{m} \times 1\ \text{m}$. The east and west sides of the aquifer are constant head boundaries with prescribed head values of 0 m, and the other two sides are no-flow boundaries (see Fig. 1). Porosity and specific storage are assumed to be constant and set equal to 0.3 and $0.02\ \text{m}^{-1}$, respectively. An injection well is located at the center of aquifer with a constant injection rate $Q = 25\ \text{m}^3/\text{day}$. The total simulation time is 30 days, discretized into ten time steps with varying time step size following a geometric sequence of ratio 1.2. The head data collected from nine wells in the first five time steps will be used for conditioning.

3 Results

Figure 2 shows two individual realizations before and after conditioning to head data using the EnKF and the EnPAT. The updated realization using the EnKF tends to lose the channel features of the training image. Partly, this is expected because in the EnKF each updated realization can be interpreted as a linear combination of the ensemble of prior models. For the EnPAT, the individual conductivity realizations retain the channel features of the training image and end with similar channel structures as the ones in the reference field.

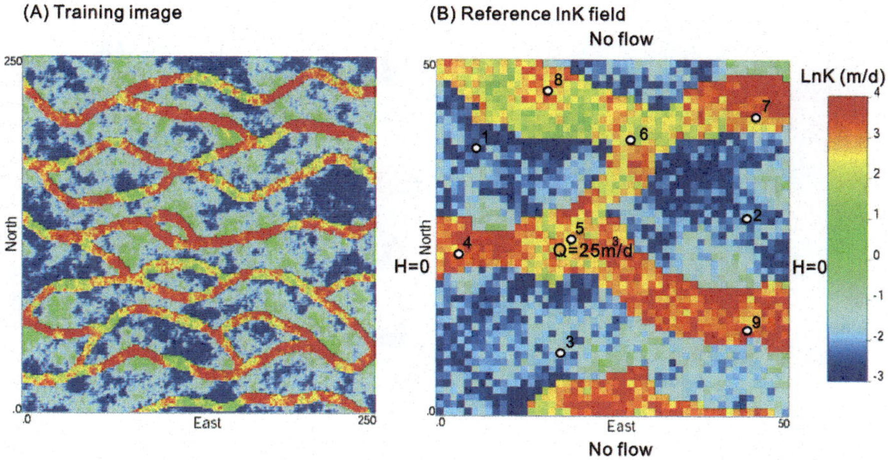

Fig. 1 Training image and reference conductivity field

Figure 3 shows the concentration distribution, after 30 days, for the reference field, and the ensemble mean and variance of the predicted concentrations for the prior models and for the updated models using the EnKF and the EnPAT. The results show that the concentration profile is heavily impacted by the connectivity of the conductivity channel structures.

4 Conclusion

The EnKF and the EnPAT are compared for inverse modeling of a conductivity field characterized by curvilinear channel features. The EnKF has been widely used in petroleum engineering and hydrogeology over the past decade. The remarkable advantages of the EnKF are the capability to handle multiple parameters, computational efficiency, and real-time data assimilation. However, one significant drawback of the EnKF is that, because it is based on two-point statistics, it is optimal only for linear state functions and parameters following multi-Gaussian distribution. As an alternative to the EnKF, the EnPAT method was proposed to condition to dynamic data without the limitation that state variables should follow a multi-Gaussian distribution. The application of the EnKF and the EnPAT to a synthetic case shows that the EnPAT yields a more accurate characterization of conductivity than EnKF, and more importantly, the updated models obtained using the EnPAT honor the prior geologic features exhibited in the training image.

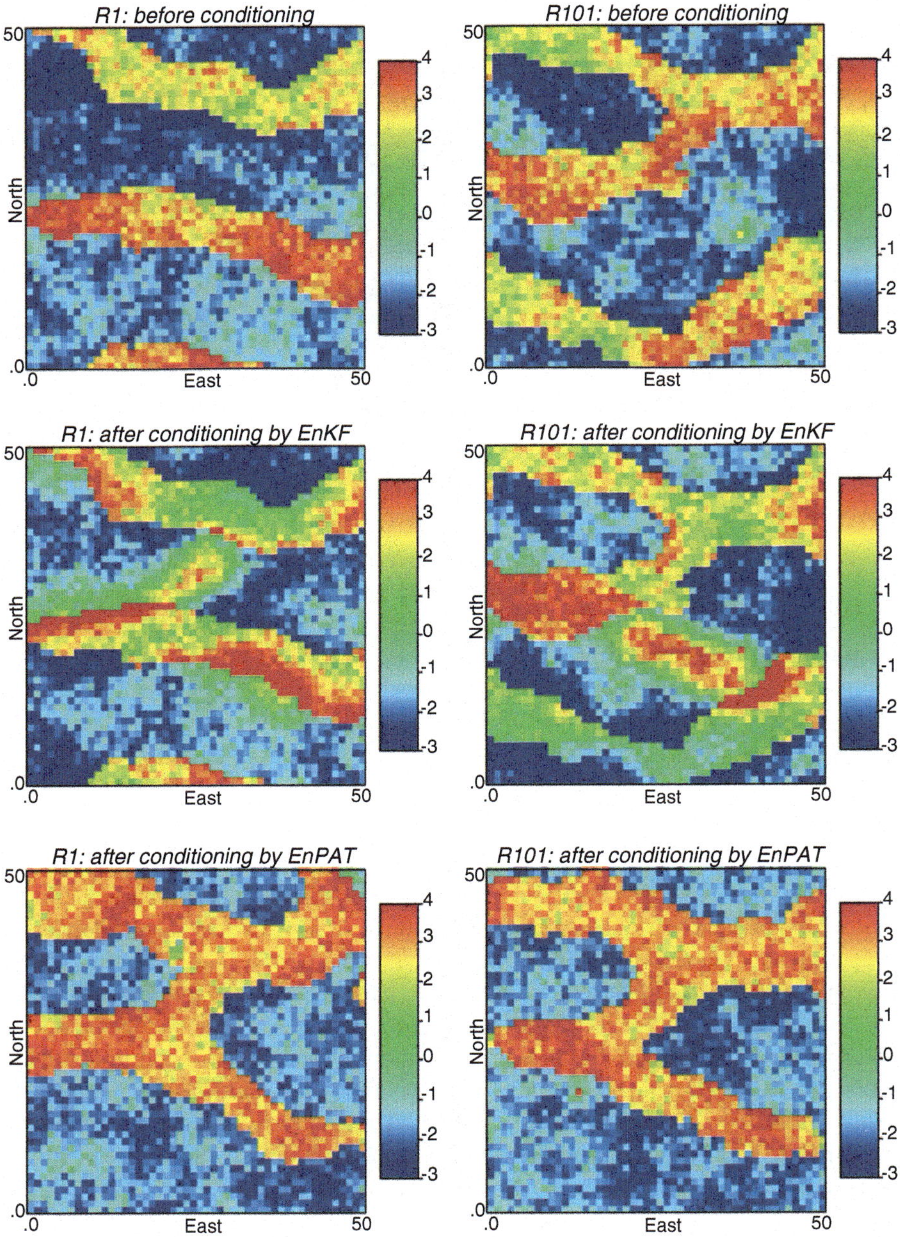

Fig. 2 Randomly selected two individual realizations before and after head data conditioning using the EnKF and the EnPAT

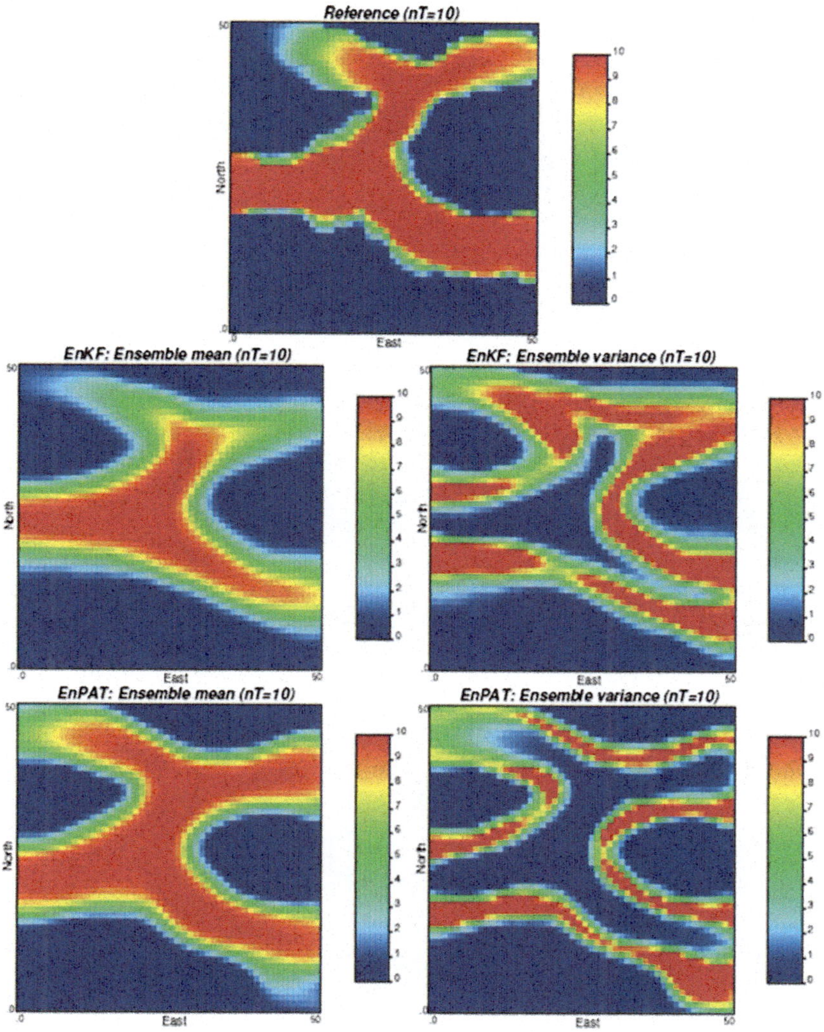

Fig. 3 Ensemble mean and variance of concentration before and after head conditioning using the EnKF and the EnPAT

Bibliography

Evensen G (2003) The ensemble Kalman filter: theoretical formulation and practical implementation. Ocean Dyn 53(4):343–367

Zhou H, Gomez-Hernandez JJ, Li L (2011) A pattern-search-based inverse method. Water Resour Res 48(3):W03505

Calibration of Land Subsidence Model Using the EnKF

Liangping Li, Meijing Zhang, and Haiyan Zhou

Abstract Land subsidence modeling has been developed for reliable modeling and prediction in the last several decades. Calibration of hydraulic properties such as transmissivity and elastic and inelastic specific storages using observation data is a challenge because of the strong nonlinearity of groundwater flow equation especially when it accounted for the interbed drainage process. The ensemble Kalman filter is applied to calibrate hydraulic properties in a synthetic land subsidence model. The characterization of transmissivity and specific storages and prediction of land subsidence are improved after the drawdown and subsidence observation data are conditioned.

1 Introduction

As a global scale problem, land subsidence has been studied by many researchers (Galloway et al. 1999; Bell et al. 2002, 2008; Zhang and Burbey 2015). Due to declining water levels, decreasing pore water pressures within the aquifer system have led to significant increases in effective stress, which accounted for large-scale compaction of sediments (Terzaghi 1925; Poland and Davis 1969; Holzer 2010). Subsidence data, when combined with groundwater drawdown data, can be used to improve groundwater model calibration of the hydrologic parameters such as elastic and inelastic skeletal specific storage, transmissivity, the compaction time constant, and others (Burbey 2001; Hoffmann et al. 2001, 2003b; Zhang et al. 2013; Zhang and Burbey 2015). Hoffman et al. (2003a, b) used a regional groundwater flow and subsidence model in conjunction with UCODE (Poeter et al. 2005) to

L. Li (✉) • H. Zhou
Geology and Geological Engineering, South Dakota School of Mines and Technology, 501 E. Saint Joseph St., Rapid City, SD 57701, USA
e-mail: Liangping.Li@sdsmt.edu

M. Zhang
Department of Agricultural and Biological Engineering, University of Florida, Gainesville, FL 32611, USA

J.J. Gómez-Hernández et al. (eds.), *Geostatistics Valencia 2016*, Quantitative Geology and Geostatistics 19, DOI 10.1007/978-3-319-46819-8_58

estimate spatially varying compaction time constants and inelastic specific skeletal storage coefficients in Antelope Valley, California (Hoffmann et al. 2003a). Zhang et al. (2013) presented a discrete adjoint algorithm for identifying suitable zonations of elastic and inelastic skeletal specific storage coefficients and hydraulic transmissivity from hydraulic head and subsidence measurements (Zhang et al. 2013). The ensemble Kalman filter (EnKF) based on the sequential Bayesian updating rule can be used to obtain results similar to those obtained by Monte Carlo (MC)-type inverse methods but with reduced CPU time (Burgers et al. 1998; Evensen 2003). The EnKF method is able to jointly map the hydraulic conductivity and porosity fields accurately and efficiently by assimilating dynamic piezometric head and multiple concentration data (Li et al. 2012). However, no research has been done to joint estimation of spatially distributed hydraulic transmissivity and elastic and inelastic specific skeletal storage coefficients using land subsidence and groundwater level measurements. We will demonstrate the capability of the EnKF to jointly map hydraulic transmissivity and elastic and inelastic specific skeletal storage coefficients in a synthetic land subsidence model with MODFLOW 2005 (Harbaugh 2005).

2 Example

A synthetic aquifer has 38×58 grid blocks with block size of 1×1 km. Aquifer thickness is 200 m. The aquifer is assumed to be confined and simulated as transient flow. The total simulation time of flow and land subsidence is 15 years, with each year being divided into two 6-month periods. A poorly permeable but highly compressible clay interbed of constant thickness of 70 m is distributed within the permeable aquifer. All the boundaries are assumed to be no-flow boundary, and the initial hydraulic head is 800 m over the domain. The preconsolidation head is set as 795 m. Aquifer storage coefficient is set as 0.002. Five wells are pumped at a constant rate in 6-month intervals (6 months on during the summer and 6 months off during the winter) (see Fig. 1). The vertical hydraulic conductivity of interbed is 0.00006 m/day.

Three cases are considered. For the first case, no conditioning data are considered. Like the "true" fields, the initial models are generated using GCOSIM3D, without conditioning any data. For the second case, only the measured static data at well locations will be considered. For the case three, besides the static data, dynamic data (drawdown and subsidence data) will be included as well; the initial models of case 2 will be used in this case, and a further conditioning on dynamic data is conducted using the EnKF such that a best characterization of both transmissivity and specific storages is achieved and a smallest uncertainty would be observed since all the data are integrated into the models.

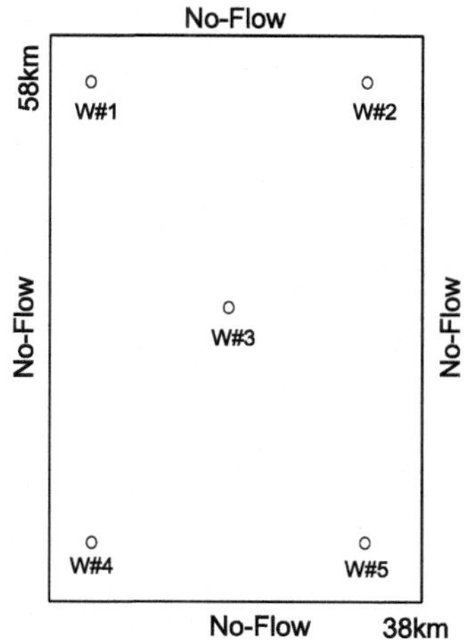

Fig. 1 Boundary conditions of hypothetic example

3 Results

The predictions of land subsidence using calibrated transmissivity and specific storages are shown in Fig. 2. The first 10 years of drawdown and subsidence observation are used for conditioning, whereas the last 5 years of simulations are used to validate the models. Like the characterization of hydraulic properties, the uncertainty of predictions for subsidence is reduced if more data are considered. The value of drawdown and subsidence data for inverse modeling is evident.

4 Conclusion

We have presented an application of the EnKF for calibration of transmissivity and interbed specific storages using drawdown and subsidence data in a synthetic example. This work is distinguished from other subsidence modeling studies. The EnKF is successfully applied for calibration in subsidence modeling for the first time, and a further application of real case studies is expected. The significance of drawdown and subsidence data is patent for the characterization of transmissivity and elastic and inelastic specific storages and for reliable predictions of land subsidence.

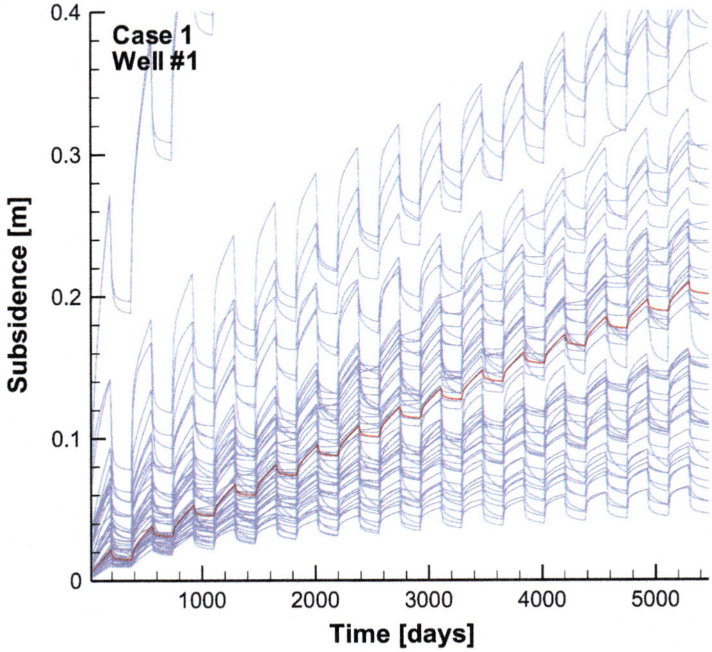

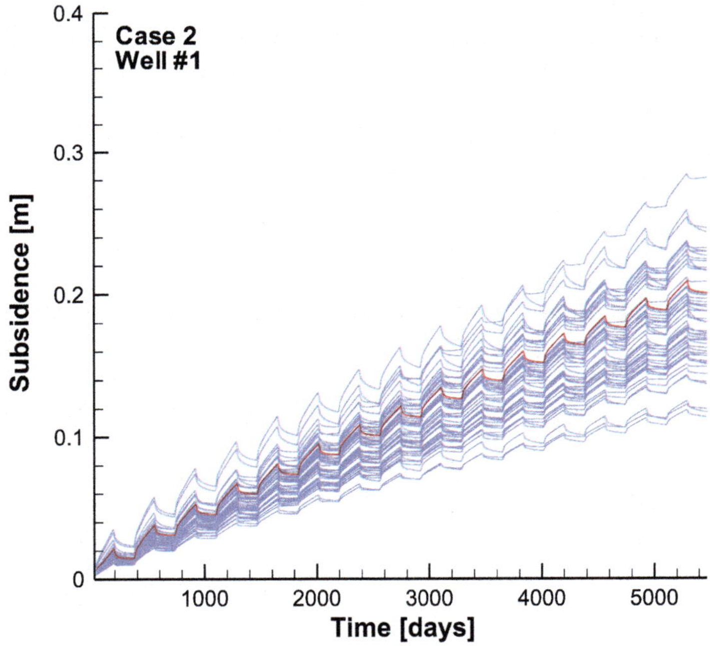

Fig. 2 Simulated subsidence for Well #1 for three cases. The *red line* indicates the observation data and the *gray lines* denote the simulation results using calibrated properties

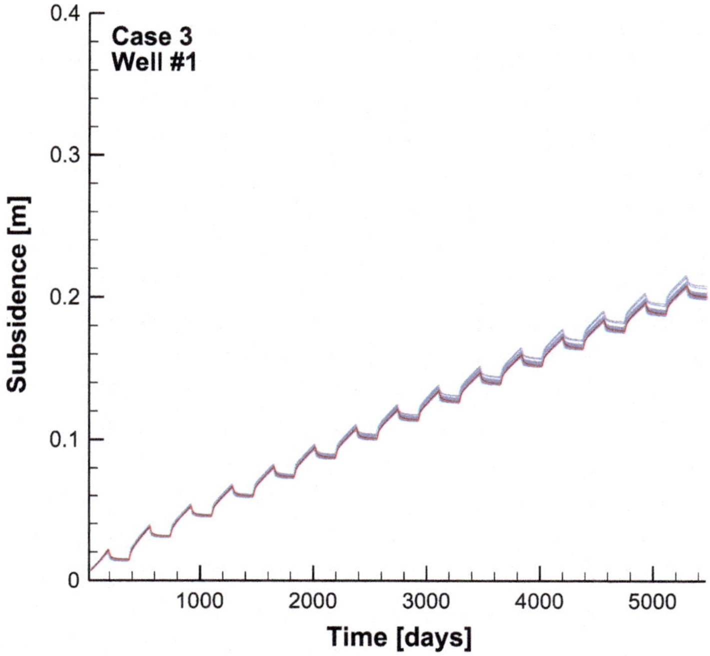

Fig. 2 (continued)

Bibliography

Bell JW, Amelung F, Ramelli AR, Blewitt G (2002) Land subsidence in Las Vegas, Nevada, 1935–2000: new geodetic data show evolution, revised spatial patterns, and reduced rates. Environ Eng Geosci 8:155–174

Bell JW, Amelung F, Ferretti A, Bianchi M, Novali F (2008) Permanent scatterer InSAR reveals seasonal and long-term aquifer-system response to groundwater pumping and artificial recharge. Water Resour Res 44:W02407

Burbey TJ (2001) Stress-strain analyses for aquifer-system characterization. Groundwater 39:128–136

Burgers G, Jan van Leeuwen P, Evensen G (1998) Analysis scheme in the ensemble Kalman filter. Mon Weather Rev 126:1719–1724

Evensen G (2003) The ensemble Kalman filter: theoretical formulation and practical implementation. Ocean Dyn 53:343–367

Galloway D, Jones DR, Ingebritsen SE (1999) Land subsidence in the United States. US Geological Survey, Reston

Harbaugh AW (2005) MODFLOW-2005, the US Geological Survey modular ground-water model: the ground-water flow process. US Department of the Interior, US Geological Survey

Hoffmann J, Zebker HA, Galloway DL, Amelung F (2001) Seasonal subsidence and rebound in Las Vegas Valley, Nevada, observed by synthetic aperture radar interferometry. Water Resour Res 37:1551–1566

Hoffmann J, Galloway DL, Zebker HA (2003a) Inverse modeling of interbed storage parameters using land subsidence observations, Antelope Valley, California. Water Resour Res 39:5

Hoffmann J, Leake S, Galloway D, Wilson AM (2003b) MODFLOW-2000 ground-water model – user guide to the subsidence and aquifer-System compaction (SUB) package. DTIC Document

Holzer T (2010) Implications of ground-deformation measurements across earth fissures in subsidence areas in the southwestern USA. Proceedings of EISOLS

Li L, Zhou H, Gómez-Hernández JJ, Franssen H-JH (2012) Jointly mapping hydraulic conductivity and porosity by assimilating concentration data via ensemble Kalman filter. J Hydrol 428:152–169

Poeter EP, Hill MC, Banta ER, Mehl S, Christensen S (2005) Ucode_2005 and six other computer codes for universal sensitivity analysis, calibration, and uncertainty evaluation: US geological survey techniques and methods 6-A11. US Geological Survey, Reston

Poland JF, Davis GH (1969) Land subsidence due to withdrawal of fluids. Rev Eng Geol 2:187–270

Terzaghi K (1925) Principles of soil mechanics, IV—settlement and consolidation of clay. Eng News Rec 95:874–878

Zhang M, Burbey T (2015) A comparison of three hydraulic parameter optimization schemes for Las Vegas Valley. Environ Eng Geosci, 1078–7275 EEG-1672

Zhang M, Burbey TJ, Nunes VDS, Borggaard J (2013) A new zonation algorithm with parameter estimation using hydraulic head and subsidence observations. Groundwater

Influence of Heterogeneity on Heat Transport Simulations in Shallow Geothermal Systems

Javier Rodrigo-Ilarri, Max Reisinger, and J. Jaime Gómez-Hernández

Abstract The influence of parameter heterogeneity, such as permeability, porosity, and thermal conductivity, over results of heat transport simulation is studied. A set of synthetic aquifer simulations considering different degrees of heterogeneity in the hydraulic conductivity, porosity, and thermal conductivity fields were created by sequential Gaussian simulation techniques. Heterogeneity of the hydraulic conductivity showed to have a significant influence on the evaluation of a cold plume in the porous media. Higher variances in the hydraulic conductivity distributions cause an important rise in the variability of the simulated temperature fields and a considerable increase of uncertainty in the simulated heat distribution in the aquifer system. Results show that considering heterogeneity on the permeability fields induces more impact on the model results than considering heterogeneity of both porosity and thermal conductivity.

1 Introduction

Shallow geothermal system uses the energy stored in the first approximately 400 m under the earth surface (Llopis Trillo and López Jimeno 2009). From about 10–20 m in depth, temperature is considered to be constant during the year. Deeper below the surface, temperatures increase according to the geothermal gradient (3 °C for each 100 m of depth on average) (Sanner 2001).

Due to low temperatures (10 °C to max. 30 °C) in the shallow zone, the so-called low enthalpy energy is obtained. Low enthalpy energy cannot be used directly and geothermal systems have to be applied to make use of it.

J. Rodrigo-Ilarri (✉) • J.J. Gómez-Hernández
Institute for Water and Environmental Engineering, Universitat Politècnica de València, Valencia, Spain
e-mail: jrodrigo@upv.es; jgomez@upv.es

M. Reisinger
REVITAL Integrative Naturraumplanung, 9990, Nußdorf-Debant, Nußdorf 7, Austria
e-mail: M.Reisinger@revital-ib.at

J.J. Gómez-Hernández et al. (eds.), *Geostatistics Valencia 2016*, Quantitative Geology and Geostatistics 19, DOI 10.1007/978-3-319-46819-8_59

The most common system to extract heat from the underground is the ground source heat pump (GSHP) system. A GSHP system extracts thermal energy from a cold zone to transport it to a warmer zone. The natural form of heat transport would be in the opposite direction (from warm to cold) according to the second law of thermodynamics. To invert the natural heat flow, it is necessary to supply the system with energy, normally with a compressor. In these systems for each kWh of electric energy used for the compressor, up to 4.5 kWh of thermal energy can be provided (Conde Lázaro and Ramos Millán 2009). Another advantage of GSHPs is the reversibility which allows to obtain heating and cooling with the same system.

Natural groundwater or collectors installed in the underground in which a fluid circulates are used as heat sources. In the first case, the natural groundwater is used directly; it is pumped up with a well and transported to the heat exchanger. After extracting energy it is reinjected to the ground. These systems are called open-loop systems. In the second case, a fluid circulates through the collectors which are installed in the underground. The fluid is heated up on this way in the collectors and transports the energy to the GSHP system. This type of systems is called closed-loop systems (Llopis Trillo and López Jimeno 2009).

2 Scope and Objectives

Different investigations on heat transport in the subsurface have been made so far, most of them assuming homogenous aquifer conditions. Kupfersberger (2009) developed a 2D numerical groundwater model to simulate the impact of groundwater heat pumps on groundwater temperature in the Leibnitzer Feld aquifer, Austria. He validated the simulated results comparing them to field site measurements. A 3D density-dependent groundwater flow and thermal transport model was developed and validated using the results of the thermal injection experiment by Molson (1992).

The effect of heterogeneity on heat transport simulation was the object of several investigations over the last few years. Ferguson (2007) presented a study on the topic, using stochastic modeling on two aquifers with low and high degrees of heterogeneity. He concluded that there is considerable uncertainty in the distribution of heat associated with injection of warm water into an aquifer. Bridger and Allen (2010) developed a model to evaluate the influence of aquifer heterogeneity as a result of geologic layering on heat transport and storage in an aquifer used for thermal energy storage. Bridger and Allen (2010) used FEFLOW to create a three-dimensional groundwater flow and heat transport model. All these investigations considered only the heterogeneity of the permeability, porosity, and thermal conductivities that were assumed to be constant.

The present work has been made in order to get more information about heat transport modeling in aquifer systems. Based on the results obtained by Shuang (2009), further investigation on how heterogeneity affects heat transport simulation has been made. Synthetic aquifers with different grades of heterogeneity were created using the Stanford Geostatistical Modeling Software (SGeMS) (Remy et al. 2009).

A set of heat transport simulations were performed using MT3DMS (Zheng and Wang 1999) as heat transport code.

To evaluate the importance of heterogeneity in permeability as well as heterogeneity in porosity and thermal conductivity, different simulations with homogeneous and heterogeneous parameters were made and compared to each other.

Therefore, the main objectives of this work are:

- To analyze the influence of heterogeneous distributions of permeability on heat transport simulations in shallow geothermal systems
- To analyze the influence of heterogeneous distributions of porosity, bulk density, and thermal conductivity on heat transport simulations in shallow geothermal systems

Mean values of permeability and porosity, injection rates, initial temperature distributions, and the model dimensions and well layout were taken from Shuang (2009).

3 Heat Transport Modeling Using MT3DMS

MT3DMS code (Zheng and Wang 1999) was originally written to simulate solute transport. The comparison of the solute transport equation (Eq. 1) and the heat transport equation (Eq. 2) shows the similarities of these two processes. Table 1 shows the nomenclature used hereafter:

$$\left(1+\frac{\rho_b K_d}{n}\right) n \frac{\partial C}{\partial t} = \nabla[n(D_m + \alpha_s \nu_a)\nabla C] - \nabla(\nu_a n C) + q_s C_s - \lambda n C \quad (1)$$

$$\left(\frac{(\rho c)_e}{n \cdot \rho_f c_f}\right) n \frac{\partial T}{\partial t} = \nabla\left[\left(\frac{\lambda_e}{n \cdot \rho_f c_f} + \alpha_h \nu_a\right)\nabla T\right] - \nabla(n \cdot \nu_a T) + \frac{q_h}{\rho_f c_f} - \frac{\lambda_u}{\rho_f c_f H F}\Delta T' \quad (2)$$

MT3DMS was verified for heat transport by Mendez Hecht (2008) and Molina Giraldo (2008). Therefore, MT3DMS can be used for the simulation of heat transport with some adaptations on the equation coefficients. Further information about heat transport modeling with MT3DMS can be found in other studies such as Molina (2009). In order to perform heat transport modeling, the following adaptations of the original mass transport parameters were performed:

Heat exchange between solid and liquid phase:

Table 1 Nomenclature

Parameter	Unit	Description
C	[kg m^{-3}]	Dissolved mass concentration
c_s	[J kg^{-1} K^{-1}]	Specific heat capacity solid
C_s	[kg m^{-3}]	Concentration sources and sinks
D_m	[m^2 s^{-1}]	Coefficient of molecular diffusion
F	[m]	Depth of the water table
H	[m]	Thickness of the aquifer
k	[m s^{-1}]	Hydraulic conductivity
K_d	m^3 kg^{-1}	Distribution coefficient
n	[–]	Effective porosity
Q	[m^3 s^{-1}]	Water injection rate
q_h	[W m^{-3}]	Heat injection or extraction
q_s	[m^3 s^{-1} m^{-3}]	Flow rate of sources and sinks solute transport
R	[–]	Retardation factor
t	[s]	Simulated time period
T_0	[K]	Initial temperature
T_f	[K]	Temperature of the water
T_{in}	[K]	Temperature of the injected water
T_s	[K]	Temperature of the solid
v_a	[m s^{-1}]	Seepage velocity
α_h	[m]	Horizontal transverse dispersivity coefficient
α_v	[m]	Vertical transverse dispersivity coefficient
α_L	[m]	Longitudinal dispersivity coefficient
α_s	[m]	Dispersivity coefficient
γ	[–]	Unit weight of the fluid
λ	[W m^{-1} K^{-1}]	Thermal conductivity
λ_e	[W m^{-1} K^{-1}]	Overall thermal conductivity of the saturated aquifer
λ_f	[W m^{-1} K^{-1}]	Thermal conductivity of the fluid
λ_s	[W m^{-1} K^{-1}]	Thermal conductivity of the solid
λ_u	[W m^{-1} K^{-1}]	Thermal conductivity of the unsaturated soil
μ	[kg s^{-1} m^{-1}]	Viscosity of the fluid
ν_s	[m s^{-1}]	Velocity
ρ_b	[kg m^{-3}]	Bulk density
$\rho_e c_e$	[J m^{-3}K^{-1}]	Volumetric heat capacity of the saturated aquifer
ρ_f	[kg m^{-3}]	Density of water
$\rho_f c_f$	[J m^{-3}K^{-1}]	Volumetric heat capacity of the fluid
ρ_s	[kg m^{-3}]	Density of the solid
$\rho_s c_s$	[J m^{-3}K^{-1}]	Volumetric heat capacity of the solid
σ^2	[–]	Variance

$$\left(1 + \frac{\rho_b K_d}{n}\right) = \left(\frac{(\rho c)_e}{n \cdot \rho_f c_f}\right) \tag{3}$$

Heat exchange between solid and liquid phase is implemented in MT3DMS in the chemical reaction package. The type of sorption has to be set to linear isotherm sorption. The input parameters required by MT3DMS to calculate the retardation factor are the bulk density ρ_b and the distribution coefficient K_d.

Conductive heat transport:

$$n(D_m + \alpha_s v_a) = \left(\frac{\lambda_e}{n \cdot \rho_f c_f} + \alpha_h v_a\right) \tag{4}$$

The conductive heat transport is implemented in MT3DMS in the dispersion package. The dispersivity coefficient α_h can be introduced without adaptations. However, the molecular diffusion coefficient D_m for heat conduction has to be calculated. Previous computations of bulk density (ρ_b) and thermal conductivity for the saturated aquifer (λ_e) are also required:

$$\rho_b = (1 - n) \cdot \rho_s \tag{5}$$

$$D_m = \frac{\lambda_e}{n \cdot \rho_f c_f} \tag{6}$$

Convective heat transport:

To simulate convective heat transport, the advection package of MT3DMS has to be activated. MT3DMS provides different solution schemes for the advection term. In this study, simulations using different solution schemes were made. The results and the simulation time were compared to evaluate the most efficient solution method. The most satisfying results were made with the hybrid MOC/MMOC (HMOC) solution scheme.

Sources and sinks:

The sources and sinks term is introduced in the well package of MODFLOW and MT3DMS. Temperature [°K] is treated like a concentration [kg/m^3] and the recharge rate is constant [kg/m^3]:

$$\frac{q_h}{\rho_f c_f}\left[\frac{K}{s}\right] = q_s C_s \left[\frac{kg}{m^3 s}\right] \tag{7}$$

4 Model Setup and Input Parameters

The model layout and dimensions were assumed considering the field site data of the Esseling site (Shuang 2009). It consists of a grid with 100 × 100 cells (1 m × 1 m x 1 m) and 40 layers. The same layout as Shuang (2009) was chosen to compare results. Following Rasouli (2008), the aquifer is assumed to be confined. The left and the right boundaries of the model are considered as constant head boundaries. The upper and lower boundary and the bottom of the model are no-flow boundaries. All other cells are assigned as active flow cells.

Flow simulations were performed assuming a hydraulic gradient $i = 0.02$. Prescribed constant heads are 45 m on the left boundary and 43 m on the right boundary. Horizontal and the vertical hydraulic conductivities were set equal to make the aquifer isotropic.

The injection well is located in the cell (30, 50) from layer 10 to layer 15 of the model. The total injection rate was divided and assigned to each layer. The recharge was assumed to be constant over 360 days. The total simulation time was divided into 12 stress periods and a steady-state simulation was performed. As the definition of the boundary conditions is required, a constant temperature value was assigned to the left-hand boundary of the model. All other cells are active temperature cells. The advection term of the heat transport equation was solved with the hybrid MOC/MMOC (HMOC) solution scheme as it runs faster than the ultimate TVD scheme and it is free of numerical dispersion (Zheng and Wang 1999). The type of sorption in the chemical reaction package was set to linear isotherm sorption. No first-order reaction was simulated.

Table 2 summarizes the flow and heat transport model input parameters that were considered to be known throughout the simulation process.

5 Results and Discussion

The stochastic simulation of the hydraulic conductivity fields was done with the Stanford Geostatistical Modeling Software (SGeMS). The sequential Gaussian simulation algorithm was used. Based on the set of hard data and a target histogram, ten synthetic conductivity fields were created for the scenarios shown in Table 3. Scenarios 1–3 correspond to heterogeneous permeability distributions and constant n and λ values, and Scenarios 4–6 correspond to heterogeneous permeability distributions and heterogeneous n and λ values.

Figure 1 shows the logK histogram and the conductivity field view of the first simulated field (Simulation #0) for Scenario 3 $\left(\sigma^2{}_{\text{logk}} = 1\right)$.

Figure 2 shows the heat transport model results and the position of the heat plume after 360 days of injection for realizations #0 of Scenarios 0, 1, 2, and 3.

Table 2 Fixed flow and heat transport model input parameters

Parameter	Symbol	Value	Unit
Total recharge rate	Q	1.84E−4	[m^3s^{-1}]
Pumping rate		3.05E−5	[m^3s^{-1}]
Initial temperature	To	283.15	[K]
Longitudinal dispersivity	α_L	0.5	[m]
Horizontal transverse dispersivity	α_h	0.05	[m]
Vertical transverse dispersivity	α_v	0.05	[m]
Effective molecular diffusion coefficient	D_m	1.838E−6	[m^2s^{-1}]
Bulk density	ρ_b	1961	[kgm^{-3}]
Distribution coefficient	K_d	1.983E−4	[m^3kg^{-1}]
Temperature of injected water	T_{in}	278.15	[K]
Retardation factor	R	2.5	
Density of the solid	ρ_s	2.65	[kgm^{-3}]
Solid specific heat capacity	c_s	830	[$Jkg^{-1}K^{-1}$]
Solid volumetric heat capacity	$\rho_s c_s$	2,200,000	[$Jm^{-3}K^{-1}$]
Water volumetric heat capacity	$\rho_f c_f$	4,185,000	[$Jm^{-3}K^{-1}$]
Saturated aquifer volumetric heat capacity	$\rho_e c_e$	2,716,000	[$Jm^{-3}K^{-1}$]
Solid thermal conductivity	λ_s	3	[$Wm^{-1}K^{-1}$]
Water thermal conductivity	λ_f	0.6	[$Wm^{-1}K^{-1}$]
Saturated aquifer thermal conductivity	λ_e	2	[$Wm^{-1}K^{-1}$]

Table 3 Overview of the simulated scenarios

Scenario	σ^2_{logK}	Mean logK	Nr. Simul.	Description
0	0	−3.86	1	K homogeneous
1	0.1	−3.86	10	k heterogeneous/n, λ constant
2	0.5	−3.86	10	k heterogeneous/n, λ constant
3	1	−3.86	10	k heterogeneous/n, λ constant
4	0.1	−3.86	10	k, n, ρ_b, λ heterogeneous
5	0.5	−3.86	10	k, n, ρ_b, λ heterogeneous
6	1	−3.86	10	k, n, ρ_b, λ heterogeneous

Results show that heterogeneity has an important effect on the shape of the temperature plume, which decreases both in length and width when heterogeneity increases.

To demonstrate the influence of heterogeneous hydraulic conductivity on the cold plume development, simulated temperatures along the observation line (from the injection well downstream, 12 m depth) were plotted.

Figure 3 shows the temperatures along the observation line of Simulation #0 for each scenario. The deviation of the cold plume in this model is mainly caused by predominating advective transport.

To visualize the differences between each of the simulations for a specific scenario, Fig. 4 shows the results obtained for the ten simulations of Scenario 1

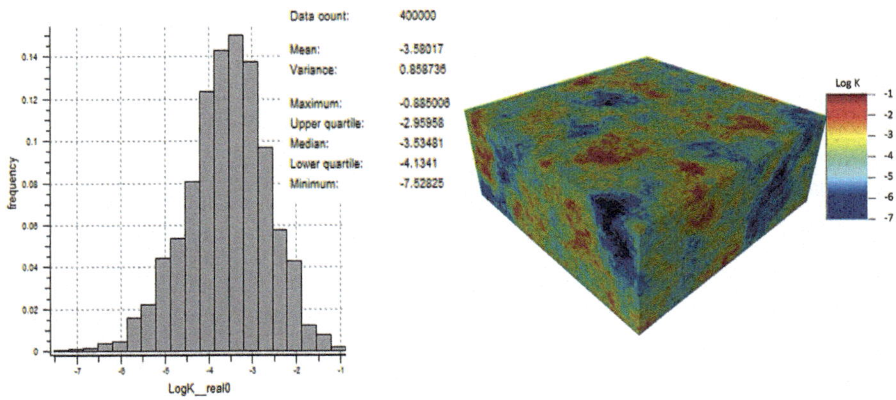

Fig. 1 Histogram and logK field – Simulation #0 of Scenario 3 $\left(\sigma^2{}_{\text{logk}} = 1\right)$

$\left(\sigma^2{}_{\text{logk}} = 0.1\right)$ and Scenario 3 $\left(\sigma^2{}_{\text{logk}} = 0.1\right)$ (n, λ constant) together with their correspondent mean values.

In order to estimate the influence of heterogeneity of all parameters over the heat transport simulation results, a set of ten new realizations for each one of the Scenarios 4–6 were obtained. These scenarios consider heterogeneous distributions of permeability, porosity, and thermal conductivity.

There are some known empirical relationships between hydraulic conductivity and porosity such as those proposed by Kozeny (1927), Carman (1937), Carrier (2003), Schneider (2003), Regalado and Carpena (2004), and Mohnke (2008). In this work, the relation of Busch and Luckner (1993) was used, as it provides a simple linear relationship between hydraulic conductivity and porosity.

Following Eqs. 5 and 6, these new porosity fields were afterward used to compute every other needed parameter, like bulk density (ρ_b) and thermal conductivity (λ) and in a further step the effective molecular diffusion coefficient (D_m).

In order to summarize the results, Fig. 5 shows the simulated temperatures along the observation line for ten realizations of Scenarios 1, 3, 4, and 6.

To quantify the uncertainty in the prediction of temperature distribution caused by heterogeneity of the parameters, the variance of the simulated temperature plumes for the ten simulations was computed for all the scenarios. Figure 6 shows the temperature variances of the ten simulations of Scenarios 3 and 6.

When comparing results obtained between simulations of corresponding scenarios (Scenarios 1 and 4 or Scenarios 3 and 6), it has been found that the temperature differences are higher as heterogeneity increases. Some simulations made for Scenario 3 $\left(\sigma^2{}_{\text{logk}} = 1\right)$ show temperature differences up to 1°K.

Results show that an increasing variance of permeability causes increasing variance in the expected temperature distribution. Scenario 3 $\left(\sigma^2{}_{\text{logk}} = 1\right)$ shows considerably higher variances than Scenario 1 $\left(\sigma^2{}_{\text{logk}} = 0.1\right)$. The highest variances have been found in the first 20 m from the injection well. The changes in flow direction generate an uncertainty in the prediction of the temperature plume.

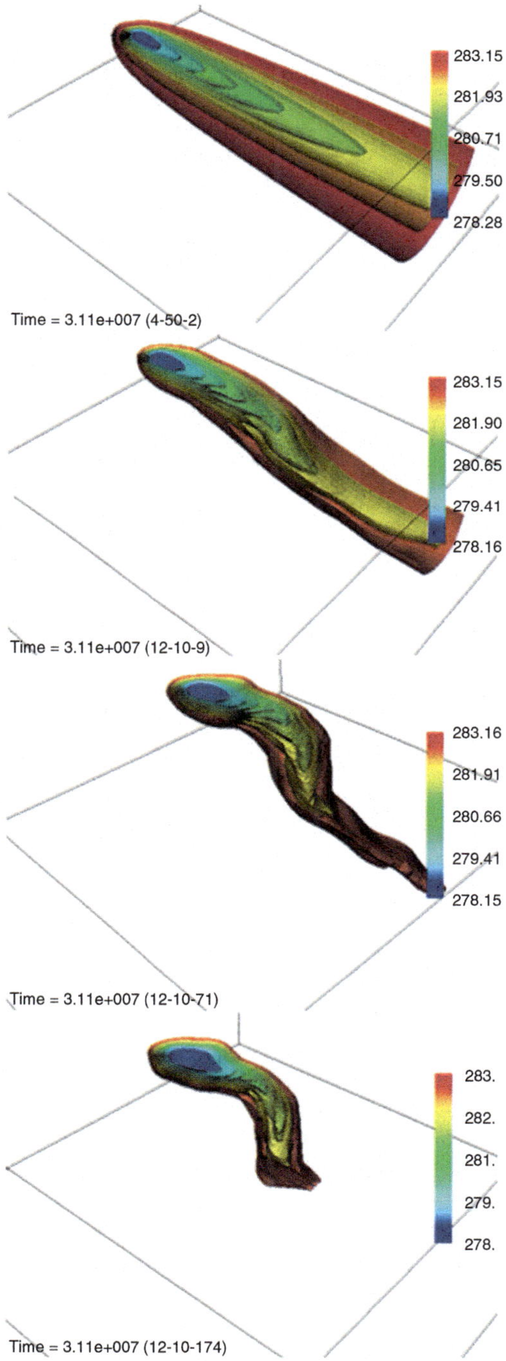

Fig. 2 Heat plume after 360 days of injection for Simulations #0 of Scenarios 0, 1, 2, and 3

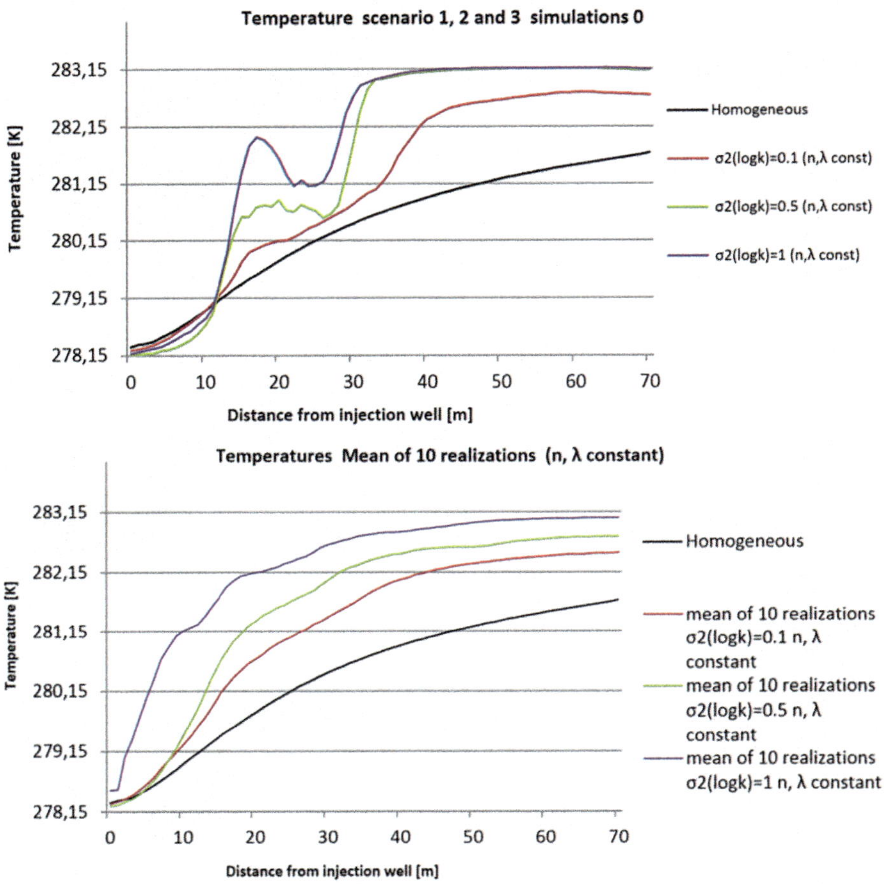

Fig. 3 Simulated temperatures along the observation line for Simulation #0 of Scenarios 0, 1, 2, and 3 (n, λ constant) (*top*) and mean value of ten realizations for each scenario (*bottom*)

If hydraulic gradients were lower and consequently flow velocities decrease, the influence of dispersion would be higher, and changes in thermal conductivity and porosity could have a major effect on the temperature distribution. Further simulations should be made using different hydraulic gradients to investigate the influence of heterogeneity in combination with the hydraulic gradient.

This shows that the effect of the heterogeneity of the porosity n and thermal conductivity λ can be important for heat transport simulation in highly heterogeneous systems. It is possible that, in combination with a low hydraulic gradient, high heterogeneity of the porosity n and thermal conductivity λ could cause a considerable increase in the uncertainty of the predictions of the temperature plume.

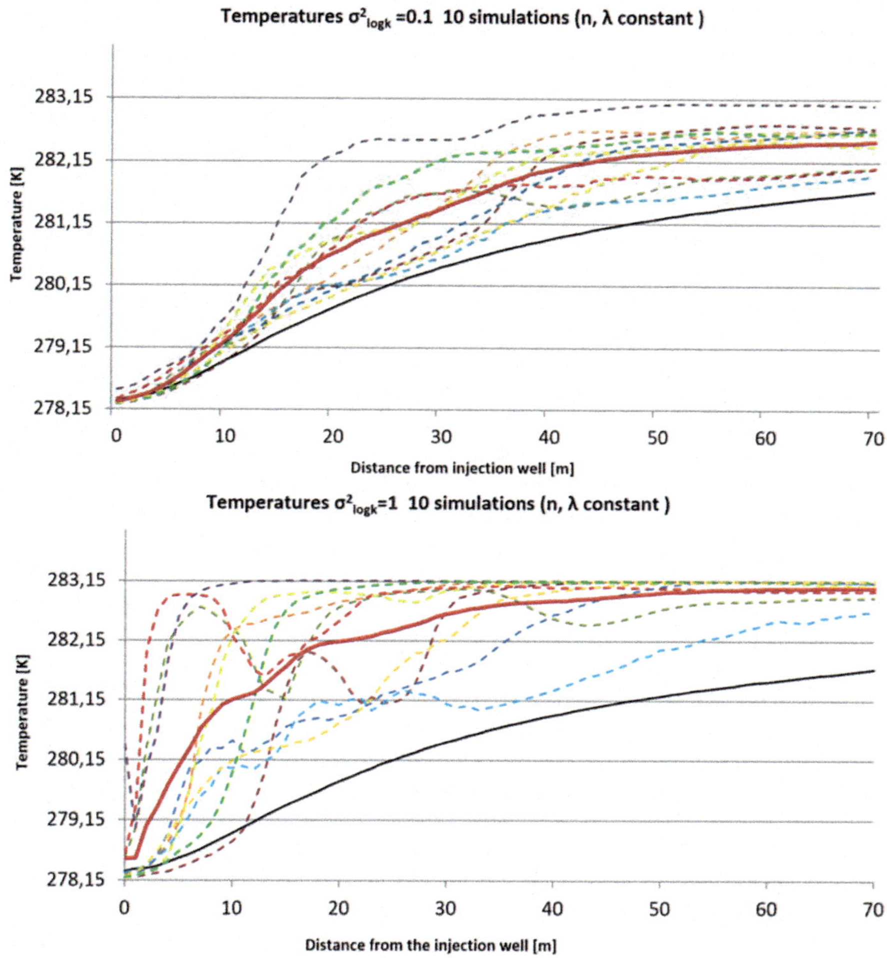

Fig. 4 Temperatures on observation line for ten realizations of Scenarios 1 (*top*) and 3 (*bottom*)

6 Conclusions

This work has been performed in order to analyze the influence of parameter heterogeneity distributions on heat transport in shallow geothermal systems. Heterogeneous distributions of permeability, porosity, bulk density, and thermal conductivity were considered on a set of simulations obtained by sequential Gaussian simulation techniques. The following conclusions can be taken:

- Heterogeneity in the hydraulic conductivity field causes significant changes in the hydraulic head distribution. This affects the flow velocity field which is used for heat transport simulation.

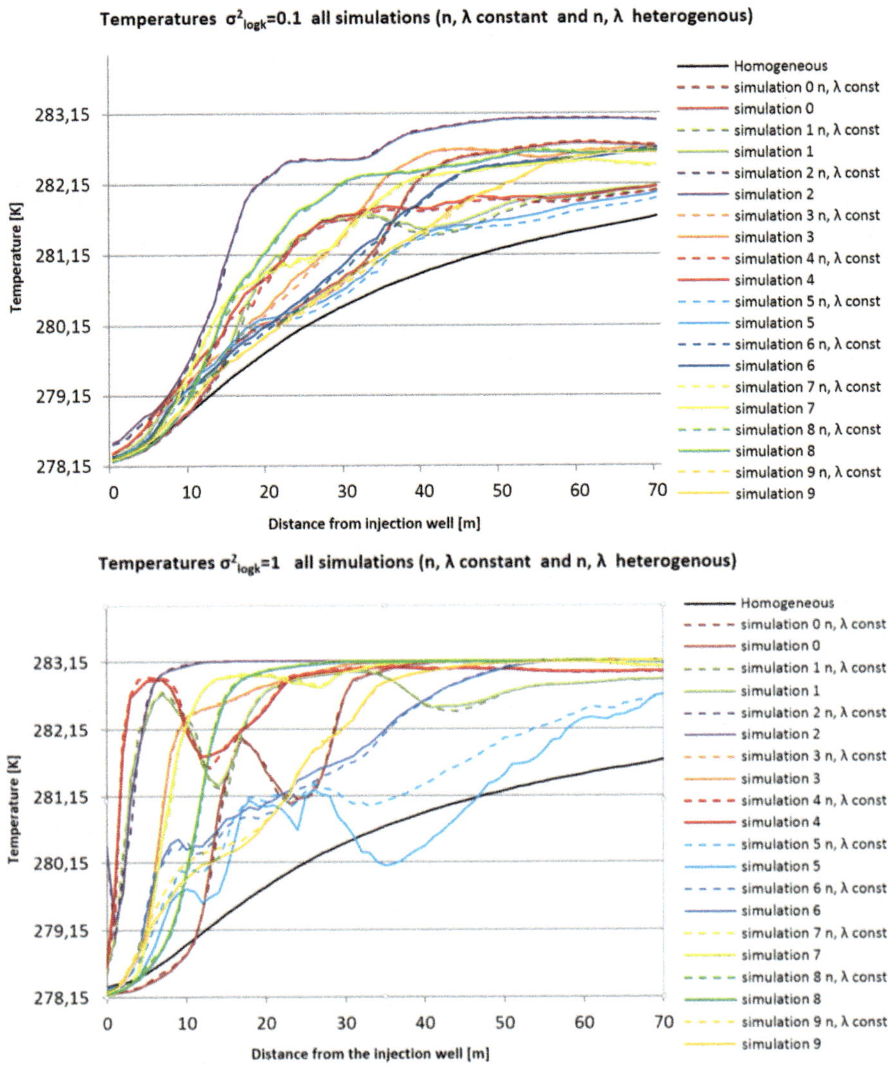

Fig. 5 Temperatures on observation line for ten realizations of Scenarios 1–4 (*top*) and 3–6 (*bottom*)

- Heterogeneity of hydraulic conductivity has a major influence on the shape and development of a temperature plume in a porous media. A high degree of variance in the logarithmic hydraulic conductivity distribution results in a rising variability of the simulated temperature fields and a considerable uncertainty in the prediction of the temperature distribution in an aquifer system. The calculated variances of the simulated temperature fields between are rising significantly with increasing degree of variance in the permeability field.

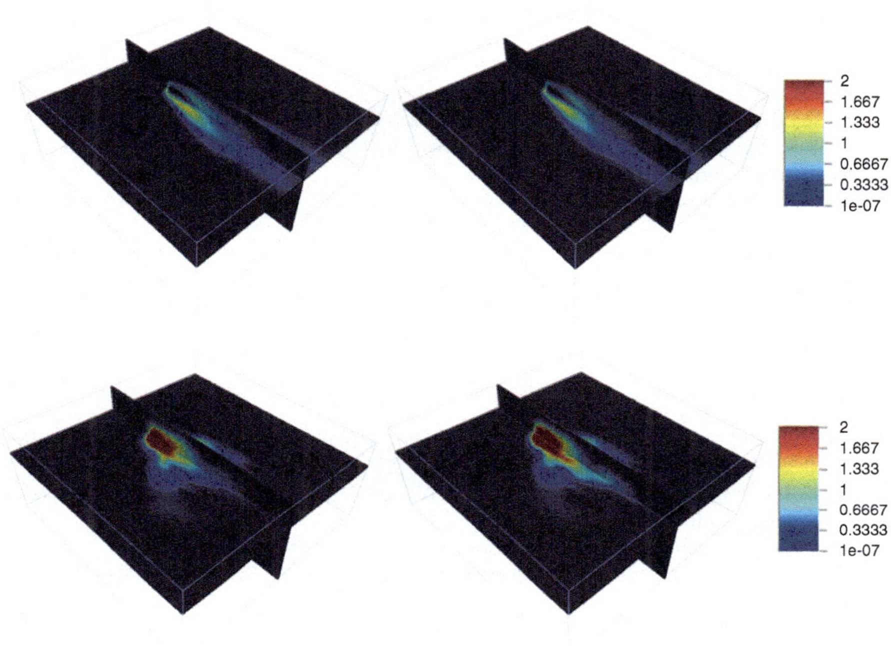

Fig. 6 Temperature variances (layer 13/row 50) of Scenarios 1–3 (*left*) and 4–6 (*right*)

- Heterogeneity in permeability distribution causes changes in the shape and configuration of the temperature plume. The length and width of the plume decreased as the variance of the permeability increases.
- The zones of cold water seem to be more concentrated when the dispersion effect gets less important due to the higher flow velocities in the pore channels. This phenomenon is widely known as "channeling effect."
- The heterogeneity of porosity and thermal conductivity seems to have less impact on modeling results than the heterogeneity of permeability. Low heterogeneity degrees in the porosity and thermal conductivity distribution do not cause important changes in shape and development of the simulated temperature plume. The calculated temperature variances in these scenarios are very small.
- However, in the most heterogeneous case $\left(\sigma^2_{\mathrm{logk}} = 1\right)$, the calculated variance of the simulated temperatures increases significantly. This effect can be even more significant when the hydraulic gradient gets lower and consequently the flow velocities are lower too. In these cases, highly heterogeneous distributions of the porosity n and thermal conductivity λ could cause a considerable increase in the uncertainty of the predictions of the temperature plume.

Acknowledgments This work has been partially funded by the Spanish *Ministerio de Economía y Competitividad* through research project ¿QUIÉN HA SIDO? – Ref: CGL2014-59841-P.

Bibliography

Bridger D, Allen D (2010) Heat transport simulations in a heterogeneous aquifer used for aquifer thermal energy storage (ATES). Can Geotechnik J 47:96–115

Busch K-F, Luckner L (1993) Geohydraulik Band 3 von Lehrbuch der Hydrogeologie. Gebr. Borntraeger, Berlin

Carman P (1937) Fluid flow through a granular bed. Trans Inst Chem Eng 15:150–167

Carrier DW (2003) Good by Hazen; Hello Kozeny-Carman. J Geotech Geoenviron 129:1054–1056

Conde Lázaro E, Ramos Millán A (2009) Guía Técnica de Bombas de Calor Geotérmicas. Gráficas Arias Montano, S.A, Madrid

Ferguson G (2007) Heterogeneity and thermal modeling of ground water. Ground Water 45:485–490

Kozeny J (1927) Über kappilare Leitung des Wassers im Boden. Wien: Hölder-Pichler-Tempsky, A.-G. [Abt.:] Akad. d. Wiss

Kupfersberger H (2009) Heat transfer modelling of the Leibnitzer Feld aquifer, Austria. Environ Earth Sci 59:561–571

Llopis Trillo G, López Gimeno C (2009) Guía Técnica de Sondeos Geotérmicos Superficiales. Gráficas Arias Montano S.A, Madrid

Mendez Hecht J (2008) Implementation and verification of the USGS solute transport code MT3DMS for groundwater heat transport modelling. Eberhard Karls Universität Tübingen, Tübingen

Mohnke O (2008) Pore size distributions and conductivities of rocks derived from magnetic resonance sounding relaxation data using multi-exponential delay time inversion. J Appl Geophys 66:73–81

Molina Giraldo NA (2008) Verification of MT3DMS as heat transport code using analytical solutions. Eberhard Karls Universität Tübingen, Tübingen

Molson JW (1992) Thermal energy storage in an unconfined aquifer: 1. Field injection experiment. Water Resour Res 28–10:2845–2856

Rasouli P (2008) Numerical verification of shallow geothermal models using FEFLOW. Eberhard Karls Universität Tübingen, Tübingen

Regalado C, Carpena R (2004) Estimating the saturated hydraulic conductivity in a spatially variable soil with different permeameters: a stochastic Kozeny-Carman relation. Soil Tillage Res 77:189–202

Remy N, Boucher A, Wu J (2009) Applied geostatistics with SGeMS. A user's guide. Cambridge University Press, Cambridge

Sanner B (2001) Shallow geothermal energy. Justus-Liebig University, Giessen

Schneider JH (2003) New least squares model used for development of permeability-porosity correlation. Poteet Texas

Shuang J (2009) Geostatistical modeling of shallow open geothermal systems. Eberhard Karls Universität Tübingen, Tübingen

Zheng C, Wang PP (1999) MT3DMS documentation and user guide. U.S. Army Corps of Engineers, Washington

Part VI
Environmental Engineering and Sciences

Building a Geological Reference Platform Using Sequence Stratigraphy Combined with Geostatistical Tools

Bernard Bourgine, Éric Lasseur, Aurélien Leynet, Guillaume Badinier, Carole Ortega, Benoît Issautier, and Valentin Bouchet

Abstract This paper presents a methodology that is currently tested at the French geological survey in order to validate drill holes interpretation. Validated drill holes are intended to be included in the future French geological reference platform which is under construction. To validate drill holes, a first subset of high-quality holes is selected. This data is interpreted in terms of geology and a geostatistical analysis is performed. A 3D geological model is built to assess the overall geological consistency. Then the rest of the drill holes is progressively and iteratively validated by geostatistical cross validation. As several thousands of drill holes are to be validated, specific software and workflows have been developed and are presented here.

1 The French Geological Reference Platform

The objective of this work is the setup of a methodology for the validation of drill holes that will be included in the French geological reference platform.

The French geological reference platform (Fig. 1) is one of the major scientific programs of the BRGM (BRGM is the French geological survey). It aims at delivering 3D validated geological data to public institutes, to the scientific community or to private companies, as well as to every citizen.

The data to be validated is of various types: geological maps, well logs, field data, seismic profiles, etc.

B. Bourgine (✉) • É. Lasseur • A. Leynet • G. Badinier • C. Ortega • B. Issautier
BRGM (French Geological Survey), 36009, 45060 Orléans Cedex 2, France
e-mail: b.bourgine@brgm.fr; e.lasseur@brgm.fr; a.leynet@brgm.fr; g.badinier@brgm.fr; c.ortega@brgm.fr; b.issautier@brgm.fr

V. Bouchet
ISTO, Orléans, France
e-mail: valentin.bouchet1@gmail.com

J.J. Gómez-Hernández et al. (eds.), *Geostatistics Valencia 2016*, Quantitative Geology and Geostatistics 19, DOI 10.1007/978-3-319-46819-8_60

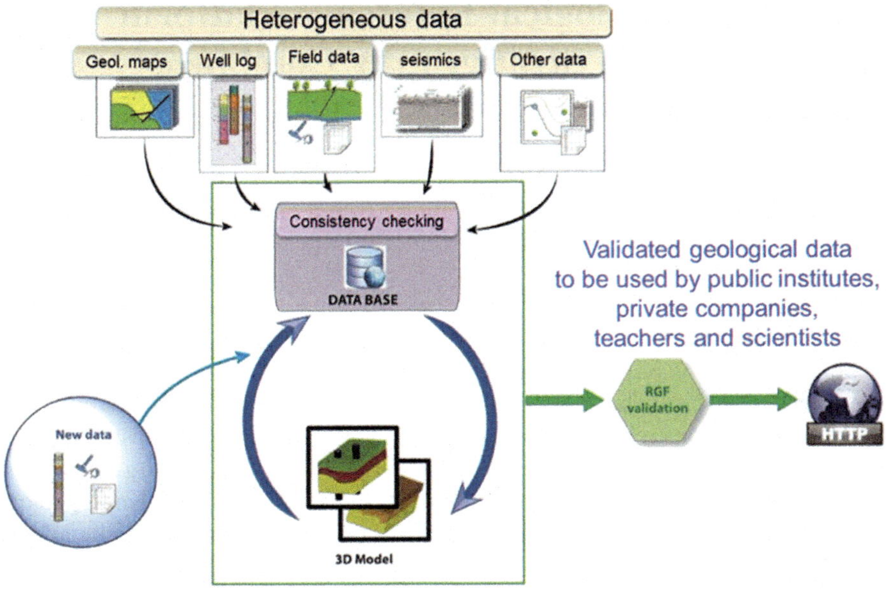

Fig. 1 The French geological reference platform

It is heterogeneous data, stored presently in independent databases. Moreover, it has often not been homogenized, verified, or compared to other data. For this reason this data cannot always be used directly and is not always reliable.

In the geological reference platform that will be built, every type of data will be stored in a normalized database and will share the same data model.

This will allow data comparison and data checking and will facilitate data interpretation and data use.

Regarding data validation, several levels of validation will be defined, but one of the key components will be the validation through the construction of 3D geological models or other types of numerical models such as flow simulation or geophysical inversion.

For new data acquired later, tools will be developed in order to check its consistency toward other existing data and 3D models.

In the end, validated data and models will be made accessible on the Internet.

2 Definition and Objective of Drill Holes Validation

Drill holes are one of the most important subsurface data to be validated in the RGF (french geological reference platform). These drill holes are currently stored in the French national drill holes database (BSS), which actually includes more than 800,000 drill holes. For most of the drill holes, the only provided information is the

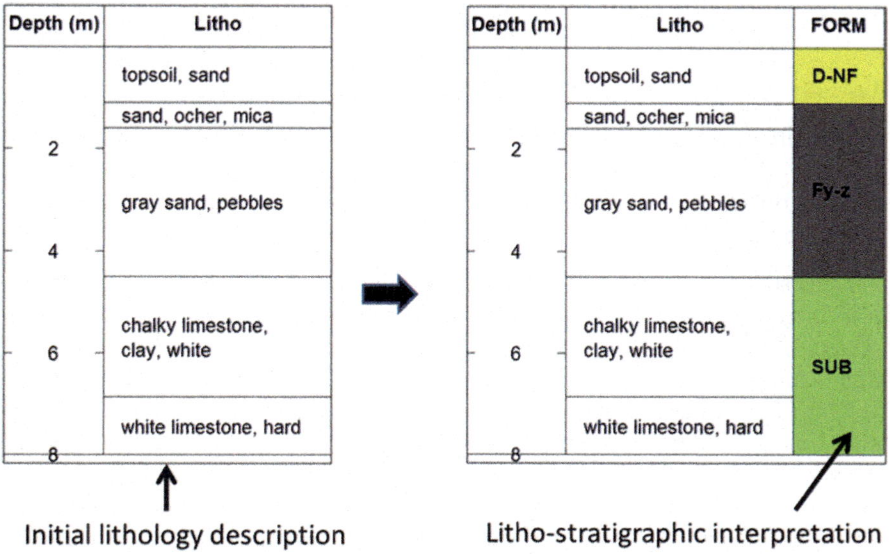

Fig. 2 First litho-stratigraphic interpretation from initial lithology description of drill holes (*D-NF* marine dunes, *Fy-z* quaternary alluviums, *SUB* bedrock)

location of the hole, as well as the description of the lithology encountered at different depths along the hole.

This information in itself is not sufficient to enable direct use of the drill holes to build a geological model. For this it is necessary at least to associate each lithology encountered along the hole with a known geological formation, a formation age, or a geological body.

This is why a preliminary litho-stratigraphic interpretation was made in the past, for a subset of 10 % of the database, approximately 90,000 holes (Fig. 2). This first level of interpretation includes the definition of lexicons concerning standardized formation names, their age, and their lithology, as well as their graphical representation.

However, the quality of this preliminary interpretation can range from very good to very low, depending on the quality of the original drill holes description.

Moreover, the drill holes were often interpreted independently, without taking into account other drill holes around, geological maps, or other information.

This is why a next step of the process is planned in the framework of the RGF. The objective is to reprocess the 90,000 drill holes in order to validate or not the preliminary interpretation in terms of litho-stratigraphy and to record and validate the main geological interfaces crossed by drill holes.

As many drill holes are to be processed and as each drill hole can intersect frequently 10–20 interfaces, it is of course necessary to automatize the work as much as possible, in order not to spend too much time and money.

3 Overview of Drill Holes Validation Methodology

Basically, the methodology is divided into two steps (Fig. 3):

1. A reference set (set *A*) is created using a limited number of high-quality drill holes.
2. Other drill holes of lower quality (set *B*) are validated progressively by comparison to set *A* and are iteratively added to set *A*.

A geostatistical analysis of set *A* data enables data characterization as well as a quantification of uncertainty and helps validating interpretation.

In the first step, the reference set (set *A*) is created by selecting a loose network of high-quality drill holes, i.e., holes owning a well log, generally a gamma ray. These well logs are correlated using sequence stratigraphy (Homewood et al. 1999), which allows an accurate litho-stratigraphic interpretation. A geostatistical analysis of interpreted data is performed. The objective of this analysis is to spatially characterize this data, detect potential errors (by cross validation), and establish the variogram model then used. 3D models are built in order to quantify uncertainty and validate the set *A* by checking the overall consistency of geological bodies with respect to geological knowledge. Isopach maps derived from the model contribute to this validation procedure.

During step 2 all other drill holes (of lower and unknown quality) are checked. A geostatistical cross validation technique is used to check their consistency as compared to the reference set *A*. Drill holes that are in good agreement with reference data are considered as validated and added to the reference set. Generally, the geostatistical model itself is not challenged because it is supposed to have been validated at step 1. However, in case of high inconsistency, it can be reviewed. Then the process is iterated to look for other drill holes to validate. 3D models are updated consequently and results verified (Fig. 3).

4 Application of the Methodology on Real Data

The methodology of validation of drill holes will be now presented in more details and illustrated on two real datasets, one in the Paris basin and the other in the North Aquitanian basin (France).

4.1 First Step: Analysis of the Reference Set

The first example shown in Fig. 4 is a 120 * 80 km area around Paris. In this area, approximately 7000 drill holes are to be validated.

Fig. 3 Overall methodology

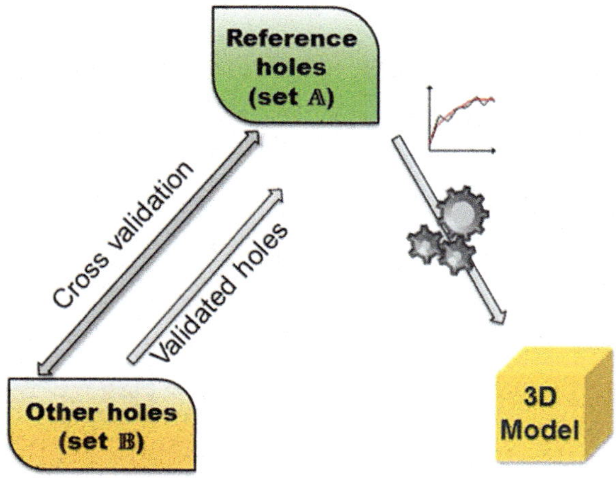

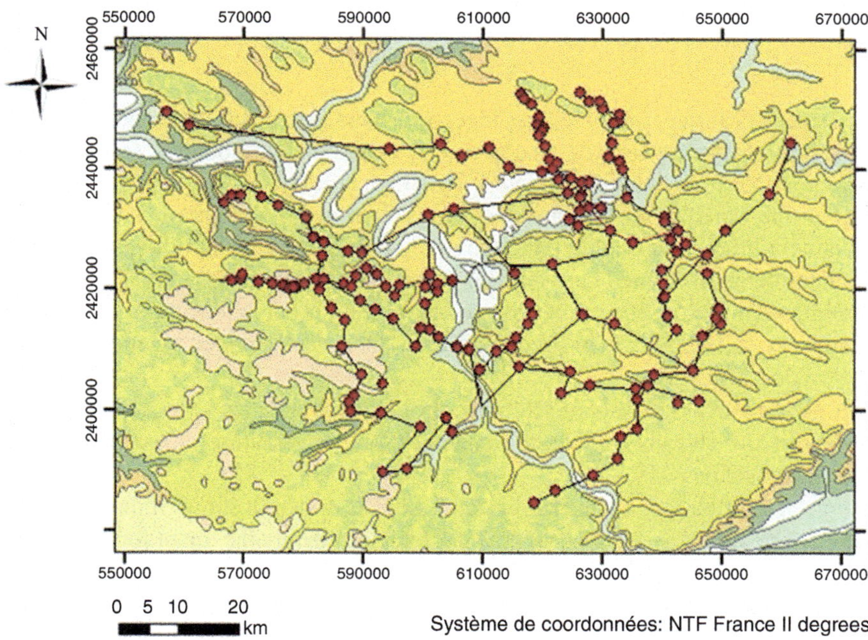

Fig. 4 Dataset in the tertiary basin around Paris, France

Among them, 168 reference drill holes with a gamma ray are chosen for the reference set (2.5 %).

These reference drill holes are grouped along vertical transects (Fig. 4) that are aligned along the major geological axes. Drill holes and well logging are displayed along these transects (e.g., Fig. 5 in the Aquitanian basin). The geologist can interpret well logs using sequence stratigraphy concepts. Basically, the geologist

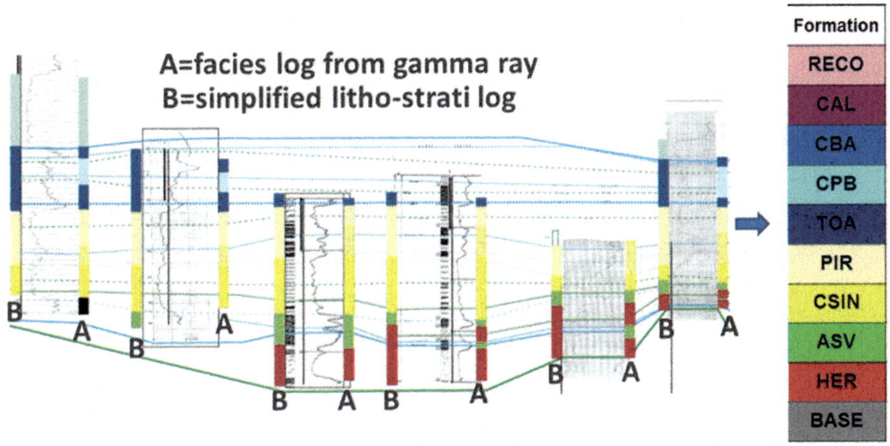

Fig. 5 Interpretation of drill holes using gamma ray logging and sequence stratigraphy concepts. Example in the Aquitanian basin

has to identify surfaces or time lines that correspond to depositional context change. This allows restoring correctly the geometry of sedimentary bodies. This will also make possible facies simulation in each geological body.

Once the well logs are interpreted in terms of sequence stratigraphy, it is possible to define a litho-stratigraphic pile along the wells (Fig. 5). The litho-stratigraphic pile interpretation defined here (shown at the left of each gamma ray (Fig. 5)) is a simplification of sequence stratigraphy interpretation (shown at the right of each gamma ray on Fig. 5), where several different facies are grouped together between two "major" time lines.

A geostatistical analysis of reference drill holes is then performed. The studied variable is the elevation of the top or base of each formation. It can also be the formation thickness.

For each formation, a "formation status map" is drawn. It shows if the formation is present or absent and if the drill hole has intersected only the top of formation, the base of formation, or both top and base (Fig. 6).

For example, in Fig. 6 the green "#" symbols correspond to a formation gap. When isolated and surrounded by holes intersecting the formation, it can indicate an error of interpretation or a coordinate error. Otherwise, these "#" symbols show where the formation has not deposited and help the geologist to check his interpretation.

From a geostatistical point of view, the variograms of tops and bottoms are computed and fitted. Generally, a polynomial drift (linear or quadratic) is considered, and the fitted variogram is that of the residual after removal of a global drift (fitted by least squares to all data – which is an approximation of the true residual). Figure 7 shows an example of variogram of the residual computed from high-quality data. The error in the interpretation of logging is of the order of few

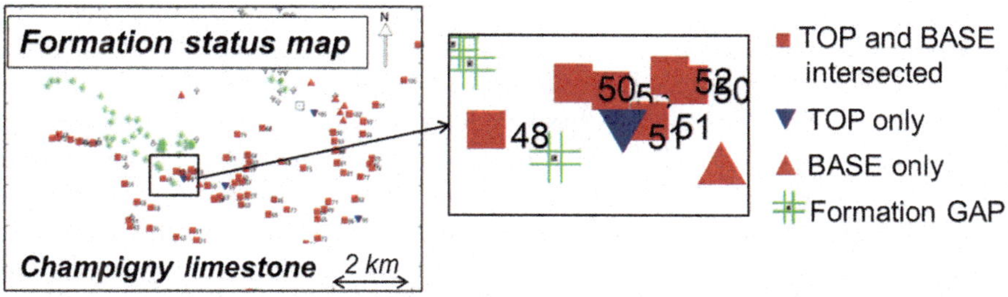

Fig. 6 Formation status map. Example for Champigny limestone. Paris basin

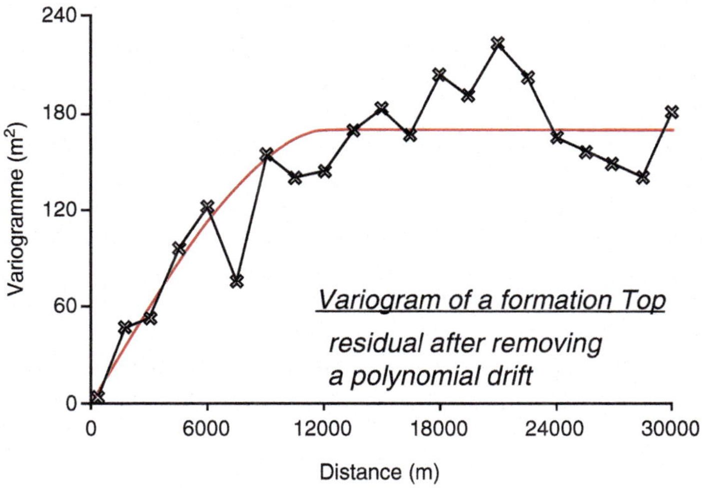

Fig. 7 Example of the variogram of the top of a formation (residual after removing a polynomial drift)

centimeters, and the nugget effect can be neglected. It would not be the case with destructive holes for which the identification of formation change, based on cuttings, can be affected by an error of 1–2 m.

Then a cross validation (based on kriging) is then performed. At this step, the coefficient of the drift function is computed locally by kriging using data of the neighborhood, so the local drift (and then the local residual) may differ from the global one used to fit the variogram of the residual. Ideally, the new residual should be recomputed and the process should be iterated as shown by Hengl et al. (2007). This is not done in our case. Consequently, there is a risk of error due to a nonoptimal geostatistical model. However, in practice the cross validation of the reference data gives very satisfactory results in terms of error and of normalized error, and outliers found in data can generally be attributed to errors in the data. In our case the approximation seems acceptable and the geostatistical model is validated.

During this cross validation, inequality constraints given by the drill holes that have not crossed or reached a given formation are taken into account. For example, if a drill hole does not reach a given formation, this gives an upper bound value for the elevation of the top of this formation. When the top of this formation is interpolated from other drill holes where it has been observed, the interpolated value at the location of the drill hole is compared to the upper bound. A warning is sent if the interpolated value is above the upper bound and data can be verified. Note that at this step a standard kriging is used instead of a "kriging taking into account inequalities" (Freulon and De Fouquet 1993; Abrahamsen and Benth 2001) because the objective is first to check data consistency and detect potential error (even in the inequality constraint themselves).

Other information like the DTM and the geological map are also taken into account. For example, the geological map, if correct, virtually gives infinity of inequality constraints: if at a given point the formation *A* is outcropping that gives constraints on the elevation of other geological formations.

The whole litho-stratigraphic pile has to be processed. If the litho-stratigraphic pile in the area includes 20 formations, this procedure has to be repeated for all the formations.

To make this work possible in a consistent and largely automated way, specific software has been developed by the BRGM: GDM MultiLayer (Geological Data Management, http://gdm.brgm.fr) (Bourgine et al. 2008; Bourgine 2015). This software and associated algorithms include the management of gaps due to erosion or formation pinching.

At last a preliminary 3D geological model is built using the reference dataset. This helps checking the overall interpretation and possibly to correct it.

4.2 Second Step: Validation of Remaining Drill Holes

Once the reference dataset is validated for all the formations, other drill holes (dataset *B*) are compared to the reference dataset.

For this, we again apply a cross validation technique. For example, if we consider the top of a formation, and if we denote *TA* as the top of the formation known from reference drill holes (set *A*) and *TB* the top of the same formation measured on drill holes dataset *B*, we estimate *TB* from drill holes belonging to dataset *A* and compare estimated value *TB** to the true value *TB*. Thus, we can compute the estimation error (*TB**-*TB*) and the normalized error if we divide by the kriging standard deviation.

The *B* holes, where we do not know the true value *TB*, but an inequality constraint on *TB*, can also be tested.

At last the software we have developed displays automatic maps highlighting outliers and potential errors. By clicking on the map, the geologist can get vertical cross sections showing the potential anomaly.

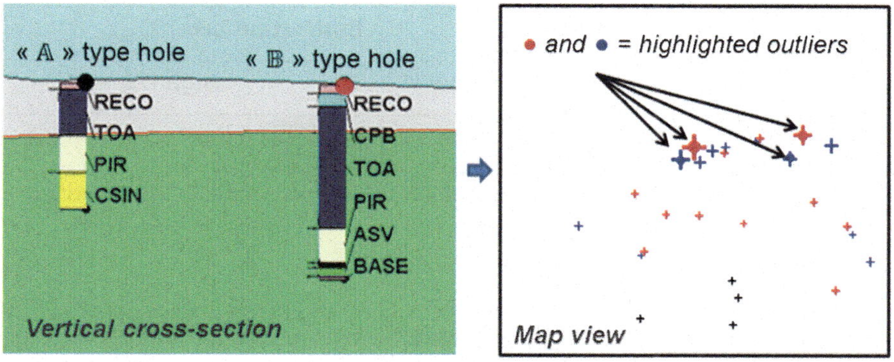

Fig. 8 Validation of B holes along a cross section. Example for the top of "TOA" formation (dark-blue formation along holes)

For example, on the cross section in Fig. 8 (data from the Aquitanian basin), the interpolation of the base of the dark-blue formation (named *TOA*) using reference holes *A* is not consistent with *B* hole where the base of dark-blue formation is intersected much lower. Either drill hole *B* is correct (e.g., in the case of a fault or a fold between the two holes) or drill hole *B* is not located here (not GPS-tagged drill hole, coordinate input error, database error) or has not been correctly interpreted. Typically, this hole has to be validated or corrected by manual check.

In fact all the *B* holes are not checked systematically, but only when necessary. In a first step, *B* holes which are consistent with reference holes and with the geological model are validated automatically. These new validated holes are then added to the reference set *A*, and the process is reiterated with nonvalidated holes remaining in set *B*. The iterative process ends when there are no more automatically validated holes in set *B*. The holes remaining in set *B* can then be checked and corrected manually. At each step, the variogram can (on demand, if necessary) be recomputed and remodeled, especially to improve fitting for the short-range components.

For the selection of automatically validated holes, we use the table shown in Fig. 9. Drill holes for which the estimation error is low and the normalized error is lower than 2 (green cell in Fig. 9) are validated automatically and added to the reference set *A*. The threshold between "low" and "high" estimation error is fixed arbitrarily and depends on geological context, as well as on the lithology contrast between successive formations, which can make the drill holes interpretation easy or difficult.

The threshold value of 2 for the normalized error corresponds to a 95 % confidence interval if we assume that the kriging error distribution is Gaussian. It is well known that kriging error is generally not Gaussian. However, we have verified that it is a reasonable approximation in our case.

		Estimation error	
		Low $\lvert T^*_{\mathbb{B}} - T_{\mathbb{B}}\rvert$	**High** $\lvert T^*_{\mathbb{B}} - T_{\mathbb{B}}\rvert$
Deviation to the model	$\lvert T^*_{\mathbb{B}} - T_{\mathbb{B}}\rvert/\sigma_K < 2$	hole $\mathbb{B}$ validated	hole $\mathbb{B}$ to be verified
	$\lvert T^*_{\mathbb{B}} - T_{\mathbb{B}}\rvert/\sigma_K > 2$	hole $\mathbb{B}$ to be verified	hole $\mathbb{B}$ to be verified

Fig. 9 Criterion for validating or not drill holes of type B

All the holes that are not automatically validated are kept for further checking, and the process is iterated using reference *A* holes plus the validated *B* holes, as soon as there is no new auto-validated *B* hole.

The *B* holes that are not automatically validated can be checked manually one by one. As it is not always possible to check all drill holes, we check preferentially drill holes that are located in critical zones, for example, that intersect important aquifers. Outliers of the cross validation are also checked because often they reveal a major error in the interpretation or in the data.

Another criterion is to select the next candidate for validation in areas where the kriging error map obtained from reference dataset shows large errors and where present *B* holes are that have not yet been validated.

For example, on the kriging standard deviation map in Fig. 10, reference holes *A* are in black and *B* holes waiting for validation in red. We will try to validate first *B* holes located in the yellow and orange areas in order to get a maximum standard deviation of 10–12 m.

At each time new holes are validated, it is possible to rebuild the whole geological model automatically.

Cross sections and isopach maps can be derived from the model and are used for verification and for the final validation.

4.3 Automatization

As many drill holes and many formations are to be validated, it is necessary to automatize the work as much as possible.

This is why R scripts are first used for methodology testing.

Once the methodology seems to work well, it is implemented in an in-house software (GDM software, http://gdm.brgm.fr) to be used by geologists who are not specialists in geostatistics.

One of the strengths of this software is that it is coupled with the geological litho-stratigraphic pile and includes many consistency checks. It includes tools for managing the results of cross validation (*set A* with *set A*, *set B* against *set A*), building the model, generating and updating cross section or isopach maps, and

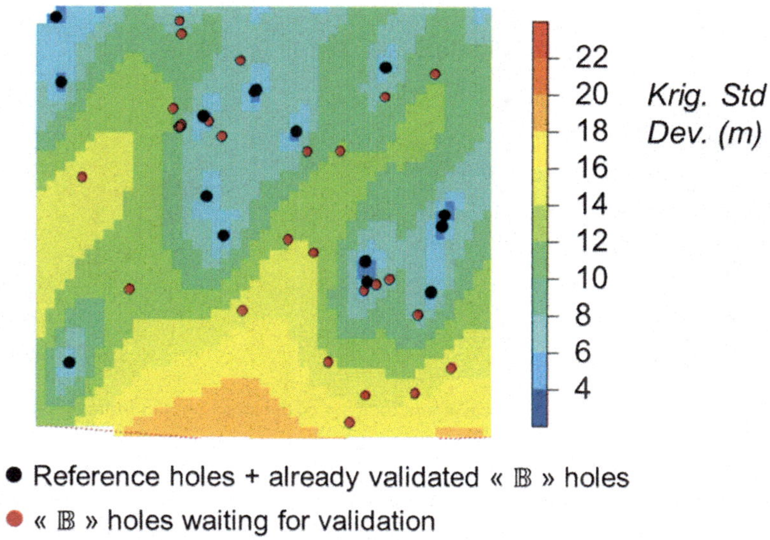

Fig. 10 Use of the kriging standard deviation map to define the next B holes to be validated

managing gaps in case of erosion or formation pinching. However, it does not yet handle variogram automatic fitting nor drift degree identification. This work is still left to the geologist and requires a short training.

This software is able to handle several thousands of drill holes and work on large areas. But in practice, it is preferable to work on a limited area, not for performance reasons, but rather to consider relatively homogeneous areas where the drift degree is constant or where the geological context does not change much.

4.4 Results

Examples of results are given for two areas: the North Aquitanian basin and the Paris basin.

In the North Aquitanian basin (Fig. 11), the geological context is a carbonate ramp, and we looked at Jurassic formations. In this area, 117 reference holes were selected (set *A*), and 10 formations of the Jurassic were considered.

For the Toarcian, which is one of those ten formations, 60 drill holes were to be validated (set *B*).

Two thirds of these drill holes could be automatically validated by the procedure, so only one third was left for manual verification.

For control purposes, they were all checked: four were discarded, seven were validated, and nine were erroneous but could be reinterpreted.

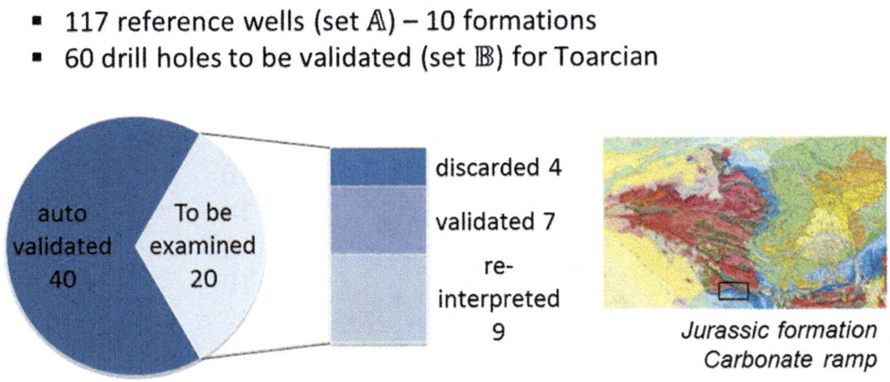

Fig. 11 Result in North Aquitanian basin and location of the study area in France

With the new validated holes, we obtained a reduction of 30 % of the average kriging standard deviation, compared to the situation with only the set *A* holes.

In such simple geological context, drill holes are of good quality and can be easily validated.

The second area where we tested the methodology is in the Paris basin, next to Paris city (Fig. 12). Here we looked more precisely at Eocene formations which deposited in shallow sea or lacustrine context. In this area there are 7000 holes to validate, represented by the black dots in the map. 168 drill holes with gamma ray were interpreted and were used as the reference set (Fig. 4). They are shown in red on the map.

Among the 7000 holes, only 137 intersect the Champigny formation that was selected for the test, and nearly 50 % of holes could be validated automatically. They are shown in blue on the map. These holes are close to reference drill holes because we introduced a low value for the maximum allowable estimation error.

In this case where the geological context is more complex, we can auto-validate only 50 % of holes for the Champigny limestone formation. As there are 19 other formations to validate, the work takes more time, but is facilitated by a semi-automatized workflow.

5 Conclusion

A methodology has been set up in order to enable drill holes validation. It is based on geological concepts mixed with basic geostatistical tools.

This methodology helps finding quickly consistent drill holes and discarding erroneous ones.

As we are considering large datasets and many geological formations, automatic and semiautomatic tools have been developed to save time and ensure repeatability.

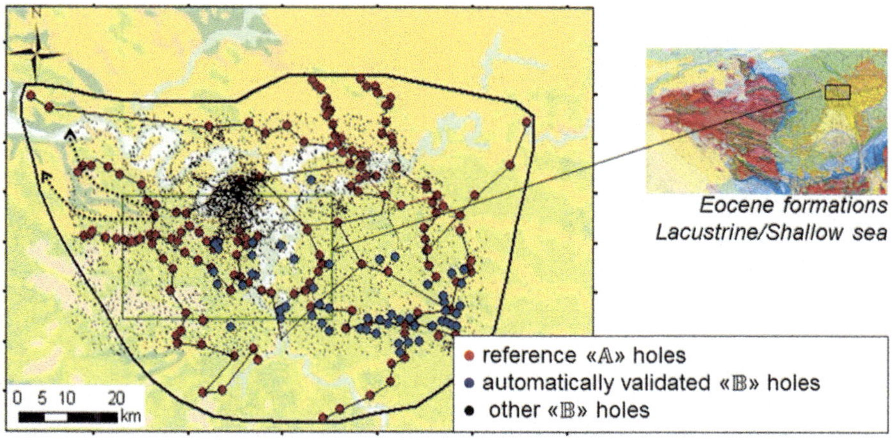

Fig. 12 Result in Paris basin. Case of the Champigny limestone formation

These tools have been made accessible to the geologists in charge of validation and who are not geostatisticians.

Further work to be done concerns (1) improvement in the automation (e.g., automatic looping on formations and automatic database management when adding validated *B* holes to dataset *A*), (2) automatic variogram fitting and drift identification, and (3) finding a way to manage drill holes that have not been automatically validated nor discarded by the present procedure. These holes are kept as "not yet validated" but are not rejected. They are candidate for a further validation, but it would be useful to assign to these drill holes some kind of probability index of being consistent with present knowledge and develop other methods to assess their quality.

References

Abrahamsen P, Benth FE (2001) Kriging with inequality constraints. Math Geol 33(6):719–744

Bourgine B (2015) MultiLayer, Manuel de référence, version 2014. BRGM Report RP64115-FR

Bourgine BLM, Lembezat C, Thierry P, Luquet C, Robelin C (2008) Tools and methods for constructing 3D geological models in the urban environment. The Paris case. In: Ortiz JM, Emery X. GEOSTATS 2008. Santiago, pp 951–960

Freulon X, De Fouquet C (1993) Conditioning a Gaussian model with inequalities. In: Soares A (ed) Geostatistics Troia 92, vol 1. Kluwer Academic, Dordrecht, pp 33–51

Hengl T, Heuvelink G, Rossiter D (2007) About regression-kriging: from equations to case studies. Comput Geosci 33:1301–1315

Homewood P, Mauriaud P, Lafont F (1999) Best practices in sequence stratigraphy for explorationists and reservoir engineers. Bull centre Rech Elf Explor Prod Mem 25:81, pp 50

Constrained Spatial Clustering of Climate Variables for Geostatistical Reconstruction of Optimal Time Series and Spatial Fields

Peter Dowd, Hong Wang, Eulogio Pardo-Igúzquiza, and Yongguo Yang

Abstract The purpose of this work is to present a new methodology for identifying geographical regions within which the climatic behaviour of a meteorological variable is coherent. We have chosen temperature as the variable of interest, and thermal coherence is defined here in the sense of having a strong (negative) correlation between terrain altitude and temperature. An improved method of constrained spatial cluster analysis is described in the form of a new constrained clustering algorithm. The methodology includes spatial bootstrap statistical tests to provide a more realistic measure of the uncertainty of the coefficient of correlation together with a spatial test of the correlation of residuals. The results are used as optimal estimates of areal temperature averages. The methodology is illustrated by applying it to the annual mean temperature measured at 1220 temperature stations across Spain.

1 Introduction

Changes in temperatures are perhaps the most common and reliable indicator of climate change or global warming (Morice et al. 2012; Robeson 1994). Temperature is usually measured at a finite number of sampling locations over regional or national networks of weather stations. Temperature measurements are affected by

P. Dowd (✉)
University of Adelaide, Adelaide, Australia
e-mail: peter.dowd@adelaide.edu.au

H. Wang • Y. Yang
School of Resources and Geosciences, China University of Mining and Technology, Xuzhou, Jiangsu Province 221116, China
e-mail: wanghongcumt@hotmail.com; ygyang88@hotmail.com

E. Pardo-Igúzquiza
Geological Survey of Spain, Ríos Rosas 23, 28003 Madrid, Spain
e-mail: e.pardo@igme.es

J.J. Gómez-Hernández et al. (eds.), *Geostatistics Valencia 2016*, Quantitative Geology and Geostatistics 19, DOI 10.1007/978-3-319-46819-8_61

many factors, including latitude, altitude, continentality, atmospheric circulation and proximity to the sea. For a geographical area of relatively small size, altitude is the most significant variable and explains most of the structural variance in spatial temperature. Thus, temperature and altitude are usually very well correlated. The correlation is usually linear and negative so that temperature decreases as altitude increases with, in general, a mean gradient of 0.6 °C per 100 m of altitude (Viers 1975). However, for large areas (several degrees of latitude), the many other factors listed above may affect the temperature in such a way that the relationship between altitude and temperature is much weaker because, for example, different climate patterns and climate zones are merged within the area. This will reduce the accuracy and reliability of estimated mean temperature values using geostatistical interpolation estimators, such as regression kriging or kriging with an external drift (Carrega 1995). It is thus useful to be able to identify zones in which the thermal behaviour is coherent in the sense that the correlation between altitude and temperature is as strong as possible. Cluster analysis is highly suited to this purpose.

Clustering algorithms (Thacker and Lewandowicz 1997; Gerstengarbe et al. 1999; DeGaetano 2001; Unal et al. 2003; Hoffman et al. 2005; Fereday et al. 2008; Zhou et al. 2009; Mahlstein and Knutti 2010; Tang et al. 2012; Zscheischler et al. 2012; Stooksbury and Michaels 1991) have been used for similar, but not identical, problems to the one dealt with here. In this work a new clustering method is proposed in which there are two novel aspects. The first is the recognition that, although the problem is outside the framework of classical cluster analysis, it is a particular form of a constrained cluster analysis problem. The second is accounting for the spatial correlation of the data when calculating the correlation statistics and in testing the spatial correlation of the residuals of the regression of temperature on altitude for the clusters. The regions resulting from the cluster analysis are not interpreted climatologically and are used solely to obtain optimal estimates of mean areal temperatures. In addition, the regional clusters implicitly take account of secondary variables such as latitude, longitude and proximity to the sea.

The remainder of this paper proceeds as follows. A brief description of approaches employed in this study is given in Sect. 2. A real case study is presented in Sect. 3. Section 4 discusses the experimental results, and finally Sect. 5 summarises the conclusions of the work.

2 Methodology

Classical cluster analysis identifies groups of objects that are similar. It does so by maximising the similarity of objects (for the purposes of this paper, temperature measurements from weather stations) within a group and maximising the dissimilarity of different groups of objects (Gordon 1981). There are two broad types of clustering methods: hierarchical clustering and non-hierarchical clustering. Among the non-hierarchical clustering algorithms, the most widely used is the *k*-mean algorithm. The similarity of objects is usually defined in terms of a distance (e.g. Euclidean, Mahalanobis) and defined according to the measured characteristics of the objects.

For the problem addressed in this paper, the first difference with respect to classical clustering is that, instead of defining the similarity measure as a distance between the objects of a group, it is an objective function to be maximised or minimised. The second difference is that the problem in this paper is a case of constrained clustering in which a contiguity constraint restricts the sets of solutions that are allowed (Gordon 1996), i.e. the objects in each group must comprise a spatially contiguous set. Thus, given a number of groups, an object can change its membership from group A to group B if two requirements are met: (*i*) groups A and B are contiguous and (*ii*) the value of the objective function is improved. Clustering temperatures into thermally homogeneous regions can thus be seen as a contiguity-constrained optimisation problem.

The first issue is the definition of clusters and contiguity. The locations of the weather stations are used as the seeds of a Voronoi tessellation of the geographic space covered by the stations. A cluster, or group, of weather stations (or of the corresponding temperature measurements) is a union of contiguous Voronoi cells, and the boundary of the cluster is the outermost sequence of its constituent cell boundaries. Two clusters are contiguous if they share a boundary or a point on their boundaries. A member, or object, belonging to cluster A is contiguous with cluster B if its Voronoi cell shares a boundary with the Voronoi cell of any member of cluster B. These definitions are used in the application of the contiguity constraint.

In the proposed algorithm for contiguity-constrained classification of a set of N objects (weather stations), the algorithm starts with an exhaustive classification into M groups. The manner in which this starting classification is obtained is described below. The classification is exhaustive in the sense that the N objects have been classified, and each belongs to one of the M groups.

For any given configuration of groups $(G_1, \ldots, G_M)$, the objective function, OF $(G_1, \ldots, G_M)$, of the configuration is defined by

$$OF(G_1, \ldots, G_M) = \sum\nolimits_{i=1}^{M} n_i \rho_i^* \tag{1}$$

where n_i is the number of objects that belong to the ith group and ρ_i^* is the value of the modified Pearson product-moment correlation coefficient:

$$\rho_i^* = \rho_i + 1.96{*}\mathrm{SE}(\rho_i) \tag{2}$$

where ρ_i is the estimated Pearson correlation coefficient of the ith group and $\mathrm{SE}(\rho_i)$ is the associated standard error. The correlation between altitude and temperature is negative, and, thus, from Eq. 2, $\rho_i < \rho_i^*$. The modified coefficient, ρ_i^*, can be used instead of ρ_i as an experimental measure of correlation between altitude and temperature that accounts for the size of the group, i.e. the uncertainty of the estimated value of the correlation coefficient as represented by the standard error in Eq. 2. The value $1.96 * \mathrm{SE}(\rho_i)$ is the lower bound of the 95 % confidence interval and has been chosen as a conservative value for including cluster size in the comparison of the correlation coefficient of two different clusters. The estimated Pearson product-moment correlation coefficient, ρ_i, for the ith group is

$$\rho_i = \sum\nolimits_{j=1}^{n_i} \left(T_{ij} - \bar{T}_i\right)\left(H_{ij} - \bar{H}_i\right) / \left(\sqrt{\sum\nolimits_{j=1}^{n_i} \left(T_{ij} - \bar{T}_i\right)^2} \sqrt{\sum\nolimits_{j=1}^{n_i} \left(H_{ij} - \bar{H}_i\right)^2}\right) \tag{3}$$

where T_{ij} is the temperature at the jth station of the ith group, H_{ij} is the altitude of the jth station of the ith group and $\bar{T}_i$ and $\bar{H}_i$ are the respective means of the ith group. Weather stations are thus defined by their pair (T, H) of temperature and altitude. Stations are also defined by their geographical co-ordinates (X, Y), which are implicitly included in the proposed methodology by the contiguity constraint.

It is obvious that

$$OF(G_1, \ldots, G_M) \geq -N \tag{4}$$

with the minimum value $-N$ being unattainable in practice because it would imply the unlikely case of a perfect (negative) correlation between altitude and temperature for a given partition ($G_1, \ldots, G_M$).

The standard error, SE(ρ_i), of the estimated Pearson correlation coefficient may be calculated by a parametric method such as Student's t-distribution or by using a non-parametric method such as the bootstrap. The advantage of the latter is that it works when the sampling distribution of the correlation coefficient is asymmetrical and the data are (spatially) correlated, as is the case in the application described here.

The clustering process is applied to each set of annual temperatures. In the clustering process, there are two permitted operations: coalescence of two groups and moving an element from one group to another. Both operations use the definitions of contiguity given above.

Two groups G_i and G_j will coalesce to form a new group, G_k, if:

1. The two groups, G_i and G_j, are contiguous.
2. The value of the objective function improves.

An object, o_k, that belongs to group G_i can move to group G_j if:

1. The object o_k is contiguous with the group G_j.
2. The value of the objective function improves.

Finally, the constrained clustering algorithm comprises the following steps:

(i) Start with a partition of M clusters, where M is greater than the expected optimal number of clusters. For example, $M = 100$ is used in the case study. Select at random M stations from the total N stations ($N > M$). Call these locations the seed stations. Next, each of the N stations is assigned to the nearest seed station, and then M groups that comply with the contiguity constraint are formed. This is the initial random partition with M groups:

$$\{G_1, G_2, \ldots, G_M\} \tag{5}$$

for which the initial objective function value is given by Eq. 1.

(ii) Each group, $\{G_i\}$, is taken in turn, and the closest contiguous group (in terms of geographical distance) $\{G_j\}$ is found. The closest group is the one that contains an object that has the shortest geographical distance (calculated using geographical co-ordinates (X, Y)) to a member of the group $\{G_i\}$. Obviously, the two objects and the two groups are contiguous. Three operations are then tried. Operation 1 ($O1$): if either of the two groups has less than n_{min} objects, the groups are merged; Operation 2 ($O2$): try to merge both groups; Operation 3 ($O3$): try to move an object from one group to the other. The purpose of $O1$ is that if either of the two groups has a small number of elements, defined by the threshold value n_{min} (e.g. $n_{min} = 20$ is used in the case study), there is no point in calculating a very unreliable correlation coefficient, and thus the groups are merged to create a larger group. The meaning of $O2$ is that of merging two groups if the resulting group is better than the worse of the two groups. In other words, Operation 2 consists in merging group $\{G_i\}$ and group $\{G_j\}$ into a single group if the following criterion is met:

$$\rho_{ij}^* \leq \max\left(\rho_i^*, \rho_j^*\right) \tag{6}$$

where ρ_{ij}^* is the modified Pearson correlation coefficient, defined in Eq. 2, for the merged group $\{G_{ij}\} = \{G_i\} \cup \{G_i\}$. Note that maximum operator is used in Eq. 6 because the correlations are negative, and groups with the largest possible negative correlation are required.

(iii) If the operation of merging the two groups fails (because Eq. 6 is not satisfied), then the operation of moving an element to the closest group is tried. If the pair of elements $\{o_{ik}, o_{jl}\}$ is the two closest elements between groups $\{G_i\}$ and $\{G_j\}$, such that $\{o_{ik}\} \in \{G_i\}$ and $\{o_{jl}\} \in \{G_j\}$, there are two possibilities to try: (1) the station $\{o_{ik}\}$ leaves group $\{G_i\}$ and joins group $\{G_j\}$, and (2) the station $\{o_{jl}\}$ leaves group $\{G_j\}$ and joins group $\{G_i\}$. Note that the possibility of both element swapping groups is not allowed because it violates the contiguity constraint. Thus, for possibility (1), let ρ_i^* and ρ_j^* be the correlation coefficients of $\{G_i\}$ and $\{G_j\}$, respectively, and let ρ_{i-}^* and ρ_{j+}^* be the correlation coefficients of $\{G_{i-}\}$ and $\{G_{j+}\}$ where $\{G_{i-}\} = \{G_i\} - \{o_k\}$, i.e. group $\{G_i\}$ without object $\{o_k\}$, and $\{G_{j+}\} = \{G_j\} + \{o_k\}$ is group $\{G_j\}$ with object $\{o_k\}$ added. The proposal to move an object from group i to group j is accepted if

$$\max\left(\rho_{i-}^{*},\rho_{j+}^{*}\right) < \max\left(\rho_{i}^{*},\rho_{j}^{*}\right) \tag{7}$$

where max(A, B) is the operator that selects the maximum of A and B.

Similarly, for possibility (2), the proposal to move an object from group *j* to group *i* is accepted if

$$\max\left(\rho_{i+}^{*},\rho_{j-}^{*}\right) < \max\left(\rho_{i}^{*},\rho_{j}^{*}\right) \tag{8}$$

Combining (7) and (8), and noting that both conditions cannot hold simultaneously, gives

$$\min\left\{\max\left(\rho_{i-}^{*},\rho_{j+}^{*}\right),\max\left(\rho_{i+}^{*},\rho_{j-}^{*}\right)\right\} < \max\left(\rho_{i}^{*},\rho_{j}^{*}\right) \tag{9}$$

(iv) Go to (ii) to operate on the next group until all N groups have been visited in turn.

3 Case Study

3.1 Study Area and Research Material

The study area is mainland Spain as shown in Fig. 1. It is bordered to the north and northeast by France, to the south and east by the Mediterranean Sea and to the west and northwest by Portugal and the Atlantic Ocean. A DEM of mainland Spain with a resolution of 751 × 728 m is used in this study. We used the Universal Transverse Mercator (UTM) projection, and we selected UTM-30N as the reference system. The DEM is a re-projection from the original resolution of 752 × 752 m. The experimental data are 1220 mean annual temperatures for year 2010, the locations of which are also shown in Fig. 1.

3.2 Results of Constrained Spatial Clustering

Scatterplots of altitude and mean temperature for the year 2010 in Fig. 2 show a clear linear relationship but with significant dispersion within a broad band. In Fig. 2 the dispersion of temperatures is approximately 7 °C, which is most evident for altitudes close to zero. Stations at zero altitude are at, or near, the coast and thus belong to different climate regions than those at higher altitudes. The total correlation coefficient is −0.729 with a 95 % confidence interval of −0.755, −0.702. Student's *t*-statistic was used to calculate this interval as, for such a large sample, the non-parametric bootstrap evaluation gives virtually the same value. The clusters retained by the spatial clustering algorithm are shown in Fig. 3.

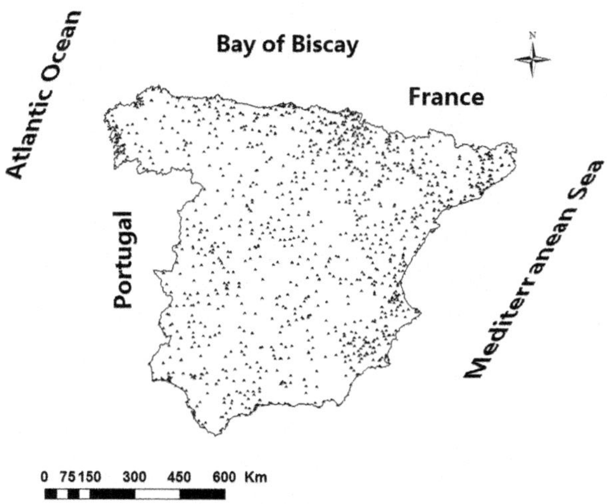

Fig. 1 Locations of the 1220 weather stations over mainland Spain

The scatterplots between altitude and mean annual temperature for each spatially constrained cluster are shown in Fig. 4 in which stations with different symbols and colours represent the different clusters.

The 13 clusters are clearly distinguishable in these plots, and the stations in each cluster display a strong (negative) correlation between altitude and mean annual temperature. In particular, the weakest correlation coefficient among all clusters is −0.848, which is significantly higher than the correlation coefficient (−0.729) for all temperatures taken as a single group, and the strongest correlation coefficient is −0.959. The value of the objective function (formula (1)) for the 13 retained clusters is −1087.55, which is higher than the value of −889.38 when all stations are considered as one group. These results demonstrate the ability of the constrained spatial clustering approach to identify geographical regions that are thermally coherent.

3.3 Hypothesis Test for Residuals of Annual Mean Temperature

In this section, we assess the ability of the spatial clustering method to retain clusters that explain most of the structural variance in temperature. If the retained clusters are thermally coherent, then the regressions of temperature on altitude within each cluster should explain almost all of the structural variance in temperature within that cluster. If this is so, then the semi-variogram of the residuals should indicate more or less spatial randomness, i.e. a pure nugget effect. The semi-variogram of the residuals from all 13 retrained clusters taken together displays some spatial variability up to a range of ~115,000 m, with a nugget variance of 0.37 and a structural variance of 0.24 as shown by the semi-variogram model fitted in

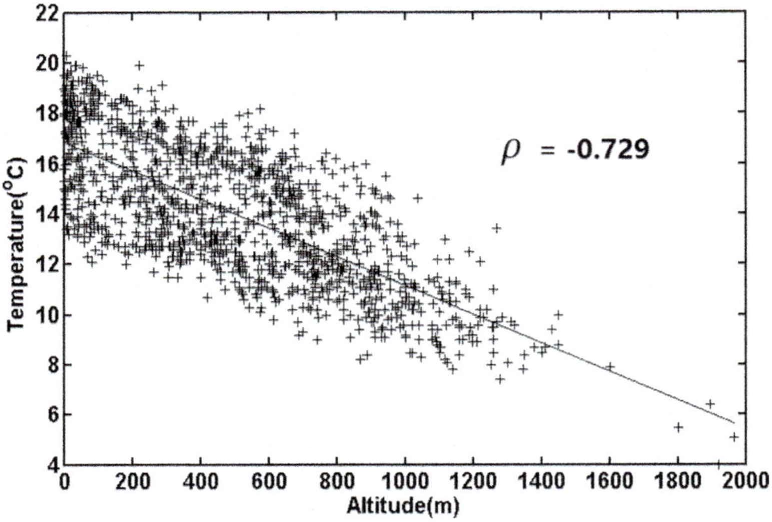

Fig. 2 Linear relationship between altitude and mean annual temperature for Spain, for all 1220 weather stations in the year 2010

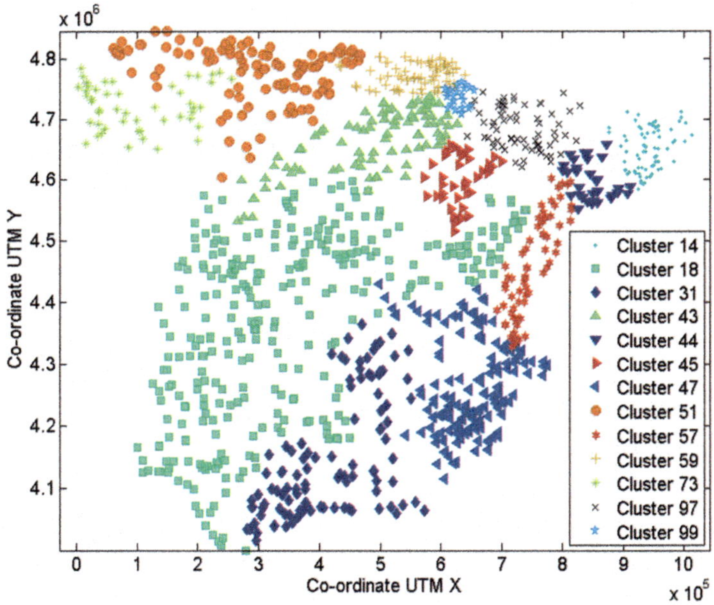

Fig. 3 Constrained spatial clusters based on temperatures measured at 1220 stations

Fig. 5. Thus, 61 % of the total variability is nugget (or random) variance, and only 39 % is spatially structured variance. By way of comparison, the model fitted to the semi-variogram of the original temperatures also has a range of ~115,000 m, but the

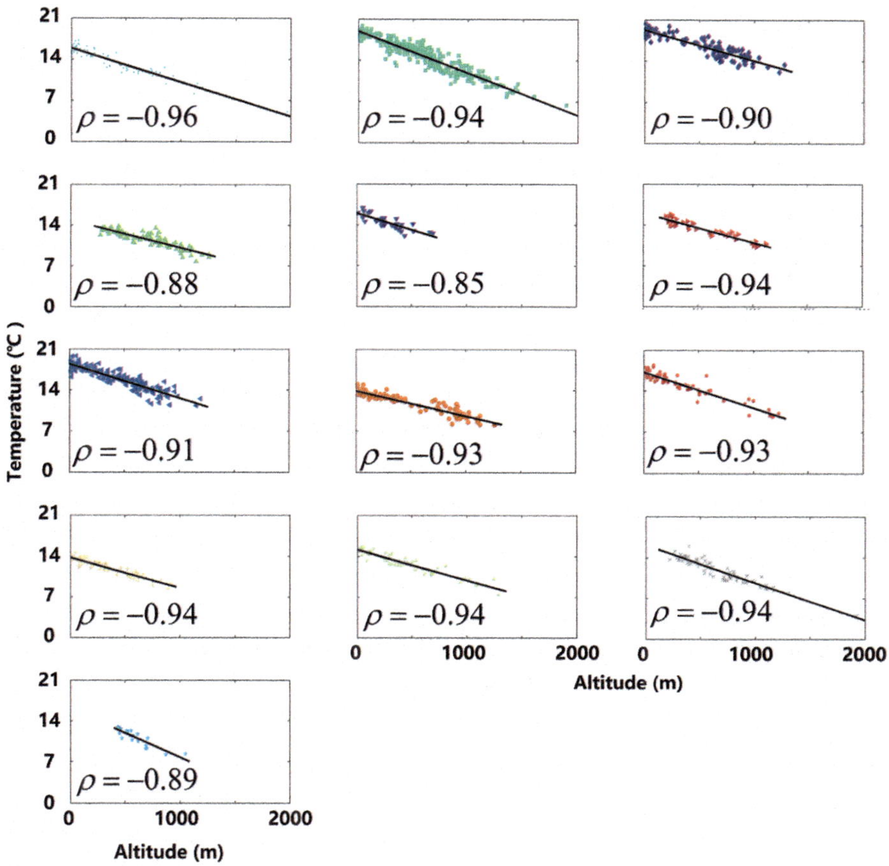

Fig. 4 Scatterplots of altitude and mean annual temperature based on constrained spatial clusters for year 2010

nugget variance of 0.60 accounts for only 8 % of the total variance with spatially structured variance accounting for the remaining 92 %. This confirms that the bulk of the structural variance has been removed by spatial clustering. Some residual structural variance might be expected from border effects of the clusters; even so, the structured variance of the residuals (0.24) represents only 3 % of the total variance of temperature (7.5).

3.4 Hypothesis Test for Correlation Coefficients

In this section we test the significance of the estimated correlation coefficients using a bootstrap statistical test and the Fisher transformation. The estimated correlation coefficients for all other clusters are inside their confidence intervals. Thus, all

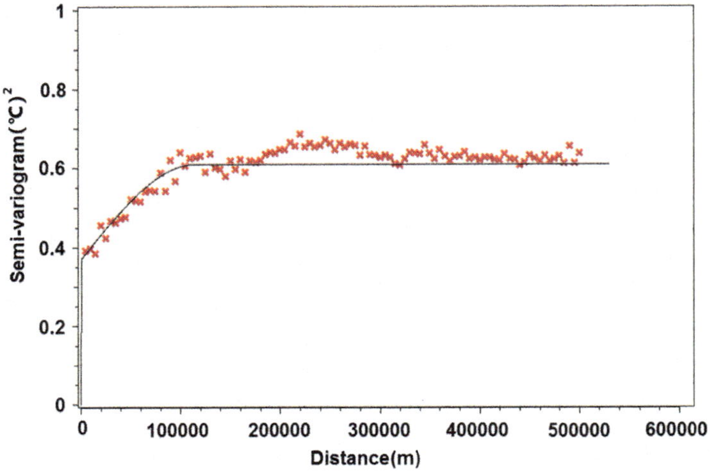

Fig. 5 Omnidirectional experimental variogram (*red crosses*) of the regression residual for the year 2010 and the fitted model (*black line*): a spherical model with a nugget variance of 0.37(°C)^2, a structural variance of 0.24(°C)^2 and a range of 115,000 m

estimated correlation coefficients are significant. For the sake of brevity, we show in Fig. 5 only two bootstrap histograms (clusters 57 and 18). With respect to the estimated confidence limits, when the distribution is symmetrical, such as in B in Fig. 5, both estimators give very similar 95 % confidence intervals. Differences between the two estimators increase as the skewness of the sampling distribution of the correlation coefficient increases as can be seen in A in Fig. 6.

3.5 Optimal Estimation of Mean Annual Temperature

The regression kriging estimator, which is optimally determined by the previous cluster analysis, is applied to estimate the mean annual temperature and the corresponding standard error. The results, using a moving window of 23 × 23 km, are shown in Fig. 7a, b. The mean standard error is 0.68 and the maximum and minimum standard errors are 0.85 and 0.63, respectively. For comparison, we have also used regression kriging to interpolate the temperature and calculate the associated standard error taking all 1220 stations as a single group. In this case, the mean standard error is 1.04, which is significantly higher than the value obtained by cluster analysis.

4 Discussion

We highlight three main points from the work presented here:

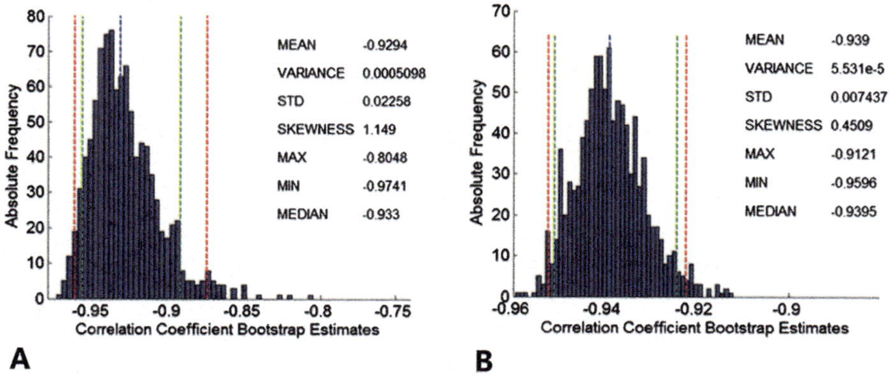

Fig. 6 Bootstrap histogram of two clusters: 95 % confidence limits of correlation coefficient from Student's t-test (*green dashed lines*) and bootstrap percentile estimates (*red dashed lines*); *blue dashed lines* are estimated correlation coefficients

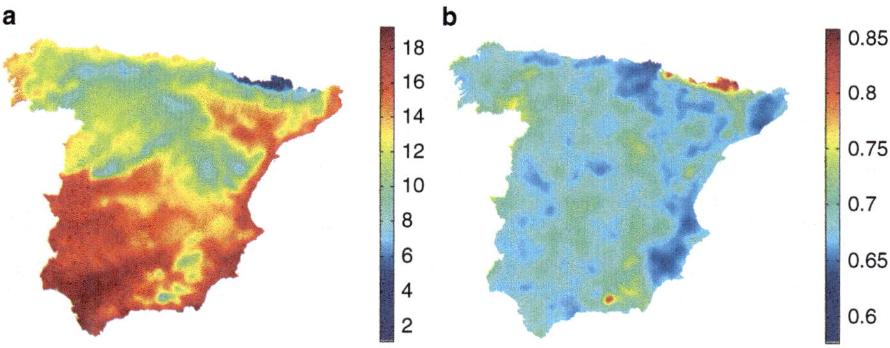

Fig. 7 (**a**) Estimated annual mean temperature; (**b**) estimated standard error

1. Firstly, the spatial clustering algorithm delivers a stable solution. Provided steps (*i*) to (*iv*) in the methodology section are repeated at least 1000 times, all results will have stabilised, in particular, the number of groups, the values of the correlation coefficients and the value of the objective function in formula (1). In addition, the objective function will have converged to an acceptable approximation of its minimum value.
2. Secondly, the role of the modified correlation coefficient. In the case study, the modified correlation coefficient outperforms the Pearson correlation coefficient because the inclusion of the standard error reduces the effects of the varying cluster sizes. In formula (2) the modified correlation coefficient is defined as the estimated correlation coefficient plus 1.96 times the corresponding standard error. This modified value is equal to the upper limit of the 95 % confidence interval using the Fisher transformation. Although the coefficient 1.96 is required for a 95 % confidence interval, it could be set to any other value depending on the requirements of the application.

3. Thirdly, the method used to quantify the uncertainty of the correlation coefficient. Provided there are sufficient data, the bootstrap will give the same result as the Fisher transformation. However, we prefer to use the bootstrap procedure because there may be significant differences when the sampling distribution of the correlation coefficient is asymmetrical as, for example, in A in Fig. 6.

5 Conclusions

The improved constrained spatial clustering analysis presented in this paper has proved to work well when applied to our experimental data. The spatial clusters that it generates are reasonable and stable, and the correlation between annual mean temperature and altitude in each cluster is more significant than the equivalent correlation for all sample stations taken as a single group. In addition, by using our modified correlation coefficient, the constrained spatial clustering approach reduces the effect of the uncertainty caused by different cluster sizes.

In testing the spatial residual of mean annual temperature, the semi-variograms of the residuals display near-random spatial variability for the complete data set. This indicates that, within each cluster, altitude explains almost all of the structural variation in mean annual temperature. For all stations taken together, it is possible that cluster (region) border effects may introduce a small amount of structural spatial variability in the residuals.

In testing the significance of the correlation coefficient, the bootstrap procedure is preferred to the Fisher transformation. The reason is that for asymmetrical distributions, which are common in practice, the bootstrap performs better.

Finally, the areal temperature average, estimated from the cluster results, provides an effective means of identifying climate zones across Spain and of detecting climate patterns.

These results may assist in identifying effects of climate change and detecting evidence of global warming in Spain especially when our approaches are applied to more extensive data sets, such as a series of annual average temperature data over a long sequence of years. We intend to investigate these possibilities in future work.

References

Carrega P (1995) A method for reconstruction of mountain air temperatures with automatic cartographic application. Theor Appl Climatol 52:69–84

DeGaetano AT (2001) Spatial grouping of United States climate stations using a hybrid clustering approach. Int J Climatol 21:791–807

Fereday DR, Knight JR, Scaife AA, Folland CK (2008) Cluster analysis of North Atlantic-European circulation types and links with tropical Pacific sea surface temperatures. J Clim 21:3687–3703

Gerstengarbe F-W, Werner PC, Fraedrich K (1999) Applying non-hierarchical cluster analysis algorithms to climate classification: some problems and their solution. Theor Appl Climatol 64:143–150

Gordon AD (1981) Classification: methods for the exploratory analysis of multivariate data. Chapman and Hall, London. ISBN 0412228505

Gordon AD (1996) A survey of constrained classification. Comput Stat Data Anal 21:17–29

Hoffman FM, Hargrove JR, Erickson DJ, Oglesby RJ (2005) Using clustered climate regimes to analyse and compare predictions from fully coupled general circulation models. Earth Interact 9:1–27

Mahlstein I, Knutti R (2010) Regional climate change patterns identified by cluster analysis. Clim Dyn 35:587–600

Morice CP, Kennedy JJ, Rayner NA, Jones PD (2012) Quantifying uncertainties in global and regional temperature change using an ensemble of observational estimates: the HadCRUT4 data set. J Geophys Res 117:1–22

Robeson SM (1994) Influence of spatial sampling and interpolation on estimates of air temperature change. Clim Res 4:119–126

Stooksbury DE, Michaels PJ (1991) Cluster analysis of South-Eastern U.S. climate stations. Theor Appl Climatol 44(3):143–150

Tang L, Su X, Shao G, Zhang H, Zhao J (2012) Clustering-assisted regression(CAR) approach for developing spatial climate data sets in China. Environ Model Softw 38:122–128

Thacker WC, Lewandowicz R (1997) Partitioning the North Atlantic into regions of similar seasonal sea-surface temperature anomalies. Int J Climatol 17:3–23

Unal Y, Kindap T, Karaca M (2003) Redefining the climate zones of Turkey using cluster analysis. Int J Climatol 23:1045–1055

Viers G (1975) Éléments de climatologie. Fernand Nathan, Paris, 309p

Zhou D, Khan S, Abbas A, Rana T, Zhang H, Chen Y (2009) Climatic regionalisation mapping of the Murrumbidgee irrigation area, Australia. Proc Natl Acad Sci U S A 19:1773–1779

Zscheischler J, Mahecha MD, Harmeling S (2012) Climate classification: the value of unsupervised clustering. Procedia Comp Sci 9:897–906

Constraining Geostatistical Simulations of Delta Hydrofacies by Using Machine Correlation

Peter Dowd, Eulogio Pardo-Igúzquiza, Sara Jorreto, Antonio Pulido-Bosch, and Francisco Sánchez-Martos

Abstract In some hydrogeology applications, the only subsurface geological information available comes from a small number of boreholes from which hydrofacies have been intersected and identified. Geostatistical simulation is a widely used stochastic technique for generating a set of possible hydrofacies images that cover the range of the complexity and heterogeneity of the structures. However, the uncertainty due to the very sparse data may be significant to the extent that the simulated images cover an unrealistically large range of possibilities for the hydrofacies characteristics. In such cases it may be desirable to constrain the simulations so as to provide a more realistic, or plausible, set of simulations. In the absence of wireline logging, outcrops, geophysics, production data or any other types of hard data, we propose the use of machine numerical correlation between hydrofacies at the boreholes as a means of constraining the range of plausible simulations. The procedure is used to simulate delta hydrofacies in a coastal aquifer in Almería (Southern Spain) where the variability of the hydrofacies is critical for managing problems related to seawater intrusion.

P. Dowd (✉)
University of Adelaide, Adelaide, Australia
e-mail: peter.dowd@adelaide.edu.au

E. Pardo-Igúzquiza
Geological Survey of Spain, Ríos Rosas 23, 28003 Madrid, Spain
e-mail: e.pardo@igme.es

S. Jorreto • A. Pulido-Bosch • F. Sánchez-Martos
University of Almería, Almería, Spain
e-mail: s.jorreto@ual.es; a.pulido@ual.es

J.J. Gómez-Hernández et al. (eds.), *Geostatistics Valencia 2016*, Quantitative Geology and Geostatistics 19, DOI 10.1007/978-3-319-46819-8_62

1 Introduction

The spatio-temporal patterns of groundwater quality in coastal aquifers are determined by the spatial heterogeneity and spatial distribution of hydrofacies (Eaton 2006). A general way of accommodating the stochastic character of the geology consists of generating a set of numerical aquifer models and applying deterministic equations of flow and transport to each of them; this approach provides a stochastic solution to the seawater intrusion problem. Although the techniques are general for different sedimentary environments, the focus in this paper is on deltaic environments. There are three main techniques for generating 3D geological models (in this case, 3D models of hydrofacies) of deltas. These methodologies are sequential stratigraphy (Cabello et al. 2007), geophysical techniques (Barakat 2010) and geostatistical models (dell'Arciprete et al. 2011). In an ideal situation, all of these techniques could be used for the integration of all the available information to provide as realistic a model as possible. However, the choice of method depends on the available data. Of the three methodologies, geostatistical methods are the least demanding with respect to data requirements and can be applied even when only a few sparsely located boreholes are available. Dell'Arciprete et al. (2011) compare sequential indicator simulation, transition probability geostatistical simulation and multiple-point simulation. We disregard multiple-point simulation because of the requirement for 3D training images or at least orthogonal 2D training images. We also disregard transition probability simulation because, as can be seen in Dell'Arciprete et al. (2011), it generates unrealistic images of spatial heterogeneity. Dowd et al. (2015) extended sequential indicator simulation to include the uncertainty of the proportions of the facies.

A problem that often remains, however, is that the data are so sparse that they cannot constrain the simulated realisations to realistic ranges. In the absence of wireline logging, outcrops, geophysics and production data (e.g. pumping tests, responses to tidal fluctuations), we propose to use machine correlation to rank the realisations by their plausibility. In the absence of the required hard data, we propose the use of the hydrofacies themselves together with the philosophy of the CORRELATOR software (Olea 2004) to build probabilistic sections of correlation between each pair of boreholes. We include a procedure to check the accuracy of the automatic correlation. Finally, the planar sections between pairs of boreholes are used to classify the simulations by assigning to them a score that quantifies their plausibility. We explain the methodology in the following section.

2 Methodology

Machine correlation, or computer-assisted correlation, of litho-stratigraphic sequences and wireline logging has been the subject of intensive research in geology in general and in petroleum exploration in particular (Fang et al. 1992).

Among the available software implementations of machine correlation, one of the best known is CORRELATOR (Olea 2004), which emulates the visual inspection used by experienced subsurface geologists in manual correlation. It uses two properties at each borehole, shale content and the similarity in the patterns of vertical variation of a specified petrophysical property. The automatic correlation can be done between any pair of boreholes; one is the reference well and the other is the matching well. We employ the same concept of a weighted correlation coefficient $w_{1,2}(i,k;n)$, which is defined as the product of a standardised similarity coefficient $\alpha_{1,2}(i,k;n)$, and the Pearson correlation coefficient, $r(i,k;n)$, (Olea 2004):

$$w_{1,2}(i,k;n) = \alpha_{1,2}(i,k;n) \cdot r_{1,2}(i,k;n) \tag{1}$$

where i is the depth index, z_i is the centre of an interval in the reference well, k is the offset between the centres of the intervals in the two wells being compared and n is the number of data either side of the central interval value; the number of data in the interval is thus $2n+1$.

However, while in CORRELATOR the similarity index is based on the concept of shale content as estimated from gamma logs, we have defined a similarity index for hydrofacies:

$$\alpha_{1,2}(i,k;n) = 1 - \frac{\sum_{s=1}^{K}\left|\sum_{j=i-n}^{i+n} I_s(j) - \sum_{j=i-n}^{i+n} I_s(j+k)\right|}{2(2n+1)} \tag{2}$$

where $I_s(j)$ is the indicator of the sth hydrofacies at the jth location in the interval in well 1, $I_s(j+k)$ is the indicator of the sth hydrofacies at the $(j+k)^{\text{th}}$ location in the interval in well 2 and K is the total number of hydrofacies. For example, if $i=5$, $k=13$, $n=3$ and $K=4$, then the total number of each of the four hydrofacies in well 1 in the interval from depth 2 to depth 8 is compared with the total number in the interval from depth 15 to depth 21 in well 2. The value of $\alpha_{1,2}(i,k;n)$ ranges between 0 and 1 in the same way as the similarity function based on the shale content in CORRELATOR.

The Pearson correlation coefficient is defined in the standard way (Olea 2004):

$$r_{1,2}(i,k;n) = \frac{\mathrm{cov}_{12}(i,j;n)}{s_1(i;n)s_2(i,k;n)} \tag{3}$$

where $\mathrm{cov}_{12}(i,j;n)$ is the covariance between the two intervals (one in each borehole) and $s_1(i;n)$and $s_2(i,k;n)$ are the respective standard deviations. The covariance term is

$$\mathrm{cov}_{12}(i,j;n) = \mathrm{E}\{C(i)C(j+k)\} - \mathrm{E}\{C(i)\}\mathrm{E}\{C(j+k)\} \tag{4}$$

where E{.} is the mathematical expectation operator and $C(i)$ is the category number at i^{th} location $C(i) \in \{1, 2, \ldots, K\}$.

The final correlation function in Eq. (1) is used to establish the likelihood of correlations between different parts of the boreholes, which, in turn, can be used to assess the plausibility of different geostatistically simulated realisations of the spatial distribution of hydrofacies.

3 Case Study

The study area (Fig. 1) is the detrital aquifer of the Andarax river delta in the province of Almería (Southern Spain). The aquifer comprises deltaic deposits from the Pleistocene overlain by fluvial and deltaic deposits from the Quaternary (Sanchez-Martos et al. 1999). The Andarax river is ephemeral, with flow usually resulting from big storms, and is typical of rivers in the semiarid coastal regions of the Mediterranean. Within the study area, there are 19 boreholes and three clusters of four piezometers each (Jorreto-Zaguirre et al. 2005), the locations of which are shown in plan view in Fig. 1. The borehole cores were classified into five types of hydrofacies according to their permeability: very permeable (category 1), permeable (category 2), low permeability (category 3), impermeable (category 4) and very impermeable (category 5).

Figure 2 shows the original 15 boreholes that were drilled along the coast. Their absolute locations are shown in Fig. 1, whereas in Fig. 2 relative spatial locations are used so as to show the distribution of the facies along the boreholes as well as the relationships between them. A visual inspection shows that the most abundant of the hydrofacies is number 2, and the least abundant is number 1. The proportions of the hydrofacies are 2.65 %, 43.70 %, 27.10 %, 18.50 % and 7.80 % for hydrofacies 1 to 5, respectively.

It is also apparent from Fig. 2 that there are no clear patterns of spatial distribution of the hydrofacies across the boreholes. This could, however, be a subjective evaluation, whereas an evaluation by machine correlation could provide a quantitative assessment that would be the same for everyone independent of individual experience. The three functions given in Eq. 1, i.e. similarity, Pearson correlation and weighted final correlation, are shown for boreholes 1 and 3 in Figs. 3, 4 and 5, respectively. In these graphs if the maximum correlations are on the 1:1 line, the implication is that the maximum correlation is along the horizontal between the boreholes, while deviations from the 1:1 line give the likely dip of the layer according to machine correlation. The similarity function has the effect of screening large correlations between intervals with different hydrofacies but with the same correlation pattern. Negative correlations are of no interest for hydrofacies correlation, so the final image of interest is the weighted correlation using positive correlations (Fig. 5).

Figure 5 shows that there is a relatively high probability (>0.6) that the hydrofacies are correlated in the interval between depths of 36 m and 62 m in

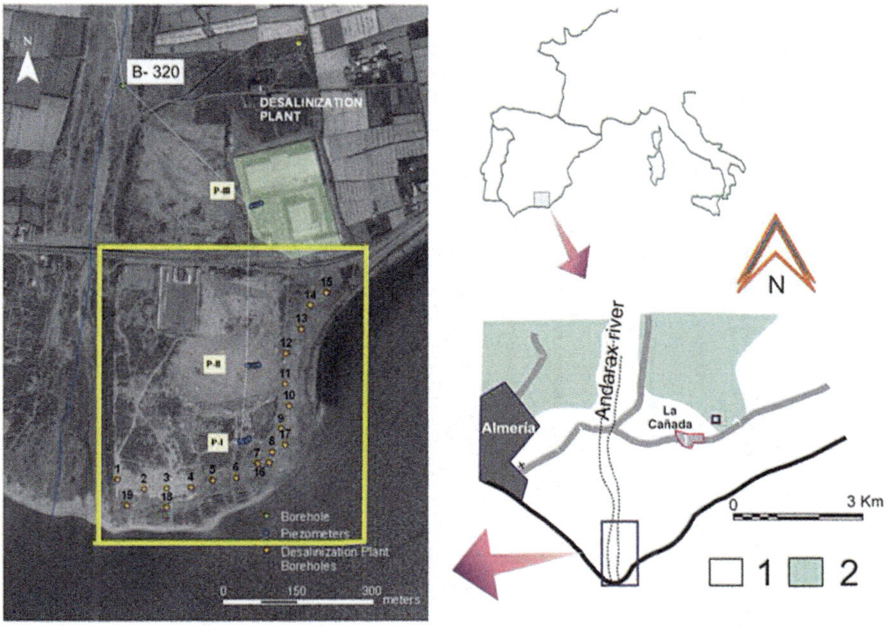

Fig. 1 Location of the study area (*yellow square*) in the Andarax river delta. *1* Quaternary detrital material. *2* Pliocene detrital material

borehole 1 and in the interval between 28 m and 54 m in borehole 3. This is an expression of the plausibility of the continuity of hydrofacies 2 between those two intervals and can be verified in Fig. 1 by looking at borehole 2, which is located between boreholes 1 and 3. To provide a better understanding of these diagrams, we have done a random permutation of the hydrofacies along boreholes 1 and 3 as shown on the left-hand side of Fig. 6.

The number and frequency of the hydrofacies in borehole 1 with random permutation are identical to the hydrofacies in the original borehole 1 (Figs. 1, 2, 3, 4, and 5). The same is true for borehole 3. However, given the random permutation, any continuity has been lost, and there is now no correlation between the two boreholes. This can be seen on the right of Fig. 6. The similarity will be high because the number and frequency of facies between the intervals will be similar, but any continuity has been lost, and the raw correlation and final weighted correlation will be low (<0.5) for virtually all the area. Another illustrative example is given in Fig. 7, which shows the final machine correlation between boreholes 10 and 11. Despite these two boreholes being close in space, the correlation is high only along the horizontal for the 8 first metres and for the last 12 metres along the boreholes, which in this case have the same length.

The ultimate goal of machine correlation in this work is the ranking of conditional geostatistical simulations and the screening of realisations using that ranking to constrain the set of plausible realisations. Figure 8 shows three conditional

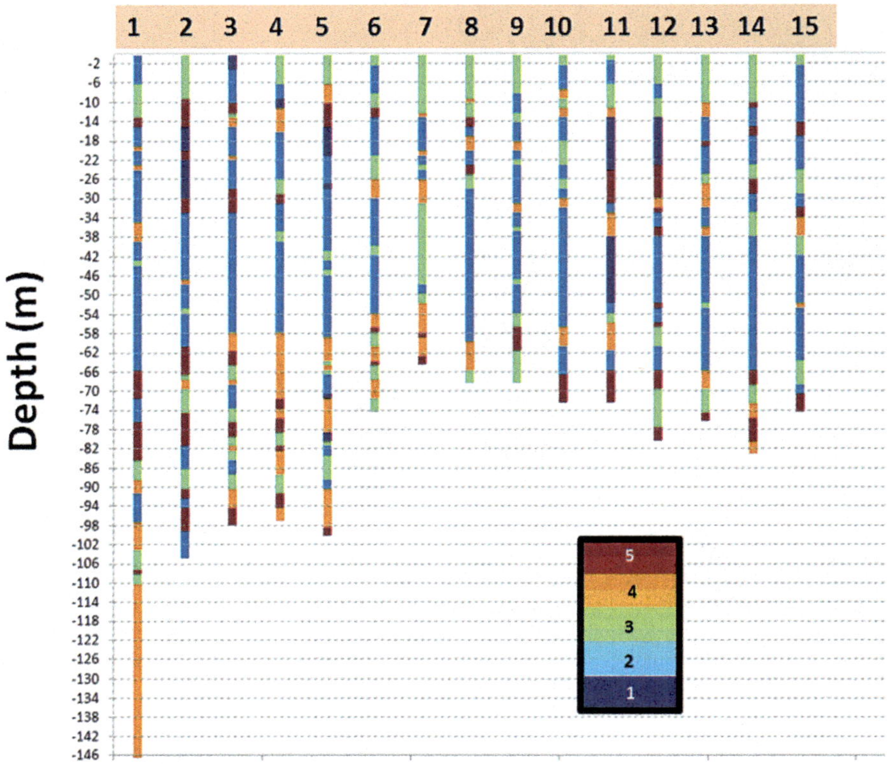

Fig. 2 Distribution of categories (from 1 to 5) along the 15 boreholes (from 1 to 15, Fig. 1) that are located parallel to the coast

simulations of the hydrofacies. Although the machine correlation between all the boreholes should be considered, for illustration purposes we concentrate on boreholes 10 and 11 for which the machine correlation is given in Fig. 7. These two boreholes are shown in 3D in Fig. 9.

Figures 10, 11 and 12 show panel 74, from the 3D simulations in Fig. 8b–d, respectively, and which passes close to the line that joins boreholes 10 and 11. The idea is to rank the plausibility of the three simulations of the panels between boreholes 10 and 11 represented by the yellow square in Figs. 10, 11 and 12 by using the machine correlation in Fig. 7.

This task could be accomplished in several ways using the fact that the conditional simulation is a 3D grid of voxels so that each column of voxels (keeping x and y constant) can be assimilated to a simulated or synthetic borehole. Thus, the equidistant "synthetic borehole" (Fig. 13) could be correlated with boreholes 10 and 11 in turn (Fig. 14), and the weighted correlation function could be compared with Fig. 7. The procedure is repeated for each simulation; it would then be possible to rank the simulations in Figs. 10 and 11 according to this rank.

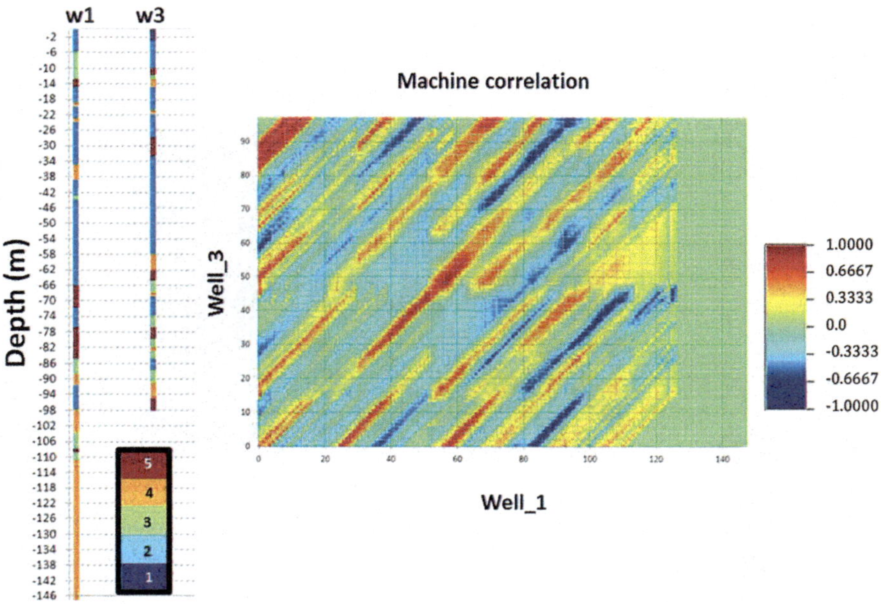

Fig. 3 Raw correlation between boreholes 1 and 3 (Figs. 1 and 2)

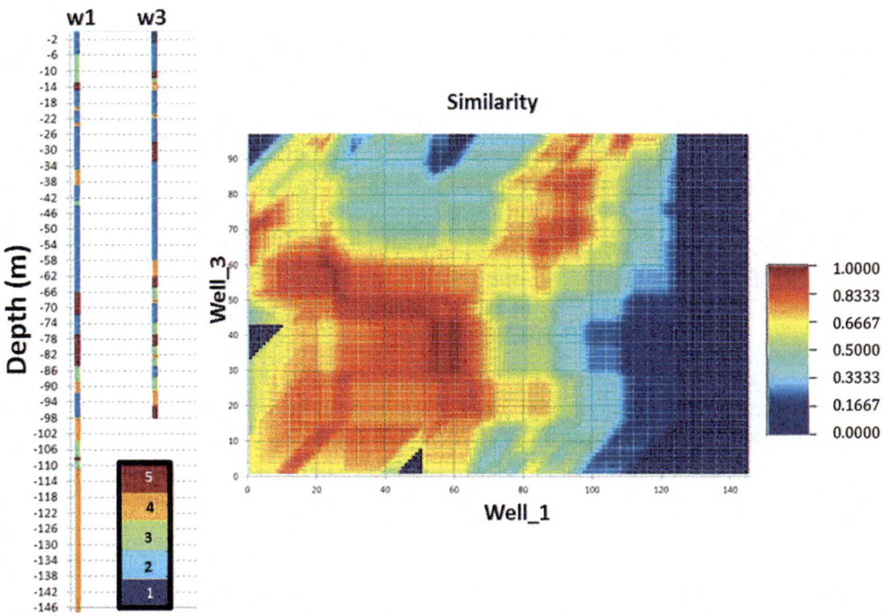

Fig. 4 Similarity function between boreholes 1 and 3 (Figs. 1 and 2)

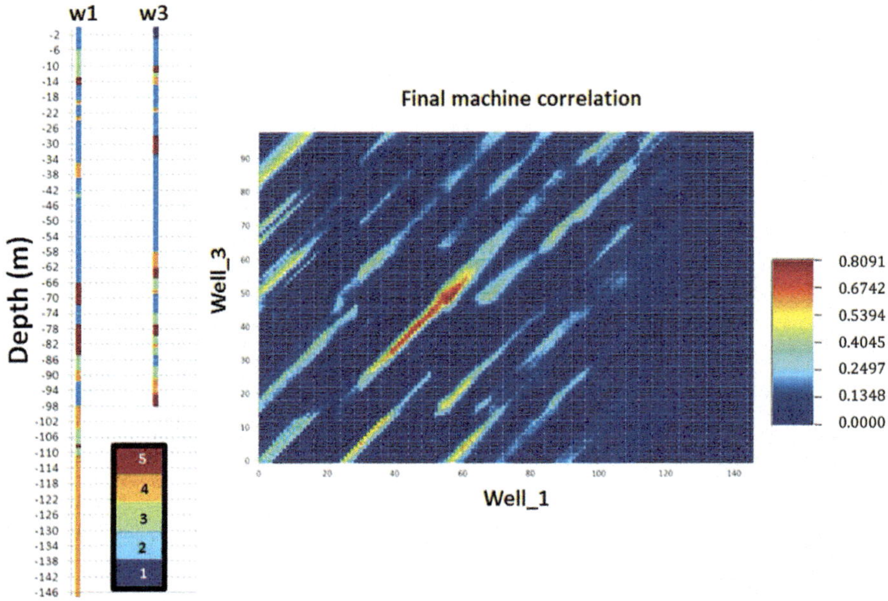

Fig. 5 Weighted correlation between boreholes 1 and 3 (Figs. 1 and 2)

A comparison of Figs. 14 and 7 shows that, of the three simulations, simulation B is the closest to the machine correlation between boreholes and would thus be the preferred choice. A quantitative assessment of the choice is left open for future work. The simulation for which the final correlation shows the highest similarity with the final correlation in Fig. 7 would be ranked first. The same procedure could be applied to any other pair of boreholes, and the final scoring would be accumulated to provide a final ranking. This procedure is computationally intensive and is still under development, but the outline of the method has been established in this work. The procedure described here is not the only possibility, and alternatives can be developed and implemented. The probability of machine correlation between each pair of boreholes could be integrated and mapped. This map could provide information on spatial variation of the continuity of the hydrofacies. It could also assist in the inference of the 3D semi-variogram by suggesting preferential directions of the anisotropy between the vertical and a non-vertical direction (horizontal or dipping). Intervals with strong connectivity could be identified and their reproduction in the simulation investigated.

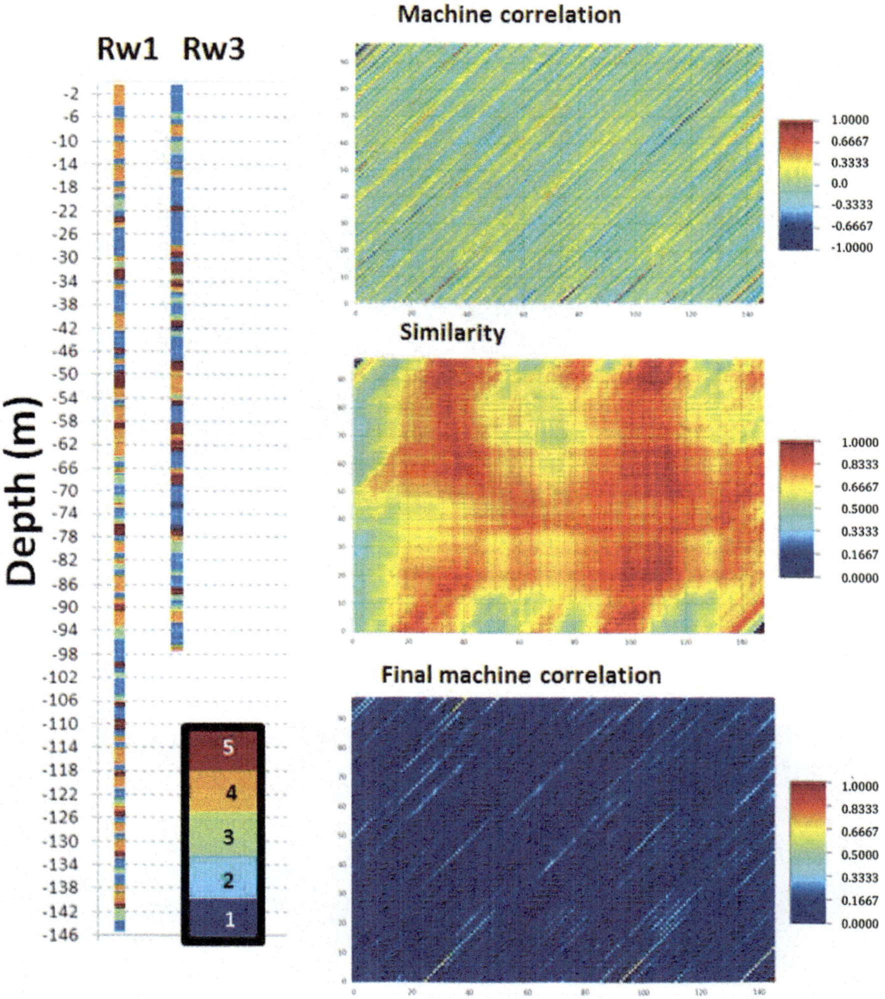

Fig. 6 *Left* random permutation of hydrofacies of boreholes 1 and 3. *Right* raw correlation (*upper*), similarity function (*middle*) and weighted correlation (*bottom*)

4 Discussion and Conclusions

The spatial distribution of hydrofacies in sedimentary delta environments is an important factor in solving practical hydrogeological problems such as seawater intrusion. Often, in these environments, as in the case study presented here, subsurface data can only be acquired either directly from boreholes or indirectly by using geophysical methods. Although both types of information are desirable and complementary, it is often the case that only a small number of boreholes are

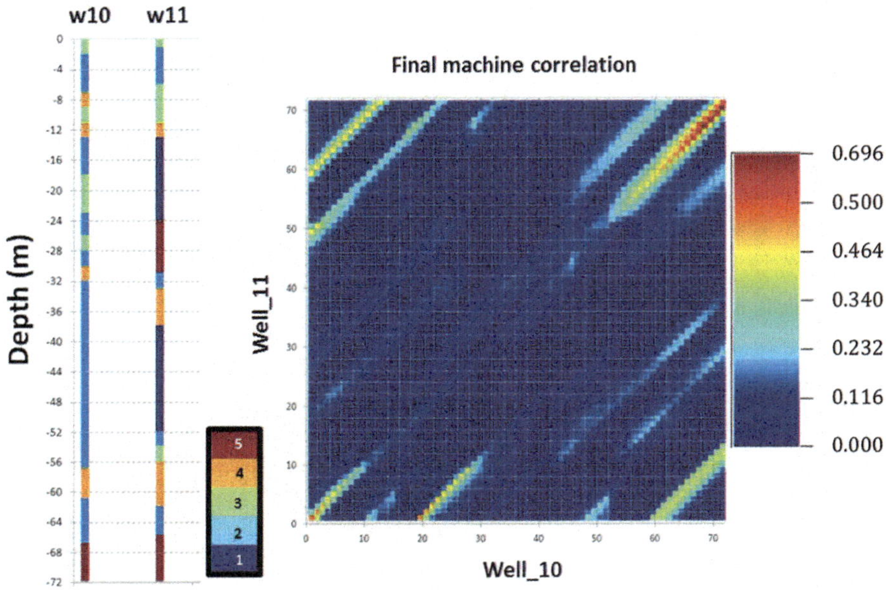

Fig. 7 Weighted correlation between boreholes 1 and 3 (Figs. 10 and 11)

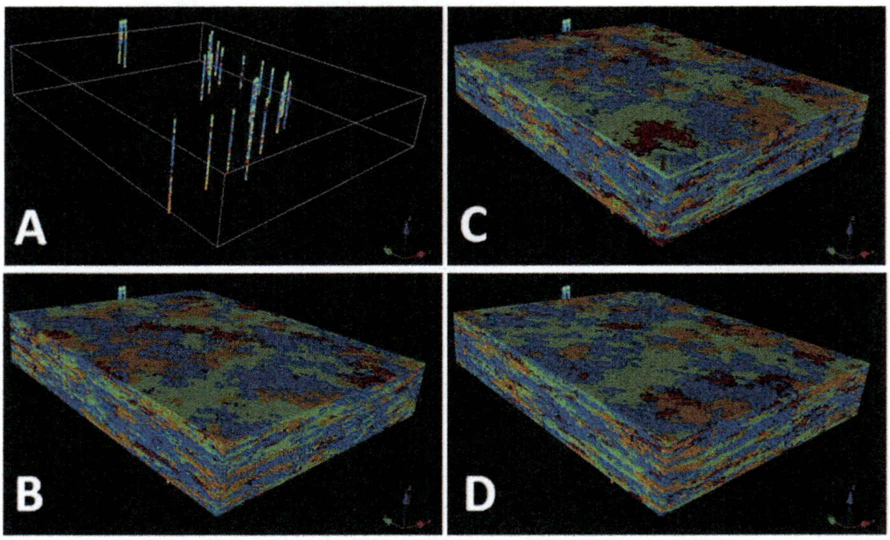

Fig. 8 (**a**) 3D view of boreholes and (**b**, **c** and **d**) three different realisations of conditional simulations of the geologic delta mediums by simulating the hydrofacies

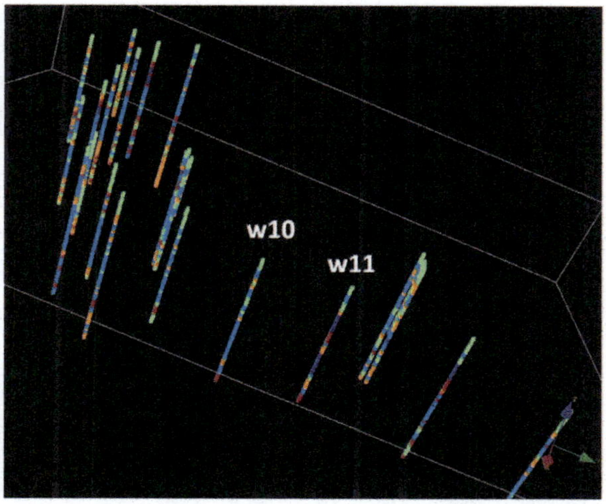

Fig. 9 3D location of boreholes 10 and 11

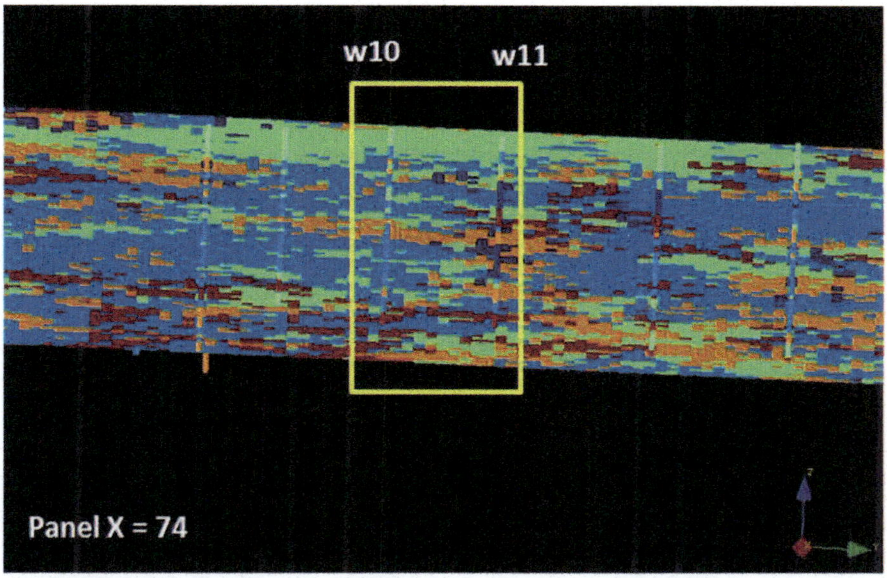

Fig. 10 Boreholes 10 and 11

available and geophysical information is absent. The uncertainty of the geological media (as described by a set of hydrofacies) can be quantified by conditional geostatistical simulation, which generates a set of plausible realisations of the spatial distribution of hydrofacies across the complete range of possible realities. This set of scenarios can then be used with flow and transport models to simulate seawater intrusion and assess the dispersion of the interface in a probabilistic way.

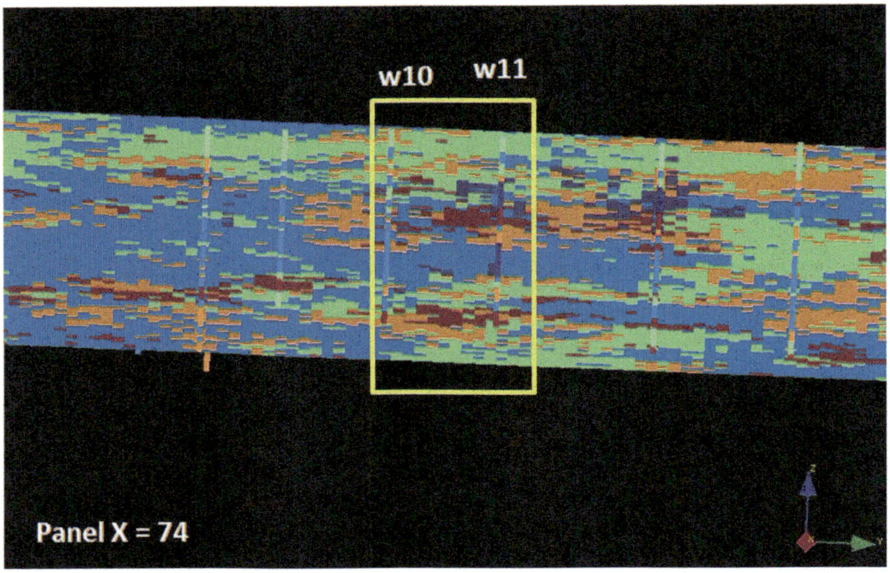

Fig. 11 Boreholes 10 and 11

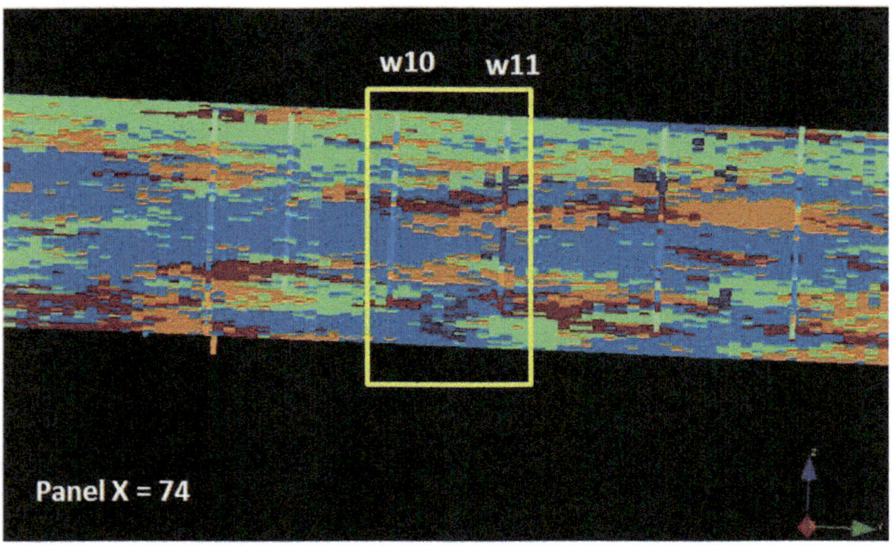

Fig. 12 Boreholes 10 and 11

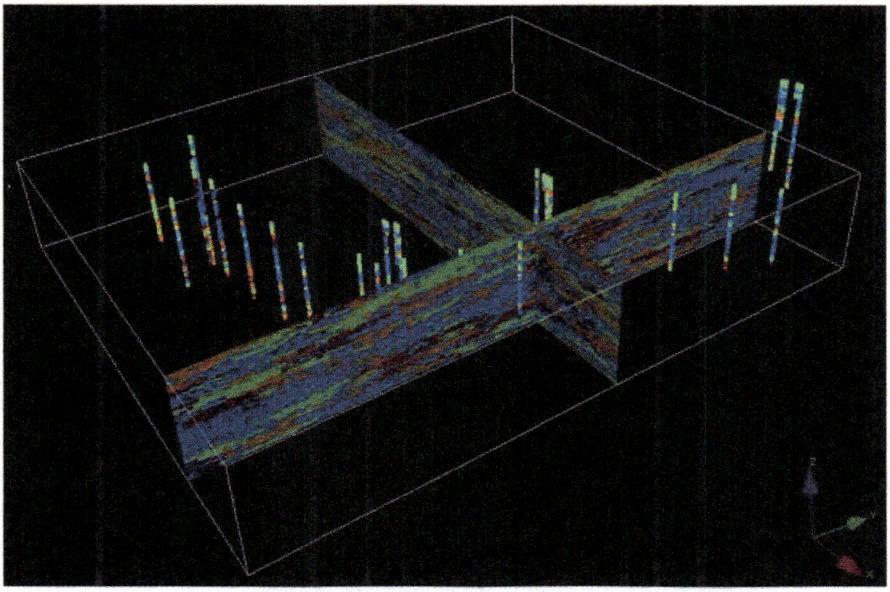

Fig. 13 Synthetic borehole equidistant from experimental boreholes 10 and 11 created by the column of voxels at the intersection of two panels from the 3D simulation

In practice, however, data are often so sparse that the range of generated realisations is unrealistically wide. In the absence of sufficient data or other information, machine correlation offers one possibility for constraining the range of realisations by determining the possible continuity of the horizons of the hydrofacies. In this work a method of machine correlation for hydrofacies has been proposed as an extension of the CORRELATOR approach (Olea 2004) and which does not require any additional information. The procedure has been validated with real data, and we have shown how it could be used in constraining the 3D realisations of hydrofacies such as those given in Dowd et al. (2015). The work presented here is just a first step, and further work is needed to rank realisations by a plausibility index, based on machine correlation, which could then be used for screening the realisations.

In future work we will also explore a variation to the approach described in this paper by conducting the well-to-well correlation before doing the geostatistical simulations. The well-to-well correlation would define the stratigraphic grid on which the simulations would be generated. As a reviewer pointed out, this would avoid mixing correlation uncertainty and geostatistical uncertainty. Nevertheless, for the particular application described in this paper, correlation after simulation is a useful approach because underground geology data are sparse and there are no 3D geophysics data available; this lack of data makes correlation before simulation far too restrictive.

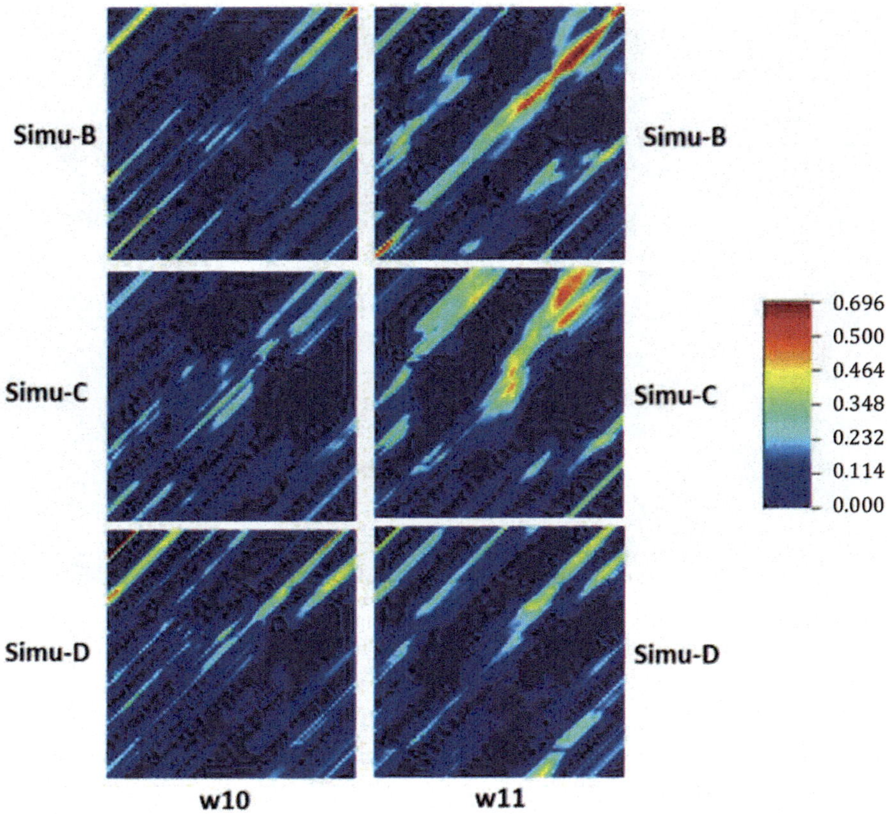

Fig. 14 Machine correlation between the synthetic borehole in Fig. 13 for the three simulations in Fig. 8b–d and boreholes 10 and 11

Acknowledgement This work was supported by research project CGL2015-71510-R from the *Ministerio de Economía y Competitividad* of Spain.

We are grateful to an anonymous reviewer for the comments on the alternative approach of conducting the well-to-well correlation before doing the geostatistical simulations.

References

Barakat MKA (2010) Modern geophysical techniques for constructing a 3D geological model on the Nile Delta, Egypt. Thesis dissertation. Technische Universität Berlin, 158 p

Cabello P, Cuevas JL, Ramos E (2007) 3D modelling of grain size distribution in quaternary deltaic deposits (Llobregat Delta, NE Spain). Geol Acta 5(3):231–244

dell'Arciprete D, Bersezio R, Felleti F, Giudici M, Comunian A, Renard P (2011) Comparison of three geostatistical methods for hydrofacies simulation: a test on alluvial sediments. Hydrogeol J 20:299–311

Dowd PA, Pardo-Igúzquiza E, Jorreto S, Pulido-Bosch A (2015) Stochastic simulation of spatial hydrofacies accounting for the uncertainty of hydrofacies proportions. In: Schaeben H, Tolosana Delgado R, van den Boogaart KG, van den Boogaart R (eds) Proceedings of IAMG 2015: the 17th annual conference of the International Association of Mathematical Geosciences. 1000-1009, ISBN 978-3-00-050337-5 (DVD)

Eaton TT (2006) On the importance of geological heterogeneity for flow simulation. Sediment Geol 184:187–201

Fang JH, Chen HC, Schultz AW, Mahmoud W (1992) Computer-aided well log correlation. Am Assoc Pet Geol Bull 76:307–317

Jorreto-Zaguirre S, Pulido-Bosch A, Gisbert-Gallego J, Sánchez-Martos F (2005) Las diagrafías y la caracterización de la influencia de los bombeos de agua de mar sobre el acuífero del delta del Andarax (Almería). Ind Minería 362:15–21

Olea RA (2004) CORRELATOR 5.2—a program for interactive lithostratigraphic correlation of wireline logs. Comput Geosci 30:561–567

Sánchez-Martos F, Pulido-Bosch A, Calaforra JM (1999) Hydrogeochemical processes in an arid region of Europe (Almería, SE Spain). Appl Geochem 14:735–745

Assessing the Performance of the Gsimcli Homogenisation Method with Precipitation Monthly Data from the COST-HOME Benchmark

S. Ribeiro, J. Caineta, and A.C. Costa

Abstract Nowadays, climate data series are used in so many different studies that their importance implies the essential need of good data quality. For this reason, the process of homogenisation became a hot topic in the last decades, and many researchers have focused on developing efficient methods for the detection and correction of inhomogeneities in climate data series. This study evaluates the efficiency of the gsimcli homogenisation method, which is based on a geostatistical simulation approach. For each instant in time, gsimcli uses the direct sequential simulation algorithm to generate several equally probable realisations of the climate variable at the candidate station's location, disregarding its values. The probability density function estimated at the candidate station's location (local probability density functions (PDF)), for each instant in time, is then used to verify the existence of inhomogeneities in the candidate time series. When an inhomogeneity is detected, that value is replaced by a statistical value (correction parameter) derived from the estimated local PDF. In order to assess the gsimcli efficiency with different implementation strategies, we homogenised monthly precipitation data from an Austrian network of the COST-HOME benchmark data set (COST Action ES0601, Advances in homogenization methods of climate series: an integrated approach – HOME). The following parameters were tested: grid cell size, candidate order in the homogenisation process, local radius parameter, detection parameter and correction parameter. Performance metrics were computed to assess the efficiency of gsimcli. The results show the high influence of the grid cell size and of the correction parameter in the method's performance.

S. Ribeiro (✉) • J. Caineta • A.C. Costa
NOVA IMS, Universidade Nova de Lisboa, Campus de Campolide, 1070-312 Lisbon, Portugal
e-mail: sribeiro@novaims.unl.pt; jcaineta@novaims.unl.pt; ccosta@novaims.unl.pt

J.J. Gómez-Hernández et al. (eds.), *Geostatistics Valencia 2016*, Quantitative Geology and Geostatistics 19, DOI 10.1007/978-3-319-46819-8_63

1 Introduction

As defined by the Intergovernmental Panel on Climate Change (IPCC), climate change refers to a change in the state of the climate that can be identified by changes in the statistical characteristics of its properties and that persists for an extended period, typically decades or longer (Bernstein et al. 2007). In order to assess climate change and to develop impact studies, it is imperative that climate signals are clean from any external factors. Hence, two steps must be performed in climate time series: quality control and homogenisation. Quality control relates to the verification and treatment of extremely high and low values (outliers). The second includes an analysis of the time series that is focused on the detection and correction of inhomogeneities caused by non-climatic factors (Bližňák et al. 2014; Vertacnik et al. 2015).

The non-climatic factors include stations' relocations, changes in the environment, instrumentation, time and the methods of measurement (Aguilar et al. 2003). Since these artificial discontinuities have often the same magnitude as the usual variability of climate data series, they can erroneously influence the analysis of natural climate variations (Hannart et al. 2014).

Homogenisation methods usually depend on the type of climate variable, the temporal resolution of the observations, the weather station network density and also the availability of metadata (Costa and Soares 2009). Indeed, metadata plays a very important role in the homogenisation of climate data and should be documented and treated with the same care as the data themselves (World Meteorological Organization 2010). Direct homogenisation methods employ metadata in order to assess the presence of a breakpoint; however, whenever metadata is absent, the indirect homogenisation methods justify the presence of a breakpoint only with the result of homogenisation tests (Ribeiro et al. 2016). The relative homogeneity principle (Conrad and Pollack 1962) assumes that neighbouring series reveal the same climate variations apart from the inhomogeneities integrated in one of the series (Hannart et al. 2014). Based on this principle, homogenisation methods can be classified in regard to the use of reference stations: absolute and relative homogenisation methods. The use of absolute methods (without using reference stations) should be used with care and always accompanied with metadata (Venema et al. 2012), since they may introduce more inhomogeneities in the data series.

Several authors have prepared reviews of homogenisation methods (Aguilar et al. 2003; Costa and Soares 2009; Domonkos 2013; Ribeiro et al. 2015). The European initiative COST Action ES0601, Advances in homogenisation methods of climate series: an integrated approach (HOME), is intended to review and improve common homogenisation methods and to assess their impact in climate time series (Chair of the Management Committee of the Action 2011). In order to achieve such goals, HOME has executed a blind intercomparison and validation study for homogenisation methods. The methods were tested against a realistic benchmark data set, which included temperature and precipitation data (Venema et al. 2012). This benchmark data set has three different groups of data: real, surrogate and synthetic. The first group contains real inhomogeneous data, while the other two enclose simulated data with inserted inhomogeneities, outliers, missing data periods, local station trends and a global trend, per network (Venema et al. 2012). Fifteen simulated networks were prepared, and they are located in different places within Europe. The networks comprise 5, 9 and 15 stations.

The submitted methods were evaluated by the calculation of performance metrics. Based on the performance metrics, the best homogenisation contributions were ACMANT (Domonkos et al. 2011), MASH (Szentimrey 1999, 2007, 2008), PRODIGE (Caussinus and Mestre 1996, 2004) and USHCN (Menne and Williams 2009; Menne et al. 2009). Recently, some of the homogenisation methods were transformed into software packages, in order to become fully automatic procedures, and they are available in (http://www.climatol.eu/tt-hom/).

This study assesses the efficiency of the gsimcli homogenisation method, which is based on a geostatistical simulation approach. To do so, we homogenised monthly precipitation data from an Austrian network of the HOME benchmark data set. The following parameters were tested: grid cell size, candidate order in the homogenisation process, local radius parameter, detection parameter and correction parameter. Performance metrics were computed to assess the efficiency of gsimcli. Precipitation is the focus on this study, since it is one of the most important variables for climate and hydrometeorology studies. Changes in precipitation pattern may lead to floods, to droughts and consequentially to the loss of biodiversity and agricultural productivity (Sayemuzzaman and Jha 2014).

This work is organised as follows. The following section depicts the study domain and data. The methodological framework includes the description of the gsimcli method and the set of performed homogeneity tests. In the results and discussion section, the performance metrics are scrutinised. Finally, the conclusion section brings a summary of the lessons learned and recommendations for future work.

2 Study Domain and Data

This study analyses monthly surrogate precipitation data that are part of the HOME data set, namely, the network 16. This network comprises 15 stations and it is located in Austria (Fig. 1). The data series include 100 years of precipitation values, between 1900 and 1999. It covers a rectangular area of approximately 24,640 km^2 (220 $\times$ 112 km). Considering the statistics for the annual series (Table 1), the lowest values were recorded in stations 4313302 (northeast corner) and 4315421 (west area), with 374.3 mm and 384.6 mm, respectively. The maximum value was recorded in station 4319710 (southwest corner). The variability of both annual and monthly series is very high. For example, the standard deviation of the annual series varies between 131.4 mm (station 4315515) and 287.7 mm (station 4320123).

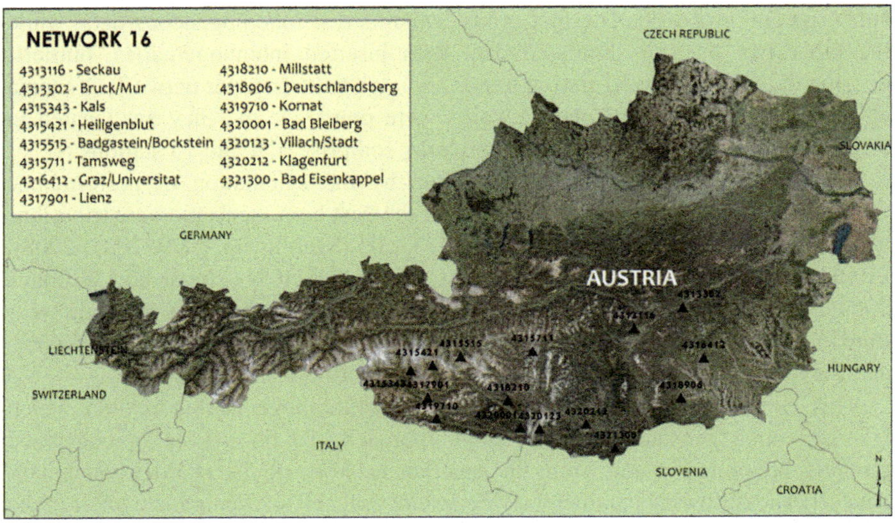

Fig. 1 Location of 15 stations from the network 16

Table 1 Summary statistics of the annual precipitation series from network 16

Station ID	Mean	Median	Std. dev.	Variance	Range	Min.	Max.
4313116	868.9	812.2	211.2	44,582	974.7	415.5	1,390
4313302	769.5	759.9	131.7	17,343	780.8	374.3	1,155
4315343	903.0	885.2	152.4	23,217	834.1	563.0	1,397
4315421	786.0	795.3	194.4	37,777	749.6	384.6	1,134
4315515	1,051.4	1051.5	131.4	17,275	639.8	726.4	1,366
4315711	793.5	773.3	136.1	18,521	733.5	465.8	1,199
4316412	824.3	825.5	188.8	35,626	994.9	422.7	1,418
4317901	1,010.5	1,015.0	206.8	42,770	942.5	597.9	1,540
4318210	916.5	908.8	178.4	31,825	866.6	470.1	1,337
4318906	1,150.4	1,158.1	210.39	44,265	1,132.1	550.5	1,683
4319710	1,290.6	1237.4	279.8	78,299	1,498.6	618.2	2,117
4320001	1,247	1,270.5	244.4	59,720	1,096.4	685.5	1,782
4320123	1,164.0	1,100.1	287.7	82,792	1,408.2	510.1	1,918
4320212	1,000.5	1,012.4	190.8	36,409	881.2	538.1	1,419
4321300	1,282.4	1297.6	203.3	41,330	1,000.2	705.7	1,706

Variography analysis is required to perform kriging interpolation, in which the gsimcli method is based. Hence, before executing the homogenisation procedure, the semivariogram model must be studied and its parameters defined. Data was divided by month and then by decade. Due to the missing periods in the beginning of the century (1900–1929) and in the fifth decade (1940–1945), the first three decades were joined into a data set, as well as the fourth and fifth decade, for the purpose of the variography study. Seven semivariograms were modelled for each of the monthly series, in a total of 84.

3 Methodological Framework

3.1 *Gsimcli Method*

This study evaluates the gsimcli homogenisation method, which is based on the direct sequential simulation (DSS) algorithm (Soares 2001). The gsimcli method uses the DSS in the calculation of the local probability density functions (PDF) (Costa and Soares 2009) at the location of the candidate station. Such calculation is prepared solely with the temporal and spatial observations of nearby reference stations. A breakpoint is identified whenever the interval of a specified probability p, centred in the local PDF, does not include the real observation of the candidate station. The detected irregular value is then replaced by a statistic value of the local PDF formerly computed (e.g. mean, median or a given percentile).

This method turned into a software package, which allowed the homogenisation process to become direct and quasi-automatic (Caineta et al. 2015). Two subsets of parameters must be defined before starting the homogenisation procedure: the simulation parameters and the homogenisation parameters. The former define the number of simulations, the kriging type of the geostatistical method, the maximum number of nodes to be found, the number of CPU cores, the simulation grid size and the semivariogram parameters. The latter depict the candidate order, the detection parameter, the local radius and the correction parameter. The simulation grid describing the area where the stations are located must also be analysed, in order to ascertain the cell size, the number of columns and the number of rows. These values take into account that the bordering stations of the network must be surrounded by a number of cells at least equal to the value of the local radius parameter. The gsimcli method is freely available at http://iled.github.io/gsimcli/.

3.2 *Homogeneity Tests*

In order to start the homogenisation procedure, the monthly precipitation surrogate data set was divided in 12 folders, 1 per month. Each of these folders included the variography file with the semivariogram parameters per decade, the grid settings file containing the grid cell size, and a subfolder with the ten decadal data files. The monthly folders are homogenised separately.

A sensitivity analysis comprising 16 different strategies was implemented for the following parameters (Table 2): grid cell size, the detection parameter (the probability value to build the detection interval centred in the local PDF), the correction

Table 2 Different homogenisation strategies (grid cell size, detection parameter, correction parameter and local radius parameter)

Test #	Grid cell size (metres)	Detection parameter (p)	Correction parameter (percentile p)	Local radius parameter (*r*)
1	1,000	0.95	0.975	0
2	1,000	0.95	0.95	0
3	5,000	0.95	0.90	1
4	5,000	0.95	0.90	2
5	5,000	0.95	0.90	3
6	5,000	0.95	0.975	0
7	5,000	0.95	0.975	1
8	5,000	0.95	0.975	2
9	10,000	0.95	0.975	0
10	10,000	0.95	0.975	1
11	10,000	0.95	0.975	2
12*	10,000	0.95	0.90	0
13*	10,000	0.95	0.90	0
14	10,000	0.95	0.90	1
15	10,000	0.95	0.90	2
16	10,000	0.975	0.975	0

In all Tests the candidate order was based on the stations' data variance, except in Test #13 (*) that was based on the network deviation

parameter (the statistic value used for the inhomogeneities correction: the inhomogeneities, outliers or missing values can be replaced by the mean, median, skewness and percentile), the local radius (sets the radius of a circle centred at the candidate station location where the simulated values of the nodes located within the circle are considered in the calculation of the local PDF) and the candidate order (the order by which the candidate stations are homogenised).

The analysed grid cell sizes are 1000 m, 5000 m and 10,000 m, which correspond to grids of 27,709 cells (229 × 121 cells), 2088 (58 × 36 cells) and 792 cells (36 × 22 cells), respectively. The values of the detection parameter analysed are 0.95 and 0.975. The values considered for the correction parameter are the percentiles of 0.90, 0.95 and 0.975. The investigated local radii are 0, 1, 2 and 3. It is noteworthy to mention that the area of the circle centred in the candidate station depends on the grid cell size and the local radius. Regarding the candidate order, Tests #1 to #12 and #14 to #16 used the descending value of the stations' data variance, while Test # 13 used the network deviation (difference between the station and the network average values) to define the sequence of the candidate stations to be homogenised (Table 2).

The values defined for the remaining simulation parameters are common to the 16 strategies and correspond to default values proposed by Ribeiro et al. (2016):

- Number of simulations: 500
- Kriging type: ordinary kriging

- Maximum number of nodes to be found: 16
- Number of CPU cores: 4

4 Results and Discussion

For each homogenisation strategy, four performance metrics are automatically calculated by gsimcli software (Table 3). Tests with the lowest values of performance metrics correspond to tests with the best set of parameters.

Those metrics are the station centred root mean square error (CRMSE), the network CRMSE, the station improvement and the network improvement, as defined by Venema et al. (2012). The CRMSE was chosen by the HOME project since the main aim of the homogenisation is not to improve the absolute values but rather the temporal consistency. The station CRMSE quantifies the homogenisation efficiency for each station individually, and it is obtained by the mean CRMSE, by station. The network CRMSE measures the efficiency of the homogenisation of the network, as a whole.

The improvement metrics assess the enhancement over the inhomogeneous data. Station (network) improvement metrics will reflect the quality of the procedure also shown in the station (network) CRMSE. The improvement metrics are computed as the ratio of the station (network) CRMSE of the homogenised networks and the station (network) CRMSE of the same inhomogeneous networks.

Analysing the station CRMSE of the 16 strategies, it is possible to note that the highest values belong to the strategies where the correction parameter was defined

Table 3 Performance metrics of the 16 homogenisation strategies

Test #	Station CRMSE	Network CRMSE	Station improvement	Network improvement
1	13.11	5.11	1.10	1.17
2	13.56	5.96	1.13	1.37
3	15.05	7.73	1.26	1.77
4	15.01	7.71	1.26	1.77
5	14.99	7.69	1.25	1.76
6	13.09	5.20	1.09	1.19
7	13.01	5.13	1.09	1.17
8	12.99	5.10	1.09	1.17
9	13.03	5.25	1.09	1.20
10	12.90	5.10	1.08	1.17
11	12.92	5.09	1.08	1.17
12*	15.02	7.71	1.26	1.77
13*	15.17	7.59	1.27	1.74
14	14.97	7.65	1.25	1.75
15	14.96	7.65	1.25	1.75
16	12.94	5.33	1.08	1.22

as the percentile of 0.90, regardless of the other parameters. Therefore, it is less appropriate for the correction of irregularities (inhomogeneities, outliers and missing values). These strategies (Tests #3, #4, #5, #12, #13, #14 and #15) also have the highest values of the network CRMSE.

The most appropriate value for the correction parameter seems to be the percentile of 0.975, because the performance metrics exhibit smaller values.

Comparing the candidate order parameter, focus must be given to Tests #12 and #13. The evaluation of their performance metrics is ambiguous, since the station CRMSE is better for the Test #12 (candidate order by variance), while Test #13 performs better in the network CRMSE (candidate order by network deviation).

Tests #9 and #16 differ in the detection parameter, and their station and network CRMSE vary in opposite directions. In this case, as it happens with the candidate order parameter, it is not possible to make any judgement about the best value for the detection parameter. The value of 0.95 is set as default.

The CRMSE metrics decrease with the increase of the cell size (e.g. Tests #1, #6 and #9); thus, increasing the cell size improves the method performance.

The best metrics are provided by Tests #10 and #11, where the cell size is 10,000 m, the detection and correction parameters are set to 0.95 and the percentile of 0.975, respectively. The difference between these two Tests is the local radius (1 and 2, respectively) parameter. Their performance metrics are very similar. The larger the cell size and the local radius are, the greater the quality of the homogenisation results. This fact relates to the area that is considered for the computation of the local PDF.

5 Conclusion

This study aimed at investigating several parameters in the homogenisation of monthly precipitation surrogate series with the gsimcli approach. The analysed parameters were the grid cell size, the detection parameter, the correction parameter and the local radius parameter.

The analysis has emphasised the importance of the grid cell size and the local radius parameters in the performance of gsimcli. The knowledge of the surrounding area of a candidate station is essential to the improvement of its series quality. The principle of relative homogeneity (Conrad and Pollack 1962) is again validated.

Acknowledgements The authors gratefully acknowledge the financial support of *Fundação para a Ciência e Tecnologia* (FCT), Portugal, through the research project PTDC/GEO-MET/4026/2012 ("GSIMCLI – Geostatistical simulation with local distributions for the homogenization and interpolation of climate data").

References

Aguilar E, Auer I, Brunet M, Peterson T, Wieringa J (2003) Guidelines on climate metadata and homogenization. In Llansó P (ed) WMO/TD No. 1186, WCDMP No. 53. World Meteorological Organization, Geneva

Bernstein L, Bosch P, Canziani O, Chen Z, Christ R, Davidson O et al (2007) Climate change 2007: synthesis report. Cambridge University Press, Cambridge

Bližňák V, Valente M, Bethke J (2014). Homogenization of time series from Portugal and its former colonies for the period from the late 19th to the early 21st century. Int J Climatol 2400–2418

Caineta J, Ribeiro S, Soares A, Costa AC (2015) Workflow for the homogenisation of climate data using geostatistical simulation, vol 1, Informatics, geoinformatics and remote sensing. International Multidisciplinary Scientific GeoConference – SGEM, Albena, pp 921–929

Caussinus H, Mestre O (1996) New mathematical tools and methodologies for relative homogeneity testing. In: Service HM (ed) Proceedings of the first seminar for homogenization of surface climatological data. World Meteorological Organization, Budapest, pp 63–82

Caussinus H, Mestre O (2004) Detection and correction of artificial shifts in climate series. Appl Stat 53:405–425

Chair of the Management Committee of the Action (2011) Monitoring progress report 03/05/2007-01/06/2011. HOME – advances in homogenisation methods of climate series: an integrated approach (COST Action ES0601)

Conrad V, Pollack LW (1962) Methods in climatology. Havard University Press, Cambridge, MA

Costa AC, Soares A (2009) Homogenization of climate data: review and new perspectives using geostatistics. Math Geosci 117(1):91–112

Domonkos P (2013) Measuring performances of homogenization methods. Időjárás Q J Hung Meteorol 117(1):91–112

Domonkos P, Poza R, Efthymiadis D (2011) Newest developments of ACMANT. Adv Sci Res 6:7–11

Hannart A, Mestre O, Naveau P (2014) An automatized homogenization procedure via pairwise comparisons with application to Argentinean temperature series. Int J Climatol 34:3528–3545

Menne MJ, Williams CN Jr (2009) Homogenization of temperature series via pairwise comparisons. J Clim 22(7):1700–1717

Menne MJ, Williams CN Jr, Vose RS (2009) The U. S. historical climatology network monthly temperature data, version 2. Bull Am Meteorol Soc 90:993–1007

Ribeiro S, Caineta J, Costa AC (2015) Review and discussion of homogenisation methods. Phys Chem Earth, (in press)

Ribeiro S, Caineta J, Costa A, Henriques R, Soares A (2016) Detection of inhomogeneities in precipitation time series in Portugal using direct sequential simulation. Atmos Res 171:147–158

Sayemuzzaman M, Jha MK (2014) Seasonal and annual precipitation time series trend analysis of North Carolina, United States. Atmos Res 137:183–194

Soares A (2001) Direct sequential simulation and cosimulation. Math Geol 33(8):911–926

Szentimrey T (1999) Multiple analysis of series for homogenization (MASH). In: Proceedings of the second seminar for homogenization of surface climatological data. WCDMP-No. 41, WMO-TD No. 1962. World Meteorological Organization, Budapest, pp 27–46

Szentimrey T (2007).Manual of homogenization software MASH v3.02. Hungarian Meteorological Service

Szentimrey T (2008) Development of MASH homogenization procedure for daily data. In: Lakatos M, Szentimrey T, Bihari Z, Szalai S (eds) Proceedings of the fifth seminar for homogenization and quality control in climatological databases. WCDMP-No. 71. World Meteorological Organization, Budapest, pp 123–130

Venema V, Mestre O, Aguilar E, Auer I, Guijarro J, Domonkos P et al (2012) Benchmarking homogenization algorithms for monthly data. Clim Past 8(1):89–115

Vertacnik G, Dolinar M, Bertalanic R, Klancar M, Dvorsek D, Nadbath M (2015) Ensemble homogenization of Slovenian monthly air temperature series. Int J Climatol 35:4015–4026
World Meteorological Organization (2010) Guide to climatological practices. WMO No. 100, 3rd edn. World Meteorological Organization, Geneva

Ecological Risk Evaluation of Heavy Metal Pollution in Soil in Yanggu

Yingjun Sun

Abstract Soil pollution caused by heavy metals was studied. The potential ecological risks posed by seven heavy metals (Cu, Pb, Cd, Cr, Ni, Zn, and As) in the agricultural soils of Yanggu, Shandong Province, China, were analyzed. The spatial variation of the seven heavy metals was used to develop a probability map of the heavy metals based on sequential simulation methods. The ecological risks to the region from the heavy metals were then assessed using the Hakanson potential danger index. The result showed that Cd was the main problem in the Yanggu area. The potential ecological risk from Cd was nearly 80 and the risk classification was medium. The total potential ecological risk from all seven heavy metals was 114.656. This indicated that heavy metal pollution in the entire study area posed only a slight ecological risk.

1 Introduction

Ecological risk assessment (ERA) is performed to evaluate the potential adverse effects of physical or chemical stressors on the environment. Risk assessment provides a way to develop, organize, and present scientific information so that it is meaningful in the consideration of environmental decisions. Soil is a basic element of ecological systems. Soil conditions affect economic development and are directly related to agricultural product safety and human health. Many studies have evaluated soil pollution caused by heavy metals and their potential ecological risks (Liu 2010; Muyessar Trudi 2013). The multivariate statistical analysis, pollution index, and potential ecological risk index methods were used to study the spatial distribution of heavy metals (Zheng 2012; Li 2014; Lu 2012).

Y. Sun (✉)
College of Geomatics, Shandong University of Science and Technology, Qingdao 266510, China

Department of Civil Engineering, Shandong Jianzhu University, Jinan 250014, China
e-mail: sdjzusyj@126.com

J.J. Gómez-Hernández et al. (eds.), *Geostatistics Valencia 2016*, Quantitative Geology and Geostatistics 19, DOI 10.1007/978-3-319-46819-8_64

Geostatistics has played an important role in soil environment studies. Geostatistics includes description of spatial patterns, quantitative modeling of spatial continuity, spatial prediction, and uncertainty assessment (Goovaerts 1998). The "bin" variogram has been used to quantify the spatial distribution of heavy metal concentrations. There are two kinds of prediction methods in geostatistics: kriging and simulation. The methods of kriging tend to produce more "accurate" and "smoothed" values for the unsampled points. The results are based on the condition of the minimum local error variance (Sun 2008). Simulation methods focus on the spatial pattern of the sample points. Simulated maps will reproduce the statistics drawn from the sample points, like the histogram or the semivariogram model (Goovaerts 1997). Therefore, the maximum and the minimum values will be retained in the map, and this is important for the discovery of pollution sources. The sequential Gaussian simulation method was used to represent the possible pollution status of heavy metals in the Yanggu study area. In advanced research, uncertainty in the spatial distribution of attribute values can be reflected in the uncertainty of risk assessments. This, in turn, may compromise the successful performance of remediation processes.

2 Experiment

2.1 Study Area

Yanggu (35°55′–36°19′ N, 115°39′–116°06′ E) is located in the west of Shandong Province (Fig. 1). It covers an area of 1,064 km^2 and the population size is 794,800. It has a typical continental monsoon climate, with warm temperatures, semi-humid conditions, and well-defined seasons. The mean annual temperature is 13.9 °C, and the mean precipitation is 523.9 mm. The soil of Yanggu is typically moist; the soil types are mainly sandy soil, loam soil, and clay soil. Yanggu is an important cotton-producing area for both the Shandong Province and China.

2.2 Sampling

A digital map of Yanggu was used to establish uniform 2 × 2 km grids for the entire study area. Each grid center represented a sampling point. Topsoil samples (0–20 cm) from Yanggu were collected from 226 different grids in 2008 (Fig. 1). Some sampling point locations varied slightly from the grid center. These changes were based on specific topography, land use, and soil type provided by 1:50,000 scale topographic maps, 1:100,000 scale soil maps, and 1:10,000 scale aerial photos. Soil samples were obtained by mixing five subsamples from each site within a 20 × 20 m area and were recorded for the central point position using

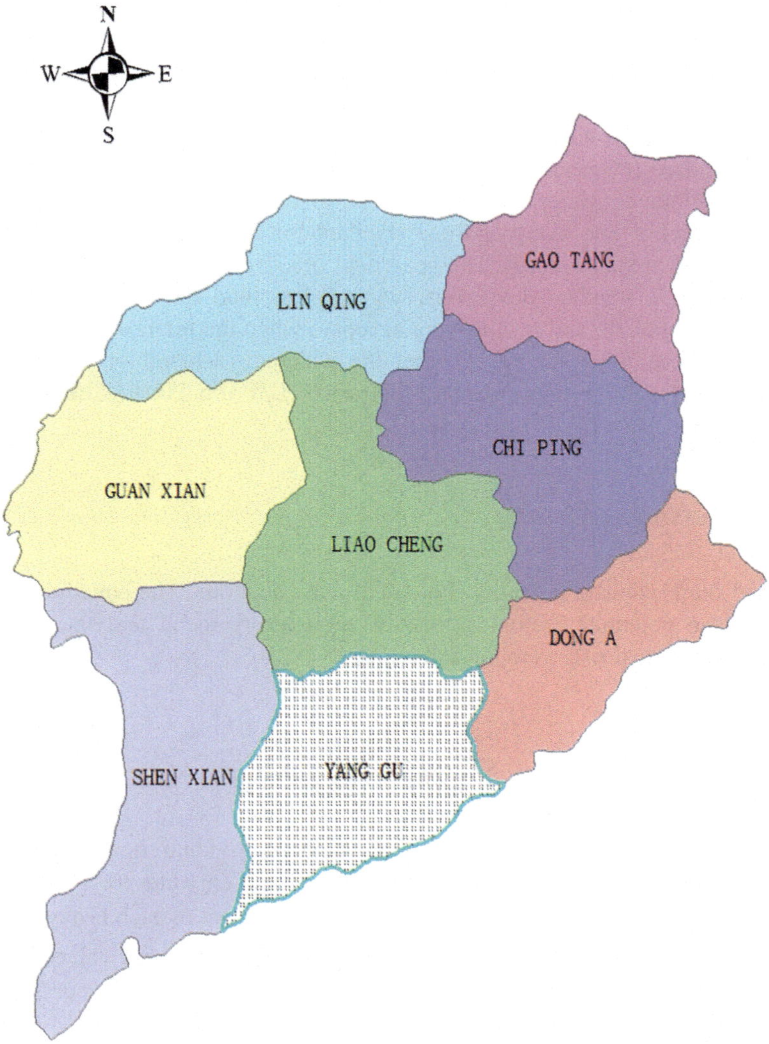

Fig. 1 Study area

GPS. Approximately 1 kg of soil sample was collected at each location using a stainless steel spade. Samples were stored in self-sealing plastic bags.

2.3 Chemical Analysis

All soil samples were air-dried, ground, and sieved through a 2-mm nylon sieve to remove coarse materials and debris. Then a portion of each sample was ground with a mortar and pestle until all particles passed through a 0.15-mm nylon sieve. For analysis of total heavy metal concentrations, 1 g of each dry soil sample was digested in a Teflon tube using a mixture of perchloric acid (HClO4), nitric acid (HNO3), and hydrogen fluoride (HF). The solution of each digested sample was analyzed by inductively coupled plasma atomic absorption spectrometry (ICP/AAS) for the following heavy metals: As, Cd, Cr, Cu, Ni, Pb, and Zn (Meng 2014). Standard reference materials were obtained from the Center of National Standard Reference Materials of China, and blank samples were used with each batch of samples (one blank and one standard for each ten samples) for quality assurance and quality control. The analytical precision values for the tests, measured as relative standard deviations, were all less than 10 %. All samples were analyzed in duplicate and results were accepted when the relative standard deviation was within 5 %. The results met the accuracy demand of the Technical Specification for Soil Environmental Monitoring HJ/T 166–2004 ((SEPAC) 2004).

2.4 Assessment Method

We used the Hakanson potential ecological risk equations. The equations used to compute the potential ecological risk of each heavy metal and the integrated potential ecological risk were the following:

$$E_r^i = T_r^i \times C_f^i, \tag{1}$$

$$\mathrm{IR} = \sum\nolimits_{i=1}^{N} E_r^i = \sum\nolimits_{i=1}^{N} T_r^i \times C_f^i, \tag{2}$$

where T_r^i is the toxic response factor of the individual heavy metal. The corresponding values of Cd, As, Pb, Ni, Cu, Cr, and Zn were 30, 10, 5, 5, 5, 2, and 1, respectively. $C_f^i = C_i/C_r^i$ is the pollution coefficient of each heavy metal; C_i is the concentration of each heavy metal; C_r^i is the recommended values of heavy metal concentration in soils. The paper took the recommended values of Shandong Province (Table 1).

A combination of different heavy metals can result in a relatively greater total potential ecological risk. The Hakanson classification is shown in Table 2.

3 Results and Discussion

3.1 Histogram and Transformation

To obtain the spatial distribution of a heavy metal using the simulation method, we initially tested for satisfaction of the assumption of multivariate normality. The histograms of each heavy metal were generated (Figs. 2 and 3). Cd and Cr had

Table 1 Recommended values of heavy metal concentrations in soil (mg/kg)

	Cu	Zn	Pb	Cd	Cr	Ni	As
Value	24	63.5	25.8	0.084	66.0	25.8	9.36

Table 2 Classification of potential ecological risk

Potential ecological risk					
	Slight	Medium	High	Higher	Highest
E_r^i	<40	40–80	80–160	160–320	>320
IR	<150	150–300	300–600		≥600

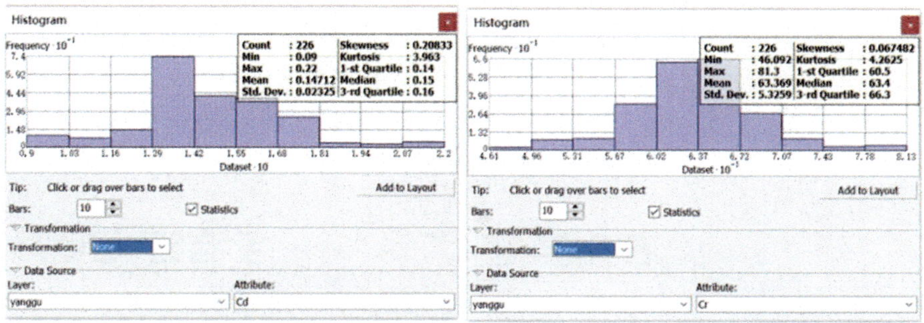

Fig. 2 Histograms of Cd and Cr

almost normal distributions, with low skewness and kurtosis values close to 3. The other five heavy metal values were log transformed to satisfy requirements for the normal distribution.

3.2 *The Semivariance Function Models*

The semivariance function models included the exponential model, Gaussian model, spherical model, and linear model (Fig. 4). The main parameters included nugget (C_0), sill (C_0+C), range (Range), nug/sill ratios ($C_0/(C_0+C)$), coefficient of determination (R^2), and RSS. Among these parameters, we focused on the nug/sill ratio. This parameter is used for spatial heterogeneity and reflects the influence of natural regional factors and the role of man-made non-regional factors. When $C_0/(C_0+C) <0.25$, it indicates that the variable space mutation has been given priority to the structural variation (nature), and the variables have strong spatial correlation. When $0.25 \leq C0/(C0 + C) < 0.75$, the variables have moderate spatial correlation. When $C0/(C0 + C) \geq 0.75$, the variables are random and the spatial correlations are weak . The result of Table 3 shows that the nug/sill ratios of As, Cd, Cr, Ni, Zn, and Cu are between 0.25 and 0.75, indicating a medium spatial correlation, while the nug/sill ratios $C_0/ (C_0+C)$ of Pb are smaller than 0.25.

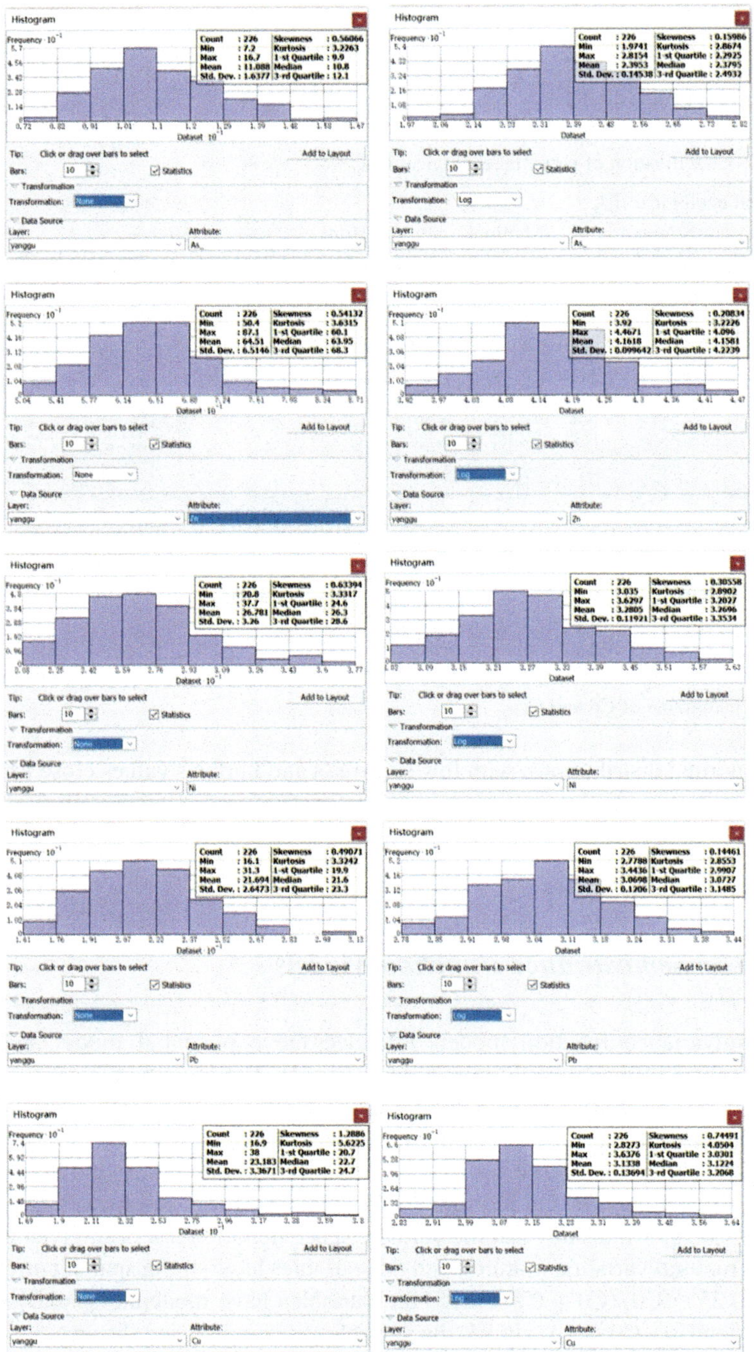

Fig. 3 Histograms of As, Zn, Ni, Pb, and Cu

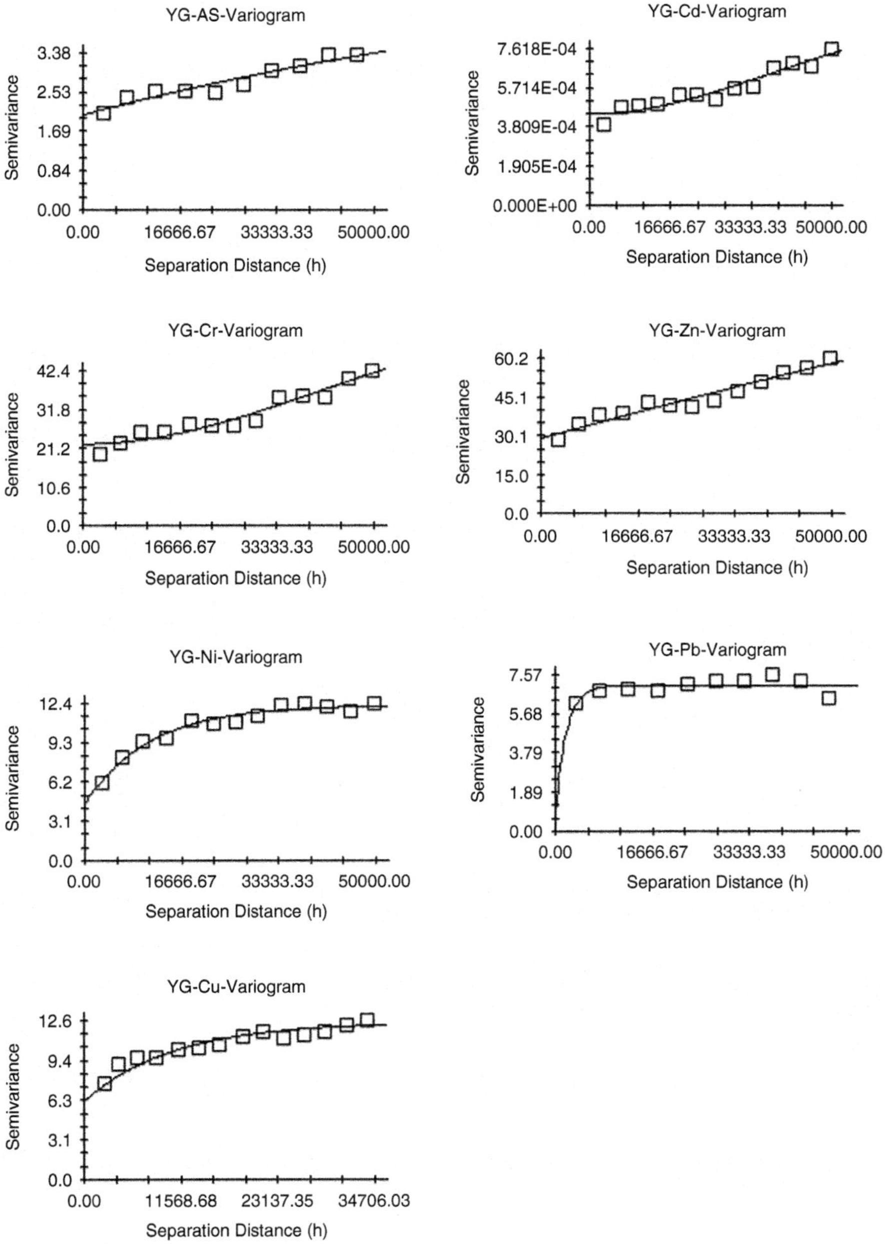

Fig. 4 The semivariogram function diagram of heavy metal in Yanggu

Table 3 Semivariogram fitting of heavy metals in soils from Yanggu

	Model	C_0	Sill	Range	$C_0/(C_0+C)$
As	Exponential	2.04	5.564	314,700	0.366643
Cd	Gaussian	0.000445	0.000934	90,239.847	0.476445
Cr	Gaussian	22.4	59.68	10,1151.767	0.375335
Ni	Exponential	4.53	12.27	34,200	0.369193
Pb	Exponential	0.83	7.068	5,700	0.117431
Zn	Spherical	29.1	83	133,000	0.350602
Cu	Exponential	6.19	12.39	31,650	0.499596

3.3 Spatial Distribution of Heavy Metals in Yanggu

The spatial distributions of the seven heavy metals were based on 1,000 simulations (shown in Fig. 5). We used the upper limit of the background value of heavy metals in soil established by the local government as the standard (Table 1). To present spatial distribution details, the legend of each heavy metal had ten or more intervals.

We did not find an obvious concentration point of Pb, Cr, and Cu in the simulation results. The areas contaminated with Zn, As, and Ni represented 25–50 % of the entire area. Cd pollution was widely spread throughout the area with higher accumulations in eastern and southern locations.

3.4 Probability Map of Heavy Metals in Yanggu

The probability map of each heavy metal was based on the 1,000 realizations, and the corresponding recommended threshold values listed in Table 1. Cr, Pb, and Cu had little probability of exceeding the threshold value (Fig. 6). Some areas for As, Zn, and Ni had probabilities >50 % for exceeding the threshold value. For Cd, almost 80 % of the total area had a probability >90 % of exceeding the threshold value. These results were the same as those from the spatial distribution analysis in Sect. 3.3.

3.5 Potential Ecological Risks of Heavy Metals in Yanggu

An evaluation of the Hakanson potential ecological risk was completed for each heavy metal and the total risk using the contamination sum. The results indicated that Cd was the main heavy metal contaminant problem in Chiping. The potential ecological risk from Cd was nearly 80 and the risk classification was medium (Fig. 7).

The total potential ecological risk from all seven heavy metals was 114.656, and this level of heavy metal pollution poses a slight potential ecological risk (Fig. 8).

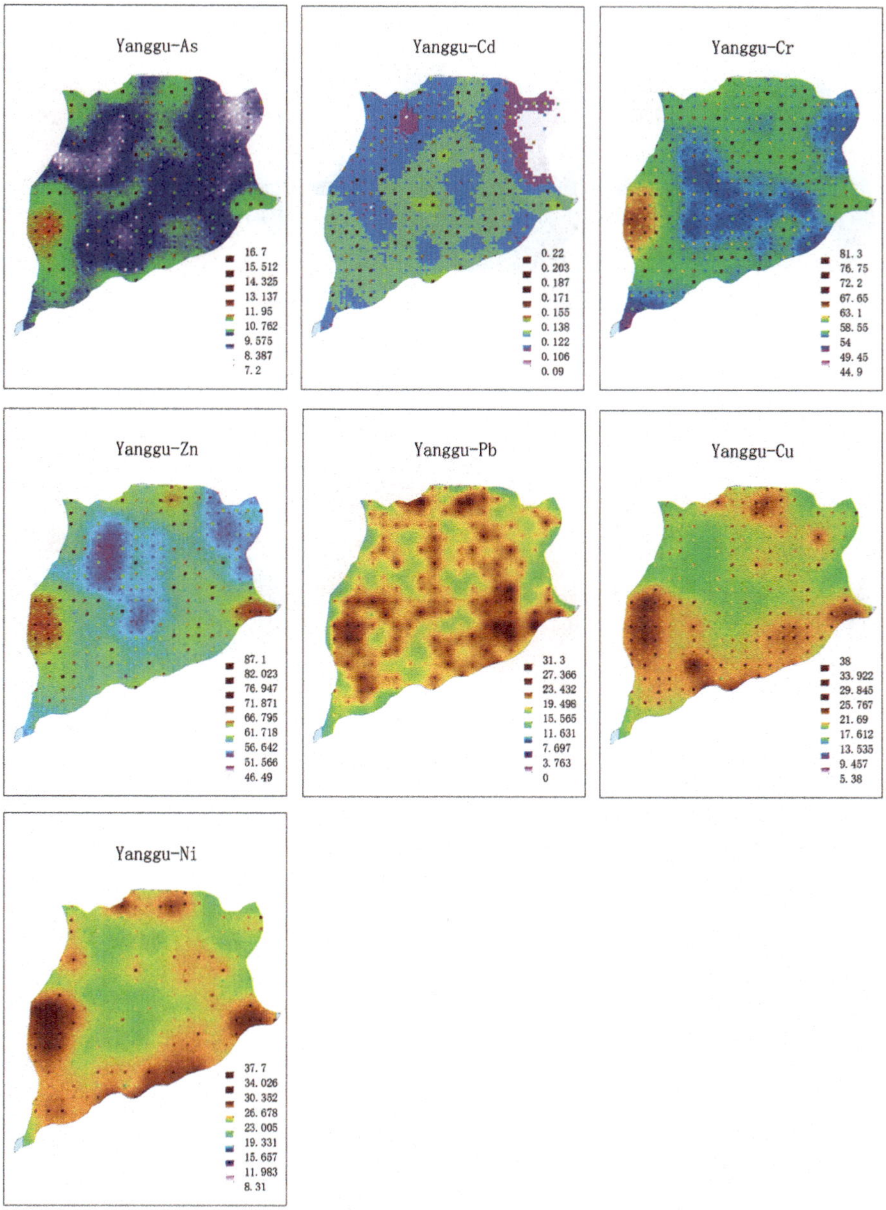

Fig. 5 Spatial distribution of heavy metals in Yanggu

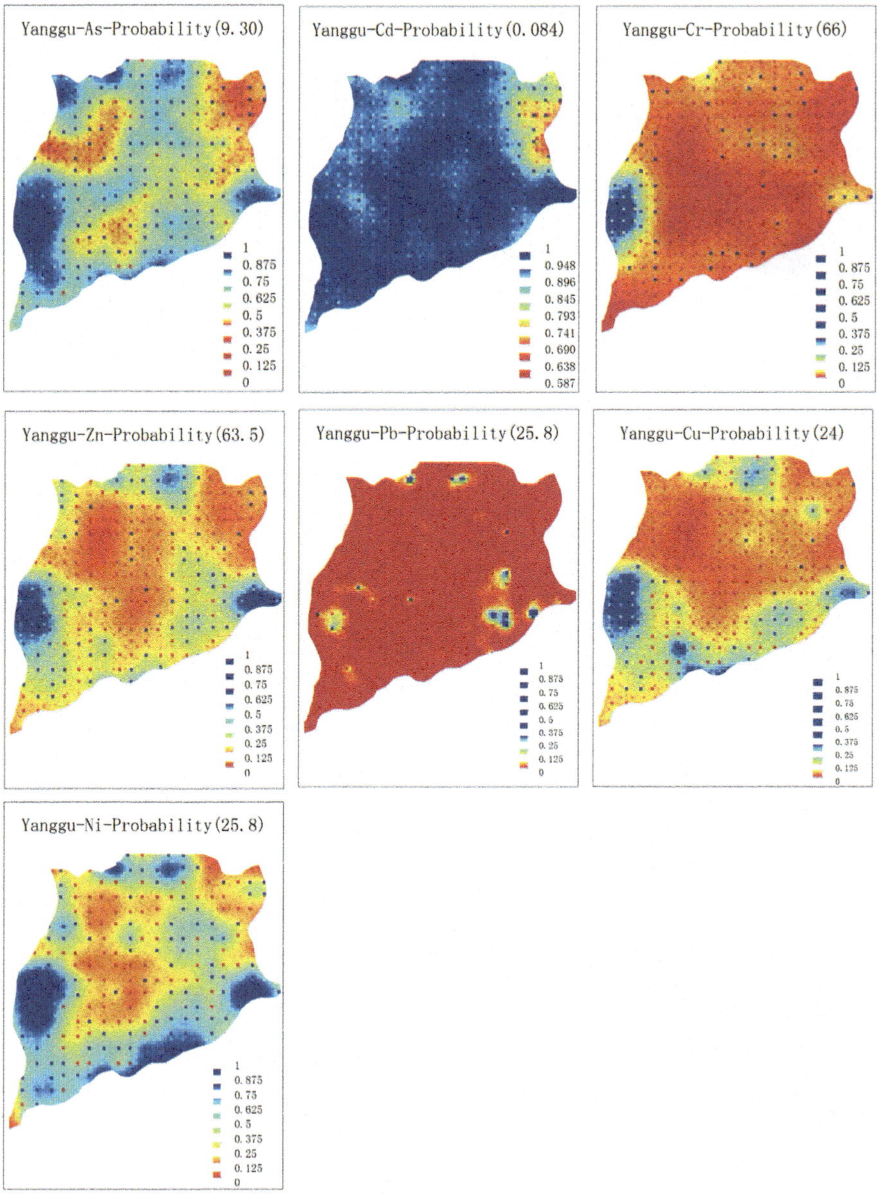

Fig. 6 Probability map of heavy metals in Yanggu

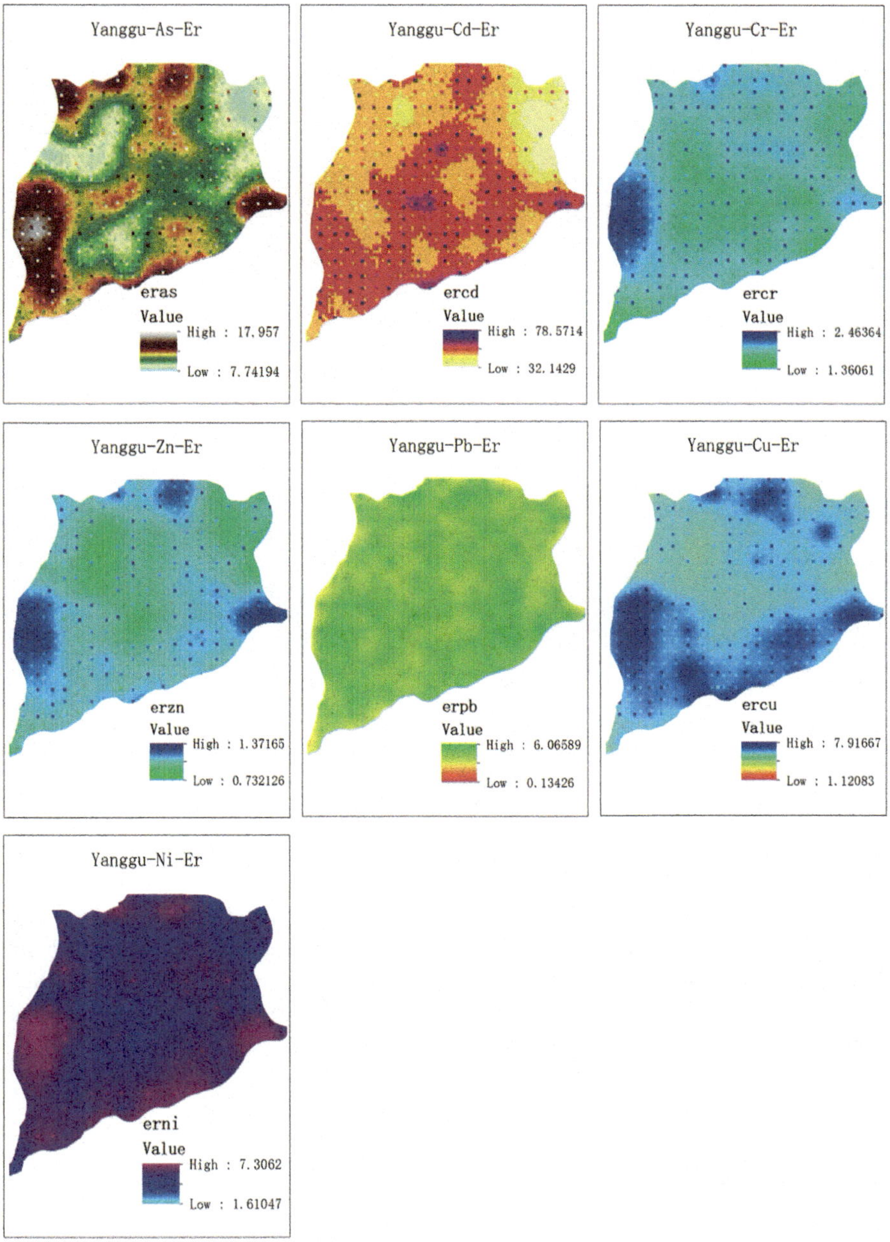

Fig. 7 Potential ecological risks of heavy metals in Yanggu

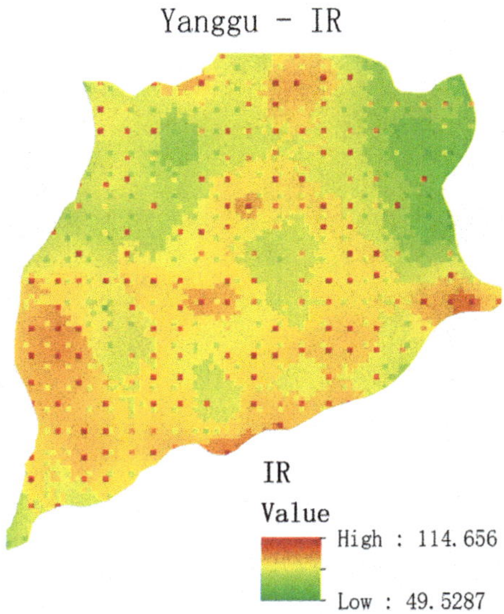

Fig. 8 Integrated potential ecological risk of heavy metal in Yanggu

3.6 Discussion

Analysis of the seven heavy metals showed that the sampling schedule was important. The goal of simulation is to get the attribution of unsampled locations. But the uniform 1-km grid sampling was limited. No information was available for distances smaller than 2 km. This meant that short-distance spatial variation was poorly estimated. For example, the nugget of Pb is forced to zero in Fig. 4, where this was not supported by the data.

In addition, cross-correlations between heavy metals should be taken into account. We found a strong positive relationship between Cu and Zn with the Pearson coefficient 0.714. Thus, a co-kriging approach could be used which estimates values of cross-variograms. This might provide an improved linear unbiased prediction of intermediate values.

4 Conclusions

Levels of seven heavy metals (Cu, Pb, Cd, Cr, Ni, Zn, and As) were studied in agricultural soils. Spatial variation of these heavy metals was determined based on sequential simulation methods. The ecological risk of the heavy metals was assessed using the Hakanson potential danger index. The results indicated that Cd was the main problem in Yanggu. The potential ecological risk from Cd contamination had a value of 80 and a medium risk classification. The total potential ecological risk in the study area from all seven heavy metals had a value of 114.656, and this total constitutes a slight ecological risk.

Acknowledgments This work was supported by the following projects: (1) the National Natural Science Foundation of China (No. 41,301,509, 41,271,413), (2) China Postdoctoral Science Foundation funded project (No. 2014 M561950), and (3) Qingdao Municipal Bureau of Human Resource and Social Security (No. 201,409). The authors thank LetPub (www.letpub.com) for its linguistic assistance during the preparation of this manuscript.

References

(SEPAC), S. E. (2004) The technical specification for soil environmental monitoring HJ/T 166–2004. China Environmental Press, Beijing

Goovaerts P (1997) Geostatistics for natural resource evaluation. Oxford University Press, New York

Goovaerts P (1998) Geostatistics in soil science: state-of-the-art and perspectives. Geodema 89:1–45

Li SQ, Yang JL et al (2014) Atmospheric deposition of heavy metals and their impacts on soil environment in typical urban areas of Nanjing. China Environ Sci 34(1):22–29

Liu HL, Yin CQ et al (2010) Distribution and speciation of heavy metals in sediments at a littoral zone of Meiliang Bay of Taihu Lake. China Environ Sci 30(3):389–394

Lu JS, Zl Z et al (2012) Sources identification and hazardous risk delineation of heavy metals contamination in Rizhao city. Acta Geograph Sin 67(7):971–984

Meng F, Ding N et al (2014) Assessment of heavy metal pollution in Chinese suburban farmland. Pol J Environ Stud 23(6):447–454

Muyessar Trudi JA (2013) Distribution characteristics of soil heavy metal content in northern slope of Tianshan Mountains and its source explanation. Chin J Eco-Agric 21(7):883–890

Sun YJ, DN (2008) Spatial analysis of heavy metals in surface soils based on GeoStatistics. Geoinformatics 2008 and joint conference on GIS and built environment: monitoring and assessment of natural resources and environments, pp 714514-714514-8

Zheng J, Jl Z (2012) Assessment of farmland soil heavy metal pollution in Urumqi Midong sewage irrigation region. Arid Environ Monit 26(1):17–21

Comparison of Trend Detection Approaches in Time Series and Their Application to Identify Temperature Changes in the Valencia Region (Eastern Spain)

Hong Wang, Eulogio Pardo-Igúzquiza, Peter Dowd, and Yongguo Yang

Abstract The identification of systematic small- and intermediate-scale temperature changes (trends) in a time series is of significant importance in the analysis of climate data. This is particularly so in the analysis of local climate change trends and their potential impact on local hydrological cycles. Although many statistical tests have been proposed for detecting these trends their effectiveness is often affected by the presence of serial correlation in the time series. Hence, it is of both interest and necessity to compare the performances of these tests by applying them under a representative range of conditions. In this study, we use Monte Carlo experiments to compare and explore six commonly used tests for detecting trend. For this purpose, we use the confidence level and power to assess the ability to detect trend in two groups of simulated time series with and without serial correlation. The statistical tests are also applied to mean annual temperature measured at 13 weather stations located in the Valencia region (Eastern Spain).

H. Wang (✉) • Y. Yang
School of Resources and Geosciences, China University of Mining and Technology, XuZhou, JiangSu Province 221116, China
e-mail: wanghongcumt@hotmail.com; ygyang88@hotmail.com

E. Pardo-Igúzquiza
Geological Survey of Spain, Ríos Rosas 23, 28003 Madrid, Spain
e-mail: e.pardo@igme.es

P. Dowd
University of Adelaide, Adelaide, Australia
e-mail: peter.dowd@adelaide.edu.au

J.J. Gómez-Hernández et al. (eds.), *Geostatistics Valencia 2016*, Quantitative Geology and Geostatistics 19, DOI 10.1007/978-3-319-46819-8_65

1 Introduction

The identification of systematic small-scale and intermediate-scale temperature changes (trends) in a time series is a significant issue in the analysis of climate data. It becomes especially crucial in the analysis of local climate change and its potential impact on local hydrological cycles. Over the past few decades, a number of studies have been undertaken to evaluate temperature trends in multi-scale climatic and hydrological domains, for example (Gilbert 1987; Lins and Slack 1999; Zhang et al. 2001). The trend detection tests used in these and other similar studies can be grouped into two broad categories: parametric and non-parametric approaches. Despite the success of the parametric approaches, non-parametric techniques are more popular in many applications because they require fewer assumptions. There is a substantial body of published work, particularly in hydrology, that reviews parametric and non-parametric approaches including, among others (Khaliq et al. 2006, 2009; Sonali and Kumar 2013). It is beyond the scope of this paper to explore all statistical tests included previously published studies and we limit our study to six commonly used and representative approaches.

The rank-based Mann-Kendall (MK) and Spearman rank correlation (SRC) tests are frequently used nonparametric tests. Study Yu et al. (1993) indicated that the MK test, supported by a large number of case studies, is more widely used than the SRC test. In contrast, the classical parametric approach of least squares linear regression (LR) has also has been successful in many applications. We group these three methods together because in time series applications they generally assume serial independence. However, this assumption does not always hold. Certain time series, for example, annual mean temperature, frequently display serial correlation. The existence of positive serial correlation has been shown in Von Storch (1995) to increase the possibility of a null hypothesis of no trend being rejected when the null hypothesis is actually true. Various statistical tests have been proposed to address the issues caused by serial correlation, the most well-known being the pre-whitening Mann-Kendall (PW-MK) (Von Storch 1995), the Trend-Free Pre-whitening Mann-Kendall (TFPW-MK) (Yue et al. 2002b) and the Variance Correction (VC) (Yue and Wang 2004) tests. These three tests are modified MK tests and are included in the work presented here.

Because of the significant diversity of climatic and geographical phenomena there is general agreement that there is no one test that surpasses the others in all cases. A suboptimal choice is to find a test that works well for a number of specific representative cases. To do so, we compare the performances of these six statistical tests under a representative range of conditions with the objective of providing useful guidelines. The comparison of MK and SRC tests in Yue et al. (2002a) indicates that the two tests have the same power. In Yue and Pilon (2004) the authors compared the MK test and LR and found that, for normally distributed data, the power of LR is slightly higher than that of the MK test and, for non-normally distributed data, the power of the MK test is higher than that of LR. It should be noted that the work reported in Yue and Pilon (2004) and Yue et al. (2002a) was conducted using serially independent simulation data and did not address confidence levels and power for serially dependent data. In Yue et al. (2002b) the authors proved that the TFPW-MK test has a larger power than that of PW-MK test with AR (1) simulated time series; performances were not assessed for correlated time series with a skewed distribution.

In spite of the very large volume of literature comparing the performances of various tests, there has been very little exploration of these six statistical tests and especially of the assumptions under different models. In this paper, we use Monte Carlo experiments to conduct a comprehensive comparison of the performances of rank-based, LR and modified MK tests with respect to the confidence level and power in order to assess the ability to detect trend in the presence and absence of serial correlation. To do so we use two groups of simulated time series – one with and one without serial correlation.

The remainder of this paper proceeds as follows. A brief description of approaches employed in this study is given in Sect. 2. Section 3 compares the performance of statistical tests using a simulation study. A real case study is presented in Sect. 4. Section 5 provides conclusions from the study.

2 Methodologies

The statistical tests employed in this study are introduced in Sects. 2.1 and 2.2.

2.1 *Statistical Tests for Trend Detection*

The MK test (Kendall 1975; Mann 1945) is a rank-based non-parametric test. The MK statistic S is calculated as:

$$S = \sum_{i=1}^{n-1} \sum_{j=i+1}^{n} \operatorname{sgn}\left(x_j - x_i\right) \tag{1}$$

where $j>i$, x_j and x_i are the values at i and j, respectively; n is the length of the time series, and

$$\operatorname{sgn}(\theta) = \begin{cases} 1 & \text{if } \theta > 0 \\ 0 & \text{if } \theta = 0 \\ -1 & \text{if } \theta < 0 \end{cases} \tag{2}$$

when $n \geq 10$, S is approximately normally distributed (Gilbert 1987) with variance

$$\mathrm{Var}(S) = 1/18\left[n(n-1)(2n+5) - \sum_{k=1}^{m} t_k(t_k-1)(2t_k+5)\right] \tag{3}$$

where m is the number of tied values and t_k is the number of observations in the kth value.

The standardized test statistic Z is computed by

$$Z = \begin{cases} \sqrt{(S-1)}/\sqrt{\mathrm{Var}(S)} & \text{if } S > 0 \\ 0 & \text{if } S = 0 \\ (S+1)/\sqrt{\mathrm{Var}(S)} & \text{if } S < 0 \end{cases} \tag{4}$$

At the α significance level the null hypothesis of no trend is rejected if the absolute value of Z is greater than $Z_{1-1/\alpha}$ which is obtained from standard normal cumulative distribution tables.

LR is a parametric approach. Assuming that there is a linear relationship in the time series, the test statistic t is defined by

$$t = \hat{b}/s(\hat{b}) \tag{5}$$

where $s(\hat{b})$ is the standard error and $\hat{b}$ is the estimated slope. The test statistic t follows a student's t-distribution with n-2 degree of freedom; n is the length of the time series. The null hypothesis of zero slope will be rejected if $|t| > t_{(1-\alpha/2),n-2}$ at the α significance level.

The SRC test (Dahmen and Hall 1990) is a rank-based nonparametric approach used for trend analysis. Spearman's rank correlation coefficient, R_s, and the test statistic, t_s, are defined by

$$R_s = 6\sum_{i=1}^{n} d_i^2/n(n^2-1) \tag{6}$$

$$t_s = R_s\sqrt{(n-2)/(1-R_s^2)} \tag{7}$$

where $d_i = rx_i - ry_i$; rx_i and ry_i are the rank numbers of x_i and y_i, respectively; and n is the length of the time series. When there are ties in rx_i and ry_i the average rank is used. Positive (negative) t_s indicates an increasing (decreasing) trend. At an α significance level the null hypothesis will be rejected if $|t_s| > t_{(1-\alpha/2),n-2}$.

2.2 Approaches for Trend Detection in Serially Correlated Time Series

The PW-MK test is proposed by Von Storch (1995) and implemented in the following steps.

i. Compute the lag-1 correlation coefficient (r_1) (Salas et al. 1980) of the time series ($x_1, x_2, \ldots\ldots, x_n$) with a significance level of α.
ii. If the lag-1 correlation coefficient is significantly different from zero at the specified significance level of α, the original time series will first be pre-whitened as $(x_2 - r_1x_1, x_3 - r_1x, \ldots\ldots, x_n - r_1x_{n-1})$, then MK test is applied to the pre-whitened time series. Otherwise, the MK test is directly applied to the original time series.

The TFPW-MK procedure (Yue et al. 2002b) is applied in the following manner to identify a trend in a time series.

i. Estimate the slope of the original time series using the approaches of Sen (1968) and Theil (1950).
ii. Compute the lag-1 correlation coefficient (Yue and Pilon 2004) of the time series with a significance level of α. If the lag-1 correlation coefficient is not significant, the time series is treated as serially independent and the MK test is applied to the original time series. Otherwise, it is treated as serially correlated, and the pre-whitening technique is applied to the de-trended time series.
iii. The MK test is applied to the blended series obtained by combining the de-trended series and the estimated slope of the trend.

In Yue and Wang (2004) a procedure is proposed for correcting the variance of the MK test statistic S by using an effective sample size (ESS) that takes into account the impact of serial correlation on the variance of S. The variance $V(S)^*$ modified by ESS is

$$\mathrm{Var}(S)^* = \mathrm{Var}(S)n/n^* \tag{8}$$

where n is the actual sample size (ASS) of the actual sample data; n^* is the ESS; *Var* (S) is the variance of S which can be calculated using Eq. 3. The formula for computing n^* for the lag-1 autoregressive process is given in (Matalas and Langbein 1962) as:

$$n^* = n/\left(1 + 2\left(\rho_1^{n+1} - n\rho_1^2 + (n-1)\rho_1\right)/\left(n(\rho_1 - 1)^2\right)\right) \tag{9}$$

The modified standard MK statistic Z^* can be expressed, by Z in Eq. 4 based on the relationship between Var(S) and Var(S)* in Eq. 8, as

$$Z^* = Z\sqrt{n^*/n} \tag{10}$$

3 Simulation Study

3.1 Simulation Data

To provide a comprehensive assessment of the performances of the statistical tests, groups of data were simulated according to the following requirements:

(i) Two types of independent time series: Gaussian and Chi-square time series. The realisations of the Gaussian time series were generated using Sequential Gaussian Simulation and the realisations of the Chi-square time series were obtained by squaring the corresponding Gaussian time series. The Gaussian time series is normally distributed. The Chi-square time series is characterized by a highly skewed probability density with a long right-hand tail.
(ii) Auto-correlated random functions best describe the variability of natural phenomena in time. In this work, the serial correlation is quantified by an exponential semi-variogram. We use two models defined by semi-variogram ranges equal to 10 % and 50 % of the side-length of the geometric field respectively.
(iii) A linear drift (trend) defined by: $m(x) = \beta_0 + \beta_1 x$, where $\beta_0 = 0$.
(iv) Five values of the scales of variability of the residual and the drift given by the ratio σ_L/σ_H equal to 0.1, 0.2, 0.3, 0.4 and 0.5; where σ_L and σ_H denote the standard deviation of the drift and the standard deviation of the residual of the time series, respectively.
(v) Four sizes (lengths) of the time series: 40, 60, 80, 100.

Six types of time series were obtained by combining (i), (ii) with (v): independent Gaussian (denoted Gaussian1), Gaussian with short range (Gaussian2), Gaussian with long range (Gaussian3), independent Chi-square (Chi-square1), Chi-square with short range (Chi-square2), Chi-square with long range (Chi-square3). A total of 1000 of each of the six types of time series without drift were generated and used to estimate the confidence level. Similarly, 1000 of each of the six types of time series with drifts specified by (iii) and (iv) were generated to evaluate the power. Examples of simulated data are shown in Fig. 1.

3.2 Simulation Results

At the significance level of $\alpha = 0.05$ for a two-tailed test, the six statistical approaches were applied to each time series with and without drift. The confidence level and the power were then computed. For the sake of brevity, as the test results are insensitive to different sample sizes, we show in this study the experimental results for sample size 60.

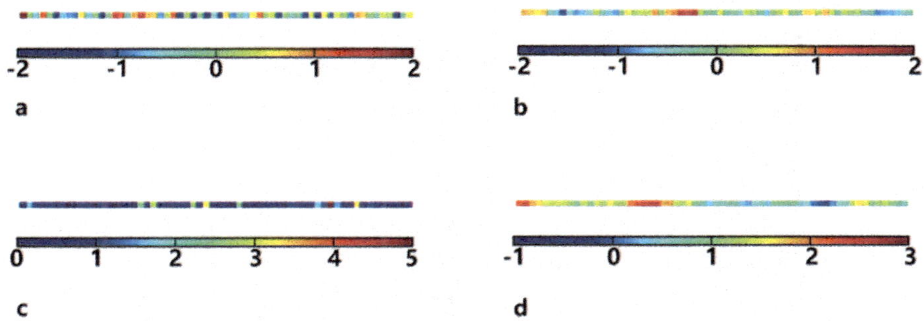

Fig. 1 Four realizations of simulated data used in the simulation study. (**a**) A realization of Gaussian1; (**b**) A realization of Gaussian3 with drift specified by σ_L/σ_H which is equal to 0.3; (**c**) A realization of Chi-square1; (**d**) A realization of Chi-square1 with drift specified by σ_L/σ_H which is equal to 0.3

3.2.1 Confidence Level of Statistic Tests for Six Types of Time Series Without Drift

In this subsection, we assess the influence of different time series in the trend detection analysis. For this purpose, we employ the simulated time series data without drift. The confidence interval is calculated for measuring purposes. Results are summarized in Fig. 2, which shows the estimated confidence levels of the six statistical tests for Gaussian1, Gaussian2, Gaussian3, Chi-square1, Chi-square2 and Chi-square3, respectively, in the first, second, third, fourth, fifth and sixth columns. In particular, Gaussian1 and Chi-square1 are simulated time series without serial correlation while the other four are time series with serial correlation. In this Monte Carlo simulation, the nominal confidence level is 95 % as the nominal significance level is set to 0.05. Using these experimental results, we now compare the performance of the six tests with respect to the presence or absence of time series correlation.

Comparing the first (Gaussian1) and fourth (Chi-square1) columns in Fig. 2, it can be seen that the six statistical tests obtain almost the same confidence level (around 95 %) and achieve the nominal confidence level for independent time series (Guassian1 and Chi-square1). Thus, for time series without serial correlation, irrespective of the type of distribution (Gaussian or Chi-square), the performances of the six tests for trend detection are very similar.

For the simulated data for correlated time series, shown in the second, third, fifth and sixth columns in Fig. 2, the performances of all the six tests are similar. In particular, (1) two of the modified MK tests, PW-MK and TFPW-MK, have almost the same confidence level, which is slightly lower than that of the VC test; (2) the confidence level of the VC test is similar to each of the other four correlated time series and achieves the nominal value (95 %); (3) the confidence level of the three modified MK tests is obviously higher than that of the rank-based tests and the LR test when the time series is serially correlated.

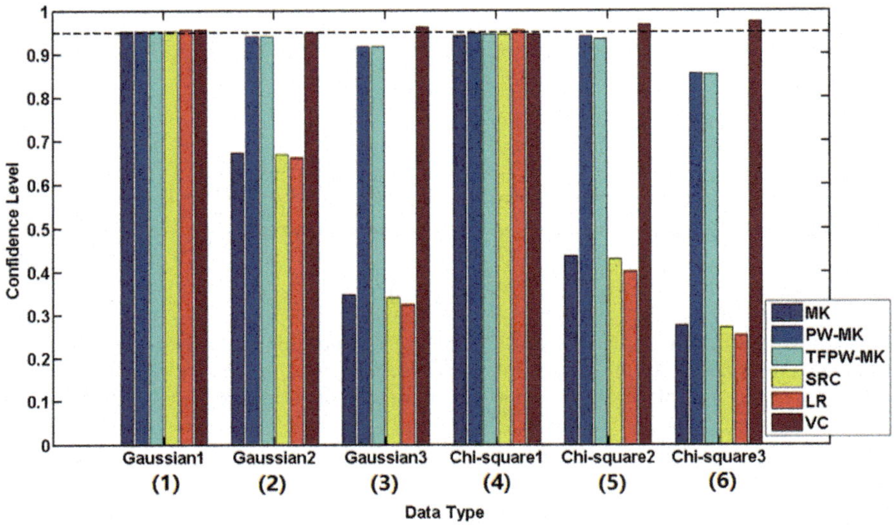

Fig. 2 Confidence levels of the six statistical tests for time series without drift at the significance level of $\alpha = 0.05$

On the other hand, the statistical tests perform differently for the correlated time series. With the exception of the VC test, the second and third columns show that the confidence levels of the six statistical tests for time series with short correlated range (Gaussian2) are higher than the corresponding confidence levels for time series with long correlated range (Gaussian3). This is also the case for the fifth (Chi-square2) and sixth (Chi-square3) columns. In particular, the confidence levels of the PW-MK and TFPW-MK tests for time series with long correlated range is slightly lower than the corresponding values of these two tests for time series with short correlated range. The differences are small and can be accepted. Similar results were obtained in Yue et al. (2002b).

Thus, the three modified MK tests demonstrate almost the same ability to detect no trend in a correlated time series without drift.

3.2.2 Powers of Statistical Tests for Independent Time Series with Drift

In this section, we summarize trend detection with the six tests in order to evaluate their performances on time series without serial correlation and with drift. The power criterion is calculated for measuring purposes. Rows 1–5 of Table 1 present the power of the six tests for uncorrelated normally distributed time series (Gaussian1) with different drifts specified by σ_L/σ_H. It can be seen that the six statistical tests have almost the same power. Thus, for independent normally time series with drift, the six tests demonstrate similar abilities to detect trend.

Rows 6–10 of Table 1 show the power for uncorrelated skewed distributed time series (Chi-square1) with different drifts specified by σ_L/σ_H. It can be seen that the

Table 1 Power of six statistic tests for Gaussian1 and Chi-square1 with different drifts specified by ratios (σ_L/σ_H) at the significant level of $\alpha = 0.05$

Data type	Ratio	MK	PW-MK	TFPW-MK	SRC	LR	VC
Gaussian1	0.1	0.14	0.14	0.14	0.14	0.14	0.14
	0.2	0.37	0.37	0.37	0.37	0.39	0.37
	0.3	0.65	0.65	0.65	0.65	0.67	0.65
	0.4	0.85	0.84	0.84	0.84	0.89	0.85
	0.5	0.94	0.94	0.94	0.94	0.97	0.94
Chi-square1	0.1	0.58	0.58	0.58	0.58	0.15	0.59
	0.2	0.93	0.92	0.93	0.93	0.43	0.93
	0.3	0.99	0.99	0.99	0.99	0.73	0.99
	0.4	1.00	1.00	1.00	1.00	0.89	1.00
	0.5	1.00	1.00	1.00	1.00	0.96	1.00

rank-based tests and the modified MK tests have almost identical powers. However, the LR test has the lowest power among the three groups of tests when drift is present in time series. Thus, for uncorrelated skewed time series with drift, LR has the lowest ability to detect trend.

Table 1 indicate that the LR test is sensitive to the distribution of uncorrelated time series with drift. The other five statistical tests have almost the same ability to detect trend irrespective of whether the time series is Gaussian or Chi-square.

3.2.3 Powers of Statistical Tests for Correlated Time Series with Drift

We conducted the trend detection with the six statistical tests to investigate their performance on correlated simulated time series and with drift. The powers were computed for four types of correlated time series (Gaussian2, Gaussian3, Chi-square2, Chi-square3) with drift and are shown in Table 2.

For the four types of correlated time series with drift in Table 2, common experimental results for the six statistical tests are (1) The LR test and rank-based tests have significantly higher powers than those of the modified MK tests. This is because positive serial correlation within a time series can increase the possibility of a trend (Von Storch 1995). Thus, although the LR and rank-based tests have more power than the modified MK tests, they are not our choices when correlated time series data with drift; (2) the power of the TFPW-MK test is higher than that of other two modified-MK tests (PW-MK and VC). This agrees with the studies in Von Storch (1995) and Yue and Wang (2004) that show that the PW-MK and VC tests underestimated the drift when both trend and positive serial correlation are present in a time series. However, in our case, the power of PW-MK is higher than that of VC for four correlated time series. This implies that, for correlated time series with drift, the VC test is more likely to underestimate the drift than the PW-MK test; (3) powers of the PW-MK and VC tests for time series with short serial correlation (Gaussian2 and Chi-square2) are greater than the corresponding

Table 2 Power of six statistic tests for Gaussian2, Gaussian3, Chi-square2 and Chi-square3 with different drifts specified by ratios (σ_L/σ_H) at the significant level of $\alpha = 0.05$

Date type	Ratio	MK	PW-MK	TFPW-MK	SRC	LR	VC
Gaussian2	0.1	0.37	0.08	0.08	0.38	0.40	0.07
	0.2	0.47	0.12	0.14	0.48	0.50	0.11
	0.3	0.63	0.18	0.20	0.63	0.65	0.17
	0.4	0.75	0.25	0.29	0.76	0.77	0.22
	0.5	0.85	0.32	0.39	0.85	0.87	0.29
Gaussian3	0.1	0.66	0.09	0.10	0.67	0.69	0.04
	0.2	0.71	0.10	0.12	0.72	0.75	0.04
	0.3	0.74	0.11	0.21	0.74	0.77	0.05
	0.4	0.78	0.13	0.30	0.79	0.80	0.06
	0.5	0.82	0.16	0.40	0.83	0.86	0.07
Chi-square2	0.1	0.60	0.07	0.09	0.62	0.63	0.04
	0.2	0.68	0.09	0.14	0.69	0.72	0.05
	0.3	0.79	0.13	0.21	0.79	0.79	0.09
	0.4	0.84	0.17	0.30	0.85	0.86	0.12
	0.5	0.91	0.21	0.42	0.91	0.92	0.17
Chi-square3	0.1	0.73	0.14	0.17	0.74	0.76	0.03
	0.2	0.76	0.16	0.21	0.77	0.79	0.04
	0.3	0.82	0.17	0.28	0.83	0.84	0.06
	0.4	0.86	0.20	0.37	0.87	0.88	0.08
	0.5	0.91	0.23	0.48	0.91	0.92	0.09

values for time series with long serial correlation (Gaussian3 and Chi-square3). This is because the estimated lag1 in a time series with long range is greater than that in a time series with short range. A clear linear positive relationship between lag1 and drift is reported in Yue and Wang (2002).

On the other hand, different performances are observed for the statistical tests for correlated time series with drift. The results show that the power of TFPW-MK for a time series with short range (Gaussian2) is similar to that of a time series with long range (Gaussian3). This is also the case for Chi-square2 and Chi-square3. These results for a Gaussian time series are similar to those reported in Yue et al. (2002b). The results show that the TFPW-MK test can effectively eliminate the serial correlation in the Chi-square long range time series. This can be considered as a supplement to the results in Yue et al. (2002b) as the authors did not consider Chi-square time series.

Thus, we can conclude that the TFPW-MK test has a greater ability than the PW-MK and VC tests to detect trend in correlated time series.

4 Case Study

In this study the six statistical tests were used to assess the significance of trend in mean annual temperatures measured at 13 weather stations located in the Valencia region in Eastern Spain. The spatial distribution of the stations is illustrated in Fig. 3. Of 13 time series, four mean annual temperature time series for stations 1, 5, 8 and 11 are shown in Fig. 4. Weather stations identified by number, X (east-west) coordinate, Y (north-south) coordinate, altitude, time series period, skewness and kurtosis are shown in rows (1–7) of Table 3 (Note that actual Y co-ordinate is Y + 4,000,000). For a normally distributed series, the skewness and kurtosis should be approximately 0 and 3, respectively. From Table 3 it can be seen that the experimental data are negatively skewed and are poorly described by a normal distribution.

The lag-1 correlation coefficient and its lower and upper confidence interval limits at the 0.05 significance level are shown in rows (8–10) of Table 3. All mean annual temperature time series are serially correlated except that for station 2. The magnitude of the slope, computed using Theil-Sen (Sen 1968; Theil 1950), is shown in column (11) of Table 3. The P-values (Yue and Pilon 2004) of MK, SRC, LR, PW-MK, TFPW-MK, VC tests are shown in column (14–17). The

Fig. 3 Locations of 13 weather stations in the Valencia region (Eastern Spain). In the figure, *triangles* are station locations and the *red circle* is the city of Valencia

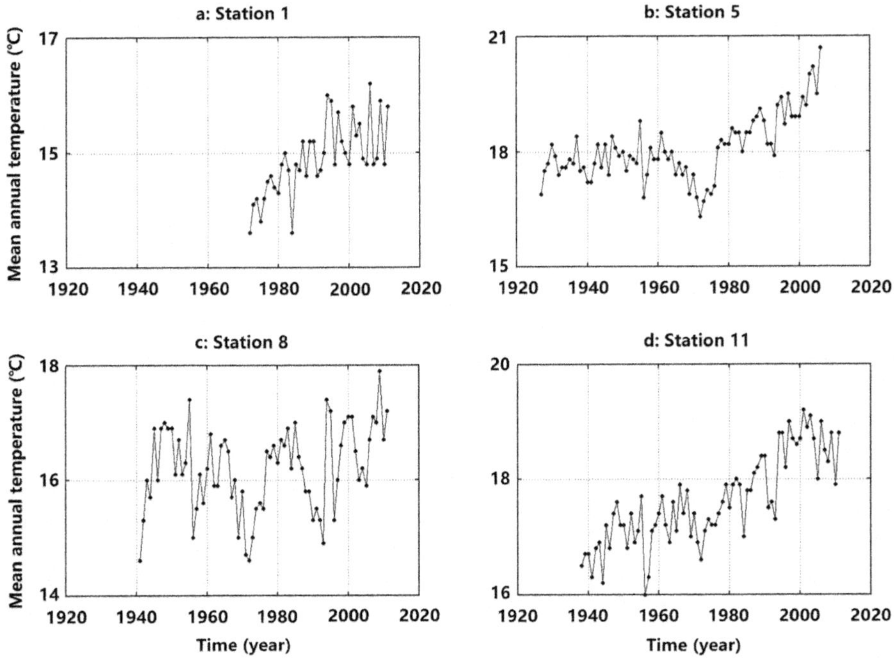

Fig. 4 Mean annual temperature time series for stations 1, 5, 8 and 11

P-values of the modified MK tests are obviously higher than those of the rank-based and LR tests for all stations except station 2. In addition, the VC test has the highest P-values for almost all stations. We conclude that the results of this real case study are entirely consistent with the simulation results.

5 Conclusions

This study investigated the performances of six commonly used trend detection approaches (two rank-based, LR and three modified MK tests) on two groups of simulated time series with and without serial correlation. The theoretical component of the study used Gaussian and Chi-square distributions as representative symmetrical and skewed distributions; used semi-variogram models with different ranges to quantify correlation in time series; and used linear forms of drift to asses trends. We used confidence level and power to quantify the ability of the six tests to detect trend in serially correlated, and in uncorrelated, time series. The Monte Carlo simulation experiments document the performances of the different statistical tests performances on different simulated time series. As an outcome of this work we recommend the following strategy for choosing a suitable statistical test: (a) assess the distribution of the time series (Gaussian or otherwise); (b) determine whether

Table 3 Basic information and P-values of mean annual temperature at 13 stations

Station (1)	1	2	3	4	5	6	7	8	9	10	11	12	13
X Co-ord. (2)	680807	684514	712468	718904	701067	700857	629220	707758	726991	717204	726514	714909	750113
Y Co-ord. (3)	283759	248681	239983	250120	205830	212364	377457	300798	340568	373677	373419	414204	427117
Altitude (4)	515	500	43	81	1	2	652	350	12	69	11	364	35
Period (5)	1972–2011	1977–2011	1967–2011	1946–2011	1927–2006	1947–2006	1966–2010	1941–2011	1987–2011	1966–2011	1938–2010	1945–2001	1976–2009
Skewness (6)	0.06	−0.56	−0.11	−0.29	0.62	−0.11	0.26	−0.24	−1.4	−0.05	0.23	−0.32	−0.2
Kurtosis (7)	2.75	2.7	1.81	2.45	3.55	2.4	2.85	2.44	4.89	2.19	2.28	2.9	2.05
Lag-1 (8)	0.37	0.21	0.71	0.49	0.72	0.65	0.58	0.47	0.51	0.76	0.73	0.39	0.75
Upper limit (9)	0.28	0.3	0.27	0.23	0.21	0.24	0.27	0.22	0.35	0.27	0.21	0.28	0.31
Lower lim. (10)	−0.34	−0.36	−0.31	−0.26	−0.23	−0.27	−0.31	−0.25	−0.43	−0.31	−0.24	−0.33	−0.37
Slope (11)	0.036	0	0.027	0.007	0.021	0.021	0.016	0.009	0.03	0.04	0.03	0.023	0.069
MK (12)	0	0.455	0	0.039	0	0	0.062	0.031	0.126	0	0	0.003	0
SRC (13)	0	0.401	0	0.039	0	0	0.029	0.022	0.081	0	0	0.05	0
LR (14)	0	0.292	0	0.067	0	0	0.072	0.020	0.042	0	0	0.015	0
TFPW-MK (15)	0.026	0.455	0.282	0.301	0.013	0.022	0.396	0.351	0.491	0.167	0.017	0.03	0.229
PW-MK (16)	0.028	0.455	0.292	0.315	0.018	0.025	0.388	0.335	0.491	0.182	0.023	0.032	0.234
VC (17)	0.001	0.463	0.079	0.148	0.008	0.025	0.209	0.129	0.251	0.042	0.001	0.032	0.010

the time series is correlated; and (c) choose a test from the analysis in Sect. 3 based on (a) and (b). We also remark here that our case study could form the basis of a feasible way of exploring proper test techniques for detecting trend.

References

Dahmen ER, Hall MJ (1990) Screening of hydrological data. International institute for land reclamation and improvement (ILRI), Netherlands

Gilbert RO (1987) Statistical methods for environmental pollution monitoring. Van Nostrand Reinhold Company, New York

Kendall MG (1975) Rank correlation methods. Griffin, London

Khaliq MN, Ouarda TBMJ, Gachon P, Sushama L, St-Hilaire A (2009) Identification of hydrological trends in the presence of serial and cross correlation: a review of selected method and their application to annual flow regimes of Canadian rivers. J Hydrol 368:117–130

Khaliq MN, Ouarda TBMJ, Ondo JC, Boee B (2006) Frequency analysis of a sequence of dependent and/or non-stationary hydrometeologyical observations: a review. J Hydrol 329:534–552

Lins HF, Slack JR (1999) Streamflow trends in the United States. Geophys Res Lett 26:227–230

Mann HB (1945) Nonparametric tests against trend. Econometrica 13:245–259

Matalas NC, Langbein WB (1962) Information content of the mean. J Geophys Res 67:3441–3448

Salas JD, Delleur JW, Yevjevich V, Lane WL (1980) Applied modelling of hydrologic time series. Water Resources Publications, Littleton

Sen PK (1968) Estimates of the regression coefficient based on Kendall's tau. J Am Stat Assoc 63:1379–1389

Sonali P, Kumar DN (2013) Review of trend detection methods and their application to detect temperature changes in India. J Hydrol 476:212–227

Theil H (1950) A rank-invariant method of linear and polynomial regression analysis. Proc K Ned Acad Wet 53:1397–1412

Von Storch VH (1995) Misuses of statistical analysis in climate research. In: von Storch H, Navarra A (eds) Analysis of climate variability: applications of statistical techniques. Springer, Berlin, pp 11–26

Yu YS, Zou S, Whittemore D (1993) Non-parametreic trend analysis of water quality data of river in Kansas. J Hydrol 150:61–80

Yue S, Pilon P (2004) A comparation of the power of the t test, Mann-Kendall and bootstrap tests for trend detection. Hydrol Sci J Sci Hydrol 49:21–37

Yue S, Pilon P, Cavadias G (2002a) Power of the Mann-Kendall and Spearman's rho tests for detecting monotonic trends in hydrological series. J Hydrol 259:254–271

Yue S, Plion P, Phinney B, Csvadias G (2002b) The influence of autocorrelation on the ability to detect trend in hydrological series. Hydrol Process 16:1807–1829

Yue Y, Wang CY (2002) Applicability of prewhitening to eliminate the influence of serial correlation on the Mann-Kendall test. Water Resour Res 38

Yue S, Wang CY (2004) The Mann-Kendall test modified by effective sample size to detect trend in serially correlated hydrological series. Water Resour Manag 18:201–218

Zhang X, Harvey KD, Hogg WD, Yunzy TR (2001) Trend in Canadian streamflow. Water Resour Res 37:987–998

Part VII
Big Data

Urban Dynamic Estimation Using Mobile Phone Logs and Locally Varying Anisotropy

Oscar F. Peredo, José A. García, Ricardo Stuven, and Julián M. Ortiz

Abstract In telecommunications, the billing data of each telephone, denoted call detail records (CDRs), are a large and rich database with information that can be geo-located. By analyzing the events logged in each antenna, a set of time series can be constructed measuring the number of voice and data events in each time of the day. One question that can be addressed using these data involves estimating the movement or flow of people in the city, which can be used for prediction and monitoring in transportation or urban planning. In this work, geostatistical estimation techniques such as kriging and inverse distance weighting (IDW) are used to numerically estimate the flow of people. In order to improve the accuracy of the model, secondary information is included in the estimation. This information represents the locally varying anisotropy (LVA) field associated with the major streets and roads in the city. By using this technique, the flow estimation can be obtained with a better quantitative and qualitative interpretation. In terms of storage and computing power, the volume of raw information is extremely large; for that reason big data technologies are mandatory to query the database. Additionally, if high-resolution grids are used in the estimation, high-performance computing techniques are necessary to speed up the numerical computations using LVA codes. Case studies are shown, using voice/data records from anonymized clients of Telefónica Movistar in Santiago, capital of Chile.

1 Introduction

The usage and ubiquity of mobile phones in all countries and social groups are generating an unprecedented amount of behavioral data. According to the International Telecommunication Union (ITC 2015), by the end of 2015, there are more

O.F. Peredo (✉) • J.A. García • R. Stuven
Telefónica R & D, Chile, Av. Manuel Montt 1404 3er piso, 46022 Santiago, Chile
e-mail: oscar.peredoa@telefonica.com

J.M. Ortiz
Department of Mining Engineering, University of Chile, Santiago, Chile

J.J. Gómez-Hernández et al. (eds.), *Geostatistics Valencia 2016*, Quantitative Geology and Geostatistics 19, DOI 10.1007/978-3-319-46819-8_66

than seven billion mobile cellular subscriptions, corresponding to a penetration rate of 97 %, up from 738 million in the year 2000. Accompanying this large pool of users comes an even larger dataset, commonly known as call detail records (CDRs), which has been used for billing purposes since the early days of telecommunications. Each record stores the activity of a user, storing the origin and destination phone number, the amount of minutes or kilobytes consumed in the event, the antenna index in which the event happened, and several other fields. Each time a user starts a phone call or consumes a fixed amount of kilobytes, a CDR event is logged in the mobile phone operator servers. The richness of this dataset has attracted researchers from different fields, all of them interested in understanding the human behavior at individual and aggregate levels. Notable contributions have been developed in the fields of transportation (Lima et al. 2016; Colak et al. 2015; Toole et al. 2015), economic and public health development (Soto et al. 2011; Mao et al. 2013; Oliver et al. 2015), and human urban mobility (Song et al. 2010; Gonzalez et al. 2008).

Regarding human urban mobility inference using CDR datasets, the simplest model uses a tower-based resolution where all devices are mapped to the geographical positions of the underlying tower or *base transceiver station* where their events are registered (Song et al. 2010; Gonzalez et al. 2008; Candia et al. 2008). This approach, although simple and straightforward, doesn't reflect the true position of the devices and also uses a comparatively small number of sample points in the map where events are being placed. An improvement on the later method is based in Voronoi diagrams calculated using the tower positions, and posteriorly the antennas attached to each tower can be decoupled and placed in new virtual locations (Horanont and Shibasaki 2008; Horanont 2012). Specifically, each Voronoi polygon is divided in several pie chart sections centered in the tower, where each section is oriented using the azimuth of the corresponding antenna. A rather different approach can be applied using probabilistic simulations of each device position, by estimating an a priori cumulative density function on the device location related with the tower location (Traag et al. 2011). In order to use this approach, a parameter that controls the speed of signal decay must be inferred from field data in each region of interest. If real-time network monitoring is available for the mapping estimation, like GPS or network package monitoring (Leontiadis et al. 2014; Görnerup 2012; Saravanan et al. 2011), in addition to an antenna radiation and propagation models (Calabrese et al. 2011), higher resolution location maps can be obtained at a higher technological complexity.

In the previous approaches, the location of people's devices is inferred without considering the urban landscape, particularly the influence of highways, railroads, and primary and secondary streets, among many other routes followed by pedestrians, cars, buses, and taxicabs. In the present work, we propose the introduction of street-level information into estimations of the urban dynamics, particularly using the geostatistical tools that allow the usage of locally varying anisotropy (LVA) fields (Boisvert and Deutsch 2011). LVA field-based methods have been developed to reduce the underlying uncertainty in natural resource estimation by adding additional sources of information to assess non-stationarity, like geophysical remote sensing data, discrete point measurements of orientation, or geological solid models. We postulate that this approach will serve to build better models for prediction and monitoring in any process related with urban mobility that uses population estimates based in CDR datasets.

The next section presents the problem of urban dynamic estimation in the context of telecommunications. Section 3 presents the proposed methodology to include the primary and secondary streets of a city into the urban estimation based in telecommunication billing data. In Sect. 4 we explain the technical difficulties associated to the extraction, aggregation, and estimation of the databases with observations in the city map. A sample estimation using data from the city of Santiago, capital of Chile, is presented in Sect. 5. Finally, conclusions and future ideas to improve and validate the proposed methodology are presented in the last section.

2 Urban Dynamic Estimation Using CDRs

Call detail records are the essential data stored for billing purposes by all telecommunication companies. In Table 1 we can observe some of the fields stored in this database, with sample data attached. Telefónica Movistar stores these records using a large number of column fields (more than 60) and keeps each day of records in a long-term storage (more than 1 year of data). In this study only voice and data records are collected. The voice CDR contains the duration, origin antenna, and destination antenna, storing the antenna IDs registered at the beginning of the call. The data CDR contains the size in kilobytes of the transaction and the origin antenna, but in this case the timestamp of the event can be registered with a delay fluctuating between 0 and 60 min on average. Although this delay can affect the final results, theoretical studies (Song et al. 2010) have shown that even using events with timestamp errors of 60 min approximately, the patterns of movement of each mobile phone can be predicted with 93 % of accuracy. The CDR dataset, in voice and data, has inherently low spatial (at antenna/tower scale) and temporal resolution (not enough to estimate velocities or other short-range properties), but contains the mobility patterns of all active cellular subscribers without support of

Table 1 Call detail records schema and a sample register

Fields	Sample value
Origin number (anonymized)	6957491b18b029f3f95e666c651ba7
Destination number (anonymized)	bd48d4c6c086d985da6a2efa698fa0
Date (yymmdd)	140823
Time (hhmmss)	084533
Duration (seconds or KB)	447
Origin antenna ID	CURZFU3
Destination antenna ID	ATLBAL1

additional data, like GPS or network monitoring tools. This feature makes the CDR dataset very attractive to perform studies on mobility for urban planning and transportation policies, among many other use cases.

Regarding the antennas and towers, each antenna is attached to a specific tower, colloquially known as base transceiver station (BTS). In Fig. 1, top, we can observe several towers located in a central neighborhood of Santiago, Chile. Each tower can have attached several antennas of different technologies (2G, 3G, or 4G), as shown in Fig. 1, bottom. As mentioned in the previous section, there are algorithms that can decouple the location of each antenna from the tower, increasing the number of geographical positions with valuable data in the map. However, in this work only the tower-based approach is used, where all antennas attached to the same tower have the same geographical positions. Future versions of the method will explore estimations increasing the number of locations for each antenna in the map, including the antenna orientation and other geometrical and physical parameters.

With both data sources, the CDRs and the tower/antenna positions, we can count the number of events registered in each tower/antenna in each second of the day. For a fixed timestamp, we can define $Z(\mathbf{x})$ as the random field that represents the number of events in the position $\mathbf{x}$. If $\{\mathbf{x}_a\}$ is the set of positions where the towers/antennas are located (without repetition, in case of erroneous tabulation of the tower/antenna positions) and are registering CDR events, we can estimate the number of events in the geographical zone of study (the complete city or some specific regions) using geostatistical tools like inverse distance weighting (IDW) and simple or ordinary kriging (Chilès and Delfiner 2012). In the related literature, typically inverse distance weighting (IDW) is applied for the estimation (Horanont and Shibasaki 2008; Horanont 2012), since no variographic analysis is involved in the workflow. As a drawback, the estimation is highly smooth and does not reflect the landscape features, such as empty lots or large open spaces, where the probability of finding a mobile device is relatively low compared to a street or a highway (Fig. 2).

As the populations of interest are devices in movement around the city, only outdoor antennas are used and a mobility filter is applied, where a minimum different tower indicator is calculated for each device (if the device does not register events in at least N different antennas on different towers, it is not included in the mobility analysis). The indoor antennas capture with higher probability the events of devices that are not in movement in the streets; for this reason they are not included in the study (the inverse is not true, since an outdoor antenna can capture events from indoor devices, but there are no methods to infer this using CDRs). Since the average distance between two towers fluctuates between 400 and 2,500 m approximately, the filter value is set to $N = 2$, meaning that each device must register a movement of at least 400 m between two different towers (approximately 20 % of the devices in this study can be dropped). Although the mobility filter works on the majority of cases, there are some special behaviors of the cellular network where a device can be "jumping" between different antennas located in the same or different towers, even if the device is not in movement. This behavior is triggered by network congestion and antenna misconfiguration, and according to recent

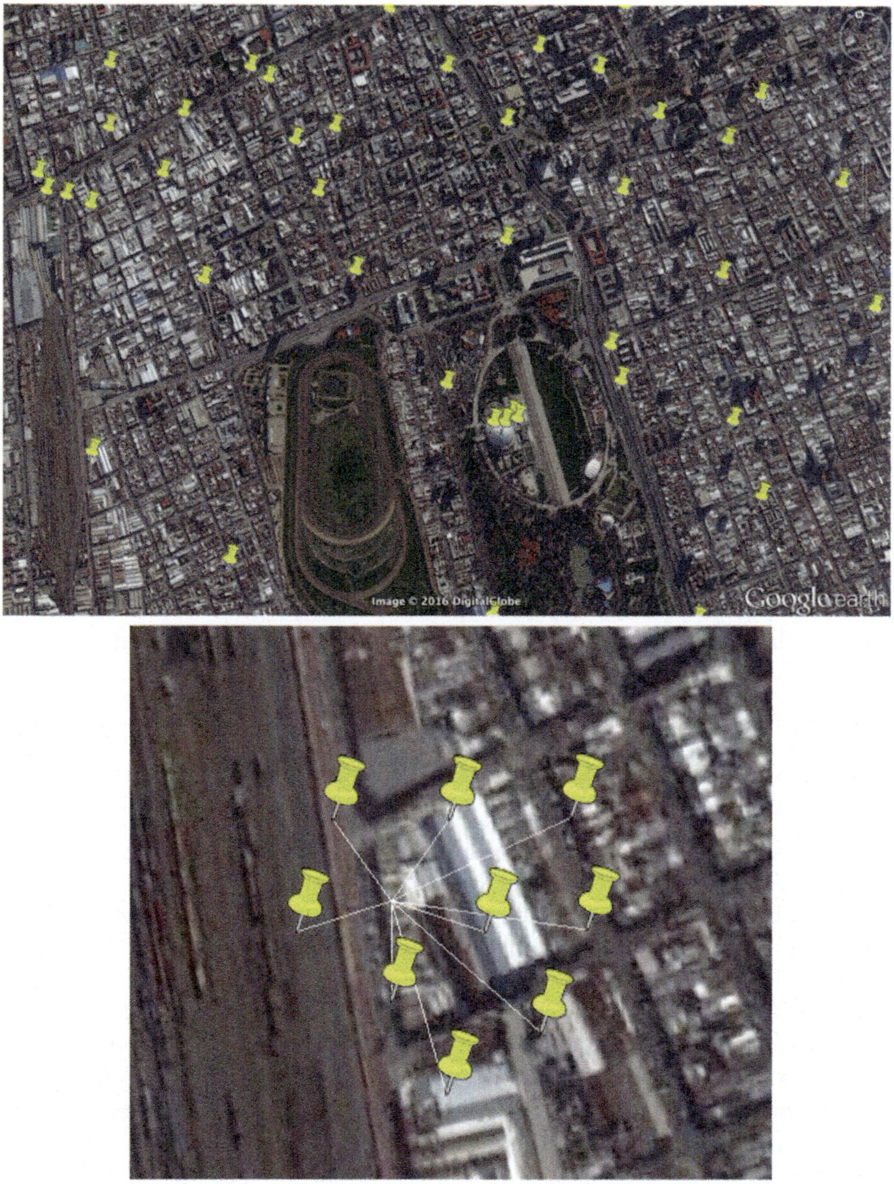

Fig. 1 *Top*: towers of antennas located in a neighborhood of Santiago, Chile. *Bottom*: zoom in a tower with several antennas attached

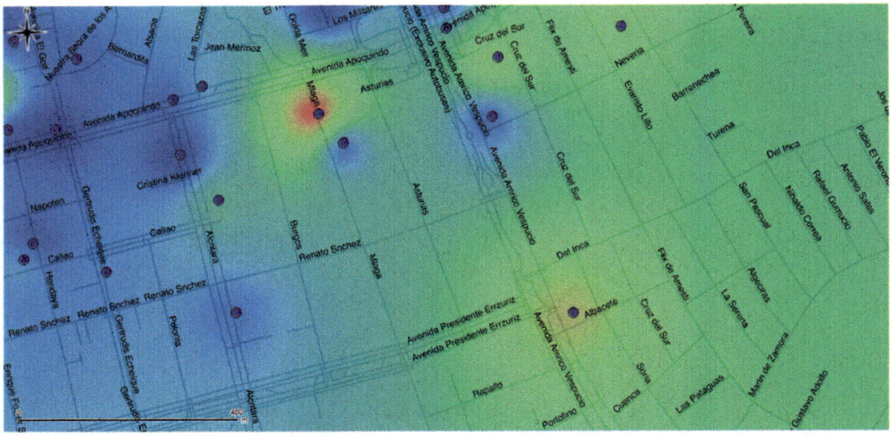

Fig. 2 IDW applied to the samples collected in towers/antennas located in a neighborhood of Santiago, Chile (*Red*, high number of events; *blue*, low number of events). The urban landscape, such as the streets and highways, is not included in the estimation

studies (Li et al. 2016), it represents less than 3 % of the observed records in network data and an order of magnitude less in the aggregated CDR dataset.

3 LVA Field Inference from City Streets

Locally varying anisotropy can be included in geostatistical estimation or simulation processes in several forms (Boisvert et al. 2009). The inference of a 2D LVA field has been studied in Maksuda and Boisvert (2015), with success stories in natural resource estimation, particularly in the mining and petroleum industries. In the specific scenario of urban dynamic estimation and using a proper scale, the streets and highways can be viewed in a similar way as iso-surfaces of geological models, exhaustive geophysical seismic images, or discrete point measurements in the context of natural resource estimation. The kind of secondary information will depend on the quality of the underlying maps and metadata descriptions of the objects in the map. In the case of Santiago, Chile, an exhaustive database of volunteered geo-located objects can be extracted from the OpenStreetMap platform (Haklay and Weber 2008). Figure 3 shows all streets registered in the platform until March 2016 (total area is approximately 50×50 km^2). In Fig. 4 we can observe a specific gridded zone of the city, used as background in Fig. 2 (squares of 50×50 m). The total number of objects labeled as streets in the platform is approximately 300,000, with the most important categories as *bridleway*, *construction*, *cycleway*, *escape*, *footway*, *living_street*, *motorway*, *path*, *pedestrian*, *primary*, *residential*, *road*, *secondary*, *service*, *tertiary*, *track*, and *unclassified*. From these categories, the ones that are representative of streets with a considerable

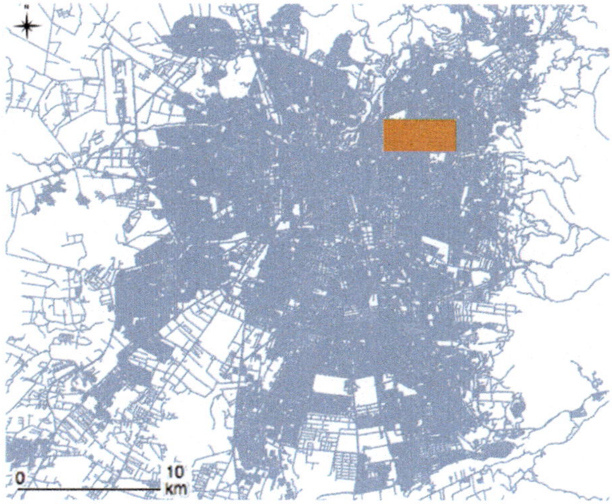

Fig. 3 Geo-located streets from Santiago, Chile, extracted from the OpenStreetMap platform

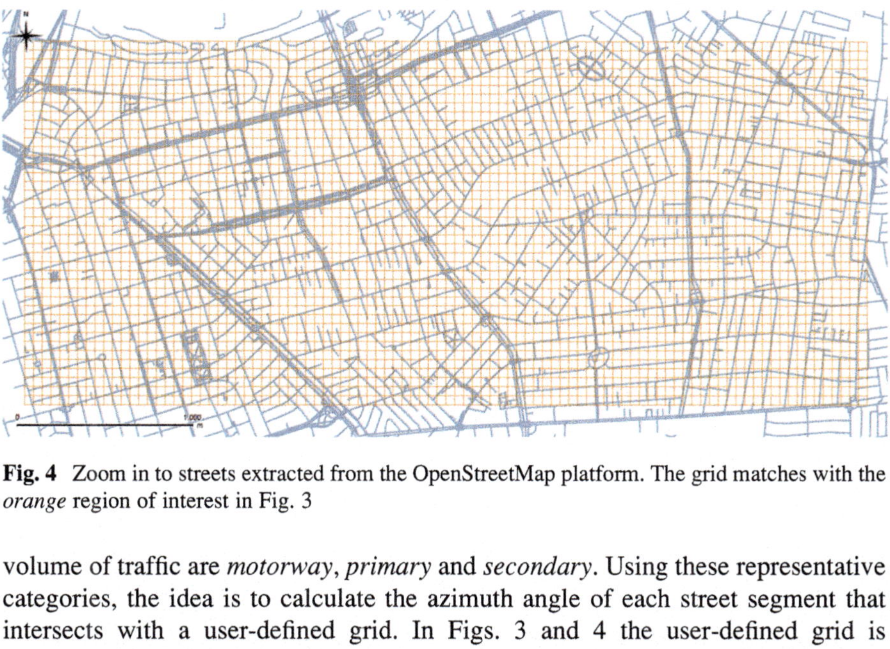

Fig. 4 Zoom in to streets extracted from the OpenStreetMap platform. The grid matches with the *orange* region of interest in Fig. 3

volume of traffic are *motorway*, *primary* and *secondary*. Using these representative categories, the idea is to calculate the azimuth angle of each street segment that intersects with a user-defined grid. In Figs. 3 and 4 the user-defined grid is represented by orange regular squares. In Fig. 5 we can observe all the intersections with their corresponding azimuth values (arrow orientation w.r.t. north) and also a zoom to a particular region of the same grid, where the intersection of the street segments and the grid square sides can be observed.

A standard IDW interpolation is applied to the azimuth values of the sparse 2D field from the previous section. After that, a full 2D LVA field is reconstructed as shown in Fig. 6. We can observe that the streets with wide azimuth are colored in white, and the streets with narrow azimuth in black. With the full 2D LVA field, we

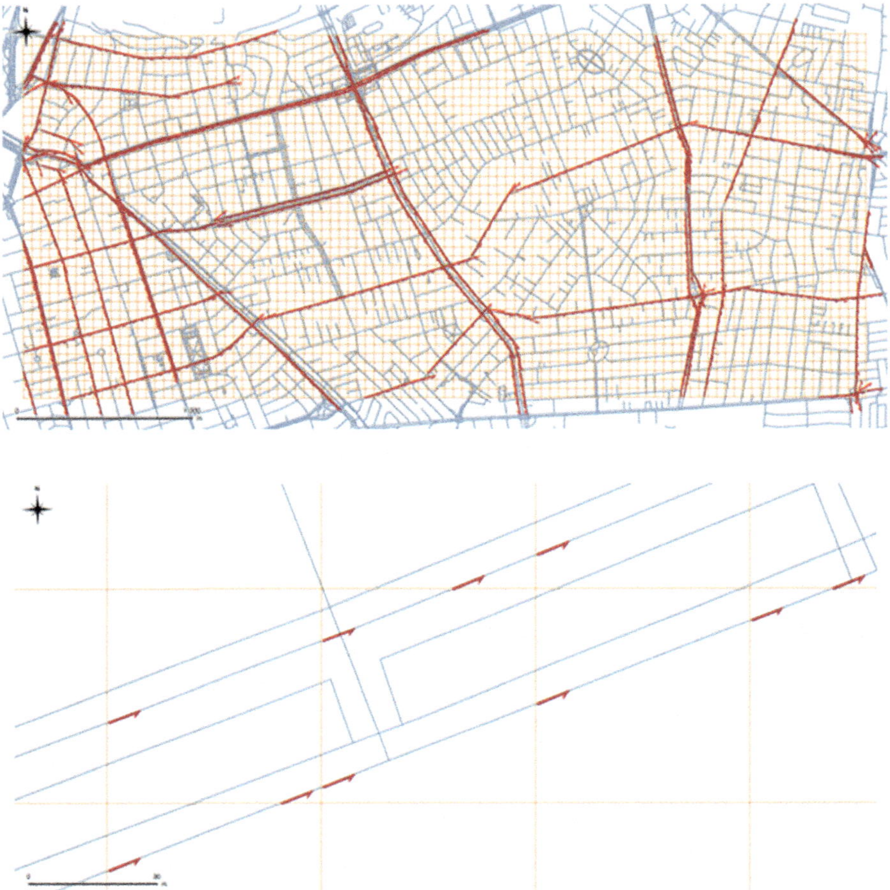

Fig. 5 *Top*: primary and secondary streets in the zone of interest (*red arrows*). *Bottom*: zoom in to observe the intersections between the street segments and the grid square sides (*arrows* with orientation)

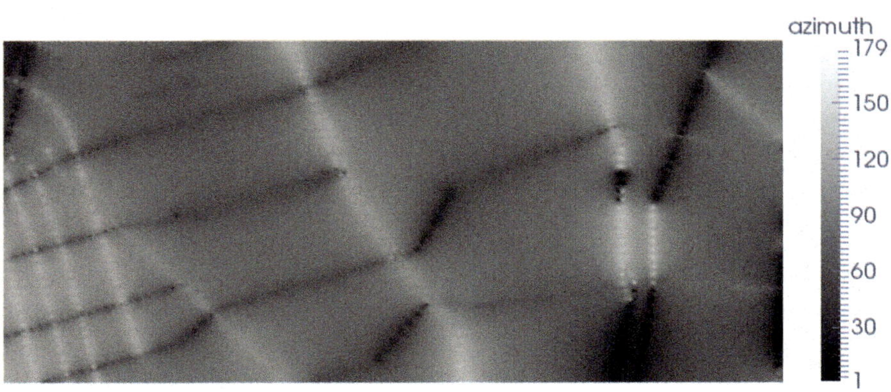

Fig. 6 IDW interpolation (*degrees*) of a 2D LVA sparse field

can use LVA-based geostatistical codes (Boisvert and Deutsch 2011) to obtain the estimation, adding the ratio between the X and Y axes.

4 Technical Aspects

4.1 Big Data

Up to this point, the CDR data have been assumed to exist in an aggregate form that can be used to extract the event number $Z(\mathbf{x}_a)$ in each sampled position $\mathbf{x}_a$ corresponding to tower/antenna locations. However, it is not straightforward to obtain these values using standard tools for data management, like text-plain or relational databases. For example, on October 6th of 2014, Telefónica Movistar generated approximately 91,088,956 CDR registers only in data traffic (24 GBytes of plain text) for all contract users located in Chile. Just in Santiago, the capital of Chile, there are approximately one million mobile devices with Movistar contracts, which are producing 38 % of the total amount of data traffic CDRs. If several days must be aggregated, a large storage facility must be used to handle the data. For instance, each period of 6 months of data recording can increase an estimated of 4 TBytes the total space in the storage devices. Using data replication, in order to backup the data in a different storage, at least 8 TBytes of space are needed to handle the new 6-month CDR dataset. To solve the previous problem, we use Hadoop technologies (White 2009), mainly because of the cheap pricing and the on-demand features delivered by cloud computing providers, like Amazon Web Services (Amazon 2008) or Microsoft Azure (Microsoft 2010). The CDRs are first collected with an Extract-Transform-Load automatic process, which selects only the columns of interest, as shown in Table 1. Those trimmed registers are stored in a Hadoop distributed filesystem (HDFS) for long-term persistency using data redundancy in different geographical locations. Finally, a Hadoop-based query is launched periodically over the last days stored in the HDFS in order to collect the event traces (timestamp and antenna ID) of each device, ordered by day and time. The aggregated data can occupy space in the storage that is one or two order of magnitude less than the raw CDRs (from 24 GBytes to 100 Mbytes in 1 day of CDRs geo-located in antennas based in Santiago, Chile).

4.2 High-Performance Computing

In terms of computing capacity, the query that aggregates the CDRs and the estimation using the LVA process are the most challenging. The query is implemented in Pig (Gates 2011), a high-level data processing language that allows defining complex data flow patterns using syntax similar to SQL. The advantage of this processing language is that it can use different back end distributed/parallel computing technologies without changing the data flow specification or significant portions of the code. Among the available back end technologies, we can mention MapReduce (Dean and Ghemawat 2008), Tez (Saha et al. 2015) and Spark (Zaharia et al. 2010). By using HDFS, the Hadoop distributed file system as storage technology, combined with a data processing language like Pig, we can launch instances of clusters of computers using on-demand services in order to execute the query and store the aggregated results in a storage container. For instance, Microsoft Azure provides the service HDInsight (Microsoft 2013), which consists in a cluster of computers, masters, and workers, all of them ready to use with an implementation of Hadoop and the data processing tools. To allow the HDInsight instance to query the raw CDR dataset, the access credentials to the corresponding HDFS storage account must be previously loaded in the cluster management services.

Regarding the LVA process, we use a modified version of the LVA kriging code developed by Jeff Boisvert (Boisvert and Deutsch 2011). This version uses multi-core support (Chandra et al. 2001) allowing multiple threads to execute algebraic operations related with the L-ISOMAP multidimensional scaling. Although the multi-core support accelerates the computation of the distance matrix in the multidimensional space, it is only useful the first time the code is executed. For large estimation grids, the parallelization of the estimation loop as shown in Peredo et al. (2015) is not yet developed. The distributed capacities of the on-demand cluster can be exploited further to accelerate the estimation of different aggregated dataset corresponding to different times of the day. We use MPI to distribute the data preparation and execution of several LVA estimations of different timestamps for each day. Using a time granularity of 1 min, a maximum of 1,440 LVA estimations can be executed efficiently in parallel up to the total number of CPU cores and depending on the free memory space in each node of the cluster. With this schema, we can use an on-demand service to query and aggregate the raw CDRs and with the same computational resources execute all the LVA estimations using the CPU cores of the nodes in parallel. The final results, expressed as GEO-EAS or CSV files containing the LVA estimations for each timestamp of the day, can be stored in the HDFS persistent storage for further usage.

5 Case Study: Santiago, Chile

The proposed methodology is applied to the city of Santiago, Chile. In the first scenario, the region of interest is described in Fig. 4, which corresponds to a zone with 83 towers of antennas in a total area of 10 km^2. The total number of event traces registered on 12 August of 2015 in this zone is 129,598.

The northeast part in this zone is a business area with high density of tower/antennas; it includes many office buildings, shops, and retail stores. The west part corresponds to a residential area with low density of tower/antennas. The middle

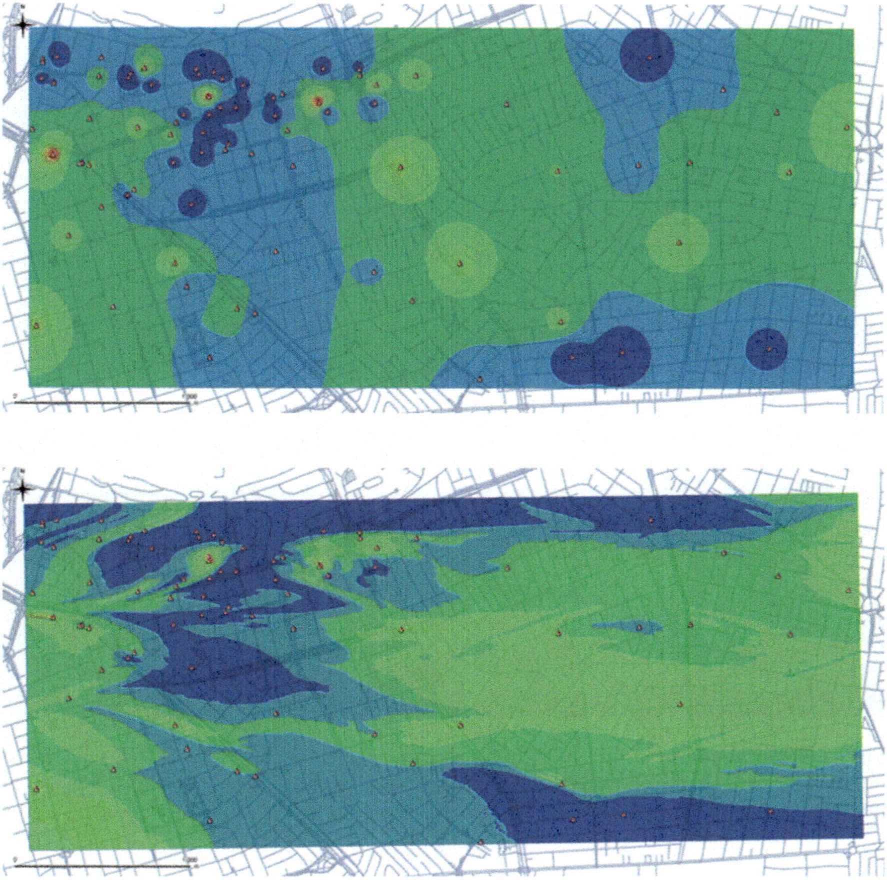

Fig. 7 Comparison of estimation contours between the IDW (*top*) and ordinary kriging using the LVA field with ratio r1 = 0.01 (*bottom*). The *small triangles* represent the towers of antennas with event numbers in the range [0,125] in each case. The time of these events is 6:08 PM

Table 2 Ordinary LVA kriging parameters using normal-scored data

Parameter	Value
Estimation/LVA grid	$576 \times 200 \times 1$
Offset number	4
Landmark points grid	$20 \times 20 \times 1$
Min/max data for kriging	2,60
Variographic parameters	Spherical, nugget 0.0, range 100

and southeast parts are mixed areas where the population can be traveling to other zones of the city. Figure 7 shows two estimations, inverse distance weighting (top) and ordinary kriging using the LVA field (bottom) generated with the primary and secondary streets. The kriging parameters are depicted in Table 2.

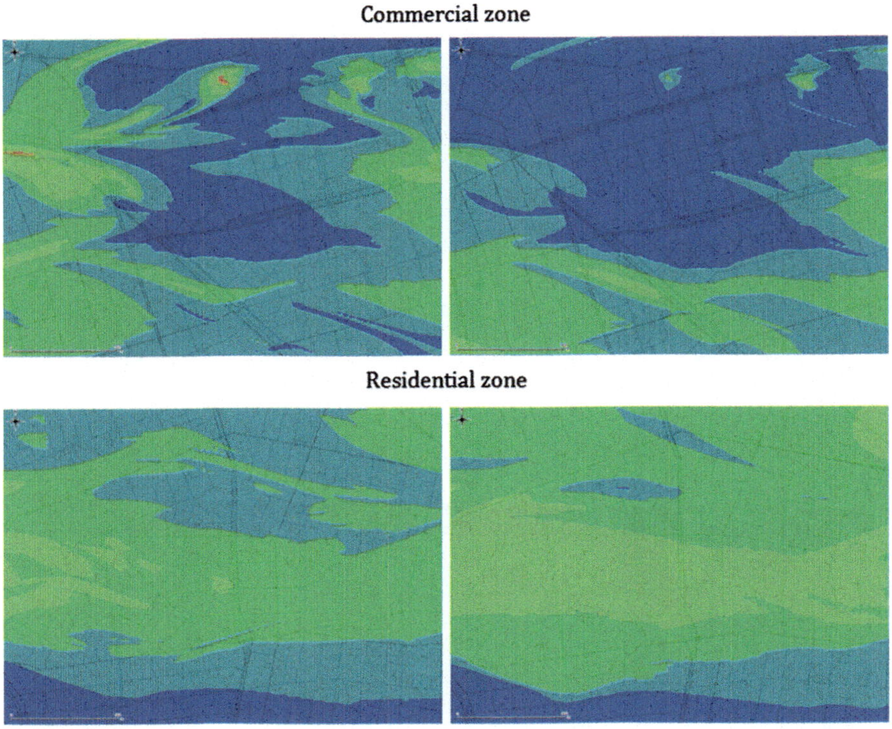

Fig. 8 Street-based anisotropy in the LVA-based estimation, comparing different zones at different times of the day. *Left column*: 3:11 pm, before rush hour. *Right column*: 8:55 pm, after rush hour

We can observe that the local anisotropy of the underlying urban landscape is qualitatively more accurate using the LVA method. In Fig. 8, we can observe different zones of the LVA estimation at different times. At 3:11 pm, more events are registered in the commercial zone, and at 8:55 pm more events are registered in the residential zone.

In order to compare quantitatively both estimations, IDW and LVA, a ground-truth data must be necessary, for example, GPS or network monitoring data. With the ground-truth data, we can locate with high precision each device at the specific time of study. According to Calabrese et al. (2011) and Leontiadis et al. (2014), extensive GPS campaigns and large volumes of network data are needed to build the ground-truth model data. In our case, we have developed a special Android application which collects the GPS position of the mobile phone each 60 s, together with the local area code (LAC) and cell ID (CI) which are the basic network parameters to identify the antenna sector of influence. Our idea is to measure a specific day in the region of interest, using several devices at our disposal, tracking their positions and antennas where the connections are made. With that ground-truth data, we can approximately pair the sparse CDR events with some of the GPS

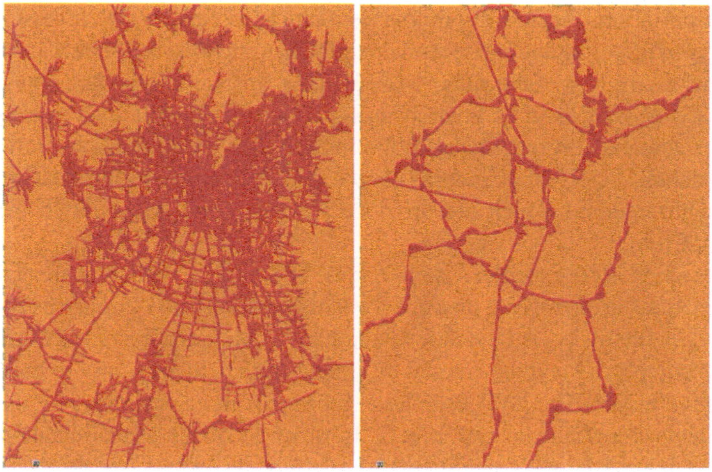

Fig. 9 Azimuth angles in Santiago metropolitan area. Using *primary* and *secondary* objects (*left*) and using the *motorway* object (*right*)

events according to their timestamps (with tolerated delays of 60 min), building an initial truth model. Due to the complexity of the validation process, we decided to exclude it from the present work.

The previous scenario was analyzed in a region of relative small area, 10 km^2. The total size of Santiago metropolitan area is approximately 1,800,000 km^2. As a consequence, the estimation grid and variographic parameters necessary to execute the full estimation are difficult to infer. Additionally, the LVA field inference is not straightforward, since the OpenStreetMap objects *primary* and *secondary* streets are densely sampled in the large region of interest (Fig. 9, top). A possible solution is to extract a coarser object from the platform, namely, *motorway*, which represents large highways and motorways in the city. Using the new object, we can calculate the azimuth angles in the intersections of a gridded area and the motorways present in the map (Fig. 9, bottom). The estimation in the large region is left for future versions of the application, since our primary interest is to include the street-based anisotropy in low-range estimations. Another possible inclusion in future versions is a velocity estimation filter, which can discriminate between different modes of transport in the highway. However, the CDR dataset by itself is not sufficiently fine grained to estimate this kind of short-range statistic.

6 Conclusions and Future Work

Two-dimensional locally varying anisotropy based in street orientation (azimuth) has been included in the geostatistical estimation of mobile events registered in different towers of antennas around a city. In related literature from urban planning and transportation, the most common estimation method is the inverse distance weighting, without including secondary information from the streets in the urban landscape. To the best of the author's knowledge, this work is the first application of LVA-based geostatistical estimation in the context of mobile data geo-located in towers of antennas. In our scenario, the mobile data corresponds to billing information, denoted call detail records (voice and data traffic), and the network of towers with antennas which are provided by Telefónica Movistar in the city of Santiago, Chile.

A case study is presented in a specific zone of Santiago, with total area of 10 km^2. The estimations using inverse distance weighting and ordinary kriging with LVA are compared qualitatively. An extended quantitative comparison is left as future work due to its complexity and the amount of ground-truth data necessary to build a pivot comparative model. The LVA estimation represents in an accurate way the flow of mobile devices around the city streets. In order to obtain the orientations of the streets, we have used the OpenStreetMap platform to extract objects from the city map, like primary and secondary streets. After that, a gridded sampling of the orientations (azimuths) is extracted and posteriorly interpolated using standard methods. With the full 2D LVA field estimated, standard LVA-based geostatistical tools can be used to obtain the final results in each timestamp of the day. The process can be repeated each day, and further analysis can be done with the urban dynamic estimation provided by the current methodology. LVA estimation parameters, such as the landmark grid, offset number, or variographic parameters, must be inferred previously using data from similar past days (working days, weekends, or holidays).

The presented work is the first step in a long-term effort with the final objective of highly accurate dynamic position estimation of mobile devices in the city using indirect geo-localization data, such as the CDR dataset. Although there are other technologies that can provide directly this information, such as GPS or network monitoring systems, the advantage of the CDRs is that no extra applications need to be active in the client (possible battery exhaustion of the mobile device) or server side (possible saturation of the network). The proposed methodology is flexible enough to allow other types of data sources, by just changing the geo-location and dataset passed to the LVA-based codes. The main feature is the inclusion of the streets into the geostatistical estimation. If stochastic simulations must be obtained to quantify uncertainty, sequential Gaussian simulation can be executed following the same steps described in this work, using the LVA-based corresponding code. Nonconventional sources of information, like antenna radiation and power, or signal propagation models for wireless communications, can be added to the models in the form of variographic parameters or local definitions of the ratio and azimuth values of the LVA field (small ratios for small influence zones and large ratios for large influence zones). This topic will be studied in future versions of the proposed methodology.

Acknowledgments The authors would like to acknowledge the project CORFO 13CEE2-21592 (2013-21592-1-INNOVA_PRODUCCION2013-21592-1), *Telefónica Investigación y Desarrollo SPA* and *Telefónica Chile SA* for the support and data provided in this study.

References

Amazon Inc (2008) Amazon elastic compute cloud. URL: http://aws.amazon.com

Boisvert J, Deutsch CV (2011) Programs for kriging and sequential Gaussian simulation with locally varying anisotropy using non-Euclidean distances. Comput Geosci 37(4):495–510

Boisvert JB, Manchuk JG, Deutsch CV (2009) Kriging in the presence of locally varying anisotropy using non-Euclidean distances. Math Geosci 41(5):585–601

Calabrese F, Colonna M, Lovisolo P, Parata D, Ratti C (2011) Real-time urban monitoring using cell phones: a case study in Rome. IEEE Trans Intell Transp Sys 12:141–151

Candia J, González MC, Wang P, Schoenharl T, Greg M, Barabási A-L (2008) Uncovering individual and collective human dynamics from mobile phone records. J Phys A: Math Theor 41:224015

Chandra R, Dagum L, Kohr D, Maydan D, McDonald J, Menon R (2001) Parallel programming in openMP. Morgan Kaufmann Publishers, San Francisco

Chilès JP, Delfiner P (2012) Geostatistics: modeling spatial uncertainty, 2nd edn. Wiley, New York, p 734, Wiley series in probability and statistics

Colak S, Alexander LP, Alvim BG, Mehndiretta SR, González MC (2015) Analyzing cell phone location data for urban travel: current methods, limitations and opportunities. Transportation Research Board 94th Annual Meeting, Issue, 15–5279

Dean J, Ghemawat S (2008) MapReduce: simplified data processing on large clusters. Commun ACM 51(1):107–113

Gates A (2011) Programming pig, 1st edn. O'Reilly Media, Sebastopol

Gonzalez MC, Hidalgo CA, Barabási A-L (2008) Understanding individual human mobility patterns. Nature 453:779–782

Görnerup O (2012) Scalable mining of common routes in mobile communication network traffic data. In: Proceedings of the 10th international conference on Pervasive Computing, Pervasive'12, pp 99–106

Haklay M, Weber P (2008) OpenStreetMap: user-generated street maps. Pervasive Comput IEEE 7:12–18

Horanont TA (2012) Study on urban mobility and dynamic population estimation by using aggregate mobile phone sources. CSIS discussion paper no. 115. Technical report, Center for Spatial Information Science, The University of Tokyo, 2012

Horanont T, Shibasaki R (2008) An implementation of mobile sensing for large-scale urban monitoring. Proceedings of Urbansense'08 2008

ITC (2015) The world in 2015 - ICT facts and figures. International Telecommunications Union, ICT Data and Statistics Division, Geneva

Leontiadis I, Lima A, Kwak H, Stanojevic R, Wetherall D, Papagiannaki K (2014) From cells to streets: estimating mobile paths with cellular-side data. In: Proceedings of the 10th ACM international on conference on emerging Networking Experiments and Technologies (CoNEXT '14). ACM, New York, pp 121–132

Lima A, Stanojevic R, Papagiannaki D, Rodriguez P, González MC (2016) Understanding individual routing behaviour. J R Soc Interface 13:20160021

Li Y, Deng H, Li J, Peng C, Lu S (2016) Instability in distributed mobility management: revisiting configuration management in 3G/4G mobile networks. ACM SIGMETRICS 2015, Antibes Juan-les-pines

Maksuda L, Boisvert JB (2015) Inference of locally varying anisotropy fields from diverse data sources. Comput Geosci 82:170–182

Mao H, Shuai X, Ahn YY, Bollen J (2013) Mobile communications reveal the regional economy in cote D'Ivoire. Cambridge, MA. NetMob 2013: data for development challenge

Microsoft Corporation (2010) Microsoft azure. URL: https://azure.microsoft.com

Microsoft Corporation (2013) Azure HDInsight. URL: https://azure.microsoft.com/en-us/services/hdinsight/

Oliver N, Matic A, Frias-Martinez E (2015) Mobile network data for public health: opportunities and challenges. Front Public Health 3:189

Peredo O, Ortiz JM, Herrero JR (2015) Acceleration of the Geostatistical Software Library (GSLIB) by code optimization and hybrid parallel programming. Comput Geosci 85:210–233, PA (December 2015)

Saha B, Shah H, Seth S, Vijayaraghavan G, Murthy A, Curino C (2015) Apache Tez: a unifying framework for modeling and building data processing applications. In: Proceedings of the 2015 ACM SIGMOD International Conference on Management of Data (SIGMOD '15). ACM, New York, pp 1357–1369

Saravanan M, Pravinth S, and Holla P (2011) Route detection and mobility based clustering. In: Proceedings of IEEE 5th international conference on Internet Multimedia Systems Architecture and Application (IMSAA), pp 1–7

Song C, Qu Z, Blumm N, Barabási A-L (2010) Limits of predictability in human mobility. Science 327(5968):1018–1021

Soto V, Frias-Martinez V, Virseda J, Frias-Martinez E (2011) Prediction of socioeconomic levels using cell phone records. Proceedings of User Modeling, Adaption and Personalization: 19th International Conference, UMAP 2011, Girona, Spain, July 11–15, 2011. Ed. Springer, Berlin, pp 377–388

Traag VA, Browet A, Calabrese F, Morlot F (2011) Social event detection in massive mobile phone data using probabilistic location inference. Privacy, Security, Risk and Trust (PASSAT) and 2011 I.E. Third International Conference on Social Computing (SocialCom), 2011, pp 625–628

Toole JL, Colak S, Sturt B, Alexander LP, Evsukoff A, González MC (2015) The path most traveled: travel demand estimation using big data resources. Transp Res Part C: Emerg Technol 58(Part B):162–177

White T (2009) Hadoop: the definitive guide, 1st edn. O'Reilly Media, Sebastopol

Zaharia M, Chowdhury M, Franklin MJ, Shenker S, Stoica I (2010) Spark: cluster computing with working sets. In: Proceedings of the 2nd USENIX conference on Hot topics in cloud computing (HotCloud'10). USENIX Association, Berkeley, pp 10–10

Part VIII
Health

Using Classical Geostatistics to Quantify the Spatiotemporal Dynamics of a Neurodegenerative Disease from Brain MRI

Robert Marschallinger, Mark Mühlau, Paul Schmidt, Peter Atkinson, Eugen Trinka, Stefan Golaszewski, and Johann Sellner

Abstract We present a novel approach to characterize multiple sclerosis (MS) from brain magnetic resonance imaging (MRI) with geostatistics. Brain MRI provides excellent, exhaustive input data to geostatistical analysis, typically several million voxels per MRI scan. A dataset of 259 spatially normalized binary MS white matter lesion (WML) patterns covering very mild to extremely severe MS cases was subject to directional variography. Using an exponential variogram model function, the observed spatial variability in x,y,z directions can be expressed by geostatistical parameters range and sill which perfectly correlate with WML pattern surface complexity and lesion volume. A scatterplot of ln(range) vs. ln(sill), classified by pattern anisotropy, enables a consistent and clearly arranged presentation of MS-lesion patterns based on their geometry. The geostatistical approach and the graphical representation of results are considered efficient exploratory data analysis tools for longitudinal, cross-sectional, and medication impact studies.

R. Marschallinger (✉)
Interfaculty Department of Geoinformatics Z_GIS, Universität Salzburg, Schillerstr. 30, 5020, Salzburg, Austria

Department of Neurology, Christian Doppler Medical Centre, Paracelsus Medical University, Ignaz Harrer-Straße 79, 5020 Salzburg, Austria
e-mail: robert.marschallinger@sbg.ac.at

M. Mühlau
Department of Neurology, Klinikum rechts der Isar, TU München, Munich, Germany

P. Schmidt
Department of Statistics, Ludwig-Maximilians-University München, Munich, Germany

P. Atkinson
Faculty of Science and Technology, Engineering Building Technology, Lancaster University, Lancaster LA14YR, UK

E. Trinka • S. Golaszewski • J. Sellner
Department of Neurology, Christian Doppler Medical Centre, Paracelsus Medical University, Ignaz Harrer-Straße 79, 5020 Salzburg, Austria

J.J. Gómez-Hernández et al. (eds.), *Geostatistics Valencia 2016*, Quantitative Geology and Geostatistics 19, DOI 10.1007/978-3-319-46819-8_67

1 Introduction

Multiple sclerosis (MS) is an autoimmune disease of the central nervous system (CNS) that causes inflammatory demyelinating lesions in the brain and in the spine. MS affects over 2.5 million people worldwide and is the primary nontraumatic cause of neurologic disability in young adults, with an associated socioeconomic impact. The individual course of MS is highly variable – while some patients may acquire severe and irreversible disability within a few years, others may follow a mild course with almost no disability even after decades (Compston and Coles 2008).

2 MRI Data

The introduction of magnetic resonance imaging (MRI) for MS diagnosis and monitoring the disease course was a milestone in MS patient's care (Filippi and Rocca 2011). Recent diagnostic criteria allow setting MS diagnosis on the basis of MRI (Polmann et al. 2011). The hallmark of MS is hyperintense sclerotic lesions within cerebral white matter, as indicated in Fig. 1 which is a T2-weighted (FLAIR) axial slice through an MS patient's brain.

Traditionally, MRI scanners yield stacks of parallel images from which 3D models can be derived; more recent MRI scanners provide voxel models which directly represent the volume of interest. Before single-patient follow-up MRI scans (longitudinal studies) or multiple-patient MRI scans (cross-sectional studies) can be

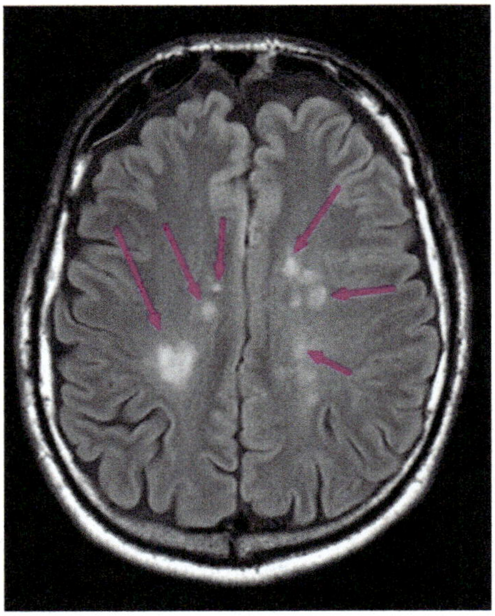

Fig. 1 MRI of MS-affected brain. *Arrows* indicate bright, hyperintense MS lesions (FLAIR MRI sequence, axial slice at *top* of ventricle system)

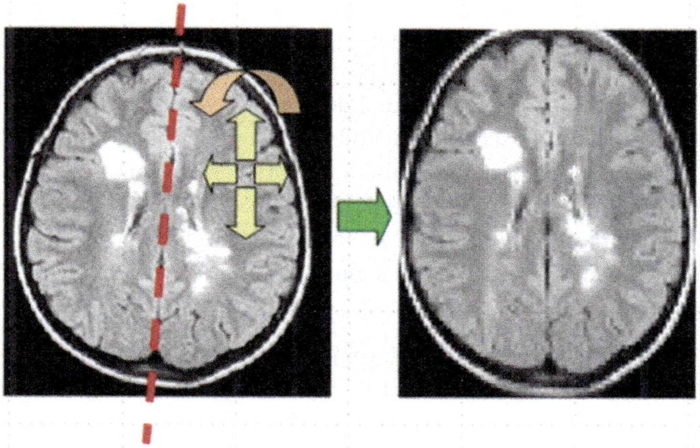

Fig. 2 Geometrical transformation of individual brain geometry (*left*) to MNB geometry (*right*). Axial slice of MS-affected brain at *top* of ventricle level

compared, the individual brain scans need to be geometrically matched to a standard brain template like the Montreal Normal Brain ("MNB," Collins et al. 1998; Penny et al. 2007). This approach compensates for varying head positions in follow-up MRI scans and ensures reference brain geometry when working with group data (Fig. 2).

From MRI data, 3D MS-lesion patterns can be extracted with semiautomatic or automatic lesion extraction software that outputs binary MS-lesion models (Garcia-Lorenzo et al. 2012).

3 The Geostatistical Approach

WML presents rather heterogeneously across patients not only with regard to the number and overall volume but also with regard to spatial pattern, predilection sites, and shape of single lesions. Routine radiological evaluation reports the number of WML, the total lesion load (volume), and the associated changes as from follow-up investigations. Our goal is to describe the spatial structure of a WML pattern in MNB geometry with standard geostatistical parameters (Marschallinger et al. 2014, 2016).

$$\text{Variogram function} \qquad \gamma(\mathbf{h}) = \frac{1}{2n(\mathbf{h})} \cdot \sum_{i=1}^{n} ((z(\mathrm{x}_i) - z(\mathrm{x}_i + \mathbf{h}))^2 \qquad (1)$$

$$\text{Exponential variogram model} \qquad \gamma(\mathbf{h}) = c \cdot \left(1 - \frac{e^{(-3\cdot|\mathbf{h}|)}}{a}\right) \qquad (2)$$

Since the MNB can be considered a Euclidean space which is then dissected into equally sized voxels, the constant extent and support of the MNB enable rapid, grid-based variography (Eq. 1), variogram modeling (Eq. 2), and the sensitive comparison of longitudinal and cross-sectional studies.

For efficient characterization of WML from MRI, we developed the following, fully automatic processing pipeline that builds on a pair of T1 and FLAIR MRI images, acquired on a 3-tesla scanner:

1. From FLAIR and T1 sequences of a single MRI investigation, automatically derive a binary 3D model of the WML pattern in MNB geometry (e.g., Fig. 3 right). Software used is Lesion Segmentation Tool (LST) (Schmidt et al. 2012).
2. Compute directional empirical variograms (Eq. 1) in x,y,z directions (dextral-sinistral, caudal-rostral, dorsal-ventral orientations). Variograms are confined to lag distances from 0 to 15 mm, because this area holds the most relevant correlation information and a variogram model can be fitted straightforwardly. Software used is Geostatistical Software Library (GSLIB) (Deutsch and Journel 1997).
3. Fit an exponential variogram model function (Eq. 2) to each directional empirical variogram to yield associated range and sill parameters. These characterize overall surface smoothness, total lesion volume (TLL, total lesion load), and preferred spatial continuity of a white matter lesion (WML) pattern. Software used is R (R development core team 2008).
4. Derive scatterplots of ln(range) vs. ln(sill) to portray WML patterns in a space defined by total pattern surface smoothness vs. total lesion volume. For the current purpose, "surface smoothness" is defined as the ratio of total WML pattern volume/total WML pattern surface. This space, making up the so-called

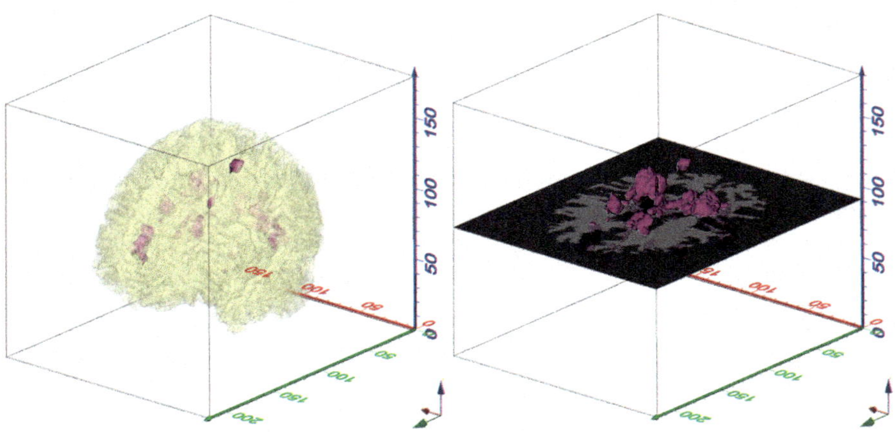

Fig. 3 *Left*: 3D model of white matter of MS-affected brain, normalized to MNB geometry (white matter visualized in translucent style to show some MS lesions for orientation in 3D). *Right*: associated WML model, white matter striped off, and axial slicing plane at ventricle level indicated for orientation. MNB geometry

MS-Lesion Pattern Discrimination Plot (LDP), enables a clear presentation of the spatial characteristics of a WML pattern. Scatterplots involving individual x,y,z directions account for anisotropy in WML geometry. Calculating mean center (e.g., De Smith et al. 2007) for a WML from x,y,z directional ln(range) and ln (sill) data, the overall x,y,z geometry of a WML can be expressed. Since all parameters are derived from geometrically standardized data (i.e., MNB geometry), in the course of longitudinal and cross-sectional studies, WML patterns can be immediately related: On the one hand, data can be compared synoptically; on the other hand, they are ready for automatic processing, e.g., for deriving evolution paths in single-patient follow-up from mean center data (Fig. 6).

The LDP conveniently combines class symbols indicating the number of individual lesions (Fig. 4) or standard distance (De Smith et al. 2007) to express overall

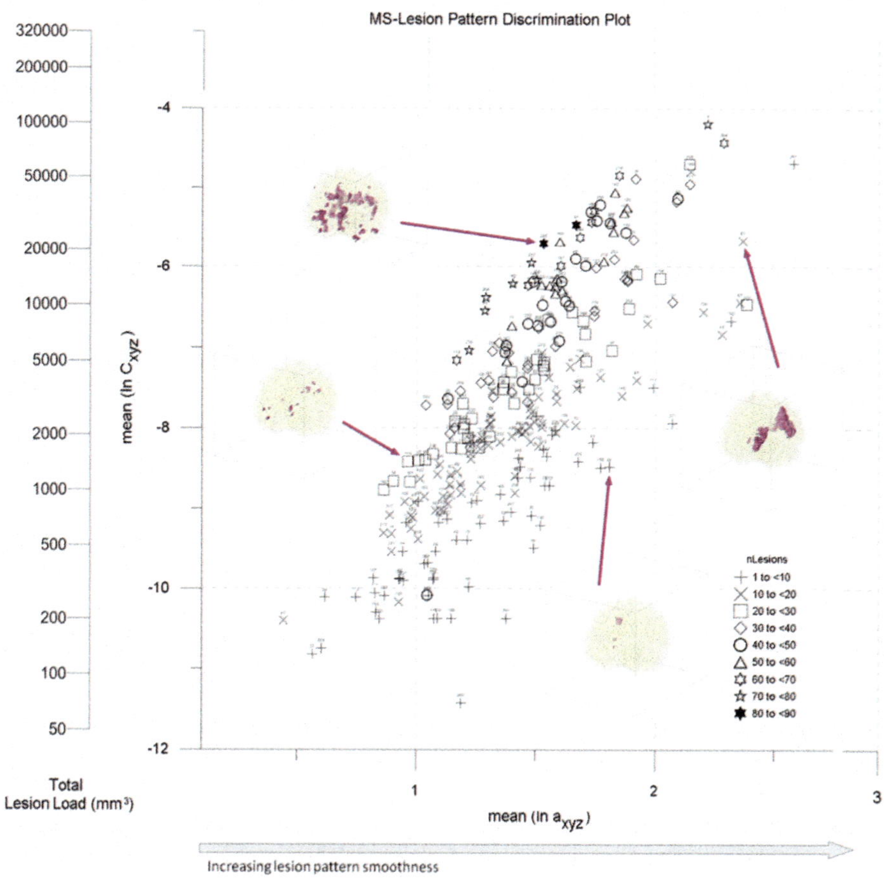

Fig. 4 Combining geostatistical Range ln(a) and geostatistical Sill ln(C) of 259 MS-affected, geometrically normalized brains in the MS-Lesion Pattern Discrimination Plot (LDP). Four WML pattern 3D models and their associated positions in the LDP are indicated. nLesions: number of individual lesion objects per WML pattern, coded by symbols. See text for details (After Marschallinger et al. 2016)

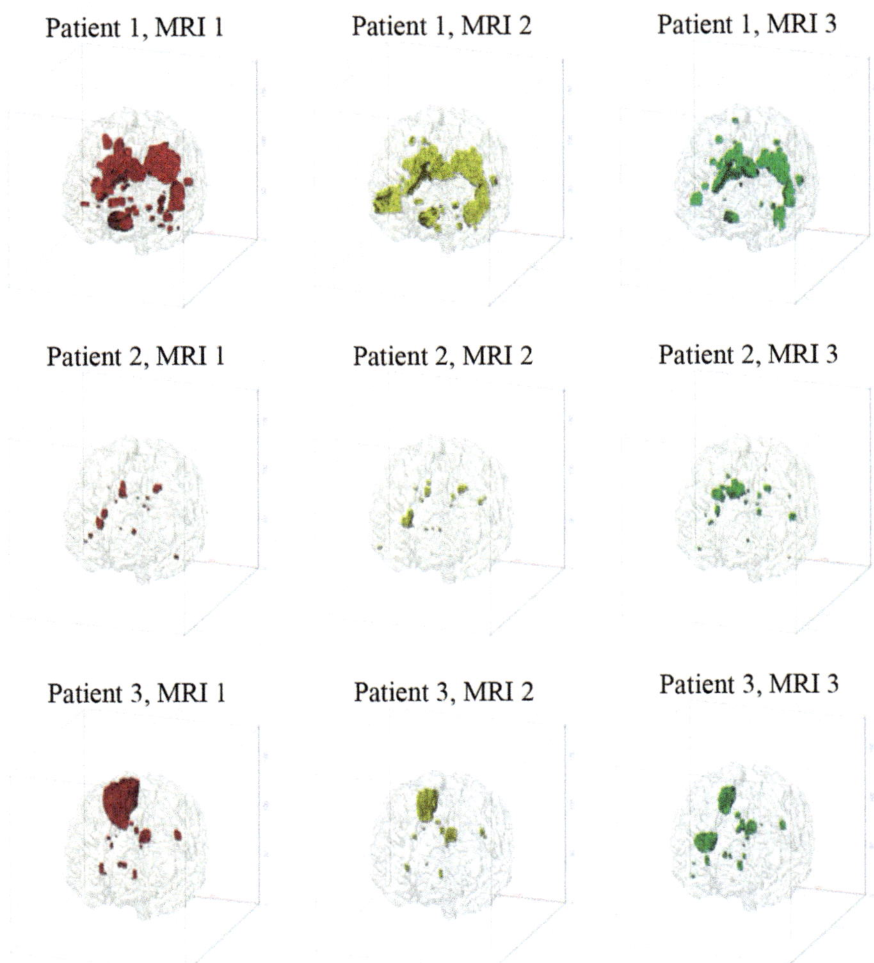

Fig. 5 WML evolution in three MS patients, three time steps each chronology from *left* to *right*, time interval 6 months each. (See text for details, compare with Fig. 6)

WML anisotropy. Working through Fig. 4 shows that complex patterns with many lesions or patterns with a "rough"/"complex" surface generally are positioned at the left fringe of the point cloud while patterns with few, big, and "smooth" lesions are placed toward the right border. Patterns around the long axis of the elliptic cloud mediate between rough and smooth extremes.

More clinically relevant, the LDP can be used to portray the evolution of WML during single-patient follow-up. Figure 5 shows WML patterns from MRI follow-ups of three patients. Each patient was scanned three times at 6-month intervals, yielding a total observation period of 1.5 years per patient. Patients are arranged in rows, and columns represent MRI investigations (time series advancing from left to right, also indicated by WML colors red-yellow-green). Using the abovementioned processing pipeline, the evolution of the three individual WML patterns can be

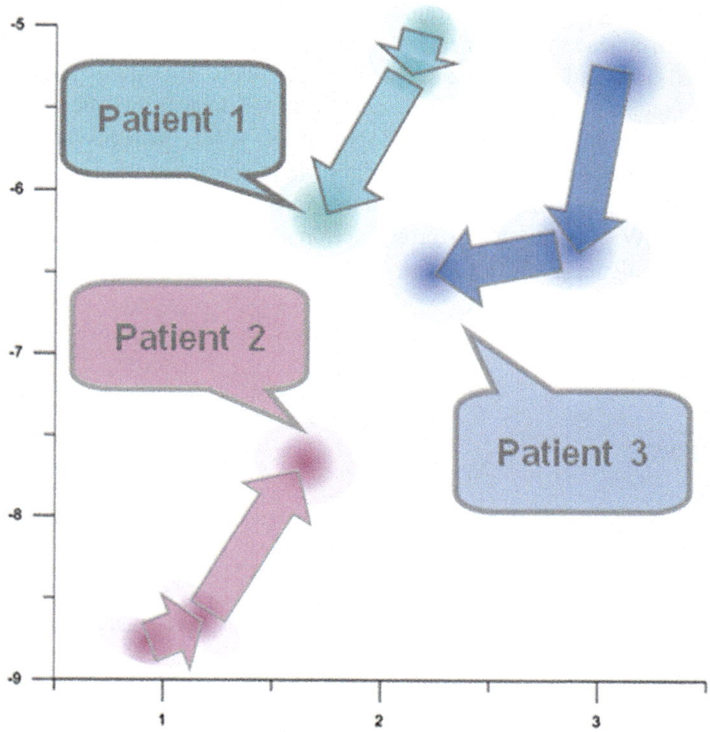

Fig. 6 WML pattern evolution paths of three MS patients in the LDP. *Arrows* indicate evolution direction. Compare Fig. 5 (See text for details)

visualized and quantitatively described by means of the LDP (Fig. 6). In Fig. 5, Patient 1 shows a decrease in MS lesion sizes and concurrent decrease in surface smoothness due to decomposition of big MS lesions into smaller ones; accordingly, in the LDP, the evolution path arrows point to lower volumes and a less smooth overall surface (Fig. 6). About the same is due for Patient 3; the WML pattern of which shows volume loss at the cost of lesion pattern smoothness. This is more pronounced between scans two and three where the biggest lesion continues to decrease but new small lesions show up, yielding approximately constant volume. Patient 2 shows increasing lesion volume at increasing WML pattern surface smoothness caused by confluence of small lesions into bigger ones.

4 Conclusions

A workflow and processing pipeline has been established that enables the characterization of MS-related white matter lesion (WML) patterns from MRI data by estimation of geostatistical parameters' range and sill. These are the bases for representing a WML pattern in the MS-Lesion Pattern Discrimination Plot

(LDP). The LDP is a versatile framework that combines WML pattern volume, WML pattern surface smoothness, and geometrical anisotropy information in a single, well-arranged plot. Major changes as well as subtle fluctuations in MS-lesion pattern geometry can be visualized straightforwardly. The LDP provides precise insight into the spatial development of WML patterns (i.e., selective growth/shrink in specific directions) without requiring object-based characterization. The LDP is considered an EDA tool that informs on the spatial/spatiotemporal properties of WML patterns in cross-sectional and longitudinal studies and in monitoring medication efficacy.

Acknowledgments We thank two anonymous reviewers for their constructive ideas.

References

Collins DL, Zijdenbos AP, Kollokian V et al (1998) Design and construction of a realistic digital brain phantom. IEEE Trans Med Imag 17(3):463–468

Compston A, Coles A (2008) Multiple sclerosis. Lancet 372:1502–1517

De Smith M, Goodchild M, Longley PA (2007) Geospatial analysis. Matador, Leicester, 491 pp

Deutsch CV, Journel AG (1997) GSLIB Geostatistical software library and user's guide. Oxford University Press, 380pp

Filippi M, Rocca MA (2011) MR imaging of multiple sclerosis. Radiology 259(3):659–681

Garcia-Lorenzo D, Francis S, Narayanan S, Arnold DL, Collins L (2012) Review of automatic segmentation methods of multiple sclerosis white matter lesions on conventional magnetic resonance imaging. Med Image Anal 17:1–18

Marschallinger R, Golaszewski S, Kunz A et al (2014) Usability and potential of geostatistics for spatial discrimination of multiple sclerosis lesion patterns. J Neuroimaging 4(3):278–286. doi:10.1111/jon.12000

Marschallinger R, Schmidt P, Hofmann P, Zimmer C, Atkinson P, Sellner J, Trinka E, Mühlau M (2016) A MS-lesion pattern discrimination plot based on geostatistics. Brain Behav 6(3), doi: 10.1002/brb3.430

Penny WD, Friston JK, Ashburner JT, Kiebel SJ, Nichols TE (2007) Statistical parametric mapping: the analysis of functional brain images. Academic Press, London, 625pp

Polmann CH, Reingold SC, Banwell B et al (2011) Diagnostic criteria for multiple sclerosis: 2010 revisions to the McDonald criteria. Ann. Neurology 69(2):292–302

R Development Core Team (2008) R: a language and environment for statistical computing. R Foundation for Statistical Computing, Vienna, ISBN 3-900051-07-0

Schmidt P, Gaser C, Arsic M, Buck D, Forschler A (2012) An automated tool for detection of FLAIR-hyperintense white-matter lesions in multiple sclerosis. Neuroimage 59:3774–3783

Printed by Printforce, the Netherlands